May 8–10, 2013
Hangzhou, China

Association for
Computing Machinery

Advancing Computing as a Science & Profession

ASIA CCS'13

Proceedings of the 8th ACM SIGSAC Symposium on
Information, Computer and Communications Security

Sponsored by:

ACM SIGSAC

In-cooperation with:

Hangzhou Normal University, Shanghai Jiaotong University, Zhejiang University and Nanjing University of Information Science & Technology

Association for
Computing Machinery

Advancing Computing as a Science & Profession

The Association for Computing Machinery
2 Penn Plaza, Suite 701
New York, New York 10121-0701

ISBN: 978-1-4503-2148-8 (Digital)
ISBN: 978-1-4503-2280-5 (Print)

Additional copies may be ordered prepaid from:

ACM Order Department
PO Box 30777
New York, NY 10087-0777, USA

Phone: 1-800-342-6626 (USA and Canada)
+1-212-626-0500 (Global)
Fax: +1-212-944-1318
E-mail: acmhelp@acm.org
Hours of Operation: 8:30 am – 4:30 pm ET

Printed in the USA

Welcome to ASIA CCS 2013

It is our great pleasure to welcome you to the 8th ACM Symposium on Information, Computer and Communications Security - ASIACCS 2013, which will be held in Hangzhou, China, May 7-10, 2013. The inaugural ASIACCS was held in Taipei (2006). Since then ASIACCS have been held in Singapore (2007), Tokyo (2008), Sydney (2009), Beijing (2010), Hong Kong (2011), and Seoul (2012).

ASIACCS is a major international forum for information security researchers, practitioners, developers, and users to explore and exchange the newest cyber security ideas, breakthroughs, findings, techniques, tools, and experiences. We welcome subm issions from academia, government, and industry presenting novel research on all theoretical and practical aspects of computer and network security.

The conference received 216 paper submissions, and accepted 35 regular papers and 27 short papers for presentation at the conference. The final versions of the accepted papers, which the authors finalized on the basis of comments from the reviewers, were included in the proceedings. Each accepted paper was carefully reviewed by at least three program committee members on the basis of its significance, novelty, and technical quality. The reviewing process was double-blind. It was followed by a Web-based discussion. Based on the comments and scores given by reviewers, the final decisions on acceptance were made.

As always, the success of an international conference was made possible through the contributions from many individuals and organizations. We are thankful to Dawn Song and Wenbo Mao for delivering the keynote addresses. We would like to thank all the authors who submitted papers. We are indebted to our program committee members and the external reviewers for the great job they did. We sincerely thank all members of the Steering Committee for their support and encouragement. Special thanks go to Adrienne Griscti from ACM. Adrienne is used to organizing the paperwork at ACM to initiate the ACM-Sheridan Proceedings Service. Assisted by Lisa Tolles, she has taken care of processing the papers in a timely manner. Last, but not least, we are grateful to Hangzhou Normal University and Shanghai Jiaotong University for sponsoring the conference.

Kefei Chen, Qi Xie, Weidong Qiu
ASIACCS'13 General Co- Chairs
Shanghai Jiaotong University, China
Hangzhou Normal University, China

Ninghui Li, Wen-Guey Tzeng
ASIACCS'13 Program Co-Chairs
Purdue University, USA
National Chiao Tung University,Taiwan

Table of Contents

Session 4: Short Papers I: Social Network

Keynote Address

Session 5: Data Outsourcing

Session 6: Applied Cryptography II

ASIA CCS 2013 Symposium Organization

General Co-Chairs: Kefei Chen *(Shanghai Jiao Tong University, China)*
Qi Xie *(Hangzhou Normal University, China)*
Weidong Qiu *(Shanghai Jiao Tong University, China)*

Program Co-Chairs: Ninghui Li *(Purdue University, USA)*
Wen-Guey Tzeng *(National Chiao Tung University, Taiwan)*

Honorary Chair: Xiuyun Yu *(Hangzhou Normal University, China)*

Workshop Chair: Xiaofeng Wang *(Indiana University at Bloomington, USA)*

Publicity Chair: Lujo Bauer *(Carnegie Mellon University, USA)*

Steering Committee Chair: Shiuhpyng Shieh *(NCTU, Taiwan)*

Steering Committee: Elisa Bertino *(Purdue University, USA)*
Mike Reiter *(University of North Carolina at Chapel Hill, USA)*
Ninghui Li *(Purdue University, USA)*
Eiji Okamoto *(University of Tsukuba, Japan)*
Li Gong *(Mozilla Online Ltd., USA)*
Vijay Varadharajan *(Macquarie University, Australia)*
David Basin *(ETH, Switzerland)*
Robert Deng *(SMU, Singapore)*
Virgil Gligor *(CMU, USA)*
Hideki Imai *(AIST, Japan)*
Sushil Jajodia *(GMU, USA)*
Pierangela Samarati *(University of Milan, Italy)*

Program Committee: Gail-Joon Ahn *(Arizona State University, USA)*
Michael Backes *(Saarland University, Germany)*
Lujo Bauer *(Carnegie Mellon University, USA)*
Elisa Bertino *(Purdue University, USA)*
Ioana Boureanu *(EPFL, Switzerland)*
Kefei Chen *(Shanghai Jiao Tong University, China)*
Liqun Chen *(Hewlett Packard Bristol Labs, UK)*
Sherman S. M. Chow *(Chinese University of Hong Kong, Hong Kong)*
Robert Deng *(Singapore Management University, Singapore)*
Wenliang Du *(Syracuse University, USA)*
Chun-I Fan *(National Sun Yat-sen University, Taiwan)*
Philip Fong *(University of Calgary, Canada)*
Debin Gao *(Singapore Management University, Singapore)*
Deepak Garg *(Max Planck Institute for Software Systems, Germany)*

Program Committee (continued):

Virgil Gligor *(Carnegie Mellon University, USA)*
Ragib Hasan *(University of Alabama at Birmingham, USA)*
Trent Jaeger *(Penn State University, USA)*
Sushil Jajodia *(George Mason University, USA)*
Xuxian Jiang *(North Carolina State University, USA)*
Yongdae Kim *(KAIST, South Korea)*
Ponnurangam Kumaraguru *(Indraprastha Institute of Information Technology, India)*
Xuejia Lai *(Shanghai Jiao Tong University, China)*
Ruby Lee *(Princeton University, USA)*
Chin-Laung Lei *(National Taiwan University, Taiwan)*
Zhichun Li *(NEC Labs, USA)*
Yingjiu Li *(Singapore Management University, Singapore)*
Jiangtao Li *(Intel, USA)*
Zhengkai Liang *(National University of Singapore, Singapore)*
David Lie *(University of Toronto, Canada)*
Dongdai Lin *(Chinese Academy of Sciences, China)*
Zhiqiang Lin *(University of Texas at Dallas, USA)*
Peng Liu *(Penn State University, USA)*
Javier Lopez *(University of Malaga, Spain)*
Z. Morley Mao *(University of Michigan, USA)*
Ian Molloy *(IBM Research, USA)*
David Naccache *(Ecole Normale Superieure, France)*
Charalampos Papamanthou *(UC Berkeley, USA)*
Kui Ren *(University at Buffalo, USA)*
Ryan Riley *(Qatar University, Qatar)*
Ahmad-Reza Sadeghi *(Technische Universitaet Darmstadt and Intel-CRISC, Germany)*
Rei Safavi-Naini *(University of Calgary, Canada)*
Pierangela Samarati *(University of Milan, Italy)*
Kouichi Sakurai *(KyuShu University, Japan)*
Shiuhpyng Shieh *(National Chiao Tung University, Taiwan)*
Willy Susilo *(University of Wollongong, Australia)*
Mahesh Tripunitara *(University of Waterloo, Canada)*
Wen-Guey Tzeng *(National Chiao Tung University, Taiwan)*
Vijay Varadharajan *(Macquarie University, Australia)*
Huaxiong Wang *(Nanyang Technological University, Singapore)*
Rui Wang *(Microsoft Research, USA)*
Guoliang Xue *(Arizona State University, USA)*
Danfeng Yao *(Virginia Tech University, USA)*
Moti Yung *(Google Inc. and Columbia University, USA)*
Fangguo Zhang *(Sun Yat-sen University, China)*
Kehuan Zhang *(Chinese University of Hong Kong, Hong Kong)*
Xiangyu Zhang *(Purdue University, USA)*

Additional reviewers:

Yousra Aafer
Massimiliano Albanese
Claudio Agostino Ardagna
Man Ho Au
Sonia Bogos
Sven Bugiel
Jianneng Cao
Yinzhi Cao
Cheng Chen
Jianhong Chen
Jie Chen
Po-Kai Chen
Si Chen
MichaelCheng-Yi Cho
Kai-Min Chung
Shuaifu Dai
Ting Dai
Luca Davi
Sabrina DeCapitani Di Vimercati
Yu Ding
Alexandra Dmitrienko
Xinshu Dong
Lei Duan
Alexandre Duc
Amit Dutta
Sebastian Faust
Sara Foresti
Siddharth Garg
Hao Hao
Md. Haque
Urs Hengartner
Chia-Wei Hsu
Yun Huang
Sungjae Hwang
Frank Imeson
Pramod Jamkhedkar
Asim Jamshed
Karthick Jayaraman
Ik Rae Jeong
Xing Jin
Patil Kailas
Seny Kamara
Min Suk Kang
Junpei Kawamoto
Rasib Khan
Junzuo Lai
Markus Miettinen
Mainul Mizan
Petros Mol
Divya Muthukumaran
Khoa Nguyen
Takashi Nishide
Miyako Ohkubo
Olga Ohrimenko
Diego Perez-Botero
Joe Politz
Raluca Ada Popa
Wahbeh Qardaji
Vineet Rajani
Paul Ratazzi
Junghwan Rhee
Bodhisatta Roy
Pouyan Sepehrdad
Bilal Shebaro
Seonghan Shin
Emil Stefanov
Chunhua Su
Jinyuan Stella Sun
Kun Sun
Yuqiong Sun
David Swasey
Jakub Szefer
Nirupama Talele
Uday Tupakula
Serge Vaudenay
Hayawardh Vijayakumar
Christian Wachsmann
Chi-Wei Wang
Jun Wang
Liangliang Wang
Lusha Wang
Tianze Wang
Tielei Wang
Yifei Wang
Zhenyu Wu
Xiang Xie
Hong Xu
Hui Xue
Weijia Xue
Guomin Yang
Kenji Yasunaga
Xun Yi

Additional reviewers: (Continued)

Adeline Langlois
Meixing Le
Su Le
Bing-Han Li
Lingjun Li
Xiangxue Li
Xiaolei Li
Yan Li
Hoon Wei Lim
Tingting Lin
Giovanni Livraga
Swee Won Lo
Yu Long
Long Lu
Tongbo Luo
Xianping Mao
Shams Zawoad
Chao Zhang
Cong Zhang
Huilin Zhang
Jiang Zhang
Lei Zhang
Lei Zhang
Mingwu Zhang
Xiao Zhang
Yun Zhang
Zhi-Kai Zhang
Fangming Zhao
Xingwen Zhao
Xinxin Zhao
Lan Zhou
Xuhua Zhou

ASIACCS 2013 Sponsor & Supporters

Sponsor:

In-cooperation:

AppInk: Watermarking Android Apps for Repackaging Deterrence

Wu Zhou
North Carolina State University
Raleigh, NC
wzhou2@ncsu.edu

Xinwen Zhang
Huawei Research Center
Santa Clara, CA
xinwen.zhang@huawei.com

Xuxian Jiang
North Carolina State University
Raleigh, NC
xjiang4@ncsu.edu

ABSTRACT

With increased popularity and wide adoption of smartphones and mobile devices, recent years have seen a new burgeoning economy model centered around mobile apps. However, app repackaging, among many other threats, brings tremendous risk to the ecosystem, including app developers, app market operators, and end users. To mitigate such threat, we propose and develop a watermarking mechanism for Android apps. First, towards automatic watermark embedding and extraction, we introduce the novel concept of *manifest app*, which is a companion of a target Android app under protection. We then design and develop a tool named *AppInk*, which takes the source code of an app as input to automatically generate a new app with a transparently-embedded watermark and the associated manifest app. The manifest app can be later used to reliably recognize embedded watermark with zero user intervention. To demonstrate the effectiveness of AppInk in preventing app repackaging, we analyze its robustness in defending against distortive, subtractive, and additive attacks, and then evaluate its resistance against two open source repackaging tools. Our results show that AppInk is easy to use, effective in defending against current known repackaging threats on Android platform, and introduces small performance overhead.

Categories and Subject Descriptors K.6.5 [**Management of Computing and Information Systems**]: Security and protection – *Invasive software*

General Terms Security; Algorithms

Keywords: Mobile Application; App Repackaging; Software Watermarking; App Protection; Smartphone Security

1. INTRODUCTION

With the unprecedented adoption of smartphones in consumer and enterprise users, a large number (and a wide variety) of mobile applications (apps) have been developed and installed to extend the capability and horizon of mobile devices. These apps in return foster an emerging app-centric business model and drive innovations across personal, social, and enterprise fields. At the same time, the wild proliferation of mobile apps has introduced serious risks to the stakeholders in the ecosystem. Particularly, app repackaging has been considered as a major threat to both app developers and end users. Through app repackaging, malicious users can breach the revenue stream and intellectual property of original app authors, and plant malicious backdoors or payloads to infect unsuspecting mobile users. Recent studies have shown that app repackaging is a real threat to both official and third-party Android markets [52,53], and regarded as one of the most common mechanisms leveraged by Android malware to spread in the wild [54]. Investigation has also shown that app repackaging presents serious vulnerability to mobile banking [27].

Facing the prevalent risks brought by app repackaging, we are in a desperate need of a reliable, efficient, and easy-to-use mechanism to detect repackaged apps and prevent their propagation. Android package obfuscation tools such as Proguard [29] and DexGuard [25] have been provided by Google and other companies to confuse attackers when they are in the process of repackaging an app. However obfuscation can only increase the difficulty of reverse engineering Android apps, and cannot stop determined attackers from achieving their purposes through manual analysis, laborious experiments, and strong persistence. To defend against Android app piracy and repackaging, Google has introduced a tool library named license verification library to protect app developers from having their apps stolen by third parties [17]. Other app market operators (such as Amazon and Verizon) also provide their own digital right management (DRM) options for app developers which can be applied to prevent apps from being copied and pirated [41]. However, these mechanisms are hard to deploy correctly [13] and also easy to crack [30–32]. Recently, researchers have introduced various techniques to detect repackaged apps on a large scale [8,16, 42,52,53]. However, these mechanisms usually cannot detect app repackaging online and in real-time. For these reasons, repackaged apps have usually been widely distributed before being detected.

To overcome the above weaknesses, we propose to embed software watermarks dynamically into the running state of an Android app to represent the author or developer's ownership. For verification, an authorized party can extract the embedded watermark by running the app with a specific input in a dedicated environment, e.g., a customized emulator. When the extracted watermark matches the one provided by the developer, the verifying party (e.g., an arbitrator) can confirm the ownership of the original developer even when the app is repackaged by another publisher. The proposed watermarking mechanism should be resistant to manipulation by common static and automated attacks, therefore making it hard for an attacker to remove the original embedded watermark or embed his own watermark.

ASIA CCS'13, May 8–10, 2013, Hangzhou, China.

A desirable solution needs to meet two requirements: watermark embedding should be readily integrated into current app development practice, and watermark extraction should be convenient for the authorized verifying party to perform. To fulfill these two requirements, we introduce the concept of *manifest app*, which is a companion app to the original app under protection. Basically the manifest app encapsulates a specific input to drive the watermark-protected app automatically, and thus eliminates the user interventions needed in traditional watermark extraction. Based on the manifest app, we design and implement a practical tool named *AppInk*, which consists of four components: watermarking code generation, automatic manifest app generation, watermark embedding, and watermark extraction. By seamlessly integrating these four components, AppInk presents an effective app protection solution for both developers and other authorized verifying parties. Specifically, to leverage AppInk to protect her own app, a developer applies the first three components of AppInk to the app's source code to generate two apps: the watermark-protected app to be released to the public; and the manifest app which is presented on demand to an authorized party to verify the originality of the watermark-protected app. Upon request, the verifying party will run the manifest app in the watermark extractor (the fourth component of AppInk), which automatically launches the app under review and extracts the originally embedded watermark.

Two typical scenarios can leverage this process to detect unauthorized repackaged apps and prevent their propagation. The first scenario has a central authority (e.g., the app market operator) to review each submitted app to verify its originality before accepting it for publication. For that purpose, each app publisher submits the manifest app along with the app under review (publishers who cannot submit a companion manifest app are most likely not original developers). The app market operator then runs the watermark extracting algorithm, using the manifest app provided by the publisher to drive the app under review.

The second scenario has a third-party arbitrator, who inspects the evidence of app ownership to resolve dispute upon request. In this scenario, when an app author suspects that one app is a repackaged version of her own, she can run the suspect app inside the watermark recognizer, using her own manifest app to feed input. If the watermark extracted from the suspect app is the same as the watermark from her own app, she can submit this as evidence to prove that the suspect app is a repackaged version of her own app. Within this scenario, AppInk provides an effective mechanism to prevent the propagation of repackaged apps across different app markets.

To demonstrate the effectiveness of AppInk in deterring app repackaging, we analyze its robustness against general watermark-targeted attacks, including distortive, subtractive, and additive attacks. We also study its resistance against two open source repackaging tools (Proguard [29] and ADAM [51]). Our results show that AppInk is effective in defending against common automatic repackaging attacks. Our performance evaluation indicates that an embedded watermark introduces only a small overhead for end users.

In summary, this paper makes the following contributions::

- First, we design a dynamic graph based watermarking mechanism for Android apps, which can be used to detect and deter further propagation of repackaged apps. To the best of our knowledge, it is the first watermarking mechanism for Android apps.
- Second, we introduce the concept of manifest app and design a series of automatic processes to make the watermarking mechanism integratable into current app development practice and also conveniently deployable for arbitrators.
- Third, we implement a prototype tool named AppInk and evaluate it against two open source repackaging tools, and demonstrate that it is effective in defending against commonly available automatic attacks.

The rest of this paper is organized as follows. We present the paper overview in Section 2, describe the AppInk design in Section 3, and present its prototype implementation in Section 4. After that, we present the robustness analysis of AppInk and evaluate it against real world repackaging attacks in Section 5, and discuss the system's limitations and suggest possible improvements in Section 6. Lastly, we describe related work in Section 7 and conclude this paper in Section 8.

2. OVERVIEW

2.1 Problem Statement

App repackaging refers to disassembling one app, making some changes (to the code, data, or simply the signing key inside the original app), and rebuilding the modified components into a new app. As a technical method, it can be used for benign purposes. For example, ADAM [51] uses app repackaging to tweak malware samples for the purpose of stress testing various Android anti-virus tools. Aurasium [50] uses app repackaging to intercept an app's interaction with its underlying OS, with the goal of enforcing user-specified security policies for the app. However, app repackaging is more commonly used for surreptitious and malicious purposes. For example, greedy publishers use app repackaging to replace existing in-app advertisements or embed new ones to steal advertisement revenues [8,52,53]. Malicious attackers use app repackaging [42, 53,54] to plant malicious backdoors or payloads into benign apps.

Because it is relatively easy to reverse engineer Android apps (which are mainly written in Java), app repackaging has been identified as a widespread practice in current diversified app distribution channels [8,16,23,42,53]. As a result, it not only brings a lot of damages to app authors (in terms of losing their monetary income and intellectual property), but also causes tremendous risks to the large community of mobile users and affects the burgeoning innovative app economy. As concrete examples, severe vulnerabilities have been found in mobile banking apps through app repackaging, and serious doubt is cast on mobile banking security and feasibility in general [27]. More recently, researchers have identified that about 10% of Android apps available in popular third-party markets are repackaged [53]. The latest investigation from the industry [48] has reported that most of the popular mobile apps are beset by app repackaging threat: 92 of the top 100 paid apps for Apple iOS, and all of top 100 paid apps for Android were found to be hacked.

From another perspective, effective security defenses are seriously lagging behind. The current industrial practices are either too weak to deter determined attackers from conducting repackaging attacks, or too complex to be deployed properly. For example, app developers are encouraged to use obfuscation to protect their apps, but the introduced confusion is usually not strong enough to prevent determined attackers from achieving their goals [29]. App market operators (e.g., Google) have provided license verification or DRM service to apps submitted to their stores, but automatic repackaging tools can work around them easily [30–32]. Recently, researchers have begun to tackle this problem, but most of the proposed solutions so far focus on feasible mechanisms to detect repackaged apps after their propagation [8,16,42,53]. Considering the wide and severe impact of app repackaging, an effective and robust mechanism is urgently demanded to efficiently prevent and deter app repackaging in the first place.

2.2 Software Watermarking

Software watermarking has been studied extensively to defend against the piracy of desktop software [4, 6, 35, 37, 39]. Since the processes of mobile app repackaging and desktop software piracy are similar, we believe that software watermarking can be a promising technique in deterring app repackaging. In general, watermarking software involves two steps: first a watermark, typically a number or a message string known only to the author or publisher, is embedded into the target software in a specific way such that it does not affect the running behaviors of the original app and is difficult to remove without modifying the original app semantic; then, a recognition technique is used to extract the original watermark from the software. The matched watermarks verify that this software package belongs to the original developer or publisher. Depending on how the watermark is embedded and extracted, there are static and dynamic watermarking methods. Static watermarking embeds the watermark into the code or data of a package and extracts the watermark without executing the code; dynamic watermarking embeds the watermark into the execution state of the target software, and extracts the watermark during runtime.

Regardless of which method is used, it is desirable that a watermark embedded in a software package should be robust to various well-known attacking techniques, especially distortive attacks – to apply semantic preserving transformation on the watermarked code to modify the embedded watermark, subtractive attacks – to remove complete or partial watermark, and additive attacks – to add attacker's own watermark and confuse the arbitrator on resolving ownership dispute. Typically, dynamic watermarking has a stronger resistance against these attacks than static watermarking, and it is used in our approach to embed and verify the ownership of a specific Android app package.

2.3 Challenges of Watermarking Android Apps

There are several key challenges to incorporate dynamic watermarking into current Android app development practice and make it easily deployable by arbitrating parties.

Firstly, the state of the art dynamic Java watermarking techniques need extensive intervention from developers to embed a watermark. For example, SandMark [5] requires developer to manually annotate source code to indicate where watermark *can* be inserted, and to manually give input to drive the software when embedding a watermark. These manual interventions make it cumbersome to apply this technique in real practice and hard to be made right.

Secondly, it is desirable to have automatic watermark recognition so that it can handle thousands of apps with little human effort in an online and realtime manner. To recover embedded watermarks, SandMark leverages programmable Java Debug Interface [24] to access memory objects on the heap in order to infer object reference relationships. However, there is no known programmable debugging interface available on Android. Even worse, manually providing input is required for watermark extraction phase as well. Obviously, this cannot scale to handle the large number of watermark recognition requests for thousands of Android apps submitted to current app markets on a daily basis.

Thirdly, Android apps, although mainly developed in Java language, have significant difference from desktop Java software. For example, Android apps depend more heavily on event-driven mechanisms and the underlying execution environment to work correctly. Unlike legacy Java applications that have a single entry point named `main`, Android apps may have multiple entry points.

2.4 Solution Overview

To overcome these challenges, we introduce an entity named *manifest app*, which is a companion app for a target app under protection. Basically, the manifest app encapsulates a sequence of input events to drive the watermark-protected app automatically, and thus eliminates the user intervention needed in traditional watermark extraction. Based on the manifest app, we design and implement a practical tool named AppInk to automatically generate the manifest app, embed the watermark, and execute the dynamic watermark extraction with zero user intervention. As an input during watermark embedding, the manifest app encodes the event sequence and accordingly indicates the event handlers of the target app where watermarking code segments can be inserted, which addresses the first challenge. As an input for watermark extraction, the manifest app automatically launches the original app and feeds the input event sequence to it, which triggers all the inserted watermarking code segments and thus recovers the watermark object embedded. By doing so, we effectively overcome the second challenge. Based on the insight that each event in Android platform is uniquely mapped to a well-known system API, we propose a conservative method to automatically generate an event flow model for Android app, and leverage model-based test generation to automatically create a suboptimal manifest app for watermarking purpose, thus resolving the third challenge.

3. APPINK DESIGN

Figure 1 depicts the overall AppInk architecture. At the app developer (left) side, AppInk consists of three components: manifest app generation, watermark code generation, and watermark embedding based on source code instrumentation. The input of the manifest app generation component is the source code of the target app including its resource files. The watermark code generation takes a watermark object (e.g., a number or a string) and outputs watermarking code segments. The watermark embedding component takes the manifest app and the code segments as inputs, and generates a watermarked Android package that can be released to app markets. At the arbitrator (right) side, the watermark recognizer takes the inputs of the released Android package and the manifest app from the app developer side, and extracts the embedded watermark.

Watermarking code generation: Given a watermark value specified by app developer, this component encodes the watermark value into a special graph structure and transforms the graph into watermarking code. In order to improve the stealthiness of the embedded watermark, AppInk splits the watermarking code into a variety of segments, each of which will be inserted into different locations of the original app's source code. The execution states of these code segments collaboratively present specific object reference relationships and thus can be leveraged to reveal the original watermark value. Section 3.1 explains the detailed design of this component.

Manifest app generation: The main function of the manifest app is to feed pre-determined user inputs to the app under review, which trigger the executions of embedded code segments and thus recover the watermark value with the help of the watermark recognizer. To ease the burden of writing manifest apps by developers, AppInk leverages the event-driven nature of Android apps and the latest

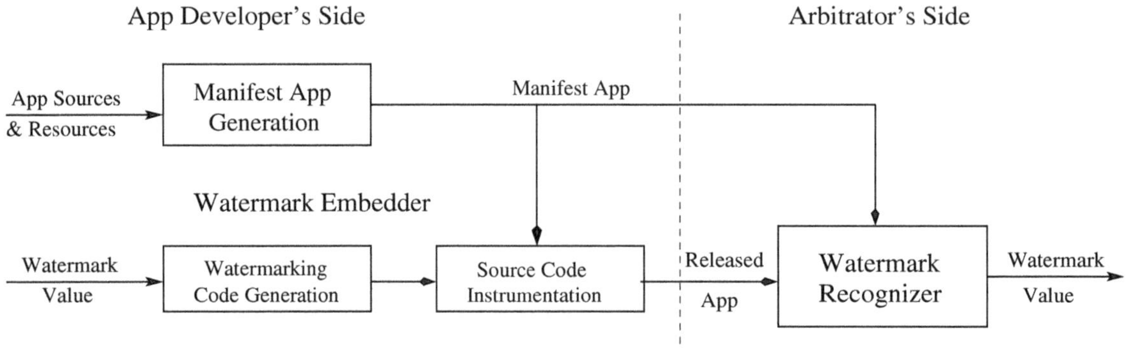

Figure 1: The overall AppInk architecture.

model-based test case generation to automatically generate these input events and makes this process totally transparent to app developers. Section 3.2 presents the detailed design of this component.

Source code instrumentation: By parsing the files in the manifest app, this component first identifies encoded user input events, and then determines their corresponding event handlers based on the source code of the original app. Next, it inserts the watermark code segments under the path of these identified event handlers. After that, AppInk packages the modified app source into a released app, which is for both public publication and arbitrating purposes, and the manifest app into another executable package, which is not released to the public but rather dedicated to arbitrator use. Section 3.3 presents the detailed design of this process.

Watermark recognizer: This component is a modified Android emulator on x86 [19], which is invoked by a shell script. The script first installs both the Android app and the manifest app in the emulator, then starts the manifest app which feeds a sequence of input events to the Android app, and then calls the extended Dalvik Virtual Machine (DVM) [10] to export all object reference information in the runtime heap. From this information, AppInk uses a special pattern to match potential watermarking structure. If such a watermarking structure is identified, a reversed process of the watermarking code generation is invoked to recover the embedded watermark value. Section 3.4 illustrates the design of this component.

3.1 Watermarking Code Generation

Different from static watermarking which embeds a secret watermark object (e.g., a numerical value or a message string) into the code or data section of a target application, dynamic watermarking embeds a watermark object into special structures that present themselves only in the runtime of the target application. AppInk adopts graph-based data structure, which is hard to be reversed by attackers due to the inherent difficulty of analyzing point-to relationship in graphs [14,45].

There are different ways to use graph to encode a watermark object. AppInk uses permutation graph, which adopts a special graph structure to encode a permutation mapping to the watermark object. As depicted in Figure 2, the graph includes 5 nodes, each of which has two outgoing edges, one in solid line and one in dotted line. Through the solid line edges, the graph forms a cycle. If we can further identify one unique node (e.g., the one which is referenced by any other object outside the figure), a specific order is defined. Suppose that only the first node is referenced by another object not in the figure, we can assign number 0 to this node, which is called root node, and number 1 to 4 for the other four along the circle. A dotted line edge of a node is then associated with a number counting the distance from this node to its target node

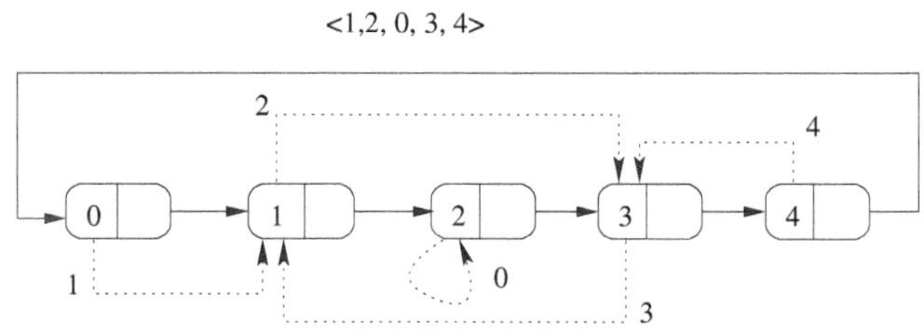

Figure 2: Example permutation graph. This graph encodes numerical value of 116 as a watermark.

```
1  class WatermarkNode {
2    WatermarkNode solid;
3    WatermarkNode dotted;
4    ......
5  }
6  ......
7  node0 = new WatermarkNode();
8  ......
9  node4 = new WatermarkNode();
10
11 node0.solid  = node1;    // Points to next node
12 node0.dotted = node1;    // Encode number 1
13 ......
14 node4.solid  = node0;
15 node4.dotted = node3;    // Encode number 4
```

Figure 3: Watermarking code for the permutation graph in Figure 2.

along the solid line edges. For example, the dotted outgoing edge from `node0` to `node1` encodes a number of 1 since the distance from `node0` to `node1` is 1 along the solid line edges. Similarly the dotted outgoing edge from `node3` to `node1` encodes a number of 3. In Figure 2, the 5 dotted edges encodes the numbers of 1, 2, 0, 3, and 4 respectively, which is a perfect form of permutation. According to the permutation-to-number algorithm in [28], `<1, 2, 0, 3, 4>` is mapped to `116`, which is the watermark value encoded by this graph.

In Java language, the permutation graph depicted in Figure 2 can be represented by a doubly linked list, as shown by the skeleton code in Figure 3. The class `WatermakrNode` (lines 1 to 5) represents the node in the permutation graph. The following initialization code (lines 7 to 9) creates five instance nodes, and the later code (lines 11 to 15) defines the object reference relationship, from which we can reconstruct the permutation graph at runtime. As commented in the list, the member field `solid` points to the next node in the list, and all these fields form a cycle. The member field `dotted` encodes the permutation distance for each node in the permutation graph, which jointly encodes the watermark value specified by the app developer.

```
1  public class TestAndroidCalculator
2    extends ActivityInstrumentationTestCase2<Main> {
3      protected void setUp() { ... }
4      protected void tearDown() { ... }
5      public void testEventSequence() {
6        enterText(0, "10");
7        enterText(1, "20");
8        clickOnButton("Multiply");
9      }
10 }
```

Figure 4: Example manifest app based on Robotium.

When the above code is executed on Android platform, memory space will be allocated for each instance object (`node0` to `node4`) at lines 7 to 9. At lines 11 to 15, the member fields are assigned, which results in the establishment of the object reference relationship among these `WatermarkNode` instances. Through analyzing the runtime heap, this object reference relationship can be extracted and decoded to recover the original watermark value. Section 3.4 presents the details of this process in the watermark recognizer.

Because linked structures are very commonly used in Java applications, it is hard to distinguish these watermarking code from other code. The *stealthy* nature of these graph structures, combined with the inherent difficulty of point-to analysis in graph [14, 45], makes it very challenging for attackers to succeed in reverse engineering the watermarked code. (More detailed analysis on the robustness of this technique is presented in Section 5.1.) To further improve the stealthiness of the watermarking code, AppInk splits the watermarking code into a number of segments and inserts them into a variety of places in an app. This is especially helpful when the watermark value is large and thus has to be represented by a large number of code segments. Section 3.3 presents more details of this technique with the help of manifest app.

3.2 Manifest App Generation

3.2.1 Manifest App Based on Robotium

Working as a companion app to drive the execution of a released app inside watermark recognizer, a manifest app functions in similar way as test cases. However, unlike common unit tests which only provide component specific tests [21] and special Android UI/application exerciser [22] which sends random stream of events to apps under test, a manifest app needs programmable event delivery within the entire target app, so that watermarking code can be scattered to different places, thus achieving better stealthiness. For this purpose, AppInk generates manifest app based on Robotium [46], which extends Android app instrumentation framework and provides precise UI element locations and event delivery. Figure 4 shows an example of the Robotium test case.

In Figure 4, the method `setUp` starts the main activity of the app under test, `tearDown` clears the execution environment and stops its execution, and `testEventSequence` (lines 5 to 9) sends specific sequence of events to the app. To automatically generate manifest app and later extract watermark by the recognizer, AppInk needs to decide a proper input sequence and fill them into the method `testEventSequence`. Specifically, AppInk has two requirements for the input sequence: its execution must deterministically trigger the watermarking code segments, and it must be diversified enough so that the watermarking code can be scattered into a large enough scope. AppInk leverages the event-driven nature of Android apps and model-based automatic test case generation to achieve these purposes.

3.2.2 Manifest App Generation

Different from desktop applications, the control flow of Android apps depends heavily on the diversified Android events, including user generated events (e.g., key presses and screen taps) and system generated events (e.g., short messages received, incoming phone calls, and various sensor events). Each event is handled by a well-defined Android API. For example, a menu item click is handled by the method `onOptionsItemSelected` in the corresponding activity, a button press is handled by the `onClick` method of the listener object registered for the button, and a short message received event is handled by the method `onReceive` of the activity registered for `SMS_RECEIVED` intent. By issuing these events in a well-defined order, the app under test invokes these event handlers in order, and responds in a deterministic manner.

For the purpose of automatic test generation, model based methods have been well studied for event-driven software in general, and actively investigated for Android apps in particular [26, 34, 38, 47]. But to use it in AppInk, we need to have an app's event flow model as input. One option is to ask developers to provide the model. But this puts an extra burden on them, and it is also prone to error. Another option is to infer the event flow model through reverse engineering. We note that generating a complete app model is hard, but unlike common test case generation (whose task is to exhaustively generate test sets to cover as many paths as possible), AppInk only needs one test case if it can trigger as large set of code segments as possible to achieve stealthy watermark embedding. Therefore we only need to have a partial model for the app event flow.

AppInk uses static method to infer a partial event flow model for Android apps in a conservative but safe way. This is achieved through parsing app source files, including `AndroidManifest.xml`, UI layout, Java and other resource files. The generated model is fed to an existing model-based test generator for Android [11] to generate a set of test cases, from which we pick the test input that covers most code segments.

We now use an example app to describe how AppInk generates the event flow model. Figure 5 shows the user interface elements and relevant events for the app `NotePad`. The first screen (5a) pops right after the app starts. By analyzing the layout and Java source files, we infer that the event `Add note` is handled by method `onOptionsItemSelected` in file `NotesList.java`. The second screen (5b) shows the UI elements for the action of adding a note, including a text input box and two menu items (`Save` and `Discard`), whose event handlers are the method `onOptionsItemSelected` in file `NoteEditor.java`. The third screen (5c) shows the UI elements for editing a note, including a test input and three menu items (`Save`, `Delete`, and `Edit Title`), each of which has its own handler. The last screen (5d) shows the UI elements for editing title, including a test input and one button (`Ok`), whose handler is the `onClick` method in file `TitleEditor.java`.

At this point, each screen shows only individual events. Through analyzing the handler for the event of clicking `Add note` (method `onOptionsItemSelected` in file `NotesList.java`), AppInk determines that it starts an activity with the intent of `ACTION_INSERT`, which is found later to be defined in file `NoteEditor.java` by parsing file `AndroidManifest.xml`. So the event `Add Note` connects screens 5a and 5b. Through a similar analysis, we determine that the event of clicking list item connects screens 5a and 5c, and the event of clicking `Edit`

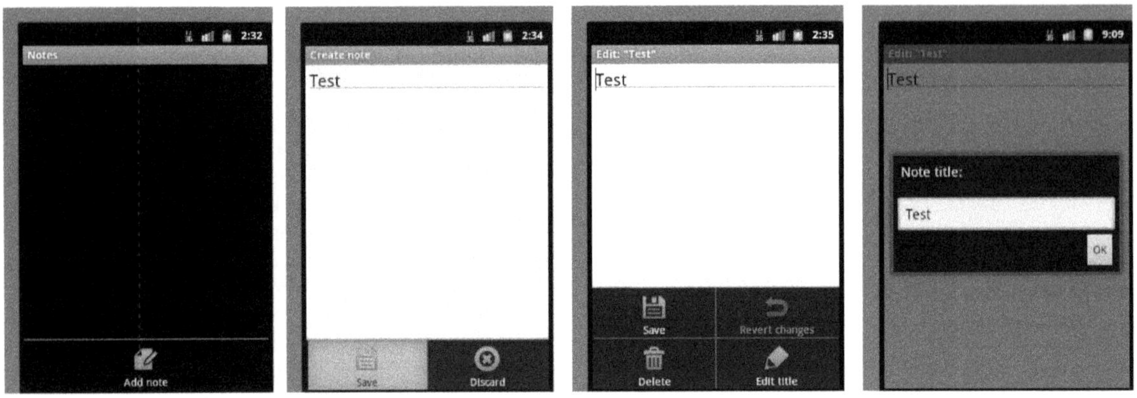

(a) Main Activity - NotesList (b) Note Editor - Create Note (c) Note Editor - Edit Note (d) Note Editor - Edit Title

Figure 5: User interface elements in app NotePad.

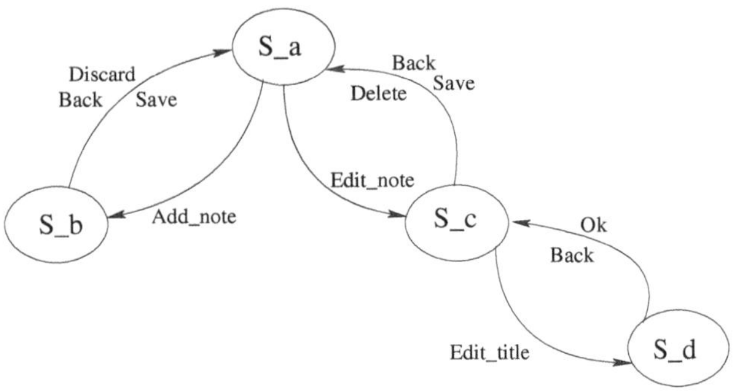

Figure 6: Event flow graph for NotePad.

```
public void testEventSequence() {
  clickOnMenuItem("Add note");
  enterText(0, "Test");
  clickOnMenuItem("Save");
  goBackToActivity("NotesList");
  clickInList(1);
  clickOnMenuItem("Edit Title");
  clickOnButton("Ok");
  clickInList(1);
  clickOnMenuItem("Delete");
}
```

Figure 7: Skeleton code to drive NotePad.

title connects screens 5c and 5d. Furthermore, there is a back button below the display screen of the phone, which connects the current screen and the one before it. Having these connecting events, AppInk generates the event flow model as depicted in Figure 6.

After feeding the above event flow graph into M[agi]C [11] — a test input generator tool, we obtain a test case with the skeleton shown in Figure 7. This skeleton encodes the event sequence of Add note, Enter text, Save, Back, Click list item 1, Edit Title, Ok, Click list item 1, and Delete. This sequence covers all the activity classes in the app, thus presenting an optimal test for watermarking purpose.

3.3 Source Code Instrumentation

Having the manifest app source, together with the original app source and the generated watermarking code, AppInk uses source code instrumentation to perform the watermark embedding. The choice of source code instrumentation is reasonable since AppInk is used by app developers, who already have the app source code at hand. This also helps integrate AppInk with the well-established app development environment for Android. We note that AppInk can be supported by bytecode level instrumentation as well.

Source code instrumentation uses three steps to embed developer-provided watermarks. First, AppInk fetches all control events (including clicking button, menu and list items) from the manifest app, each of which is mapped to a single event handler in the original app (such as onOptionsItemSelected and onClick). Next, AppInk splits the watermarking code into the same number of code segments as the number of the event handlers, and generates a configuration file to record the one-to-one mapping from the watermarking code segments to the event handlers. Last, AppInk parses the source code of the original app, generates its abstract syntax tree, identifies nodes for the event handlers, and inserts the watermarking code segments into their corresponding event handlers.

After the above instrumentation, AppInk automatically builds and generates an executable app package and signs it [20], which can be used for public release. The manifest app is built into another executable package, which is not released to public. Instead it will be submitted upon request to the arbitrator for the verification purpose. All these steps are serialized with an automated script, which is seamlessly integrated into the app-building process.

3.4 Watermark Recognizer

The watermark recognizer takes both apps as input: a released app for reviewing and the associated manifest app as the driver. The core part of the recognizer (Figure 8) is an extended Dalvik virtual machine (DVM), which is the execution engine for Android app code and maintains the runtime heap. Unlike traditional watermarking tool for Java that uses the Java debug interface to access object reference information and reconstruct the watermarking object, AppInk leverages the customized DVM to fetch object reference information from the runtime heap directly. Further, with the help of the manifest app generated in Section 3.2, the watermark recognizer enables automatic watermark extraction without any user intervention, which is highly desirable for scalable handling of a large number of apps submitted to app markets.

Just like Java virtual machine executing Java code by interpreting bytecode, DVM executes Dalvik bytecode, which is the main body of Android apps. Therefore it has access to all the needed information for watermark extraction purpose. Particularly, DVM manages memory space for Android apps, and maintains relevant information for memory reclaim (garbage collection). All object

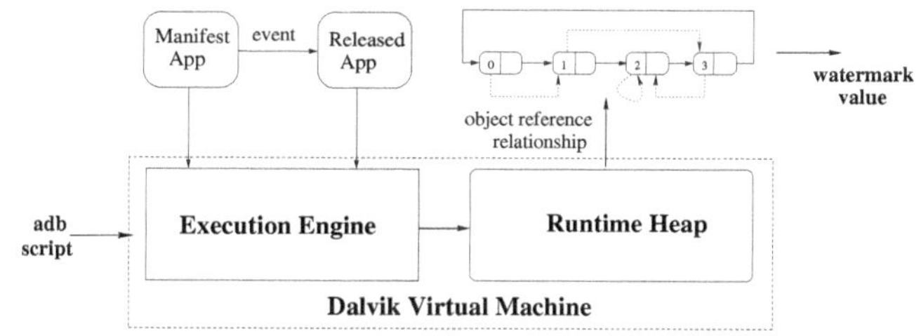

Figure 8: Work flow of watermark recognizer.

```
1 adb install -r releasedApp.apk
2 adb install -r appTest.apk
3 adb shell am instrument -w InstrumentTestRunner
4
5 pid=`adb shell ps|grep appName| awk '{print \$2}'`
6 ### Send USR2 signal to trigger GC
7 adb shell kill -10 $pid
8 adb logcat -d | grep $pid > $pid.log
9 java appink.wmGraphRecognizer $pid.log
```

Figure 9: Shell script to drive watermark extraction.

reference information is maintained, so AppInk only needs to extend the garbage collector to record and export this information, among which a later module will search for the watermarking graph. The identified watermarking graph is then decoded to recover its corresponding watermark object and to verify whether it is the same as what the author claims.

More specifically, DVM uses mark-and-sweep algorithm [49] to execute the task of garbage collection, which scans all allocated objects and their member fields (a reference relationship forms between an object and its member fields) and determines if any object is not needed any more and thus can be reclaimed. We develop a module in DVM to record these object reference relationships in the scanning phase and export them into a log file. From these reference relationships, a reverse process of that in the watermarking code generation (cf. Section 3.1) is applied to recover the watermark value.

Towards automatic operations, we create a shell script based on Android debug bridge [18] to link all these steps, as shown in Figure 9. The script first installs both apps – the released app and the manifest app (lines 1 and 2), and then starts the manifest app through the instrumentation command of Android activity manager (line 3). This will feed the event sequence to the released app in a specified order. The command at line 5 gets the process identifier of the running DVM, and line 7 sends a `SIGUSR2` signal to trigger the object reference recording module inside the extended DVM. The commands at lines 8 and 9 fetch these recorded messages, search for reference relationship pattern among them, and try to extract the embedded watermark.

4. IMPLEMENTATION

We have implemented an AppInk prototype on Ubuntu 10.04. The watermarking code generation component is implemented in Java, which accepts an integer or a string as input and outputs its corresponding watermarking code. In the manifest app generation component, the parsing of Java source files is based on ANTLR [40] – a language parser generator. More concretely, we input a Java language grammar [15] into ANTLR, which generates a Java AST (abstract syntax tree) parser. By iterating through the AST, AppInk can locate the nodes for all event handlers, and identify the connecting events for different UI states. The parsing of `AndroidManifest.xml`, UI layouts, and resource files is written in Python. Another Python script glues the output results from these parsing modules, generates the event flow graph, feeds it to the test case generator named M[agi]C, and picks up the test case which has the largest coverage of watermarking code.

The source code instrumentation component includes three steps. It first parses the manifest app source to identify all these event handlers to insert the watermarking code, and then splits the watermarking code into segments with the same number as that of events to be delivered. Last it extends the Java source parser generated in the manifest app generation component to insert the watermarking code segments into the execution path for their corresponding event handlers. To automate the watermark extraction on the arbitrating side, this component also generates the shell script to drive watermark extraction (as presented in Figure 9). Basically, this script only needs relevant information for a released app and its manifest app, which is readily available after the completion of the first three components.

Having this watermark extraction script at hand, the watermark recognizer needs two modules to achieve the final watermark recognition task. The first module implements the extended DVM to record and export object reference information when the app receives a `SIGUSR2` signal. This is achieved by modifying the garbage collector code (in C language) in DVM and rebuilding the Android open source project. The second module is written in Java, which searches through these object reference relationships to match any potential watermarking graph, and decodes the graph to recover the corresponding watermark value.

5. ANALYSIS AND EVALUATION

While AppInk aims to embed strong ownership verification mechanism into Android apps, attackers always strive to defeat the protections in any way they can think of. In this section, we first analyze the robustness of AppInk against three common attacks toward watermarking, namely distortive, subtractive, and additive attacks. We then evaluate it against two open source repackaging tools to demonstrate its effectiveness. Finally we evaluate the runtime performance overhead for watermarked apps.

5.1 Robustness Analysis

AppInk adopts dynamic graph based watermarking as its key technique to defend against app repackaging. Therefore its robustness depends heavily on that of dynamic graph watermarking, which is highly resistant against distortive attacks, subtractive attacks, and additive attacks according to our analysis.

Distortive attacks: This type of attacks applies semantic-preserving transformations on target apps, trying to make it hard or impossible to extract original watermarks from the modified apps. Many static watermarking mechanisms are highly susceptible to distortive attacks since they leverage the code or data syntax to encode the watermark, which is very sensitive to semantic-preserving transformations. Dynamic watermarking, however, never depends on any syntax structure in application code, but instead encodes watermark object into the execution state of the application. Furthermore, the semantic of runtime graph data structures is usually hard to analyze without executing it in real environment, because of the inherent difficulty in analyzing point-to relationships [14,45]. These factors make it very hard for any static transformation to change these graph structures without changing the application semantics. For these reasons, most semantic-preserving transformations cannot affect the execution states of apps, and theoretically dynamic graph watermarking is resistant to distortive attacks. To further confirm AppInk's robustness in this aspect, we evaluate AppInk against a series of semantic-preserving transformations available in two open source tools [1], and report our results later in this section.

Subtractive attacks: This type of attacks tries to remove watermarking relevant code segments in an application, and usually needs manual analysis to identify the location of these code segments in the first place. The dynamic graph based watermarking mechanism adopted in AppInk makes it relatively easier to defend against subtractive attacks. First, since the arbitrators are in general trustworthy, we can assume the manifest app is kept as secret. So although the watermarking mechanism used in AppInk can become public knowledge, the secrecy of manifest app provides one layer of protection for the watermarking code segments. Second, the data structures used in AppInk are commonly used in normal Java applications, which makes it hard to separate these watermarking code segments from other functional code.

What is more, we can also leverage the inherent difficulty of alias analysis [14,45] to add another layer of protection against subtractive attacks. Since the runtime graph data structures in AppInk have reference relationships among themselves, an app developer can easily know the correct reference relationships among the inserted graphic nodes. Instead, without this pre-knowledge, attackers have great difficulty in identifying these reference relationships through reverse engineering. Therefore we can create bogus dependency relationships between the original code and the newly inserted code [7]. The attempt to remove or modify the watermarking graph code segments will have high probability to damage the original application logic, making it useless after repackaging.

Additive attacks: This type of attacks tries to add another watermark on a watermarked app, with the assumption that attackers somehow understand and implement the same watermark embedding algorithm as presented in AppInk. In the defense model applied in centralized app markets (cf. Section 1), there is an inherent timing gap between the submission of the original app and the repackaged app. That is, the original author always submits her AppInk-protected app before an attacker succeeds in executing an additive attack on that app. Therefore when the attacker submits his repackaged app to the app market, the app market can detect that there is an earlier app which has the same functionality but has a different watermark extracted. The operator can then launch another watermark extracting session on the app under review, using the manifest app for the earlier app as the watermark extraction driver. If the same watermark is extracted as the earlier published app, it is derived that the second is a repackaged one. However, this does not prevent the attacker from downloading an app from one store and publishing in another one where the original app has not been published yet.

In the postmortem arbitrary model, when an app author suspects that one app is a repackaged version of her own, she can apply the watermark recognizer with her own manifest app on the suspected app. If the extracted watermark is identical with the one from her own app, she can submit this as evidence to prove that the suspected app is a repackaged version of her own one. In case that an attacker can somehow embed his own watermark into the AppInk-protected app and generate his own manifest app, two watermarks can be extracted by using their corresponding manifest apps. Under this confusion, the original author can present another evidence to show that her watermark is original, and the other is additional. The evidence is that her manifest app can extract the same watermark from both her original app and the repackaged app, but the attacker's manifest app can only extract his own watermark from the repackaged app.

Next we evaluate the AppInk's robustness against real repackaging tools. With our best effort we do not find any available tool for subtractive and additive attacks. Therefore our evaluation is for distortive attacks only.

5.2 Evaluation with Repackaging Tools

We evaluate AppInk against two open-source tools, which apply semantic-preserving transformations on Android app code and therefore can simulate the aforementioned distortive attacks. We have five Android apps under evaluation, one named `AndroidCalculator` from Robotium, and the other four from Android SDK samples (including `ContactManager`, `NotePad`, `HoneycombGallery`, and `SearchableDictionary`). We first apply AppInk to embed watermarks on these apps, and then apply the available transformations present in the above tools to the watermarked apps. Last we feed the modified apps to the AppInk watermark recognizer to see if the originally embedded watermarks can be extracted.

ADAM: We first evaluate AppInk against an automatic Android app repackaging tool named ADAM [51], which operates on Android apps (`.apk` files) directly and automatically repackages apps with different code transformation techniques. Figure 10 shows the above three-step evaluation process for the app named `NotePad`. The first snapshot (Figure 10a) shows the embedding session on `NotePad.apk`, with a string of `1234567890abcdef` as the watermark value. It clearly demonstrates the working process of the three AppInk components.

The second snapshot (Figure 10b) shows the repackaging session, where we apply seven semantic-preserving transformations from ADAM on `NotePad.apk`. Among these transformations, three do not modify the app code but change other phases in the app packaging process. For example, *resign* transformation disassembles the app and re-signs the app using attacker's signing key, *rebuild* transformation disassembles the app and re-assembles the components into a new one with the open-source tool named apktool [44], and *zipalign* transformation realigns the locations of different data in the app package in a different and pre-determined way. Four other techniques apply various code obfuscation transformations on the app code, including defunct code insertion, identifier renaming (include packages, classes, methods, and fields), control flow obfuscation, and string encryption. To conduct the evaluation, we apply these seven transformations on the watermarked code with a bash script, and output the repackaged apps into a directory.

The third snapshot (Figure 10c) shows the recognizing session. We create a script to feed each of these seven repackaged apps into the watermark recognizer, and check the extracted watermark value. The first attempt shows that AppInk recognizes the watermark value of `1234567890abcdef` correctly from six of these repackaged apps. The app repackaged with *identifier renaming* obfuscation fails the first attempt. Further analysis shows that ADAM incorrectly renames one Android API method, which results in the incorrect execution of the repackaged app. After fixing this bug in ADAM, AppInk recognizes the correct watermark from all these repackaged apps (Figure 10c shows the result after fixing the ADAM bug).

Proguard: Our second evaluation is against a popular Android app obfuscation tool named Proguard [29]. Different from ADAM which works directly on final `.apk` files, Proguard operates on class files generated in Android app building process. To conduct

[1]They can be used for app repackaging as well.

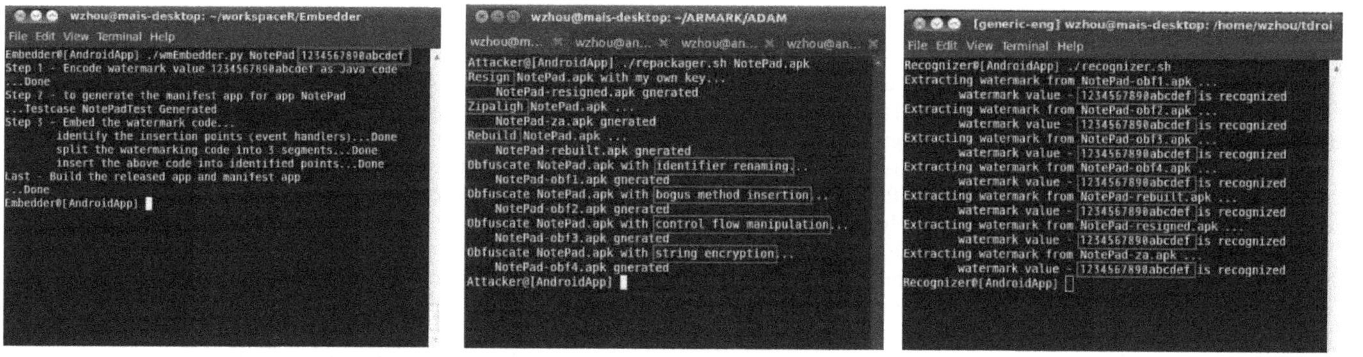

(a) Snapshot of AppInk watermark embedding (b) Snapshot of ADAM repackaging (c) Snapshot of AppInk watermark recognizing

Figure 10: Snapshots for watermark embedding, app repackaging, and watermark recognizing.

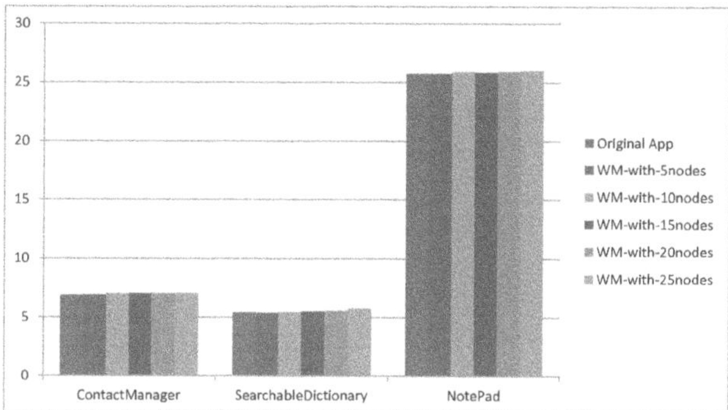

Figure 11: Execution time of watermarked app.

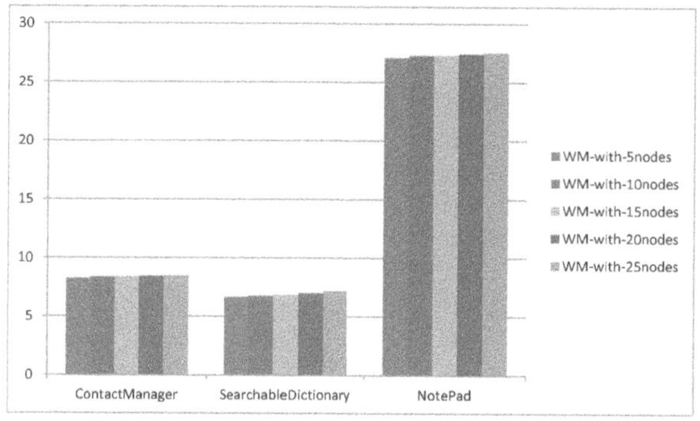

Figure 12: Extraction time of watermarked app.

this evaluation, we modify the watermarking embedding process as presented in Section 3, by adding Proguard obfuscation as a post-compilation action into the app building process [2]. With this extra action, the generated class files are optimized and obfuscated first, and then packaged into the final released apps. Last we feed these obfuscated apps into the AppInk recognizer. Our experiments show that AppInk recognizer can extract the correct watermarks embedded into all of these transformed apps successfully.

These two sets of evaluations demonstrate that AppInk has high resistance against currently available repackaging and transformation tools, and thus is very robust against distortive attacks.

5.3 Performance Evaluation

We conduct performance evaluation in two aspects with three different Android apps (`ContactManager`, `SearchableDictionary`, and `NotePad`). As the size of the permutation graph is the main factor to decide the extra code size and thus the final performance, we watermark each of these three apps with five different watermark values, which encode permutation graphs with sizes of 5, 10, 15, 20, and 25, respectively. These values can encode a number from `24` to $1.6 * 10^{25}$. Our experiments show that even the longest watermark value only introduces trivial performance overhead.

First, we measure how much extra time is required to execute these watermarked apps, which affects the user experience of an end mobile user when running these AppInk protected apps on their devices. To reduce the undecidability of human input and also exercise all these watermarking code, we use the manifest apps to drive these apps in a normal Android emulator. Figure 11 shows the times in seconds to finish each watermarked app. Please note that for each app, the first column shows the time to execute the original (un-watermarked) app. The small differences between the five watermarked apps and the original apps show that AppInk causes very small runtime overhead (2.4% at most in our evaluation).

Second, we measure how much time is required to recognize a watermark. This is the time that an arbitrating party needs to verify an app's originality. For that purpose, we feed these watermarked apps into the extended Android emulator as presented in Section 3.4, and measure how much time elapses when AppInk recognizes the watermark values. Figure 12 shows the measurement results in seconds. As shown, it takes 7 seconds to 28 seconds to verify these apps. Additionally, a longer watermark value in general requires more time to be recognized, but the difference is small. Compared with the data in Figure 11, we find that most time for watermark recognition is spent on the app execution itself. The time differences for the same watermarked apps in these two figures show the execution time dedicated to the watermark extraction, is from 1.3 seconds to 1.6 seconds. With this scale of time requirement, AppInk's watermark recognition can be deployed at current largest app market to handle thousands of app submissions every day.

6. DISCUSSION

Our prototype implementation and evaluation have demonstrated the effectiveness of AppInk for preventing the propagation of repackaged Android apps and deployable capability for general Android app development practice. In this section, we examine possible limitations in the current prototype and discuss future improvements.

[2]Concretely, we add a new Proguard configuration file and Proguard action into one ant [20] building script.

First, AppInk uses a conservative model-based test generation algorithm to generate manifest apps, which may not be the optimum for watermarking purpose. One possible enhancement is to investigate the latest automatic test case generation methods that have been studied by researchers in software engineering field. For example, the concolic execution based and GUI ripping based techniques [1, 2, 33] are actively investigated in software engineering community to enable automatic generation of high-coverage test inputs for Android apps. We plan to study these methods to see if they can be leveraged by AppInk for watermarking purpose.

Second, our AppInk prototype supports user input events only, which are the primary driver for app functionality, but ignores possible discrete system events, such as short messages received, incoming phone calls, and various sensor events. We plan to study the working mechanisms of all these events and explore ways to incorporate them into AppInk.

7. RELATED WORK

Software watermarking: Static watermarking embeds watermarks into the code or data of applications [35, 37], which usually involves syntax transformation and is vulnerable to semantic-preserving transformations. A variety of dynamic watermarking mechanisms have been proposed to overcome these attacks, including graph based [6, 39], thread based [36], and path based watermarking [4]. AppInk does not claim any novel contribution in this aspect. We instead leverage existing dynamic graph based watermark to improve Android app's capability in preventing and defending against common app repackaging attacks.

Java software protection: Android apps are mainly written in Java. Due to its high-level expressiveness, Java code is relatively easier to be decompiled and reversed [43] than native code. To protect software written in Java, various solutions are pursued since its inception. One popular solution is to apply different levels of obfuscation to Java code, such as code or data layout obfuscation, control flow obfuscation, and string encryption [3, 9, 12]. Watermarking is also used to prove the ownership of Java code and to discourage Java software piracy [6, 35]. SandMark [5] is one popular research platform to study how well different obfuscation and watermarking mechanisms work in protecting Java software.

Android app protection: To protect Android apps from piracy and foster the healthy development of Android app economy, Google has introduced several mechanisms. For example, Google recommends developers to leverage ProGuard [29] to optimize and obfuscate apps. Google also provides licensing verification library [17] to query a server to verify if an app running on a mobile device has been properly downloaded from Google's app market. There are also attempts from other parties in this aspect. For example, Amazon and Verizon have introduced their own digital right management solutions for Android apps available in their app markets [41]. These mechanisms can increase the difficulty of reverse engineering Android apps, but are not strong enough to deter determined attackers from repackaging through more laborious manual analysis. For example, open source tools are available to automatically crack these protections [30–32].

More recently, there are a series of work studying the app repackaging problem in Android platform [8, 16, 42, 52, 53]. Different from the dynamic watermarking mechanism proposed by AppInk, all these systems attack the app repackaging problem from the point of view of measuring the apps similarity. DroidMOSS [53] uses fuzzy hashing to speed up the pair-wise similarity comparison at the opcode level. Potharaju et al. [42] compares each app pair using different syntactic fingerprinting schemes, and can handle different levels of obfuscation used by the attacker. Juxtapp [16] collects static code features and represents them as bit vectors to improve the efficiency of pairwise comparison. It also supports incremental update and distributed analysis. DNADroid [8] uses program dependency graph (PDG) to characterize Android app and compares PDGs between methods in app pairs, showing resistance to several evasion techniques. By proposing a new distance metric design and an associated nearest neighbor search algorithm, PiggyApp [52] overcomes the scalability limitation from the pairwise comparison as presented in previous systems, and achieves a better scalability with $O(n * logn)$ complexity. None of these similarity based methods need modification to the released apps, but their results can only indicate possible repackaging, the final decision of which is left for human review. AppInk, although requiring some modification from the app developer side, does not have this drawback. The successful extraction of the author's watermark from the suspected app clearly derives the fact that it is repackaged. What is more, AppInk automates the work at both the app developer and arbitrator sides, enables the online and realtime detection of repackaged apps, and thus provides stronger deterrence towards app repackaging threat.

8. CONCLUSION

App repackaging is a serious threat to Android ecosystem including app developers, app store operators, and end users. To prevent the propagation of unauthorized repackaged apps, we propose to adopt a dynamic graph based watermarking mechanism and discusses two scenarios where this mechanism is mostly useful. To make the watermarking mechanism readily integratable into current app development practice and conveniently deployable by relevant parties, we introduce the concept of manifest app, which is a companion app for an Android app under protection. We then design and implement a tool named AppInk to generate manifest apps, embed watermarks to apps, and extract watermarks without any user intervention. Our robustness analysis and practical evaluation against currently available open source tools demonstrate that AppInk is effective in defending against common automatic repackaging threats while introducing trivial performance overhead.

Acknowledgements The authors would like to thank the anonymous reviewers for their insightful comments that helped improve the presentation of this paper. We also want to thank Yajin Zhou, Michael Grace, Chiachih Wu, Minh Q. Tran, Lei Wu and Kunal Patel for the helpful discussion. This work was supported in part by the US National Science Foundation (NSF) under Grants 0855297, 0855036, 0910767, and 0952640. Any opinions, findings, and conclusions or recommendations expressed in this material are those of the authors and do not necessarily reflect the views of the NSF.

9. REFERENCES

[1] Domenico Amalfitano, Anna Rita Fasolino, Porfirio Tramontana, Salvatore De Carmine, and Atif M. Memon. Using GUI Ripping for Automated Testing of Android Applications. In *Proceedings of the 27th IEEE/ACM International Conference on Automated Software Engineering*, ASE 2012, pages 258–261, New York, NY, USA, 2012. ACM.

[2] Saswat Anand, Mayur Naik, Hongseok Yang, and Mary Jean Harrold. Automated Concolic Testing of Smartphone Apps.

In *ACM International Symposium on Foundations of Software Engineering*, FSE, 2012.

[3] Jien-Tsai Chan and Wuu Yang. Advanced Obfuscation Techniques for Java Bytecode. *J. Syst. Softw.*, 71(1-2):1–10, April 2004.

[4] Christian Collberg, Edward Carter, Saumya Debray, Andrew Huntwork, John Kececioglu, Cullen Linn, and Michael Stepp. Dynamic Path-based Software Watermarking. In *Proceedings of the ACM SIGPLAN 2004 Conference on Programming Language Design and Implementation*, PLDI '04, pages 107–118, New York, NY, USA, 2004. ACM.

[5] Christian Collberg, Ginger Myles, and Andrew Huntwork. SandMark - A Tool for Software Protection Research. IEEE Security and Privacy, Vol. 1, Num. 4, July/August 2003.

[6] Christian Collberg and Clark Thomborson. Software Watermarking: Models and Dynamic Embeddings. In *Proceedings of the 26th ACM SIGPLAN-SIGACT Symposium on Principles of Programming Languages*, POPL '99, pages 311–324, New York, NY, USA, 1999. ACM.

[7] Christian Collberg, Clark Thomborson, and Douglas Low. Manufacturing Cheap, Resilient, and Stealthy Opaque Constructs. In *Proceedings of the 25th ACM SIGPLAN-SIGACT Symposium on Principles of Programming Languages*, POPL '98, pages 184–196, New York, NY, USA, 1998. ACM.

[8] Jonathan Crussell, Clint Gibler, and Hao Chen. Attack of the Clones: Detecting Cloned Applications on Android Markets. In *17th European Symposium on Research in Computer Security*, ESORICS 2012, September 2012.

[9] Mila Dalla Preda and Roberto Giacobazzi. Control Code Obfuscation by Abstract Interpretation. In *Proceedings of the Third IEEE International Conference on Software Engineering and Formal Methods*, SEFM '05, pages 301–310, Washington, DC, USA, 2005. IEEE Computer Society.

[10] DalvikVM.com. Dalvik Virtual Machine - Brief Overview of the Dalvik Virtual Machine and Its Insights. `http://www.dalvikvm.com/`. Online; accessed at Nov 30, 2012.

[11] FITTEST. M[agi]C Tool: M*C Test Generation Tool. `http://selab.fbk.eu/magic/`. Online; accessed at Dec 1, 2012.

[12] Kazuhide Fukushima and Kouichi Sakurai. A Software Fingerprinting Scheme for Java Using Classfiles Obfuscation. In Ki-Joon Chae and Moti Yung, editors, *Information Security Applications*, volume 2908 of *Lecture Notes in Computer Science*, pages 303–316. Springer Berlin Heidelberg, 2004.

[13] Dan Galpin and Trevor Johns. Evading Pirates and Stopping Vampires Using License Verification Library, In-App Billing, and App Engine. `http://www.google.com/events/io/2011/sessions/evading-pirates-and-stopping-vampires-using-license-verification-library-in-app-billing-and-app-engine.html`. Online; accessed at Nov 30, 2012.

[14] Rakesh Ghiya and Laurie J. Hendren. Is It a Tree, a DAG, or a Cyclic Graph? A Shape Analysis for Heap-directed Pointers in C. In *Proceedings of the 23rd ACM SIGPLAN-SIGACT Symposium on Principles of Programming Languages*, POPL '96, pages 1–15, New York, NY, USA, 1996. ACM.

[15] Dieter Habelitz. Java 1.5 Grammar for ANTLR v3 That Builds Trees. `http://www.antlr.org/grammar/1207932239307/Java1_5Grammars`. Online; accessed at Nov 30, 2012.

[16] Steve Hanna, Ling Huang, Saung Li, Charles Chen, and Dawn Song. Juxtapp: A Scalable System for Detecting Code Reuse Among Android Applications. In *9th Conference on Detection of Intrusions and Malware & Vulnerability Assessment*, DIMVA 2012, July 2012.

[17] Google Inc. Android Application Licensing. `http://developer.android.com/guide/google/play/licensing/index.html`. Online; accessed at Nov 30, 2012.

[18] Google Inc. Android Debug Bridge. `http://developer.android.com/tools/help/adb.html`. Online; accessed at Nov 30, 2012.

[19] Google Inc. Android Emulator. `http://developer.android.com/tools/help/emulator.html`. Online; accessed at Nov 30, 2012.

[20] Google Inc. Building and Running Android App from Command Line. `http://developer.android.com/tools/building/building-cmdline.html`. Online; accessed at Nov 30, 2012.

[21] Google Inc. Testing Fundamental | Android Developers. `http://developer.android.com/tools/testing/testing_android.html`. Online; accessed at Nov 30, 2012.

[22] Google Inc. UI/Application Exerciser Monkey. `http://developer.android.com/tools/help/monkey.html`. Online; accessed at Nov 30, 2012.

[23] Lookout Inc. App Genome Report: February 2011. `https://www.mylookout.com/appgenome/`. Online; accessed at Nov 30, 2012.

[24] Oracle Inc. Java Debug Interface. `http://docs.oracle.com/javase/1.5.0/docs/guide/jpda/jdi/`. Online; accessed at Nov 30, 2012.

[25] Saikoa Inc. A Specialized Optimizer and Obfuscator for Android. `http://www.saikoa.com/dexguard`. Online; accessed at Nov 30, 2012.

[26] Yiming Jing, Gail-Joon Ahn, and Hongxin Hu. Model-Based Conformance Testing for Android. In Goichiro Hanaoka and Toshihiro Yamauchi, editors, *Advances in Information and Computer Security*, volume 7631 of *Lecture Notes in Computer Science*, pages 1–18. Springer Berlin Heidelberg, 2012.

[27] Seolwoo Joo and Changyeon Hwang. Mobile Banking Vulnerability: Android Repackaging Threat. Virus Bulletin, May 2012.

[28] Donald Knuth. *Fundamental Algorithms, Volume 1 of The Art of Computer Programming, Third Edition.* Addison-Wesley, 1997.

[29] Eric Lafortune. ProGuard. `http://proguard.sourceforge.net/`. Online; accessed at Nov 30, 2012.

[30] Lohan+. AntiLVL - Android License Verification Library Subversion. `http://androidcracking.blogspot.com/p/antilvl_01.html`. Online; accessed at Nov 30, 2012.

[31] Lohan+. Cracking Amazon DRM. `http://androidcracking.blogspot.com/2011/04/cracking-amazon-drm.html`. Online; accessed at Nov 30, 2012.

[32] Lohan+. Cracking Verizon's V Cast Apps DRM. `http://androidcracking.blogspot.com/2011/06/`

cracking-verizons-v-cast-apps-drm.html. Online; accessed at Nov 1, 2012.

[33] Riyadh Mahmood, Naeem Esfahani, Thabet Kacem, Nariman Mirzaei, Sam Malek, and Angelos Stavrou. A Whitebox Approach for Automated Security Testing of Android Applications on the Cloud. In *Proceedings of the 7th International Workshop on Automation of Software Test*, AST 2012, 2012.

[34] Atif M. Memon. An event-flow model of GUI-based applications for testing: Research Articles. *Softw. Test. Verif. Reliab.*, 17(3):137–157, September 2007.

[35] Akito Monden, Hajimu Iida, Ken-ichi Matsumoto, Koji Torii, and Katsuro Inoue. A Practical Method for Watermarking Java Programs. In *24th International Computer Software and Applications Conference*, COMPSAC '00, pages 191–197, Washington, DC, USA, 2000. IEEE Computer Society.

[36] Jasvir Nagra and Clark Thomborson. Threading Software Watermarks. In *Proceedings of the 6th International Conference on Information Hiding*, IH'04, pages 208–223, Berlin, Heidelberg, 2004. Springer-Verlag.

[37] Jasvir Nagra, Clark Thomborson, and Christian Collberg. A Functional Taxonomy for Software Watermarking. In *Proceedings of the Twenty-fifth Australasian Conference on Computer Science - Volume 4*, ACSC '02, pages 177–186, Darlinghurst, Australia, Australia, 2002. Australian Computer Society, Inc.

[38] Cu D. Nguyen, Alessandro Marchetto, and Paolo Tonella. Combining Model-based and Combinatorial Testing for Effective Test Case Generation. In *Proceedings of the 2012 International Symposium on Software Testing and Analysis*, ISSTA 2012, pages 100–110, New York, NY, USA, 2012. ACM.

[39] Jens Palsberg, Sowmya Krishnaswamy, Minseok Kwon, Di Ma, Qiuyun Shao, and Yi Zhang. Experience with Software Watermarking. In *Proceedings of the 16th Annual Computer Security Applications Conference*, ACSAC '00, pages 308–, Washington, DC, USA, 2000. IEEE Computer Society.

[40] Terence Parr. ANTLR - ANother Tool for Language Recognition. http://www.antlr.org/. Online; accessed at Nov 30, 2012.

[41] Android Police. [Updated: Amazon Provides Clarifications] Amazon App Store's DRM To Be More Restrictive Than Google's? http://www.androidpolice.com/2011/03/07/amazon-app-stores-drm-to-be-more-restrictive-than-googles/. Online; accessed at Nov 30, 2012.

[42] Rahul Potharaju, Andrew Newell, Cristina Nita-Rotaru, and Xiangyu Zhang. Plagiarizing Smartphone Applications: Attack Strategies and Defense Techniques. In *Proceedings of the 4th International Conference on Engineering Secure Software and Systems*, ESSoS'12, pages 106–120, Berlin, Heidelberg, 2012. Springer-Verlag.

[43] Todd A. Proebsting and Scott A. Watterson. Krakatoa: Decompilation in Java (Does Bytecode Reveal Source?). In *In Third USENIX Conference on Object-Oriented Technologies and Systems (COOTS)*, 1997.

[44] Google Code Project. Android-apktool - Tool for Reengineering Android apk Files. http://code.google.com/p/android-apktool/. Online; accessed at Nov 30, 2012.

[45] G. Ramalingam. The Undecidability of Aliasing. *ACM Trans. Program. Lang. Syst.*, 16(5):1467–1471, September 1994.

[46] Renas Reda. Robotium. http://code.google.com/p/robotium/. Online; accessed at Nov 30, 2012.

[47] Tommi Takala, Mika Katara, and Julian Harty. Experiences of System-Level Model-Based GUI Testing of an Android Application. In *Proceedings of the 4th IEEE International Conference on Software Testing, Verification, and Validation (ICST 2011)*, pages 377–386, Los Alamitos, CA, USA, March 2011. IEEE Computer Society.

[48] Arxan Technologies. State of Security in the App Economy: Mobile Apps Under Attack. http://www.arxan.com/assets/1/7/state-of-security-app-economy.pdf, 2012.

[49] Paul R. Wilson, Mark S. Johnstone, Michael Neely, and David Boles. Dynamic Storage Allocation: A Survey and Critical Review. In *Proceedings of the International Workshop on Memory Management*, IWMM '95, pages 1–116, London, UK, UK, 1995. Springer-Verlag.

[50] Rubin Xu, Hassen Saïdi, and Ross Anderson. Aurasium: Practical Policy Enforcement for Android Applications. In *Proceedings of the 21st USENIX Conference on Security Symposium*, Security'12, pages 27–27, Berkeley, CA, USA, 2012. USENIX Association.

[51] Min Zheng, Patrick P.C. Lee, and John C.S. Lui. ADAM: An Automatic and Extensible Platform to Stress Test Android Anti-Virus Systems . In *Proceedings of the 9th Conference on Detection of Intrusions and Malware & Vulnerability Assessment*, 2012.

[52] Wu Zhou, Yajin Zhou, Michael Grace, Xuxian Jiang, and Shihong Zou. Fast, Scalable Detection of Piggybacked Mobile Applications. In *Proceedings of the 3nd ACM Conference on Data and Application Security and Privacy*, CODASPY '13, February 2013.

[53] Wu Zhou, Yajin Zhou, Xuxian Jiang, and Peng Ning. DroidMOSS: Detecting Repackaged Smartphone Applications in Third-Party Android Marketplaces. In *Proceedings of the 2nd ACM Conference on Data and Application Security and Privacy*, CODASPY '12, February 2012.

[54] Yajin Zhou and Xuxian Jiang. Dissecting Android Malware: Characterization and Evolution. In *Proceedings of the 33rd IEEE Symposium on Security and Privacy*, Oakland 2012, May 2012.

PSiOS: Bring Your Own Privacy & Security to iOS Devices

Tim Werthmann[1], Ralf Hund[1], Lucas Davi[2,3],
Ahmad-Reza Sadeghi[2,3] and Thorsten Holz[1]

[1]Ruhr-Universität Bochum, Germany [2]Intel CRI-SC at TU Darmstadt, Germany
[3]Technische Universität Darmstadt, Germany

ABSTRACT

Apple iOS is one of the most popular mobile operating systems. As its core security technology, iOS provides *application sandboxing* but assigns a generic sandboxing profile to *every* third-party application. However, recent attacks and incidents with benign applications demonstrate that this design decision is vulnerable to crucial privacy and security breaches, allowing applications (either benign or malicious) to access contacts, photos, and device IDs. Moreover, the dynamic character of iOS apps written in Objective-C renders the currently proposed static analysis tools less useful.

In this paper, we aim to address the open problem of *preventing* (not only detecting) privacy leaks and simultaneously strengthening security against runtime attacks on iOS. Compared to similar research work on the open Android, realizing such a system for the closed-source iOS is highly involved.

We present the design and implementation of *PSiOS*, a tool that features a novel policy enforcement framework for iOS. It provides fine-grained, application-specific, and user/administrator defined sandboxing for each third-party application *without* requiring access to the application source code. Our reference implementation deploys control-flow integrity based on the recently proposed MoCFI (Mobile CFI) framework that only protects applications against runtime attacks. We evaluated several popular iOS applications (e.g., Facebook, WhatsApp) to demonstrate the efficiency and effectiveness of *PSiOS*.

Categories and Subject Descriptors

D.4.6 [**Operating Systems**]: Security and Protection

General Terms

Security

Keywords

iOS; Application Sandboxing; Software Security

1. INTRODUCTION

Smartphones and tablet computers are becoming ubiquitous, and the sales figures for both types of devices are growing rapidly. Probably the most important factor behind this popularity is the availability of a large number of mobile applications, ranging from simple games over messaging apps to office applications. Consumers can easily install these apps via so-called *app stores* and then use them. Privacy and security concerns arise because an application can also access personal/sensitive information on the device, such as for example contact details, messages, location information, or private photos. Current mobile operating systems are following slightly different approaches to protect a user's personal information and we focus in this paper on the closed-source Apple iOS since it is after Android the most popular mobile OS on the market. Further, most security extensions for smartphones target the *open* Android, while security extensions for the *closed* iOS are rarely available.

iOS assigns a generic sandboxing profile to every third-party application allowing access to contacts, location, photos, recent searches, and device information. This design decision contradicts the least-privilege principle, and easily allows an application (either benign or malicious) to access sensitive information. Developers must adhere to certain guidelines for their applications to be accepted in the Apple App Store [7]. According to these guidelines, apps must ask for the user's permission to read out sensitive data. In the recent past, however, several (legitimate) apps were detected that abused these privileges and for instance uploaded the complete contact list to the app developers without the user's consent [26]. These incidents even lead to an investigation of the US Congress regarding data collection practices on mobile devices [21].

To mitigate this problem, one of the first approaches is to *detect* privacy leaks in iOS apps by using *static* analysis, as recently proposed by Egele et al. [14]. Their static analysis tool, called PiOS, generates the control-flow graph of an application to identify code regions that potentially steal sensitive information. PiOS focuses on a *static* analysis approach and we identified shortcomings of the actual analysis phase which we detail in Section 8. Up to now, there is no solid privacy framework for iOS that enables a user to *protect* his personal information. Compared to Android, for which such tools have been proposed [19, 24], our main challenges concern the proprietary, closed-source nature of iOS as well the usage of the programming language Objective-C. The latter significantly complicates the implementation due to the highly dynamic nature (e.g., dynamic binding and

Objective-C class clusters). Moreover, while on Android all applications access the main phone functionalities through the Java Android API, iOS apps may use either Objective-C, standard C/C++, or direct system calls to use the main phone services. This raises a significant technical challenge, since all possible access techniques have to be considered to ensure accurate access control on privacy-related information and operations.

Contribution.

In this paper, we aim to address the open problem of not only detecting privacy leaks for iOS apps, but actually *preventing* them. The key idea of our approach is to assign specific sandboxing profiles to each application to enforce a given fine-grained privacy policy. Such a profile may either be defined by a user at installation time, or centrally provided by a system administrator or an enterprise. Logically, we generate a protection layer between applications and the iOS Objective-C Runtime environment. Further, we monitor applications at execution time and ensure that they only perform actions that adhere to the given sandboxing profiles. Our solution operates *directly* on the application binary, and hence, neither requires recompilation nor access to the source code, which enables enforcement of policies for arbitrary applications.

We have implemented a fully-working and efficient prototype of our framework in a tool called *PSiOS* that is based on MoCFI [11], a tool that enforces control-flow integrity on iOS devices. *PSiOS* enables user-driven and fine-grained application sandboxing: on the one hand, a user can dynamically update sandboxing profiles, without the need to recompile or reinstall the application. Hence, end-users can easily revoke or assign privileges. On the other hand, our framework enables very fine-grained policies in which the user or system administrator can precisely specify which privileges are assigned to an application. This is possible, since our sandboxing profiles cover the *entire* Objective-C runtime and allows argument validations for each API call. Since our framework is based on a complete control-flow integrity tool, we also prevent attackers from exploiting vulnerabilities in the application code to hijack its assigned rights [20, 22]. We demonstrate that *PSiOS* effectively prevents privacy breaches by testing our tool with SpyPhone [23], an iOS app specifically designed to steal sensitive information from an iOS device. Furthermore, we successfully instrumented many real-world, complex applications like Facebook or WhatsApp and performance measurements indicate that the overhead introduced by our tool is reasonable.

Note that our approach differs from Apple's recently introduced entitlement keys (i.e., permissions). Specifically, Apple provides 25 keys to confine the privileges of apps [6]. However, these keys are specified by the app developer or directly by Apple. Hence, neither the end-user nor an enterprise can apply custom sandboxing policies to their apps. Instead, one needs to rely on Apple to apply the appropriate entitlements to each app, while malware writers will avoid to confine their apps for obvious reasons. Moreover, as we will elaborate in Section 2.2, these entitlements are enforced on the basis of the built-in iOS sandboxing framework, which (in contrast to *PSiOS*) cannot enforce fine-grained sandboxing rules.

Independent from our research work on *PSiOS*, Apple has very recently introduced privacy settings options (starting from iOS 6) where end-users can disable or enable access to private information on an app-by-app basis [8]. We believe that this new feature is a step in the right direction since iOS devices suffer from lack of privacy protection. However, we stress that our tool *PSiOS* does not only cover the same access rules, but also features argument validation (e.g., to allow access to a subset of private information), enables fine-grained access control beyond access to private information (e.g., any system call an application may invoke), and at the same time prevents runtime attacks.

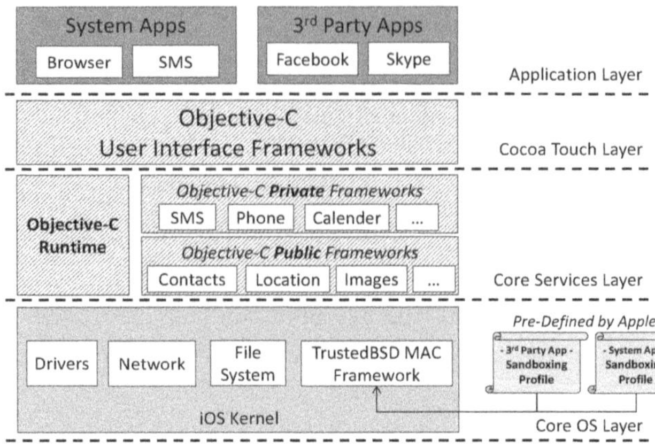

Figure 1: iOS Software Architecture

2. BACKGROUND AND PROBLEM DESCRIPTION

In this section, we recall the basics of the iOS architecture and describe why iOS sandboxing suffers from severe security and privacy problems.

2.1 iOS Background

Apple devices such as iPhone, iPod touch, and iPad are based on the iOS operating system that provides several security features, e.g., mandatory code signing, data encryption, or memory randomization. In particular, iOS enforces *application sandboxing* to isolate applications from each other, i.e., an application cannot access files in another application's directory. Moreover, it constrains third-party applications from accessing the underlying operating system kernel.

Figure 1 shows an abstract representation of the iOS software architecture. Note that the original iOS architecture also includes a Media layer, which we did not include for the sake of brevity. Basically, iOS features four software layers that are relevant for our analysis: (1) an application layer, (2) the so-called Cocoa Touch layer which mainly provides objects for the application display, (3) the Core Services layer which provides frameworks to access main phone facilities, and (4) the Core OS layer (the iOS kernel) that provides basic OS facilities such as a system call wrapper, device drivers, and the file-system.

Objective-C Runtime.

iOS applications and the main iOS system libraries are implemented in the object-oriented language Objective-C. Since iOS defers many decisions from compile-time to runtime, a *runtime system* is required in order to use Objective-C [4]. The Objective-C runtime is linked to every process and interacts with the so-called Objective-C *frameworks*. A framework is Apple's notion for a directory containing a shared library along with resources to support the framework (e.g., images, header files). The Objective-C core frameworks are provided in the Cocoa Touch and Core Services layer, while iOS distinguishes between *public* and *private* frameworks. The private frameworks are only accessible by system applications, while the public frameworks can be accessed by every third-party application. Of particular interest are the frameworks located in the Core Services layer as they provide access to main phone facilities (such as Location, SMS, Calendar, Contacts, etc.).

2.2 Problem Description

iOS sandboxing is realized by a kernel module which has been adopted from the TrustedBSD kernel. This module mediates and validates every system call and its arguments according to sandboxing profiles already *pre-defined* by Apple. As already mentioned, iOS assigns a generic sandboxing profile to *every* third-party application which enables every application to access the public frameworks and specifically grants access to contacts, location, device information, call history, keyboard cache, recent searches, e-mail account configurations, and photos. Recently, several attacks were reported where applications abused their privilege set to steal, for instance, the user's address book [26]. Moreover, whenever an application is exploited by a runtime attack, the adversary can misuse the application's privileges to steal sensitive information as well [20, 22].

Note that iOS already supports sandboxing at the kernel-level, but not within the Objective-C runtime. The design decision taken by Apple leads to coarse-grained sandboxing, because the Core OS layer misses the semantics of the Objective-C runtime. Instead of enforcing access-control on a specific API call, iOS has to enforce access-control based on invoked system calls. Hence, it cannot provide a fine-grained access control. In particular, the Core OS layer cannot enforce access control on the main Objective-C constructs such as used classes, objects, variables, and methods, which are extensively used by iOS applications and involve a chain of diverse system calls, files, and memory structures.

Individual iOS sandboxing rules can be bound to mobile apps using *entitlements* [6]. These entitlements assign certain rights to applications, which are in turn enforced by the iOS sandbox described previously. Apple currently supports 25 different OS X entitlements for that purpose, of which only a subset is available on iOS. However, the system has a major drawback: entitlements are requested by the developer and included in the digital signature of the application by Apple's app store. This means that they cannot be changed afterwards by the end user since this would break the entire signature.

3. HIGH-LEVEL IDEA

In this paper, we address the mentioned security and design weaknesses of the current iOS sandboxing realization (see Section 2). In particular, we aim towards a framework that allows access-control for the Objective-C runtime and the enforcement of the least-privilege principle. Note that realizing such a system for iOS is highly involved, since iOS is closed-source. Hence, we cannot simply replace or extend existing modules as typically performed in recent Android security research proposals, e.g., Kirin [16] or TaintDroid [15] to name a few.

The high-level idea of *PSiOS* (**P**rivacy and **S**ecurity for **iOS** devices) is shown in Figure 2. In contrast to Apple's approach (where sandboxing profiles are generic and pre-defined), *PSiOS* allows a different and user-defined sandboxing profile for each application. Basically, we add a new module that operates between the Application and Cocoa Touch Layer (see Section 2), which we call *policy enforcement* component. As shown in Figure 2, this component mediates every access request to the Objective-C runtime, the frameworks, and the system call wrapper. It enforces access control rules on each access request based on the user-defined sandboxing profiles. Only when the policy has not been violated, we forward the request to its original destination. Note that the current iOS system does not enforce *any* access control mechanism to the Objective-C runtime and frameworks. On the other hand, iOS already enforces access control on system calls, but our approach allows an individual enforcement policy for each iOS application.

To monitor if an application adheres to the given sandboxing profile, we instrument the application so that all access requests are redirected to the policy enforcement. Alternatively, one could directly extend the Objective-C runtime and frameworks with dedicated interfaces that validate whether the caller has the appropriate privileges (similar to the Android permission system [18]). However, the iOS Objective-C runtime and frameworks are closed-source. Hence, extending them directly is infeasible.

Furthermore, the Objective-C runtime operates on the same level as the application code and checks within the runtime could thus easily be bypassed, either intentionally by the application author or by a control-flow attack against a software vulnerability. Our framework does not suffer from this shortcoming since we have full control over the application code. We also protect the application from being compromised by control-flow attacks (such as code injection [2] and return-oriented programming [28]). Otherwise an adversary could leverage such attack techniques to circumvent our instrumentation points and the policy enforcement component. To tackle this problem, we ensure the integrity of the application's execution flow by enforcing control-flow integrity (CFI) which is a general countermeasure to defeat such attacks by validating if an application always follows the application's control-flow graph (CFG). Recently, we introduced MoCFI (Mobile CFI) [11] which enforces CFI on mobile devices, where iOS has been used for the proof-of-concept implementation. We leverage MoCFI for our prototype, but it is noteworthy to mention that in contrast to MoCFI, we replace PiOS [14] with a novel Objective-C analyzer, to tackle shortcomings of PiOS (see Section 8).

4. DESIGN OF PSIOS

In this section we introduce the design of *PSiOS* (**P**rivacy and **S**ecurity for **iOS** devices) which enforces our high-level idea presented in Section 3.

4.1 Architecture

The general architecture of *PSiOS* is depicted in Figure 3. Basically, our design can be divided into three distinct phases: (1) static analysis (*offline*), (2) binary rewriting at *load-time*, and (3) runtime CFI and policy enforcement at

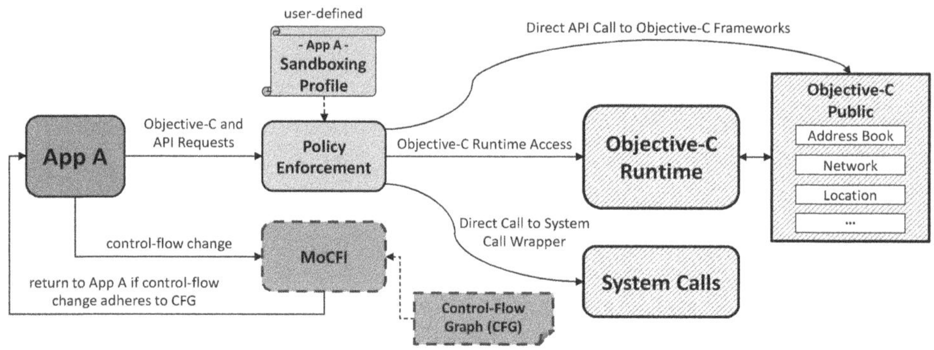

Figure 2: High-Level Idea of *PSiOS*

execution-time. While the static analysis phase needs to be performed only once, the binary rewriting and runtime enforcement phase are performed whenever the application is launched by the user.

The general workflow is as follows: First, we reverse-engineer the application binary by using automated tools to derive the application's structure. In particular, we leverage MoCFI [11] to derive the application's control-flow graph (CFG), which is required to extract all valid execution paths (step 1). Further, we implemented a static Objective-C analyzer that reuses techniques of existing tools such as PiOS [14] and Objective-C helper scripts [13] to identify used Objective-C classes and methods (step 2). Note that we extended these tools to cover all calls to the system call wrapper as well. When the application is launched by the user, we first perform binary rewriting to integrate control-flow checks into the binary (step 3).

Second, we leverage binary rewriting to insert checkpoints into the application that will be reached whenever an application aims to access the Objective-C runtime, the public frameworks, or the system call wrapper (step 4).

At execution-time, we first use a novel runtime Objective-C analyzer that tackles the incompleteness of the static analysis and retrieves important runtime information on Objective-C constructs (step 5). Afterwards, CFI ensures that the control-flow of an application always follows the legitimate paths of the CFG (step 6). Further, for all access requests to the Objective-C environment and the system call wrapper, our policy enforcement component validates if the request adheres to the given policy rules (step 7). In the following, we elaborate in more detail on the individual components and phases.

Static Analysis.

Since iOS applications are encrypted by default, we first obtain the decrypted version of an iOS binary by using a technique called *process dumping* [14], which automatically creates memory snapshots of the unencrypted code at runtime. Afterwards, we use MoCFI to derive the CFG, i.e., to resolve the targets of indirect branches (e.g., indirect calls/jumps, and function returns). In the original design of MoCFI, PiOS [14] is used to resolve Objective-C calls. However, as our experiments revealed, the static analysis performed by PiOS is not always able to retrieve the used Objective-C structures. Other existing approaches to parse Mach-O Objective-C files [13] are outdated and only process a subset of the required constructs. Hence, we developed a novel static Objective-C analyzer which is capable of identifying all relevant Objective-C structures describing classes, methods, and inheritance relationships. Further, our tool provides a more fine-grained analysis than the existing ones, since it completely parses the iOS Mach-O File Header and accurately resolves all calls to the system call wrapper. Moreover, note that the existing tools [14, 13] do not take the dynamic nature of the Objective-C runtime environment into account, which is necessary to enforce fine-grained policy enforcement and handle cases that cannot be resolved statically. The CFG and the Objective-C information are stored in separate configuration files.

Binary Rewriting and Runtime Enforcement.

To preserve the application signature, we perform binary rewriting after the iOS application loader has verified the application signature. For this *PSiOS* employs a rewriting engine to patch all indirect branch instructions with a control-flow check. Furthermore, *PSiOS* rewrites all access requests to the Objective-C runtime to insert checkpoints. After the binary rewriting, our runtime Objective-C analyzer requests crucial runtime information on Objective-C constructs (such as registered parent and child classes, and runtime addresses of invoked methods) which cannot be derived in the static analysis phase. Afterwards, the application starts executing and whenever a checkpoint has been reached, *PSiOS* ensures that the call follows a valid CFG path and adheres to the given sandboxing profile. Our approach ensures that the application cannot escape from the newly established sandbox, which may occur by either directly calling the Objective-C runtime or frameworks, or indirectly, by launching a runtime attack.

Enforcement Policies.

PSiOS supports three different policy enforcement types: *Log*, *Exit*, and *Replace*.

Each enforcement type can be used to achieve a different objective. The *Log* option ensures that all policy violations are recorded by the system. However, the application is allowed to continue executing, but the system's log file can be inspected afterwards. This option is in particular useful when applied in a learning phase (for instance in case of enterprises) in order to help system administrators to identify which Objective-C calls are performed and required by the application. Hence, it facilitates the specification of

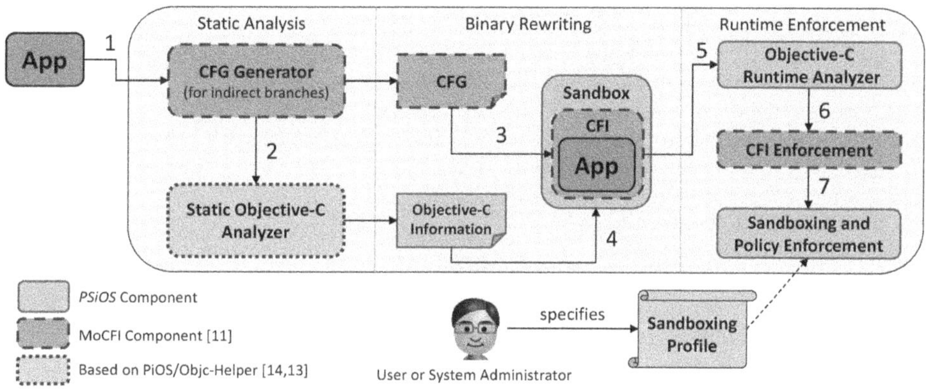

Figure 3: Architecture of *PSiOS*

sandboxing profiles. The *Exit* option is more restrictive, because it immediately terminates the process whenever a policy violation occurs. Due to usability reasons, we also introduced the *Replace* option which allows the application to continue executing, but *PSiOS* replaces the return values of the Objective-C runtime with shadow data. For instance, if the sandboxing profile prohibits access to the address book, the return value will be either fake contacts or an empty data set indicating to a user that a policy violation occurred.

Our policy enforcement can be used to achieve different goals. For example, an end user can be enabled to specify which mobile app may access which privacy critical resources. A graphical interface can then manage the assignment of policies to applications in a user-friendly way, similar to Android's permission system [18]. Furthermore, policies could also be deployed by a central instance (e.g. an enterprise) in a *bring your own device* environment. This allows the company to mitigate the risk of sensitive corporate data leakage (e-mails, text message, address book entries, etc.) through insecure private applications. Our current prototype only provides the underlying framework for policy enforcement and does not come with a frontend. However, it is straightforward to implement such a component in the future, we focus on showing the feasibility of the approach.

Note that providing complete policies is tedious work. Many tasks can be achieved by using different APIs. For instance, files can be accessed by using a variety of function calls (e.g., through Objective-C, the C standard lib, etc.). Identifying all functions that can be used to access a privacy critical ressource takes considerable effort, but is possible nevertheless.

4.2 Format of Sandboxing Profiles

The sandboxing profile consists of blacklisted Objective-C and API Calls, while each policy rule follows the following XML-based format:

```
<rule type="objc" class="<classname>"
  selector="<selector>" mode="log|exit|replace">
    <arg number="<arg number>" type=
      "int|string|..." operator="=|!|<|>|<=|>="
      value="<argument value>"/>
    <arg ... />
</rule>
<rule type="api" function="<functionname>"
   mode="log|exit|replace">
    <arg ... />
     ...
</rule>
```

The first attributes of an Objective-C rule refer to the class and selector (i.e., the Objective-C method name). If the policy targets an API call, the API call name is stored in the `function` field and the `type` is set to `api`. The `mode` field indicates the enforcement type. The `number` field of an `arg` mode indicates the argument number and the `type` field identifies the argument's type, e.g., **int** refers to an integer. The `operator` indicates the compare method (=, !, <, >, >=, <=) and the `value` field holds the compare value. For better understanding, consider the following Objective-C message:

```
[NSUserDefaults valueForKey:@"
     SBFormattedPhoneNumber"]
```

This Objective-C message initializes a new instance of the `NSUserDefaults` class and uses the `valueForKey:` selector (i.e., method) with the argument `SBFormattedPhoneNumber` to retrieve the device's phone number. Note that this call can be invoked by every application. Hence, a policy rule to strictly deny this Objective-C message would look as follows:

```
<rule type="objc" class="NSUserDefaults"
   selector="valueForKey:" mode="exit" />
```

This rule is always triggered if the class and selector are set to `NSUserDefaults` and `valueForKey:`. If only those messages that use the `SBFormattedPhoneNumber` as method argument should be denied, the rule has to be changed as follows:

```
<rule type="objc" class="NSUserDefaults"
   selector="valueForKey:" mode="exit">
    <arg number="1" type="string" operator="="
        value="SBFormattedPhoneNumber"/>
</rule>
```

This allows us to define fine-grained policy rules on each Objective-C method and API call. Moreover, the argument validation of our policy language supports logical operations such as *AND* and *OR*.

5. BACKGROUND ON OBJECTIVE-C AND CHALLENGES

In this section, we briefly recall the basics of the Objective-C runtime and elaborate on the challenges to be tackled when implementing *PSiOS*. Further, we describe why existing static tools [14, 13] miss important information that is only available at runtime.

In general, Objective-C is an object-oriented, dynamically typed language, and mainly operates on objects. In particular, Objective-C provides three basic constructs for encapsulating data with methods: metaclasses, classes, and objects. Objects are instances of classes, whereby classes are instantiated from their metaclasses, and metaclasses are instantiated from their root's metaclass. Further, every class can have an arbitrary amount of child classes.

Messaging.

A distinct feature of Objective-C is its messaging engine. Instead of calling a method directly, applications send a message to the Objective-C runtime, which deploys a dynamic message dispatcher to interpret the message. The generic syntax of an Objective-C message is as follows:

```
[object method:arg1 param_name1:arg2 ... param_
    nameN-1:argN]
```

It consists of the target object (usually called *receiver*) and the method name along with its parameter names and argument values. In particular, the method name and the parameter names form the so-called Objective-C *selector*. The Objective-C runtime interprets these messages and forwards them to the receiving objects which execute the desired methods.

For better understanding, consider the example Objective-C message used in Section 4.2, where `valueForKey:` forms the selector of the message:

```
[NSUserDefaults valueForKey:@"
    SBFormattedPhoneNumber"]
```

The selector is unique to the runtime system and is used to derive a *unique identifier* of type `SEL` (a compiled selector) that is registered by the runtime system. However, since each object can implement its own version of a method, the same selector might refer to different implementations. Thus, the Objective-C runtime determines the appropriate method implementation based on the class of the receiver at runtime. This technique is referred to as *dynamic binding* and is realized by compiling every Objective-C message into a default call to the Objective-C runtime messaging function `objc_msgSend`, which has the following format: *objc_msgSend(receiver, selector, arg1, arg2, ...)*

Challenge: Due to dynamic binding, a compiled iOS binary will consists of a high number of calls to the generic dispatcher function `objc_msgSend`. Hence, straight-forward disassembling (as performed by IDA Pro) will only output a number of calls to the dispatcher function rather than providing the real target object and method. Although existing tools [14, 13] partly address this problem, they cannot determine the actual implementation of the method since the same selector might reference different implementations.

Class Clusters.

Objective-C class clusters allow the grouping of related or dependent objects under a single public abstract superclass. A typical example is the Objective-C class cluster for numbers: while the *Number* class is the public abstract superclass of the cluster, the *Integer* or *Float* classes are private subclasses of the *Number* class. Hence, the private subclasses are not visible to the developer, who only uses the public *Number* class to instantiate a new number. In turn, the *Number* class will itself perform the correct instantiation of the respective integer or float.

Challenge: The challenge in this context is that private subclasses are not visible at static analysis time, and hence, existing static tools [14, 13] can neither identify malicious behavior through these classes nor provide accurate control-flow information with regard to the control-flow graph.

6. IMPLEMENTATION

We implemented the design of *PSiOS* (see Figure 3) in a prototype that supports iOS version 4.3.2, 4.3.3, 5.0.1, and 5.1.1. The static Objective-C analyzer is realized as a new Python module for the reverse-engineering tool IDA Pro 6.x. For rewriting the application binary and enforcing CFI, we base our implementation on our MoCFI framework [11]. However, we need to extend MoCFI to introduce the policy enforcement and the runtime Objective-C analyzer. In particular, the latter component enables runtime analysis of Objective-C constructs to tackle the challenges mentioned in Section 5. The entire runtime tools are realized as one shared iOS library that is developed in the Objective-C++ language. Since Apple prohibits any user from installing a new shared library, we had to jailbreak our test devices to inject our library to every iOS application and to enable binary rewriting at runtime, as discussed in Section 7.4. In the following, we present selected implementation details of the core components of the *PSiOS* framework, while we refer to [11] for the implementation details of MoCFI.

6.1 Static Objective-C Analyzer

Basically, the static Objective-C analyzer performs binary analysis of an iOS application to identify all implemented and referenced Objective-C classes and selectors. For this we reuse techniques from existing tools such as PiOS [14] and Objective-C scripts [13]. However, we extend the existing tools to resolve all direct calls to the system call wrapper. Note that static analysis of iOS binaries is a challenging task because most internals have not been documented, and required high reverse-engineering efforts on our side.

Analysis of iOS File Header.

In the first step, our analyzer parses the entire iOS file header (Mach-O) header, which is necessary to locate all code and data sections and to identify important side information on used Objective-C constructs and the binding information referenced at runtime by the iOS linker. For this, we examine the so-called load commands included in the iOS Mach-O header [3]. For the sake of completeness, the load commands inspected by *PSiOS* are listed in Table 2 in Appendix A. Of particular interest are the `LC_DYLD_INFO` and the `LC_DYLD_INFO_ONLY` load commands as they provide binding information that is used at load-time by the iOS loader to resolve API calls and Objective-C classes.

Objective-C Classes and Selectors.

After analyzing the iOS header, we identify all Objective-C classes and selectors used by the application. This includes new Objective-C classes the developer implemented himself and classes that have been already pre-defined by the Objective-C runtime. In particular, we reuse the existing tools [14, 13] to record for each call to the `objc_msgSend` dispatcher function the referenced Objective-C class and selector.

API Calls.

Direct API calls to the public frameworks and the system call wrapper can be resolved similarly to imported Objective-C constructs since the symbol section (`__lazy_symbol`) holds 4-byte addresses which identify API calls used in the code. Backtracking the references to the symbol section reveals that each address is referenced by an address in the section `__symbolstub1`[1]. The addresses of the `__symbolstub1` section are then used throughout the code to invoke API calls, which allows us to identify the location and type of all API calls.

6.2 Objective-C Runtime Analyzer

Our novel Objective-C runtime analyzer tackles the incompleteness of the static analysis phase. In particular, it starts its operation after the iOS loader has loaded the application into memory. The basic information our analyzer derives are the *runtime addresses* of all selectors and the missing parent and child classes (which are necessary to support class clustering).

First, we retrieve the runtime addresses of the used selectors by referencing the `__objc_selrefs` section. However, the selectors can be only available as textual strings (rather than runtime addresses). To tackle this problem, we use the string representation of the selector and issue the Objective-C Call `NSSelectorFromString` to retrieve the selector's runtime address. In particular, we create a hashmap *objcHashmap* that stores all selector runtime addresses.

Afterwards, we add all Objective-C classes to the *objcHashmap* by using their runtime addresses. We retrieve the runtime addresses from the `__objc_classrefs` section and in case only the class name has been derived at static analysis time, we use the `objc_getClass` function to retrieve the corresponding runtime address. Note that we complete this step by requesting all registered classes from the Objective-C runtime. This comprises all parent and child classes which allows us to support class clustering and tackles the challenge mentioned in Section 5.

6.3 Policy Enforcement

The policy enforcement is the core component of *PSiOS* as it enforces access control on each Objective-C message and direct call to the system call wrapper.

To initialize the policy enforcement we first use the API call name (derived from the static tools) and leverage the dynamic iOS loader to retrieve the runtime address for each API call. Afterwards, we open and read the sandboxing profile file of the application. Since each rule in the sandboxing profile targets a particular Objective-C selector, class, and API call, we validate which of these constructs are actually used by the application. This can be determined by comparing the constructs of the sandboxing profile to the *objcHashmap* and the derived API calls. Only for those rules that are relevant for the application, we add the class, selector, API call name, and the policy rule to a second hashmap, we refer to as *polHashmap*.

At runtime we enforce a policy check whenever a function call to the Objective-C runtime, the public frameworks, or the system call wrapper occurs. First, MoCFI validates that control-flow integrity is preserved for the call. Afterwards, our policy enforcement checks if the call adheres to the given sandboxing profile.

For all Objective-C messages invoked at runtime, our policy enforcement looks up the class and the selector of the involved message in the *polHashmap*. When the *polHashmap* contains a rule for this message we validate whether the rule matches the message and (if specified) validate the arguments of the message with regard to a policy violation. If a rule completely matches the invoked Objective-C message, then a policy violation has occurred. Based on the enforcement option we either log the violation, terminate the program, or return shadow data (see also Section 4.1). The same policy enforcement is applied to every API call invoked at runtime, while the difference is that we look up the API call name in the *polHashmap* rather than the class and selector.

While the implementation of the *Log* and *Exit* enforcement option are straightforward (i.e., log the violation in a file, or terminate the program), it is involved to realize the *Replace* option for Objective-C messages. In particular, we achieve this by *replacing* the method implementation at runtime. For this, we request a pointer to the indicated method by contacting the Objective-C runtime. Afterwards, we force the Objective-C runtime to replace the method's implementation with the implementation of a novel and already prepared method that simply returns an empty or arbitrary data structure. Note that we roll back the original method implementation after the shadow data has been returned to ensure a new validation when the same Objective-C message but with a different argument set is invoked.

7. EVALUATION

In this section, we analyze the effectiveness and efficiency of *PSiOS* by using the SpyPhone application [23] as proof-of-concept since it demonstrates which private data can be accessed by every iOS application. Afterwards, we apply *PSiOS* to several popular iOS applications (e.g., Facebook and WhatsApp). Finally, we measure the performance overhead induced by *PSiOS* and compare it to the native runtime execution of applications.

7.1 SpyPhone

SpyPhone is an open-source proof-of-concept application that demonstrates which data can be collected by iOS third-party applications. As SpyPhone is able to retrieve a significant amount of private and sensitive data about the user and the device, we thoroughly analyzed how this is achieved and how the data harvesting can be prevented using *PSiOS*. The data collected by SpyPhone includes, amongst others:

- e-mail account information, including server addresses, user names,
- information on WLANs the device was connected to,
- phone data, such as phone number, UUID, ICCID, IMSI, and call history,
- location data from the weather and the map app,
- history logs from Safari and Youtube,
- personal photos including tagged GPS information,
- address book entries,
- keyboard cache.

[1]Note that the exact section name might differ. However, our implementation does not depend on the section name, instead it uses the `__lazy_symbol` section to locate a `__symbolstub1` like section.

Hence, we have analyzed how SpyPhone collects this data and created policy rules that prevent the requested access. As enforcement type we used the *Replace* option meaning that *PSiOS* returns shadow data for each access request.

Basically, SpyPhone retrieves Wi-Fi configurations, location, call histories, and e-mail account information by accessing property lists that are stored as XML files on any iOS device at the Core OS layer (see Figure 1 in Section 2). For instance, the information about a user's email accounts is retrieved from the file *com.apple.accountsettings.plist*. In general, access to these files can be denied by restricting the Objective-C `dictionaryWithContentsOfFile` method of the `NSDictionary` class, which is used to parse the XML file into an Objective-C data structure. For each file, we defined one rule that restricts the first parameter of the method, namely the filename.

SpyPhone also accesses sensitive information by calling appropriate Objective-C methods at the Core Services Layer (see Figure 1 in Section 2). For instance, this applies to the user's address book or when requesting device information (e.g., the user's phone number). In particular, the address book is used to store private phone numbers, email addresses, and home addresses. We specified policy rules that prohibit SpyPhone from using these Objective-C methods (e.g., `ABAddressBookCopyArrayOfAllPeople` to access the user's address book).

Table 1 shows an excerpt of our policy rules to prevent SpyPhone from accessing private user information. For the sake of completeness the entire sandboxing profile (including all policy rules) for SpyPhone is shown in Appendix B.

7.2 Applying PSiOS to iOS Apps

To demonstrate the effectiveness of our approach, we applied *PSiOS* to a number of popular iOS applications such as Facebook, WhatsApp, ImageCrop, BatteryLife, Flashlight, ImageCrop, InstaGram, MusicDownloader, MyVideo, NewYork (Game), Quickscan, LinPack, Satellite TV and the Audi App. For each app we defined a specific sandboxing profile that prevents the app from accessing private user information. In particular, *PSiOS* successfully prevents access to the address book (for Quickscan, Facebook, and Whatsapp), to personal photos (for ImageCrop and InstaGram), and to the iOS universal unique identifier, short UUID (for Quickscan, BatterLife, Flashlight, MusicDownloader, MyVideo, NewYork, and Audi). As the other applications did not contain any privacy related calls we successfully applied proof of concept enforcement rules to them (access to the file system for LinPack and an URL request for Satellite TV).

7.3 Performance

We conducted a variety of performance tests to assess the practical usability of *PSiOS*. In particular, we performed runtime measurements based on the iOS gensystek benchmark app by using two configurations: (1) *PSiOS* without policy enforcement, but with CFI checks enabled, and (2) *PSiOS* with policy enforcement. For the latter configuration, we applied several policies on Objective-C selectors (such as NSString and NSBundle) which frequently trigger a policy validation for each of the individual benchmarks. The results of our measurements are shown in Figure 4.

Compared to native execution of the individual benchmarks, *PSiOS* (with policy enforcement enabled) only adds

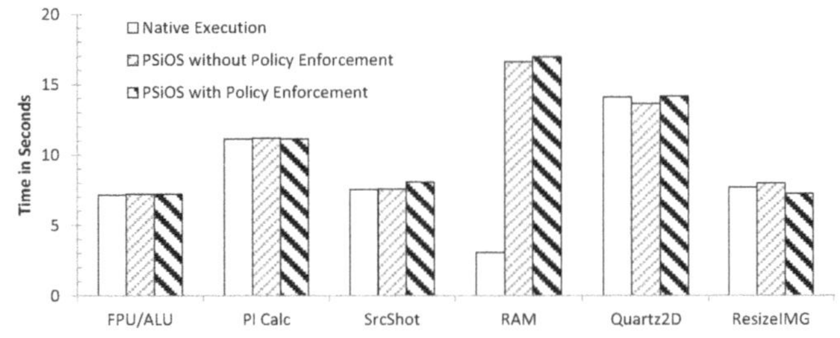

Figure 4: Performance Overhead Measurements with iOS Gensystek App

a negligible overhead ranging from 0.18% to 8%. However, the RAM speed read/write benchmarks induces a significant slowdown of approximately factor 4. Nevertheless, this slowdown is not directly caused by *PSiOS*'s policy enforcement rather by the CFI checks, which are intensively issued for this particular benchmark.

In addition, we measured the overhead of the policy enforcement based on the SpyPhone rule set presented in Section 7.1 using the Instruments application [5]. Figure 5 shows the measured time for native execution, execution with *PSiOS* but without policy enforcement, and execution with *PSiOS* and policy enforcement. Compared to native execution, the load-time overhead of *PSiOS* is only 0.2s. At runtime, the performance overhead for validating an Objective-C message or API call is in the worst-case 0.1s, while sometimes *PSiOS* speeds up the execution. This is due to the fact that we apply the *Replace* enforcement option, which returns shadow data rather than invoking the desired Objective-C call. One of the most called API functions (2163 of 2813 branches and seven percent of the overall execution time) is caused by the Objective-C dispatcher function `objc_msgSend`.

In general, the performance overhead is not noticeable for an end-user, because *PSiOS* enforces access control at the Objective-C layer. Moreover, *PSiOS* applies policy enforcement on the main application binary, while the iOS system libraries and Objective-C frameworks execute without being instrumented by *PSiOS*. Since Apple prohibits developers from using own shared libraries, an adversary cannot deploy own shared libraries to circumvent *PSiOS*.

7.4 Jailbreak Issues

The runtime components of *PSiOS* are completely implemented in one shared library. We opted for this implementation approach, because it enables a system-centric sandboxing solution, and allows every iOS app to immediately benefit from our tools. To install and push our *PSiOS* library on an iOS device, we require a jailbreak of the device, since Apple is closed-source and strictly prohibits any installation of a new shared library. For *PSiOS*, we *only* require a jailbreak for setting a single environment variable, installing a shared library, and allowing our library to rewrite the application code during load-time. Setting the variable as well as installing and signing our library can be easily done by Apple for future iOS releases. Further, for binary rewriting our library only needs to be assigned the dynamic code signing entitlement which allows an application generate code just-in-time.

Information	Policy Rule
Wi-Fi	<rule type="objc" class="NSDictionary" selector="dictionaryWithContentsOfFile:" mode="replace"> <arg number="1" type="string" operator="=" value="/Library/Preferences/SystemConfiguration/com.apple.wifi.plist"/> </rule>
Call history	<rule type="objc" class="NSDictionary" selector="dictionaryWithContentsOfFile:" mode="replace"> <arg number="1" type="string" operator="=" value="/var/mobile/Library/Preferences/com.apple.mobilephone.plist"/> </rule>
Phone number	<rule type="objc" class="NSUserDefaults" selector="valueForKey:" mode="replace"> <arg number="1" type="string" operator="=" value="SBFormattedPhoneNumber"/> </rule>
Address book	<rule type="api" function="ABAddressBookCopyArrayOfAllPeople" mode="replace" />

Table 1: An excerpt of the policy rules we applied to SpyPhone

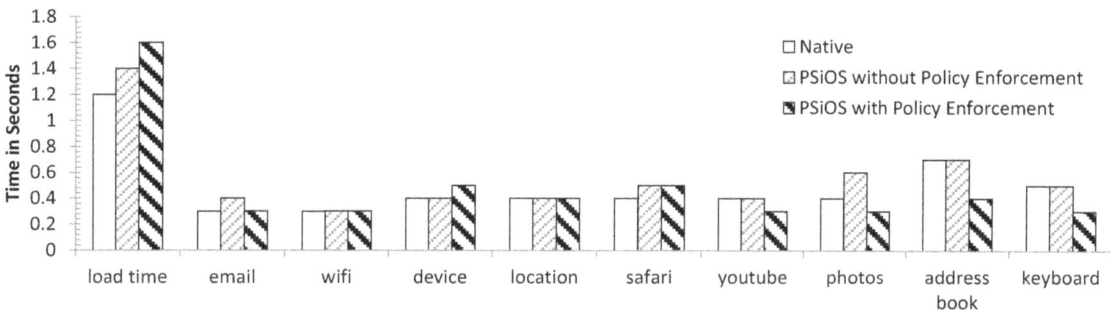

Figure 5: Runtime comparison chart.

Note that many similar approaches targeting Android (for example TaintDroid and AppFence) require the rooting of an Android-powered device as well. Moreover, *PSiOS* does not necessarily require a jailbreak. For instance, *PSiOS* could be provided as a static rewriting tool that Apple applies to all app binaries before releasing them on the App Store. The static analysis (which requires IDA Pro at the moment) could be incorporated using cloud services. Further, app developers could run *PSiOS* before submitting their applications to the App Store. Both approaches are compatible to Apple's signature scheme, since *PSiOS* rewrites the app before it is signed by Apple. We are currently working on the implementation of this approach.

8. RELATED WORK

In this section, we elaborate on related work in the area of mobile security and application sandboxing.

Research on iOS.

The closest work to our framework is PiOS [14] which is a static analysis tool to detect privacy leaks of iOS applications. PiOS generates an application's control-flow graph by backtracking all Objective-C calls and by reconstructing the class inheritance relationships. However, PiOS suffers from some shortcomings: First, its analysis does not cover embedded metaclasses (i.e., root classes of Objective-C classes), which frequently occur in iOS applications. Second, it does not support class clusters (see Section 5). Third, it mainly uses a backtracking of ARM processor registers to determine used classes and selectors. However, as our experiments have shown this approach often fails to resolve the used class. In contrast, *PSiOS* tackles this problem by including the Objective-C sections in its analysis and by retrieving runtime information. To summarize, PiOS is constrained in its analysis because it fails to cover the full picture of the Objective-C runtime, particularly when an adversary deploys obfuscation techniques to circumvent static analysis tools.

Our recent work on iOS has focused on detection of runtime attacks against mobile devices. In particular, our previous work MoCFI [11] enforces control-flow integrity (CFI) on iOS devices running on ARM processors, while CFI [1] has been originally proposed for desktop PCs. However, MoCFI focuses on runtime attacks rather than application sandboxing. Nevertheless, *PSiOS* leverages MoCFI to ensure that our policy enforcement cannot be circumvented by a control-flow attack (see also Section 3).

MobileSubstrate [25] is a framework for jailbroken iOS devices that provides run-time patching of existing programs. To this end, application code can be rewritten to install hooks for Objective-C message handlers and C/C++ functions. The rewriting engine works similar to our approach, however, MobileSubstrate merely provides hooking support. Further, it does not provide protection against runtime attacks which means that a policy framework that is built on top of it could be undermined by runtime attacks.

Research on Android.

In the last years Android has been an appealing subject of research. Kirin [16] is an extended application installer that checks application's permission combinations according to a given policy. Apex [24] goes a step further and allows end-users to choose permissions at install-time.[2] However, since iOS is closed-source, these approaches are not feasible on iOS.

TaintDroid [15] is a framework to detect data leakage attacks on Android. It uses dynamic taint analysis and warns the user whenever sensitive data leaves the device at a taint sink (e.g., the network interface). The AppFence [19] framework builds upon TaintDroid and enables fine-grained privacy rules, and enables the return of shadow data when a policy rule has been violated. However, TaintDroid does not fully cover native code, and could be subverted by runtime attacks. Moreover, it is directly implemented into Android's Java virtual machine (Dalvik). Since iOS use Objective-C rather than the interpreted Java language, it remains open how such a system could be integrated in iOS. Further, in parallel to our work, several security extensions have been proposed to enable fine-grained sandboxing rules (Aurasium [30]) or fine-grained privacy controls [10], but on Android, while we focus on iOS.

Finally, there are several works on Android that tackle privilege escalation attacks at application-level [17, 12, 9]. These attacks are based on the observation that two applications merge their permissions (either directly or indirectly) resulting in a larger sandbox. However, inter-app communication is still an exceptional event on iOS. Nevertheless, in our future work we aim to investigate the feasibility and detection of privilege escalation attacks on iOS with *PSiOS*.

Application Sandboxing Techniques.

Application sandboxing has its origins in the field of software fault isolation (SFI) [29]. The basic idea of SFI is to isolate code and data of an untrusted module in a separate fault domain. Afterwards, the untrusted module is instrumented to ensure that the code cannot jump or reference data beyond its fault domain. In particular, NativeClient [31, 27] builds upon SFI and enables a sandboxed environment for native code plugins in web browsers. Our iOS policy framework is closely related to these works. In particular, we enable SFI for iOS applications, by instrumenting all calls to the Objective-C runtime. However, note that the existing works either focus on entirely different computing platforms or are incompatible to Objective-C.

9. CONCLUSION

In this paper, we introduced *PSiOS*, a novel policy enforcement framework for the closed-source mobile operating system iOS. *PSiOS* provides fine-grained, application-specific, and user-driven sandboxing for third-party applications *without* requiring access to source code. We presented the design decisions behind the framework and discussed in detail the improvements to prior work in this area. We implemented a fully working prototype of *PSiOS* and also illustrated technical details of the implementation. In an empirical evaluation, we demonstrated that *PSiOS* effectively prevents privacy breaches by testing our reference implementation with *SpyPhone* [23], a tool specifically designed to steal data from an iOS device. The runtime overhead introduced by our tool is reasonable and we demonstrated that even complex applications can be instrumented by *PSiOS*. In our future work, we aim to provide *PSiOS* as a static rewriter that can be leveraged by app developers and App Store manufactures to allow fine-grained application sandboxing without requiring users to jailbreak their device.

[2]Note that the standard Android system follows the all-or-nothing principle.

10. ACKNOWLEDGEMENTS

This work has been supported by the Ministry of Economic Affairs and Energy of the State of North Rhine-Westphalia (Grant IV.5-43-02/2-005-WFBO-009).

11. REFERENCES

[1] M. Abadi, M. Budiu, U. Erlingsson, and J. Ligatti. Control-Flow Integrity: Principles, Implementations, and Applications. In *ACM Conference on Computer and Communications Security (CCS)*, 2005.

[2] Aleph One. Smashing the Stack for Fun and Profit. *Phrack Magazine*, 49(14), 1996.

[3] Apple Inc. Mac OS X ABI Mach-O File Format Reference. `http://developer.apple.com/library/mac/documentation/developertools/conceptual/MachORuntime/Mach-O\_File\_Format.pdf`, 2009.

[4] Apple Inc. Objective-C Runtime Programming Guide. `http://developer.apple.com/library/mac/documentation/Cocoa/Conceptual/ObjCRuntimeGuide/ObjCRuntimeGuide.pdf`, 2009.

[5] Apple Inc. Instruments user guide. `https://developer.apple.com/library/mac/\#documentation/DeveloperTools/Conceptual/InstrumentsUserGuide/Introduction/Introduction.html`, 2010.

[6] Apple Inc. Entitlement key reference. `http://developer.apple.com/library/mac/#documentation/Miscellaneous/Reference/EntitlementKeyReference/Chapters/EnablingAppSandbox.html#//apple_ref/doc/uid/TP40011195-CH4-SW1`, 2011.

[7] Apple Inc. App Store Review Guidelines. `https://developer.apple.com/appstore/guidelines.html`, 2012.

[8] Apple Inc. ios sdk release notes for ios 6. `http://developer.apple.com/library/ios/#releasenotes/General/RN-iOSSDK-6_0/index.html`, 2013.

[9] S. Bugiel, L. Davi, A. Dmitrienko, T. Fischer, A.-R. Sadeghi, and B. Shastry. Towards taming privilege-escalation attacks on android. In *Symposium on Network and Distributed System Security (NDSS)*, Feb 2012.

[10] S. Bugiel, S. Heuser, and A.-R. Sadeghi. myTunes: Semantically Linked and User-Centric Fine-Grained Privacy Control on Android. Technical Report TUD-CS-2012-0226, Center for Advanced Security Research Darmstadt (CASED), 2012.

[11] L. Davi, A. Dmitrienko, M. Egele, T. Fischer, T. Holz, R. Hund, S. Nürnberger, and A.-R. Sadeghi. MoCFI: A framework to mitigate control-flow attacks on smartphones. In *Symposium on Network and Distributed System Security (NDSS)*, 2012.

[12] M. Dietz, S. Shekhar, Y. Pisetsky, A. Shu, and D. S. Wallach. QUIRE: Lightweight provenance for smartphone operating systems. In *USENIX Security Symposium*, 2011.

[13] J. Duart. Objective-C helper script. https://github.com/zynamics/objc-helper-plugin-ida.

[14] M. Egele, C. Kruegel, E. Kirda, and G. Vigna. PiOS: Detecting Privacy Leaks in iOS Applications. In *Symposium on Network and Distributed System Security (NDSS)*, 2011.

[15] W. Enck, P. Gilbert, B.-G. Chun, L. P. Cox, J. Jung, P. McDaniel, and A. N. Sheth. TaintDroid: An information-flow tracking system for realtime privacy monitoring on smartphones. In *USENIX Symposium on Operating Systems Design and Implementation (OSDI)*, 2010.

[16] W. Enck, M. Ongtang, and P. McDaniel. On lightweight mobile phone application certification. In *ACM Conference on Computer and Communications Security (CCS)*, 2009.

[17] A. P. Felt, H. Wang, A. Moshchuk, S. Hanna, and E. Chin. Permission re-delegation: Attacks and defenses. In *USENIX Security Symposium*, 2011.

[18] Google. Security and permissions. http://developer.android.com/guide/topics/security/security.html, 2012.

[19] P. Hornyack, S. Han, J. Jung, S. Schechter, and D. Wetherall. These aren't the Droids you're looking for: Retrofitting Android to protect data from imperious applications. In *ACM Conference on Computer and Communications Security (CCS)*. ACM, 2011.

[20] V. Iozzo and R.-P. Weinmann. Pwn2Own contest. http://blog.zynamics.com/2010/03/24/ralf-philipp-weinmann-vincenzo-iozzo-own-the-iphone-at-pwn2own/, 2010.

[21] J. Lowensohn. US Congress probes iOS app makers over privacy. http://bit.ly/IuqIFu, March 2012.

[22] C. Miller and D. Blazakis. Pwn2Own contest. http://www.ditii.com/2011/03/10/pwn2own-iphone-4-running-ios-4-2-1-successfully-hacked/, 2011.

[23] N. Seriot. SpyPhone. https://github.com/nst/SpyPhone, 2011.

[24] M. Nauman, S. Khan, and X. Zhang. Apex: Extending Android permission model and enforcement with user-defined runtime constraints. In *ACM Symposium on Information, Computer and Communications Security (ASIACCS)*, 2010.

[25] saurik. MobileSubstrate. http://iphonedevwiki.net/index.php/MobileSubstrate.

[26] M. J. Schwartz. iOS Social Apps Leak Contact Data. http://www.informationweek.com/news/security/privacy/232600490, 2012.

[27] D. Sehr, R. Muth, C. Biffle, V. Khimenko, E. Pasko, K. Schimpf, B. Yee, and B. Chen. Adapting Software Fault Isolation to Contemporary CPU Architectures. In *USENIX Security Symposium*, 2010.

[28] H. Shacham. The Geometry of Innocent Flesh on the Bone: Return-into-libc Without Function Calls (on the x86). In *ACM Conference on Computer and Communications Security (CCS)*, 2007.

[29] R. Wahbe, S. Lucco, T. E. Anderson, and S. L. Graham. Efficient Software-based Fault Isolation. *ACM SIGOPS Operating Systems Review*, 27(5), 1993.

[30] R. Xu, H. Saidi, and R. Anderson. Aurasium: Practical Policy Enforcement for Android Applications. In *USENIX Security Symposium*, 2012.

[31] B. Yee, D. Sehr, G. Dardyk, J. B. Chen, R. Muth, T. Ormandy, S. Okasaka, N. Narula, and N. Fullagar. Native Client: A Sandbox for Portable, Untrusted x86 Native Code. *IEEE Symposium on Security and Privacy*, 2009.

APPENDIX

A. LOAD COMMANDS

Our static Objective-C analyzer parses every load command included in the Mach-O file header and in particular inspects the load commands listed in Table 2.

Load command	Purpose
LC_SEGMENT	Defines code and data segments, while each segment contains of a set of sections
LC_SYMTAB	Specifies the symbol table
LC_DYLD_INFO and LC_DYLD_INFO_ONLY	Define binding information

Table 2: Relevant Mach-O Load Commands

B. SANDBOXING PROFILE FOR SPYPHONE

```
<rule type="api" function="
      ABAddressBookCopyArrayOfAllPeople:" mode="
      replace" />
<rule type="objc" class="NSDictionary"
      selector="dictionaryWithContentsOfFile:" mode=
          "replace">
  <arg number="1" type="string" operator="=" value=
        "/Library/Preferences/SystemConfiguration/
        com.apple.wifi.plist"/>
</rule>
<rule type="objc" class="NSDictionary"
      selector="dictionaryWithContentsOfFile:" mode=
          "replace">
  <arg number="1" type="string" operator="=" value=
        "/var/mobile/Library/Preferences/com.apple.
        accountsettings.plist"/>
</rule>
<rule type="objc" class="NSDictionary"
      selector="dictionaryWithContentsOfFile:" mode=
          "replace">
  <arg number="1" type="string" operator="=" value=
        "/var/mobile/Library/Preferences/com.apple.
        accountsettings.plist"/>
</rule>
<rule type="objc" class="NSDictionary"
      selector="dictionaryWithContentsOfFile:" mode=
          "replace">
  <arg number="1" type="string" operator="=" value=
        "/var/mobile/Library/Preferences/com.apple.
        commcenter.plist"/>
</rule>
<rule type="objc" class="NSDictionary"
      selector="dictionaryWithContentsOfFile:" mode=
          "replace">
  <arg number="1" type="string" operator="=" value=
        "/var/mobile/Library/Preferences/com.apple.
        mobilephone.settings.plist"/>
</rule>
<rule type="objc" class="NSDictionary"
      selector="dictionaryWithContentsOfFile:" mode=
          "replace">
```

```
<arg number="1" type="string" operator="=" value=
       "/var/mobile/Library/Preferences/com.apple.
       mobilephone.plist"/>
</rule>
<rule type="objc" class="NSDictionary"
     selector="dictionaryWithContentsOfFile:" mode=
          "replace">
 <arg number="1" type="string" operator="=" value=
       "/var/mobile/Library/Preferences/com.apple.
       Maps.plist"/>
</rule>
<rule type="objc" class="NSDictionary"
     selector="dictionaryWithContentsOfFile:" mode=
          "replace">
 <arg number="1" type="string" operator="=" value=
       "/var/mobile/Library/Preferences/com.apple.
       preferences.datetime.plist"/>
</rule>
<rule type="objc" class="NSDictionary"
     selector="dictionaryWithContentsOfFile:" mode=
          "replace">
 <arg number="1" type="string" operator="=" value=
       "/var/mobile/Library/Preferences/com.apple.
       weather.plist"/>
</rule>
<rule type="objc" class="NSDictionary"
     selector="dictionaryWithContentsOfFile:" mode=
          "replace">
 <arg number="1" type="string" operator="=" value=
       "/var/mobile/Library/Preferences/com.apple.
       mobilesafari.plist"/>
</rule>
<rule type="objc" class="NSDictionary"
     selector="dictionaryWithContentsOfFile:" mode=
          "replace">
 <arg number="1" type="string" operator="=" value=
       "/var/mobile/Library/Preferences/com.apple.
       youtube.plist"/>
</rule>
<rule type="objc" class="UIDevice" selector="
       uniqueIdentifier:"
     mode="replace" />
<rule type="objc" class="NSUserDefaults" selector=
       "valueForKey:"
     mode="replace">
 <arg number="1" type="string" operator="=" value=
       "SBFormattedPhoneNumber"/>
</rule>
<rule type="objc" class="NSFileManager" selector="
       directoryContentsAtPath:" mode="replace">
 <arg number="1" type="string" operator="=" value=
       "/var/mobile/Library/Keyboard/"/>
</rule>
<rule type="objc" class="NSFileManager" selector="
       enumeratorAtPath:"
     mode="replace">
 <arg number="1" type="string" operator="=" value=
       "/var/mobile/Media/DCIM"/>
</rule>
```

Listing 1: Full rule set for SpyPhone

On the Effectiveness of API-Level Access Control Using Bytecode Rewriting in Android

Hao Hao, Vicky Singh and Wenliang Du
Dept. of Electrical Engineering & Computer Science, Syracuse University
Syracuse, New York, USA
{hahao,vsignh02,wedu}@syr.edu

ABSTRACT

Bytecode rewriting on Android applications has been widely adopted to implement fine-grained access control. It endows more flexibility and convenience without modifying the Android platform. Bytecode rewriting uses static analysis to identify the usage of security-sensitive API methods, before it instruments the bytecode to control the access to these API calls. Due to the significance of this technique, the effectiveness of its performance in providing fine-grained access control is crucial. We have provided a systematic evaluation to assess the effectiveness of API-level access control using bytecode rewriting on Android Operating System. In our evaluation, we have identified a number of potential attacks targeted at incomplete implementations of bytecode rewriting on Android OS, which can be applied to bypass access control imposed by bytecode rewriter. These attacks can either bypass the API-level access control or make such access control difficult to implement, exposing weak links in the bytecode rewriting process. Recommendations on engineering secure bytecode rewriting tools are presented based on the identified attacks. This work is the first systematic study on the effectiveness of using bytecode rewriting for API-level access control.

Categories and Subject Descriptors

D.4.6 [**Operating Systems**]: Security and Protection

Keywords

Android, Access Control, Bytecode Rewriting

1. INTRODUCTION

In the Android Operating System, application is the main unit that interacts with users. Android applications are implemented in Java, which is then converted into Dalvik bytecode that resides within DEX (Dalvik Executable) files after compilation. During execution, Dalvik bytecode is interpreted by register-based Dalvik Virtual Machine (DVM) before it is executed. Android provides rich Java API methods for applications to access privileged resources as shown in Figure 1. In addition to application isolation enforced by Linux Operating System, Android also supplies a permission system [17] to restrict operations, which an application can perform, on privileged resources. Permissions required by applications are declared in the `AndroidManifest.xml` files. During application installation, users are notified of permissions that are required by the applications. The users have the option to grant these permissions if they are satisfied; otherwise, they could choose to cancel the installation.

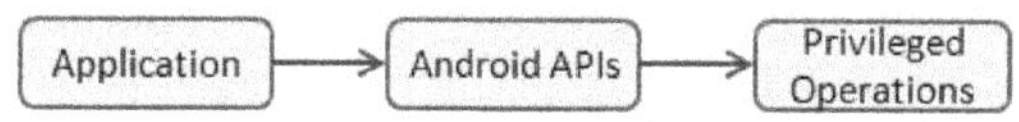

Figure 1: Android API Usage

The current Android permission system is coarse-grained, which causes the applications to be over-privileged. For example, when an application uses the `loadUrl` method of class `android.webkit.WebView` to load `www.facebook.com`, `INTERNET` permission is required. However, because `INTERNET` permission is coarse-grained which does not restrict which domain the application can access, the application is provided with more privilege than it truly needs.

Several methods have been proposed to provide fine-grained access control for Android. The first method is the bytecode rewriting technique [11, 14, 19, 22]. By instrumenting applications' bytecode, security policies could be enforced upon security-sensitive Android API methods. Bytecode rewriting can be done on both Java bytecode and Dalvik bytecode. Some tools actually convert Dalvik bytecode to Java bytecode, and rewrite on Java bytecode, because of the maturity of Java bytecode rewriting tools. Our evaluation could be applied to both Java and Dalvik bytecode rewriting. The second method is native library rewriting, which imposes fine-grained access control by intercepting the native calls to Bionic library [26]. A third way to implement fine-grained access control can be achieved by modifying the Android Operating System [15, 18, 20, 28]. Each of these methods has its own advantages as well as disadvantages. The purpose of this paper is not to compare these methods; instead, we would like to focus on one of the methods, the bytecode rewriting technique. Compared to the other methods, the bytecode rewriting endows more flexibility and convenience, and has been used in a number of existing works.

The existing work point out several places where bytecode rewriting can be attacked if it is not implemented cor-

ASIA CCS'13, May 8–10, 2013, Hangzhou, China.
Copyright 2013 ACM 978-1-4503-1767-2/13/05 ...$15.00.

rectly [14], but no work has systematically studied the potential attacks on the implementation of bytecode rewriting. It is unclear whether there are other attacks beyond what are described in the existing work. Because of the ever-increasing importance of such a technique, having a full understanding of its effectiveness in providing access control is crucial. This paper is the first to conduct such a study.

The purpose of this paper is three-fold: (1) We would like to study all possible attacks against bytecode rewriting, beyond what have been mentioned by the existing work. In fact, we did find some interesting attacks that were not documented in the literature. Without the awareness of these new attacks, the rewritten applications can be compromised by adversary. (2) For all the attacks, we would like to understand why such attacks work, and the fundamental cause for these attacks. Some attacks may appear to be different, but we found them to be quite similar upon close examination. Such observations are described in the paper to give readers a deeper insight into the attacks. (3) Based on our understanding of the attacks, we make recommendations on how the bytecode rewriting can defend against these attacks. These recommendations could help better the design of bytecode rewriting tools.

2. API-LEVEL ACCESS CONTROL USING BYTECODE REWRITING

The key idea with the API-level access control using bytecode rewriting is to provide modified behaviors, which allow fine-grained access control at the Java API level. The modified behaviors are introduced by secure wrappers which function as replacements of sensitive Android APIs. Thus, before invocations of those APIs, specified security policies are enforced. Figure 2 illustrates the approach.

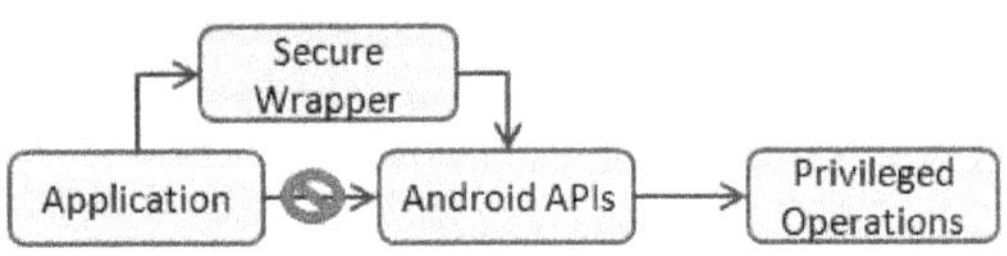

Figure 2: API-Level Access Control

The process of API-level access control using bytecode rewriting involves three steps. After the bytecode file is extracted from the APK [1] and decompiled [2,8], static analysis is performed to identify the occurrences of the security-sensitive APIs in the code. These occurrences are modified, so customized access control logic is added before these APIs are invoked during the runtime. Finally, the bytecode file and all the other resources are repackaged into a single APK file, with a new digital signature generated for the file.

2.1 Dalvik Bytecode Analysis

To identify usages of Android APIs, method-invocation instructions need to be identified in the DEX file. The instructions are listed below [4]:

- `invoke-virtual` (`invoke-virtual/range`): used to invoke a normal virtual method.
- `invoke-direct` (`invoke-direct/range`): used to invoke either a private instance method or a constructor.
- `invoke-static` (`invoke-static/range`): used to invoke a static method.
- `invoke-interface` (`invoke-interface/range`): used to invoke an interface method on an object whose concrete class is not known.
- `invoke-super` (`invoke-super/range`): used to invoke the closest superclass' virtual method.

All of the aforementioned instructions have an argument called *method index*, which represents the method to be invoked. From this index, the fully-qualified method signature can be resolved, including parameters types, return type, as well as package, class, and method name. From this method signature, simple pattern matching can be used to identify the usages of the API methods that need to be restricted.

2.2 Dalvik Bytecode Rewrite

The existing work includes several rewriting mechanisms to enforce more fine-grained access control policies, e.g., placing the reference monitor in another service or within the application. Both of these techniques revise Dalvik bytecode to inject dynamic data checking, based on the policies specified by users. For the first approach [19], i.e., encapsulating sensitive APIs in a separate service, the application's `AndroidManifest.xml` file is modified so that permissions are removed and replaced with new permissions to access the secure service. Within the secure service, more fine-grained access control to the privileged resources are provided. During the process of bytecode rewriting, the calls to sensitive API methods are substituted with a completely new set of calls to the secure service. Because of its fail-safe default property, we did not evaluate this type of code rewriting, but our evaluation may be useful for this technique.

Instead, we focused on the second approach, i.e., the reference monitor on API methods is directly added to the application. After recognizing the utilization of API calls, the rewriting tool places the reference monitor, which uses secure wrappers, on sensitive APIs within the application. Access control policies can be placed on any public interfaces. Therefore, all the Java methods on the call chains would be restricted by the policies. The API method on which to interpose policy could be a method either in a class that can be extended, or in a class that is final and cannot be extended. We are going to discuss these two scenarios in more details below. Note that all the modifications are performed on the bytecode, but for a clearer presentation, we used Java code and not the bytecode in our examples.

Bytecode rewriting for a `non-final` class. Let us use an example to illustrate bytecode rewriting in this scenario. In Android, `WebView` is a `non-final` class. Assume that we want to put some restrictions on its `loadUrl` method, so we only allow the application to load certain URLs. To achieve this goal, we can define a wrapper class for `WebView`, and it is named `SecureWebView`. In this wrapper class, we perform some access control on the URL string before passing it to `WebView`'s `loadUrl`. See the following code:

[1]Android Applications are distributed as a single archive file called APK (Android Application Package) , which holds a DEX bytecode file, `AndroidManifest.xml`, XML resources, and any other resources the applications need.

```
public class SecureWebview extends WebView {
    public void loadUrl(String str) {
        //access control implementation
        ...
        super.loadUrl(str); }}
```

Once we have this wrapper, we replace all the occurrences of `WebView` in the bytecode with `SecureWebView`. See the following examples:

```
"WebView wv;"  is replaced by
"SecureWebView wv;"

"public class MyWebView extends WebView"
    is replaced by
"public class MyWebView extends SecureWebview"
```

Bytecode rewriting for a `final` class. Unlike in the previous method, there are cases in which the API methods belong to a `final` class. In these cases, we cannot define a subclass or use it as the wrapper. Another way to write the wrapper is thus needed. For the purpose of illustration, let us assume that `WebView` is a `final` class, and the following code shows how its `loadUrl` API is used.

```
class UserClass {
public void navigatePage(String url) {
        WebView w = new Webview();
        w.loadUrl(url); }}
```

We would like to put a restriction on the URL string when `WebView`'s `loadUrl` API is called. To achieve this goal, we can create another class `SecureWebView` with a static method `loadUrl`. In this static method, instances of `WebView` are passed as a parameter. All the invocations of `WebView.loadUrl` are replaced with `SecureWebView.loadUrl`, which implements access control. See the following code:

```
public class SecureWebView {
public static void loadUrl(WebView w, String s){
        //access control implementation
        ...
        w.loadUrl(s); }}

public class userClass {
public void navigatePage(String url) {
        WebView w = new Webview();
        /* w.loadUrl(url); is replaced by the
           following */
        SecureWebView.loadUrl(w, url); }}
```

2.3 Bytecode Rewrite Assumption

Android application can introduce its own native code mainly for performance reasons. With the existence of native code, the applications can directly invoke any native library functions of Android framework. Because app introduced native code is running in the same process space as DVM. The applications also have the ability to tamper the integrity of DVM state making method signature information unreliable. Hence, bytecode rewriting can be easily circumvented. Current bytecode rewriter assumes that applications either do not have native code or their native code is blocked from being executed [14]. This assumption is reasonable. From previous work [27], only 4.52% apps have native code. Hence, in our study we evaluated bytecode rewriting technique without considering app introduced native code.

3. EFFECTIVENESS OF API LEVEL ACCESS CONTROL

To measure the effectiveness of the API-level access control using the methods described above, we need to understand what it is trying to protect and how these protected resources are accessed under the hood. As depicted in Figure 3, resources that are out of process boundaries require privileged operations for access. These resources could be hardware devices, kernel data, or data in another process space. Hardware accesses are performed through instructions only executable in the kernel. Data from another process space is retrieved through inter-process communication, which is implemented by the binder driver in the kernel [3, 10, 13]. Access to kernel data and privileged instructions within the kernel are achieved through system calls.

To initiate system calls into the kernel, Android provides standard native C libraries called Bionic libc. These native libraries are bridged to Android Java APIs, so that they can be invoked directly from Java via the Java Native Interface (JNI). For example, `loadUrl` is a Java API in `android.webkit.WebView`. It invokes the native method `nativeLoadUrl` of the shared library `libwebcore.so` via JNI, and the shared library communicates with the kernel via socket system calls in order to access internet resources.

To acquire privileged resources protected by system services, such as locations, the `LocationManager` API interacts with the service called `LocationManagerService` using inter-process communications. The service obtains the location from the GPS device and then returns the result.

As discussed in the previous section, the Android install-time permission system protects these privileged resources, but at a level that is quite coarse-grained. The API-level access control using bytecode rewriting can intercept the calls from applications' own Java code to APIs, and place fine-grained access control before the invocation of those APIs. However, the way in which applications access privileged resources are quite complicated; therefore, it is very natural to question whether the bytecode rewriting is complete enough to prevent applications from directly accessing privileged resources, bypassing the secure wrapper. More specifically, we have the following questions: (1) Can applications directly talk to kernel? (2) Can applications directly invoke native libraries? (3) Can applications directly talk to system services? (4) Can applications hide the usage of APIs to deceive the bytecode rewriting?

These four questions represent four potential paths, as indicated by the numbers in Figure 3, to access privileged resources. If an application takes advantage of any of these paths, it would lead to an attack on API-level access control using bytecode rewriting.

Access Kernel from Application Directly. The first attack depicted in Figure 3 is not possible, because for applications to interact with the kernel, they must be able to invoke system calls. To be specific, they have to be able to execute special instructions provided by the CPU, such as the SWI instruction in ARM processors or the SYSENTER, SYSEXIT instructions in Intel processors [9]. DVM does not provide opcode to directly execute these instruc-

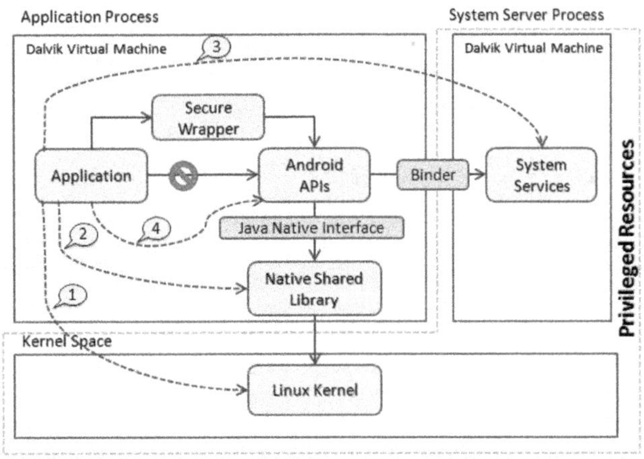

Figure 3: Potential Vulnerabilities of API-Level Access Control

tions. In other words, Java code must invoke the native code, such as the Bionic libraries, to communicate with the kernel. Therefore, the path marked with 1 in Figure 3 is not a feasible attacking path.

Disguise Usage of APIs. Because of Java language flexibility, the fourth attack which is concealing APIs usage is feasible. The most well-known case is Java *reflection*, whose troublesome aspect to bytecode rewriting has been documented by the existing work [14]. Using APIs `java.lang.reflect` package, any fields or methods of classes can be accessed or modified despite public or private at runtime. Thus, besides normal method invocation, reflection provides another way to invoke methods or change the behavior of methods. Reflection even can be used recursively (self-reflection). Hacker may make use of these powerful capabilities to disguise APIs usage. However, current bytecode rewriter [14] proposed defense mechanism against these attacks by detecting and preventing methods invocation of `java.lang.reflect` package. Other than Java reflection, other techniques such as polymorphism may also result in incomplete detection of APIs usages. Dynamic binding [21] enables methods to be bound during runtime but compilation. Any methods that adopts dynamic binding cannot be determined at compile time, and hence can escape from detection. However, researchers have discussed how to perform type inference on Dalvik bytecode effectively [23].

Invoke Native Methods & System Services Directly. Because the first attack path is infeasible and the fourth attack path has been actively discussed, our main focus is second and third attack path. In the following sections, we present our successful attacks with structure of attack mechanism, attack evaluation and recommendation on countermeasure. Attack mechanism illustrates techniques required to make attack successful. Attack evaluation demonstrates how applications can bypass current bytecode rewriter using our attacks. Our evaluation is performed based on Android 4.0.3 [1] SDK. All case studies are tested on a Android 4.0.3 ARM emulator. To help readers understand our attacks, we first present some background materials on Java Native Interface (JNI).

3.1 JNI Native Methods Resolution

To let Java code invoke native code and vice verse, there needs to be an interface. This interface is called Java Native Interface (JNI) [6]. In Android, JNI is the only interface that enables the two-way communication. Native code written in C/C++ and assembly is made available via shared objects (.so files) or shared library. For Java code to make a call into native code in a shared library, first, DVM needs to identify that the Java code tries to invoke native code. This is done through the definition of `native` method in Java, and `native` methods are the entries to native code. Second, DVM needs to identify which functions in the loaded shared library should be invoked, and then execute them. We call these functions *native library functions* to distinguish with the *native Java methods*. The following code snippet demonstrates the usage of native code in Java.

```
1 package edu.com;
2 public class MyClass {
3    native public long myFunc();
4    static {
5       System.loadLibrary("myLib"); }}
```

In the code above, the `MyClass` class from the package `edu.com` declares the `myFunc` method using the **native** modifier, which indicates that this method is actually an entry point to native library function. At line 6, the shared library `myLib` which contains the native functions is loaded into the DVM. Once the library is loaded, DVM will attempt to link the native Java method `myFunc()` to its corresponding native library function. The linking can be conducted in two different ways.

Dynamic Name Resolution. When the first invocation of the `myFunc` method happens, DVM searches in the loaded native libraries to find the corresponding native library function. The searching is performed based on names. DVM deduces the name of the corresponding native library function using the following convention:

1. The name uses `Java/` as the prefix.

2. The package name and class name are appended to the prefix. Since the package name is `edu.com` and the class name is `MyClass`, the resulting string is `Java/edu/com/MyClass`.

3. The name of the native Java method is appended to the above string. Since the method is `myFunc`, the resulting string is `Java/edu/com/MyClass/myFunc`.

4. Any "`/`" in the above string is replaced with "`_`". This results in the final string `Java_edu_com_MyClass_myFunc`.

The DVM will then search the loaded native libraries for a function called `Java_edu_com_MyClass_myFunc`. If it is found, the DVM will set up the internal data structure to direct all future invocations of `myFunc` to this native library function. In the native library, the function corresponding to `myFunc` must be defined in the following way:

```
JNIEXPORT jlong Java_edu_com_MyClass_myFunc(
    JNIEnv* env, jobject thiz);
```

Static Methods Association In the method mentioned above, linking to native functions is decided by DVM and conducted when the first invocation occurs. There is another approach, in which the linking is decided by the native library through registration, not DVM. This registration typically happens when the native library is loaded, at which time a JNI function called `JNI_onLoad` in the native library is invoked. This function registers native functions to Java methods using the `RegisterNatives`. The following code snippet from a native library illustrates the process.

```
static JNINativeMethod method_table [] = {{
"myFunc", "(J)J",
            (void *) myFunc_Implementation }};

extern "C" jint JNI_OnLoad(JavaVM* vm, ... ) {
  jclass c = env->FindClass("edu/com/MyClass");
  env->RegisterNatives(c, method_table, 1); }
```

In the above example, within the `JNI_OnLoad` function, the `myFunc_Implementation` native library function is registered to the `myFunc` native Java method in the `edu/com/MyClass` class. Note that in this case, native function `myFunc_Implementation` does not need to follow the JNI naming convention as that in the previous dynamic name resolution case. When a shared library is loaded using the JNI method `System.loadLibrary`, DVM searches for `JNI_OnLoad` in the shared library. If the function is found, it will be invoked, so DVM can link the future invocation of `myFunc` to `myFunc_Implementation`.

Another technique for static association is to delegate this responsibility to another native function within the library and invoke this function after the library is successfully loaded. This is commonly used in Android framework. The following code snippet from `android_runtime` library demonstrates this technique:

```
static JNINativeMethod camMethods[] = {
{ "native_takePicture", "(I)V",
  (void*)android_hardware_Camera_takePicture},
  ... };
int register_android_hardware_Camera(JNIEnv* env)
{ return AndroidRuntime::registerNativeMethods
  (env, "andriod/hardware/Camera", camMethods,
  ...);}
extern "C" jint
Java_com_android_internal_util_WithFramework_
  registerNatives(JNIEnv* env, ...) {
  return register_android_hardware_Camera(env); }
```

In the above example, `Java_com_android_internal_util_WithFramework_registerNatives` native library function can be invoked by `registerNatives` in `com.android.internal.util.WithFramework` class through the dynamic JNI naming resolution. This function registers other native functions to Java methods of `Camera` class.

4. INVOKE NATIVE METHOD DIRECTLY

From the discussions above, it is clear that to access privileged resources at the kernel, applications have to utilize shared libraries. Apparently, the shared libraries could be either provided by the Android platform or introduced by applications. As we have stated earlier, in this paper, we follow current bytecode rewriter assumption that app introduced native code is not considered.

Bytecode rewriting enforces access control at the API-level in Java bytecode, not in native code. Therefore, if an application can directly invoke the native code without going through those restricted APIs, the API-level access control can be bypassed. This path is illustrated in Figure 3 as path 2 and Figure 4. In this section, we analyzed whether it is possible for application's Java code to directly invoke the native code in the shared libraries provided by the Android platform. We did identify two possible ways. Attack 1 is presented in this section, and attack 2 is presented in the section 5.

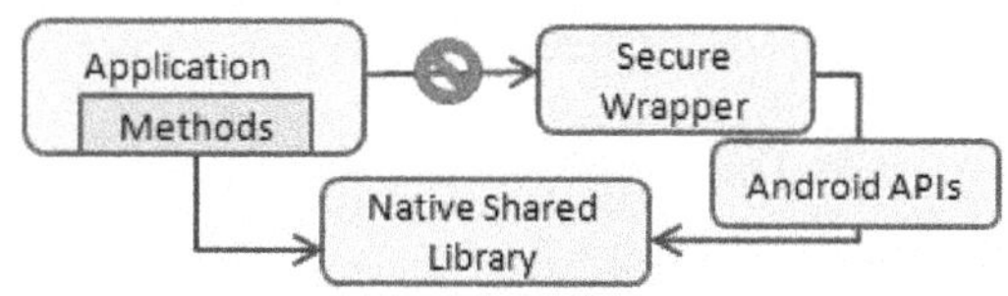

Figure 4: Invoke Native Library Directly

4.1 Exploit JNI Naming Convention

The way how native library functions are bound to native Java methods, either through the JNI naming convention or through registration, ensures that a native library function is only bound to a particular Java API, i.e., that API is the only path leading to the native library function. Therefore, if the bytecode rewriting puts a secure wrapper on that Java API, all the invocation of the corresponding native library function has to go through the wrapper, hence going through the intended access control. If we can find a way to invoke a native library function without going through its corresponding Java API, we can evade the restriction enforced by the wrapper.

Objective and Approach. The objective of our study is to evaluate whether there is a way to invoke native library functions through an unintended Java method, instead of from the one under the protection. In this subsection, we focus on the JNI naming convention, i.e., the dynamic naming resolution. This naming resolution attempts to link a native library function to a unique Java method, achieving a one-to-one mapping. Our approach is to study whether there is a loophole in the naming resolution that allows us to invoke the same native library function from two different Java methods, breaking the one-to-one mapping. Our attack strategy is illustrated in Figure 5.

Attempt 1 (Failed): As demonstrated in the last example, when DVM deduces the name of a native library function, "/" is replaced with "_". One hypothesis is *whether using "_" in the package, class, or method name can cause ambiguity in the naming convention, and thus break the one-to-one mapping.* Let us look the following two classes:

```
//Example 1:
package edu.com;
public class MyClass {
   native public long my_Func(); }

//Example 2:
package edu.com.MyClass;
public class my {
   native public long Func(); }
```

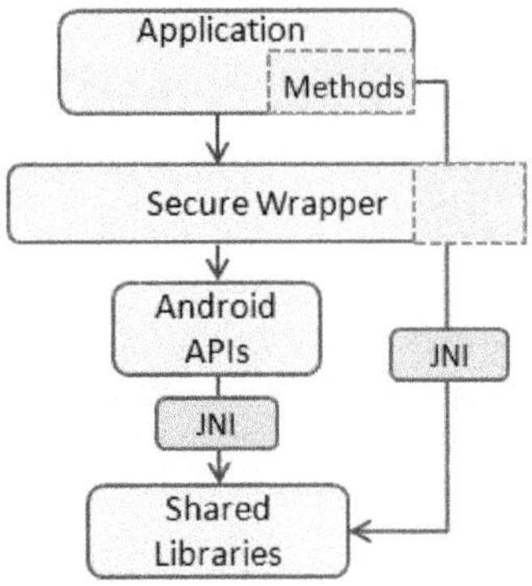

Figure 5: Exploit JNI Naming Convention

It appears that both native Java methods will map to `Java_edu_com__MyClass_my_Func`. Unfortunately (to attackers), Android has already thought about such a potential ambiguity, and its naming resolution replaces any "_" in the name with "_1". Therefore, the first method maps to `Java_edu_com_MyClass_my_1Func`, while the second maps to `Java_edu_com_MyClass_my_Func`. Other unicode characters, such as "&", ";", "[", etc., are also replaced in the same fashion. Our attempt failed.

Attempt 2 (Succeeded) Our failed attempt does lead to another hypothesis: *what if we put the number "1" in front of a package, class, or method name, can we cause ambiguity?*

The reason why the JNI naming convention can use "_1" to avoid the ambiguity caused by "_" is that in Java, no package, class, or method name can start with a digit [7]. Any attempts to do so will encounter compile-time errors. After digging into this, we realize that this naming violation is only detected at the compile time, not during the run time. Therefore, an attacker can add the digit to the beginning of a name at the bytecode level, i.e., through his/her own bytecode rewriting. We tried this, and have successfully invoked a native library function from an unintended Java API. We will show an example in the following.

Based on the JNI naming convention, the native library function `Java_edu_com_MyClass_my_1Func` is supposed to be mapped only to the `my_Func` function defined in the first example in our previous attempt. However, we can successful call this same native library function through a different Java method, i.e., the `1Func` method. It should be noted that in our Java code, we used `Func()` to pass the compiler, and then changed `Func()` to `1Func()` directly on the bytecode. Here is what the program looks like if we convert the modified bytecode back to Java code:

```
package edu.com.MyClass;
public class my {
   native long 1Func();
   static { System.loadLibrary('myLib'); }}
```

Because now two methods can invoke the same native library function, if the API-level access control only block the intended API, i.e., `my_Func`, in order to prevent applications from accessing privileged operations through the corresponding native library function, the attacker can bypass this blocking and invoke the native library function through another API.

4.2 Case Study

Our attack works on the native functions that have "_1" in their names. We searched in the shared libraries provided by Android, and found some cases. For example, the `sqlite_jni` library contains a function called `Java_SQLite_Database_error_1string`, with the "_1" pattern in the name. This function maps to the `error_string` method in the `Database` class of the `SQLite` package. We evaluated our attack based on this case. The attack objective is to invoke native library function `Java_SQLite_Database_error_1string` of `sqlite_jni` library from a different Java method `1string` of class `SQLite.Database.error`.

```
package SQLite.Database;
public class error {
   public static native String 1string(...);
   static { System.loadLibrary('sqlite_jni'); }}
```

We had to rename Java method `string` to `1string` directly on bytecode after compilation to deceive compiler. Figure 6 taken from logcat output demonstrates that by invoking `SQLite.Database.error.1string` we successfully invoked `Java_SQLite_Database_error_1string`. Because of different method signatures, any access control policies placed on the `SQLite.Database.error_string` method by current bytecode rewriter would not affect invocation of method `SQLite.Database.error.1string`. Restrictions on the application become ineffective.

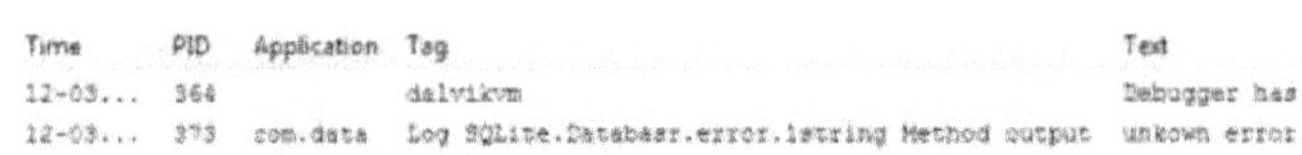

Figure 6: Successful Attack on SQLite app

Actually, if "_" appears in the name of a package, class, or native Java method, then the corresponding native library function must have "_1" in its name, creating an opportunity for our attacks. Having "_" in these names is not a common practice but at the same time is not totally absent. Although today, luckily, there are not many native library functions with this pattern, the existence of this loophole can cause many problems when more native libraries are added to Android. To ensure the security of the API-level access control, this loophole needs to be fixed.

4.3 Recommendations

To circumvent exploiting JNI name convention, bytecode rewriter can perform checking on naming convention of type descriptor and method. If any of them starts with numbers, bytecode rewriter should remove the digit as it is illegal.

5. EXPLOIT JAVA CLASS RELOADING

Another way to defeat the secure wrapper is to somehow modify the implementation of the APIs that the wrapper is trying to protect. If we can do this, the unauthorized access can be launched from the modified implementation, which is "behind the defense line" protected by the wrapper.

Objective and Approach. To achieve the above goal, we need to find ways to replace an existing trusted Java class (provided by Android) with another Java class provided by the application (untrusted). The application's Java class (and its methods) will have the same name as the existing Java class (and its methods). This way, the application's class can invoke all the native functions that the existing Java class can invoke. These invocations cannot be restricted by the API-level access control placed in the secure wrapper. We would like to exploit problems with the class loading mechanism in Android to achieve our goal. Figure 7 depicts our strategy.

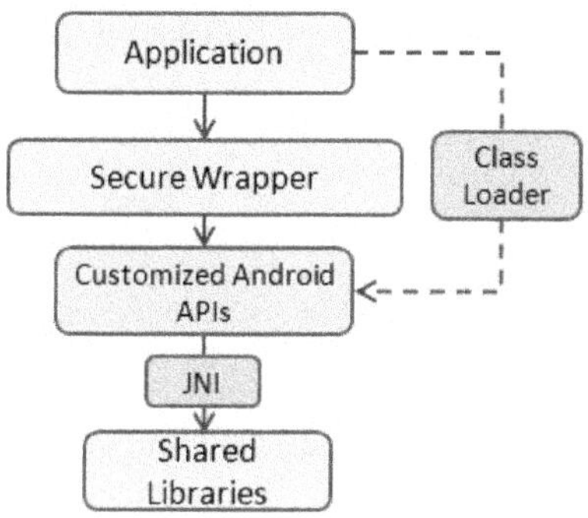

Figure 7: Exploit Reload Java Classes

In this study, we use `android.hardware.Camera` class as an example. This class is provided by Android, and is considered as trusted. A malicious application wants to redefine this class, so when this name is used, the redefined class is used, not the one provided by Android. Here is the code snippet of the redefined class:

```
package android.hardware;
public class Camera{
   final public void someFunc() {
       // Calling the privileged function
       privilegedFunc(); }
   native void privilegedFunc(); }
```

In the above code, the native function `privilegedFunc` is one of the APIs protected by the secure wrapper in the bytecode rewriting technique, but `someFunc` is not restricted (the name of this method can be arbitrary). If we use the original `android.hardware.Camera` class, our access to the `privilegedFunc` is restricted because of the wrapper placed on this class. However, if we can get DVM to load our redefined class, instead of the original Android class, and invoke `privilegedFunc` through `someFunc`, the access of the privileged function becomes unrestricted.

Attempt 1 (Failed): We attempt to load our redefined class (stored in the `Camera.apk` file) in DVM. Android provides `dalvik.system.DexClassLoader` to allow applications to dynamically load classes. We did the following:

```
DexClassLoader classLoader = new DexClassLoader(
    "Camera.apk", ..., getClassLoader());
Class mClass = classLoader.loadClass("android.
    hardware.Camera");
android.hardware.Camera c = (android.hardware.
    Camera)mClass.newInstance();
//Access the privileged native code through
    someFunc()
c.someFunc();
```

Unfortunately, the above attack does not work. Apparently, our redefined class was not loaded. We looked into the code and find out where the problem is. In the class `DexClassLoader`, we found that using `DexClassLoader`, a Java class cannot be loaded again if it has already been loaded by this class loader or by its parent class loaders [2]. This loading policy is implemented in `DexClassLoader`.

Attempt 2 (Succeeded): To make our previous attempt successful, we need to change the loading policy, which means we need to change `DexClassLoader`. This is impossible without modifying the Android operating system. Fortunately, DVM allows us to write our own customized class loader. Our idea is to use this customized class loader to load our redefined `android.hardware.Camera` class, and then somehow use this class in the application, instead of the one provided by Android. DVM does allow two classes with the same name to coexist, as long as they are in separate class loaders.

In DVM, the `DexClassLoader` class is a subclass of `BaseDexClassLoader`, and the loading policy is implemented in the `loadClass` method. We can write another subclass of `BaseDexClassLoader`, override its `loadClass` method, but we skip the loading policy enforcement logic in this method, i.e., we load the class without checking whether it is already loaded by itself or by its parent class loaders. Here is the implementation of our class loader:

```
public class MyDexLoader extends
             BaseDexClassLoader {
   // Constructor omitted
   @Override
   public Class<?> loadClass(String s) {
      Class c;
      try { c = super.findClass(s);
         return c;
      } catch (ClassNotFoundException e) {
         // handling the exceptions
      }  return null; } }
```

In our customized class loader, we use `super.findClass` to load the class. Unlike the implementation in the class `DexClassLoader`, we do not check whether the class is already loaded by its parent class loaders. Now, we can use `MyDexLoader` to load our redefined `android.hardware.Camera` class without any problem.

There is another challenge. The `privilegedFunc` native function in our redefined `Camera` class needs to be linked to its corresponding native library function. Here is the problem: In Android, a Java method can only be linked to a native library function if they are both associated with the same class loader. In our case, the redefined `Camera` class is associated with our class loader `MyDexLoader`, but the native library will be associated with another class loader if it loaded the library [3].

We tried to resolve the challenge from association of native library and class loader, but without much success, mostly

[2]Each class loader needs to have a parent class loader, unless it is the first one.

[3]A native library is associated with the class loader who loads it.

because the association policy is enforced by DVM, which cannot be overwritten by Java code. However, we investigated that if a class loader is not updated with the list of libraries that loaded in DVM, then it is possible to reload those libraries.

Our investigation revealed that the default class loader (bootstrap) of DVM is not aware of `android_runtime` library being loaded. Hence making Java classes associated with this library a candidate for attack. `registerNatives` native function in this library would link plenty of internal native functions with Android Java classes.

5.1 Case Study

Attack Background. We performed our attack on a camera application [5]. Its main functionalities are taking pictures and then save the taken photos under the external SD card directory.

```
public boolean onOptionsItemSelected(MenuItem
    item) {
/* normal method invocation */
camera.takePicture(...);

/* equivalent method invocation using reflection
   */
Class c=Class.forName("android.hardware.Camera");
Method m=c.getDeclaredMethod("takePicture", ...);
m.invoke(camera,...); */  }
```

Suppose current bytecode rewriter enforced finer access control on method `Camera.takePicture` which specifies that pictures can only be taken at daytime between 8am to 6pm.

```
public class SecureCamera{
public static void takePicture(Camera camera,
    ...){
    Time now = new Time();
    now.setToNow();
    if(now.hour > 8 && now.hour < 18) {
       camera.takePicture(...); }}}
```

Current bytecode rewriter would also put some restrictions on *reflection* method `java.lang.reflect.Method.invoke` to prevent method `android.hardware.Camera.takePicture` from being invoked through `invoke`, as discussed in section 3.

```
public class SecureMethod{
public static void invoke(Method m, ...){
    String name = m.getDeclaringClass().getName()
        + "." + m.getName());
    if(!"android.hardware.Camera.takePicture".
        equals(name)) { m.invoke(); }}
```

Attack Detail. By exploiting Java class reloading, malicious developer can still take photos despite the access control policy. The attack mainly involves three steps.

First reload redefined `android.hardware.Camera` class into a new class loader. Without the needs of registration to native functions through name mapping, non native Java method can be renamed in the user-defined class. Method `takePicture` is a non native Java method so that we can rename it to `takeMyPicture`.

```
package android.hardware;
public class Camera {
 public void takeMyPicture(...) {...}}
```

However, `takePicture` invokes `native_takePicture` native Java methods which contains actual implementation of taking pictures. Hence, we need to associate native Java methods of `Camera` class with corresponding native library functions. Refer to the code snippet in section 3.1, Java method `com.android.internal.util.WithFramework.registerNatives` in `android_runtime` library registers native library functions to `Camera` native Java methods. By reloading `android_runtime` library and invoking method `registerNatives`, native library functions are linked to redefined `Camera` class.

Then `takeMyPicture` can be invoked using reflection.

```
// Create a customized class loader
MyDexLoader ld = new MyDexLoader(...);

// Load redefined Camera class
Class c = l.loadClass("android.hardware.Camera");
Class util = l.loadClass("com.android.internal.
    util.WithFramework");
Method m = util.getDeclaredMethod("
    registerNatives", ...);
m.invoke(...);

// Invoke takeMyPicture method using reflection
m = c.getDeclaredMethod("takeMyPicture", ...);
m.invoke(...);   ... }
```

Because the method name `android.hardware.Camera.takePicture` has changed to `android.hardware.Camera.takeMyPicture`, invocation of `takeMyPicture` cannot be restricted. Figure 8 demonstrates the application can still take photos even though current bytecode rewriter enforced access control policy that pictures cannot be taken during nighttime.

To estimate the potential impact of this attack, we have identified all the native libraries that are not associated with default class loader in Android. The full list can be found in Appendix A. The functions in these native libraries are mapped to the Java methods in various Java classes. In the Appendix B, we present all Android Java classes that can be registered to internal native functions in `android_runtime` library. Several Java classes on this list seem security sensitive. Some examples are `shutdown` method of class `android.os.Power` and `getWifiState` method of class `androd.net.wifi.WifiManager`. If the API-level access control wants to restrict the access to the APIs in this class, it needs to prevent our attacks; otherwise, its access control can be circumvented.

5.2 Recommendations

To cope with the problems caused by the class reloading, one possible way is to stop application's Java code from reloading preloaded Android core classes in any class loader instance. Digging deep into the call chain of the `findClass` method, which is used to perform class loading, we found that eventually `defineClass` native method in `DexFile` is invoked, and its corresponding native library function does the actual class loading task. Therefore, if API-level access control using bytecode rewriting is applied to prevent loading of Android classes, restriction on `BaseDexClassLoader`.

(a) Picture Saved

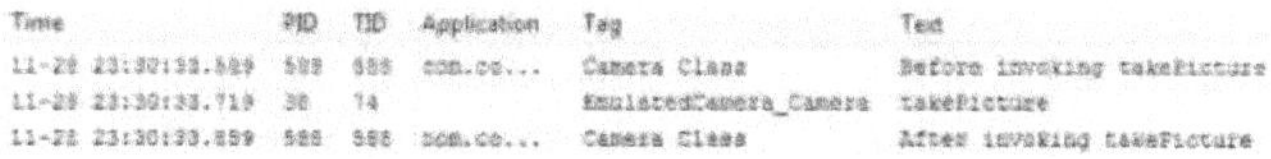

(b) Logcat Output

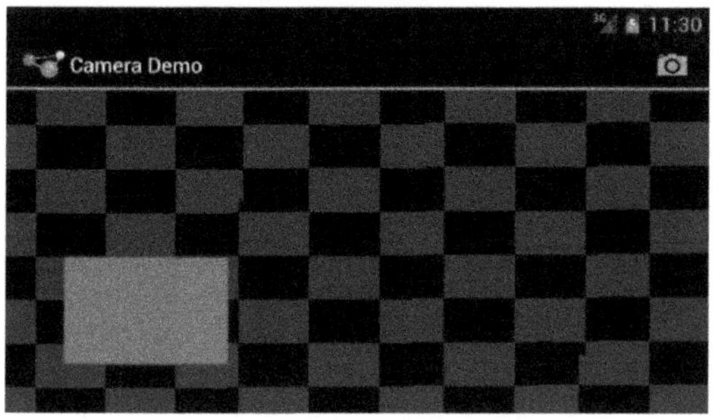

(c) Screenshot

Figure 8: Successful Attack on Camera App

`loadClass` alone is not enough. All the invocations of methods within the call chain from `findClass` in the class `BaseDexClassLoader` to `loadClass` in `DexFile` should be restricted.

6. ACCESS SYSTEM SERVICE DIRECTLY

In Android, some of the implementation of the APIs are performed through the system process via inter-process communication. APIs invoke RPC stubs which initiates inter-process requests to ask system services to perform actual actions [17]. If applications can directly communicate with the system services without going through those APIs, the access control put on those APIs can be bypassed. The objective of this study is to find ways to achieve this. Figure 9 illustrates our strategy.

6.1 Exploit Customized RPC Stubs

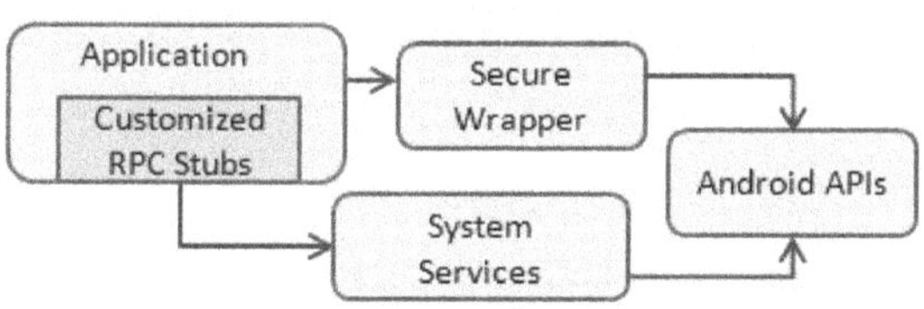

Figure 9: Access System Services Directly

Let us use location service as an example. Android provides an API `getLastKnownLocation` in the `android.location.LocationManager` class, and this API returns the last known location. The call to this API will be directed to `ILocationMangaer$Stub$Proxy`, which is the RPC stub to the system service `LocationManagerService`. If bytecode rewriting wants to put some fine-grained access control on the location access, it usually adds a layer of access control to this API. Unfortunately, there is a way to directly send a request to `LocationManagerService`, without going through the `getLastKnownLocation` API. To achieve this, the application can write its own RPC stub to communicate with `LocationManagerService`. We did a case study on geolocation app to demonstrate how exploit customized RPC stubs can bypass current bytecode rewriter.

6.2 Case Study

We evaluated our attack on a geolocation application which get last known location and display on the map.

```
import android.location.Location;
import android.location.LocationManger;

/* Return a handler to geolocation service */
LocationManager ser = (LocationManager)
    getSystemService(LOCATION_SERVICE);
/* Retrieve last know location */
Location loc = ser.getLastKnowLocation(...);
```

Suppose current bytecode rewriter enforced finer access control policy that the application can only retrieve location information when the location is within Alaska.

```
class SecureLocationManager extends
    LocationManager{
public Location getLastKnownLocation(...) {
  Location loc = super.getLastKnownLocation(...);
  if(loc.getLatitude()>60&&loc.getLatitude()<70&&
     loc.getLongtitude()>140&&loc.getLongtitude()
         <160) {
     return loc; }}
```

However, malicious app could introduce customized RPC with different method signature.

```
package my.location;
/* User-defined RPC stub class */
public interface LocMgr extends android.os.
    IInterface {
    public static abstract class Stub extends
        android.os.Binder implements
            my.location.LocMgr {...}}
```

```
1 import my.location.LocMgr;
2
3 IBinder b = android.os.ServiceManager.getService(
      LOCATION_SERVICE);
4 LocMgr sLocationManger = LocMgr.Stub.asInterface(
      b);
5 Location loc = sLocationManger
6         .getLastKnownLocation(...);
7 }
```

The above code snippet illustrates how customized RPC stub is used. Line 1 returns an IBinder instance, which is the gateway to the RPC interface. Line 2 establishes local clients by converting the IBinder instance to an instance of `LocMgr`, so that any invocation of inter-process calls are thereafter acted as regular method calls on `LocMgr`.

Note that `android.os.ServiceManager` is hidden in Android framework, which means the compiler would throw an error when an application tries to invoke any methods of it. At runtime `android.os.ServiceManager` is already loaded by DVM when application is started. However, one can cheat the compiler by writing a dummy `android.os.ServiceManager` to overcome compile time error.

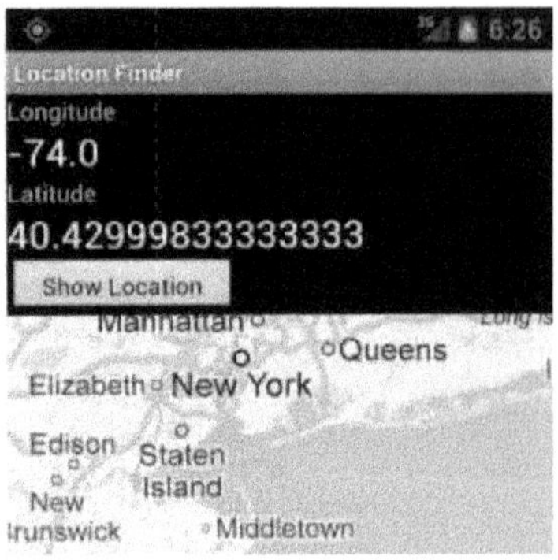

Figure 10: Successful Attack on Geolocation App

`my.location.LocMgr.getLastKnownLocation` has different method signature compared to that of `android.location.LocationManager.getLastKnownLocation`. Malicious application can escape access control check of current bytecode rewritier. Figure 10 demonstrates last known location is successfully retrieved by the application despite the location is at New York City.

6.3 Recommendations

The above problem is not difficult to fix; the easiest fix is to apply the API-level access control on `android.os.ServiceManager`'s `getService` API, so application's Java code cannot use this API to get system services. It is very rare that generic application would need to create custom client for system services, since rich android APIs are already provided to handle communications with system services. If we do not want to block this API, we can apply the bytecode rewriting technique to `android.os.Binder`.

7. RELATED WORK

Fine-grained access control using bytecode rewriting. Several researchers have proposed to implement fine-grained access control using bytecode rewriting on Android platform [11,14,19,22]. I-ARM embedded In-App Reference Monitors into Android application [14]. Via instrumenting the bytecode, security policies are interposed on all secure-sensitive API calls. Several other work [11,19,22] placed Reference Monitors in other Android services and substituted the calls to API methods with calls to other services. The work in [19,22] removes all the assigned permissions from the applications, and reassigns the applications with fine-grained permissions. The executions of all secure-sensitive APIs will be delegated to the services. In-vivo uses another approach, it only uses the services to checks whether the application can invoke sensitive API calls or not [11], if it is approved, executions of APIs still take place inside the applications. Some researchers have proposed similar work on fine-grained access control using bytecode rewriting for Java platform [12,16,24]. Ajay et. al insert runtime checks into Java application code using rewriting [12]. Rudy and Wallach built three bytecode rewriting frameworks to add security semantics to the Java Virtual Machine [24]. Erlingsson has discussed practical applications of Java bytecode rewriting [16].

I-ARM pointed out some potential attacks on bytecode rewriting, i.e., usage of Java reflection and `ClassLoader` [14]. However, the main focus of the paper is on the implementation of a bytecode rewriting framework, analysis on potential attacks is quite brief. Our study is the first one presenting a systematical evaluation on API-level access control using bytecode rewriting. We did a comprehensive study on the attacking surfaces of the API-level access control using bytecode rewriting.

Fine-grained access control through library interposition. Instead of using bytecode rewriting, Aurasium accomplishes fine-grained access control via instrumenting Android Bionic libraries [26]. It encapsulates the original Bionic libraries functions inside its interposition routines, so fine-grained security policies can be enforced. These new routines will be invoked when applications invoke the routines in the Bionic libraries. This is achieved through the dynamic linking mechanism used by the native libraries. Compared to bytecode rewriting, this technique is more secure, as it does not worry about whether applications can directly jump to native library functions. However, by conducting access control at the lower level, it loses the rich context information that is available to bytecode rewriting. Such context is beneficial for enforcing more fine-grained security policies.

Fine-grained access control through OS modification. Other researches have modified the Android OS to monitor the data flow, detect data leakage and permission exploits [15,18,20,28]. TISSA modified the Android OS to provide fine-grained protection on personal information [28]. TaintDroid performs taint analysis on applications and checks for privacy leakage [15]. Hornyack et al. modified the Android OS to replace private data with dummy data when the data are offered to applications [18]. It also blocks network transmissions if they contain data that are not supposed to be sent out. The main drawbacks of the OS-modification approach is the lack of flexibility and the need to update the phones.

Bytecode rewriting for other purposes. Several works use bytecode rewriting to instrument Android applications [23,25], although not for enforcing API-level access control. Adsplit used bytecode rewriting to remove advertisement components to another advertisement service [25]. The work in [23] used bytecode rewriting to bind applications to fake Market app instead of the genuine one, so attacks can be launched in Google In-App Billing.

8. CONCLUSION AND FUTURE WORK

API-level access control using bytecode rewriting is a common technique used by existing work to provide fine-grained access control in Android. To fully understand how secure this technique is, we have conducted a systematic study on the effectiveness of the implementation of this technique. We have identified several new attacks. Although all the problems are fixable, our work manifests the need to perform more static analysis and dynamic checking should be performed to fulfill an effective API-level access control using bytecode rewriting. Our work can be beneficial to those who make use of bytecode rewriting or those who develop bytecode rewriting tools.

9. REFERENCES

[1] Android developer. http://www.developer.android.com/about/versions/android-4.0.3.html.

[2] Android reverse engineering honeynet project. http://www.honeynet.org/node/783.

[3] Binder. http://www.developer.android.com/reference/andrdoid/os/Binder.html.

[4] Bytecode for the dalvik vm. http://www.source.android.com/tech/dalvik-bytecode.html.

[5] Commonsware camera application. http://github.com/commonsguy/cw-omnibus.

[6] Jni tips. http://www.developers.android.com.guide/practices/jni.html.

[7] Naming a package. http://www.docs.oracle.com/javase/tutorial/package/namingpkgs.html.

[8] smali: An assembler/disassembler for android's dex format. http://www.code.google.com/p/smali.

[9] Swi handlers. http://www.infocenter.arm.com/help/index.jsp?topic=/com.arm.doc.dui0040d/Cacdfeci.html.

[10] Android binder: Android interprocess communication. 2011.

[11] A. Bartel, J. Klein, K. Allix, Y. Traon, and M. Monperrus. Improving privacy on android smartphones through in-vivo bytecode instrumentation. *CoRR*, abs/1208.4536, 2012.

[12] A. Chander, J. C. Mitchell, and I. Shin. Mobile code security by java bytecode instrumentation. pages 1027–1040, 2001.

[13] E. Chin, A. P. Felt, K. Greenwood, and D. Wanger. Analyzing inter-application communication in android. *In Proceedings of the 9th International Conference on Mobile system, application and serivecs*, 2011.

[14] B. Davis, B. Sanders, A. Khodaverdian, and H. Chen. I-arm-droid: A rewriting framework for in-app reference monitors for android applications. *In IEEE Mobile Security Technologies*, 2012.

[15] W. Enck, P. Gilbert, B. Chun, L. P. Cox, J. Jung, P. McDaniel, and A. N. Sheth. Taintdroid: an information-flow tracking system for realtime privacy monitoring on smartphones. *In Proceedings of the 9th USENIX Conference on Operating Systems Design and Implementation, pp.1-6*, 2010.

[16] U. Erlingsson. The inlined reference monitor approach to security policy enforcement. 2004.

[17] A. P. Felt, E. Chin, S. Hanna, D. Song, and D. Wagner. Android permissions demystified. *In Proceedings of the 18th ACM Conference on Computer and Communication Security*, 2011.

[18] P. Hornyack, S. Han, J. Jung, S. Schechter, and D. Wetherall. "these aren't the droids you're looking for": Retrofitting android to protect data from imperious applications. *In Proceedings of the 18th ACM conference on Computer and communication security*, 2011.

[19] J. Jeon, K. K. Micinski, and J. A. Vaughan. Dr. android and mr. hide: Fine-grained security policies on unmodified android. *Technical Report, Department of Computer Science, University of Maryland*, 2011.

[20] M. Nauman, S. Khan, and X. Zhang. Apex: Extending android permission model and enforcement with user-defined runtime constraints. *in Proceedings of the 5th ACM Symposium on Information, Computer and Communications Security*, 2010.

[21] D. Poo, D. Kiong, and S. Ashok. Object-oriented programming and java, 2nd edition. 2007.

[22] N. Reddy, J. Jeon, J. Vaughan, T. Millstein, and J. Foster. Application-centric security policies on unmodified android. *UCLA Computer Science Department, Technical Report*, 2011.

[23] D. Reynaud, E. C. R. Shin, T. R. Magrino, E. X. Wu, and D. Song. Freemarket: Shopping for free in android applications. *In Proceedings of the 19th Annual Symposium on Network and Distributed System Security*, 2012.

[24] A. Rudys and D. S. Wallach. Enforceing java run-time properties using bytecode rewriting. 2002.

[25] S. Shekhar, M. Dietz, and D. S. Wallach. Adsplit: Separating smartphone advertising from applications. *CoRR,abs/102.4030*, 2012.

[26] R. Xu, H. Saidi, and R. Anderson. Aurasium: Practical policy enforcement for android applications. *In Proceedings of the 21st USENIX Security Symposium*, 2012.

[27] Y. Zhou, Z. Wang, W. Zhou, and X. Jiang. Hey, you, get off of my market: Detection malicious apps in official and alternative android markets. *In Proceedings of the 19th Annual Symposium on Network and Distributed System Security*, 2012.

[28] Y. Zhou, X.Zhang, X. Jiang, and V. Freeh. Taming information-stealing smartphone applications(on android). *Trust and Trustworthy Computing, pp.93-107*, 2011.

APPENDIX

A. LIBRARIES NOT PRELOADED

The following are shared native libraries provided by the Android platform. These libraries are stored under the directory /system/lib, and default class loader is not aware of their existence. Applications can reload them during runtime. If bytecode rewriting puts any access control on Java native APIs in order to restrict applications' access to the native library functions in these libraries, the access control can be bypassed using our attacks.

```
libEGL.so,  libETC1.so,  libFFTEm.so
libGLESv1_CM.so,  libGLESv1_enc.so
libGLESv2.so,  libGLESv2_dbg.so
libOpenSLES.so,  libOpenglSystemCommon.so
libRS.so,  libSR_AudioIn.so
libWnnEngDic.so, libWnnJpnDic.so
lib_renderControl_enc.so,  libandroid.so
libandroid_runtime.so,  libandroid_servers.so
libaudioeffect_jni.so,  libaudioflinger.so
libbcc.so,  libbcc.so.sha1
libbcinfo.so,  libbinder.so
libc.so,  libc_malloc_debug_leak.so
libc_malloc_debug_qemu.so,  libcamera_client.so
libcameraservice.so,  libclcore.bc
libcrypto.so,  libctest.so
libcutils.so,  libdefcontainer_jni.so
libdiskconfig.so,  libdl.so
libdrm1.so,  libdrm1_jni.so
```

libdrmframework.so, libdvm.so
libeffects.so, libemoji.so
libexif.so, libexpat.so
libext4_utils.so, libfilterfw.so
libfilterpack_imageproc.so
libgabi++.so, libgui.so
libhardware.so, libhardware_legacy.so
libharfbuzz.so, libhwui.so
libicui18n.so, libicuuc.so
libinput.so, libjni_latinime.so
libjni_mosaic.so, libjnigraphics.so
libjpeg.so, liblog.so
libm.so, libmedia.so
libmediaplayerservice.so
libmtp.so, libnativehelper.so
libnetutils.so, libnfc_ndef.so
libpagemap.so, libpixelflinger.so
libpower.so, libpowermanager.so
libreference-ril.so, libril.so
libsensorservice.so, libskia.so
libsonivox.so, libspeexresampler.so
libsqlite.so, libsqlite_jni.so
libsrec_jni.so, libssl.so
libstagefright.so
libstagefright_amrnb_common.so
libstagefright_foundation.so
libstagefright_omx.so
libstagefright_soft_aacdec.so
libstagefright_soft_amrdec.so
libstagefright_soft_g711dec.so
libstagefright_yuv.so
libstdc++.so, libstlport.so
libsurfaceflinger.so
libsurfaceflinger_client.so
libsystem_server.so
libsysutils.so, libthread_db.so
libttscompat.so, libttspico.so
libui.so, libusbhost.so
libutils.so, libvariablespeed.so
libvorbisidec.so
libwebrtc_audio_preprocessing.so
libwilhelm.so, libwnndict.so
libwpa_client.so, libz.so

B. CLASSES LINKED TO ANDROID_RUNTIME

The following are Android framework Java classes that link to the native library functions in the android_runtime library. If an API-level access control is put upon these classes to restrict the applications' access of their linked native library functions, our attacks can bypass this access control by directly invoking those native library functions.

/android/debug/JNITest
/com/android/internal/os/RuntimeInit
/android/os/SystemClock
/android/util/EventLog
/android/util/Log
/android/util/FloatMath
/android/text/format/Time
/android/pim/EventRecurrence
/android/content/AssetManager
/android/security/Md5MessageDigest
/android/text/AndroidCharacter
/android/text/AndroidBidi
/android/text/KeyCharacterMap
/android/os/Process
/android/os/Binder
/android/view/Display
/android/nio/utils
/android/graphics/PixelFormat
/android/graphics/Graphics
/android/view/Surface
/android/view/ViewRoot
/com/google/android/gles/jni/EGLImpl
/com/google/android/gles/jni/GLImpl
/android/opengl/jni/GLES10
/android/opengl/jni/GLES20
/android/graphics/Bitmap
/android/graphics/BitmapFactory
/android/graphics/BitmapRegionDecoder
/android/graphics/Camera
/android/graphics/Canvas
/android/graphics/ColorFilter
/android/graphics/DrawFilter
/android/graphics/Interpolator
/android/graphics/LayerRasterizer
/android/graphics/MaskFilter
/android/graphics/Matrix
/android/graphics/Movie
/android/graphics/NinePatch
/android/graphics/Paint
/android/graphics/PorterDuff
/android/graphics/Rasterizer
/android/graphics/Region
/android/graphics/Shader
/android/graphics/Typeface
/android/graphics/Xfermode
/android/graphics/YuvImage
/com/android/internal/graphics/NativeUtils
/android/database/CursorWindow
/android/database/SQLiteCompiledSql
/android/database/SQLiteDatabase
/android/database/SQLiteDebug
/android/database/SQLiteProgram
/android/database/SQLiteQuery
/android/database/SQLiteStatement
/android/os/Debug
/android/os/FileObserver
/android/os/FileUtils
/android/os/MessageQueue
/android/os/ParcelFileDescriptor
/android/os/Power
/android/os/StatFs
/android/os/SystemProperties
/android/os/UEventObserver
/android/net/LocalSocketImpl
/android/net/NetworkUtils
/android/net/TrafficStats
/android/net/wifi/WifiManager
/android/nfc/NdefMessage
/android/nfc/NdefRecord
/android/os/MemoryFile
/com/android/internal/os/ZygoteInit
/android/hardware/Camera
/android/hardware/SensorManager
/android/media/AudioRecord
/android/media/AudioSystem
/android/media/AudioTrack
/android/media/ToneGenerator
/android/opengl/classes
/android/bluetooth/HeadsetBase
/android/bluetooth/BluetoothAudioGateway
/android/bluetooth/BluetoothSocket
/android/bluetooth/ScoSocket
/android/server/BluetoothService
/android/server/BluetoothEventLoop
/android/server/BluetoothA2dpService
/android/server/Watchdog
/android/message/digest/sha1
/android/ddm/DdmHandleNativeHeap
/android/backup/BackupDataOutput
/android/backup/FileBackupHelperBase
/android/backup/BackupHelperDispatcher
/android/content/res/ObbScanner
/android/content/res/Configuration

Designing Leakage-Resilient Password Entry on Touchscreen Mobile Devices

Qiang Yan†, Jin Han‡, Yingjiu Li†, Jianying Zhou‡, Robert H. Deng†
†School of Information Systems, Singapore Management University, Singapore
‡Cryptography and Security Department, Institute for Infocomm Research, Singapore
{qiang.yan.2008, yjli, robertdeng}@smu.edu.sg, {hanj, jyzhou}@i2r.a-star.edu.sg

ABSTRACT

Touchscreen mobile devices are becoming commodities as the wide adoption of pervasive computing. These devices allow users to access various services at anytime and anywhere. In order to prevent unauthorized access to these services, passwords have been pervasively used in user authentication. However, password-based authentication has intrinsic weakness in password leakage. This threat could be more serious on mobile devices, as mobile devices are widely used in public places.

Most prior research on improving leakage resilience of password entry focuses on desktop computers, where specific restrictions on mobile devices such as small screen size are usually not addressed. Meanwhile, additional features of mobile devices such as touch screen are not utilized, as they are not available in the traditional settings with only physical keyboard and mouse. In this paper, we propose a user authentication scheme named CoverPad for password entry on touchscreen mobile devices. CoverPad improves leakage resilience by safely delivering hidden messages, which break the correlation between the underlying password and the interaction information observable to an adversary. It is also designed to retain most benefits of legacy passwords, which is critical to a scheme intended for practical use. The usability of CoverPad is evaluated with an extended user study which includes additional test conditions related to time pressure, distraction, and mental workload. These test conditions simulate common situations for a password entry scheme used on a daily basis, which have not been evaluated in the prior literature. The results of our user study show the impacts of these test conditions on user performance as well as the practicability of the proposed scheme.

Categories and Subject Descriptors

K.6.5 [**Management of Computing And Information Systems**]: Security and Protection— *Authentication*; H.5.2 [**Information Interfaces and Presentation**]: User Interfaces— *Evaluation/methodology, User-centered design, Haptic I/O*

ASIA CCS'13, May 8–10, 2013, Hangzhou, China.

Keywords

User Authentication; Leakage-Resilience; Mobile Devices

1. INTRODUCTION

Mobile devices are becoming essential tools in modern life, which seamlessly connect human beings to the cyberspace. A user can now use his smartphone or tablet to access not only general informative services but also sensitive services such as mobile banking and corporate services. In order to prevent unauthorized access to these services, user authentication is required to verify the identity of a user. Among existing user authentication mechanisms, passwords are still the most pervasive due to their significant advantage in usability over other alternatives such as smartcards and biometrics [28]. However, password-based user authentication has intrinsic weakness in password leakage, which may lead to financial loss or corporate data disclosure. This threat could be more serious in scenarios when mobile devices are involved, as mobile devices are widely used in public places.

Password leakage is a classic problem in password-based authentication. Since password leakage usually happens during authentication when a user inputs his password, we focus on the problem of improving leakage resilience of password entry in this work. Most prior research [19, 27, 38, 39, 5, 26, 33, 13, 24] on this problem focuses on desktop computers, where specific restrictions on mobile devices are usually not addressed. These restrictions mainly include: 1) a mobile device usually has a smaller screen size than a desktop computer; 2) a mobile device needs to be operable in non-stationary environments such as on public transit. On the other hand, mobile devices provide additional features such as touch screen, which may not be available in traditional settings. These new features can be utilized to support advanced security properties that were difficult to achieve before.

In this paper, we propose a *concise yet effective* authentication scheme named CoverPad, which is designed for password entry on touchscreen mobile devices. CoverPad improves *leakage resilience* of password entry while *retaining most benefits* of legacy passwords. Leakage resilience is achieved by utilizing the gesture detection feature of touch screen in forming a *cover* for user inputs. This cover is used to safely deliver hidden messages, which break the correlation between the underlying password and the interaction information observable to an adversary. From the other perspective, our scheme is also designed to retain the benefits provided by legacy passwords. This requirement is critical, as Bonneau et al. [10] conclude that any user authentication is unlikely to gain traction if it does not retain comparable benefits of legacy passwords. Our scheme approaches this requirement by involving only intuitive cognitive operations and requiring no extra devices in the design.

We implement three variants of CoverPad and evaluate them with an extended user study. This study includes additional test conditions related to *time pressure*, *distraction*, and *mental workload*. These test conditions simulate common situations for a daily-used password entry scheme, which have not been evaluated in the prior literature. We design new experiments to examine their influence based on previous work in psychology literature [23, 12, 21]. Experimental results show the influence of these conditions on user performance and the practicability of our proposed scheme.

The contributions of this paper are summarized as follows.

- We propose CoverPad to protect password entry on touch-screen mobile devices. It achieves leakage resilience and retains most benefits of legacy passwords by involving only intuitive cognitive operations and requiring no extra devices.
- We implement three variants of CoverPad to address different user preferences. Our user study shows the practicability of these variants.
- We extend user study methodology to examine the influence of various additional test conditions. Among these conditions, time pressure and mental workload are shown to have significant impacts on user performance. Therefore, it is recommended to include these conditions in the evaluation of user authentication schemes in the future.

2. THREAT MODEL

Passwords are the most pervasive user authentication that allows a human *user* to be authenticated to a (local or remote) computer *server*. *Password leakage* is a threat that a user's password is directly disclosed or indirectly inferred. It usually happens during *password entry*, when a user inputs his password in order to prove his identity. In the case of legacy passwords, a user directly enters his plaintext password so that the password may be captured via various eavesdropping attacks including key logger, hidden camera, and malware. We classify these attacks into two types, *external* or *internal*, according to whether an adversary can access the internal states of a device for password entry, such as device memory.

An external eavesdropping attack is an attack exploiting a leakage channel outside a device. This type of attacks includes *vision*-based eavesdropping such as hidden camera, *haptics*-based eavesdropping such as physical key logger, and *acoustics*-based eavesdropping such as tone analysis. Compared to traditional scenarios involving only desktop computers, an adversary has more opportunities to launch an external eavesdropping attack against mobile devices, as mobile devices are widely used in public places. In a crowded area, an adversary may observe password entry in a close distance without being noticed (see Figure 1).

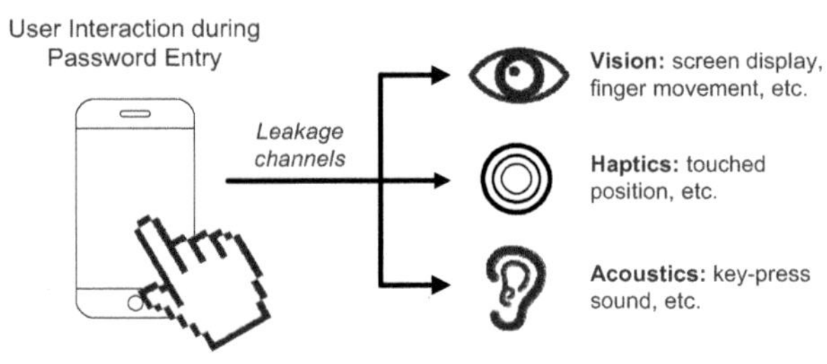

Figure 1: Attack scenarios

For vision-based attacks, an adversary may infer the actual password by observing the movement of fingers even without direct line-of-sight on the screen display. This capability is significantly enhanced with emerging augmented-reality accessory like Google Glass [18], which is a small wearable glass transferring real-time video captured by a tiny camera to a server and displaying the analyzed results received from the server.

Haptics-based attacks are most likely to happen when users use public mobile devices. Mobile devices, such as iPad, have been used as public computer kiosks as observed in museums, restaurants, and hotels [22, 20, 41]. In addition, many existing kiosks are also equipped with touch screen similar to mobile devices. This provides an incentive for an adversary to install a physical "touch" logger. Although such touch logger has not been observed in the wild, it is technically feasible to implement as other physical key loggers [36]. Considering that the thickness of touch screen in Samsung Galaxy S3 is just 1.1mm [2], it may not be noticeable to users if an extra physical touch logger is installed on a normal touch screen.

The effectiveness of acoustics-based attacks depends on whether user actions can be distinguished by their tone patterns. For example, different tones are played when a user dials different numbers on an old-style phone. Due to environmental noises, acoustics-based attacks are usually not as effective as vision-based attacks and haptics-based attacks.

The other type of attacks that cause password leakage is the internal eavesdropping attack. Such attacks exploit a leakage channel inside a device, where an adversary is allowed to access the internal states such as reading device memory. This type of attacks include *logic key logger*, *malware*, and *network eavesdropping*, which are common to all password-based user authentication schemes. Like most prior research [26, 33, 13, 15, 24, 8, 7], our scheme design does not address these attacks for the following reasons: 1) Existing solutions [30, 3, 6, 37] such as application sandbox are available to effectively defend against these attacks, though it takes time for them to replace legacy vulnerable systems; 2) these solutions are independent on user interaction during password entry so that they can be adapted to any user authentication schemes. Compared to external eavesdropping attacks, the threat from internal eavesdropping attacks can be effectively mitigated if a user uses a computer system that is properly updated and configured [16], while it is not easy to defend against external eavesdropping attacks as they are caused by *inevitable* exposure of human interaction during password entry. These external eavesdropping attacks impose realistic threats leading to password leakage. We will thus focus on external eavesdropping attacks in our scheme design.

Besides the above attacks which happen during password entry, password leakage may also be caused by other types of attacks including social engineering and phishing [28]. Although their mitigation technologies such as secure URL checker and spam filter have been widely deployed in modern computer systems, some of these attacks may not be completely preventable by technical solutions alone. Another example is the database reading attack, where the back-end databases are intruded so that all user passwords are compromised. Since these attacks are orthogonal to the password entry problem, they are out of the scope of this paper.

3. COVERPAD DESIGN

In this section, we present the design of CoverPad. First, we describe our design objectives from both security and usability perspectives. Then, we introduce the conceptual design of CoverPad. Lastly, we present three variants in implementing CoverPad.

3.1 Design Objectives

CoverPad is designed to improve leakage resilience of password entry while retaining most benefits of legacy passwords. We describe our design objectives as follows.

First, in terms of security, a scheme should minimize password leakage during password entry under realistic settings. To achieve this objective, a user should 1) input obfuscated response derived from his password, and/or 2) input his password in a protected environment. A recent study [40] shows strong evidence on the infeasibility of using obfuscated response solely based on human cognitive capabilities. Therefore, it is necessary to rely on certain protected environment to achieve this security objective. However, a fully protected environment may be difficult to establish in practice, which requires to protect all messages delivered between user and server. Therefore, we choose a hybrid solution in our scheme design, where the requirement on a protected environment is significantly reduced with the assistance of simple obfuscation. Such environment is referred to as *partially* protected environment.

In the presence of a partially protected environment, it is possible to achieve the optimal security objective – *no password leakage*. As long as the partially protected environment is not compromised, CoverPad provides the same leakage resilience as *one-time pad* [31], where the most efficient attacks for an adversary to learn the password are online dictionary attacks. We will show how this security objective is achieved in our scheme in the following sections.

Second, in terms of usability, a scheme should preserve the benefits of legacy passwords in order to gain traction [10]. The major benefits of legacy passwords include no extra devices required, and only intuitive cognitive operations performed. We further consider additional restrictions on mobile devices including that 1) a mobile device usually has a *smaller* screen size compared to a desktop computer; 2) a mobile device needs to be operable in a *non-stationary* environment such as on public transit. So we minimize the number of visual elements that are displayed simultaneously on the screen, and also simplify the involved operations to make them suitable in a non-stationary environment.

3.2 Conceptual Design

The conceptual design of CoverPad is shown in Figure 2, where a hidden transformation $T_i(\cdot)$ is a random mapping $\Omega \to \Omega$, where Ω is the set of all individual elements contained in the password alphabet.

Setup:
A server and a user agree on a k-length password $pwd = (a_1, a_2, \ldots, a_k)$, where a ***password element*** *$a_i = pwd[i]$ belongs to an alphabet with size w. It is allowed that $a_i = a_j$, for $i \neq j$.*

Password Entry:
For each i from $[1, k]$:
Step 1: *The touch screen shows a keypad with all the elements in the alphabet.*
Step 2: *The user is asked to perform a hand-shielding gesture to read the hidden transformation $T_i(\cdot)$ protected by the hand-shielding gesture. $T_i(\cdot)$ will immediately disappear if the gesture is no longer detected.*
Step 3: *The user clicks on response element e_i, where $e_i = T_i(a_i) = (a_i + r_i \mod w)$, where r_i is a random number drawn from a uniform distribution. A new random number r_i is generated for each round i. The hand-shielding gesture is not required for this step.*

Figure 2: Conceptual design of CoverPad

An example of using CoverPad is given as follows. Suppose a user has a k-length password. At the beginning of password entry, the user performs the hand-shielding gesture to view the current hidden transformation T_1 for the first character a_1 in his password. Then, he applies T_1 to a_1 and enters the transformed response e_1. This procedure repeats for each password element a_i. During the whole password entry, T_i disappears immediately once the gesture is not being detected. A user can always view T_i by performing the gesture again before inputting e_i.

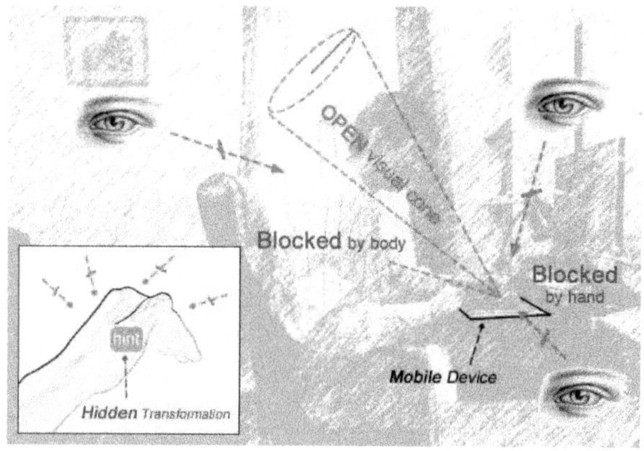

Figure 3: The hand-shielding gesture and its effectiveness

Figure 3 shows how to correctly perform a hand-shielding gesture. This gesture restricts the vision channel to a small visual cone. This visual cone is not accessible to an adversary unless the adversary's eyes are close enough to the user's head, which makes the adversary easily exposed. A hidden camera near the line of sight may help capture the hidden transformation. However, it needs to be adjusted according to the user's height and current position, which may lead to user's awareness. On the other hand, the observable responses for the same password element are uniformly randomized. Thus, CoverPad is also immune to haptics-based eavesdropping. Further analysis is provided in the next section.

Therefore, it is difficult to compromise the partially protected environment formed by the hand-shielding gesture from external eavesdropping attacks in practice, though the use of this gesture is simple. If the protective gesture is not being detected by the touch screen, the hidden transformation will not be displayed such that the hidden transformation is always protected under the required gesture. Note that a hidden transformation alone does not leak any information about the password. As long as the hidden transformation is not revealed together with the corresponding response, observed interaction provides no valuable information for an adversary to infer the actual password. A proof about this security property will be given in Section 4.

3.3 Implementation Variants

We provide three variants of CoverPad that implement different features tailored for users with various skill sets, which are described and illustrated as follows (see Figure 4).

3.3.1 NumPad-Add

In NumPad-Add, the alphabet of password consists of digits 0 to 9 only. The hidden transformation is performed by *adding* a random digit to the current password element and then $\mod 10$ if the sum is larger than 9, where the value of the random digit ranges from 0 to 9. For example, the correct response for the first round

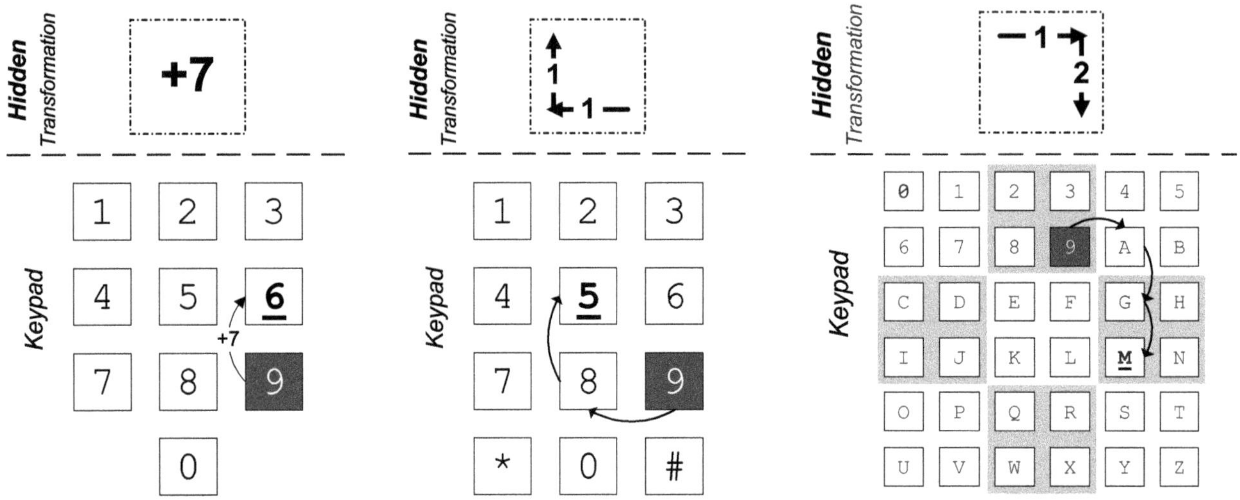

(a) NumPad-Add (b) NumPad-Shift (c) LetterPad-Shift

Figure 4: Demonstration of three implementation variants

is $6 = (9 + 7) \mod 10$ given password $\underline{9}34567$ and the hidden message '*plus 7*'.

3.3.2 NumPad-Shift

In NumPad-Shift, the alphabet of password consists of digits 0 to 9 only. The hidden transformation is performed by *shifting* the location of the current password element by X-offset and Y-offset, where the offset values are randomly taken from $\{-1, 0, 1\}$ for X-offset, and $\{-1, 0, 1, 2\}$ for Y-offset. For a 3×4 keypad design shown in Figure 4(b), the transformed response for a_i is calculated as $pad[x(a_i) + \Delta x \mod 3][y(a_i) + \Delta y \mod 4]$, where Δx is the X-offset, Δy is the Y-offset, and $x(a_i)$ is the X-index of a_i, and $y(a_i)$ is the Y-index of a_i. For example, the correct response for the first round is 5 if the password is $\underline{9}34567$ and the hidden message is '*move left by 1 step and move up by 1 step*'.

Note that two extra keys $*$ and $\#$ are added to the keypad; otherwise, the distribution of hidden transformations is not uniform on the keypad layout. The proof for the necessity of these two keys is given as follows. Assuming $*$ and $\#$ keys are removed, the keypad now contains only 10 keys for digits 0 to 9. To provide a full transformation from a secret key to a random key, the minimum value set is $\{-1, 0, 1\}$ for X-offsets and $\{-1, 0, 1, 2\}$ for Y-offsets. There are twelve combinations between X-offsets and Y-offsets, but only ten keys on the keypad. If the offset values are drawn from a uniform distribution, certain response keys for a given password element would have a higher frequency compared to others (it is similar as placing twelve balls in ten buckets in a deterministic way). The exact distribution of response keys is decided by the underlying password element, thus it discloses valuable information about the password. From the other perspective, if response keys are drawn from a uniform distribution, the offset values will not be uniformly distributed due to similar reason. Therefore, it is necessary to add these two extra keys to the NumPad-Shift keypad.

3.3.3 LetterPad-Shift

In LetterPad-Shift, the alphabet of password consists of letters a to z and digits 0 to 9 (36 elements in total). The hidden transformation is the same as NumPad-Shift. The offset values are randomly taken from $\{-2, -1, 0, 1, 2, 3\}$ for both X-offset and Y-offset for a 6×6 keypad design. The transformed response for a_i is calculated as $pad[x(a_i)+\Delta x \mod 6][y(a_i)+\Delta y \mod 6]$ in a similar way as for NumPad-Shift. A background grid is added to ease the calculation of shifting, as shown in Figure 4(c).

4. SECURITY ANALYSIS

4.1 External Eavesdropping Attacks

Common external eavesdropping attacks leading to password leakage may exploit vision, haptics, or acoustics channel as analyzed in Section 2. For CoverPad, an adversary using these attacks can observe *at most* a complete response key sequence pressed by a user, while the hidden transformation is protected by our design. From this key sequence, the adversary knows the i-th pressed key is decided by the i-th element in the password. However, the adversary cannot further infer what the i-th password element is, as proved as follows.

Proof: Given a pressed key e_i, and two password elements a_x and a_y in a w-sized password alphabet, let $Pr(e_i|a_x)$ and $Pr(e_i|a_y)$ be the probabilities for e_i being pressed when the underlying password element are a_x and a_y, respectively. We have $Pr(e_i|a_x) = Pr(e_i = a_x + r_i \mod w) = Pr(r_i = e_i - a_x \mod w) = Pr(r_i = C \mod w) = 1/w = Pr(e_i|a_y)$ for any i, x, and y, where C is a constant integer randomly drawn from a uniform distribution. Therefore, a sequence of pressed keys observed by an adversary is equivalent to a random sequence, which is similar to a ciphertext generated by a one-time pad. □

In a partially protected environment where the hidden transformation is protected by the hand-shielding gesture, our scheme achieves no password leakage. As long as the hidden transformation is not disclosed together with the corresponding response, an adversary cannot infer any information about the underlying password (except password length) even after an infinite number of observations.

4.2 Side-channel Attacks

In reality, it is possible for an adversary to exploit subtle side-channels to collect password information during password entry. These attacks are not usually considered in common threat mod-

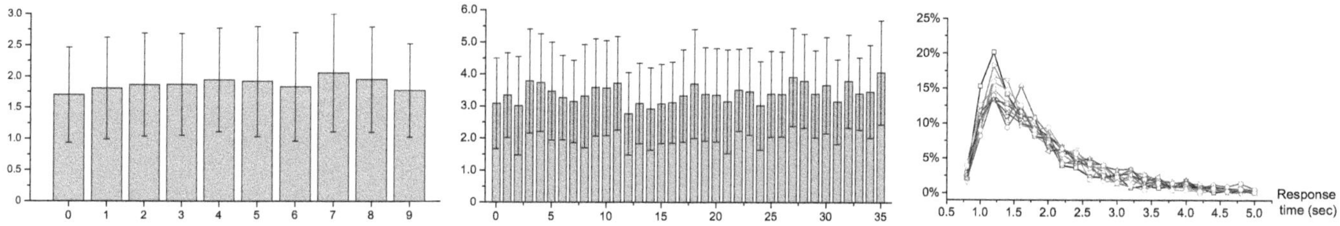

(a) Timing deviation for NumPad-Add (b) Timing deviation for LetterPad-Shift (c) Timing Distribution for NumPad-Add

Figure 5: Timing deviations and distributions for entering each password element. The results of NumPad-Shift are similar to the results of NumPad-Add shown in these figures.

els [19, 27, 38, 39, 5, 26, 33, 13, 15, 24, 8, 7]. A typical side-channel attack is timing analysis [35], which analyzes the patterns in the response time of entering individual password elements. The preliminary results of our scheme against timing analysis are given in Figure 5. For the timing deviation shown in Figure 5(a) and 5(b), each bar with x-value i represents the average response time for entering the transformed responses for a specific password element i. For the timing distribution shown in Figure 5(c), each line in the figure represents the distribution of the response time for entering the transformed responses for a specific password element. These results show the range and the distribution of the response time for entering different password elements are almost overlapped. This indicates that timing analysis is not a major concern for our scheme, though it is difficult to completely prevent such attacks due to inevitable human behavior patterns during password entry. Detailed analysis on side channel attacks is out of the scope of this paper.

5. USABILITY EVALUATION

5.1 Methodology

The participants in our user study are recruited from undergraduate students in our university. There are 61 participants in total, 30 male and 31 female, with age range between 20 and 25. These participants come from five different departments, in which 42 of them have a social science or business related background, and the remaining have a computer science or information technology related background. Each participant is paid with 10 dollars as compensation for their time. We establish a *ranking* system from which a participant can see a *performance score* representing how well he performs compared to other participants. This ranking system provides a moderate level of motivation for the participants to do their best in tests. A numerical identifier is assigned to each participant in order to protect user privacy.

The user study is conducted in a quiet room. The experiments use a within-objects design. Each participant is asked to use all three variants as three *test groups*. These variants are implemented on Apple iPad, which are referred to as *schemes* in this section. The order of the schemes is randomized to avoid the learning effect that affects the performance for a specific scheme. For each test group, a user is required to memorize a *randomly generated* password in the beginning. The password strength is set to be equivalent to 6-digit PIN, where the password length is four for LetterPad-Shift, and six for both NumPad-Shift and NumPad-Add. The same password will be used for the same test group and a "*show my password*" button is provided in case a participant forgot his password. The participants learn how to use a scheme by an interactive step-by-step tutorial. The participants are required to go through the whole tutorial for the first scheme appearing in the tests, and they may skip the tutorial for the second and third schemes after learning the basic scheme design. In the end of each tutorial, there is a short pretest for the participant to exercise. If a participant fails to pass the pretest, the researchers will provide help to him to ensure that he understands how to use the scheme before the tests start.

In each test group, there are six tests simulating additional *test conditions* that evaluate the influence of time pressure, distraction, and mental workload. The details of these test conditions are described in the next subsection. The order of these tests is also randomized in order to avoid the learning effect.

All three test groups consist of 18 tests in total. To avoid the participants from feeling exhausted and bored, each test is designed to be short and can be finished within one or two minutes. The participants are given a short break after each test group. At the end of the user study, the participants are given a questionnaire using 5-point Likert scale to collect their perception on the schemes. The whole user study takes $35 \sim 50$ minutes to complete.

5.2 Simulating Various Test Conditions

In order to simulate various test conditions related to time pressure, distraction, and mental workload, we introduce two extra experimental tools, timer and secondary task. A *timer* is used to create time pressure by showing a participant how much time is left for the current test condition. It is implemented as a progress bar whose length increases every second with a countdown text field showing how many seconds are left. *Secondary tasks* are used to simulate unexpected distraction and persistent mental workload. We use CRT (*choice reaction time*) tasks as secondary tasks, which is a standard technology in experimental psychology [23, 12, 21]. CRT tasks usually work as secondary tasks that occupy the central executive[1] in human brain when evaluating the performance of a primary task in the presence of a secondary task. CRT tasks require participants to give distinct responses for each possible stimulus. In our implementation, the participants are asked to press the correct button among N buttons, where the correct button should have the same color as the stimulus. For example, if the stimulus shows a red button, a participant should press the red button among N buttons with different colors. We use $N = 2$ for tests in the distraction condition as the major focus is to unexpectedly disrupt password entry with a CRT task. We use $N = 8$ for tests in the mental workload condition so as to create a considerable mental workload, which is the same as in the classic Jensen Box setting [23].

Based on the above experimental tools, we simulate six test conditions for each test group by combining the two modes and three statuses. Two modes related to a timer are described as follows:

- **Relaxed mode:** A participant is asked to minimize the error rate in a fixed number of login attempts where time is not

[1]The central executive is a control system that mediates attention and regulation of processes occurring in working memory [4].

considered in performance score calculation. The number of login attempts is 5 for no-extra-task status and 3 for distraction and mental workload statuses.

- **Timed mode:** A participant is asked to perform as many successful logins as possible within 1 minute, where both time and accuracy are considered in performance score calculation. The countdown of a timer creates time pressure.

Three statuses related to secondary tasks are described as follows:

- **No-extra-task status:** A participant is asked to perform the login task only.
- **Distraction status:** A simple CRT task may appear with $1/3$ probability each time when a participant presses a response key. This task is used to create unexpected distractions during password entry.
- **Mental workload status:** A relatively complex CRT task appears every time when a participant presses a response key. This task is used to create continuing mental workload during password entry.

Among six conditions, we referred to the combination of *relaxed* mode and *no-extra-task* status as the **normal condition**, which is the common condition usually tested in prior work [19, 27, 38, 39, 5, 26, 33, 13, 15, 24, 8, 7]. The short names for the other five conditions are given in Table 1.

Short name	Full specification
normal	*relaxed* mode + *no-extra-task* status
timed	*timed* mode + *no-extra-task* status
distraction	*relaxed* mode + *distraction* status
distraction+timed	*timed* mode + *distraction* status
mental workload	*relaxed* mode + *mental workload* status
mental workload+timed	*timed* mode + *mental workload* status

Table 1: Short names for test conditions

The hypotheses related to these test conditions are described as follows.

(H1) *Compared to the normal condition,* ***login time*** *will be significantly shorter when* ***time pressure*** *is present.*

(H2) *Compared to the normal condition,* ***login accuracy*** *will be significantly lower when* ***time pressure*** *is present.*

(H3) *Compared to the normal condition,* ***login time*** *will be significantly longer when unexpected* ***distraction*** *is present.*

(H4) *Compared to the normal condition,* ***login accuracy*** *will be significantly lower when unexpected* ***distraction*** *is present.*

(H5) *Compared to the normal condition,* ***login time*** *will be significantly longer when persistent* ***mental workload*** *is present.*

(H6) *Compared to the normal condition,* ***login accuracy*** *will be significantly lower when persistent* ***mental workload*** *is present.*

(H7) *Compared to a condition in* ***relaxed*** *mode with* ***secondary tasks, login time*** *will be significantly shorter for its counterpart in* ***timed*** *mode.*

(H8) *Compared to a condition in* ***relaxed*** *mode with* ***secondary tasks, login accuracy*** *will be significantly lower for its counterpart in* ***timed*** *mode.*

5.3 Learning Curve

Although our scheme design involves intuitive operations only, it requires a different process for password entry compared to legacy passwords. While we expect the participants can learn this process with the tutorial and pretests, we observed that some participants were impatient to read all instructions and keep pressing the next button. These participants proceeded to the evaluation stage before they fully understand our scheme design.

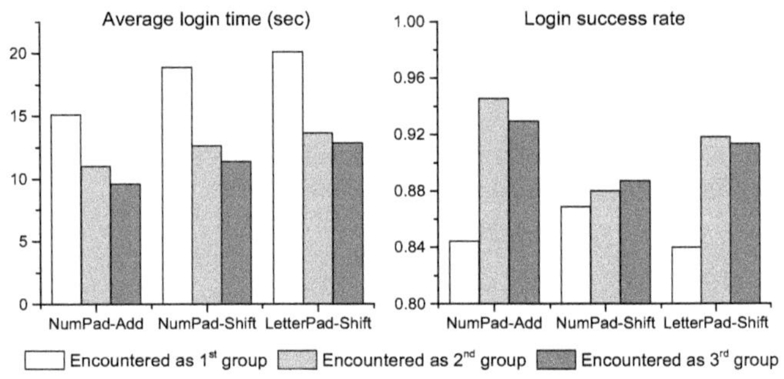

Figure 6: Learning curve of our schemes

Figure 6 compares user performance under the normal condition for different positions where a scheme appears in the study. These results show the user performance in terms of login time and login success rates is significantly worse when the tested scheme is the first scheme which a participant encountered in the user study. But the differences on user performance are not significant if a scheme is encountered as the second or third test group, as all our schemes are similar due to the fact that they are based on the same conceptual design. As shown in the learning curve in Figure 6, most participants get familiar with our scheme design after the first test group. Therefore, we consider the first test group as part of the learning process, and use the performance data collected from the second and third test groups only in the following analysis.

5.4 Experimental Results

We measure user performance with the following metrics: average login time, login success rates, round success rates, and average edit distances. A *round* success rate is the average success rate for a user to correctly input one password element by applying a hidden transformation. An *edit distance* is the minimum number of insertions, deletions, substitutions, and adjacent transpositions required to transform an input string into the correct password string so that an *average* edit distance is the average value of edit distances calculated from all login attempts of a user under a test condition. Among these metrics, login success rates, round success rates, and average edit distances are used to evaluate *login accuracy*.

We use the following statistical tools to test the significance of our experimental results, where a significance level of $\alpha = .05$ is used. For each comparison, we run an *omnibus* test across all test conditions for each scheme. Since all our performance data are quantitative, we use *Kruskal-Wallis* (KW) test for omnibus tests, which is an analogue of ANOVA but does not require normality. If the omnibus test indicates significance, we further use *Mann-Whitney* (MW) U test to perform pair-wise comparisons so as to identify specific pairs with significant differences. The detailed results of our statistical tests are given in Appendix A.

5.4.1 Performance under Normal Condition

In the normal condition, a participant is only asked to perform login tasks without any time pressure or secondary tasks. It corresponds to the combination of relaxed mode and no-extra-task status, which is used as a *baseline* in our tests.

Figure 7(a) shows the average time for a successful login attempt

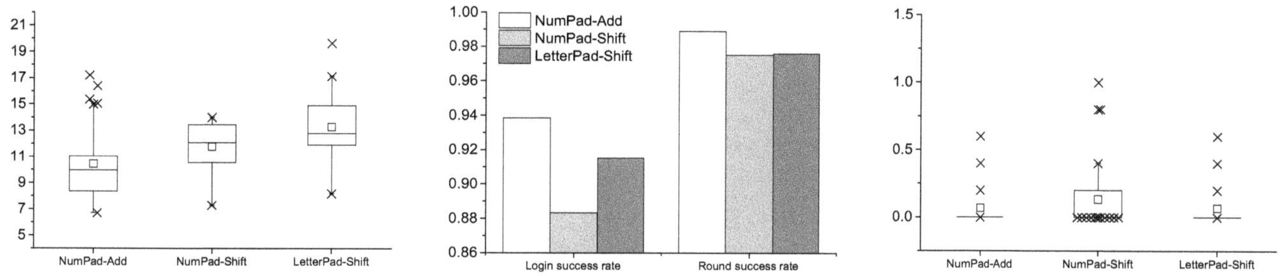

(a) Login time distribution (sec) (b) Average success rate (c) Edit distance distribution

Figure 7: Average login time, success rate, and edit distance under the normal condition

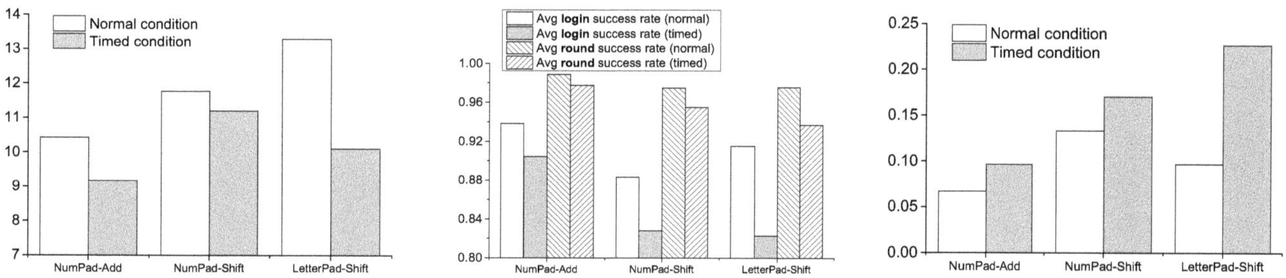

(a) Average login time (sec) (b) Average success rate (c) Average edit distance

Figure 8: Impact of time pressure

in the normal condition. For all the three schemes, most participants are able to finish the login within 13 seconds. Figure 7(b) and 7(c) show the corresponding login accuracy. Since our experiment limits the number of login attempts to 5 in order to prevent the participants from feeling exhausted or bored, even a single mistake would take the login success rate down to 80%. Our results indicate that most participants make *at most* one mistake when they use our schemes for the first time after a short training. This is shown by 97.5% average round success rate and 0.13 average edit distance in the worst case. Particularly, for the distribution of average edit distance of NumPad-Shift, 27 participants among 40 samples (after removing the experimental data when NumPad-Shift appears as the first test group) has an average edit distance of zero (i.e. no mistakes during all tests under the test condition), which are shown as a cluster of *outliers* at the bottom of the box chart. The login accuracy is expected to increase after the participants get more familiar with the schemes.

5.4.2 *Influence of Time Pressure*

Figure 8 shows the impact of time pressure without any secondary tasks. The results show that the participants behave much hastily in the presence of time pressure. The average time for a successful login attempt becomes shorter and the login accuracy is decreased. The statistical tests show the difference in login time is significant (p=.017 for NumPad-Add and p<.001 for LetterPad-Shift) but the difference in login accuracy is not. Therefore, **H1** is supported while **H2** is not.

The insignificant results in login accuracy are due to the *ceiling* effect [1], which implies the tests are not sufficiently difficult to distinguish the influence of different test conditions. This effect could be caused by our scheme design, which is not difficult for the participants to use so that the majority of the participants did not make any mistakes during all the tests. This effect will be further discussed in Appendix A. However, even without statistical significance, we still observe the average results of login accuracy become worse for all three tested schemes. Considering the simple design of our schemes, this indicates that time pressure may have a larger influence on the login accuracy of a more complex scheme.

5.4.3 *Influence of Distraction*

Figure 9 shows the impact of distraction without time pressure. Many participants made a mistake when they saw a distraction task for the first time (however, NumPad-Shift is an exception). For NumPad-Add and LetterPad-Shift shown in Figure 9(b), the round success rate returns to a comparable level as the normal condition, after the first time the distraction task appears. This indicates that the distraction task is no longer a surprise for the participants. However, even after the participants get familiar with the distraction tasks, compared to the normal condition, the success rate is still lower, the average edit distance is larger, and the average login time is longer. But the statistical tests show these differences are not significant. Therefore, **H3** and **H4** are not supported in our experiments.

5.4.4 *Influence of Mental Workload*

Figure 10 shows the impact of mental workload without time pressure. The average login time becomes significantly longer with mental workload (p=.003 for NumPad-Add) due to context switch in users' mind between password inputs and secondary CRT tasks. An extra startup time is required to release the central executive after each CRT task. Our experiment simulates the case when users cannot get rid of other thoughts during password entry. The actual effect of mental workload depends on the status of users' mind. The impact may be elevated when the actual mental workload is higher than our CRT tasks. On the other hand, the login accuracy is lower compared to the normal condition but the difference is not significant due to the same ceiling effect mentioned in Section 5.4.2. Therefore, **H5** is supported and **H6** is not. These results

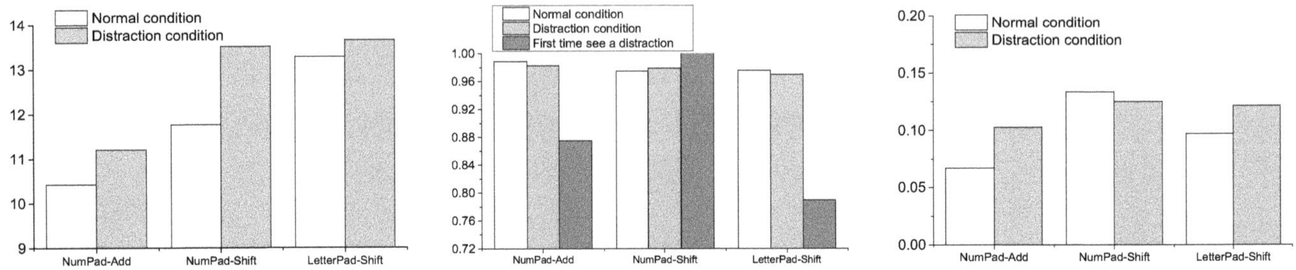

(a) Average login time (sec) (b) Average **round** success rate (c) Average edit distance

Figure 9: Impact of distraction

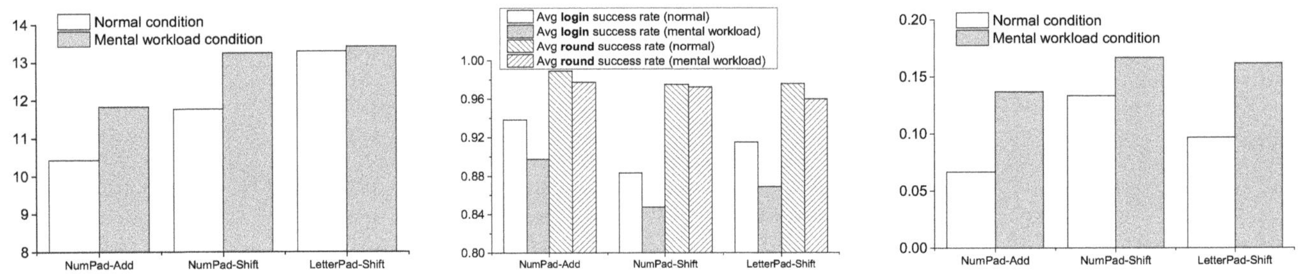

(a) Average login time (sec) (b) Average success rate (c) Average edit distance

Figure 10: Impact of mental workload

show that persistent mental workload significantly slows the process of password entry for our schemes.

5.4.5 *Performance under Combined Conditions*

We also examine the overall impact when distraction or mental workload appears together with time pressure. As expected, compared to their counterparts without time pressure, the average login time becomes shorter (from 10.3 seconds to 11.7 seconds on average), the login success rate becomes even lower (from 81.3% to 87.5%), and the average edit distance becomes larger (from 0.151 to 0.243). The statistical tests show the difference in login time is significant (p=.009 for NumPad-Add, p=.019 for NumPad-Shift, and p<.001 for LetterPad-Shift) and the difference in login accuracy is still not significant due to the ceiling effect explained in Section 5.4.2. Therefore, **H7** is supported but **H8** is not. These results show time pressure is still an effective stimulus to speed password entry even in the presence of secondary tasks.

5.4.6 *Effectiveness of Secondary Tasks*

Figure 11 shows the distribution of the accuracy rate which represents the percentage of secondary tasks being correctly performed by a participant under certain test condition. The overall average accuracy rate is 98.3% across all these test conditions. It implies that the participants did pay attention to these tasks, as they were told that the performance of these tasks also contributes to their scores in the ranking system. Therefore, these CRT tasks work as intended in disturbing participants' mind during password entry.

5.4.7 *Memory Interference by Mental Calculation*

Figure 12 shows how frequently a participant presses the "*show my password*" button during all tests in a test group. Note that the participants are not allowed to write down their assigned passwords, but they can always click that button in case they forgot their passwords. The overall average value for the total number of times to press the "*show my password*" button is only 0.31 across all three test groups. As shown in Figure 12, most users did not use

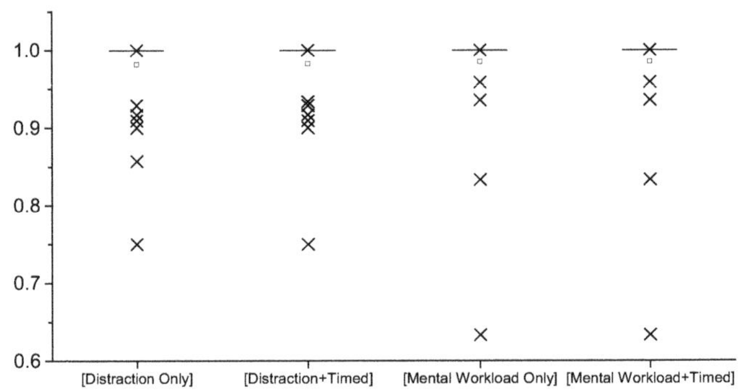

Figure 11: Accuracy rate of performing secondary tasks

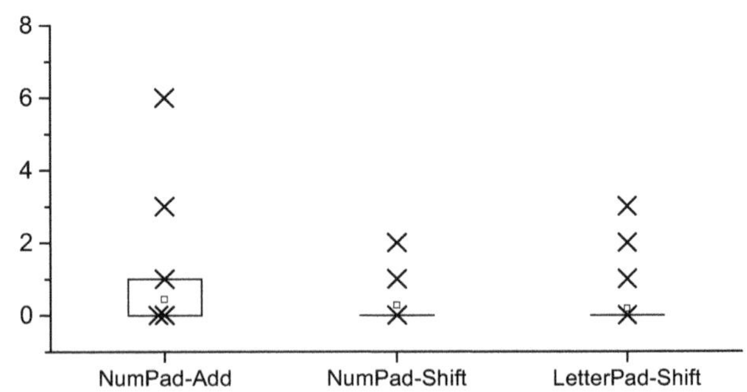

Figure 12: Total number of times for each participant to press the "*show my password*" button

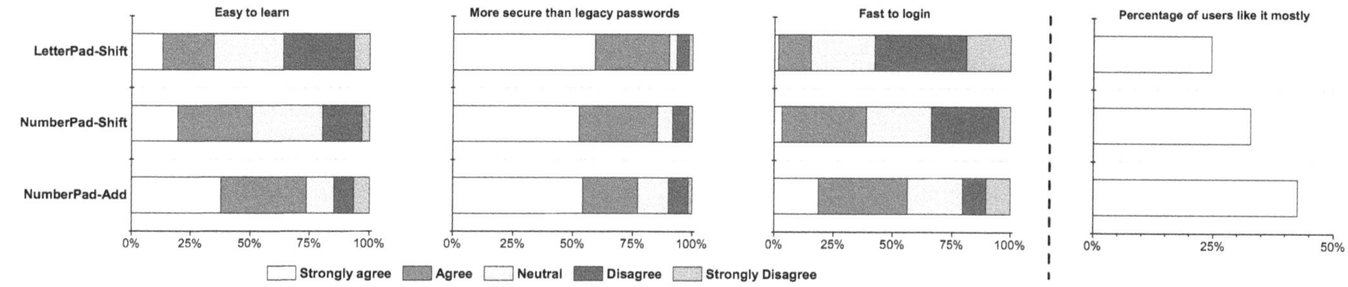

Figure 13: Perception of participants

this button during the tests. This implies that the mental calculation involved in the hidden transformation of our schemes does not pose a significant interference on participants' capability of recalling their passwords.

5.4.8 *User Perception*

Figure 13 shows the perception of participants collected from questionnaires. The results indicate that the participants generally feel that our schemes are secure and easy to use. While NumPad-Add is the most popular, the other two schemes also have their favorite users.

5.5 Comparison with Legacy Passwords

Table 2 gives a comparison between CoverPad and legacy passwords based on the *usability-deployability-security* metrics proposed in [10], where a metric is not shown if neither our schemes nor legacy passwords offer corresponding benefit. We have the following observations in comparison. 1) Our schemes are rated as not *mature* since they are just proposed and have not been widely deployed. 2) Our schemes are not *server-compatible*, as most current servers support only static and replayable passwords, which could be changed in the near future. 3) Our schemes are *quasi-resilient-to-internal-observation* in a sense that any key logger or malware which fails to capture the hidden transformation causes no password leakage. Overall, this table shows that our schemes significantly improve the security strength while retaining most benefits of legacy passwords.

	Nothing-to-Carry	Easy-to-Learn	Efficient-to-Use	Infrequent-Errors	Easy-Recovery-from-Loss	Accessible	Negligible-Cost-per-User	Server-Compatible	Browser-Compatible	Mature	Non-Proprietary	Resilient-to-Physical-Observation	Resilient-to-Targeted-Impersonation	Resilient-to-Internal-Observation	Resilient-to-Theft	No-Trusted-Third-Party	Requiring-Explicit-Consent	Unlinkable
CoverPad Schemes	•	•	○	○	•	•	•		•		•	•	○	○	•	•	•	•
Legacy Passwords	•	•	•	○	•	•	•	•	•	•	•		○		•	•	•	•

Table 2: Comparison between CoverPad and legacy passwords using usability-deployability-security metrics [10]. • = offer the benefit, ○ = almost offer the benefit, *no circle* = does not offer the benefit

6. DISCUSSION

6.1 Eavesdropping Attacks

Eavesdropping attacks such as vision-based eavesdropping may require the physical presence of an adversary, which limits the scale of their threat. However, the scale of attacks is not the only factor that determines the impact of attacks, which is also decided by the severity of potential losses. If a victim is an important person in a company, password leakage may lead to disclosing sensitive corporate data, which would provide sufficient incentives to an adversary. While internal attacks such as malware and logic key logger could be prevented by properly updating and configuring the computing system [30, 3, 6, 37] used by the victim, it is difficult to effectively mitigate the threat of external eavesdropping attacks due to inevitable exposure of human-computer interaction during the entry of legacy passwords. This threat becomes more serious in scenarios when a mobile device is used in public places.

Nevertheless, the threat of external eavesdropping attacks can be effectively mitigated with CoverPad. Besides enhanced security features, our scheme retains most benefits of legacy passwords and can be implemented on *commodity* devices. Our scheme is not only applicable to mobile devices but also other devices equipped with touch screen. For example, many ATM machines have been deployed with touch screen. Our scheme can be deployed on these machines to mitigate the threat of the ATM skimming attack [25].

6.2 Device Screen Size

Although we implement our scheme on Apple iPad, it could be easily adapted to other screen sizes, as illustrated in Figure 14. For a mobile phone with a small touch screen like Apple iPhone, a user can use a hand A to perform the hand-shielding gesture, and use the other hand B to hold the phone. The thumb on hand B can be used to press the response keys. For a mobile phone with a larger touch screen like Samsumg Galaxy Note II, a user may not be able to click all the keys with the thumb of hand B that holds the device. To deal with this situation, he only needs to use one hand A to perform the hand-shielding gesture and key pressing sequentially. Once the user raises his hand before pressing a key, the hidden transformation immediately disappears because the gesture is no longer detected by the touch screen. Meantime, the user does not need to worry about whether the actual keys pressed or the finger movements during key pressing may be observed by an adversary, as the sequence of pressed keys alone does not leak any information about the underlying password as analyzed in Section 4.

6.3 Limitations

Ecological validity is a challenging issue in any user study. Like most prior research [19, 27, 26, 15, 24], our experiments engage only university students. These participants are younger and more educated compared to the general population. Therefore, usability evaluation may vary with other populations. Our experiments are

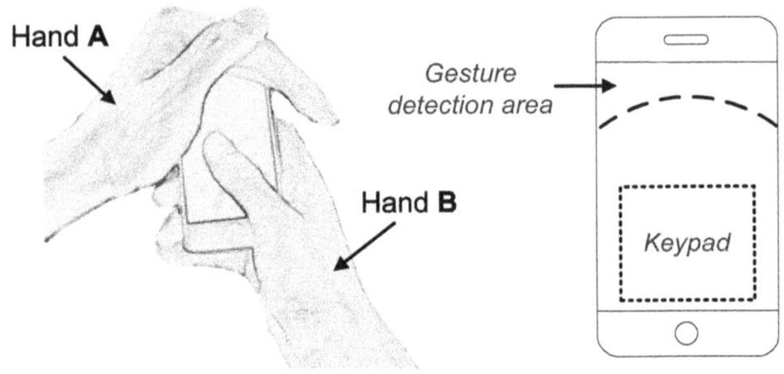

Figure 14: Conceptual demonstration on a small screen device

also restricted by the sample size, which may affect the results of statistical tests. Typical examples are the insignificant results on the login accuracy of our schemes. Moreover, our user study does not include experiments on memory effects (e.g. forgetting). Since our scheme uses the same alphabet and password composition as legacy passwords, the users may use the same coping strategies to help themselves to memorize the passwords in our scheme. The impact of memory effects on the user performance would be similar to legacy passwords as shown in the prior literature [14, 34].

7. RELATED WORK

In this section, we summarize closely related work on achieving leakage resilience of password entry in three different aspects.

Although the problem of achieving leakage resilience of password entry was proposed two decades ago [29], it is still a challenge to design a practical solution till now. Early work in this direction [19, 27, 38, 39, 5] focused on designing schemes solely rely on the cognitive capability of human beings. Unfortunately, all such schemes with acceptable usability have been broken [27, 38, 39, 5]. Recent investigations [11, 40] provided strong evidence for the necessity to construct a protected environment to hide certain user interaction during password entry in order to achieve both security and usability. The establishment of such protected environment may require the features only available from new user interface technologies. A few schemes [26, 33, 13, 15, 24, 8, 7] were designed in this strategy. Among them, our scheme design was mostly inspired by the concept of physical metaphor introduced in [24]. Our scheme distinguishes itself from prior work in the sense that it not only achieves leakage resilience but also retains most benefits of legacy passwords, while some of prior schemes [33, 15] are flawed in terms of security, and the others incur extra usability costs due to various reasons including: 1) using an uncommon device such as gaze tracker [26, 13], haptic motor [8], and large pressure-sensitive screen [24], 2) requiring an extra accessory device [7], and 3) inoperable in a non-stationary environment [8].

On the other hand, the procedure of applying random transformations on a fixed password used in our scheme design is a classic idea to prevent password leakage, but it is not easy to be realized in a *human-friendly* manner without the new user interface technologies, which are only available on modern computing devices. These new technologies give our scheme advantages when compared to recently patented schemes. Take GridCode [17] as an example, which asks users to memorize extra secrets (besides the passwords) in order to perform the transformations specified in its scheme design, while our scheme does not have such requirement. Another advantage of our scheme is that each character of the password uses a different hidden transformation during an authentication attempt, while GridCode uses the same transformation for all the characters in the password. If a hidden transformation in GridCode is disclosed, the entire password will be exposed. However, if a hidden transformation in our scheme is disclosed, only the single character associated with the transformation will be exposed. These two fundamental differences show both security and usability advantages of our scheme.

In terms of design principles, Roth et al. [32] proposed to use a cognitive trapdoor game to transform the knowledge of the underlying password into obfuscated responses. Li and Shum [27] later suggested three other principles including time-variant responses, randomness in challenges and responses, and indistinguishability against statistical analysis. Yan et al. [40] further extended the coverage by including the design principles against brute force attacks, and provided concrete guidelines against generic statistical attacks. Our proposed scheme follows all these design principles to avoid corresponding security flaws.

Bonneau et al. [10] recently proposed a generic framework for evaluating user authentication proposals and emphasized the importance of retaining the benefits of legacy passwords. Their framework introduced twenty-five benefits covering usability, deployability and security. This framework is used in our study to guide the scheme design in retaining the benefits of legacy passwords. Other research on password-based user authentication can be found in a recent survey paper [9], which summarized the development of new password schemes in the past decade.

8. CONCLUSION

In this paper, we proposed a leakage-resilient password entry scheme leveraging on the touch screen feature of mobile devices. It improves leakage resilience while preserving most benefits of legacy passwords. Three variants of this scheme were implemented. The practicability of our scheme was evaluated in an extended user study that incorporates new experiments to examine the influence of additional test conditions related to time pressure, distraction, and mental workload. These conditions were tested for the first time in the evaluation of user authentication schemes. Among these conditions, time pressure and mental workload were shown to have significant impacts on user performance. Therefore, we suggest including these conditions in the evaluation of user authentication schemes in the future research.

9. ACKNOWLEDGEMENTS

Yingjiu Li's work is supported in part by SMU Research Office under number 12-C220-SMU-006.

10. REFERENCES

[1] Ceiling effect. http://en.wikipedia.org/wiki/Ceiling_effect.

[2] Androidcommunity. Samsung galaxy siii display specs. http://androidcommunity.com/samsung-galaxy-siii-display-specs-edge-out-iphone-5-20121002/.

[3] Apple. Mac os x. http://www.apple.com/osx/.

[4] A. D. Baddeley and G. Hitch. Working memory. *The psychology of learning and motivation*, 8:47–89, 1974.

[5] X. Bai, W. Gu, S. Chellappan, X. Wang, D. Xuan, and B. Ma. Pas: Predicate-based authentication services against powerful passive adversaries. In *Proceedings of the 2008 Annual Computer Security Applications Conference*, pages 433–442, 2008.

[6] O. Begemann. Remote view controllers in ios 6. http://oleb.net/blog/2012/10/remote-view-controllers-in-ios-6.

[7] A. Bianchi, I. Oakley, V. Kostakos, and D. S. Kwon. The phone lock: audio and haptic shoulder-surfing resistant pin entry methods for mobile devices. In *Proceedings of the fifth international conference on Tangible, embedded, and embodied interaction*, pages 197–200, 2011.

[8] A. Bianchi, I. Oakley, and D.-S. Kwon. Obfuscating authentication through haptics, sound and light. In *Proceedings of the 2011 annual conference extended abstracts on Human factors in computing systems*, pages 1105–1110, 2011.

[9] R. Biddle, S. Chiasson, and P. C. van Oorschot. Graphical passwords: Learning from the first twelve years. *ACM Computing Surveys*, 44(4), 2012.

[10] J. Bonneau, C. Herley, P. van Oorschot, and F. Stajano. The quest to replace passwords: A framework for comparative evaluation of web authentication schemes. In *Proceedings of IEEE Symposium on Security and Privacy*, 2012.

[11] B. Coskun and C. Herley. Can "something you know" be saved? In *Proceedings of the 11th international conference on Information Security*, pages 421–440, 2008.

[12] F. I. Craik and J. M. McDowd. Age differences in recall and recognition. *Journal of Experimental Psychology: Learning, Memory, and Cognition*, 13(3):474–479, 1987.

[13] A. De Luca, M. Denzel, and H. Hussmann. Look into my eyes!: can you guess my password? In *Proceedings of the 5th Symposium on Usable Privacy and Security*, pages 7:1–7:12, 2009.

[14] A. De Luca, M. Langheinrich, and H. Hussmann. Towards understanding atm security: a field study of real world atm use. In *Proceedings of the Sixth Symposium on Usable Privacy and Security*, 2010.

[15] A. De Luca, E. von Zezschwitz, and H. Husmann. Vibrapass: secure authentication based on shared lies. In *Proceedings of the 27th international conference on Human factors in computing systems*, pages 913–916, 2009.

[16] M. Georgiev, S. Iyengar, S. Jana, R. Anubhai, D. Boneh, and V. Shmatikov. The most dangerous code in the world: validating ssl certificates in non-browser software. In *Proceedings of the 19th ACM Conference on Computer and Communications Security*, pages 38–49, 2012.

[17] L. Ginzburg, P. Sitar, and G. K. Flanagin. User authentication system and method. US Patent 7,725,712, SyferLock Technology Corporation, 2010.

[18] Google. Google glass. http://plus.google.com/+projectglass.

[19] N. J. Hopper and M. Blum. Secure human identification protocols. In *Proceedings of the 7th International Conference on the Theory and Application of Cryptology and Information Security: Advances in Cryptology*, pages 52–66, 2001.

[20] H. B. Hotel. ipad - free for every hotel guest. http://www.hollmann-beletage.at/en/ipad.

[21] I. Imbo and A. Vandierendonck. The role of phonological and executive working memory resources in simple arithmetic strategies. *European Journal Of Cognitive Psychology*, 19(6):910–933, 2007.

[22] A. Imran. ipads can now be used as public kiosks. http://www.redmondpie.com/ipad-public-kiosks-video/.

[23] A. R. Jensen. Process differences and individual differences in some cognitive tasks. *Intelligence*, 11(2):107–136, 1987.

[24] D. Kim, P. Dunphy, P. Briggs, J. Hook, J. W. Nicholson, J. Nicholson, and P. Olivier. Multi-touch authentication on tabletops. In *Proceedings of the 28th international conference on Human factors in computing systems*, pages 1093–1102, 2010.

[25] Krebs. Would you have spotted the fraud? http://krebsonsecurity.com/2010/01/would-you-have-spotted-the-fraud.

[26] M. Kumar, T. Garfinkel, D. Boneh, and T. Winograd. Reducing shoulder-surfing by using gaze-based password entry. In *Proceedings of the 3rd symposium on Usable privacy and security*, pages 13–19, 2007.

[27] S. Li and H. yeung Shum. Secure human-computer identification (interface) systems against peeping attacks: SecHCI. In *Cryptology ePrint Archive, Report 2005/268*, 2005.

[28] J. Long and J. Wiles. *No Tech Hacking: A Guide to Social Engineering, Dumpster Diving, and Shoulder Surfing*. Syngress, 2008.

[29] T. Matsumoto and H. Imai. Human identification through insecure channel. In *Proceedings of the 10th annual international conference on Theory and application of cryptographic techniques*, pages 409–421, 1991.

[30] Microsoft. Windows 8. http://windows.microsoft.com.

[31] F. Miller. *Telegraphic code to insure privacy and secrecy in the transmission of telegrams*. C.M. Cornwell, 1882.

[32] V. Roth, K. Richter, and R. Freidinger. A pin-entry method resilient against shoulder surfing. In *Proceedings of the 11th ACM conference on Computer and communications security*, pages 236–245, 2004.

[33] H. Sasamoto, N. Christin, and E. Hayashi. Undercover: authentication usable in front of prying eyes. In *Proceeding of the twenty-sixth annual SIGCHI conference on Human factors in computing systems*, pages 183–192, 2008.

[34] R. Shay, P. G. Kelley, S. Komanduri, M. L. Mazurek, B. Ur, T. Vidas, L. Bauer, N. Christin, and L. F. Cranor. Correct horse battery staple: exploring the usability of system-assigned passphrases. In *Proceedings of the Eighth Symposium on Usable Privacy and Security*, 2012.

[35] D. X. Song, D. Wagner, and X. Tian. Timing analysis of keystrokes and timing attacks on ssh. In *Proceedings of the 10th USENIX Security Symposium*, 2001.

[36] Spycop. Hardware keylogger detection. http://spycop.com/keyloggerremoval.htm.

[37] TCG. Trusted computing group. http://www.trustedcomputinggroup.org.

[38] D. Weinshall. Cognitive authentication schemes safe against spyware (short paper). In *Proceedings of the 2006 IEEE Symposium on Security and Privacy*, pages 295–300, 2006.

[39] S. Wiedenbeck, J. Waters, L. Sobrado, and J.-C. Birget. Design and evaluation of a shoulder-surfing resistant graphical password scheme. In *Proceedings of the working conference on Advanced visual interfaces*, pages 177–184, 2006.

[40] Q. Yan, J. Han, Y. Li, and R. H. Deng. On limitations of designing leakage-resilient password systems: Attacks, principles and usability. In *Proceedings of the 19th Annual Network and Distributed System Security Symposium*, 2012.

[41] ZDNet. More ipad love: Now hotels offer ipad to customers. http://www.zdnet.com/blog/apple/more-ipad-love-now-hotels-offer-ipad-to-customers/6850.

APPENDIX

A. STATISTICAL TEST RESULTS

In this section, we provide the detailed results of statistical tests. Table 3 shows the results for login time, which indicates that the same test condition may have difference impact on the login time of different schemes.

Average login time of NumPad-Add - *omnibus* KW χ^2_5=32.423, p<.001

normal (10.4)	timed (9.2)	U=551, p=.017 ★
	distraction (11.2)	U=679, p=.184
	distraction+timed (10.3)	U=878, p=.989
	mental workload (11.8)	U=515, p=.003 ★
	mental workload+timed (10.7)	U=696, p=.319
distraction (11.2)	distraction+timed (10.3)	U=718, p=.107
mental workload (11.8)	mental workload+timed (10.7)	U=558, p=.009 ★

Average login time of NumPad-Shift - *omnibus* KW χ^2_5=11.965, p=.034

normal (11.7)	timed (11.2)	U=666, p=.199
	distraction (13.5)	U=645, p=.137
	distraction+timed (11.7)	U=727, p=.485
	mental workload (13.3)	U=655, p=.164
	mental workload+timed (11.4)	U=644, p=.135
distraction (13.5)	distraction+timed (11.7)	U=565, p=.024 ★
mental workload (13.3)	mental workload+timed (11.4)	U=555, p=.019 ★

Average login time of LetterPad-Shift - *omnibus* KW χ^2_5=49.252, p<.001

normal (13.2)	timed (10.1)	U=294, p<.001 ★
	distraction (13.6)	U=774, p=.667
	distraction+timed (11.0)	U=413, p<.001 ★
	mental workload (13.4)	U=653, p=.116
	mental workload+timed (11.5)	U=472, p=.002 ★
distraction (13.6)	distraction+timed (11.0)	U=422, p<.001 ★
mental workload (13.4)	mental workload+timed (11.5)	U=631, p=.075

Table 3: The results of statistical tests for login time (sec). All pairwise tests are Mann-Whitney U. The statistically significant results are marked with ★.

The results of statistical tests on login accuracy are not shown as none of them indicate significance. This is caused by the ceiling effect, which can be observed from the data shown in Table 4. Even in the worst case, 50.0% participants did not make any mistakes during all tests in the test condition, which implies our tests are not sufficiently difficult to distinguish these test conditions regarding their influence on the login accuracy of our schemes. This could be caused by the simple design of our schemes such that they are easy to use even in the presence of time pressure, distraction, and mental workload. However, it does not necessarily imply that these factors will not significantly influence the login accuracy of other user authentication schemes. Since the average results of login accuracy are observed to be worse due to the presence of these factors in our tests, we expect they would have a more significant influence on other schemes with higher complexity.

	NumPad-Add	NumPad-Shift	LetterPad-Shift
normal	82.9%	67.5%	75.6%
timed	78.0%	62.5%	53.7%
distraction	80.5%	70.0%	63.4%
distraction+timed	70.7%	55.0%	58.5%
mental workload	75.6%	57.5%	65.9%
mental workload+timed	65.9%	50.0%	51.2%

Table 4: Evidence for the ceiling effect in statistical tests on login accuracy. Each cell in this table shows the percentages of the participants who did not make any mistakes in a test condition.

Your Love is Public Now: Questioning the use of Personal Information in Authentication

Payas Gupta, Swapna Gottipati, Jing Jiang and Debin Gao
{payas.gupta.2008, swapnag.2010, jingjiang, dbgao}@smu.edu.sg
School of Information Systems, Singapore Management University, Singapore

ABSTRACT

Most social networking platforms protect user's private information by limiting access to it to a small group of members, typically friends of the user, while allowing (virtually) everyone's access to the user's public data. In this paper, we exploit public data available on Facebook to infer users' undisclosed interests on their profile pages. In particular, we infer their undisclosed interests from the public data fetched using Graph APIs provided by Facebook. We demonstrate that simply liking a Facebook page does not corroborate that the user is interested in the page. Instead, we perform sentiment-oriented mining on various attributes of a Facebook page to determine the user's real interests. Our experiments conducted on over 34,000 public pages collected from Facebook and data from volunteers show that our inference technique can infer interests that are often hidden by users on their personal profile with moderate accuracy. We are able to disclose 22 interests of a user and find more than 80,097 users with at least 2 interests. We also show how this inferred information can be used to break a preference based backup authentication system.

Categories and Subject Descriptors

D.4.6 [**Security and Protection**]: Authentication; H.2.4 [**Systems**]: Textual databases

General Terms

Security

Keywords

Facebook; preference based authentication; Graph API; semantic analysis

1. INTRODUCTION

With the rise of online social networks (OSNs), more and more personal information of users is available on the web. It also forms a fertile ground for a variety of research efforts. The information shared on OSNs can be classified into two broad categories, *private* (shared with a limited set of users) and *public* (shared with the whole world). Prior research shows that the information shared with a limited set of users can leak undisclosed privacy attributes, e.g., users' interests and even sexual orientation [28, 21]. Authors of [4] crawled Facebook users' personal profiles to infer users' undisclosed interests. However, getting access to such information (without explicit permission) that is shared with a limited set of users is non-trivial as it is not available to public via any APIs. Moreover, OSNs are adopting ways to restrict crawling unless explicit permission is granted [25, 31].

In comparison to the information available only to a limited set of users, public information is readily available. In many cases APIs are provided by OSNs for anyone to efficiently download such public data. Facebook, for example, has made all the fan pages public by default. Access to the data of these pages can be conveniently obtained through Graph APIs [8]. It is generally believed that these public pages hardly contain any secret information, and mining some useful information from them is not easy mainly due to the large amount of noise contained in the heterogeneous pages, and the huge amount of unstructured data involved. For example, Facebook has little control on the titles and descriptions of fan pages; posts from users may contain text and multimedia content; users use a lot of short sentences and slang (e.g., "LOL", "LMAO", etc); off-topic discussions go on frequently (e.g., on a "Jazz" page, we found users discussing the latest soccer game). All this adds to the noise in public data on OSNs. Moreover, almost 15 percent of user-submitted content on large Facebook fan pages is spam [10]. Such noise and the huge amount of unstructured data to be processed usually makes mining interesting information not practical.

In this paper, we show that this belief might not be true in certain aspect. In particular, we show how we use *publicly available* data from Facebook to infer users' interests that are usually only on their personal profile pages. We make use of the graph APIs provided by Facebook to obtain public fan pages [7]. As these pages are public irrespective of the users' privacy settings, an attacker can grab the unique profile IDs of those who have interacted with the page. We show that by aggregating different interests of the users found across different pages, one could build users' interests profiles from the public data without gaining access to the personal profile pages of any of the users. This collective information can be used in many ways including targeted spamming, show-

ASIA CCS'13, May 8–10, 2013, Hangzhou, China.

ing ads without the consent of users, or even breaking into specific authentication systems, in our case preference based authentication [17]. A system based on user preferences was proposed in order to reduce the vulnerability to data-mining and maximize the success rate of legitimate reset attempts. The viability of such an approach is supported by findings in psychology, showing that personal preferences remain stable for a long period of time.

To demonstrate the security and privacy implication of this, we base our experiments on mining personal interests to break into Blue Moon™ [15] introduced by RavenWhite as a backup authentication system to provide better security and usability. From the dataset in our experiments involving 1.1 million different user IDs from 34,000 Facebook public pages, we detected 80,097 (6.89%) users with two or more interests. Out of these 80,097 users, there are 66 who have been found with more than 8 interests, which is enough to break their corresponding Blue Moon accounts (if they have) with reasonable accuracy under certain assumptions. In one case, we were able to build a user profile with as many as 22 interests by mining the data we collected. We also present valuable lessons we learned in our experiments, among which the most notable one being that users' sentiment orientation might not be inclined towards the sentiment orientation of the page i.e. simply liking a page does not corroborate enough that the user is really interested in the page. Therefore, we performed sentiment mining to find out the actual sentiment orientation of the user.

In summary, this paper makes the following contributions.

- We use publicly available data on Facebook to infer users' interests and aggregate this information across different pages. This differs from prior research as we do not use user's personal data posted on their profile page (e.g., gender, current location, activities, interests, etc.).

- We find that liking a page does not corroborate a user's inclination towards a page or interest category. We performed an in-depth analysis (sentiment) using text mining to find the real sentiment orientation or polarity (like or dislike) of the user towards a page and an interest.

- We use Facebook's public Graph API [8] to obtain the public pages. Unlike crawling which is usually restricted in its usage by OSNs to a small number of partners, our method could be easily used by anyone with little restriction.

- We demonstrate the severe implication of interests mining by showing that interests inferred from the public data can be used to exploit a previously proposed preference based authentication system.

The rest of the paper is structured as follows. We provide background and related work in Section 2 where we show some of the important prior work to abuse OSNs. We explain our technique to mine user interests from Facebook public pages in Section 3, and report experimental results in Section 4. We then discuss the limitations and errors that could have occurred in our technique in Section 5 and conclude the paper in Section 6.

2. RELATED WORK AND BACKGROUND

In this section, we first discuss related work in obtaining a user's private information by abusing OSNs in general. After that, we discuss the more specific interests inference techniques in social networks.

2.1 Abusing OSN data

With the increasing popularity of OSNs, people start to find ways of abusing it, e.g., illegitimate use by spammers with ad deals. In this paper, we focus on the abuse in which a user's privacy attributes are inferred from information hosted on OSNs. In general, attackers could base their attacks on two types of data obtained in different ways.

One is to use restricted pages by crawling. Prior research shows that information on restricted pages (shared with a limited set of users) can leak undisclosed privacy attributes about the users [28, 21]. Existing techniques have demonstrated that private information can be crawled to obtain attributes like mother's maiden name, date of birth, hometown, first school attended to break into backup authentication mechanisms that are based on such privacy attributes [16]. Attackers can also correlate information from different OSNs to retrieve undisclosed attributes of the users [21]. Authors of [4] crawled users' personal profiles of Facebook to infer their undisclosed interests. [3] describes how an attacker could query popular social networks for registered e-mail addresses on a large scale and information from different social networks can be aggregated to launch sophisticated and targeted attacks.

An important limitation to using restricted pages by crawling is that most OSNs restrict crawling to a small number of partners only. That is, crawling restricted pages is not a technique available to general attackers.

The other type of data to use is public pages. As compared to crawling restricted pages on OSNs, anyone can use a legitimate channel (usually by using public APIs provided by OSNs) to gather public information. Although these public pages are more readily available for anyone to analyze, as pointed out in Section 1, it is generally believed that mining interesting private information from these public pages is difficult due to the noise in it and the huge amount of unstructured data to be analyzed. In this paper, we show that mining users' otherwise undisclosed interests from public pages on OSNs is, in fact, practical.

There are strong security and privacy implications to such abuse of OSN data because the private information mined could potentially be used to break existing authentication systems, typically those that use challenge questions as a backup to the main authentication mechanism. In Table 1, we highlight noticeable differences between this work and prior research. Previous work has shown that OSN data and public databases can be used to infer or guess sensitive information about users [28, 21]. A number of incidents, e.g., in 2008 the Republican vice presidential nominee Sarah Palin's email account was compromised by an attacker who guessed her personal authentication question (where did you meet your spouse?) [26], in 2009 a vandal successfully guessed a Twitter executive's password and leaked the company's internal documents [6], have shown the severe damage such attacks could have. Personal authentication questions are usually a weaker link in authentication systems [24]. Authors show that answers to predefined questions can be easily guessed or obtained from OSNs [24]. Instead of specific

#	Paper	Datasets source	Dataset type	Count	Inferred information	Collected information	Dataset gathering technique
1	Mislove et al. [21]	Rice University Facebook network	Personal profiles	6,156	Missing attribute on personal profile page	Attributes of others (friends etc.)	Crawling
		New Orleans Facebook network	Personal profiles	90,269			Crawling
2	Chaabane et al. [4]	Facebook	Personal profiles	104,000	Gender, Relationship status, Country-level location and Age	Interests	Crawling
		Volunteers from Facebook	Personal profiles	200			Voluntarily provided
3	Lindamood et al. [20]	LiveJournal	Personal profiles	66,766	Nodes in the social graph	Friendship relations	Crawling
4	Goga et al. [13]	Twitter	Personal profiles	93,839	Correlating missing attributes	Profile and privacy information	Friend-finder
		Flicker		59,476			
		Yelp		24,176			
5	Avello et al. [12]	Twitter	Personal Profiles	4.98M	Sex, age, political orientation, religious affiliation, race	Personal details and tweets	Crawling
			Tweets	27.9M			
6	Zheleva et al. [33]	Flicker	Personal Profiles	9,179	Location	Information on profile	Crawling
		Facebook	Personal Profiles	1,598	Gender, political orientation	Information on profile	Crawling
		Dogster	Dog Profiles	2,632	Breed category	Information on profile	Crawling
		BibSonomy	Personal Profiles	31,715	Spammer	Information on profile	N/A
7	*This paper*	Facebook	*Public pages*	34,000	Interests	Public page's data	*API*
			Personal profiles	1.1M			

Table 1: Comparison with the related work

attack incidents where one or two particular accounts are compromised, our work presented in this paper shows an attack to the authentication system and evaluates the extent to which thousands of users of such a system could be attacked.

2.2 Interests mining from OSN data

This paper focuses on personal interests mining because personal interest is one of the most popular choices used in challenge questions. Authors of [22] leverage on the friendship network to mine users' interests. [11] employs feature engineering to generate hand-crafted meta-descriptors as fingerprints for a user. However, such models alone may not derive the complete interest list of any user [32]. [19, 5] resort to collaborative filtering techniques to profile user interests by collaboratively uncovering user behaviors.

The "Like" function on OSN provides a more intuitive way of estimating user interests as compared to non-direct indicators such as user-service interactions. Clicking on the "Like"/"Dislike" button associated with an object usually indicates that (s)he is highly interested/disinterested in the object [32]. Recent approaches like LikeMiner [18] assumes that clicking the "like" button demonstrates the user's liking towards the object. In this paper, we show complications in using such an assumption on large datasets and propose solutions to it.

3. INTERESTS INFERENCE

Although users understand that public pages are for everyone to view and should not contain sensitive or private information, these pages nevertheless reveal what users do and what users think. Therefore, it is probably not difficult to be convinced that such pages still contain private information, probably indirectly and to a limited degree, e.g., by reflecting what users like and dislike. This paper is not to argue this, but to rather investigate how practical it is to mine interesting personal interests from the large amount of unstructured data on public pages that contain a lot of noise.

To do this, we first introduce the data source on which our analysis is performed, i.e., the public pages on Facebook (see Section 3.1). Section 3.2 presents our methodology to fetch information from these public pages. Finally, we present our methodology to infer users' interests (i.e. likes and dislikes) for different categories (e.g., music, cars, sports) in Section 3.3.

3.1 Facebook Page Layout

A Facebook Page is a public profile where users can talk (and comment/like) about a particular topic. As shown in Figure 1, it usually contains many attributes including title, page description, profile picture, wall posts, likes, etc. Any registered user can create a page and by default all Facebook pages are public. Please note that anyone can view the page, however, to interact with a page (i.e. comment or post something), a registered Facebook user must "like" it first by pressing the like button on the page. After liking the page, the user can post a message/link/photo/video which will appear on the wall of that page. Other users who have already liked that page can post comments on a post, like the post, etc.

As an illustration, Figure 1 shows a Facebook page "LSU Football" with one post "LSU Tigers in the NFL – Week 10" from the user "LSU Football". 244 users have liked that post; a user has commented on it and 4 users have liked it.

As shown in Figure 2, in general, a public page P is a collection of many attributes. First, there is a title and description t of the page. Each page P may contain a number of posts p^1, p^2, etc. Each post p^i may have a few likes L_1^i, L_2^i, etc. and a few comments c_1^i, c_2^i, etc. Each comment c_j^i might have a few likes $l_{j,1}^i$, $l_{j,2}^i$, etc.

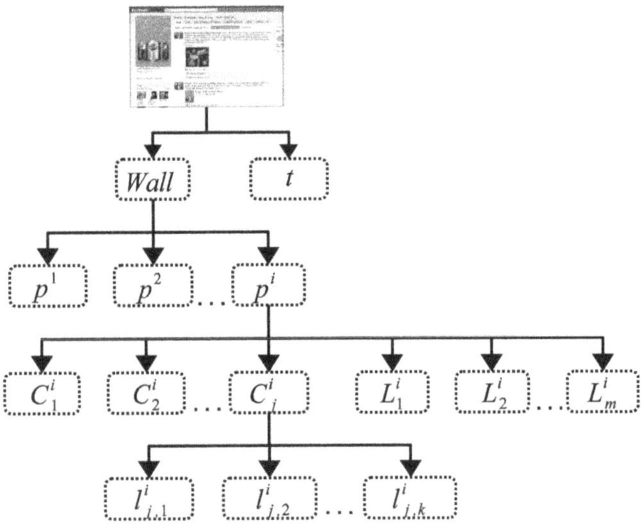

Figure 2: Structure of a Facebook page

3.2 Data collection

As discussed in Section 2, since we make use of public pages only, we can use Facebook's public Graph API to fetch data of any Facebook public page. Information that we manage to fetch for each page includes its title and description t, all posts p^i, comments c_j^i and likes L_m^i of each post, as well as likes of each comment $l_{j,k}^i$.

A small difficulty we faced was the authentication needed to use the Graph API. To fetch pages from Facebook using Graph API, one requires an authentication code. This authentication code is generally provided to Facebook applications for a limited amount of time. However, we did not build any Facebook application to obtain this authentication code. Instead, we created a php script to automatically login to Facebook and then parse the webpage returned at the following URL and look for the authentication code `https://developers.facebook.com/tools/access_token/`. Access token is only granted for a limited time, therefore this process can be repeated whenever the access token is expired. This work around was possible at the time of writing this paper.

There are limitations in using the public APIs only. For example, we were not able to obtain the IDs of those people who only liked the page and did not comment or posted anything on the page. We were also limited by the number of API calls we can make in a certain duration. At the time when the experiments were conducted, Facebook used to provide the list of users who liked a page, this feature is not supported anymore.

3.3 Automated profilin with attributes

In this subsection, we first present how we analyze an individual attribute on a Facebook page to figure out if a user has personal interests in the topic covered in that page. This might sound simple, as the user's interaction with the attributes of a page reveals some inclination towards that page. For example, if a user posts a positive message on the wall of a soccer page, it can be inferred that the user may be interested in soccer. For this we propose a technique SPM (**S**im**P**le **M**ining) to mine the information of users' interest (see Section 3.3.1). However, we also observed during our analysis that many users may "like" a page even though they are not interested in the corresponding topic, or if they have strong negative opinions on the topic. To solve this problem, in Section 3.3.2 we propose a more advanced technique called SOM (**S**entiment **O**riented **M**ining) to use sentiment analysis of the attributes to find the actual sentiment orientation of the users.

3.3.1 SPM: SimPle Mining

In SPM, we simply assume that a user's involvement in any of the attributes of P in whichever way indicates that the user has a same interest on the topic of the page, which can be inferred from t. For example, if a user likes a comment or adds a comment on a post on P, then we believe that the user is interested in P. If multiple users have interacted with P, we add all these users into the set $u_q^=$ which denotes the set of users who are interested in P about the interest q.

SPM is simple, but can easily introduce errors to $u_q^=$ because there could be a group of users of P (denoted $u_q^{\neq}$) who hold an opinion opposite to the focus of P. For example, a user who liked the cats page posted the following posts "I hate cats", "Lewis is a mad cat", "go doggies cats are crap" etc. They "liked" the page not because they really like it, but simply because Facebook does not allow them to add a post until they "like" it. To minimize this noise, we perform sentiment mining on textual data to find out the inclination of all users towards that page. See the following section for details.

3.3.2 SOM: Sentiment Oriented Mining

Sentiment analysis is the task of identifying positive and negative opinions, emotions, and evaluations [30]. Sentiment analysis has been used in many fields where users have subjective agenda such as movie reviews [23]. Intuitively, the content of the posts/comments should be accounted in deriving the users' interest. Hence, the polarity of the sentiment information of the text aids in conforming the users' interest.

We first define two sets of attributes on a Facebook page P. $\mathcal{A}_\mathcal{C} = \{t, p, c\}$ is the set of text-based attributes which consists of text, while $\mathcal{A}_\mathcal{D} = \{L, l\}$ is the set of dependent attributes which does not contain text. We separate these

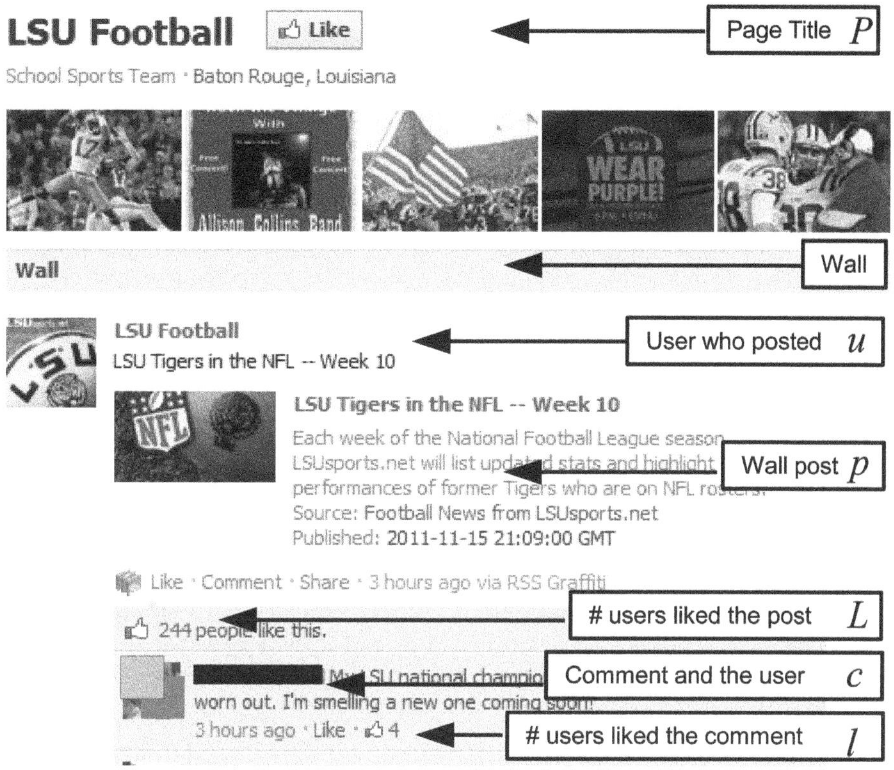

Figure 1: A public Facebook Page

attributes into two groups because those that consist of text can go through a more thorough sentiment analysis on the text, while attributes of the other set are more dependent on the post or comment upon which the "like" was applied.

Sentiment analysis on $\mathcal{A}_\mathcal{C}$.

We propose to use lexicon approach [14], which is one of the most popular methods used in sentiment analysis to detect the opinion bearing words. Lexicon approach concerns the use of lexical resources such as a dictionary of opinionated terms or opinion words. Collectively, they are called the opinion lexicon and are instrumental for sentiment analysis. Opinion words are the words that are commonly used to express positive (s^+) or negative (s^-) sentiments. For example: 'beautiful', 'wonderful', 'good', and 'amazing' are positive opinion words, and 'bad', 'poor', and 'terrible' are negative opinion words. Many opinion words are adjectives and adverbs. Sometimes, nouns such as 'rubbish', 'junk', and 'crap' and verbs such as 'hate' and 'like' also indicate opinions. Words which are neither positive nor negative are marked as neutral ($s^{\bigcirc}$).

Several opinion lexicons are available and SentiWordNet [2] is one such resource containing opinion information on terms extracted from the WordNet database and made publicly available for research purposes. SentiWordNet is a lexical resource built on top of WordNet. WordNet [9] is a thesaurus containing descriptions of terms, and relationships between terms and part-of-speech (POS) types. For example "car" is a subtype of vehicle and car has same concept as automobile. Hence, a synset (a synonym set) in WordNet comprises of all the terms with the same concept, e.g., the synset is car, automobile.

SentiWordNet assigns three sentiment scores to each synset of WordNet: positivity, negativity, objectivity/neutral. The sentiscores are in the range of [0, 1] and sum up to 1 for each triplet. For example, in SentiWordNet, the sentiscore of the term "good" is (pos, neg, obj) = (0.875, 0.0, 0.125). For our experiments, the scores are approximated with labels/part-of-speech of term in the text or sentence. First, the text is tagged using a standard POS tagger. A standard POS Tagger [27] is a piece of software that reads text in some language and assigns parts of speech to each word, such as noun, verb, adjective, etc. Then the SentiWordNet is used to get the scores for each term in the text. As our sentiment analysis is domain independent, we choose the general lexicon method as compared to the corpus-based method which is a domain dependent approach.

We now explain in detail how the sentiment is derived for the attributes in $\mathcal{A}_\mathcal{C}$. We call them text-based attributes because their sentiment orientation is based on the text of the attribute. For a particular attribute $a \in \mathcal{A}_\mathcal{C}$, we count the total number of words/phrases with positive sentiment (p_s) and that of words/phrases with negative sentiment (n_s) for all posts and comments he/she has, and use the term-counting method proposed by [29] to determine $\Psi(a)$ i.e. the sentiment orientation of a to be

$$\Psi(a) = \begin{cases} s^+, & \text{if } p_s > n_s \\ s^-, & \text{if } p_s < n_s \\ s^{\bigcirc}, & \text{otherwise} \end{cases}$$

Sentiment analysis of $\mathcal{A}_\mathcal{D}$.

Attributes in $\mathcal{A}_\mathcal{D}$ does not contain text, but they also contribute to a user's sentiment orientation. We call them dependent attributes because their sentiment orientation is dependent on other attributes. For example, if a user u_1 has a post p^i with negative opinion, and user u_2 likes that post L^i_m, then both u_1 and u_2 share negative sentiment orientation on the topic. Similarly, if a user u_1 has a comment c^i_j with positive opinion, and user u_2 likes that comment $l^i_{j,k}$, then both u_1 and u_2 share positive sentiment orientation on the topic. That is,

$$\Psi\left(L^i_m\right) = \Psi\left(p^i\right)$$

and

$$\Psi\left(l^i_{j,k}\right) = \Psi\left(c^i_j\right)$$

Aggregating interests profilin from multiple attributes on multiple pages.

A user might have multiple posts, comments, and likes on a single Facebook page, and multiple Facebook pages might be about the same interest. Therefore, we have to aggregate the sentiment analysis results on multiple attributes from multiple pages in order to figure out the sentiment orientation of the user on that interest.

Let $A = \{a_1, a_2, ..., a_k\}$ be the set of posts, comments, and likes of a user u on a page P about a particular interest q. For each $a \in A$, we compute the sentiment orientation. Then, the sentiment orientation of u towards P, S^u_P is s^+/s^- if the number of attributes with positive sentiment is greater/lesser than the number of attributes with negative sentiment respectively, otherwise $s^{\bigcirc}$. Aggregating all Facebook pages about q, sentiment orientation of u towards q (S^u_q) is s^+/s^- if the number pages with positive sentiment orientation is greater/lesser than the number of pages with the negative sentiment orientation respectively; otherwise $s^{\bigcirc}$. If the sentiment orientation of q and S^u_q is same, then, u is added to the set $u^=_q$ otherwise to $u^{\neq}_q$. That is,

$$u^=_q = \left\{ u \in U \;\middle|\; \left(\left(\Psi(q) = s^+ \;\&\&\; S^u_q \in \{s^+, s^{\bigcirc}\}\right) \;\|\; \left(\Psi(q) = s^- \;\&\&\; S^u_q = s^-\right) \right) \right\}$$

$$u^{\neq}_q = \left\{ u \in U \;\middle|\; \left(\left(\Psi(q) = s^+ \;\&\&\; S^u_q = s^-\right) \;\|\; \left(\Psi(q) = s^- \;\&\&\; S^u_q \in \{s^+, s^{\bigcirc}\}\right) \right) \right\}$$

4. EXPERIMENTAL RESULTS

To base our analysis on a concrete example, we focus on breaking Blue Moon™ [15], a backup authentication system which can be used by a user to reset his lost or forgotten credentials. For example, if a user forgets his password of an email account, he or she can use Blue Moon to reset the password. The idea is to use personal preferences as challenge questions for authentication. Figure 3 shows a screenshot of Blue Moon[1].

During enrollment, the user is asked to select 8 items which he likes and 8 items which he dislikes from a list of 76 common interests. During authentication, the user is presented with a set containing the chosen items in a randomized fashion. The user categorizes the items to like and dislike. A user is not required to pick all the interests correctly. Instead, the user just need to correctly categorize 8 items[2] to reset his password [17]. To make our analysis consistent and the evaluation comparable, in the rest of the paper we assume that a user has to correctly categorize 8 items from the entire list of 76 interests which are shown in Table 2.

4.1 Dataset Description

In order to attack the Blue Moon system, we assume a strategy taken by an attacker as follows. He first constructs a set of interests Q from Table 2, and another set Q' containing the corresponding negated items like "I hate golf" and "I hate jazz". He then leverages the Facebook's public Graph API to 1) find all public Facebook pages related to $q \in \{Q \cup Q'\}$; 2) fetch all attribute data of these pages; and 3) use the technique described in Section 3 to find $u^=_q$ and $u^{\neq}_q$ for all $q \in \{Q \cup Q'\}$.

Note that this methodology does not cover all those pages which are semantically related with the query term. For example, query term "cars" may not fetch pages of Mercedes, Hyundai or Porche which are indeed the pages of cars. [4] provides a solution to fetch these pages using semantic search with the help of an ontology build upon Wikipedia. We are sure that this could significantly increase accuracy of the attack, although that comes with a price of longer processing of a large set of pages. We leave this as a future work to increase the size of the corpus.

Table 3 summarizes the availability of the attributes with their counts. From 34K pages fetched for 152 categories we found 2.5 million posts, 7.5 million likes and 4.3 million comments on these posts, and 1.3 million likes on those comments.

Attribute	Count
Categories	152
Pages	34,738
Posts	2,538,987
Post likes	7,574,965
Comments	4,381,967
Comment Likes	1,361,361

Table 3: Number of attributes found

We apply both SPM and SOM mining techniques as discussed in Section 3.3.1 and 3.3.2 respectively to the dataset `PubProf`, and discuss the results in the next subsections.

4.2 Inferred interests using SPM

We found a total of 1,162,575 unique users whose interests can be inferred from the pages analyzed. These users

[1]This image is taken from `http://www.ravenwhite.com/iforgotmypassword.html`.

[2]The threshold where false acceptance rate and false rejection rate meets.

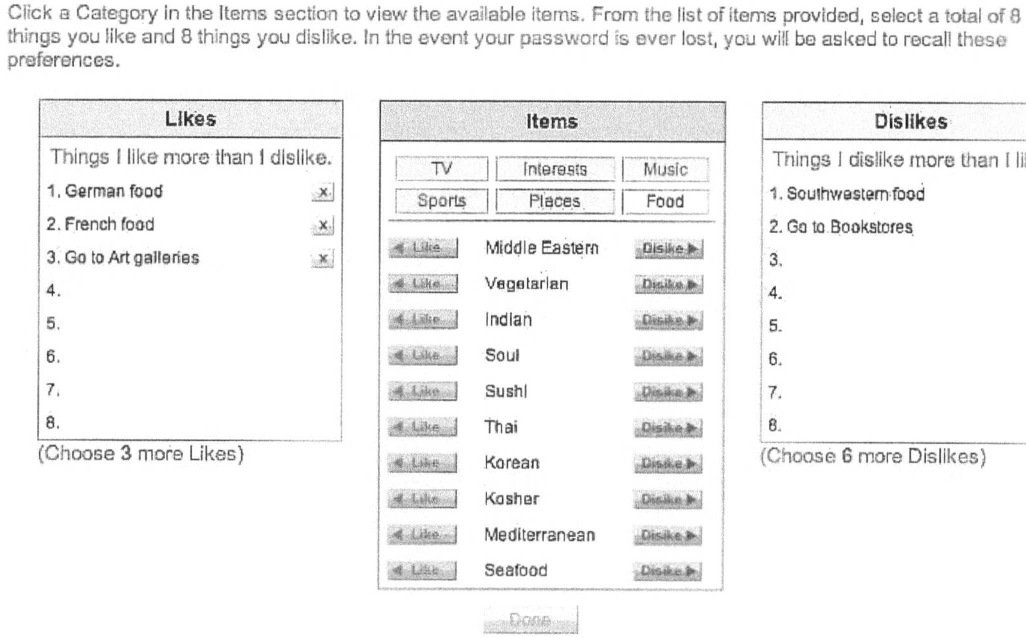

Figure 3: Blue Moon^{TM}

are the users who have either posted something on the pages, commented on the posts, or liked the posts/comments. Applying the SPM approach to our dataset, we detected 80,097 users with 2 or more interests. This amounts to 6.89% of all the user IDs collected in our dataset. Figure 4 shows breakdown of users with different number of interests found. We were able to build a user profile with as many as 22 interests.

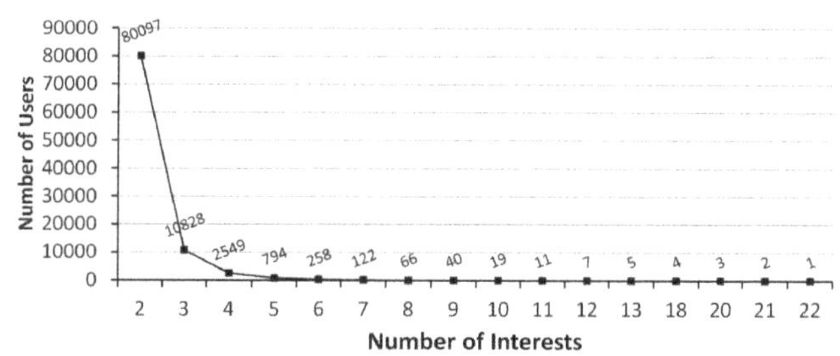

Figure 4: Inferred interests of the users using SPM

Note that although SPM might not be accurate in finding the true sentiment orientation of the user over an interest, the numbers presented here is not affected by this inaccuracy. That is, results presented in Figure 4 actually applies to SOM as well.

Results show that the number of users whose interests could be inferred from the public pages is significant, and this would have an important impact on the possibility of breaking into such users' Blue Moon account. For example, for those users found to have two interests, the search space for breaking into their Blue Moon account is reduced by about a factor of 3,000 ($^{76}C_2$). Please also note the dataset we use represents a tiny subset of the Facebook pages.

4.3 Inferred interests using SOM

In this section, we first investigate the inaccuracies when applying SPM on our data. As discussed in Section 3.3.2, these inaccuracies happen when users like a page but oppose to the topic in it. For example, if the page's title is "I hate Cats", we want to find those users who clicked on the like button on this page, however, actually like cats. Figure 5 shows the result of our sentiment analysis on all the comments, in particular, the number of negative comments different user posts. We can observe that about 10% of the comments posted are negative, meaning that the comment itself does not have the same sentiment orientation as that of the page. This suggests that more careful analysis and handling of the sentiments of posts and comments are important in order to find out the users' interests.

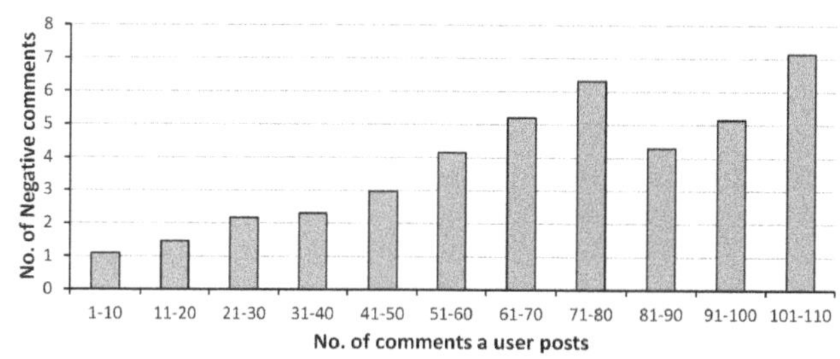

Figure 5: No. of negative comments posted by users

One interesting finding we also observe from Figure 5 is that users who comment a lot (more than 70 comments) tend to have a smaller percentage of negative comments. It is our future work to investigate whether the same is observed on a larger dataset.

To further investigate the result of our sentiment analysis, Table 4 shows 15 interests with the largest percentage of users who disagree with the corresponding topic of the page. $|u_q^{=}|$ is the total number of users whose sentiment orientation is same as that of the category, and $|u_q^{\neq}|$ is the total number

Sports	Music	Places	Interests	TV	Food
Aerobics	Instrumental	Garage sales	Cars	Watching Extreme sports	Soul food
Billiards	Symphony	Bookstores	Crafts	Documentaries	Indian food
Racing	Folk	Political events	Creative writing	Watching Auto racing	Korean food
Martial arts	Easy listening	Art galleries	Casino	Watching News	Kosher food
Baseball	Gospel	Raves	Painting	Watching Figure skating	Middle Eastern food
Soccer	Electronics	Antique stores	Gardening	Watching Diving	Southwestern food
Golf	Classical	Museums	Religion	Watching Baseball	French food
Running	Jazz	Flea markets	Politics	Watching Hockey	Seafood food
Yoga	Big Band	Libraries	Poetry	Watching Golf	Thai food
Skating	Reggae	The opera	Gaming	Game shows	Vegetarian food
Cycling	Show tunes	Politics	Reading comics	Watching Soccer	German food
Hockey	Heavy Metal		Cats	Watching Bowling	Mediterranean food
Pool			Gambling		
Motocross					
Basketball					
Football					

Table 2: User interests domain

of users whose sentiment orientation is opposite to that of the category. We see that although the inaccuracies from the SPM technique exist, most pages, especially those with a large number of users discussing, tend to have less than 10% of the users with negative sentiment orientation.

Category (q)	$\lvert u_q^{\neq} \rvert$ (%)	$\lvert u_q^{=} \rvert$ (%)	Total
Hate motocross	25.00	75.00	8
Hate skating	25.00	75.00	4
Hate heavy Metal	14.28	85.72	14
Hate poetry	12.50	87.50	8
Hate hockey	11.53	88.47	26
Watching baseball	11.12	88.88	9
Hate baseball	8.33	91.67	48
Hate basketball	5.52	94.48	181
Hate cats	5.39	94.61	260
Hate religion	5.36	94.64	56
Hate football	3.35	96.65	359
Raves	3.22	96.78	280
Gamble	3.08	96.92	195
Hate cars	3.08	96.92	130
Game shows	2.95	97.05	34

Table 4: The percentage of users whose sentiment orientation is not inclined / inclined towards the sentiment orientation of the page across all interests categories.

Another interesting observation is that most of the entries in Table 4 are of pages with a negative sentiment, i.e., q is "Hate xxx". We believe that it is because there are more people who want to voice out their disagreement with such pages than people who disagree with pages with a positive sentiment.

4.4 Comparing SPM and SOM

Table 5 shows the number of users found liking/disliking selected categories for both the SPM and SOM techniques. We only show a few categories with the largest discrepancies due to space constraint. Note that these are accumulated categories, e.g., we combined all 12 musical categories like jazz, classical, etc.

Results show that although there are users who like the page while having a different sentiment orientation as shown in the previous subsection, these users are minorities, and that is why we do not see a large discrepancy between results from SPM and SOM. In this respect results here seem to be consistent with those presented in Table 4.

Also, we observe that sports is the most popular category where 12.62% of users are inclined to the sports. These numbers could potentially be used to obtain the a priori probability for an unknown user having different interests, and subsequently used in attacking the Blue Moon system. We leave more detailed analysis on this for our future work.

Main Category	#Users using SPM	#Users using SOM
Like Sports	146756	145141
Like Music	29597	29255
Like General	65354	64571
Like Entertainment	163	158
Like Food	4031	3990
Do not like Sports	1120	1079
Do not like Music	30	28
Do not like General	825	800

Table 5: Likes and Dislikes

4.5 Errors in sentiment analysis

As text mining is prone to errors, in this section we evaluate the correctness of our SOM approach in detecting the correct sentiment orientation. We manually label 300 sentences randomly chosen from various categories including sports, music, religion, politics, cats, and food. An independent human annotator then labels each sentence to either s^+, $s^{\bigcirc}$ or s^- depending on his/her understanding of the sentence. We then evaluate SentiWordNet's accuracy using precision metrics (automated labeling against manually annotated labels). This measure have been commonly used to evaluate the accuracy of various retrieval, classification, and mining algorithms. Precision refers to the proportion of true positives over the sum of the true positives and false positives. The sentiment mining technique provided an overall accuracy of 69.33%, see Table 6 where the diagonal figures represent the accurate labeling, while off diagonal figures represent false positives.

		Estimated/Sentiwordnet		
		s^+	$s^{\bigcirc}$	s^-
Actual/Human	s^+	**66.33**	17.35	16.33
	$s^{\bigcirc}$	23.68	**63.16**	13.16
	s^-	18.25	06.35	**73.02**

Table 6: Confusion matrix for sentiment (%)

Sentiment analysis failed for conjugate and multi sentences. For example, "*As far as intelligence goes cats have a different kind of intelligence than that of dogs. They can MANIPULATE their environment to SURVIVE can hunt on their own and...*" is labeled negative and "*really no 1 can say our governance in 9ja is understood... we are just driven here and there no prosperous direction... we really don't know... Im shot of words 4 my dear country*" is labeled as positive. Sentiment approach also failed for sarcastic statements like "*GREAT.. now I can not get ONLINE...*" which has been labeled positive.

The inaccurate estimation of positive to negative or negative to positive labels has more impact on building the user profile. The neutral messages are overwritten by the page polarity and hence no impact on user profile.

4.6 Concentrated group with ground truth

We tried to get some ground truth to be compared with the results obtained. We chose 450 users from $u^=$ obtained using the SPM approach with the largest number of interests inferred (more than 4 in particular). Out of the 450 user profiles, 47 have either been deactivated or deleted. We manually sent Facebook friend requests and messages to the remaining 403 Facebook users to know more about their interests in certain categories. In particular, we provided the user with a list of all the inferred interests (using SPM) of the user and asked him/her to classify into three categories i.e. s^+, s^-, $s^{\bigcirc}$. Due to privacy settings imposed by many users, we were only able to send 334 friend request (70 accepted) and 299 messages (15 replied back).

We expected those who accepted our friend request would reply to our messages sent; however, there were only 12 (out of 70) who replied to our message. 56 did not reply and we were not able to send message to 2 users because of the privacy settings imposed by them.

Majority users (212) neither accepted our friend request nor responded to our message sent. There were 2 participants who did not accepted our friend request but still replied to the message. Please refer to Table 7 for a summary of the responses we got.

		Disallow Message	Allow Message	
			No reply	Replied
Friend req. not allowed		52	16	1
Friend req. allowed	Not added	50	212	2
	Added	2	56	12

Table 7: Users category settings in `VolProf`

We take the responses of those 15 users who replied to our messages and compared them with the corresponding

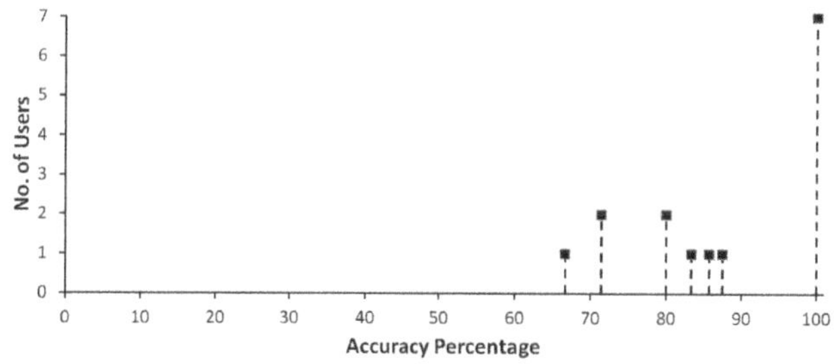

Figure 6: Users Interest from `VolProf`

interests inferred using SPM. Figure 6 shows the percentage of the correctly inferred interests of these users. It shows that we were able to infer approximately half (7 out of 15) of the users' interests with 100% accuracy. Also, we can see from the figure that at least two-third of the interests are inferred correctly for all the 15 users. Although we did not manage to get a larger pool of people with ground truth due to the manual work involved, available data seems to suggest that our technique provides reasonable accuracy in inferring users' interests from public pages on Facebook.

5. DISCUSSION AND LIMITATIONS

In this section we first discuss another threat model and results under this new threat model with our manual analysis. We then discuss other factors that could have contributed to inaccuracies and point out limitations in our results.

5.1 A different threat model

Apart from the threat model discussed so far, we discuss the strategy that can be adopted by an attacker who has access to *all* the pages the victim has liked on Facebook. The objective of the attack is still to guess the Blue Moon™ categories. This is a reasonable threat model when the attacker is a malicious app or some close friend/colleague of the victim. For example, spouses check their partners accounts without their permission, which can lead to divorces later on [1]. Partners are usually friends on Facebook who have access to most of the information including "likes" etc.

We conducted a small experiment by inviting some users to the lab and collecting their page "likes" (only after seeking their permission). We hired two research assistants to independently categorize these pages according to the Blue Moon™ categories. This manual analysis on one user having the highest (926) number of page likes shows that 71 out of 926 pages can be categorized to Blue Moon™ categories. These 71 pages fall into 15 Blue Moon™ categories. This shows that an attacker who has access to the user's like pages could possibly use them to break into the victim's Blue Moon™ system.

5.2 Limitations

One of the most important contribution to inaccuracies in our analysis is noise in the public pages on Facebook. This noise could come from advertisements, spamming in general, conversations in comments, and others.

Facebook is popular marketing media and some users are actually advertisers. For example, a person selling Nike shoes may post his ad in all the categories corresponding to sports. Our system might therefore believe that this user has an interest in sports. To have a sense on the noise level, we search for advertisements in selected categories by randomly choosing some posts and manually labeling them as advertisements. Table 8 shows the number of advertisements found in a number of posts for selected categories.

Category	No of posts scanned	% of Ads found
Sports	1,348	0.445
Food	2,312	0.346
Cats	2,312	0.216
Music	7,000	0.171
Politics	9,928	0.060

Table 8: Percentage of Advertisement posts

It appears that the noise level in our dataset is very low, however, our manual process in finding advertisements might be error prone, too. Another step we took to minimize this error was to manually check whether the users inferred with more than eight interests are real users. Our simple manual checking revealed that except a few users whose accounts had been deactivated or deleted, all of them seem to be legitimate. We also filtered off users who posted same message in more than four pages.

We have seen many people get into their personal conversations on public pages that are not related to the topic of the page. Unfortunately we do not find a scalable way of filtering out such noise, and therefore it might have contributed to errors made.

There are also limitations in the techniques that we use. First, we use a context independent text mining algorithm. This limits our capability in analyzing the sentiment of certain pages, e.g., "Lets help the dogs in the streets and kick the cats." The sentiment scoring without context will fail to identify the user interest in this example. To solve this problem, we need additional scoring models that can handle the sentiment with the context.

Second, we only managed to obtain ground truth for a small set of users. We wish we could find a better approach to obtain the ground truth, but sending out message to a large number of users had one of our Facebook account suspended, and that was why we did not go further to target a larger group.

Last but not the least, manual work was involved in a number of steps in our experiments, including evaluation of the accuracy of sentiment analysis, spam detection, etc. This manual work could potentially introduce errors into the evaluation.

6. CONCLUSION AND FUTURE WORK

In this paper we raised two important questions, as why public information is made public? And is there any information that can be mined to break preference based authentication. We present two mining based approaches to predict undisclosed users' interests. Using only simple mining approach we extracted unobservable Interest topics by analyzing the corpus of Interests obtained via legitimate use of Graph API provided by Facebook. From our experiments, we were able to disclose 22 interests of a user and found more than 80,097 users with at least more than 2 interests. We also show how this inferred information can be used to break preference based backup authentication system. We also demonstrated that simply liking a Facebook page does not imply the users' inclination towards that page. We showed that there exists many users who liked a Facebook page, however they post negative comments on the page. We also compared our work with prior research work which involves crawling personal profiles (either public or private). In future, we would like to improve our mining approach to infer more privacy attributes apart from interests.

7. REFERENCES

[1] F. L. Attorney. Maintaining privacy and starting a separate life during divorce. `http://goo.gl/Nm5Wc`.

[2] A. E. S. Baccianella and F. Sebastiani. Sentiwordnet 3.0: An enhanced lexical resource for sentiment analysis and opinion mining. In *Proceedings of the Seventh conference on International Language Resources and Evaluation (LREC'10)*, Valletta, Malta, May 2010. European Language Resources Association (ELRA).

[3] M. Balduzzi, C. Platzer, T. Holz, E. Kirda, D. Balzarotti, and C. Kruegel. Abusing social networks for automated user profiling. In *Proceedings of the 13th international conference on Recent advances in intrusion detection*, RAID'10, pages 422–441, Berlin, Heidelberg, 2010. Springer-Verlag.

[4] A. Chaabane, G. Acs, and M. A. Kaafar. You are what you like! Information leakage through users' Interests. In *Proceedings of the 19th Annual Network & Distributed System Security Symposium*, Feb. 2012.

[5] Y. Chen, D. Pavlov, and J. F. Canny. Large-scale behavioral targeting. In *Proceedings of the 15th ACM SIGKDD international conference on Knowledge discovery and data mining*, KDD '09, pages 209–218, New York, NY, USA, 2009. ACM.

[6] N. Cubrilovic. The anatomy of the twitter attack. `http://techcrunch.com/2009/07/19/the-anatomy-of-the-twitter-attack/`.

[7] Facebook. Facebook pages. `http://www.facebook.com/directory/pages/`.

[8] Facebook. Graph api. `https://www.developers.facebook.com/docs/reference/api/`.

[9] C. Fellbaum. Wordnet: An electronic lexical database. `http://wordnet.princeton.edu/`.

[10] L. Gannes. Mole-whacking: Vendor says spam is growing on facebook fan pages. `http://goo.gl/a89Aa`.

[11] S. Gauch, M. Speretta, A. Chandramouli, and A. Micarelli. The adaptive web. In P. Brusilovsky, A. Kobsa, and W. Nejdl, editors, *The Adaptive Web*, chapter User profiles for personalized information access, pages 54–89. Springer-Verlag, Berlin, Heidelberg, 2007.

[12] D. Gayo Avello. All liaisons are dangerous when all your friends are known to us. In *Proceedings of the 22nd ACM conference on Hypertext and hypermedia*, HT '11, pages 171–180, New York, NY, USA, 2011. ACM.

[13] O. Goga, H. Lei, S. H. K. Parthasarathi, G. Friedland, R. Sommer, and R. Teixeira. On exploiting innocuous user activity for correlating accounts acrossăsocial network sites & twitter & personal profiles. Technical report, International Computer Science Institute, 2012.

[14] M. Hu and B. Liu. Mining and summarizing customer reviews. In *KDD*, pages 168–177, 2004.

[15] R. Inc. The blue moon authentication system. http://www.ravenwhite.com/iforgotmypassword.html.

[16] D. Irani, S. Webb, K. Li, and C. Pu. Modeling unintended personal-information leakage from multiple online social networks. *IEEE Internet Computing*, 15(3):13–19, May 2011.

[17] M. Jakobsson, E. Stolterman, S. Wetzel, and L. Yang. Love and authentication. In *Proceedings of the twenty-sixth annual SIGCHI conference on Human factors in computing systems*, CHI '08, pages 197–200, New York, NY, USA, 2008. ACM.

[18] X. Jin, C. Wang, J. Luo, X. Yu, and J. Han. Likeminer: a system for mining the power of 'like' in social media networks. In *Proceedings of the 17th ACM SIGKDD international conference on Knowledge discovery and data mining*, KDD '11, pages 753–756, New York, NY, USA, 2011. ACM.

[19] Y. Koren, R. Bell, and C. Volinsky. Matrix factorization techniques for recommender systems, 2009.

[20] J. Lindamood, R. Heatherly, M. Kantarcioglu, and B. Thuraisingham. Inferring private information using social network data. In *Proceedings of the 18th international conference on World wide web*, WWW '09, pages 1145–1146, New York, NY, USA, 2009. ACM.

[21] A. Mislove, B. Viswanath, K. P. Gummadi, and P. Druschel. You are who you know: inferring user profiles in online social networks. In *Proceedings of the third ACM international conference on Web search and data mining*, WSDM '10, pages 251–260, New York, NY, USA, 2010. ACM.

[22] J. Owyang. The many challenges of social network sites, 2008.

[23] B. Pang, L. Lee, and S. Vaithyanathan. Thumbs up?: sentiment classification using machine learning techniques. In *Proceedings of the ACL-02 conference on Empirical methods in natural language processing - Volume 10*, EMNLP '02, pages 79–86, Stroudsburg, PA, USA, 2002. Association for Computational Linguistics.

[24] A. Rabkin. Personal knowledge questions for fallback authentication: security questions in the era of facebook. In *Proceedings of the 4th symposium on Usable privacy and security*, SOUPS '08, pages 13–23, New York, NY, USA, 2008. ACM.

[25] Robots.txt. A standard for robot exclusion. http://www.robotstxt.org/orig.html.

[26] U. T. Ted Bridis, Associated Press Writer. Hacker impersonated palin, stole e-mail password. http://www.usatoday.com/news/politics/2008-09-17-152224562_x.htm.

[27] K. Toutanova, D. Klein, C. D. Manning, and Y. Singer. Feature-rich part-of-speech tagging with a cyclic dependency network. In *Proceedings of the 2003 Conference of the North American Chapter of the Association for Computational Linguistics on Human Language Technology - Volume 1*, NAACL '03, pages 173–180, Stroudsburg, PA, USA, 2003. Association for Computational Linguistics.

[28] A. Tsotsis. Hacker proves facebook's public data is public. http://techcrunch.com/2010/07/28/hacker-proves-facebooks-public-data-is-public/.

[29] P. D. Turney. Thumbs up or thumbs down?: semantic orientation applied to unsupervised classification of reviews. In *ACL '02: Proceedings of the 40th Annual Meeting on Association for Computational Linguistics*, pages 417–424, Morristown, NJ, USA, 2002. Association for Computational Linguistics.

[30] T. Wilson, J. Wiebe, and P. Hoffmann. Recognizing contextual polarity: An exploration of features for phrase-level sentiment analysis. *Computational Linguistics*, 35(3):399–433, 2009.

[31] Yahoo. Rate limiting for yahoo! search web services. http://developer.yahoo.com/search/rate.html.

[32] S.-H. Yang, B. Long, A. Smola, N. Sadagopan, Z. Zheng, and H. Zha. Like like alike: joint friendship and interest propagation in social networks. In *Proceedings of the 20th international conference on World wide web*, WWW '11, pages 537–546, New York, NY, USA, 2011. ACM.

[33] E. Zheleva and L. Getoor. To join or not to join: the illusion of privacy in social networks with mixed public and private user profiles. In *Proceedings of the 18th international conference on World wide web*, WWW '09, pages 531–540, New York, NY, USA, 2009. ACM.

Multi-Key Leakage-Resilient Threshold Cryptography*

Cong Zhang
Department of Computer Science
The University of Hong Kong
czhang2@cs.hku.hk

Tsz Hon Yuen†
Department of Computer Science
The University of Hong Kong
thyuen@cs.hku.hk

Hao Xiong‡
Department of Computer Science
The University of Hong Kong
hxiong@cs.hku.hk

Sherman S. M. Chow
Department of Information Engineering
The Chinese University of Hong Kong
sherman@ie.cuhk.edu.hk

Siu Ming Yiu
Department of Computer Science
The University of Hong Kong
Hong Kong
smyiu@cs.hku.hk

Yi-Jun He
Hong Kong R&D Centre for Logistics and Supply Chain Management Enabling Technologies
ahe@lscm.hk

ABSTRACT

With the goal of ensuring availability of security services such as encryption and authentication, we initiate the study of leakage-resilient threshold cryptography, for achieving formal security guarantee under various key-exposure attacks. A distinctive property of threshold cryptosystems is that a threshold number of secret keys are used in the main cryptographic function such as decryption or signing. Even though some existing security models allow leakages of multiple keys of different users, these keys are not used simultaneously to decrypt a ciphertext or sign a message.

In this paper, we introduce the multi-key leakage-resilient security model for threshold cryptography. We also propose constructions with formal security guarantee with respect to our model, one is a dynamic threshold public key encryption scheme and another is a threshold ring signature scheme.

Categories and Subject Descriptors

E.3 [**Data Encryption**]: Public key cryptosystems

General Terms

Security, Algorithms, Design

Keywords

Leakage-resilience, key-exposure, dynamic threshold public key encryption, threshold ring signatures, pairing, privacy

*Supported by the National Natural Science Foundation of China under Grant No. 60970135 and 61170282
†Supported by the HKU Small Project Funding under Grant No. 201109176192
‡Corresponding author

ASIA CCS'13, May 8–10, 2013, Hangzhou, China.

1. INTRODUCTION

We consider the question of how to ensure availability of security services such as encryption and authentication. A traditional paradigm for this goal is threshold cryptography, in which the secret key is shared in a way that any threshold number of key shares is sufficient for the cryptographic function. This also helps to maintain security in the face of key-exposure attack, such that as long as the adversary cannot obtain shares up to this threshold, the cryptosystem remains secure. Putting this in another way, that means the security is only maintained when all other key shares are fully share after a certain subset of the shares are fully compromised. However, side-channel attacks like power analysis demonstrate that such assumption may not hold in the real world. Therefore, leakage-resilient cryptography is proposed to formally capture such scenarios and provides security guarantees even if the adversary can obtain partial information about the secret key.

In the security model of leakage-resilient cryptography, the leakage is modeled by a leakage function f applied on the secret key and/or the randomness used. Various restrictions are applied on f in different models to define different reasonable levels of security. The relative leakage model [2] only allows f to output at most l bits, where l is smaller than the size of the secret key. The bit size restriction is later relaxed such that f is restricted to lower the entropy of the secret key by at most l bits [14]. These two models consider l as a fraction of the secret key. On the other hand, the bounded retrieval model (e.g., [3]) considers l as a system parameter and increases the size of the secret key to accommodate l bits of leakage. The continual leakage model [11, 6] allows periodic updates on the secret key, while bounding the leakage among updates. The auxiliary input model [10] only restricts that f to be an uninvertible function and hence allows larger classes of leakage functions. Finally, the continual auxiliary leakage model [17] combines both.

Multi-Key Leakage Model. In many leakage-resilient non-threshold cryptosystems, there is only one secret key and hence the leakage function f is always applied to the same secret key. Leakage of more than one keys may be allowed in some advanced encryption primitives, such as leakage-resilient identity-based encryption (IBE), in which the attacker can choose to leak different identity-based secret keys (e.g., [8]) or even the master secret key (e.g., [13]) in an arbitrary order; however, most of them do not model the leakage of both kinds of key at the same time (a notable exception is [17]), since decryption can be done by the knowledge of any one of them. However, this kind of leakage function is simply not general enough if multiple secret keys are involved during decryption or signing algorithm. since it is possible for an attacker to obtain leakage from more than one user during a single decrypt/sign operation. In this paper, we tackle the challenge of modelling multi-key leakage-resilience in threshold cryptosystems. We demonstrate our multi-key leakage-resilient model in the contexts of dynamic threshold public key encryption (DT-PKE) [9] and threshold ring signature (TRSig) [7].

Dynamicity. In traditional threshold encryption, the threshold is often fixed during the key generation phase, if not during the setup. In dynamic threshold public key encryption, for each ciphertext, an encryptor can dynamically choose the authorized set of recipients and the threshold governing the decryption capability among this set. This is especially useful when leakage resilience is considered. In the case that a certain entity (may it be a recipient, or one of the devices of a single user storing a key share) is compromised, one may choose to take it out from the authorized set. The same holds true for threshold ring signature. For each signature, a different ad-hoc group can be formed.

Related Work. To the best of our knowledge, we give the first multi-key leakage-resilient threshold ring signature scheme. For leakage-resilient threshold cryptosystem, Dodis *et al.* [12] proposed a updatable threshold encryption scheme, which is useful for sharing a secret between leaky devices. However, their motivating scenario is that all devices are owned by a single entity who also decides the threshold (actually, a general access structure), but not the encryptor. Also, each leakage allowed by their model is restricted to a single device. On the other hand, Akavia *et al.* [1] proposed a leakage-resilient distributed public-key encryption. Since one of their motivating application is long-term secure storage on leaky hardware, they also consider the (optimal) setting of two devices and the leakage is only restricted to one of them at a time. Moreover, decryption in their scheme requires an interactive two-party protocol among the devices.

Challenges in Our Construction. For the construction of multi-key leakage-resilient DT-PKE, we extend the DT-PKE scheme of Delerablée and Pointcheval [9] to composite order group, in order to achieve leakage-resilient through dual system encryption [13]. In dual system encryption, there are some extra semi-functional (SF) keys and SF ciphertexts. An SF key can decrypt normal ciphertext and a normal key can decrypt SF ciphertext. On the other hand, an SF key cannot decrypt an SF ciphertext.

When we extend the dual system setting to DT-PKE, multiple key shares are used to decrypt a single ciphertext. Therefore, the SF definition for keys cannot be applied trivially to DT-PKE. We define a new *nominally semi-functional key set* to capture the possibility that multiple SF key shares may be jointly used in decryption and the SF property is canceled out in this set of key shares. Using this new definition of nominally semi-functional key set, we are able to achieve multi-key leakage-resilient through dual system encryption. The assumption originally used by Delerablée and Pointcheval [9] has to be changed in a non-trivial way to fit with our construction as well.

For the construction of multi-key leakage-resilient TRSig, we extend the TRSig of Yuen *et al.* [18]. In order to provide the leakage-resilient property, we use the approach of using admissible hash function and non-interactive zero-knowledge proof system [19]. However, Yuen *et al.* [19] gave a generic construction and their example of admissible hash function does not fit our underlying TRSig [18].

To deal with the problem in constructing TRSig, we define a new function F which is similar to the Waters hash function [15]. In contrast to [15], this function F is used to generate outputs in two different subgroups. We can then define the admissible hash function upon F. Furthermore, the two different subgroups also enable us to generate the hiding and binding commitments, which are used in the signing oracle query and the challenge message during the security proof.

2. BACKGROUND

We use the notation $\mathsf{poly}(\cdot)$ to refer to some polynomially bounded function and $\mathsf{negl}(\cdot)$ to refer to some negligible function which is smaller than $1/\mathsf{poly}(\cdot)$ for any positive $\mathsf{poly}(\cdot)$. We use the notation $x \in_R X$ to denote picking an element x uniformly at random from the set X. PPT stands for probabilistic polynomial-time, and WLOG stands for without loss of generality.

Composite Order Bilinear Groups. Let $\mathcal{G}$ be an algorithm that takes a security parameter 1^λ as input, outputs a description of bilinear group $(N = p_1p_2p_3, \mathbb{G}, \mathbb{G}_T, \hat{e})$, where p_1, p_2, p_3 are distinct λ-bit primes, $\mathbb{G}$ and $\mathbb{G}_T$ are cyclic groups of order N, and $\hat{e} : \mathbb{G} \times \mathbb{G} \to \mathbb{G}_T$ is a bilinear map such that $\forall g, h \in \mathbb{G}$ and $a, b \in \mathbb{Z}_N$, $\hat{e}(g^a, h^b) = \hat{e}(g, h)^{ab}$.

Denote $\mathbb{G}_{p_i}$ as the subgroup of order p_i in $\mathbb{G}$ and g_i as its generator ($i = 1, 2, 3$). For all $h_i \in \mathbb{G}_{p_i}$ and $h_j \in \mathbb{G}_{p_j}$, if $i \neq j$, $\hat{e}(h_i, h_j) = 1$. We denote $\mathbb{G}_{p_1p_2}$ as the subgroup of order p_1p_2 in $\mathbb{G}$. For all $T \in \mathbb{G}_{p_1p_2}$, T can be written uniquely as the product of an element of $\mathbb{G}_{p_1}$ and an element of $\mathbb{G}_{p_2}$. We refer to these elements as the "$\mathbb{G}_{p_1}$ part of T" and the "$\mathbb{G}_{p_2}$ part of T" respectively. We also define $\mathbb{G}_{p_1p_3}$ and $\mathbb{G} = \mathbb{G}_{p_1p_2p_3}$ similarly.

Finally, we define another algorithm $\mathcal{G}'$ which outputs a bilinear group of order $N = p_1p_2$ (instead of a product of three primes). Other properties of $\mathcal{G}'$ follow from $\mathcal{G}$.

Vector-in-the-Exponent Notation. Let $\mathbb{G}$ be a group of order q generated by an element $g \in \mathbb{G}$ and let $\vec{\gamma} = \langle \gamma_1, \ldots, \gamma_n \rangle \in \mathbb{F}_q^n$ be a vector. We use the notation $g^{\vec{\gamma}} \in \mathbb{G}^n$ to denote the vector $\langle g^{\gamma_1}, \ldots, g^{\gamma_n} \rangle$ of group elements. Given a vector of group elements $g^{\vec{\gamma}} \in \mathbb{G}^n$ and a vector $\vec{\theta} \in \mathbb{F}_q^n$ of *exponents*, one can efficiently compute $g^{\vec{\gamma} \cdot \vec{\theta}}$. Given a bilinear map $\hat{e} : \mathbb{G} \times \mathbb{G} \to \mathbb{G}_T$, $g^{\vec{\gamma}}, g^{\vec{\theta}} \in \mathbb{G}^n$, we denote $\hat{e}(g^{\vec{\gamma}}, g^{\vec{\theta}})$ as $\hat{e}(g, g)^{\vec{\gamma} \cdot \vec{\theta}} = \prod_{i=1}^{n} \hat{e}(g^{\gamma_i}, g^{\theta_i})$. We also denote $g_1^{\vec{\gamma}} * g_2^{\vec{\theta}}$ as component-wise multiplication $\langle g_1^{\gamma_1} g_2^{\delta_1}, \ldots, g_1^{\gamma_n} g_2^{\delta_n} \rangle$.

Intractability Assumptions. We make some standard assumptions that no PPT adversary can solve the following problems:

- *Discrete Logarithm (DL) Problem*: Given $y, g \in \mathbb{G}_{p_1}$, output $x = \log_g y$.
- *Computational Diffie-Hellman (CDH) Problem*: Given g, g^a, $g^b \in \mathbb{G}_{p_1}$ for random $a, b \in \mathbb{Z}_{p_1}$, output g^{ab}.

- *Subgroup Decision Problem*: Given ($N = p_1p_2, \mathbb{G}, \mathbb{G}_T, \hat{e}, w$), decide if $w \in \mathbb{G}_{p_1}$ or not.

We also use some complexity assumptions which are related to composite order groups:

- *Assumption 1.* Given $(N = p_1p_2p_3, \mathbb{G}, \mathbb{G}_T, \hat{e}), g_1 \in \mathbb{G}_{p_1}, g_3 \in \mathbb{G}_{p_3}$ and T, no PPT adversary can decide if $T \in \mathbb{G}_{p_1p_2}$ or $\mathbb{G}_{p_1}$ with non-negligible probability.

- *Assumption 2.* Given $(N = p_1p_2p_3, \mathbb{G}, \mathbb{G}_T, \hat{e}), g_1, g_3, g_1^z g_2^\nu, g_1^\mu g_3^\rho$ and T, for some unknowns $z, \nu, \mu, \rho \in \mathbb{Z}_N$, and g_i are generators of the group $\mathbb{G}_{p_i}$ for $i = 1, 2, 3$, no PPT adversary can decide if $T \in \mathbb{G}_{p_1p_3}$ or $\mathbb{G}$ with non-negligible probability.

Assumptions 1 and 2 have appeared in the literature [13]. Our Assumption 3 extends the Multi-Sequence of Exponents Diffie-Hellman assumption [9] to the subgroup setting.

DEFINITION 1. (ℓ, m, t)-Assumption 3. *Given $(N = p_1p_2p_3, \mathbb{G}, \mathbb{G}_T, \hat{e}), g_1, h \in \mathbb{G}_{p_1}, g_2 \in \mathbb{G}_{p_2}, g_3 \in \mathbb{G}_{p_3}$ and let ℓ, m and t be three integers, and two random coprime polynomials f and g, of respective orders ℓ and m, with pairwise distinct roots, as well as several sequences of exponentiations:*

$$g_1 g_2^{\theta_0}, g_1^\gamma g_2^{\theta_1}, \ldots, g_1^{\gamma^{m-1}} g_2^{\theta_{m-1}}, g_1^{\kappa\gamma f(\gamma)g(\gamma)} g_2^{\theta}$$
$$h, h^\gamma, \ldots, h^{\gamma^{m-2}}, h^\alpha, h^{\alpha\gamma}, \ldots, h^{\alpha\gamma^{2m-1}}, h^{\kappa g(\gamma)} g_2^{\omega}$$
$$g_1^{g(\gamma)}, g_1^{\gamma g(\gamma)}, \ldots, g_1^{\gamma^{\ell+t-2} g(\gamma)}, g_1^{\alpha g(\gamma)}, g_1^{\alpha\gamma g(\gamma)}, \ldots, g_1^{\alpha\gamma^{\ell+t} g(\gamma)}$$

and $T \in \mathbb{G}_T$, for random unknown $\theta_0, \ldots, \theta_{m-1}, \kappa, \alpha, \gamma, \theta, \omega \in \mathbb{Z}_N$, no PPT adversary can decide whether T is equal to $\hat{e}(g_1, h)^{\kappa \cdot f(\gamma)}$ or a random element of $\mathbb{G}_T$.

3. MULTI-KEY LEAKAGE-RESILIENT DYNAMIC THRESHOLD PKE

3.1 Syntax

A DT-PKE scheme has the following algorithms:

Setup$(1^\lambda) \to (\mathsf{EK}, \mathsf{CK}, \mathsf{MK})$. The setup algorithm takes as input the security parameter 1^λ, and outputs the public parameters: encryption key EK and combination key CK; and a master secret key, MK. Denote the message space as $\mathcal{M}$ and the identity space as $\mathcal{I}$ which we omitted from the syntax.

Join$(\mathsf{MK}, \mathsf{ID}) \to (\mathsf{usk}, \mathsf{upk})$. It takes as input the master secret key MK and an identity $\mathsf{ID} \in \mathcal{I}$ of the user joining the system, and outputs the user's key $(\mathsf{usk}, \mathsf{upk})$. The user secret key usk is sent to the user via a secure channel, while the user public key upk is widely distributed.

Encrypt$(\mathsf{EK}, M, \mathcal{S}, t) \to C$. It takes as input EK, a message $M \in \mathcal{M}$, a set $\mathcal{S}$ of user public keys and a threshold t, and outputs a ciphertext C.

Validate$(\mathsf{EK}, \mathcal{S}, t, C) \to 1/0$. It takes as input EK, $\mathcal{S}$, t and C, it outputs 1 if C is a valid ciphertext with respect to EK, $\mathcal{S}$ and t, or outputs 0 otherwise.

ShareDecrypt$(\mathsf{ID}, \mathsf{usk}, C) \to \sigma_I/\perp$. It takes as input the identity ID, its secret key usk and the ciphertext C, and outputs the decryption share σ_{ID} or $\perp$ for invalid decryption.

Combine$(\mathsf{CK}, \mathcal{S}, t, C, T, \Sigma) \to M/\perp$. It takes as input CK, a ciphertext C, a subset $T \subseteq \mathcal{S}$ of t users and $\Sigma = (\sigma_1, \ldots, \sigma_t)$ the corresponding decryption shares. It outputs the message M or $\perp$ for invalid decryption.

A DT-PKE scheme must satisfy the correctness. For any message $M \in \mathcal{M}$ associated with a set $\mathcal{S}$ of registered users and a threshold t in an encryption and resulted in a ciphertext C, then C must be validated as a correct ciphertext; and if t users from $\mathcal{S}$ correctly produce the partial decryption $(\sigma_1, \ldots, \sigma_t)$, then the message M must be recovered.

3.2 Leakage Resilience

Informally, security for leakage-resilient DT-PKE requires the following property. For any message encrypted for a set $\mathcal{S}$ of registered users with a threshold t, any attacker cannot learn any information about the message after obtained (1) collusion of less than t users from $\mathcal{S}$, (2) a bounded leakage from the uncorrupted users in $\mathcal{S}$, and (3) collusion of users that do not belong to $\mathcal{S}$.

We formally define the security notion using a game $\mathsf{CorruptLeak}$ between an adversary $\mathcal{A}$ and a challenger $\mathcal{C}$. In this game, $\mathcal{A}$ can obtain a number of usk by the $\mathsf{Corrupt}$ oracle, and obtain some leakage of the uncorrupted keys by the Leak oracle. The restrictions we impose are:

- $\mathcal{A}$ cannot get any leakage about the master secret key.

- $\mathcal{A}$ has to decide the challenge set $\mathcal{S}^*$, the challenge threshold t^*, and the identities set $\overline{\mathcal{I}}$ that he will corrupt at the beginning of the game. It follows the existing non-adaptive adversary and corruption model [9]. We restrict that $|\mathcal{S}^* \cap \overline{\mathcal{I}}| \leq t^* - 1$.

- $\mathcal{A}$ cannot get leakage of more that ℓ_{usk} bits per key, where ℓ_{usk} is a parameter of the game.

We use a handle to refer to a user secret key in different oracle queries. The security game is parameterized by a security parameter λ and a leakage bound $\ell_{\mathsf{usk}} = \ell_{\mathsf{usk}}(\lambda)$. The game $\mathsf{CorruptLeak}$ consists of the following phases:

Initial: The adversary $\mathcal{A}$ gives the challenger $\mathcal{C}$ the authorized set of identities $\mathcal{S}^*$, the threshold t^* and the identities set $\overline{\mathcal{I}}$ he wants to corrupt. Here, we reasonably assume that $\mathcal{A}$ only issues the Leak query to the uncorrupted identities from $\mathcal{S}^*$ that do not appear in the $\mathsf{Corrupt}$ oracle query, since $\mathcal{A}$ can get the entire secret key of the corrupted identities.

Setup: $\mathcal{C}$ runs $(\mathsf{EK}, \mathsf{CK}, \mathsf{MK}) \leftarrow$ **Setup**(1^λ). The public parameters $(\mathsf{EK}, \mathsf{CK})$ are given to $\mathcal{A}$. $\mathcal{C}$ maintains a list $\mathcal{T}$ that holds tuples of handles, user identities, keys and leaked bits, which is initially empty. Also, a handle counter H is initialized to 0.

Phase 1: $\mathcal{A}$ can make the following queries adaptively:

- $\mathsf{Join}(\mathsf{ID})$: On input a new identity $\mathsf{ID} \in \mathcal{I}$, $\mathcal{C}$ runs $(\mathsf{usk}, \mathsf{upk}) \leftarrow$ **Join**$(\mathsf{MK}, \mathsf{ID})$ and returns upk and the handle $(H + 1)$. $\mathcal{C}$ puts $(H + 1, \mathsf{ID}, \mathsf{usk}, \mathsf{upk}, 0)$ in the list $\mathcal{T}$ and sets $H \leftarrow H + 1$.

- $\mathsf{Corrupt}(h)$: On input a handle h, $\mathcal{C}$ scans $\mathcal{T}$ to find the tuple $(h, \mathsf{ID}, \mathsf{usk}, \mathsf{upk}, L)$. If $\mathsf{ID} \in \overline{\mathcal{I}}$, it returns usk.

- $\mathsf{Leak}(h_{i_1}, \cdots, h_{i_c}, f)$: On input the handles $h_{i_1}, \cdots h_{i_c}$ and a polynomial-time computable function f of constant output size, $\mathcal{C}$ scans $\mathcal{T}$ to find those corresponding tuples $(h_{i_j}, \mathsf{ID}_{i_j}, \mathsf{usk}_{i_j}, \mathsf{upk}_{i_j}, L_{i_j})$ for $j \in [1, c]$. It checks if $L_{i_j} + |f(\mathsf{usk}_{i_1}, \ldots, \mathsf{usk}_{i_c})| \leq \ell_{\mathsf{usk}}$ for each tuple. If this is true, it responds with $f(\mathsf{usk}_{i_1}, \ldots, \mathsf{usk}_{i_c})$ and updates each L_{i_j} in all these tuples with $L_{i_j} + |f(\mathsf{usk}_{i_1}, \ldots, \mathsf{usk}_{i_c})|$ where $j \in [1, c]$. If the check fails, it returns $\perp$ to $\mathcal{A}$.

Challenge: $\mathcal{A}$ submits two challenge messages M_0^*, M_1^* of equal length. $\mathcal{C}$ flips a uniform coin $b \in_R \{0,1\}$ and encrypts M_b^* by running $C^* \leftarrow$ **Encrypt**$(\mathsf{EK}, M_b^*, \mathcal{S}^*, t^*)$. $\mathcal{C}$ sends the challenge ciphertext C^* to $\mathcal{A}$.

Phase 2: This is the same as **Phase 1** with the restriction that the Leak queries are not allowed, and the number of different identities such that $\mathsf{Corrupt}(h)$ queries have been asked is strictly less than t. (If the adversary has corrupted $t^* - 1$ identities from $\mathcal{S}^*$, $\mathcal{A}$ can win trivially by encoding the entire decryption algorithm of C^* as a function of a uncorrupted secret key.)

Guess: $\mathcal{A}$ outputs a bit $b' \leftarrow \{0,1\}$ and wins if $b = b'$.

We use $\mathsf{Adv}_{\mathcal{A}}(\lambda, \ell_{\mathsf{usk}})$ to define the advantage of adversary $\mathcal{A}$ in game $\mathsf{CorruptLeak}$ with security parameter 1^λ and leakage parameter ℓ_{usk}, which is defined as $\mathsf{Adv}_{\mathcal{A}}(\lambda, \ell_{\mathsf{usk}}) = \left|\Pr[\mathcal{A} \text{ wins}] - \frac{1}{2}\right|$, where the probability is over all random bits used by the challenger and the adversary.

DEFINITION 2. *A DT-PKE system is ℓ_{usk}-leakage-resilient if $\mathsf{Adv}_{\mathcal{A}}(\lambda, \ell_{\mathsf{usk}}) \leq \mathsf{negl}(\lambda)$ for all PPT adversaries $\mathcal{A}$.*

3.3 Our Construction

Our new leakage-resilient DT-PKE is based on Delerablée and Pointcheval's construction [9] and its leakage-resilience is inspired from the tagging technique initially used by Chow *et al.* [8] which is generalized by Lewko *et al.* [13].

Setup(1^λ). Given the security parameter 1^λ, the algorithm runs a bilinear group generator $(N, \mathbb{G}, \mathbb{G}_T, \hat{e}) \leftarrow \mathcal{G}(1^\lambda)$, where $N = p_1p_2p_3$. Also three generators $g, g_1 \in \mathbb{G}_{p_1}$ and $g_3 \in \mathbb{G}_{p_3}$ are given. With those, another generator $h \in \mathbb{G}_{p_1}$ is chosen. This algorithm also randomly picks $\gamma, \alpha, y_1, y_2, \ldots, y_n \in \mathbb{Z}_N$. Finally, a set $\mathcal{D} = \{d_i\}_{i=1}^{m-1}$ of values in $\mathbb{Z}_N^*$ is randomly selected, where m is the maximal size of an authorized set. This corresponds to a set of dummy users, that will be used in the encryption algorithm to complete a set of authorized users.
The encryption key is $\mathsf{EK} = (m, u = g^{\alpha\gamma}, v = \hat{e}(g,h)^\alpha, g_1, \{h^{\alpha\gamma^i}, g_1^{y_j}, h^{\alpha y_j \gamma^i}\}_{i\in[0,2m-1], j\in[1,n]}, \mathcal{D})$. The combining key is $\mathsf{CK} = (m, \{h^{\gamma^i}\}_{i=0}^{m-2}, \mathcal{D})$. The master secret key is defined as $\mathsf{MK} = (g, \gamma, \alpha)$.

Join$(\mathsf{MK}, \mathsf{ID})$. Given MK and an identity ID, this algorithm chooses a random $x \in \mathbb{Z}_N^*$ which is different from the x_i's chosen in all previous execution of **Join** protocol, and the d_i's in the dummy set $\mathcal{D}$; subject to the constraints that $\gcd(\gamma + x, N) = 1$, $\gcd(x - x_i, N) = 1$ for all previously chosen x_i's, and $\gcd(x - d_i, N) = 1$ all $d_i \in \mathcal{D}$. The reason for these will become apparent in §3.5.
It randomly picks $(z_1, \ldots, z_n) \in_R \mathbb{Z}_N^n, \vec{\rho} \in_R \mathbb{Z}_N^{n+1}$ and outputs the user's key with $\mathsf{upk} = x$ and:

$$\mathsf{usk} = (g_1^{z_1}, g_1^{z_2}, \ldots, g_1^{z_n}, g^{\frac{1}{\gamma+x}} \cdot \prod_{j=1}^n g_1^{-y_j z_j}) * g_3^{\vec{\rho}}.$$

The private key usk is given to the user via a secure channel, whereas upk is widely published in an authenticated way.

Encrypt$(\mathsf{EK}, M, \mathcal{S}, t)$. When a user wants to encrypt a message $M \in \mathbb{G}_T$, for a set of users of $\mathcal{S} = (\mathsf{upk}_1 = x_1, \ldots, \mathsf{upk}_s = x_s)$ and a threshold t (with $t \leq s = |\mathcal{S}| \leq m$), it picks a random $k \in \mathbb{Z}_N^*$ and a random subset $\bar{\mathcal{D}} \subseteq \mathcal{D}$ of size $\bar{d} = (m + t - s - 1)$, and outputs the ciphertext $C = (C_1, C_2, \ldots, C_{n+3}, \bar{\mathcal{D}})$, where

$$C_j = h^{\alpha k y_j \prod_{x_i \in \mathcal{S}}(\gamma + x_i) \prod_{x \in \bar{\mathcal{D}}}(\gamma + x)} \quad \forall\, j \in [1, n],$$
$$C_{n+1} = h^{\alpha k \prod_{x_i \in \mathcal{S}}(\gamma + x_i) \prod_{x \in \bar{\mathcal{D}}}(\gamma + x)},$$
$$C_{n+2} = u^{-k}, \qquad C_{n+3} = M \cdot v^k.$$

The crucial point is that **Encrypt** includes a set of $\bar{d}$ dummy users, in order to obtain a polynomial of degree exactly $(\bar{d} + s) = (m + t - 1)$ in the exponent of h. Such a setting allows t authorized users to decrypt by their respective user private key together with the $(m-1)$ elements from $\{h^{\gamma^i}\}_{i=0}^{m-2}$ in the combining key.

Validate$(\mathsf{EK}, \mathcal{S}, t, C)$. Given EK, the authorized set $\mathcal{S}$, the threshold t and the ciphertext $C = (C_1, \ldots, C_{n+3}, \bar{\mathcal{D}})$, one can compute

$$C'_j = h^{\alpha y_j \prod_{x_i \in \mathcal{S}}(\gamma + x_i) \prod_{x \in \bar{\mathcal{D}}}(\gamma + x)} \quad \forall j \in [1, n],$$
$$C'_{n+1} = h^{\alpha \prod_{x_i \in \mathcal{S}}(\gamma + x_i) \prod_{x \in \bar{\mathcal{D}}}(\gamma + x)}, \quad C'_{n+2} = u^{-1}.$$

One can see that the ciphertext is valid with respect to $\mathcal{S}$ if and only if $\mathcal{S}$ contains at least t user public keys, and there exists a scalar k such that for $C_j = {C'_j}^k$ $\forall j \in [1, n+2]$. Consequently, **Validate** can check whether $\hat{e}(C'_j, C_{n+2}) = \hat{e}(C_j, C'_{n+2})$ $\forall j \in [1, n]$ and if $\hat{e}(C'_{n+1}, C_{n+2}) = \hat{e}(C_{n+1}, C'_{n+2})$. It outputs 1 if they both hold, and outputs 0 otherwise.

ShareDecrypt$(\mathsf{ID}, \mathsf{usk}, C)$. To decrypt a ciphertext $C = (C_1, \ldots, C_{n+3}, \bar{\mathcal{D}})$, denote $\mathsf{usk} = (D_1, \ldots, D_{n+1})$ with corresponding $\mathsf{upk} = x_i$. Compute the decryption share:

$$\sigma_i = \prod_{j=1}^{n+1} \hat{e}(D_j, C_j) = \hat{e}(g, C_{n+1})^{\frac{1}{\gamma + x_i}}$$

Since $\mathbb{G}_{p_1}$ and $\mathbb{G}_{p_3}$ are orthogonal, the result of the pairing algorithm is valid.

Combine$(\mathsf{CK}, \mathcal{S}, t, C, T, \Sigma)$. Given CK, the authorized set $\mathcal{S}$, the threshold t, the ciphertext $C = (C_1, \ldots, C_{n+3}, \bar{\mathcal{D}})$, a subset $T = (x_1, \ldots, x_t) \subseteq \mathcal{S}$ of t users and $\Sigma = (\sigma_1, \ldots, \sigma_t)$ the corresponding decryption shares, denote Δ as the set containing the difference among the public keys used in encryption and decryption, i.e. $\Delta := (\mathcal{S} \cup \bar{\mathcal{D}}) \setminus T$. Define the algorithm **Aggregate** by [9]:

$$\mathbf{Aggregate}(\Sigma) = \hat{e}(g, C_{n+1})^{\prod_{x \in T}(\frac{1}{\gamma + x})}$$
$$= \hat{e}(g, h)^{\alpha \cdot k \cdot \prod_{x \in \Delta}(\gamma + x)}.$$

The **Aggregate** algorithm will be described in §3.5. Denote $p(\gamma)$ as a polynomial of γ with degree $(m-2)$:

$$p(\gamma) = \frac{1}{\gamma} \cdot \left(\prod_{x \in \Delta}(\gamma + x) - \prod_{x \in \Delta} x\right).$$

Note that $h^{p(\gamma)}$ is computable from CK. Finally, it outputs $M = C_{n+3}/K$ where:

$$K = \left(\hat{e}(C_{n+2}, h^{p(\gamma)}) \cdot \mathbf{Aggregate}(\Sigma)\right)^{\frac{1}{\prod_{x \in \Delta} x}}.$$

3.4 Security Proof

To prove the security of our scheme, we use the dual system encryption technique.

Dual-System DT-PKE. We first define the two additional algorithms required for a dual system DT-PKE.

JoinSF(MK, ID) $\rightarrow \widetilde{K}$. The semi-functional join algorithm works in a similar way to **Join** but outputs a semi-functional (SF) keys, also this algorithm *need not be polynomial time computable*, in contrast to **Setup**, **Join**, **Encrypt**, **Validate**, **ShareDecrypt** and **Combine**.

EncryptSF(EK, $M, \mathcal{S}, t$) $\rightarrow \widetilde{C}$. The semi-functional encryption algorithm works in a similar way to **Encrypt** but outputs a semi-functional (SF) ciphertext, $\widetilde{C}$. This algorithm *also needs not be polynomial computable.*

Note that an SF-key can decrypt normal ciphertext correctly, and a normal key can decrypt SF ciphertext correctly. However, an SF-key cannot decrypt an SF ciphertext.

Semi-functionality Construction. In our concrete construction, all the ciphertexts and the secret keys holds by the users are *normal*, where by *normal* we mean that they do not have $\mathbb{G}_{p_2}$ parts. On the other hand, an SF key or ciphertext has $\mathbb{G}_{p_2}$ part. We denote g_2 as a generator of $\mathbb{G}_{p_2}$. The remaining algorithms of our dual system DT-PKE are the following:

JoinSF(MK, ID) $\rightarrow \widetilde{K}$. It calls **Join**(MK, ID) to get a normal key $\vec{K}$. Then it picks $\vec{\theta} \in_R \mathbb{Z}_N^{n+1}$ and outputs $\widetilde{K} = \vec{K} * g_2^{\vec{\theta}}$.

EncryptSF(EK, $M, \mathcal{S}, t$) $\rightarrow \widetilde{C}$. It calls **Encrypt**(EK, $M, \mathcal{S}$, t) to obtain a normal ciphertext $C = (C_1, \ldots, C_{n+3})$. Then it picks $\vec{\delta} \in_R \mathbb{Z}_N^{n+1}$, $\mu \in \mathbb{Z}_N$ and outputs

$$\widetilde{C} = \Big((C_1, \ldots, C_{n+1}) * g_2^{\vec{\delta}}, \quad C_{n+2} \cdot g_2^{\mu}, \quad C_{n+3}\Big).$$

We call the terms $\vec{\theta}$, $(\vec{\delta}, \mu)$ the SF parameters of the secret key and the ciphertext, respectively. If an SF key is used to decrypt an SF ciphertext, it is hindered by an extra element $\hat{e}(g_2, g_2)^{\vec{\theta} \cdot \vec{\delta}}$ in $\mathbb{G}_T$.

Nominally Semi-Functionality. Following previous dual system encryption schemes [16, 13], we call a key *nominally semi-functional (NSF)* if $\vec{\theta} \cdot \vec{\delta} = 0 \mod p_2$, since the **ShareDecrypt** still succeeds in this case. Otherwise, the key is truly SF. More precisely, t (the threshold of the encryption) NSF keys can correctly decrypt an SF ciphertext. It is obvious that the decryption by t NSF keys is valid, because, 1) in the **ShareDecrypt** algorithm, we get an extra term $\hat{e}(g_2, g_2)^{\vec{\theta} \cdot \vec{\delta}} = 1$ by the pairing algorithm; 2) the CK does not have the $\mathbb{G}_{p_2}$ parts, thus $\hat{e}(C_{n+2} \cdot g_2^{\mu}, h^{p(\gamma)})$ does not have the $\mathbb{G}_{p_2}$ part either.

However, this traditional definition for NSF key is not sufficient for DT-PKE. The extra terms $\hat{e}(g_2, g_2)^{\vec{\theta_i} \cdot \vec{\delta}}$ in multiple decryption shares generated by truly SF keys may cancel out each other in the **Combine** algorithm. Therefore, we have to define the nominally semi-functional property for a set of keys used in the **Combine** algorithm.

Suppose WLOG there are t users $(x_1, x_2, \ldots, x_t)$ and the first i users hold the SF key, the t users can correctly decrypt the ciphertext if and only if:

$$D_1 f_1 - D_2 f_2 + \ldots + (-1)^{i+1} D_i f_i \equiv 0 \mod p_2, \qquad (1)$$

where $D_j = \vec{\theta_j} \cdot \vec{\delta}$, $\vec{\theta_j}$ is $\mathbb{G}_{p_2}$ part of the j-th SF key, $\vec{\delta}$ is the $\mathbb{G}_{p_2}$ part of the SF ciphertext, and

$$f_j = \frac{1}{\prod_{k=t-i+j+1}^{t}(x_k - x_{t-i+j}) \cdot \prod_{\ell=1}^{t-i+j-1}(x_{t-i+j} - x_\ell)}.$$

We say that the secret keys for a set of users $(x_1, x_2, \ldots, x_t)$ form a *nominally semi-functional key set* if they satisfy equation (1). Its correctness will be analyzed in §3.5.

THEOREM 1. *Under assumptions 1, 2, 3 and for $\ell_{\mathsf{usk}} = (n - 2c - 2)\log p_2$, where $c > 0$ is a fixed positive constant, our DT-PKE scheme is ℓ_{usk}-leakage-resilient.*

Hybrid Game Definitions. Before giving the proof, we first define several hybrid games. The main part of the proof is to show computational or statistical indistinguishability among these games, denoted by $\approx$. The output of each game consists of the view of the adversary $\mathcal{A}$ as well as a bit b chosen by the challenger $\mathcal{C}$ representing its choice of which message M_0^*, M_1^* to encrypt. Denote Q as the maximum number of Join queries on the uncorrupted identities from $\mathcal{S}^*$ during the game. We describe the games in details below.

GameReal: This is the original "CorruptLeak" game.

Game 0: It is the same as **GameReal**, except that the challenge ciphertext returned by $\mathcal{C}$ is an SF ciphertext of the message M_b^*.

Game i: In Game i, for the first i Join queries on the uncorrupted identities from $\mathcal{S}^*$, $\mathcal{C}$ will create the SF key, store it in $\mathcal{T}$ and send a handle referring to the key to the adversary, and the remaining queries will create normal keys.The challenge ciphertext is still SF.

GameFinal: It is the same as **Game** Q, except that $\mathcal{C}$ just encrypts a random message $M' \in \mathcal{M}$ instead of the challenge message M_b^*. More specifically, in **GameFinal**, the secret keys used in the Leak Oracle are all SF, and the ciphertext is just encrypting a random message. Thus, the view of $\mathcal{A}$ in **GameFinal** is independent of the challenge bit b.

We now give the proof of theorem 1.

PROOF. We will build several different games to simulate the different cases and show that those games are indistinguishable under the corresponding number-theoretic assumptions. First, we review a useful lemma.

LEMMA 1 ([6]). *Let $n \in \mathbb{N}, n \geq 3$, and let p be a prime. Let $\vec{\delta} \in_R \mathbb{Z}_p^n, \vec{\tau} \in_R \mathbb{Z}_p^n$, and let $\vec{\tau'}$ be chosen uniformly randomly from the set of vectors in $\mathbb{Z}_p^n$ which are orthogonal to $\vec{\delta}$ under the dot product modulo p. Let $f : \mathbb{Z}_p^n \rightarrow W$ be some function. Then: $dist((\vec{\delta}, f(\vec{\tau'})), (\vec{\delta}, f(\vec{\tau}))) \leq \epsilon$, as long as $|W| \leq 4 \cdot (1 - \frac{1}{p}) \cdot p^{n-2} \cdot \epsilon^2$.*

This lemma allows us to set $\ell_{\mathsf{usk}} = (n - 2c - 2)\log(p_2)$ for our construction, where c is any fixed positive constant (so that $\epsilon := p_2^{-c}$ is negligible).

THEOREM 2. *If assumption 1 holds, **GameReal** $\approx$ **Game** 0.*

PROOF. We will build a PPT simulator $\mathcal{B}$ that breaks assumption 1 with the help of a PPT attacker $\mathcal{A}$ who can distinguish **GameReal** from **Game 0** with non-negligible advantage. $\mathcal{B}$ initially receives $(N, \mathbb{G}, \mathbb{G}_T, \hat{e}, g_1, X_3, T)$ from the problem instance. $\mathcal{B}$ simulates **GameReal** or the **Game 0** with $\mathcal{A}$ in the following way:

Initial Phase: $\mathcal{A}$ gives $\mathcal{B}$ the authorized set $\mathcal{S}^*$, the threshold t^* and a set $\overline{\mathcal{I}}$ of the identities he wants to corrupt, with $|\mathcal{S}^* \cup \overline{\mathcal{I}}| \leq t^* - 1$.

Setup: $\mathcal{B}$ picks $\langle \alpha, \gamma, a, b, y_1, y_2, \ldots, y_n \rangle \in_R \mathbb{Z}_N^{n+4}$, a proper value m and an eligible dummy set $\mathcal{D} = \{d_i\}_{i=1}^{m-1}$. It computes $g = g_1^a, h = g_1^b, v = e(g,h)^\alpha, u = g^{\alpha\gamma}, g_1^{y_1}, g_1^{y_2}, \ldots, g_1^{y_n}$, then gives out the public parameter $(N, \mathbb{G}, \mathbb{G}_T, \hat{e})$ and:

$$\mathsf{EK} = (m, \{h^{\alpha\cdot\gamma^i}\}_{i=0}^{2m-1}, \{h^{\alpha\cdot y_1\cdot\gamma^i}\}_{i=0}^{2m-1}, \ldots, \{h^{\alpha\cdot y_n\cdot\gamma^i}\}_{i=0}^{2m-1}, u, v, g_1, g_1^{y_1}, g_1^{y_2}, \ldots, g_1^{y_n}, \mathcal{D}), \mathsf{CK} = (m, \{h^{\gamma^i}\}_{i=0}^{m-2}, \mathcal{D}),$$

to $\mathcal{A}$. $\mathcal{B}$ keeps the master secret key $\{g, \alpha, \gamma\}$.

Phase 1: $\mathcal{B}$ can answer all of the adversary's queries, because $\mathcal{B}$ only needs $\{g, \alpha, \gamma\}$ to generate the secret key of any user. It can work according to the definition of the scheme.

Challenge: $\mathcal{A}$ gives $\mathcal{B}$ two equal-length messages M_0^* and M_1^*. $\mathcal{B}$ flips a coin $b' \in_R \{0,1\}$ and outputs the ciphertext (for simplicity, we denote $D = \prod_{x_i \in \mathcal{S}^* \cup \bar{\mathcal{D}}} (\gamma + x_i)$, where $\bar{\mathcal{D}} \subseteq \mathcal{D}$ has size $(m + t - s - 1)$):

$$C_j = T^{\alpha\cdot y_j\cdot b\cdot D} \quad \forall\, j \in [1,n], \qquad C_{n+1} = T^{\alpha\cdot b\cdot D},$$
$$C_{n+3} = e(T^a, g_1^{b\cdot\alpha}) \cdot M_{b'}^*, \qquad C_{n+2} = T^{-a\cdot\alpha\cdot\gamma},$$

where T is the challenge term from the assumption.

Phase 2: $\mathcal{B}$ works in the same way as **Phase 1** except responding $\perp$ when $\mathcal{A}$ makes the leak queries on the uncorrupted identities from $\mathcal{S}$.

If $T = g_1^z g_2^\nu$, then the ciphertext is SF, since $\forall j \in [1,n]$,

$$C_j = h^{\alpha\cdot z\cdot y_j\cdot D} g_2^{\nu\cdot b\cdot\alpha\cdot y_j\cdot D}, \qquad C_{n+1} = h^{\alpha\cdot z\cdot D} g_2^{\nu\cdot b\cdot\alpha\cdot D},$$
$$C_{n+2} = u^{-z} g_2^{\nu\cdot a\cdot\alpha\cdot\gamma}, \qquad C_{n+3} = e(g,h)^{\alpha\cdot z} \cdot M_{b'}^*.$$

This implicitly sets $\kappa = z$ and $\vec{\delta} = (\nu b\alpha y_1 D, \ldots, \nu b\alpha y_n D, \nu b\alpha D, \nu a\alpha\gamma)$. Obviously, κ is properly distributed since $z \in_R \mathbb{Z}_N$ according to the assumption. The vector $\vec{\delta}$ is properly distributed in the adversary $\mathcal{A}$'s view since the multiplying factors $b\alpha y_1 D, \ldots, b\alpha y_n D, b\alpha D$ and $a\alpha\gamma$ are only seen modulo p_1 in public parameter $(\mathsf{EK}, \mathsf{CK})$ but not modulo p_2. Thus, in $\mathcal{A}$'s view they are random modulo p_2 by Chinese Remainder Theorem. This means that $\mathcal{B}$ has properly simulated the **Game 0**.

If $T = g_1^z$, it is obvious that the challenge ciphertext is normal since it has no $\mathbb{G}_{p_2}$ part, and $\mathcal{B}$ has correctly simulated **GameReal**.

Hence, if $\mathcal{A}$ has a non-negligible difference in the advantage of **Game 0** and **GameReal**, $\mathcal{B}$ can use it to break assumption 1 with non-negligible probability. □

THEOREM 3. *If assumption 2 holds,* ***Game*** $i \approx$ ***Game*** $i+1$ *with the leakage* $\ell_{\mathsf{usk}} \leq (n - 2c - 2)\log(p_2)$.

PROOF. In this proof, we introduce an additional game, called **AltGame** i. This game is the same as **Game** i except the i-th Join query on the uncorrupted identity from $\mathcal{S}^*$; in **Game** i the challenger creates a truly SF key and in **AltGame** i the challenger creates a NSF key. Then we have the following lemma.

LEMMA 2. *We suppose the leakage is at most* $(\ell_{\mathsf{usk}} = (n - 2c - 2)\log(p_2))$*, where* $c \geq 0$ *is any fixed positive constant, then* ***Game*** $i+1 \approx$ ***AltGame*** $i+1$.

PROOF. Suppose there exists a PPT adversary $\mathcal{A}$ whose advantage changes by ϵ from **Game** $i+1$ to **AltGame** $i+1$. Using $\mathcal{A}$, we will create a PPT algorithm $\mathcal{B}$ which will distinguish between the distribution $(\vec{\delta}, f(\vec{\tau'}))$ and $(\vec{\delta}, f(\vec{\tau}))$ from **Lemma** 1 with non-negligible probability (where $p = p_2$). This will yield a contradiction, since these distributions share a negligible statistical distance.

$\mathcal{B}$ simulates either **Game** $i+1$ or **AltGame** $i+1$ with $\mathcal{A}$ as follows. It starts by running the **Setup** algorithm by itself and giving $\mathcal{A}$ the public parameters. Since $\mathcal{B}$ knows (g, α, γ) and $g_2 \in \mathbb{G}_{p_2}$, it can create normal and SF keys. Hence, it can respond to all the $\mathcal{A}$'s queries.

For the $(i+1)$-th Join query on the uncorrupted identity from $\mathcal{S}^*$, $\mathcal{B}$ just responds with a handle h^* without creating the key. When $\mathcal{A}$ issues the **Leak**(h^*, f), $\mathcal{B}$ will encode the leakage $\mathcal{A}$ asking for this key in **Phase 1** as a single polynomial time computable function f within domain $\mathbb{Z}_{p_2}^n$ and with an image of size $2^{\ell_{\mathsf{usk}}}$. It can do this by fixing the values of all other keys and fixing all other variables involved in the challenge key. $\mathcal{B}$ then receives a sample $\langle \vec{\delta}, f(\vec{\Gamma}) \rangle$ (where $\vec{\Gamma}$ is either distributed as $\vec{\tau}$ or as $\vec{\tau'}$ using the notation of the **Lemma** 1). $\mathcal{B}$ will use $f(\vec{\Gamma})$ to answer $\mathcal{A}$'s Leak queries for the $(i+1)$-th key by implicitly defining the challenge key as follows: $\mathcal{B}$ chooses two random values $r_1, r_2 \in \mathbb{Z}_{p_2}$, and implicitly sets the $\mathbb{G}_{p_2}$ components of the key to be $g_2^{\vec{\Gamma'}}$, where $\vec{\Gamma'}$ is defined to be $\langle \vec{\Gamma}, 0 \rangle + \langle 0, \ldots, 0, r_1, r_2 \rangle$. $\mathcal{B}$ defines the non-$\mathbb{G}_{p_2}$ components of the key to fit their appropriate distribution.

Eventually, $\mathcal{A}$ declares the challenge message. Then $\mathcal{B}$ chooses a element $t_2 \in \mathbb{Z}_{p_2}$ such that $\delta_n r_1 + t_2 r_2 \equiv 0 \mod p_2$. It then constructs the challenge ciphertext using $\langle \vec{\delta}, 0 \rangle + \langle 0, \ldots, 0, 0, t_2 \rangle$ as the $\mathbb{G}_{p_2}$'s exponents (recall that $\vec{\delta}$ is of length n). Now, if $\vec{\Gamma}$ is orthogonal to $\vec{\delta}$, then the $(i+1)$-th key is an NSF key. Otherwise, it is a truly SF key.

It is clear that $\mathcal{B}$ can easily answer **Phase 2** queries, since the uncorrupted keys from $\mathcal{S}^*$ cannot be queried in **Phase 2**. Hence, $\mathcal{B}$ can use the output of $\mathcal{A}$ to gain a non-negligible advantage in distinguishing $(\vec{\delta}, f(\vec{\tau'}))$ from $(\vec{\delta}, f(\vec{\tau}))$. □

LEMMA 3. *If assumption 2 holds,* ***Game*** $i \approx$ ***AltGame*** $i+1$ *with the leakage* $\ell_{\mathsf{usk}} \leq (n - 2c - 2)\log(p_2)$.

PROOF. If there exists a PPT adversary $\mathcal{A}$ that can distinguish **Game** i and **AltGame** $i+1$ with non-negligible probability, then we can build a PPT simulator $\mathcal{B}$ to break assumption 2. $\mathcal{B}$ will simulate either **Game** i or **AltGame** $i+1$. $\mathcal{B}$ receives $(N, \mathbb{G}, \mathbb{G}_T, \hat{e}, g_1, g_3, g_1^z g_2^\nu, g_2^\mu g_3^\rho)$ and a challenge term T, which is equal either to $g_1^\omega g_3^\sigma$ or $g_1^\omega g_2^\kappa g_3^\sigma$, from the problem instance. Initially, $\mathcal{A}$ sends $\mathcal{S}^*, t^*$ and $\overline{\mathcal{I}}$ to $\mathcal{B}$. $\mathcal{B}$ simulates as follows:

Setup: It is the same as described in the Setup phase of the proof of Theorem 2.

Phase 1: Since the simulator knows $\{g, \alpha, \gamma\}$, $\mathcal{B}$ can adaptively answer the following queries:

When $\mathcal{A}$ issues the Corrupt queries with an identity from $\overline{\mathcal{I}}$, $\mathcal{B}$ picks $\langle z_1, \ldots, z_n \rangle \in_R \mathbb{Z}_N^n$, and a random vector $\vec{\rho} \in_R \mathbb{Z}_N^{n+1}$. It sends the following key to $\mathcal{A}$:

$$\mathsf{usk} = \left\langle g_1^{z_1}, \ldots, g_1^{z_n}, g^{\frac{1}{\gamma+x_i}} \prod_{i=1}^{n} g_1{}^{-y_i z_i} \right\rangle * g_3^{\vec{\rho}}.$$

When $\mathcal{A}$ issues the Leak queries using the first i handles, $\mathcal{B}$ picks $\vec{\gamma}, \vec{\rho} \in_R \mathbb{Z}_N^{n+1}$ and generates an SF key

$$\widetilde{\mathsf{usk}} = \left\langle g_1^{z_1}, \ldots, g_1^{z_n}, g^{\frac{1}{\gamma+x_i}} \prod_{i=1}^{n} g_1{}^{-y_i z_i} \right\rangle * (g_2^{\mu} g_3^{\rho})^{\vec{\gamma}} * g_3^{\vec{\rho}}.$$

When $\mathcal{A}$ issues the Leak queries using the last $(Q-i-1)$ handles, $\mathcal{B}$ generates a normal key as above and sends the result of the leakage function f using this key.

When $\mathcal{A}$ issues the Leak queries using the $(i+1)$-th handle, $\mathcal{B}$ uses the challenge term T to generate the key:

$$\mathsf{usk} = \left\langle T^{z'_1}, \ldots, T^{z'_n}, g^{\frac{1}{\gamma+x_i}} \prod_{i=1}^{n} T^{-y_i z'_i} \right\rangle * g_3^{\vec{\rho}},$$

where $\langle z'_1, \ldots z'_n \rangle \in_R \mathbb{Z}_N^n$ and $\vec{\rho} \in_R \mathbb{Z}_N^{n+1}$. It is easy to see that if $T = g_1^{\omega} g_3^{\sigma}$, then the $\mathbb{G}_{p_2}$ part above is omitted and $\mathcal{B}$ has formed a properly distributed normal secret key. If $T = g_1^{\omega} g_2^{\kappa} g_3^{\sigma}$, then this key is SF with the parameter:

$$z_i = \omega z'_i \quad \forall i \in [1,n], \quad \vec{\gamma} = \kappa \cdot \langle z'_1, \ldots, z'_n, -\sum y_i z_i{}' \rangle.$$

Since all terms $z'_1, \ldots, z'_n, y_1, \ldots, y_n$ are only seen modulo p_1 in the public parameters, they appear random modulo p_2 by the Chinese Remainder Theorem. Therefore, in this case, $\mathcal{B}$ has formed a properly distributed SF key.

Challenge: $\mathcal{B}$ gets two messages M_0^* and M_1^* from $\mathcal{A}$ and flips a coin $b' \in_R \{0,1\}$. Then $\mathcal{B}$ generates the SF ciphertext using $(g_1^{z} g_2^{\nu})$ from the assumption:

$$C_j = (g_1^{z} g_2^{\nu})^{\alpha \cdot y_j \cdot b \cdot D} \quad \forall j \in [1,n], \; C_{n+1} = (g_1^{z} g_2^{\nu})^{\alpha \cdot b \cdot D},$$
$$C_{n+2} = (g_1^{z} g_2^{\nu})^{-a \cdot \alpha \cdot \gamma}, \quad C_{n+3} = e((g_1^{z} g_2^{\nu})^{a}, g_1^{b \cdot \alpha}) \cdot M_{b'}^*.$$

It is obvious that the ciphertext's SF parameters are:

$$s = z, \quad \mu = \nu \alpha a \gamma, \quad \text{and} \quad \vec{\delta} = \nu \alpha b D \cdot (y_1, \ldots, y_n, 1).$$

In this case, though the ciphertext is not well-distributed (there exists an equation maybe the adversary would find the invalidity of the simulation), according to the lemma 2, no PPT adversary can distinguish this from a truly random vector. Thus, the ciphertext seems properly distributed to the adversary.

Phase 2: Same as described in the proof of Theorem 2.

Here, we claim that if T has $\mathbb{G}_{p_2}$ term, the secret key of the $Q-i+1$ handle is NSF with respect to the aforementioned ciphertext as:

$$\kappa \langle z'_1, \ldots, z'_n, -\sum y_i z_i{}' \rangle \cdot \nu \alpha b D \langle y_1, \ldots, y_n, 1 \rangle \equiv 0 \mod p_2.$$

Thus, if $T = g_1^{\omega} g_2^{\kappa} g_3^{\sigma}$, $\mathcal{B}$ simulates the **AltGame** $i+1$. Otherwise, $\mathcal{B}$ simulates the **Game** i. □

In summary, under assumption 2, we have **Game** $i \approx$ **AltGame** $i+1$, and according to the **Lemma** 2, if the leakage from any uncorrupted identity belonging to $\mathcal{S}$ is at most $(n-2c-2)\log(p_2)$, no PPT adversary has a non-negligible probability in distinguishing **Game** i from **Game** $i+1$. □

THEOREM 4. *If (ℓ, m, t^*)-assumption 3 holds,* ***Game*** $Q \approx$ ***GameFinal****, where $\ell + t^* - 1$ is the size of the corrupt identity set $\overline{\mathcal{I}}$, m is the maximum size of an authorized set and t^* is the challenge threshold.*

PROOF. We build a PPT simulator $\mathcal{B}$ that breaks the (ℓ, m, t^*)-assumption 3 with the help of a PPT adversary $\mathcal{A}$ who can distinguish **Game** Q from **GameFinal** with non-negligible probability.

From the assumption 3, $\mathcal{B}$ is given $(N, \mathbb{G}, \mathbb{G}_T, \hat{e}, g_1, g_2, g_3)$ and can construct two coprime polynomials f and g, of respective orders ℓ and m, with their pairwise distinct roots $(-x_1, \ldots, -x_\ell)$ and $(-x_{\ell+t}, \ldots, -x_{\ell+t+m-1})$, and several sequences of exponentiations:

$$g_1 g_2^{\theta_0}, g_1^{\gamma} g_2^{\theta_1}, \ldots, g_1^{\gamma^{m-1}} g_2^{\theta_{m-1}}, g_1^{\kappa \gamma f(\gamma) g(\gamma)} g_2^{\theta}$$
$$h, h^{\gamma}, \ldots, h^{\gamma^{m-2}}, h^{\alpha}, h^{\alpha\gamma}, \ldots, h^{\alpha\gamma^{2m-1}}, h^{\kappa g(\gamma)} g_2^{\omega}$$
$$g_1^{g(\gamma)}, g_1^{\gamma g(\gamma)}, \ldots, g_1^{\gamma^{\ell+t-2} g(\gamma)}, g_1^{\alpha g(\gamma)}, g_1^{\alpha\gamma g(\gamma)}, \ldots, g_1^{\alpha\gamma^{\ell+t} g(\gamma)},$$

as well as $T \in \mathbb{G}_T$ which is either $e(g_1, h)^{\kappa \cdot f(\gamma)}$ or a random element of $\mathbb{G}_T$. For the sake of simplicity, we state that f and g are unitary polynomials, but this is not a mandatory requirement:

$$f(X) = \prod_{i=1}^{\ell} (X + x_i), \quad q(X) = \prod_{i=\ell+1}^{\ell+t-1} (X + x_i),$$
$$g(X) = \prod_{i=\ell+t}^{\ell+t+m-1} (X + x_i).$$

The polynomial f corresponds to a set of l identities that can be corrupted, not in $\mathcal{S}^*$. The polynomial g corresponds to the m users of the target set that cannot be corrupted, but they can be leaked in **Phase 1** with bound. We will simulate $(\ell+t-1)$ normal keys (which will be corrupted in both **Phase 1** and **Phase 2**), with $(t-1)$ of them, only, in the target set, and (s^*-t+1) SF keys (which will be leaked only in **Phase 1**), and the rest $(m+t-s^*-1)$ are in the dummy set. For $i \in [1, \ell+t-1]$, we set $f_i(x) = \frac{f(x) \cdot q(x)}{x + x_i}$, which is a polynomial of degree $(\ell+t-2)$.

Initial Phase: The adversary $\mathcal{A}$ outputs the challenge set $\mathcal{S}^*$ of size s^*, the challenge threshold t^* and the identities set which he wants to corrupt $\overline{\mathcal{I}}$.

Setup: To generate the system parameters, $\mathcal{B}$ implicitly sets $g = g_1^{f(\gamma) \cdot q(\gamma)}$(without computing it) and sets

$$u = g_1^{\alpha \cdot \gamma \cdot f(\gamma) \cdot q(\gamma)} = g^{\alpha \cdot \gamma},$$
$$v = e(g_1, h)^{\alpha(\gamma + x_1) \cdot f_1(\gamma)} = e(g, h)^{\alpha}.$$

The two latter formulae can be computed from the problem instance, since $f \cdot q$ is of degree $(l+t-1)$. Then $\mathcal{B}$ randomly picks $\langle y_1, \ldots, y_n \rangle \in_R \mathbb{Z}_N^n$ and computes the public parameters $\{h^{a y_1 \gamma^i}\}_{i=0}^{2m-1}, \ldots, \{h^{a y_n \gamma^i}\}_{i=0}^{2m-1}$ and $g_1^{y_1}, \ldots, g_1^{y_n}$ using the input of the assumption. WLOG, suppose $\mathcal{B}$ sets the set $\mathcal{D} = \{d_i\}_{i=1}^{m-1}$ corresponding to dummy users:

- Let $\mathcal{D}^* = \{d_i\}_{i=1}^{m+t-s^*-1}$ be a subset of $\{x_j\}_{j=\ell+t}^{\ell+t+m-1}$. This subset corresponds to the dummy users included to complete the target set in the challenge phase.
- $\{d_i\}_{i=m+t-s^*}^{m-1}$ is a set of random elements in $\mathbb{Z}_N^*$.

Note that $\mathcal{B}$ cannot compute the value of g, but it can still simulate the game.

Phase 1: Here we first show the generation of keys of different identities:

- For each $\mathsf{ID} \in \overline{\mathcal{I}}$, $\mathcal{B}$ picks $\vec{\rho} \in_R \mathbb{Z}_N^{n+1}$, computes and sends $(\overline{\mathsf{usk}}, \overline{\mathsf{upk}})$ to $\mathcal{A}$ with $\overline{\mathsf{upk}} = x_i$ and
$$\overline{\mathsf{usk}} = \langle g_1^{z_1}, \ldots, g_1^{z_n}, g_1^{g(\gamma)f_i(\gamma)} \prod_{i=1}^{n} g_1^{-y_i z_i} \rangle * g_3^{\vec{\rho}}$$
$$= \langle g_1^{z_1}, \ldots, g_1^{z_n}, g^{\frac{1}{\gamma+x_i}} \prod_{i=1}^{n} g_1^{-y_i z_i} \rangle * g_3^{\vec{\rho}}.$$
with the following constraint: if $\mathsf{ID} \in \mathcal{S}^*$, then the corresponding x_i must be taken from $\{x_j\}_{j=\ell+1}^{\ell+t-1}$. Otherwise x_i must be taken from $\{x_j\}_{j=1}^{\ell}$. Used x_i cannot be chosen again.
- For each $\mathsf{ID} \in \mathcal{S}^* \setminus (\mathcal{S}^* \cap \overline{\mathcal{I}})$, $\mathcal{B}$ sends $\mathsf{upk} = x_i$ to $\mathcal{A}$, with the constraints that x_i must be taken from $\{x_j\}_{j=\ell+t}^{\ell+t+m-1} \setminus \mathcal{D}^*$. $\mathcal{B}$ computes the corresponding SF key using $g_1 g_2^{\theta_0}, g_1^{\gamma} g_2^{\theta_1}, \ldots, g_1^{\gamma^{m-1}} g_2^{\theta_{m-1}}$ from the problem instance. $\mathcal{B}$ can easily compute a term in $\mathbb{G}_{p_1 p_2}$ with the $\mathbb{G}_{p_1}$ part as $g^{\frac{1}{\gamma+x_i}}$, thus $\mathcal{B}$ can compute the valid SF key.
- For each $\mathsf{ID} \notin \mathcal{S}^* \cup \overline{\mathcal{I}}$, $\mathcal{B}$ sends $\mathsf{upk} = x$ to $\mathcal{A}$, with $x \notin \{x_j\}_{j=1}^{\ell+t+m-1}$.

Challenge: $\mathcal{B}$ receives M_0^* and M_1^*, randomly chooses $b \leftarrow \{0,1\}$ and computes the SF ciphertext as follows:
$$C_j = (h^{\kappa \cdot g(\gamma)} g_2^{\omega})^{y_j} \quad \forall j \in [1,n], \quad C_{n+1} = h^{\kappa \cdot g(\gamma)} g_2^{\omega},$$
$$C_{n+2} = g_1^{-\kappa \cdot \gamma \cdot f(\gamma)} g_2^{\theta}, \quad C_{n+3} = T \cdot M_b^*.$$
One can verify that, if we set $k' = \frac{\kappa}{\alpha \cdot g(\gamma)}$, then the SF ciphertext can be viewed as:
$$C_j = h^{\kappa' \alpha y_j \prod_{x_i \in \mathcal{S}^*} (\gamma + x_i) \prod_{x \in \mathcal{D}^*} (\gamma + x)} g_2^{\omega y_j}, \quad \forall j \in [1,n]$$
$$C_{n+1} = h^{\kappa' \alpha \prod_{x_i \in \mathcal{S}^*} (\gamma + x_i) \prod_{x \in \mathcal{D}^*} (\gamma + x)} g_2^{\omega}, \quad C_{n+2} = u^{-\kappa'} g_2^{\theta}.$$

Phase 2: Same as Phase 1 except answering $\perp$ for the Leak query on the uncorrupted identities in $\mathcal{S}^*$.

Note that if $T = e(g_1, h)^{\kappa \cdot f(\gamma)}$, then we can get that $T = v^{\kappa'}$, and it is obvious that the ciphertext is well distributed. Hence $\mathcal{B}$ simulates the **Game** Q correctly in $\mathcal{A}$'s view. If $T \in_R \mathbb{G}_T$, the term C_{n+3} is entirely random and we get an SF ciphertext of a random message. Thus $\mathcal{B}$ properly simulates the **GameFinal**. Hence, $\mathcal{B}$ can use $\mathcal{A}$'s guess to break assumption 3 with non-negligible probability. □

Putting It All Together. With Theorems 2, 3, 4 for the sequence of games, we get **GameReal** $\approx$ **GameFinal**. In the **GameFinal**, the view of $\mathcal{A}$ is independent of the random bit b. Note that the probability of the uncorrupted users form a *nominally SF key set* is negligible ($\leq Q \cdot \ell_{\mathsf{usk}}/p_2$) and hence the Leak oracle cannot help $\mathcal{A}$ to determine if the challenge ciphertext is an encryption of a random message. Hence we have $\Pr[\mathcal{A} \text{ wins}] = \frac{1}{2}$ in **GameFinal**, which concludes the proof of Theorem 1. □

3.5 More Details of the Construction

Sub-routine. The **Aggregate** algorithm [9] aims to compute
$$L = \hat{e}(g, C_{n+1})^{\frac{1}{(\gamma+x_1)\cdots(\gamma+x_t)}},$$
from $\{\sigma_j = \hat{e}(g, C_{n+1})^{\frac{1}{\gamma+x_j}}\}_{j=1}^{t}$ and pairwise distinct $x_1, \ldots, x_t$, where for all $i, j \in [1,t], \gcd(x_i - x_j, N) = 1$. Define for any (j, ℓ) such that $1 \leq j < \ell \leq t$,
$$L_{j,\ell} = \sigma_\ell^{\frac{1}{\prod_{\kappa=1}^{j}(\gamma+x_\kappa)}} = \hat{e}(g, C_{n+1})^{\frac{1}{\gamma+x_\ell} \cdot \frac{1}{\prod_{\kappa=1}^{j}(\gamma+x_\kappa)}}.$$
The **Aggregate** algorithm computes sequentially $L_{j,l}$ for $j \in [1, t-1]$ and $\ell \in [j+1, t]$ using the induction
$$L_{j,\ell} = \left(\frac{L_{j-1,j}}{L_{j-1,\ell}}\right)^{\frac{1}{x_\ell - x_j}},$$
and posing $L_{0,\ell} = \sigma_\ell$, for $\ell = 1, \ldots, t$. The algorithm finally outputs $L_t = L_{t-1,t}$.

Eligibility of upk. We now show that a suitable x satisfying the three conditions in **Join** can be found with overwhelming probability. Let $x_1, \ldots, x_l$ be the previously eligible upk (including the dummy keys) and we randomly choose $x \in_R \mathbb{Z}_N^*$. We claim that x is an eligible one if $\forall i \in [1,n]$, $x \neq x_i$, $\gcd(x - x_i, N) = 1$ and $(\gamma + x) \nmid p_1$. We can simplify this condition as randomly choosing $l+2$ numbers from $\mathbb{Z}_N$ and they are all coprime with N. The probability of this happening is (WLOG we assume $p_1 \leq p_2 \leq p_3$):
$$\left(\frac{\Phi(N)}{N}\right)^{l+2} = \left(\frac{(p_1-1)(p_2-1)(p_3)}{p_1 p_2 p_3}\right)^{l+2}$$
$$\geq \left(1 - \frac{3}{p_1}\right)^{l+2} = \left(1 - \frac{1}{\frac{p_1}{3}}\right)^{\frac{p_1}{3} \cdot \frac{3(l+2)}{p_1}} \geq \left(\frac{1}{e}\right)^{\frac{3(l+2)}{p_1}}.$$
Hence, a random x is eligible with overwhelming probability, since $l \lll p_1$.

Correctness of NSF Key Set. We show the correctness of the equation (1) by induction. First, in the case that $i = 1$, the *NSF key set* is the same as the NSF key, $D_1 = 0$. So the equation is correct.

If in the case that there are j SF keys of the t users, by induction, the t keys forms a *NSF key set*, then:
$$D_1 f_1 - D_2 f_2 + D_3 f_3 - \ldots (\pm) D_j f_j \equiv 0 \mod p_2.$$

Note that increasing (or decreasing) t only adds (or removes) a user with normal key to the decryption. Since normal keys do not have any $\mathbb{G}_{p_2}$ part, it will not change the equation modulo p_2. So we can just change from t to $t-1$ without affecting the equation.

Next, we consider the case that there are $(j+1)$ SF keys. According to the **Aggregate** algorithm, we have
$$L_{t-1,t} = \left(\frac{L_{t-2,t-1}}{L_{t-2,t}}\right)^{\frac{1}{x_t - x_{t-1}}}.$$
For the numerator, the $\mathbb{G}_{p_2}$ part of $L_{t-2,t-1}$ is in the induction by changing t to $t-1$. More precisely, for each $s \in [1,j]$
$$f_s = \frac{1}{\prod_{k=t-i+s}^{t-1}(x_k - x_{t-1-i+s}) \prod_{q=1}^{t-i+s-2}(x_{t-1-i+s} - x_q)}.$$
For the denominator $L_{t-2,t}$, the $\mathbb{G}_{p_2}$ part is the same as $L_{t-2,t-1}$'s except changing from D_{t-1} to D_t and changing from x_{t-1} to x_t in every f_s. By combining $L_{t-2,t-1}$ with $L_{t-2,t}$, we can get the $\mathbb{G}_{p_2}$ part of $L_{t-1,t}$ for t users as
$$D_1 f_1 - D_2 f_2 + D_3 f_3 - \ldots (\pm) D_{j+1} f_{j+1}.$$

4. MULTI-KEY LEAKAGE-RESILIENT THRESHOLD RING SIGNATURES

4.1 Syntax

A TRSig scheme consists of four PPT algorithms:

Setup: On input a security parameter 1^λ, it generates the set of system parameters param. We denote by $\mathcal{M}$ the message space.

KeyGen: On input param, it generates a pair of signing key sk_i and verification key vk_i.

Sign: On input param, a set of n public keys $\mathcal{Y}$, a set of d secret keys $\mathcal{X}$ (with the corresponding public keys in $\mathcal{Y}$) and a message $m \in \mathcal{M}$, it outputs a signature σ.

Verify: On input param, a set of n public keys $\mathcal{Y}$, a threshold d, a message m and a signature σ, it outputs a bit 1 or 0 symbolizing the validity of the signature.

For correctness, we require that $\forall m \in \mathcal{M}$, we have $1 \leftarrow \mathbf{Verify}(\mathsf{param}, \mathcal{Y}, t, m, \mathbf{Sign}(\mathsf{param}, \mathcal{Y}, \mathcal{X}, m))$, where $(\mathsf{sk}_i, \mathsf{vk}_i) \leftarrow \mathbf{KeyGen}(\mathsf{param})$ for $i \in [1, t]$, $\mathcal{X} = \{sk_1, \ldots, sk_t\}$ and $\{\mathsf{vk}_1, \ldots, \mathsf{vk}_t\} \subset \mathcal{Y}$.

4.2 Leakage-Resilience

We consider the following existential unforgeability game against adaptive chosen message attacks (EUF-CMA), in the multi-key leakage-resilient model. We allow the leakage function to take all secret keys (and the past randomness used) as the input and output a length-bounded leakage.

Denote $\mathcal{R}$ as a set of random coins used. To make it explicit, we write $(\mathsf{sk}_i, \mathsf{vk}_i) \leftarrow \mathbf{KeyGen}(\mathsf{param}; r_i')$, $\sigma \leftarrow \mathbf{Sign}(\mathsf{param}, \mathcal{Y}, \mathcal{X}, m_i; r_i)$, where r_i' and r_i are the randomness used in **KeyGen** and **Sign** respectively. The game mLR-TRSig is defined as follows:

1. Setup. The challenger samples $r \leftarrow \{0,1\}^*$ and runs $\mathsf{param} \leftarrow \mathbf{Setup}(1^\lambda)$. The challenger gives param to the adversary $\mathcal{A}$ and sets $\mathcal{R}$ as an empty set.

2. Query. The following oracles can be queried by $\mathcal{A}$:

   - Join Oracle $\mathcal{JO}$: It samples $r_i \leftarrow \{0,1\}^*$ and runs $(\mathsf{sk}_i, \mathsf{vk}_i) \leftarrow \mathbf{KeyGen}(\mathsf{param}; r_i)$. It sets $\mathcal{R} := \mathcal{R} \cup \{r_i\}$ and returns vk_i.
   - Corrupt Oracle $\mathcal{CO}(\mathsf{vk}_i)$: On input vk_i which was output by $\mathcal{JO}$, it returns the corresponding sk_i.
   - Signing Oracle $\mathcal{SO}(\mathcal{Y}, \mathcal{V}, m_i)$: On input a set of n public keys $\mathcal{Y}$, a set of d signer's public key $\mathcal{V} \subset \mathcal{Y}$ and a message $m_i \in \mathcal{M}$, it samples $r_i' \leftarrow \{0,1\}^*$. It returns the signature $\sigma_i \leftarrow \mathbf{Sign}(\mathsf{param}, \mathcal{Y}, \mathcal{X}, m_i; r_i')$ where $\mathcal{X}$ is the set of signing keys corresponding to $\mathcal{V}$. It sets $\mathcal{R} := \mathcal{R} \cup \{r_i'\}$.
   - Leak Oracle $\mathcal{LO}(f_j)$: On input a polynomial-time computable function $f_j : \{0,1\}^* \to \{0,1\}^{\ell_j}$, it returns $f_j(\mathcal{R})$.

3. Output. $\mathcal{A}$ returns a message-signature pair (m^*, σ^*) for a set of n^* public keys $\mathcal{Y}^*$ and a threshold d^*.

$\mathcal{A}$ wins if $1 \leftarrow \mathbf{Verify}(\mathsf{param}, \mathcal{Y}^*, d^*, m^*, \sigma^*)$, m^* was not queried to the $\mathcal{SO}$, the sum of all ℓ_j is at most ℓ, all of the public keys in $\mathcal{Y}^*$ are query outputs of the $\mathcal{JO}$, and there are at most $(d^* - 1)$ queries to the $\mathcal{CO}$.

DEFINITION 3. *A threshold ring signature scheme is ℓ-mLR-EUF-CMA (multi-key leakage-resilient) secure if no PPT $\mathcal{A}$ wins the game above with non-negligible probability.*

We use the anonymity model of threshold ring signature under full key exposure attack [18], so there is no need to explicitly allow secret key leakage. We left it as an open problem to allow leakage of the randomness used in the challenge signature.

4.3 Our Construction

Our construction is based on the TRSig of Yuen *et al.* [18]. There is a commitment C_i which commits to 1 or 0 to indicate whether the user at index i is a real signer or not. The corresponding zero-knowledge proof is π_i. Then we split the secret key into q pieces, and generate a (partial) signature $(A_{i,j}, B_j)$ for them.

In order to prove the security of the scheme in the multi-key leakage-resilient model, we add the admissible hash function [4] ensemble H and implicitly define the function:

$$F(m_1, \ldots, m_\tau) = v' \prod_{k=1}^{\tau} v_k^{m_k}.$$

The function F is similar to the Waters hash function [15]. However in the security proof, the admissible hash function ensemble helps us to divide the output of F into two subgroups: in $\mathbb{G}_{p_2}$ and not in $\mathbb{G}_{p_2}$ (which is different from the Waters identity-based encryption [15]). These two groups are indistinguishable under the subgroup decision assumption. In the subgroup $\mathbb{G}_{p_2}$, it helps us to extract the secret keys from the adversary's forgery. In the group $\mathbb{G}$, it allows us to simulate the signing oracle queries using some *fake* witness, and then leak the fake witness later on. Hence, the leakage will not help the attacker. Admissible hash function was also used in some leakage-resilient signatures [5, 19].

Finally, since we are now working in the composite order group, we need the extra elements (D, E) to ensure that the signature falls in the desired subgroup.

Setup(1^λ): On input 1^λ, it samples $(\mathbb{G}, \mathbb{G}_T, N = p_1 p_2, \hat{e}) \leftarrow \mathcal{G}'(1^\lambda)$. In addition, it also obtains the generators of the subgroups $\mathbb{G}$ and $\mathbb{G}_{p_2}$. It randomly picks $u, v, g_1, \ldots, g_q, Z \in \mathbb{G}, h, v', v_1, \ldots, v_\tau \in \mathbb{G}_{p_2}, \alpha \in \mathbb{Z}_N$ and sets $\hat{h} = h^\alpha$ and $\hat{u} = u^\alpha$. Observe that $\hat{e}(u, \hat{h}) = \hat{e}(h, \hat{u})$. Let $H : \{0,1\}^* \to \{0,1\}^\tau$ be an admissible hash function ensemble. It outputs $\mathsf{param} =$

$$(\mathbb{G}, \mathbb{G}_T, N, \hat{e}, u, \hat{u}, \hat{h}, h, H, g_1, \ldots, g_q, v, v', v_1, \ldots, v_\tau, Z).$$

KeyGen(param): On input param, it samples $s_1, \ldots s_q \in \mathbb{Z}_N$ and computes $\mathsf{sk} = (u^{s_1}, \ldots, u^{s_n}), \mathsf{vk} = \prod_{j=1}^{q} g_j^{s_j}$.

Sign$(\mathsf{param}, \mathcal{Y}, \mathcal{X}, M)$: Suppose $\mathcal{Y} = \{\mathsf{vk}_1, \ldots, \mathsf{vk}_n\}$ is the user ring, $\mathcal{X}$ is the set of private keys of d participating signers to sign for the message M. WLOG, suppose that $\{1, 2, \ldots, d\}$ is the indices of participating signers, and $\{d+1, \ldots, n\}$ is the indices of non-signers. Define f_i such that $f_i = 1$ if $i = 1, \ldots, d$ and $f_i = 0$ if $i = d+1, \ldots, n$.

1. One of the signers picks $x_i \in_R \mathbb{Z}_N$ for $i \in [1, n]$ and sets

$$C_i = \Big(\frac{\mathsf{vk}_i}{Z}\Big)^{f_i} \hat{h}^{x_i}, \quad \pi_i = \Big(\big(\frac{\mathsf{vk}_i}{Z}\big)^{2f_i - 1} \hat{h}^{x_i}\Big)^{x_i}.$$

   Let $C = \prod_{i=1}^{n} C_i$. Then we have $Z^d C = \hat{h}^x \prod_{i=1}^{d} \mathsf{vk}_i$ where $x = \sum_{i=1}^{n} x_i$.

2. Each signer i (with secret key $\mathsf{sk}_i = (u^{s_{i,1}}, \ldots, u^{s_{i,q}})$) computes $W = v' \prod_{k=1}^{\tau} v_k^{m_k}$ where $(m_1, \ldots, m_\tau) = H(d, \mathcal{Y}, M)$. For $j \in [1, q]$, he picks a random $r_{i,j} \in \mathbb{Z}_N$ and sets

$$A_{i,j} = u^{s_{i,j}}(hW)^{r_{i,j}}, \quad B_i = \prod_{j=1}^{q} g_j^{r_{i,j}}.$$

Signer i sends $(\{A_{i,1}, \ldots, A_{i,q}\}, B_i)$ to the signer in step 1.

3. After collecting $(\{A_{i,1}, \ldots, A_{i,q}\}, B_i)$ from the t signers, calculates

$$\bar{A}_j = \prod_{i=1}^{d} A_{i,j}, \quad B = \hat{u}^{-x} \prod_{i=1}^{d} B_i, \quad D = W^x, \quad E = v^x.$$

The signature is $(\{\bar{A}_j\}_{j\in[1,q]}, B, \{C_i, \pi_i\}_{i\in[1,n]}, D, E)$.

Verify$(\mathsf{param}, \mathcal{Y}, t, M, \sigma)$: On input $(n, d, \mathcal{Y}, M, \sigma)$, first computes $(m_1, \ldots, m_\tau) = H(d, \mathcal{Y}, M)$ and $W = v' \prod_{k=1}^{\tau} v_k^{m_k}$. For $i = 1, \ldots, n$, checks if

$$\hat{e}(C_i, C_i) = \hat{e}(\hat{h}, \pi_i) \cdot \hat{e}(C_i, \frac{\mathsf{vk}_i}{Z}), \quad \text{and} \quad \hat{e}(D, v) = \hat{e}(W, E).$$

If they are true, computes $C = \prod_{i=1}^{n} C_i$ and checks if:

$$\prod_{j=1}^{n} \hat{e}(\bar{A}_j, g_j) = \hat{e}(u, Z^d C) \cdot \hat{e}(hW, B) \cdot \hat{e}(D, \hat{u}).$$

THEOREM 5. *Our scheme is* $(\mathsf{negl}(\lambda))$*-mLR-EUF-CMA if the DL, CDH and subgroup decision assumptions hold. Our scheme is anonymous against full key exposure if the subgroup decision assumption holds.*

The proof is omitted due to the space limit.

5. CONTINUAL LEAKAGE EXTENSIONS

The *continual leakage* model [11, 6, 13, 17] allows periodic updates on the secret key, while bounding the leakage between updates. This model offers an overall unbounded leakage throughout the life time of the key. It is possible to extend our DT-PKE and TRSig schemes in the continual leakage model. Similar to the continual leakage-resilient IBE [13, 17], we can update our keys by re-randomizing the $\mathbb{G}_{p_3}$ element of the private key shares for DT-PKE. For TRSig, we have to extend the scheme to the group of order $N = p_1 p_2 p_3$ first, set the secret keys to be in the group of $\mathbb{G}_{p_1 p_3}$ and update the keys in $\mathbb{G}_{p_3}$.

6. CONCLUSION

Key-exposure attacks constitute threat to cryptosystems. We initiate the study of leakage-resilient threshold cryptography, for achieving formal security guarantee under various key-exposure attacks. We propose a dynamic threshold public key encryption scheme and a threshold ring signature scheme which are secure in our multi-key leakage model.

7. REFERENCES

[1] A. Akavia, S. Goldwasser, and C. Hazay. Distributed public key schemes secure against continual leakage. In *PODC*, pages 155–164, 2012.

[2] A. Akavia, S. Goldwasser, and V. Vaikuntanathan. Simultaneous hardcore bits and cryptography against memory attacks. In *TCC*, volume 5444 of *LNCS*, pages 474–495. Springer, 2009.

[3] J. Alwen, Y. Dodis, M. Naor, G. Segev, S. Walfish, and D. Wichs. Public-key encryption in the bounded-retrieval model. In *EUROCRYPT*, volume 6110 of *LNCS*, pages 113–134. Springer, 2010.

[4] D. Boneh and X. Boyen. Secure identity based encryption without random oracles. In *CRYPTO*, volume 3152 of *LNCS*, pages 443–459. Springer, 2004.

[5] E. Boyle, G. Segev, and D. Wichs. Fully leakage-resilient signatures. In *EUROCRYPT*, volume 6632 of *LNCS*, pages 89–108. Springer, 2011.

[6] Z. Brakerski, Y. T. Kalai, J. Katz, and V. Vaikuntanathan. Overcoming the hole in the bucket: Public-key cryptography resilient to continual memory leakage. In *FOCS*, pages 501–510. IEEE, 2010.

[7] E. Bresson, J. Stern, and M. Szydlo. Threshold ring signatures and applications to ad-hoc groups. In *CRYPTO*, volume 2442 of *LNCS*, pages 465–480. Springer, 2002.

[8] S. S. M. Chow, Y. Dodis, Y. Rouselakis, and B. Waters. Practical leakage-resilient identity-based encryption from simple assumptions. In *CCS*, pages 152–161. ACM, 2010.

[9] C. Delerablée and D. Pointcheval. Dynamic threshold public-key encryption. In *CRYPTO*, volume 5157 of *LNCS*, pages 317–334. Springer, 2008.

[10] Y. Dodis, S. Goldwasser, Y. T. Kalai, C. Peikert, and V. Vaikuntanathan. Public-key encryption schemes with auxiliary inputs. In *TCC*, volume 5978 of *LNCS*, pages 361–381. Springer, 2010.

[11] Y. Dodis, K. Haralambiev, A. López-Alt, and D. Wichs. Cryptography against continuous memory attacks. In *FOCS*, pages 511–520. IEEE, 2010.

[12] Y. Dodis, A. B. Lewko, B. Waters, and D. Wichs. Storing secrets on continually leaky devices. In *FOCS*, pages 688–697, 2011.

[13] A. B. Lewko, Y. Rouselakis, and B. Waters. Achieving leakage resilience through dual system encryption. In *TCC*, volume 6597 of *LNCS*, pages 70–88. Springer, 2011.

[14] M. Naor and G. Segev. Public-key cryptosystems resilient to key leakage. In *CRYPTO*, volume 5677 of *LNCS*, pages 18–35. Springer, 2009.

[15] B. Waters. Efficient identity-based encryption without random oracles. In *EUROCRYPT*, volume 3494 of *LNCS*, pages 114–127. Springer, 2005.

[16] B. Waters. Dual system encryption: Realizing fully secure IBE and HIBE under simple assumptions. In *CRYPTO*, volume 5677 of *LNCS*, pages 619–636. Springer, 2009.

[17] T. H. Yuen, S. S. M. Chow, Y. Zhang, and S. M. Yiu. Identity-based encryption resilient to continual auxiliary leakage. In *EUROCRYPT*, volume 7237 of *LNCS*, pages 117–134. Springer, 2012.

[18] T. H. Yuen, J. K. Liu, M. H. Au, W. Susilo, and J. Zhou. Threshold ring signature without random oracles. In *ASIACCS*, pages 261–267. ACM, 2011.

[19] T. H. Yuen, S. M. Yiu, and L. C. K. Hui. Fully leakage-resilient signatures with auxiliary inputs. In *ACISP*, volume 7372 of *LNCS*, pages 294–307. Springer, 2012.

Privacy-preserving Multi-keyword Text Search in the Cloud Supporting Similarity-based Ranking

Wenhai Sun[†,††]
whsun@xidian.edu.cn

Bing Wang[††]
bingwang@vt.edu

Ning Cao[§]
ncao@ece.wpi.edu

Ming Li[‡]
ming.li@usu.edu

Wenjing Lou[††]
wjlou@vt.edu

Y. Thomas Hou[††]
thou@vt.edu

Hui Li[†]
lihui@mail.xidian.edu.cn

[†]The State Key Laboratory of Integrated Services Networks, Xidian University, China
[††]Virginia Polytechnic Institute and State University, USA
[§]Worcester Polytechnic Institute, USA
[‡]Utah State University, USA

ABSTRACT

With the increasing popularity of cloud computing, huge amount of documents are outsourced to the cloud for reduced management cost and ease of access. Although encryption helps protecting user data confidentiality, it leaves the well-functioning yet practically-efficient secure search functions over encrypted data a challenging problem. In this paper, we present a privacy-preserving multi-keyword text search (MTS) scheme with similarity-based ranking to address this problem. To support multi-keyword search and search result ranking, we propose to build the search index based on *term frequency* and the *vector space model* with *cosine similarity measure* to achieve higher search result accuracy. To improve the search efficiency, we propose a tree-based index structure and various adaption methods for *multi-dimensional (MD) algorithm* so that the practical search efficiency is much better than that of linear search. To further enhance the search privacy, we propose two secure index schemes to meet the stringent privacy requirements under strong threat models, i.e., known ciphertext model and known background model. Finally, we demonstrate the effectiveness and efficiency of the proposed schemes through extensive experimental evaluation.

Categories and Subject Descriptors

E.3 [**Data Encryption**]; H.3.3 [**Information Storage and Retrieval**]: Information Search and Retrieval

Keywords

Cloud Computing; Privacy-preserving Search; Multi-keyword Search; Similarity-based Ranking

1. INTRODUCTION

Cloud computing is a new model of enterprise IT infrastructure that enables ubiquitous, convenient, and on-demand network access to a shared pool of configurable computing resources (e.g., networks, servers, storage, applications, and services) [8]. Due to the centralized management of elastic resources, all players in this emerging X-as-a-service (XaaS) model, including the cloud provider, application developers, and end-users, can reap benefits. Especially, for the end-users, they can outsource large volumes of data and workloads to the cloud and enjoy the virtually unlimited computing resources in a pay-per-use manner. Indeed, many companies, organizations, and individual users have adopted the cloud platform to facilitate their business operations, research, or everyday needs [24].

Despite the tremendous business and technical advantages, privacy concern is one of the primary hurdles that prevents the widespread adoption of the cloud by potential users, especially if their sensitive data are to be outsourced to and computed in the cloud. Examples may include financial and medical records, and social network profiles. Cloud service providers (CSPs) usually enforce users' data security through mechanisms like firewalls and virtualization. However, these mechanisms do not protect users' privacy from the CSP itself since the CSP possesses full control of the system hardware and lower levels of software stack. There may exist disgruntled, profiteered, or curious employees that can access users' sensitive information for unauthorized purposes [14, 26]. Although encryption before data outsourcing [15, 33] can preserve data privacy against the CSP, it also makes the effective data utilization, such as search over encrypted data, a very challenging task. Without being able to extract useful information from the outsourced data in a secure and private manner, the cloud will merely be a remote storage which provides limited value to all parties.

One fundamental and common form of data utilization is the search operation, i.e., to quickly sort out information of interest from huge amount of data. The information retrieval community has the state-of-the-art techniques that are readily available to achieve rich search functionalities, such as result ranking and multi-keyword queries, on plaintext. For example, *cosine measure* in the *vector space model* [30] is a state-of-the-art similarity measure widely used in plaintext information retrieval community, which incorporates the "term frequency (TF) × inverse document frequency (IDF)" weight to evaluate the similarity between a document and a particular query, and yield accurate ranked search result. However, implementing a secure version of such techniques over outsourced encrypted data in the cloud is not straightforward, and is susceptible to privacy breach [29]. Although inverted index (a.k.a. inverted file) is the most popular and efficient index data structure used in document retrieval systems, it is not directly applicable in TF-based multi-keyword encrypted text search environment [28,29,34].

In the literature, searchable encryption (SE) techniques can partially address the need for secure outsourced data search. Many researchers have developed SE schemes that allow searches over encrypted keyword indexes, either based on public key cryptography (PKC) [3,4,12,13] or symmetric key cryptography (SKC) [6,10,11,16,27]. In general, although the PKC-based schemes allow more expressive queries than SKC-based ones, they are much less efficient. Thus, there has been significant interest in developing efficient SKC-based SE mechanisms. Curtmola et al. were the first to propose a symmetric SE scheme with security guarantees under rigorous definitions [10]. Other works [28, 29, 34] targeted on providing ranked search. These schemes only support single-keyword queries which is too restrictive for practical use. To enrich search functionality, Cao et al. [5] attempted privacy-preserving multi-keyword ranked search over encrypted cloud data. Nevertheless, the search complexity is linear to the number of documents in the dataset, which becomes undesirable and inefficient when a huge amount of documents are present. In addition, the heuristic ranking function, i.e., "coordinate matching", failed to yield more accurate search result, compared to the state-of-the-art multi-keyword search over plaintext. Thus, the quest for secure data search mechanisms that can simultaneously achieve high efficiency and functionality (such as expressive/usable queries) still remains open up to date.

In this paper, we address the challenges of constructing practically efficient and flexible encrypted search functionalities that support result ranking and multi-keyword queries. In particular, to support multi-keyword queries and search result ranking functionalities, we propose to build the search index based on the vector space model , i.e., cosine measure, and incorporate the *TF* × *IDF* weight to achieve high search result accuracy. To improve the search efficiency, we propose a tree-based index structure, where each value in a node is a vector of term frequency related information. We then apply the search algorithm, adapted from the MD-algorithm [19], so as to realize efficient search functionality. Our basic scheme for multi-keyword text search with similarity-based ranking (BMTS) is secure under the known ciphertext model. In order to further enhance the search privacy, we propose another enhanced secure index scheme (EMTS) against sensitive frequency information leakage to meet more stringent privacy requirements under a stronger threat model, i.e., known background model. Finally, we demonstrate the effectiveness and efficiency of the proposed schemes through extensive experimental evaluation.

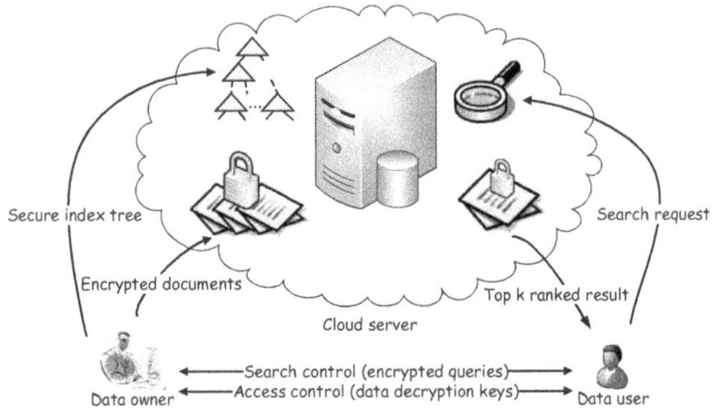

Figure 1: Framework of the search over outsourced encrypted cloud data

Our contributions are summarized as follows:

1. By incorporating the state-of-the-art information retrieval techniques, we propose a privacy-preserving multi-keyword text search scheme supporting similarity-based ranking, which enjoys the same flexibility and search result accuracy as the existing state-of-the-art multi-keyword search over plaintext.

2. We propose a randomization (phantom terms) approach in the enhanced scheme to prevent sensitive frequency information leakage thus achieving better privacy of keywords. We show that with the proposed methods, user can balance between search precision and privacy.

3. With improved security guarantee, EMTS is still comparable in search time to BMTS. In addition, we investigated various index building methods to speed up the search of common cases. The results demonstrate much improved search efficiency compared with [5].

2. PROBLEM FORMULATION

2.1 System Model

The system model considered in this paper consists of three entities: the *data owner*, the *data user*, and the *cloud server*, as illustrated in Fig. 1. The data owner outsources a huge size of document collection $\mathcal{DC}$ in the encrypted form $\mathcal{C}$, together with an encrypted searchable index tree $\mathcal{I}$ generated from $\mathcal{DC}$, to the cloud server. We assume that the data user has the mutual authentication capability with the data owner. As such, search control mechanisms can be applied here, e.g., broadcast encryption [10], through which the data user obtains the encrypted search query $\widetilde{Q}$. Upon the receipt of $\widetilde{Q}$, the cloud server starts searching the index tree $\mathcal{I}$ and will return the corresponding set of encrypted documents, which have been well-ranked by our frequency based similarity measures (as will be introduced shortly).

An additional feature is that the data user may not want to receive all the relevant documents. Instead, the data user may send a search parameter k along with the search query $\widetilde{Q}$ such that the cloud server only returns the top-k most relevant documents. The capability of the user to decrypt the received documents [15,33] is a separate issue and is out of the scope of this paper.

2.2 Threat Model

We assume that the data user is honest but that the cloud server acts in an "honest-but-curious" manner, which is also employed by related works on secure cloud data search [5, 29]. In other words, the cloud server honestly follows the protocol execution, but curiosity propels him/her to the speculation and analysis over the data and searchable index tree available at the server. Depending on the available information to the cloud server, two threat models are considered here.

Known Ciphertext Model: Only the encrypted document set $\mathcal{C}$, searchable index tree $\mathcal{I}$ and encrypted query vector $\widetilde{Q}$, all of which are outsourced from the data owner, are available to the cloud server. Specifically, we intend to protect the plaintext query/index information against the cloud server and keep the dictionary $\mathcal{T}$ as secret that was used to build the searchable index tree $\mathcal{I}$.

Known Background Model: In this stronger model, the cloud server is equipped with more knowledge than what can be accessed in the known ciphertext model. In particular, the attacker may extract the statistical information from a known comparable dataset which bears the similar nature to the targeting dataset, e.g., the TF distribution information of a specific keyword. Given such statistical information, the cloud server is able to launch statistical attack to deduce/identify specific keywords in the query [28,29,34].

2.3 Design Goals

To enable effective, efficient and secure multi-keyword ranked search over encrypted cloud data under the aforementioned models, our mechanism is aiming to achieve the following design goals.

Accuracy-improved Multi-keyword Ranked Search: To design an encrypted cloud data search scheme which not only supports the effective *multi-keyword search* functionality, but also, by adoption of the vector space model, achieves the accuracy-improved *similarity-based search result ranking*.

Search Efficiency: Instead of linear search [5], we explore a tree-based index structure and an efficient search algorithm to achieve better practical search efficiency.

Privacy Goals: The general goal is to protect user privacy by preventing the cloud server from learning information of the document set, the index tree, and the queries. In particular, search privacy requirements that we are concerned with are 1) *Index Confidentiality*: the underlying plaintext information pertaining to the encrypted index tree, e.g., keywords and TF of keywords; 2) *Query Confidentiality*: the plaintext information regarding the encrypted query, e.g., keywords in the query and document frequency (DF) of these keywords; 3) *Query Unlinkability*: whether two or more encrypted queries are from the same search request; 4) *Keyword Privacy*: the identification of specific keyword in the index tree, in the query or in the document set. Note

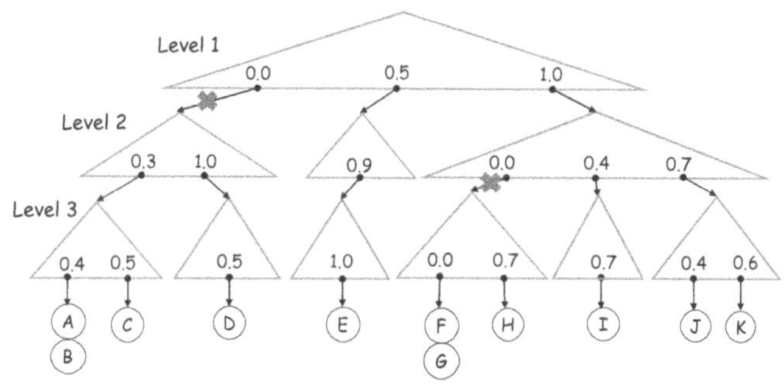

Figure 2: Illustration of the MD-algorithm on the MDB-tree

that protecting *access pattern*, i.e., the sequence of returned documents, is extremely expensive since the algorithm has to "touch" the whole document set [7]. We do not aim to protect it in this work for efficiency concerns.

2.4 Notations

For the sake of clarity, we introduce the main notations used in this paper.

- $\mathcal{DC}$ – the plaintext document collection, denoted as a set of m documents $\mathcal{DC} = \{d|d_1, d_2, \ldots, d_m\}$.
- $\mathcal{C}$ – the encrypted form of $\mathcal{DC}$ stored in the cloud server, denoted as $\mathcal{C} = \{c|c_1, c_2, \ldots, c_m\}$.
- $\mathcal{T}$ – the dictionary, composed of n keywords, denoted as $\mathcal{T} = \{t|t_1, t_2, \ldots, t_n\}$.
- $\bar{\mathcal{T}}$ – a subset of $\mathcal{T}$, indicating the keywords in a search request.
- $\mathcal{I}$ – the h-level searchable index tree for the whole document set $\mathcal{DC}$. Each level corresponds to an exclusive subset of keywords. Documents c_i's are associated with the leaf nodes.
- $\mathcal{T}_i$ – a subset of $\mathcal{T}$, which constitutes the ith level of $\mathcal{I}$, $i = 1, \ldots, h$.
- D_d – the index vector of document d for all the keywords in $\mathcal{T}$.
- Q – the query vector for the keyword set $\bar{\mathcal{T}}$.
- $\widetilde{D_d}$ – the encrypted form of D_d.
- $\widetilde{Q}$ – the encrypted form of Q.

2.5 Preliminaries

Vector Space Model: Among many similarity measures in plaintext information retrieval, vector space model [30] is the most popular one, supporting both conjunctive search and disjunctive search. Specifically, document rankings are realized by comparing the deviation of angles, i.e., cosine values, between each document vector and the query vector. The cosine measure allows accurate rankings due to the "TF×IDF rule", where TF denotes the occurrence count of a term within a document (it is used to measure how important a specific term is to a particular document), and IDF is obtained by dividing the total number of documents in the collection by the number of documents containing the term (it implies that this frequency of a term tends to be inversely proportional to its ranking). We adopt the similarity evaluation function for cosine measure from [30], where

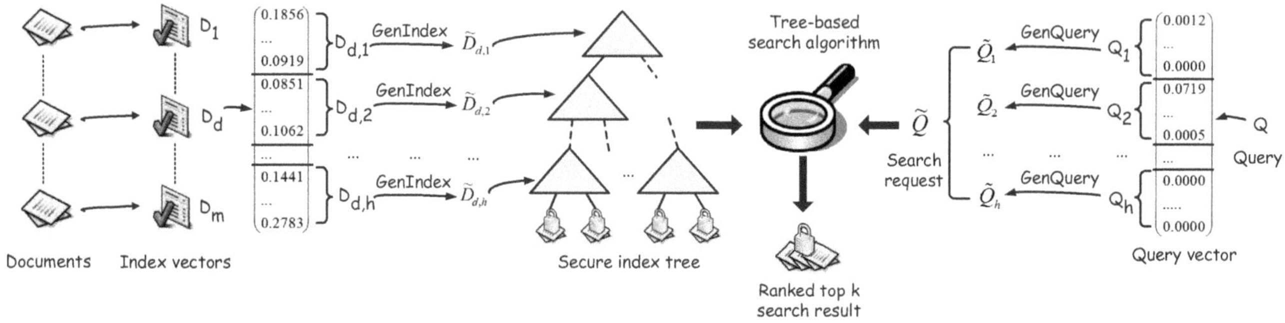

Figure 3: Overview of secure index scheme

the calculation methods for TF weight and IDF weight are equally effective compared to others [35]. The following statistical values are used in our similarity evaluation function:

- $f_{d,t}$, the TF of the keyword t within the document d;
- f_t, the number of documents containing the keyword t;
- N, the total number of documents in the document set;
- $w_{d,t}$, the TF weight for $f_{d,t}$;
- $w_{q,t}$, the IDF weight (query weight);
- W_d, the Euclidean length of $w_{d,t}$;
- W_q, the Euclidean length of $w_{q,t}$.

The definition of the similarity function is as follows:

$$Cos(D_d, Q) = \frac{1}{W_d W_q} \sum_{t \in Q \cap D_d} w_{d,t} \cdot w_{q,t} \quad (1)$$

where $W_d = \sqrt{\sum_{t \in Q \cap D_d} w_{d,t}^2}$, $W_q = \sqrt{\sum_{t \in Q \cap D_d} w_{q,t}^2}$, $w_{d,t} = 1 + \ln(f_{d,t})$, $w_{q,t} = \ln(1 + \frac{N}{f_t})$, and hence, the index vector D_d and query vector Q are both unit vectors.

MD-algorithm: The MD-algorithm [19] is used to find the k-best matches in a database that is structured as an MDB-tree [22], as shown in Fig. 2. In the database scenario, each level of the MDB-tree represents an attribute domain and each attribute in that domain is assigned an attribute value. All the attributes sharing the same value in the upper domain forms a child node. As such, a set of objects is allowed to be indexed in one data structure. An important search parameter, the prediction threshold value $\hat{P}_i$ for each level i, is obtained from the maximum attribute value P_i at each level, for example, in Fig. 2, $\hat{P}_i = P_i = 1.0$. In a depth-first manner, MD-algorithm starts from the root node with a recursive procedure upon this tree. Specifically, search process selects the unused maximum attribute value when it enters a node, and based on $\hat{P}_i$'s below this level, predicts the maximum possible final score to be obtained. The criteria for node selection is that if this predicted final score is less than or equal to the minimum score of the top-k objects which have been selected, search process returns to the parent node, otherwise, it goes down to the child node at the next level. This procedure is executed recursively until the objects with top-k scores are selected. The search can be done very efficiently due to the relatively accurate final score prediction, and thus only part of the objects in the tree are accessed. Fig. 2 shows an example that, when $k = 3$, the set of objects, E, K, and J, are returned to the user and the cross signs in the figure indicate that it is not necessary to access the nodes below. More details of the MD-algorithm and MDB-tree can be found in [19].

3. SECURE INDEX SCHEME

To achieve accurate multi-keyword ranked search, we adopt the cosine measure to evaluate similarity scores. In particular, we divide the original long document index vector D_d into multiple sub-vectors such that each sub-vector $D_{d,i}$ represents a subset of keywords $\mathcal{T}_i$ of $\mathcal{T}$, and becomes a part of the ith level of the index tree $\mathcal{I}$, as shown in Fig. 3. The query vector Q is divided in the same way D_d is done. Let Q_i be the query sub-vector at the ith level. As such, the final similarity score for document d can be obtained by summing up the scores from each level. Based on these similarity scores, the cloud server determines the relevance of document d to the query Q and sends the top-k most relevant documents back to the user. By using the level-wise secure inner product scheme, similar to the techniques applied in [5, 31], the document index vector $D_{d,i}$ and the query vector Q_i are both well protected, and we show that this basic scheme is secure in the known ciphertext model. To further protect the sensitive frequency information from leakage, we also propose an enhanced scheme in the known background model.

3.1 BMTS in Known Ciphertext Model

In order to facilitate the relevance rankings, the similarity scores, i.e., cosine values, are revealed to the cloud server, which differs from the schemes adopted in [5, 31]. In other words, we do not apply the dimension extension technique to our basic scheme in the known ciphertext model. For each level i of $\mathcal{I}$, our basic secure index scheme can be described as follows:

- **Setup** In this initialization phase, the secret key SK_i is produced by the data owner, including: 1) a $|\mathcal{T}_i|$-bit randomly generated vector S_i, where $|\mathcal{T}_i|$ is the length of $\mathcal{T}_i$; 2) two $(|\mathcal{T}_i| \times |\mathcal{T}_i|)$ invertible random matrices $\{M_{1,i}, M_{2,i}\}$. Hence, SK_i can be denoted as a 3-tuple $\{S_i, M_{1,i}, M_{2,i}\}$.
- **GenIndex** $(\mathcal{DC}, SK_i)$ For each document d, the data owner generates an index vector $D_{d,i}$ according to $\mathcal{T}_i$, and each dimension is a normalized TF weight $w_{d,t}$.

Next, the splitting procedure is applied to $D_{d,i}$, which splits $D_{d,i}$ into two random vectors as $\{D_{d,i}', D_{d,i}''\}$. Specifically, with the $|\mathcal{T}_i|$-bit vector S_i as a splitting indicator, if the jth bit of S_i is 0, $D_{d,i}'[j]$ and $D_{d,i}''[j]$ are set as the same as $D_{d,i}[j]$; if the jth bit of S_i is 1, $D_{d,i}'[j]$ and $D_{d,i}''[j]$ are set to two random numbers so that their sum is equal to $D_{d,i}[j]$. Finally, the encrypted index vector $\widetilde{D_{d,i}}$ is built as $\{M_{1,i}^T D_{d,i}', M_{2,i}^T D_{d,i}''\}$.

- **GenQuery**$(\mathcal{T}, SK_i)$ With the keywords of interest in $\mathcal{T}$, the query vector Q_i is generated, where each dimension is a normalized IDF weight $w_{q,t}$ ($w_{q,t} = 0$ for any keyword t not present in Q_i). Subsequently, Q_i is split into two random vectors as $\{Q_i', Q_i''\}$ with the similar splitting procedure. The difference is that if the jth bit of S_i is 0, $Q_i'[j]$ and $Q_i''[j]$ are set to two random numbers so that their sum is equal to $Q_i[j]$; if the jth bit of S_i is 1, $Q_i'[j]$ and $Q_i''[j]$ are set as the same as $Q_i[j]$. Finally, the encrypted query vector $\widetilde{Q_i}$ is yielded as $\{M_{1,i}^{-1} Q_i', M_{2,i}^{-1} Q_i''\}$.
- **SimEvaluation** $(\widetilde{D_{d,i}}, \widetilde{Q_i})$ The cloud server executes similarity evaluation with query vector $\widetilde{Q_i}$ as in Eq. 2.

The similarity score at the ith level is computed as follows:

$$\begin{aligned}
&Cos(\widetilde{D_{d,i}}, \widetilde{Q_i}) \\
=&\{M_{1,i}^T D_{d,i}', M_{2,i}^T D_{d,i}''\} \cdot \{M_{1,i}^{-1} Q_i', M_{2,i}^{-1} Q_i''\} \\
=&D_{d,i}' \cdot Q_i' + D_{d,i}'' \cdot Q_i'' \\
=&D_{d,i} \cdot Q_i.
\end{aligned} \quad (2)$$

Hence, the final similarity score for document d is $\sum_{i=1}^{h} D_{d,i} \cdot Q_i = D_d \cdot Q$.

Security Analysis We analyze BMTS concerning the search privacy requirements as described in section 2.

1) Index confidentiality and Query confidentiality: In BMTS, $\widetilde{D_{d,i}}$ and $\widetilde{Q_i}$ are obfuscated vectors. As long as the secret key SK_i is kept confidential, the cloud server cannot infer the original vectors $D_{d,i}$ or Q_i. Neither can it deduce the keywords nor the TF and IDF information included in the documents or queries from the result similarity scores, which appear to be random values to the server. This has been proven in the known ciphertext model in [31]. Therefore, *index confidentiality* and *query confidentiality* are well protected.

2) Query unlinkability: The adopted vector encryption method provides non-deterministic encryption, in light of the random vector splitting procedure. Thus the same search request (e.g. same search keywords) will be encrypted to different query vector $\widetilde{Q}$. The non-linkability of search requests can be provided to this extent. However, if a cloud server is capable of tracking the nodes visited and the intermediate similarity results, it is possible for the cloud server to link the same search request based on the same similarity scores. In this case the *search pattern* or the *access pattern* will be leaked even in the known ciphertext model.

3) Keyword privacy: In the known background model, the cloud server may have the knowledge of not only the TF distributions, but also the normalized TF distributions of some sensitive keywords from a known comparable dataset.

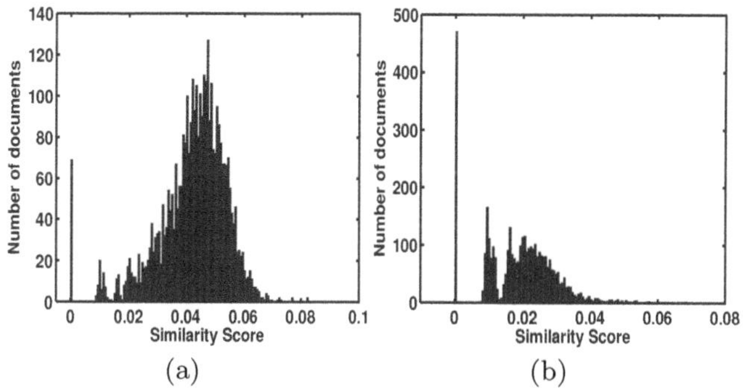

Figure 4: Distribution of similarity score when a single keyword in a query vector with our basic scheme. (a) For keyword "network". (b) For keyword "search".

It is worth noting that these distributions are keyword specific respectively, as shown in Fig. 4, such that the corresponding keywords can be differentiated by the slope and value range of these distributions [28, 29, 34]. In the worst case, where only one keyword t appears in the query vector Q (the normalized $w_{q,t}$ is 1), the normalized TF distribution of this keyword is exposed directly.

In order to enhance security and boost search efficiency, search evaluation may be only executed at certain levels where the user-intended keywords reside; for the other levels, we can render these similarity scores some fixed values, e.g., 0, during the execution of both similarity score prediction and evaluation in the search process.

3.2 EMTS in Known Background Model

The previous security analysis shows that in the known background model, *keyword privacy* breach is possible, due to the distance-preserving property of BMTS, i.e., the cosine value calculated from $\widetilde{D_{d,i}}$ and $\widetilde{Q_i}$ is equal to the one from $D_{d,i}$ and Q_i. In order to break such equality, we introduce some tunable randomness into the similarity score evaluation, by which the cloud server cannot differentiate keywords from the particular similarity score distributions. In addition, this randomness can be calibrated by the user to represent the user's preference for the more accurate ranked search result versus better-protected *keyword privacy*. Specifically, U_i phantom terms are added into the query vector Q_i, and we extend the index vector $D_{d,i}$ from $|\mathcal{T}_i|$ dimensions to $|\mathcal{T}_i| + U_i$ dimensions. We denote the subset of h levels where the keywords of interest reside as w and its size $|w| \leq h$.

Our EMTS scheme is performed almost the same as BMTS, except that at the ith level: 1) in Setup phase, S_i becomes $(|\mathcal{T}_i| + U_i)$-bit long. $M_{1,i}$ and $M_{2,i}$ are $(|\mathcal{T}_i| + U_i) \times (|\mathcal{T}_i| + U_i)$ dimensional matrices; 2) in GenIndex phase, by choosing V_i out of U_i phantom terms, the corresponding entries in the $(|\mathcal{T}_i| + U_i)$-dimensional index vector $D_{d,i}$ are set to 1; 3) when generating encrypted query, the $(|\mathcal{T}_i| + j)^{th}$ entry in Q_i where $j \in [1, U_i]$ is set to a random number $\varepsilon_{i,j}$; 4) The cloud server executes similarity evaluation and obtains the final similarity score for document d equal to $(D_d \cdot Q + \sum_{i \in w} \sum_{j \in \bar{V}_i} \varepsilon_{i,j})$, where $\bar{V}_i$ is the set of the V_i selected phantom terms, and it is different for each index vector at level i.

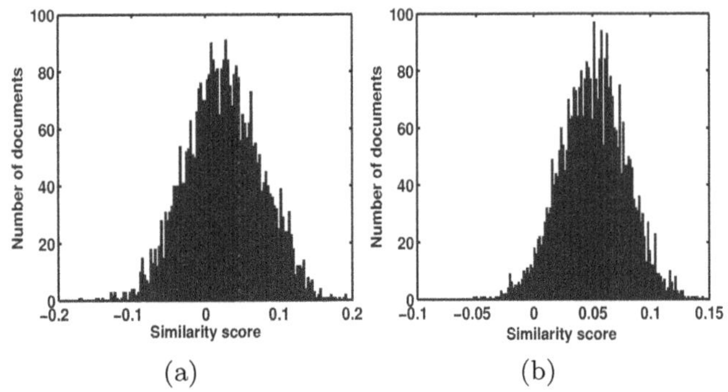

Figure 5: Distribution of similarity score for keyword "network" with different standard deviations, 3500 documents, only one keyword "network" in the query, in our enhanced scheme. (a) $\sigma = 0.05$. (b) $\sigma = 0.03$.

Security Analysis We analyze EMTS again with respect to the aforementioned search privacy requirements.

1) Index confidentiality and Query confidentiality: EMTS can protect *index confidentiality* and *query confidentiality* in both the known ciphertext model and the known background model, which is inherited from BMTS.

2) Query unlinkability: The introduction of randomly generated $\varepsilon_{i,j}$ will allow EMTS to produce different similarity scores even for the same search request. The value of $\varepsilon_{i,j}$ can be adjusted to control the level of variance thus the level of unlinkability. It is worth noting that this query-side randomization technique significantly differs from [5], where randomization occurs on the index vector side and is not possible to be tweaked as an effective privacy-preserving parameter for users. Query unlinkability is thus much enhanced compared with BMTS to the extent that there is no easy way for the attacker to link the queries. However, since we do not intend to protect access pattern for efficiency reasons, the returned results from the same request will always bear some similarity which could be exploited with powerful statistical analysis by the very motivated cloud server. This is a trade-off that one has to make between efficiency and privacy.

3) Keyword privacy: At level i, the number of the index vectors is denoted as l_i. On one hand, to render separate $\sum_{j\in\bar{V}_i} \varepsilon_{i,j}$ of this level the different values for one search request, we set $\binom{U_i}{V_i} \geq l_i$. On the other hand, $\sum_{j\in\bar{V}_i} \varepsilon_{i,j}$ for the same $D_{d,i}$ is different with multiple search requests, since $\varepsilon_{i,j}$ is generated uniformly at random upon each search request. The cloud server cannot eliminate the impact of these phantom terms from the final similarity scores without the exact values of them. Furthermore, every $\varepsilon_{i,j}$ follows the same uniform distribution $M(\mu' - c, \mu' + c)$, where the mean is μ' and the variance as σ'^2 is $c^2/3$. According to the central limit theorem, the $\sum_{i\in w} \sum_{j\in\bar{V}_i} \varepsilon_{i,j}$ follows the Normal distribution $N(\mu, \sigma^2)$, where the mean as μ is $\sum_{i\in w} V_i\mu'$ and the variance as σ^2 is $\sum_{i\in w} V_i c^2/3$. Therefore, we may generate $\varepsilon_{i,j}$ with the value of μ' as $\mu/\sum_{i\in w} V_i$ and the value of c as $\sqrt{3/\sum_{i\in w} V_i} \cdot \sigma$. As shown in Fig. 5, with larger σ selected by the user, it is more difficult for the cloud server to infer the corresponding statistical information, and further reverse-engineer the keyword, from the well-obfuscated distribution of the similarity score. However, it does not suffice to protect *keyword privacy*. For simplicity, we assume that there are 2 levels, i.e., only one keyword t at level 1 and two or more other keywords at level 2. At level 1, $\sum_{j\in\bar{V}_1} \varepsilon_{1,j}$ does not follow the Normal distribution with σ selected by the user, in that with the smaller V_1, the value of ${\sigma_1}^2$ as $V_1 c^2/3$ is smaller than the value of σ^2 as $(V_1 + V_2)c^2/3$. It is possible that σ_1 is too small to obfuscate the distribution of the similarity score, so that the cloud server may identify the keyword t at level 1. For better protecting *keyword privacy*, the user chooses an appropriate σ_1, i.e., large enough to obfuscate the distribution, to generate $\varepsilon_{1,j}$ accordingly, while σ remains as the overall search parameter. Hence, the variance ${\sigma_2}^2$ for level 2 can be set as $\sigma^2 - {\sigma_1}^2$ and $\varepsilon_{2,j}$ is generated accordingly, as the normal random variables $\sum_{j\in\bar{V}_1} \varepsilon_{1,j}$ and $\sum_{j\in\bar{V}_2} \varepsilon_{2,j}$ are independent to each other. Finally, *keyword privacy* can be well protected by these phantom keywords.

Remarks Recently Yao et al. [32] find that this underlying encryption method [31] is susceptible to chosen plaintext attack. However, it is not applicable under our defined threat models, since in order to launch such attack, the cloud server has to acquire plaintext query information, i.e., the normalized IDF weights, which are *only* possessed by the data owner and protected by BMTS and EMTS.

4. EFFICIENCY OF THE TREE-BASED SEARCH ALGORITHM

In the plaintext information retrieval community, many well-developed techniques have been adopted to accelerate the search process, e.g., inverted index [18], B-tree [9], etc. However, in the ciphertext scenario, they cannot be implemented in a straightforward manner. In [10, 28, 29, 34], the inverted index based search methods are employed to achieve an extremely efficient search process. However, these schemes are only designed for single keyword search. Efficient range search in database [17] can be realized by using B^+-tree, but it is not applicable to the text search scenario. The similarity score in our scheme is a value depending on the query and has to be evaluated in the runtime, which makes the fixed tree structures, such as B-tree or B^+-tree, not suitable here. In this paper, we propose a tree-based search algorithm, which is adapted from MDB-tree based MD-algorithm, to enable efficient multi-keyword ranked search. In what follows, we briefly introduce our tree-based search algorithm and present some experimental results from our implementation of the proposed tree-based search algorithm on a real-world document set: the recent ten years' INFOCOM publications. We identify key factors that affect the search efficiency and propose strategies in building the index tree that effectively speed up the search process.

4.1 Tree-based Search Algorithm

The MD-algorithm is originally designed for plaintext database search. In the case of privacy-preserving similarity-based multi-keyword ranked text search, it cannot be applied in a straightforward manner. Instead of a numerical "attribute value" for each attribute in the MDB-tree, our index tree structure has to be built on vectors. The secure index scheme described in section 3 is for this purpose and it enables the

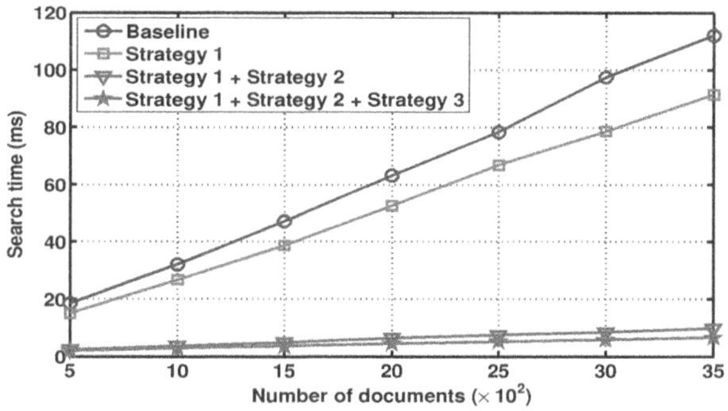

Figure 6: Comparison of search efficiency with different efficiency-improving strategies

search algorithm to take the inputs of the encrypted searchable index tree and the encrypted query, and ensures that the search algorithm is conducted in a secure way to protect important search privacy in the whole search process.

Another remarkable difference between our search algorithm and MD-algorithm is that we cannot set $\hat{P}_i$ to P_i as running the MD-algorithm in database scenario, since P_i varies for queries in our scenario and has to be securely evaluated (as described in section 3) in the runtime. The pseudo code for our tree-based search algorithm is presented in Appendix.

4.2 Impact of Prediction Threshold Value

An important factor that affects the search efficiency is the prediction threshold value $\hat{P}_i$ at each level i. To ensure the search precision, $\hat{P}_i \geq P_i$ should hold where P_i is the maximum similarity score at level i. As shown in Tab. 1, the tighter the prediction value of $\hat{P}_i$, the higher the search efficiency. The reason is that the search process can be terminated earlier without going into unnecessary nodes. On the other hand, when $\hat{P}_i < P_i$, the search precision (a quantitative measure for search accuracy, cf. section 5) drops below 100% while the rank privacy (a privacy measure. cf. section 5) increases.

Strategy 1 Based on this observation, our first efficiency enhancement aims to produce a better estimation of $\hat{P}_i$ that approximates to its ideal value P_i. We propose the following strategy to achieve this. During the index tree generation phase, the data owner retains a vector E_i for each level i. This vector consists of the maximum values at each dimension among all the indexes at this level. Subsequently, during the query generation phase, $\hat{P}_i$ is equal to the inner product of E_i and Q_i, and $\hat{P}_i$ will be set to 1 if it is greater than 1, thus $P_i \leq \hat{P}_i \leq 1$. $\hat{P}_i$ can be taken as an additional search parameter to be sent with $\widetilde{Q_i}$ to the cloud server. As for EMTS, we add the maximum $\sum_{j \in \bar{V}_i} \varepsilon_{i,j}$ from Q_i to $\hat{P}_i$, and refer to this sum as the final prediction threshold value. In Fig. 6, it is shown that the search efficiency is improved with this strategy, compared to the baseline search, in which $\hat{P}_i$ is always set to 1, the upper bound of cosine function.

4.3 Impact of Intended Keyword Position

Another factor we observed that affects the search efficiency is the position of the search keywords on the index tree. As shown in Tab. 2, the higher level the intended keywords reside, the higher the search efficiency. This is very different from using the MD-algorithm in database scenario where all the attributes are involved in searching the relevant objects. In the text search scenario, people are likely to complete a search with a query only comprising five keywords or less [1]. Consequently, the search algorithm needs to go through a larger number of nodes to evaluate an intended keyword if it resides at a lower level.

Strategy 2 The insight from this observation is that the average search time can be improved by strategically arranging keyword position in the index tree – the most frequently searched keywords on the top levels. In our experiment, we collected a set of 100 search requests from the volunteering users of this prototype secure search system. We then build the index tree where keywords are re-ordered by their search popularity. In practice, the information on the search keyword distribution can be extracted from the user search history. As shown in Fig. 6, the efficiency of the search algorithm is ameliorated significantly when applying this strategy.

Table 1: Impact of prediction threshold

$\hat{P}_i$	Time (ms)	# of accessed nodes	Precision	Rank privacy
1	33	17007	100%	0%
0.05	32	15012	100%	0%
0.02	28	14326	90.3%	6.5%
0.01	11	6410	10.6%	87.2%

Table 2: Impact of keyword position

Keyword position	Time (ms)	# of accessed nodes
1st	5	491
15th	32	13304
30th	33	23844
50th	36	49001

Table 3: Impact of clustering

Ed	Time (ms)	# of accessed nodes	Precision	Rank privacy
0.02	7.4	7062	100%	0%
0.05	6.24	6728	97.00%	2.1%

4.4 Impact of Index Vector Clustering

Another idea for improving the search efficiency is to cluster "similar" index vectors, as shown in Tab. 3. The improved efficiency comes from the reduced number of accessed nodes in the index tree, but at the expense of lower search precision. The bigger each cluster is, the higher the search efficiency, but the lower the search precision.

Strategy 3 To maximize the possibility of clustering, the length of the index vector at each level should be as short as possible (but at least achieve 80-bit symmetric key security [31]) in order to group the "similar" indexes. Inspired by the k-means method, which is the most widely used clustering technique in the data mining community [21], we use Euclidean distance (Ed) as a metric to cluster "close enough" vectors, e.g., when $Ed < 1$. For EMTS, we may first cluster original index vectors, and then execute dimension extension. The time cost for the search scheme after combining all the three efficiency-improving strategies, where $Ed = 0.02$, is shown in Fig. 6 as well.

Remarks The original combination of the MD-algorithm

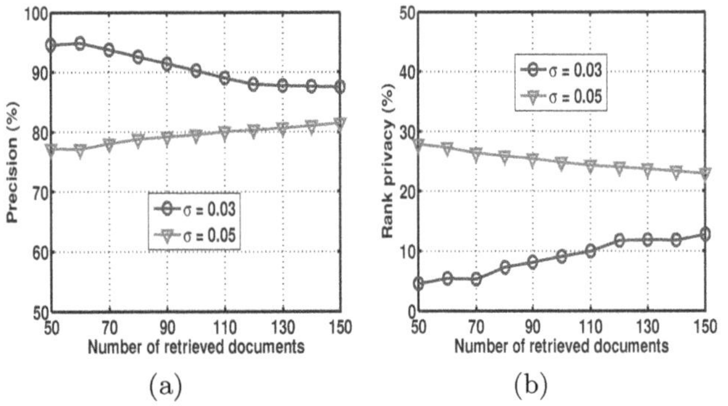

Figure 7: By choosing different standard deviation σ, the trade-off, between (a) Precision, and (b) Rank privacy, can be achieved.

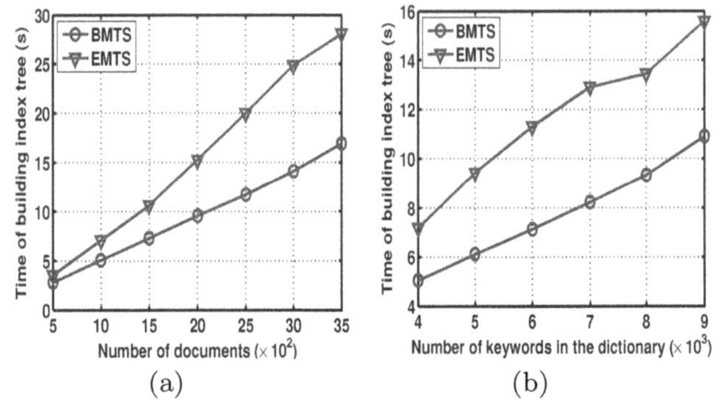

Figure 8: Time cost for building index tree. (a) For the different size of document set with the same dictionary, $n = 4000$. (b) For the different size of dictionary with the same document set, $m = 1000$.

and the MDB-tree is not directly applicable for efficient search over vector indexes. From the extensive experiments on our prototype implementation of the search algorithm, we identified three efficiency-crucial factors and proposed effective strategies to improve the practical search efficiency with our vector indexes. Although the worst case search complexity is no better than linear search that is the state-of-the-art search efficiency in the multi-keyword encrypted text search scenario, this much less time-consuming tree-based search algorithm represents a solid step forward on the utilization of encrypted cloud data in practice. From the security point of view, the entire search process does not introduce new privacy vulnerability when used with our secure index scheme. In particular, our scheme is secure against search time analysis, i.e., the cloud server cannot infer specific keyword by the difference of search time, even if he/she knows that the keyword resides at a certain level. In fact, for efficiency, the cloud server performs level-selected similarity evaluation (see section 3), such that he/she has already possessed the keyword position information. There are at least 80 dimensions in an index vector at each level, and all the words falling into this level have almost the same search time. This effectively blinds one keyword within at least 79 other keywords at the same level. In addition, the cloud server has no knowledge about which concrete set of keywords are selected to build this level without the dictionary $\mathcal{T}$. Therefore, it is not possible to differentiate these keywords or identify a particular keyword of interest.

5. PERFORMANCE ANALYSIS

To evaluate the overall performance of our proposed techniques, we implemented the entire secure search system using JAVA on a Linux Server with Intel Core i3 Processor 3.3GHz. The document set is built from the recent ten years' IEEE INFOCOM publications, including about 3600 publications, from which we extract about 9000 keywords. In this section we present the detailed performance result. The documents and keywords used in the evaluation are selected randomly from the created document sets.

5.1 Precision and Privacy

To evaluate the impact on the accuracy of search result introduced by phantom terms in EMTS, we adopt the definition of "precision" in [5]. Namely, the "precision" of a top-k search is defined as $P_k = k'/k$ where k' is the number of the real top-k documents that are returned by the cloud server. Fig. 7(a) shows that with a small σ, the effectiveness of the search scheme is not affected much. The user can still enjoy almost the same search result as BMTS. On the other hand, we evaluate the "rank privacy" obtained from introducing phantom terms. The definition of "rank privacy" is also adopted from [5], i.e., the rank privacy at point k is calculated as $\widetilde{P_k} = \sum \widetilde{p_k}/k^2$. For every document d in the returned top-k documents, let the rank perturbation $\widetilde{p_k}$ be $|u_d - u_d'|$, where u_d is the rank number of document d in the returned top-k documents and it is set to k if greater than k, and u_d' is its rank number in the real ranked documents. As shown in Fig. 7(b), large σ provides better protection of rank information in EMTS.

It is worth noting that σ is a tunable search parameter at the discretion of the user. The selection of different σ reflects his/her predilection for the better effectiveness of the search scheme or the better protected rank privacy and *keyword privacy* (see section 3.2).

Table 4: Size of index tree

Size of dictionary	2000	4000	6000	8000
BMTS (MB)	23.16	46.34	69.51	92.68
EMTS (MB)	26.87	53.77	80.66	107.54

5.2 Construction for Index Tree

Apparently, the time cost for generating the proposed index tree structure mainly depends on the size of the document set $\mathcal{DC}$ and dictionary $\mathcal{T}$. At a particular level i, the major computation is the encryption of the divided index vectors, which involves the splitting process and two multiplications of an $n_i \times n_i$ matrix and an n_i-dimensional vector where $n_i = |\mathcal{T}_i|$ in BMTS and $n_i = |\mathcal{T}_i| + U_i$ in EMTS. Fig. 8(a) shows that the time cost for building the index tree is nearly linear to the number of the documents, given the same dictionary, i.e., n_i is fixed at each level i when the tree structure has a predefined h levels. Fig. 8(b) shows that with the same document set, the index construction time is proportional to the number of keywords in the dictionary. Besides, building the index tree for EMTS is slightly more time-consuming than BMTS due to the dimension extension, which is shown in both Fig 8(a) and Fig 8(b). Note

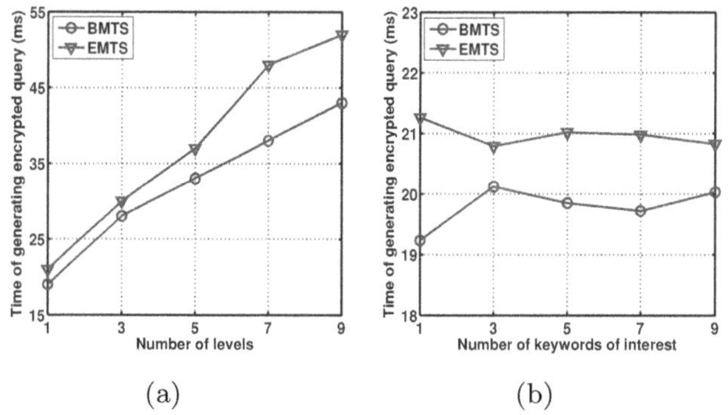

Figure 9: Time cost for generating encrypted query, when $|\mathcal{T}_i| = 80$. (a) For the different number of levels where user intended keywords reside. (b) For the different number of keywords of interest in one level.

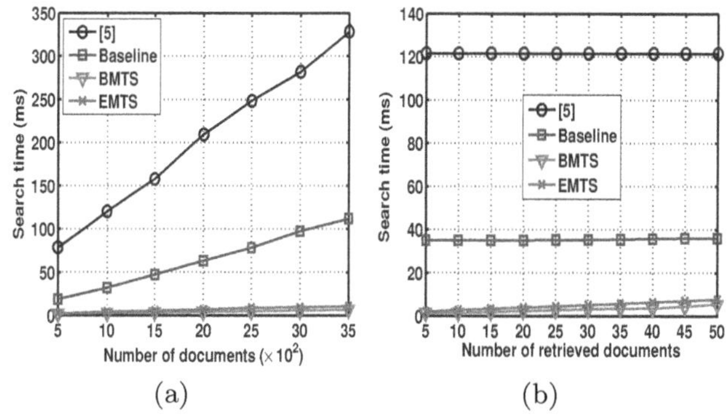

Figure 10: Search efficiency, with the same 10 keywords of interest. (a) For the different size of document set with the same dictionary, $n = 4000$, $k = 10$. (b) For the different number of retrieved documents with the same document set and dictionary, $m = 1000$, $n = 4000$.

that this computation is a one-time cost on the data owner side. On the other hand, the storage overhead, listed in Tab. 4, of the index tree with the fixed size of document set m=3000 is determined by the number of keywords in the dictionary. EMTS is very close to BMTS in the size of the index tree, and considering the massive storage capacity and low storage cost in the cloud, it is practical and completely affordable.

5.3 Query Generation

Typical search requests only consist of a few keywords [1]. Therefore, it is inefficient to generate a query that involves all the keywords in the dictionary, which not only makes the search process costly (similarity evaluation needs to be carried out on long vectors) but also brings some unnecessary computation overhead into query generation procedure. Fig. 9(a) demonstrates that when $|\mathcal{T}_i|$ is fixed, the time cost for generating an encrypted query is only linear to the number of levels where the searched keywords reside. Furthermore, the strategy to place the most frequently searched keywords on the top levels of the index tree will lead to the result that a good portion of queries are only generated for a few limited levels. As such, when $|\mathcal{T}_i|$ is chosen properly, the average query generation can be extremely efficient, as shown in Fig. 9(b). The major computation overhead stems from the vector encryption process, and due to the dimension extension, the time cost in EMTS is a bit more expensive than that in BMTS.

5.4 Search Efficiency

The search process executed at the cloud server is comprised of computing and ranking the similarity scores of relevant documents. The search algorithm terminates after the top-k documents have been selected. We evaluate the search efficiency of BMTS and EMTS with our proposed efficiency-improving strategies. Fig. 10(a) shows the search time for BMTS and EMTS, compared with [5] and baseline search with respect to the size of document set. In the baseline search, all the keywords in the dictionary are dispersed uniformly at random within the index tree, but it is still far efficient than [5] due to our proposed search algorithm and the tree-based index structure. It is noteworthy that the time cost of our proposed encrypted cloud data search is much more efficient than [5] and baseline search. In addition, with the increased size of document set, our two proposed schemes enjoy almost the same and nearly constant search time. Fig. 10(b) demonstrates that when user requests more relevant documents, our search algorithm is still extremely efficient.

6. RELATED WORK

6.1 Searchable Encryption with Single Keyword

Song et al. [27] propose the first SE scheme, where, to search a certain keyword in a document, user has to go through the whole document. After this work, many improvements and novel schemes [6,10,11,16] have been proposed. Curtmola et al. [10] propose an inverted index based SE scheme with extremely efficient search process, but the *keyword privacy* will be revealed if the corresponding keywords have been searched. Frequency information is not involved in the similarity evaluation processes of the above techniques to provide accurate search functionality. In [28, 29,34], the order-preserving techniques are utilized to protect the rank order. Due to the index and query built from frequency related information and the inverted index as the underlying index structure, they can achieve accurate and efficient search at the same time. Boneh et al. [3] propose the first PKC-based SE scheme, where anyone with public key can write to the data stored on server but only authorized users with private key can search. However, all of the aforementioned solutions only support single keyword search.

6.2 Searchable Encryption with Multiple Keywords

In the public key setting, a lot of works have been done to realize the conjunctive keyword search, subset search, or range queries [4,12,13], but they are too computationally intensive to be implemented for practical use. Predicate encryption is another promising technique to fulfill the search over encrypted data [2,23,25]. In [17], a logarithmic-time search scheme is presented to support range queries, which is orthogonal to our text search scenario. Furthermore, no similarity measure is adopted in these works to provide multi-keyword ranked search functionality. In text retrieval scenario, Pang et al. [20] also propose a vector space model based secure search scheme. An access manager is supposed to exist in their protocol except for a document server, and additional overhead occur on the user side, which is apparently not applicable to cloud environment. Without the security analysis for frequency information in their scheme, it is not clear whether such sensitive information disclosure could lead to *keyword privacy* infringement. Besides, the practical search performance is absent from the demonstration of their experiment. Cao et al. [5] propose a privacy-preserving multi-keyword ranked search scheme. Although with "coordinate matching", this scheme can produce the ranked search result by the number of matched keywords, more accurate ranked search result is not considered there, and the search complexity is constant in that the cloud server has to traverse all the indexes of the document set for each search request.

7. CONCLUSION

In this paper, as an initial attempt to achieve practical and effective multi-keyword text search over encrypted cloud data, we make contributions in two major aspects, supporting similarity-based ranking for more accurate search result and a tree-based search algorithm that achieves better-than-linear search efficiency. For the accuracy aspect, we first exploit the popular similarity measure, i.e., vector space model with cosine measure, to effectively procure the accurate search result. We propose two secure index schemes to meet various privacy requirements in the two threat models. Eventually, the leakage of sensitive frequency information can be avoided. To boost search efficiency, we propose a tree-based index structure for the whole document set. From the utilization of the prototype of our secure search system, we identify three essential efficiency-related factors, by which the efficiency of the search algorithm upon our index tree can be significantly improved. Finally, thorough analysis on the real-world document set demonstrates the performance of BMTS and EMTS in terms of search effectiveness, efficiency and privacy.

Acknowledgment We would like to thank Matúš Ondreička for his helpful comments and sharing the source code of MD-algorithm. This work was supported in part by the NSFC 61272457, the FRFCU K50511010001, the PCSIRT 1078, and the NSF grants CNS-1217889, CNS-1155988 and CNS-1218085.

8. REFERENCES

[1] Keyword and search engines statistics. `http://www.keyworddiscovery.com/keyword-stats.html?date=2013-01-01`, 2013.

[2] N. Attrapadung and B. Libert. Functional encryption for inner product: Achieving constant-size ciphertexts with adaptive security or support for negation. In *Proc. of PKC*, pages 384–402, 2010.

[3] D. Boneh, G. D. Crescenzo, R. Ostrovsky, and G. Persiano. Public key encryption with keyword search. In *Proc. of EUROCRYPT*, pages 506–522, 2004.

[4] D. Boneh and B. Waters. Conjunctive, subset, and range queries on encrypted data. In *Proc. of TCC*, pages 535–554, 2007.

[5] N. Cao, C. Wang, M. Li, K. Ren, and W. Lou. Privacy-preserving multi-keyword ranked search over encrypted cloud data. In *Proc. of IEEE INFOCOM*, pages 829–837, 2011.

[6] Y.-C. Chang and M. Mitzenmacher. Privacy preserving keyword searches on remote encrypted data. In *Proc. of ACNS*, pages 391–421, 2005.

[7] B. Chor, E. Kushilevitz, O. Goldreich, and M. Sudan. Private information retrieval. *Journal of the ACM*, 45(6):965–981, 1998.

[8] Cloud Security Alliance. Security guidance for critical areas of focus in cloud computing v3.0. `http://www.cloudsecurityalliance.org`, 2011.

[9] D. Comer. Ubiquitous b-tree. *ACM computing surveys*, 11(2):121–137, 1979.

[10] R. Curtmola, J. A. Garay, S. Kamara, and R. Ostrovsky. Searchable symmetric encryption: improved definitions and efficient constructions. In *Proc. of ACM CCS*, pages 79–88, 2006.

[11] E.-J. Goh. Secure indexes. Cryptology ePrint Archive. `http://eprint.iacr.org/2003/216`, 2003.

[12] P. Golle, J. Staddon, and B. R. Waters. Secure conjunctive keyword search over encrypted data. In *Proc. of ACNS*, pages 31–45, 2004.

[13] Y. Hwang and P. Lee. Public key encryption with conjunctive keyword search and its extension to a multi-user system. In *Pairing*, pages 2–22, 2007.

[14] B. Krebs. Payment processor breach may be largest ever. `http://voices.washingtonpost.com/securityfix/2009/01/payment_processor_breach_may_b.html`, 2009.

[15] M. Li, S. Yu, K. Ren, and W. Lou. Securing personal health records in cloud computing: Patient-centric and fine-grained data access control in multi-owner settings. In *Proc. of SecureComm*, pages 89–106, 2010.

[16] P. Liesdonk, S. Sedghi, J. Doumen, P. Hartel, and W. Jonker. Computationally efficient searchable symmetric encryption. *Secure Data Management*, pages 87–100, 2010.

[17] Y. Lu. Privacy-preserving logarithmic-time search on encrypted data in cloud. In *Proc. of NDSS*, 2012.

[18] NIST. NIST's dictionary of algorithms and data structures: inverted index. `http://xlinux.nist.gov/dads/HTML/invertedIndex.html`.

[19] M. Ondreička and J. Pokorný. Extending fagin's algorithm for more users based on multidimensional b-tree. In *Proc. of ADBIS*, pages 199–214, 2008.

[20] H. Pang, J. Shen, and R. Krishnan. Privacy-preserving similarity-based text retrieval. *ACM Transactions on Internet Technology*, 10(1):4, 2010.

[21] A. Rajaraman and J. D. Ullman. Mining of massive datasets. Cambridge University Press, Dec. 2011.

[22] P. Scheuermann and M. Ouksel. Multidimensional b-trees for associative searching in database systems. *Information systems*, 7(2):123–137, 1982.

[23] E. Shen, E. Shi, and B. Waters. Predicate privacy in encryption systems. In *Proc. of TCC*, pages 457–473, 2009.

[24] J. Sheridan and C. Cooper. Defending the cloud. `http://www.reactionpenetrationtesting.co.uk/Defending%20the%20Cloud%20v1.0.pdf`, 2012.

[25] E. Shi, J. Bethencourt, H. Chan, D. Song, and A. Perrig. Multi-dimensional range query over encrypted data. In *Proc. of S&P*, pages 350–364, 2007.
[26] Z. Slocum. Your google docs: Soon in search results? `http://news.cnet.com/8301-17939_109-1035713%207-2.html`, 2009.
[27] D. Song, D. Wagner, and A. Perrig. Practical techniques for searches on encrypted data. In *Proc. of S&P*, pages 44–55, 2000.
[28] A. Swaminathan, Y. Mao, G.-M. Su, H. Gou, A. L. Varna, S. He, M. Wu, and D. W. Oard. Confidentiality-preserving rank-ordered search. In *Proc. of the 2007 ACM Workshop on Storage Security and Survivability*, pages 7–12, 2007.
[29] C. Wang, N. Cao, K. Ren, and W. Lou. Enabling secure and efficient ranked keyword search over outsourced cloud data. *IEEE Transactions on Parallel and Distributed Systems*, 23(8):1467–1479, 2012.
[30] I. H. Witten, A. Moffat, and T. C. Bell. Managing gigabytes: Compressing and indexing documents and images. Morgan Kaufmann Publishing, San Francisco, May 1999.
[31] W. K. Wong, D. W. Cheung, B. Kao, and N. Mamoulis. Secure knn computation on encrypted databases. In *Proc. of SIGMOD*, pages 139–152, 2009.
[32] B. Yao, F. Li, and X. Xiao. Secure nearest neighbor revisited. `http://www.cs.utah.edu/~lifeifei/papers/snnicde.pdf`, to appear in *ICDE*, 2013.
[33] S. Yu, C. Wang, K. Ren, and W. Lou. Achieving secure, scalable, and fine-grained data access control in cloud computing. In *Proc. of IEEE INFOCOM*, pages 1–9, 2010.
[34] S. Zerr, D. Olmedilla, W. Nejdl, and W. Siberski. Zerber+r: Top-k retrieval from a confidential index. In *Proc. of EDBT*, pages 439–449, 2009.
[35] J. Zobel and A. Moffat. Exploring the similarity space. *SIGIR Forum*, 32(1):18–34, 1998.

APPENDIX

The Proposed Tree-based Search Algorithm

The following notations are used in the pseudo code of the algorithm:

- O – it is denoted as a subtree of the searched index tree, or as a document array residing in the leaf node. For the subtree O at the ith level, it can be defined by the tree identifier $(\widetilde{D_{d,1}}, \widetilde{D_{d,2}}, ..., \widetilde{D_{d,i-1}})$.
- x – an encrypted document in a document array O. For simplicity, it could be a document ID.
- S_i – the similarity score that is equal to $Cos(\widetilde{D_{d,i}}, \widetilde{Q_i})$ as Eq. 2.
- $PF(\widetilde{D_{d,i}})$ – the function for predicting the maximum possible final similarity score from index vector $\widetilde{D_{d,i}}$. $PF(\widetilde{D_{d,i}}) = \sum_{t=1}^{i} S_t + \sum_{j=i+1}^{h} \hat{P_j}$, where h is the total number of levels of the index tree.
- $F(\cdot)$ – the final similarity score of a document, e.g., $F(x) = \sum_{i=1}^{h} S_i$.
- L_k – the list for selected top-k documents that are stored in descending order according to $F(x)$.
- DL_i – the list for index vectors $\widetilde{D_{d,i}}$'s stored in a node at the ith level of a subtree of the index tree.
- M_k – the similarity score of the kth document in L_k.

Algorithm 1 Proposed Search Algorithm for Top-k Ranking on Index Tree

```
begin
for (i = 1 → h) do
    if (Q̃_i does not exist in the search request) then
        P̂_i ← 0 and all S_i ← 0 at the i^th level;
    else
        use the received P̂_i with the search request;
    end if
end for
findTopK(I, (∅), F(.), k);
return L_k;
end

procedure findTopK(IndexTree I, Identifier (D̃_{d,1},
        D̃_{d,2}, ..., D̃_{d,i−1}), Score F(·), int k);
if (O is a document array) then
    while (there is next document in O) do
        if (|L_k| < k) then
            insert document x into L_k according to F(x);
        else
            if (F(x) > M_k) then
                delete k^th document from L_k;
                insert x into L_k according to F(x);
            end if
        end if
    end while
else
    DL_i = getIndexList(I, (D̃_{d,1}, D̃_{d,2}, ..., D̃_{d,i−1}));
    counter ← 0;
    while (there is next index vector in the subtree) do
        if (|L_k| = k and FP(DL_i(counter)) ≤ M_k) then
            return;
        else
            findTopK(I, (D̃_{d,1}, ..., D̃_{d,i−1}, D̃_{d,i}), F(·), k);
        end if
        counter ← counter + 1;
    end while
end if
end procedure

procedure getIndexList(IndexTree I, Identifier (D̃_{d,1},
        D̃_{d,2}, ..., D̃_{d,i−1}));
if (Q̃_i exists in the search request) then
    for (each D̃_{d,i} within the node at the i^th level in
        subtree of I with identifier (D̃_{d,1}, D̃_{d,2}, ..., D̃_{d,i−1}))
do
        compute S_i = Cos(D̃_{d,i}, Q̃_i);
    end for
    sort DL_i in descending order according to S_i;
    return DL_i;
else
    return the original DL_i;
end if
end procedure
```

Practical and Post-Quantum Authenticated Key Exchange from One-Way Secure Key Encapsulation Mechanism

Atsushi Fujioka, Koutarou Suzuki, Keita Xagawa, Kazuki Yoneyama
NTT Secure Platform Laboratories
3-9-11, Midori-cho Musashino-shi
Tokyo 180-8585 Japan

ABSTRACT

This paper discusses how to realize practical *post-quantum* authenticated key exchange (AKE) with strong security, i.e., CK^+ security (Krawczyk, CRYPTO 2005). It is known that strongly secure post-quantum AKE protocols exist on a generic construction from IND-CCA secure key encapsulation mechanisms (KEMs) in the standard model. However, when it is instantiated with existing IND-CCA secure post-quantum KEMs, resultant AKE protocols are far from practical in communication complexity. We propose a generic construction of AKE protocols from OW-CCA secure KEMs and prove CK^+ security of the protocols in the random oracle model. We exploit the random oracle and instantiate AKE protocols from various assumptions; DDH, gap DH, CDH, factoring, RSA, DCR, (ring-)LWE, McEliece one-way, NTRU one-way, subset sum, multi-variate quadratic systems, and more. For example, communication costs of our lattice-based scheme is approximately 14 times lower than the previous instantiation (for 128-bit security). Also, in the case of code-based scheme, it is approximately 25 times lower.

Categories and Subject Descriptors

C.2.0 [**Computer-Communication Networks**]: General—*Security and protection*; E.3 [**Data**]: Data Encryption—*Public key cryptosystems*

General Terms

Security

Keywords

authenticated key exchange, CK^+ model, key encapsulation mechanism, random oracle model, post-quantum cryptography, ring-LWE assumption.

1. INTRODUCTION

1.1 Background

Following the celebrated Diffie-Hellman key exchange [17], researchers proposed several cryptographic schemes based on various problems. The problems are classified into two classes. The one is number-theoretical problems such as factoring [50], RSA [53], and the decisional Diffie-Hellman (DDH) problem [21]. The other is combinatorial problems such as subset-sum problem [42] and code problem [41]. After Shor [55] proposed *quantum* polynomial-time algorithm for factorization and the discrete logarithm problem, cryptographic researchers have proposed a great number of cryptographic primitives based on combinatorial problems, since combinatorial problems seem hard even for quantum algorithms. As examples of *post-quantum cryptography*, we have *public-key encryption* (PKE) and *key-encapsulation mechanism* (KEM) with provable security based on subset-sum problem [39], lattice problems [3, 27, 52], code problem [41], and problems on multi-variate quadratic (MQ) systems [30].

However, we only know a few results on *authenticated key exchange* (AKE) based on the combinatorial problems [10, 23], which is often employed to construct a secure channel between two parties in a public channel for secrecy and authenticity. When the parties share secret information (called *a session key*), they can construct a secure channel by symmetric-key encryptions and message authentication codes with the shared session key. AKE provides a solution to share a session key even if parts of secret information are exposed to an adversary. Formally speaking, in an AKE protocol, each party has public information, called a *static public key*, which is authorized by a trusted third party, e.g., certification authority (CA), and the corresponding secret information, called a *static secret key*. A user who wants to share information with an entity exchanges *ephemeral public keys*, generated from the corresponding *ephemeral secret keys*, and computes some *session state* from their static public keys, the corresponding static secret keys, the exchanged ephemeral public keys, and the corresponding ephemeral secret keys. Both parties then derive a *session key* from these values including session state with a function called the *key derivation function*. The desirable security notion of AKE is formulated as CK^+ security [34, 23]. Here, CK^+ security guarantees the Canetti-Krawczyk (CK) security [12], weak perfect forward secrecy (wPFS), resilience to key compromise impersonation (KCI), and resilience to maximal exposure attacks (MEX) which may exploit exposed secret information of the target session.

Recently, generic constructions of AKE from KEM have been investigated [10, 23]. Boyd et al. [10] and Fujioka et al. [23] gave constructions of CK and CK^+ secure AKE protocols from IND-CCA secure KEM schemes with pseudo-random functions (PRFs), respectively. Plugging the existing IND-CCA secure KEM schemes based on lattice problems and code problems in the standard model (StdM) into the generic constructions, we obtain AKE protocols CK and CK$^+$ secure against quantum adversaries.

Unfortunately, KEM schemes secure in the StdM are often inefficient on time and memory. In the case of the KEM schemes based number-theoretic assumptions, we have less problem on efficiency. However, when we consider post-quantum cryptography, we have unsatisfactory inefficiency. For example, on the IND-CCA secure KEM scheme based on the ring-LWE problem [40], our rough estimation says that the sizes of encapsulation key and ciphertext are about 540 kbits. Consequently, the AKE protocol from the KEM scheme also are inefficient on the communication costs, 1.12 Mbits (see 7-th rows in Table 1).

1.2 Our Contribution

In this paper, we investigate how efficiency of AKE is improved by relaxing the security model, i.e., adapting the random oracle model (ROM) [5]. We answer the following natural questions:

1. Which level of security in KEM is required to construct CK$^+$ secure AKE?
2. Which (post-quantum) assumptions can be used to construct CK$^+$ secure AKE?
3. How is efficiency of AKE improved?

As the answer of Question 1, we show that CK$^+$ secure AKE protocols can be constructed from OW-CCA secure KEM schemes with hash functions (regarded as the random oracles (RO)).

Regarding to Question 2, we show the followings:

- Rabin-KEM [16] and RSAP-H [15] can be used since they are OW-CCA secure. Rabin-KEM and RSAP-H are secure under the factoring and RSA assumptions in the ROM, respectively.
- Regarding to lattice problems, which seem to be hard even for quantum machines, we have efficient KEM schemes [40, 56]. They are secure under the ring-learning with errors (LWE) assumption, however, they achieve only IND-CPA security. Thus, we apply the Fujisaki-Okamoto (FO) conversion [24] to enhance their security to IND-CCA one, and then, construct CK$^+$ secure AKE protocols using the converted schemes. It may be worth noting that although applying the FO conversion requires the ROM in their security proofs, it is no problem since the security proof of the proposed construction needs the ROM, also.
- For a code-based KEM scheme, we can use an IND-CCA secure KEM scheme under the McEliece one-way assumption [33] as it is OW-CCA secure in the ROM, also.
- NTRU encryption scheme [27, 31] is also IND-CCA secure under the NTRU one-way assumption in the ROM. We can construct an efficient AKE from it.
- Similarly, we have a KEM scheme IND-CCA secure in the ROM under the subset-sum assumption [39]. Thus, the proposed construction generates an AKE protocol CK$^+$ secure under the subset-sum assumption with help of the FO conversion.
- Under the new assumption on multi-variate quadratic systems [30], we have an IND-CCA secure KEM scheme in the ROM. Thus, we also obtain a CK$^+$ secure AKE protocol from the assumption in the ROM.

As the answer of Question 3, we show that adapting the ROM in the security proofs, when we set the security parameter as $\kappa = 128$, each protocol becomes efficient on the communication costs as approximately half under the factoring assumption (from 22.74 kbits to 13.00 kbits), approximately 1/14 under the ring-LWE assumption (from 1.12 Mbits to 80.65 kbits), and approximately 1/25 under the code assumption (from 1.31 Mbits to 52.32 kbits). Thus, adapting ROM brings much benefit in efficiency. The details are in Section 5.4 and Section 6.

Our contributions are summarized as follows:

- We propose a generic CK$^+$ secure AKE construction from a OW-CCA secure KEM in the ROM.
- We achieve efficient CK$^+$ secure AKE protocols based on the hardness of integer factoring, the ring-LWE problem, and the code problem in the ROM.
- We achieve new CK$^+$ secure AKE protocols based on the hardness of the NTRU problem, subset-sum problem, and MQ problem.
- Especially, we have *quasi-linear* time AKE protocols based on the hardness of the ring-LWE problem in the ROM.

For the summary, see Table 1.

Table 1: Comparison between the previous schemes and instantiations of our scheme.

	Assumption	StdM?	Comm. Costs (bits)
[23]	DDH	StdM	$2,048$
[23]	Factoring	StdM	$22,736$
HMQV [34]	GapDH+KEA	ROM	512
Ours	GapDH/CDH	ROM	$2,048$
Ours	RSA	ROM	$16,240$
Ours	Factoring	ROM	$12,992$
[23]	Ring-LWE	StdM	$\approx 1,117,000$
[23]	Codes	StdM	$\geq 1,312,672$
Ours	Ring-LWE	ROM	$\approx 80,600$
Ours	Codes	ROM	$52,320$
Ours	NTRU	ROM	$19,756$
Ours	Subset Sum	ROM	$1,295,960$
Ours	MQ [30]	ROM	$\approx 11,440,000$
Ours	MQ [30]	ROM	$\approx 7,672,000$

Open problem.

We left an open problem to show our construction is secure in "quantum accessible" random oracle model. In this paper, we successfully showed that our construction is secure in ROM if the underlying KEM is secure. In addition, our proof does not require rewinding. Hence, one would consider our construction also quantumly secure. However, what we show is that if underlying problem is hard then there are no quantum adversary violating CK$^+$ security with *classical access to the random oracles*. In the real world, the random oracle is instantiated by a hash function H and a quantum adversary can evaluate on a quantum superposition of input. A quantum-accessible random oracle model (QAROM) [9, 60] captures this ability. Therefore, it would be interesting to show our construction is secure in the QAROM.

2. CK+ SECURITY MODEL

In this section, we quote the CK+ model that was introduced by [34, 23]. We show the model for two-pass protocols. It can be trivially extended to n-pass protocols for any $n > 2$.

U_i denotes a party indexed by i. Parties are modeled as probabilistic polynomial-time (PPT) interactive Turing machines w.r.t. security parameter $\kappa \in \mathbb{N}$. For party U_i, we denote static secret (public) key by s_i (S_i) and ephemeral secret (public) key by x_i (X_i), respectively. Party U_i generates its own keys, s_i and S_i, and the static public key S_i is linked with U_i's identity in some systems like PKI.[1]

Session.

We call an invocation of a protocol *session*. A session is activated by an incoming message of the forms $(\Pi, \mathcal{I}, U_A, U_B)$ or $(\Pi, \mathcal{R}, U_B, U_A, X_A)$, where Π is a protocol identifier, $\mathcal{I}$ and $\mathcal{R}$ are role identifiers, and U_A and U_B are user identifiers. If U_A is activated with $(\Pi, \mathcal{I}, U_A, U_B)$, then U_A is called the session *initiator*. If U_B is activated with $(\Pi, \mathcal{R}, U_B, U_A, X_A)$, then U_B is called the session *responder*. The initiator U_A outputs X_A on activation, then may receive an incoming message of the forms $(\Pi, \mathcal{I}, U_A, U_B, X_A, X_B)$ from the responder U_B, then computes the session key SK if U_A received the message. On the contrary, the responder U_B outputs X_B, and computes the session key SK.

If U_A is the initiator of a session, the session is identified by $\mathsf{sid} = (\Pi, \mathcal{I}, U_A, U_B, X_A)$ or $\mathsf{sid} = (\Pi, \mathcal{I}, U_A, U_B, X_A, X_B)$. If U_B is the responder of a session, the session is identified by $\mathsf{sid} = (\Pi, \mathcal{R}, U_B, U_A, X_A, X_B)$. We say that U_A is the *owner* of session sid, if the third coordinate of session sid is U_A. We say that U_A is the *peer* of session sid, if the fourth coordinate of session sid is U_A. We say that a session is *completed* if its owner computes the session key. The *matching session* of $(\Pi, \mathcal{I}, U_A, U_B, X_A, X_B)$ is session $(\Pi, \mathcal{R}, U_B, U_A, X_A, X_B)$ and vice versa.

Adversary.

The adversary $\mathcal{A}$, which is modeled as a PPT interactive Turing machine, controls all communications between parties including session activation by performing the following adversary query.

- Send(message): The message has one of the following forms: $(\Pi, \mathcal{I}, U_A, U_B)$, $(\Pi, \mathcal{R}, U_B, U_A, X_A)$, or $(\Pi, \mathcal{I}, U_A, U_B, X_A, X_B)$. The adversary $\mathcal{A}$ obtains the response from the party.

To capture exposure of secret information, the adversary $\mathcal{A}$ is allowed to issue the following queries.

- $\mathsf{SessionKeyReveal}(\mathsf{sid})$: The adversary $\mathcal{A}$ obtains the session key SK for the session sid if the session is completed.
- $\mathsf{SessionStateReveal}(\mathsf{sid})$: The adversary $\mathcal{A}$ obtains the session state of the owner of session sid if the session is not completed (the session key is not established yet). The session state includes all ephemeral secret keys and intermediate computation results except for immediately erased information but does not include the static secret key.
- $\mathsf{Corrupt}(U_i)$: This query allows the adversary $\mathcal{A}$ to obtain all information of the party U_i. If a party is corrupted by a $\mathsf{Corrupt}(U_i)$ query issued by the adversary $\mathcal{A}$, then we call the party U_i *dishonest*. If not, we call the party *honest*.

[1] Static public keys must be known to both parties in advance. They can be obtained by exchanging them before starting the protocol or by receiving them from a certificate authority. This situation is common for *all* PKI-based AKE schemes.

Freshness.

For the security definition, we need the notion of freshness.

Definition 1. Let $\mathsf{sid}^* = (\Pi, \mathcal{I}, U_A, U_B, X_A, X_B)$ or $(\Pi, \mathcal{R}, U_A, U_B, X_B, X_A)$ be a completed session between honest users U_A and U_B. If the matching session exists, then let $\overline{\mathsf{sid}^*}$ be the matching session of sid^*. We say session sid^* is *fresh* if none of the following conditions hold:

1. $\overline{\mathsf{sid}^*}$ exists and the adversary $\mathcal{A}$ makes either of the following queries;
   - $\mathsf{SessionKeyReveal}(\mathsf{sid}^*)$ or
   - $\mathsf{SessionKeyReveal}(\overline{\mathsf{sid}^*})$:
2. $\overline{\mathsf{sid}^*}$ exists and the adversary $\mathcal{A}$ makes either of the following queries;
   - $\mathsf{SessionStateReveal}(\mathsf{sid}^*)$ or
   - $\mathsf{SessionStateReveal}(\overline{\mathsf{sid}^*})$:
3. $\overline{\mathsf{sid}^*}$ does not exist and the adversary $\mathcal{A}$ makes the following query; $\mathsf{SessionStateReveal}(\mathsf{sid}^*)$.

Security experiment.

For the security definition, we consider the following security experiment. Initially, the adversary $\mathcal{A}$ is given a set of honest users and makes any sequence of the queries described above. During the experiment, the adversary $\mathcal{A}$ makes the following query.

- $\mathsf{Test}(\mathsf{sid}^*)$: sid^* must be a fresh session. Select random bit $b \in_U \{0, 1\}$, and return the session key held by sid^* if $b = 0$, and return a random key if $b = 1$.

The experiment continues until the adversary $\mathcal{A}$ makes a guess b'. The adversary $\mathcal{A}$ *wins* the game if the test session sid^* is still fresh and if the guess of the adversary $\mathcal{A}$ is correct, i.e., $b' = b$. The advantage of the adversary $\mathcal{A}$ in the AKE experiment with the PKI-based AKE protocol Π is defined as

$$\mathsf{Adv}^{\mathsf{ake\text{-}ck^+}}_{\Pi,\mathcal{A}}(\kappa) = \Pr[\mathcal{A} \text{ wins}] - \frac{1}{2}.$$

We define the security as follows.

Definition 2. We say that a PKI-based AKE protocol Π is *secure in the* CK+ *model* if the following conditions hold:

1. (Completeness:) if two honest parties complete matching sessions, then, except with negligible probability, they both compute the same session key.
2. (Soundness:) for any PPT bounded adversary $\mathcal{A}$, its advantage $\mathsf{Adv}^{\mathsf{ake\text{-}ck^+}}_{\Pi,\mathcal{A}}(\mathcal{A})$ is negligible in security parameter κ for the test session sid^*,
   (a) if $\overline{\mathsf{sid}^*}$ does not exist, and the static secret key of the owner of sid^* is given to $\mathcal{A}$.
   (b) if $\overline{\mathsf{sid}^*}$ does not exist, and the ephemeral secret key of sid^* is given to $\mathcal{A}$.
   (c) if $\overline{\mathsf{sid}^*}$ exists, and the static secret key of the owner of sid^* and the ephemeral secret key of $\overline{\mathsf{sid}^*}$ are given to $\mathcal{A}$.
   (d) if $\overline{\mathsf{sid}^*}$ exists, and the ephemeral secret key of sid^* and the ephemeral secret key of $\overline{\mathsf{sid}^*}$ are given to $\mathcal{A}$.

(e) if $\overline{\mathsf{sid}^*}$ exists, and the static secret key of the owner of sid^* and the static secret key of the peer of sid^* are given to $\mathcal{A}$.

(f) if $\overline{\mathsf{sid}^*}$ exists, and the ephemeral secret key of sid^* and the static secret key of the peer of sid^* are given to $\mathcal{A}$.

Note that the item 2.a corresponds to resistance to KCI, item 2.e corresponds to wPFS, and items 2.b, 2.c, 2.d and 2.f correspond to resistance to MEX.

3. GENERIC AKE CONSTRUCTION FROM KEM

In this section, we propose a generic construction (GCwR) of CK^+ secure AKE from one-way KEMs in the random oracle model. Before describing our GCwR, we recall the notion of KEM.

3.1 Security Notions of KEM Schemes

Let us recall the model and security notions of KEM.

Definition 3. A KEM scheme KEM consists of the following 3-tuple (KeyGen, EnCap, DeCap):

$(ek, dk) \leftarrow \mathsf{KeyGen}(1^\kappa; r_g)$: a key-generation algorithm which on inputs 1^κ and $r_g \in \mathcal{RS}_G$, where κ is the security parameter and $\mathcal{RS}_G$ is a randomness space, outputs a pair of an encapsulation key and a decapsulation key (ek, dk).

$(K, C) \leftarrow \mathsf{EnCap}_{ek}(r_e)$: an encapsulation algorithm which takes as inputs encapsulation key ek and $r_e \in \mathcal{RS}_E$, outputs session key $K \in \mathcal{KS}$ and ciphertext $C \in \mathcal{CS}$, where $\mathcal{RS}_E$ is a randomness space, $\mathcal{KS}$ is a session key space, and $\mathcal{CS}$ is a ciphertext space.

$K/\perp \leftarrow \mathsf{DeCap}_{dk}(C)$: a decapsulation algorithm which takes as inputs decapsulation key dk and ciphertext $C \in \mathcal{CS}$, outputs session key $K \in \mathcal{KS}$ or a rejection symbol $\perp$.

We next recall the definitions of one-wayness under chosen-ciphertext and chosen-plaintext attacks (denoted by OW-CCA and OW-CPA) for key encapsulation, respectively.

Definition 4. A KEM scheme KEM is (t, ϵ)-OW-CCA secure for KEM if the following property holds for security parameter κ; For any adversary $\mathcal{A} = (\mathcal{A}_1, \mathcal{A}_2)$ with a time complexity at most t,

$$\mathsf{Adv}^{\mathsf{ow\text{-}cca}}_{\mathsf{KEM},\mathcal{A}}(\kappa) = \Pr\left[\begin{array}{l}(ek, dk) \leftarrow \mathsf{KeyGen}(1^\kappa; r_g);\\ state \leftarrow \mathcal{A}_1^{\mathcal{DO}_\perp(dk,\cdot)}(ek);\\ (K^*, C^*) \leftarrow \mathsf{EnCap}_{ek}(r);\\ K'^* \leftarrow \mathcal{A}_2^{\mathcal{DO}_{C^*}(dk,\cdot)}(ek, C^*, state);\\ K'^* = K^*\end{array}\right] \leq \epsilon$$

where $\mathcal{DO}_a(dk, C)$ is the decapsulation oracle that returns $K = \mathsf{DeCap}_{dk}(C)$ if $C \neq a$ and returns a reject symbol otherwise, *state* is state information which $\mathcal{A}$ wants to preserve from $\mathcal{A}_1$ to $\mathcal{A}_2$, and $\mathcal{A}$ runs in at most t steps.

We say a KEM scheme is OW-CPA secure for key encapsulation if $\mathcal{A}$ does not access $\mathcal{DO}$.

3.2 Construction from One-way KEM

We propose a generic construction (GCwR), which is based on a OW-CCA secure KEM and the ROM. (GCwR stands for generic construction with random oracles.) At the price of using ROs, we can weaken the assumption on the KEM from IND-CCA to OW-CCA. Since a OW-CCA secure KEM is efficiently obtained in the ROM (e.g., RSAP-H, RSA-KEM, and ECIES-KEM) and we do not need PRFs and a strong randomness extractor, we can efficiently instantiate GCwR. The details of instantiations are given in Section 5 and Section 6. To achieve ephemeral secret key exposure resistance, the ordinary NAXOS technique [36] is known and used in some protocols [36, 43, 23]. The NAXOS technique means that the DH-exponent is set as an output of a RO on inputs static and ephemeral secret keys. Since adversaries are not allowed to expose both static and ephemeral secret keys, the DH-exponent is not exposed if one of secret keys are exposed. In GCwR, we also apply the NAXOS technique to generating randomness of EnCap instead of generating the DH-exponent.

Generic construction GCwR.

Let KEM = (KeyGen, EnCap, DeCap) be a OW-CCA secure KEM and wKEM = (wKeyGen, wEnCap, wDeCap) be a OW-CPA secure KEM.

The protocol of GCwR is as follows.

Public Parameters: Let κ be the security parameter and let $H_1 : \{0,1\}^* \rightarrow \mathcal{RS}_E$ and $H_2 : \{0,1\}^* \rightarrow \{0,1\}^\kappa$ be hash functions modeled as ROs. These are provided as a part of the public parameters.

Secret and Public Keys: User U_I randomly selects $r_I \in \mathcal{RS}_G$, and runs the generation algorithm $(dk_{I,1}, ek_{I,1}) \leftarrow \mathsf{KeyGen}(r_I)$. User U_I's static secret and public keys are $(dk_{I,1}, ek_{I,1})$.

Key Exchange: User U_A with secret and public keys $(dk_{A,1}, ek_{A,1})$, who is the initiator, and user U_B with secret and public keys $(dk_{B,1}, ek_{B,1})$, who is the responder, perform the following two-pass key exchange protocol.

1. To initialize the protocol, user U_A selects ephemeral secret keys $r_{A,1} \in \{0,1\}^\kappa$ and $r_{A,2} \in \mathcal{RS}_G$ randomly. User U_A computes $(C_{A,1}, K_{A,1}) \leftarrow \mathsf{EnCap}_{ek_{B,1}}(H_1(r_{A,1}, dk_{A,1}))$ and $(dk_{A,2}, ek_{A,2}) \leftarrow \mathsf{wKeyGen}(r_{A,2})$, and sends $(U_A, U_B, C_{A,1}, ek_{A,2})$ to user U_B.

2. Upon receiving $(U_A, U_B, C_{A,1}, ek_{A,2})$, user U_B chooses ephemeral secret keys $r_{B,1} \in \{0,1\}^\kappa$ and $r_{B,2} \in \mathcal{RS}_E$ uniformly at random, computes $(C_{B,1}, K_{B,1}) \leftarrow \mathsf{EnCap}_{ek_{A,1}}(H_1(r_{B,1}, dk_{B,1}))$ and $(C_{B,2}, K_{B,2}) \leftarrow \mathsf{wEnCap}_{ek_{A,2}}(r_{B,2})$, and sends $(U_A, U_B, C_{B,1}, C_{B,2})$ to user U_A.

   User U_B computes $K_{A,1} \leftarrow \mathsf{DeCap}_{dk_{B,1}}(C_{A,1})$, sets the session identity $\mathsf{sid} = (U_A, U_B, ek_{A,1}, ek_{B,1}, C_{A,1}, ek_{A,2}, C_{B,1}, C_{B,2})$ and the session key $SK = H_2(K_{A,1}, K_{B,1}, K_{B,2}, \mathsf{sid})$, and completes the session.

3. Upon receiving $(U_A, U_B, C_{B,1}, C_{B,2})$, user U_A computes $K_{B,1} \leftarrow \mathsf{DeCap}_{dk_{A,1}}(C_{B,1})$, $K_{B,2} \leftarrow \mathsf{wDeCap}_{dk_{A,2}}(C_{B,2})$, sets the session identity $\mathsf{sid} = (U_A, U_B, ek_{A,1}, ek_{B,1}, C_{A,1}, ek_{A,2}, C_{B,1}, C_{B,2})$ and the session key $SK = H_2(K_{A,1}, K_{B,1}, K_{B,2}, \mathsf{sid})$, and completes the session.

The session state of a session owned by U_A contains ephemeral secret keys $(r_{A,1}, r_{A,2})$ and a KEM key K_{A_1}. Similarly, the session state of a session owned by U_B contains ephemeral secret keys $(r_{B,1}, r_{B,2})$ and KEM keys $(K_{B,1}, K_{B,2})$.

Security.

We show the following theorem.

THEOREM 1. *If a KEM scheme* KEM = (KeyGen, EnCap, DeCap) *is* OW-CCA *secure, and a KEM scheme* wKEM = (wKeyGen, wEnCap, wDeCap) *is* OW-CPA *secure, then the AKE scheme with* GCwR *is* CK^+ *secure where* H_1 *and* H_2 *are modeled as ROs.*

The precise proof of Theorem 1 is in Appendix A.

Here, we give an overview of the security proof for the case that the test session has a matching session.

We have to consider the following four exposure patterns in the CK$^+$ security model:

2-(c) the static secret key of the initiator and the ephemeral secret key of the responder

2-(d) both ephemeral secret keys

2-(e) both static secret keys

2-(f) the ephemeral secret key of the initiator and the static secret key of the responder

Intuitively speaking, in case 2-(c), $K_{A,1}$ is protected by the security of $CT_{A,1}$ because $H_1(r_{A,1}, dk_{A,1})$ is hidden and $dk_{B,1}$ is not exposed. In case 2-(d), $K_{A,1}$ and $K_{B,1}$ are protected by the security of $C_{A,1}$ and $C_{B,1}$ because $H_1(r_{A,1}, dk_{A,1})$ and $H_1(r_{B,1}, dk_{B,1})$ are hidden, and $dk_{A,1}$ and $dk_{B,1}$ are not exposed. In case 2-(e), $K_{B,2}$ is protected by the security of $C_{B,2}$ because $dk_{A,2}$ and $r_{B,2}$ are not exposed. In case 2-(f), $K_{B,1}$ is protected by the security of $C_{B,1}$ because $H_1(r_{B,1}, dk_{B,1})$ is hidden and $dk_{A,1}$ is not exposed.

Then, we construct a reduction from the OW-CCA game (for cases 2-(c), 2-(d), and 2-(f)) or the OW-CPA game (for 2-(e)) to the CK$^+$ security game. The simulator embeds the challenge ciphertext in the OW-CCA (OW-CPA) game into the hash list of H_2 corresponding to the test session. If an adversary in the CK$^+$ security game succeeds with non-negligible probability, then the simulator in the OW-CCA (OW-CPA) game also succeeds with non-negligible probability. The simulation of SessionStateReveal queries is done by the power of the decryption oracle in the OW-CCA game. For case 2-(e), the simulator does not need the decryption oracle because he knows all static secret keys. We can show a similar proof in non-matching cases.

4. FUJISAKI–OKAMOTO CONVERSION, RECONSIDERED

In this section, we briefly review the Fujisaki–Okamoto (FO) conversion [24] and discuss that we can slightly weaken the conditions.

Let us review the model and security notions of public key encryption (PKE) scheme. The model for PKE schemes is summarized as follows:

Definition 5. A PKE scheme PKE consists of the following 3-tuple (Gen, Enc, Dec):

$(pk, sk) \leftarrow \mathsf{Gen}(1^\kappa; r_g)$: a key-generation algorithm which on input 1^κ, where κ is the security parameter, outputs a pair of keys (pk, sk). pk and sk are called public key and secret key, respectively.

$C \leftarrow \mathsf{Enc}_{pk}(M; r_e)$: an encryption algorithm which takes as input public key pk and plaintext M, outputs ciphertext C.

$M/\perp \leftarrow \mathsf{Dec}_{sk}(C)$: a decryption algorithm which takes as input secret key sk and ciphertext C, outputs plaintext M or a rejection symbol $\perp$.

We say PKE has perfect correctness if for any (pk, sk) generated by Gen, $M \in \mathcal{MS}$, and $r_e \in \mathcal{RS}_E$, we have that $\mathsf{Dec}_{sk}(\mathsf{Enc}_{pk}(M; r_e)) = M$.

The security of PKE schemes is defined by several notions like one-wayness and indistinguishability. Here, we recall the definition of indistinguishability under chosen-ciphertext and chosen-plaintext attacks (denoted by IND-CCA and IND-CPA) for PKE, respectively.

Definition 6. A PKE scheme is (t, ϵ)-IND-CCA secure if the following property holds for security parameter κ; For any adversary $\mathcal{A} = (\mathcal{A}_1, \mathcal{A}_2)$,

$$\mathsf{Adv}^{\text{ind-cca}}_{\mathcal{A}}(\kappa) = \left| \Pr\left[\begin{array}{l} b \leftarrow \{0,1\}; \\ (pk, sk) \leftarrow \mathsf{Gen}(1^\kappa); \\ (M_0, M_1, state) \leftarrow \mathcal{A}_1^{\mathcal{DO}_\perp(sk,\cdot)}(pk); \\ C^* \leftarrow \mathsf{Enc}_{pk}(M_b); \\ b' \leftarrow \mathcal{A}_2^{\mathcal{DO}_{C^*}(sk,\cdot)}(pk, C^*, state); \\ b' = b \end{array} \right] - \frac{1}{2} \right| \leq \epsilon,$$

where $\mathcal{DO}_a(sk, C)$ is the decryption oracle that returns a message $M = \mathsf{Dec}_{sk}(C) \in \mathcal{MS} \cup \{\perp\}$ if $C \neq a$ and returns a reject symbol otherwise, *state* is state information (possibly including pk, M_0 and M_1) which $\mathcal{A}$ wants to preserve, and $\mathcal{A}$ runs in at most t steps. We say a PKE scheme is IND-CPA secure, if $\mathcal{A}$ does not access $\mathcal{DO}$.

Review of the FO conversion.

The conversion transforms an IND-CPA secure PKE scheme, which has perfect correctness and γ-uniformity (defined later) into an IND-CCA secure PKE scheme in the ROM.

Let wPKE = (wGen, wEnc, wDec) be a PKE scheme with a message space $w\mathcal{MS}$, randomness spaces $w\mathcal{RS}_G$ and $w\mathcal{RS}_E$, and a ciphertext space $w\mathcal{CS}$. We define γ-uniformity, which essentially says that ciphertext has a min-entropy at least $-\lg \gamma$.

Definition 7. We say a PKE scheme wPKE = (wGen, wEnc, wDec) is *γ-uniform* if for any $(pk, sk) \leftarrow \mathsf{Gen}(1^n; r_g)$, $M \in w\mathcal{MS}$, and $C \in w\mathcal{CS}$, the inequality $\Pr_{r_e}\left[\mathsf{Enc}_{pk}(M; r_e) = C\right] \leq \gamma$ holds.

We decompose $w\mathcal{MS}$ into two finite sets $\mathcal{MS}$ and $\mathcal{RS}_E$, that is, $w\mathcal{MS} = \mathcal{MS} \times \mathcal{RS}_E$. Let $H : w\mathcal{MS} \rightarrow w\mathcal{RS}_E$ be a hash function modeled as the random oracle. The FO conversion converts wPKE into an encryption scheme PKE = FO(wPKE) = (Gen, Enc, Dec) with a message space $\mathcal{MS}$, a randomness space $\mathcal{RS}_E$, and a ciphertext space $\mathcal{CS} = w\mathcal{CS}$ defined as follows:

$\mathsf{Gen}(1^n; r_g)$**:** Output $(pk, sk) \leftarrow \mathsf{wGen}(1^n; r_g)$.

$\mathsf{Enc}_{pk}(M; r_e)$**:** Output $C \leftarrow \mathsf{wEnc}_{pk}((M, r_e); H(M, r_e))$.

$\mathsf{Dec}_{sk}(C)$**:** Compute $M' \leftarrow \mathsf{wDec}_{sk}(C)$. If $M' = \perp$ then output $\perp$. Otherwise parse $(M, r_e) \leftarrow M'$ and verify C by checking $C = \mathsf{wEnc}_{pk}(M'; H(M'))$. If the verification is not passed then output $\perp$. Otherwise output M.

They showed that PKE is IND-CCA secure if wPKE is IND-CPA secure, γ-uniform, and perfectly correct.

Relaxation.

We define semi-uniformity in order to relax the requirements on a PKE scheme. We only require that the min-entropy of the ciphertext is at least $-\lg \gamma$ for all but δ fraction of key pairs rather than any key pairs.

Definition 8. We say a PKE scheme wPKE = (wGen, wEnc, wDec) is *(δ, γ)-semi-uniform* if for all $M \in w\mathcal{MS}$ and $C \in w\mathcal{CS}$,

$$\Pr_{r_g}\left[\Pr_{r_e}[\mathsf{wEnc}_{pk}(M; r_e) = C] \leq \gamma : (pk, sk) \leftarrow \mathsf{wGen}(1^n; r_g)\right] \geq 1 - \delta.$$

It is obvious that $(0, \gamma)$-semi-uniformity implies γ-uniformity.

This relaxation does not harm the security of the converted scheme. Without modifications, we can adapt the proof of the security by introducing δ into the bound below.

Theorem 2 (Theorem 5.4 in [24], adapted). *Suppose that* wPKE *is a* (δ, γ)*-semi-uniform PKE scheme. Let* PKE = FO(wPKE) *be the converted scheme from* wPKE. *If* wPKE *is* $(t_{\mathsf{cpa}}, \epsilon_{\mathsf{cpa}})$*-IND-CPA secure then* PKE *is* $(t_{\mathsf{cca}}, \epsilon_{\mathsf{cca}})$*-IND-CCA secure, where*

$$t_{\mathsf{cca}} \le t_{\mathsf{cpa}} - Q_H \cdot (T_{\mathsf{wEnc}} + O(\kappa)),$$
$$\epsilon_{\mathsf{cca}} \le \frac{1}{(1-\gamma)^{Q_D}} \cdot \epsilon_{\mathsf{cpa}} + \frac{2Q_H}{|w\mathcal{RS}_E|} + \delta,$$

Q_D *and* Q_H *denote the numbers of the queries to the decryption oracle and the random oracle* H*, and* T_{wEnc} *denotes the computational running time of* wEnc.

5. INSTANTIATIONS FROM RING-LWE

We have new concrete AKE protocols based on the worst-case hardness of the (ring-)LWE problems derived from our generic construction, GCwR. We can employ IND-CCA secure PKE schemes in the standard model [47, 45, 13, 1, 2, 57, 40] as OW-CCA secure KEM schemes, as Fujioka et al. did [23]. Unfortunately, the obtained AKE protocols are inefficient since these PKE schemes require huge keys, say, the quadratic or cubic order of the security parameter.

In order to instantiate practical AKE protocols with CK$^+$ security, we construct efficient OW-CCA KEM schemes in the ROM from IND-CPA secure PKE scheme under the ring-LWE assumption. Formally speaking, we construct IND-CCA secure PKE schemes by applying the Fujisaki–Okamoto conversion [24] to IND-CPA secure PKE schemes [40, 56]. Since IND-CCA secure PKE schemes with sufficiently large plaintext space are OW-CCA secure, we have instantiated OW-CCA secure KEM schemes from them. An obstacle to applying the Fujisaki–Okamoto conversion is that the (ring-)LWE-based PKE schemes do not have perfect correctness but negligible decryption errors. To eliminate this obstacle, i.e., eliminate the decryption errors, we carefully choose parameters and slightly strengthen the assumptions. We should note that one can also eliminate errors by using the technique proposed by Dwork, Naor, and Reingold [20]. We do not adopt it since it uses pseudorandom generators.

Remark 1. We note that our CCA-secure schemes are based on the worst-case hardness of the ring-LWE problem and *not* based on the worst-case hardness of the approximating shortest vector problem. This is because we employ the Fujisaki-Okamoto conversion and the proof of the conversion in quantum setting is still open [9, 60].

5.1 Ring-LWE Assumption

We next review the ring-LWE assumption [40]. For simplicity, practical issues, and limitation of space, we adapt a simplified notion of the ring-LWE problem by Ducas and Durms [19] with parameter $n = 2^z$.

Hereafter, an element $c \in \mathbb{Z}_q$ is represented by a corresponding integer in $[-(q-1)/2, (q-1)/2]$. For an element $c \in \mathbb{Z}_q$, $|c|$ denotes the Lee value, which is defined by $|c| = c$ if $0 \le c < q/2$ and $-c$ if $-q/2 < c < 0$. For a real $x \in \mathbb{R}$, $\lfloor x \rceil$ denotes the nearest integer, that is, $\lceil x - 1/2 \rceil$. In what follows, $\lg(x)$ denotes $\log_2(x)$.

Let $\mathcal{R}_q = \mathbb{Z}_q[X]/(X^n+1)$ and $\mathcal{R}_q^* = \{\mathbf{x} \in \mathcal{R}_q : \exists \mathbf{y}, \mathbf{xy} = 1\}$. We identify a polynomial $\mathbf{f} = f_0 + f_1 X + \cdots + f_{n-1}X^{n-1} \in \mathcal{R}_q$ with a vector $\vec{f} = (f_0, \ldots, f_{n-1})^T \in \mathbb{Z}_q^n$. The ℓ_p and maximum norm of a polynomial $\mathbf{f}$ is denoted by $\|\mathbf{f}\|_p$ and $\|\mathbf{f}\|_\infty$, which is $(\sum_i |f_i|^p)^{1/p}$ and $\max_i |f_i|$, respectively.

The Gaussian distribution with mean 0 and variance σ^2 is denoted by $N(0, \sigma)$. For a real s and $q \in \mathbb{N}$, $\bar{\Psi}_s$ denotes the folded and discretized Gaussian distribution over $\mathbb{Z}_q$, that is, $\lfloor N(0, s^2/2\pi) \rceil \bmod q$. For $s \in \mathbb{R}$, we define n-dimensional Gaussian function ρ_s as $\rho_s(\vec{x}) = \exp(-\pi \cdot \|\vec{x}\|_2^2/s^2)$. For countable set $L \subseteq \mathbb{R}^n$, the discrete Gaussian distribution over L with parameter s is $D_{L,s}(\vec{x}) = \rho_s(\vec{x})/(\sum_{\vec{y}\in L} \rho_s(\vec{y}))$ for $\vec{x} \in L$. Identifying $\mathcal{R}_q$ and $\mathbb{Z}_q^n$, we may choose a polynomial in $\mathcal{R}_q^n$ from the distribution $\bar{\Psi}_s^n$ or $D_{\mathbb{Z}^n,\sigma}$ over $\mathbb{Z}^n$.

For a polynomial $\mathbf{s} \in \mathcal{R}_q$ and a distribution χ over $\mathcal{R}_q$, we define the oracle $A_{\mathbf{s},\chi}$ as follows: (1) take samples $\mathbf{a} \leftarrow \mathcal{R}_q$ and $\mathbf{e} \leftarrow \chi$ and (2) output $(\mathbf{a}, \mathbf{as} + \mathbf{e})$. Let $U(\mathcal{R}_q^2)$ be a oracle which returns two random elements in $\mathcal{R}_q$.

Let us define the $\mathrm{RLWE}_{q,\chi}$ problem and an assumption related to the problem.

Definition 9. Let $n = 2^z$, let q be a prime congruent to 1 modulo $2n$, and let χ be a distribution over $\mathcal{R}_q$. The $\mathrm{RLWE}_{q,\chi}$ *problem* is distinguishing $A_{\mathbf{s},\chi}$ from $U(\mathcal{R}_q^2)$, where $\mathbf{s}$ is chosen at random from $\mathcal{R}_q$. The advantage of $\mathcal{A}$ is defined as

$$\mathsf{Adv}_{\mathcal{A}}^{\mathrm{RLWE}}(n) = \left|\Pr[\mathcal{A}^{A_{\mathbf{s},\chi}}(1^n) = 1] - \Pr[\mathcal{A}^{U(\mathcal{R}_q^2)}(1^n) = 1]\right|,$$

where the probability is defined by $\mathcal{A}$'s random coins, choices of $\mathbf{s} \in \mathcal{R}_q$, and randomness of the oracles. We say that the $\mathrm{RLWE}_{q,\chi}$ *assumption* holds if for any PPT adversary $\mathcal{A}$ its advantage $\mathsf{Adv}_{\mathcal{A}}^{\mathrm{RLWE}}(n)$ is negligible in n.

Lemma 1 ([4, Lemma 2], adapted, and [56]). *Let* $n = 2^z$ *and* q *be a prime with* $q \equiv 1 \bmod 2n$*. Then, for any* $k \ge 1$*, there is a randomized reduction from distinguishing* $A_{\mathbf{s}',\chi}$ *from* $U(\mathcal{R}_q^2)$ *with* $k+1$ *samples where* $\mathbf{s}' \leftarrow \chi$ *to distinguishing* $A_{\mathbf{s},\chi}$ *from* $U(\mathcal{R}_q^2)$ *with* k *samples where* $\mathbf{s} \leftarrow U(\mathcal{R}_q)$.

We finally note that the $\mathrm{RLWE}_{q,\chi}$ problem is reduced to the approximating shortest vector problem in any ideal of the polynomial ring. The details are in [40, 19].

5.2 The Lyubashevsky–Peikert–Regev Scheme

We here describe the modified version of the PKE scheme which appeared in the presentation of [40]. The differences are error checking steps to wGen and wEnc, which make the scheme perfectly correct.

Let us define the parameters as the follows:

$$q = \mathrm{poly}(n): \text{prime}, \quad p = 2, \quad t = \omega(\sqrt{\lg n}),$$
$$s \le \sqrt{\frac{q-p}{2(2n+1)pt^2}}, \quad \chi = \bar{\Psi}_s^n, \quad \lambda = n\lg(s),$$
$$\gamma = 2^{-\lambda/2}, \quad \delta = 2^{-\lambda/2},$$

where t is a threshold parameter.

The following is a description of the PKE scheme $\mathsf{wPKE}_{\mathrm{LPR}}$ = (wGen, wEnc, wDec) with $w\mathcal{MS} = \mathbb{Z}_p^n$, $w\mathcal{RS}_E = \{0,1\}^*$, $w\mathcal{CS} = \mathcal{R}_q^2$. The public parameter contains the security parameter and a polynomial $\mathbf{a}$ chosen from $\mathcal{R}_q^*$ uniformly at random.

$\mathsf{wGen}(1^n; r_g)$**:** (1) Generate two polynomials $\mathbf{s}, \mathbf{e} \leftarrow \chi$. (2: error checking) If $\|\mathbf{s}\|_\infty, \|\mathbf{e}\|_\infty \le st$, then go to next step; otherwise go to step (2). (3) Compute $\mathbf{b} \leftarrow \mathbf{as} + \mathbf{e} \in \mathcal{R}_q$. (4) The encryption key is $pk = (\mathbf{a}, \mathbf{b})$ and the decryption key is $sk = \mathbf{s}$.

$\mathsf{wEnc}_{pk}(\mathbf{k}; r_e)$**:** (1) Generate three polynomials $\mathbf{t}, \mathbf{e}_1, \mathbf{e}_2 \leftarrow \chi$. (2: error checking) If $\|\mathbf{t}\|_\infty, \|\mathbf{e}_1\|_\infty, \|\mathbf{e}_2\|_\infty \le st$, then go to next step; otherwise go to step (1). (3) Compute $\mathbf{u} \leftarrow \mathbf{at} + \mathbf{e}_1$ and $\mathbf{v} \leftarrow \mathbf{bt} + \mathbf{e}_2$. (4) Compute $\mathbf{c} \leftarrow \mathbf{v} + \lfloor (q/p)\mathbf{k} \rceil$. (5) The ciphertext is $(\mathbf{u}, \mathbf{c})$.

$\mathsf{wDec}_{sk}(\mathbf{u}, \mathbf{c})$**:** (1) Compute $\mathbf{d} \leftarrow \mathbf{c} - \mathbf{us} \in \mathcal{R}_q$. (2) Output $\mathbf{k}' \leftarrow \lfloor (p/q)\mathbf{d} \rceil \bmod p$.

We can show the following lemmas from our parameter settings.

Lemma 2 (Perfect Correctness). *There are no decryption errors in* $\mathsf{wPKE}_{\mathrm{LPR}}$.

Lemma 3 (IND-CPA Security). *Under the* $\mathrm{RLWE}_{q,\chi}$ *assumption, the scheme* $\mathsf{wPKE}_{\mathrm{LPR}}$ *is* IND-CPA *secure.*

Lemma 4 (Semi-Uniformity). $\mathsf{wPKE}_{\mathrm{LPR}}$ *is* (δ,γ)*-semi-uniform.*

The proofs of the above lemmas will appear in the full version.

Applying the FO conversion, we obtain an IND-CCA secure PKE scheme, $\mathsf{PKE}_{\mathrm{LPR}}$, based on the $\mathrm{RLWE}_{q,\chi}$ problem.

Theorem 3. *Let* $\mathsf{PKE}_{\mathrm{LPR}} = \mathrm{FO}(\mathsf{wPKE}_{\mathrm{LPR}})$ *where we employ the random oracle* $H : \mathbb{Z}_p^n \to w\mathcal{RS}_E$ *with* $\mathcal{MS} = \mathcal{RS}_E = \mathbb{Z}_p^{n/2}$. *Then* $\mathsf{PKE}_{\mathrm{LPR}}$ *is* IND-CCA *secure if the* $\mathrm{RLWE}_{q,\chi}$ *assumption holds.*

5.3 The Stehlé and Steinfeld Scheme

Moreover, we can employ a secure variant of the NTRU encryption scheme [27] by Stehlé and Steinfeld [56] with the simplified variant of Ring-LWE by Ducas and Durms [19].

Comparing with the LPR scheme, there are additional parameters. A parameter σ and a constant ϵ defines the uniformity of the public key. We define the parameters as follows:

$$\begin{array}{lll} q = \mathrm{poly}(n) : \text{prime}, & p = 2 \in \mathcal{R}_q^*, & t = \omega(\sqrt{\lg n}), \\ s \leq \dfrac{q}{4n\sigma t p(4p+1)}, & \epsilon \in (0, 1/4), & \sigma \geq 2n\sqrt{\ln(8nq)} \cdot q^{\frac{1}{2}+2\epsilon}, \\ \chi = \bar{\Psi}_s^n, & \lambda = n\lg(s), & \\ \gamma = 2^{-\lambda/2}, & \delta = 2^{-\lambda/2}. & \end{array}$$

We modify the original scheme by adding error checking steps to wGen and wEnc. The following is a description of the PKE scheme $\mathsf{wPKE}_{\mathrm{SS}} = (\mathsf{wGen}, \mathsf{wEnc}, \mathsf{wDec})$ with $w\mathcal{MS} = \mathcal{R}/p\mathcal{R}$, $w\mathcal{RS}_E = \{0,1\}^*$, $w\mathcal{CS} = \mathcal{R}_q$.

$\mathsf{wGen}(1^n; r_g)$: (1) Generate a random polynomial $\mathbf{f}' \leftarrow D_{\mathbb{Z}^n,\sigma}$ and compute $\mathbf{f} = p\mathbf{f}' + 1$; if $\mathbf{f} \notin \mathcal{R}_q^*$, re-sample; if $\|\mathbf{f}\|_2 \geq 4p\sqrt{n}\sigma$, re-sample. (2) Generate a random polynomial $\mathbf{g} \leftarrow D_{\mathbb{Z}^n,\sigma}$; if $\mathbf{g} \notin \mathcal{R}_q^*$, re-sample; if $\|\mathbf{g}\|_2 \geq \sqrt{n}\sigma$, re-sample. (3) Compute $\mathbf{h} \leftarrow p\mathbf{g}/\mathbf{f} \in \mathcal{R}_q^*$. (4) The encryption key is $pk = \mathbf{h}$ and the decryption key is $sk = \mathbf{f}$.

$\mathsf{wEnc}_{pk}(\mathbf{k}; r_e)$: (1) Generate three polynomials $\mathbf{s}, \mathbf{e} \leftarrow \chi$. (2: error checking) If $\|\mathbf{s}\|_\infty, \|\mathbf{e}\|_\infty \leq st$ then go to next step; otherwise, go to step (1). (3) Compute $\mathbf{c} \leftarrow \mathbf{hs} + p\mathbf{e} + \mathbf{k} \in \mathcal{R}_q$. (4) The ciphertext is $\mathbf{c}$.

$\mathsf{wDec}_{sk}(\mathbf{c})$: (1) Compute $\mathbf{d} \leftarrow \mathbf{cf} \in \mathcal{R}_q$. (2) Output $\mathbf{k}' \leftarrow \mathbf{d} \bmod p$.

The properties of the scheme are summarized as follows:

Lemma 5 (Perfect Correctness). *There are no decryption errors in* $\mathsf{wPKE}_{\mathrm{SS}}$.

Lemma 6 (IND-CPA Security). *Under the* $\mathrm{RLWE}_{q,\chi}$ *assumption, the scheme* $\mathsf{wPKE}_{\mathrm{SS}}$ *is* IND-CPA *secure*

Lemma 7 (Semi-Uniformity). $\mathsf{wPKE}_{\mathrm{SS}}$ *is* (δ,γ)*-semi-uniform.*

The proofs will appear in the full version.

Applying the FO conversion, we again obtain an IND-CCA secure PKE scheme based on the $\mathrm{RLWE}_{q,\chi}$ assumption.

Theorem 4. *Let* $\mathsf{PKE}_{\mathrm{SS}} = \mathrm{FO}(\mathsf{wPKE}_{\mathrm{SS}})$ *employing the random oracle* $H : \mathbb{Z}_p^n \to w\mathcal{RS}_E$ *with* $\mathcal{MS} = \mathcal{RS}_E = \mathbb{Z}_p^{n/2}$. *Then* $\mathsf{PKE}_{\mathrm{SS}}$ *is* IND-CCA *secure if the* $\mathrm{RLWE}_{q,\chi}$ *assumption holds.*

The proof is obtained by combining the previous lemmas.

Remark 2. We note that Steinfeld et al. [58, Sect. 4] proposed an IND-CCA secure PKE scheme based on $\mathsf{wPKE}_{\mathrm{SS}}$. From their parameter settings [58, Sect. 4.2], a public key and a ciphertext includes at least 8 polynomials, while ours consists of one polynomial.

5.4 Efficiency of the Ring-LWE-based Construction

As a concrete example, we consider the AKE protocol instantiated from $\mathsf{PKE}_{\mathrm{LPR}}$. In the protocol, the initiator sends $C \in \mathcal{R}_q^2$ and $ek \in \mathcal{R}_q$ and the responder sends $C \in \mathcal{R}_q^2$ twice. Hence, the communication costs are seven polynomials, which are $7n\lg q$ bits.

Next, we take parameters chosen by Rückert and Schneider [54, Table 12], which are more conservative than these chosen by Lindner and Peikert [38, Section 1.2]. We let $n = 512$ and $q = 5941249$ by following them. We additionally set $s = 8.0$ and $t = 4.5$ in order to make scheme perfectly correct. Such parameters yield $7n\lg q \approx 80.65$ kbits as the concrete communication costs.

On the above parameter $(n,q) = (512, 5941249)$, Pöppelmann and Güneysu [49] reported FPGA implementation of the computations in $\mathbb{Z}_q[X]/(X^{2^z} + 1)$. Their implementation requires 53 μs per polynomial multiplication on a Spartan-6 LX100.

We compare the existing AKE protocol with our instantiated protocol under the Ring-LWE assumption. Using the generic construction by Fujioka et al. [23] with IND-CCA secure KEM scheme under the ring-LWE assumption, we obtain an AKE protocol based on the ring-LWE problem. There are CCA secure PKE schemes based on the ring-LWE assumption in the StdM: the ring-LWE version of the Peikert PKE scheme [45] over polynomial ring, and the scheme [37] obtained by applying the CHK conversion [8] to the ring-LWE version of the ABB IBE scheme [1]. Those schemes require larger keys and ciphertext, $\kappa \cdot O(\lg q)$ polynomials in the former scheme and $O(\lg q)$ polynomials in the latter scheme, and q should be larger than ours from technical reasons. As a concrete example, we can set $2\lceil\lg q\rceil + 1$ as the number of polynomials in keys and ciphertexts. If we adopted $(n,q) = (512, 5941249)$ as the parameters, then a ciphertext consists from 47 polynomials and the length of it results in approximately 541.50 kbits. In the protocol, the initiator sends a ciphertext of IND-CCA KEM and an encapsulation key of IND-CPA KEM, and the responder sends a ciphertext of IND-CCA KEM and that of IND-CPA KEM. Adopting $\mathsf{wPKE}_{\mathrm{LPR}}$ as IND-CPA KEM, we obtain 97 polynomials ≈ 1.12 Mbits as the communication costs. For comparison, see Table 1.

6. OTHER INSTANTIATIONS

Instantiations from Diffie-Hellman problems.

We can realize various AKE protocols as concrete instantiations of our generic construction GCwR in Section 3 from OW-CCA KEM schemes based on the hardness of the DH problem and its variants. For instance, we can adopt the following efficient KEM schemes based on computational DH (CDH) assumption and gap DH (GDH) assumption.

From the CDH assumption, we have PSEC-KEM [32] that is IND-CCA secure in the ROM under the CDH assumption [24].

From the GDH assumption, we have ECIES-KEM [32] that is IND-CCA secure in the ROM under the GDH assumption [16].

In addition, we have a KEM scheme ($ek = g^x$, $K = H(g^r)$, and $C = (g^x)^r$) that is already IND-CCA secure in the ROM under the GDH assumption, where y is a public key, as Boyen [11] pointed out.

Instantiations from factoring or RSA.

We have new AKE protocols as concrete instantiations of our generic construction GCwR in Section 3 based on the hardness of integer factoring, which are slightly more efficient than those in the standard model in [23], say, from 2.27kbits to 1.30kbits.

From the factoring assumption, we have Rabin-SAEP$^+$ [7] or Rabin-KEM [16] that are IND-CCA secure in the ROM under the factoring assumption.

For the RSA assumption, we have the REACT PKE scheme [44] that was shown to be OW-CCA secure PKE scheme under the RSA assumption.

If we employ the Rabin and RSA PKE schemes as wKEM, we may face an efficiency issue because these schemes require prime generation in the key-generation algorithm, which is executed in the key-generation algorithms of wKEM in our generic construction GCwR. In order to address this issue, we introduce the following simple OW-CPA secure KEM based on Wee's IND-CPA PKE scheme [59], which requires no prime generation in the key-generation algorithm and has ciphertext of length $2|N|$. Let us briefly recall the properties of the signed quadratic residues [28, 29] Fix a Blum integer $N = pq$ for safe primes $p, q \equiv 3 \pmod 4$. The signed quotient group $\mathbb{QR}_N^{\prime} = \mathbb{QR}_N/(\pm 1)$ is a cyclic group of order $(p-1)(q-1)/4$ and efficiently recognizable by computing Jacobi symbol. Let g be a random generator of $\mathbb{QR}_N^{\prime}$. The public parameter is (N, g) and the key pair is (ek, dk), where $dk \leftarrow [1, (N-1)/4]$ and $ek = g^{2dk}$. The encapsulation algorithm outputs $K = g^r$ and $C = (u, \tau) = (g^{2r}, (ek \cdot g)^r)$, where $r \leftarrow [1, (N-1)/4]$. The decapsulation algorithm outputs $K = \tau \cdot u^{-dk}$. The one-wayness follows from the hardness of factoring the Blum integer with safe primes.

Instantiations from codes.

We can instantiate new AKE protocols as concrete instantiations of our generic construction GCwR in Section 3 from code-based problems, though the protocols are not so efficient.

We have Dowsley et al.'s PKE [18] scheme that is IND-CCA secure in the StdM under the McEliece and LPN assumptions. (See Ref. [18] for definitions of these assumptions.)

We also have a padding for the McEliece PKE scheme [33] that is IND-CCA secure in the ROM under the McEliece one-way assumption. Cayrel, Hoffmann, and Persichetti [14] provided a variant of McEliece PKE scheme, converted it into IND-CCA secure schemes by the FO conversion, and reported implementation result.

On the key size, Bernstein et al. [6] estimated the size of a public key of the McEliece using binary Goppa codes at about 1.8 Mbits for 128-bit security. Since the key is huge, there are several attempts to reduce the size of keys. Cayrel et al. [14] reported an implementation result based on very structured code in [48].

If we adapt the scheme in [14] with parameter set in [48, Table 3][2], then we obtain an AKE protocol with communication costs about $|ek| + 3|C| = 52.32$ kbits. If we adopt the IND-CCA secure PKE scheme in [18] with parameter set in [48, Table 3] and a one-time signature scheme whose verification key and signature are of length 128 bits, then the communication costs results in approximately $|ek| + |C| + 2 \cdot 128 \cdot |C| \approx 1.31$ Mbits.

[2] Persichetti [48] proposed the parameters $(q, m, n, k, s, t) = (2^5, 2{,}992, 415, 2^5, 9)$ for 128-bit security. We have $|ek| = kmt \lg q = 37440$ and $|C| = n \lg q = 4960$ bits.

Instantiation from NTRU.

We have already a padding scheme for NTRU encryption [27, 31]. Hence, we can have an AKE protocol based on the NTRU encryption scheme. The scheme enjoys fast computation and light communication costs, as the ring-LWE based AKE does. For example, the communication costs is only $|ek| + 3|C| \approx 19.76$ kbits if we employ the parameter set [31, ees449ep1] for 128-bit security and optimized size. [3]

Instantiation from subset sum.

As a bonus of the ROM, we also have a new AKE protocol from the subset-sum problem. Lyubashevsky, Palacio, and Segev [39] proposed an IND-CPA secure PKE scheme based on the subset-sum problem. To obtain IND-CCA PKE scheme by applying the FO conversion [24], we have to modify the parameters of the scheme slightly, and the details will appear in the full version of this paper. For example, if we set the security parameter n as 256, then we have 1.30 Mbits for the communication cost. [4]

Instantiation from multi-variate quadratic systems.

It is known that solving multi-variate quadratic equations over $\mathbb{F}_q$ is hard in the worst case. There were several PKE schemes exploiting its hardness and several attacks against them.

Huang, Liu, and Yang [30] proposed a new PKE scheme based on the hardness of solving a special case of multi-variate quadratic equations over $\mathbb{F}_q$. A slight modification on the parameters makes their scheme perfectly correct as in the case of subset-sum based construction (the details appear in the full version). When we adapt their estimation [30, Case 1], the communication cost is 11.44 Mbits. [5] Meanwhile, we can hide the quadratic part from public parameters. Employing the idea in the case of LWE-based PKE [46], we have $pp \in \mathbb{Z}_q^{m\times n}$, $ek \in \mathbb{Z}_q^{m\times \ell}$, and $C \in \mathbb{Z}_q^{n+\ell}$. In this case, we have $|ek| = m\ell \lg q = 7.57$ Mbits and $|C| = (n+\ell)\lg q = 33.71$ kbits. They result in 7.67 Mbits as the communication cost.

7. REFERENCES

[1] Agrawal, S., Boneh, D., and Boyen, X. Efficient lattice (H)IBE in the standard model. In Gilbert [25], pp. 553–572.

[2] Agrawal, S., Boneh, D., and Boyen, X. Lattice basis delegation in fixed dimension and shorter-ciphertext hierarchical IBE. In Rabin [51], pp. 98–115.

[3] Ajtai, M., and Dwork, C. A public-key cryptosystem with worst-case/average-case equivalence. In *STOC '97* (1997), ACM, pp. 284–293. See also ECCC TR96-065.

[4] Applebaum, B., Cash, D., Peikert, C., and Sahai, A. Fast cryptographic primitives and circular-secure encryption based on hard learning problems. In Halevi [26], pp. 595–618.

[3] In the parameter set [31, ees449ep1], the parameters are $(N, q, p) = (449, 2048, 3)$. Since $ek, C \in \mathbb{Z}_q[X]/(X^N - 1)$, we have $|ek| = |C| = N \lg q = 4939$ bits.

[4] We set parameters as $(n, \ell, q) = (256, 256, 10n^2)$ instead of $q = 10n\log^2 n$ to make a scheme perfectly correct. We have $pp \in \mathbb{Z}_q^{n\times n}$, $ek \in \mathbb{Z}_q^{n\times \ell}$, and $C \in \mathbb{Z}_q^{n+\ell}$. Hence, we have $|ek| = n\ell \lg q = 1266.28$ kbits and $|C| = (n+\ell)\lg q = 9.89$ kbits.

[5] Huang et al. proposed the parameters $(n, m) = (200, 400)$ and $q \approx 2^{73.93292}$ [30, Case 1]. We additionally set $\ell = 256$, which defines the length of the plaintext. We have $pp \in \mathbb{Z}_q^{m\times n} \times (\mathbb{Z}_q^{n\times n})^m$, $ek \in \mathbb{Z}_q^m$, and $C \in (\mathbb{Z}_q^{n+1})^\ell$. Hence, we have $|ek| = m \lg q = 29.57$ kbits and $|C| = \ell(n+1)\lg q = 3.80$ Mbits. Consequently, we have 11.44 Mbits as the communication cost.

[5] Bellare, M., and Rogaway, P. Random oracle are practical: A paradigm for designing efficient protocols. In *CCS '93* (1993), ACM, pp. 62–73.

[6] Bernstein, D. J., Lange, T., and Peters, C. Smaller decoding exponents: Ball-collision decoding. In *CRYPTO 2011* (2011), P. Rogaway, Ed., vol. 6841 of *LNCS*, Springer, Heidelberg, pp. 743–760.

[7] Boneh, D. Simplified OAEP for the RSA and Rabin functions. In *CRYPTO 2001* (2001), J. Kilian, Ed., vol. 2139 of *LNCS*, Springer, Heidelberg, pp. 275–291.

[8] Boneh, D., Canetti, R., Halevi, S., and Katz, J. Chosen-ciphertext security from identity-based encryption. *SIAM Journal on Computing 36*, 5 (12 2006), 1301–1328.

[9] Boneh, D., Dagdelen, Ö., Fischlin, M., Lehmann, A., Schaffner, C., and Zhandry, M. Random oracles in a quantum world. In *ASIACRYPT 2011* (2011), D. H. Lee and X. Wang, Eds., vol. 7073 of *LNCS*, Springer, Heidelberg, pp. 41–69.

[10] Boyd, C., Cliff, Y., González Nieto, J. M., and Paterson, K. G. One-round key exchange in the standard model. *International Journal of Applied Cryptography (IJACT) 1*, 3 (2009), 181–199. A preliminary version appeared in *ACISP 2008*, 2008.

[11] Boyen, X. Miniature CCA2 PK encryption: Tight security without redundancy. In Kurosawa [35], pp. 485–501.

[12] Canetti, R., and Krawczyk, H. Analysis of key-exchange protocols and their use for building secure channels. In *EUROCRYPT 2001* (2001), B. Pfitzmann, Ed., vol. 2045 of *LNCS*, Springer, Heidelberg, pp. 453–474.

[13] Cash, D., Hofheinz, D., Kiltz, E., and Peikert, C. Bonsai trees, or how to delegate a lattice basis. In Gilbert [25], pp. 523–552.

[14] Cayrel, P.-L., Hoffmann, G., and Persichetti, E. Efficient implementation of a CCA2-secure variant of McEliece using generalized Srivastava codes. In Fischlin et al. [22], pp. 138–155.

[15] Coron, J.-S., Gouget, A., Paillier, P., and Villegas, K. SPAKE: A single-party public-key authenticated key exchange protocol for contact-less applications. In *FC 2010 Workshops* (2010), R. Sion, R. Curtmola, S. Dietrich, A. Kiayias, J. M. Miret, K. Sako, and F. Sebé, Eds., vol. 6054 of *LNCS*, Springer, Heidelberg, pp. 107–122.

[16] Dent, A. W. A designer's guide to KEMs. In *IMA 2003* (2003), K. G. Paterson, Ed., vol. 2898 of *LNCS*, Springer, Heidelberg, pp. 133–151.

[17] Diffie, W., and Hellman, M. E. New directions in cryptography. *IEEE Transactions on Information Theory 22* (November 1976), 644–654.

[18] Dowsley, R., Müller-Quade, J., and Nascimento, A. C. A. A CCA2 secure public key encryption scheme based on the McEliece assumptions in the standard model. In *CT-RSA 2009* (2009), M. Fischlin, Ed., vol. 5473 of *LNCS*, Springer, Heidelberg, pp. 240–251.

[19] Ducas, L., and Durmus, A. Ring-LWE in polynomial rings. In Fischlin et al. [22], pp. 34–51.

[20] Dwork, C., Naor, M., and Reingold, O. Immunizing encryption schemes from decryption errors. In *EUROCRYPT 2004* (2004), C. Cachin and J. Camenisch, Eds., vol. 3027 of *LNCS*, Springer, Heidelberg, pp. 342–360.

[21] ElGamal, T. A public key cryptosystem and a signature scheme based on discrete logarithms. *IEEE Transaction on Information Theory 31*, 4 (1985), 469–472.

[22] Fischlin, M., Buchmann, J., and Manulis, M., Eds. *Public Key Cryptography - PKC 2012 - 15th International Conference on Practice and Theory in Public Key Cryptography, Darmstadt, Germany, May 21-23, 2012, Proceedings* (2012), vol. 7293 of *LNCS*, Springer, Heidelberg.

[23] Fujioka, A., Suzuki, K., Xagawa, K., and Yoneyama, K. Strongly secure authenticated key exchange from factoring, codes, and lattices. In Fischlin et al. [22], pp. 467–484.

[24] Fujisaki, E., and Okamoto, T. How to enhance the security of public-key encryption at minimum cost. *IEICE transactions on fundamentals of electronics, communications and computer sciences 83*, 1 (2000), 24–32. A preliminary version appeared in *PKC '99*, 1999.

[25] Gilbert, H., Ed. *Advances in Cryptology - EUROCRYPT 2010, 29th Annual International Conference on the Theory and Applications of Cryptographic Techniques, French Riviera, May 30-June 3, 2010. Proceedings* (2010), vol. 6110 of *LNCS*, Springer, Heidelberg.

[26] Halevi, S., Ed. *Advances in Cryptology - CRYPTO 2009, 29th Annual International Cryptology Conference, Santa Barbara, CA, USA, August 16-20, 2009. Proceedings* (2009), vol. 5677 of *LNCS*, Springer, Heidelberg.

[27] Hoffstein, J., Pipher, J., and Silverman, J. H. NTRU: A ring-based public key cryptosystem. In *ANTS-III* (1998), J. Buhler, Ed., vol. 1423 of *Lecture Notes in Computer Science*, Springer-Verlag, pp. 267–288.

[28] Hofheinz, D., and Kiltz, E. The group of signed quadratic residues and applications. In Halevi [26], pp. 637–653.

[29] Hofheinz, D., and Kiltz, E. Practical chosen ciphertext secure encryption from factoring. In *EUROCRYPT 2009* (2009), A. Joux, Ed., vol. 5479 of *LNCS*, Springer, Heidelberg, pp. 313–332.

[30] Huang, Y.-J., Liu, F.-H., and Yang, B.-Y. Public-key cryptography from new multivariate quadratic assumptions. In Fischlin et al. [22], pp. 190–205.

[31] IEEE. *IEEE P1363.1/D12 Draft Standard for Public-Key Cryptographic Techniques Based on Hard Problems over Lattices*, October 2008. Available at `http://grouper.ieee.org/groups/1363/lattPK/`.

[32] ISO/IEC. *ISO/IEC 18033-2 Information technology—Security techniques—Encryption algorithms—Part 2: Asymmetric ciphers.* Geneva, 2006.

[33] Kobara, K., and Imai, H. Semantically secure McEliece public-key cryptosystems –conversions for McEliece PKC–. In *PKC 2001* (2001), K. Kim, Ed., vol. 1992 of *LNCS*, Springer, Heidelberg, pp. 19–35.

[34] Krawczyk, H. HMQV: A high-performance secure Diffie-Hellman protocol. In *CRYPTO 2005* (2005), V. Shoup, Ed., vol. 3621 of *LNCS*, Springer, Heidelberg, pp. 546–566.

[35] Kurosawa, K., Ed. *Advances in Cryptology - ASIACRYPT 2007, 13th International Conference on the Theory and Application of Cryptology and Information Security, Kuching, Malaysia, December 2-6, 2007, Proceedings* (2007), vol. 4833 of *LNCS*, Springer, Heidelberg.

[36] LaMacchia, B. A., Lauter, K., and Mityagin, A. Stronger security of authenticated key exchange. In *ProvSec 2007* (2007), W. Susilo, J. K. Liu, and Y. Mu, Eds., vol. 4784 of *LNCS*, Springer, Heidelberg, pp. 1–16.

[37] Langlois, A., and Stehlé, D. Hardness of decision (R)LWE for any modulus. Cryptology ePrint Archive, Report 2012/091, 2012. Available at `http://eprint.iacr.org/2012/091`.

[38] Lindner, R., and Peikert, C. Better key sizes (and attacks) for LWE-based encryption. In *CT-RSA 2011* (2011), A. Kiayias, Ed., vol. 6558 of *LNCS*, Springer, Heidelberg, pp. 319–330.

[39] Lyubashevsky, V., Palacio, A., and Segev, G. Public-key cryptographic primitives provably as secure as subset sum. In *TCC 2010* (2010), D. Micciancio, Ed., vol. 5978 of *LNCS*, Springer, Heidelberg, pp. 382–400.

[40] Lyubashevsky, V., Peikert, C., and Regev, O. On ideal lattices and learning with errors over rings. In Gilbert [25], pp. 1–23.

[41] McEliece, R. J. A public key cryptosystem based on algebraic coding theory. Tech. rep., DSN progress report, 1978.

[42] Merkle, R. C., and Hellman, M. E. Hiding information and signatures in trap door knapsacks. *IEEE Transactions on Information Theory 24*, 5 (September 1978), 525–530.

[43] Okamoto, T. Authenticated key exchange and key encapsulation in the standard model. In Kurosawa [35], pp. 474–484.

[44] Okamoto, T., and Pointcheval, D. REACT: Rapid enhanced-security asymmetric cryptosystem transform. In *CT-RSA 2001* (2001), D. Naccache, Ed., vol. 2020 of *LNCS*, Springer, Heidelberg, pp. 159–175.

[45] Peikert, C. Public-key cryptosystems from the worst-case shortest vector problem: extended abstract. In *STOC 2009* (2009), M. Mitzenmacher, Ed., ACM, pp. 333–342.

[46] Peikert, C., Vaikuntanathan, V., and Waters, B. A framework for efficient and composable oblivious transfer. In *CRYPTO 2008* (2008), D. Wagner, Ed., vol. 5157 of *LNCS*, Springer, Heidelberg, pp. 554–571.

[47] Peikert, C., and Waters, B. Lossy trapdoor functions and their applications. In *STOC 2008* (2008), C. Dwork, Ed., ACM, pp. 187–196.

[48] Persichetti, E. Compact McEliece keys based on quasi-dyadic Srivastava codes. *Journal of Mathematical Cryptology 6*, 2 (2012), 149–169.

[49] Pöppelman, T., and Güneysu, T. Towards efficient arithmetic for lattice-based cryptography on reconfigurable hardware. In *LATINCRYPT 2012* (2012), A. Hevia and G. Neven, Eds., vol. 7533 of *LNCS*, Springer, Heidelberg, pp. 139–158.

[50] Rabin, M. O. Digitalized signatures and public-key functions as intractable as factorization. Tech. rep., MIT, January 1979.

[51] Rabin, T., Ed. *Advances in Cryptology - CRYPTO 2010, 30th Annual International Cryptology Conference, Santa Barbara, CA, USA, August 15-19, 2010. Proceedings* (2010), vol. 6223 of *LNCS*, Springer, Heidelberg.

[52] Regev, O. On lattices, learning with errors, random linear codes, and cryptography. *Journal of the ACM 56*, 6 (2009), Article 34. A preliminary version appeared *STOC 2005*, 2005.

[53] Rivest, R. L., Shamir, A., and Adleman, L. M. A method for obtaining digital signatures and public-key cryptosystems. *Communications of the ACM 21*, 2 (Febrary 1978), 120–126.

[54] Rückert, M., and Schneider, M. Estimating the security of lattice-based cryptosystems. Cryptology ePrint Archive, Report 2010/137, 2010. Available at `http://eprint.iacr.org/2010/137`.

[55] Shor, P. W. Polynomial-time algorithms for prime factorization and discrete logarithms on a quantum computer. *SIAM Journal on Computing 26*, 5 (1997), 1484–1509.

[56] Stehlé, D., and Steinfeld, R. Making NTRU as secure as worst-case problems over ideal lattices. In *EUROCRYPT 2011* (2011), K. G. Paterson, Ed., vol. 6632 of *LNCS*, Springer, Heidelberg, pp. 27–47.

[57] Stehlé, D., Steinfeld, R., Tanaka, K., and Xagawa, K. Efficient public key encryption based on ideal lattices. In *ASIACRYPT 2009* (2009), M. Matsui, Ed., vol. 5912 of *LNCS*, Springer, Heidelberg, pp. 617–635.

[58] Steinfeld, R., Ling, S., Pieprzyk, J., Tartary, C., and Wang, H. NTRUCCA: How to strengthen NTRUEncrypt to chosen-ciphertext security in the standard model. In Fischlin et al. [22], pp. 353–371.

[59] Wee, H. Efficient chosen-ciphertext security via extractable hash proofs. In Rabin [51], pp. 314–332.

[60] Zhandry, M. Secure identity-based encryption in the quantum random oracle model. In *CRYPTO 2012* (2012), R. Safavi-Naini and R. Canetti, Eds., vol. 7417 of *LNCS*, Springer, Heidelberg, pp. 758–775.

APPENDIX

A. PROOF OF THEOREM 1

In the experiment of CK^+ security, we suppose that sid^* is the session identity for the test session. We will show that if a PPT bounded adversary $\mathcal{A}$ can distinguish the session key of a fresh session from a randomly chosen session key, we can construct a OW-CCA or OW-CPA adversary $\mathcal{S}$. Let κ be the security parameter, and let $\mathcal{A}$ be a polynomially (in κ) bounded adversary. We use adversary $\mathcal{A}$ to construct $\mathcal{S}$ that succeeds with non-negligible probability. Suc denotes the event that $\mathcal{A}$ wins. Let AskH be the event that adversary $\mathcal{A}$ poses $(K_{A,1}, K_{B,1}, K_{B,2}, U_A, U_B, ek_{A,1}, ek_{B,1}, C_{A,1}, ek_{A,2}, C_{B,1}, C_{B,2})$ to H_2. Let $\overline{\textsc{AskH}}$ be the complement of event AskH. Let sid be any completed session owned by an honest party such that $\mathsf{sid} \neq \mathsf{sid}^*$ and sid is non-matching to sid^*. Since sid and sid^* are distinct and non-matching, the inputs to the key derivation function H_2 are different for sid and sid^*. Since H_2 is a random oracle, $\mathcal{A}$ cannot obtain any information about the test session key from the session keys of non-matching sessions. Hence, $\Pr[\textsc{Suc} \wedge \overline{\textsc{AskH}}] \leq \frac{1}{2}$ and $\Pr[\textsc{Suc}] = \Pr[\textsc{Suc} \wedge \textsc{AskH}] + \Pr[\textsc{Suc} \wedge \overline{\textsc{AskH}}] \leq \Pr[\textsc{Suc} \wedge \textsc{AskH}] + \frac{1}{2}$. Henceforth, the event $\textsc{Suc} \wedge \textsc{AskH}$ is denoted by $\textsc{Suc}^*$.

For party P, we denote the static secret key by $dk_{P,1}$, the ephemeral secret key for an initiator by $r_{P,1} \in \{0,1\}^\kappa$ and $r_{P,2} \in \mathcal{RS}_G$ and the ephemeral secret key for a responder by $r_{P,1} \in \{0,1\}^\kappa$ and $r_{P,2} \in \mathcal{RS}_E$. Assume that $\mathcal{A}$ succeeds in an environment with N users, activates at most ℓ sessions within a party.

We consider the following events.

- Let AskS be the event that $\mathcal{A}$ poses the static secret key $dk_{A,1}$ to H_1 when U_A is the owner of sid^*.
- Let $\overline{\textsc{AskS}}$ be the complement of event AskS.

More specifically, we consider the following events that cover all cases of the behavior of $\mathcal{A}$.

- Let E_1 be the event that the test session sid^* has no matching session $\overline{\mathsf{sid}}^*$, the owner of sid^* is the initiator and the static secret key of the initiator is given to $\mathcal{A}$.
- Let E_2 be the event that the test session sid^* has no matching session $\overline{\mathsf{sid}}^*$, the owner of sid^* is the initiator and the ephemeral secret key of sid^* is given to $\mathcal{A}$.

- Let E_3 be the event that the test session sid^* has no matching session $\overline{\mathsf{sid}}^*$, the owner of sid^* is the responder and the static secret key of the responder is given to $\mathcal{A}$.
- Let E_4 be the event that the test session sid^* has no matching session $\overline{\mathsf{sid}}^*$, the owner of sid^* is the responder and the ephemeral secret key of sid^* is given to $\mathcal{A}$.
- Let E_5 be the event that the test session sid^* has matching session $\overline{\mathsf{sid}}^*$, and both static secret keys of the initiator and the responder are given to $\mathcal{A}$.
- Let E_6 be the event that the test session sid^* has matching session $\overline{\mathsf{sid}}^*$, and both ephemeral secret keys of sid^* and $\overline{\mathsf{sid}}^*$ are given to $\mathcal{A}$.
- Let E_7 be the event that the test session sid^* has matching session $\overline{\mathsf{sid}}^*$, and the static secret key of the owner of sid^* and the ephemeral secret key of $\overline{\mathsf{sid}}^*$ are given to $\mathcal{A}$.
- Let E_8 be the event that the test session sid^* has matching session $\overline{\mathsf{sid}}^*$, and the ephemeral secret key of sid^* and the static secret key of the owner of $\overline{\mathsf{sid}}^*$ are given to $\mathcal{A}$.

To finish the proof, we investigate events $\text{AskS} \wedge \text{Suc}^*$ and $E_i \wedge \overline{\text{AskS}} \wedge \text{Suc}^*$ $(i = 1, \ldots, 8)$ that cover all cases of event Suc^*.

A.1 Event $\text{AskS} \wedge \text{Suc}^*$

$\mathcal{S}$ receives the challenge ciphertext C^* for the key K^*. In the event AskS, $\mathcal{A}$ poses the static secret key $dk_{A,1}$ to H_1. $\mathcal{S}$ embeds the instance as $C^*_{B,1} = C^*$ and decrypts C^* with $dk_{A,1}$. Then, $\mathcal{S}$ obtains K^*.

A.2 Event $E_1 \wedge \overline{\text{AskS}} \wedge \text{Suc}^*$

In the event E_1, the test session sid^* has no matching session $\overline{\mathsf{sid}}^*$, the static secret key of U_A is given to $\mathcal{A}$. In the case of event $E_1 \wedge \overline{\text{AskS}} \wedge \text{Suc}^*$, OW-CCA adversary $\mathcal{S}$ performs the following steps.

Initialization.

$\mathcal{S}$ receives the public key ek^* as a challenge. Also, $\mathcal{S}$ receives the challenge ciphertext C^* for the key K^*.

Setup.

$\mathcal{S}$ randomly selects two parties U_A, U_B and $i \in [1, \ell]$ that becomes a guess of the test session with probability $1/N^2\ell$. $\mathcal{S}$ sets all N users' static secret and public keys except U_B. $\mathcal{S}$ selects $r_I \in \mathcal{RS}_G$ randomly, runs the key generation algorithm $(dk_{I,1}, ek_{I,1}) \leftarrow \mathsf{KeyGen}(r_I)$, and sets U_I's static secret and public key as $(dk_{I,1}, ek_{I,1})$. $\mathcal{S}$ sets ek^* as the static public key of U_B.

Also, $\mathcal{S}$ sets the ephemeral public key of i-th session of U_A as follows: $\mathcal{S}$ generates $ek_{A,2}$ obeying the protocol and sets $(C^*, ek_{A,2})$ as the ephemeral public key.

Simulation.

$\mathcal{S}$ simulates oracle queries by $\mathcal{A}$ as follows. $\mathcal{S}$ maintains the lists $\mathcal{L}_{H_1}$ and $\mathcal{L}_{H_2}$ that contains queries and answers of the H_1 and H_2 oracles respectively, and the list $\mathcal{L}_{SK}$ that contains queries and answers of $\mathsf{SessionKeyReveal}$.

1. $H_1(r_i, dk_i)$: If there exists a tuple $(r_i, dk_i, *) \in \mathcal{L}_{H_1}$, $\mathcal{S}$ returns the registered value [6]; otherwise, $\mathcal{S}$ chooses $h_i \in \mathcal{RS}_E$ randomly, returns h_i and records it to $\mathcal{L}_{H_1}$.

[6] $H_1(r_{A,1}, dk_{A,1})$ is not registered in $\mathcal{L}_{H_1}$. However, $\mathcal{A}$ does not pose $\mathsf{SessionStateReveal}(\mathsf{sid}^*)$ by the security definition; thus, $\mathcal{A}$ cannot know information about $r_{A,1}$. Thus, $\mathcal{A}$ cannot distinguish the real experiment from the simulation by such queries.

2. $H_2(K_{P,1}, K_{Q,1}, K_{Q,2}, U_P, U_Q, ek_{P,1}, ek_{Q,1}, C_{P,1}, ek_{P,2}, C_{Q,1}, C_{Q,2})$:
   (a) If $P = A$, $Q = B$, $C_{A,1} = C^*$, and $(\Pi, \mathcal{I}, U_A, U_B, ek_{A,1}, ek^*, (C^*, ek_{A,2}), (C_{B,1}, C_{B,2}))$ is i-th session of U_A, then $\mathcal{S}$ outputs $K_{A,1}$ as the answer of OW-CCA game (i.e., K^*);
   (b) else if there exists a tuple $(K_{P,1}, K_{Q,1}, K_{Q,2}, U_P, U_Q, ek_{P,1}, ek_{Q,1}, C_{P,1}, ek_{P,2}, C_{Q,1}, C_{Q,2}, h) \in \mathcal{L}_{H_2}$, $\mathcal{S}$ returns registered value h;
   (c) else if $P = B$ and there exists a tuple $(U_B, U_B, ek^*, ek_{Q,1}, (C_{B,1}, ek_{B,2}), (C_{Q,1}, CT_{Q,1}), h) \in \mathcal{L}_{SK}$, and $K_{B,1} = K'_{B,1}$, $K_{Q,1} = K'_{Q,1}$ and $K_{Q,2} = K'_{Q,2}$, where $\mathcal{S}$ computes $K'_{B,1} = \mathsf{DeCap}_{dk_{Q,1}}(C_{B,1})$ and $K'_{Q,2} = \mathsf{wDeCap}_{dk_{P,2}}(C_{Q,2})$, poses $C_{Q,1}$ to the decryption oracle, and the oracle outputs $K'_{Q,1}$, then $\mathcal{S}$ returns recorded value h, and records $(K_{B,1}, K_{Q,1}, K_{Q,2}, U_B, U_Q, ek^*, ek_{Q,1}, (C_{B,1}, ek_{B,2}), (C_{Q,1}, C_{Q,2}), h)$ in the list $\mathcal{L}_H$;
   (d) else if $Q = B$ and there exists a tuple $(U_P, U_B, ek_{P,1}, ek^*, (C_{P,1}, ek_{P,2}), (C_{B,1}, C_{B,2}), h) \in \mathcal{L}_{SK}$, and $K_{P,1} = K'_{P,1}$, $K_{B,1} = K'_{B,1}$ and $K_{B,2} = K'_{B,2}$, where $\mathcal{S}$ computes $K'_{B,1} = \mathsf{DeCap}_{dk_{P,1}}(C_{B,1})$ and $K'_{B,2} = \mathsf{wDeCap}_{dk_{P,2}}(C_{B,2})$, poses $C_{P,1}$ to the decryption oracle, and the oracle outputs $K'_{P,1}$, then $\mathcal{S}$ returns recorded value h, and records $(K_{P,1}, K_{B,1}, K_{B,2}, U_P, U_B, ek_{P,1}, ek^*, (C_{P,1}, ek_{P,2}), (C_{B,1}, C_{B,2}), h)$ in the list $\mathcal{L}_H$;
   (e) else if there exists a tuple $(U_P, U_Q, ek_{P,1}, ek_{Q,1}, (C_{P,1}, ek_{P,2}), (C_{Q,1}, C_{Q,2}), h) \in \mathcal{L}_{SK}$, and $K_{P,1} = K'_{P,1}$, $K_{Q,1} = K'_{Q,1}$ and $K_{Q,2} = K'_{Q,2}$, where $\mathcal{S}$ computes $K'_{Q,1} = \mathsf{DeCap}_{dk_{P,1}}(C_{Q,1})$, $K'_{P,1} = \mathsf{DeCap}_{dk_{Q,1}}(C_{P,1})$ and $K'_{Q,2} = \mathsf{DeCap}_{dk_{P,2}}(C_{Q,2})$, than $\mathcal{S}$ returns recorded value h and record $(K_{P,1}, K_{Q,1}, K_{Q,2}, U_P, U_Q, ek_{P,1}, ek_{Q,1}, (C_{P,1}, ek_{P,2}), (C_{Q,1}, C_{Q,2}), h)$ in the list $\mathcal{L}_{H_2}$;
   (f) otherwise, $\mathcal{S}$ returns a random value $SK \in \{0,1\}^\kappa$ and records it in the list $\mathcal{L}_{H_2}$.
3. $\mathsf{Send}(\Pi, \mathcal{I}, U_P, U_Q)$: If $P = A$ and the session is i-th session of U_A, $\mathcal{S}$ returns the ephemeral public key $(C^*, ek_{A,2})$ computed in the setup. Otherwise, $\mathcal{S}$ computes the ephemeral public key $(C_{P,1}, ek_{P,2})$ obeying the protocol, returns it and records $(\Pi, U_P, U_Q, (C_{P,1}, ek_{P,2}))$.
4. $\mathsf{Send}(\Pi, \mathcal{R}, U_Q, U_P, (C_{P,1}, ek_{P,2}))$: $\mathcal{S}$ computes the ephemeral public key $(C_{Q,1}, C_{Q,2})$ obeying the protocol, and the session key SK as follows:
   (a) If $Q = B$ and there exists a tuple $(K_{P,1}, K_{B,1}, K_{B,2}, U_P, U_B, ek_P, ek^*, (C_{P,1}, ek_{P,2}), (C_{B,1}, C_{B,2}), h) \in \mathcal{L}_H$ for $(K_{P,1}, K_{B,1}, K_{B,2})$, where $K_{B,1}$ and $K_{B,2}$ are known for $\mathcal{S}$, and $\mathcal{S}$ poses $C_{P,1}$ to the decryption oracle and the oracle outputs $K_{P,1}$, then $\mathcal{S}$ sets $SK = h$;
   (b) else if there exists a tuple $(K_{P,1}, K_{Q,1}, K_{Q,2}, U_P, U_Q, ek_{P,1}, ek_{Q,1}, (C_{P,1}, ek_{P,2}), (C_{Q,1}, C_{Q,2}), h) \in \mathcal{L}_{H_2}$ for $(K_{P,1}, K_{Q,1}, K_{Q,2})$, where $K_{P,1} = \mathsf{DeCap}_{dk_{Q,1}}(C_{P,1})$, and $K_{Q,1}$ and $K_{Q,2}$ are known for $\mathcal{S}$, then $\mathcal{S}$ sets $SK = h$;
   (c) otherwise, $\mathcal{S}$ chooses $SK \in \{0,1\}^\kappa$ randomly.

   Finally, $\mathcal{S}$ records $(\Pi, U_P, U_Q, (C_{P,1}, ek_{P,2}), (C_{Q,1}, C_{Q,2}))$ as the completed session and SK in the list $\mathcal{L}_{SK}$.
5. $\mathsf{Send}(\Pi, \mathcal{I}, U_P, U_Q, (C_{P,1}, ek_{P,2}), (C_{Q,1}, C_{Q,2}))$:
   (a) If $P = B$ and there exists a tuple $(K_{B,1}, K_{Q,1}, K_{Q,2}, U_B, U_Q, ek^*, ek_Q, (C_{B,1}, ek_{B,2}), (C_{Q,1}, C_{Q,2}), h) \in \mathcal{L}_H$ for $(K_{B,1}, K_{Q,1}, K_{Q,2})$, where $K_{B,1} = \mathsf{DeCap}_{dk_{Q,1}}(C_{B,1})$ and $K_{Q,2} = \mathsf{wDeCap}_{dk_{B,2}}(C_{Q,2})$, and $\mathcal{S}$ poses $C_{Q,1}$ to the decryption oracle and the oracle outputs $K_{Q,1}$, then $\mathcal{S}$ sets $SK = h$;
   (b) else if $Q = B$ and there exists a tuple $(K_{P,1}, K_{B,1}, K_{B,2}, U_P, U_B, ek_P, ek^*, (C_{P,1}, ek_{P,2}), (C_{B,1}, C_{B,2}), h) \in \mathcal{L}_H$ for $(K_{P,1},$

$K_{B,1}, K_{B,2})$, where $K_{B,1} = \mathsf{DeCap}_{dk_{P,1}}(C_{B,1})$ and $K_{B,2} = \mathsf{wDeCap}_{dk_{P,2}}(C_{B,2})$, and $\mathcal{S}$ poses $C_{P,1}$ to the decryption oracle and the oracle outputs $K_{P,1}$, then $\mathcal{S}$ sets $SK = h$;

(c) else if there exists a tuple $(K_{P,1}, K_{Q,1}, K_{Q,2}, U_P, U_Q, ek_{P,1}, ek_{Q,1}, (C_{P,1}, ek_{P,2}), (C_{Q,1}, C_{Q,2}), h) \in \mathcal{L}_{H_2}$ for $(K_{P,1}, K_{Q,1}, K_{Q,2})$, $K_{P,1} = \mathsf{DeCap}_{dk_{Q,1}}(C_{P,1})$, $K_{Q,1} = \mathsf{DeCap}_{dk_{P,1}}(C_{Q,1})$ and $K_{Q,2} = \mathsf{DeCap}_{dk_{P,2}}(C_{Q,2})$, then $\mathcal{S}$ sets $SK = h$;

(d) otherwise, $\mathcal{S}$ chooses $SK \in \{0,1\}^\kappa$ randomly.

Finally, $\mathcal{S}$ records $(\Pi, U_P, U_Q, (C_{P,1}, ek_{P,2}), (C_{Q,1}, C_{Q,2}))$ as the completed session and SK in the list $\mathcal{L}_{SK}$.

6. SessionKeyReveal(sid):
   (a) If the session sid is not completed, $\mathcal{S}$ returns an error message;
   (b) else if sid is recorded in the list $\mathcal{L}_{SK}$, then $\mathcal{S}$ returns the recorded value SK;
   (c) otherwise, $\mathcal{S}$ returns a random value $SK \in \{0,1\}^\kappa$ and records it in the list $\mathcal{L}_{SK}$.
7. SessionStateReveal(sid): $\mathcal{S}$ responds the ephemeral secret key and intermediate computation results of sid as the definition. If the owner of sid is U_B, $\mathcal{S}$ poses ciphertexts received by U_B to the decryption oracle and can simulate all intermediate computation results. Note that the SessionStateReveal query is not posed to the test session from the freshness definition.
8. Corrupt(U_P): $\mathcal{S}$ responds the static secret key of U_P as the definition.
9. Test(sid): If sid is not i-th session of U_A, then $\mathcal{S}$ aborts with failure. Otherwise, $\mathcal{S}$ responds to the query as the definition.
10. If $\mathcal{A}$ outputs a guess b', $\mathcal{S}$ aborts with failure.

A.2.1 Analysis.

The simulation for $\mathcal{S}$ is perfect except with negligible probability. The probability that $\mathcal{A}$ selects the session as the test session sid* is at least $\frac{1}{N^2\ell}$.

Under the event Suc*, $\mathcal{A}$ poses correctly formed $K_{P,1}, K_{Q,1}, K_{Q,2}$ to H_2. Therefore, $\mathcal{S}$ is successful and does not abort.

Thus, $\mathcal{S}$ is successful with non-negligible probability.

A.3 Other Events

A.3.1 Event $E_2 \wedge \overline{\text{AskS}} \wedge \text{Suc}^*$.

In the event E_2, the test session sid* has no matching session $\overline{\text{sid}}^*$, the ephemeral secret key of sid* is given to $\mathcal{A}$. Thus, $\mathcal{A}$ cannot obtain any information about $dk_{A,1}$ except negligible guessing probability, since H_1 is the random oracle. Hence, $\mathcal{S}$ performs the reduction same as in the case of event $E_1 \wedge \overline{\text{AskS}} \wedge \text{Suc}^*$.

A.3.2 Event $E_3 \wedge \overline{\text{AskS}} \wedge \text{Suc}^*$.

In the event E_3, the test session sid* has no matching session $\overline{\text{sid}}^*$, the static secret key of U_B is given to $\mathcal{A}$. Thus, $\mathcal{A}$ cannot obtain any information about $dk_{B,1}$ except negligible guessing probability, since H_1 is the random oracle. Hence, $\mathcal{S}$ performs the reduction same as in the case of event $E_1 \wedge \overline{\text{AskS}} \wedge \text{Suc}^*$.

A.3.3 Event $E_4 \wedge \overline{\text{AskS}} \wedge \text{Suc}^*$.

In the event E_4, the test session sid* has no matching session $\overline{\text{sid}}^*$, the ephemeral secret key of sid* is given to $\mathcal{A}$. Thus, $\mathcal{A}$ cannot obtain any information about $dk_{B,1}$ except negligible guessing probability, since H_1 is the random oracle. Hence, $\mathcal{S}$ performs the reduction same as in the case of event $E_2 \wedge \overline{\text{AskS}} \wedge \text{Suc}^*$.

A.3.4 Event $E_5 \wedge \overline{\text{AskS}} \wedge \text{Suc}^*$.

In the event E_5, the test session sid* has the matching session $\overline{\text{sid}}^*$, both static secret keys of U_A and U_B are given to $\mathcal{A}$. OW-CPA adversary $\mathcal{S}$ embeds ek^* to $ek_{A,2}$ and C^* to $C_{B,2}$. Then, $\mathcal{S}$ obtains K^* by the hash list $\mathcal{L}_{H_2}$.

A.3.5 Event $E_6 \wedge \overline{\text{AskS}} \wedge \text{Suc}^*$.

In the event E_6, the test session sid* has the matching session $\overline{\text{sid}}^*$, both the ephemeral secret keys of *sid** and $\overline{\text{sid}}^*$ are given to $\mathcal{A}$. Then, $\mathcal{A}$ cannot obtain any information about $dk_{A,1}$ except negligible guessing probability because H_1 is the random oracle. Hence, $\mathcal{S}$ performs the reduction same as in the case of event $E_2 \wedge \overline{\text{AskS}} \wedge \text{Suc}^*$.

A.3.6 Event $E_7 \wedge \overline{\text{AskS}} \wedge \text{Suc}^*$.

In the event E_7, the test session sid* has the matching session $\overline{\text{sid}}^*$, the static secret keys of U_A and the ephemeral secret key of $\overline{\text{sid}}^*$ are given to $\mathcal{A}$. Then, $\mathcal{A}$ cannot obtain any information about $r_{A,1}$ except negligible guessing probability because H_1 is the random oracle. Hence, $\mathcal{S}$ performs the reduction same as in the case of event $E_1 \wedge \overline{\text{AskS}} \wedge \text{Suc}^*$.

A.3.7 Event $E_8 \wedge \overline{\text{AskS}} \wedge \text{Suc}^*$.

In the event E_8, the test session sid* has the matching session $\overline{\text{sid}}^*$, the static secret keys of U_B and the ephemeral secret key of sid* are given to $\mathcal{A}$. $\mathcal{S}$ embeds ek^* to $ek_{A,1}$ and C^* to $C_{B,1}$. Then, $\mathcal{S}$ obtains K^* by the hash list $\mathcal{L}_{H_2}$.

□

Blank Digital Signatures

Christian Hanser
Institute for Applied Information Processing and Communications (IAIK)
Graz University of Technology (TUG)
Inffeldgasse 16a, 8010 Graz, Austria
christian.hanser@iaik.tugraz.at

Daniel Slamanig
Institute for Applied Information Processing and Communications (IAIK)
Graz University of Technology (TUG)
Inffeldgasse 16a, 8010 Graz, Austria
daniel.slamanig@iaik.tugraz.at

ABSTRACT

In this paper we present a novel type of digital signatures, which we call *blank digital signatures*. The basic idea behind this scheme is that an originator can define and sign a *message template*, describing fixed parts of a message as well as multiple choices for exchangeable parts of a message. One may think of a form with blank fields, where for such fields the originator specifies all the allowed strings to choose from. Then, a proxy is given the power to sign an *instantiation* of the template signed by the originator by using some secret information. By an instantiation, the proxy commits to one allowed choice per blank field in the template. The resulting *message signature* can be publicly verified under the originator's and the proxy's signature verification keys. Thereby, no verifying party except the originator and the proxy learn anything about the "unused" choices from the message template given a message signature. Consequently, the template is hidden from verifiers.

We discuss several applications, provide a formal definition of *blank digital signature schemes* and introduce a security model. Furthermore, we provide an efficient construction of such a blank digital signature scheme from any secure digital signature scheme, pairing-friendly elliptic curves and polynomial commitments, which we prove secure in our model. We also provide a detailed efficiency analysis of our proposed construction supporting its practicality. Finally, we outline several open issues and extensions for future work.

Categories and Subject Descriptors

[**Security and privacy**]: Digital signatures

General Terms

Algorithms, Design, Security, Theory

Keywords

Digital signature scheme, blank digital signatures, elliptic curves, pairings, polynomial commitments

1. INTRODUCTION

Digital signatures provide the means to achieve source authentication and data integrity for digital messages in a publicly verifiable way meaning that at signing time a signer commits himself to a concrete message. In this paper, we propose the novel concept of a *blank digital signature scheme*. Here, an originator can define and sign a *message template*, describing fixed parts of a message as well as several choices for exchangeable parts of a message (one may think of a form with blank fields, where for such fields the originator specifies all the allowed strings to choose from), for which he can delegate signing permissions to a proxy. This proxy is given the power to sign *template instantiations* of the template given by the originator by using some secret information. The resulting *message signature* can be publicly verified under the originator's and the proxy's signature verification keys. Thereby, no verifying party except the originator and the proxy learn anything about the "unused" choices from the message template and, consequently, about the template when given a message signature. In order to construct such a scheme it is helpful to look at existing variants of digital signature schemes to figure out, whether they can be used to instantiate blank digital signatures.

Conventional digital signatures require the signer to be available during signature creation, e.g., when a contract is signed. To overcome this limitation, the concept of proxy signatures [16] has been introduced. Basically, a proxy signature scheme allows an entity (the delegator) to delegate his signing capabilities to another entity (the proxy) that can then construct signatures on behalf of the delegator. This concept has seen a considerable amount of interest since then [7]. Surprisingly, only quite recently a suitable security model for proxy signatures has been introduced [6], and been extended to multi-level and identity-based proxy signature schemes later on [21]. Since in a practical application, the delegator may not want to give the proxy the power to sign *any* message on behalf of the signer, the *delegation by warrant* [16] approach was proposed. Here, a signed *warrant* is used to describe the delegation. Thereby, any type of security policy may be included in the warrant to describe the restrictions under which the delegation is valid. This approach seems to be particularly attractive and received the most attention, since the designator can define a message space for which he delegates his signing rights. In state of the art schemes [21, 7], a *warrant* consists of the description ω of the message space for which signing is being delegated, together with a "certificate", which is a signature on ω under the delegators private signing key. We are given a similar requirement and, consequently, could ask whether proxy signatures can be used in this setting. In proxy signatures, this warrant is an abstract description, which could, for instance, be a context-free grammar, a regular expression, or as in [6], the description of a polynomial-time Turing machine computing the characteristic function of all potential messages, i.e., given a message to decide, whether the message is covered by ω or not. However, in proxy signatures the proxy is allowed to sign arbitrary messages from this abstract message space with the downside that the verifier learns the entire message space. Consequently, our requirement that the proxy can sign instantiations of a template without a verifier learning the corresponding template can not be realized by using existing proxy signature schemes.

Conventional digital signature schemes do not allow alterations of a signed document without invalidating the signature. Since it may be valuable to have the possibility to replace or remove (specified) parts of a message after signature creation such that the original signature stays valid (and no interaction with the original signer is required), redactable [23, 13] as well as sanitizable signature schemes [3] have been introduced. Signature schemes, which allow *removal* of content (replacement by some special symbol $\perp$) by *any* party are called redactable [23, 13], while signature schemes allowing (arbitrary) *replacements* of *admissible* parts by a *designated* party are called sanitizable signature schemes [3], cf. [20] for a comparison. As in our setting, the proxy should be allowed to choose from a list of predefined replacements for designated parts of the message, one could ask whether redactable or sanitizable signatures can be used in this setting. Since in redactable signature schemes any party is allowed to modify signed messages by removing message parts, such signature schemes are obviously not compatible with our requirements. The original concept of sanitizable signatures [3] allows designated sanitizers to replace designated parts of a message. However, here the sanitizer does not have the role of a proxy meaning that it does not sign the modified message. Furthermore, a sanitizer can replace the designated parts with arbitrary strings, which is clearly not meeting our requirements. The concept of sanitizable signatures was later on extended to allow only permitted replacements [15], yet, the Bloom filter [5] based construction does not meet cryptographic security requirements and the cryptographic accumulator [4] based approach [15, 9] allows to securely restrict replacements. Yet, both approaches are not designed and also do not support the hiding of the set of accumulated values (allowed replacements) and, thus, are not suitable for our construction.

To sum this up, our concept has more in common with proxy signatures than with sanitizable signatures. This is due to our requirements that the signature of the originator is not publicly verifiable as it is the case in sanitizable signatures and only instantiations can be publicly verified as it is the case for proxy signatures.

1.1 Contribution

Since, however, none of the existing concepts covers all our requirements, we propose the novel concept of a *blank digital signature scheme*. Here, an originator, i.e., the signer delegating signing permissions, can define and sign a *message template*, describing fixed parts of a message as well as several choices for exchangeable parts of a message. One may think of a form with blank fields, where for such fields the originator specifies all the allowed strings to choose from. Then, a proxy is given the power to sign *template instantiations* of the template given by the originator by using some secret information. The resulting *message signature* can be publicly verified under the originator's and the proxy's signature verification keys. Thereby, no verifying party except the originator and the proxy learn anything about the "unused" choices from the message template and, consequently, about the template given a message signature. Since this setting is quite different from the security requirements of proxy signatures and sanitizable signatures, most importantly, the template should be hidden from verifiers, we define a novel type of signature scheme along with a suitable security model. Similar to proxy signatures and sanitizable signatures, we require a public key infrastructure meaning that the originator and proxy are in possession of authentic signing keys. Moreover, since we use polynomial commitments in our construction, we need the parameters to be generated by a trusted third party.

A naive approach to realize blank digital signatures is that the originator produces n signatures for all n possible instantiations together with the public key of the proxy using a standard digital signature scheme, whereas the proxy simply signs the originator's signature for the chosen instantiations. However, the number of signatures issued by the originator would then be $O(n)$, which gets impractical very soon with increasing number of choices in exchangeable parts. By using randomized Merkle hash trees [17] as in redactable signatures, the number of signatures of the originator could be reduced to $O(1)$, whereas the signature of the proxy would then, however, be of size $O(\log n)$. At first glance, this may seem attractive, yet in Section 5.4 we illustrate that this approach also becomes soon impractical with an increasing number of choices. In our construction, the number of signatures of the originator is $O(1)$, whereas the size of both signatures, of the originator and the proxy, are also $O(1)$ and, in particular, very small and constant. Clearly, this is far more appealing than the aforementioned naive approaches.

1.2 Outline

In Section 2, we sketch some application scenarios for blank digital signatures. Section 3 discusses the mathematical and cryptographic preliminaries. Then, in Section 4 we introduce the notion of blank digital signatures and the corresponding security model. A construction of a blank digital signature scheme along with its security proof, an efficiency analysis and a comparison to the naive approaches are given in Section 5. Finally, Section 6 concludes the paper and lists open issues for future work.

2. APPLICATIONS

Here, we sketch some application scenarios which we envision for this novel type of digital signatures.

Partially blank signed contracts: Suppose a person is willing to sign a contract under certain predefined conditions, e.g., set of potential prices, range of possible contract dates, but is not able to sign the contract in person. Then, this person can elegantly delegate this task to another semi-trusted party, e.g., his attorney, by using blank digital signatures. The third party is then able to conclude the contract on behalf of his client. The client can do so by defining a contract template thereby leaving certain positions "blank", i.e., defining certain potential choices for the position without committing to one, and signing the template. Then, at a later point in time, the attorney is able to "fill in the gaps" by choosing from predefined choices, whereas the original signature of the client remains valid, and then signing the resulting contract as a proxy.

"Sanitizable" signatures: We may interpret exchangeable parts of message templates as replacements (with a potentially empty string) and, thus, can achieve a scheme with similar capabilities, but different meaning and strength as a non-interactive publicly accountable sanitizable signature scheme [8][1], which supports controlled replacements [9, 15]. Note that such a sanitizable signature scheme does not yet exist. However, there are some key differences. In contrast to sanitizable signatures, our template signature is not intended to be publicly verifiable, i.e., can only be verified by the proxy and, thus, the originator does not commit to a concrete instantiation of the template. Furthermore, in blank digital signatures, the allowed replacements are hidden, which is not supported by sanitizable signatures allowing such replacements [9, 15]. Consequently, blank digital signatures may be seen as signature schemes supporting sanitizing capabilities, but are a different concept as it is clear from the differences mentioned above.

3. PRELIMINARIES

In this section we firstly provide an overview of required mathematical and cryptographic preliminaries.

3.1 Mathematical Background

An elliptic curve over the finite field $\mathbb{F}_q$ is a plane, smooth curve described by the Weierstrass equation:

$$E : Y^2 + a_1XY + a_3Y = X^3 + a_2X^2 + a_4X + a_6, \quad (1)$$

where $a_1, a_2, a_3, a_4, a_6 \in \mathbb{F}_q$. The set $E(\mathbb{F}_q)$ of points $(x, y) \in \mathbb{F}_q^2$ satisfying Equation (1) plus the point at infinity ∞, which is the neutral element, forms an additive Abelian group, whereas the group law is determined by the *chord-and-tangent* method [22].

Let G be a cyclic group and p be a divisor of its group order, then there exists a subgroup of order p, which we subsequently denote by $G[p]$.

Definition 1 (Bilinear Map). *Let G_1, G_2 and G_T be three cyclic groups of the same prime order p, where G_1, G_2 are additive groups and G_T is a multiplicative group. We call the map $e : G_1 \times G_2 \to G_T$ a bilinear map or pairing, if the following conditions hold:*

Bilinearity: *For all $P_1, P_2 \in G_1$ and $P_1', P_2' \in G_2$ we have:*

- $e(P_1 + P_2, P') = e(P_1, P') \cdot e(P_2, P')$ *for all* $P' \in G_2$,
- $e(P, P_1' + P_2') = e(P, P_1') \cdot e(P, P_2')$ *for all* $P \in G_1$.

Non-degeneracy: *If P is a generator of G_1 and P' a generator of G_2, then $e(P, P')$ is a generator of G_T, i.e., $e(P, P') \neq 1_{G_T}$.*

Efficiently computable: *e can be computed efficiently.*

If $G_1 = G_2$, then e is called *symmetric* and *asymmetric* otherwise. The former type is also called *Type-1* pairing, whereas in case of the latter we distinguish between *Type-2* and *Type-3* pairings. For Type-2 pairings there is an efficiently computable isomorphism $\Psi : G_2 \to G_1$ [10] and for Type-3 pairings such an efficiently computable isomorphism does not exist. In our setting, G_1 and G_2 are p-order elliptic curve group over $\mathbb{F}_q$ and $G_T = \mathbb{F}_{q^k}^*[p]$, which is an order p subgroup of $\mathbb{F}_{q^k}^*$. Note that k, the so called *embedding degree*, is defined as $k = \min\{\ell \in \mathbb{N} : p \mid q^\ell - 1\}$.

3.2 Digital Signatures

Here, we briefly recall the definition of a standard digital signature scheme.

Definition 2 (Digital Signature Scheme). *A digital signature scheme is a tuple* $(\mathsf{KeyGen}^{\mathsf{DSS}}, \mathsf{Sign}^{\mathsf{DSS}}, \mathsf{Verify}^{\mathsf{DSS}})$ *of polynomial-time algorithms:*

$\mathsf{KeyGen}^{\mathsf{DSS}}(\kappa)$: *Is a key generation algorithm that takes as input a security parameter $\kappa \in \mathbb{N}$ and outputs a private (signing) key $\mathsf{sk}^{\mathsf{DSS}}$ and a public (verification) key $\mathsf{pk}^{\mathsf{DSS}}$.*

$\mathsf{Sign}^{\mathsf{DSS}}(M, \mathsf{sk}^{\mathsf{DSS}})$: *Is a (probabilistic) algorithm taking input a message $M \in \{0,1\}^*$, a private key $\mathsf{sk}^{\mathsf{DSS}}$ and outputs a signature σ.*

$\mathsf{Verify}^{\mathsf{DSS}}(\sigma, M, \mathsf{pk}^{\mathsf{DSS}})$: *Is a deterministic algorithm taking input a signature σ, a message $M \in \{0,1\}^*$, a public key $\mathsf{pk}^{\mathsf{DSS}}$ and outputs a single bit $b \in \{\mathtt{true}, \mathtt{false}\}$ indicating whether σ is a valid signature for M.*

Furthermore, we require the digital signature scheme to be correct, i.e., for all $(\mathsf{sk}^{\mathsf{DSS}}, \mathsf{pk}^{\mathsf{DSS}}) \in \mathsf{KeyGen}(\kappa)$ and all $M \in \{0,1\}^*$, $\mathsf{Verify}^{\mathsf{DSS}}(\mathsf{Sign}^{\mathsf{DSS}}(\mathsf{M}, \mathsf{sk}^{\mathsf{DSS}}), \mathsf{M}, \mathsf{pk}^{\mathsf{DSS}}) = \mathtt{true}$ must hold. A digital signature scheme is secure, if it is existentially unforgeable under adaptively chosen-message attacks (UF-CMA) [12]. Note that in practice, the sign and verify algorithms will typically use a hash function to map input messages to constant size strings, which is also known as the *hash-then-sign paradigm*.

3.3 Polynomial Commitments

In [14], Kate et al. introduced the notion of constant-size polynomial commitments. The authors present two distinct commitment schemes, whereas one is unconditionally binding and computationally hiding ($\mathsf{PolyCommit}_{\mathsf{DL}}$) and the other is unconditionally hiding as well as computationally binding ($\mathsf{PolyCommit}_{\mathsf{Ped}}$). For our scheme, we are using $\mathsf{PolyCommit}_{\mathsf{Ped}}$, which is based on Pedersen commitments [19]. The constructions of [14] use an algebraic property of polynomials $f(X) \in \mathbb{Z}_p[X]$. Namely, that $(X - \lambda)$ perfectly divides the polynomial $f(X) - f(\lambda)$ for $\lambda \in \mathbb{Z}_p$: Now, we briefly present the $\mathsf{PolyCommit}_{\mathsf{Ped}}$ construction of [14].

$\mathsf{Setup}(\kappa, t)$: Pick two groups G, G_T of the same prime order p (with p being a prime of bitlength κ) having a symmetric pairing $e : G \times G \to G_T$. Choose two

[1]In such a sanitizable signature scheme, when given a signature anybody can verify, whether a modification has been conducted by the original signer or the sanitizer without interacting with any party.

generators $P, Q \in G$ and $\alpha \in_R \mathbb{Z}_p^*$ and output $\mathsf{pk} = (G, G_T, p, e, P, \alpha P, \ldots, \alpha^t P, Q, \alpha Q, \ldots, \alpha^t Q)$ as well as $\mathsf{sk} = \alpha$.

$\mathsf{Commit}(\mathsf{pk}, f(X))$: Given $f(X) \in \mathbb{Z}_p[X]$ with $\deg(f) \leq t$, choose a random polynomial $r(X) \in \mathbb{Z}_p[X]$ with $\deg(f) \leq \deg(r) \leq t$, compute the commitment $\mathcal{C} = f(\alpha)P + r(\alpha)Q \in G$ and output $\mathcal{C}$.

$\mathsf{Open}(\mathsf{pk}, \mathcal{C}, f(X), r(X))$: Output $(f(X), r(X))$.

$\mathsf{VerifyPoly}(\mathsf{pk}, \mathcal{C}, f(X), r(X))$: Verify whether

$$\mathcal{C} = \sum_{i=0}^{\deg(f)} f^{(i)}(\alpha^i P) + \sum_{i=0}^{\deg(r)} r^{(i)}(\alpha^i Q)$$

holds and output `true` on success and `false` otherwise.

$\mathsf{CreateWit}(\mathsf{pk}, f(X), r(X), \lambda)$: Compute $\phi(X) = \frac{f(X)-f(\lambda)}{X-\lambda}$, $\hat{\phi}(X) = \frac{r(X)-r(\lambda)}{X-\lambda}$ and $W_\lambda = \phi(\alpha)P + \hat{\phi}(\alpha)Q$ and output $(\lambda, f(\lambda), r(\lambda), W_\lambda)$.

$\mathsf{VerifyEval}(\mathsf{pk}, \mathcal{C}, \lambda, f(\lambda), r(\lambda), W_\lambda)$: Verify that $f(\lambda)$ is the evaluation of f at point λ. This is done by checking whether

$$e(\mathcal{C}, P) = e(W_\lambda, \alpha P - \lambda P) \cdot e(f(\lambda)P + r(\lambda)Q, P)$$

holds. Output `true` on success and `false` otherwise.

This scheme can be proven secure under the t-SDH assumption in G, as long as $t < \sqrt{2^\kappa}$. For the proof we refer the reader to [14]. Notice that α must remain unknown to the committer (and thus the setup has to be run by a trusted third party), since, otherwise, it would be a trapdoor commitment scheme.

Moreover, we note that we do not need the algorithms $\mathsf{CreateWit}$ and $\mathsf{VerifyEval}$ in our construction, since we do not need to prove valid evaluations of the polynomial $f(X)$ without revealing the polynomial itself. For the same reason, we simply use a random point R instead of $\sum_{i=0}^{\deg(r)} r^{(i)}(\alpha^i Q)$, which obviously also yields an unconditionally hiding Pedersen commitment.

4. BLANK DIGITAL SIGNATURES

In this section we introduce the notion of blank digital signatures as well as the according security model. As a prerequisite we first need to introduce representations and encodings for message templates and template instantiations.

4.1 Template and Message Representation

In the following we introduce a representation for message templates. A message template $\mathcal{T}$ describes all potential template instantiations that correspond to a single template. More formally, a message template is defined as follows.

Definition 3 (Message Template). *A message template $\mathcal{T}$ is a sequence of non-empty sets $T_i = \{M_{i_1}, \ldots, M_{i_l}\}$ of bitstrings M_{i_j} and uniquely identified by $id_\mathcal{T}$. If the size of T_i is one, then the set T_i is called* fixed element *of $\mathcal{T}$ and* exchangeable element *otherwise. The set of all message templates is denoted by $\mathbb{T}$.*

An exchangeable element T_i represents allowed substitutions, i.e., T_i can be replaced by any of its elements M_{i_j} in order to obtain an *instantiation* of the template. Let n be the sequence length of $\mathcal{T}$, then n is called *length of template $\mathcal{T}$*. Furthermore, with $|\mathcal{T}|$ we denote the *size of template $\mathcal{T}$*, that is $|\mathcal{T}| = \sum_{i=1}^n |T_i|$. Finally, we call $\max_{1\leq i \leq n} |T_i|$ the *depth of template $\mathcal{T} = (T_i)_{i=1}^n$*.

Definition 4 (Template Instantiation). *A* template instantiation *$\mathcal{M}$ of some template $\mathcal{T} = (T_i)_{i=1}^n$ is derived from $\mathcal{T}$ as follows. For each $1 \leq i \leq n$ choose exactly one element $M_i \in T_i$ and set $\mathcal{M} = (M_i)_{i=1}^n$. A template instantiation $\mathcal{M}$ is called* valid, *which we denote by $\mathcal{M} \preceq \mathcal{T}$, if it represents choices that were intended by the originator of template $\mathcal{T}$. Furthermore, we use $\mathbb{M}_\mathcal{T} = \{\mathcal{M} : \mathcal{M} \preceq \mathcal{T}\}$ to denote the set of all possible template instantiations of a template $\mathcal{T}$.*

A message template $\mathcal{T}$ is called

- *trivial* if it does not contain any exchangeable elements. Note that this implies $|\mathcal{T}| = n$, and
- *minimal* if no two fixed elements are adjacent.

The *minimal* property guarantees that the number of fixed elements is kept minimal. The *complement* of the template instantiation $\mathcal{M}$ denoted as $\overline{\mathcal{M}}$ is a sequence of sets of bitstrings and represents all unused choices in the exchangeable elements, that is $\overline{\mathcal{M}} = (T_i \setminus \{M_i\})_{i=1}^n$ for an instantiation $\mathcal{M} = (M_i)_{i=1}^n$ of $\mathcal{T} = (T_i)_{i=1}^n$.

Now, we give a short example to illustrate our concept.

Example 1. *Let $\mathcal{T} = (T_1, T_2, T_3)$ with*

- *$T_1 = \{$"I, hereby, declare to pay "$\}$,*
- *$T_2 = \{$"100\$", "120\$", "150\$"$\}$ and*
- *$T_3 = \{$" for this tablet device."$\}$.*

Here, T_1 and T_3 are fixed elements and T_2 is an exchangeable element with three choices. A template instantiation could, for instance, be $\mathcal{M} = ($"I, hereby, declare to pay ", "120\$", " for this tablet device."$)$. The complement of template instantiation $\mathcal{M}$ is then $\overline{\mathcal{M}} = (\emptyset, \{$"100\$", "150\$"$\}, \emptyset)$.

In the following, we define encodings of templates and template instantiations, for which we use polynomials in the Euclidean ring $\mathbb{Z}_p[X]$. This allows us to perform polynomial division with remainder, which is essential to our construction.

Definition 5 (Template Encoding). *Let $\mathcal{T} = (T_i)_{i=1}^n$ be a message template and $H : \{0,1\}^* \to \mathbb{Z}_p$ be a full-domain cryptographic hash function. A* template encoding function *$t : \mathbb{T} \to \mathbb{Z}_p[X]$ is defined as follows:*

$$\mathcal{T} \mapsto \prod_{i=1}^{n} \prod_{M \in T_i} (X - H(M \| id_\mathcal{T} \| i)).$$

The evaluation $t(\mathcal{T})$ results in a so-called template encoding polynomial *$t_\mathcal{T} \in \mathbb{Z}_p[X]$ of degree $|\mathcal{T}|$.*

Note that the degree of the resulting polynomial needs to be bounded by $\sqrt{2^\kappa}$, as otherwise the security of the polynomial commitment scheme is no longer guaranteed. However, this has no impact in practice, since the polynomial can only be created by a polynomial time algorithm.

Definition 6 (Message Encoding). *Similar to Definition 5, a* message encoding function $m_{\mathcal{T}} : \mathbb{M}_{\mathcal{T}} \rightarrow \mathbb{Z}_p[X]$ *with respect to a message template* $\mathcal{T}$ *is defined as follows:*

$$\mathcal{M} \mapsto \prod_{i=1}^{n} (X - H(M_i \| id_{\mathcal{T}} \| i)),$$

where $\mathcal{M} = (M_i)_{i=1}^n$ *is an instantiation of* $\mathcal{T} = (T_i)_{i=1}^n$*. We call* $m_{\mathcal{M}} = m_{\mathcal{T}}(\mathcal{M}) \in \mathbb{Z}_p[X]$ message encoding polynomial. *Furthermore, we define the* complementary message encoding function $\overline{m}_{\mathcal{T}} : \mathbb{M}_{\mathcal{T}} \rightarrow \mathbb{Z}_p[X]$*:*

$$\mathcal{M} \mapsto \prod_{i=1}^{n} \prod_{M \in (T_i \setminus \{M_i\})} (X - H(M \| id_{\mathcal{T}} \| i)).$$

We call $m_{\overline{\mathcal{M}}} = \overline{m}_{\mathcal{T}}(\mathcal{M}) = (\frac{t_{\mathcal{T}}}{m_{\mathcal{M}}}) \in \mathbb{Z}_p[X]$ *the* complementary message encoding polynomial *of* $m_{\mathcal{M}}$ *with respect to template* $\mathcal{T}$*.*

In the following, we consider all polynomials to be expanded. To do so, we assume that an algorithm Exp, which carries out the polynomial expansion, is applied implicitly to all polynomials.

Typically, a template instantiation $\mathcal{M} = (M_i)_{i=1}^n \preceq \mathcal{T}$ will be mapped to a single bitstring $M = M_1 \| \ldots \| M_n$. We denote the mapping leading from M to $\mathcal{M}$ by $\lambda(M, \mathcal{I}) = \mathcal{M}$, where $\mathcal{I} = (|M_i|)_{i=1}^n$ is a descriptional sequence holding the lengths of the n elements of $\mathcal{M}$ as given by template $\mathcal{T}$. For sake of simplicity, we consider all templates to be non-trivial as well as minimal and do not differentiate between a template instantiation $\mathcal{M}$ and its corresponding bitstring M.

4.2 Blank Digital Signature Scheme

Now, we are able to formally define what we mean by a blank digital signature scheme.

Definition 7 (Blank Digital Signature Scheme). *A blank digital signature scheme* BDSS *consists of a tuple* (KeyGen, Sign, $\mathsf{Verify}_{\mathcal{T}}$, Inst, $\mathsf{Verify}_{\mathcal{M}}$) *of polynomial-time algorithms:*

$\mathsf{KeyGen}(\kappa, t)$**:** *This probabilistic algorithm gets the security parameter* $\kappa \in \mathbb{N}$ *and a value* $t \in \mathbb{N}$ *specifying the maximum template size. It generates private and public keys* $(\mathsf{pk_{orig}}, \mathsf{sk_{orig}})$ *for the originator and returns them.*

$\mathsf{Sign}(\mathcal{T}, \mathsf{sk_{orig}}, \mathsf{pk_{orig}}, \mathsf{sk^{DSS}_{orig}}, \mathsf{pk^{DSS}_{proxy}})$**:** *This probabilistic algorithm takes a message template* $\mathcal{T}$*, the originator's private key* $\mathsf{sk_{orig}}$*, the originator's public key* $\mathsf{pk_{orig}}$*, the originator's signing key* $\mathsf{sk^{DSS}_{orig}}$*, the proxy's verification key* $\mathsf{pk^{DSS}_{proxy}}$ *and outputs a template signature* $\sigma_{\mathcal{T}}$ *and a template dependent private key for the proxy* $\mathsf{sk^{\mathcal{T}}_{proxy}}$*.*

$\mathsf{Verify}_{\mathcal{T}}(\mathcal{T}, \sigma_{\mathcal{T}}, \mathsf{pk_{orig}}, \mathsf{pk^{DSS}_{orig}}, \mathsf{sk^{\mathcal{T}}_{proxy}}, \mathsf{pk^{DSS}_{proxy}})$**:** *This deterministic algorithm takes a template* $\mathcal{T}$*, its signature* $\sigma_{\mathcal{T}}$*, the originator's public key* $\mathsf{pk_{orig}}$*, the originator's signature verification key* $\mathsf{pk^{DSS}_{orig}}$*, the private key of the proxy* $\mathsf{sk^{\mathcal{T}}_{proxy}}$*, and the proxy's verification key* $\mathsf{pk^{DSS}_{proxy}}$*. It outputs a bit* $b \in \{\mathtt{true}, \mathtt{false}\}$ *indicating whether* $\sigma_{\mathcal{T}}$ *is a valid signature for* $\mathcal{T}$*.*

$\mathsf{Inst}(\mathcal{T}, \mathcal{M}, \sigma_{\mathcal{T}}, \mathsf{pk_{orig}}, \mathsf{sk^{\mathcal{T}}_{proxy}}, \mathsf{sk^{DSS}_{proxy}}, \mathsf{pk^{DSS}_{proxy}})$**:** *This probabilistic algorithm takes a template* $\mathcal{T}$*, an instantiation* $\mathcal{M}$*, a template signature* $\sigma_{\mathcal{T}}$*, the public key of the originator* $\mathsf{pk_{orig}}$*, the private key of the proxy* $\mathsf{sk^{\mathcal{T}}_{proxy}}$*, the signing key pair* $(\mathsf{sk^{DSS}_{proxy}}, \mathsf{pk^{DSS}_{proxy}})$ *of the proxy and outputs a message signature* $\sigma_{\mathcal{M}}$*.*

$\mathsf{Verify}_{\mathcal{M}}(\mathcal{M}, \sigma_{\mathcal{M}}, \mathsf{pk_{orig}}, \mathsf{pk^{DSS}_{proxy}}, \mathsf{pk^{DSS}_{orig}})$**:** *This deterministic algorithm takes a template instantiation* $\mathcal{M}$ *of* $\mathcal{T}$*, the signature* $\sigma_{\mathcal{M}}$*, the public key of the originator* $\mathsf{pk_{orig}}$*, the signature verification key of the proxy* $\mathsf{pk^{DSS}_{proxy}}$*, and the signature verification key of the originator* $\mathsf{pk^{DSS}_{orig}}$*. It outputs a bit* $b \in \{\mathtt{true}, \mathtt{false}\}$ *indicating whether* $\sigma_{\mathcal{M}}$ *is a valid signature for* $\mathcal{M} \preceq \mathcal{T}$*.*

4.3 Security Definitions

In the following, we define the security properties a blank digital signature scheme needs to satisfy in order to be secure. Therefore, we start with a brief overview of the required properties.

Correctness: The scheme must be correct in terms of signature correctness, signature soundness and instantiation correctness, i.e., both template and message signatures are accepted when valid and template signatures of the originator are binding.

Unforgeability: No entity without knowledge of the private key $\mathsf{sk_{orig}}$ and the signing keys $\mathsf{sk^{DSS}_{orig}}$ and $\mathsf{sk^{DSS}_{proxy}}$ should be able to forge a valid template or message signature. This is analogous to the security of traditional digital signatures. In particular, this means that it is infeasible to produce forgeries of the following types:

- **T1** finding a template $\mathcal{T}^*$ yielding the same template signature $\sigma_{\mathcal{T}^*} = \sigma_{\mathcal{T}_i}$ for one queried template $\mathcal{T}_i$.
- **T2** forging a valid template signature $\sigma_{\mathcal{T}^*}$ for some unqueried template $\mathcal{T}^*$.
- **M1** finding another message $\mathcal{M}^* \neq \mathcal{M}_{i_j}$ yielding message signature $\sigma_{\mathcal{M}^*}$ such that $\sigma_{\mathcal{M}^*} = \sigma_{\mathcal{M}_{i_j}}$ for some previously queried message $\mathcal{M}_{i_j}$.
- **M2** forging a valid message signature $\sigma_{\mathcal{M}}$ for some non-queried $\mathcal{M}$ from template signature $\sigma_{\mathcal{T}_i}$ for some previously queried template $\mathcal{T}_i$. There are two cases:
  - $\mathcal{M} \preceq \mathcal{T}_i$ for some queried template $\mathcal{T}_i$,
  - $\mathcal{M} \npreceq \mathcal{T}_i$ for some queried template $\mathcal{T}_i$.

Immutability: The proxy having access to $\mathsf{sk^{\mathcal{T}}_{proxy}}$ and $\mathsf{sk^{DSS}_{proxy}}$, when given a template signature $\sigma_{\mathcal{T}}$ for template $\mathcal{T}$ should not be able to forge signatures. Here, cases **T1**, **T2** and **M2** from above apply.

Privacy: No entity should be able to restore template $\mathcal{T}$ from a message signature $\sigma_{\mathcal{M}}$ representing a signature for an instantiation $\mathcal{M}$ of template $\mathcal{T}$.

4.3.1 *Correctness*

For a blank digital signature scheme the usual correctness properties are required to hold, i.e., genuinely signed templates and message signatures are accepted. Furthermore, we require template signatures to be sound, i.e., the originator commits to exactly one template by creating a template signature.

Signature correctness: For any key pairs $(\mathsf{sk^{DSS}_{orig}}, \mathsf{pk^{DSS}_{orig}}) \in \mathsf{KeyGen^{DSS}}(\kappa)$ and $(\mathsf{sk^{DSS}_{proxy}}, \mathsf{pk^{DSS}_{proxy}}) \in \mathsf{KeyGen^{DSS}}(\kappa)$, any BDSS key $(\mathsf{pk}_{\mathsf{orig}}, \mathsf{sk}_{\mathsf{orig}}) \in \mathsf{KeyGen}(\kappa, \mathsf{t})$, any template $\mathcal{T}$ and any honestly computed template signature

$$\sigma_{\mathcal{T}} = \mathsf{Sign}(\mathcal{T}, \mathsf{sk}_{\mathsf{orig}}, \mathsf{pk}_{\mathsf{orig}}, \mathsf{sk}^{\mathsf{DSS}}_{\mathsf{orig}}, \mathsf{pk}^{\mathsf{DSS}}_{\mathsf{proxy}}),$$

we require that the verification

$$\mathsf{Verify}_{\mathcal{T}}(\mathcal{T}, \sigma_{\mathcal{T}}, \mathsf{pk}_{\mathsf{orig}}, \mathsf{pk}^{\mathsf{DSS}}_{\mathsf{orig}}, \mathsf{sk}^{\mathcal{T}}_{\mathsf{proxy}}, \mathsf{pk}^{\mathsf{DSS}}_{\mathsf{proxy}}) = \mathtt{true}$$

holds.

Signature soundness: For any key pairs $(\mathsf{sk}^{\mathsf{DSS}}_{\mathsf{orig}}, \mathsf{pk}^{\mathsf{DSS}}_{\mathsf{orig}}) \in \mathsf{KeyGen}^{\mathsf{DSS}}(\kappa)$ and $(\mathsf{sk}^{\mathsf{DSS}}_{\mathsf{proxy}}, \mathsf{pk}^{\mathsf{DSS}}_{\mathsf{proxy}}) \in \mathsf{KeyGen}^{\mathsf{DSS}}(\kappa)$, any BDSS key $(\mathsf{pk}_{\mathsf{orig}}, \mathsf{sk}_{\mathsf{orig}}) \in \mathsf{KeyGen}(\kappa, \mathsf{t})$, any template $\mathcal{T}$ and any honestly computed template signature

$$\sigma_{\mathcal{T}} = \mathsf{Sign}(\mathcal{T}, \mathsf{sk}_{\mathsf{orig}}, \mathsf{pk}_{\mathsf{orig}}, \mathsf{sk}^{\mathsf{DSS}}_{\mathsf{orig}}, \mathsf{pk}^{\mathsf{DSS}}_{\mathsf{proxy}}),$$

we require that for any $\mathcal{T}^* \neq \mathcal{T}$ the probability that the verification

$$\mathsf{Verify}_{\mathcal{T}}(\mathcal{T}^*, \sigma_{\mathcal{T}}, \mathsf{pk}_{\mathsf{orig}}, \mathsf{pk}^{\mathsf{DSS}}_{\mathsf{orig}}, \mathsf{sk}^{\mathcal{T}}_{\mathsf{proxy}}, \mathsf{pk}^{\mathsf{DSS}}_{\mathsf{proxy}}) = \mathtt{true}$$

holds is negligibly small as a function of the security parameter κ.

Instantiation correctness: For any key pairs $(\mathsf{sk}^{\mathsf{DSS}}_{\mathsf{orig}}, \mathsf{pk}^{\mathsf{DSS}}_{\mathsf{orig}}) \in \mathsf{KeyGen}^{\mathsf{DSS}}(\kappa)$ and $(\mathsf{sk}^{\mathsf{DSS}}_{\mathsf{proxy}}, \mathsf{pk}^{\mathsf{DSS}}_{\mathsf{proxy}}) \in \mathsf{KeyGen}^{\mathsf{DSS}}(\kappa)$, any BDSS key $(\mathsf{pk}_{\mathsf{orig}}, \mathsf{sk}_{\mathsf{orig}}) \in \mathsf{KeyGen}(\kappa, t)$, any template $\mathcal{T}$, any honestly computed signature $\sigma_{\mathcal{T}}$ and corresponding $\mathsf{sk}^{\mathcal{T}}_{\mathsf{proxy}}$ such that

$$\mathsf{Verify}_{\mathcal{T}}(\mathcal{T}, \sigma_{\mathcal{T}}, \mathsf{pk}_{\mathsf{orig}}, \mathsf{pk}^{\mathsf{DSS}}_{\mathsf{orig}}, \mathsf{sk}^{\mathcal{T}}_{\mathsf{proxy}}, \mathsf{pk}^{\mathsf{DSS}}_{\mathsf{proxy}}) = \mathtt{true},$$

any honestly computed message signature

$$\sigma_{\mathcal{M}} = \mathsf{Inst}(\mathcal{T}, \mathcal{M}, \sigma_{\mathcal{T}}, \mathsf{pk}_{\mathsf{orig}}, \mathsf{sk}^{\mathcal{T}}_{\mathsf{proxy}}, \mathsf{sk}^{\mathsf{DSS}}_{\mathsf{proxy}}, \mathsf{pk}^{\mathsf{DSS}}_{\mathsf{proxy}}),$$

we require that the verification

$$\mathsf{Verify}_{\mathcal{M}}(\mathcal{M}, \sigma_{\mathcal{M}}, \mathsf{pk}_{\mathsf{orig}}, \mathsf{pk}^{\mathsf{DSS}}_{\mathsf{proxy}}, \mathsf{pk}^{\mathsf{DSS}}_{\mathsf{orig}}) = \mathtt{true}$$

holds.

4.3.2 Unforgeability

Unforgeability in the context of blank digital signatures resembles the notion of existential unforgeability against adaptive chosen message attacks (UF-CMA) in classic digital signature schemes. We adapt the classical notion to our setting in Game 1. Unforgeability is a protection against attacks mounted by parties not having access to any secret information.

Definition 8 (Unforgeability). *A* BDSS *is called* unforgeable, *if for any polynomial-time algorithm A the probability of winning Game 1 is negligible as a function of security parameter κ.*

4.3.3 Immutability

Immutability guarantees that no malicious proxy can compute message templates or template instantiations not intended by the signer. In contrast to unforgeability, immutability deals with malicious insiders.

The immutability game differs only slightly from the unforgeability game. Therefore, we only briefly discuss the differences. Here, the adversary additionally obtains $\mathsf{sk}^{\mathcal{T}}_{\mathsf{proxy}}$ and $\mathsf{sk}^{\mathsf{DSS}}_{\mathsf{proxy}}$ from the challenger in the setup phase. In the query phase, A has access to a template signing oracle and can query signatures for arbitrary templates. A can compute valid template instantiations locally. In this game, A wins if he outputs valid forgeries of type **T1, T2** or **M2**.

Setup: The challenger B runs $\mathsf{KeyGen}(\kappa, t)$ to obtain $(\mathsf{pk}_{\mathsf{orig}}, \mathsf{sk}_{\mathsf{orig}})$. Furthermore, B runs $\mathsf{KeyGen}^{\mathsf{DSS}}(\kappa)$ of a secure digital signature scheme twice to generate $(\mathsf{sk}^{\mathsf{DSS}}_{\mathsf{orig}}, \mathsf{pk}^{\mathsf{DSS}}_{\mathsf{orig}})$ and $(\mathsf{sk}^{\mathsf{DSS}}_{\mathsf{proxy}}, \mathsf{pk}^{\mathsf{DSS}}_{\mathsf{proxy}})$. It gives the adversary A the resulting public keys $\mathsf{pk}_{\mathsf{orig}}, \mathsf{pk}^{\mathsf{DSS}}_{\mathsf{orig}}$ and $\mathsf{pk}^{\mathsf{DSS}}_{\mathsf{proxy}}$ and keeps the private keys $\mathsf{sk}_{\mathsf{orig}}, \mathsf{sk}^{\mathsf{DSS}}_{\mathsf{orig}}$ and $\mathsf{sk}^{\mathsf{DSS}}_{\mathsf{proxy}}$ to itself.

Query: The adversary A has access to an oracle $\mathcal{O}$, which is simulated by the challenger and answers to queries of the form $(\mathcal{T}_i, \mathcal{M}_{\mathcal{T}_i})$ as follows. For each template $\mathcal{T}_i$ the challenger chooses $id_{\mathcal{T}_i}$, computes and stores $(\sigma_{\mathcal{T}_i}, \mathsf{sk}^{\mathcal{T}}_{\mathsf{proxy}}) = \mathsf{Sign}(\mathcal{T}_i, \mathsf{sk}_{\mathsf{orig}}, \mathsf{pk}_{\mathsf{orig}}, \mathsf{sk}^{\mathsf{DSS}}_{\mathsf{orig}}, \mathsf{pk}^{\mathsf{DSS}}_{\mathsf{proxy}})$. Next, B computes $\sigma_{\mathcal{M}_{i_j}} = \mathsf{Inst}(\mathcal{T}_i, \mathcal{M}_{i_j}, \sigma_{\mathcal{T}}, \mathsf{pk}_{\mathsf{orig}}, \mathsf{sk}^{\mathcal{T}}_{\mathsf{proxy}}, \mathsf{sk}^{\mathsf{DSS}}_{\mathsf{proxy}}, \mathsf{pk}^{\mathsf{DSS}}_{\mathsf{proxy}})$ and returns $\sigma_{\mathcal{M}_{i_j}}$. A can issue these queries for different templates as well as template instantiations in an adaptively interleaved manner.

Output: The adversary A outputs either a triple $(\mathcal{T}^*, \sigma_{\mathcal{T}^*}, \mathsf{sk}^{\mathcal{T}^*}_{\mathsf{proxy}})$ or a pair $(\mathcal{M}^*, \sigma_{\mathcal{M}^*})$. A wins if either $\mathsf{Verify}_{\mathcal{T}}(\mathcal{T}^*, \sigma_{\mathcal{T}^*}, \mathsf{pk}_{\mathsf{orig}}, \mathsf{pk}^{\mathsf{DSS}}_{\mathsf{orig}}, \mathsf{sk}^{\mathcal{T}^*}_{\mathsf{proxy}}, \mathsf{pk}^{\mathsf{DSS}}_{\mathsf{proxy}})$ or $\mathsf{Verify}_{\mathcal{M}}(\mathcal{M}^*, \sigma_{\mathcal{M}^*}, \mathsf{pk}_{\mathsf{orig}}, \mathsf{pk}^{\mathsf{DSS}}_{\mathsf{proxy}}, \mathsf{pk}^{\mathsf{DSS}}_{\mathsf{orig}})$ accepts, whereas the forgeries must be of type **T1, T2** or **M1, M2**.

Game 1: Unforgeability Game

Definition 9 (Immutability). *A* BDSS *is called* immutable, *if for any polynomial-time algorithm A the probability of winning the immutability game is negligible as a function of security parameter κ.*

4.3.4 Privacy

Privacy captures that any verifier except for the originator and the proxy, which is given a signature for a template instantiation $\mathcal{M}$ of a non-trivial template $\mathcal{T}$, can not gain any information about $\overline{\mathcal{M}}$ and thereby learn about $\mathcal{T}$. This means that even if all but one choice of a single exchangeable element has been revealed no verifier should be able to gain complete knowledge of $\mathcal{T}$. More precisely, let k be the depth of template $\mathcal{T}$, which is the smallest number of instantiations necessary to reveal all choices of $\mathcal{T}$, then no verifier should be able to derive information about the remaining unseen choices inside $\mathcal{T}$ by knowing $k-1$ instantiations.

Definition 10 (Privacy). *A* BDSS *is called* private, *if for any polynomial-time algorithm A the probability of winning Game 2 is negligible as a function of security parameter κ.*

4.3.5 Security

Now, we can define what constitutes a secure blank digital signature scheme.

Definition 11. *We call a* BDSS secure, *if it is correct, unforgeable, immutable and private.*

5. CONSTRUCTION

In this section we detail our construction and present an analysis of its efficiency in terms of computational effort as well as key and signature sizes. Moreover, we prove its security, i.e., we show that our presented construction is correct, unforgeable, immutable and private.

KeyGen: On input (κ, t), choose an elliptic curve $E(\mathbb{F}_q)$ with a subgroup of large prime order p generated by $P \in E(\mathbb{F}_q)[p]$, such that the bitlength of p is κ. Choose a pairing $e : E(\mathbb{F}_q)[p] \times E(\mathbb{F}_q)[p] \to \mathbb{F}^*_{q^k}[p]$, a full-domain cryptographic hash function $H : \{0,1\}^* \to \mathbb{Z}_p$ for use with the encoding functions and a full-domain cryptographic hash function $H' : \mathbb{F}^*_{q^k}[p] \to \{0,1\}^\kappa$. Let $P \in E(\mathbb{F}_q)[p]$ be a generator and let $g = e(P,P)$ be a generator of $\mathbb{F}^*_{q^k}[p]$. Pick two elements $\alpha, \delta \in_R \mathbb{Z}^*_p$, compute $(\alpha P, \dots, \alpha^t P)$ as well as $(\delta P, \alpha\delta P, \dots, \alpha^t \delta P)$ and output $\mathsf{sk}_{\mathsf{orig}} = \delta$, $\mathsf{pk}_{\mathsf{orig}} = (H, H', E(\mathbb{F}_q), e, p, P, g, \alpha P, \dots, \alpha^t P, \delta P, \alpha\delta P, \dots, \alpha^t \delta P)$.

Sign: Given $\mathcal{T}, \mathsf{sk}_{\mathsf{orig}}, \mathsf{pk}_{\mathsf{orig}}, \mathsf{sk}^{\mathsf{DSS}}_{\mathsf{orig}}$ and $\mathsf{pk}^{\mathsf{DSS}}_{\mathsf{proxy}}$, where $\mathcal{T}$ is a template of size $|\mathcal{T}| = \ell$ and length n with $\ell > n$, this algorithm picks a unique $id_\mathcal{T} \in_R \{0,1\}^\kappa$, computes $t_\mathcal{T} = t(\mathcal{T}) \in \mathbb{Z}_p[X]$, picks a random point $R \in E(\mathbb{F}_q)[p]$, computes $R_0 = \delta R$,

$$\mathcal{C} = H'(e(\sum_{i=0}^{\ell} t^{(i)}_\mathcal{T}(\alpha^i P) + R, \delta P)) \quad \text{and} \quad \tau = \mathsf{Sign}^{\mathsf{DSS}}(id_\mathcal{T} \| \mathcal{C} \| R_0 \| R_2 \| n \| \mathsf{pk}^{\mathsf{DSS}}_{\mathsf{proxy}}, \mathsf{sk}^{\mathsf{DSS}}_{\mathsf{orig}}).$$

It picks a point $R_1 \in_R E(\mathbb{F}_q)[p]$, computes $R_2 = R - R_1$, and returns the template signature $\sigma_\mathcal{T} = (id_\mathcal{T}, \mathcal{C}, R_0, R_2, n, \tau)$ as well as $\mathsf{sk}^{\mathcal{T}}_{\mathsf{proxy}} = R_1$.

Verify$_\mathcal{T}$: Given $\mathcal{T}, \sigma_\mathcal{T}, \mathsf{pk}_{\mathsf{orig}}, \mathsf{pk}^{\mathsf{DSS}}_{\mathsf{orig}}, \mathsf{sk}^{\mathcal{T}}_{\mathsf{proxy}}$ and $\mathsf{pk}^{\mathsf{DSS}}_{\mathsf{proxy}}$, where $\mathcal{T}$ is a template of size $|\mathcal{T}| = \ell$ and length n with $\ell > n$, this algorithm checks whether $|\mathcal{T}| \le t$. If not, it returns `false`. Otherwise, it computes $t_\mathcal{T} = t(\mathcal{T})$ and checks whether

$$\mathsf{Verify}^{\mathsf{DSS}}(\tau, id_\mathcal{T} \| \mathcal{C} \| R_0 \| R_2 \| n \| \mathsf{pk}^{\mathsf{DSS}}_{\mathsf{proxy}}, \mathsf{pk}^{\mathsf{DSS}}_{\mathsf{orig}}) = \mathtt{true} \ \wedge \ e(R_0, P) = e(R_1 + R_2, \delta P) \ \wedge \ H'(e(\sum_{i=0}^{\ell} t^{(i)}_\mathcal{T}(\alpha^i \delta P) + R_0, P)) = \mathcal{C}$$

where $t^{(i)}_\mathcal{T}$ is the i'th coefficient of $t_\mathcal{T}$. If so, return `true` and `false` otherwise.

Inst: Given $\mathcal{T}, \mathcal{M}, \sigma_\mathcal{T}, \mathsf{pk}_{\mathsf{orig}}, \mathsf{sk}^{\mathcal{T}}_{\mathsf{proxy}}$ and $\mathsf{sk}^{\mathsf{DSS}}_{\mathsf{proxy}}, \mathsf{pk}^{\mathsf{DSS}}_{\mathsf{proxy}}$, where $\mathcal{T}$ is a template of size $|\mathcal{T}| = \ell$ and length n with $\ell > n$, this algorithm, chooses $\gamma \in_R \mathbb{Z}^*_p$, and computes $m_{\overline{\mathcal{M}}} = \overline{m}_\mathcal{T}(\mathcal{M}) \in \mathbb{Z}_p[X]$. Then, it computes

$$\mathcal{C}_{\overline{\mathcal{M}}} = \sum_{i=0}^{\ell-n} m^{(i)}_{\overline{\mathcal{M}}}(\alpha^i P) + \gamma R_1 \quad \text{and} \quad \mu = \mathsf{Sign}^{\mathsf{DSS}}(\tau \| \mathcal{C}_{\overline{\mathcal{M}}} \| \gamma \| \mathcal{I}, \mathsf{sk}^{\mathsf{DSS}}_{\mathsf{proxy}}),$$

where $m^{(i)}_{\overline{\mathcal{M}}}$ is the i'th coefficient of $m_{\overline{\mathcal{M}}}$. It returns $\sigma_\mathcal{M} = (\mu, \mathcal{C}_{\overline{\mathcal{M}}}, \gamma, \mathcal{I}, \sigma_\mathcal{T})$.

Verify$_\mathcal{M}$: Given $\mathcal{M}, \sigma_\mathcal{M} = (\mu, \mathcal{C}_{\overline{\mathcal{M}}}, \gamma, \mathcal{I} = (|M_i|)^n_{i=1}, \sigma_\mathcal{T}), \mathsf{pk}_{\mathsf{orig}}$, $\mathsf{pk}^{\mathsf{DSS}}_{\mathsf{proxy}}$ and $\mathsf{pk}^{\mathsf{DSS}}_{\mathsf{orig}}$ this algorithm verifies whether

$$\mathsf{Verify}^{\mathsf{DSS}}(\tau, id_\mathcal{T} \| \mathcal{C} \| R_0 \| R_2 \| n \| \mathsf{pk}^{\mathsf{DSS}}_{\mathsf{proxy}}, \mathsf{pk}^{\mathsf{DSS}}_{\mathsf{orig}}) = \mathtt{true} \wedge \mathsf{Verify}^{\mathsf{DSS}}(\mu, \tau \| \mathcal{C}_{\overline{\mathcal{M}}} \| \gamma \| \mathcal{I}, \mathsf{pk}^{\mathsf{DSS}}_{\mathsf{proxy}}) = \mathtt{true} \wedge |\mathcal{I}| = n \wedge \sum_{i=1}^{n} |M_i| = |\mathcal{M}|$$

On failure return `false`, otherwise evaluate $m_\mathcal{M} = m_\mathcal{T}(\mathcal{M})$, let $m^{(i)}_\mathcal{M}$ be the i'th coefficient of $m_\mathcal{M}$, and compute

$$\mathcal{C}_\mathcal{M} = \sum_{i=0}^{n} m^{(i)}_\mathcal{M}(\alpha^i P) \text{ and } \mathcal{C}'_\mathcal{M} = \sum_{i=0}^{n} m^{(i)}_\mathcal{M}(\alpha^i \delta P)$$

and check whether

$$H'(e(\mathcal{C}_{\overline{\mathcal{M}}} + \gamma R_2, \mathcal{C}'_\mathcal{M}) \cdot e((1-\gamma)R_0, \mathcal{C}_\mathcal{M}) \cdot e(P - \mathcal{C}_\mathcal{M}, R_0)) = \mathcal{C}$$

On success return `true` and `false` otherwise.

Scheme 1: Blank Digital Signature Scheme

5.1 Intuition

Before we present the detailed construction, we provide some intuition in order to make our design choices comprehensible. As already noted, we use standard digital signatures, such as ECDSA [11], as a building block assuming the respective signature keys to be available to every participant in an authentic fashion. Note that this requires the availability of public key infrastructures, which are, however, commonly used in practice today. In our construction, DSS signatures provide authenticity of template and message signatures.

As already discussed in Section 4, we use polynomials to represent templates and template instantiations. The intuition is that the originator commits to a template polynomial. By construction every allowed template instantiation is represented by a message encoding polynomial that perfectly divides the template polynomial. A proxy can now commit to a message polynomial, by computing and signing a commitment to the complementary message encoding polynomial. However, he can not choose arbitrary divisors of the template polynomial, as the indexes of message elements are incorporated into the encoding and the length of the message, i.e., the degree of the message polynomial, is fixed by the originator.

In the verification, the verifier computes a commitment to the message polynomial and checks whether the computed commitment and the commitment given by the proxy re-

We assume that the originator as well as the proxy both own an authentic key pair for a secure digital signature scheme $(\mathsf{sk}^{\mathsf{DSS}}_{\mathsf{orig}}, \mathsf{pk}^{\mathsf{DSS}}_{\mathsf{orig}})$ and $(\mathsf{sk}^{\mathsf{DSS}}_{\mathsf{proxy}}, \mathsf{pk}^{\mathsf{DSS}}_{\mathsf{proxy}})$, respectively.

$\mathsf{Setup}_\mathsf{T}$: The trusted third party T chooses a suitable security parameter κ and a value $t \in \mathbb{N}$ representing the maximum template length, runs $\mathsf{KeyGen}(\kappa, t)$, sends $\mathsf{sk}_{\mathsf{orig}}$ to the originator O and publishes $\mathsf{pk}_{\mathsf{orig}}$ in an authentic fashion.

$\mathsf{Issue}_\mathsf{O}$: O defines a message template $\mathcal{T}$, runs $\mathsf{Sign}(\mathcal{T}, \mathsf{sk}_{\mathsf{orig}}, \mathsf{pk}_{\mathsf{orig}}, \mathsf{sk}^{\mathsf{DSS}}_{\mathsf{orig}}, \mathsf{pk}^{\mathsf{DSS}}_{\mathsf{proxy}})$ and gives $\sigma_\mathcal{T} = (id_\mathcal{T}, \mathcal{C}, n, \tau)$ as well as $\mathsf{sk}^{\mathcal{T}}_{\mathsf{proxy}}$ to P.

$\mathsf{Issue}_\mathsf{P}$: P runs $\mathsf{Verify}_\mathcal{T}(\mathcal{T}, \sigma_\mathcal{T}, \mathsf{pk}_{\mathsf{orig}}, \mathsf{pk}^{\mathsf{DSS}}_{\mathsf{orig}}, \mathsf{sk}^{\mathcal{T}}_{\mathsf{proxy}}, \mathsf{pk}^{\mathsf{DSS}}_{\mathsf{proxy}})$ to check whether $\sigma_\mathcal{T}$ is a valid signature for $\mathcal{T}$ issued by O. On success, P, on behalf of O, defines a template instantiation $\mathcal{M} \preceq \mathcal{T}$ and runs $\mathsf{Inst}(\mathcal{T}, \mathcal{M}, \sigma_\mathcal{T}, \mathsf{pk}_{\mathsf{orig}}, \mathsf{sk}^{\mathcal{T}}_{\mathsf{proxy}}, \mathsf{sk}^{\mathsf{DSS}}_{\mathsf{proxy}}, \mathsf{pk}^{\mathsf{DSS}}_{\mathsf{proxy}})$ and publishes $(\mathcal{M}, \sigma_\mathcal{M})$.

Verify: Anybody in possession of the public keys can now take $(\mathcal{M}, \sigma_\mathcal{M})$ and run $\mathsf{Verify}_\mathcal{M}(\mathcal{M}, \sigma_\mathcal{M}, \mathsf{pk}_{\mathsf{orig}}, \mathsf{pk}^{\mathsf{DSS}}_{\mathsf{proxy}}, \mathsf{pk}^{\mathsf{DSS}}_{\mathsf{orig}})$ to check whether $\sigma_\mathcal{M}$ is a valid signature for $\mathcal{M}$ issued by O and P.

Protocol 1: Blank Digital Signature Protocol

Setup: The challenger B runs $\mathsf{KeyGen}(\kappa, t)$ to obtain $(\mathsf{pk}_{\mathsf{orig}}, \mathsf{sk}_{\mathsf{orig}})$. Furthermore, B runs $\mathsf{KeyGen}^{\mathsf{DSS}}(\kappa)$ of a secure digital signature scheme twice to generate $(\mathsf{sk}^{\mathsf{DSS}}_{\mathsf{orig}}, \mathsf{pk}^{\mathsf{DSS}}_{\mathsf{orig}})$ and $(\mathsf{sk}^{\mathsf{DSS}}_{\mathsf{proxy}}, \mathsf{pk}^{\mathsf{DSS}}_{\mathsf{proxy}})$. It gives the adversary A the resulting public keys $\mathsf{pk}_{\mathsf{orig}}$, $\mathsf{pk}^{\mathsf{DSS}}_{\mathsf{orig}}$ and $\mathsf{pk}^{\mathsf{DSS}}_{\mathsf{proxy}}$ and keeps the private keys $\mathsf{sk}_{\mathsf{orig}}$, $\mathsf{sk}^{\mathsf{DSS}}_{\mathsf{orig}}$ and $\mathsf{sk}^{\mathsf{DSS}}_{\mathsf{proxy}}$ to itself.

Query 1: The adversary A issues template signature queries $\mathcal{T}_1, \ldots, \mathcal{T}_q$. To each query $\mathcal{T}_i$ the challenger responds by running $\mathsf{Sign}(\mathcal{T}_i, \mathsf{sk}_{\mathsf{orig}}, \mathsf{pk}_{\mathsf{orig}}, \mathsf{sk}^{\mathsf{DSS}}_{\mathsf{orig}}, \mathsf{pk}^{\mathsf{DSS}}_{\mathsf{proxy}})$ to generate a signature $\sigma_{\mathcal{T}_i}$ for $\mathcal{T}_i$ and sending it together with $\mathsf{sk}^{\mathcal{T}_i}_{\mathsf{proxy}}$ to the adversary. For each template signature query $\mathcal{T}_i$, A can issue an arbitrary number of signature instantiation queries of the form $(\mathcal{T}_i, \mathcal{M}_{i_j})$. To every such query B responds by running $\mathsf{Inst}(\mathcal{T}_i, \mathcal{M}_{i_j}, \sigma_\mathcal{T}, \mathsf{pk}_{\mathsf{orig}}, \mathsf{sk}^{\mathcal{T}_i}_{\mathsf{proxy}}, \mathsf{sk}^{\mathsf{DSS}}_{\mathsf{proxy}}, \mathsf{pk}^{\mathsf{DSS}}_{\mathsf{proxy}})$ and returning $\sigma_{\mathcal{M}_{i_j}}$. All of these queries can be made adaptively interleaved.

Challenge: At some point A signals B that he is ready to be challenged. B constructs a non-queried random template $\mathcal{T}$ of depth k and produces a template signature for $\mathcal{T}$, but keeps $\mathcal{T}$ and $\mathsf{sk}^{\mathcal{T}}_{\mathsf{proxy}}$ to itself.

Query 2: The adversary A is allowed to issue queries as in query phase 1 and at most $k-1$ instantiation queries to the unknown template $\mathcal{T}$. All of these queries can be made adaptively interleaved.

Output: The adversary A outputs $\mathcal{T}^*$ and wins if $\mathcal{T}^* = \mathcal{T}$.

Game 2: Privacy Game

late to the commitment given by the originator, whereas the originator's commitment is given as hash value in order to prevent arithmetics. We need a trusted third party, as the originator should not know the value α. Otherwise, he could exchange the template polynomial after signature generation for another polynomial having the same evaluation at the point α. Note that in the context of polynomial commitments the setup must always be run by a trusted third party, as otherwise these commitments represent trapdoor commitments, i.e., the knowledge of α allows to open the commitment to arbitrary polynomials.

We use polynomial Pedersen commitments of the form $\mathcal{C} = f(\alpha)P + R$ to hide the committed polynomials. Pedersen commitments provide computational binding as long as the discrete logarithm between the values R and P is unknown. Typically, this is achieved by letting a third party compute these parameters. In our scenario, R is chosen by the originator, which requires us to additionally commit to a multiple of R and to use an additive decomposition of R in order to fix it.

5.2 Scheme

In Scheme 1, we present the detailed construction of our proposed BDSS. Moreover, in Protocol 1, we illustrate a typical scenario for the interaction of the originator, the proxy and the verifier.

We note that Scheme 1 can easily be turned into a scheme using asymmetric pairings giving flexibility in the choice of curves and pairings as well as improved efficiency. In case of Type-2 pairings there are only minor modifications necessary, as there is an efficiently computable isomorphism between G_1 and G_2, whereas in the Type-3 setting this comes at the costs of doubling the size of $\mathsf{pk}_{\mathsf{orig}}$. This is because the values $\alpha P, \ldots, \alpha^t P, \delta P, \alpha\delta P, \ldots, \alpha^t \delta P \in G_1$ also need to be mapped to elements of group G_2, i.e., we need to put the additional points $\alpha P', \ldots, \alpha^t P', \delta P', \alpha\delta P', \ldots, \alpha^t \delta P' \in G_2$ into $\mathsf{pk}_{\mathsf{orig}}$, where P' is a generator of G_2.

5.3 Efficiency

In the following, we give a detailed analysis of the efficiency of our scheme in terms of computational effort as well as key and signature sizes. We use $\mathtt{P}, \mathtt{S}, \mathtt{A}$ and $\mathtt{m}$ to denote the costs of a pairing evaluation, the costs of an elliptic curve scalar multiplication, the costs of a point addition and the costs of one multiplication in $\mathbb{Z}_p$, respectively. Moreover, we use $\mathtt{Exp(d)}$, $\mathtt{s}$ and $\mathtt{v}$ for the costs of the expansion of a polynomial of degree d, the costs of a signing operation and the costs of a signature verification, respectively. Note that $\mathtt{Exp(d)}$ is $\mathtt{d(d+1)/2m}$ when ignoring shifts and field additions, which are inexpensive. The costs of the algorithms of Scheme 1 are summarized in Table 1. Figure 1 illustrates the performance of the algorithms in Scheme 1 for varying template size $|\mathcal{T}| = \ell$ and template length n set to $\ell/4$, meaning that on average there are three choices per exchangeable element. We conducted the experiments on an Intel Core i5-2540M equipped with 8GB RAM running Ubuntu 12.10/amd64 and OpenJDK 6/amd64. For ECDSA and basic elliptic curve arithmetics we were using the Java™ library ECCelerate™ 2.0 [1], which is freely available for

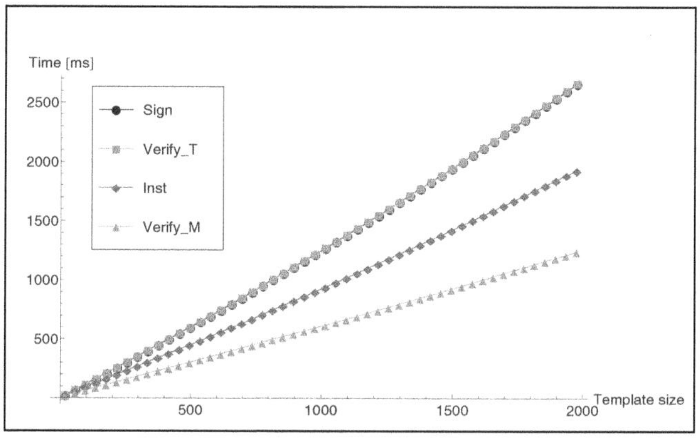

Figure 1: Performance Evaluation of the Algorithms in Scheme 1.

purposes of research, as well as the jPBC library [2] version 1.2.1 for the pairing evaluations. Thereby, we are using the NIST P-224 curve [11] for ECDSA and the Tate pairing on an MNT curve [18] with a group size of 224 bits and embedding degree $k = 6$ in order to perform our benchmarks, respectively. Note that we have chosen the asymmetric version of Scheme 1 to gain improved efficiency.

Algorithm	Costs
KeyGen	$(2t+1)S$
Sign	$P+(\ell+1)S+(\ell+2)A+Exp(\ell)+s$
$\mathsf{Verify}_{\mathcal{T}}$	$3P+(\ell+1)S+(\ell+2)A+Exp(\ell)+v$
Inst	$3P+(\ell-n+2)S+(\ell-n+1)A+Exp(\ell-n)+s+v$
$\mathsf{Verify}_{\mathcal{M}}$	$3P+(2n+4)S+(2n+2)A+Exp(n)+2v$

Table 1: Costs of the Algorithms of Scheme 1 for templates of size ℓ and length n.

In Table 2, we analyze the sizes of keys and signatures of Scheme 1. We use a pairing-friendly curve of bitsize κ, assume all points to be compressed and for the DSS we use ECDSA with key sizes of κ bits. Consequently, the public and private keys of the DSS are of size κ bits and the signature is of size 2κ. Note that we only count values that are related to the security parameter κ and, consequently, no small values, such as integers and the like.

Component	Size
$\mathsf{sk}_{\mathsf{orig}}$	κ
$\mathsf{pk}_{\mathsf{orig}}$	$2(t+1)\kappa$
$\mathsf{sk}_{\mathsf{proxy}}$	κ
$\sigma_{\mathcal{T}}$	6κ
$\sigma_{\mathcal{M}}$	10κ

Table 2: Sizes of Keys and Signatures in Scheme 1.

Observe that only $\mathsf{pk}_{\mathsf{orig}}$ depends on the maximum degree ℓ of the template polynomial $t_{\mathcal{T}}$. For instance, given a template $\mathcal{T}$ with 20 fixed elements and 25 exchangeable elements with 5 choices each, we obtain a template of size $\ell = 145$ yielding a template polynomial $t_{\mathcal{T}}$ with $\deg(t_{\mathcal{T}}) = 145$. Consequently, we have $t \geq 145$ and the size of $\mathsf{pk}_{\mathsf{orig}}$ would be approximately 8.1kB for $\kappa = 224$, which is absolutely reasonable.

5.4 Comparison to the Naive Approaches

Recall that the first naive approach given in Section 1.1 would require the originator to produce one signature for every possible template instantiation. Let us look at the above example, where we have 20 fixed elements and 25 exchangeable elements with 5 choices each. Note that this is an absolutely reasonable example, which is far from being overstated. Then, the originator would have to compute $5^{25} \approx 298 \cdot 10^{15}$ signatures, which is obviously impractical.

The second naive approach we have mentioned is the use of Merkle hash trees to reduce the number of signatures that need to be computed by the originator at the expense of higher computational costs and an increased size of the signature. This means that the originator needs to build a complete binary tree, where the number of leaves equals the number of possible template instantiations. Furthermore, each leaf would need to include a random string as additional input to the hash function in order to hide the instantiations from a verifier as it is done in redactable signatures [13]. In our above example, the number of leaves would then be $5^{25} \approx 298 \cdot 10^{15}$. In order to build the hash tree, the originator would need to perform one hash evaluation per node in the tree. Note that for a complete binary tree with n leaves there would be at most $2n - 1$ nodes in the tree. For our above example, this would yield at most $2 \cdot 5^{25} - 1$ hash evaluations and the same number of PRF evaluations to randomize the tree (see [13] for more details on how to compute the random strings using a PRF). Although the verification of a signature for an instantiation in this construction would be quite efficient, as it can be carried in logarithmic time in the number of possible instantiations, the $\mathsf{Sign}, \mathsf{Verify}_{\mathcal{T}}$ and Inst algorithms all require the computation of the full hash tree rendering this approach impractical.

We emphasize that in our approach the signature size stays constant, regardless of the number of possible template instantiations. This is due to the fact that the template polynomial, whose degree grows only linearly in the template size, is mapped to a point on the curve, which is further mapped to a field element and then hashed. Notice that in our construction, the computational effort is independent of the number of potential template instantiation. Instead, it grows only linearly with the template size, i.e., with the number and the cardinality of exchangeable elements and the number of fixed elements.

5.5 Security

Subsequently, we investigate the security of our construction in the proposed security model by considering all the required security properties.

THEOREM 1. *Assuming the hardness of the* ECDLP *as well as the existence of second preimage resistant hash functions, Scheme 1 is correct with respect to Definition 8.*

PROOF. See Appendix A.2. □

THEOREM 2. *Assuming the hardness of the* ECDLP *as well as the existence of second preimage resistant hash functions and secure digital signature schemes, Scheme 1 is unforgeable with respect to Definition 8.*

PROOF. See Appendix A.1. □

THEOREM 3. *Assuming the hardness of the* ECDLP *as well as the existence of second preimage resistant hash functions and secure digital signature schemes, Scheme 1 is immutable with respect to Definition 9.*

PROOF. See Appendix A.3. □

THEOREM 4. *Scheme 1 is private with respect to Definition 10.*

PROOF. See Appendix A.4. □

Taking Theorem 1-Theorem 4 together, we obtain the following corollary.

COROLLARY 1. *Scheme 1 is a secure* BDSS.

6. CONCLUSIONS

In this paper we have introduced a new notion of digital signatures, namely so-called blank digital signatures. We have provided the abstract scheme, a security model and a concrete construction of such a scheme from any secure digital signature scheme, pairing-friendly elliptic curves and polynomial commitments. Moreover, we have proven the security of our construction, have analyzed its efficiency supporting its practicality and have given several use cases, such as delegated contract signing.

6.1 Future Work

Since blank digital signatures are a novel concept, there are several open issues for future work, which we outline subsequently. One issue for future work is to get rid of the trusted third party for key generation. Furthermore, it would be desirable to generalize the blank digital signature scheme and its security model to multiple designated proxies, which seems to be straight-forward by inclusion of multiple proxy signature verification keys into the template signature. However, in this naive construction every proxy and every verifier can determine the set of designated proxies. This may not be desirable in practice, whereas to achieve this goal does not seem to be that straight-forward. Another issue is to find alternative designated constructions for blank signatures potentially without relying on standard digital signature schemes. Additionally, it would be desirable to prove the security of our construction under weaker assumptions and to impose further restrictions on allowed template instantiations, i.e., to further limit the allowed combinations of choices over all exchangeable elements of templates. Also, allowing blank fields, which can be substituted with arbitrary strings, would be desirable. Finally, it may be interesting to investigate concepts applied in the construction of blank digital signatures in the proxy signature setting.

7. ACKNOWLEDGEMENTS

We would like to thank Jiangtao Li for his many valuable suggestions for improving the presentation of this paper. The work of both authors has been supported by the European Commission through project FP7-FutureID, grant agreement number 318424.

8. REFERENCES

[1] ECCelerate 2.0. `http://jce.iaik.tugraz.at/sic/Products/Core-Crypto-Toolkits/ECCelerate`, 2012.

[2] jPBC 1.2.1. `http://gas.dia.unisa.it/projects/jpbc`, 2012.

[3] G. Ateniese, D. H. Chou, B. de Medeiros, and G. Tsudik. Sanitizable Signatures. In *ESORICS*, volume 3679 of *LNCS*, pages 159–177. Springer, 2005.

[4] J. C. Benaloh and M. de Mare. One-way accumulators: A decentralized alternative to digital sinatures (extended abstract). In *Advances in Cryptology - EUROCRYPT '93*, volume 765 of *Lecture Notes in Computer Science*, pages 274–285, 1993.

[5] B. H. Bloom. Space/time trade-offs in hash coding with allowable errors. *Commun. ACM*, 13(7):422–426, 1970.

[6] A. Boldyreva, A. Palacio, and B. Warinschi. Secure proxy signature schemes for delegation of signing rights. *IACR Cryptology ePrint Archive*, 2003:96, 2003.

[7] A. Boldyreva, A. Palacio, and B. Warinschi. Secure Proxy Signature Schemes for Delegation of Signing Rights. *J. Cryptology*, 25(1):57–115, 2012.

[8] C. Brzuska, H. C. Pöhls, and K. Samelin. Non-interactive public accountability for sanitizable signatures. In *Proc. of the 9th European PKI Workshop: Research and Applications (EuroPKI 2012)*, LNCS. Springer-Verlag, 2012.

[9] S. Canard and A. Jambert. On Extended Sanitizable Signature Schemes. In *Topics in Cryptology - CT-RSA 2010*, volume 5985 of *LNCS*, pages 179–194, 2010.

[10] S. Chatterjee and A. Menezes. On cryptographic protocols employing asymmetric pairings - the role of ψ revisited. *Discrete Applied Mathematics*, 159(13):1311–1322, 2011.

[11] P. Gallagher and C. Furlani. FIPS PUB 186-3 federal information processing standards publication digital signature standard (dss), 2009.

[12] S. Goldwasser, S. Micali, and R. L. Rivest. A Digital Signature Scheme Secure Against Adaptive Chosen-Message Attacks. *SIAM J. Comput.*, 17(2):281–308, 1988.

[13] R. Johnson, D. Molnar, D. X. Song, and D. Wagner. Homomorphic Signature Schemes. In *Topics in Cryptology - CT-RSA 2002*, volume 2271 of *LNCS*, pages 244–262, 2002.

[14] A. Kate, G. M. Zaverucha, and I. Goldberg. Constant-size commitments to polynomials and their applications. In *Advances in Cryptology - ASIACRYPT 2010*, pages 177–194, 2010.

[15] M. Klonowski and A. Lauks. Extended sanitizable signatures. In *ICISC*, pages 343–355, 2006.

[16] M. Mambo, K. Usuda, and E. Okamoto. Proxy Signatures for Delegating Signing Operation. In *ACM Conference on Computer and Communications Security (CCS 1996)*, pages 48–57, 1996.

[17] R. C. Merkle. A digital signature based on a conventional encryption function. In *CRYPTO*, pages 369–378, 1987.

[18] Miyaji, Nakabayashi, and Takano. New Explicit Conditions of Elliptic Curve Traces for FR-Reduction. *TIEICE: IEICE Transactions on Communications/Electronics/Information and Systems*, 2001.

[19] T. P. Pedersen. Non-interactive and information-theoretic secure verifiable secret sharing. In *CRYPTO*, pages 129–140, 1991.

[20] K. Samelin, H. C. Poehls, J. Posegga, and H. de Meer. Redactable vs. Sanitizable Signatures. Technical Report Number MIP-1208, Department of Informatics and Mathematics, University of Passau, 2012.

[21] J. C. N. Schuldt, K. Matsuura, and K. G. Paterson. Proxy Signatures Secure Against Proxy Key Exposure. In *11th International Workshop on Practice and Theory in Public-Key (PKC 2008)*, volume 4939 of *LNCS*, pages 141–161, 2008.

[22] J. Silverman. *The Arithmetic of Elliptic Curves*, volume 106 of *Graduate Texts in Mathematics.* Springer, 1986.

[23] R. Steinfeld, L. Bull, and Y. Zheng. Content Extraction Signatures. In *ICISC 2001*, volume 2288 of *LNCS*, pages 285–304. Springer, 2001.

APPENDIX

A. PROOFS

This section contains the proofs of Theorem 1-Theorem 4.

A.1 Proof of Theorem 2 (Sketch)

The proof consists of two parts. The first part covers unforgeability of template signatures, whereas the second part covers unforgeability of message signatures. Both parts consist of two cases covering the reuse of queried signatures and existential forging of signatures as detailed in Section 4.3.

During the query phase, A is allowed to issue an arbitrary number of template instantiation queries for every previously queried template $\mathcal{T}_i$.

Case T1. This case covers the infeasability of finding some $\mathcal{T}^* \neq \mathcal{T}_i$ for all previously queried $\mathcal{T}_i$ such that $\mathcal{T}^*$ verifies under some queried template signature $\sigma_{\mathcal{T}_i}$. If A was able to find - with non-negligible probability - a $\mathcal{T}^* \neq \mathcal{T}_i$ for all $1 \leq i \leq q$ such that

$$\mathcal{C}_i = H'(e(\sum_{i=0}^{\ell} t^{(i)}_{\mathcal{T}^*}(\alpha^i P) + R_i), \delta P)),$$

for some $1 \leq i \leq q$, then A has either

1. found a template $\mathcal{T}^* \neq \mathcal{T}_i$ such that $t_{\mathcal{T}_i}(X) = t_{\mathcal{T}^*}(X)$, or
2. found a second polynomial $t_{\mathcal{T}^*}$ with $\deg(t_{\mathcal{T}^*}) \leq t$ with $t_{\mathcal{T}_i}(\alpha) = t_{\mathcal{T}^*}(\alpha)$ and the corresponding template $\mathcal{T}^*$.

In case one, A has found second preimages in H with non-negligible probability. More precisely, $\mathsf{sk}^{\mathcal{T}}_{\mathsf{proxy}}$ and $\sigma_{\mathcal{T}_i}$ are fixed. If A is able to generate a $\mathcal{T}^* \neq \mathcal{T}_i$ with $t_{\mathcal{T}_i}(X) = t_{\mathcal{T}^*}(X)$ with non-negligible probability, then he must have found second preimages in H with non-negligible probability for ℓ_i roots of $t(\mathcal{T}_i)$, whereas the suffix of each preimage must be of the form $id_{\mathcal{T}^*}\|j$ and only the value of M is arbitrary.

In the second case, if A is able to do so, then A must have found a polynomial $t_{\mathcal{T}^*}$ with $t_{\mathcal{T}_i}(\alpha) = t_{\mathcal{T}^*}(\alpha)$, i.e., having the same evaluation at the unknown point α as $t_{\mathcal{T}_i}$ for some $1 \leq i \leq q$. If A is able to extract the value α from αP by breaking the ECDLP, for which the probability is negligibly small, then he could construct a polynomial $t_{\mathcal{T}^*}$ with $t_{\mathcal{T}_i}(\alpha) = t_{\mathcal{T}^*}(\alpha)$. However, then A would still need to find second preimages in H as in case 1 to construct a corresponding template $\mathcal{T}^*$.

Taking all together, gives us the desired contradiction.

Case T2. This case covers the infeasability of computing a valid signature $\sigma_{\mathcal{T}^*}$ for some $\mathcal{T}^*$ giving $\mathcal{C}^*$ (which may be computed by A), which differs from all previously queried signatures. If A was able to find a pair $(\mathcal{T}^*, \sigma_{\mathcal{T}^*}) \neq (\mathcal{T}_i, \sigma_{\mathcal{T}_i})$ for all $1 \leq i \leq q$ such that

$$\mathsf{Verify}^{\mathsf{DSS}}(\tau, id_{\mathcal{T}^*}\|\mathcal{C}^*\|R^*_0\|R^*_2\|n^*\|\mathsf{pk}^{\mathsf{DSS}}_{\mathsf{proxy}}, \mathsf{pk}^{\mathsf{DSS}}_{\mathsf{orig}}) = \mathtt{true}$$

then A must be able to forge signatures of the digital signature scheme DSS under $\mathsf{sk}^{\mathsf{DSS}}_{\mathsf{orig}}$. This gives us the desired contradiction.

Case M1. This case covers the infeasability of finding some $\mathcal{M}^* \neq \mathcal{M}_{i_j}$ for all previously queried $\mathcal{M}_{i_j}$ such that $\mathcal{M}^*$ verifies under some issued message signature $\sigma_{\mathcal{M}_{i_j}}$. If A was able to find - with non-negligible probability - an $\mathcal{M}^* \neq \mathcal{M}_{i_j}$ for all $1 \leq i \leq q$ and $1 \leq j \leq q_i$ such that

$$|\mathcal{M}^*| = |\mathcal{M}_{i_j}| \ \wedge \ \deg(m_{\mathcal{M}^*}) = n \ \wedge$$
$$H'(e(\mathcal{C}_{\overline{\mathcal{M}}_{i_j}} + \gamma_{i_j} R_{2,i}, \mathcal{C}'_{\mathcal{M}^*}) \cdot e((1-\gamma_{i_j})R_{0,i}, \mathcal{C}_{\mathcal{M}^*}) \cdot$$
$$\cdot e(P - \mathcal{C}_{\mathcal{M}^*}, R_{0,i})) = \mathcal{C}_i$$

for some $1 \leq i \leq q$ and $1 \leq j \leq q_i$ then A has found second preimages in H with non-negligible probability. More precisely, since due to μ all values in the verification relation are fixed, the only way for A to output an $\mathcal{M}^*$ that passes the signature verification for an existing signature, is to find an $\mathcal{M}^*$ such that $m(\mathcal{M}^*) = m(\mathcal{M}_{i_j})$ and $|\mathcal{M}^*| = |\mathcal{M}_{i_j}|$, i.e.,

$$H(M^*_l\|id_{\mathcal{T}_i}\|l) = H(M_{i_{j_l}}\|id_{\mathcal{T}_i}\|l)$$

for all n roots of the polynomial $m_{\mathcal{M}_{i_j}}$, whereas the suffix of each preimage must be of the form $id_{\mathcal{T}_i}\|l$ and only the value of M^*_l is arbitrary.

Case M2. This case covers the infeasability of computing a valid signature $\sigma_{\mathcal{M}^*}$ for some $\mathcal{M}^*$, which differs from all previously queried signatures. A needs to find a pair $(\mathcal{M}^*, \sigma_{\mathcal{M}^*})$ $\neq (\mathcal{M}_{i_j}, \sigma_{\mathcal{M}_{i_j}})$ for all $1 \leq j \leq q_i$ such that $\deg(m_{\mathcal{M}^*}) = n$ and

$$H'(e(\mathcal{C}_{\overline{\mathcal{M}}^*} + \gamma^* R^*_2, \mathcal{C}'_{\mathcal{M}^*}) \cdot e((1-\gamma^*)R^*_0, \mathcal{C}_{\mathcal{M}^*}) \cdot$$
$$\cdot e(P - \mathcal{C}_{\mathcal{M}^*}, R^*_0)) = \mathcal{C}_i$$

First we consider finding $\mathcal{M}^* \preceq \mathcal{T}_i$ for some i, for which $\sigma_{\mathcal{M}^*} \neq \sigma_{\mathcal{M}_{i_j}}$ for all $1 \leq j \leq q_i$. In order for this to hold for a given $\mathcal{C}_i$, A must construct $\mathcal{C}_{\overline{\mathcal{M}}^*}$. Since A knows all templates, A can choose $\mathcal{M}^* \preceq \mathcal{T}_i$ and compute $\mathcal{C}_{\overline{\mathcal{M}}^*} = \sum_{l=0}^{\ell_i - n_i} m^{(l)}_{\overline{\mathcal{M}}^*}(\alpha^l P) + \gamma^* R_1$, whereas R_1 can be extracted from previous queries as γ_{i_j} and $m_{\overline{\mathcal{M}}_{i_j}}$ are known. Still, A must be able to forge signatures of the digital signature scheme DSS under $\mathsf{sk}^{\mathsf{DSS}}_{\mathsf{proxy}}$.

Secondly, we consider finding $\mathcal{M}^* \npreceq \mathcal{T}_i$, for which $\sigma_{\mathcal{M}^*} \neq \sigma_{\mathcal{M}_{i_j}}$ for all $1 \leq j \leq q_i$. Finding such an $\mathcal{M}^*$ implies that A has found a polynomial $m_{\mathcal{M}^*}$ such that $m_{\mathcal{M}^*}$ does not perfectly divide $t_{\mathcal{T}_i}$. This means that $t_{\mathcal{T}_i} = m_{\overline{\mathcal{M}}^*} \cdot m_{\mathcal{M}^*} + \xi$ with $\xi \neq 0$. Note that the only way for A to include the remainder ξ for the verification relation is to put $\xi(\alpha)$ into $\mathcal{C}_{\overline{\mathcal{M}}^*}$. However, since the only place where $\mathcal{C}_{\overline{\mathcal{M}}^*}$ is input to the verification is the first pairing $e(\mathcal{C}_{\overline{\mathcal{M}}^*} + \gamma R_2, \mathcal{C}'_{\mathcal{M}^*})$, this will result in $\mathcal{C}_{\overline{\mathcal{M}}^*} = \sum_{l=0}^{\ell_i - n_i} m_{\overline{\mathcal{M}}^*}^{(l)}(\alpha^l P) + \frac{\xi(\alpha)}{m_{\mathcal{M}^*}(\alpha)} R_1$. Furthermore, A then needs to compute γ as $\frac{\xi(\alpha)}{m_{\mathcal{M}^*}(\alpha)}$, as these are the only values A can influence. However, this requires A to have knowledge of α, which requires A to break the ECDLP. Another strategy, A can follow is to use non-intended perfect divisors of $t_{\mathcal{T}_i}$, i.e., constructing an $\mathcal{M}^*$ such that $\deg(m_{\mathcal{M}^*}) \neq n$ and/or the elements of $\mathcal{M}^*$ are not consecutive. However, the degree n of all valid message polynomials as well as the index values i in the message encoding are fixed by the originator. Thus, the verification can never be satisfied, unless A has the power to compute second preimages.

In all cases, A must be able to forge signatures of the digital signature scheme DSS under $\mathsf{sk}_{\mathsf{proxy}}^{\mathsf{DSS}}$. This gives us the desired contradictions.

Since all cases gave us the desired contradictions, the proof is complete. □

A.2 Proof of Theorem 1 (Sketch)

We show the signature and instantiation correctness, where we omit showing that the DSS verification works, since it is clear from the construction. The signature correctness is obvious, and we now show the signature soundness. This means that we have to show that the originator commits to exactly one template polynomial by issuing a template signature. Similarly to case **T1** in the proof of unforgeability, the originator would need to find second preimages or another polynomial having the same evaluation at the unknown point α. Moreover, since he commits to R by signing R_0 along with R_2 and gives R_1 to the proxy, the proxy can check this commitment by verifying whether $e(R_0, P) = e(R_1 + R_2, \delta P)$ holds, which prevents the originator from changing R after signature generation.

It remains to show that the message signature verification works, i.e., we have to prove the correctness of

$$H'(e(\mathcal{C}_{\overline{\mathcal{M}}} + \gamma R_2, \mathcal{C}'_{\mathcal{M}}) \cdot e((1-\gamma)R_0, \mathcal{C}_{\mathcal{M}}) \cdot e(P - \mathcal{C}_{\mathcal{M}}, R_0)) = \mathcal{C},$$

whereas $\mathcal{C} = H'(e(t_{\mathcal{T}}(\alpha)P + R, \delta P))$. The left-hand-side yields

$$g^{(\overline{m} + \gamma r_1 + \gamma r_2)\cdot \delta m + ((1-\gamma)\cdot \delta r)\cdot m + (1-m)\cdot \delta r} =$$
$$g^{(\delta m \overline{m} + \gamma \delta r m) + (\delta r m - \gamma \delta r m) + (\delta r - \delta r m)} =$$
$$g^{\delta t + \delta r} = e(t_{\mathcal{T}}(\alpha)P + R, \delta P),$$

whereas we assume R_1, R_2 and R to be r_1P, r_2P and rP for some unknown values r_1, r_2 and r, respectively. Furthermore, we write t, m and $\overline{m}$ instead of $t_{\mathcal{T}}(\alpha), m_{\mathcal{M}}(\alpha)$, and $m_{\overline{\mathcal{M}}}(\alpha)$, respectively. This completes the proof. □

A.3 Proof of Theorem 3 (Sketch)

Case T1 & T2. The proofs for these two cases are analogous to the proof in Appendix A.1, since A has exactly the same knowledge as in the previous proof. More precisely, this is due to the knowledge of all templates in the unforgeability game.

Case M2. The only difference to the power of the adversary A in the proof in Appendix A.1 is that A knows R_1 and $\mathsf{sk}_{\mathsf{proxy}}^{\mathsf{DSS}}$ beforehand. However, as we have shown in the proof in Appendix A.1, due to our strong adversary model, A can obtain R_1 anyway. Therefore, the proof is analogous to the proof of case **M2** above. □

A.4 Proof of Theorem 4 (Sketch)

Let $\mathcal{T}$ be a template and let k be the depth of $\mathcal{T}$. Now, we assume that A only knows at most $k-1$ instantiations and, thus, A has not full knowledge of $\mathcal{T}$. What A can try to do is to guess the remaining choices based on $\mathcal{C}$ and the instantiation dependent values $\mathcal{C}_{\overline{\mathcal{M}}}$. However, note that γR_1 as well as R are random values with unknown discrete logarithms, which can not be computed using any values from signatures or public keys. Thus, the values $t_{\mathcal{T}}(\alpha)P + R$ and δP are uniformly random and, consequently, $e(t_{\mathcal{T}}(\alpha)P + R, \delta P)$ as well as $\mathcal{C} = H'(e(t_{\mathcal{T}}(\alpha)P + R, \delta P))$ are also uniformly random. Moreover, in the same fashion also $\mathcal{C}_{\overline{\mathcal{M}}}$ is a uniformly random value. As a consequence, $\mathcal{C}$ and $\mathcal{C}_{\overline{\mathcal{M}}}$ constitute unconditionally hiding Pedersen commitments. Note that the instantiation dependent value γ guarantees that the values γR_1 in the commitments $\mathcal{C}_{\overline{\mathcal{M}}}$ of different template instantiations differ. Thus, an adversary can gain no knowledge by using differences of distinct values $\mathcal{C}_{\overline{\mathcal{M}}}$. Note also that since $\mathcal{C}$ is a hash value this prevents A from performing arithmetic with values $\mathcal{C}_{\overline{\mathcal{M}}}$ and thereby gaining knowledge of $t_{\mathcal{T}}$. From this follows the privacy of Scheme 1. □

Pseudorandom Signatures

Nils Fleischhacker
Cryptographic Algorithms Group
Saarland University, Germany
fleischhacker@cs.uni-saarland.de

Franziskus Kiefer, Mark Manulis
Department of Computing
University of Surrey, United Kingdom
f.kiefer@surrey.ac.uk, mark@manulis.eu

Felix Günther
Cryptographic Protocols Group
Technische Universität Darmstadt, Germany
guenther@cs.tu-darmstadt.de

Bertram Poettering
Information Security Group, Royal Holloway
University of London, United Kingdom
bertram.poettering@rhul.ac.uk

ABSTRACT

We develop a three-level hierarchy of privacy notions for (unforgeable) digital signature schemes. We first prove mutual independence of existing notions of anonymity and confidentiality, and then show that these are implied by higher privacy goals. The top notion in our hierarchy is *pseudorandomness*: signatures with this property hide the entire information about the signing process and cannot be recognized as signatures when transmitted over a public network. This implies very strong unlinkability guarantees across different signers and even different signing algorithms, and gives rise to new forms of private public-key authentication.

We show that one way towards pseudorandom signatures leads over our mid-level notion, called *indistinguishability*: such signatures can be simulated using only the public parameters of the scheme. As we reveal, indistinguishable signatures exist in different cryptographic settings (e.g. based on RSA, discrete logarithms, pairings) and can be efficiently lifted to pseudorandomness deploying general transformations using appropriate encoding techniques. We also examine a more direct way for obtaining pseudorandomness for any unforgeable signature scheme. All our transformations work in the standard model. We keep public verifiability of signatures in the setting of system-wide known public keys. Some results even hold if signing keys are disclosed to the adversary — given that signed messages have high entropy.

Categories and Subject Descriptors

E.3 [**Data Encryption**]: Public key cryptosystems

General Terms

Security, Theory, Algorithms, Design

Keywords

Digital signatures; privacy; hierarchy; pseudorandomness

1. INTRODUCTION

A digital signature σ on a message m is generated using a private key sk and is verified in respect to the corresponding public key pk. Digital signatures shall be unforgeable and offer authenticity of signers and integrity of signed messages. In the 90's, however, with the advent of public key infrastructures (PKIs), digital signatures were criticized for being a threat to user's *privacy* [20]. For instance, with system-wide known (PKI-certified) public keys, and due to the public verifiability of signatures, any transmission of (m, σ) over a public network such as the Internet implicitly reveals to all intermediate parties the identity of the signer, i.e. owner of (certified) pk — and not only to the intended recipients/verifiers. Considering public availability of both σ and pk, we can hope to obtain privacy only by restricting the amount of publicly available information about message m. Indeed, messages might be delivered through a different communication channel (e.g. in an online banking scenario with two-factor authentication) or at some earlier or later point in time (e.g. in anonymous auctions), and thus still remain out of reach of the adversary that obtains signatures and public keys. Further on, verifiers might expect signatures on messages that need not be transmitted at all: for example, private outputs computed with secure multi-party computation techniques or in (anonymous) key exchange protocols can be viewed as messages for which parties may wish to exchange signatures.

Current Privacy Notions and Open Issues. Privacy of digital signatures, where signatures σ are revealed but associated messages m are not disclosed, has found attention in definitions and security models of *anonymous signatures* by Yang et al. [29] and Fischlin [17], and in the notion of *confidential signatures* by Dent et al. [16]. These schemes aim at either hiding the identities/public keys of signers (anonymity) or the content of signed messages (confidentiality). Both privacy notions were defined for high-entropy message spaces, which is a necessary requirement, presuming the system-wide knowledge of public keys and signatures. This conceptual similarity raises a question on the possible relationship between anonymity and confidentiality, and triggers interest in a deeper study on the limits of privacy achievable with traditional signature schemes: Are there schemes whose signatures σ hide signer's identity/public key and simultaneously keep signed messages secret? What are the differences between deterministic and probabilistic schemes in terms of these goals? Is the property of 'message recovery' damaging for privacy? Answers to these questions would clarify the relationship among the existing notions of privacy for signature schemes, shed light on their privacy-enabling properties, and possibly pave the way for more sophisticated privacy guarantees. We give answers to many such questions.

Pseudorandom Signatures and Applications. In fact, the most interesting question in respect to privacy of signature schemes is whether signatures can look (to observers) completely random. On the one hand, this property would repel attention of intermediate parties, possibly performing traffic analysis, to the transmission of signatures. Those parties could not learn whether a given datagram represents something potentially valuable (in this case a signature) or not. On the other hand, and more importantly, signatures that cannot be distinguished from random strings (of some fixed length) also hide which signing algorithm its signer was using — this knowledge alone is often sufficient to identify signers (even if the signature scheme itself is anonymous).

We give some examples where signers, or groups of signers, naturally use different parameters settings for (potentially) the same signing algorithm.

For instance, in the new European travel documents, the selection of specific (elliptic curve) parameters is the priority of respective states [9]. Many banks and health insurance companies issue smart cards to their customers, initialized with different signing algorithms and parameters. We observe that distinguishing among different settings, implementations, or instantiations of the same signature scheme S can be seen as a privacy problem, e.g. if signatures can be used to derive which citizenship or which customer relationship to which bank or insurance company the signer has. Now assume that different signature schemes $S_1, S_2, \ldots$ (unforgeable, possibly under different hardness assumptions or with varying levels of security) output signatures $\sigma_1, \sigma_2, \ldots$, respectively, of fixed length L. If *all* these signatures looked random to observers, then σ_i would hide the applied scheme S_i, i.e. only the possession of m and (the expected signer's) pk would allow to verify σ_i. In contrast, any other party (even with knowledge of the keys of the whole system) would remain totally clueless whether σ_i represents a signature or not, and, if so, which signing algorithm was used.

Furthermore, pseudorandom signatures give rise to *covert public-key authentication*, offering cryptographic protection to covert channels, i.e. channels that appear random to any entity other than the communication partner (as defined, e.g. in [10,21,28]). With pseudorandom signatures, it would be possible to perform public-key authentication and execute authenticated key exchange over covert channels without loosing covertness. For example, parties could first run an unauthenticated key exchange protocol that has random-looking messages (those can easily be constructed using the Diffie-Hellman approach in various groups based on techniques used in our work (cf. Lemma 13)) and then exchange pseudorandom signatures on the protocol transcript and an additional high-entropy confirmation token derived from the established shared key material. Since all exchanged messages are random-looking, the established secure channel between the two mutually authenticated parties would remain covert. Moreover, if pseudorandomness of signatures can be preserved even in the unfortunate case where the signing key sk is leaked, then the above protocol would guarantee 'forward covertness' (akin to forward secrecy, e.g. [13]). More generally, pseudorandom signatures seem to offer very strong and useful privacy guarantees in communication protocols and applications where authentication should remain unobservable by traffic analysis [23].

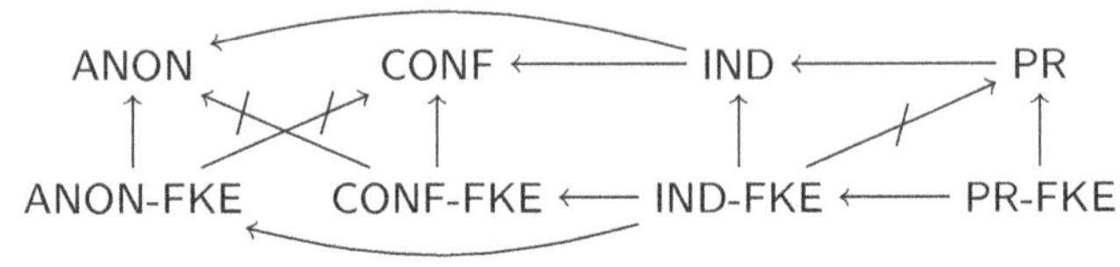

Figure 1: Privacy Hierarchy for Digital Signatures. Notation: $X \rightarrow Y$ means that X is a strictly stronger privacy notion than Y. $X \nrightarrow Y$ means that X does not imply Y. X-FKE stands for X defined with of full key exposure.

1.1 Our Results and Techniques

1.1.1 *Privacy Hierarchy for Digital Signatures*

We develop a three-level hierarchy of privacy notions for digital signatures, starting our work with the investigation of the relationship between anonymity and confidentiality, two previously established privacy notions for signatures. Definitions of anonymity were first given by Yang et al. [29], with later refinements by Fischlin [17]. These definitions presume messages with high entropy (unlike Bellare and Duan [2] and Saraswat and Yun [25], who regard a more restrictive form of signatures[1]). For confidentiality, we use the original definitions by Dent et al. [16] (in its strongest variant). We show that anonymous signatures and confidential signatures are independent privacy notions[2]. For this reason, *anonymity* (ANON) and *confidentiality* (CONF) are located at the lowest level of our privacy hierarchy.

At the mid-level of our hierarchy we have *indistinguishability* (IND), expressing that signatures can be simulated from the public parameters of the scheme. We prove that such signatures offer stronger privacy guarantees than purely anonymous and purely confidential schemes. In particular, any transmission of an IND signature simultaneously hides both the signer *and* the associated message. The IND prop-

[1]Anonymity definitions in [2, 25] assume that σ consists of two parts — signature σ_1 and a value σ_2 (called *de-anonymizer* in [2] or *verification token* in [25]), both of which are needed to perform the verification. Transmission of the entire $\sigma = (\sigma_1, \sigma_2)$, that is amenable to verification, becomes a two-stage process, with σ_1 (together with m) being disclosed in the first stage, and σ_2 in the second stage. Anonymity of the signer is then defined with respect to an adversary that knows σ_1 but not σ_2. As discussed in [2,25], this definition essentially implies anonymity from [17, 29], allowing also low-entropy message spaces. Since we look on privacy from a more general perspective (e.g. consider confidentiality as another privacy goal), it appears more advisable to work with high-entropy messages and use conceptually simpler anonymity definitions from [17,29].

[2]Interestingly, Dent et al. [16] mention similarities between the notions but do not formally investigate their relationship. They only expect it to be similar to the relationship between anonymous (key-private) encryption schemes [1] and traditional public-key encryption schemes. Arguably, these notions do not seem to be independent, unless anonymous encryption schemes that do not offer secrecy of encrypted messages are interesting on their own.

Setting	Example scheme	IND with FKE	PR with FKE	Model
RSA	FDH-RSA [4]	randomized hash with padding	IND-to-PR compiler	ROM
DL	Schnorr [26]	✓	IND-to-PR compiler	ROM
Pairings	Boneh-Boyen [5]	for random ℓ-bit m / hash-then-sign	IND-to-PR compiler	STD / ROM
any unforgeable signature scheme		inherited from PR compiler	direct PR compiler	STD

Table 1: Indistinguishable and Pseudorandom Signatures (Settings, Techniques, Compilers)

erty is thus a generalization of both ANON and CONF and is sufficient for obtaining privacy in anticipated applications of both schemes. IND signatures may, however, leak some recognizable structure about the signature scheme in use.

The strongest privacy notion in our arsenal lets signatures appear to privacy adversaries as (structureless) strings of random bits, and is hence termed *pseudorandom* (PR). We show why this property is sufficient to hide *all* information about the signing process, including the signature scheme itself. The introduced privacy hierarchy with the intuitively strongest notion of pseudorandomness on top is thus likely to close the subject of privacy-preserving signatures.

Each of our privacy notions is defined in two variants, reflecting that adversaries might of might not have a copy of the signing key. That is, we also address security with *full key exposure* (FKE), which has been formalized for anonymous schemes by Fischlin [17] (and is also considered in [2, 25], but did not find formal treatment in the definitions of confidentiality in [16]).

The universe of privacy notions considered in this paper and their relationships is illustrated in Figure 1.

1.1.2 Constructions and Transformations

For a selection of existing signature schemes, we investigate indistinguishability and pseudorandomness properties and, where appropriate, propose modifications and generic transformations to achieve these goals. We build privacy-preserving signatures in three different cryptographic settings, namely using RSA parameters on the example of the full-domain hash RSA (FDH-RSA) scheme [4], using cyclic prime-order groups on the example of Schnorr's scheme [26], and using pairings on the example of the Boneh-Boyen (BB) scheme [5]. These results are summarized in Table 1 and detailed in the following.

Indistinguishable Signatures. We formalize the notion of *information recovering signatures* (as a generalization of the known concept of 'message recovery') and discuss its negative impact on confidentiality, and hence on indistinguishability and pseudorandomness. We show that not only known message recovering schemes (e.g. 'text-book' RSA and Nyberg-Rueppel [22], together with their 'hash-then-sign'-based transformations), but also several other schemes (in different cryptographic contexts) fall under our more general notion of 'information recovery' and thus do not offer indistinguishability. Examples include schemes by Cramer and Shoup (CS) [15], Camenisch and Lysyanskaya (CL) [11], and Boneh, Lynn, and Shacham (BLS) [7]. That is, signatures in these schemes leak information even if they do not allow full recovery of signed messages.

We then focus on three cryptographic settings — RSA, discrete logarithms (DL), and pairings — and provide examples of indistinguishable schemes. In particular, we show that, using appropriate randomization and padding techniques, FDH-RSA becomes indistinguishable. In DL setting we prove that the (generalized) signature scheme by Schnorr [26] offers indistinguishability 'off the shelf', yet assuming that its cyclic group is shared among all signers. We notice that anonymity of FDH-RSA and Schnorr's schemes was previously analyzed in [29], their confidentiality in [16]. Our analysis essentially shows that both schemes admit much stronger privacy guarantees. Finally, we prove that the pairing-based Boneh-Boyen (BB) [5] scheme is indistinguishable in the standard model, yet for uniformly distributed (fixed-length) messages only. As suggested in [5], the 'hash-then-sign' approach can be used in standard model to sign longer messages. We show that in the random oracle model this method readily offers indistinguishability for arbitrary long high-entropy messages. We note that indistinguishability of all analyzed schemes holds in the presence of full key exposure (FKE). Bottom line, we show existence of IND schemes in different cryptographic settings, paving the way towards pseudorandom signature schemes (based on different hardness assumptions).

Pseudorandom Signatures. PR signatures, which cannot be distinguished from random bit strings of the same length, offer the highest form of privacy that signatures can provide.

Our first result on PR signatures is a generic transformation that strengthens IND signatures to obtain the PR property (we call it IND-to-PR compiler in Table 1). It uses *admissible encodings*, introduced in a different context by Boneh and Franklin [6] for elliptic curves, generalized later by Brier et al. [8], and also used to preserve privacy in the password-based authentication protocol by Bringer, Chabanne and Icart [9] (we thus show another interesting application of this primitive). By finding appropriate admissible encodings for different types of sets, we can immediately obtain the PR property for the IND versions of FDH-RSA, Schnorr, and BB schemes. We also prove that the obtained PR property holds in the presence of full key exposure.

Our next result is a second generic transformation that achieves the PR property directly for any (unforgeable) digital signature scheme. This PR compiler is powerful enough to guarantee the PR property also for information recovering schemes and works irrespective of whether the original scheme is probabilistic or deterministic. Moreover, the PR property is guaranteed even if secret signing keys are exposed (FKE). This compiler uses randomness extractors and its techniques have been previously developed by Fischlin [17] to construct anonymous signatures. We thus prove formally that Fischlin's anonymous signature admits a general transformation, which is strong enough to convert any unforgeable signature scheme into a scheme satisfying the strongest[3] privacy property. This direct PR compiler, however, is slightly less efficient than our IND-to-PR compiler.

[3]Fischlin [17] mentioned informally that his anonymous signature scheme offers pseudorandomness, although this no-

1.2 Related Work on Signature Privacy

Anonymity of signers assuming high-entropy messages was initially defined by Yang et al. [29], who analyzed anonymity of 'text-book' RSA, PSS, and Schnorr signatures, after applying some necessary modifications. Their definition was simplified by Fischlin [17] (and relaxed to full key exposure), who showed how to obtain anonymity using randomness extractors. Anonymity definitions for arbitrary messages, yet with specific restrictions on the format of disclosed signatures, were formulated independently by Bellare and Duan [2], Saraswat and Yun [25], and Zhang and Imai [30]. Using the 'sign-then-commit' approach, Bellare and Duan [2] gave four constructions: Their first scheme (also presented in [25]) uses commitments as black-box to produce anonymous signatures in the standard model. The second scheme uses randomized hash functions and can be applied to deterministic signature schemes. Their third solution relies on deterministic hash functions and can be used with probabilistic signature schemes. The fourth scheme from [2], termed 'splitting construction' follows closely the design of Schnorr signatures in the discrete logarithm setting. Saraswat and Yun [25] proved anonymity of the signature scheme by Boneh and Boyen [5]. Furthermore, both [2] and [25] formalize another requirement (called unambiguity in [2] and unpretendability in [25]) that prevents the adversary from claiming the ownership of an anonymous signature at a later stage. Note that this requirement is orthogonal to privacy and can be handled separately. The anonymous scheme by Zhang and Imai [30] uses what they call 'collision-resistant exposure-free functions' which are instantiated with randomized hash functions in the random oracle model. Confidentiality of signature schemes has been considered so far formally by Dent et al. [16] for messages with high entropy, inspired by the definitional treatment of confidentiality for deterministic public-key encryption [3]. They defined three flavors (weak, mezzo, and strong) with increasing strength and addressed both deterministic and probabilistic schemes (without full key exposure). Confidentiality of several schemes used in practice was analyzed as well, including those that use full-domain hash constructions (for which [16] defined confidentiality of hash functions, following earlier ideas from [12]), those obtained from Fiat-Shamir transformation, and those based on randomness extractors. Manifold solutions for obtaining privacy were also proposed with more advanced signing techniques. For example, in group signatures [14], users obtain membership certificates from the manager of a group and issue signatures that identify the signer as a valid group member without revealing its actual identity. The latter can be recovered from the signature only by the group manager. Ring signatures [24] allow the signer to form 'ad-hoc' groups and so hide its own identity (in an unrevocable way) from a potential verifier, who only learns that the signer belongs to the formed group. As discussed in [17], ring signatures differ substantially from anonymous signatures as, in the former, anonymity is bound to a (presumably small) group, and in the latter it is guaranteed as long as some information needed for the public verification of signatures remains secret. Anonymity notions have also been considered for other signature types, e.g. undeniable and confirmer signatures by Galbraith and Mao [19]; as discussed in [25], these notions differ from anonymity in ordinary signature schemes.

tion was not yet defined. Our hierarchy clarifies his intuition formally and further implies indistinguishability and confidentiality of his scheme (in presence of full key exposure).

$\mathsf{Exp}^{\mathsf{ANON[-FKE]},b}_{\mathsf{S},\mathcal{A}}(\lambda)$:

$(sk_0, pk_0) \leftarrow \mathsf{KGen}(1^\lambda)$

$(sk_1, pk_1) \leftarrow \mathsf{KGen}(1^\lambda)$

$m \leftarrow \mathcal{A}_1^{\mathsf{Sign}(sk_0,\cdot),\mathsf{Sign}(sk_1,\cdot)}(pk_0, pk_1[, sk_0, sk_1])$

$\sigma \leftarrow \mathsf{Sign}(sk_b, m)$

$d \leftarrow \mathcal{A}_2^{\mathsf{Sign}(sk_0,\cdot),\mathsf{Sign}(sk_1,\cdot)}(pk_0, pk_1, \sigma[, sk_0, sk_1])$

output d

Figure 2: Anonymity Experiments (without and with Full Key Exposure)

2. PREVIOUS PRIVACY NOTIONS

We recall the syntax of digital signatures in Definition 1. Note that all schemes used in this work are also unforgeable (cf. Appendix A). We focus on privacy of probabilistic[4] schemes in a setting with system-wide known public keys.

DEFINITION 1 (DIGITAL SIGNATURE SCHEME). *A digital signature scheme* $\mathsf{S} = (\mathsf{KGen}, \mathsf{Sign}, \mathsf{Ver})$ *is given by three algorithms: The key generation algorithm* KGen, *on input security parameter* 1^λ, *generates a key pair* $(sk, pk) \leftarrow \mathsf{KGen}(1^\lambda)$, *the signing algorithm* Sign, *on input a secret key sk and a message* $m \in \{0,1\}^*$, *outputs a signature* $\sigma \leftarrow \mathsf{Sign}(sk, m)$, *and the (deterministic) verification algorithm* Ver, *on input a public key pk, a message m, and a candidate signature* σ, *outputs a bit* $d \leftarrow \mathsf{Ver}(pk, m, \sigma)$. *The scheme* S *is* correct *if for all* $\lambda \in \mathbb{N}$, $(sk, pk) \leftarrow \mathsf{KGen}(1^\lambda)$, $m \in \{0,1\}^*$, *and* $\sigma \leftarrow \mathsf{Sign}(sk, m)$, *we have* $\mathsf{Ver}(pk, m, \sigma) = 1$. *The scheme* S *is* deterministic *if every two invocations of* $\mathsf{Sign}(sk, \cdot)$ *on the same input message m result in the same signature* σ.

2.1 Anonymous Signatures

Anonymity of signatures for high-entropy messages [17,29] hides which signer (presuming the system-wide knowledge of public keys) generated a given challenge signature σ. Definition 2 is essentially from [17], where we also let the adversary $\mathcal{A}$ choose the challenge message m (similar to [2]). In case of full key exposure, $\mathcal{A}$ receives secret keys, which obsoletes the signing oracle.

DEFINITION 2 (ANONYMOUS SIGNATURES). *A signature scheme* S *is* anonymous (ANON), *possibly* with full key exposure (ANON-FKE), *if for all PPT adversaries* $\mathcal{A} = (\mathcal{A}_1, \mathcal{A}_2)$ *the advantage function*

$$\mathsf{Adv}^{\mathsf{ANON[-FKE]}}_{\mathsf{S},\mathcal{A}}(\lambda) := \Big| \Pr\Big[\mathsf{Exp}^{\mathsf{ANON[-FKE]},0}_{\mathsf{S},\mathcal{A}}(\lambda) = 1\Big] - \Pr\Big[\mathsf{Exp}^{\mathsf{ANON[-FKE]},1}_{\mathsf{S},\mathcal{A}}(\lambda) = 1\Big] \Big|$$

is negligible in λ, *where* $\mathsf{Exp}^{\mathsf{ANON[-FKE]},b}_{\mathsf{S},\mathcal{A}}(\lambda)$, $b \in \{0,1\}$, *are the* anonymity experiments *from Figure 2, and where the following* high entropy constraint *holds: The function* $\mu(\lambda) =$

[4] Due to space limitations, please refer to the full version [18] of this paper for all (impossibility) results on privacy for *deterministic* schemes.

$\mathsf{Exp}_{\mathsf{S},\mathcal{A}}^{\mathsf{CONF[-FKE]},b}(\lambda)$:

$(sk, pk) \leftarrow \mathsf{KGen}(1^\lambda)$

$(\vec{m}_0, t_0) \leftarrow \mathcal{A}_1^{\mathsf{Sign}(sk,\cdot)}(pk[, sk])$

$(\vec{m}_1, t_1) \leftarrow \mathcal{A}_1^{\mathsf{Sign}(sk,\cdot)}(pk[, sk])$

if $|\vec{m}_0| \neq |\vec{m}_1|$ then output 0

$\vec{\sigma}^* \leftarrow \mathsf{Sign}(sk, \vec{m}_b)$

$t' \leftarrow \mathcal{A}_2^{\mathsf{Sign}(sk,\cdot)}(pk, \vec{\sigma}^*[, sk])$

if $t' = t_0$ then output 1, else output 0

Figure 3: Confidentiality Experiments (without and with Full Key Exposure).[5]

$\max_{M\in\{0,1\}^*} \Pr[M = m : m \leftarrow \mathcal{A}_1]$ *is negligible in* λ, *assuming that* $\mathcal{A}_1$ *is provided with all admissible inputs and oracles as specified in the respective anonymity experiment.* The minimum entropy *of* $\mathcal{A}$ *is then given by* $-\log_2 \mu(\lambda)$.

2.2 Confidential Signatures

Confidentiality of digital signatures, formalized by Dent et al. in [16], hides information about the message m that was signed. Definition 3 corresponds to *strong confidentiality* from [16], the strongest among the three notions (weak, mezzo, strong) proposed there.

DEFINITION 3 (CONFIDENTIAL SIGNATURES). *A signature scheme* S *is* confidential (CONF), *possibly* with full key exposure (CONF-FKE), *if for all PPT adversaries* $\mathcal{A} = (\mathcal{A}_1, \mathcal{A}_2)$ *the following advantage function*

$$\mathsf{Adv}_{\mathsf{S},\mathcal{A}}^{\mathsf{CONF[-FKE]}}(\lambda) := \Big|\Pr\left[\mathsf{Exp}_{\mathsf{S},\mathcal{A}}^{\mathsf{CONF[-FKE]},0}(\lambda) = 1\right] - \Pr\left[\mathsf{Exp}_{\mathsf{S},\mathcal{A}}^{\mathsf{CONF[-FKE]},1}(\lambda) = 1\right]\Big|$$

is negligible in λ, *where* $\mathsf{Exp}_{\mathsf{S},\mathcal{A}}^{\mathsf{CONF[-FKE]},b}(\lambda)$, $b \in \{0,1\}$, *are the* confidentiality experiments *from Figure 3, and where the following* high entropy constraint *holds: The function* $\mu(\lambda) = \max_{M\in\{0,1\}^*} \Pr[M \in \vec{m} : (\vec{m}, t) \leftarrow \mathcal{A}_1]$ *is negligible in* λ, *assuming* $\mathcal{A}_1$ *is provided with all admissible inputs and oracles as specified in the resp. confidentiality experiment.* The minimum entropy *of* $\mathcal{A}$ *is then given by* $-\log_2 \mu(\lambda)$.

Observe that, in the confidentiality experiments from Figure 3, the first-stage adversary $\mathcal{A}_1$ outputs a vector $\vec{m}$ of messages, each of high entropy, and an additional token t. This token models the intuition that confidential signatures shouldn't leak 'any information' about signed messages.[6]

2.3 Independence of ANON and CONF

Lemmas 1 and 2, proven in Appendix B, separate the two notions ANON and CONF. Their independence follows then from the fact that some unforgeable schemes are neither anonymous nor confidential and that for other schemes both notions hold simultaneously (cf. Section 4).

LEMMA 1 (CONF[-FKE] $\not\Rightarrow$ ANON). *Confidential signature schemes (with full key exposure) are not necessarily anonymous.*

LEMMA 2 (ANON[-FKE] $\not\Rightarrow$ CONF). *Anonymous signature schemes (with full key exposure) are not necessarily confidential.*

3. DIGITAL SIGNATURES WITH INFORMATION RECOVERY

The following definition of 'information recovery', where in the verification procedure signer's public key is used together with the signature to compute some information that is then compared to information determined by the public key and the message, generalizes the known concept of 'message recovery'.

DEFINITION 4 (INFORMATION RECOVERING SIGNATURES). *A digital signature scheme* S = (KGen, Sign, Ver) *is called* information recovering *if there exist two polynomial-time algorithms* Inf *and* Rec *such that for all* $(sk, pk) \leftarrow \mathsf{KGen}(1^\lambda)$, $m \in \{0,1\}^*$, *and* $\sigma \leftarrow \mathsf{Sign}(sk, m)$, *the verification algorithm* $\mathsf{Ver}(pk, m, \sigma)$ *outputs 1 if and only if* $\mathsf{Inf}(pk, m) = \mathsf{Rec}(pk, \sigma)$.

REMARK 1. *Note that information recovering signature schemes with* $\mathsf{Inf}(pk, \cdot) = pk = \mathsf{Rec}(pk, \cdot)$ *might be correct, but are necessarily forgeable. More precisely, if* $\mathsf{Inf}(pk, m_0) = \mathsf{Inf}(pk, m_1)$ *happens with noticeable probability (for independently drawn* $m_0, m_1 \in_R \{0,1\}^\lambda$*), an adversary* $\mathcal{A}$ *against unforgeability can be constructed by letting* $\mathcal{A}$ *request a signature* σ_0 *on a random message* m_0 *and output* (m_1, σ_0)*, for random* m_1*, as a forgery. To see that* $\mathcal{A}$ *has non-negligible success probability, observe that* σ_0 *will verify successfully for* m_1 *if we have* $\mathsf{Rec}(pk, \sigma_0) = \mathsf{Inf}(pk, m_1) = \mathsf{Inf}(pk, m_0)$*, what happens with non-negligible probability by assumption.*

Examples of Schemes with Information Recovery. Classical examples of information recovering schemes include "text-book" RSA and Nyberg-Rueppel [22], whose signatures can be used to recover messages. Observe that if the "hash-then-sign" approach is used, e.g. FDH-RSA [4], then signatures computed on hash values $H(m)$ (rather than on messages m) are still information recovering, even if H is modeled as a random oracle, i.e. the corresponding algorithm $\mathsf{Inf}(pk, m)$ would simply output $H(m)$. This property can also be found amongst signature schemes that are not message recovering or where messages need not be hashed to compute (unforgeable) signatures (as shown in Appendix C).

Information Recovery Limits Privacy. The property of information recovery of a scheme can be immediately used to break the scheme's confidentiality by including information derived via algorithm Inf from messages in $\vec{m}$ into t, as shown in the following lemma.

LEMMA 3. *There is no unforgeable information recovering signature scheme that provides confidentiality.*

PROOF. Let S = (KGen, Sign, Ver) be an information recovering signature scheme. Consider the following adversary $\mathcal{A} = (\mathcal{A}_1, \mathcal{A}_2)$ against confidentiality of S: $\mathcal{A}_1$ on input pk picks a random λ-bit message m, computes $t \leftarrow$

[5]Values $\vec{m}$ output by (stateless) $\mathcal{A}_1$ are *vectors* of messages in $\{0,1\}^*$ and $|\vec{m}|$ denotes the number of elements in $\vec{m}$. Accordingly, by $\vec{\sigma} \leftarrow \mathsf{Sign}(sk, \vec{m})$ we denote the process of signing the messages in $\vec{m}$ *element-wise* and *independently* of each other, resulting in a vector $\vec{\sigma}$ of corresponding signatures.

[6]We stick here to the confidentiality definition introduced by Dent et al. in [16] using two separate calls of $\mathcal{A}_1$.

$\mathsf{Inf}(pk, m)$, and outputs (m, t). $\mathcal{A}_2$ on input (pk, σ^*) outputs $t' \leftarrow \mathsf{Rec}(pk, \sigma^*)$. In this setting, we observe that correctness of S and the construction of Ver from Inf and Rec implies $\Pr[\mathsf{Exp}^{\mathsf{CONF},0}_{\mathsf{S},\mathcal{A}}(\lambda) = 1] = 1$. We now consider $\mathsf{Exp}^{\mathsf{CONF},1}_{\mathsf{S},\mathcal{A}}(\lambda)$, where we use σ_1 to denote the challenge signature $\sigma^* \leftarrow \mathsf{Sign}(sk, m_1)$. Note that by construction of $\mathcal{A}_2$ we have $t' = \mathsf{Rec}(pk, \sigma_1)$. We see that

$$\Pr[\mathsf{Exp}^{\mathsf{CONF},1}_{\mathsf{S},\mathcal{A}}(\lambda) = 1] = \Pr[t' = t_0] = \Pr[\mathsf{Rec}(pk, \sigma_1) = \mathsf{Inf}(pk, m_0)] = \Pr[\mathsf{Inf}(pk, m_1) = \mathsf{Inf}(pk, m_0)],$$

which, as S is unforgeable, is negligible (cf. Remark 1). For $\mathcal{A}$ we thus proved non-negligible advantage $\mathsf{Adv}^{\mathsf{CONF[-FKE]}}_{\mathsf{S},\mathcal{A}}(\lambda)$ against confidentiality of signature scheme S. □

4. INDISTINGUISHABLE SIGNATURES

The independence of the notions of anonymity and confidentiality of digital signature schemes places these concepts at the bottom level of our privacy hierarchy and puts forward the question on the existence of a more general privacy property that implies both of them. We call this property *indistinguishability* and formalize it in Definition 5. We use a simulation-based approach, following the intuition that anonymity and confidentiality are implied if all information that can be extracted from a real signature can also be extracted from a 'signature' that was *simulated* without knowledge of keys and messages.[7]

DEFINITION 5 (INDISTINGUISHABLE SIGNATURES). *A signature scheme* S *is* indistinguishable (IND), *possibly* with full key exposure (IND-FKE), *if there exists a PPT* simulator Sim *such that for all PPT adversaries* $\mathcal{A} = (\mathcal{A}_1, \mathcal{A}_2)$ *the following advantage function*

$$\mathsf{Adv}^{\mathsf{IND[-FKE]}}_{\mathsf{S,Sim},\mathcal{A}}(\lambda) := \Big|\Pr\Big[\mathsf{Exp}^{\mathsf{IND[-FKE]},0}_{\mathsf{S,Sim},\mathcal{A}}(\lambda) = 1\Big] - \Pr\Big[\mathsf{Exp}^{\mathsf{IND[-FKE]},1}_{\mathsf{S,Sim},\mathcal{A}}(\lambda) = 1\Big]\Big|$$

is negligible in λ, *where* $\mathsf{Exp}^{\mathsf{IND[-FKE]},b}_{\mathsf{S,Sim},\mathcal{A}}(\lambda)$, $b \in \{0,1\}$, *are the* indistinguishability experiments *from Figure 4, and where the following* high entropy constraint *holds: The function* $\mu(\lambda) = \max_{M \in \{0,1\}^*} \Pr[M \in \vec{m} : (\vec{m}, t) \leftarrow \mathcal{A}_1]$ *is negligible in* λ, *assuming* $\mathcal{A}_1$ *is provided with all admissible inputs and oracles as specified in the resp. indistinguishability experiment. The* minimum entropy *of* $\mathcal{A}$ *is given by* $-\log_2 \mu(\lambda)$.

Lemmas 4 and 5, proven in Appendix D, confirm the intuition that indistinguishable signatures are also anonymous and confidential (even in presence of full key exposure).

LEMMA 4 (IND ⇒ ANON, IND-FKE ⇒ ANON-FKE). *Every indistinguishable signature scheme is anonymous. The same implication holds in presence of full key exposure.*

LEMMA 5 (IND ⇒ CONF, IND-FKE ⇒ CONF-FKE). *Every indistinguishable signature scheme is confidential. The same implication holds in presence of full key exposure.*

[7]Since our IND definition involves a simulator Sim that generates signatures in $\vec{\sigma}$ without knowledge of (sk, m), one may ask about the relationship to zero-knowledge proofs, which also, by definition, are simulatable without knowledge of the secret. We observe that indistinguishable signatures are not zero-knowledge proofs — simulated IND signatures need not to be convincing (as opposed to simulated proofs), i.e. they do not need to pass the regular signature verification.

$\mathsf{Exp}^{\mathsf{IND[-FKE]},b}_{\mathsf{S,Sim},\mathcal{A}}(\lambda)$:
- $(sk, pk) \leftarrow \mathsf{KGen}(1^\lambda)$
- $(\vec{m}, t) \leftarrow \mathcal{A}_1^{\mathsf{Sign}(sk,\cdot)}(pk[, sk])$
- $\vec{\sigma}_0 \leftarrow \mathsf{Sign}(sk, \vec{m})$
- $\vec{\sigma}_1 \leftarrow \mathsf{Sim}(1^\lambda, |\vec{m}|)$
- $t' \leftarrow \mathcal{A}_2^{\mathsf{Sign}(sk,\cdot)}(pk, \vec{\sigma}_b[, sk])$
- if $t' = t$ then output 1, else output 0

Figure 4: Indistinguishability Experiments (without and with Full Key Exposure).

4.1 Techniques and Examples

We now exemplify IND constructions using three known signature schemes: FDH-RSA [4], Schnorr [26], and Boneh-Boyen [5]. That is, we show that indistinguishable schemes can be obtained in different cryptographic setting, i.e. RSA, discrete logarithms (DL), and pairings. We notice that our techniques can be applied to many existing schemes that either fulfill this privacy notion directly or can be slightly modified to become indistinguishable.

4.1.1 RSA-based Construction

On the example of FDH-RSA [4], which is neither confidential nor anonymous, we demonstrate two techniques to obtain indistinguishability. First, we apply a *randomized hash* [2], where a message m is hashed together with some randomness r, which is chosen within the signing procedure. The hash value $H(m, r)$ is then used in the signing algorithm (instead of m), and r is appended to the resulting signature. This method eliminates information recovery since the output of $\mathsf{Inf}(pk, m)$ depends now on $H(m, r)$ (and not only on m), and the probability that the first stage IND adversary learns information about $H(m, r)$ is negligible (given that r is sufficiently long and chosen in the challenge phase). We can then apply *padding* to hide the length of signature components that are elements of $\mathbb{Z}_N$, and by this protect anonymity [29]. These methods turn out to be sufficient for the indistinguishability of the scheme.

Randomized FDH-RSA with Padding. Let $\mathsf{GenRSA}(1^\lambda)$ denote an algorithm that outputs tuples (N, e, d) where N is an RSA modulus, i.e. $N = pq$ for two prime numbers p and q of length $\lambda/2$, and $e, d \in \mathbb{Z}^\times_{\varphi(N)}$ with $ed = 1 \bmod \varphi(N)$, where $\varphi(N) = (p-1)(q-1)$. Let $H_N : \{0,1\}^* \to \mathbb{Z}_N$ be a hash function modeled as random oracle and Z_λ be a fixed number of 2λ bits, independently of N. The randomized FDH-RSA scheme with padding is defined as follows.

$\mathsf{FDH\text{-}RSA.KGen}(1^\lambda)$: Let $(N, e, d) \leftarrow \mathsf{GenRSA}(1^\lambda)$, $pk := (N, e)$, and $sk := d$. Output (sk, pk).

$\mathsf{FDH\text{-}RSA.Sign}(sk, m)$: Choose $r \in_R \{0,1\}^\lambda$ and $k \in_R [0, \lfloor Z_\lambda / N \rfloor - 1]$. Let $\sigma' := H_N(m \| r)^d \bmod N$. Output $\sigma := (\sigma' + kN, r)$.

$\mathsf{FDH\text{-}RSA.Ver}(pk, m, \sigma)$: Parse σ as (σ', r). Let $h' \leftarrow H_N(m \| r)$ and $h := (\sigma')^e \bmod N$. Output $h = h'$.

Observe that on each signature σ' a probabilistic padding is applied, computing $k \in_R [0, \lfloor Z_\lambda/N \rfloor - 1], \sigma := \sigma' + kN$, which can be reversed by computing $\sigma' := \sigma \bmod N$. It maps uniformly distributed integers from $[0, N-1]$ to (nearly) uniformly distributed integers in $[0, Z_\lambda - 1]$ (cf. Lemma 13).

LEMMA 6. *The probabilistic FDH-RSA scheme with padding is indistinguishable with full key exposure, in the random oracle model.*

PROOF. We will consider the simulator Sim that, on input security parameter 1^λ and message number ℓ, outputs a vector of ℓ integers in $[0, Z_\lambda - 1]$, drawn uniformly at random. Let $\mathcal{A} = (\mathcal{A}_1, \mathcal{A}_2)$ be any indistinguishability adversary against the modified FDH-RSA scheme. Consider experiments $\mathsf{Exp}^{\mathsf{IND},0,j}_{\mathsf{FDH\text{-}RSA},\mathsf{Sim},\mathcal{A}}(\lambda)$, similar to $\mathsf{Exp}^{\mathsf{IND},0}_{\mathsf{FDH\text{-}RSA},\mathsf{Sim},\mathcal{A}}(\lambda)$, except that the first j signatures in $\vec{\sigma}^*$ are simulated using $\mathsf{Sim}(1^\lambda, j)$, i.e. $\mathsf{Exp}^{\mathsf{IND},0}_{\mathsf{FDH\text{-}RSA},\mathsf{Sim},\mathcal{A}}(\lambda) = \mathsf{Exp}^{\mathsf{IND},0,0}_{\mathsf{FDH\text{-}RSA},\mathsf{Sim},\mathcal{A}}(\lambda)$ and $\mathsf{Exp}^{\mathsf{IND},1}_{\mathsf{FDH\text{-}RSA},\mathsf{Sim},\mathcal{A}}(\lambda) = \mathsf{Exp}^{\mathsf{IND},0,\ell}_{\mathsf{FDH\text{-}RSA},\mathsf{Sim},\mathcal{A}}(\lambda)$ for some ℓ polynomially bounded by λ. It will suffice to prove that

$$\left|\Pr\left[\mathsf{Exp}^{\mathsf{IND},0,j}_{\mathsf{FDH\text{-}RSA},\mathsf{Sim},\mathcal{A}}(\lambda) = 1\right] - \Pr\left[\mathsf{Exp}^{\mathsf{IND},0,j+1}_{\mathsf{FDH\text{-}RSA},\mathsf{Sim},\mathcal{A}}(\lambda) = 1\right]\right|$$

is negligible for all j.

Let Exp be the experiment which is like $\mathsf{Exp}^{\mathsf{IND},0,j}_{\mathsf{FDH\text{-}RSA},\mathsf{Sim},\mathcal{A}}(\lambda)$ except that, in the generation of the $(j+1)$th signature of $\vec{\sigma}^*$, we replace the output of hash function H_N by a value $h \in_R \mathbb{Z}_N$, picked uniformly at random (or, equivalently, we use the value h^d, for $h \in_R \mathbb{Z}_N$). By the Random Oracle Model, this change can only be detected by adversaries that correctly guess both: message m and randomizer r. But this will happen only with negligible probability, since $\mathcal{A}_1$ would have to guess $r \in \{0,1\}^\lambda$, and $\mathcal{A}_2$ would have to guess m, which has large entropy. Consider now the hop to experiment $\mathsf{Exp}^{\mathsf{IND},0,j+1}_{\mathsf{FDH\text{-}RSA},\mathsf{Sim},\mathcal{A}}(\lambda)$, which is exactly like Exp, except that the returned $(j+1)$th 'signature' is not computed via the padding, i.e. $\sigma = (h + kN, r)$, but instead via $\sigma = (h', r)$, where $h' \in_R [0, Z_\lambda - 1]$. We will show in Lemma 13 (1) that this introduces only a negligible statistical difference between the output distributions of Exp and $\mathsf{Exp}^{\mathsf{IND},0,j+1}_{\mathsf{FDH\text{-}RSA},\mathsf{Sim},\mathcal{A}}(\lambda)$. □

4.1.2 DL and Pairing-Based Constructions

We now move to the DL and pairing-based settings and focus on the signature schemes by Schnorr [26] and by Boneh and Boyen [5], respectively.

Schnorr Signature Scheme. Let $G = \langle g \rangle$ be a cyclic group of prime order q, where $|q| = \lambda$, and $H : \{0,1\}^* \to \mathbb{Z}_q$ be a hash function modeled as random oracle. Schnorr's signature scheme is specified as follows.

$\mathsf{SCH.KGen}(1^\lambda)$:	Choose $x \in_R \mathbb{Z}_q$. Output $(sk, pk) := (x, g^x)$.
$\mathsf{SCH.Sign}(sk, m)$:	Choose $r \in_R \mathbb{Z}_q$. Let $c \leftarrow H(g^r \parallel m)$ and $s := sk \cdot c + r \bmod q$. Output $\sigma := (c, s)$.
$\mathsf{SCH.Ver}(pk, m, \sigma)$:	Parse σ as (c, s). Compute $c' \leftarrow H(pk^{-c} \cdot g^s \parallel m)$. Output $c = c'$.

In the indistinguishability analysis (akin to prior work on anonymity and confidentiality of the scheme [16, 29]) we assume that all signers use the same group G.

LEMMA 7. *The (generalized) Schnorr signature scheme is indistinguishable with full key exposure, in the random oracle model.*

PROOF. Consider the following simulator Sim: On input security parameter 1^λ and message number ℓ, Sim independently samples ℓ random pairs $(c'_1, s'_1), \ldots, (c'_\ell, s'_\ell) \in_R \mathbb{Z}_q \times \mathbb{Z}_q$ and outputs $\vec{\sigma}^* = (\sigma_i)_{1 \le i \le \ell}$, where $\sigma_i = (c'_i, s'_i)$. Let $\mathcal{A} = (\mathcal{A}_1, \mathcal{A}_2)$ be any indistinguishability adversary against the Schnorr signature scheme. Denote by $\mathsf{Exp}^{\mathsf{IND},0,j}_{\mathsf{SCH},\mathsf{Sim},\mathcal{A}}(\lambda)$ the experiment that is like $\mathsf{Exp}^{\mathsf{IND},0}_{\mathsf{SCH},\mathsf{Sim},\mathcal{A}}(\lambda)$, except that the first j signatures in $\vec{\sigma}^*$ are simulated using $\mathsf{Sim}(1^\lambda, j)$, i.e. $\mathsf{Exp}^{\mathsf{IND},0}_{\mathsf{SCH},\mathsf{Sim},\mathcal{A}}(\lambda) = \mathsf{Exp}^{\mathsf{IND},0,0}_{\mathsf{SCH},\mathsf{Sim},\mathcal{A}}(\lambda)$ and $\mathsf{Exp}^{\mathsf{IND},1}_{\mathsf{SCH},\mathsf{Sim},\mathcal{A}}(\lambda) = \mathsf{Exp}^{\mathsf{IND},0,\ell}_{\mathsf{SCH},\mathsf{Sim},\mathcal{A}}(\lambda)$ for some ℓ polynomially bounded by λ. It suffices to prove that, for all j, $\left|\Pr\left[\mathsf{Exp}^{\mathsf{IND},0,j}_{\mathsf{SCH},\mathsf{Sim},\mathcal{A}}(\lambda) = 1\right] - \Pr\left[\mathsf{Exp}^{\mathsf{IND},0,j+1}_{\mathsf{SCH},\mathsf{Sim},\mathcal{A}}(\lambda) = 1\right]\right|$ is negligible.

Denote by Exp the experiment which is like $\mathsf{Exp}^{\mathsf{IND},0,j}_{\mathsf{SCH},\mathsf{Sim},\mathcal{A}}(\lambda)$ except that, in the generation of the $(j+1)$th signature of $\vec{\sigma}^*$, we replace the output of hash function H by a value $c' \in_R \mathbb{Z}_q$, picked uniformly at random. By the Random Oracle Model, this change can only be detected by adversaries that correctly guess both: group element g^r and message m. This will happen only with negligible probability, since $\mathcal{A}_1$ would have to guess g^r (with success probability $\frac{1}{q} \approx 2^{-\lambda}$), and $\mathcal{A}_2$ would have to guess m, which has large entropy. Note that c is now independent of r, i.e., in the calculation of $s := sk \cdot c + r$, variable r acts like a one-time pad on $sk \cdot c$. It follows that also s can be replaced by a uniformly random value $s' \in_R \mathbb{Z}_q$, without $\mathcal{A}$ noticing it. We have just shown that $|\Pr[\mathsf{Exp}^{\mathsf{IND},0,j}_{\mathsf{SCH},\mathsf{Sim},\mathcal{A}}(\lambda) = 1] - \Pr[\mathsf{Exp} = 1]|$ is negligible, and that $|\Pr[\mathsf{Exp} = 1] - \Pr[\mathsf{Exp}^{\mathsf{IND},0,j+1}_{\mathsf{SCH},\mathsf{Sim},\mathcal{A}}(\lambda) = 1]| = 0$. This concludes the proof. □

Boneh-Boyen Signature Scheme.

The signature scheme by Boneh and Boyen [5] based on pairings works over cyclic groups G_1, G_2, G_T of prime order q (with $|q| = \lambda$) for which an efficient bilinear map $e : G_1 \times G_2 \to G_T$ is known. Let g_1 and g_2 be generators of G_1 and G_2, respectively. The scheme is specified for message space $\mathbb{Z}_q$ as follows.

$\mathsf{BB.KGen}(1^\lambda)$:	Choose $x, y \in_R \mathbb{Z}_q \setminus \{0\}$. Let $sk := (x, y)$ and $pk := (u, v) = (g_2^x, g_2^y)$. Output (sk, pk).
$\mathsf{BB.Sign}(sk, m)$:	Choose $r \in_R \mathbb{Z}_q \setminus \{-\frac{x+m}{y}\}$. Let $\sigma' := g_1^{1/(x+m+yr)}$. Output $\sigma := (\sigma', r)$.
$\mathsf{BB.Ver}(pk, m, \sigma)$:	Parse σ as (σ', r). Output $e(\sigma, u g_2^m v^r) = e(g_1, g_2)$.

LEMMA 8. *The "hash-then-sign" version of the signature scheme by Boneh and Boyen is indistinguishable with full key exposure, in the random oracle model.*

We first prove that this construction is indistinguishable (in the standard model), yet for uniform message distributions only, and then generalize this result to arbitrary distributions in $\{0,1\}^*$ (in the random oracle model).

In order to prove Lemma 8 we first prove the following lemma (which holds in the standard model).

LEMMA 9. *Signature scheme* BB *is (perfectly) indistinguishable with respect to full key exposure, for* uniformly distributed messages.

PROOF. Consider the following simulator Sim: On input security parameter 1^λ and message number ℓ, Sim independently samples ℓ random pairs $(\sigma', r) \in_R (G_1 \setminus \{1\}) \times \mathbb{Z}_q$. Let $\mathcal{A} = (\mathcal{A}_1, \mathcal{A}_2)$ be any indistinguishability adversary against the BB signature scheme such that $\mathcal{A}_1$ outputs uniformly distributed messages in $\mathbb{Z}_q$. Denote by $\mathsf{Exp}^{\mathsf{IND},0,j}_{\mathsf{BB},\mathsf{Sim},\mathcal{A}}(\lambda)$ the experiment that is like $\mathsf{Exp}^{\mathsf{IND},0}_{\mathsf{BB},\mathsf{Sim},\mathcal{A}}(\lambda)$, except that the first j signatures in $\vec{\sigma}^*$ are simulated using $\mathsf{Sim}(1^\lambda, j)$, i.e. $\mathsf{Exp}^{\mathsf{IND},0}_{\mathsf{BB},\mathsf{Sim},\mathcal{A}}(\lambda) = \mathsf{Exp}^{\mathsf{IND},0,0}_{\mathsf{BB},\mathsf{Sim},\mathcal{A}}(\lambda)$ and finally $\mathsf{Exp}^{\mathsf{IND},1}_{\mathsf{BB},\mathsf{Sim},\mathcal{A}}(\lambda) = \mathsf{Exp}^{\mathsf{IND},0,\ell}_{\mathsf{BB},\mathsf{Sim},\mathcal{A}}(\lambda)$ for some ℓ polynomially bounded by λ.

Note that in BB.Sign the mapping $r \mapsto \sigma' = g_1^{1/(x+m+yr)}$ is one-to-one between domain $\mathbb{Z}_q \setminus \{-\frac{x+m}{y}\}$ and range $G_1 \setminus \{1\}$. Basically, this is due to the fact that $\mathbb{Z}_q$ is a finite field, in which all elements but zero can be multiplicatively inverted. It follows that, if r is picked uniformly at random from the given domain, then r acts like a one-time pad on m and makes σ' uniformly distributed, in $\mathcal{A}_1$'s eyes. The analog holds for $\mathcal{A}_2$: This time, it is uniformly distributed message m that makes σ' look uniform. This shows that $\Pr[\mathsf{Exp}^{\mathsf{IND},0,j}_{\mathsf{BB},\mathsf{Sim},\mathcal{A}}(\lambda) = 1] = \Pr[\mathsf{Exp}^{\mathsf{IND},0,j+1}_{\mathsf{BB},\mathsf{Sim},\mathcal{A}}(\lambda) = 1]$ for all j, and hence concludes the proof. □

In general, digital signature schemes are expected to support arbitrary message spaces, i.e. messages $m \in \{0,1\}^*$. As pointed out by Boneh and Boyen [5], their scheme can be converted into an unforgeable signature scheme for arbitrary long messages by using the "hash-then-sign" approach, for a suitable hash function $H : \{0,1\}^* \to \mathbb{Z}_q$. Such 'hybrid' version of BB can still be proven unforgeable in the standard model, as the only condition posed on H is that of collision-resistance. If, on the other hand, the hash function H is additionally modeled as a random oracle (that smoothes the entropy in the message space to a uniform distribution) the proof of Lemma 8 follows directly from Lemma 9.

5. PSEUDORANDOM SIGNATURES

Although indistinguishability is already a strong privacy notion it still has one important limitation: The simulator Sim used to define the IND property depends on the signature scheme S; in particular, it simulates signatures using public parameters of S. Although both the signer and the message are successfully hidden in IND signatures, the very scheme that was used to create a given signature might not be, e.g. IND signatures may have characteristic lengths or follow specific formats, like element representation of the components of Schnorr signatures, and so on. In practice, usage of some S in an application or network protocol can be prescribed via standards. However, instantiations with concrete parameters (e.g. prime modulus p in a Schnorr group $G \subseteq \mathbb{Z}_p^\times$) is often left unspecified. Different parameter choices may introduce a unique pattern that can be (mis)used to distinguish among the signing algorithms and by this obtain more information about signers and about the context in which the signatures were produced.

We address this limitation in Definition 6 with the property of *pseudorandomness*, where we require that signatures output by S are indistinguishable from randomly chosen binary strings of length $L(\lambda) = L_{\mathsf{S}}(\lambda)$. This simpler definition

$$
\begin{aligned}
&\mathsf{Exp}^{\mathsf{PR[\text{-}FKE]},b}_{\mathsf{S},\mathcal{A}}(\lambda):\\
&\quad (sk, pk) \leftarrow \mathsf{KGen}(1^\lambda)\\
&\quad (\vec{m}, t) \leftarrow \mathcal{A}_1^{\mathsf{Sign}(sk,\cdot)}(pk[, sk])\\
&\quad \vec{\sigma}_0 \leftarrow \mathsf{Sign}(sk, \vec{m})\\
&\quad \vec{\sigma}_1 \in_R \{0,1\}^{L(\lambda)\times|\vec{m}|}\\
&\quad t' \leftarrow \mathcal{A}_2^{\mathsf{Sign}(sk,\cdot)}(pk, \vec{\sigma}_b[, sk])\\
&\quad \text{if } t' = t \text{ then output } 1, \text{ else output } 0
\end{aligned}
$$

Figure 5: Pseudorandomness Experiments (without and with Full Key Exposure).[8]

is sufficient to obtain pseudorandom signatures of some fixed length L^*, viewed as a global upper bound on the individual lengths $L_{\mathsf{S}}(\lambda)$ for all signature schemes S in the system, using a simple *padding with random bits*. That is, *all* signatures in the system would be L^* bits long and look completely random, no matter how they were produced. It would hence become impossible, for some given signature σ to derive any information about the scheme S that was used to generate it. This seems to be the highest level of privacy that can be offered by a signature scheme.

DEFINITION 6 (PSEUDORANDOM SIGNATURES). *A signature scheme* S *is* pseudorandom (PR), *possibly* with full key exposure (PR-FKE), *if there is a polynomially bounded function $L(\lambda)$ such that for all PPT adversaries $\mathcal{A}=(\mathcal{A}_1, \mathcal{A}_2)$ the advantage function*

$$
\begin{aligned}
\mathsf{Adv}^{\mathsf{PR[\text{-}FKE]}}_{\mathsf{S},\mathcal{A}}(\lambda) := \Big| &\Pr\left[\mathsf{Exp}^{\mathsf{PR[\text{-}FKE]},0}_{\mathsf{S},\mathcal{A}}(\lambda) = 1\right]\\
&- \Pr\left[\mathsf{Exp}^{\mathsf{PR[\text{-}FKE]},1}_{\mathsf{S},\mathcal{A}}(\lambda) = 1\right]\Big|
\end{aligned}
$$

is negligible in λ, where $\mathsf{Exp}^{\mathsf{PR[\text{-}FKE]},b}_{\mathsf{S},\mathcal{A}}(\lambda)$, $b \in \{0,1\}$, are the pseudorandomness experiments *from Figure 5, and where the following* high entropy constraint *holds: The function $\mu(\lambda) = \max_{M\in\{0,1\}^*} \Pr[M \in \vec{m} : (\vec{m}, t) \leftarrow \mathcal{A}_1]$ is negligible in λ, assuming that $\mathcal{A}_1$ is provided with all admissible inputs and oracles as specified in the respective pseudorandomness experiment. The* minimum entropy *of $\mathcal{A}$ is then given by $-\log_2 \mu(\lambda)$.*

Pseudorandomness is as a special case of indistinguishability, where simulator Sim draws at random from $\{0,1\}^{L(\lambda)}$. However, PR is strictly stronger than IND, i.e. Lemmas 10 and 11, proven in Appendix E, settle the PR notion at the top of the privacy hierarchy:

LEMMA 10 (PR $\Rightarrow$ IND, PR-FKE $\Rightarrow$ IND-FKE). *Every pseudorandom signature scheme is indistinguishable. The same implication holds with full key exposure.*

LEMMA 11 (IND[-FKE] $\not\Rightarrow$ PR). *Indistinguishable signature schemes (with full key exposure) are not necessarily pseudorandom.*

[8]We denote by $\vec{\sigma} \in_R \{0,1\}^{L\times|\vec{m}|}$ the process of picking $|\vec{m}|$ strings independently at random from $\{0,1\}^L$. We comprehend $L(\lambda)$ as the fixed length of signatures conforming to security level λ.

5.1 Two Pseudorandomness Compilers

We present two compilers for pseudorandomness of digital signatures. Our first compiler assumes that the underlying scheme is indistinguishable (with some additional constraints), while our second compiler offers pseudorandomness for arbitrary (unforgeable) signature schemes. Both transformations work *without* random oracles.

5.1.1 IND-*to*-PR *Compiler: From Indistinguishability to Pseudorandomness*

Our IND-to-PR compiler converts any indistinguishable signature scheme into a pseudorandom one. This is done by considering the different elements that form the signature component-wise (e.g., in case of Schnorr signatures, the elements $c \in \mathbb{Z}_q$ and $s \in \mathbb{Z}_q$), and encoding them as binary strings via appropriate *admissible encodings*. Resulting strings are concatenated to obtain the pseudorandom signature. The concept of admissible encodings was created for the main purpose of hashing into elliptic curves in the IBE scheme of Boneh and Franklin [6]. Their definition was later generalized to arbitrary sets by Brier et al. [8], and used recently in the construction of a privacy-preserving authentication protocol by Bringer, Chabanne and Icart [9].

Definition 7 (Admissible Encoding [8]). *Let S, R denote finite sets with $|S| > |R|$. A function $F : S \to R$ is called* ϵ-admissible encoding *for (S, R) if it satisfies the following properties:*

1. *Computable: F is computable in deterministic polynomial time.*
2. *Invertible: There exists a PPT algorithm $\mathcal{I}_F$ such that $\mathcal{I}_F(r) \in F^{-1}(r) \cup \{\bot\}$ for all $r \in R$, and for r uniformly distributed in R the distribution of $\mathcal{I}_F(r)$ is ϵ-statistically indistinguishable from the uniform distribution in S.*

If ϵ is a negligible function of the security parameter then F is called an admissible encoding.

Intuitively, an admissible encoding $F : S \to R$ shifts the process of picking elements uniformly at random in S to the process of picking elements uniformly at random in R, and vice versa. Not surprisingly, the following aggregation lemma holds.

Lemma 12 (Aggregation of Admissible Encodings). *Let $\mathcal{S} = S_1 \times \ldots \times S_n$ and $\mathcal{R} = R_1 \times \ldots \times R_n$ denote Cartesian products of finite sets. For each $1 \le i \le n$, let $F_i : S_i \to R_i$ denote an ϵ_i-admissible encoding. Then $F : \mathcal{S} \to \mathcal{R}$; $(s_1, \ldots, s_n) \mapsto (F_1(s_1), \ldots, F_n(s_n))$ is an ϵ-admissible encoding, for $\epsilon = \sum_{i=1}^{n} \epsilon_i$.* □

In our compiler, we will use admissible encodings $F : S \to R$, where $S = \{0,1\}^{L(\lambda)}$ and R is the 'signature space' of the scheme. Note that w. l. o. g. it would even suffice to have $S = \{0,1\}^{\ell}$ for $\ell < L(\lambda)$ as we can always pad[9] with $L(\lambda) - \ell$ random bits at the end in $\mathcal{I}_F$, and ignore the last $L(\lambda) - \ell$ bits when evaluating F. Thus, if σ is indistinguishable then $\sigma' := \mathcal{I}_F(\sigma) \in \{0,1\}^{L(\lambda)}$ is pseudorandom. This admissible encoding-based compiler works as follows, where the input signature scheme S is assumed to be indistinguishable with a regular simulator and $(F, \mathcal{I}_F)$ denotes an appropriate admissible encoding that maps strings in $\{0,1\}^{L(\lambda)}$ into the signature space.

$\mathsf{AEC.KGen}(1^\lambda)$: Output $(sk, pk) \leftarrow \mathsf{S.KGen}(1^\lambda)$.

$\mathsf{AEC.Sign}(sk, m)$: Let $\sigma \leftarrow \mathsf{S.Sign}(sk, m)$. Output $\sigma' \leftarrow \mathcal{I}_F(\sigma)$.

$\mathsf{AEC.Ver}(pk, m, \sigma')$: Let $\sigma \leftarrow F(\sigma')$. Output $\mathsf{S.Ver}(pk, m, \sigma)$.

The pseudorandomness of the resulting scheme AEC is proven in Theorem 1. This proof requires the following notion of *regular simulators* that can be found in our proofs of IND signature schemes from Section 4.1 and exist for many other schemes.

Definition 8 (Regular Simulators). *A simulator* Sim *in the indistinguishability experiment (cf. Figure 4) is called* regular *if it samples uniformly at random from the 'signature space' $\mathcal{S}(\lambda)$, i.e. the range of the* Sign *algorithm. That is, for regular simulators* Sim*, running $\vec{\sigma}^* \leftarrow \mathsf{Sim}(1^\lambda, \ell)$ and $\vec{\sigma}^* \in_R \mathcal{S}(\lambda)^\ell$ are identical.*

Theorem 1. *If* S *is an indistinguishable signature scheme with a regular simulator* Sim *and F is an ϵ-admissible encoding that maps $\{0,1\}^{L(\lambda)}$ into the signature space of* S *then the* AEC *signature scheme, obtained via* IND-*to*-PR *compiler, is pseudorandom.*

Proof. Let $\mathcal{A} = (\mathcal{A}_1, \mathcal{A}_2)$ be any pseudorandomness adversary against AEC signature scheme. Denote by Exp the experiment that is like $\mathsf{Exp}^{\mathsf{PR},0}_{\mathsf{AEC},\mathcal{A}}(\lambda)$, except that the signatures in $\vec{\sigma}^*$ are not computed individually as $\sigma := \mathcal{I}_F(\mathsf{S.Sign}(sk, m))$, but as $\vec{\sigma}^* := \mathcal{I}_F(\mathsf{Sim}(1^\lambda, |\vec{m}|))$, where $\mathcal{I}_F$ is executed component-wise. As S is indistinguishable by assumption, we know that $\left|\Pr\left[\mathsf{Exp}^{\mathsf{PR},0}_{\mathsf{AEC},\mathcal{A}}(\lambda) = 1\right] - \Pr\left[\mathsf{Exp} = 1\right]\right|$ is negligible. Consider now experiment Exp^j, which is like Exp except that the first j values in $\vec{\sigma}^*$ are randomly picked strings in $\{0,1\}^{L(\lambda)}$, i.e. $\mathsf{Exp} = \mathsf{Exp}^0$ and $\mathsf{Exp}^{\mathsf{PR},1}_{\mathsf{AEC},\mathcal{A}}(\lambda) = \mathsf{Exp}^\ell$ for some ℓ polynomially bounded by λ. As Sim is regular, the distribution of $\mathcal{I}_F(\mathsf{Sim}(1^\lambda, 1))$ is ϵ-close to the uniform distribution over $\{0,1\}^{L(\lambda)}$. That is, we proved that $\left|\Pr\left[\mathsf{Exp}^j = 1\right] - \Pr\left[\mathsf{Exp}^{j+1} = 1\right]\right| \le \epsilon$ is negligible for all j. All in all, we showed $\mathsf{Adv}^{\mathsf{PR}}_{\mathsf{AEC},\mathcal{A}}(\lambda) = \left|\Pr\left[\mathsf{Exp}^{\mathsf{PR},0}_{\mathsf{AEC},\mathcal{A}}(\lambda) = 1\right] - \Pr\left[\mathsf{Exp}^{\mathsf{PR},1}_{\mathsf{AEC},\mathcal{A}}(\lambda) = 1\right]\right|$ is negligible. Note that AEC preserves the unforgeability of scheme S. □

As shown above, general transformation of indistinguishable signatures into pseudorandom becomes straightforward — once appropriate admissible encodings are identified. If signatures are formed by tuples of elements of certain sets then by the aggregation lemma it will suffice to identify encodings for these particular sets. Lemma 13 shows existence of admissible encodings for a variety of algebraic sets that are often used in practical cryptography, including sets behind the indistinguishable versions of FDH-RSA, Schnorr, and Boneh-Boyen schemes from Section 4.1.

Lemma 13 (Sets with Admissible Encodings). *For the following sets R there exist polynomials $\ell(\lambda)$ and admissible encodings $F : \{0,1\}^{\ell(\lambda)} \to R$:*

[9] Such pad/ignore steps can also be seen as the aggregation of F with the canonical admissible encoding for $\{0,1\}^{L(\lambda)-\ell} \to \{0,1\}^0$ (where $\{0,1\}^0$ denotes the language that contains only the empty word).

(1) Ranges $R = \{0, \ldots, N-1\} = \mathbb{Z}_N$ of natural numbers, for arbitrary $N \in \mathbb{N}$.

(2) The set of quadratic residues modulo safe primes p, i.e. $R = QR(p) \subseteq \mathbb{Z}_p^\times$.

(3) Arbitrary subgroups $G_q \subseteq \mathbb{Z}_p^\times$ of prime order q.

(4) The set $R = E(\mathbb{F})$ of rational points on (certain) elliptic curves, defined over a finite field.

PROOF. The set $\{0,1\}^{\ell(\lambda)}$ can be canonically identified with $T^{\ell(\lambda)} := \mathbb{Z}_{2^{\ell(\lambda)}}$. It will hence, for all considered sets R, suffice to indicate admissible encodings $T^{\ell(\lambda)} \to R$.

(1) Let $N \in \mathbb{N}$ be a natural number, and $\lambda = |N|$ its length. Consider polynomial $\ell(\lambda) = 2\lambda$, function $F\colon T^{\ell(\lambda)} \to \mathbb{Z}_N; a \mapsto a \bmod N$, and the probabilistic mapping $\mathcal{I}_F\colon \mathbb{Z}_N \to T^{\ell(\lambda)}$; $b \mapsto b + kN$ for random $k \in_R [0, \lfloor 2^{\ell(\lambda)}/N \rfloor - 1]$. It is easy to see that $\mathcal{I}_F$ inverts F. According to [27, Section 8.8], the statistical distance ϵ between $\mathcal{I}_F(r)$ for $r \in_R \mathbb{Z}_N$ and the uniform distribution in $T^{\ell(\lambda)}$ is bounded by $\epsilon < N/2^{\ell(\lambda)} \approx 2^\lambda/2^{2\lambda} = 2^{-\lambda}$, and hence negligible.

(2) and (3) Let $p = \alpha q + 1$ with primes p, q such that $\gcd(\alpha, q) = 1$ and $|p| = \lambda$ (if $\alpha = 2$ we have the safe prime setting). Let g be a generator of $G = \mathbb{Z}_p^\times$. Consider the probabilistic $\mathcal{I}_F\colon G_q \to \mathbb{Z}_p^\times$; $a \mapsto (g^q)^r a^{\alpha^{-1} \bmod q} \bmod p$ for $r \in_R \mathbb{Z}_\alpha$ together with its inversion $F\colon \mathbb{Z}_p^\times \to G_q$; $b \mapsto b^\alpha \bmod p$. As g^q and also $(g^q)^r$ have order α, it is easy to see that F perfectly 'inverts' $\mathcal{I}_F$. The encoding is 0-admissible as for every $h \in G$ we have $h = g^t = (g^q)^x (g^\alpha)^y$ for some t, x, y (by CRT or Euclid), i.e. every element in G is the (unique) product of the power of an order-α and the power of an order-q element. In above construction, the role of the former is taken by $(g^q)^r$, while element $a^{\alpha^{-1}} \in G_q$ corresponds to the latter. This encoding can be composed with (1) for $N = p$ to obtain the desired $F\colon \{0,1\}^{\ell(\lambda)} \to R$.

(4) We refer to Brier et al. [8] for an overview of (pairing-friendly) curves with suitable admissible encodings. □

5.1.2 *Direct* PR *Compiler*

The PR compiler introduced below outputs pseudorandom signatures (with full key exposure) for any signature scheme S; underlying techniques were proposed in [17] for building anonymous signatures. Its main building block is a pair of *associated randomness extractor $\mathcal{E}$ and hash function H*. Here, we only recall the properties of such a pair $(\mathcal{E}, H)$ and refer to [17] for a rigorous treatment. Basically, a randomized hash function H takes a message m and some randomness $r \in_R \{0,1\}^{t(\lambda)}$ and outputs $h = H(m; r)$. H is called *collision-intractable* if it is difficult to find $m \neq m'$ and $r \in \{0,1\}^{t(\lambda)}$ with $H(m; r) = H(m', r)$, for the same randomness r. The task of the randomness extractor $\mathcal{E}$ is to distill uniformly distributed strings of fixed length from inputs $x \leftarrow \mathcal{X}$ whose distribution is unknown but where a certain minimum level of entropy is assumed. As auxiliary input, $\mathcal{E}$ gets a uniformly distributed randomness $u \in \{0,1\}^{d(\lambda)}$. The extracted value is denoted by $\mathcal{E}(m; u)$. A pair $(\mathcal{E}, H)$ is called *pseudorandom* if tuples (r, y, u, e) and (r, y, u, v) are computationally indistinguishable, where $r \in_R \{0,1\}^{t(\lambda)}$, $y \leftarrow H(x; r)$, $u \in_R \{0,1\}^{d(\lambda)}$, $e \leftarrow \mathcal{E}(x; u)$, and $v \in_R \{0,1\}^{|e|}$, for $x \leftarrow \mathcal{X}$. Fischlin [17] offers an efficient instantiation for such primitive in the standard model[10]. Our analysis shows that this primitive gives rise to the following compiler for pseudorandom signatures, which are also indistinguishable and confidential (by Lemmas 10 and 5). We notice that this compiler results in somewhat less efficient schemes as opposed to our IND-to-PR compiler (e.g. when used with our IND examples).

$\mathsf{DPRC.KGen}(1^\lambda)$:	Output $(sk, pk) \leftarrow \mathsf{S.KGen}(1^\lambda)$.
$\mathsf{DPRC.Sign}(sk, m)$:	Choose $r \in_R \{0,1\}^{t(\lambda)}$ and $u \in_R \{0,1\}^{d(\lambda)}$. Let $h \leftarrow H(m; r)$ and $\sigma \leftarrow \mathsf{S.Sign}(sk, h)$. Compute $\tau := \sigma \oplus \mathcal{E}(m; u)$ and output $\sigma' := \tau \,\|\, r \,\|\, u$.
$\mathsf{DPRC.Ver}(pk, m, \sigma')$:	Parse σ' as $\sigma' = \tau \,\|\, r \,\|\, u$. Let $\sigma := \tau \oplus \mathcal{E}(m; u)$ and $h \leftarrow H(m; r)$. Output $\mathsf{S.Ver}(pk, h, \sigma)$.

THEOREM 2. *If $(\mathcal{E}, H)$ is a pseudorandom pair of an associated randomness extractor $\mathcal{E}$ and a hash function H then for any signature scheme* S *the* DPRC *signature scheme, obtained using our direct* PR *compiler, is pseudorandom with respect to full key exposure (in the standard model).*

PROOF SKETCH. Anonymity of DPRC is shown in [17] by presenting a game-hopping proof that, in the first hop, modifies ANON game such that challenge signature σ^* is computed as specified in DPRC, except that τ is replaced by $\tau := \sigma \oplus v$ for random $v \in_R \{0,1\}^{|\mathcal{E}(m;u)|}$. As v acts as a one-time pad on σ, component τ is uniformly distributed in $\{0,1\}^{|\sigma|}$. Obviously, the concatenation $\sigma' = \tau \,\|\, r \,\|\, u$ is uniformly distributed as well, in $\{0,1\}^{|\sigma| + t(\lambda) + d(\lambda)}$. This shows pseudorandomness of DPRC. Note that the transformed scheme inherits its unforgeability from S under standard assumptions as proven [17]. □

6. CONCLUSION

In this paper we gave a detailed account on the privacy hierarchy for ordinary signature schemes, taking into account earlier definitions of anonymity and confidentiality in the setting of high-entropic message spaces and system-wide known public keys. Our major result are pseudorandom signatures that cannot be distinguished from random strings and thus hide the entire information about message, signer, and signing algorithm. To obtain such fully private signatures we gave two compilers: the more efficient one adds pseudorandomness to indistinguishable signature schemes and we have shown that such schemes exist in different cryptographic settings; our second compiler, based on Fischlin's work [17], adds pseudorandomness to any unforgeable signature scheme but is less efficient (though in the standard model). In summary, with our hierarchy of privacy notions and generic transformations we showed how to efficiently achieve an ultimate form of privacy for arbitrary signature schemes, both in the random oracle and the standard model.

Acknowledgments

The majority of work on this paper was performed while all its authors were in the Cryptographic Protocols Group at TU Darmstadt. This research was supported by the German Science Foundation (DFG) through the project PRIMAKE (MA 4957).

[10] In the random oracle model $H(x; r) := H^\#(0 \,\|\, x \,\|\, r)$ and $\mathcal{E}(x; u) := H^\#(1 \,\|\, x \,\|\, r)$ for a hash function $H^\#$ is an efficient instantiation of a pseudorandom associated pair $(\mathcal{E}, H)$.

7. REFERENCES

[1] M. Bellare, A. Boldyreva, A. Desai, and D. Pointcheval. Key-Privacy in Public-Key Encryption. In *ASIACRYPT 2001*, *LNCS* 2248, pp. 566–582. 2001.

[2] M. Bellare and S. Duan. Partial Signatures and their Applications. Cryptology ePrint Archive, Report 2009/336, 2009. `http://eprint.iacr.org/2009/336`.

[3] M. Bellare, M. Fischlin, A. O'Neill, and T. Ristenpart. Deterministic Encryption: Definitional Equivalences and Constructions without Random Oracles. In *CRYPTO 2008*, *LNCS* 5157, pp. 360–378. 2008.

[4] M. Bellare and P. Rogaway. Random Oracles are Practical: A Paradigm for Designing Efficient Protocols. In *CCS 1993*, pp. 62–73. 1993.

[5] D. Boneh and X. Boyen. Short Signatures Without Random Oracles and the SDH Assumption in Bilinear Groups. *J. of Cryptology*, 21(2):149–177, 2008.

[6] D. Boneh and M. Franklin. Identity-Based Encryption from the Weil Pairing. In *CRYPTO 2001*, *LNCS* 2139, pp. 213–229. 2001.

[7] D. Boneh, B. Lynn, and H. Shacham. Short Signatures From the Weil Pairing. *J. of Cryptology*, 17(4):297–319, 2004.

[8] E. Brier, J.-S. Coron, T. Icart, D. Madore, H. Randriam, and M. Tibouchi. Efficient Indifferentiable Hashing into Ordinary Elliptic Curves. In *CRYPTO 2010*, *LNCS* 6223, pp. 237–254. 2010.

[9] J. Bringer, H. Chabanne, and T. Icart. Password Based Key Exchange Protocols on Elliptic Curves Which Conceal the Public Parameters. In *ACNS 2010*, *LNCS* 6123, pp. 291–308. 2010.

[10] C. Cachin. An Information-Theoretic Model for Steganography. In *Information Hiding 1998*, *LNCS* 1525, pp. 306–318. 1998.

[11] J. Camenisch and A. Lysyanskaya. A Signature Scheme with Efficient Protocols. In *Security in Communication Networks 2002*, *LNCS* 2576, pp. 268–289. 2003.

[12] R. Canetti. Towards Realizing Random Oracles: Hash Functions That Hide All Partial Information. In *CRYPTO 1997*, *LNCS* 1294, pp. 455–469. 1997.

[13] R. Canetti and H. Krawczyk. Security Analysis of IKE's Signature-Based Key-Exchange Protocol. In *CRYPTO 2002*, *LNCS* 2442, pp. 143–161. 2002.

[14] D. Chaum and E. van Heyst. Group Signatures. In *EUROCRYPT 1991*, *LNCS* 547, pp. 257–265. 1991.

[15] R. Cramer and V. Shoup. Signature Schemes Based on the Strong RSA Assumption. *ACM TISSEC*, 3(3):161–185, 2000.

[16] A. W. Dent, M. Fischlin, M. Manulis, M. Stam, and D. Schröder. Confidential Signatures and Deterministic Signcryption. In *PKC 2010*, *LNCS* 6056, pp. 462–479. 2010.

[17] M. Fischlin. Anonymous Signatures Made Easy. In *PKC 2007*, *LNCS* 4450, pp. 31–42. 2007.

[18] N. Fleischhacker, F. Günther, F. Kiefer, M. Manulis, and B. Poettering. Pseudorandom Signatures. Cryptology ePrint Archive, Report 2011/673, 2013. `http://eprint.iacr.org/2011/673`.

[19] S. D. Galbraith and W. Mao. Invisibility and Anonymity of Undeniable and Confirmer Signatures. In *CT-RSA 2003*, *LNCS* 2612, pp. 80–97. 2003.

[20] G. Greenleaf and R. Clarke. Privacy Implications of Digital Signatures. IBC Conference on Digital Signatures, 1997. Available at `http://www.anu.edu.au/people/Roger.Clarke/DV/DigSig.html`.

[21] N. J. Hopper, J. Langford, and L. von Ahn. Provably Secure Steganography. In *CRYPTO 2002*, *LNCS* 2442, pp. 77–92. 2002.

[22] K. Nyberg and R. Rueppel. Message Recovery for Signature Schemes Based on the Discrete Logarithm Problem. In *EUROCRYPT 1994*, *LNCS* 950, pp. 182–193. 1995.

[23] J.-F. Raymond. Traffic Analysis: Protocols, Attacks, Design Issues, and Open Problems. In *Workshop on Design Issues in Anonymity and Unobservability 2000*, *LNCS* 2009, pp. 10–29. 2000.

[24] R. L. Rivest, A. Shamir, and Y. Tauman. How to Leak a Secret. In *ASIACRYPT 2001*, *LNCS* 2248, pp. 552–565. 2001.

[25] V. Saraswat and A. Yun. Anonymous Signatures Revisited. In *Provable Security 2009*, *LNCS* 5848, pp. 140–153. 2009.

[26] C.-P. Schnorr. Efficient Identification and Signatures for Smart Cards. In *CRYPTO 1989*, *LNCS* 435, pp. 239–252. 1990.

[27] V. Shoup. *A Computational Introduction to Number Theory and Algebra*. Cambridge Uni Press, 2008.

[28] G. J. Simmons. The Prisoners' Problem and the Subliminal Channel. In *CRYPTO 1983*, pp. 51–67. 1983.

[29] G. Yang, D. S. Wong, X. Deng, and H. Wang. Anonymous Signature Schemes. In *PKC 2006*, *LNCS* 3958, pp. 347–363. 2006.

[30] R. Zhang and H. Imai. Strong Anonymous Signatures. In *Inscrypt 2008*, *LNCS* 5487, pp. 60–71. 2009.

APPENDIX

A. DEFINITION OF UNFORGEABILITY

DEFINITION 9 (EXISTENTIAL UNFORGEABILITY). *A signature scheme* S *is* existentially unforgeable under adaptive chosen-message attacks ($\mathsf{EUF\text{-}CMA}$) *if all PPT adversaries* $\mathcal{A}$ *have negligible probability to output* $(m^*, \sigma^*) \leftarrow \mathcal{A}^{\mathsf{Sign}(sk,\cdot)}(pk)$ *(where* $(sk, pk) \leftarrow \mathsf{KGen}(1^\lambda)$ *and* $\mathsf{Sign}(sk,\cdot)$ *is a signature oracle) such that* $\mathsf{Ver}(pk, m^*, \sigma^*) = 1$*, provided that* m^* *was not queried to* $\mathsf{Sign}(sk,\cdot)$ *oracle.*

B. PROOFS OF LEMMAS 1 AND 2

Proof of Lemma 1 (CONF[-FKE] $\not\Rightarrow$ ANON)**.** Let S be a confidential signature scheme. We construct a confidential but not anonymous signature scheme S' from S as follows.

$\mathsf{KGen}'(1^\lambda)$: Output $(sk, pk) \leftarrow \mathsf{KGen}(1^\lambda)$.

$\mathsf{Sign}'(sk, m)$: $\sigma \leftarrow \mathsf{Sign}(sk, m)$. Output $\sigma' := (\sigma, pk)$.

$\mathsf{Ver}'(pk, m, \sigma')$: Parse σ' as (σ, pk'). Return 1 iff $\big(\mathsf{Ver}(pk, m, \sigma) = 1 \wedge pk = pk'\big)$.

As pk is extractable from σ', S' is not anonymous. Yet, S' is confidential (as the CONF adversary knows pk anyway). □

Proof of Lemma 2 (ANON[-FKE] $\not\Rightarrow$ CONF)**.** Let S be an anonymous signature scheme and $\mathsf{L}(x)$ denote the the last bit of x. We construct an anonymous but not confidential signature scheme S' from S as follows.

$\mathsf{KGen}'(1^\lambda)$: Output $(sk, pk) \leftarrow \mathsf{KGen}(1^\lambda)$.
$\mathsf{Sign}'(sk, m)$: $\sigma \leftarrow \mathsf{Sign}(sk, m)$. Output $\sigma' := (\sigma, \mathtt{L}(m))$.
$\mathsf{Ver}'(pk, m, \sigma')$: Parse σ' as (σ, b). Return 1 iff $(\mathsf{Ver}(pk, m, \sigma) = 1 \wedge \mathtt{L}(m) = b)$.

To see that S' is not confidential, consider the following confidentiality adversary $\mathcal{A} = (\mathcal{A}_1, \mathcal{A}_2)$. $\mathcal{A}_1$ outputs $(\vec{m}_0, t_0)$ (resp. $(\vec{m}_1, t_1)$), where $\vec{m}_i = (M_i)$ for $M_i \in_R \{0,1\}^\lambda$, and $t_i = \mathtt{L}(M_i)$. $\mathcal{A}_2$ parses $\vec{\sigma}^*$ as $\vec{\sigma}^* = ((\sigma, b))$ and outputs b. So $\mathsf{Adv}^{\mathsf{CONF}}_{\mathsf{S}',\mathcal{A}}(\lambda) = |\Pr[\mathsf{Exp}^{\mathsf{CONF},0}_{\mathsf{S}',\mathcal{A}}(\lambda) = 1] - \Pr[\mathsf{Exp}^{\mathsf{CONF},1}_{\mathsf{S}',\mathcal{A}}(\lambda) = 1]| = |1 - \frac{1}{2}| = \frac{1}{2}$. To show anonymity of S', consider any adversary $\mathcal{A} = (\mathcal{A}_1, \mathcal{A}_2)$ against ANON of S'. Flip a coin $\beta \in_R \{0,1\}$ and define ANON adversary $\mathcal{B} = (\mathcal{B}_1, \mathcal{B}_2)$ against S as follows: On input pk_0, pk_1, $\mathcal{B}_1$ runs $\mathcal{A}_1(pk_0, pk_1)$ as a black-box and receives a message m. If $\mathtt{L}(m) = \beta$, m is output, otherwise the simulation aborts which is modeled by letting $\mathcal{B}$'s simulator output 0, i.e. $\Pr[\mathsf{Exp}^{\mathsf{ANON},b}_{\mathsf{S},\mathcal{B}}(\lambda) = 1|\mathtt{A}] = 0$ for $b \in \{0,1\}$. The "abort" event $\mathtt{A}$ will occur with probability $\frac{1}{2}$. Now, on input pk_0, pk_1, σ, algorithm $\mathcal{B}_2$ appends β to σ, runs $d \leftarrow \mathcal{A}_2(pk_0, pk_1, (\sigma, \beta))$ and outputs d. Now

$$\Pr[\mathsf{Exp}^{\mathsf{ANON},b}_{\mathsf{S},\mathcal{B}}(\lambda) = 1] = \Pr[\mathsf{Exp}^{\mathsf{ANON},b}_{\mathsf{S},\mathcal{B}}(\lambda) = 1|\neg\mathtt{A}]\Pr[\neg\mathtt{A}] + \Pr[\mathsf{Exp}^{\mathsf{ANON},b}_{\mathsf{S},\mathcal{B}}(\lambda) = 1|\mathtt{A}]\Pr[\mathtt{A}] = \Pr[\mathsf{Exp}^{\mathsf{ANON},b}_{\mathsf{S}',\mathcal{A}}(\lambda) = 1]\cdot\tfrac{1}{2} + 0\cdot\tfrac{1}{2}$$

and hence $\mathsf{Adv}^{\mathsf{ANON}}_{\mathsf{S},\mathcal{B}}(\lambda) = \frac{1}{2}\cdot\mathsf{Adv}^{\mathsf{ANON}}_{\mathsf{S}',\mathcal{A}}(\lambda)$. Thus S' is anonymous, as the left hand side is negligible by assumption. □

C. INFORMATION RECOVERY

Cramer-Shoup (CS) [15]. The Strong RSA-based CS scheme outputs signatures of the form $\sigma = (e, s, \sigma'_1, \sigma'_2)$ and its verification algorithm checks whether e is an odd integer of certain length, followed by two checks of the form $\sigma_1'^{e} \equiv th^{H(s)} \bmod N$ and $\sigma_2'^{\tilde{e}} \equiv sh^m \bmod N$ with $\tilde{e}$, t, h, and N being part of the public key pk. These equation can be rewritten to $\mathsf{Inf}(pk, m) = \mathsf{Rec}(pk, \sigma)$ using $\mathsf{Inf}(pk, m)$ that outputs a pair $(t, h^m \bmod N)$ and $\mathsf{Rec}(pk, \sigma)$ returning a pair $(\sigma_1'^{e}\cdot h^{-H(s)}, \sigma_2'^{\tilde{e}}\cdot s^{-1} \bmod N)$ after verifying the appropriate form for e. The equality of the outputs of Inf and Rec can then be tested component-wise to verify the signature.

Camenisch-Lysyanskaya (CL) [11]. The Strong RSA-based CL scheme outputs signatures of the form $\sigma = (e, s, \sigma')$ and its verification algorithm checks if e is in the appropriate range and $\sigma'^{e} \equiv a^m b^s c \bmod N$ with a, b, c, and N being part of the public key pk. By rewriting the verification equation to $a^m \equiv \sigma'^{e}/(b^s c) \bmod N$ we can define algorithm $\mathsf{Inf}(pk, m)$ to output $a^m \bmod N$ and $\mathsf{Rec}(pk, \sigma)$ to output $\sigma'^{e}\cdot(b^s c)^{-1} \bmod N$ if e is in the appropriate range.

Boneh-Lynn-Shacham (BLS) [7]. The pairing-based BLS scheme, which can be initialized for example in cyclic groups $G = \langle g\rangle$ of prime order q with a suitable bilinear map $e : G\times G \mapsto G_T$, outputs signatures of the form $\sigma = H(m)^x$ where $H : \mathbb{Z}_q^* \mapsto G$ is a random oracle and x is a secret key. Its verification equation $e(H(m), y) = e(\sigma, g)$ with g, y belonging to pk, immediately defines $\mathsf{Inf}(pk, m)$ and $\mathsf{Rec}(pk, \sigma)$.

D. PROOFS OF LEMMAS 4 AND 5

Proof of Lemma 4 (IND/IND-FKE⇒ANON/ANON-FKE). Let S be an indistinguishable signature scheme, with simulator Sim. For an ANON-adversary $\mathcal{A}$ consider experiment $\mathsf{Exp}^{*}_{\mathsf{S},\mathsf{Sim},\mathcal{A}}(\lambda)$, which is like $\mathsf{Exp}^{\mathsf{ANON},0}_{\mathsf{S},\mathcal{A}}(\lambda)$, except that challenge signature σ is computed as $\sigma \leftarrow \mathsf{Sim}(1^\lambda, 1)$. Construct IND-adversary $\mathcal{B}$ by generating random $(sk', pk') \leftarrow \mathsf{KGen}(1^\lambda)$ and defining $\mathcal{B}_1$ and $\mathcal{B}_2$ as follows: $\mathcal{B}_1$, on input pk and having oracle access to $\mathsf{Sign}(sk,\cdot)$, runs $m \leftarrow \mathcal{A}_1^{\mathsf{Sign}(sk,\cdot),\mathsf{Sign}(sk',\cdot)}(pk, pk')$ as a black-box, relaying oracle queries to $\mathsf{Sign}(sk,\cdot)$, and answering $\mathsf{Sign}(sk',\cdot)$ queries itself. After receiving m, $\mathcal{B}_1$ outputs $(\vec{m}, t) = ((m), 1)$ and stops. $\mathcal{B}_2$, on input pk and challenge signature $\vec{\sigma} = (\sigma)$, feeds σ into $\mathcal{A}_2$ (together with pk and pk'). Value d output by $\mathcal{A}_2$ is used as return value t' of $\mathcal{B}_2$. Careful inspection results in $\left|\Pr[\mathsf{Exp}^{\mathsf{ANON},0}_{\mathsf{S},\mathcal{A}}(\lambda) = 1] - \Pr[\mathsf{Exp}^{*}_{\mathsf{S},\mathsf{Sim},\mathcal{A}}(\lambda) = 1]\right| = \left|\Pr[\mathsf{Exp}^{\mathsf{IND},0}_{\mathsf{S},\mathsf{Sim},\mathcal{B}}(\lambda) = 1] - \Pr[\mathsf{Exp}^{\mathsf{IND},1}_{\mathsf{S},\mathsf{Sim},\mathcal{B}}(\lambda) = 1]\right| = \mathsf{Adv}^{\mathsf{IND}}_{\mathsf{S},\mathsf{Sim},\mathcal{B}}(\lambda)$. Equality $|\Pr[\mathsf{Exp}^{\mathsf{ANON},1}_{\mathsf{S},\mathcal{A}}(\lambda) = 1] - \Pr[\mathsf{Exp}^{*}_{\mathsf{S},\mathsf{Sim},\mathcal{A}}(\lambda) = 1]| = \mathsf{Adv}^{\mathsf{IND}}_{\mathsf{S},\mathsf{Sim},\mathcal{B}}(\lambda)$ can be shown similarly (by swapping pk and pk' in the construction of $\mathcal{B}$). All in all, $\mathsf{Adv}^{\mathsf{ANON}}_{\mathsf{S},\mathcal{A}}(\lambda) \leq 2\cdot\mathsf{Adv}^{\mathsf{IND}}_{\mathsf{S},\mathsf{Sim},\mathcal{B}}(\lambda)$, which is negligible by assumption. The implication IND-FKE ⇒ ANON-FKE is shown analogously. □

Proof of Lemma 5 (IND/IND-FKE⇒CONF/CONF-FKE). Let S be an indistinguishable signature scheme, with simulator Sim. For a CONF-adversary $\mathcal{A}$ consider experiment $\mathsf{Exp}^{*}_{\mathsf{S},\mathsf{Sim},\mathcal{A}}(\lambda)$, which is like $\mathsf{Exp}^{\mathsf{CONF},0}_{\mathsf{S},\mathcal{A}}(\lambda)$, except challenge signatures $\vec{\sigma}^*$ are computed as $\vec{\sigma}^* \leftarrow \mathsf{Sim}(1^\lambda, |\vec{m}_0|)$. Construct IND-adversary $\mathcal{B}$ defining $\mathcal{B}_1, \mathcal{B}_2$ as follows: $\mathcal{B}_1$, on input pk and with oracle $\mathsf{Sign}(sk,\cdot)$, runs $\mathcal{A}_1^{\mathsf{Sign}(sk,\cdot),\mathsf{Sign}(sk',\cdot)}(pk)$ twice as a black-box, relaying oracle queries to $\mathsf{Sign}(sk,\cdot)$ and answering $\mathsf{Sign}(sk',\cdot)$ queries itself, to obtain $(\vec{m}_0, t_0)$ and $(\vec{m}_1, t_1)$, respectively. Algorithm $\mathcal{B}_1$ aborts if $|\vec{m}_0| \neq |\vec{m}_1|$. Else it outputs $(\vec{m}_0, t_0)$ and stops. $\mathcal{B}_2$, on input pk and challenge signatures $\vec{\sigma}$, feeds $\vec{\sigma}$ into $\mathcal{A}_2$ (together with pk). Value t' output by $\mathcal{A}_2$ is used as return value t' of $\mathcal{B}_2$. Now we have $\left|\Pr[\mathsf{Exp}^{\mathsf{CONF},0}_{\mathsf{S},\mathcal{A}}(\lambda) = 1] - \Pr[\mathsf{Exp}^{*}_{\mathsf{S},\mathsf{Sim},\mathcal{A}}(\lambda) = 1]\right| = \left|\Pr[\mathsf{Exp}^{\mathsf{IND},0}_{\mathsf{S},\mathsf{Sim},\mathcal{B}}(\lambda) = 1] - \Pr[\mathsf{Exp}^{\mathsf{IND},1}_{\mathsf{S},\mathsf{Sim},\mathcal{B}}(\lambda) = 1]\right| = \mathsf{Adv}^{\mathsf{IND}}_{\mathsf{S},\mathsf{Sim},\mathcal{B}}(\lambda)$. Equality $|\Pr[\mathsf{Exp}^{\mathsf{CONF},1}_{\mathsf{S},\mathcal{A}}(\lambda) = 1] - \Pr[\mathsf{Exp}^{*}_{\mathsf{S},\mathsf{Sim},\mathcal{A}}(\lambda) = 1]| = \mathsf{Adv}^{\mathsf{IND}}_{\mathsf{S},\mathsf{Sim},\mathcal{B}}(\lambda)$ can be shown similarly (by letting $\mathcal{B}_1$ output $(\vec{m}_1, t_0)$ instead of $(\vec{m}_0, t_0)$). All in all, $\mathsf{Adv}^{\mathsf{CONF}}_{\mathsf{S},\mathcal{A}}(\lambda) \leq 2\cdot\mathsf{Adv}^{\mathsf{IND}}_{\mathsf{S},\mathsf{Sim},\mathcal{B}}(\lambda)$, which is negligible by assumption. The implication IND-FKE ⇒ CONF-FKE is shown analogously. □

E. PROOFS OF LEMMAS 10 AND 11

The proof of Lemma 10 (PR ⇒ IND, PR-FKE ⇒ IND-FKE) follows directly from Definitions 5 and 6.

Proof of Lemma 11 (IND[-FKE] ⇏ PR). Let S be an indistinguishable signature scheme, with simulator Sim. We construct an indistinguishable but not pseudorandom signature scheme S' from S as follows.

$\mathsf{KGen}'(1^\lambda)$: Output $(sk, pk) \leftarrow \mathsf{KGen}(1^\lambda)$.
$\mathsf{Sign}'(sk, m)$: $\sigma \leftarrow \mathsf{Sign}(sk, m)$. Output $\sigma' := \sigma \,\|\, 1$.
$\mathsf{Ver}'(pk, m, \sigma')$: Parse σ' as $\sigma \,\|\, b$. Return 1 iff $(\mathsf{Ver}(pk, m, \sigma) = 1 \wedge b = 1)$.

It is easy to prove that Scheme S' is not pseudorandom as the last bit of a signature is always 1. However, S' is still indistinguishable: An appropriate simulator $\mathsf{Sim}'(1^\lambda)$ is given by $\mathsf{Sim}(1^\lambda) \,\|\, 1$. It is straightforward to show that any successful indistinguishability adversary for S' can be turned into a successful adversary against S. □

Looking at the Bag is not Enough to Find the Bomb: An Evasion of Structural Methods for Malicious PDF Files Detection

Davide Maiorca
Department of Electrical and Electronic Engineering
University of Cagliari
Piazza d'Armi, 09123
Cagliari, Italy
davide.maiorca@diee.unica.it

Igino Corona
Department of Electrical and Electronic Engineering
University of Cagliari
Piazza d'Armi, 09123
Cagliari, Italy
igino.corona@diee.unica.it

Giorgio Giacinto
Department of Electrical and Electronic Engineering
University of Cagliari
Piazza d'Armi, 09123
Cagliari, Italy
giacinto@diee.unica.it

ABSTRACT

PDF files have proved to be excellent malicious-code bearing vectors. Thanks to their flexible logical structure, an attack can be hidden in several ways, and easily deceive protection mechanisms based on file-type filtering. Recent work showed that malicious PDF files can be accurately detected by analyzing their *logical structure*, with excellent results. In this paper, we present and practically demonstrate a novel evasion technique, called *reverse mimicry*, that can easily defeat such kind of analysis. We implement it using real samples and validate our approach by testing it against various PDF malware detectors proposed so far. Finally, we highlight the importance of developing systems robust to adversarial attacks and propose a framework to strengthen PDF malware detection against evasion.

Categories and Subject Descriptors

D.4.6 [**Software**]: Operating Systems—*Security and Protection*; I.2.6 [**Computing Methodologies**]: Artificial Intelligence—*Learning*

Keywords

Machine Learning; PDF Malware Detection; Detection Evasion; Reverse Mimicry

1. INTRODUCTION

PDF files bearing malicious content have been harming computer systems for more than four years, and in 2010 they have been considered one of the most dangerous threats [15]. From an attacker point of view, the PDF format is an extremely flexible code-bearing vector. It is possible to embed different types of attacks, such as sophisticated `JavaScript` and `ActionScript` code [28, 27], as well as malicious files of different type, such as `EXE` files, and malicious `PDF` files themselves [37]. Moreover, the structure of the PDF format allows an attacker to hide these attacks in a number of different ways, e.g., by employing *encryption*, or *obfuscation* techniques [36, 12]. Finally, malicious PDF files may also exploit specific vulnerabilities of the reader application, such as `Adobe Reader` [2] or `Foxit` [13]. As a consequence, PDF documents constitute a natural, simple way of devising polymorphic attacks. To cope with attack polymorphism, *machine learning* approaches have attracted researchers for their ability to learn from examples. To date, machine learning techniques have been also proposed for the analysis of PDF files. State-of-the-art machine learning approaches for malicious PDF detection basically rely on static detection techniques. The analysis is typically performed by measuring features related either to the JavaScript code embedded in the PDF file, or to the *structure* of PDF files, such as the number of specific keywords, the number of lowercase/uppercase characters, the number of objects and streams, etc. [26, 33, 20, 24, 40]. Such techniques claim to detect different kinds of maliciousness, because they are agnostic with respect to the way an attack is embedded.

As a matter of fact, malicious PDFs typically do not contain any readable document, and their structure is quite different from the one observed in benign PDF files, created for exchanging information in a standard format. This allows attackers to devise lightweight PDFs that can easily spread through the Internet (e.g., by means of `spam` techniques), infecting personal computers as well as mobile devices. For this reason, structural detection techniques showed excellent performance in detecting malicious PDF files. However, we observe that, as soon as such techniques are adopted in real-world deployment, the behavior of attackers may rapidly change to evade them. Now, a natural question arises. Do malicious PDFs always show structural differences with respect to benign files? Is it possible to evade detection by creating malicious PDFs that mimic the structure of benign files (i.e. *mimicry* attacks)? A theoretical study of such mimicry attacks has been presented in recent works [33, 40]. However, it is still not clear how to devise such evasion attacks in practice, through real PDF samples. In addition, it is not clear how difficult this task would be, and how much such attacks can be effective against state-of-the-art PDF malware detectors.

In this work, we investigate the above aspects. To this

ASIA CCS'13, May 8–10, 2013, Hangzhou, China.

end, we develop a new attack methodology designed to evade structural PDF malware detection systems, that we call *reverse mimicry*. Our methodology is simple, yet effective. We inject malicious content on a legitimate PDF so as to introduce minimum differences within its structure. Our attack can be easily automated and grants the possibility of a fast and effective deployment. Moreover, it does not require knowledge about the learning algorithm attacked, nor about the structural features employed to characterize a PDF document.

In particular, we present three different approaches to implement this attack, that allow creating *real* samples of malicious PDF files. Reported experiments show that these files evade detection by state-of-the-art malicious PDF detectors based on learning-by-example techniques. Quite surprisingly, we found that even those which employ non-structural features may be evaded, but for different reasons. Finally, we suggest some approaches to make PDF malware detection more robust against evasion attacks.

1.1 Contributions

The contributions of this paper can be summarized as follows:

- we present a novel attack methodology called *reverse mimicry*, designed to evade malicious PDF detectors based on structural analysis;
- we present and experimentally evaluate different techniques to create malicious PDF files according to the proposed methodology;
- we propose some approaches to improve robustness against evasion of malicious PDF detection.

1.2 Paper Organization

There are seven more sections beyond this one. In Section 2, we provide the basics of the PDF file format. In Section 3, we discuss the related works. In Section 4, we provide a description of how structural systems work, along with the differences with the other tools. In Section 5, we describe the *reverse mimicry* attack, as well as three different methodologies to implement this attack. Furthermore, we illustrate how certain types of detectors can be immediately evaded by this attack. In Section 6 we test our attack and its variants against recently developed systems. In Section 7 we provide an explanation of possible countermeasures against evasion attacks, such as those described in this paper. Section 8 closes the paper with the conclusions.

2. PDF BASICS

A PDF file is a hierarchy of objects logically connected to each other. For the sake of the following discussion, we will model the PDF file structure as composed by four basic parts [11]: Objects, File Structure, Document Structure, and Content Streams.

2.1 Objects

Objects are divided into *indirect objects*, i.e., objects referenced by a number (and that are used by the reader to build its *logical* structure), and *direct objects*, i.e., objects that are not referenced by a number. Basically, PDF objects can be of eight types:

boolean an object whose value can be `True` or `False`.

numeric an object represented by a real or integer number.

string a sequence of literal characters enclosed by parenthesis `( )` or hexadecimal data enclosed by angle brackets `< >`.

name a literal sequence of characters starting with `/`.

array a sequence of objects, between square brackets `[ ]`.

dictionary a sequence of pairs made up of a keyword (name object) and a value (it could be boolean, numeric, another keyword, or an array). They are enclosed between << and >>.

stream a special dictionary object between the keywords `stream` and `endstream`. It is used to store stream data such as images, text, script code, and it can be compressed using special filters.

null an empty object represented by the keyword `null`.

2.2 File Structure

The File Structure determines how objects are accessed and updated inside the PDF file. Each PDF file is composed by four parts:

header a line which gives information on the PDF version used by the file.

body it is the main portion of the file, and contains all the PDF objects.

cross-reference table it indicates the position of every *indirect* object in memory.

trailer it gives relevant information about the root object and number of revisions made to the document.

2.3 Document Structure

The Document Structure specifies how objects are used to represent several parts of the PDF document, such as pages, font, animations and so on. It describes the hierarchy of the objects in the *body* of the PDF file. The main object in the hierarchy is the `catalog` object, represented by a dictionary. Most of the indirect objects in a PDF file are dictionaries. Each page of the document is a `page` object, which contains also the references to the other objects that are part of that page. The position of the catalog dictionary is marked by the `/Root` name object located in the *trailer*.

2.4 Content Streams

Content Streams are *stream objects* containing a sequence of instructions which describe the appearance of the page and the graphical entity. Although they are defined as objects, they are conceptually different from the objects representing the document structure. The instructions can also refer to other *indirect* objects which contain information about the *resources* adopted by the stream.

The logical structure can be really complex, since there are a number of degrees of freedom in establishing references between objects. Moreover, with the exception of *linearized* files, the order of objects inside the file is fully arbitrary. Figure 1 shows an example of the PDF structure.

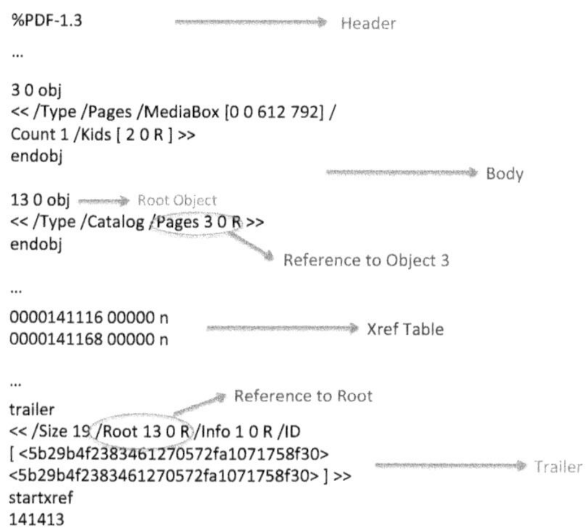

Figure 1: An example of the PDF structure

Usually, it is not possible to modify objects within the file, once they got their memory reference inside the cross-reference table. In order to do so, a new version of an object must be created and added after the trailer, together with a new trailer and a new cross-reference table. That is, original objects are preserved inside the file. This procedure is also called *version update*.

3. RELATED WORK

The first approaches based on machine learning for malicious PDF detection proposed static analysis on the raw (byte-level) document, by means of *n-gram* analysis [25, 32] and *decision trees* [38]. Anyway, these approaches were not really tailored to PDF files, but designed to analyze files in very different formats such as `doc`, `exe`, *etc.* Whereas raw analysis may detect many malware "implementations" besides malicious PDFs, it is not robust against modern obfuscation techniques, such as AES encryption [12], or different attack techniques, such as Return Oriented Programming, Heap Spraying or JIT Spraying [17, 29, 16].

To address the evolution in malware propagation through documents in PDF format, subsequent works focused on the analysis of embedded JavaScript code. A number of solutions for the detection of malicious JavaScript code have been proposed to date in the context of web security. For instance, **Jsand** [19], **Cujo** [31], **Zozzle** [21], **Prophiler** [18] are well-known tools for the dynamic and static analysis of JavaScript code. These tools are often employed by systems designed to identify threats embedded in different document formats.

`Wepawet` [10], a framework for the analysis of web-based threats, relies on **JSand** to analyze JavaScript code within PDF files. **Jsand** [19] adopts `HtmlUnit`[1], a Java-based browser simulator, and Mozilla's **Rhino**[2] to extract behavioral features related to the execution of JavaScript code. A statistical classifier is trained on a representative sample of benign code, and malicious code is spot by detecting anomalous patterns.

[1]`http://htmlunit.sourceforge.net`
[2]`http://www.mozilla.org/rhino/`

A similar approach is adopted by **MalOffice** [22]. **MalOffice** uses **pdftk**[3] to extract JavaScript code, and **CWSandbox** [41] to analyze the code behavior: code classification is carried out by a set of rules. **CWSandbox** has also been used to classify general malware behavior [30]. **MDScan** [39] follows a different approach as malicious behavior is detected through **Nemu**, a tool able to intercept memory-injected shellcode. A very similar idea, but with a different implementation, has been developed in **ShellOS** [34].

Dynamic detection by the execution of JavaScript code in a virtual environment may be time consuming and computationally expensive, and it is prone to evasion by a clever attacker that leverage on different implementations of the JavaScript engine used by the PDF reader, and by the code analyzer [39]. To reduce computational costs, **PJScan** [24] proposed a fully static lexical analysis of JavaScript code by training a statistical classifier on malicious files. Latest research work developed **Malware Slayer** [26] and **PDFRate** [33, 6], two malware detectors based on machine learning for PDF structural analysis. Such solutions do not specifically analyze malicious code within PDF files, but they rely on structural patterns to discriminate malicious from benign files. While **Malware Slayer** mainly focuses on *name objects*, **PDFRate** tries to retrieve as much information as possible on the PDF structure (number of objects, streams, lowercase/uppercase letters, *etc.*). Both tools have shown very high detection rate with low false positives rate. In 2013, an approach similar to **Malware Slayer** has been proposed by adopting the hierarchical sequence of *name objects* as features [40]. It is worth noting that a number of tools are also available for the analysis of PDF documents, such as **Pdf Tools** [5], **PeePDF** [7] and **Origami** [4]. Although they are not designed to detect malicious PDF files, they extract useful information on the file structure and content that can be used for classification. For example, the number of indirect objects, how objects are distributed and connected to each other, *etc.*

4. TECHNIQUES FOR DETECTING MALICIOUS PDF FILES

As mentioned in Section 3, in the past years there has been an increased interest in developing machine learning approaches for malicious PDF files detection. Basically, the two main approaches to generic malware detection, namely, *dynamic* and *static* analysis, are also applied to the detection of malicious PDF files.

`Wepawet` [10, 19] implements a dynamic analysis. It extracts JavaScript code from PDF files, executes it in a sandbox, and extracts specific features from the run-time execution. This tool is also able to analyze JavaScript code extracted from other sources, such as malicious web pages.

In order to improve accuracy and time response, recent tools focus on the *static analysis* of PDF files. These tools can be subdivided into two categories:

1. malicious JavaScript detectors. These tools look for specific PDF objects that contain JavaScript code and analyze their content. The features adopted by these tools are related to characteristics of JavaScript code. For example, the frequency of specific tokens, or the presence of specific functions, such as `unescape`.

[3]`http://www.pdflabs.com/tools/pdftk-the-pdf-toolkit`

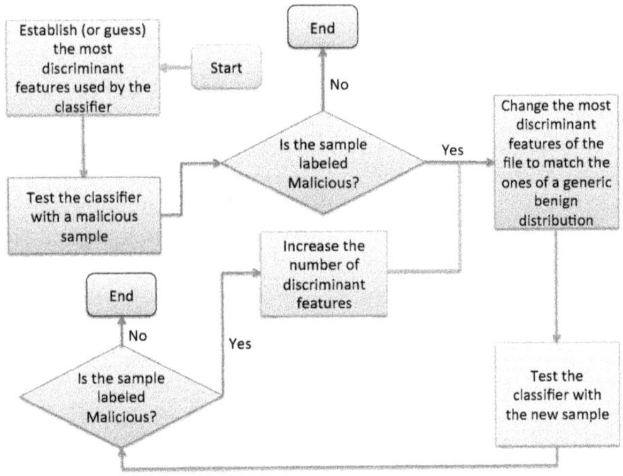

Figure 2: A conceptual structure of the mimicry attack

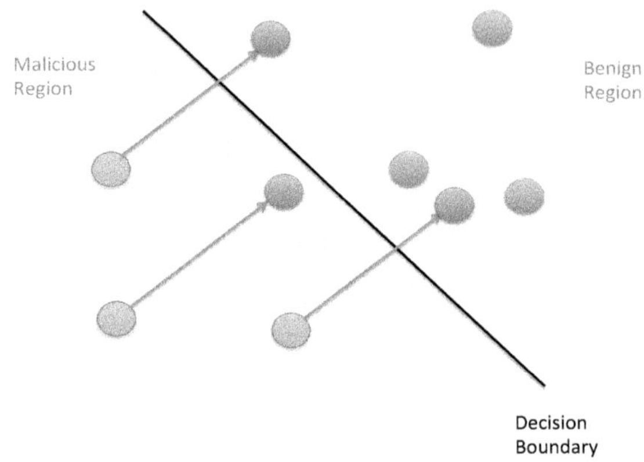

Figure 3: An example of mimicry attack

2. malicious structure detectors. These tools analyze the internal structure of a PDF file without analyzing executable code within the file. For instance, features can be related to raw object keywords within the PDF or to lowercase and uppercase characters. These tools assume structural differences between malicious and benign files, caused by the presence of malicious content.

`PjScan` [24] is a tool which applies the first methodology. It extract *lexical* features from JavaScript code. For example, it analyzes the frequency of specific tokens such as `+`, `(`, *etc.*, and looks for suspicious functions such as `unescape`, `eval`. Using these features, a one-class `SVM` classifier is trained.

`Malware Slayer` [26], on the contrary, adopts the second methodology. It selects, thanks to a clustering process, the most frequent `name objects` in malicious and benign files. It then adopts their frequencies as features to train a `random forest` classifier. This tool relies on the assumption that the frequency of the `name objects` is somewhat related to the maliciousness of the file. A similar approach has been proposed in 2013 [40] by considering, as features, the presence of specific *sequences* of `name objects` obtained by parsing the *logical* tree of the file.

`PDFRate` [33, 6], like `Malware Slayer`, performs a structural analysis of a `PDF` file, but it has a more extensive number of features, compared to `Slayer`. An example of these features can be the number of `stream` markers, the number of dot characters, and so on. Besides doing a distinction between malicious and benign, this tool also distinguishes malicious samples between *targeted*, whose payload directly implements the attack that is executed on the victim system, and *opportunistic*, whose payload downloads other malicious content from the Internet.

5. EVADING MALICIOUS PDF DETECTORS BASED ON STRUCTURAL ANALYSIS

5.1 Overview

Machine learning systems aimed at detecting malicious PDF files by analyzing the file structure, have been recently proposed. They rely on extracting a set of structural features according to the `Adobe` PDF standard [11]. Such features are not directly related to specific vulnerabilities or characteristics of the embedded malicious code, but, more in general, they are related to *elements that are part of the general PDF file structure.* Examples of structural features are the number of keywords [26], their distribution based on the *logical* tree of the file [40], or more general properties such as the number of lowercase or uppercase characters [33]. The reason why these systems are exceptionally good at detecting malicious samples is that malware in-the-wild shows an *evident* structural difference with respect to benign files.

5.2 The Mimicry Attack

While a benign PDF file is usually generated by an *off-the-shelf* production tool, which creates a series of objects that are not particularly useful to an attacker (for example, the `/Font` related objects that describe how the characters will appear on the screen), a malicious PDF sample can be obtained by creating specific objects through low-level tools such as `Python` libraries [14, 8]. Of course, the latter approach is more feasible for an attacker, as it grants a better control of the PDF contents.

Smutz and Šrndić [33, 40] theoretically investigated the possibility of a mimicry attack. In the analysis performed by Smutz et al., an attacker *exactly* knows what are the most N (where N is a number much lower than the total number of structural features) *discriminant* features used by the classifier. They extract an average *estimate* of the features values used in a *benign* file set. Finally, they modify the specific features of a malicious sample to match those determined before. In the approach adopted by Šrndić, the malicious sample is modified to match the most "benign" sample in the attacker dataset (i.e. the sample with the lowest *classification score*). The feature values can only be *incremented* and the choice of the feature to be changed depends on the *type* of classifier adopted (assuming that the attacker *perfectly* knows its model). Figures 2 and 3 show a graphical structure of the mimicry attack.

Let us assume, for the sake of simplicity, that patterns representing PDF files are represented as points in a 2D plane, and that malicious PDF (represented as red dots) are separated from benign PDF files (blue dots) by a line. Per-

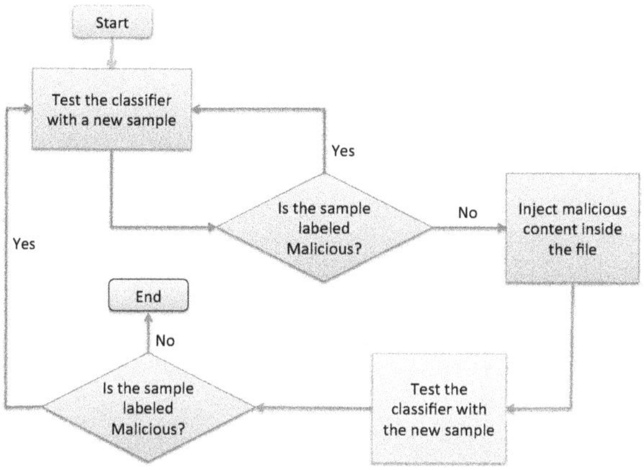

Figure 4: Conceptual structure of the reverse mimicry attack

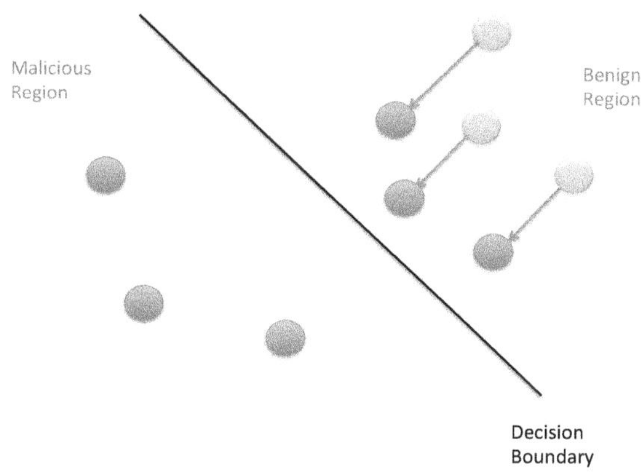

Figure 5: An example of reverse mimicry attack

forming a mimicry attack translates into moving red dots in the direction of the arrows, so that malicious samples are represented in a way similar to the benign ones. The length and direction of the arrow depends on the effort needed to transform a malicious PDF into a benign one from the point of view of the learning algorithm. This approach, though, does not guarantee an effective attack, as an attacker has to guess a reasonable model of benign samples based on the knowledge of the learning algorithm, and the features used. This means that, if an attacker makes a wrong guess, the sample can go farther from the benign region. Moreover, the changes that the attacker should do to the feature values might be impossible to be done concretely, due to some limitations in operating with `PDF data`. For example, *removing* data from the file might be a non feasible operation, as it can, for example, completely break the `tree` or the `Xref table`, making the file not readable anymore. For this reason, the approach proposed by Smutz et al. has been kept theoretical only.

5.3 Reverse Mimicry

In the case of attacks carried by PDF files, we believe

```
Benign Sample                              Malicious Sample

75 0 obj                                   75 0 obj
<<                                         <<
/Type /Catalog                             /Type /Catalog
/Pages 72 0 R                  ------->    /Pages 72 0 R/Names 178 0 R
/PageLabels 71 0 R                         /PageLabels 71 0 R
/Outlines 69 0 R                           /Outlines 69 0 R
/PageLayout /SinglePage                    /PageLayout /SinglePage
/OpenAction 176 0 R                        /OpenAction 182 0 R
/Metadata 177 0 R                          /Metadata 177 0 R
>>                                         >>
endobj                                     endobj

0005878978 00000 n             ------->    0006475864 00000 n
```

Figure 6: Changes in `root` object structure and memory allocation from a benign to a malicious sample

that a structural detection model can be evaded by *reverse mimicry* attacks. The rationale behind the reverse mimicry attack is the following. Instead of manipulating a malicious sample to mimic benign patterns, *we propose to manipulate a benign file to make it malicious*, with minimum structural differences. Figures 4 and 5 show an example analogous to Figures 3 and 2. The recognized benign samples are poisoned by the introduction of a malicious payload (the initial benign samples are in clear blue). This operation may determine a variation of some features that make the sample closer to the malicious region. However, we will show that this variation can be very limited: the new malicious samples may not cross the boundary of the decision region, thus bypassing the detection system.

This process is relatively easy to be implemented in PDF files, because of their particular standard. In the following, we will describe three possible ways of implementing this attack.

5.3.1 EXE Embedding

As described in Section 2, when an existing PDF object is edited without rewriting the entire file, a new `version` is added after its trailer. This version has *a new trailer* which defines the main object (`root`) of the PDF tree. In other words, with a new version, it is possible to completely redraw the tree of a PDF file. However, if there are compressed data, removing the previous version or its objects from the file can be a difficult operation: there are strict boundaries indicated by the `Xref table`. Indeed, when new objects are added, their related `Xref table` values are included as well. This means that *adding* structural features is really easy but, on the contrary, *removing* them can be quite complex. We added a new version of the file containing a malicious embedded `EXE` payload, by using the `Social Engineering Toolkit (SET)` [9]. In this version, a new `root` object is added, so that the new trailer will point to this new object. Figure 6 shows the changes between the `root` object in the benign sample and the one in the malicious sample (embedded in the second version) and the *trailer*, respectively, for the benign and the malicious file. As it can be seen from the picture, a new object called `Names` is added and the `OpenAc-`

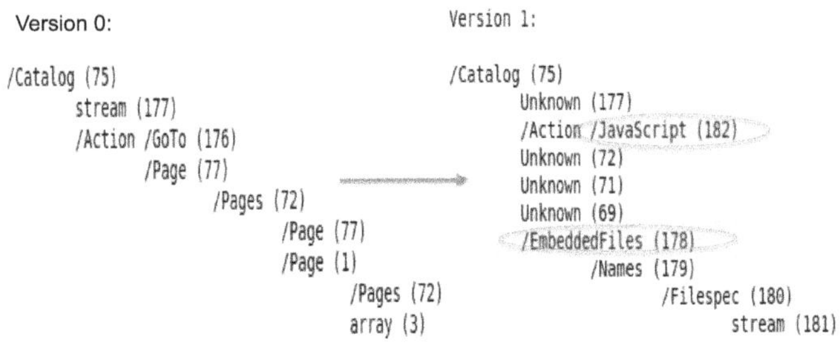

Figure 7: Differences between the beginning of the trees in benign and malicious sample

```
65 0 obj
<< /Length 121365
/Type /EmbeddedFile
/Filter /FlateDecode
/Params << /Size 123679
/Checksum <24c0c2b54da6b50a4e7ee3c5e8a9b3eb> >>
/Subtype /application#2Fpdf >>
stream
xúåªy<TÔ˝˝? ä±À:FYÚb≤¬√
  Ÿ¯e≤ÕëA˙Rñ±K NHñ»^&[e'dãÏ[∂flÒ˙º üflÁw˚~˛¯˛·ÃôÀôsùsû◊Û¯∏?Œ9◊5
```

Figure 8: An example of a PDF embedded object

tion object is changed from 176 to 182. Those two objects are usually related to *actions* such as code execution or, for benign files, the filling of a form. Needless to say, the presence of these objects make the file more suspicious, but it does not constitute a proof of maliciousness by itself. Figure 7 better explains the changes introduced by the new version to the logical tree.

Object 182 contains the `JavaScript` keyword, which *automatically*[4] triggers a launch action on the embedded file, described by the keyword `EmbeddedFiles`, which is finally contained inside the `stream` 181. This is also known as the `CVE-2010-1240` vulnerability, and it was discovered by Didier Stevens. Mr. Stevens has also implemented a way to make an application be launched inside a PDF *without* using JavaScript [35]. It is not important to explain any single detail of how the vulnerability can be exploited. The reason why this vulnerability is interesting, although not being very recent (indeed, this attack will not work on `Adobe Reader X`), is that it is a clear (and concrete, as we have generated *real samples*) proof-of-concept of the effectiveness of the reverse mimicry attack.

5.3.2 PDF Embedding

A limit of the attack we have just described is that it is related to an old vulnerability, and therefore, even if it is able to evade structural analysis techniques, it might not work in patched `Adobe Reader` versions (specifically, after the 9.3). However, the PDF standard supports the embedding of other file formats apart from the `EXE` one (whose automatic execution is currently banned by Adobe). In particular, the most interesting format is the PDF one itself. Indeed, it is possible to embed a PDF file inside another one, so that *the embedded file is automatically opened without user interaction.* Therefore, it is possible to embed a malicious file inside a benign one. Interestingly for our purposes, there are no restrictions on the file that can be embedded: it can contain JavaScript, Flash code, *etc.* Such samples can be automatically built using the embedding function provided by the `PeePDF` tool [7]. To improve the obfuscation of an embedded file, `PeePDF` can embed the file in a compressed `stream`. This means that, in order to retrieve the features of the embedded file, the parser should be able to *decompress the stream* by applying the correct *filter*. This can give a strong advantage against *raw* parsers which do not decompress objects contained inside a PDF file. Figure 8 shows an example of a typical PDF file embedded object. As it can be seen, the keyword `EmbeddedFile` (along with some others) is added to the carrier file. This is a clear marker that an embedded file is contained inside the carrier. Again, the presence of such a keyword may make the file more suspicious, but it does not constitute a proof of maliciousness by itself. The couple of keys `Filter` and `FlateDecode`, followed by the `stream`, means that the embedded object is compressed with the `FlateDecode` filter. This simple approach may completely bypass simple parsers such as `PdfID` [5], that are not able to decompress object streams. Therefore, accurate parsers are fundamental in order to retrieve useful information about embedded content. Even if a parser is able to decompress object streams, embedding a PDF file allows a fine-grain control on the structural features of the carrier file. For example, we can carefully choose what structural features are injected, depending also on the vulnerability that is exploited by the malicious sample. In other terms, PDF embedding allows devising reverse mimicry attacks that exploit a wide spectrum of vulnerabilities, even those that affect the latest versions of PDF readers.

5.3.3 JavaScript Injection

Most of malicious PDF files adopt JavaScript code [24] to exploit vulnerabilities in the reader application. Smutz et al. showed that there are basically two types of JavaScript code that can be used: one type that, along with the code to exploit the vulnerability, *includes the payload used for the attack* and the other one that relies to other objects in the file, or to external malicious links, to download malicious code [33]. The best way to perform a reverse mimicry attack is to encapsulate a malicious JavaScript code that does not contain references to other objects. This is because only one object will be added to the PDF tree and, consequently, a minimum variation over the structure of the `root` (benign) file is introduced. Furthermore, a single object addition is much faster and more feasible for an attacker than a complicated obfuscation process. An interesting characteristic of JavaScript injection is that no new version is added: the whole tree is completely rewritten. This causes little modifications in the reference to the tree structure: the whole structure is substantially the same, and this leads to little variations of the PDF pattern in the feature space. Figure 9 shows an example of this specific attack, that can be easily built using specific `Python` libraries [14]. As it can be seen, a JavaScript object is added, and the JavaScript code

[4] That is, without user interaction.

```
93 0 obj
<<
/Type /Action
/S /JavaScript
/JS (app.alert({cMsg: 'Hello from PDF JavaScript', cTitle:
>>
endobj
94 0 obj
<<
/Type /Catalog
/Pages 1 0 R
/OpenAction 93 0 R
>>
```

Figure 9: An example of an injected JavaScript object and how the Catalog object is changed consequently

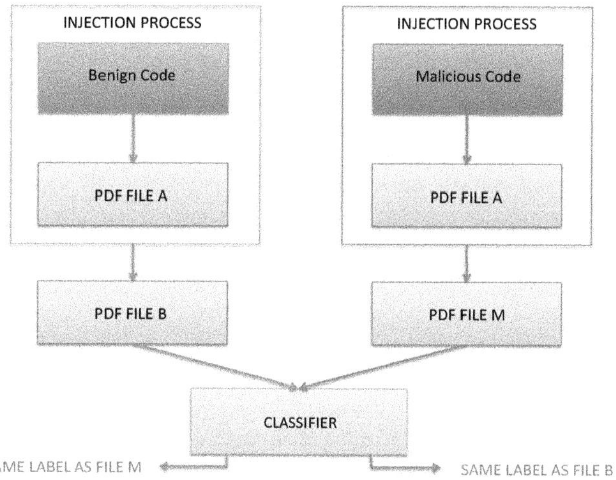

Figure 11: An example of evasion of systems that extract features from STO and SUO layers

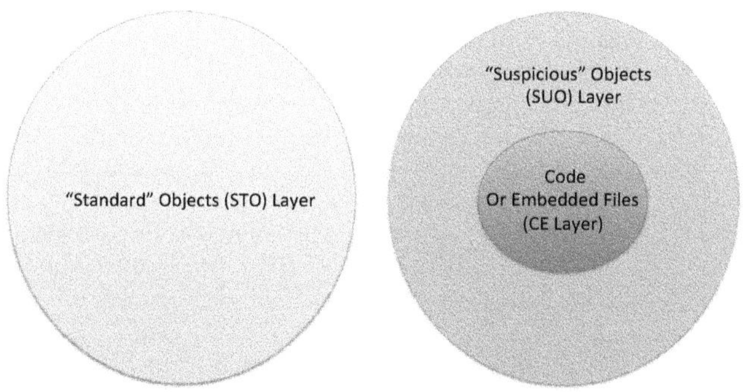

Figure 10: A conceptual representation of the layers

is inserted after the JS keyword (the one in the Figure was cut for space reason). It is also possible to observe how the Catalog object is changed. The reference marked by 93 0 R has been added in order for the code to be executed *independently* from the presence of other objects within the PDF file.

5.4 Systems Weaknesses

The reason *why* the reverse mimicry, along with its variants, is so effective against Machine Learning systems, is easily deducible by observing the *parts of the PDF file* on which the detection systems focus. Figure 10 shows a *conceptual* representation of the layers on which systems such as PJScan, Malware Slayer and PDFRate calibrate their analysis.

Generally speaking, we can distinguish three types of layers:

- Standard Objects (STO) Layer: objects that are not related to external actions (such as execution of code). They can, however, contain references to suspicious or malicious objects.
- Suspicious Objects (SUO) Layer: objects that are related to external actions (such as code execution, forms, etc.). In this category there are objects that can contain, for example, keywords such as Acroform, Names, JavaScript, etc.
- Malicious code or Embedded Files (CE) Layer: this layer describes *codes or embedded files contained inside suspicious objects*. It is the point in which the *real* attack is contained, therefore, in this representation, it is considered the most internal layer.

Malware Slayer [26] *only* extracts information from the STO and the SUO layers. The injection, for example, of JavaScript, introduces small changes in both layers. The classifier can see this variation in terms of suspicious objects. However, objects in SUO layer do *not* necessarily contain malicious code. Indeed, if there is a code inside the suspicious object, it might also be benign. Therefore, there might be a *benign* file with a *suspicious* distribution of the objects in the STO and SUO layers. Now, let us consider a malicious and a benign JavaScript code: suppose that these codes are injected inside the *same* PDF file, therefore obtaining two files, one malicious and one benign. Given a set of features related to STO and SUO layers only, the two files would be indistinguishable from each other. Figure 11 shows a graphical representation of this behavior.

Of course, the same approach is valid when an embedded file is used instead of code. PJScan [24], on the contrary, does not suffer from this issue, as its features are mainly extracted from the CE layer. However, the tool *only* analyzes JavaScript code, and therefore it could be easily evaded by a sample carrying other malicious content, such as ActionScript code, EXE Payloads, *etc.*

6. EXPERIMENTAL EVALUATION

To assess the efficacy of reverse mimicry attacks, we implemented the three attack variants presented in Section 5.3,

Table 1: Efficacy of EXEembed, PDFembed, JSinject evasion techniques against six PDF malware detection tools

Detector	EXEembed	PDFembed	JSinject
PJScan	**evaded** (-/.317)	**evaded** (-/-)	detected (-/.905)
PDFRate Contagio	**evaded** (.002/.069)	**evaded** (0/.008)	**evaded** (0/.148)
PDFRate George Mason	**evaded** (0/.162)	**evaded** (0/0)	**evaded** (0/.013)
PDFRate Community	**evaded** (.001/.2)	**evaded** (.008/.024)	**evaded** (0/.125)
Malware Slayer	**evaded** (0/.08)	**evaded** (0/0)	**evaded** (0/.08)
Wepawet	-	-	-

namely, EXE Payload embedding —**EXEembed**—, PDF embedding —**PDFembed**—, JavaScript injection —**JSinject**— and tested them against various PDF malware detectors proposed so far, namely, **Wepawet** (online tool) [10], **PJScan** [24], **Malware Slayer** [26] and **PDFRate** [33]. More precisely, we tested three *online* versions of **PDFRate** differing each other on the data source employed for training: **Contagio**, **George Mason** and **Community** version[5].

PJScan and **Malware Slayer** have been trained on a dataset composed by **5993** malicious samples retrieved from the Contagio repository [1], that is, the same data source employed by **PDFRate Contagio**. For **Malware Slayer**, **5591** benign files collected automatically using the Yahoo search API [42] have also been adopted. Each attack variant has been built starting from a randomly-chosen benign file[6], that we call as **root**. In particular we built the attacks as follows:

- **EXEembed**: we embedded a **Zeus EXE** payload which implements a simple type of compression;
- **PDFembed**: we embedded a **PDF** file containing malicious JavaScript code that implements the **CVE-2009-4324** vulnerability. The embedded file is opened automatically (without user interaction) as soon as the **root** file is opened.
- **JSinject**: we injected the same **JavaScript** code employed for **PDFembed** in the **root** file; we recall that this task does *not* require an addition of a version on the file.

For each PDF malware detector we verified that such a **root** file was indeed correctly labelled as benign. For those detectors that provide also a *maliciousness* score, we computed how much such score changed before and after the attack implementation, i.e., we computed the score of the **root** file, as well as the score of the PDF attack instance built upon it. This task allowed us to evaluate the *sensitivity* of each detector with respect to a certain evasion technique.

Table 1 shows a summary of evasion results. For each detector we indicate whether an evasion attack is successful (evaded, in bold) or not (detected). Between parentheses, when available, we indicate **root** and attack score, separated by a /. The maliciousness score ranges from 0 to 1, and an ideal detector should assign high values for attack instances and very low values to benign samples. As it can be seen from Table 1, the reverse mimicry attacks are incredibly effective against all PDF detectors. Almost all PDF detectors assign a relatively low score to the three attack instances, and are thus successfully evaded. More precisely, all detectors which adopt structural analysis, i.e., **PDFRate** and **Malware Slayer** are successfully evaded, whereas **PJScan** is evaded by both **EXEembed** and **PDFembed**, but it is able to detect **JSinject**. In fact, this behavior is quite reasonable, because as explained in Section 5, **PJScan** operates at the **CE** layer, and therefore its features are not influenced by the structural injection operated by reverse mimicry attacks. More in detail, **PJScan** does not provide a score for the **root** file, since it does not contain JavaScript. Moreover, **PJScan** does not provide a score for **PDFembed**, because such detector is not able to analyze JavaScript code within embedded PDF files, due to a limitation of its file parsing mechanism.

An analogous result has been obtained with **Wepawet**. From **Wepawet** we were unable to get any detection result, because our attacks systematically raised *internal errors*. On the other hand, the detector was indeed active, because we were able to analyze benign PDFs[7] as well as malicious PDFs taken from the Contagio repository [1]. As a consequence, we speculate that **Wepawet** was unable to analyze our attacks due to some parsing error caused by our deliberate PDF manipulation. In a recent work, Jana and Shmatikov [23] showed that such kind of errors are pretty common in malware detectors, and actually represent the weakest link of malware defense.

Finally, we would like to discuss a particularly interesting finding, related to the **JSinject** attack. An attack instance containing the same JavaScript code can be built automatically through the **Metasploit** framework [3]. However, doing so would make this attack detectable by *all* PDF malware detectors. There is a good reason for this, since this attack instance contains *both* structural and code differences with respect to benign files. This aspects highlights once more the efficacy of the reverse mimicry evasion strategy.

7. COUNTERMEASURES

Experimental results presented in Section 6 clearly show that, even if PDF malware detectors are able to accurately detect malware in-the-wild, they may miserably fail against simple, yet effective evasion techniques such as the reverse mimicry attack described in Section 5. While such an attack has been specifically developed against detectors based on

[5]More details are available at **https://www.csmutz.com/pdfrate/data**.

[6]We performed this operation multiple times, with different benign files. In all runs, we were able to get the same results in terms of evasion success (with slightly different score). For simplicity, we report only the results attained by attacks built in the first run. Such reverse mimicry attacks, together with the associated **root** files, can be downloaded at **https://dl.dropbox.com/u/65430791/ASIACCSSamples.zip**. Open them at your own risk!

[7]**Wepawet** classified correctly the **root** file, but no maliciousness score was available.

PDF structural analysis, we found it to be effective against tools that operate at CE layer, due to errors in the parsing process. In other terms, current PDF malware detectors are weak against adversarial attacks. We identify two main key points for improving state-of-the-art solutions: (1) accuracy of the file parsing process; (2) robustness of the feature set against evasion.

The first of issue is directly related to all embedding attacks. The parsing process should (a) be aware of the PDF versioning mechanism, (b) identify and extract all embedded content. This way, embedded content can be thoroughly analyzed as a different entity by a content-specific detector. For instance, embedded PDF files can be analyzed by the same tools used to analyze the `root` file. Other type of content, such as `SWF` or `EXE`, can be analyzed by other, specialized tools (e.g., `Wepawet`). Then, the maliciousness of the original PDF file can be evaluated as the result of the correlation between the outputs of all these specialized detectors (e.g., through OR rule).

On the other hand, improving the robustness of the feature set against evasion is very useful when facing *code injection*. In particular, features that involve all layers described in Section 5, namely, `STO`, `SUO` and `CE`, should be extracted. `CE` features should be specifically related to the *type of code*. For example, combining a structural system such as `Malware Slayer` or `PDFRate` with a code analyzer such as `PjScan` may be a first, simple solution, to improve the robustness of the analysis.

According to the above discussion, Figure 12 shows a possible framework that is able to cope with adversarial attacks against PDF malware detectors. We have realized a simple implementation of this framework. We have empowered the parsing process by extracting any embedded content inside the main PDF, and we have combined the output of `Malware Slayer` and `PDFRate` (structural analysis), `PJScan` (JS analysis), `Wepawet` (JS, EXE and SWF analysis) according to an OR rule[8]. This allows to thoroughly analyze PDF content and combine structural and code analysis. Table 2 shows that such a new framework is able to spot *all* attacks presented in Section 6. In the following, for each attack, we provide a detailed explanation of the operations performed by the framework:

- `EXEembed`: our parser successfully extracts the executable file, which in turn is analyzed by `Wepawet`. This tool raises an alert and correctly identifies the related exploit.

- `PDFembed`: our parser successfully extracts the embedded PDF file, as well as the JavaScript within this file. `PJScan`, `Malware Slayer` and `PDFRate` (all versions) correctly raise an alert when analyzing the embedded PDF file. `Wepawet` correctly raises an alert when analyzing the JavaScript code.

- `JSinject`: our parser successfully extracts the JavaScript within this file. `PJScan` correctly raises an alert when analyzing the main PDF. `Wepawet` correctly raises an alert when analyzing the JavaScript code.

Indeed, the correlation of different detectors can improve robustness against evasion. However, we observe that it may also increase the rate of false alarms. So, there may be a tradeoff between overall accuracy of the framework and robustness against evasion. To this end, the choice of base detectors and correlation rules should be tailored according to the target tradeoff. A more detailed study of this aspect is outside the scopes of this work, but it is indeed interesting for future research activity.

[8]A file is classified malicious if at *least one* tool classifies the main PDF file or one of its embedded content, as malicious.

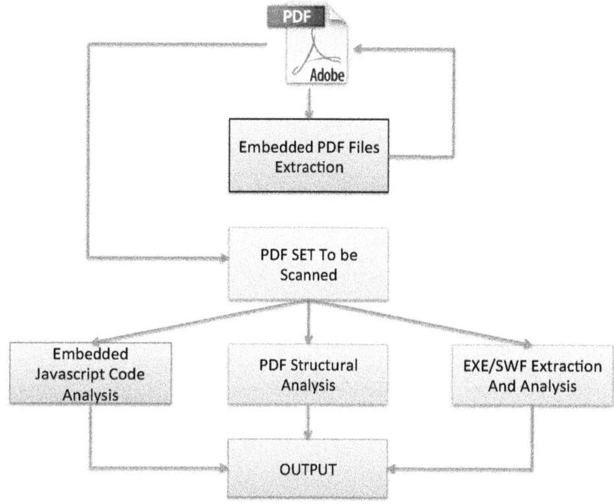

Figure 12: A possible framework to make ML systems more robust against adversarial attacks

Table 2: Efficacy of `EXEembed`, `PDFembed`, `JSinject` evasion techniques against the proposed framework, which combines different PDF malware detectors using a OR rule

Attack	Alerts	Result
`EXEmbed`	`Wepawet`	detected
`PDFEmbed`	`PJScan`, `Wepawet`, `Malware Slayer`, `PDFRate`	detected
`JSinject`	`PJScan`, `Wepawet`	detected

8. CONCLUSIONS

Nowadays, malicious PDF files represent a concrete and relevant threat. Recent work showed that malicious PDF files can be accurately detected by analyzing their *logical structure*, with excellent results. However, structural analysis can be easily defeated by a skilled attacker, by developing PDF malware that mimic the structure of benign PDF documents. In this paper, we practically demonstrated this claim through the development and the implementation of a novel attack methodology called *reverse mimicry*. We implemented it using real PDF malware and validated our approach by testing it against various PDF malware detectors proposed so far. Surprisingly, we found that even detectors which employ non-structural features may be evaded, due to errors in the file parsing mechanism.

This case study highlights once more the importance of developing malware detection systems robust to adversarial attacks. As a countermeasure, we proposed a framework based on (a) the development of an accurate PDF parsing mechanism and (b) the correlation of outputs from different PDF malware detectors. This framework allows detecting all attack instances implemented in this work, and it is a promising solution for future research activity.

Acknowledgements

This work is supported by the Regional Administration of Sardinia, Italy, within the project "Advanced and secure sharing of multimedia data over social networks in the future Internet" (CUP F71J11000690002). Davide Maiorca gratefully acknowledges Sardinia Regional Government for the financial support of his PhD scholarship (P.O.R. Sardegna F.S.E. Operational Programme of the Autonomous Region of Sardinia, European Social Fund 2007-2013 - Axis IV Human Resources, Objective l.3, Line of Activity l.3.1.).

9. REFERENCES

[1] Contagio. `http://contagiodump.blogspot.it`.

[2] Malware tracker. `http://www.malwaretracker.com/pdfthreat.php`.

[3] Metasploit framework. `Http://www.metasploit.com/`.

[4] Origami framework. `Http://esec-lab.sogeti.com/pages/Origami`.

[5] Pdf tools. `http://blog.didierstevens.com/programs/pdf-tools/`.

[6] Pdfrate. `http://pdfrate.com`.

[7] Peepdf. `Http://eternal-todo.com/tools/peepdf-pdf-analysis-tool`.

[8] Pypdf. `http://pybrary.net/pyPdf/`.

[9] Social engineering toolkit. `Https://www.secmaniac.com/`.

[10] Wepawet. `Http://wepawet.iseclab.org/index.php`.

[11] *PDF Reference. Adobe Portable Document Format Version 1.7.* Adobe, November 2006.

[12] *Adobe Supplement to ISO 32000.* Adobe, June 2008.

[13] Foxit reader stack overflow exploit. `http://www.exploit-db.com/foxit-reader-stack-overflow-exploit-egghunter/`, November 2010.

[14] Add javascript to existing pdf files (python). `http://blog.rsmoorthy.net/2012/01/add-javascript-to-existing-pdf-files.html`, 2012.

[15] *Internet Security Threat Reports. 2011 Trends.* Symantec, April 2012.

[16] P. Bania. Jit spraying and mitigations. *CoRR*, `http://www.piotrbania.com/all/articles/pbania-jit-mitigations2010.pdf`, 2010.

[17] E. Buchanan, R. Roemer, S. Sevage, and H. Shacham. Return-oriented programming: Exploitation without code injection. In *Black Hat '08*, 2008.

[18] D. Canali, M. Cova, G. Vigna, and C. Kruegel. Prophiler: a fast filter for the large-scale detection of malicious web pages. In *Proceedings of the 20th international conference on World wide web*, WWW '11, pages 197–206, New York, NY, USA, 2011. ACM.

[19] M. Cova, C. Kruegel, and G. Vigna. Detection and analysis of drive-by-download attacks and malicious javascript code. In *Proceedings of the 19th international conference on World wide web*, WWW '10, pages 281–290, New York, NY, USA, 2010. ACM.

[20] J. S. Cross and M. A. Munson. Deep pdf parsing to extract features for detecting embedded malware. Technical report, Sandia National Laboratories, 2011.

[21] C. Curtsinger, B. Livshits, B. Zorn, and C. Seifert. Zozzle: fast and precise in-browser javascript malware detection. In *Proceedings of the 20th USENIX conference on Security*, SEC'11, pages 3–3, Berkeley, CA, USA, 2011. USENIX Association.

[22] M. Engleberth, C. Willems, and T. Holz. Detecting malicious documents with combined static and dynamic analysis. Technical report, Virus Bulletin, 2009.

[23] S. Jana and V. Shmatikov. Abusing file processing in malware detectors for fun and profit. In *IEEE Symposium on Security and Privacy*, pages 80–94, 2012.

[24] P. Laskov and N. Šrndić. Static detection of malicious javascript-bearing pdf documents. In *Proceedings of the 27th Annual Computer Security Applications Conference*, ACSAC '11, pages 373–382, New York, NY, USA, 2011. ACM.

[25] W.-J. Li, S. Stolfo, A. Stavrou, E. Androulaki, and A. D. Keromytis. A study of malcode-bearing documents. In *Proceedings of the 4th international conference on Detection of Intrusions and Malware, and Vulnerability Assessment*, DIMVA '07, pages 231–250, Berlin, Heidelberg, 2007. Springer-Verlag.

[26] D. Maiorca, G. Giacinto, and I. Corona. A pattern recognition system for malicious pdf files detection. In *Proceedings of the 8th international conference on Machine Learning and Data Mining in Pattern Recognition*, MLDM'12, pages 510–524, Berlin, Heidelberg, 2012. Springer-Verlag.

[27] S. Porst. A brief analysis of a malicious pdf file which exploits this week's flash 0-day. `http://blog.zynamics.com/`, 2010.

[28] M. A. Rahman. Getting owned by malicious pdf - analysis. Technical report, SANS Institute, 2008.

[29] P. Ratanaworabhan, B. Livshits, and B. Zorn. Nozzle: a defense against heap-spraying code injection attacks. In *Proceedings of the 18th conference on USENIX security symposium*, SSYM'09, pages 169–186, Berkeley, CA, USA, 2009. USENIX Association.

[30] K. Rieck, T. Holz, C. Willems, P. Düssel, and P. Laskov. Learning and classification of malware behavior. In *Proceedings of the 5th international conference on Detection of Intrusions and Malware, and Vulnerability Assessment*, DIMVA '08, pages 108–125, Berlin, Heidelberg, 2008. Springer-Verlag.

[31] K. Rieck, T. Krueger, and A. Dewald. Cujo: efficient detection and prevention of drive-by-download attacks. In *Proceedings of the 26th Annual Computer Security Applications Conference*, ACSAC '10, pages 31–39, New York, NY, USA, 2010. ACM.

[32] M. Z. Shafiq, S. A. Khayam, and M. Farooq. Embedded malware detection using markov n-grams. In *Proceedings of the 5th international conference on Detection of Intrusions and Malware, and*

Vulnerability Assessment, DIMVA '08, pages 88–107, Berlin, Heidelberg, 2008. Springer-Verlag.

[33] C. Smutz and A. Stavrou. Malicious pdf detection using metadata and structural features. In *Proceedings of the 28th Annual Computer Security Applications Conference*, ACSAC '12, 2012.

[34] K. Z. Snow, S. Krishnan, F. Monrose, and N. Provos. Shellos: enabling fast detection and forensic analysis of code injection attacks. In *Proceedings of the 20th USENIX conference on Security*, SEC'11, 2011.

[35] D. Stevens. Escape from pdf. `http://blog.didierstevens.com/2010/03/29/escape-from-pdf/`, 2010.

[36] D. Stevens. *Free Malicious PDF Analysis.* `http://didierstevens.com/files/data/malicious-pdf-analysis-ebook.zip`, 2010.

[37] D. Stevens. Malicious pdf documents explained. *IEEE Security and Privacy*, 9(1):80–82, Jan. 2011.

[38] S. M. Tabish, M. Z. Shafiq, and M. Farooq. Malware detection using statistical analysis of byte-level file content. In *Proceedings of the ACM SIGKDD Workshop on CyberSecurity and Intelligence Informatics*, CSI-KDD '09, pages 23–31, New York, NY, USA, 2009. ACM.

[39] Z. Tzermias, G. Sykiotakis, M. Polychronakis, and E. P. Markatos. Combining static and dynamic analysis for the detection of malicious documents. In *Proceedings of the Fourth European Workshop on System Security*, EUROSEC '11, pages 4:1–4:6, New York, NY, USA, 2011. ACM.

[40] N. Šrndić and P. Laskov. Detection of malicious pdf files based on hierarchical document structure. In *Proceedings of the 20th Annual Network & Distributed System Security Symposium*, 2013.

[41] C. Willems, T. Holz, and F. Freiling. Toward automated dynamic malware analysis using cwsandbox. *IEEE Security and Privacy*, 5(2), March 2007.

[42] Yahoo. Search api. `http://developer.yahoo.com`, December 2012.

Efficient User-Space Information Flow Control

Ben Niu
Department of Computer Science and Engineering
Lehigh University
19 Memorial Drive West
Bethlehem, PA, 18015
ben210@lehigh.edu

Gang Tan
Department of Computer Science and Engineering
Lehigh University
19 Memorial Drive West
Bethlehem, PA, 18015
gtan@cse.lehigh.edu

ABSTRACT

The model of Decentralized Information Flow Control (DIFC [17]) is effective at improving application security and can support rich confidentiality and integrity policies. We describe the design and implementation of duPro, an efficient user-space information flow control framework. duPro adopts Software-based Fault Isolation (SFI [22]) to isolate protection domains within the same process. It controls the end-to-end information flow at the granularity of SFI domains. Being a user-space framework, duPro does not require any OS changes. Since SFI is more lightweight than hardware-based isolation (e.g., OS processes), the inter-domain communication and scheduling in duPro are more efficient than process-level DIFC systems. Finally, duPro supports a novel checkpointing-restoration mechanism for efficiently reusing protection domains. Experiments demonstrate applications can be ported to duPro with negligible overhead, enhanced security, and with tight control over information flow.

Categories and Subject Descriptors

D.4.6 [**Software**]: Operating Systems—*Security and Protection*

General Terms

Security

Keywords

Secure information flow, software-based fault isolation, checkpointing

1. INTRODUCTION

Decentralized information flow control (DIFC [17]), which is evolved from classic information flow control models ([7, 9]), is an effective way of improving application security. In DIFC, components of an application are put into separate protection domains; information flow between domains is restricted by a trusted reference monitor (e.g., the OS) to enforce application-wide data confidentiality and integrity. For example, information in system password files should not flow to untrusted domains without explicit declassification. As another example, a web browser might forbid network data from contaminating user profiles.

There has been substantial research progress to apply DIFC at the OS-process level to improve application security, including Asbestos [10], HiStar [26], and Flume [14]. In this line of work, each protection domain is realized by an OS process and a reference monitor tracks inter-process information flow. Information flow between processes is forbidden by the reference monitor if that flow would violate the policy implied by the security labels of the sending and receiving processes. These systems' practicality in terms of security benefits has been demonstrated on real-world applications.

On the other hand, existing process-level DIFC systems have some drawbacks that hinder their adoption in practice. Since many process-level DIFC systems such as Asbestos and HiStar are implemented as new OSes, porting them to other OSes is nontrivial; they also do not benefit from OS kernel updates or new device driver support. To address the challenge of OS portability, Flume [14] is proposed as a user-space DIFC implementation (on Linux and OpenBSD). However, it still requires the root privilege to inject an OS kernel module that performs system-call interposition. Furthermore, performance overhead of process-level DIFC is significant when there is heavy inter-process communication (IPC) in an application. IPC involves heavy process-level context switches (including the cost of switching the page table). As a result, Flume imposes over 30% performance overhead on real applications.

We propose duPro, an efficient framework that allows applications to control information flow between components to enhance their security. One key starting point of duPro is that protection domains are put into the same process, but are isolated through software-based fault isolation (SFI [22]). Because of this design, duPro provides the following advantages over previous DIFC systems.

- duPro requires no OS changes. Separation between domains is achieved via a pure user-space implementation. Therefore, applications running in duPro can benefit from new OS kernel updates and driver support. In addition, duPro can be deployed as a user utility program; non-root users can use it without the root privilege.

- duPro is efficient. duPro domains reside in the same address space, enabling efficient context switches between domains. The state of an SFI-enforced domain is much smaller than a process state and there is no need to switch page tables when performing context switches. This makes SFI especially suitable for situations where domain communications are frequent. Moreover, duPro applies SFI to intercept system calls. It incurs less overhead compared to a kernel-level

ASIA CCS'13, May 8–10, 2013, Hangzhou, China.

system-call interposition mechanism such as Flume's kernel module or a `ptrace`-based solution.

A user-space DIFC framework is not without its challenges. One challenge we have encountered during experiments is to provide an efficient user-space *domain-reuse* mechanism. The need can be seen through the example of a web server. To isolate between client connections, the server could allocate a new domain to handle each connection. When connection requests are frequent, this would be inefficient in duPro because SFI domain creation is relatively expensive. A more efficient design is to reuse a domain for future connections after a previous connection has been handled. However, the challenge is that the domain has been tainted by data of previous connections and reusing it without caution would allow information flows between connections, violating the connection-isolation policy. To address the challenge, duPro implements a novel *checkpointing-restoration* mechanism based on signals. An untrusted domain can checkpoint its state before performing actions that can taint the domain. At a later point, the domain can be restored to an early state for reuse by rolling back all the changes to memory and DIFC labels between checkpointing and restoration. This mechanism provides a safe and efficient way of reusing a domain.

Another challenge is how to minimize changes to source code when porting applications to duPro. Similar to process-level DIFC systems, duPro requires changes to source code to adapt to its programming model. Since processes are not supported in duPro, duPro provides a set of APIs with similar semantics and interfaces as the process-related abstractions (e.g., pipes, socketpairs) in order to reduce the effort spent on porting applications to duPro. Our evaluation demonstrates that complex applications such as Apache can be ported to duPro without many changes to the source code.

At a high level, duPro's technical contributions are mostly two-fold. First, it provides an efficient, pure user-space DIFC enforcement tool at the granularity of SFI protection domains. This includes adapting an existing DIFC model (the Flume model) to a new context and building the infrastructure. Second, it provides an effective and efficient domain-reuse mechanism based on domain checkpointing-restoration.

The remainder of the paper is organized as follows. Section 2 discusses related work. Section 3 presents the background of two pieces of work that duPro builds upon. Section 4 describes the design and implementation of duPro. Evaluation of duPro is presented in section 5. We conclude in section 6.

2. RELATED WORK

duPro adopts SFI to isolate protection domains. SFI is a special kind of Inlined Reference Monitors (IRM), which statically weave security checks into the application code to enforce a security policy. Specifically, an SFI security policy specifies valid code and data regions in which at runtime the program's control flow and memory accesses are restricted. According to the policy, the branch and memory access instructions are instrumented with checks to assure the destinations of such instructions fall within the valid regions. IRMs have had a long tradition in improving software security (e.g., [22, 21, 15, 12, 23, 5, 11, 6, 8]).

Mandatory access control (MAC) [20] enforces mandatory instead of discretionary data access by applications. Example MAC systems include SELinux [4], AppArmor [3] and `systrace` [19]. SELinux and AppArmor are Linux kernel extensions, which enforce MAC based on application-specific configuration files. In both SELinux and AppArmor, untrusted application components are forbidden to define or update the security policy. This design impacts flexibility and performance if there exist many such untrusted components. The system `systrace` intercepts Linux system calls and enforces MAC by checking the system calls of an application.

Classic information-flow-control models [7, 9] are special MAC models. In such models, the security levels (with a partial order) are predefined and managed by a centralized security officer. All the protection domains in the system are granted with certain security levels and the information is only allowed to flow from domains with low security levels to domains with higher levels. Example systems include TightLip [24], IX [16] and LOMAC [13]. These systems are also inflexible because the security levels are defined and maintained by a security administrator, who acts as the only data owner on the system and whose policy may be unsuitable for individual applications.

DIFC relaxes the restriction of centralized information flow control to allow an application to define its own custom policy. In such systems, data owners can create and maintain new security labels and grant capabilities to other principals discretionarily. Therefore, a data owner is capable of declassifying data. DIFC can be enforced at the language level or the protection-domain level. Example language-level DIFC enforcement tools (e.g., [17, 25, 27]) add DIFC related annotations in the source code and analyze the information flow mostly statically. Language-level DIFC enforcement provides flexibility, portability and fine granularity (at byte level). However, porting legacy applications to these systems is time consuming. Protection-domain level DIFC systems include Asbestos [10], HiStar [26], and Flume [14], all enforcing end-to-end information flow at the granularity of processes. As discussed before, duPro adopts DIFC on SFI-enforced protection domains, which provides better inter-domain communication performance and requires no OS changes.

Asbestos provides an abstraction called event processes, which has the same motivation as duPro's checkpointing-restoration mechanism: to reuse protection domains. In Asbestos, a base process can spawn multiple event processes that share most of the memory pages with the base process. As a result, an event process is lightweight. In this way, resources are reused across multiple event processes. To provide efficient sharing between an event process and the base process, Asbestos's OS kernel manipulates page tables. Being a user-space mechanism, duPro cannot manipulate page tables directly and instead relies on a mechanism that allows one domain to return to its previous memory and security state. We believe duPro's domain-reuse mechanism is conceptually simpler and can handle most of the practical situations when a domain needs to be reused.

3. BACKGROUND

duPro builds upon two pieces of previous work: (1) uPro, a user-space framework for isolating domains using SFI; and (2) Flume's DIFC model.

3.1 uPro

uPro [18] is a capability-based framework for enforcing privilege separation on applications. In uPro, applications are separated into modules that are loaded into separate protection domains. Each domain is given minimal privileges (e.g., allowed communications) specified by a configuration file. Communications between domains and system-call support are provided by uPro's runtime. We next discuss the major design choices in uPro.

First, it adopts SFI to isolate domains in a single process. The uPro toolchain compiles an application's source code and emits SFI-compliant object code. The toolchain inserts checks into the object code to enforce SFI domains. Thanks to SFI, context switches between domains are more efficient than the case when domains are implemented as OS processes. Moreover, SFI provides an efficient mechanism for intercepting system calls. In particular, each SFI-enforced domain is provided with a set of safe exits, called *trampolines*, which are the only ways that code in a protection domain can escape the protection domain to uPro's runtime. Trampolines allow a domain to invoke OS system calls (after security checks succeed) or invoke uPro's API functions for inter-domain communication.

Second, uPro provides a configuration language that allows explicit modeling of an application's security architecture. An application's configuration file declares protection domains, privileges of domains (e.g., allowed inter-domain calls), execution units, the initial configuration, and other information.

Third, uPro separates protection domains from execution units. A protection domain is a *protection unit*, whose functionality is isolation. In contrast, an *execution unit* is for control flow and is realized by an OS thread. An OS process combines protection and execution in one abstraction. In uPro, multiple execution units can execute in one protection domain and one execution unit can cross multiple protection domains during its lifetime.

Finally, two kinds of communications are provided: cross-domain calls (CDCs) and communicators. CDCs have similar semantics to remote procedure calls and allow one domain to call functions provided by another domain. Allowed CDCs are statically declared in the configuration file of an application. CDCs involve lightweight context switches between domains within the same process; their performance is much better than inter-process communication mechanisms. Communicators are either unidirectional or bidirectional and have similar semantics to OS pipes and socketpairs. When a communicator is dynamically created, two handles representing the two ends are returned to the two domains involved in the communication; each domain has one handle. Handles are passed to uPro's `read` and `write` API functions to receive and send data.

3.2 Flume's DIFC Model

A DIFC model assigns security labels to protection domains and includes rules for allowing or forbidding inter-domain information flow and rules for updating security labels when flow is allowed. Many DIFC models have been formulated in the literature and they are equivalent in terms of supported security policies. duPro adopts Flume's DIFC model for its succinct and intuitive formulation. We next give a brief introduction to the model and leave example uses of the model to the Flume paper [14].

Each domain is associated with a *security label*: (S, I, Cap), in which S is the *secrecy label*, I the *integrity label* and Cap the *capability set*. A secrecy/integrity label is a set of tags, which are drawn from an opaque token set. A tag is associated with some category of secrecy or integrity. As one example, t_1 may be associated with data in a secret file and a domain having t_1 in its secrecy label implies that the domain contains data in the file. As another example, t_2 might label data that is endorsed by a trusted entity and a domain having t_2 in its integrity label implies that the domain contains only data endorsed by the trusted entity.

The model allows decentralized control of tags. Any domain can create new tags and can dynamically adjust its secrecy label S and integrity label I if it has a corresponding capability. The capability set, Cap, of a domain is the set of tag operations that the domain can discretionarily perform to change S or I. If $t^+ \in Cap$, then the domain can dynamically add tag t to either S or I. If $t^- \in Cap$, the domain can dynamically drop t from S or I. Adding a tag t to S allows itself to receive secret t data and dropping t from S declassifies secret t data in the domain. In an opposite way, adding an integrity tag t to I endorses its data as having high-t-integrity and removing t from I allows it to receive low-t-integrity data.

As notational convenience, when a domain's capability set is empty, we omit Cap and write just (S, I) for its security label. We write C for the set of global capabilities owned by all domains. When C is clear from the context, we omit those capabilities from Cap when writing (S, I, Cap). For a domain d, we write D_d for the set $\{t \mid \{t^+, t^-\} \subseteq Cap_d\}$, where Cap_d is the domain's capability set. Those tags in D_d are owned by domain d.

DEFINITION 1 (SAFE MESSAGES). *Domain p is allowed to send to domain q messages iff the following two conditions are satisfied:*

- *Secrecy:* $S_p - D_p \subseteq S_q \cup D_q$
- *Integrity:* $I_q - D_q \subseteq I_p \cup D_p$

When D_p and D_q are empty sets, the rule says that information can flow from domains with less secret data and higher integrity to domains with more secret data and lower integrity. With non-empty D_p and D_q, the sending and receiving domains can first adjust their secrecy/integrity labels using capabilities in D_p and D_q, send the message, and adjust the labels back to their original values.

External objects such as files, network connections, or the user's terminal also have security labels. However, they always have empty capability sets and their labels are immutable once set. Before a domain communicates with an external object, an *endpoint* must be created; the endpoint serves as an intermediary between the domain and the external object. An endpoint e also has a security label. Information can flow between an endpoint and its associated external object if their labels obey rules in Definition 1. By default, an endpoint e has the same label as its creator domain p, but its label can be adjusted dynamically by p to a new security label (S_e, I_e, Cap_e) as long as one of the following conditions is met.

DEFINITION 2 (SAFE ENDPOINTS).

- *e is a safe readable endpoint iff* $(S_e - S_p) \cup (I_p - I_e) \subseteq D_p$
- *e is a safe writeable endpoint iff* $(S_p - S_e) \cup (I_e - I_p) \subseteq D_p$
- *e is a safe readable and writable endpoint iff the above two conditions are met.*

Two endpoints, e and f, may be two ends of a communication channel (e.g., the two ends of a pipe), and the information flow from e to f is allowed iff the following rule is obeyed.

DEFINITION 3 (SAFE ENDPOINT COMMUNICATION). *e is writeable, f is readable,* $S_e \subseteq S_f \wedge I_f \subseteq I_e$

4. DUPRO

As a first step, duPro combines Flume's DIFC model with uPro's user-space protection domains. Each domain is associated with a security label (S, I, Cap) and information flow between domains is regulated by the DIFC model. External objects such as files, network, and terminal are also labeled. Domains communicate with external objects through endpoints. Furthermore, the configuration language of uPro is extended to allow specification of initial security labels of domains and external objects. Labels of domains may change dynamically because of new data received, explicit declassification, or newly created categories of tags. The duPro runtime monitors inter-domain information flow and prevents forbidden information flow.

The combination of DIFC and SFI domains yields immediate advantages over uPro and Flume. One drawback of uPro is that privileges of domains are assigned statically and it does not support dynamic creation/revocation of privileges and transferring of

```
Trusted_PD_Type CM {
  export = {next_submission};
};
PD_Type GM {
  import = CM:{next_submission};
};
Labels {
  GM = ({}, {}, {});
};
```

Figure 1: Configuration of the duPro-protected grading application.

privileges from one domain to another; furthermore, there is no notion that treats data owned by a protection domain at different security levels. These features are useful in many practical situations, but the challenge is how to prevent privileges from being propagated arbitrarily. This question is elegantly addressed by the DIFC model. Compared to Flume, duPro maintains the strength of user-space protection domains: low context-switch overhead and requiring no OS changes.

Next we discuss a few unique design points in duPro when combining DIFC with user-space protection domains. We first introduce a running example, which will be used to illustrate concepts in duPro.

4.1 A Running Example

We consider a grading application that runs on a college server shared by users including professors and students. Each student submits his/her homework to the professor of a course. After the submission deadline, the professor launches a grading application to grade submissions. The grading application consists of two modules: the control module (CM) and the grading module (GM). The CM module launches GM and iterates a list of students. For each student, it sends the file name of the submission to GM, waits for the completion of grading before proceeding to the next student. GM receives the file name of a student submission, grades the submission, and stores the grading result in a file.

For this application, we assume CM is simple and trusted, while GM is untrusted and may contain bugs that are exploitable by malicious student submissions. We desire a security policy that the grading application should not leak any student's submission or grade to other students even if the GM module has been taken over by an attacker. When both CM and GM are put into one protection domain, the policy is not enforced as an attacker can hijack the application using a malicious submission and leak other students' submissions and files.

We show how this policy can be enforced within duPro. As discussed before, each duPro-protected application comes with a configuration file, which declares the number of domains, their communication channels, and initial security labels. Figure 1 presents (part of) the configuration of the duPro-protected grading application. CM runs in a trusted domain, declared in a `Trusted_PD_Type` block; GM is in an untrusted domain, declared in a `PD_Type` block. CM exports a CDC function `next_submission`, which is invoked by the GM domain. The `Labels` block declares initial labels for domains and external objects. The GM domain has an initial label $(\{\},\{\},\{\})$.

Being a trusted domain, the CM domain is blessed with the ability to dynamically adjust security labels of other domains and external objects. A duPro application often needs a small trusted domain for controlling the setting of security labels. Figure 2 visualizes the grading application and its label settings at the moment when the application is grading the submission of student s_1. For simplicity,

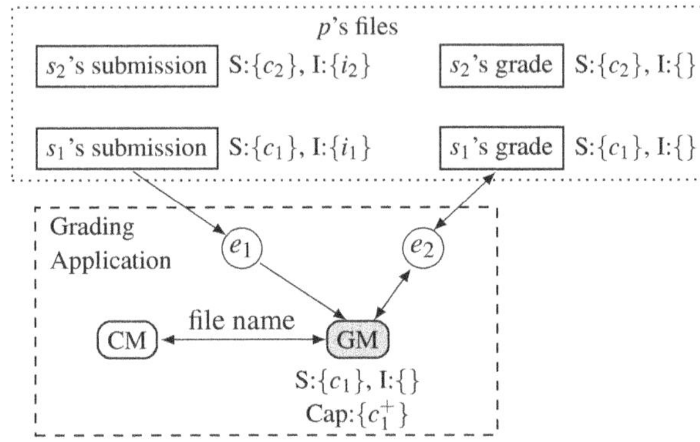

Figure 2: duPro-protected grading application.

only two students are included in the figure: student s_1 and s_2. In this graph and others, we use circles for endpoints, square boxes for external objects such as files, and rounded boxes for domains. An endpoint has the same security label as the domain it is attached to when there is no label for the endpoint in a graph.

The CM domain assigns a submission-specific label to each submission. The submission of student s_1 has label $(\{c_1\},\{i_1\})$ and the submission of student s_2 has label $(\{c_2\},\{i_2\})$. Intuitively, tag c_k stands for s_k's secret submission and tag i_k stands for data endorsed by s_k. Before the CM domain sends the GM domain the submission file name of s_1, it creates one grade report file with label $(\{c_1\},\{\})$ and grants capability c_1^+ to the GM domain. When grading the submission of s_1, the GM domain adds c_1 to its secrecy label so that it can read the submission file; it then performs the grading and outputs the grade to the grade report. When grading s_1's submission, the security label of the GM domain prevents it from touching s_2's submission. Therefore, even if the GM domain is hijacked, the attacker cannot read or write other students' files.

There is one issue that the above setup has not addressed. After grading s_1's submission, the GM module is supposed to grade s_2's submission. However, the GM module has already been tainted with data from s_1. Simply giving the GM module the capability of c_2^+ would allow a malicious submission from s_2 to possibly learn information from the submission of s_1. One solution is to allocate a new GM domain from scratch for grading s_2's submission. This ensures separation between submissions, but performance would be a concern as multiple GM domains have to be launched when many students' submissions need to be graded. Another solution is to ask the CM domain to perform erasure of sensitive information from GM, downgrade GM's security label, and then grant capability c_2^+ to GM for grading s_2's submission. This would work and reuse the GM domain. However, its drawback is that what information to erase is highly application-specific and the CM domain may forget erase part of the sensitive information. duPro's solution is to introduce a checkpointing-restoration mechanism, which is a principled way of performing erasure and downgrading.

4.2 Checkpointing and Restoration

Two API functions are implemented in duPro: one for checkpointing and one for restoration.

(1) `ST_HANDLE checkpoint();`

(2) `void restore(ST_HANDLE s);`

The `checkpoint` function remembers the current state of the calling domain. The domain state includes the state of its memory pages, its security label, context, and its control data. It returns a

```
1:   init();
2:   ch = checkpoint();         // ({}, {}, {})
3:   f = next_submission();  // ({}, {}, {c_k^+})
4:   cap = get_ownership();
5:   change_label(Secrecy, cap.plus[0]);
6:   grading(f);              // ({c_k}, {}, {c_k^+})
7:   restore(ch);
```

Figure 3: Pseudocode of the reusable GM domain. For understanding, we add security labels as comments at certain program states.

handle, which is later used by the `restore` function to restore the domain to its last checkpointed state.

It is important to note that these two functions are not privileged and can be used even by an untrusted domain. The reason is that checkpointing not only remembers the memory pages, but also the security label of the domain. If the domain is in a *consistent state* during checkpointing, meaning its security label accurately reflects the taint levels of data in those memory pages, then it is safe to go back to this state after the domain has evolved into another consistent state. It is safe even if there are side effects such as writing to external files between checkpointing and restoration because the interaction between the domain and external files is already governed by the DIFC model.

With the checkpointing-restoration mechanism, the GM domain in the grading application can be reused safely. The pseudocode of the GM domain is presented in Figure 3.

After initialization (line 1), the code calls `checkpoint` (line 2) to take a snapshot of its state, including the security label, memory pages and others. Since no grading work has been performed at this point, it is in a clean state with label $(\{\},\{\},\{\})$. Then the GM domain calls `next_submission` (line 3) exported by CM to retrieve next student's submission file name from the CM domain. The CM domain also changes GM's capability set to add c_k^+, assuming c_k is next student's secrecy tag. After GM receives the file name, it queries its own capability set (line 4), adds c_k to its secrecy label (line 5), and then performs grading (line 6). After grading, it calls `restore` (line 7) to revert to the last checkpointed state. Consequently, the memory pages tainted when grading the student's submission are rolled back to the clean state. Note that the `restore` operation also restores the program counter to one that follows the `checkpoint` operation and therefore GM's control continues at line 3.

The attacker can hijack the GM domain when grading (line 6), but the GM domain can neither read nor write other students' files. The CM domain checks whether the security label of the GM domain is restored before granting next student's secrecy tag, and hence if the GM domain does not restore itself after grading, the CM domain detects this and rejects the capability granting for the next student.

We next discuss how checkpointing is implemented in duPro. One way is of course to back up the entire domain state, including all memory pages. However, it performs unnecessary copying when the domain modifies only a subset of the memory pages between checkpointing and restoration. duPro adopts lazy copying. When checkpointing a domain, all domain state (the security label, context, and control data) except for memory pages is copied. Memory pages of the domain are set to read only through page protection, but are not copied at the moment of checkpointing. The domain continues execution and a `SIGSEGV` signal is generated when it writes data to a memory page. The signal is handled by duPro's runtime to copy the memory page and set the page to be writable, and then the domain code resumes performing the write to the original page. Restoration is implemented by restoring those copied memory pages. Note that only one signal is generated for an entire page when the page is first written. Therefore, if the domain writes a small number of pages between checkpointing and restoration, only those written pages are copied and restored.

The checkpointing-restoration mechanism does not forbid concurrency in a domain. Note that only the control of the thread that calls the `restore` is reverted and therefore the thread synchronization in a domain is left to the application.

4.3 Application-Controlled Label Setting

Process-level DIFC systems such as Flume use system-wide persistent labels for external objects such as files. Objects have security labels shared by all applications. The advantage is that the security policy can be enforced even after the system is shut down or crashes and later recovered. This implies that (1) the system provides persistent labels so that it is fail-safe; (2) the system also provides a mechanism to help rebooted processes to regain their security labels.

On the other hand, system-wide persistent labels have disadvantages. First, persistent labels cost extra storage. Second, system-wide persistent labels are inflexible and complicates the label setting of applications. We believe each application should be responsible for how it sets labels of objects to best suit its security goal. A persistent label setting is unlikely to work for all applications.

Instead of persistent labels, duPro allows applications to set their own labels of domains and external objects such as files and the network. Those labels are stored in duPro's memory. They are temporal and only durable during the execution of an application. Two applications running concurrently have two instances of duPro and can therefore have different labels for the same object. As a result, one duPro instance may forbid information flow between two files while another may allow such flow. This design is justified by duPro's goal, which is to regulate information flow during the run of an application according to an application-controlled policy; it is not to regulate information flow of the whole system. Though duPro does not maintain persistent labels, an application can still achieve application-level persistent labels. Its trusted domain can store in-memory labels to non-volatile storage before the application exits, and restores the labels when the application reloads.

All objects, including domains, files and the network have a default label. By default, a file f has label $(\{f_S\},\{f_I\})$, where f_S and f_I are tags unique to the file. Similarly, a domain d has label $(\{d_S\},\{d_I\})$ as well as an empty capability set, where d_S and d_I are unique tags. The network and terminal are with the default label $(\{\},\{\})$. The default labels can be overridden in multiple ways. The configuration file can declare initial labels for domains and other objects. A trusted domain can also set labels of other domains.

4.4 duPro API

duPro provides API functions to domains to interact with other domains and the duPro runtime. Table 1 presents a subset of the API functions. There are two groups: OS system-call wrappers and duPro-specific API functions. For untrusted domains, some API functions such as those performing I/O are related to information flow and their invocation has to be *checked* according to the DIFC rules by the duPro runtime. Some APIs are *unchecked* because they are unrelated to information flow; for example, `gettimeofday` queries the current time and it is unchecked. Other APIs such as `fork` are *forbidden* since processes are not allowed. Instead, duPro provides `pd_dup`, which allows the creation of a child domain that duplicates the parent domain. It also provides

Category	API functions	
	System-call wrappers	duPro-specific
Unchecked	gettimeofday, dup, dup2, exit, fstat, ftruncate, fcntl, flock, fsync, getuid, setsockopt, lseek, bind, close, select, mmap, socket, ...	eu_create, com_create, create_tag*, checkpoint, restore, ...
Checked	open, read, write, recv, send, stat, accept, connect, readlink, access, link, mkdir, rmdir, getcwd, chdir, chmod, ...	get_label*, com_open, get_ownership*, change_label*, pd_dup, reduce_ownership*, get_fd_label*, change_fd_label*, pd_create, ...
Forbidden	fork, execve, ...	trusted_change_label, setgid, seteuid, setuid, setpgid, setgroups, ...

Table 1: duPro API functions. Those marked with * provide the same semantics as their Flume counterparts.

`pd_create` for creating a new protection domain and loading code from an object file; in essence, `pd_create` is similar to `fork` followed by `execve` except that an SFI domain is created instead of a process.

All API functions are available to a trusted domain and no check is performed. In addition, a trusted domain can change the security label of an object by invoking `trusted_change_label`. We explain part of duPro-specific API functions next.

- `int trusted_change_label(TYPE t, OBJECT_ID obj_id,` $\{(S,I,Cap)\}$`)`
  Change the security label of an object of type `t`, identified by `obj_id`, with the specified security label. This API function can be used to change labels of protection domains, files, directories, and other objects.

- `PD_HANDLE pd_create(PD_Type pt, char* exec, char* argv[], char *env[], COM_HANDLE ch,` (S_c,I_c,Cap_c)`)`
  Create a domain c of type `pt` with the object code file `exec`, the initial arguments, the environment variables, the created communicator, and optionally the specified security label. Suppose the parent p's security label is (S_p,I_p,Cap_p), it is required that $S_c \subseteq S_p \cup D_p$, $I_c \subseteq I_p \cup D_p$ and $Cap_c \subseteq Cap_p$. If c's security label is not specified, the default one (defined either by the application or the duPro runtime) is granted.

- `PD_HANDLE pd_dup()`
  Create a child domain with the same type and state as the parent. The context, memory pages, security label and file descriptors are copied to the child. The program counter of both the parent and the child is set to the instruction right after this call.

- `COM_HANDLE com_create(FLAG f)`
  Create a communicator and return a handle to it. The argument `f` indicates whether the communicator is *unidirectional* or *bidirectional*.

- `int com_open(COM_HANDLE ch, END e)`
  Return the logical read or write end of a communicator, depending on the second argument `e`, which is either *read* or *write*. The first argument `ch` is a created communicator handle.

4.5 Covert Channel Control

Besides explicit information flow through communication channels, information can also implicitly flow between domains through covert channels. Complete elimination of covert channels is difficult (if not impossible) because any resource sharing between domains might form covert channels. duPro limits covert channels in similar ways as previous process-level DIFC systems. For completeness, we list how certain covert channels are controlled and also list those uncontrolled covert channels.

Inter-domain communication. In duPro, communication can be either synchronous or asynchronous. Cross-domain calls (CDC) provide synchronous communication while communicators provide asynchronous unidirectional or bidirectional communication. A CDC can go through only when the security labels of the caller and callee domains allow bidirectional information flow, even if the callee returns no result. It is because the callee can control the response time—a timing channel. When using communicators, the operation of writing data to a communicator returns immediately with a success return code. However, communicators are unreliable in that data may fail to reach the reader domain and there is no failure notification; a notification would allow the reader domain to transmit one bit to the writer domain. Finally, if domains involved in inter-domain communication change their security labels during runtime, the change may result in the shutdown of communication channels as the new labels may no longer allow communication according to the DIFC model.

File descriptors. duPro protection domains reside in the same process. All domains share a single process-level file descriptor table. Without control, file descriptors could be used as covert channels between domains. File descriptors are allocated sequentially in OSes such as Linux. In addition, Linux sets a small limit on the size of the file descriptor table (e.g., 1024) and as a result the table entries can be exhausted in a short period of time. Both the predictability of file descriptors and possible exhaustion of the file descriptor table can allow one domain to observe the behavior of other domains. To limit such covert channels, duPro adds a layer of indirection. It assigns to each protection domain a domain-specific, logical file descriptor table, which records the correspondence between logical file descriptors and real file descriptors. Logical file descriptors are provided to a domain as results of API functions such as `open`. The duPro runtime performs translation from logical to real file descriptors before invoking OS system calls that require file descriptors. Logical file descriptors of a domain are allocated sequentially but they are not observable to other domains.

Tags/Handles. Tags and handles are created for entities such as protection domains and communication channels. As the tag/handle space is shared by all domains, predicable allocation of tags and handles would allow one domain to observe the number and the rate of tag/handle creation by a different domain. In duPro, tags and handles are 64-bit integers and are generated by using a block cipher (AES specifically) to encrypt a counter.

Memory/CPU/Storage/Network. duPro's protection domains do not share memory and one domain cannot observe other domains' memory. However, one covert channel may arise when an attacker domain tries to exhaust memory of duPro's runtime. In this scenario, one domain might dynamically create many domains, which requires memory allocation from duPro's runtime. It can then use creation and destruction of protection domains to exhaust and release duPro runtime memory to transmit one bit at a time to another domain. This channel is uncontrolled in duPro.

Covert channels may also be possible because of shared resources such as CPU, storage, and network. duPro is a user-space implementation so that it has no control over CPU scheduling or cache usage, for which the OS kernel is responsible. In addition, the CPU cache is shared by all duPro domains and it can be used as timing channels. duPro also does not control covert timing channels that are possible because of the shared network and the shared storage (e.g., the disk). Since duPro intercepts system calls, it would be possible to impose quotas on network and storage usage of a domain to limit covert channels.

4.6 Implementation

duPro is implemented on top of uPro. Its current implementation targets 32-bit Linux and is implemented in C++. Security and integrity labels are represented by C++ STL sets. The capability set is represented by two sets of tags: the plus set contains the set of tags that can be mounted and the minus set contains those tags that can be dropped. Each domain's control structure has a field representing its security label. External objects such as a logical file descriptor are also associated with security labels when they are created.

The implementation has around 1,500 lines of code for the DIFC model and around 500 lines of code for implementing checkpointing and restoration. The total runtime of duPro is about 26,000 lines of C code, together with around 6,500 lines of wrappers of duPro trampolines, which facilitates porting application programs.

5. EVALUATION

In evaluating duPro we consider whether duPro improves application security and how much performance penalty is imposed on ported applications. For evaluation we ported wget and Apache to run them in duPro.

To evaluate performance, we first carried out a set of micro-benchmarks to test the latency for a domain to invoke OS system calls and inter-domain communication. We also measured the performance of the duPro-protected wget and Apache and compared with their native counterparts.

To evaluate security benefits, we searched known vulnerabilities for wget and Apache and analyzed whether attackers could still succeed in exploiting such vulnerabilities in duPro-protected wget and Apache. Successful exploits mean that given the same threat model and security policy an attacker can perform the same attacks and cause the same damage on duPro-protected applications as on native applications. If the same attacks fail on duPro-protected applications, the vulnerability is contained.

All evaluation was performed on a system running 32-bit Ubuntu 12.04 with the Linux kernel version 3.2.0, the Ext4 file system, an Intel Core i7 870 CPU, and 4GB of memory. When testing Apache's performance, a client machine was set up with 64-bit CentOS based on the Linux kernel version 2.6.32, an Intel Xeon X5550 CPU, and 12GB of memory. The average latency between the server and the client is 0.23 ms.

5.1 Micro-benchmarking

As discussed before, duPro runs application code inside SFI-enforced domains. Application code can invoke allowed OS system calls through trampolines. Trampolines responsible for system calls are mapped to system-call wrappers, which perform DIFC checks and invoke OS system calls if checks succeed. Both the trampoline calls and the wrapper functions incur overhead.

Operations	duPro multiplier	Flume multiplier
`open`		
`-create`	1.4	16
`-exists`	2.1	12.5
`-non-exists`	5.5	9.5
`close`	2.7	1.3
`stat`	2.6	13.7
`unlink`	1.2	7.2
`readlink`	2.4	33.0
`mkdir`	1.0	4.3
`rmdir`	1.1	7.7
`IDC latency`	1.2	8.2

Table 2: Normalized operation latency in uPro and Flume.

We used a set of micro-benchmarks to measure duPro system-call overhead. In the experiment, we also compared duPro's system-call overhead with Flume's system-call overhead. Table 2 presents the results. The column of duPro multiplier presents the latency of performing a system call in duPro, normalized by allowing code to invoke Linux system calls directly (that is, without going through SFI and DIFC checks, which is unsafe). The `open` system call was tested in three modes: in the `create` mode, a new file is created; in the `exists` mode, an existing file is opened; in the `non-exists` mode, it tries to open a non-existent file and fails. All the system calls were performed 10K times and the average is presented.

Also presented in the table is Flume's multipliers for those system calls, extracted from the Flume paper. In general, duPro imposes much less system-call overhead than Flume, due to the implementation differences. Flume's protection domains are based on processes. It injects a kernel module to intercept system calls and redirects certain system call requests to a reference-monitor process. Handling a system call in Flume involves multiple process-level context switches: for instance, from the user process to the reference-monitor (RM) process, from the RM process to its file-server process, from the file-server process back to the RM process, and from the RM process back to the user process. The procedure involves multiple heavyweight process context switches. By contrast, duPro runs each module in a user-level sandbox and only two lightweight domain-level switches are needed to invoke a system call. Therefore, duPro's support of protection domains is lightweight and its handling of system calls is more efficient.

We also measured the inter-domain communication latency using two unidirectional communicators by transferring one byte back and forth between two domains one million times. The average is then calculated and normalized by the case of sending one byte using pipes between processes. The result and its comparison with Flume is presented in the last row of the table.[1] duPro performs better than Flume because the Flume RM process mediates inter-process communication. In Flume, a sending process sends data to the RM process and the RM then sends the data to the receiving process. This involves two process context switches. Furthermore, the RM process needs to queue the data-sending requests, which further increases the latency.

[1]duPro communicators are implemented on top of OS pipes and socketpairs, which is why the communicator performance is slower than pipe performance. An alternative implementation, which implements communicators in the duPro runtime, would be more efficient than pipes as it would not go through the OS for communication. The uPro paper [18] shows that cross-domain calls, which is implemented by the uPro runtime without going through the OS, is an order of magnitude faster than remote procedure calls.

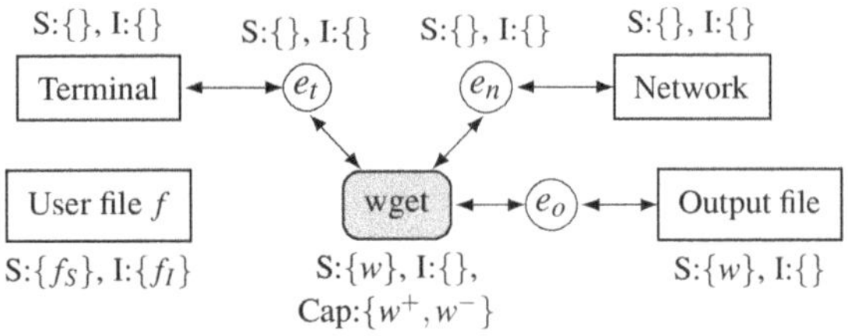

Figure 4: wget's label setting at runtime.

5.2 wget

wget is a command-line tool and commonly used to download files stored on a remote HTTP or FTP server. It parses the input URL of a remote file, connects to the server, creates a local file, retrieves the file contents, and writes the contents to the local file. We assume a threat model in which the remote server may hijack wget by sending malicious files. The desired security policy is that wget should neither taint existing user files nor upload any user data to the Internet, even if wget's implementation contains bugs. Obviously, the native wget fails to meet the security policy.

We ported wget version 1.14 to run in duPro and enforced the above security policy by restricting information flow. A process in the native wget was mapped to a duPro protection domain and an execution unit; this required an appropriate configuration file. wget's build scripts were also changed so that the default building toolchain was replaced by the duPro toolchain. We call this version duPro-wget.

No trusted domain is needed to enforce the desired security policy. Figure 4 shows the label setting of duPro-wget at runtime. The terminal, network, and other user files are associated with their default labels: the terminal and network are associated with label $(\{\}, \{\})$, and each user file f is associated with $(\{f_S\}, \{f_I\})$. In the configuration file, the initial label of the wget domain is $(\{\}, \{\})$, and so is the directory `dir` that holds the output file. The execution steps of duPro-wget are as follows:

1. The wget domain is given the initial label $(\{\}, \{\})$. The domain then dynamically creates a tag w, which is associated with data downloaded from the remote server. Since wget owns the tag, duPro adds $\{w^+, w^-\}$ to wget's capability set. Then wget adds w to its secrecy label and creates an output file with the same security label as its label in directory `dir`.

2. wget connects to the network, retrieves the remote file, and writes to the output file. After finishing the writing, wget reports success to the terminal and exits.

During the execution, the wget domain cannot read or write other user files. A user file f has label $(\{f_S\}, \{f_I\})$. Its secrecy label prevents wget from reading the user file and its integrity label prevents wget from writing to the user file. Therefore, even if wget is hijacked, the security policy is not violated.

To evaluate the security benefits of duPro-wget, we searched for past vulnerabilities of wget and investigated whether these vulnerabilities would be mitigated. We found 9 vulnerabilities [1] in Linux wget. Our analysis concluded that 6 out of 9 vulnerabilities would be contained. For example, vulnerability CVE-2002-1565 allows an attacker to execute arbitrary code and read/write arbitrary files with respect to wget. With duPro-wget, it would be contained and could not touch arbitrary files. The vulnerabilities that would not be contained are irrelevant to information flow control. For instance, vulnerability CVE-2006-6719 allows a remote attacker to perform denial-of-service attacks by sending a response message of large size to crash wget. Exploits of this vulnerability does not violate information flow control and are not contained.

File Size	50MB	100MB	150MB	200MB
Native wget	0.56s	1.13s	1.74s	2.29s
duPro-wget	0.58s	1.15s	1.77s	2.32s
Slowdown	4.4%	1.7%	1.5%	1.4%

Table 3: wget performance on files of different sizes.

To evaluate the performance overhead of duPro-wget, we compared the remote-file retrieval time between the native wget and duPro-wget. Results are shown in Table 3. Data files of various sizes were stored on a remote HTTP server. For each file, five tests were performed and the average time is presented. The results show that duPro imposes negligible overhead on wget.

5.3 Apache

Apache is an HTTP server running as a privileged program. On Linux, Apache can be configured with various Multi-Processing Modules (MPM) managed by a running process called Apache kernel at runtime. In the Prefork MPM, the Apache kernel forks a set of worker processes and each of them handles one connection at a time. When a worker handles a request, it logs the connection information (e.g., the IP address), sends web pages to the client, and processes the next one. If an attacker hijacks a worker, information of previous or subsequent connections handled by the same worker is leaked. The Prefork MPM can be configured so that a worker handles exactly one connection (called the Fork MPM). However, the Fork-MPM mode incurs frequent worker process creation and destruction, which impacts the performance.

We ported Apache (version 2.2.21) with the Prefork MPM to duPro. We call this version duPro-Apache. In duPro-Apache, both workers and the Apache kernel are untrusted. Instead, there is a trusted wrapper (around 90 lines), which is responsible for setting and adjusting labels of the workers and the kernel. In native Apache, a worker is created in the Apache kernel by invoking the system call `fork`. As `fork` is forbidden in duPro, it was replaced by the API function `pd_dup`. Besides, code for communication between a worker and the trusted wrapper via CDCs was added (sending log entries and retrieving labels for the next connection). In total, the porting required us to add around 140 lines of code and adjust the build scripts. The following security policy is enforced in duPro-Apache.

1. Client connections are isolated. This implies: (1) information should not flow between different workers; (2) if a worker handles multiple connections, the information of one connection should not flow to another connection.

2. Workers are only allowed to read web pages. Low-integrity data from the network should not flow to the web pages.

The policy is not enforced by the native Apache. duPro-Apache can enforce this policy with appropriate label settings. Figure 5 shows the label setting of duPro-Apache when handling requests. The initial labels of web page files, logs, Apache kernel and worker domains are specified in a configuration file. The trusted wrapper domain performs dynamic label changes. All the web pages have label $(\{p_S\}, \{p_I\})$. The logs have label $(\{l_S\}, \{l_I\})$. Each domain has label $(\{p_S\}, \{\})$ when created. Each connection is modeled as an individual object. A secrecy tag is allocated for data received from a connection. For instance, connection C_x's secrecy tag is x and the connection's label is $(\{p_S, x\}, \{\})$.

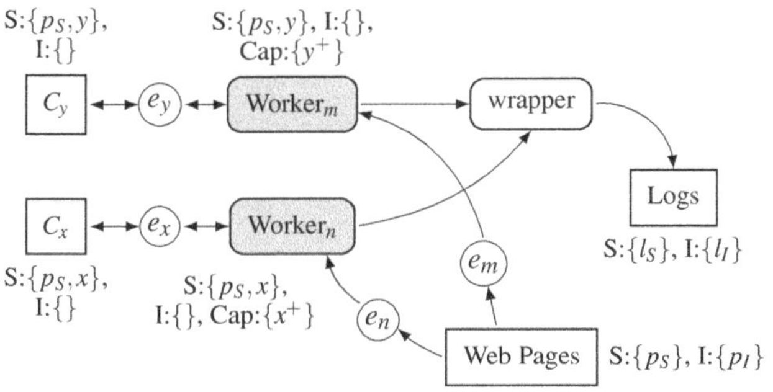

Figure 5: Label setting for Apache at runtime. The Apache kernel is not shown because, after worker domain creation, it enters into an idle state and does not handle connections.

```
1:   init();
2:   s = checkpoint();              // ({ps}, {})
3:   next_connection();             // ({ps, x}, {x+})
4:   d = accept();
5:   handle_conn(d);
6:   do_log();
7:   restore(s);
```

Figure 6: The pseudocode showing how an duPro-Apache worker domain handles connections.

In the Prefork MPM mode, each worker domain handles multiple connections. To ensure isolation between connections, duPro-Apache's worker code was modified to use duPro's checkpointing-restoration mechanism. Figure 6 shows the pseudocode for demonstrating how a duPro-Apache worker domain handles connections. The following steps happen during a worker's execution:

1. A worker domain is created with labels $(\{p_S\}, \{\})$. The `init` function (line 1) initializes the worker. The worker then calls `checkpoint` (line 2) to checkpoint itself.

2. The worker calls `next_connection` (line 3) to send a capability change request to the wrapper domain and the wrapper domain grants x^+ to the worker's capability set. Afterwards, the worker adds x to its secrecy label.

3. The worker accepts connection C_x by calling `accept` (line 4) and processes it by calling `handle_conn` (line 5).

4. The worker finishes processing and calls `restore` (line 7) to revert to the last checkpointed state; it then starts to handle the next connection.

In this setup, a worker cannot communicate information of its connections to another worker that has handled other connections. As Figure 5 shows, once two different workers have learned information from their connections, they are tainted with labels of different secrecy tags. This prevents communication between workers.

If two connections are consecutively processed by the same domain, shown in Figure 7, no information should flow between the two connections. In the figure, the worker handles C_{x_1} after handling C_x. The worker is supposed to call `restore` to restore its state so that C_x-related data are purged by the duPro's restoration mechanism and cannot be mixed with data from C_{x_1}. If the worker has been hijacked by C_x and it calls `next_connection` without first invoking `restore`, its secrecy label remains $\{p_S, x\}$. In this case, the trusted wrapper checks the worker's security label and refuses to grant the capability x_1^+ to the worker.

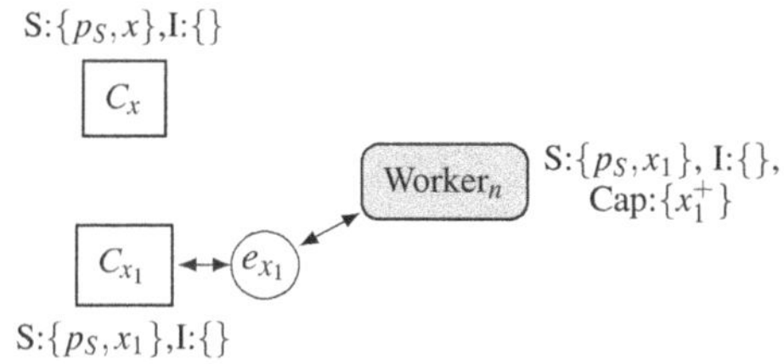

Figure 7: Label setting for an Apache domain handling consecutive connections. The worker can communicate with the current connection C_{x_1}. However, it cannot communicate with the previous connection C_x.

Native Apache's worker processes write logs to two files (for error logging and connection logging). The sharing of the log files forms covert channels between workers and between connections based on log-written speed and size. A possible approach to limiting such covert channels is that each worker domain writes a separate log file for each connection. However, this is impractical because a large number of small log files would be generated. This would waste the disk space and burden the file system. We moved the Apache logging code (only the part performing the I/O) to the wrapper domain (by adding around 60 lines) and exported CDCs for worker domains to invoke. The function `do_log` at line 6 sends a log entry to the wrapper domain, which truncates the log entry to at most 256 bytes and writes it to the log files. This limits the covert channels based on disk space exhaustion.

To evaluate duPro-Apache's security benefits, we inspected 36 publicly known Apache vulnerabilities posted on the Apache website [2]. All vulnerabilities are with respect to the Prefork-MPM Apache on Linux. After analyzing each vulnerability, we concluded duPro-Apache can contain the damages of 10 out of 36 vulnerabilities. Most of the vulnerabilities that cannot be contained are Denial-of-Service (DoS), which are out of the capability of duPro.

We measured and compared the performance of native Apache and duPro-Apache, in terms of the throughput and per-request latency. Native Apache was configured with the Prefork model and the Fork model. We configured all servers so that the number of workers at runtime was fixed. For duPro-Apache, the number of workers is the number of SFI domains created during initialization. For native-Apache-Prefork, the number of workers is the number of processes forked during initialization of Apache. for native-Apache-Fork, the number of workers refers to the maximum number of live worker processes at a given time. Since the Fork MPM creates one separate process to handle a connection, processes get created and destroyed dynamically.

Given a fixed number of workers, we tested the performance of the three versions by requesting a static 10KB page 10,000 times with different numbers of concurrent requests and took the geometric mean. Figure 8 shows the performance results of three Apache servers, including throughput and per request latency. The performance of duPro-Apache increases as the number of worker domains grow and maintains a relatively stable throughput. Its performance is comparable to native-Apache-Prefork. It even outperforms native-Apache-Prefork when the number of workers is large. The performance of native-Apache-Prefork has a slight decrease when the number of processes grow; we believe that is because of the process scheduling. Native-Apache-Fork has the same trend as duPro-Apache, but due to the expensive process creation, it performs much worse than the other two versions.

To evaluate the performance benefits of reusing workers in duPro-

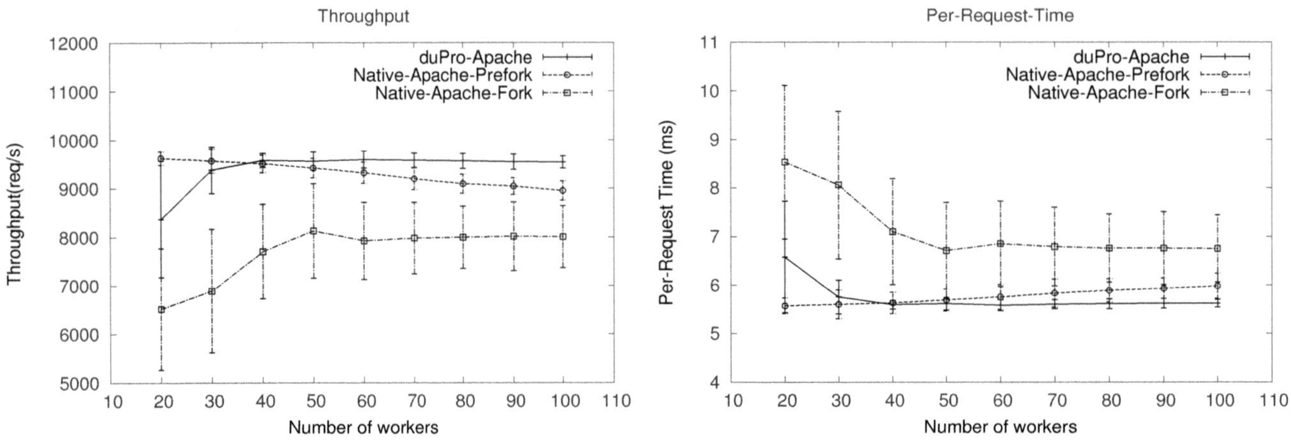

Figure 8: The performance comparison between duPro-Apache and native Apache. The left figure shows the throughput information and the right figure shows the per-request latency.

Apache, we also tested a version that recreates a worker SFI domain for each connection. We found the throughput of this version is three orders of magnitude worse than duPro-Apache because in the non-reuse version duPro spends almost all the time on creating and collecting protection domains. This demonstrates that reusing domains is critical to acceptable performance in duPro-Apache.

5.4 Summary of experiments

The experimental results of duPro are encouraging. The results demonstrate the efficiency of protecting user programs using duPro. In both applications, the performance of duPro-protected applications is close to its native counterpart, and with the enhanced security. Though duPro requires porting of application code, experiments demonstrate porting is straightforward to perform in practice.

6. CONCLUSIONS

duPro is an efficient user-level DIFC enforcement framework. duPro applies SFI to isolate protection domains in the user space. Each protection domain is associated with a DIFC label in order to track information flow. duPro neither requires OS changes nor root privileges when used, and thus benefits from the OS updates and new driver support. Thanks to SFI, duPro provides high performance IDC mechanisms. duPro also provides a checkpointing-restoration mechanism for efficiently reusing protection domains. The evaluation shows that the effort spent on porting applications to duPro is minimum and the performance overhead imposed by duPro on ported applications is negligible. At the same time, duPro-protected applications can use DIFC to provide end-to-end security guarantee on information flow.

7. ACKNOWLEDGMENTS

This research is supported by US NSF grants CCF-0915157, CCF-1149211, CCF-1217710, and two research awards from Google.

8. REFERENCES

[1] `http://web.nvd.nist.gov/view/vuln/search`. [Online; accessed on 21-October-2012].

[2] `http://httpd.apache.org/security/vulnerabilities_22.html`. [Online; accessed on 22-October-2012].

[3] AppArmor. `http://wiki.apparmor.net`.

[4] SELinux. `http://selinuxproject.org`.

[5] ABADI, M., BUDIU, M., ERLINGSSON, Ú., AND LIGATTI, J. Control-flow integrity. In *12th ACM Conference on Computer and Communications Security (CCS)* (2005), pp. 340–353.

[6] AKRITIDIS, P., CADAR, C., RAICIU, C., COSTA, M., AND CASTRO, M. Preventing memory error exploits with wit. In *IEEE Symposium on Security and Privacy (S&P)* (2008), pp. 263–277.

[7] BELL, D., AND LAPADULA, L. Secure computer system: Unified exposition and multics interpretation. Tech. Rep. ESD-TR-75-306, MITRE Corp., Mar. 1976.

[8] CASTRO, M., COSTA, M., MARTIN, J.-P., PEINADO, M., AKRITIDIS, P., DONNELLY, A., BARHAM, P., AND BLACK, R. Fast byte-granularity software fault isolation. In *ACM SIGOPS Symposium on Operating Systems Principles (SOSP)* (2009), pp. 45–58.

[9] DENNING, D. E. A lattice model of secure information flow. *Communications of The ACM 19* (1976), 236–243.

[10] EFSTATHOPOULOS, P., KROHN, M., VANDEBOGART, S., FREY, C., ZIEGLER, D., KOHLER, E., MAZIÈRES, D., KAASHOEK, M. F., AND MORRIS, R. Labels and event processes in the Asbestos operating system. In *ACM SIGOPS Symposium on Operating Systems Principles (SOSP)* (2005), pp. 17–30.

[11] ERLINGSSON, Ú., ABADI, M., VRABLE, M., BUDIU, M., AND NECULA, G. XFI: Software guards for system address spaces. In *USENIX Symposium on Operating Systems Design and Implementation (OSDI)* (2006), pp. 75–88.

[12] FORD, B., AND COX, R. Vx32: Lightweight user-level sandboxing on the x86. In *USENIX Annual Technical Conference* (2008), pp. 293–306.

[13] FRASER, T. LOMAC: Low water-mark integrity protection for COTS environments. In *Proceedings of the 2000 IEEE Symposium on Security and Privacy* (2000), pp. 230–245.

[14] KROHN, M., YIP, A., BRODSKY, M., CLIFFER, N., KAASHOEK, M. F., KOHLER, E., AND MORRIS, R. Information flow control for standard OS abstractions. In *ACM SIGOPS Symposium on Operating Systems Principles (SOSP)* (2007), pp. 321–334.

[15] McCamant, S., and Morrisett, G. Evaluating SFI for a CISC architecture. In *15th Usenix Security Symposium* (2006).

[16] McIlroy, M. D., and Reeds, J. A. Multilevel security in the UNIX tradition. *Software Practice and Experience 22* (1992), 673–694.

[17] Myers, A., and Liskov, B. Protecting privacy using the decentralized label model. *ACM Transactions on Software Engineering Methodology 9* (Oct. 2000), 410–442.

[18] Niu, B., and Tan, G. Enforcing user-space privilege separation with declarative architectures. In *Proceedings of the seventh ACM workshop on Scalable trusted computing* (New York, NY, USA, 2012), STC '12, ACM, pp. 9–20.

[19] Provos, N. Improving host security with system call policies. In *12th Usenix Security Symposium* (2003), pp. 257–272.

[20] Saltzer, J., and Schroeder, M. The protection of information in computer systems. *Proceedings of The IEEE 63*, 9 (Sept. 1975), 1278–1308.

[21] Small, C. A tool for constructing safe extensible C++ systems. In *COOTS'97: Proceedings of the 3rd conference on USENIX Conference on Object-Oriented Technologies (COOTS)* (1997), pp. 174–184.

[22] Wahbe, R., Lucco, S., Anderson, T., and Graham, S. Efficient software-based fault isolation. In *ACM SIGOPS Symposium on Operating Systems Principles (SOSP)* (New York, 1993), ACM Press, pp. 203–216.

[23] Yee, B., Sehr, D., Dardyk, G., Chen, B., Muth, R., Ormandy, T., Okasaka, S., Narula, N., and Fullagar, N. Native client: A sandbox for portable, untrusted x86 native code. In *IEEE Symposium on Security and Privacy (S&P)* (May 2009).

[24] Yumerefendi, A. R., Mickle, B., and Cox, L. P. Tightlip: keeping applications from spilling the beans. In *Proceedings of the 4th USENIX conference on Networked systems design & implementation* (Berkeley, CA, USA, 2007), NSDI'07, USENIX Association, pp. 12–12.

[25] Zdancewic, S., Zheng, L., Nystrom, N., and Myers, A. Secure program partitioning. *ACM Transactions on Compututer Systems (TOCS) 20*, 3 (2002), 283–328.

[26] Zeldovich, N., Boyd-Wickizer, S., Kohler, E., and Mazières, D. Making information flow explicit in HiStar. In *USENIX Symposium on Operating Systems Design and Implementation (OSDI)* (2006), pp. 263–278.

[27] Zheng, L., Chong, S., Myers, A., and Zdancewic, S. Using replication and partitioning to build secure distributed systems. In *IEEE Symposium on Security and Privacy (S&P)* (2003), pp. 236–250.

SocialWatch: Detection of Online Service Abuse via Large-Scale Social Graphs

Junxian Huang[1] Yinglian Xie[2] Fang Yu[2] Qifa Ke[2]
Martín Abadi[2*] Eliot Gillum[3] Z. Morley Mao[1]
[1]University of Michigan [2]Microsoft Research Silicon Valley [3]Microsoft Corporation

ABSTRACT

In this paper, we present a framework, SocialWatch, to detect attacker-created accounts and hijacked accounts for online services at a large scale. SocialWatch explores a set of social graph properties that effectively model the overall social activity and connectivity patterns of online users, including degree, PageRank, and social affinity features. These features are hard to mimic and robust to attacker counter strategies. We evaluate SocialWatch using a large, real dataset with more than 682 million users and over 5.75 billion directional relationships. SocialWatch successfully detects 56.85 million attacker-created accounts with a low false detection rate of 0.75% and a low false negative rate of 0.61%. In addition, SocialWatch detects 1.95 million hijacked accounts—among which 1.23 million were not detected previously—with a low false detection rate of 2%. Our work demonstrates the practicality and effectiveness of using large social graphs with billions of edges to detect real attacks.

Categories and Subject Descriptors

G.2.2 [**Graph Theory**]: Graph algorithms; K.6.5 [**Security and Protection**]: Authentication

Keywords

SocialWatch, social graph, spam, PageRank, security

1. INTRODUCTION

Virtually all large-scale online services today (*e.g.,* Hotmail, Facebook) have become popular platforms for attackers to conduct a wide range of nefarious activities. Although the precise methods or scales of these attacks may differ, the common thread among them is the requirement of a large number of malicious user accounts. These accounts can be either created anew or obtained via compromising real user accounts. They serve as the channels to propagate spam, malware, social scam, or other malicious contents.

*Martín Abadi is also affiliated with the University of California, Santa Cruz.

ASIA CCS'13, May 8–10, 2013, Hangzhou, China.

Many existing security mechanisms to deal with malicious accounts have extensively studied the attack characteristics [1, 2, 3, 4]. However, as attacks evolve over time, one has to constantly adapt features for detection in the arms race. Thus, two important questions naturally arise: Are there robust features for distinguishing malicious accounts from legitimate ones? If so, are they practical to compute?

In this paper, we explore the use of *social graph features* for detecting both attacker-created accounts and compromised accounts at a large scale. The motivation is that a social graph describes not just the behaviors of individual user accounts, but also the interactions among them and the structures of their relationships. Although attackers can easily adjust or adapt the behaviors of the accounts that they control, they cannot change the behaviors of other legitimate accounts. For example, most legitimate users do not respond to spam emails and attackers cannot force them to reply. Therefore, attackers will have limited effect on the local graph features of legitimate users. Furthermore, it is fundamentally difficult for them to change the overall communication patterns involving other legitimate users. The larger the legitimate user population is, the harder it is for attackers to influence the global social graph structures.

We design and implement a framework called SocialWatch to detect attacker-controlled accounts using social graphs. Unlike previous work that focuses on a specific feature, SocialWatch explores a set of graph properties. Some of them are local graph features based on degree and PageRank. However, we show that directly applying degree and PageRank to our context is not effective as legitimate users have diversified behaviors and attackers can be stealthy. Therefore, we modify these well-known features and use them in combination.

In addition, to address the challenges of detecting hijacked accounts with mixed behaviors of normal and malicious activities, we propose a set of *social affinity features* that not only describe individual account behaviors, but also the behaviors of its contacts and the community structures. We leverage these fine-grained social features and use a Bayesian decision framework for detection. We show that, despite the presence of noise and missing information in the data set, the selected social affinity features can still effectively differentiate hijacked accounts from normal ones. In summary, SocialWatch puts attack detection in a large social context and hence is robust to attacker counter strategies.

To make our solution practical, all the graph features that we select can be efficiently computed using the cloud computing techniques today. We implement SocialWatch on a cluster of 240 machines using Dryad/DryadLINQ [5] and evaluate it using a large dataset with more than 682 million users and 5.75 billion directional relationships. We demonstrate the practicality and effectiveness of our approach by detecting real, large-scale attacks from the dataset. We summarize our key results as follows:

- Using algorithms that combine degree and PageRank features together, SocialWatch detects 56.85 million attacker-created accounts, with a 0.75% false detection rate and a 0.61% false negative rate.
- Using the fine-grained social affinity features, at a false detection rate of 2%, SocialWatch identifies 1.95 million hijacked accounts, among which 1.23 million were not detected previously.

The rest of the paper is organized as follows. We discuss related work in §2. We present our problem formulation in §3 and describe the detection methodology in §4. We evaluate the evaluation results in §5, before concluding in §6.

2. RELATED WORK

Detecting malicious accounts for online services has been an important problem in recent years. Existing studies have proposed to identify correlated abnormal behaviors for detecting such attacks at a large scale [4, 1]. Although these defense systems are demonstrated to work effectively for existing attacks, attackers may evade by modifying the behaviors of multiple malicious accounts simultaneously to make them look different. The goal of our study is to design a more robust social framework by looking at not only individual user behaviors, but also the interactions among all users.

Sybil attacks and defense in social networks have been extensively studied [6, 7, 8, 9]. Most existing Sybil-defense solutions are designed for distributed environments. Further, recent studies show that some of the key assumptions required by Sybil solutions may not hold on real social graphs [10]. In contrast, our approach takes a centralized view of the entire social graph, allowing us to explore more efficient and practical solutions. As we focus on attack detection, our work is also different from the recent work that focuses on identifying legitimate user populations early using social graphs [11].

With Online Social Networks (OSN) growing rapidly, there exist extensive studies for understanding their graph properties [12, 13, 14]. Their findings are important and complementary to our work, where we focus on leveraging their findings and various graph properties for improving security.

There also exist previous studies that explore social graph features for attack detection [15, 16, 17, 18]. The early success of these proposals motivate us to further explore richer graph properties and use them in combination. In addition, we also explore more fine-grained graph features that capture community structures and user social distances for detecting hijacked accounts.

3. PROBLEM FORMULATION

We leverage social graphs to detect both attacker-created accounts and hijacked accounts. Although SocialWatch is a general solution that is applicable to different types of online services, in this paper, we use an email application to drive our presentation due to the use of our dataset.

3.1 Input Dataset

We have access to a large data set collected from Microsoft Hotmail. The dataset contains coarse-grained communication information of anonymized, sampled user accounts from October 2007 to April 2010. Based on whether the email address is registered with Hotmail, we classify accounts into *internal accounts* and *external accounts*. Each entry in the data set contains the communication history summary between a sampled internal user account and another user (either internal or external), including the anonymized email addresses of both users and the number of emails that the two users have mutually sent to each other.

Note that the sampling method is user-based. In other words, if an internal user is sampled, the dataset records all other users that this sampled user has ever communicated with during the data collection period. Thus for each sampled account, we have a complete view of its communication patterns to construct graphs. For other internal accounts and external accounts, we have only a partial view of their communications, derived from the records of the set of sampled internal accounts that they have communicated with.

In total, our dataset contains 5.745 billion directional records involving more than 682 million unique accounts, among which 269.5 million are internal accounts.

3.2 Graph Construction

Given the input, SocialWatch builds two social graphs. On both graphs, a node v corresponds to a unique user, but the edges are defined differently.

The first graph is a *directed email-communication graph* $G_d = (V, E_d)$, which records the detailed email activities between users. Each directed edge $e(v_1, v_2) \in E_d$ represents that node v_1 has sent emails to v_2, and the weight w of this edge indicates the number of emails that v_1 has ever sent to v_2 during the data collection period.

The second graph is an *undirected friendship graph* $G_u = (V, E_u)$, which records only the mutual relationship between users. The edges in E_u are extracted from the friendship patterns of E_d, *i.e.,* if both $e(v_1, v_2) \in E_d$ and $e(v_2, v_1) \in E_d$, and the weight of the two edges are at least 2, we define an undirected edge e' that connects v_1 and v_2 in G_u. By requiring mutual email exchanges on the undirected graph, SocialWatch attempts to exclude connections from malicious accounts to legitimate accounts, as legitimate users usually do not reply to spam emails. The edge weight represents the strength of the connection between two accounts. By requiring an edge to have a weight of at least 2, SocialWatch prunes weak connections due to occasional or accidental email exchanges (*e.g.,* accidentally reply to a malicious account). The constructed G_d has 682 million nodes with degree ≥ 1 and 5.745 billion edges, and G_u includes 255 million nodes and 436.9 million edges.

3.3 Problem Statements and Challenges

Given both the directed email-communication graph and the undirected friendship graph, our goal is to develop methods to detect the set of nodes tat correspond to either attacker-created accounts or hijacked accounts. In [19], we investigate in detail the adversary behaviors of these two types of accounts on the social graphs that we created. We aim to maximize detection coverage while minimizing the false detection rates, *i.e.,* the percentage of legitimate users that are mistakenly marked as malicious over the total number of users that are marked as malicious by our methods. Selecting robust and effective graph properties for malicious account detection is challenging, especially under our practical settings:

- **Incomplete graph.** Any large graphs constructed in practice have only a partial view based on the specific set of accounts (either due to sampling or the limited view from one provider). Therefore, the graph-based detection algorithms must be robust to missing data.
- **Large-scale graph.** As our graph is huge with billions of edges, the selected graph properties need to be computed efficiently on large graphs.
- **Legitimate accounts have diverse behaviors.** Given our graphs have hundreds of million users, the behaviors of legitimate accounts can be highly diversified. For example, some

legitimate accounts, such as mailing-lists, may have thousands of contacts. On the other hand, attackers could choose to send spam to only tens of victims from each malicious account to evade detection. Our method needs to distinguish both aggressive and stealthy malicious accounts from the remaining ones.

- **Hijacked accounts have mixed behaviors**. Traditional graph properties such as PageRank and degrees may not able to provide strong signals. Hence, more fine-grained graph properties are required to detect the subtle differences between hijacked accounts and the remaining legitimate ones.

Due to the above challenges, there is no single graph property that can distinguish malicious accounts from legitimate ones. Therefore, SocialWatch resorts to a set of complementary graph properties. For each account, we are interested in not only its own behaviors, but also the behaviors of its contacts, or even all the other accounts to some extent. Putting the detection in a large social context is the key for the social graph based detection.

4. DETECTION METHODOLOGY

In this section, we present our detection methodology for attacker-created accounts and hijacked accounts.

4.1 Detecting Attacker-created Accounts

There are two most prominent and easy-to-compute graph properties that are widely used for detection: node degree and PageRank. Degree is a *local* graph feature that captures the aggressiveness of an account. PageRank is a *global* graph feature that calculates the weights of the nodes on the overall graph. We next discuss how we modify these two properties in the social context for detection.

4.1.1 *Defense Using Degree*

The straightforward degree-based method of disallowing any account sending emails to more than N recipients may affect the usability of active legitimate users. Therefore, we introduce the *response rate* of an account, *i.e.,* the ratio between the number of replied recipients to the total number of recipients. Note that we use the number of recipients rather than the number of emails because a friend would typically reply to *at least one* emails sent over the entire history, but may not respond to *all* emails. Therefore, the number of recipients is a better metric to quantify friendship than using the number of emails. For spam emails, most users do not reply to *any* of them, so spammers typically have low response rates.

4.1.2 *Defense Using PageRank*

PageRank is widely used for ranking Web pages and recently has been applied to social graphs to detect spammers [17]. Each node is assigned a uniform reputation score initially. In each iteration of the PageRank computation, each node propagates its reputation to neighbors. After the ith iteration, the node A's new reputation score $R_{A,i+1}$ is given as

$$R_{A,i+1} = 1 - d + d \sum_{\{X:e_{XA} \in E\}} \frac{R_{X,i}}{outdegree(X)} \qquad (1)$$

where d is the damping factor usually set to 0.85 [20]. $R_{X,i}$ is the reputation score of node X after the previous iteration, and $\{X : e_{XA} \in E\}$ is the set of nodes in the graph that have directed edges pointing to A.

One method is to directly apply PageRank to the directed G_d, where the PageRank scores propagate along directed edges. Intuitively, the PageRank of a node models the *goodness* of the user, and we call it the *goodness score*. However, even with a response rate as low as 1%, an aggressive spammer who sent thousands of emails may still receive emails from legitimate accounts (*e.g.,* accidental replies) and get a high goodness score. On the other hand, inactive, legitimate accounts who do not receive many emails may have a low goodness score.

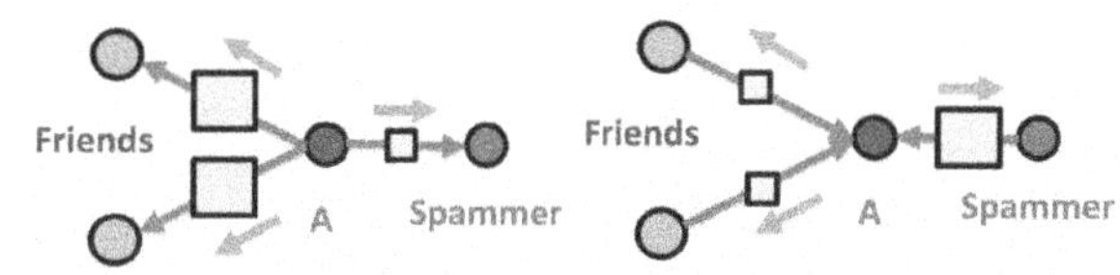

(a) goodness score (b) badness score

Figure 1: Goodness and badness score propagation. The box size is in proportion to the edge weight[1].

To mitigate this problem, we introduce two modifications to this approach. The first is that we adjust edge weights based on email exchange patterns. We use the number of emails exchanged between two accounts to infer this relation. Assume that an account A has n contacts (C_i, $1 \leq i \leq n$), and A sends s_i emails to C_i, while receiving r_i emails from C_i. The percentage ω_i of the goodness score that A should propagate to C_i is defined by

$$\omega_i = \left(\frac{s_i + 1}{r_i + 1}\right) \Big/ \sum_{k=1}^{n} \left(\frac{s_k + 1}{r_k + 1}\right) \qquad (2)$$

Here, we add 1 to s_i and r_i to avoid the divided-by-zero problem. As in Figure 1(a), A would propagate more fraction of her score to friends who have "balanced" mutual email exchanges with her, while less to the spammers who send multiple emails to A with few or zero replies.

The second modification we adopt is the reverse PageRank [21], calculated by running the PageRank algorithm on the social graph obtained by reversing the directions of all edges as illustrated in Figure 1. We name the score of a node obtained in this way a *badness* score. The intuition behind the badness score is that bad accounts normally send many emails, so if we propagate badness scores from recipients to senders, bad accounts would receive high badness scores. Similarly, the edge weight is adjusted based on email exchange patterns so that a node would propagate more of her badness score to spammers and less to her friends.

We compute the ratio of goodness to badness scores by PageRank computation and use this ratio to detect spammers. We further combine the modified PageRank and degree-based approaches to achieve better detection results using the following steps:

- Identify aggressive spamming accounts with high out degrees ($\geq N'$) and low response rates ($\leq \theta$).
- Identify less aggressive spamming accounts using the badness-goodness PageRank ratio ($\geq \alpha$).

The settings of N', θ, α are based on a small set of known good and bad accounts [19].

4.2 Methods for Hijacked Account Detection

The communication patterns of hijacked accounts are different from attacker-created ones, *e.g.,* hijacked accounts will continue to receive emails from their contacts, while attacker-created accounts often do not receive emails from legitimate users. The degree and PageRank-based approaches in §4.1 thus may not catch hijacked accounts. Instead, we explore fine-grained graph properties and use two social-affinity features to detect hijacked accounts.

[1] Circles in the graph represent user nodes. Boxes denote the propagation of goodness/badness score and the box size is proportional to the edge weight.

4.2.1 *Social-Affinity Features*

We introduce social-affinity features defined for each user v, aiming to capture the subtle differences between hijacked accounts and legitimate accounts.

Recipient connectivity r: A legitimate user usually belongs to one or a few communities in the real world based on geographic locations, working connections, or interest, and so are her recipients. For example, we have observed strong evidence of geographic locality in communication patterns—80.7% of communication edges are between users within a same country.

In contrast, spam emails are often sent to users who are not socially connected. It is difficult for attackers to identify community patterns and mimic normal user behaviors. Thus, how well the email recipients of a given user v are socially connected can be defined as a social affinity feature for hijacked account detection.

The recipient connectivity for a user v is defined based on how well $R(v)$, the recipients of v, are connected in the undirected friendship graph G_u. SocialWatch computes the fraction of such connected recipients among $R(v)$. For each account v, $R(v)$ forms a subgraph $G_u(v)$. Note that $G_u(v)$ does not include v. SocialWatch then identifies all the connected components $C_1 \cdots C_k$ with size of at least two on the subgraph $G_u(v)$. Let $R_i(v)$ denote the set of recipients in C_i ($1 \leq i \leq k$). Then the recipient connectivity feature $r(v)$, the fraction of socially connected recipients, is defined as:

$$r(v) = \frac{\sum_{i=1}^{k} |R_i(v)|}{|R(v)|} \tag{3}$$

Social distance l: In practice, the email communication graph is only a sub-graph of the real-world communication graph. For example, we may not be able to observe the connections among accounts whose communication records are not included in our dataset (e.g., accounts external to the service that we are protecting). Such incomplete graph may lead to incorrect measurements in the recipient connectivity feature $r(v)$, because two external recipients of v that are socially connected in the real world are now disconnected (not belonging to the same connected component) in our observed graph G_u.

We use a social distance $d(v_1, v_2)$ on the friendship graph G_u to describe how close two users v_1 and v_2 are on the social graph. It is defined by the *distance of the shortest path* between v_1 and v_2 on G'_u, where G'_u is defined by removing v (and edges connected with v) from G_u.

We define the second social feature $l(v)$ for user v as the mean of all pairwise social distances between any two users in $R(v)$:

$$l(v) = \frac{1}{N} \sum_{v_i, v_j \in R(v), i<j} d(v_i, v_j) \tag{4}$$

where $N = |R(v)| * |R(v) - 1|/2$ is the number of all possible user pairs.

In the case of missing observations mentioned above, the social distance defined by the shortest-path distance is an upper-bound approximation of the real social distance on the complete graph. $l(v)$ is robust to missing data and complementary to the recipient-connectivity feature $r(v)$.

The social affinity features $r(v)$ and $l(v)$ exploit the structure of the subgraph defined by the email recipients of the user v. We use $r(v)$ and $l(v)$, combined with node out-degree (denoted as $D(v)$), for hijacked account detection. Previous work has leveraged clustering coefficients for similar purposes [16]. We do not use clustering coefficients because they only account for direct connections between pair-wise accounts and are thus sensitive to missing data.

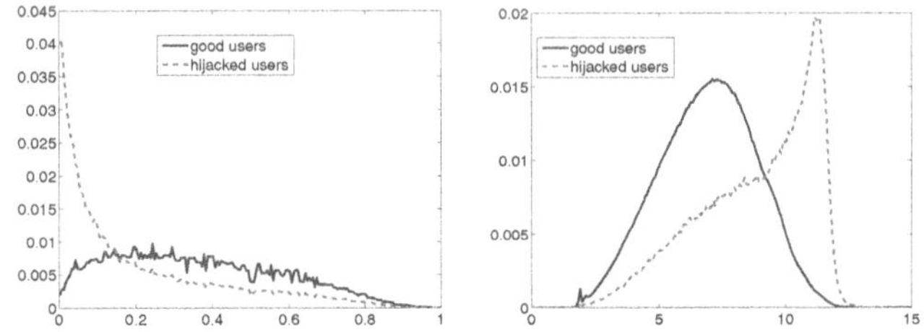

(a) $r(v)$ (b) $l(v)$

Figure 2: Social affinity features for detecting hijacked accounts.

4.2.2 *Computing Social Affinity Features*

The recipient connectivity feature $r(v)$ is relatively easy to compute in parallel for all accounts. However, the social-distance feature $l(v)$ requires computing shortest-path distances on the large graph G_u for all user pairs in $R(v)$, which is computational expensive and hard to parallelize. SocialWatch adopts a sketch-based approach to compute an approximate shortest-path distance between a pair of nodes using sampled landmark nodes [22]. Using this algorithm, the estimated path length is slightly larger than the actual value with provable bounds. We find that the approximation works well in practice for our purpose of detecting hijacked accounts.

Figure 2 shows the distributions of the two social-affinity features for a set of known hijacked accounts and another set of sampled legitimate accounts, both labeled by the email provider. As we can see, for each feature, the distribution of hijacked accounts is significantly different from that of legitimate accounts. The distribution of $r(v)$, the recipient-connectivity feature, is relatively even between 0 and 1 for legitimate users, but is heavily peaked at zero for hijacked users, which means that the email recipients of a hijacked user are mostly disconnected. The average pairwise shortest-path distance $l(v)$ follows a normal distribution with a mean of about 7 for legitimate accounts, but is peaked at about 12 for hijacked accounts.

These two social-affinity features are complementary to each other as they capture different social graph properties of a user. In the following we present two scenarios for using these features together to detect hijacked accounts. The first scenario is when we do not have pre-labeled hijacked accounts, i.e., we want to detect hijacked accounts based on the properties of legitimate accounts. The second scenario is a refined approach when we are provided with a set of known hijacked accounts.

4.2.3 *Detection Without Known Hijacked Accounts*

For legitimate accounts, the distributions of their graph features are usually stable over time and can be estimated accurately. Without known hijacked accounts, we can use one-tailed hypothesis testing [23] to detect hijacked accounts. The null hypothesis is:

$$H_0 : \omega = W_0 \tag{5}$$

where ω denotes the unknown account label, and W_0 denotes the normal account label. Given the observed features $\mathbf{x}^0$ of a user v (in our case $\mathbf{x}^0 = \{D(v), r(v), l(v)\}$), and assuming the null hypothesis is true, we can estimate the one-tailed critical p-value (the statistical significance level):

$$P_0 = P(\mathbf{x} > \mathbf{x}^0 | \omega = W_0) = \int p(\mathbf{x} > \mathbf{x}^0 | \omega = W_0) d\mathbf{x} \tag{6}$$

For simplicity, we assume that the social features are independent. Equation 6 becomes:

$$P_0 = \prod_{i=1}^{3} \int p(x_i > x_i^0 | \omega = W_0) dx_i \tag{7}$$

Symbol	N	N'	θ	α
Description	Inactive limit	Aggressive limit	Response rate	Badness/ goodness ratio
Value	5	500	5%	4.5

Table 1: Selected values for detection thresholds.

If P_0 is less than a predefined statistical significance level t, then we reject the hypothesis. The significance level t is a parameter to control the potential false detection rate of such detection. Specifically, given t, SocialWatch computes a threshold along each feature dimension based on data. We classify an account as a hijacked account if one of its feature values violates the computed threshold.

4.2.4 Detection With Known Hijacked Accounts

In practice, a service provider can often identify a subset of hijacked accounts based on user reports, or manual examination, *etc.* Although the distribution of hijacked accounts is not as stable as normal accounts, we can still leverage the knowledge of known hijacked accounts to detect remaining ones. SocialWatch uses a Bayesian decision framework to detect additional hijacked accounts using the known ones as training data.

Given the learned probability distributions of both normal and hijacked accounts, the Bayesian decision theory [24] gives the following decision boundary:

$$\frac{p(\mathbf{x}|W_1)}{p(\mathbf{x}|W_0)} > \frac{c_0}{c_1}\frac{P(W_0)}{P(W_1)} \tag{8}$$

where W_1 denotes the class of hijacked accounts. Here c_0 represents the cost (*i.e.*, penalty) of labeling a normal account as a hijacked account, and c_1 the cost of labeling a hijacked account as a normal account. Assuming that features are independent, the decision ratio can be directly computed from the learned probability density functions shown in Figure 2(a)-(c). If the above inequality holds, $\mathbf{x}$ is classified as hijacked, otherwise it is classified as normal. $P(W_0)$ is the prior of a given account to be a normal account and $P(W_1)$ the prior of being a hijacked accounts. Both $P(W_0)$ and $P(W_1)$ can be estimated by the fraction of normal and hijacked accounts in the training set. The only parameter to set is the ratio of $\frac{c_0}{c_1}$, which is independent of the observed data, and is intuitive thus easier to understand and control by the service providers.

5. EVALUATION

We implement SocialWatch using Dryad/DryadLINQ [5] and it processes data in parallel on a 240-machine cluster. In this section, we evaluate SocialWatch on the large graphs that we build (see Section 3.2). We describe how we label the ground-truth of data for validation in the longer version of this paper [19].

5.1 Degree and PageRank Based Detection

After filtering inactive users, we first detect all the active users using degree and PageRank based method as described in §4.1.

We pick a small set of labeled accounts as training data to derive the threshold parameters (listed in Table 1) and we exclude them from our evaluation.

Using the degree-based method alone, SocialWatch detects a vast majority of the attacker-created accounts with a low false detection rate (1.06%) and a low false negative rate (1.19%). Indeed, legitimate and malicious accounts exhibit very different patterns in their email response rates. In our data set, more than 95% of the bad accounts have 0 in-degrees on the directed graph, meaning they receive no replies. In contrast, the median response rate of good accounts is over 50% and only 5% of them receive no replies (most of them are actually inactive users with small out-degrees).

We further apply PageRank-based detection to increase the detection coverage and reduce the false detection rate. It takes about

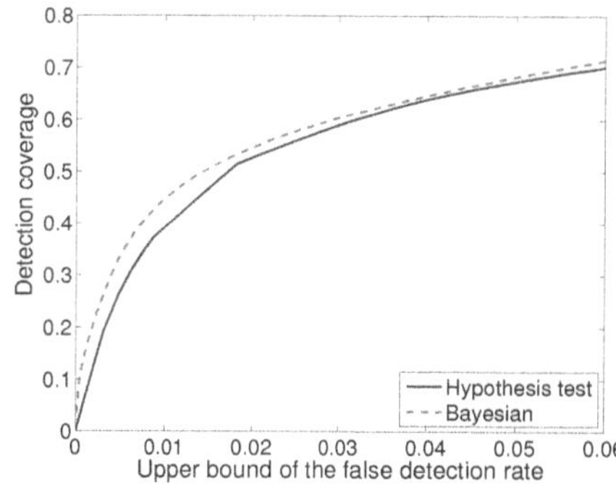

Figure 6: Detection coverage vs. the upper bound of false detection rate for hijacked users (the newly detected users are all regarded as false positives in this figure).

30 iterations for the goodness and badness scores of all the nodes to converge ($\leq 1\%$ difference between two consecutive iterations).

After obtaining the badness-goodness PageRank ratio, we need to select a cutoff threshold. To do so, we leverage the training data mentioned earlier. Figure 3 shows the CDF of their badness-goodness ratio. We find good and bad accounts in the training data have clear distinctions. Accordingly, we select $\alpha = 4.5$ as the threshold for detection on the entire graph. In total, among all 269.5 million internal accounts, we detect 56.85 million bad accounts by combining degree and PageRank in detection and reduce the false detection rate to 0.75%. Among the falsely detected accounts, we find some of them have a high badness score because they sent emails to a few very large mailing-list addresses (which typically have high badness scores). Thus we can potentially improve our detection further by removing large mailing lists from detection. Our false negative rate is as low as 0.61%. Among those false negative users, 38.88% are actually known hijacked accounts that we fail to detect using degree and PageRank.

The above results suggest that degree and PageRank can effectively detect attacker-created accounts accurately. For the set of detected accounts, service providers can apply more strict policies to further validate them (*e.g.*, giving more CAPTCHA or asking security questions) while throttling their attack behaviors substantially (*e.g.*, by limiting the email-sending rates aggressively).

5.2 Social Affinity Based Detection

After applying the combined degree and PageRank algorithm to the email graph to detect attacker-created accounts, we apply the social-affinity based method on the remaining users to detect hijacked accounts. In particular, we leverage the two detection methods, namely *hypothesis testing* and *Bayesian decision* introduced in §4.2.1. The hypothesis testing does not need known hijacked accounts, so we use the labeled good accounts to build normal account distributions and use the remaining users as testing data. For the Bayesian decision method, we use half labeled good accounts and half known hijacked accounts for training, and the remaining data for testing.

Figure 6 shows the detection coverage vs. false detection rate from hypothesis testing (solid curve) and Bayesian decision (dashed curve). We vary the false detection rate from 0 to 6%. The detection results are compared to the labels provided by the email provider. We observe that Bayesian decision performs slightly better. At a 6% false detection rate, hypothesis testing detects about 70% of the hijacked accounts, while Bayesian decision detects about 72%. At a false detection rate of 2%, using hypothesis testing, SocialWatch detects 53.3% hijacked accounts. One advantage of using the Bayesian approach is that the parameter has an intuitive meaning, and is easier to understand and control.

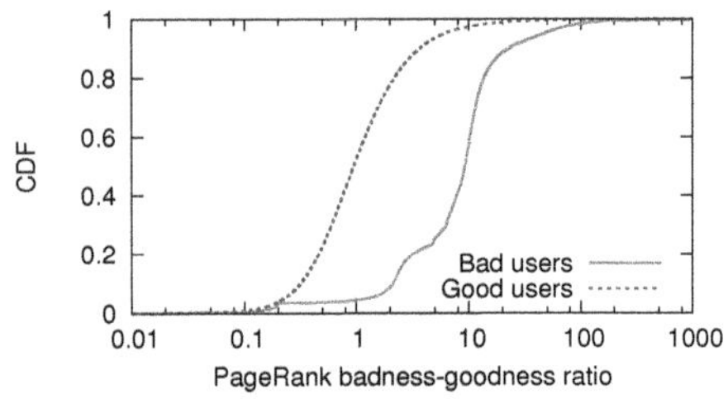

Figure 3: PageRank **badness-goodness ratio.**

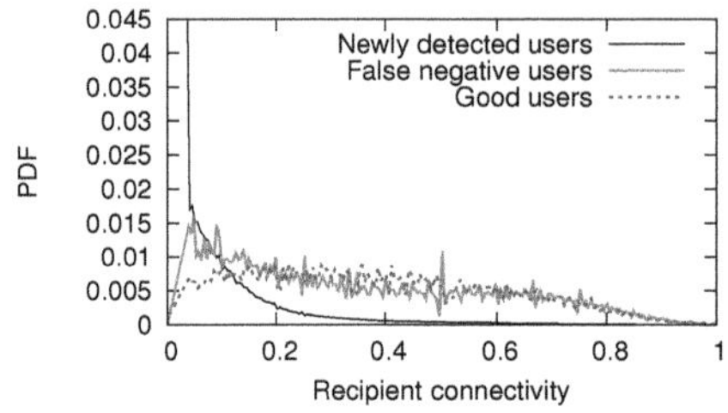

Figure 4: Recipient connectivity Eq 3[2]**.**

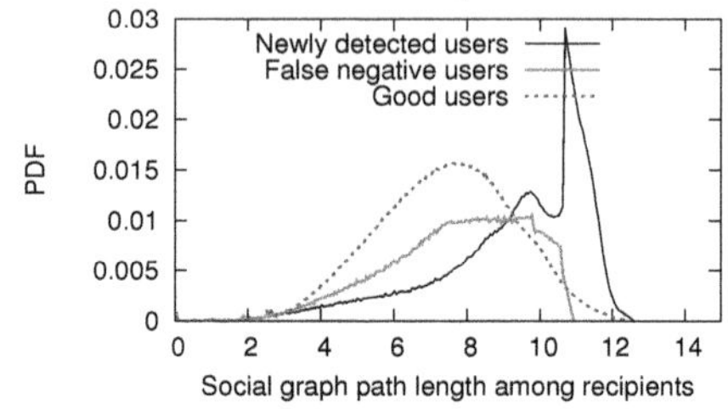

Figure 5: Social distance among recipients Eq 4.

Our method also flags 1.53 million additional accounts as being hijacked. These accounts are our new findings, which have not been reported by the existing system yet. Similarly, using Bayesian decision, SocialWatch detects 54.6% known hijacked accounts as well as 1.23 million additional ones.

For the newly detected accounts, the question is "are they indeed hijacked users that were not detected yet?" We study them using two approaches. First, we submitted these accounts to the email provider for manual verification based on sampling. They confirmed that the sampled accounts exhibit suspicious email activities. After a year, most of them were either reported by the users or captured by content-based detection. Second, we compare two social-affinity features of these accounts with the overall good account population (Figure 4, 5). These two features describe the community structures of the accounts. We observe that the newly detected accounts have clearly different feature distributions from the good account population, *i.e.,* with smaller ratios of recipient connectivity $r(v)$ and larger social distances $l(v)$. In fact, their distributions are similar to the set of known hijacked accounts. These accounts thus may already have started spamming activities at the time of our detection.

On the other hand, for the false negative accounts, i.e., the set of hijacked accounts that we fail to detect, we find that they have similar social feature distributions to the known good ones. These may be more stealthy accounts or the set of accounts not actively misbehaving yet, making it inherently challenging to detect them.

6. CONCLUSION

In this paper, we present an online service protection framework, SocialWatch, that uses social connectivity features to detect attacker-created accounts and hijacked accounts at a large scale. We explore a rich set of graph properties including degree and PageRank that are effective for detecting aggressive attack behaviors, as well as social affinity metrics that capture the subtle differences between the stealthy hijacked accounts and the remaining legitimate users. We show that SocialWatch can effectively detect tens of millions of malicious accounts in the wild. Since the social connectivity features that we rely on can be derived from a wide class of applications, we believe that SocialWatch is general and promising for protecting both the providers and the online users of many services.

7. REFERENCES

[1] Y. Xie, F. Yu, K. Achan, R. Panigrahy, G. Hulten, and I. Osipkov, "Spamming Botnets: Signatures and Characteristics," in *SIGCOMM*, 2008.

[2] S. Hao, N. A. Syed, N. Feamster, A. G. Gray, and S. Krasser, "Detecting Spammers with SNARE: Spatio-temporal Network-level Automatic Reputation Engine," in *USENIX Security*, 2009.

[2]False negative accounts are those detected by the email provider but not by SocialWatch. Newly detected accounts are those classified as hijacked by SocialWatch but not by the email provider. They may actually be hijacked accounts that the current system fails to detect.

[3] Z. Qian, Z. M. Mao, Y. Xie, and F. Yu, "On Network-level Clusters for Spam Detection," in *NDSS*, 2010.

[4] Y. Zhao, Y. Xie, F. Yu, Q. Ke, Y. Yu, Y. Chen, and E. Gillum, "BotGraph: Large Scale Spamming Botnet Detection," in *NSDI*, 2009.

[5] Y. Yu, M. Isard, D. Fetterly, M. Budiu, U. Erlingsson, P. K. Gunda, and J. Currey, "DryadLINQ: A System for General-Purpose Distributed Data-Parallel Computing Using a High-Level Language," in *OSDI*, 2008.

[6] H. Yu, M. Kaminsky, P. B. Gibbons, and A. Flaxman, "SybilGuard: Defending Against Sybil Attacks via Social Networks," in *SIGCOMM*, 2006.

[7] H. Yu, P. B. Gibbons, M. Kaminsky, and F. Xiao, "SybilLimit: A Near-Optimal Social Network Defense against Sybil Attacks," in *IEEE Symposium on Security and Privacy (Oakland)*, 2008.

[8] N. Tran, B. Min, J. Li, and L. Subramanian, "Sybil-Resilient Online Content Voting," in *NSDI*, 2009.

[9] B. Viswanath, K. P. Gummadi, A. Post, and A. Mislove, "An Analysis of Social Network-Based Sybil Defenses," in *SIGCOMM*, 2010.

[10] A. Mohaisen, A. Yun, and Y. Kim, "Measuring the Mixing Time of Social Graphs," in *IMC*, 2010.

[11] Y. Xie, F. Yu, Q. Ke, M. Abadi, E. Gillum, K. Vitaldevaria, J. Walter, J. Huang, , and Z. M. Mao, "Innocent by Association: Early Recognition of Legitimate Users," in *CCS*, 2012.

[12] Y.-Y. Ahn, S. Han, H. Kwak, S. Moon, and H. Jeong, "Analysis of Topological Characteristics of Huge Online Social Networking Services," in *WWW*, 2007.

[13] A. Bonato, J. Janssen, and P. Pralat, "A Geometric Model for On-line Social Networks," in *WOSN*, 2010.

[14] J. Leskovec, J. Kleinberg, and C. Faloutsos, "Graphs over Time: Densification Laws, Shrinking Diameters and Possible Explanations," in *KDD*, 2005.

[15] M. Sirivianos, K. Kim, and X. Yang, "SocialFilter: Introducing Social Trust to Collaborative Spam Mitigation," in *INFOCOM*, 2011.

[16] V. P. R. P. Oscar Boykin, "Leveraging Social Networks to Fight Spam," in *IEEE Computer Vol 38*, 2005.

[17] P. A. C. andJoerg Diederich and W. Nejdl, "MailRank: Global Attack-Resistant Whitelists for Spam Detection," in *CIKM*, 2005.

[18] S. Garriss, M. Kaminsky, M. J. Freedman, B. Karp, D. Mazieres, and H. Yu, "RE: Reliable Email," in *NSDI*, 2006.

[19] J. Huang, Y. Xie, F. Yu, Q. Ke, M. Abadi, E. Gillum, and Z. M. Mao, "SocialWatch: Detection of Online Service Abuse via Large-Scale Social Graphs," in *Tech Report MSR-TR-2013-24*, 2013.

[20] S. Brin and L. Page, "The Anatomy of a Large-Scale Hypertextual Web Search Engine," in *WWW*, 1998.

[21] Z. Bar-Yossef and L.-T. Mashiach, "Local Approximation of PageRank and Reverse PageRank," in *WWW*, 2008.

[22] A. D. Sarma, S. Gollapudi, M. Najork, and R. Panigrahy, "A Sketch-Based Distance Oracle for Web-Scale Graphs," in *WSDM*, 2010.

[23] E. Lehmann and J. P. Romano, *Testing Statistical Hypotheses*. Springer, 2005.

[24] R. O. Duda, P. E. Hart, and D. G. Stork, *Pattern Classification*. Wiley-Interscience, 2001.

Privacy Settings in Social Networking Systems: What You Cannot Control

Amirreza Masoumzadeh
School of Information Sciences
University of Pittsburgh
Pittsburgh, PA, USA
amirreza@sis.pitt.edu

James Joshi
School of Information Sciences
University of Pittsburgh
Pittsburgh, PA, USA
jjoshi@pitt.edu

ABSTRACT

In this paper, we propose a framework to formally analyze what privacy-sensitive information is protected by the stated policies of a Social Networking System (SNS), based on an expression of ideal protection policies for a user. Our ontology-based framework can capture complex and fine-grained privacy-sensitive information in SNSs, and find out missing policies, given a user's ideal policies, and SNS's privacy settings and described system policies. We propose notions of policy completeness for SNSs to facilitate such an analysis. Our case study of using this approach on Facebook shows that we can effectively identify important missing policies.

Categories and Subject Descriptors

K.4.1 [**COMPUTERS AND SOCIETY**]: Public Policy Issues—*Privacy*

Keywords

Social Networking Systems, Privacy Settings, Privacy Control Policies

1. INTRODUCTION

In a typical Social Networking System (SNS) such as Facebook, there exist several privacy settings that can be configured by a user in order to control others' access to the information related to her. These settings are in fact equivalent to access control policies that are expressed by the users for the respective digital objects. Such privacy settings are by no means complete in the sense that they do not control access to all the potentially privacy-sensitive information about a user. That is the case even for Facebook, which has fairly the most extensive set of privacy settings among SNSs. Access to the rest of the information related to the user is governed by a set of fixed rules set by the SNS itself. We call these system defined policies and the user-configurable privacy settings together as *privacy control policies* in an SNS.

ASIA CCS'13, May 8–10, 2013, Hangzhou, China.

There is a major issue with the current practice of privacy control policies in SNSs such as Facebook. The current privacy settings do not provide users with adequate power to protect their privacy-sensitive information. Moreover, the system defined policies enforced by an SNS are not clearly described to the users. Therefore, users are unsure about what to expect from the system. Users need to learn about them either by harvesting help pages in the SNS or by observing the system's behavior. Worse is that, since the system defined policies are not well documented, SNSs can modify them without users noticing it and put them at great risk of privacy violations.

In order to identify the privacy risks users are dealing with in such an abovementioned ecosystem, we propose an approach to formally reason about *completeness* of privacy control policies in SNSs. Our notion of completeness ensures that privacy control policies are applicable to every piece of information related to a user. Therefore, the user can clearly expect to know that how her information is protected. Most of the recent literature on policy analysis focuses on XACML policies [5, 6], which are not suitable for representing complex policies in SNSs such as "*Who can see posts you've been tagged in on your timeline?*". A separate body of literature has focused on modeling access control policies for SNSs. Most of those models formalize specification of access subjects based on the type and distance of their relationships to an object's owner [2, 3]. However, such models cannot consider protection of complex privacy-sensitive resources in SNSs as required for our analysis purpose. A few recent works approach the access control modeling problem in SNSs by using Semantic Web technologies [1, 7, 8], which shows great promise in capturing the complex and fine-grained access control policy requirements in these systems. Like in [7, 8] we approach this issue by modeling the information contained in an SNS using an ontology. However, we employ a more rigorous approach with regards to modeling data constraints in SNSs, as they are vital to provide sound analysis of privacy control policies.

In this work, we propose a framework to formally reason about the completeness of privacy control policies in an SNS. To the best of our knowledge, we provide the first of its kind policy analysis approach for SNSs that can theoretically reason about the missing pieces of policies and controls in SNSs. Such a systematic approach can tell the SNS users if their expectations about privacy control (or system defined policies in place) is provided by an SNS, and ultimately help the developers of such systems to resolve these issues. In summary, we make the following contributions:

- We propose a fine-grained ontology-based approach to model information in SNSs, including data integrity constraints, that are necessary for the reasoning tasks in our framework.
- We present an approach to specify permissions in such an ontology-based knowledge base, using which we reason about completeness of privacy control policies provided by an SNS, and detect missing policies.
- Using our framework, we analyze Facebook's privacy control policies as a case-study, and present our findings with regards to its completeness.

The rest of the paper is organized as follows. We propose a fine-grained model of information in SNSs, and an approach to specify privacy-sensitive permissions in these systems in Sections 2 and 3, respectively. In Section 4, we formally define our notion of completeness, and show how our framework can analyze policies in an SNS based on that. In Section 5, we demonstrate the applicability of our approach by providing a case study, and conclude the paper in Section 6.

2. MODELING SNS INFORMATION

In this section, we propose a model of SNS information using OWL [4] as our modeling language. We first discuss a generic information model for SNSs and then explain in more detail an ontology for Facebook. Clearly, a similar ontology can be developed for other SNSs.

2.1 Basic Concepts

We model SNS information as a set of users, digital objects, and data values, which are related to each other by relationships. In OWL terminology, we model the first two concepts using *classes* as abstract objects. The relationships between objects are captured using the following *object properties*: between users to define the social network (e.g., friendship, following, etc.), between digital objects (e.g., a comment that is related to a photo), or between users and digital objects (e.g., a user may own a photo). Moreover, class objects can be related to data values using *data properties*, e.g., the relationship between a photo and its binary content. We model annotations as a special class of digital objects. Annotations are objects that annotates one object with another object. Comments and photo tags are two common ways of annotation. For instance, a photo tag can annotate a photo with a user. As another example, a check-in in Foursquare is an annotation that annotates a venue (place) with a user, which may also be associated with a time stamp. These are the minimum concepts and relationship types that we capture for an SNS; they can be easily extended depending on the needs of an SNS.

Figure 1 depicts our proposed ontology for Facebook. Note that many details, especially restrictions, are not captured in the figure or in the following description. Entity is the root class to our ontology (a subclass of owl:Thing, which is OWL's built-in most general class). It is specialized by User and DigitalObject. There is a predefined individual in class User, i.e., me, for whom policy analysis is performed. The isFriendOf object property expresses friendship relationship between instances of User. Data properties such as hasFullname associate data values to a User. The owns object property defines a User as the owner of a DigitalObject. DigitalObject is the union of four major subclasses: Content, Wall, Event, and Annotation. Content represents an object that has data content such as a photo or a text (specified using hasContent data property). Classes Wall and Event correspond to profile wall page and events that users can attend in Facebook. Annotation, as mentioned before, represents objects that instead of directly carrying content, annotate a DigitalObject (e.g., a wall or a photo) with an Entity (e.g., a wallpost or a comment), using object properties annotates and annotatesWith, respectively. Actual forms of annotations are defined as its subclasses: Comment, UserTag, and WallPost. A Comment annotates a Commentable (a class which is the union of Photo and WallPost) with a Text. A UserTag annotates a UserTaggable (a class which is the union of Photo and Text) with a User.

2.2 Restrictions

It is crucial to accurately model the data constraints in an SNS knowledge base, in order to facilitate meaningful policy analysis. Any inaccuracy will result in false positives/ negatives in our analysis. We employ the following restriction features in OWL to model such data constraints:

Disjoint Union of Subclasses: We ensure our class hierarchy is captured completely by defining each class as disjoint union of its subclasses, i.e., union of subclasses which are pair-wise disjoint. For instance, Annotation is equivalent to the disjoint union of Comment, UserTag, and WallPost.

Property Domain/Range: The domain and range of a property can be restricted to specific classes, e.g., owns has User and DigitalObject as its domain and range, respectively.

Property Characteristics: OWL supports several restrictions to accurately model property constraints: functional, inverse functional (i.e., the inverse of the relation is a function), transitive, symmetric, asymmetric, reflexive, and irreflexive. For instance, isFriendOf is defined as irreflexive; hence, no user can be a friend of herself.

Class Property Restriction: OWL supports existential, universal, cardinality, and value constraints for properties when applied to a class. For instance, WallPost is defined equivalent to a class that annotates exactly one Wall.

3. MODELING PRIVACY-SENSITIVE PERMISSIONS

For our analysis purpose, we do not aim at analyzing the expressive power of the SNS privacy settings in terms of characterizing the access subjects. Instead, we focus on capturing the fine-granularity of the protected objects in an SNS, as described in this section.

3.1 Properties as Protected Resources

The ontology modeling approach proposed in Section 2 captures SNS information in the forms of classes and properties. We argue that an individual of a class does not represent any privacy-sensitive information unless its properties are considered. In other words, the essential knowledge is captured by the triples that represent properties between two individuals (object properties), or between an individual and a data value (data properties). Based on this observation, we consider such triples as privacy-sensitive information that needs to be protected.

We elaborate on this further using the example ontology in Figure 2. Assume Alice has a photo, in which Bob has been tagged by Carol: Alice owns Photo1, and Carol owns PhotoUserTag1 which annotates Photo1 with Bob. By con-

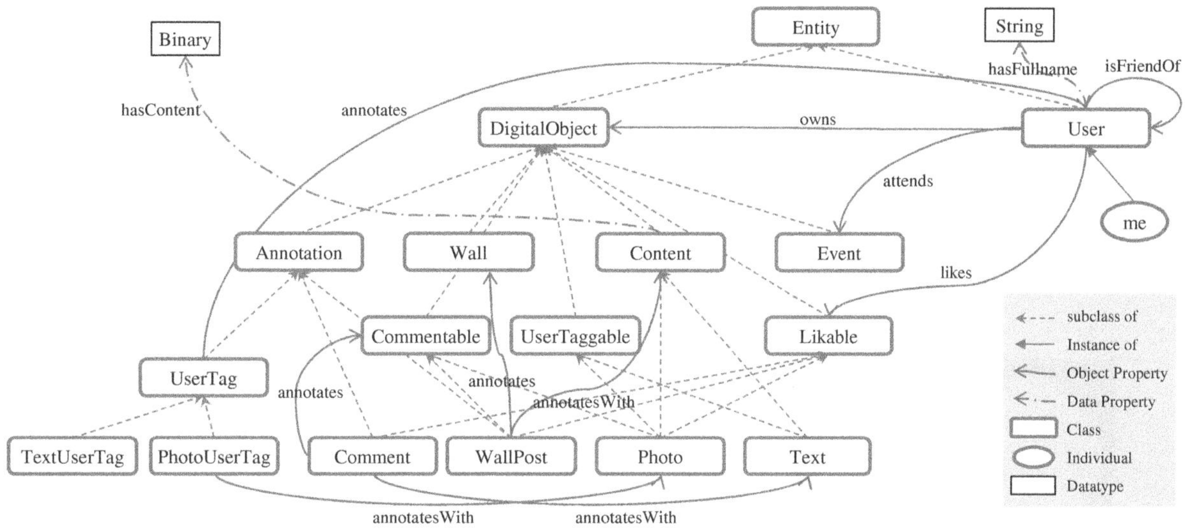

Figure 1: A Model of Concepts and Properties for Facebook

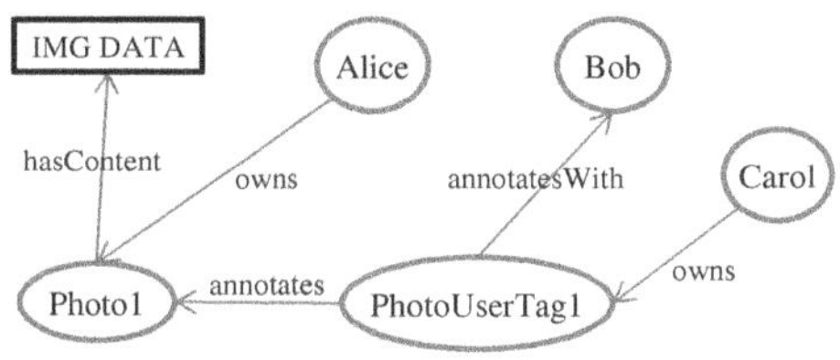

Figure 2: A PhotoUserTag Example

sidering the object and data properties in this example as protected resources, we can control the access to privacy-sensitive information. We assume a simple and effective policy authority scheme: the owners of the endpoints of each property are eligible to define policy for that property. Therefore, for instance, visibility of the tag can be controlled using relationships PhotoUserTag1 annotates Photo1 (by Alice or Carol) and PhotoUserTag1 annotatesWith Bob (by Bob or Carol). Carol can control revealing the fact that she has created this tag by controlling Carol owns PhotoUserTag1. Finally, Alice can control the association of the actual binary content of the photo using relationship Photo1 hasContent IMG-DATA. Various permissions can be specified with respect to the three basic possible actions on these resources: selection, insertion, and deletion. For example, the insertion of a tag can be controlled in a similar fashion.

3.2 Reification of Properties

In order to specify permissions, we need to characterize classes of relationships in an ontology based on certain restrictions. However, OWL does not support such expressions about relationships. We use the concept of reification to overcome this limitation. We reify each object/data property in our SNS ontology as a class in OWL. Figure 3 depicts the policy ontology corresponding to the ontology presented in Section 2. Class ReifiedProperty is the root to all such classes, with the two subclasses ReifiedObjectProperty and ReifiedDataProperty, which are the disjoint union of the corresponding reified property classes. We define two special object properties, ropSbj and ropObj, that relate a reified object property to its subject and object, respectively. Analogously, the special object and data properties rdpSbj and rdpData relate a reified data property to its subject and data value, respectively.

The reified property classes will essentially replace SNS ontology properties that were described in Section 2. Therefore, they must conform to the information domain constraints that we considered in designing the ontology. We demonstrate with an example how a property restriction presented in 2.2 can be translated for reified properties. Let p be an object property in the SNS ontology, and RPp be its corresponding reified class. If p is a functional property, such a characteristic can be ensured for RPp by the following restriction: owl:Thing $\sqsubseteq$ inverse ropSbj max 1 RPp.

3.3 Representing Permissions

A permission can be represented by a set of protected resources and corresponding action(s). The set of possible actions on a resource include viewing, creating, or removing it from an SNS knowledge base. We define class Action which consists of three individual members select, insert, and delete, respectively. We represent permissions by characterizing a subset of class Permission. The class Permission is related to class RefieidProperty using object property pResource to specify the resources. It is also related to class Action using object property pAction in order to specify the considered action(s) in the permission. As an example, consider the privacy setting in Facebook that states *"who can post to your timeline?"*. The corresponding permission can be represented using the following specification:

S2 ≡(pAction value insert)
and (pResource some (RPannotates
and (ropObj some (Wall
and (inverse ropObj some (RPowns
and (ropSbj value me)))))))

In the above conjunctive expression, the first clause specifies the action as insertion. The second complex clause indicates that the resource should be of a reified property class RPannotate, where its object is a Wall, and that wall is

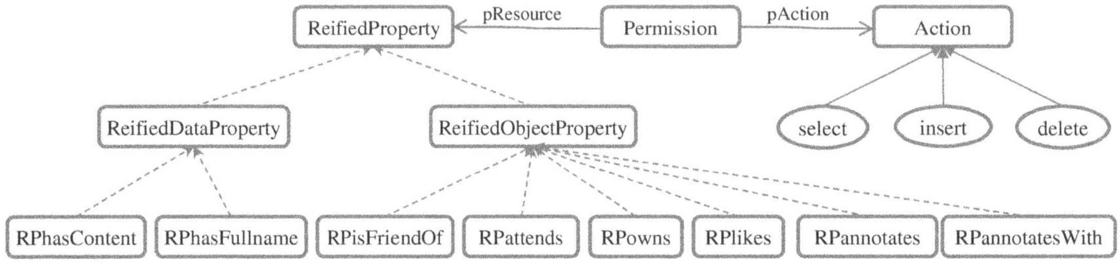

Figure 3: Representing Permission Specification in Ontology

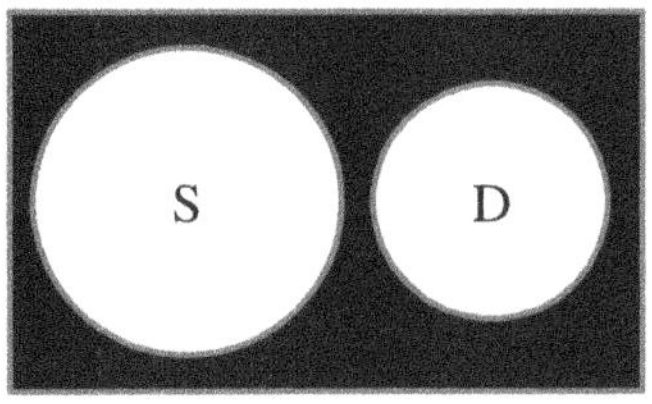

Figure 4: Permission Categories Space for an SNS

owned by me. In other words, the resource is the annotates relationship that annotates the Wall belonging to me.

4. ANALYZING PRIVACY CONTROL PERMISSIONS

We identify three categories of permissions in an SNS:

Privacy Setting Permissions These are captured by user-configurable privacy settings, which usually have dedicated control elements in the system.

Described System Permissions These consists of the permissions that are not configurable by users and their corresponding system-defined policy is well documented.

Ideal Permissions These refer to the permissions that a user ideally considers as privacy-sensitive, which may depend on the information model of the SNS.

Let S, D, and I, represent the classes of privacy setting, described system, and ideal permissions in an SNS, respectively. Figure 4 depicts these permission classes in relation to each other. The privacy setting permissions and described system permissions are disjoint by definition. The ideal permissions (the rectangle) is likely to cover both these classes. However, we note that there may exist parts of ideal permissions that are covered by neither privacy settings nor described system permissions. In other words, there may exist some privacy-sensitive permissions which are not mediated by a transparent policy. Our framework can reason about existence and nature of such permissions.

Our permission specification, as proposed in Section 3.3, is expressed in terms of classes in the ontology. Hence, we can leverage ontological reasoning power to compare and analyze such permissions. Let P and Q be two permission classes. An OWL DL reasoner can efficiently evaluate the following permission compositions:

- not P: denotes what P does not cover.
- P and Q: denotes what both P and Q cover.
- P or Q: denotes what either P or Q (or both) cover.

Using the above compositional semantics and subsumption re we can analyze permissions in an SNS. The class M $\equiv$ I and not (S or D) represents the permissions that users are missing from the picture. We further elaborate on such analysis in our case study in Section 5.

We propose two completeness privacy properties for an SNS based on this analysis. The following property ensures that users' ideal permissions are part of the privacy settings and can be controlled by them.

DEFINITION 1 (COMPLETELY CONTROLLABLE). *An SNS with its privacy setting permissions* S *is* completely controllable *with regards to ideal permissions* I, *if and only if* I $\sqsubseteq$ S.

In practice, with a reasonable assumption about ideal permissions, the above notion of completeness cannot be satisfied in SNSs. The users may be overwhelmed by the complexity of the options, and the system may choose not to provide such controls due to various design considerations. The next notion of completeness is more practical.

DEFINITION 2 (COMPLETELY KNOWN). *An SNS with its privacy setting and described system permissions* ⟨S, D⟩, *is* completely known *with regards to ideal permissions* I, *if and only if* I $\sqsubseteq$ (S or D).

The above notion of completeness verifies if the collection of privacy settings and described system permissions by an SNS covers the ideal permissions. If not, we can employ subsumption reasoning to find permissions that the user misses.

5. CASE STUDY: FACEBOOK PRIVACY CONTROL PERMISSIONS

In order to evaluate our framework, we analyze the privacy control policies in Facebook. We leverage the ontology described in Section 2, in which we refrain from completely modeling every detail in Facebook for brevity reasons. We believe the current ontology is representative enough to describe our methodology, and other features can be captured in a similar fashion.

Facebook provides a centralized dashboard for controlling privacy settings such as the visibility of the statuses, tagging, etc. Moreover, a user can determine the visibility of her profile information such as education, contact info, etc. We collected those settings and formulated their corresponding permissions. In Table 1, we list those settings either worded as exactly as in the settings' page or representative enough, and corresponding permissions. Note that the listed privacy settings are limited to the information that can be captured by our proposed ontology in Section 2. For instance, the setting "*Who can send you Facebook messages?*" is not listed since we did not include messages in our ontology. Our current ontology also does not model apps, or

	Privacy Setting	Corresponding Permission
S1	Profile attributes privacy (for each item)	(pAction value select) and (pResource some (rdpSbj value me))
S2	Who can post on your timeline?	(pAction value insert) and (pResource some (RPannotates and (ropObj some (Wall and (inverse ropObj some (RPowns and (ropSbj value me)))))))
S3	Who can see what others post on your timeline?	(pAction value select) and (pResource some (RPannotates and (ropObj some (Wall and (inverse ropObj some (RPowns and (ropSbj value me))))) and (ropSbj some (WallPost and (inverse ropObj some (RPowns and (ropSbj value me)))))))
S4	Who can see what you post? (per item)	(pAction value select) and (pResource some (RPhasContent and (rdpSbj some (inverse ropObj some (RPowns and (ropSbj value me))))))
S5	Review tags friends add to your own posts	(pAction value insert) and (pResource some (RPannotates and (ropObj some (inverse ropObj some (RPowns and (ropSbj value me)))) and (ropSbj some PhotoUserTag)))
S6	Visibility of photos (managed per album)	(pAction value select) and (pResource some (RPhasContent and (rdpSbj some (inverse ropObj some (RPowns and (ropSbj value me))))) or (RPowns and (ropSbj value me) and (ropObj some Photo)))

Table 1: Privacy Setting Permissions in Facebook

	Described System Policy	Corresponding Permission
D1	Who can see tags I make?	(pAction value select) and (pResource some (RPannotatesWith and (ropSbj some (UserTag and (inverse ropObj some (RPowns and (ropSbj value me)))))))
D2	Who can see that I'm tagged in a post?	(pAction value select) and (pResource some (RPannotatesWith and (ropObj value me)))
D3	Who can see a tag that someone added to my post?	(pAction value select) and (pResource some (RPannotates and (ropObj some (UserTaggable and (inverse ropObj some (RPowns and (ropSbj value me)))))))

Table 2: Described System Permissions in Facebook

the posts that are automatically made by the system on one's timeline when she is tagged in another post. Moreover, in order to avoid enumerating a tedious list of settings, we provide one privacy setting that accounts for all profile attributes (S1). That setting can cover permissions corresponding to a setting like "*Who can look you up using the email address or phone number you provided?*" as well as many other single visibility controls in front of every profile item under a user's *about page*.

Unfortunately, Facebook is not very transparent about its system defined policies. Some of the system defined policies are documented in the help pages, in the format of questions and answers, which is not easily accessible, unless one goes through all the pages. For instance, the following question is provided in the help page for tagging (and not under privacy!): "*When I tag someone in a photo, post or app activity, who can see it?*". The answer provided to this question explains that the audience selected for the post, the user tagged, etc., can see it. Table 2 lists some of the described system policies that we were able to harvest from Facebook's help pages, and that were related to the information captured in our ontology.

Based on our model of SNS information, our intuition is that a user should be able to control relationships that are about her. In terms of our proposed Facebook ontology, those includes properties that directly relate to the user, and the properties that relate to some digital objects owned by the user. We consider these as our ideal permissions for a user in an SNS, as shown in Table 3. Note that this is a fairly conservative ideal model in the sense that every related information to a user is considered as potentially privacy-sensitive. However, we see it as a safe and reasonable assumption.

Based on the abovementioned permissions, it is clear that Facebook does not satisfy *completely controllable* property (Definition 1) with regards to ideal policies listed in Table 3. The permissions in Table 2 are clearly not covered by the privacy settings, and we were able to verify this using our framework. For instance, D1 $\sqsubseteq$ I and not S is satisfiable.

Our framework also shows negative result for Facebook with regards to the *completely known* property (Definition 2), i.e., I $\sqsubseteq$ (S or D) is not satisfiable. In order to identify some of the missing policies, we formulated several permission classes (subclasses of ideal permissions), and tested against formula M $\equiv$ I and not (S or D), as explained in Section 4. We report in Table 4 some permissions that are missing in current Facebook privacy control policies according to our analysis. The table lists only a sample of permissions with selection as action. As expected, many missing permissions can be found for insertion and deletion permissions, that we did not include.

	Ideal Privacy Setting	Corresponding Permission
I1	Control whatever that relates to you	(pAction some Action) and (pResource some (ReifiedProperty and ((rdpSbj value me) or (ropObj value me) or (ropSbj valueuj me))))
I2	Control whatever that relates to something belonging to you	(pAction some Action) and (pResource some (ReifiedProperty and ((rdpSbj some (inverse ropObj some (RPowns and (ropSbj value me)))) or (ropObj some (inverse ropObj some (RPowns and (ropSbj value me)))) or (ropSbj some (inverse ropObj some (RPowns and (ropSbj value me)))))))

Table 3: Ideal Permissions for Facebook

	Missing Policy	Corresponding Permission
M1	Who can see that I have tagged someone?	(pAction value select) and (pResource some (RPowns and (ropObj some PhotoUserTag) and (ropSbj value me)))
M2	Who can see that I have liked something?	(pAction value select) and (pResource some (RPlikes and (ropSbj value me)))
M3	Who can see my comment on someone else's post?	(pAction value select) and (pResource some (RPannotates and (ropSbj some (Comment and (inverse ropObj some (RPowns and (ropSbj value me))))) and (ropObj some (inverse ropObj some (RPowns and (ropSbj some Others)))))
M4	Who can see if I am friend with someone?	(pAction value select) and (pResource some (RPisFriendOf and (ropSbj value me)))

Table 4: Sample Missing Permissions in Facebook

6. CONCLUSIONS

The privacy settings in existing SNSs such as Facebook do not provide users with adequate power to control their privacy-sensitive information. Furthermore, SNS system-defined policies are not typically clearly described and hence may not be easily understood by the users. To enable assessment and analysis of privacy protection achieved within an SNS environment, we proposed an analysis framework based on the ontology-based formulation of permissions in privacy control policies. We formulated completeness properties based on a given ideal specification of permissions and policies by an SNS in terms of privacy settings and described system policies. Our framework is capable of formally reasoning about such completeness properties and identifying undocumented and unclear practices of an SNS with regards to protecting privacy-sensitive information. We demonstrated the analysis power of our framework by performing a case study on analyzing Facebook's privacy control policies.

As future work, we plan to investigate efficient and effective algorithms to automate the process of identifying the missing policies. Moreover, we plan to extend our approach towards a comprehensive framework that can consider other components in the privacy control policies (e.g., subjects), and perform variety of privacy analysis tasks to support users and system developers.

Acknowledgments

This research has been partly supported by the U.S. National Science Foundation awards IIS-0545912 and DUE-0621274. We would like to thank Professor Elisa Bertino and our anonymous reviewers for their helpful comments.

7. REFERENCES

[1] B. Carminati, E. Ferrari, R. Heatherly, M. Kantarcioglu, and B. Thuraisingham. A semantic web based framework for social network access control. In *Proc. 14th ACM Symposium on Access Control Models and Technologies*, pages 177–186. ACM, 2009.

[2] B. Carminati, E. Ferrari, and A. Perego. Enforcing access control in Web-based social networks. *ACM Trans. Inf. Syst. Secur.*, 13(1):1–38, Nov. 2009.

[3] P. W. Fong. Relationship-based access control: protection model and policy language. CODASPY '11, pages 191–202, San Antonio, TX, USA, 2011. ACM.

[4] P. Hitzler, M. Krötzsch, B. Parsia, P. F. Patel-Schneider, and S. Rudolph. OWL 2 Web Ontology Language - Primer. `http://www.w3.org/TR/owl2-primer/`, 2009.

[5] V. Kolovski, J. Hendler, and B. Parsia. Analyzing web access control policies. In *Proc. 16th Int'l Conference on World Wide Web*, pages 677–686, 2007.

[6] D. Lin, P. Rao, E. Bertino, N. Li, and J. Lobo. EXAM: a comprehensive environment for the analysis of access control policies. *International Journal of Information Security*, 9(4):253–273, Aug. 2010.

[7] A. Masoumzadeh and J. Joshi. OSNAC: An Ontology-based Access Control Model for Social Networking Systems. In *Proc. 2nd IEEE Int'l Conference on Information Privacy, Security, Risk and Trust (PASSAT 2010)*, pages 751–759, 2010.

[8] A. Masoumzadeh and J. Joshi. Ontology-based access control for social network systems. *International Journal of Information Privacy, Security and Integrity (Special Issue: Selected Papers from PASSAT 2010)*, 1(1):59–78, Jan. 2011.

Trustworthy Distributed Computing on Social Networks

Abedelaziz Mohaisen
Verisign Labs, VA, USA

Huy Tran
KAIST, Daejeon, South Korea

Abhishek Chandra
University of Minnesota, MN, USA

Yongdae Kim
KAIST, Daejeon, South Korea

ABSTRACT

We investigate a new computing paradigm, called SocialCloud, in which computing nodes are governed by social ties driven from a bootstrapping trust-possessing social graph. We investigate how this paradigm differs from existing computing paradigms, such as grid computing and the conventional cloud computing paradigms. We show that incentives to adopt this paradigm are intuitive and natural, and security and trust guarantees provided by it are solid. We propose metrics for measuring the utility and advantage of this computing paradigm, and using real-world social graphs and structures of social traces; we investigate the potential of this paradigm for ordinary users. We study several design options and trade-offs, such as scheduling algorithms, centralization, and straggler handling, and show how they affect the utility of the paradigm. Interestingly, we conclude that whereas graphs known in the literature for high trust properties do not serve distributed trusted computing algorithms, such as Sybil defenses—for their weak algorithmic properties, such graphs are good candidates for our paradigm for their self-load-balancing features.

Categories and Subject Descriptors

C.2.0 [**Computer Communication Networks**]: General – *Security and Protection*; C.4 [**Performance of Systems**]: Design studies

General Terms

Security, Design, Experimentation

Keywords

Distributed computing, Trust, Social Computing.

1. INTRODUCTION

Cloud computing is a paradigm that overcomes restrictions of conventional computing systems by enabling elasticity and pay-as-you-go, which free users from long-term commitments and obligation towards providers. Cloud computing is beneficial for both consumers and cloud service providers. Despite such benefits, this paradigm also poses several challenges, including the need for architectures to support various potential applications, programming models to address large scale data-centric computing, and the need for strong security and data privacy protection guarantees. Indeed, both outsider and insider threats to security and privacy of data in cloud systems are unlimited. Also, there are many incentives for providers to use of users' data residing in cloud for their own benefits, for the lack of regulations and enforcing policies.

In this paper, we oversee a new type of computing paradigm, called SocialCloud, that enjoys parts of the merits provided by the conventional cloud. Imagine the scenario of a computing paradigm where users who collectively construct a pool of resources perform computational tasks on behalf of their social acquaintance. Our paradigm and model are similar in many aspects to the conventional grid-computing paradigm. It exhibits such similarities in that users can outsource their computational tasks to peers, complementarily to using friends for storage, which is extensively studied in literature. Our paradigm is, however, very unique in many aspects as well. Most importantly, our paradigm exploits the trust exhibited in social networks as a guarantee for the good behavior of other "workers" in the system. Accordingly, the most important ingredient to our paradigm is the social bootstrapping graph, a graph that is used for recruiting workers for SocialCloud. Most important to the context of SocialCloud is the aggregate computational power provided by users willing to share their idle time compute cycles [3]. In SocialCloud, owners of these computing resources are willing to share their computing resources for their friends, and for a different economical model than in the conventional cloud computing paradigm—fully altruistic one. This behavior makes our work share commonalities with an existing stream of work on creating computing services through volunteers [20, 5], although by enabling trust. Our results hence highlight technical aspects of this direction and pose challenges for designs options when using social networks for recruiting such workers and enabling trust.

The contribution of this paper is twofold. First, we investigate the potential of the social cloud computing paradigm by introducing a design that bootstraps from social graphs to construct distributing computing services. We advocate the merits of this paradigm over existing ones such as the grid computing paradigm. Second, we verify the potential of our paradigm using simulation set-up and real-world social graphs with varying social characteristics that reflect different, and possibly contradicting, trust models. Both graphs and the simulator are made public to the community to make use of them, and improve by additional features [14].

Organization. In §2 we review preliminaries followed by the design in §3. In §4, we describe our simulator followed in §5 by preliminary results, analyses and discussion. In §6, we summarize some of the related work followed by concluding remarks in §7.

ASIA CCS'13, May 8–10, 2013, Hangzhou, China.

2. ASSUMPTIONS AND SETTINGS

In this section, we review the preliminaries required for understanding the rest of this paper. In particular, we elaborate on the social networks, their popularity, and their potential for being used as bootstrapping tools for systems, services, and protocols. We describe the social network formulation at a high level, the economical aspect of our system, and finally, the attacker model.

Social Graphs—High Level Description. We view the social network as an undirected and unweighted graph $G = (V, E)$, where $V = \{v_1, \ldots, v_n\}$ is the set of vertexes, representing the set of nodes in the social graph, and correspond to users (or computing machines), and $E = \{e_{ij}\}$ (where $1 \leq i \leq n$ and $1 \leq j \leq n$) is the set of edges connecting those vertices. $|V| = n$ denotes the size of G and $|E| = m$ denotes the number of edges in G.

Economics of SocialCloud. In our design we assume an altruistic model, which simplifies the behavior of users and arguments on the attacker model. In this model, users *donate* their *computing resources* while not using them. One can further improve this model by incorporating a differential trust in scheduling [13]. In this work, and in order to make use of and confirm this model, we limit outsourced computations at 1-hop.

Use Model and Applications. For our paradigm, we envision compute intensive applications for which other systems have been developed in the past using different design principles, but lacking trust features. These systems include ones with resources provided by volunteers, as well as grid-like systems, like in Condor [11], MOON [10], Nebula [5], and SETI@Home [1]. Specific examples of applications built on top of these systems, that would as well fit to our use model, include blog analysis [20], web crawling and social apps (collaborative filtering, image processing, etc) [4], scientific computing [19], among others.

Attacker Model. In this paper, as it is the case in many other systems built on top of social networks, we assume that the attacker is restricted in many aspects. For example, the attacker has a limited capability of creating arbitrarily many edges between himself and other nodes in the social graph. For understanding the rationale of this model, see the related literature; e.g., [13]

Comparison with Trust in Grid Computing Systems. While there has been a lot of research on characterizing and improving trust in the conventional grid computing paradigm [2]—which is the closest paradigm to compare to ours, trust guarantees in such paradigm are less strict than what is expressed by social trust. For that, it is easy to see that some nodes in the grid computing paradigm may act maliciously by, for example, giving wrong computations, or refusing to collaborate; which is even easier to detect and tolerate, as opposed to acting maliciously.

3. THE DESIGN OF SOCIALCLOUD

The main design of SocialCloud is very simple, where complexities are hidden in design choices and options. In SocialCloud, the computing overlay is bootstrapped by the underlying social structure. Accordingly, nodes in the social graph act as workers to their adjacent nodes (i.e., nodes which are one hop away from the outsourcer of computations). An illustration of this design is depicted in Figure 1. In this design, nodes in the social graph, and those in the SocialCloud overlay, use their neighbors to outsource computational tasks to them. For that purpose, they utilize local information to decide on the way they schedule the amount of computations they want each and every one of their neighbors to take care of. Accordingly, each node has a scheduler which she uses for deciding the proportion of tasks that a node wants to outsource to any given worker among her neighbors. Once a task is outsourced to the given worker, and assuming that both data and code for processing the task are transferred to the worker, the worker is left to decide how to schedule the task locally to compute it. Upon completion of a task, the worker sends back the computations result to the outsourcer.

3.1 Design Options: Scheduling Entity

In SocialCloud two schedulers are used; one for determining the proportion of task outsourced to each worker and the second scheduler is used at each worker to determine how tasks outsourced by outsourcers are computed and in which order. While the latter scheduler can be easily implemented locally without impacting the system complexity, the decision used for whether to centralize or decentralize the former scheduler impacts the complexity and operation of the entire system. In the following, we elaborate on both design decisions, their characteristics, and compare them.

Decentralized Scheduler. In our paradigm, we limit selection of workers to 1-hop from the outsourcer. This makes it possible, and perhaps plausible, to incorporate scheduling of outsourcing tasks at the side of the outsourcer in a decentralized manner—thus each node takes care of scheduling its tasks. On the one hand, this could reduce the complexity of the design by eliminating the scheduling server in a centralized alternative. However, on the other hand, this could increase the complexity of the used protocols and the cost associated with them for exchanging *states*—such as availability of resources, online and offline time, among others. All of such states are exchanged between workers and outsourcers in our paradigm. These states are essential for building basic primitives in any distributed computing system to improve efficiency (see below for further details). An illustration of this design option is shown in Figure 1. In this scenario, each outsourcer and worker has its own separate scheduler.

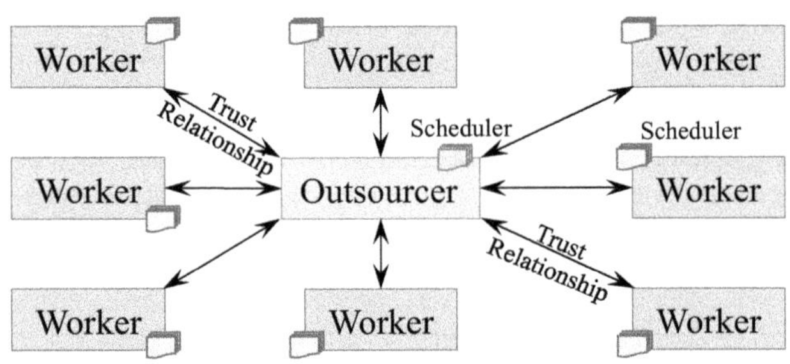

Figure 1: A depiction of the main SocialCloud paradigm as viewed by an outsourcer of computations.

Centralized Scheduler. Despite that nodes may only require their neighbors to perform the computational tasks on behalf of them and that may require only local information—which could be available to these nodes in advance, the use of a centralized scheduler might be necessitated to reduce communication overhead at the protocol level. For example, in order to decide upon the best set of nodes to which to outsource computations, a node needs to know which of its neighbors are available, among other statistics. For that purpose, and given that the underlying communication network topology may not necessarily have the same proximity of the social network topology, the protocol among nodes needs to incur back and forth communication cost. One possible solution to the problem is to use a centralized server that maintains states of the different nodes. Instead of communicating directly with neighbor nodes, an outsourcer would request the best set of candidates among its neighbors to the centralized scheduling server. In response, the server

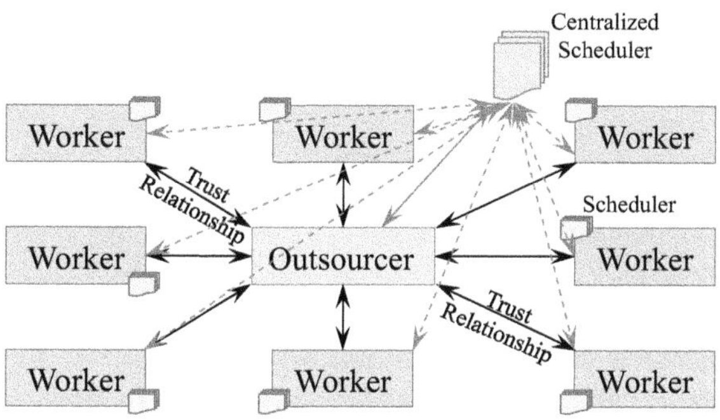

Figure 2: Decentralized model of scheduling in SocialCloud.

Table 1: A comparison of design options

Option	Failure	Comm.	Hardware	Trust
Centralized	✖	$O(n)$	✖	✖
Decentralized	✔	$O(m)$	✔	✔

will produce a set of candidates, based on the locally stored states. Such candidates would typically be those that would have the most available resources to handle outsourced computation tasks.

An illustration is shown in Figure 2. In this design, each node in SocialCloud would periodically send states to a centralized server. When needed, an outsourcer node contacts the centralized server to return to it the best set of candidates for outsourcing computations, which the server would return based on the states of these candidates. Notice that only states are returned to the outsourcer, upon which the outsourcer would send tasks to these nodes on its own—Thus, the server involvement is limited to the control protocol.

The communication overhead of this design option to transfer states between a set of d nodes is $2d$, where d messages are required to deliver all nodes' states and d messages are required to deliver states of all other nodes to each node in the set. On the other hand, $d(d-1)$ messages are required in the decentralized option (which requires pairwise communication of states update). When outsourcing of computations is possible among all nodes in the graph, this translates into $O(n)$ for the centralized versus $O(n^2)$ communication overhead for the decentralized option. To sum up, Table 1 shows a comparison between both options.

3.2 Tasks Scheduling Policy

While using distributed or centralized scheduler resolves scheduling at the outsourcer, two decisions remain untackled: how much computation to outsource to each worker, and how much time a worker should spend on a given task for a certain outsourcer. We address these two issues separately.

Any off-the-shelf scheduling algorithm can be used to schedule at outsourcer's side, which can be further improved by incorporating trust models for weighted job scheduling [13]. On the other hand, we consider several scheduling algorithms for workers scheduling, as follows. (*i*) Round Robin (RR) Scheduling Policy This is the simplest policy to implement, in which a worker spends an equal share of time on each outsourced task in a round robin fashion among all tasks he has. (*ii*) Shortest First (SF) Scheduling Policy The worker performs shortest task first. (*iii*) Longest First (LF) Scheduling Policy The worker performs longest task first.

Notice that we omit a lot of details about the underlying computing infrastructure, and abstract such infrastructure to "time sharing machines", which further simplifies much of the analysis in this work. However, in a working version of this paradigm, all of these aspects are addressed in a similar manner is in other distributed systems and paradigms. See §5 for details on limitations of this approach and possible extensions in the future work.

3.3 Handling Outliers

The main performance criterion used for evaluating SocialCloud is the time required to finish computing tasks for all nodes with tasks in the system. Accordingly, an outlier (also called a computing straggler) is a node with computational tasks that take a long time to finish, thus increasing the overall time to finish and decreasing the performance of the overall system. Detecting outliers in our system is simple: since the total time is given in advance, outliers are nodes with computing tasks that have longer time to finish when other nodes participating in the same outsourced computation are idle. Our method for handling outliers is simple too: when an outlier is detected, we outsource the remaining part of computations on all idle nodes neighboring the original outsourcer. For that, we use the same scheduling policy used by the outsourcer when she first outsourced this task. In the simulation part, we consider both scenarios of handled and unhandled outliers, and observe how they affect the performance of the system.

3.4 Deciding Workers Based on Resources

In real-world deployment of SocialCloud, we expect heterogeneity of resources, such as bandwidth, storage, and computing power, in workers. This heterogeneity would result in different results and utilization statistics of a system like SocialCloud, depending on which nodes are used for what tasks. While our work does not address this issue, and leaves it as a future work. We further believe that simple decisions can be made in this regard so as to meet the design goals and achieve the good performance. For example, we expect that nodes would select workers among their social neighbors that have resources and link capacities exceeding a threshold, thus meeting an expected performance outcome.

4. SIMULATOR OF SOCIALCLOUD

To demonstrate the potential of SocialCloud as a computing paradigm, we implement a batch-based simulator [16] that considers a variety of scheduling algorithms, an outlier handling mechanism, job generation handling, and failure simulation. A flow diagram of the simulator is in Figure 3.

The flow of the simulator is in Figure 3. First, a node factory uses the bootstrapping social graph to create nodes and workers. Each node then decides on whether she has a task or not, and if she has a task she schedules the task according to her scheduling algorithm. If needed, each node then transfers code on which computations are to be performed to the worker along with the chunks of the data for these codes to run on. Each worker then performs the computation according to her scheduling algorithm and returns results to the outsourcer.

Timing. In SocialCloud, we use *virtual time* to simulate computations and resources sharing. We scale down the simulated time by 3 orders of magnitude of that in reality. This is, for every second worth of computations in real-world, we use one millisecond in the simulation environment. Thus, units of times in the rest of this paper are in virtual seconds.

5. RESULTS AND ANALYSIS

In this section, and in order to derive insight on the potential of SocialCloud, we experiment with the simulator described above. Before getting into the details of the experiments, we describe the data and evaluation metric used in this section.

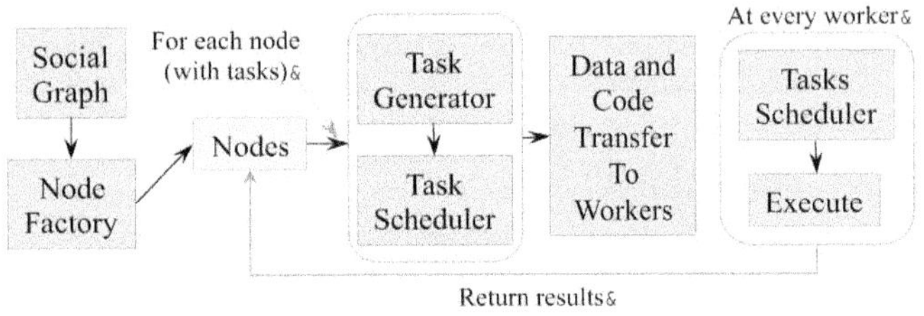

Figure 3: The flow diagram of SocialCloud.

5.1 Evaluation Metric

To demonstrate the potential of operating SocialCloud, we use the "normalized finishing time" of a task outsourced by a user to other nodes in the SocialCloud as the performance metric. For that, we use the empirical commutative distribution function (CDF) as an aggregate measure among all nodes with tasks to compute. For a random variable X, the CDF is $F_X(x) = P_r(X \leq x)$. In our experiments, the CDF measures the fraction of nodes that finish their tasks before time x, as part of the overall number of tasks. We define x as the factors of time of normal operation per dedicated machines, if they were to be used instead of outsourcing computations. This is, suppose that the overall time of a task is T_{tot} and the time it takes to compute the subtask by the slowest worker is T_{last}, then x for that node is defined as T_{last}/T_{tot}.

5.2 Tasks Generation

To demonestrate the operation of our simulator and the trade-off our system provides, we consider two different approaches for the tasks generated by each user. The size of each generated task is measured by virtual units of time, and for our demonstration we use two different scenarios. (*i*) **Constant task weight.** each outsourcer generates tasks with an equal size. These tasks are divided into equal shares and distributed among different workers in the computing system. The size of each task is $\bar{T}$. (*ii*) **Variable task weight.** each outsourcer has a different task size. We model the size of tasks as a uniformly distributed random variable in the range of $[\bar{T} - \ell, \bar{T} + \ell]$ for some $\bar{T} > \ell$. Each worker receives an equal share of the task from the outsourcer.

5.3 Deciding Tasks Outsourcers

Not all nodes in the system are likely to have tasks to outsource for computation at the same time. Accordingly, we denote the fraction of nodes that have tasks to compute by p, where $0 < p < 1$. In our experiments we use p from 0.1 to 0.5 with increments of 0.1. We further consider that each node in the network has a task to compute with probability p, and has no task with probability $1 - p$—thus, whether a node has a task to distribute among its neighbors and compute or not follows a binomial distribution with a parameter p. Once a node is determined to be among nodes with tasks at the current round of run of the simulator, we fix the task length. For tasks length, we use both scenarios mentioned in §5.2; with fixed or constant and variable tasks weights.

5.4 Social Graphs

To derive insight on the potential of SocialCloud, we run our simulator on several social graphs with different size and density, as shown in Table 2. The graphs used in these experiments represent three co-authorship social structures (DBLP, Physics 1, and Physics 2), one voting network (of Wiki-vote for wikipedia administrators election), and one friendship network (of the consumer review website, Epinion). Notice the varying density of these graphs, which also reflects on varying topological characteristics. Also, notice the nature of these social graphs, where they are built in different social contexts and possess varying qualities of trust that fits to the application scenario mentioned earlier. The proposed architectural design of SocialClould, however, minimally depends on these graphs, and other networks can brought instead of them. As these graphs are widely used for verifying other applications on social networks, we believe they enjoy a set of representative characteristics to other networks as well.

Table 2: Social graphs used in our experiments.

Dataset	# nodes	# edges	Description
DBLP	614981	1155148	CS Co-authorship
Epinion	75877	405739	Friendship network
Physics 2	11204	117649	Co-authorship
Wiki-vote	7066	100736	Voting network
Physics 1	4158	13428	Co-authorship

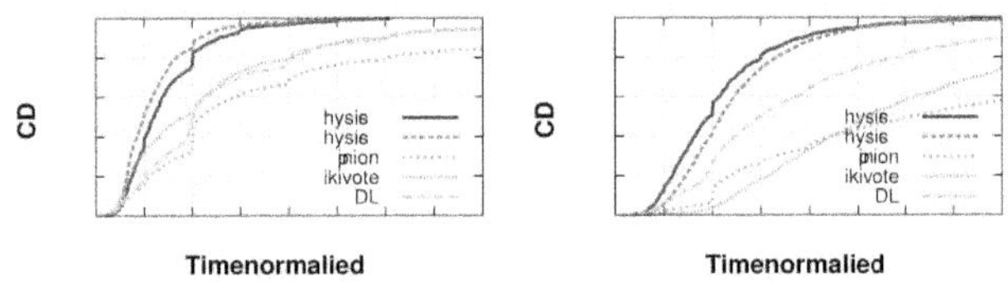

(a) Handled outliers (b) Unhandled outliers

Figure 5: The performance of SocialCloud on the different social graphs. Both figures use $p = 0.3$ and the RRS.

5.5 Main Results

Number of Outsourcers. In the first experiment, we run our SocialCloud simulator on the different social graphs discussed earlier to measure the evaluation metric when the number of the outsourcers of tasks increases. We consider $p = 0.1$ to 0.5 with increments of 0.1 at each time. The results of this experiment are in Figure 4. On the results of this experiment we make several observations. First, we observe the potential of SocialCloud, even when the number of outsourcers of computations in the social network is as high as 50% of the total number of nodes, which translates into a small normalized time to finish even in the worst performing social graphs (about 60% of all nodes with tasks would finish in 2 normalized time units). However, this advantage varies for different graphs: we observe that sparse graphs, like co-authorship graphs, generally outperform other graphs used in the experiments (by observing the tendency in the performance in figures 4(a) through 4(b) versus figures 4(c) and 4(d)). In the aforementioned graphs, for example, we see that when 10% of nodes in each case is used, and by fixing x, the normalized time, to 1, the difference of performance is about 30%. This difference of performance can be observed by comparing the Physics co-authorship graphs—where 95% of nodes finish their computations—and the Epinion graph—where only about 65% of nodes finish their computations.

Second, we observe that the impact of p, the fraction of nodes with tasks in the system, would depend greatly on the underlying graph rather than p alone. For example, in Figure 4(a), we observe that moving from $p = 0.1$ to $p = 0.5$ (when $x = 1$) leads to a decrease in the fraction of nodes that finish their computations from 95% to about 75%. On the other hand, for the same settings, this would lead to a decrease from about 80% to 40%, a decrease from about 65% to 30%, and a decrease from 70% to 30% in DBLP, Epinion, and Wiki-vote, respectively. This suggests that the de-

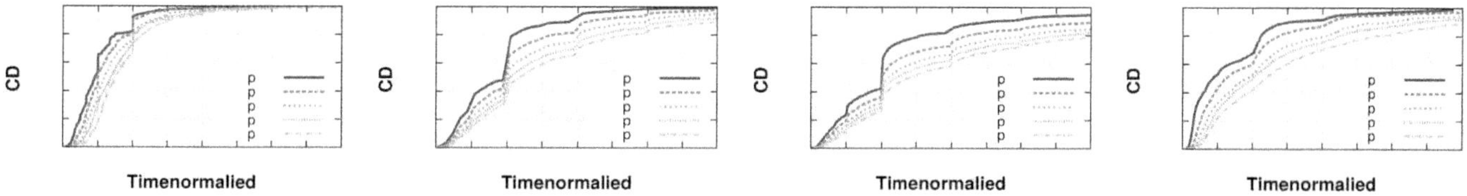

(a) Physics 1. (b) DBLP. (c) Epinion. (d) Wiki-vote.

Figure 4: The normalized time it takes to perform outsourced computations in SocialCloud.

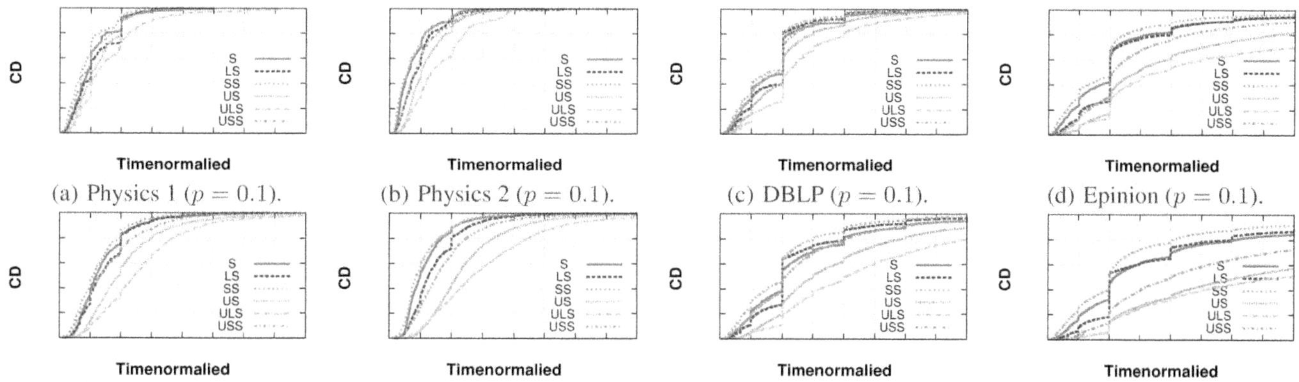

(a) Physics 1 ($p = 0.1$). (b) Physics 2 ($p = 0.1$). (c) DBLP ($p = 0.1$). (d) Epinion ($p = 0.1$).

(e) Physics 1 ($p = 0.3$). (f) Physics 2 ($p = 0.3$). (g) DBLP ($p = 0.3$). (h) Epinion ($p = 0.3$).

Figure 6: The normalized time it takes to perform outsourced computations in SocialCloud for different scheduling policies; U and stand for unhandled and handled (balanced) outlier. RRS, SFS, and LFS are the scheduling policies.

creases in the performance are due to an inherit property of each graph. The inherit property of each graph and how it affects the performance of SocialCloud is further illustrated in Figure 5. We find that even when DBLP's size is two orders of magnitude the size of Wiki-vote, it outperforms Wiki-vote when not using outlier handling, and gives almost the same performance when using it.

Scheduling Policy. To understand the impact of policies on the performance, we use $p = 0.1$ to 0.5 with 0.2 increments (partly shown in Figure 6). The observed consistent pattern in all figures in this experiment tells that shortest first policy always outperforms the round robin policy, whereas the round robin policy outperforms the longest first. This pattern is consistent regardless of p and the outlier handling policy. The difference in the performance when using different policies can be as low as 2% (when $p = 0.1$ in physics co-authorship; shown in Figure 6(f)) and as high as 70% (when using $p = 0.5$ and outlier handling as in wiki-vote; not shown for the lack of space). The patterns are made clearer in Figure 6 by observing combinations of parameters and policies.

Performance with Outliers Handling. Outliers drag the performance of the entire system down. However, handling outliers is quite simple in SocialCloud if accurate timing is used. Here we consider the impact of the outlier handling policy in §3.3. In this figure 6, we see that the simple handling policy we proposed improves the performance of the system greatly in all cases. The improvement differs depending on other parameters, such as p, and the scheduling policy. As with the scheduling policy, the improvement can be as low as 2% and as high as more than 60%. When p is large, the potential for improvement is high—see, for example, $p = 5$ in Physics 2 with the round robin scheduling policy where almost 65% improvement is due to outlier handling when $x = 1$.

Variable Task Size. In all experiments so far, we considered tasks of fixed size; 1000 of virtual time units in each of them. Whether the same pattern would be observed in tasks with variable size is unclear. Here we experimentally address this concern by using variable duty size that is uniformly distributed in the interval of $[500, 1500]$ time units. The results are shown in Figure 7. Comparing these results to the middle row of Figure 6 (for the fixed size tasks), we make two observations. (i) While the average task size in both scenarios is same, we observe that the performance with variable task size is worse. This performance is anticipated as our measure of performance is the time to finish that would be definitely increased as some tasks with longer time to finish are added. (ii) The same patterns advantaging a given scheduling policy on another are maintained as in earlier with fixed task length.

Structure and Performance. We note that the performance of SocialCloud is quite related to the underlying structure of the social graph. We see that sparse graphs, like co-authorship graphs, have a performance advantage in SocialCloud. These graphs are shown to possess a nice trust value that can be further utilized in SocialCloud. Furthermore, this trust value is unfound in online social networks which are prone to infiltration, making the case for trust-possessing graphs even stronger, as they achieve performance guarantees as well. This, indeed, is an interesting finding by itself, since it shows contradicting outcomes to what is known in the literature on the usefulness of these graphs—see §2 for more details and the work in [13] for prior literature that agrees with our findings.

5.6 Additional Features and Limitations

Our simulator of SocialCloud omits a few details concerning the way a distributed system behaves. For example our measurements do not consider failure. Furthermore, our simulator considers a simplistic scenario by abstracting the hardware, and does not consider additional resources consumed, such as memory and I/O resources. We also do not consider the heterogeneity of resources, such as bandwidth and memory. Last, we did not consider how this affects the usability of our system and what decision choices this particular aspect of distributed computing systems would have on the utility of our paradigm. While this work would be pursued in the future, we expect that nodes would select workers among their

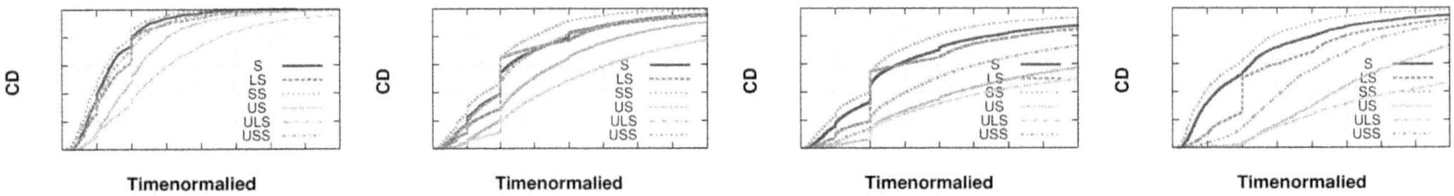

(a) Physics 1 ($p = 0.3$). (b) DBLP ($p = 0.3$). (c) Epinion ($p = 0.3$). (d) Wiki-vote ($p = 0.3$).

Figure 7: The normalized time to perform outsourced computations in SocialCloud, for variable task size.

social neighbors that have resources and link capacities exceeding a threshold, thus meeting an expected performance outcome.

6. RELATED WORK

Systems built on top of social networks include file sharing [8, 13], anonymous communication [18], Sybil defenses [21, 12, 15], routing [7], and content distribution [22], among many others. Most of these systems use social networks' trust and connectivity for their operation. Concurrent to our work, and following their work in [6], Chard et al. suggested the use of social networks to build a resource sharing system. Whereas their main realization was still a social storage system as in [6], they also suggested that the same vision can be used to build a distributed computing service as we advocate in this work. Recent realizations of this vision have been reported in [17] and [9]. In [17], Thaufeeg et al. devised an architecture where "individuals or institutions contribute the capacity of their computing resources by means of virtual machines leased through the social network".

In [9] Koshy et al. further explored the motivations of users to enable social cloud systems for scientific computing. With similar flavor of distributed computing services design, there has been prior works in literature on using volunteers' resources for computations exploiting locality of data [5, 20], examination of programing paradigms on such system [10]. Finally, our work shares several commonalities with grid and volunteer computing systems [11, 10, 5, 20, 1], of which many aspects are explored in the literature. Trust of grid computing and volunteer-based systems is explored in [2]. Applications built on top of these systems, that would fit to our use model, are reported in [20], among others.

7. SUMMARY AND FUTURE WORK

In this paper we have introduced the design of SocialCloud, a distributed computing service that recruits computing workers from friends in social networks and use such social networks that characterize trust relationships to bootstrap a computing service. To demonstrate the potential of our design, we used several social graphs to show that the majority of nodes in most graphs would benefit from outsourcing their computations to such service. We considered several basic distributed system characteristics and features, such as outlier handling, scheduling decisions, and scheduler design, and show advantages in each of these features and options when used in our system.

In the future we aim to complete the missing features of the simulator and enrich it by further scenarios of deployment of our design, under failure, with different scheduling algorithms at both sides of the outsourcer and workers. We will consider other overhead characteristics that might not be in line with topological characteristics in the social graph. These characteristics may include the uptime, downtime, communication overhead, and I/O overhead consumption. Second, we will implement a proof-of-concept application by utilizing design options discussed in this paper. We anticipate hidden complexities in the design to arise, and significant findings to come out of the deployment.

Acknowledgement. The last author was supported by the KCC (Korea Communications Commission), Korea, under the R&D program supervised by the KCA (Korea Communications Agency), KCA- 2013-12911-05003.

8. REFERENCES

[1] D. Anderson, J. Cobb, E. Korpela, M. Lebofsky, and D. Werthimer. Seti@ home: an experiment in public-resource computing. *Communications of the ACM*, 45(11):56–61, 2002.

[2] F. Azzedin and M. Maheswaran. Towards trust-aware resource management in grid computing systems. In *Proc. of CCGRID*, 2002.

[3] L. Barroso and U. Holzle. The case for energy-proportional computing. *Computer*, 40(12):33–37, 2007.

[4] M. Cardosa, C. Wang, A. Nangia, A. Chandra, and J. Weissman. Exploring mapreduce efficiency with highly-distributed data. In *Proc. of ACM MapReduce*, 2011.

[5] A. Chandra and J. Weissman. Nebulas: Using distributed voluntary resources to build clouds. In *Proc. of HotCloud*, 2010.

[6] K. Chard, S. Caton, O. Rana, and K. Bubendorfer. Social cloud: Cloud computing in social networks. In *CLOUD*. IEEE, 2010.

[7] E. M. Daly and M. Haahr. Social network analysis for routing in disconnected delay-tolerant manets. In *Proc. of ACM Mobihoc*, 2007.

[8] T. Isdal, M. Piatek, A. Krishnamurthy, and T. Anderson. Privacy-preserving p2p data sharing with oneswarm. In *ACM SIGCOMM*, 2010.

[9] J. Koshy, K. Bubendorfer, and K. Chard. A social cloud for public eresearch. In *eScience*, pages 363–370. IEEE, 2011.

[10] H. Lin, X. Ma, J. Archuleta, W.-c. Feng, M. Gardner, and Z. Zhang. Moon: Mapreduce on opportunistic environments. In *Proc. of HPDC*, pages 95–106, New York, NY, USA, 2010. ACM.

[11] M. Litzkow, M. Livny, and M. Mutka. Condor-a hunter of idle workstations. In *Proc. of ICDCS*, pages 104–111. IEEE, 1988.

[12] P. Mittal, M. Wright, and N. Borisov. Pisces: Anonymous communication using social networks. In *NDSS*, 2013.

[13] A. Mohaisen, N. Hopper, and Y. Kim. Keep your friends close: Incorporating trust into social network-based sybil defenses. In *Proc. of INFOCOM*, pages 1943–1951, 2011.

[14] A. Mohaisen, H. Tran, A. Chandra, and Y. Kim. Socialcloud: distributed computing on social networks. Technical report, University of Minnesota, 2011.

[15] A. Mohaisen, H. Tran, N. Hopper, and Y. Kim. On the mixing time of directed social graphs and security implications. In *ASIACCS*, 2012.

[16] SCloud Team. Anonymized for submission. Anonymized for submission, 2012.

[17] A. M. Thaufeeg, K. Bubendorfer, and K. Chard. Collaborative eresearch in a social cloud. In *eScience*, pages 224–231. IEEE, 2011.

[18] E. Vasserman, R. Jansen, J. Tyra, N. Hopper, and Y. Kim. Membership-concealing overlay networks. In *Proc. of ACM CCS*, pages 390–399, 2009.

[19] L. Wang, J. Tao, M. Kunze, A. Castellanos, D. Kramer, and W. Karl. Scientific cloud computing: Early definition and experience. In *Proc. of IEEE HPCC*, pages 825–830. Ieee, 2008.

[20] J. B. Weissman, P. Sundarrajan, A. Gupta, M. Ryden, R. Nair, and A. Chandra. Early experience with the distributed nebula cloud. In *Proc. of ACM DIDC*, pages 17–26, 2011.

[21] H. Yu, P. B. Gibbons, M. Kaminsky, and F. Xiao. Sybillimit: A near-optimal social network defense against sybil attacks. In *Proc. of IEEE S&P*, pages 3–17, 2008.

[22] F. Zhou, L. Zhang, E. Franco, A. Mislove, R. Revis, and R. Sundaram. WebCloud: Recruiting social network users to assist in content distribution. In *IEEE NCA*, 2012.

On the Feasibility of Inference Attacks by Third-Party Extensions to Social Network Systems

Seyed Hossein Ahmadinejad
Department of Computer Science
University of Calgary
Calgary, Alberta, Canada
shahmadi@ucalgary.ca

Philip W. L. Fong
Department of Computer Science
University of Calgary
Calgary, Alberta, Canada
pwlfong@ucalgary.ca

ABSTRACT

Social Network Systems (SNSs) providers allow third-party extensions to access users' information through an Application Programming Interface (API). Once an extension has been authorized by a user to access data in a user's profile, there is no more control on how that extension uses the data. This raises serious concerns about user privacy because a malicious extension may infer some private information based on the legitimately accessible information. This information leakage is called an *inference attack*. In addition, inference attacks are not only a privacy violation, they could also be used as the building blocks for more dangerous security attacks, such as identity theft. In this work, we conduct a comprehensive empirical study to assess the feasibility and accuracy of inference attacks that are launched from the extension API of SNSs. We also discuss an attack scenario in which inference attacks are employed as building blocks. The significance of this work is in thoroughly discussing how inference attacks could happen in practice via the extension API of SNSs, and highlighting the clear and present danger of even the naively crafted inference attacks.

Categories and Subject Descriptors

K.4.1 [**Computers and Society**]: Public Policy Issues—*Privacy*; K.6.5 [**Management of Computing and Information Systems**]: Security and Protection—*Invasive software*

Keywords

Social networks; Privacy; Inference attacks

1. INTRODUCTION

Third-party applications are a popular feature of SNSs. For instance, there are applications with more than 50M monthly active users on the Facebook platform. Third-party developers host their Facebook applications on their own (untrusted) servers. Such applications then interact with users and access their personal information through an API. Every access request sent by an application via an API call is guarded by a permission check. If the permission is granted by the user, the data will be sent to the application.

Problem definition.

Not all third-party applications are benign. For example, a malicious application may sell user information to marketing companies. As a result, delivering user data via the extension API to such applications puts the data at risk. While this problem has to do with the misuse of *legitimately accessible* information, this work is instead about a more challenging problem: i.e., through the extension API, malicious applications may gain access to private information for which they are not authorized. The following is an example.

Example 1. *In a Facebook user profile is a "*wall*" on which the user or her friends may post status updates, photos, messages, etc. Now, a malicious third-party application asks for the permission to access the wall. Say the permission is granted by the user. The application accesses the user's wall through the extension API, and then scans the wall for a day when a considerable number of birthday greetings were posted. The application can thus infer the user's birthday.*

In the above example, if the individual considers her birthday as private information (and thus should be inaccessible), then the malicious application has effectively inferred some supposedly inaccessible information about the user from the information that is accessible through the extension API.

Note that an application may also have access to some information about an individual outside the SNS. So it might utilize such information in inferring the individual's inaccessible information. We use the term ***SNS API inference attacks*** to refer to the inference of inaccessible information from both accessible information and background information. The term highlights the fact that this work focuses on inference attacks that are launched by extensions to SNSs through the extension API. Our emphasis on the role of third-party extensions in launching inference attacks differentiates us from other related work on generic inference attacks over social network data sets [6, 5, 3].

Significance of the problem.

A naive interlocutor may argue that the above issue has already been addressed by the permission-based access control mechanism, in that third-party extensions cannot access user information without seeking the required permissions. If a user does not trust a third-party application, then she shall not authorize it. This argument presumes that ordinary users have the necessary information and expertise to judge whether the applications they subscribe to are benign. In reality, most of the third-party applications are developed by developers who are not widely known to the user community. It is therefore not always possible for a user to assess if she can trust an application.

One may also claim that SNS API inference attacks are but another minor privacy violation. We disagree for two reasons. First, analyzing the threats of any security concern must be accompanied by assessing the number of potential victims. When the number of potential victims reaches, for example, 50M, then we are facing a trouble with costly consequences. Popular Facebook applications have 50M monthly users, implying that an inference attack with a meagre success rate of 10% leads to privacy violations of 5M victims.

Second, SNS API inference attacks can be employed as building blocks for conducting more dangerous security attacks. For instance, an alternative authentication mechanism is to ask users security questions (e.g., "what is the name of your youngest sibling?"). Answers to these security questions can usually be harvested systematically by launching SNS API inference attacks. The ability to answer a victim's security questions is the first step of identity theft. Therefore, inference attacks could be an initial step in the launching more dangerous attacks. Now, who is best positioned to launch covert inference attacks? The answer is third-party extension developers.

Contributions.

The specific contributions of this work are the following:

- By way of a comprehensive empirical study (Section 3), we develop deep insight into the problem of SNS API inference attacks, and demonstrate the growing threat of such attacks to user privacy.
- We devised an analytical framework for evaluating the risk of SNS API inference attacks. A key component of our framework is the classification of user profiles into four categories from the perspective of an adversary (Section 4). Based on this classification, we devised a scoring technique for assessing the success rate of inference algorithms (Section 6). The insights we gained can inform the design of protection mechanisms for mitigating the threat of SNS API inference attacks. Our analytical methodology can be applied for analyzing SNS API inference attacks not covered in this work.
- Eight (8) realistic inference algorithms are devised for our empirical study, covering a wide variety of inference techniques and inference channels. This diversity of inference techniques improves our understanding of the threat of SNS API inference attacks (Section 5).
- We examine a scenario, namely, identity theft, in which SNS API inference attacks are used as building blocks of more dangerous security attacks (Section 7). We formally model the success rate of this attack, with the success rates of the component SNS API inference attack algorithms as model parameters. We then feed our empirical data into the model to obtain the success rate of the identity theft attack. This modelling exercise offers an innovative means for demonstrating the threat of SNS API inference attacks.

In this work, we mainly focus on Facebook, the most popular SNS. However, our observations regarding Facebook third-party applications also apply to other extensible SNSs (e.g., OpenSocial).

2. SNS API INFERENCE ATTACKS

In a typical SNS, every user owns a profile consisting of attributes such as birthdate, education information, etc. Permission-based authorization schemes allow users to restrict access to their attributes. Therefore, a third-party application that needs to access a user's attributes needs to request the user for granting the required permissions.

A user can also permit an application to access certain information that she does not own. Specifically, given that the required conditions are satisfied, a user can allow an application to access her friend's photo albums. In such cases, the user who runs the application assumes the role of a ***proxy*** through whom the application accesses the friend's profile. An application can access the friend's information if the following conditions are met: (i) the proxy user grants indirect access to the application; (ii) the friend grants permissions to the proxy user for accessing the requested information; (iii) the friend explicitly allows her information to be accessible by applications through proxy users.

Suppose a user u subscribes to a third-party application π in an SNS. As a result, in her interactions with π, the SNS makes some information accessible to π (aka ***accessible information***), given that the required permissions are granted. Such accessible information may contain information about users other than u (e.g., u's friends). Moreover, there might be some personal information about u (aka ***inaccessible information***) that she intends not to share with π. Such an intention is sometimes declared explicitly in her privacy settings, but sometimes implicitly willed by her. π may also have some ***background information*** regarding u. Now a successful SNS API inference attack launched by π against u is defined as follows: *π infers some inaccessible information about u from its background information as well the accessible information it could obtain in its interaction with u from the SNS API. The inferred information must not be inferable solely from π's background information.*

As we want to focus on the role of third-party extensions in launching inference attacks, we give the above definition to emphasize that the user's accessible information through the SNS API must be utilized for inferring information. Note also that π complies with the protection mechanisms of the target SNS, by accessing nothing but the legitimately accessible information. Yet, the protection mechanisms fail to prevent inference of inaccessible information.

The definition above does not cover inference attacks that infer information about friends of the user who runs the application. Our goal is to show that, even by considering only such limited attacks, the damages can still be significant.

3. EMPIRICAL STUDY: DESIGN

This section reports the design of an empirical study we conducted to assess the feasibility of SNS API inference attacks. For the purpose of this study, we identified eight sample inference tasks and devised an inference algorithm for each task. Each inference algorithm takes as input some accessible information, carries some background information that it employs for inference, and infers some targeted inaccessible information. As mentioned previously, inference attacks can be employed as building blocks for other attacks such as identity theft. We therefore evaluated the success rate of these inference algorithms with this application in mind. Specifically, an attacker is usually allowed to make some α attempts for each security question, before the authentication mechanism blocks off the attacker (where $\alpha > 1$). Therefore, each of our algorithms will return α answers, corresponding to the number of attempts the attacker is allowed to make. We set $\alpha = 4$ in our study.

A third-party Facebook application (coded in JavaScript) was developed as an environment for simulation and data collection. Embedded in the application are our inference algorithms that were executed on the participants' profiles. To achieve 95% confidence level and 5% margin of error, we recruited 424 participants (according to [4], 384 participants are required) from 150 universities across North America.

An execution of this application involves the following steps. First, the participant selects a subset of the eight sample inference algorithms to be included in the simulation. Then, the application asks the participant to grant the permissions needed for setting up the accessible information of the selected inference algorithms. The application simulates the accessing of the accessible information, and the running of the inference algorithms. Each algorithm infers up to α answers for each of the inference tasks. Next, for each selected inference task, if the corresponding inference algorithm is able to infer at least one answer, then the answers is presented to the participant, in the order of confidence of the algorithm. The participants are to confirm which answer is the right answer, or to declare that none of the answers is correct. In summary, the execution of an inference algorithm could lead to three possible outcomes: 1) no answer is returned, 2) none of the returned answers is correct, or 3) one of the returned answers is correct. The first two outcomes simply mean algorithm failure.

4. ANALYTICAL FRAMEWORK

In this section, we describe our analytical framework based on which we analyze the behaviour of our inference algorithms. When we assess the success rate of inference algorithms, there is more than one reason for an algorithm to fail. For example, sometimes the data needed by an algorithm to make inference is not available, and at other times the data is available but the algorithm cannot gain access to that data. For each inference algorithm, we classify user profiles based on the level of availability of the information that the algorithm needs to extract from the profile in order to make correct inference. In essence, this classification corresponds to the different reasons for inference failure. For each algorithm, we will define a different success rate when each of the failure conditions is ruled out.

In Facebook, there are two types of access permissions involved in user-application interactions:

1. ***Type-1 permissions***: These are permissions that are granted or denied by the user who is running the application. They include the permissions to access the user's attributes, as well as the permissions to access the attributes of the user's friends.

2. ***Type-2 permissions***: These are permissions that are granted or denied by the friends of the user who is running the application. They include the permissions that a friend grants to the user running the application, as well as the permissions that friends can specify regarding which attributes can be accessed by applications through proxy users[1].

We specify below a scheme for classifying user profiles. Corresponding to each class of user profiles is a predicate.

Definition 1. *A user profile j is* **type-1 accessible** *to algorithm i ($accessible1(i,j) = 1$) iff the type-1 permissions requested by algorithm i have been granted by user j.*

Definition 2. *A user profile j is* **type-2 accessible** *to algorithm i ($accessible2(i,j) = 1$) iff $accessible1(i,j) = 1$ and the type-2 permissions required by algorithm i are granted by at least one of the friends of user j.*

For example, suppose algorithm i, executing on user j's profile, needs to access user j's friends' photo albums. User j's profile is type-2 accessible for algorithm i if photo albums of at least one of her friends is accessible to algorithm i.

Definition 3. *A user profile j is* **available** *to algorithm i ($available(i,j) = 1$) iff $accessible2(i,j) = 1$ and algorithm i could at least return one answer when it is executed on user profile j. In other words, the required data for making a guess is both accessible and available.*

Definition 4. *A user profile j is* **applicable** *for algorithm i ($applicable(i,j) = 1$) iff $available(i,j) = 1$ and the information algorithm i attempts to infer is logically defined for user j. That is, it is not a logical impossibility for algorithm i to infer the target information from the user j's profile.*

For instance, if an individual is single, then it is a logical impossibility for an inference algorithm to infer his/her spouse's name. So this individual's profile is not applicable for this specific algorithm.

We define below a predicate pertinent to the success scoring of an inference algorithm.

- $succeed(i,j,\alpha)$: This binary predicate evaluates to 1 iff, when algorithm i is executed on user profile j, the correct answer of the inference task is among the first α candidate answers returned by algorithm i.

We propose here four different success rates for evaluating the effectiveness of an inference algorithm in targeting the different classes of user profiles. Table 1 shows the formulas for the success rates of algorithm i. Here, N is the total number of user profiles on which algorithm i was executed. For every profile class and for every α, we define a different success rate. For example, $P^i_{ava}(\alpha)$ is the ratio of available profiles for which algorithm i successfully returned the right answer in at most α attempts. Note that the four success rates are totally ordered: i.e. $P^i_{acc1}(\alpha) \leq P^i_{acc2}(\alpha) \leq P^i_{ava}(\alpha) \leq P^i_{app}(\alpha)$. Note also that the success rates in Table 1 are conditional probabilities.

5. INFERENCE ALGORITHMS

This section presents a brief description of the inference strategy in the eight sample inference algorithms. In all of

[1]Recall from Section 2 that when an application accesses a friend attribute, both type-1 and type-2 permissions are needed.

Success rate	Formula
Type-1 Accessibility	$P^i_{acc1}(\alpha) = \frac{\sum_{j=1}^{N} succeed(i,j,\alpha)}{\sum_{j=1}^{N} accessible1(i,j)}$
Type-2 Accessibility	$P^i_{acc2}(\alpha) = \frac{\sum_{j=1}^{N} succeed(i,j,\alpha)}{\sum_{j=1}^{N} accessible2(i,j)}$
Availability	$P^i_{ava}(\alpha) = \frac{\sum_{j=1}^{N} succeed(i,j,\alpha)}{\sum_{j=1}^{N} available(i,j)}$
Applicability	$P^i_{app}(\alpha) = \frac{\sum_{j=1}^{N} succeed(i,j,\alpha)}{\sum_{j=1}^{N} applicable(i,j)}$

Table 1: Success rate formulas for algorithm i

our algorithms, the victim is the participant: i.e., the user who runs the application. If a participant does not grant the type-1 permissions required for executing the selected inference algorithms, no data point is collected for this participant. In other words, all the data points we collected correspond to type-1 accessible profiles.

We devised inference algorithms that use different types of information as the basis of inference, and resemble real security questions. Our inference algorithms are described in the following. Note that in the description of the algorithms, the basis of inference that is accessed in the victim's profile for inferring her inaccessible information is underlined.

1. Birthday (`birthday`): A day when the participant receives a considerable number of birthday greetings on her wall is inferred as her birthday.
2. Partner's First Name (`partner`): A user who has the highest number of appearances in the participant's photo albums (specially albums with captions containing "marriage", "wedding", etc.), and is of the opposite gender, is inferred to be the partner of the participant. By partner we mean spouse, boy friend or girl friend.
3. Favorite Author (`author`): An author who authors the majority of the books on the participant's list of favorite books is inferred to be her favorite author.
4. Favorite Movie Genre (`genre`): A genre that accounts for the majority of the movies on the participant's list of favorite movies is inferred to be her favorite movie genre.
5. Youngest Sibling's First Name (`sibling`): Access the participant's friends' family information. The youngest friend who has listed the participant in her family information as a sibling is inferred to be the youngest sibling. If no friend has identified the participant to be a sibling, the youngest friend in the participant's friends list who shares the same family name as the participant is inferred to be her youngest sibling. In both cases above, if the friends' birthdays are not accessible, break tie randomly.
6. Hometown (`hometown`): The participant's hometown is inferred to be in the same town as her high school is. If the participant does not list her high school in her education information, then her college or university are used for inference.
7. Oldest Friend (`oldestF`): Access the participant's and her friends' education information. A friend who went to the same high school as the participant did and, is in around the same age as the participant is inferred to be her oldest friend.
8. Political View (`polView`): Access the participant's family information. The prominent political view of the participant's close relatives (e.g., spouse) is inferred to be the political view of the participant. If `polView` could not access the political view of the participant's relatives, then the political views of all friends are consulted.

We do not claim that the above algorithms are particularly sophisticated, but rather we aim at showing even such simple algorithms can yield unintentional information disclosure.

6. ANALYSIS

In this section, we apply the analytical framework of Section 4 to the data points we collected from our participants.

6.1 Classification of User Profiles

Figure 1 depicts the distribution of the various classes of profiles for each algorithm. All profiles are type-1 accessible for all algorithms, i.e., if participant j does not grant the type-1 permissions requested by algorithm i, then algorithm i will not be executed on participant j's profile.

Only three algorithms require type-2 permissions: `sibling`, `polView` and `oldestF`. Thus, all profiles are vacuously type-2 accessible for the rest of the algorithms. The `sibling` algorithm has a secondary inference rule that serves as a fall back when the required type-2 permissions are not in place. That is why to `sibling` all profiles are type-2 accessible. In the case of `oldestF`, the required type-2 permissions happen to be always granted in our empirical data due to the Facebook default privacy settings. A similar reason is behind the low number of type-2 accessible profiles for `polView`.

The higher number there are available profiles, the easier the algorithm can find an answer. For example, only 34.5% of user profiles were available for `author`, whereas the figure is more than doubled for `hometown`. It means the chance that a user has at least one school added to her profile is much higher than the chance that she has at least one entry in her list of favorite books. As a result, the number of potential victims of `hometown` is larger than that of `author`.

Applicability is a more realistic metric for evaluating vulnerability of a profile to inference algorithms. For instance, despite `partner` returns at least one answer for 73.2% of the participants, only 30.2% of the profiles are applicable, i.e., owners of the 69.8% of profiles are single and do not have a significant other. Hence, it is impossible for `partner` to find the right answer in such cases. Analyzing this class of profiles gives us valuable insights on how many users are potentially vulnerable to an inference attack. Figure 1 illustrates that except for `polView`, all the other algorithms have quite a high number of applicable profiles.

6.2 Success Rates of Inference Algorithms

6.2.1 Success rate

Figure 2 depicts the success rates of the inference algorithms as computed using the formulas in Table 1 when $\alpha = 1$. The result shows surprisingly high success rates for some of our algorithms, even though their designs are relatively straightforward. For example, the success rate of `birthday` is 77.4% even for type-1 accessible user profiles. This means, for 77.4% of the users who grant type-1 permissions to `birthday` (i.e., permission to access wall posts), their birthdays can be successfully inferred. In addition, if

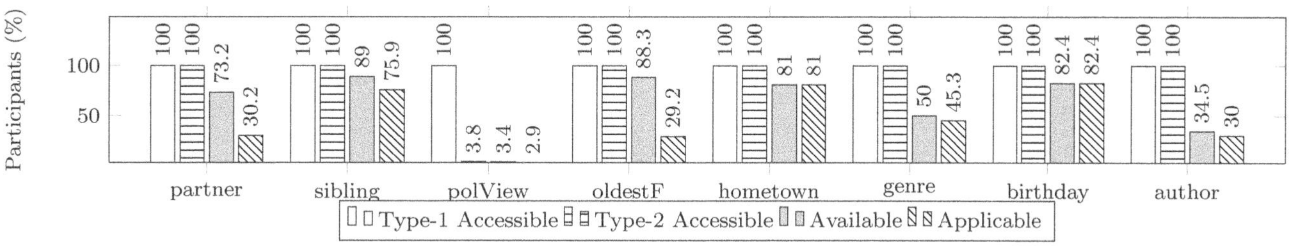

Figure 1: Classification of participants' profiles

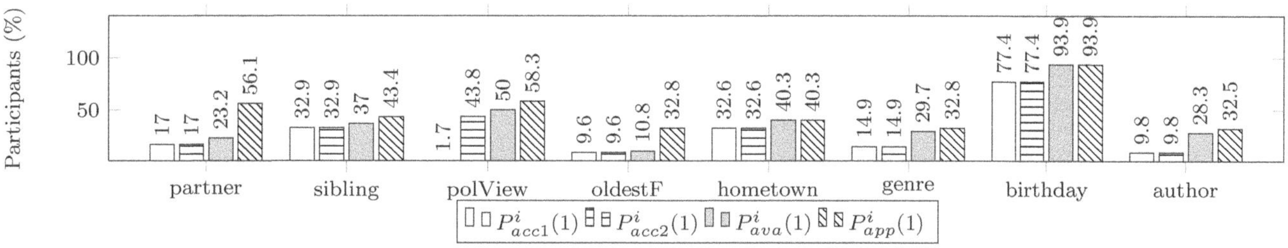

Figure 2: Success rates of inference algorithms when $\alpha = 1$

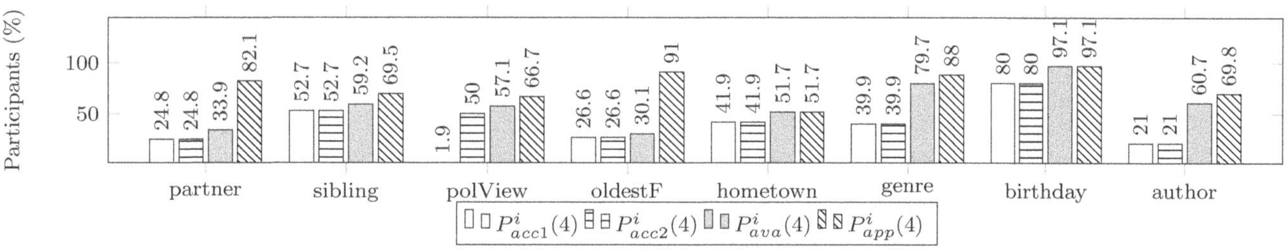

Figure 3: Success rates of inference algorithms when $\alpha = 4$

there is at least one birthday greeting on a user profile, then the success probability ($P^i_{ava}(1)$) is even higher, at 93.9%.

The `polView` had the lowest number of available profiles (3.4%), but $P^i_{ava}(1)$ is a moderate 0.5. This implies, although it is unlikely that an application can access and find the required information for inferring a user's political view, if such an access could be gained and the data is present, then the probability of success is 0.5. This moderately high success rate compared to the small number of available profiles for `polView` shows that the inference strategy behind this algorithm is a reasonable one, but the conditions for the availability of data is rarely satisfied.

6.2.2 *Unexpected disclosure*

Prior to the experiment, there were a number of algorithms that we expected a low success rate. One of them was `partner` because its inference strategy did not appear for us to be universally applicable. However, the experiment shows that if a user has a partner, algorithm `partner` can identify his/her partner with a non-trivial probability ($P^i_{app}(1)$) of 56.1%.

The difference between $P^i_{ava}(1)$ and $P^i_{app}(1)$ in `partner` is significant (23.2% versus 56.1%) due to the large difference between the number of its available and applicable profiles. The algorithm `partner` has a large number of available profiles because finding at least one photo containing a tagged user of the opposite gender is not difficult. But many of our participants were single. As a result, the number of applicable profiles is much lower. The high applicability success rate of `partner` confirms the effectiveness of its simple inference strategy, contrary to our original expectation.

Although information about one's partner is sensitive on its own, revealing an individual's ex-partner is an even more dangerous kind of privacy violation. People typically hide information about their former relationships. Based on the feedbacks we received from the participants, in at least 7% of cases when all inferred answers by `partner` were wrong, the participant's ex-partner (e.g., ex-wife) was identified instead. More importantly, a few participants told us that algorithm `partner` has identified a person whom they are looking forward to date. Identifying such information through other information sources is not straightforward. In other words, inference attacks through SNS extension API could result in accessing highly sensitive information that are not easy to achieve through other adversarial techniques.

6.2.3 *Multiple attempts*

By increasing the value of α from one to four, some algorithms showed much higher success rates (Figure 3). For example, $P^i_{app}(4)$ for `genre` increases to 88% (compared with $P^i_{app}(1) = 32.8\%$). On the contrary, success rate of some algorithms like `birthday` and `hometown` do not change significantly: i.e., they are likely to either return the right answer in their first attempt or fail.

7. INFERENCE ATTACKS AS BUILDING BLOCKS: IDENTITY THEFT

Inference attacks are not only privacy violations, they are building blocks for launching other security attacks. For instance, an individual who wants to register for an online banking access account is usually asked to select an alternative authentication mechanism. One of such mechanisms is security questions. The idea is that, a user first configures k supposedly private questions out of a set of N questions prepared by the service provider, together with their answers, and then when it is necessary, she is challenged to answer those k security questions (e.g., *"What is the first name of your youngest sibling?"*). Inference attacks can be employed by adversaries to find the answers to such security questions. An adversary who knows an individual's username, claims that she forgot her password. The pre-configured security questions will now be presented to the adversary. Then, the adversary launches inference attacks against the victim to find the answers to these security questions. As SNS users disclose a vast amount of private information in their profiles, SNS API inference attacks launched by third-party applications could yield an alarming success rate in identity theft attacks, as we shall see in the following.

In the following, we estimate the probability of success for an identity theft attack when the above authentication strategy is used by the service provider. We assume that the type-1 permissions required for launching inference attacks are indeed granted. Let Z be the set of questions supported by the service provider. Assume every subset of k questions is equally likely to be adopted by a user. The probability that an adversary can answer all k questions is the following:

$$\pi_k(Z) = \sum_{S \in [Z]^k} \frac{\prod_{i \in S} P^i_{acc1}(4)}{\binom{N}{k}} \quad (1)$$

where $[Z]^k$ is the set of all size-k subsets of Z, and $N = |Z|$. We use $P^i_{acc1}(4)$ in (1) because a user is typically allowed to make more than one attempt to answer his security questions. Note that we assume the successful execution of an algorithm is independent from the other algorithms.

If Z contains exactly the 8 inference algorithms that we designed (i.e., $N = 8$) and $k = 3$, then the value of $\pi_k(Z)$ equals 0.04, i.e. for 4% of the potential victims, all three selected security questions can be answered correctly. The impression that this success probability is low and thus the expected number of victims is low is deceptive. One should be reminded that there are applications in Facebook with around 50M, 28M, or 15M monthly active users. With a success probability of 4%, an application with 15M users means an adversary can successfully impersonate 600,000 users by first launching inference attacks, and then following up with identity theft attacks. This analysis testifies to the unfortunate effectiveness of SNS API inference attacks as a stepping stone for dangerous security attacks.

8. RELATED WORK

Inference attacks in social networks is a new research problem. [6, 5, 3] proposed inference techniques that mainly use two types of information as the basis of inference: (1) friendship information, and (2) group membership. The main idea is that the value of an attribute in an individual's profile is likely to be the same as its value in the majority of her friends' profiles and/or her fellow group members' profiles. Such works assume they have full access to the social network data set (i.e., profiles), and then they suppose half of the profiles are public and the other half are private. However, they do not specify a realistic mechanism (crawling, SNS API, etc.) through which certain profiles come to become visible (or private) to the adversary. On the contrary, we make the more realistic assumption that an application only has access to a profile as well as the information that is accessible via that profile.

Inference attacks can use different inference channels and target different types of users. In [1], inference attacks are classified along two dimensions: (1) inference channel, and (2) victim. Knowing all possible inference channels and all potential victims of inference attacks is of great use to researchers for proposing the required protection mechanisms.

Felt and Evans [2] proposed the first work on protecting SNS users against threats specifically posed by third-party applications. But, they did not relate their work to inference attacks. In other words, they present a protection mechanism, called privacy-by-proxy, to prevent third-party applications from accessing original information in user profiles.

9. CONCLUSION

In this work, we took the first step to understand the feasibility of SNS API inference attacks, as well as to assess their privacy impacts. Future work includes the redesign of SNS APIs that would mitigate the threat of inference attacks.

10. ACKNOWLEDGMENTS

This work is funded in part by a Google Research Award, stipends from ISSNet – an NSERC Strategic Research Network, and an NSERC Discover Grant.

11. REFERENCES

[1] S. Ahmadinejad, M. Anwar, and P. Fong. Inference attacks by third-party extensions to social network systems. In *Proc. of IEEE 9th International Conference on Pervasive Computing and Communications Workshops*, pages 282 –287, 2011.

[2] A. Felt and D. Evans. Privacy protection for social networking APIs. *Web 2.0 Security and Privacy*, 2008.

[3] J. He, W. Chu, and Z. Liu. Inferring Privacy Information from Social Networks. In *Intelligence and Security Informatics*, volume 3975 of *Lecture Notes in Computer Science*, pages 154–165. 2006.

[4] J. Kotrlik and C. Higgins. Organizational research: Determining appropriate sample size in survey research appropriate sample size in survey research. *Information Technology, Learning, and Performance Journal*, 19(1):43, 2001.

[5] W. Xu, X. Zhou, and L. Li. Inferring privacy information via social relations. In *Proc. of IEEE 24th International Conference on Data Engineering Workshop*, pages 525–530, 2008.

[6] E. Zheleva and L. Getoor. To join or not to join: the illusion of privacy in social networks with mixed public and private user profiles. In *Proc. of the 18th international conference on World wide web*, pages 531–540, 2009.

Dynamix: Anonymity on Dynamic Social Structures

Abedelaziz Mohaisen
Verisign Labs, VA, USA

Yongdae Kim
KAIST, Daejeon, South Korea

ABSTRACT

In this paper we advance communication using social networks in two directions by considering dynamics of social graphs. First, we formally define the problem of routing on dynamic graphs and show an interesting and intuitive connection between graph dynamics and random walks on weighted graphs; graphs in which weights summarize history of edge dynamics and allow for future dynamics to be used as weight adjustment. Second, we present several measurements of our proposed model on dynamic graphs extracted from real-world social networks and compare them to static structures driven from the same graphs. We show several interesting trade-offs and highlight the potential of our model to capture dynamics, enrich graph structure, and improves the quantitative sender anonymity when compared to the case of static graphs.

Categories and Subject Descriptors

C.2.0 [**Computer Communication Networks**]: General – *Security and Protection*; C.4 [**Performance of Systems**]: Design studies

Keywords

Social networks, Anonymity, Dynamics, P2P communication.

1. INTRODUCTION

Social networks provide rich algorithmic and structural properties that can be used for building certain classes of application benefiting from these properties. When considered along with the main characteristic of social networks, trust, the potential of these networks becomes very promising in solving real-world problems. For example, social networks have been proposed as a building block to defend against the Sybil attack [19], to enable routing in delay tolerant networks[3, 6], and to provide peer-to-peer and private communication [14, 9, 4]. However, social network-based systems make certain assumptions towards achieving their goals; trust among nodes is assumed binary [10], associations are bidirectional [12], and static [14]. For example, insight is brought on the potential of these designs by experimenting with static social graphs, and by ignoring the dynamic nature of social graphs. Ignoring this nature might be due to unavailability of tools to capture the dynamic nature of social graphs, or the unavailability of measures to quantify the performance of the proposed designs on such dynamic social graphs. However, the limited nature of the static social graphs prohibits us from making a concrete insight of these designs in reality when considered for deployment settings in which social graphs exhibit a dynamic behavior [7, 15]. Such behavior greatly alters graphs structure, which is an essential determining factor of the performance of these designs on social networks.

In this paper we proceed further to understand dynamic social graphs for another family of applications; anonymous communication systems [14, 9, 4]. On the one hand, we extend and utilize earlier findings in [14] and [4] of using social graphs as mixers for anonymity. On the other hand, we improve on these results by formalizing the use of dynamic social structures for anonymity, and establishing a relationship between dynamic and weighted graphs. We show how our new design improves anonymous communication and stands against possible attacks by empowering a richer social structure. We validate our model using empirical studies on two dynamic social structures driven from real-world networks.

The rest of this paper is organized as follows. Preliminaries are outlined in §2, theoretical formalism is introduced in§3, and results and discussion in §4. We review the related work in §5, and draw concluding remarks in §6.

2. PRELIMINARIES

In this section we review preliminaries of the prior literature on the problem, which are required for understanding the rest of this paper. This known literature assumes a static graph. Unless otherwise is mentioned, this formalism follows from [14], which is to the best of our knowledge the first work that directly touches upon the problem (other literature work use the same model [9, 4]).

2.1 System Settings and Application Scenario

The idea of building mixers over social links is very simple. Im such model, users recruit their social social acquaintance to provide anonymity to their traffic. In a nutshell, each node (user) forwards her own traffic to her friends, and friends forward that traffic to their friends, and so on, for a certain number of hops, say ℓ. The number of hops ℓ used for forwarding the traffic is a system-wide parameter, which is determined by the security level desired in the system. For simplicity, and without losing generality, let n be the number of users in the system. Accordingly, the anonymity is defined for two parties; the sender and the receiver of traffic. For the receiver, the *anonymity set* is n, and the entropy of the probability distribution for a certain node being the sender is $H_s = \log_2(n)$. On the other hand, the anonymity of a sender is determined by

ASIA CCS'13, May 8–10, 2013, Hangzhou, China.

the probability distribution achieved after the fixed number of hops used in the system. Let the distribution of the final node selected in a *random walk* after ℓ hops starting from node v_j be π_j^ℓ, where $\pi_j^\ell = [\pi_{ij}^\ell]^{1\times n}$, then the anonymity of the sender of the traffic (at the last hop in the walk) is H_r, which is given as:

$$H_r = -\sum_{i=1}^{n} \pi_{ij}^\ell \log_2 \pi_{ij}^\ell \quad (1)$$

Using (1), we define the *anonymity set* A^ℓ as

$$A^\ell = 2^{H_r} \quad (2)$$

Every random walk on a graph with certain properties—see §2.3 for details—has a unique bounding or stationary distribution which captures the maximum achieved entropy.

In the rest of this paper, and to simplify the notation, we omit the index j, which is understood implicitly. Furthermore, the entropy and anonymity set for the sender v_j is obtained from the $j-$th column in the matrix P, after ℓ hops (multiplications).

2.2 Threat Model and Design Goals

In this paper we use the classical model of a colluding adversary with the capability of launching Byzantine attacks against the system built on top of social networks [9]. The adversary has the capability of logging end-to-end information, perform active attacks, and passive attacks. We also assume that the adversary has a limited capability of launching a Sybil attack in reality by inserting large number of Sybil identities [19]. In this work, however, we evaluate the performance of our system under an ideal setting where there is no Sybil identities are injected into the system.

As for the design goals, our design aims to provide a scalable and efficient solution, to provide natural incentives of participation, and to limit the attackers capabilities.

2.3 Formalization: The Case of Static Graphs

Let $G = (V, E)$ be an undirected and unweighted graph where $|V| = n$, $|E| = m$, $V = \{v_1, v_2, \ldots, v_n\}$, and $e_{ij} \in E$ iff $v_i \sim v_j \in V$. We define $\mathbf{A} = [a_{ij}]^{n\times n}$ as the adjacency matrix of G where $a_{ij} = 1$ iff $e_{ij} \in E$ or 0 otherwise. Define the Markov chain on the graph G following the transition matrix $\mathbf{P}$ which is defined according to $\mathbf{P} = [p_{ij}]^{n\times n}$ where:

$$p_{ij} = \begin{cases} \frac{1}{\deg(v_i)}, & v_i \sim v_j \\ 0 & \text{otherwise} \end{cases} \quad (3)$$

A unique stationary distribution is defined for the Markov chain over the transition probabilities defined above if the Markov chain is ergodic—requiring it to be both *irreducible* and *aperiodic* [13]. Theorem 1 states such distribution.

THEOREM 1. *(Stationary distribution on static graph) For an undirected and unweighted graph G, the stationary distribution of the Markov chain defined over G according to transitions in (3) is the probability vector, given as* $\pi = [\pi_i]^{1\times n}$, *where*

$$\pi_i = \deg(v_i)/2m \quad (4)$$

PROOF. The proof is a special case of the weighted graph case discussed in section 3 and follows from Theorem 2 □

Using the model in (1) and the distribution in (4), we define the maximal (in size) anonymity set following the same model as in (2) as $A^\infty = 2^{H_r^\infty}$, where:

$$H_r^\infty = -\sum_{i=1}^{n} (\frac{\deg(v_i)}{2m}) \log_2(\frac{\deg(v_i)}{2m}) \quad (5)$$

2.4 Lower-bound of Anonymity

In [14] Nagaraja considered the average distribution achieved after ℓ hops from any potential source in the social graph as the anonymity achieved of every potential source. While this captures the average performance in the system, it simply does not show the worst case scenario observed at the lower-bound of the achieved anonymity for sender. Here, we revise Nagaraja's definition in [14] and outline a straightforward fix for the measure of the anonymity provided in a system that uses walks on the social graph.

Without losing generality, let ℓ be a system-wide parameter, which represents the number of hops from the source to the destination (or receiver) in the graph, and each node between them is chosen uniformly at random from its predecessor. For each source v_j (for $1 \leq j \leq n$), we define the probability distribution after ℓ hops as $\pi^\ell(v_j) = [\pi_i^\ell(v_j)]^{1\times n}$ for $(1 \leq i \leq n)$. The anonymity achieved in the system is bounded below by the entropy achieved in the probability distribution obtained by walking from the worst source in the graph:

$$H_r \geq \inf_{v_j}\{-\sum_{i=1}^{n} \pi_i^\ell(v_j) \log_2 \pi_i^\ell(v_j)\} \quad (6)$$

By extending (2) to the case in (6), we get the following

$$A^\ell = 2^{H_r} \geq 2^{\inf_{v_j}\{-\sum_{i=1}^{n} \pi_i^\ell(v_j) \log_2 \pi_i^\ell(v_j)\}} \quad (7)$$

The intuition of this lower bound is very simple, practical, and follows from the definition. Technically, this lower bound follows the classical theoretical trend in security: proving lower bounds of security (or anonymity as it is the case in hand) would enable us to guarantee, in the worst time, that our system would perform better than this bound for every user. On the other hand, considering the average case for achieved entropy might be very deceiving since many senders are not likely to achieve this average bound.

3. DYNAMIC GRAPHS FOR ANONYMITY

We extend findings in the literature on static graphs as mixers for anonymous communication to the case of the dynamic graphs. Such dynamic graphs arise naturally in many contexts due to social churn imposed by node and edge dynamics (joining and leaving social networks). It is worth noting that this is the first work of its own type to consider extending such results for building anonymous communication systems on top of dynamic social graphs.

3.1 Formalization: Dynamic Graphs

The dynamic graph is a simple generalization of the static graph used in literature. In particular, $G = \{G^{(i)}\}$ for $1 \leq i \leq t$ is a dynamic graph over t time periods. Let $G^{(i)} = (V^{(i)}, E^{(i)})$ for $1 \leq i \leq t$, where $|V^{(i)}| = n^{(i)}$ and $|E^{(i)}| = m^{(i)}$, be an unweighted and undirected graph (later we extend that to the weighted graph case). Let $V^{(i)} = \{v_1^{(i)}, v_2^{(i)}, \ldots, v_{n^{(i)}}^{(i)}\}$ and $E^{(i)}$ be the set of pairs of vertices $v_j^{(i)}$-$v_k^{(i)}$ if both nodes $v_j^{(i)}$ and $v_k^{(i)}$ in $V^{(i)}$ are connected to each other. For $G^{(i)}$, we define $\mathbf{A}^{(i)}$ where $\mathbf{A}^{(i)} = [a_{jk}^{(i)}]^{n^{(i)}\times n^{(i)}}$ (the superscription is used as part of the notation, and does not mean power), where:

$$a_{jk}^{(i)} = \begin{cases} 1 & v_j^{(i)} \sim v_k^{(i)} \in G^{(i)} \\ 0 & \text{otherwise} \end{cases} . \quad (8)$$

For the same graph $G^{(i)}$, we define the transition probability matrix

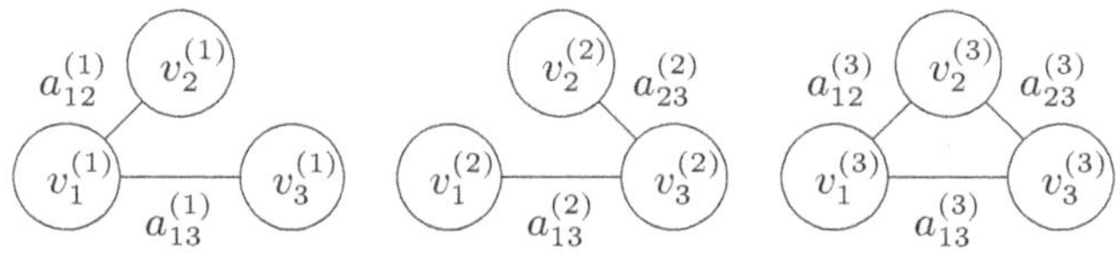

Figure 1: Simple example of dynamic graph. $a_{12}^{(i)} = w(v_1^{(i)}, v_2^{(i)})$ is as per the definition.

$\mathbf{P}^{(i)}$ such that $\mathbf{P}^{(i)} = [p_{jk}^{(i)}]^{n^{(i)} \times n^{(i)}}$, where:

$$p_{jk}^{(i)} = \begin{cases} 1/\deg(a_{jk}^{(i)}) & v_j^{(i)} \sim v_k^{(i)} \in G^{(i)} \\ 0 & \text{otherwise} \end{cases}. \quad (9)$$

Extending and generalizing (8) and (9) to the weighted case is easy if weights are given on edges in the graph. We define

$$a_{jk}^{(i)} = \begin{cases} w(v(i)_j, v(i)_k) & v_j^{(i)} \sim v_k^{(i)} \in G^{(i)} \\ 0 & \text{otherwise} \end{cases}, \quad (10)$$

where $w : E^{(i)} \to \mathbb{R}$ is a weight function that assigns real-valued weights to edges in $G^{(i)}$. Using (10), we define the degree of a node to be in terms of weights associated with edges for which that node is an end-vertex, as

$$\deg^w(v_j^{(i)}) = \sum_k w(v_j^{(i)}, v_k^{(i)}), \quad v_j^{(i)} \sim v_k^{(i)} \in G^{(i)}. \quad (11)$$

Notice that (11) can also be written as $\deg^w(v_j^{(i)}) = \sum_k a_{jk}^{(i)}$—where $a_{jk}^{(i)}$ is defined in (10). Using (11), we can compute $\mathbf{P}^{(i)} = [p_{jk}^{(i)}]$ for weighted graphs, where

$$p_{jk}^{(i)} = \begin{cases} w(v_j^{(i)}, v_k^{(i)})/\deg^w(v_j^{(i)}) & v_k^{(i)} \sim v_k^{(i)} \in G^{(i)} \\ 0 & \text{otherwise} \end{cases}. \quad (12)$$

In a matrix form, $\mathbf{P}^{(i)}$ can be defined as $\mathbf{P}^{(i)} = (\mathbf{D}^{(i)})^{-1}\mathbf{A}^{(i)}$ where $\mathbf{D}^{(i)}$ is a diagonal matrix computed from $\mathbf{A}^{(i)}$, where the diagonal element $d_{jj}^{(i)}$ in $\mathbf{D}^{(i)}$ is the sum of ones in the j-th row in $\mathbf{A}^{(i)}$ (that is, the degree of node v_{ix} in $G^{(i)}$). At any time slot i, we define the bounding distribution of the Markov chain on the graph G_i as in literature defined $[\deg(v_j^{(i)})/2m^{(i)}]$. It is, however, unclear how to proceed with the different snapshots of the same graphs at different times.

As shown in Figure 1, both nodes $v_1^{(1)}$ and $v_2^{(1)}$ are connected, but not with their future images—$v_1^{(2)}$ or $v_1^{(2)}$ and $v_2^{(2)}$ or $v_2^{(3)}$, respectively. This also applies to states in the future not connected to the past images. In the following, we investigate several techniques for modeling the dynamic social graph as a graph where transitions from future states to past states is possible. Techniques utilize here are generic, and can be used to any graph with multiple labels.

Prior work in the literature has tried to model dynamic graphs as *3-mode tensor* [1] or *union multigraph* [5]. However, while the first uses high dimensionality—making computations on the tensor computationally expensive, the second technique reduces dimensionality and loses some information about the graph. Indeed, the second technique computes the union between multiple graphs (edge- and node-wise) and omits any potential multiple edges between two nodes in the union. While this is meaningful to understand a union snapshot of multiple graphs, demonstrate connectivity characteristics of the union graph driven from multiple attributes, and potentially other benefits, it does not capture the "depth" of edges and does not differentiate between different edges based on their "real value". For example, while edges in the union multigraph are all the same, some in reality might be the result of multiple edges whereas others could be the result of a single edge.

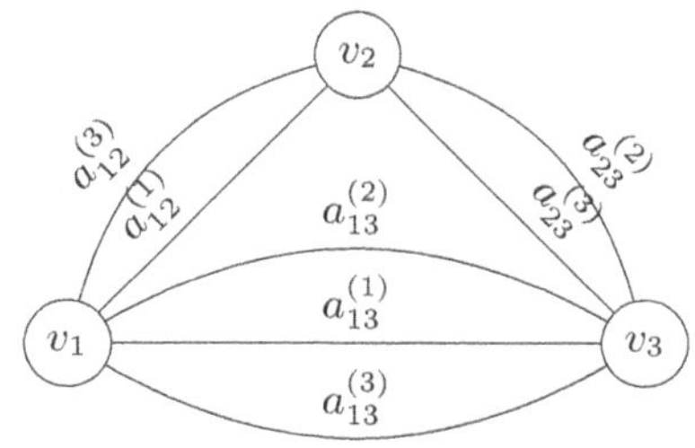

Figure 2: Simple example of converting a dynamic graph into multigraph by collapsing all images of a node to the node itself.

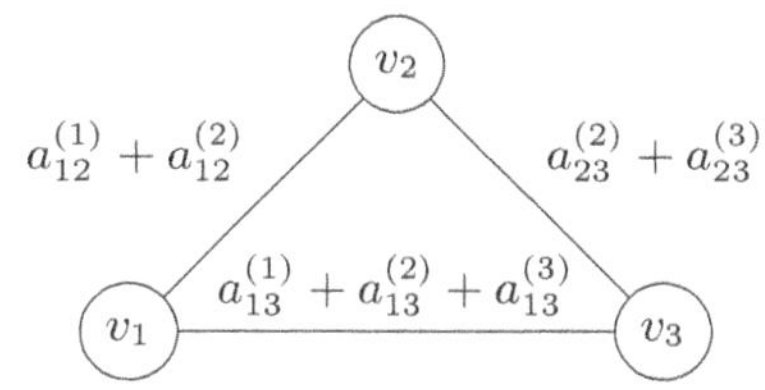

Figure 3: An example of multigraph conversion into weighted graph by summing weights of edges between pairs of nodes.

3.2 Dynamic Graphs as Multigraphs

Formally, for the dynamic graph $G = \{G^{(i)}\}$ described in section 3.1, we define a multigraph $\mathbb{G}$ as $\mathbb{G} = (\mathbb{V}, \mathbb{E})$, where

$$\mathbb{V} = \bigcup_{i=1\ldots t} \{V^{(i)}\}, \text{ and } \mathbb{E} = \biguplus_{i=1\ldots t} \{E^{(i)}\}. \quad (13)$$

Notice that $\cup$ is a *set union*, which does not allow repetition of vertices, whereas $\uplus$ is a *multiset union*, which allows edge repetition. When $\mathbb{E}$ is computed, the index that corresponds to the time of the edges in E_i can be removed for simplicity. A simple toy example of transforming the multiple snapshots of the dynamic graph in Figure 1 into a multigraph is in Figure 2.

Our formalization above of the graph as a multigraph (rather than union multigraph as per the way defined in [5]) follows the intuition of what a dynamic graph could yield of associations at any time. At a time i, where $1 \leq i \leq t$, constructing the proper graph for operating a potential system, like mixing-based anonymous communication system, and maintaining the same information driven from the original multiple snapshots of the graph is possible.

3.3 Dynamic Graphs as Weighted Graphs

Now, we convert the dynamic graph model represented as a multigraph, as in (13), into a weighted graph. We generalize formalizations in section 3.1. In particular, the model in (10) can be rewritten (for weighted undirected graph) as $\mathbf{A} = [a_{jk}]^{n \times n}$—here, $n = |\mathbb{V}|$—where

$$a_{jk} = \sum_{i=1\ldots t} w(v_j^{(i)}, v_k^{(i)}), \quad v_j^{(i)} \sim v_k^{(i)} \in G^{(i)} \forall i. \quad (14)$$

Similarly, we extend the model in (11) into

$$\deg^w(v_j) = \sum_{i=1\ldots t} \deg^w(v_j^{(i)}) = \sum_{\forall k} a_k^{(j)} \quad (15)$$

$$= \sum_{\forall k} \sum_{i=1\ldots t} w(v_j^{(i)}, v_k^{(i)}), v_j^{(i)} \sim v_k^{(i)} \in G^{(i)}. \quad (16)$$

We can further extend the transition probability formulation to cover the weighted graph by plugging both (14) and (15) into a similar

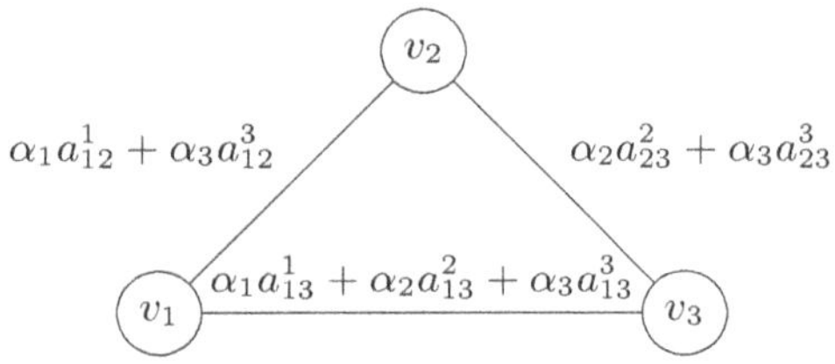

Figure 4: A weighted graph model to express dynamic graphs.

model to that of (12), to get $\mathbf{P} = [p_{jk}]^{n \times n}$, where

$$p_{jk} = a_{jk} / \deg^w(v_j) \quad (17)$$

For a random walk defined on $\mathbb{G}$ according to the transition probability defined in (17), the following theorem states the stationary distribution. This theorem (and the proof herein) are essential for latter results on characterizing and operating on dynamic graphs. Also, the proof of Theorem 1 follows similarly as in below.

THEOREM 2. *Let* $\mathbb{G} = (\mathbb{V}, \mathbb{E})$ *be a* connected, undirected, *and* weighted *graph defined as in (13). For a random walk following transition probabilities as in (17), the stationary distribution is defined as* $\pi = [\pi_i]^{1 \times n}$ *(for* $n = |\mathbb{V}|$*), where:*

$$\pi_i = \deg^w(v_i) / \sum_{k=1\ldots n} \deg^w(v_k) \quad (18)$$

PROOF. See the appendix. The proof is also in [2]. □

3.4 Generalized weighted graphs

In many natural social contexts, recent associations are more valued than older ones, or vice-versa. Accordingly, a general framework for quantifying the potential of any system on top of social networks should consider implicit social network characteristics, such as link age, in addition to the explicit differences among links captured by the topological structure. We generalize the model in section 3.3 to accommodate for implicit values of associations over time. Without losing generality, let α_i (for $1 \leq i \leq t$) be a set of parameters that take numerical values. An extension of the social graph model in (14) is as follows:

$$a_{jk} = \sum_{i=1\ldots t} \alpha_i w(v_j^{(i)}, v_k^{(i)}),\ v_j^{(i)} \sim v_k^{(i)} \in G^{(i)} \forall i. \quad (19)$$

The rest of the model in section 3.3, particularly in (15) onward, holds for this generalization after adjusting a_{jk} as in (19). A toy example demonstrating the adjustment of weights in Figure 3 is shown in Figure 4.

4. RESULTS

4.1 Datasets and Data Preprocessing

Our sources of data are the Facebook social network dataset [18] and the DBLP [8] co-authorship graph, which are explained below.

The DBLP Dataset. The DBLP dataset represents co-authorship graph, where nodes are authors and a link between two authors implies that the authors have co-authored a paper. The original DBLP dataset consists of $943,316$ nodes and $6,379,554$ edges between them, for publication records until May 2011 in computer science areas. The largest connected component consists of $769,642$ and $3,051,127$ undirected edges. To generate dynamic graphs from that component, we select the period of 2006 to 2010 inclusive, by selecting authors who have publications in each and every of these years. The result is a multigraph where two nodes would have an edge if they co-authored a paper in a given year, and number of such papers per year is used as weight for weighted graphs. Multiple edges could be created between two authors if they co-authored over multiple years. Multiple edges are labeled with respect to the year of publication. The final multigraph has $46,994$ nodes and $458,736$ edges. We decompose each multigraph to multiple-graphs with respect to the edge label. Finally, as some nodes who published in the given period could be isolated in a certain year, we remove these nodes so as each resulting graph is connected. Statistics of the different resulting graphs are shown in Table 1. For our study, we consider several cases of the same graph including both weighted and unweighted, with respect to the time.

Table 1: Statistics of DBLP time-varying graphs. Metrics of comparison are number of nodes (n), number of edges (m), average clustering coefficient, diameter, and radius.

	nodes	edges	clustering	diameter	radius
DBLP (1)	31704	71994	0.483	26	14
DBLP (2)	33012	79475	0.480	27	14
DBLP (3)	33923	84125	0.467	24	13
DBLP (4)	33071	82282	0.453	23	12
DBLP (5)	26150	62161	0.419	24	13

Graphs used in our experiments are as follows. (i) Unweighted graph with respect to each year (5 graphs). (ii) Unweighted single graph representing the entire dataset (1 graph). (iii) Weighted single graph representing the entire dataset. The weight on an edge connecting two nodes is the sum of all weights of edges between these nodes over time from the beginning to the end of recording the graph structure. (iv) Weighted multiple snapshot graphs (up to each year; 5); a graph G_{1i} combines all nodes in G_1 to G_i and edges between them. (v) Unweighted multiple snapshot (up to each year; 5). These graphs are obtained using the same method as in the previous step but without weights. (vi) Weighted graphs with weights assigned based on the age of the link. We use geometrical (2^{1-x}; $x = 1$ newest) and reciprocal ($1/x$) decay distributions.

In Table 1 and Table 2, the basic structural properties are as follows. (i) Graph size: the number of nodes and number of edges in the social graph (denoted as n and m). (ii) Clustering coefficient: is the average (thus in [0, 1]) of local clustering coefficient for all nodes. The local clustering coefficient for a node is the fraction of possible triangles that go through that node. (iii) Diameter: the longest of eccentricities among all nodes in the graph. The eccentricity of a node is defined as the longest shortest path from that node to other nodes in the graph. (iv) Radius: shortest eccentricity.

The Facebook Dataset. The Facebook dataset [18] is for wall posts in New Orleans regional network from 2004 to 2009. A link between two nodes indicates that the first node has interacted with the second node. Further details on statistics of the entire dataset is in [18]. To obtain a dynamic graph from this dataset, we limit ourselves to the last 30 months of interactions, with each graph obtained over 6 months of interaction. The resulting five graphs are shown in Table 2. The same variations used above are also used for Facebook.; we omit details for the lack of space.

4.2 Results

In the following we outline the results of utilizing the different social graphs obtained in section 4.1, using the techniques described earlier in this work. Our main measurement metric is the achieved anonymity in terms of the total entropy in the distribution of the last hop in a random walk, as the length of the random walk increases; computed as in (1). We keep in mind that potential utilization of social graphs for anonymity systems would be sub-

Table 2: **Statistics of Facebook time-varying graphs. Metrics of comparison are number of nodes (n), number of edges (m), average clustering coefficient, diameter, and radius.**

	nodes	edges	clustering	dimeter	radius
Facebook (1)	9154	23245	0.102	19	10
Facebook (2)	13288	37908	0.101	18	10
Facebook (3)	16540	42427	0.092	19	10
Facebook (4)	23879	59190	0.085	21	11
Facebook (5)	35665	86525	0.084	18	10

ject to the performance of this systems, which necessitate to a short random walk length. We consider walk lengths varying from 1 to 20 steps, where walk lengths of $1-12$ are demonstrated in most experiments. As the entropy varies depending on the source of the random walk, we are interested in the maximum, minimum (advocated in section 2.4 and expressed in (6)), and mean entropies for a given dataset as the walk length increases.

Original Unweighted Graphs. We first consider operating the anonymity system on top of the graphs shown in Table 1 and Table 2. First, we observe that lower-bound on the achieved entropy or the entire system is much smaller than that of the average and maximum entropy, for any walk length. This tells that the measure of the lower-bound on the entropy, while theoretically appealing for the guarantees advocated earlier, do not provide a representative measure for the whole set of nodes in the system or graph. Second, we observe that both the mean and the maximum of the entropy in each of the graphs stabilizes, and reaches its potential maximum entropy within a relatively smaller number of steps, corresponding to shorter random walk length. This indeed interesting, and agrees with prior work in [14], despite that the results in the prior work have been on relatively a faster mixing social graph [13].

Dynamics as Weights. We consider modeling dynamics of social graphs as weights on edges. We use the method in 3.3 for generating these graph with weights. The method of obtaining the graphs is explained in 4.1. Similar observations on the results for the tendency of weights is made as on the previous measurements. Furthermore, the *general* tendency of improvement of the entropy value as the time goes is made clear in both measurements.

Unweighted Dynamic Graphs. We consider removing weights from the different dynamic graphs to observe how this affects the entropy as walks length increases. For the same experiment above, with the only difference being deletion of weights. The most important point made clear in this experiment is that unweighted graph generated from these weighted dynamic graphs provide higher entropy for the same walk length (over the same number of nodes).

Different Weighting Scenarios. We consider the potential ways of assigning the different weights on the social graph, based on the age of the link, and how this impacts the achieved anonymity on these graphs. We consider the graphs constructed from the multiple-snapshots, according to the way described in section 3.3 and section 3.4. We use the result in section 3.3 to generate a "linear" weighting factor (the coefficient is 1), and thus the weight of an edge is the number of interactions between the two nodes over all years. We consider the same graph generated in this step without weights as well. Finally, we use the model in section 3.4 to generate generalized weighted graphs, where weights are formed according to the reciprocal or geometrical decay distributions explained in 4.1. In these measurements, and somewhat counter-intuitively, we observe that the unweighted graph model results in best entropy (in all three categories: min, mean, and max)—Further details are in the discussion section. On the other hand, we also observe that in all of these graphs, the achieved anonymity is good enough (as a portion of the maximal) even with a walk length of 10, suggesting the usefulness of this design. Both remarks apply to both datasets, though bias is a lot higher in Facebook than in DBLP.

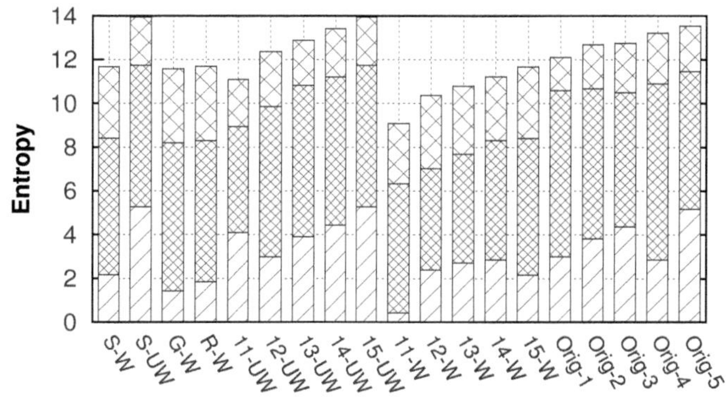

Figure 5: Average entropy for walk length of 10; all Facebook graphs (S. is single, G and R are geometrical reciprocal distributions of weights, W is weighted, and UW is unweighted, and numbers are to indicate which graph is used: 1-5 are original graphs whereas 11-15 are the dynamic graph model).

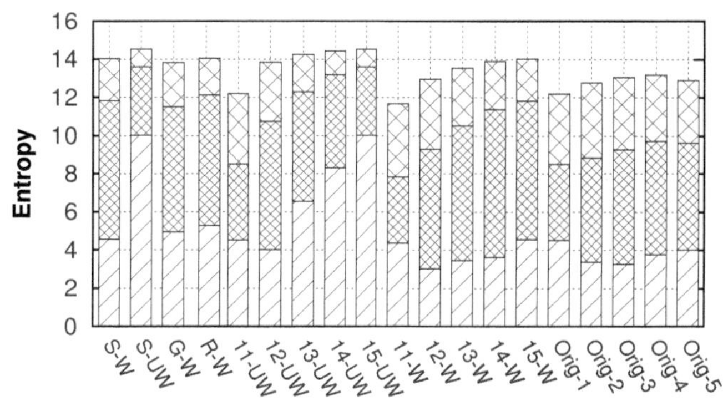

Figure 6: Average entropy for walk length of 10; all DBLP graphs (S. is single, G and R are geometrical reciprocal distributions of weights, W is weighted, and UW is unweighted, and numbers are to indicate which graph is used: 1-5 are original graphs whereas 11-15 are the dynamic graph model).

Combining All Scenarios. We consider all scenarios mentioned above, for each of the datasets we had, and for a fixed random walk length to compare them relatively and draw final conclusions on the impact of the underlying social structure on the achieved anonymity. We consider the random walk length $\ell = 10$, and experiment for both datasets to compute the entropy—both mean and max. The results are shown in Figure 5 through Figure 6.

4.3 Analysis and Discussion

In most of the measurements of the entropy on the distribution of the random walk after ℓ hops, we observe a relatively good entropy which supports the claimed efficiency advocated in this work and [14]. However, this entropy, for example is not as high in some graphs, especially those the single snapshots that consider the graph at one time period, and those resulting from assigning weights on the graphs corresponding to the richness of the edges. This pattern is shown in both datasets, which call for explanations.

One potential explanation of the relative difference between the achieved entropy in the individual graphs and that obtained from the graphs computed using our dynamic graph model is the inherit increment in the size of the resulting final graph. For example, while the largest DBLP graph is about 36,000 nodes, the final graph of the DBLP after using our model that considers graph dynamics would result into about 44,000 nodes in the largest connected components. This, and the fact that the graph becomes richer of more edges that connect multiple components in the graph, improves the mixing characteristics of the graph, which ultimately improves the achieved entropy as ℓ increases.

This does not explain the difference between the achieved entropy in both weighted and unweighted graphs, even for the graph with the same number of nodes and edges. For example, when $\ell = 10$, the achieved entropy on DBLP-15 when weighted is 14 while it's 14.5 for the same graph when it is unweighted—while the difference in entropy is small, i.e., 0.5, the 14.5 bits of entropy provide about $23,170$ anonymity set whereas only $16,384$ are provided for the other case, which translates to more than $6,786$ of difference in anonymity set. One possible explanation of this behavior is the intuitive meaning of weighting graphs: by assigning weights on edges, we are biasing the random walk on such graphs and favoring a node over another of being reaching at any time when running the random walk. This is, some nodes are more likely reached whereas other nodes while less likely reaching by the random walk which definitely decreases the potential set of nodes being used as a last hop in the random walk. This intuitive meaning explains the difference in the entropy in both cases.

Unexpectedly, both entropy and anonymity sets are greatly decreased when using the weighted graphs that model dynamic structure. One possible explanation is that these weights are obtained by favoring some edges over others, which is more meaningful from an anonymity point of view, whereas edges in the unweighted graph simply make all relationships over time equal. In a realistic scenario, where potential insider attacker could exist to penetrate the system to get communicated messages the social overlay, the model, which considers links to be equal independent of their history or time of creation, could be problematic. Given this intuitive explanation of weights associated with edges, one would anticipate the use of weighted graphs in real-world scenarios despite this degradation in the achieved entropy and anonymity set sizes given their potential for minimizing harms due to edge infiltration.

5. RELATED WORK

Exploiting static social networks for anonymous communication has been explored in [14, 9, 4], some of which has been discussed earlier, and all did not consider the dynamic graph case. Modeling of dynamic social graph and extracting a definition for the mixing time is done in [1], whereas sampling multigraph defined as node set union graphs is done in [5]. To the best of our knowledge, no prior work considered dynamic graphs in the context of the problem in hand and as per our method. On the other hand, other assumptions in social network-based systems, like binary trust [10], edge directionality [12], and expansion properties [11] are previously consider and challenged in separate studies.

Dynamic social graphs have been studied in [7, 17, 15, 16]. Most of these studies, however, considered mining known simple properties of social graphs, but not the mixing time and patterns used for anonymity. Finally, observing dynamics of social networks as set of static graphs over time has been most recently used in [20].

6. CONCLUDING REMARKS

In this paper we considered the problem of building anonymous communication systems on (unstructured) dynamic social graphs. We have pointed out an interesting relationship between dynamic structures and weighted graphs, and formalized the anonymity achieved under dynamics as a random walk on weighted graphs. We formulated the problem in hand, and shown the bounding distribution, which captures the maximal achieved entropy of a random walk on an anonymous communication system, which uses these dynamic structures. Through experiments on real-world datasets, we have shown the potential of these dynamic structures, and despite their numerical disadvantage over unweighted versions, we have pointed out their benefits for anonymity for that they capture more meaningful structure that represents stronger ties.

This work has considered "unstructured" social graphs, which a non-constant (and likely power-law) degree distribution. On one hand, the potential of structured graphs for anonymity is well studied, and beautiful theoretical results are already provided. On the other hand, suitability of these results for real-world social structures, especially when considering dynamics, is unclear. In the future, we will look at creating structured graphs from unstructured social graphs, and explore their potential for anonymity systems.

Acknowledgement. Yongdae Kim was supported by the KCC (Korea Communications Commission), Korea, under the R&D program supervised by the KCA (Korea Communications Agency), KCA- 2013-12911-05003. Part of this work was done while A. Mohaisen was at the University of Minnesota.

7. REFERENCES

[1] U. Acer, P. Drineas, and A. Abouzeid. Random walks in time-graphs. In *WMON*. ACM, 2010.
[2] B. Bollobás. *Modern graph theory*, volume 184. Springer, 1998.
[3] E. M. Daly and M. Haahr. Social network analysis for routing in disconnected delay-tolerant manets. In *MobiHoc*, 2007.
[4] G. Danezis, C. Díaz, C. Troncoso, and B. Laurie. Drac: An architecture for anonymous low-volume communications. In *PETS*, 2010.
[5] M. Gjoka, C. T. Butts, M. Kurant, and A. Markopoulou. Multigraph sampling of online social networks. *IEEE JSAC*, 2011.
[6] P. Hui, J. Crowcroft, and E. Yoneki. Bubble rap: Social-based forwarding in delay-tolerant networks. *IEEE TMC*, 2011.
[7] R. Kumar, J. Novak, and A. Tomkins. Structure and evolution of online social networks. *Link Mining: Algorithms and Apps*, 2010.
[8] M. Ley. The DBLP computer science bibliography: Evolution, research issues, perspectives. In *SPIR*, 2009.
[9] P. Mittal, M. Wright, and N. Borisov. Pisces: Anonymous communication using social networks. In *NDSS*, 2013.
[10] A. Mohaisen, N. Hopper, and Y. Kim. Incorporating trust into social network-based sybil defenses. In *INFOCOM*, 2011.
[11] A. Mohaisen, H. Tran, N. Hopper, and Y. Kim. Understanding social network properties for trustworthy computing. In *SIMPLEX*, 2011.
[12] A. Mohaisen, H. Tran, N. Hopper, and Y. Kim. On the mixing time of directed social graphs and security implications. In *ASIACCS*, 2012.
[13] A. Mohaisen, A. Yun, and Y. Kim. Measuring the mixing time of social graphs. In *IMC*, pages 383–389. ACM, 2010.
[14] S. Nagaraja. Anonymity in the wild. In *PETS*, 2007.
[15] L. Tang, H. Liu, J. Zhang, and Z. Nazeri. Community evolution in dynamic multi-mode networks. In *KDD*, 2008.
[16] C. Tantipathananandh, T. Berger-Wolf, and D. Kempe. A framework for community ident. in dynamic social networks. In *KDD*, 2007.
[17] H. Tong, S. Papadimitriou, J. Sun, P. Yu, and C. Faloutsos. Colibri: fast mining of large static and dynamic graphs. In *KDD*, 2008.
[18] B. Viswanath, A. Mislove, M. Cha, and K. P. Gummadi. On the evolution of user interaction in facebook. In *WOSN*, 2009.
[19] H. Yu, M. Kaminsky, P. Gibbons, and A. Flaxman. SybilGuard: defending against sybil attacks via social networks. In *SIGCOMM*, 2006.
[20] X. Zhao, A. Sala, C. Wilson, X. Wang, S. Gaito, H. Zheng, and B. Y. Zhao. Multi-scale dynamics in a massive online social network. In *IMC*, 2012.

Protecting Access Privacy of Cached Contents in Information Centric Networks

Abedelaziz Mohaisen
Verisign Labs, VA, USA

Xinwen Zhang
Huawei Technologies, CA, USA

Max Schuchard
University of Minnesota, MN, USA

Haiyong Xie
Huawei Technologies and USTC, China

Yongdae Kim
KAIST, Daejeon, South Korea

ABSTRACT

In recently proposed information centric networks (ICN), a user issues "interest" packets to retrieve contents from network by names. Once fetched from origin servers, "data" packets are replicated and cached in all routers along routing and forwarding paths, thus allowing further interests by other users to be fulfilled quickly. However, the way ICN caching works poses a great privacy risk: the time difference between responses for an interest of cached and uncached content can be used as an indicator to infer whether or not a near-by user has previously requested the same content as that requested by an adversary. This work introduces the extent to which the problem is applicable in ICN and provides several solutions that try to strike a balance between their cost and benefits, and raise the bar for the adversary to apply such attack.

Categories and Subject Descriptors

C.2.0 [**Computer Communication Networks**]: General – *Security and Protection*; C.4 [**Performance of Systems**]: Design studies

Keywords

Information centric networks, privacy, side channel attacks, caching.

1. INTRODUCTION

Information centric networks (ICNs) have been proposed as new Internet architectures towards secure and efficient content dissemination. In several ICNs such as content centric network (CCN) [11] and named data network (NDN) [19], contents are fetched by their names from caches deployed in the network or from origin servers—servers that serve the contents if they are not cached in the network. In such ICN architectures, once a content data packet is fetched from an origin server, it is replicated and cached in all routers along the routing and forwarding path—starting from the router that connects user who issues the interest to the one that connects the origin server to the ICN—thus allowing further interests with the same content name to be fulfilled quickly [11]. For example, when another user issues an interest in these contents that have been previously served to a user on the same path, the interest is fulfilled from the near-by cache. This design choice—as advocated by many ICN designs and architectures—is considered a great advantages in reducing overall content retrieval latency [11, 19]. However, this universal caching mechanism poses a great privacy risk: the time difference between a response of cached, when compared to uncached content, can be used as a side channel to infer whether a near-by user has previously requested that content.

ASIA CCS'13, May 8–10, 2013, Hangzhou, China.

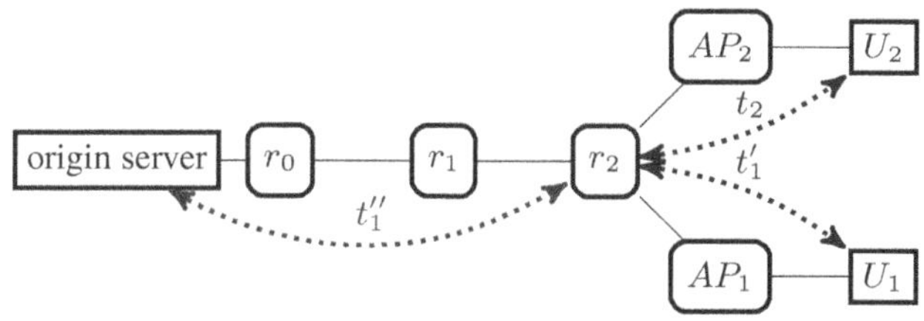

Figure 1: Toy example of timing attack in ICN.

For example, consider the topology in Figure 1, which depicts users U_1 and U_2, and a set of routers r_0 to r_2 (each with its own cache) connecting both users to an origin server that holds content with name n. Suppose that user U_2 is the adversary, whereas user U_1 is honest. If U_1 issues an interest in content n that resides behind r_0, the interest traverses the path $U_1 \rightarrow AP_1 \rightarrow r_2 \rightarrow r_1 \rightarrow r_0$, from which it retrieves a requested packet of the content. The packet is then sent back over the returning path $r_0 \rightarrow r_1 \rightarrow r_2 \rightarrow AP_1 \rightarrow U_1$. In total, the path from U_1 to the source of the content and the returning path to U_1 have four-hop each. The total round trip time required for sending the request until starting to receive data packets on the returning path is t_1. On the other hand, if U_2 is to request the same content by its name, n, the path that the interest would traverse is $U_2 \rightarrow AP_2 \rightarrow r_2$, and the contents would return on the reversed path ($r_2 \rightarrow AP_2 \rightarrow U_2$), which is two-hop in each direction, and would require a time t_2. Obviously, the time t_1 is greater than t_2, which an adversary U_2 can use to infer that user U_1 has accessed the content n.

Although pinpointing U_1 precisely may require additional side information, an attack like the one described above is still critical since it reduces the anonymity set of that user greatly. Such scenario is equal in value to identifying individual users when combined with easily available information. For example, an adversary launching a business intelligence attack might be interested in knowing what contents are being retrieved by a competing company, rather than by individual users. This attack would be possible if the adversary is co-located with that company behind an edge router, and using the above technique.

Shortcomings of Simple Solutions. Simple solutions cannot prevent the attack. For example, a user can mark a content object with a privacy flag to disable in-network caching. However, this will degrade the quality of experience of other users. Also, an adversary can send two consecutive requests of the same name to infer routers' behavior. In the first request, and assuming caching is enabled, the requested data will result in data caching The second request in that case will result in cache-hit, and the content will be served to the adversary quickly. However, if caching is disabled, the second request will result in a delay close to the delay in the first request, from which the adversary can infer that caching is disabled, and that another user has likely used the privacy flag.

Other solutions to the attack by "intelligent caching" based on router-to-user distance have their limitations. To do that, a router has to either know the user's location in advance or have the location provided at the time of caching. While the first approach requires partial topology information that can easily exceed typical routers' resources, the latter one is vulnerable to misuse; to negatively impact the users' experience the adversary can flag any content from an arbitrary location so as to disable caching.

1.1 Contributions

We make the following contributions:

- we demonstrate timing attacks on the universal caching mechanism proposed in ICN designs like CCN and NDN. For that, we make use of fine-grain per-hop timing measurements of cached and uncached contents using real-world time measurements with CCNx, a prototype implementation of the CCN [2]. We disclaim the originality of the attack in its general form but claim its suitability and applicability to ICN.
- we propose three protocols, each with different levels of complexity, cost, and privacy guarantees that prevent an adversary co-located with benign users to infer whether they have accessed certain contents or not by relying on the timing. Each and every of these protocols tries to strike a balance between the privacy provided to legitimate users from potential adversary, and the overhead added for requests performed by other legitimate users to the privacy-related contents.

1.2 Organization

The organization of the rest of this paper is as follows. In Section 2 we review the preliminaries and terminologies used in this paper. In Section 3 we introduce three protocols to solve the problem and maintain the privacy of users access, where each protocol comes at different cost and privacy guarantees. In Section 4 we present our simulation results to validate the attack and evaluate the performance of our defense protocols. In Section 5 we highlight several discussion points, including potential attacks and their applicability to our protocols. Section 6 reviews related work and Section 7 concludes this work and point out our future work.

2. PRELIMINARIES AND TERMINOLOGY

In ICN, contents are fetched by their names [11]. An ICN consists of routers, where each router has a cache, and edge routers are connected to users and origin servers. An *Interest* in ICN encapsulates a request for a content packet by its name. An *origin server* is a server that originates contents to be served in the network, thus fulfilling interests. The contents (data packets) may or may not be cached in the network. In the rest of this work, we use total Round Trip Time (RTT) to denote the time from the start of sending the first interest until the start of receiving a content packet fulfilling it (also known in the literature as Time to First Byte; TTFB). Similarly, we define RTT per hop. In ICN, contents are forwarded back to a user on the same path as they are requested by that user, thus PIT (*pending interest table*) at each ICN router records which interest is not fulfilled yet. A *face* in ICN is the port at which data is sent or received in a router. In our protocols we make use of an access point (AP), which is the closest connecting point of the user to the ICN (not to be confused with a wireless access point). Each router maintains a set of states to record the number of times that a privacy-sensitive content object has been fetched by each user or face. pmode is a flag to indicate that the privacy of a content name being accessed need to be preserved in future access and requests.

2.1 Attack Model

We consider an adversary co-located with an honest user who tries to access contents from ICN . To this end, we assume that the adversary has the capability to perform fine-grained time measurements to perform attacks. We also assume that the attacker has a list of potential "names", where he wants to verify whether the benign user has accessed such names or not. We do not assume any insider attacks, since such names are easy to infer given that domain-specific names are common among people working in that domain, and are easy to infer. From this assumption it follows that the adversary has no control over which path interests are sent, and cannot be geographically distributed to perform an intersection attacks by combining several measurements at different network locations (see Section 5). Finally, for the operation of our attack, we assume that the adversary has enough time to perform the attack, which implies that the content caching lifetime is long enough that the adversary would have a cache hit for contents previously cached by the benign user's requests.

In this paper we assume that the underlying infrastructure used by both adversaries and benign users is honest. In particular, a common router that holds traffic of the adversary and the honest user cannot collude to perform an attack against the benign user (e.g., r_2 in Figure 1). On the other hand, the adversary, if at the scale of a subdomain, may control a router where no traffic of the benign user passes through (e.g., r_3 could replace AP_2 in Figure 1). This assumption can be further used by the adversary to enumerate in real-time what contents are being consumed by other users in his domain, and to help him improve the inference attack on other users within proximity but in other domains (see §1 for such scenario). Finally, the attack discussed in this paper is applicable to CCN, and to a lesser extent to other proposals [5, 10, 16].

2.2 Design Goals and Privacy Definition

The design goals in this paper is to protect the privacy of the users fetching contents using the ICN at reasonable cost. We use the classical definition and meaning of the privacy as "anonymity": the adversary should not be able with his reasonable resources (as assumed in section 2.1) to pinpoint the user fetching such contents among a finite number of users within his proximity. To end, we definite the anonymity set of a user fetching the contents as the number of users at an less than or equal distance from the adversary to that user, who could potentially be fetching that same contents. To this end, we outline the following design goals. (i) *Protect the user privacy*: the main design goal in this work is to defend an adversary from inferring users' access patterns to contents fetched and distributed using ICN (details below). The privacy is defined as anonymity, and increasing the anonymity set of the requested traffic so that to make the inference as less accurate as possible would suffice the purpose. (ii) *Cost effective:* the modifications and overhead for providing privacy to the access pattern of the users should not represent a great overhead in relation with the operational overhead of the ICN (used for routing or caching). Furthermore, the protection mechanisms should not generate an excessive amount of communication overhead (in the form of bits on wire). (iii) *Minimal change to existing ICN protocols:* ideally, we want our solutions not to alter the way caching and routing operate in the ICN.

3. PROTECTION MECHANISMS

As mentioned earlier, simple solutions cannot prevent the timing attacks for privacy while they greatly degrade the benefits of ICN architectures. Also, intelligent caching requires a topology knowledge that is beyond a router's resources. To this end, we propose several solutions without requiring such knowledge.

Before going into the details of the protocols, we introduce the time (delay) generation procedure performed by an edge router, and takes several parameters based on the specific protocol, and the number of hops to be added as noise to prevent the timing attack. For a content name $n \in N$, the total number of hops h, RTT td_x, and the time delay for the first hop td_0, $td(n)$ is chosen as follows to balance privacy and loss in performance. For a given n, the same value of $td(n)$ is used for subsequent requests.

$$td(n) = \begin{cases} 0 & h = 1 \\ 2td_0 < td(n) < td_x & h > 1 \end{cases} \quad (1)$$

3.1 The "Vanilla" Approach

The vanilla algorithm to prevent timing attacks on privacy in ICN is described in Algorithm 1. The main ingredient of the algorithm is a carefully chosen delay added to subsequent responses to make them similar to the responses that fall back on the origin servers to ensure that the contents that are sent to an adversary do not expose timing patterns—such patterns could be used to infer if other users have requested the same contents. For that, the protocol relies on states stored by each edge router to name the contents that are of privacy-value to users, the number of times the contents are being served to each user, and the user id.

Particularly, for a user u (U_1 in Fig. 1), its edge router (r_2 in Fig. 1) maintains $\varphi(u,n) : U \times N \rightarrow INT$, where U, N, and INT are the sets of users, content names, and integers, respectively. $\varphi(u,n)$ indicates the number of times that user u has accessed the content name n. At the beginning, assuming benign user U_1 first generates interest *Ints* $= (U_1, n, \mathsf{pmode}, ts_0)$ with pmode = 1, where ts_0 is the timestamp of when the interest is issued. When r_2 receives this, it follows the ICN protocol [11] to retrieve a data packet $Data$ from the origin server, and records ts_2 upon the arrival of the first packet in response of the interest. Following Eq. 1, r_2 computes expected number of hops from the user U_1 to the origin server as $h = td_x(N)/(2td_0)+1$, and then records td_x along with (U_1, n), and updates the φ to indicate the times that the user has accessed the content. r_2 then serves the content to U_1. When another interest for n is issued by user U_2, who is a potential attacker, the router r_2 acts in response to this interest as follows: If U_2 has previously requested n, r_2 responses directly and serves contents from the cache. Else r_2 applies the random delay and returns $Data$ to U_2.

3.2 An Efficient Approach

While the vanilla algorithm preserves the privacy of user's access history from attackers in the same domain, it consumes significant resources in edge routers, especially when threats from different domains are concerned, where each domain may have large number of users. In order reduce the states stored in each router, a more efficient way is to maintain per-face state instead of per-user ones. The main observation made here is that interests from different (sub-)domains traverse different faces at an edge router, while interests coming from same (sub-)domain would traverse the same face. Accordingly, per-face states are stored and maintained in each router, and decisions to preserve privacy are made upon those states. Unlike Algorithm 1, each router stores $\varrho : F \times N \rightarrow INT$, where F is the set of faces. $\varrho(f,n)$ indicates the number of times that content

Algorithm 1: The "vanilla" approach to preserving the privacy of cache access. The description makes use of the toy example in Figure 1.

Input: n - a content name, u - a user, φ - access state,
Ints $= (u, n, \mathsf{pmode}, ts_0)$
Output: A data packet to u in a privacy-preserving manner.

When R receives *Ints* from u, it records ts_1, the timestamp of interest arrival, and computes $td_0 = ts_1 - ts_0$ as a one-hop time delay.

```
if pmode == 0 then
    if td(n) == 0 then
        // default value td(n) = 0
        R follows ICN protocol to obtain data packet Data
        from the origin server;
        R returns Data to u;
    else
        R follows ICN protocol to obtain data packet Data;
        R delays td(n);
        R returns Data to u;
    end
else
    if φ(u,n) == 0 then
        R follows the ICN protocol to obtain data packet
        Data from the origin server;
        R records ts2 upon the arrival of Data, and computes:
            tdx = ts2 − ts1; // RTT from R to origin server
            h = tdx/(2td0) + 1; // expected # of hops from u
        to the origin server
            Generate td(n) according to Eq. 1;
            φ(u,n) + +;
        R returns retrieved Data to u;
    else
        R returns cached Data to u;
    end
end
```

name n is requested from face f. The protocol can be illustrated in Figure 1, where router r_2, for example, keeps track of the faces connecting it to other routers and access points (e.g., r_3 and r_4), and the times each face has requested content names that have been previously marked as privacy-related contents. After that, r_2 follows the protocol by adding random delays when fulfilling interests that could potential thwart the privacy of other users' access.

3.3 Low Granularity Approach

The main shortcoming of the approach described in §3.2 is that it does not enable lower granularity of the preserved privacy—which is especially required when both the adversary and honest users us the same AP—unlike the protocol described in §3.1. To enable lower granularity in the protocol described in §3.1, we maintain several states in the router, which result high overhead that can be misused, whereas the protocol in §3.2 reduces this overhead at the cost of reduced granularity. We propose a new algorithm aiming to maintain the advantage of both protocols, by maintaining and distributing these states concerning access patterns of individual users at the APs, located closer to but not controlled by end users.

The main idea of the protocol is to distribute state $\varphi(u,n)$ on the AP associated with users generating such requests, and to store the face state $\varrho(f,n)$ in the router. Decisions for access privacy are made at the router with the help of the AP. When the AP receives a request from the user, it checks if the user requested the content before. If not, the pmode value is discarded (to eliminate possible cheating attack about pmode), and the AP forwards the request to the router. Otherwise, the AP directly sends the interest to the router. Upon receiving the interest from a given face, the router initially looks if the content is in the cache or not. If not, it retrieves the content from the origin server and serves it to the requesting user through that face; otherwise, the router checks the face state $\varrho(f, n)$: if it is zero, which implies that no user on that face has requested the content, the router returns the content after a delay $td(n)$ expires; otherwise, it looks at the flag generated by the AP: if it is true, which means that the user has already requested the content before, the router fulfills the interest immediately; otherwise, the interest is fulfilled after a delay $td(n)$ is expired.

4. RESULTS AND ANALYSIS

To understand the potential of the attack and how our designs impact the performance, we perform several measurements on the CCNx prototype [2] in simulated setting. To derive an accurate representation of real-world timings, we feed the simulator with topologies and per-hop RTT traces driven from the current Internet using traceroute [3] to request several websites (origin servers).

4.1 Settings and Timing Data-sets

Our measurements are based on CCNx, an open source system that implements CCN. CCNx implements both the communication operations (naming and data conventions) and security operations (using OpenSSL), for signature generation and verification.

Because no ICN design is deployed yet, we lack real-world traces of RTTs for real ICN. However, designs like CCN suggest operating on top of IP, making IP timings relevant for experiments. To this end, we instrument the CCNx simulator with per-hop round trip delays when issuing interests from within our campus (connected directly to the Internet backbone) to reach each of the Alexa top-100 sites [1]. We use traceroute to obtain per-hop RTT delay to each of these sites—each of these sites is an origin server.

We notice that traceroute has several limitations that prevent direct use of its measurements in our study. First, as the hop count increases, there is no guarantee to have larger RTT than previous RTT for smaller hop count. Second, path to origin servers may change at any time, making two measurements for the same route greatly different. Last, traceroute provides cumulative RTT as the hop count increases, but not the per-hop time delay needed in our study. To address the first issue, we run many traceroute requests at different times of the day to account for different network conditions (which is the main reason that raises this issue), and record different readings for a fixed path to the requested site. Then, for each hop we consider the median RTT among all RTTs given for that hop. We observe that as we increase the number of measurements of the traceroute for the same site we get ordered set of (median) readings: the closer to destination the hop count is, the larger the RTT. To address the second issue of traceroute, we only consider the path that is most popular in the returned traceroute results, and discard all other paths. Once both issues are addressed, we compute the per-hop delay RTT_i as:

$$RTT_i = \begin{cases} RTT_i^t - RTT_{i-1}^t & i > 1 \\ RTT_1^t & i = 1 \end{cases},$$

where RTT_i^t is the i-th returned record by traceroute for the given site. A CDF of the per-hop RTT on the path to each origin server is shown in Figure 2. Complementary CDF for smaller range of RTT ($RTT_i \leq 1$) is shown as a small graph within the CDF in

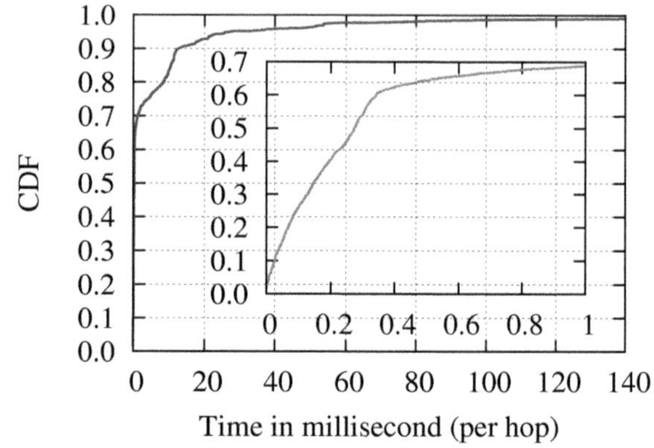

Figure 2: An empirical CDF of the per-hop RTT for alexa-100.

Figure 2 (70% of the hops' RTT in all per-hop measurements are less 1 ms). Notice that the per-hop RTT are smaller than expected on the Internet, which might be due to that two hops are in the same router, same datacenter, or same CDN. However, the results in this study are less significantly affected by other than the total RTT and the first hop delay (used for validating the attack).

We feed the per-hop RTT to a dummy CCNx topology corresponding to the toy example in Figure 1 for each of the hop counts and the per-hop RTT to request these sites. That is, in each case we control the number of hops between router r_2 and r_1 in Figure 1 to correspond to the number of hops returned by traceroute for the given site. We then add the delay incurred over that hop as measured by traceroute using the method explained earlier.

Because the hop count to reach different sites varies, we consider 24 sites that had exactly 16 returned valid hops in traceroute to unify our analysis and discussion in this section. We only limit our attention to those sites where traceroute returned 16 unmasked hops and discard timed-out hops, if any. A boxplot of the normalized per-hop RTT (defined as $RTT_i/\max\{RTT_k\}$ for $1 \leq k \leq h$ and h is the hop-count, where RTT_i is the i-th hop—notice that $RTT_1 = 2td_0$ in our protocols) for each of the 24 sites is shown in Figure 3. Finally, we define the RTT up to each hop as the sum of the per-hop RTT normalized by the total RTT to the origin server. This is, RTT_k is defined as $RTT_k^t = \sum_{i=1}^{k} RTT_i/RTT_h$, where $RTT_h = 2td_0 + td_x$ is the total RTT up to the origin server, and RTT_i is the the i-th hop RTT. Notice that RTT_k^t is returned by traceroute for each k, and can be used immediately in this study. A plot of the normalized RTT up to the origin server is in Figure 4.

4.2 Results

Attack validation. First, we examine whether an adversary co-located one-hop away from a legitimate user is able to exploit the timing attack explained earlier to infer whether some contents are being retrieved by that user or not. We note that as per the ICN caching policy in CCN, contents are replicated and cached at each hop, thus future requests are fulfilled immediately from the closest router to the user. From Figure 4, we observe that an adversary who is co-located with the user who has requested these sites benefit from the caching, and would ideally reduce the total RTT for fulfilling a request by a cache hit at the first hop by around 98% for the most conservative sites (and more than 99% for the median site). Even when a cache-miss happens, an RTT by a cache hit at the sixth hop away from the user, for example, would be 40 times at average (and about 25 times at worst) less than the RTT when retrieving contents directly from the origin server—although this scenario may not breach the privacy of user access patterns since a 6-hop network has a large anonymity set. By feeding the timing

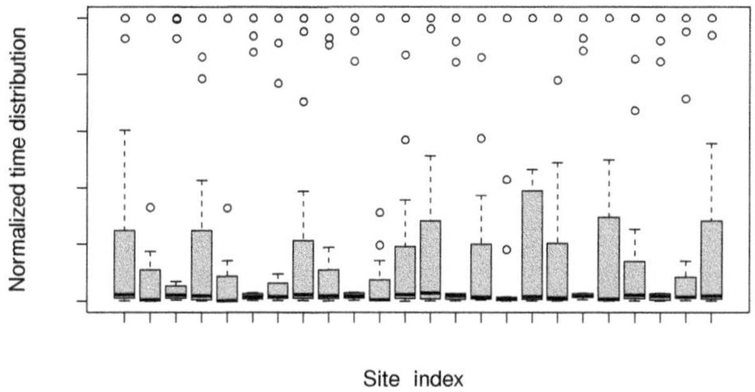

Figure 3: The normalized per-hop total RTT for 24 sites.

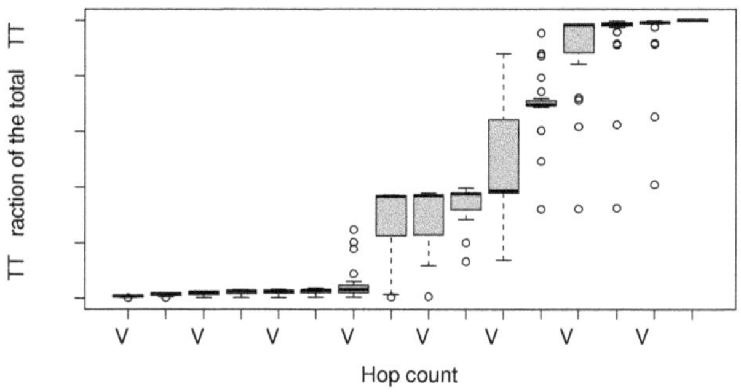

Figure 4: The normalized RTT up to the given hop count.

profiles in Figure 3 in CCNx we observe that the network latency is the dominating part of the RTT in CCN, and other ICN-related delay is negligible. From that, we conclude that an adversary that relies only on the timing information can easily and successfully infer that the contents are being cached in a near-by router due to their access be a potentially co-located user with him.

How defenses impact the performance. One critical parameter for our designs is $td(n)$, which corresponds to the number of hops d that an edge router estimates and according to which he generates noise and uses it to fulfill pending interests issued by end users while maintaining privacy of prior requests. This parameter is generated and used in the three different protocols proposed in this work. Given that we have access to the per-hop delays (as shown in §4.1), we use $d \leq h$ directly to compute $td(n)$ instead of the approximation in Eq. 1. To understand the impact of different values of d we define the maintained RTT gain metric as the difference between the gain in RTT due to caching for subsequent interest fulfillments (when contents are cached 1-hop away from the requesting host) and the the incurred delay due to the added noise in our protocols at a given d. This maintained gain is especially significant to benign users requesting the contents in the future. By observing that the first hop's RTT is negligible, we consider the maintained RTT gain (normalized) as $1 - (td(n)/td_x) \approx 1 - RTT_d^t$. We compute this quantity for the min, max, mean, and median RTT_d^t of the different sites, for different d values.

Even when the router has the capability to record a per-hop RTT and add a given number of hops as noise—not an estimate as described in the protocols, the overhead as additional time delay added to the RTT of fulfilling requests to users still maintains the benefits of ICN as shown in Figure 5. For example, when $d = 6$ (which is one-third of the hop count to the origin server thus providing high anonymity set), a request to an average site would be fulfilled about 40 times faster than retrieving contents from the origin server (0.975 gain). Even for the site with the longest RTT, it would be 25 times (0.96 gain) faster than getting contents from the origin server. Even when d increases the results are not affected greatly: for $d = 7$, the mean, median, and max gain are 0.965, 0.97, 0.75,

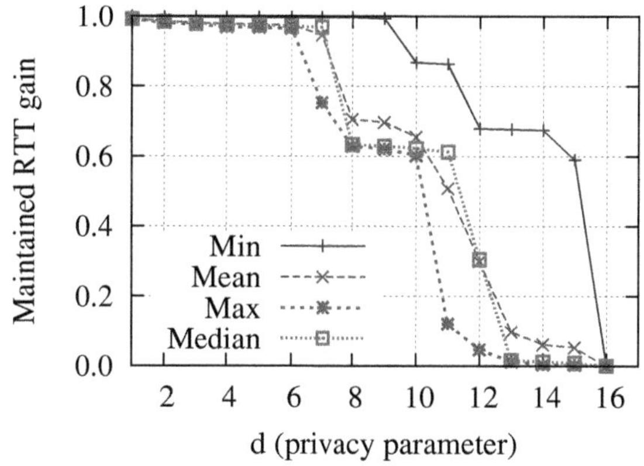

Figure 5: Maintained RTT gain for varying d values.

respectively. Similarly, for $d = 8$, we obtain 0.7, 0.633, and 0.62, respectively. However, as d reaches a value that makes the path traverse the core congested network with high TTL, this result degrades greatly: the performance worsen to reach an average gain of 0.5 at $d = 11$. As before, RTT is dominated by network latencies, whereas CCNx delays are negligible, supporting our claim that our designs maintain ICN's gain in RTT, and that the performance is tunable depending on the desirable privacy to provide to users.

How network conditions affect the performance. Both of the previous sections make conclusions that are network-dependent. Accordingly, we perform similar requests from another commercial campus network that is separated from the Internet backbone by several hops, where several middle boxes are used for security purpose (the average total RTT has increased in these measurements by 300%). In these measurements we observe that the first hop would at average constitute 1% of the overall RTT, making the attack easily applicable, and the maintained gain for $d = 6$ in sites that have 16 returned hops by traceroute is 0.88 at average (8 times faster than retrieving contents from the origin server). We further make similar measurements by performing those requests from a residential network, and find a similar RTT for the first hop, although the gain for $d = 6$ for similar set of sites is 0.92 at average.

Overhead evaluation. The overhead depends greatly on how often contents are flagged for privacy. Since we assume that a user who uses the pmode with requests is trusted, the overhead is a good estimate of real privacy needs. Misuses that try to exploit that and generate excessive overhead on routers can be penalized by feedbacks from other users. We notice that the last protocol, which outperforms all others, have limited overhead on routers. Also, we emphasize that there is no overhead on the network, since the delay generated would not affect the location of contents in the cache, but the time at which an interest is fulfilled.

5. DISCUSSION

We assume that the adversary and the benign user are residing behind the same router, and are 1-hop away from each other. Thus, if both users are 2-hops away, the adversary will still be able to infer some information about the co-location of the benign user who has requested the contents. We address this issue in two ways. First, given that the first few hops (as shown in Figure 4) have small RTTs, the adversary must have sensitive measurements capability at the microsecond level to be able to tell if the user is 2, 3, or 4 hops away with confidence. Second, even in current networks which have many subscribers to the same routing infrastructure, 2-hop away users could likely be hidden in a large enough anonymity set. This makes it hard for the adversary to pinpoint a smaller set of users who could be potentially the requesters of the contents.

We also assume that the adversary cannot collude with routers. However, two users acting as adversaries may collude with each other and try to bypass our defenses. For example, each of the colluding malicious users could issue an interest for a certain content, and compare their timings to infer whether the content has been cached or not. We notice that such collusion is only applicable to the first protocol. In the two latter protocols, requests have to go through the same face, thus they will be considered as if they are from the same entity, regardless to the users who issued them.

A final attack is what we coin the "intersection attack", in which two geographically distributed attackers collude to infer if a piece of contents is cached or not. For example two nodes that are 3 hops away on the path of contents receiving the same time measurements can infer the manipulation by the router by contradicting their measurements . However, in order for this attack to work, the attackers need to: 1) be geographically distributed, and 2) know in advance the path benign requests have traversed. While the first requirement would violate one of our attacker model assumptions, we believe that the second requirement would require collusion of the underlying infrastructure (routers) or much larger number of attackers to make a good estimate of the path. Even though the attack is possible in theory, our defenses and privacy protection mechanisms raise the bar greatly for the attack in practice.

6. RELATED WORK

Concurrent to our work, [14] pointed out the attack under different caching policies, but falls short in not providing any workable solutions to it. Caching has been widely investigated, although motivated by the performance rather than privacy. Examples of the prior literature include the work in [8, 13, 15].

Security and privacy in ICN have been discussed in several recent works. In [18], secure naming system has been proposed. Named-based trust and security protection mechanisms are introduced in [20]. Different naming conventions in ICN architectures and their security features are discussed in [9]. A privacy-preserving contents retrieval in ICN (that assumes the origin server is dishonest) is proposed in [6]. A diverse array of security mechanisms for ICN is introduced in [12]. A closely related architecture that makes accountability as a first-order property, named AIP, is introduced in [4] (which shares similarities with the naming in [7]). Arguments for ICN and future Internet design in general are in [17].

7. CONCLUSION

We have introduced an attack on content access privacy that is applicable to several ICN architectures, including the NDN. We show that an adversary with the capability to perform timing measurements can infer whether contents have been fetched by other users by exploiting the universal caching mechanism deployed in such architecture. We verify such attack theoretically and empirically using real-world per-hop time measurements. To withstand such attack, we introduce three protocols, each of which comes at varying cost and benefits to the network. In these protocols, we make use of carefully chosen time delay to responses given by routers to fulfill requests by users. The delay is chosen to strike a balance between the amount of privacy provided to users—which is determined by the delay added to increase a number of virtual hops away from the user requesting privacy-related contents, the overhead on routers, and the degradation of service to benign users. Our future work will include looking into how different caching polices will affect our attack and countermeasures.

Acknowledgement

H. Xie is supported in part by Grant No. WK0110000014, by NSFC Grant No. 61073192 and by 973 Program Grant No. 2011CB302905. The last author was supported by the KCC (Korea Communts Commission), Korea, under the R&D program supervised by the KCA (Korea Communications Agency), KCA- 2013-12911-05003.

8. REFERENCES

[1] Alexa's top sites. `http://bit.ly/2FzvdZ`, 2012.

[2] CCNx. `https://www.ccnx.org/`, July 2012.

[3] Traceroute. `www.traceroute.org`, 2012.

[4] D. G. Andersen, H. Balakrishnan, N. Feamster, T. Koponen, D. Moon, and S. Shenker. Accountable internet protocol (AIP). In *ACM SIGCOMM*, 2008.

[5] T. Anderson, K. Birman, R. Broberg, M. Caesar, D. Comer, C. Cotton, M. Freedman, A. Haeberlen, Z. Ives, et al. Nebula-a future internet that supports trustworthy cloud computing. *White Paper*, 2010.

[6] S. Arianfar, T. Koponen, B. Raghavan, and S. Shenker. On preserving privacy in content-oriented networks. In *ACM ICN*, 2011.

[7] H. Balakrishnan, K. Lakshminarayanan, S. Ratnasamy, S. Shenker, I. Stoica, and M. Walfish. A layered naming architecture for the internet. In *ACM SIGCOMM*, 2004.

[8] W. Chai, D. He, I. Psaras, and G. Pavlou. Cache "less for more" in icns. *NETWORKING*, 2012.

[9] A. Ghodsi, T. Koponen, J. Rajahalme, P. Sarolahti, and S. Shenker. Naming in content-oriented architectures. In *ACM ICN*, 2011.

[10] D. Han, A. Anand, F. Dogar, B. Li, H. Lim, M. Machado, A. Mukundan, W. Wu, A. Akella, D. G. Andersen, J. W. Byers, S. Seshan, and P. Steenkiste. Xia: efficient support for evolvable internetworking. In *NSDI*, 2012.

[11] V. Jacobson, D. K. Smetters, J. D. Thornton, M. F. Plass, N. H. Briggs, and R. Braynard. Networking named content. In *ACM CoNEXT*, 2009.

[12] J. Jeong, T. T. Kwon, and Y. Choi. Host-oblivious security for content-based networks. In *ACM CFI*, 2010.

[13] K. Katsaros, G. Xylomenos, and G. Polyzos. A hybrid overlay multicast and caching scheme for information-centric networking. In *IEEE INFOCOM*, 2010.

[14] T. Lauinger, N. Laoutaris, P. Rodriguez, T. Strufe, E. Biersack, and E. Kirda. Privacy risks in named data networking: what is the cost of performance? *Computer Communication Review*, 42(5):54–57, 2012.

[15] S. Salsano, A. Detti, M. Cancellieri, M. Pomposini, and N. Blefari-Melazzi. Transport-layer issues in information centric networks. In *ACM ICN*, 2012.

[16] I. Seskar, K. Nagaraja, S. Nelson, and D. Raychaudhuri. Mobilityfirst future internet architecture project. In *Proc. of ACM AINTEC*, 2011.

[17] D. Trossen, M. Sarela, and K. Sollins. Arguments for an information-centric internetworking architecture. *ACM CCR*, 40(2), 2010.

[18] W. Wong and P. Nikander. Secure naming in information-centric networks. In *ACM ReARCH*, 2010.

[19] L. Zhang, D. Estrin, J. Burke, V. Jacobson, J. Thornton, D. Smetters, B. Zhang, G. Tsudik, D. Massey, C. Papadopoulos, et al. Named data networking (ndn) project. Technical report, PARC, 2010.

[20] X. Zhang, K. Chang, H. Xiong, Y. Wen, G. Shi, and G. Wang. Towards name-based trust and security for content-centric network. In *IEEE ICNP*, 2011.

The Role and Effectiveness of Cryptography in Network Virtualization

Wenbo Mao
DaoliCloud Company
Room 2329, Building 2
ZhongGuanCun Software Park
Haidian District
Beijing 100193, China
www.daolicloud.com

Abstract

Communications of IT boxes need control. For IT boxes standing on floors, the control is done by wiring the boxes to some machines that specialize in controlling communications. Since through the wires the controlling machines can see the addresses of the wired IT boxes, the control of communications is done by the controlling machines working on the address information. Today in a paradigm shift, IT boxes are more and more standing on intelligent and distributed software: hypervisors. However, the industry remains in a legacy momentum of controlling communications by hypervisors still working only on the address information. This is not only under using the intelligence of hypervisors in capturing the identities of the IT boxes standing on them, but worse, making some infamous problems associated with working on addresses to get inhered into the new way of life of networking.

We identify problems of today's secure network virtualization due to its negligence in using the intelligence and distributed power of hypervisors, and propose to use hypervisors to work on the identities of IT boxes to increase the controlling power. The result is a new technology for network virtualization with many useful applications including secure multi-tenancy in cloud computing. This work will manifest the powerful role of cryptography in leveraging the intelligence and distributed power of hypervisors.

Categories and Subject Descriptors

D.1.0 Software, PROGRAMMING TECHNIQUES -- General

Keywords

Cloud computing, Network virtualization, Secure multi-tenancy

Introduction

Cloud computing is multi-tenancy: compute, storage and network resources of a cloud datacenter are shared by multiple client organizations called tenants. Virtualization is the enabling technology for resource sharing multi-tenancy, and secure multi-tenancy requires enhancing security in virtualization. In this position paper we shall focus our discussions on secure network virtualization.

To ease a bit for the introduction to the problem of secure network virtualization, let us begin by discussing something probably easier, definitely much smaller, and closer to our daily life reality, than secure multi-tenancy in cloud computing is: BYOD, short for "Bring Your Own Device" (to office). More and more young people, the so-called "net-natives", are entering their employment age. The net-natives grow up accustomed to bringing their personal mobile computers with them all the time and to wherever they are, in particular including when they are in office as employees. More and more employers (the smarter ones in particular) are embracing BYOD, permitting employees to connect their mobile computers to the office intranet. In fact, permitting BYOD can let an employer not only make use of free-cost IT resources which are contributed by employees, but also attract smarter employees who consider their mobile computers being their always available knowledge extension. Of course and inevitably, most employers will also be concerned of information security issues when they adopt a BYOD scheme. Security concerns include: BYOD may cause data leakage from the organization, and virus or malware infection, and even intrusion, to the organization's internal IT infrastructure. There have been many proposals for secure BYOD, scanning and inspecting the software environment of employees' mobile platforms are among very common approaches. We believe that such approaches, while their effectiveness is in doubt, seem to constitute a privacy intrusion to the employee.

The CPU virtualization technology can clean-cut divide a software platform into different computing environments. Let us consider the well-known application of VMware Workstation which is widely used by professionals in technical communities: it runs on common commercial operating systems and virtualizes a guest OS which is a completely independent computing environment from that of the underlying host OS. A natural BYOD solution can be as follows: let the native OS of the employee's platform be the host OS which the employer should not stick nose into; instead, let the employer pay attention to, due to having ownership of, a guest OS which is virtualized on top of the host OS using VMware Workstation. For instance, the employer has constructed the image file of the guest OS, can keep scanning and inspecting the guest OS (this does not form an intrusion since the employer owns the guest OS), and keep monitoring the network traffics to and from the

ASIA CCS'13, May 8–10, 2013, Hangzhou, China.

guest OS. Provided that the employer can differentiate the network packets which are generated by the guest OS from those which are generated by the host OS, then it is very easy for the employer to place its guest OS inside its inner IT infrastructure, and place the employee's host OS outside. If the differentiation can be done, then this employer does have a nice and effective secure BYOD scheme. However unfortunately, such a differentiation is not as straightforward as we may think in a first glance. We notice that all the network traffics from the employee's platform are through the network card of the hardware platform, which are completely under the control of the host OS since the network card is a device of the host OS. Consequently, purely inspecting the metadata part of the IP packets from this hardware platform there would be no way to differentiate the two cases; the host OS can even setup a relay service to make all packets from the guest OS exactly the same as those from itself, or even worse forge packets as if they were from the guest OS. Inspecting the payload data of IP packets cannot be effective as doing so is just the same as scanning and inspecting the host OS, yet not to mention forming a privacy intrusion to the employee. The essential problem we have met here is: VMware Workstation has very well virtualized the CPU, however has not virtualized the network. By saying that VMware Workstation has very well virtualized the CPU, we mean that the guest OS can protect its computing environment from intrusion, e.g., by the host OS, to a quite satisfactory quality. However, since the network traffics of the guest OS completely route via the host OS, the lack of secure network virtualization in VMware Workstation means that the secure BYOD scenario we are sketching here cannot be achieved by directly applying VMware Workstation.

Now let us see the power of cryptography in enabling secure network virtualization for our VMware Workstation based secure BYOD scheme. We notice that the guest OS whose image file is stored in the organization's internal infrastructure can contain sufficiently large amount of the organization defined and controlled data which can be used for the organization to identify the guest OS. As we have discussed above, the well-done CPU virtualization can have these in-guest-OS data well protected from comprehension by the host OS as well as from the employee user of the guest OS. So protected guest-OS identity data can be used to setup a secure communication channel between the guest OS and the organization. For instance, the hypervisor (Workstation) can easily make use of such guest-OS identity data to run IPsec with the organization. Thus, the payload data from the guest OS is protected by IPsec which the host OS can do nothing about them, not forgeable, or not even visible by the host OS, however, the organization running IPsec with the guest OS hypervisor can know for 100% sure guest-OS sent packets, and thereby accept the guest OS inside its internal firewall. What has happened here? Of course secure network virtualization: the physical network resource on the employee's hardware platform is securely virtualized into clear-cut two virtual networks: one for the guest OS, and the other for the host OS. The secure network virtualization in this BYOD use-case is only possibly and practically enabled by the use of cryptography.

Network Virtualization for Secure Multi-tenancy in Cloud Computing: Problem Identification

Cloud computing is a service, everything as a service! A tenant will not pay attention to how computing is done, and therefore should be hidden away from responsibilities of dealing with technical problems itself. This new "lifestyle" for a tenant using a secure multi-tenancy cloud datacenter, the phrase is "Security as a Service": "As for how to configure my firewall, how to control virus for me, how to detect and prevent intrusion to my rental resource, how to prevent data loss from my cloud, …, etc, these must all be services done by you the cloud". The tenant should not care how these are done, as long as they are properly done to a quality by an expert.

Today we are in a well connected era, and thus cloud computing should also have a luxury of wide distribution. This second new "lifestyle" for a tenant should be: "If you the cloud cannot do services well enough, or if the services will later turn out to be too expensive, I shall go to other competitors. To ease, just in case, going away, I won't put all my eggs in bucket. Rent a bit of my IT from this cloud, and a bit of from that cloud. Therefore, my cloud shall be a "hybrid" one."

Technologies to achieve secure multi-tenancy in cloud datacenter inevitably use secure virtualization technologies. Secure CPU virtualization and secure storage virtualization are relatively more mature technologies than secure network virtualization. This statement is even truer for the case of realizing CPU and storage virtualization technologies in a distributed manner, even in trans-multiple datacenter fashion. A virtual machine (VM) of a tenant can run in this server or that server, and even run in this datacenter or that datacenter, and so is a virtual disk (VD) of a tenant. Quality of service, such as efficiency of a VM, size of a VD, their backup copies, etc, should not be very different if the services are offered by this datacenter or that datacenter, or even offered by co-working of multiple datacenters. Moreover, the notion of "in" and "out" for a tenant in the case of its rental VMs or VDs can be very clear defined. Clearly, the notion of "in" and "out" for a VM or for a VD is straightforward. Thus, labeling the VMs and VDs, which are rented by a tenant, with the identity of the tenant, then "in" and "out" for the tenant's rental VMs and VDs also become clearly defined. The case for network virtualization is very different.

First of all, the notion of "in" and "out" of a network is traditionally defined by a firewall. A firewall for IT as an asset (i.e., IT infrastructure which is privately owned) is defined on the metadata of network packets—IP/MAC addresses; a firewall filters IP/MAC packets on their addresses, lets the correctly addressed packets pass, and blocks (drops) the incorrectly addressed ones. Because IT boxes used to stand on floors which do not think, the control of communications of the floor-standing IT boxes is done by wiring them to central control points which specialize in network control, and in our topic of discussion, these control points include firewall control for "in" and "out" for a network of floor-standing IT boxes. In the traditional network technology for IT as an asset, the wiring essentially only needs to pass the IP/MAC addresses to the firewall as the identities of the floor-standing IT boxes, and that is why a traditional firewall doesn't see, and hence cannot work on, the identities of floor-standing IT boxes other than IP/MAC addresses.

In today's server virtualization multi-tenancy cloud datacenters, IT boxes are VMs standing on hypervisors. As we have discussed for CPU virtualization, in order to identify VMs which are rented by different tenants, VMs have machine identities and the datacenter does manage these identities for multiple of tenants. However, in network virtualization (let's say today's cloud datacenter does do network virtualization), the hypervisors still only work on the IP/MAC addresses of the VMs, and so does the firewall, whether the so-called "virtual firewall" in hypervisors (e.g., VMware's vShield technology), or real and conventional ones outside. Although a hypervisor can so easily see machine identities (e.g., UUID) of each VM on it, in terms of working on network, it chooses to only work on IP/MAC addresses of the VMs. Thus, "in" and "out" for a tenant's rental LAN is still defined by the IP/MAC-address working firewall, exactly the same working principle of firewall for IT boxes on floors. This non-virtualized style of networking has many problems for secure network virtualization. In this position paper, we shall only point out one which is responsible for a very poor quality of service for the second new "lifestyle" we listed in the beginning of this section: if a tenant chooses to distribute its cloud in trans-datacenter fashion (very reasonable), the LAN distributes with a very poor quality of service due to tremendous network traffics which are flowing inter-datacenters.

Why? It's the firewall which cannot be virtualized nor distributed if only working on IP/MAC addresses. Working on addresses, "in" and "out" of a trans-datacenter "LAN" for a tenant is defined by an address filtering firewall. Such a firewall can only be defined at a single point of control, or else there would be loss of control. Distribution of an address filtering firewall is pointless unless the distributed firewalls are identical; if two firewalls for one tenant are different in filtering policy, the packets will choose to pass the one with less controlled policy. Thus, if a tenant chooses to have its cloud in trans-datacenter distribution, then today's address filtering firewall technology requires that a VPN technique is used: the firewall is deployed on the master or main datacenter of the VPN, and all VMs in the slave or minor datacenters must wire their network traffics through the VPN connection to the firewall in the master datacenter for firewall policy control. This is obviously a non-virtualized network. Let's only look at the following couple of.

Problem 1 (Poor QoS, minor)

Communications of two VMs in a slave datacenter cannot enjoy the LAN bandwidth; their packets must be wired through the VPN cross the WAN to the master datacenter for firewall control.

Problem 2 (Lack of distribution, very serious)

If the tenant later turns out to be unhappy on the service of the master datacenter, then moving out of the master datacenter means to redeploy and reconfigure the VPN of its complete distributed cloud. In essence and in fact, the tenant has not distributed its cloud at all.

Network Virtualization for Secure Multi-tenancy in Cloud Computing: Solution Proposal

Recall that in the "BYOD" scheme using an in-organization guest OS on VMware Workstation, only using IP addresses the organization cannot distinguish IP packets sent by the guest OS from those sent by the host OS. In other words, the network provided by the physical NIC card of the employee's platform is not virtualized. There we propose to use of cryptography to form an effective solution to have virtualized the network. The use of cryptography in that solution must use the identity of the guest OS.

Here we meet the same problem in network virtualization for secure multi-tenancy cloud datacenter: only working on IP addresses, network virtualization cannot be properly done. The solution is the same: working on VM identities.

In server virtualization datacenters, IT boxes are VMs standing on distributed and powerfully intelligent hypervisors. For CPU and storage virtualization, hypervisors already work on managing VM identities for multiple of tenants. Then why not also let hypervisors also work on VM identities for network virtualization?

Let us highlight below a scheme for a completely distributed and virtualized firewall for a tenant.

1) A hypervisor proves the identity (e.g., UUID) for a VM sitting atop. The proof can be done using public-key cryptography to make the proof verifiable globally.
2) A hypervisor verifies identities of VMs sitting on top of other hypervisors. Again, using public key cryptography, the verification is a computation in a global fashion, i.e., identities of VMs in remote datacenters can be verified.
3) A tenant composes a list of global VM identities for its rental VMs and specifies communications policy for these identities. For example, the tenant can specify that VM-Global-ID-i can or cannot communicate with VM-Global-ID-j, etc. This list is in fact the virtual firewall policy which is defined by the tenant. If the tenant chooses to have its cloud in trans-datacenter distribution, then this list is also in trans-datacenter distribution.
4) Hypervisors, whether they are in one datacenter, or are in trans-datacenter distribution, consult the tenant's distributed virtual firewall list to allow or disallow communications between any two VMs according to the tenant's instructions in its virtual firewall list.

Obviously, a firewall defined in this method is completely virtualized, not only within one datacenter, but also trans-datacenter. We can list below a number of advantages of the completely virtualized firewall which is defined and working on the VMs global identities:

- Motion of a VM, in particular when motion is in trans-datacenter, may cause VM IP address change, however, the globally defined VM identity will not change, and hence the new firewall which is defined and working on the unchanging VM identities will not be affected. We notice that, motion of VMs is not only a need of a tenant, it is more a need of a datacenter, e.g., for load-balance, server maintenance, continuity of services, etc.

- Complete distribution of the cloud for a tenant. There is not a need of a master VPN datacenter and slave ones. The traffics between any two VMs are completely controlled by the hypervisors serving them, and when two such VMs are in one datacenter, their firewall policy is controlled locally, and no trans-datacenter traffic whatsoever.
- Working with cryptography, traffics between VMs within a firewall can be encrypted, and hence secure multi-tenancy is protected strongly by cryptography.
- For VMs having public-key infrastructure (PKI) based globally verifiable IDs, a Domain Name Service (DNS) can verify the genuineness of a web server on users' behalf (as a service) without involving in content liability of a web server; anti-phishing will become much eased
- Hypervisors working with a database of the tenant's defined firewall, this algorithm can be done in high-level language programming. It is much easier than configurations of IP addresses filtering policy of a traditional firewall. The tenant's firewall can be programmed to work in complete automation, and for tenant self-serve.

Of course, it will be a long journey for this new network virtualization technology to be getting mature. Many things in this technology will involve standards work, and hence can only be done in joint efforts with the industry, academia and R&D communities.

Let's work together to make this new and useful network virtualization technology for secure multi-tenancy cloud computing toward a reality.

Efficient Dynamic Provable Possession of Remote Data via Balanced Update Trees

Yihua Zhang and Marina Blanton
Department of Computer Science and Engineering
University of Notre Dame
{yzhang16,mblanton}@nd.edu

ABSTRACT

The emergence and availability of remote storage providers prompted work in the security community that allows a client to verify integrity and availability of the data she outsourced to an untrusted remove storage server at a relatively low cost. Most recent solutions to this problem allow the client to read and update (insert, modify, or delete) stored data blocks while trying to lower the overhead associated with verifying data integrity. In this work we develop a novel and efficient scheme, computation and communication overhead of which is orders of magnitude lower than those of other state-of-the-art schemes. Our solution has a number of new features such as a natural support for operations on ranges of blocks, and revision control. The performance guarantees that we achieve stem from a novel data structure, termed *balanced update tree*, and removing the need to verify update operations.

Categories and Subject Descriptors

K.6 [**Management of Computing and Information Systems**]: Security and Protection; E.1 [**Data Structures**]: Trees; H.3.4 [**Information Storage and Retrieval**]: Systems and Software—*distributed systems*

General Terms

Security, Algorithms, Verification

Keywords

Authentication, dynamic provable data possession, balanced update tree, outsourced storage, proof of retrievability

1. INTRODUCTION

Cloud computing and storage services today enable convenient on-demand access to computing and data storage resources, which make them attractive and economically sensible for clients with limited computing or storage resources. Security and privacy, however, have been suggested to be the top impediment on the way of harnessing full benefits of these services (see, e.g., [1]). For that reason, there has been an increased interest in the research community in securing outsourced data storage and computation, and in particular, in verification of remotely stored data.

The line of work on proofs of retrievability (POR) or provable data possession (PDP) was initiated in [4, 20] and consists of many results [11, 25, 5, 24, 19, 10, 29, 6, 8, 9, 16, 26, 27, 13, 28, 30, 21, 17, 22] that allow for integrity verification of large-scale remotely stored data. At high level, the idea consists of partitioning a collection of data into data blocks and storing the blocks together with metadata at a remote storage server. Periodically, the client issues integrity verification queries (normally in the form of challenge-response protocols), which allow the client to verify a number of data blocks independent of the overall number of outsourced blocks using the metadata to ensure with high probability that all stored blocks are intact and available. Schemes that support dynamic operations [5, 19, 16, 26, 27, 30, 21, 17, 22], DPDP, additionally allow the client to issue modify, insert, and delete requests, after each of which the integrity of the newly stored data is verified.

The motivation for this work comes from (i) improving the performance of the existing schemes when modifications to the data are common, and (ii) extending the available solutions with new features such as support for revision control and multi-user access to shared data[1]. Toward this end, we design and implement a novel mechanism for efficient verification of remotely stored data with support for dynamic operations. Our solution uses a new data structure, which we call a balanced *update tree*. The size of the tree is independent of the overall size of the outsourced storage, but rather depends on the number of updates to the remote blocks. The data structure is designed to provide a natural support for handling ranges of blocks (as opposed to always processing individual blocks) and is balanced allowing for very efficient operations. Unlike all prior work with support for dynamic operations where each dynamic operation is followed by verification of its correct execution by the server, our scheme eliminates such checks. Instead, verification is performed only at the time of retrieving the data or through periodic challenge queries (both of which are also present in prior work). This distinctive feature of our scheme therefore results in substantial communication and computation savings.

Today many services outsource their storage to remote servers or the cloud, which can include web services, blogs, and other applications in which there is a need for multiple users to access and update the data, and modifications to the stored data are common. For example, many subscribers of a popular blog hosted by a cloud-based server are allowed to upload, edit, or remove blog content ranging from a short commentary to a large video clip. This demands support for multiple user access while maintaining data

ASIA CCS'13, May 8–10, 2013, Hangzhou, China.

[1]Due to space constraints, multi-user access is described in the full version of this work.

consistency and integrity, which current schemes do not provide. In addition to supporting this feature, our solution provides support for revision control which can be of value for certain applications as well. Because in the existing solutions the server maintains only the up-to-date values of each data block, support for revision control can be added by means of additional techniques (such as [3]), but they result in noticeable overhead per update. In our solution, on the other hand, there is no additional cost for enabling retrieval and verification of older versions of data blocks beyond the obvious need for the server to store them with small metadata. Finally, because the size of the maintained data structure grows with the number of dynamic operations, by issuing a commit command, the client will be able to keep the size of the maintained update tree below a desired threshold if necessary.

To summarize, our solution enjoys the following features:

- *Improved efficiency in handling dynamic operations.* In our solution there is no need to verify integrity of updates (e.g., a data block which is modified a large number of times and is subsequently deleted), while prior schemes invest resources in verifying correct implementation of each user's action by the storage server.
- *Support for range operations.* The natural support and use of range operations allows for additional performance improvement compared to the existing solutions.
- *Balanced data structure.* The update tree used for verifying correctness of the stored data blocks is always balanced regardless of the number and order of dynamic operations on the storage. This results in similar performance for locating information about each data block in the tree and is logarithmic in the size of the tree.
- *Size of the maintained data structure.* In our solution the size of the maintained update tree is independent of the outsourced data size, while it is linear for other solutions that support dynamic operations. The size of the update tree grows with the number of dynamic operations, but can be reduced by issuing a commit command.
- *Support for revision control.* We provide natural support for revision control and allow clients to retrieve previous versions of data and efficiently verify their integrity. There is no additional overhead for either the client or the server for enabling this feature (besides the obvious need to maintain different versions of data by the server).
- *Public verifiability.* Our scheme can be easily modified to support public verifiability, which allows the client to outsource periodic verification of storage integrity to a third party auditor (who is different from the server).

These features come at the cost of increased storage (compared to other schemes) at the client who in our solution maintains the update tree locally. Because the data structure size is not large (and is independent of the outsourced data size), we believe it is a reasonable tradeoff for other improvements that we achieve. In particular, any PC-based client will not be burdened by the local storage even if it reaches a few MB. Other weaker clients (such as mobile users) and battery-operated devices in particular are power-bound and benefit from the reduced computation in our scheme while still are able to store the data structure locally. A detailed comparison of the performance of our and other DPDP schemes is given in section 8, which shows that both computation and communication overhead of our scheme is orders of magnitude lower than those of other existing solutions.

2. RELATED WORK

We next review selected PDP/POR schemes from prior literature and their difference with the proposed solution. In particular, we are interested in the schemes that support dynamic operations on outsourced storage.

One line of research [20, 5, 26] relies on so-called sentinels which are outsourced together with the client's data and are used to verify remotely stored blocks. Such scheme, however, poorly scale when blocks need to be updated and allow only for a limited number of audits.

Another line of research with support for dynamic operations utilizes specialized data structures such as Merkle hash trees or chained hashes [27, 21, 22] or skip lists [16, 19, 17] to organize the data blocks outsourced to a server. When a Merkle hash tree is used, each leaf node corresponds to the hash of a data block, and the client locally keeps the root value of the tree. Correctness of a read request on the ith block is verified in a standard way by reconstructing the root value from the ith block and hashes of the sibling nodes on the path from the block to the root. For an update request on the ith block, the client retrieves and verifies the same information as that of the read request. The client then computes a new root value based on the new ith block, substitutes it for the previously stored root value, and uses it afterwards.

A disadvantage of Merkle hash tree based solutions is that the tree becomes unbalanced after a series of insert and delete requests. That is, a block insert request at position i is handled by replacing the $(i-1)$th block's node with a new node that has two children: a node for the previously stored $(i-1)$th and a node for the newly inserted ith block. Similarly, a deletion request is handled by removing the corresponding node from the tree and making its sibling take the place of its parent. As access patterns normally do not uniformly span across the stored data, inserting multiple blocks at the same position will result in the height of the tree growing for each inserted data block. This will result in a large variance in the time to locate different blocks in the tree. Using an advanced tree structure (e.g., red-black tree) can help mitigate this problem by balancing the tree when necessary, but it comes with an additional cost of recomputing the hash of nodes affected by the balancing process.

To support the dynamic operations, [16] develops a scheme based on a skip list, which extends the original skip list [23] by incorporating *label* [18] and *rank* information to enable efficient authentication of client's updates. This allows for each update to be verified in expected $O(\log n)$ time with high probability, where n is the size of the skip list. The skip list remains balanced regardless of the client's access or update patterns. The authors also propose support for variable-sized blocks, which is achieved by handling a number of fixed size blocks as a single block and performing standard operations on it. While this approach guarantees the integrity of variable-sized data blocks in their entirety, it becomes impossible to verify an individual block upon receiving a request on it. Furthermore, the time to locate a fixed size block is linear in the number of blocks stored in a node, which may dominate the overall time when a node contains many blocks.

The original Merkle hash tree and skip list schemes maintain only the most recent copy of data. To incorporate revision control capabilities, [3] used a persistent authenticated data structure, which adds $O(\log n)$ extra space for each update, where n is the number of nodes in the data structure.

The data structure that we build has three properties that make it favorably compare to the existing schemes. First, each node in our update tree corresponds to a range of block indices (instead of a specific index as in prior work) defined by a dynamic operation performed on a range of consecutive blocks. The reason for assigning a range of block indices to a node is motivated by a study [15] on user's file access patterns that observed that a large number of file accesses are sequential. Second, unlike maintaining a data structure of size linear in the number of outsourced data blocks, in our solution it is independent of the size of the stored data. Previously, the large size required the client to outsource the data structure to the cloud while locally maintaining only a constant-size data for integrity verification. In our update tree, on the other hand, a node represents a user-triggered update, and multiple updates issued on the same range of blocks can also be condensed into a single node. Due to its moderate size, the client can maintain the data structure locally, which makes the verification process more efficient. Third, we can specify requirements that define when the data structure should be rebalanced. Once the requirement is violated, the tree is re-organized to satisfy the constraint. As an example, the constraint of AVL trees [2] can be used that requires that the heights of a node's subtrees must differ by at most 1.

Prior to this work, the notion of a range tree was used in the databases to deal with range queries [7]. The range tree data structure, however, is majorly dissimilar to our update trees. For instance, range trees store one record per node as opposed to a range, are static as opposed to be dynamically updated and balanced throughout system operation, etc. Similarly, interval trees [12] cannot support insertions and deletion of block ranges which require partitioning of existing ranges/intervals in the tree and index changes. The operational details of update trees are therefore very different from those of interval trees. One of most significant challenges of this work was to design a dynamic update tree that can be rebalanced at low cost after arbitrary changes to it. A balanced update tree is therefore one of the novel aspects of this work.

3. PROBLEM DEFINITION

We consider the problem in which a resource-limited client is in possession of a large amount of data partitioned into blocks. Let N denote the initial number of blocks and m_i denote the data block at index i, where $1 \leq i \leq N$. The client C outsources her data to a storage or cloud server S and would like to be able to update and retrieve her data in a way that integrity of all returned data blocks can be verified. If the data is sensitive and its secrecy is to be protected from the server, the client should encrypt each data block using any suitable encryption mechanism prior to storing it at the remote server. In that case, each data block m_i corresponds to encrypted data, and the solution should be oblivious to whether data confidentiality is protected or not. We assume that the client and the server are connected by (or establish) a secure authenticated channel for the purposes of any communication.

The primary feature that we would like a scheme to have is *support for dynamic operations*, which include modifying, inserting, or deleting one or more data blocks. We also consider minimal-overhead *support for revision control* which allows the client to access and verify previous versions of its data, as a desirable feature to have, but it is not strictly necessary for a PDP scheme. Our scheme achieves this property at no extra cost beyond maintaining previous versions by the server, and we defer any additional discussion of this feature to the full version of this work.

We define a dynamic provable data possession scheme (DPDP) in terms of the following procedures:

- $\mathsf{KeyGen}(1^\kappa) \rightarrow \{\mathsf{sk}\}$ is a probabilistic algorithm run by C that on input a security parameter 1^κ produces key sk.
- $\mathsf{Init}(\langle \mathsf{sk}, m_1, \ldots, m_N \rangle, \langle \perp \rangle) \rightarrow \{\langle \mathsf{M}_C \rangle, \langle \mathsf{M}_S, D \rangle\}$ is a protocol run between C and S during which C uses sk to encode the initial data blocks $m_1, \ldots, m_N$ and store them at S who maintains all data blocks outsourced by the client in D. C's and S's metadata are maintained in M_C and M_S, resp.
- $\mathsf{Update}(\langle \mathsf{sk}, \mathsf{M}_C, \mathsf{op}, \mathsf{ind}, \mathsf{num}, m_{\mathsf{ind}}, \ldots, m_{\mathsf{ind+num}-1} \rangle, \langle \mathsf{M}_S, D \rangle) \rightarrow \{\langle \mathsf{M}'_C \rangle, \langle \mathsf{M}'_S, D' \rangle\}$ is a protocol run between C and S, during which C prepares num blocks starting at index ind and updates them at S. The operation type op is either modification (0), insertion (1), or deletion (-1), where no data blocks are communicated for deletion.
- $\mathsf{Retrieve}(\langle \mathsf{sk}, \mathsf{M}_C, \mathsf{ind}, \mathsf{num} \rangle, \langle \mathsf{M}_S, D \rangle) \rightarrow \{\langle m_{\mathsf{ind}}, \ldots, m_{\mathsf{ind+num}-1} \rangle, \langle \perp \rangle\}$ is a protocol run between C and S, during which C requests num data blocks starting from index ind, obtains them from S and verifies their correctness.
- $\mathsf{Commit}(\langle \mathsf{sk}, \mathsf{M}_C, \mathsf{ind}, \mathsf{num}, m_{\mathsf{ind}}, \ldots, m_{\mathsf{ind+num}-1} \rangle, \langle \mathsf{M}_S, D \rangle) \rightarrow \{\langle \mathsf{M}'_C \rangle, \langle \mathsf{M}'_S, D' \rangle\}$ is a protocol run between C and S, during which C re-stores metadata of num data blocks starting from index ind at S. S erases all previous copies of the data blocks in the range as well as previously deleted by C blocks that fall into the range if they were kept for revision control.

Our formulation of the scheme has minor differences with prior definitions of DPDP, e.g., as given in [16]. First, update and retrieve operations are defined as interactive protocols rather than several algorithms run by either the client or the server. Second, in addition to using the Retrieve protocol for reading data blocks, in the current formulation it is also used to execute periodic audits. That is, verification of each read is necessary to ensure that correct blocks were received even if the integrity of the overall storage is assured through periodic challenges, and the verification is performed similar to periodic audits. In particular, because the Retrieve protocol is executed on a range of data blocks and can cover a large number of blocks, verification is performed probabilistically by checking a random sample of blocks of sufficient (but constant) size to guarantee the desired confidence level. (And if the number of requested blocks is below the constant, all of them are verified.) This protocol can then be easily adapted to implement periodic audits denoted as $\mathsf{Challenge}(\langle \mathsf{sk}, \mathsf{M}_C \rangle, \langle \mathsf{M}_S, D \rangle) \rightarrow \{\langle m_{i_1}, \ldots, m_{i_c} \rangle, \langle \perp \rangle\}$ during which a random subset of blocks at indices $i_1, \ldots, i_c$ is verified. To implement Challenge, we simply call Retrieve on the entire storage with the difference that data blocks which are not being verified are not returned. In other words, during each Challenge query, the client receives and verifies correctness of c data blocks. We stress that defining audit Challenge queries in terms of Retrieve requests is the matter of notational preference: the functionality and probabilistic nature of verification of both Retrieve and Challenge requests is the same in our and other DPDP schemes. We obtain that prior work requires verification for each block update operation and a constant number of verifications per audit or read request. In our proposed scheme, only read and audit requests need to be verified by checking a constant number of blocks per request.

This constant c is computed in our and prior work by using detection probability of $1 - ((\mathsf{num} - t)/\mathsf{num})^c$, where num is the number of blocks being checked, from which the server tampers with t. Then, say, using $c = 460$ the client can detect the problem with 99% probability if the server tampers with 1% or more of the data regardless of the data size. This means that during Retrieve or Challenge calls, $\min(c, \mathsf{num})$ data blocks need to be verified.

To show security, we follow the definition of secure dynamic PDP from prior literature. In particular, we base our definition on the original definition of secure DPDP from [16] and introduce logical changes to account for the slightly different setting. In this context, the client should be able to verify the integrity of any data block returned by the server. This includes the verification that the returned data block corresponds to the most recent version of it (or, when revision control is used, a specific previous version, including deleted content, as requested by the client). The server is considered fully untrusted and can modify the stored data in any way it wishes (including deleting the data). Our goal is to design a scheme in which any violations of data integrity or availability will be detected by the client. More precisely, in the single-user setting the security requirements are formulated as a game between a challenger (who acts as the client) and any probabilistic polynomial time (PPT) adversary $\mathcal{A}$ (who acts as the server):

Setup: the challenger runs $\mathsf{sk} \leftarrow \mathsf{KeyGen}(1^\kappa)$. $\mathcal{A}$ specifies the data blocks $m_1, \ldots, m_N$ and their number N for the initialization and obtains initial transmission from the challenger.

Queries: The adversary $\mathcal{A}$ specifies what type of a query to perform and on what data blocks. The challenger prepares the query and sends it to $\mathcal{A}$. If the query requires a response, $\mathcal{A}$ sends it to the challenger, who informs $\mathcal{A}$ about the result of verification. The adversary can request any polynomial number of queries of any type, participate in the corresponding protocols, and be informed of the result of verification.

Challenge: At some point, $\mathcal{A}$ decides on the content $m_1, \ldots, m_R$ on which it wants to be challenged. The challenger prepares a query that replaces the current storage with the requested data blocks and interacts with $\mathcal{A}$ to execute the query. The challenger and adversary update their metadata according to the verifying updates (non-verifying updates are considered not to have taken place), and the challenger and $\mathcal{A}$ execute $\mathsf{Challenge}(\langle \mathsf{sk}, \mathsf{M}_C \rangle, \langle \mathsf{M}_S, D \rangle)$. If verification of $\mathcal{A}$'s response succeeds, $\mathcal{A}$ wins. The challenger has the ability to reset $\mathcal{A}$ to the beginning of the Challenge query a polynomial number of times with the purpose of data extraction. The challenger's goal is to extract the challenged portions of the data from $\mathcal{A}$'s responses that pass verification.

DEFINITION 1. *A DPDP scheme is secure if for any PPT adversary $\mathcal{A}$ who can win the above game with a non-negligible probability, there exists an extractor that allows the client to extract the challenged data in polynomial time.*

The existence of an extractor in this definition means that the adversary that follows any strategy can win the game above with probability negligibly larger than the probability with which the client is able to extract correct data. In our case, the probability of catching a cheating server is the same as in prior literature, and its security is analyzed in section 6.

Besides security, efficient performance of the scheme is also one of our primary goals. Toward that goal, we would like to minimize all of the client's local storage, communication, and computation involved in using the scheme. We also would like to minimize the server's storage and computation overhead when serving the client's queries. For that reason, the solution we develop has a natural support for working with ranges of data blocks which is also motivated by users' sequential access patterns in practice.

4. PROPOSED SCHEME

Building blocks. In this work we rely on a message authentication code (MAC) scheme, defined by three algorithms:

1. The key generation algorithm Gen, which on input a security parameter 1^κ produces a key k.
2. The tag generation algorithm Mac, which on input key k and message $m \in \{0,1\}^*$, outputs a fixed-size tag t.
3. The verification algorithm Verify, which on input a key k, message m, and tag t outputs a bit b, where $b = 1$ iff verification was successful.

For compactness, we write $t \leftarrow \mathsf{Mac}_k(m)$ and $b \leftarrow \mathsf{Verify}_k(m, t)$. The correctness requirement is such that for every κ, every $k \leftarrow \mathsf{Gen}(1^\kappa)$, and every $m \in \{0,1\}^*$, $\mathsf{Verify}_k(m, \mathsf{Mac}_k(m)) = 1$. The security property of a MAC scheme is such that every PPT adversary $\mathcal{A}$ succeeds in the game below with at most negligible probability in κ:

1. A random key k is generated by running $\mathsf{Gen}(1^\kappa)$.
2. $\mathcal{A}$ is given 1^κ and oracle access to $\mathsf{Mac}_k(\cdot)$. $\mathcal{A}$ eventually outputs a pair (m, t). Let Q denote the set of all of $\mathcal{A}$'s queries to the oracle.
3. $\mathcal{A}$ wins iff both $\mathsf{Verify}_k(m, t) = 1$ and $m \notin Q$.

Overview of the scheme. To mitigate the need for performing verifications for each update on the outsourced data, in our solution both the client and the server maintain metadata in the form of a binary tree of moderate size. We term the new data structure a *block update tree*. In the update tree, each node corresponds to a range of data blocks on which an update (i.e., insertion, deletion, or modification) has been performed. The challenge with constructing such a tree was to ensure that (i) a data block or a range of blocks can be efficiently located within the tree and (ii) we can maintain the tree to be balanced after applying necessary updates caused by client's queries. With our solution, we obtain that all operations on the remote storage (i.e., retrieve, insert, delete, modify, and commit) involve only work logarithmic in the tree size.

Each node in the update tree contains several attributes, one of which is the range of data blocks $[\mathsf{L}, \mathsf{U}]$. Each time the client requests an update on a particular range, the client and the server first need to find all nodes in the update tree with which the requested range overlaps (if any). Depending on the result of the search and the operation type, either 0, 1, or 2 nodes might need to be added to the update tree per single-block request. Operating on ranges helps to lower the size of the tree. For any given node in the update tree the range of its left child always covers data blocks at strictly lower indices than L, and the range of the right child always contains a range of data blocks with indices strictly larger than U. This allows us to efficiently balance the tree using standard algorithms such as that of AVL trees [2]. Furthermore, because insert and delete operations affect indices of the existing data blocks, in order to quickly determine (or verify) the indices of the stored data blocks after a sequence of updates, we store an offset value R with each node of the update tree which indicates how the ranges of the blocks stored in the subtree rooted at that node need to be adjusted. Lastly, for each range of blocks stored in the update tree, we record the number of times the blocks in that range have been updated. This information will allow the client to verify that the data she receives corresponds to the most recent version and integrity of the data (or, alternatively, to any previous version requested by the client).

At the initialization time, the client computes a MAC of each data block together with its index and version number (which is initially set to 0). The client stores the blocks and their corresponding MACs at the server. If no updates take place, the client will be able to retrieve a data block by its index number and verify its integrity. To support dynamic operations, the update tree is first initialized to empty. To modify a range of existing blocks, we insert a node in the tree that indicates that the version of the blocks in the range has increased. To insert a range of blocks, the client creates a node in the tree with the new blocks and also indicates that the indices of the blocks that follow need to be increased by the number of inserted blocks. A node's offset affects its entire subtree, which removes the need to touch many nodes. To delete a range of blocks, the deleted blocks are marked with operation type "-1" in the tree and the offset of blocks that follow is adjusted accordingly. Then to perform an update (insert, delete, or modify), the client first modifies the tree, computes the MACs of the blocks to be updated, and communicates the blocks (for insertion and modification only) and the MACs to the server. Upon receiving the request, the server also modifies the tree according to the request and stores the received data and MACs. If the server behaves honestly, the server's update tree will be identical to the client's update tree (i.e., all changes to the tree are deterministic). To retrieve a range of blocks, the client receives a number of data blocks and their corresponding MACs from the server and verifies their integrity by using information stored in the tree.

Update tree attributes. Before we proceed with the description of our scheme, we outline the attributes stored with each node of the update tree, as well as global parameters. Description of the update tree algorithms is deferred to Section 5.

With our solution, the client and the server maintain two global counters together with the update tree, GID and CID, both of which are initially set to 0. GID is incremented for each insertion operation to ensure that each insert operation is marked with a unique identifier. This allows the client to order the blocks that have been inserted into the same position of the file through different operations. CID is incremented for each commit operation and each commit is assigned a unique identifier. For a given data block, the combination of its version number and commit ID will uniquely identify a given revision of the block. In addition, each node in the update tree stores several attributes:

Node type Op represents the operation type associated with the node, where values -1, 0, and 1 indicate deletion, modification, and insertion, respectively.

Range L, U specifies the start and end indices of the data blocks, information about which is stored at the node.

Version number V indicates the number of modifications performed on the data blocks associated with the node. The version number is initially 0 for all data blocks (which are not stored in the update tree), and the version is also reset to 0 during a commit operation for all affected data blocks (at which point information about them is combined into a single node).

Identification number ID of a node has a different meaning depending on the node type. For a node that represents an insertion, ID denotes the value of GID at the time of the operation, and for a node that represents a modification or deletion, ID denotes the value of CID at the time of the last commit on the affected data blocks (if no commit operations were previously performed on the data blocks, the value is set to 0). In order to identify the type of ID (i.e., GID or CID) by observing its value, we use non-overlapping ranges for the values from which IDs for the two different types are assigned.

Offset R indicates the number of data blocks that have been added to, or deleted from, the range of data block indices that precede the range of the node (i.e., $[0, \mathsf{L}-1]$). The offset value affects all data blocks information about which is stored directly in the node as well as all data blocks information about which is stored in the right child subtree of the node.

Pointers $\mathsf{P_l}$ and $\mathsf{P_r}$ point to the left and right children of the node, respectively, and $\mathsf{P_p}$ points to the node's parent.

In addition to the above attributes, each node in the server's update tree also stores pointers to the data blocks themselves (and tags used for their verification).

Construction. We next provide the details of our construction. Because the solution relies on our update tree algorithms, we outline them first, while their detailed description is given in Section 5.

- $\mathsf{UTInsert}(\mathsf{T}, \mathsf{s}, \mathsf{e})$ inserts a range of new blocks into the update tree T, where the range starts from index s and consists of $(\mathsf{e}-\mathsf{s}+1)$ data blocks. It returns a node v that corresponds to the newly inserted block range.
- $\mathsf{UTDelete}(\mathsf{T}, \mathsf{s}, \mathsf{e})$ marks blocks in the range $[\mathsf{s}, \mathsf{e}]$ as deleted in the update tree T and adjusts the indices of the data blocks that follow. It returns an array of nodes C from T that correspond to the deleted data blocks.
- $\mathsf{UTModify}(\mathsf{T}, \mathsf{s}, \mathsf{e})$ updates the version of the blocks in the range $[\mathsf{s}, \mathsf{e}]$ in the tree T. If some of blocks in the range have not been modified in the past (and therefore are not represented in the tree), the algorithm inserts necessary nodes with version 1. The function returns all the nodes in T that correspond to the modified data blocks.
- $\mathsf{UTRetrieve}(\mathsf{T}, \mathsf{s}, \mathsf{e})$ returns the nodes in T that correspond to the data blocks in the range $[\mathsf{s}, \mathsf{e}]$.
- $\mathsf{UTCommit}(\mathsf{T}, \mathsf{s}, \mathsf{e})$ replaces nodes in T that correspond to the data blocks in the range $[\mathsf{s}, \mathsf{e}]$ with a single node and balances the remaining tree.

The protocols that define our solution are as follows:

1. $\mathsf{KeyGen}(1^\kappa) \to \{\mathsf{sk}\}$: C calls $\mathsf{sk} \leftarrow \mathsf{Gen}(1^\kappa)$.

2. $\mathsf{Init}(\langle \mathsf{sk}, m_1, \ldots, m_N \rangle, \langle \perp \rangle) \to \{\langle \mathsf{M}_C \rangle, \langle \mathsf{M}_S, D \rangle\}$: C and S initialize the update tree T to empty and set $\mathsf{M}_C = \mathsf{T}$ and $\mathsf{M}_S = \mathsf{T}$, respectively. For $1 \le i \le N$, C computes $t_i = \mathsf{Mac_{sk}}(m_i||i||0||0||0)$, where "$||$" denotes concatenation and the three "0"s indicate the version number, CID, and operation type, resp. C sends each $\langle m_i, t_i \rangle$ to S who stores this information in D.

3. $\mathsf{Update}(\langle \mathsf{sk}, \mathsf{M}_C, \mathsf{op}, \mathsf{ind}, \mathsf{num}, m_{\mathsf{ind}}, \ldots, m_{\mathsf{ind}+\mathsf{num}-1} \rangle, \langle \mathsf{M}_S, D \rangle) \to \{\langle \mathsf{M}'_C \rangle, \langle \mathsf{M}'_S, D' \rangle\}$: the functionality of this protocol is determined by the operation type op and is defined as:
   (a) **Insert** $\mathsf{op} = 1$: C executes $u \leftarrow \mathsf{UTInsert}(\mathsf{M}_C, \mathsf{ind}, \mathsf{ind}+\mathsf{num}-1)$.
   Delete $\mathsf{op} = -1$: C executes $U \leftarrow \mathsf{UTDelete}(\mathsf{M}_C, \mathsf{ind}, \mathsf{ind}+\mathsf{num}-1)$.
   Modify $\mathsf{op} = 0$: C executes $U \leftarrow \mathsf{UTModify}(\mathsf{M}_C, \mathsf{ind}, \mathsf{ind}+\mathsf{num}-1)$.
   C stores the updated update tree in M'_C.
   (b) For each $u \in U$ (or a single u in case of insertion), C locates the data blocks corresponding to the node's range from the m_i's, for $\mathsf{ind} \le i \le \mathsf{ind}+\mathsf{num}-1$, and computes $t_i \leftarrow \mathsf{Mac_{sk}}(m_i||u.\mathsf{L}+j||u.\mathsf{V}||u.\mathsf{ID}||\mathsf{op})$, where $j \ge 0$ indicates the position of the data block within the node's blocks. C sends op, ind, and num together with the $\langle m_i, t_i \rangle$ pairs to S, except that for deletions the data blocks themselves are not sent.
   (c) **Insert** $\mathsf{op} = 1$: S executes $u \leftarrow \mathsf{UTInsert}(\mathsf{M}_S, \mathsf{ind}, \mathsf{ind}+\mathsf{num}-1)$.
   Delete $\mathsf{op} = -1$: S executes $U \leftarrow \mathsf{UTDelete}(\mathsf{M}_S, \mathsf{ind}, \mathsf{ind}+\mathsf{num}-1)$.
   Modify $\mathsf{op} = 0$: S executes $U \leftarrow \mathsf{UTModify}(\mathsf{M}_S, \mathsf{ind}, \mathsf{ind}+\mathsf{num}-1)$.
   S stores the updated tree in M'_S and combines D with received data (using returned u or U) to obtain D'.

Recall that the protocol does not involve integrity verification for each dynamic operation, which removes a round of interaction between the client and server. Instead, the server records the operation in its metadata, which will be used for proving the integrity of returned blocks at retrieval time.

4. $\mathsf{Retrieve}(\langle \mathsf{sk}, \mathsf{M}_C, \mathsf{ind}, \mathsf{num}\rangle, \langle \mathsf{M}_S, D\rangle) \rightarrow \{\langle m_{\mathsf{ind}}, \ldots, m_{\mathsf{ind}+\mathsf{num}-1}\rangle, \langle \perp \rangle\}$:
   (a) C sends ind and num to S.
   (b) S executes $U \leftarrow \mathsf{UTRetrieve}(\mathsf{M}_S, \mathsf{ind}, \mathsf{ind}+\mathsf{num}-1)$. For each $u \in U$, S retrieves the attributes (L, U, and pointer to the data blocks) from u, locates the blocks and their tags $\langle m_i, t_i\rangle$ in D, and sends them to C.
   (c) Upon receiving $\langle m_i, t_i\rangle$, C executes $U \leftarrow \mathsf{UTRetrieve}(\mathsf{M}_C, \mathsf{ind}, \mathsf{ind}+\mathsf{num}-1)$. C chooses a random subset of data blocks of size $\min(c, \mathsf{num})$. For each chosen data block m_i, C locates the corresponding $u \in U$ and computes $b_i \leftarrow \mathsf{Verify}_{\mathsf{sk}}(m_i || u.\mathsf{L}+j || u.\mathsf{V} || u.\mathsf{ID} || u.\mathsf{Op}, t_i)$, where $j \geq 0$ is the data block position within the node's data blocks. If $b_i = 1$ for each verified m_i, C is assured of integrity of returned data.

5. $\mathsf{Challenge}(\langle \mathsf{sk}, \mathsf{M}_C\rangle, \langle \mathsf{M}_S, D\rangle) \rightarrow \{\langle m_{i_1}, \ldots, m_{i_c}\rangle, \langle \perp \rangle\}$:
   (a) C chooses c distinct indices $i_1, \ldots, i_c$ at random from the range $[1, N]$ and sends them to S.
   (b) If none of the indices are adjacent, S executes $U_j \leftarrow \mathsf{UTRetrieve}(\mathsf{M}_S, i_j, i_j)$ for each $j \in [1, c]$. Otherwise, S combines adjacent indices in ranges and executes UTRetrieve for each range.
   (c) For each U_j, S retrieves the attributes (L, U, and pointer to the data block) from M_S, locates the blocks and their tags $\langle m_{i_j}, t_{i_j}\rangle$ in D, and sends them to C.
   (d) Upon the receipt of c data blocks and their corresponding tags $\langle m_{i_1}, t_{i_1}\rangle, \ldots, \langle m_{i_c}, t_{i_c}\rangle$, C executes $U_j \leftarrow \mathsf{UTRetrieve}(\mathsf{M}_C, i_j, i_j)$ for each j (or for each range when some indices are adjacent). For each data block m_{i_j}, C verifies its tag using the same computation as in Retrieve.

6. $\mathsf{Commit}(\langle \mathsf{sk}, \mathsf{M}_C, \mathsf{ind}, \mathsf{num}, m_{\mathsf{ind}}, \ldots, m_{\mathsf{ind}+\mathsf{num}-1}\rangle, \langle \mathsf{M}_S, D\rangle) \rightarrow \{\langle \mathsf{M}'_C\rangle, \langle \mathsf{M}'_S, D'\rangle\}$:
   (a) C executes $u \leftarrow \mathsf{UTCommit}(\mathsf{M}_C, \mathsf{ind}, \mathsf{ind}+\mathsf{num}-1)$ and stores updated metadata in M'_C. C next computes $t_{\mathsf{ind}+i} \leftarrow \mathsf{Mac}_{\mathsf{sk}}(m_{\mathsf{ind}+i} || \mathsf{L}+i || 0 || \mathsf{CID} || 0)$ for $0 \leq i \leq \mathsf{num}-1$, and sends the tags and parameters ind and num to S (the blocks are assumed to be unchanged).
   (b) S executes $u \leftarrow \mathsf{UTCommit}(\mathsf{M}_S, \mathsf{ind}, \mathsf{ind}+\mathsf{num}-1)$ and updates its metadata to M'_S. S updates the affected blocks' tags in D to obtain D'.

Public verifiability. To enable outsourcing of periodic challenge queries to a third party auditor, instead of choosing a private key sk, the client creates a public-private key pair (pk, sk). The client then replaces MAC computation with a signature produced using sk. An auditor with an up-to-date copy of the update tree will be able to perform challenge queries on behalf of the client and verify them using pk.

5. UPDATE TREE OPERATIONS

In this section we describe all operations on the new type of data structure, balanced update tree, that allow us to achieve attractive performance of the scheme. The need to maintain several attributes associated with a dynamic operation and the need to keep the tree balanced add complexity to the tree algorithms. Initially, the tree is empty and new nodes are inserted upon dynamic operations triggered by the client. All data blocks information about which is not stored in the tree have not been modified and their integrity can be verified by assuming version number and commit ID to be 0.

When traversing the tree with an up-to-date range [s, e] of data blocks, the range will be modified based on the R value of the nodes lying on the traversal path. By doing that, we are able to access the original indices of the data blocks (prior to any insertions or deletions) to either correctly execute an operation or verify the result of a read request. We illustrate the tree operations on the example given in Figure 1, in which the leftmost tree corresponds to the result of three modify requests with the ranges given in the figure. We highlight modifications to the tree after each additional operation.

The first operation is an insertion, the range of which falls on left side of node A's range and overlaps with the range of node B. To insert the blocks, we partition B's range into two (by creating two nodes) and make node D correspond to an insertion (Op = 1). Note that the offset R of node A is updated to reflect the change in the indices for the blocks that follow the newly inserted blocks. Furthermore, for the insertion operation, only the position at which the new blocks are inserted matters, not its range. For instance, executing an insertion with a block range [61, 300] or [61, 70] will have the same impact on the tree structure. The second operation is a modification, the range of which lies on the right to node A's range. When going down the tree, we modify the block range contained in the original request based on A's offset R (for the right child only), which now overlaps with node C's range. To accommodate the request, we increment the version of C's blocks and insert two new nodes with ranges before and after C's range. The last operation is a deletion, the range of which likewise fall on the right to A's range and the indices in the original request are adjusted. Because the adjusted range falls before all ranges in C's subtree, it is inserted as the left child of E_1 with type Op $= -1$ and the offset R of both C and E_1 is adjusted to reflect the change in block indices for these nodes and their right children.

We first list sub-routines called by the main algorithms followed by an outline of main operations. Due to space constraints, pseudo-code and detailed explanations are given in the full version.

Sub-routines:

UTInsertNode(u, w, dir) inserts a node w into a (sub-)tree rooted at node u. The routine is called only in the cases when after the insertion, w becomes either the leftmost (dir = left) or the rightmost (dir = right) node of the subtree.

When the node w is inserted into the left subtree of node u, the offset R of each node on the path should be updated according to the range of indices of w when the operation is insertion or deletion, because the new range lies to the left of the blocks of the current u. When the node w is inserted into the right subtree of node u, the range of node w should also be modified as it traverses the tree, since we need to store the original indices of data blocks .

UTFindNode(u, s, e, op) searches the tree rooted at node u for a node corresponding to block range [s, e] for the purposes of executing operation op on that range.

After the function is invoked on range [s, e] and that range does not overlap with the ranges of any of the existing nodes, the function creates a new node and returns it. Otherwise, the function needs to handle the case of range overlap, defined as follows: (i) op is insertion and the index s lies within the range of a tree node or (ii) op is modification or deletion and the range [s, e] overlaps with the range of at least one existing tree node.

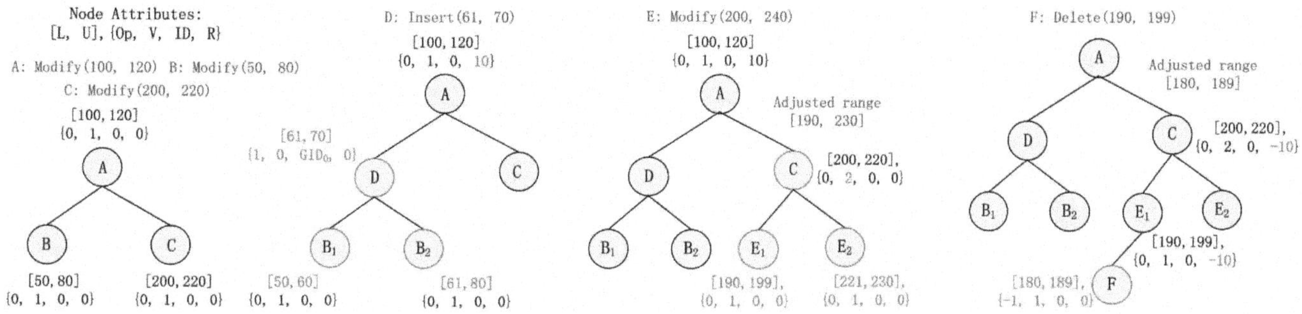

Figure 1: Example of update tree operations.

The tricky part of the algorithm is to avoid returning nodes that correspond to deleted block ranges. If such a node is found, we should ignore it and keep searching until we find a node that represents either an insertion or modification operation. The function can be invoked for any dynamic operation and takes $O(\log n)$ time.

$\mathsf{UTUpdateNode}(\mathsf{u}, \mathsf{s}, \mathsf{e}, \mathsf{op})$ is called by a modification or deletion routine on a sub-tree rooted at node u when the range $[\mathsf{s}, \mathsf{e}]$ of data blocks needs to be updated and falls into the range of u.

The function handles four different situations based on the type of intersection of ranges $[\mathsf{s}, \mathsf{e}]$ and $[\mathsf{u.L}, \mathsf{u.U}]$. If the two ranges are identical several attribute of u (i.e., V, ID and Op) will be reset with values that depend on the operation type. If only the lower (only the upper) bound of the two ranges coincide, we reset the range of the current root node to $[\mathsf{s}, \mathsf{e}]$, fork a new node corresponding to the remaining range, and insert it into the right (resp., left) subtree of current root node. If neither the lower nor the upper bound matches with each other, we fork two child nodes corresponding to the head and tail remaining ranges, and insert each of them into the left or right subtree of current root node respectively. As can be expected, the node generated with the remaining range will become either a leftmost or a rightmost node of the subtree, and we use UTInsertNode sub-routine to achieve it. The time complexity of the sub-routine is $O(\log n)$.

$\mathsf{UTBalance}(\mathsf{u})$ balances the tree rooted at node u using AVL trees method and returns the root of a balanced structure. This function will only be called on trees both direct child sub-trees of which are already balanced rather than on arbitrarily unbalanced trees. The time complexity of this function is linear in the height difference of u's child sub-trees.

Main routines:

$\mathsf{UTInsert}(\mathsf{T}, \mathsf{s}, \mathsf{e})$ updates the tree T for an insert request with the block range $[\mathsf{s}, \mathsf{e}]$ by inserting a node in the tree.

The main functionality of the routine is (i) to find a position for node insertion (using UTFindNode sub-routine), and (ii) to insert a new node into the tree. When the range $[\mathsf{s}, \mathsf{e}]$ does not overlap with any existing nodes, UTFindNode inserts a new node into the tree and no other action is necessary. Otherwise, an existing node u' that overlaps with $[\mathsf{s}, \mathsf{e}]$ is returned and determines the number of nodes that need to be created. That is, if the (adjusted) insertion position s equals to the lower bound of u', u' is substituted with a new node and is inserted into the right subtree of the new node. Otherwise, u' is split into two nodes, which are inserted into the left and right subtrees of u', respectively while u' itself is set to correspond to the insertion. The insert request in Figure 1 corresponds to the scenario.

$\mathsf{UTModify}(\mathsf{u}, \mathsf{s}, \mathsf{e})$, when called with $\mathsf{u} = \mathsf{T}$, updates the tree T based on a modification request with the block range $[\mathsf{s}, \mathsf{e}]$ and returns the set of nodes that correspond to the range. The algorithm creates a node for the range if T is empty, and otherwise invokes UTFindNode to locate the positions of nodes to be modified. After finding them, the algorithm distinguishes between three cases based on how the (adjusted) range $[\mathsf{s}, \mathsf{e}]$ overlaps with the range of a found node u_i (i.e., $[\mathsf{u}_i.\mathsf{L}, \mathsf{u}_i.\mathsf{U}]$):

1. If the adjusted range is contained in u_i's range, u_i is the only node to be modified, and this is handled by UTUpdateNode. The modify request in Figure 1 corresponds to the scenario.
2. If the adjusted range overlaps with the ranges of u_i and its one subtree, the algorithm updates the range of u_i and then recursively calls itself to update the remaining nodes.
3. If the adjusted range overlaps with the range of u_i and both of its subtrees, the algorithm updates u_i and calls UTModify twice to handle changes to its child subtrees.

$\mathsf{UTDelete}(\mathsf{u}, \mathsf{s}, \mathsf{e})$, when called with $\mathsf{u} = \mathsf{T}$, updates the update tree T based on a deletion request with the block range $[\mathsf{s}, \mathsf{e}]$. It does not delete any node from T, but rather finds all nodes whose ranges fall into $[\mathsf{s}, \mathsf{e}]$, sets their operation types to -1, and returns them to the caller. UTDelete works similar to UTModify.

$\mathsf{UTRetrieve}(\mathsf{u}, \mathsf{s}, \mathsf{e})$, when called with $\mathsf{u} = \mathsf{T}$, returns the nodes whose ranges fall into $[\mathsf{s}, \mathsf{e}]$. Its high-level structure follows that of UTModify.

$\mathsf{UTCommit}(\mathsf{T}, \mathsf{s}, \mathsf{e})$ replaces all nodes in tree T whose ranges falls into $[\mathsf{s}, \mathsf{e}]$ with a single node with the range $[\mathsf{s}, \mathsf{e}]$. The goal of a commit operation is to reduce the tree size, but in the process it may become unbalanced or even disconnected. Thus, to be able to maintain the desired performance guarantees, we must restructure and balance the remaining portions of the tree. To achieve that, we first search for two nodes that contain the lower and upper bounds s and e, respectively, and make the adjusted s and e (denoted s' and e', respectively) become the left or right bound of the nodes that contain them. We then traverse T from the two nodes to their least common ancestor T', remove the nodes with ranges falling into the range $[\mathsf{s}', \mathsf{e}']$, and apply UTBalance sub-routine to balance the tree if necessary. Lastly, we traverse T from T' to the root, and balance the tree if necessary. We also add a node with $[\mathsf{s}, \mathsf{e}]$ and new CID. The routine returns adjusted lower bound s' and updated CID.

To illustrate how the tree is being traversed and balanced in the process, let us consider an example in Figure 2. In the figure, u_1 and u_2 correspond to the tree nodes that incorporate the smallest and largest block indices falling in the commit range (i.e., s and e), respectively, and T' is their lowest common ancestor. The nodes and their subtrees shown using dotted lines corresponds to the nodes whose entire subtrees are to be removed. To remove all nodes in the subtree of T' with block indices larger than adjusted s (located in the left child's subtree), we traverse the path from u_1 to T'. Ev-

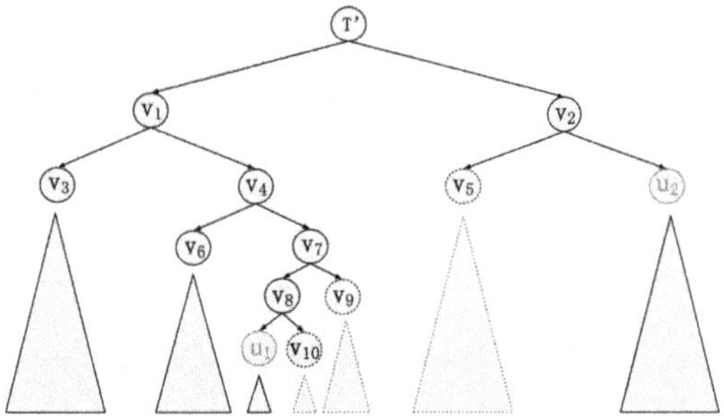

Figure 2: Illustration of the commit algorithm.

ery time u_1 is the left child of its parent, we remove u_1's right sibling and its subtree, remove u_1's parent node, and make u_1 take the place of its parent. For the example in the figure, it means that nodes v_{10} and v_9 are removed together with their subtrees, nodes v_8 and v_7 are also removed, and u_1 takes the place of v_7. At this point u_1 becomes the right child of its parent, and we balance the subtree rooted at u_1's parent and make u_1 point to its parent node (by calling UTBalance(u) sub-routine). This rebalancing procedure is continued by traversing up to v_7, v_4, and v_1, until the left subtree of T' is completely balanced.

The same process applies to the right child's tree of T' that contain u_2 with the difference that node removal is performed when u_2 is the right child of its parent and rebalancing is performed when u_2 is the left child of its parent. For the example in Figure 2, we obtain that node v_5 is removed together with its subtree, node v_2 is removed, and u_2 takes the place of v_2.

The last step that remains is to rebalance the subtree rooted at T' and the subtrees of all other nodes on the path from T' to the root. This is accomplished by making T' point to its parent after each rebalancing procedure. We obtain a balanced tree T with all nodes in the range [s, e] removed and insert one single node corresponding to this range that indicates that the commit number CID of all blocks in the range [s, e] has been increased.

6. ANALYSIS OF THE SCHEME

Complexity analysis. In what follows, we analyze the complexity of main update tree algorithms and the protocols that define the scheme. Each UTInsert adds one or two nodes to the tree, and all operations are performed during the process of traversing the tree. Therefore, its time complexity is $O(\log n)$, where n is the current number of nodes in the tree. Both UTModify and UTDelete can add between 0 and $O(\min(n, \mathsf{e}-\mathsf{s}))$ nodes to the tree, but as our experiments suggest, a constant number of nodes is added per range on average. Their time complexity is $O(\log n + \min(n, \mathsf{e}-\mathsf{s}))$, and both the size of the block range and the number of nodes in the tree form the upper bound on the number of returned nodes. UTRetrieve does not add nodes to the tree and its complexity is also $O(\log n + \min(n, \mathsf{e}-\mathsf{s}))$. Lastly, UTCommit removes between 0 and $O(\min(n, \mathsf{e}-\mathsf{s}))$ nodes from the tree and its time complexity is $O(\log n + \min(n, \mathsf{e}-\mathsf{s}))$. While the function calls UTBalance, the worst case complexity of which is $O(\log n)$, at most $O(\log n)$ number of times, due to the careful construction of the tree and the commit function, we are able to achieve $O(\log n)$ node rearrangement time (plus, node deallocation time). This is due to the fact that balancing a tree, both subtrees of which are themselves balanced, but their heights differ by a constant, requires only a constant number of operations. A detailed proof can be found in the full version.

Next, we analyze the complexity of the protocols themselves. It is clear that Init has time and communication complexity of N, i.e., the number of transmitted blocks. Update for any operation type has time complexity of $O(\log n + \mathsf{num})$ and communication complexity of $O(\mathsf{num})$. Retrieve has the same complexities as Update, unless it is used for integrity verification rather than block retrieval. In the latter case, its computation and communication complexities become $O(\log n + \min(\mathsf{num}, c))$ and $O(\min(\mathsf{num}, c))$, respectively, where constant c bounds the number of $\langle m_i, t_i \rangle$ pairs used for the purpose of probabilistic verification. Lastly, the complexities of Commit are $O(\log n + \mathsf{num})$ and $O(\mathsf{num})$, because the client needs to communicate num MACs to the server.

Security analysis. Security of our scheme can be shown according to the definition of DPDP in Section 3.

THEOREM 1. *The proposed update tree scheme is a secure DPDP scheme assuming the security of MAC scheme.*

PROOF SKETCH. Suppose that the adversary $\mathcal{A}$ wins the data possession game with a non-negligible probability. Then the challenger can either extract the challenged data blocks (i.e., if $\mathcal{A}$ has not tampered with them) or break the security of the MAC scheme (i.e., if $\mathcal{A}$ tampered with the data). In particular, in the former case, the challenger can extract the genuine data blocks from $\mathcal{A}$'s response. In the latter case, if the adversary tampers with a data block (by possibly substituting it with a previously stored data for the same or a different block), it will have to forge a MAC for it, which the challenger can use to win the MAC forgery game. This is because our solution is designed to ensure that any two MACs communicated by the client to the server are computed on unique parameters. That is, two different versions of the same data block i will have either their version, CID, or operation type differ, while two different blocks that at different points in time assume the same index i (e.g., a deleted block and a block inserted in its place) can be distinguished by the value of their ID (i.e., at least one of them will have a GID, and two GIDs or a GID and CID are always different). □

The probability that a cheating server is caught on a Retrieve or Challenge request of size $\mathsf{num} < c$ is 1, and otherwise the probability is $1 - ((\mathsf{num} - t)/\mathsf{num})^c$, where t is the number of tampered blocks among the challenged blocks.

7. ENABLING REVISION CONTROL

In this section, we sketch how our scheme can be extended to support revision control. The exact algorithms are omitted and can be found in the full version. To enable versioning functionality, we need to specify (i) how a user can retrieve a specific version of a data block or range and (ii) how a user can retrieve deleted data blocks (prior to a commit on them, which permanently removes them from the server).

Specific version retrieval can be realized by modifying the Retrieve protocol to add version number V to the set of parameters sent to the server with the request. After receiving the request, the server executes UTRetrieve as usual, but returns to the client the data blocks and their tags that correspond to version V. To verify the response, the client verifies the returned tags using the intended version V and other attributes obtained from UTRetrieve.

Deleted data retrieval can be realized by extending the Retrieve protocol's interface with a flag that indicates that deleted data is to be returned and modifying UTRetrieve that currently skips all deleted data. The difficulty in specifying what deleted data to retrieve is caused by the fact that deleted blocks no longer have indices associated with them. To remedy the problem, we propose to specify in the request a range [s, e] that contains the desired deleted range and includes one or more non-deleted blocks before and after the deleted range being requested. We also need to use an alternative UTRetrieve function for retrieval of deleted data, which instead of ignoring nodes that represent deletions will return them as the output and will allow the server to locate the deleted blocks and their tags.

8. PERFORMANCE EVALUATION

To evaluate performance of our scheme and provide a comparison with prior solutions, we designed experiments which measure the computation, communication, and storage requirements of three different schemes. The schemes that we compare are: (i) our update tree (UTree) solution, (ii) the solution based on Merkle hash tree (MHT) [27], and (iii) the solution based on skip lists (SL) [16, 19]. The asymptotic complexities of the schemes being compared are given in Table 1. The table provides storage complexities for both the server (in addition to the data blocks themselves, i.e., space for maintaining metadata) and the client, as well as computation complexities per operation. For a retrieve, update, insert, or delete operation, it is assumed that the operation is executed on a range consisting of t blocks. For MHT and SL schemes, the integrity of a dynamic operation on t blocks is verified by executing the appropriate verification procedure for each of the t blocks. Verification of retrieve and challenge operations is assumed to be probabilistic in all schemes. For a retrieve operation, $\min(c, t)$ blocks are verified for constant c, while transmitting all t blocks to the client. The complexity of the challenge query of the *entire outsourced storage* is that of verifying c data blocks for constant c. In the table, N denotes the number of data blocks stored at the server, and M denotes the number of dynamic operations on the stored blocks. The worst-case complexity of an operation with the MHT-based solutions is linear in the size of the repository because after arbitrary insertions and deletions the height of the tree is $O(N)$ in the worst case.

We evaluate the performance of the schemes in three different settings: the first uses 1GB of outsourced storage with 4KB data blocks, the second uses 256GB of storage with 4KB blocks, and the third uses 256GB of storage with 64KB blocks. The first 1GB+4KB setting was chosen for consistency with experiments in prior work and the other two allow us to examine the systems' behavior when one parameter remains fixed while the other changes (i.e., 4KB block size in the first two settings and 256GB overall storage in the last two settings). As another important observation about the chosen settings, notice that the number of blocks are 2^{18}, 2^{26}, and 2^{22}, respectively, which allows us to test the performance with respect to its dependence on the number of outsourced data blocks. In situations when the amount of outsourced storage is significantly larger than in our experiments, we expect the data block size to become larger as well leaving the number of blocks within the range that we evaluate. Lastly, we evaluate the performance of the schemes in the realistic scenarios by utilizing the file traces gathered by Microsoft Research data centers for a week time period. We believe the access patterns we observed in the trace are representative of a large number of small to medium size enterprise data centers.

We implement our and MHT schemes in C, while the SL scheme was implemented as in [16] in Java. Despite the programming language difference, the time to compute a hash is similar in both implementations. Because the overall computation of the SL scheme is dominated by hash function evaluation, we consider the performance of all implementations to be comparable. We use SHA-224 for hash function evaluation and HMAC with SHA-224 for MAC computation. The experiments were run on 2.4GHz Linux machines (both the client and the server).

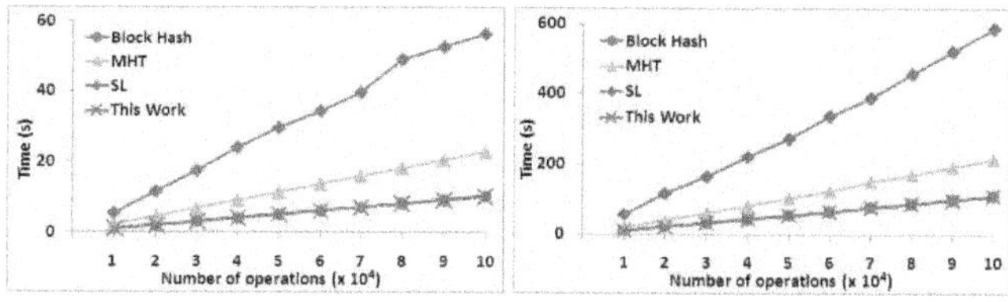

(a) single-block mixed operations (b) multiple-block mixed operations

Figure 3: Aggregate client's computation time after n operations with 1GB outsourced storage and 4KB block.

Computation. To evaluate computation, we measure the client's time after executing n client's requests for n between 10^4 and 10^5. The server's overhead in all schemes is similar to that of the respective client's overhead. The initial cost of building the data structures in MHT and SL schemes or computing MACs in our solution is not included in the measured times.

In the first experiment, we choose one of four operations (insert, delete, modify, and retrieve) at random and execute it on a randomly selected single data block. From the schemes that we compare, only our solution provides a natural support for querying ranges of blocks (the SL solution in [16] can provide a limited support for block ranges as previously described). Then because in practice accesses are often consecutive in their nature (see, e.g., [15]), the performance of our scheme is expected to be even better in practice. For all of the above operations except deletion, the client needs to compute a hash (or MAC) of the data block used in the request. Because this computation is common to all three schemes[2], we separately measure it and also provide the times for the remaining processing that the client needs to do.

Because of drastic differences in performance of the schemes, we present many results in tables instead of displaying them as plots. This allows us to convey information about the growth of each function. For that reason, Figure 3(a) plots the aggregate and Figure 4 plots average performance of all schemes for the first 1GB+4KB setting, while Table 2 provides average computation for all three settings. Notice that in the figure the overhead of each scheme is added to the common hash computation work, while in the table the common and additional computation are shown separately. We were unable to complete 256GB+4KB experiments for SL due to its extensive computation (primarily to build the data structure) and memory requirements. As can be seen from the results, the overhead of UTree scheme is 2 to 3 orders of magnitude lower for the settings used in the experiment. The majority of the total work in a UTree scheme comes from MAC computation, while in MHT and SL the proof often dominates the cost. Also note that the overhead of SL is larger than that of MHT due to the use of longer proofs and commutative hashing in the former, where the majority of the difference comes from the hashing. As expected, the number of data blocks in the storage affects performance of MHT and SL schemes (the proof sizes of which are logarithmic in the total number of blocks), while the average time per operation remains near a constant for each setting. In our scheme, on the other hand, the time grows slowly with the number of operations, but does not increase with the total storage size.

[2]In MHT and SL schemes the client needs to compute the hash of a data block, while in our scheme the client computes a MAC of it. Because MAC computation is slightly more expensive than the hash, we include MAC's additional overhead into the performance of our scheme.

Scheme	Cost per operation (Server and Client)								Storage	
	Update or Insert		Delete		Retrieve		Challenge		Server	Client
	Computation	Communic.	Computation	Communic.	Computation	Communication	Computat.	Communic.		
MHT [27]	$O(Nt)$	$O(Nt)$	$O(Nt)$	$O(Nt)$	$O(\min(c,t)N+t)$	$O(\min(c,t)N+t)$	$O(cN)$	$O(cN)$	$O(N)$	$O(1)$
SL [16, 19]	$O(t\log N)$	$O(t\log N)$	$O(t\log N)$	$O(t\log N)$	$O(\min(c,t)\log N+t)$	$O(\min(c,t)\log N+t)$	$O(c\log N)$	$O(c\log N)$	$O(N)$	$O(1)$
This work	$O(\log M+t)$	$O(t)$	$O(\log M+t)$	$O(1)$	$O(\log M+t)$	$O(t)$	$O(c\log M)$	$O(c)$	$O(N)$	$O(M)$

Table 1: Asymptotic complexities of DPDP schemes.

Setting	File size	Block size	Scheme	Total number of operations n					Block hash
				20000	40000	60000	80000	100000	
Single-block mixed random operations	1GB	4KB	UTree	1.34	1.51	1.62	1.70	1.77	134
			MHT	127	127	127	127	127	
			SL	481	502	473	512	462	
	256GB	64KB	UTree	1.19	1.33	1.43	1.49	1.57	1972
			MHT	148	147	148	147	148	
			SL	619	595	633	652	703	
	256GB	4KB	UTree	1.20	1.32	1.43	1.50	1.56	134
			MHT	180	179	179	178	183	
Multi-block mixed random operations	1GB	4KB	UTree	1.68	2.14	2.48	2.78	2.99	1340
			MHT	944	978	1000	1050	1090	
			SL	4750	4460	4570	4640	4800	
Single block same position insertions	1GB	4KB	UTree	1.13	1.17	1.21	1.26	1.76	134
			MHT	77400	155000	232000	308000	466000	
			SL	250	235	242	258	212	

Table 2: Average client's computation time per operation measured after n operations for various schemes in μsec.

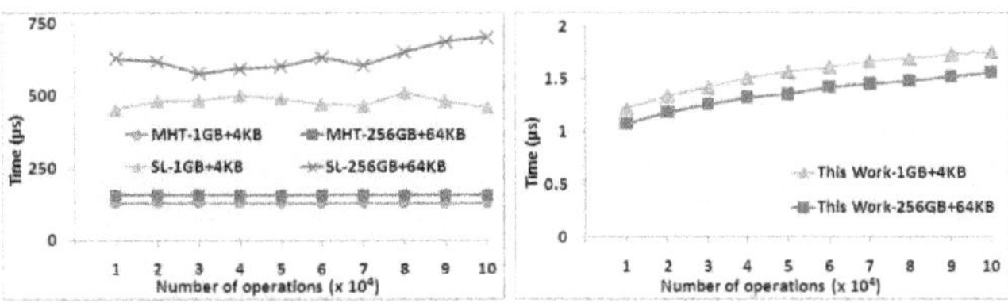

(a) MHT and SL (b) UTree

Figure 4: Average client's computation time measured after n single-block randomly chosen operations with 1GB storage and 4KB block (without block hash computation).

File size	Block size	Scheme	Total number of operations n				
			20000	40000	60000	80000	100000
1GB	4KB	UTree	0.4	0.8	1.2	1.6	2
		MHT	9.6	19.3	28.9	38.6	48.3
		SL	21	39.6	62.2	77.5	98.3
		Data	60	120	180	240	300
256GB	64KB	UTree	0.4	0.8	1.2	1.6	2
		MHT	11.7	23.5	35.3	47.0	58.8
		SL	25.1	54.7	79.2	101	126
		Data	960	1920	2880	3840	4800
256GB	4KB	UTree	0.4	0.8	1.2	1.6	2
		MHT	13.9	27.8	41.7	55.5	69.4
		Data	60	120	180	240	300

Table 3: Aggregate communication size after n operations for various schemes measured in MB.

For the second experiment, we changed the first experiment to execute each operation on a range of data blocks of size between 1 and 20. To be able to verify correctness of individual blocks, we do not implement the variable-sized data block approach suggested in [16], but rather repeat the single-block operation multiple times for each operation. However, to improve efficiency of MHT scheme, for a range insertion operation we construct an independent tree from the blocks specified in the request first and then insert it into the MHT in the same way as a single node. The performance results are given in Table 2 and Figure 3(b).

Compared to the single-block operations, performance of MHT and SL schemes deteriorates by a factor of 9–10, while for our UTree, which was designed to work on ranges, there is only a modest (20%–70%) increase in the performance. The increase in the tree size can be explained by using more nodes to partition an existing node. We expect that in the other settings with a larger number of blocks (256GB+4KB and 256GB+64KB), the tree will have a smaller size as the ranges are spread out over a larger space.

For the third experiment, we used a new access pattern that inserts data blocks at the same position in a file. The goal is to demonstrate the effect of such operation on an unbalanced data structure. The results are shown in Table 2. As we can see, UTree and SL schemes exhibit stable performance due to their balanced data structures, while performance of MHT grows significantly and linearly with the number of operations (as its height in that case exhibits linear in the number of operations growth). There is no easy way to remedy the situation by balancing MHT in a similar way to our solution, as the client does not have complete information about the MHT.

Communication. To evaluate communication, we measure the amount of data exchanged between the client and the server after executing a number of single-block requests. The data transferred in each operation consists of a data block (except for deletion) and corresponding auxiliary data. The former cost is common to all schemes, while the latter varies in its format and size. In particular, for UTree the auxiliary data consists of a single MAC, while for MHT and SL it is the proof linear in the height of the data structure. Another difference is that UTree involves a unidirectional communication for all except one operation (i.e., Retrieve which returns a response), while all operations in MHT and SL require bidirectional communication. For that reason, we measured the aggregated data exchanged for each operation, without considering the direction of data transfer. The results are given in Table 3.

Because deletion does not involve data block transfer in all three schemes, the average size of data block communication per operation is $3/4$ of the block size. As can be observed from Table 2, UTree scheme's communication is independent of the data structure size or the setting and is constant per operation. For MHT and SL scheme, on the other hand, performance depends on the data structure size. For data blocks of small size, the proof overhead of MHT and SL schemes constitutes a significant portion of the overall communication volume (14–30%), which could be a fairly large burden for a user constrained by a limited network bandwidth. The overhead of UTree scheme, on the other hand, is no more than 0.6%. Lastly, the difference in performance of MHT and SL schemes can be explained by the length of the proof and the size of elements within the proof.

Storage. To evaluate storage, we measure the size of data structures after executing client's requests on single blocks as well as ranges. In both MHT and SL schemes, the server maintains a data structure while the client keeps only constant-sized data for integrity verification. In our scheme, both the server and the client maintain a data structure, but it should be moderate in size for a variety of operating environments (and can be reduced using commit).

The data structures maintained in the schemes consists of a static portion that corresponds to the initially uploaded data and a dynamic portion that corresponds to dynamic operations issued afterwards. In MHT and SL schemes, the static component is linear in the number of outsourced blocks and is expected to be fairly large. In our scheme, on the other hand, there is no static component.

As far as dynamic component goes, the size of the data structure in our solution grows upon executing dynamic operations according to the analysis in Section 6. The growth is always constant per single-block operation, and the use of ranges allows us to reduce the overall growth. With MHT and SL schemes, the size of the data structure remains at the same level as long as the number of insertions is similar to the number of deletions. Block modifications do not affect the data structure size. Lastly, because MHT and SL scheme do not support versioning functionality, to enable it, they can be upgraded using persistent authenticated data structure [3]. The use of persistent data structures increase the data structure size by $O(\log n)$ per single-block update, where n is the number of nodes within the data structure. Therefore, considering both static and dynamic components, UTree inevitably leads to a more compact data structure, and its size is also the reason why the client can store the data structure locally.

The results of our experiments are given in Table 4. For each setting, the first row for UTree corresponds to single-block mixed dynamic operations at random locations, while the second row corresponds to similar range operations (1–20 blocks per operations). The performance is estimated based on the number of nodes (measured using UTree, MHT, and SL implementations) in the data structures and the approximate node size of 50 bytes for each scheme. It does not correspond to the memory measurement at the run time. Clearly, there is a large difference in the performance of our scheme and other approaches for the tested settings.

The experiments correspond to the original solutions, without the support of versioning functionality. This means that a deletion operation actually deletes a node and a modification does not contribute additional nodes to the data structure, and the size of the data structure remains constant after executing an equal number of different types of updates. As expected, the size of UTree grows linearly with the number of dynamic operations. Another observation that aligns with our experiments above is that the additional UTree size of range operations compared to the single-block case is significantly reduced with a larger number of outsourced blocks (compare, e.g., 112%, 0.8%, and 1.1% overhead at 100000 operations). As before, it is caused by fewer range overlaps, which results in fewer node partitioning and smaller tree size.

File size	Block size	MHT any n	SL any n	UTree for n operations				
				20000	40000	60000	80000	100000
1GB	4KB	25	24	0.70	1.37	2.03	2.66	3.28
				0.97	2.26	3.74	5.32	6.94
256GB	64KB	400	391	0.71	1.43	2.15	2.85	3.57
				0.71	1.46	2.15	2.89	3.60
256GB	4KB	6400	6206	0.71	1.43	2.15	2.86	3.57
				0.72	1.44	2.16	2.88	3.61

Table 4: The size of data structures for various schemes after n single-block or range mixed operations in MB.

Volume	Max offset	Operations ($\times 10^6$)	MHT	SL	UTree
Proj-0	170 GB	4.2	6400	21000	8.4
Proj-1	880 GB	24	29000	95000	1200
Proj-2	880 GB	29	48000	120000	2200
Proj-3	240 GB	2.2	670	2300	3
Proj-4	240 GB	6.5	3400	12000	150

Table 5: Client's aggregate computation time measured in seconds for each volume (without block hash computation).

Real life data. In the above experiments, we evaluated the efficiency of the schemes on synthetic randomly generated data. To provide more convincing results, we also conduct experiments on real life data sets from [14], which consist of file traces gathered from Microsoft data centers for a period of a week. We believe that the access patterns in the traces are representative of data usage types seen in practice. The traces were collected per volume below the file system cache and capture all block-level reads and writes performed on 36 volumes. Out of these volumes, for our experiments we select five that belong to a single server (a research project server) and contain 66 million events. Each event contains information such as a timestamp, a disk number, the start logical block number (i.e., offset), the number of blocks transferred, and its type (i.e., read or write).

For the purposes of MHT and SL schemes, we find the maximum offset that appears in the trace of a volume, consider it as the size of "outsourced data," and use it construct the corresponding data structure. (In contrast, the operation of our scheme does not need that information.) We then map a "read" or "write" operation in a event will to a respective "retrieve" or "modify" operation in all three schemes. Because there are no insertions or deletions, each operation takes the same amount of time in MHT and SL schemes.

Because the communication overhead of our scheme is always smaller than those of MHT and SL schemes in all cases, here we concentrate on computation and storage overhead. As before, we leave out the initial cost of building the data structures in MHT and SL schemes and exclude the time for computing MACs for our solution. We also do not measure the time of computing a hash (or a MAC) of a data block in the request. The results are presented in table 5 and assume a block size of 4KB. As can be observed from the table, out of five volumes our solution is almost two orders of magnitude faster than the other two schemes. For the storage overhead, we measure the data structure size after executing all operations appeared in a file trace. In case of MHT and SL schemes, the data structures depend on the size of outsourced data and range from 4 to 22GB. In our scheme, after executing all requests in a volume, the data structure size ranges from 1 to 150MB.

9. CONCLUSIONS

We propose a novel solution to provable data possession with support for dynamic operations, access to shared data by multiple users, and revision control. Our solution utilizes a new type of data structure that we term balanced update tree. Unique features of our scheme include orders of magnitude faster than in other schemes data verification and removing the need for the server to maintain data structures linear in the size of the outsourced data. The advantages come at the cost of requiring the client to maintain a data structure of modest, but non-constant size.

10. ACKNOWLEDGMENTS

This work was supported in part by grants FA9550-09-1-0223 from the Air Force Office of Scientific Research and CNS-1223699 from the National Science Foundation. Any opinions, findings, and conclusions or recommendations expressed in this publication are those of the authors and do not necessarily reflect the views of the funding agencies.

11. REFERENCES

[1] IT cloud services user survey, pt. 2: Top benefits & challenges. http://blogs.idc.com/ie/?p=210.

[2] G. Adelson-Velskii and E.M. Landis. An algorithm for the organization of information. In *Proceedings of the USSR Academy of Sciences*, pages 263–266, 1962.

[3] A. Anagnostopoulos, M. Goodrich, and R. Tamassia. Persistent authenticated dictionaries and their applications. In *International Conference on Information Security (ISC)*, pages 379–393, 2001.

[4] G. Ateniese, R. Burns, R. Curtmola, J. Herring, L. Kissner, Z. Peterson, and D. Song. Provable data possession at untrusted stores. In *ACM Conference on Computer and Communications Security (CCS)*, pages 598–609, 2007.

[5] G. Ateniese, R. Di Pietro, L. Mancini, and G. Tsudik. Scalable and efficient provable data possession. In *Security and Privacy in Communication Networks (SecureComm)*, 2008.

[6] G. Ateniese, S. Kamara, and J. Katz. Proofs of storage from homomorphic identification protocols. In *Advances in Cryptology – ASIACRYPT*, pages 319–333, 2009.

[7] J. Bentley. Decomposable searching problems. *Information Processing Letters*, 8(5):244–251, 1979.

[8] K. Bowers, A. Juels, and A. Oprea. HAIL: A high-availability and integrity layer for cloud storage. In *ACM Conference on Computer and Communications Security (CCS)*, pages 187–198, 2009.

[9] K. Bowers, A. Juels, and A. Oprea. Proofs of retrievability: Theory and Implementation. In *ACM Workshop on Cloud Computing Security (CCSW)*, pages 43–54, 2009.

[10] E. Chang and J. Xu. Remote integrity check with dishonest storage server. In *European Symposium on Research in Computer Security (ESORICS)*, pages 223–237, 2008.

[11] R. Curtmola, O. Khan, R. Burns, and G. Ateniese. MR. PDP: Multiple-replica provable data possession. In *International Conference on Distributed Computing Systems (ICDCS)*, pages 411–420, 2008.

[12] M. de Berg, M. van Kreveld, M. Overmars, and O. Schwarzkopf. Interval trees. In *Computational Geometry*, chapter 10.1, pages 212–217. Springer-Verlag, second edition, 2000.

[13] Y. Dodis, S. Dadhan, and D. Wichs. Proofs of retrievability via hardness amplification. In *Theory of Cryptography Conference (TCC)*, pages 109–127, 2009.

[14] N. Dushyanth, D. Austin, and R. Antony. Write off-loading: Practical power management for enterprise storage. *Transactions on Storage*, 4(3):10:1–10:23, 2008.

[15] D. Ellard, J. Ledlie, P. Malkani, and M. Seltzer. Passive NFS tracing of email and research workloads. In *USENIX Conference on File and Storage Technologies (FAST)*, 2003.

[16] C. Erway, A. Kupcu, C. Papamanthou, and R. Tamassia. Dynamic provable data possession. In *ACM Conference on Computer and Communications Security (CCS)*, pages 213–222, 2009.

[17] M. Goodrich, C. Papamanthou, R. Tamassia, and N. Triandopoulos. Athos: Efficient authentication of outsourced file systems. In *International Conference on Information Security*, pages 80–96, 2008.

[18] M. Goodrich, R. Tamassia, and A. Schwerin. Implementation of an authenticated dictionary with skip lists and commutative hashing. In *DARPA Information Survivability Conference and Exposition*, pages 68–82, 2001.

[19] A. Heitzmann, B. Palazzi, C. Papamanthou, and R. Tamassia. Efficient integrity checking of untrusted network storage. In *ACM International Workshop on Storage Security and Survivability (StorageSS)*, pages 43–54, 2008.

[20] A. Juels and B. Kaliski. PORs: Proofs of retrievability for large files. In *ACM Conference on Computer and Communications Security (CCS)*, pages 584–597, 2007.

[21] A. Oprea and M. Reiter. Integrity checking in cryptographic file systems with constant trusted storage. In *USENIX Security Symposium*, pages 183–198, 2007.

[22] R. Popa, J. Lorch, D. Molnar, H. Wang, and L. Zhuang. Enabling security in cloud storage SLAs with CloudProof. In *USENIX Annual Technical Conference*, pages 355–368, 2011.

[23] W. Pugh. Skip lists: a probabilistic alternative to balanced trees. *Communications of the ACM*, 33:668–676, 1990.

[24] F. Sebe, J. Domingo-Ferrer, A. Martinez-Belleste, Y. Deswarte, and J.-J. Quisquater. Efficient remote data possession checking in critical information infrastructures. *IEEE Transactions on Knowledge and Data Engineering (TKDE)*, 20(8):1034–1038, 2008.

[25] H. Shacham and B. Waters. Compact proofs of retrievability. In *Advances in Cryptology – ASIACRYPT*, pages 90–107, 2008.

[26] C. Wang, Q. Wang, K. Ren, and W. Lou. Ensuring data storage security in cloud computing. In *International Workshop on Quality of Service*, pages 1–9, 2009.

[27] Q. Wang, C. Wang, J. Li, K. Ren, and W. Lou. Enabling public verifiability and data dynamics for storage security in cloud computing. In *European Symposium on Research in Computer Security (ESORICS)*, pages 355–370, 2009.

[28] L. Wei, H. Zhu, Z. Cao, W. Jia, and A. Vasilakos. SecCloud: Bringing secure storage and computation in cloud. In *ICDCSW*, pages 52–61, 2010.

[29] K. Zeng. Publicly verifiable remote data integrity. In *International Conference on Information and Communications Security (ICICS)*, pages 419–434, 2008.

[30] Q. Zheng and S. Xu. Fair and dynamic proofs of retrievability. In *ACM Conference on Data and Application Security and Privacy (CODASPY)*, pages 237–248, 2011.

Weak Leakage-Resilient Client-side Deduplication of Encrypted Data in Cloud Storage*

Jia Xu
Institute for Infocomm Research
1 Fusionopolis Way
Singapore, 138632
xuj@i2r.a-star.edu.sg

Ee-Chien Chang
National University of Singapore
21 Lower Kent Ridge Road
Singapore, 119077
changec@comp.nus.edu.sg

Jianying Zhou
Institute for Infocomm Research
1 Fusionopolis Way
Singapore, 138632
jyzhou@i2r.a-star.edu.sg

ABSTRACT

Recently, Halevi *et al.* (CCS '11) proposed a cryptographic primitive called *proofs of ownership* (PoW) to enhance security of client-side deduplication in cloud storage. In a proof of ownership scheme, any owner of the same file F can prove to the cloud storage that he/she owns file F in a robust and efficient way, in the *bounded leakage* setting where a certain amount of efficiently-extractable information about file F is leaked. Following this work, we propose a secure client-side deduplication scheme, with the following advantages:

- our scheme protects data confidentiality (and *some* partial information) against both outside adversaries and honest-but-curious cloud storage server, while Halevi *et al.* trusts cloud storage server in data confidentiality;
- our scheme is proved secure w.r.t. any distribution with sufficient min-entropy, while Halevi *et al.* (the last and the most practical construction) is particular to a specific type of distribution (a generalization of "block-fixing" distribution) of input files.

The cost of our improvements is that we adopt a weaker leakage setting: We allow a bounded amount *one-time* leakage of a target file *before* our scheme starts to execute, while Halevi *et al.* allows a bounded amount *multi-time* leakage of the target file *before and after* their scheme starts to execute. To the best of our knowledge, previous works on client-side deduplication prior Halevi *et al.* do not consider any leakage setting.

Categories and Subject Descriptors

H.3.5 [**Information Systems**]: Information storge and retrieval—*On-line information services*; E.3 [**Data Encryption**]

*Funded by A*Star project SecDC-112172014.

General Terms

Algorithms, Security

Keywords

Cloud Storage, Client-side Deduplication, Proofs of Ownership, Privacy, Leakage-Resilient, Universal Hash

1. INTRODUCTION

Cloud storage service is gaining popularity in recent years. To reduce resource consumption in both network bandwidth and storage, many cloud storage services including Dropbox [1] and Wuala [2] employs client-side deduplication [21, 39]. That is, when a user tries to upload a file to the server, the server checks whether this particular file is already in the cloud (uploaded by some user previously), and saves the uploading process if it is already in the cloud storage. In this way, every single file will have only one copy in the cloud (i.e. Single Instance Storage). SNIA white paper [34] reported that the deduplication technique can save up to 90% storage, dependent on applications.

According to Halevi *et al.* [20] and Dropship [17], an existing implementation of client-side deduplication is as below: Cloud user Alice tries to upload a file F to the cloud storage. The client software of the cloud storage service installed on Alice's computer, will compute and send the hash value $\mathsf{hash}(F)$ to the cloud server. The cloud server maintains a database of hash values of all received files, and looks up the value $\mathsf{hash}(F)$ in this database. If there is no match found, then file F is not in the cloud storage yet. Alice's client software will be required to upload F to the cloud storage, and the hash value $\mathsf{hash}(F)$ will be added into the look-up database. If there is a match found, then file F is already in the cloud storage, uploaded by other users or even by the same user Alice before. In this case, uploading of file F from Alice's computer to the cloud storage is saved, and the cloud server will allow Alice to access the file F in its cloud storage. We may refer to the above client-side deduplication method as "hash-as-a-proof" method. In this method, the hash value $\mathsf{hash}(F)$ serves two purposes: (1) it is an index of file F, used by the cloud server to locate information of F among a huge number of files; (2) it is treated as a "proof" that Alice owns file F. Previously,

[1] http://www.dropbox.com/
[2] http://www.wuala.com/

Dropbox[3] applied the above "hash-as-a-proof" method on block-level cross-users deduplication [20][17]. If the client software of the cloud storage service is trusted and the hash function $\mathsf{hash}(\cdot)$ is collision-resistant, then the "hash-as-a-proof" method is sufficiently secure. However, malicious users may develop their own version of client software using public API[4] of the cloud service, so that they can send any manipulated messages (e.g. manipulated hash output) to the cloud server. Therefore, a more sophisticated solution without trusting the client software is required.

1.1 Security Concerns

Different users may possess some identical sensitive files for many reasons, even if they have no knowledge on each other. For example they may receive a classified or copyright-protected file directly or indirectly from the same source. Partial information of these sensitive files could be leaked via various channels [20, 21] by some owners intentionally or unintentionally. Despite its significant benefits in saving resource, client-side deduplication may bring in new security vulnerability and lead to leakage of users' sensitive files, especially when a certain amount of partial information of these files have already been leaked.

1.1.1 Data Privacy against Outside Adversaries

Recently, an attack on "hash-as-a-proof" method in popular cloud storage service like Dropbox and MozyHome is proposed [20, 17]: If the adversary *somehow* has the short hash value of a file stored in the cloud storage, he/she could fool the cloud server that he has the file by presenting only the hash value as "proof" in the client-side deduplication process, and thus gain access to that file via the cloud. This attack is practical and does not require the adversary to find a collision of the hash function, since client software of cloud service could be bypassed.

1.1.2 Data Privacy against Inside Adversaries (Cloud Storage Servers)

Confidentiality of users' sensitive data against the cloud storage server itself is another important security concern that is not addressed by Halevi *et al.* [20]. As long as it is possible, prudent users hope to ensure that the cloud storage server is technically unable to access their data. Dropbox claims they protect users' data with AES encryption. However, the encryption keys are chosen and kept by Dropbox itself. It is reported that, Dropbox mistakenly kept all user accounts unlocked for almost 4 hours, due to a new bug in their software [40]. If users' data are encrypted on client side and the encryption keys are kept away from Dropbox, then there will be no such single point of failure of privacy protection of all users' data, even if Dropbox made such mistakes or was hacked in. Very recently, a bug in Twitter's client software is discovered [37], which allows adversary to access users' private data.

It is worth noting that, cloud storage service providers, including Amazon (S3), Apple (iCloud), Dropbox, Google (Drive) and Microsoft (SkyDrive), explicitly or implicitly declare that they reserve rights to access users' files, in their official statements of privacy policy [10, 2, 22, 16, 19, 25].

1.1.3 Divide and Conquer Attack

Let us consider an example: A classified document consists of many pages. Although the whole document has sufficient min-entropy to the view of adversaries, the first page has very low min-entropy, say 1 bit min-entropy which indicates "Acceptance" or "Rejection". Suppose this classified document is stored in a cloud storage, which supports block-level cross-user deduplication. Then the adversary could recover the 1 bit unknown information in the first page, through the block-level deduplication[5]. This is because: (1) deduplication inevitably provides adversaries a way to do brute force search for unknown information, and (2) block-level deduplication that divides a file into blocks and applies deduplication on each block, will isolate min-entropy of each block, and allow adversaries to do brute force search in a much smaller search space. It is not unusual that a file with high min-entropy contains some part, which has very low min-entropy compared to its bit-length. Deterministic encryption scheme also need resolve this issue [26]. We emphasize that block-level cross-user client-side deduplication should not be applied over sensitive files.

Very recently, Ng *et al.* [28] proposed a scheme, which aimed to support PoW over encrypted data. However, their work [28] did not address the above issue and consequently is secure only if every block of the file of interest has sufficient min-entropy. We will discuss this work later in the related work section.

1.1.4 Poison Attack

When a file F is encrypted on client side with randomly chosen encryption key, the cloud server may not be able to verify consistency between the ciphertext and meta-data of file F uploaded by a user. For example, given a tuple (H_F, C_F), the cloud server is unable to verify whether there exists a file F and an encryption key k such that $H_F = \mathsf{hash}(F)$ and $C_F = \mathsf{Enc}_k(F)$. A malicious user may substitute the valid ciphertext C_F with an equal size poisoned file before uploading it to the cloud. Suppose a subsequent user Carol tries to upload the same file F to the cloud. She will be told that F is already in cloud and uploading of F is saved. She may delete her local copy of F to save local storage, and will retrieve file F from the cloud when required in the future. However, what she can retrieve from the cloud is a poisoned file—her file F is lost! This attack is also known as *Target Collision attack* [36]. We remark that this attack does not require to find a collision of the underlying hash function, and could be practical if the client-side deduplication over encrypted data is not properly implemented. For example, the "hash-as-a-proof" method suffers from such poison attack, if the cloud server

[3]In Feb 2012, we noticed that Dropbox disabled the deduplication across different users, probably due to recent vulnerabilities discovered in their original cross-user client-side deduplication method. This also indicates the importance and urgency in the study of security in client-side deduplication.

[4]Dropbox provides public API. Furthermore, this issue can not be eliminated just by hidding API, since the adversary could perform reverse-engineering attack to guess the communication protocol of the cloud service. Note the effect of obfuscating is limited [6].

[5]Users can find whether deduplication occurs by timing the uploading time or monitoring communication packet between the cloud client software and cloud server, or develop a custom cloud client software using public API. Here we assume each block contains only one page.

does not verify the hash computation in order to save computation burden.

1.1.5 Plausible Approaches

Convergent Encryption. Intuitively, convergent encryption [14, 15] together with PoW might provide a solution for client-side deduplication of encrypted files: Encrypt file F to generate ciphertext C_F with hash value $\mathsf{hash}(F)$ as encryption key and then apply PoW scheme over C_F. Indeed, cloud storage service provider Wuala adopts convergent encryption to encrypt users files on client side and supports cross-user deduplication. However, the threat models of PoW and convergent encryption are incompatible. In the setting of PoW where a bounded amount of efficiently-extractable information about the file F can be leaked, convergent encryption is insecure, since its short encryption key is generated from the input file in a deterministic way and could be leaked. Roughly speaking, convergent encryption is as insecure as "hash-as-a-proof" method (i.e using hash value $\mathsf{hash}(F)$ as a proof of ownership of file F), in the presence of leakage. Therefore, all existing works on applying convergent encryption method to implement deduplication of encrypted data (e.g. [36, 3, 24]) are insecure in the bounded leakage setting.

Per-User Encryption Key. Another approach is that each cloud user chooses his/her own per-user encryption key, and all files uploaded to the cloud by the same user will be deterministically encrypted under this user's encryption key. If a single user uploads the same file more than once to the cloud, the subsequent upload will be saved. This approach only allows deduplication of files that belong to the same user, which will severely whittle down the effect of deduplication. In this paper, we are interested in the deduplication cross different users, that is, identical duplicated files from different uses will be detected and removed safely.

1.1.6 Current states of various Cloud Storage Services

We collect some technique information about various cloud storage services (Dropbox, SpiderOak [6] and Wuala) in Table 1. All information comes from public official blogs, white papers, private communication with these cloud storage service providers, or through simple experiments with their public service. We notice that Microsoft SkyDrive and Google Drive do not provide client-side deduplication function, even within a single user account. We conjecture that all cloud storage service with simple web access support (i.e. without requiring special browser plug-in) either do not encrypt users' data or encrypt users' data on server side only.

Table 1: Comparison of various cloud storage services.

Name	Deduplication	Cross-User	Encryption
Dropbox	Yes	No (See footnote 1)	Server side enc
SpiderOak	Yes	No [35]	Client side enc
Wuala	Yes	Yes	Convergent enc

[6]https://spideroak.com/

1.2 Our results and contribution

1.2.1 Overview of proposed scheme

We briefly describe the proposed client-side deduplication scheme over encrypted files as below.

First Upload of File F. Suppose Alice is the first user who uploads a sensitive file F to the cloud storage. She will independently choose a random AES key τ, and produces two ciphertexts as below (See Figure 1): The first ciphertext C_F is generated by encrypting file F with encryption key τ using AES method, and the size of C_F is almost equal to the size of F; the second ciphertext C_τ is generated by encrypting the short AES key τ with file F as the encryption key using some *custom encryption method*, and the size of C_τ is in $\mathcal{O}(|\tau|)$ which is very small.

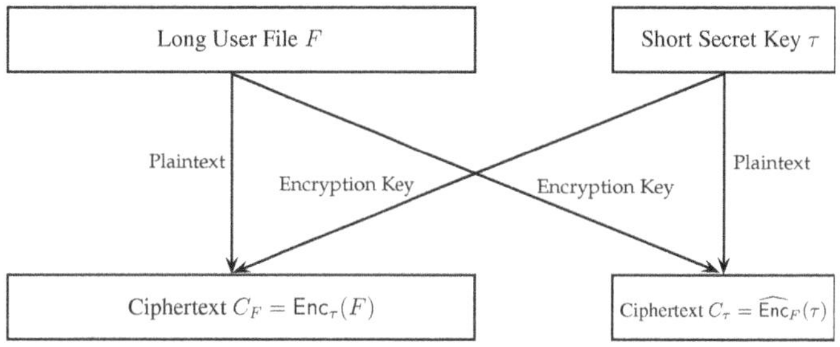

Figure 1: The generation of large ciphertext C_F and short ciphertext C_τ.

Finally, Alice will send a hash value $\mathsf{hash}(F)$ and two ciphertexts (C_F, C_τ) to the cloud storage server. The cloud storage server will compute the hash value $\mathsf{hash}(C_F)$, insert a *short* entry ($\mathtt{key} = \mathsf{hash}(F)$; $\mathtt{value} = (\mathsf{hash}(C_F), C_\tau)$) into its lookup database, and store the potentially *large* ciphertext C_F separately.

Subsequent Upload of File F. Suppose another user Carol tries to upload the same file F into the cloud, after Alice has already uploaded F. Carol sends hash value $\mathsf{hash}(F)$ to the cloud storage server, and the cloud storage server finds a matched entry in its lookup database: ($\mathtt{key} = \mathsf{hash}(F)$; $\mathtt{value} = (\mathsf{hash}(C_F), C_\tau)$). Next, the cloud storage server will send the short ciphertext C_τ to Carol. Carol can decrypt C_τ using file F as decryption key and obtain the secret AES key τ. Carol can encrypt her file F with AES key τ to generate C_F and send the newly computed hash value $\mathsf{hash}(C_F)$ to the cloud storage server. The cloud will compare Carol's version of hash value $\mathsf{hash}(C_F)$ with the one computed by itself. If the two hash values are equal, then Carol is allowed to download C_F from the cloud storage from now on. If the two hash values are different, then with overwhelming high probability[7], either Alice has launched a poison attack w.r.t. file F, or Carol is cheating, or both. If Alice is honest, she can recover file F from the cloud, and present file F as a proof; if Carol is honest, she can present her local copy of F as a proof.

After this, assuming that both Alice and Carol are honest, Carol may remove the local copy of file F if she likes and keeps the AES key τ safely in local storage. Carol can always recover file F by downloading the ciphertext C_F from the cloud and decrypting it with key τ.

[7]Except the negligibly rare case that a collision of the hash function is found.

1.2.2 Our Contributions

In this paper, we focus on cross-user client-side deduplication over users' sensitive data files, and protect data privacy from both outside adversaries and the honest-but-curious cloud storage server. Our contributions in this paper can be summarized as below:

- In Section 3, we propose a formulation for client-side deduplication of encrypted files. Our formulation protects confidentiality of users' sensitive files against both malicious outside adversaries and honest-but-curious inside adversaries. Furthermore, our formulation also protects an important type of partial information (particularly, any physical bit $F[i]$ at position i of file F) of users' sensitive files, although the nature of deduplication implies that semantic-security is unachievable. In contrast, the recent work by Ng *et al.* [28] does not protect partial information and suffers from divide and conquer attack.
- In Section 4, we propose the first secure (Definition 2) client-side deduplication scheme of encrypted files in the bounded leakage setting. We prove its security against malicious outside adversaries and honest-but-curious inside adversaries and with respect to *any* distribution of user file in Theorem 1. In contrast, the PoW schemes by Halevi *et al.* [20] only deals with outside adversaries, and their most practical construction is only proved secure against a particular type of distribution of user file and their proof (in random oracle) relies on an untypical assumption (More details are given later in Section 2).

Our scheme can be applied for deduplication of sensitive files that have very low min-entropy due to a one-time leakage. It is worth pointing out our leakage setting is weaker than Halevi *et al.* [20], so that our scheme achieves a weaker goal in this aspect and allows exposure of the whole file by leaking a short string (rather than account id/password) in the strong leakage setting of Halevi *et al.* This weakness has been resolved in our full paper [42].

The next Section 2 briefs the background and discusses related works. Experiment result is reported in Section 5. Section 6 concludes this paper.

2. RELATED WORKS

2.1 Pairwise-Independent Hash Family

A hash family $\{\mathsf{H}_k : \mathbb{M} \to \{0,1\}^L\}$ is *pairwise-independent* (or say *universal hash* [8, 38]), if for any two distinct inputs $x_1, x_2 \in \mathbb{M}$, $\mathsf{Pr}_k[\mathsf{H}_k(x_1) = \mathsf{H}_k(x_2)] = 2^{-L}$.

2.2 Works on Secure Deduplication before PoW

Deduplication of encrypted files have been studied since the design of convergent encryption by Douceur *et al.* [14]. To the best of our knowledge, all existing works (e.g. [14, 36, 3, 24]) before Halevi *et al.* [20] do not consider leakage setting and most of them focus on server-side deduplication.

2.3 Proofs of Ownership

To prevent leakage of private data to outside adversary, Halevi *et al.* [20] proposed a notion of "proofs of ownership" (PoW). In a PoW scheme, any owner of a file F, without necessarily knowing other owners of F, can efficiently prove to the cloud storage server that he/she owns the file F; any outside adversary cannot prove that he/she has the file F with probability larger than a predefined threshold, even if a certain amount of efficiently-extractable information of file F is leaked to the adversary. Such leakage may occur at any time except during the course of interactive proof between the cloud storage server and the cloud user.

Halevi *et al.* [20] proposed three constructions. The first construction encodes a file using some error erasure code, and then applies the standard Merkle Hash Tree proof method over the encoded file. The second construction is a generic framework. Let $\mathsf{H}_k : \{0,1\}^M \to \{0,1\}^L$ be *any* pairwise independent hash family. Given a file F of M bits long, the second construction computes the hash value $\mathsf{H}_k(F)$ with a public randomness k as hash key and applies the standard Merkle Hash Tree proof method over the L bits value $\mathsf{H}_k(F)$. The third construction is the most practical one. It designs an efficient hash family $\mathsf{H}_k^{'} : \{0,1\}^M \to \{0,1\}^L$ and applies the standard Merkle Hash Tree proof method over $\mathsf{H}_k^{'}(F)$. However, their construction of $\mathsf{H}_k^{'}$ is not pairwise-independent (even if in the random oracle model). Consequently, the generic framework in the second construction cannot apply and a new security proof is required. As the authors explicitly mentioned, Halevi *et al.* [20]'s security proof for their third construction has some limitations: (1) the proof assumes that the file F is sampled from a *particular* type of distribution (a generalization of "block-fixing distribution"); (2) the proof is given "*under the unproven assumption that their scheme will generate a good code*" (See text around Theorem 3 in their paper [20]); (3) the proof is given in random oracle model, where `SHA256` is treated as a random function.

2.4 Extremely Efficient but Less Secure "PoW"

Very recently, Pietro and Sorniotti [32] proposed an efficient "PoW" scheme: They use the projection of the file F onto K [8] randomly selected bit-position $i_1, \ldots, i_K$ as the "proof" of ownership of the file F, that is, the knowledge of bit-string $F[i_1]\|\ldots\|F[i_K]$ is a "proof" of ownership of file F.

This scheme is extremely efficient. However, this work [32] has at least these limitations: (1) it does not protect privacy against honest-but-curious cloud storage server; (2) it is secure only if the min-entropy of file F to the view of adversaries is close to the bit-length of file F, after the leakage occurs.

2.5 Existing Plausible Attempt for Privacy Preserving PoW

Recently, Ng *et al.* [28] made an attempt to support PoW over encrypted files. Their method encrypted files on client side and shared the encryption key among a group of users who know each other. Their method applies existing scheme [12] to do key management within the group, and focus on formulating and devising proofs of ownership scheme in a privacy preserving manner.

Here we brief their PoW scheme as below: A file is divided into many blocks x_i's, and a commitment c_i is computed from each x_i under a secret key. Then the standard Merkle Hash Tree method applies over the commitments $(c_1, c_2, \ldots)$.

[8] K is a system parameter. In their experiment [32], K takes values in the range $[100, 2000]$.

After the completion of Merkle Hash Tree proof protocol, the verifier knows some commitment value c_i, and the prover has to show that he/she has the knowledge of some secret value x_i whose commitment is c_i, without revealing information on x_i to the verifier.

We observe that their proof of knowledge of x_i against c_i is similar to the generalized Okamoto-Identification scheme [30], given by Alwen *et al.* [1]. This proof of identification scheme allows the verifier to efficiently decide whether the secret value x_i is equal to any given candidate value x, thus allows brute-force search of x.

In summary, Ng *et al.* [28]'s PoW scheme has the following limitations: (1) it is very slow in computation: in every execution of the proof protocol, to generate all commitment values c_i's, $|F|/1024$ number of exponentiations[9] in a modulo group of size $\approx 2^{1024}$ are required where $|F|$ denotes the bit-length of file F; (2) the encryption key is shared among a group of "friends", which is not suitable for client-side deduplication over encrypted files, since in current typical client-side duplication setting, owners of the same file will be anonymous to each other; (3) it suffers from "divide and conquer attack" mentioned previously in Section 1.1.3, although (i) it is provably privacy-preserving under their formulation [28] and (ii) it is applied in file-level instead of block-level.

In an extreme example, a large file F with $L(\geq \lambda)$ bits min-entropy to the view of the curious cloud server is divided into L blocks x_i's where each x_i has exactly 1 bit min-entropy to the view of the curious cloud server. If Ng *et al.* [28]'s method is applied over such a file F, then the honest-but-curious cloud server could learn everything of the file efficiently in time $\mathcal{O}(L)$ instead of $\mathcal{O}(2^L)$ during the proof process, by brute-force searching of the value of each x_i independently. In contrast, if the proposed scheme in this paper applies over the same file F, any efficient curious cloud server or outside adversary cannot recover the file and cannot obtain the bit value $F[i]$ at any bit-position i in file F, assuming $F[i]$ was unknown to adversaries before the execution of the proof protocol.

Another recent work [43] combines proofs of storage (i.e POR [23, 9, 33, 41] and PDP [4]) with proofs of ownership. However, this work [43] fails to fulfill the efficiency requirement on server side given by Halevi *et al.* [20], not to mention privacy protection of user data against the curious server.

It is worth noting that, after our work (the full version [42] of this paper), a very recent independent work by Bellare *et al.* [7] contains a scheme called "Randomized Convergent Encryption" (RCE), which has encryption part essentially identical to the encryption part of our scheme. However, Bellare *et al.* does not consider any leakage setting.

3. SECURITY MODEL

In this section, we propose a security formulation for client-side deduplication of encrypted files.

[9]One such group exponentiation requires more than 3 milliseconds in a model PC, which means that it requires more than 3000 seconds to generate all commitments for a file of size 1 giga-bits. Such expensive operation will be executed every time when a user tries to upload the same file to the cloud.

3.1 System Model and Trust Model

3.1.1 Cloud Storage Server

Cloud Storage Server (Cloud Server or Cloud for short) is the entity who provides cloud storage service to various users. Cloud Storage Server has a small and fast primary storage and a large but slow secondary storage. Although the computation power (CPU, I/O, network bandwidth, etc) of Cloud Storage Server is much stronger than a single average user, the average computation power per each online user is usually very limited. We assume that the small and fast primary storage is well-protected from outside adversaries, and the large but slow secondary storage could be visible to outside adversaries.

An example of cloud storage server is Dropbox. Users' files uploaded to Dropbox are actually stored in Amazon's S3 data center (i.e. Dropbox's secondary storage) and Dropbox only runs relatively small server to manage meta data (i.e. Dropbox's primary storage). Another cloud storage service provider Wuala stored users' files in P2P network (the secondary storage) in the early stage of the company.

3.1.2 Cloud Users

Many cloud users may upload their files to the cloud storage and possibly remove their local copies. These users may download files, which are uploaded by themselves, from cloud storage. File sharing among users is not the focus of this paper, although it can be achieved along with our solution for encrypted data.

3.1.3 Adversaries

We consider two types of adversaries: Malicious outside adversary and honest-but-curious Cloud server.

Malicious Outside Adversary. The outside adversary may obtain some knowledge (e.g. a hash value) of the file of interest via some channels, and plays a role of cloud user to interact with the cloud server.

Semi-honest Inside Adversary (Honest-but-Curious Cloud Server). This honest but curious cloud storage server (also known as inside adversary) will maintain the integrity of users' files and availability of the cloud service, but is curious about users' sensitive files. This could capture at least the following cases in real world applications:

1. Some technical employee or even the owner of the cloud tries to access user data due to some reason.
2. The company, which provide the cloud storage service, made careless technical mistakes which may leak users' private data, e.g. introducing a software bug. It is reported that Dropbox [40] made users' data open to public for almost 4 hours due a new software bug. Very recently, a bug is discovered in one of Twitter's official client software, which allows attackers to access users' accounts [37].
3. The cloud storage server is hacked in.

3.2 Syntax Definition

A Client-side Deduplication (called $\mathcal{CSD}$ for short) scheme $(\mathcal{E}, \mathcal{D}, \mathcal{P}, \mathcal{V})$ consists of four PPT algorithms $\mathcal{E}$, $\mathcal{D}$, $\mathcal{P}$ and $\mathcal{V}$, which are explained as below:

- $\mathcal{E}(F, 1^\lambda) \rightarrow (\tau, C_0, C_1)$: The probabilistic encoding algorithm $\mathcal{E}$ takes as input a data file F and a security parameter λ, and outputs a short secret per-file encryption key τ, a short encoding C_0 which contains $\mathsf{hash}(F)$ as a part, and a long encoding C_1. C_0 will be stored in cloud server's small and secure primary storage and C_1 will be stored in cloud server's large but potentially insecure secondary storage. The lengths of τ and C_0 should be both in $\mathcal{O}(\lambda)$, and the length of C_1 should be in $|F| + \mathcal{O}(poly(\lambda))$.

- $\mathcal{D}(\tau, C_1) \to F$: The deterministic decoding algorithm takes as input a secret key τ and the long encoding C_1, and outputs a file F.
- $\langle \mathcal{P}(F), \mathcal{V}(C_0) \rangle \to (y_0;\ y_1, y_2)$: The prover algorithm $\mathcal{P}$, which takes a file F as input, interacts with the verifier algorithm $\mathcal{V}$, which takes a short encoding C_0 as input. At the end of interaction, the prover algorithm $\mathcal{P}$ gets output $y_0 \in \{\tau, \perp\}$ and the verifier algorithm $\mathcal{V}$ gets output (y_1, y_2) where $y_1 \in \{\mathtt{Accept}, \mathtt{Reject}\}$ and $y_2 \in \{\mathsf{hash}(C_1), \perp\}$.

We point out, the efficiency requirement in the above description follows Halevi *et al.* [20]. Such efficiency requirement excludes some straightforward secure methods: For example, both prover and verifier have access to the file F during the interactive proof and compute a key-ed hash value over F with a randomly chosen fresh nonce as hash key per each proof session. Another negative example is Zheng *et al.* [43] which turns a proof of storage scheme into a proof of ownership scheme.

DEFINITION 1 (CORRECTNESS). *We say a $\mathcal{CSD}$ scheme $(\mathcal{E}, \mathcal{D}, \langle \mathcal{P}, \mathcal{V} \rangle)$ is* correct, *if the following conditions hold with overwhelming high probability (i.e. $1 - negl(\lambda)$): For any data file $F \in \{0,1\}^*$ and any positive integer λ, and $(\tau, C_0, C_1) := \mathcal{E}(F, 1^\lambda)$,*

- $\mathcal{D}(\tau, C_1) = F$.
- $\langle \mathcal{P}(F), \mathcal{V}(C_0) \rangle = (\tau;\ \mathtt{Accept}, \mathsf{hash}(C_1))$.

Here the hash value $\mathsf{hash}(C_1)$ is required, in order to defend poison attack. In case that the cloud does not have plaintext of file F, the cloud storage server *alone* is not able to decide whether a given tuple $(\mathsf{hash}(F), C_0, C_1)$ is consistent or inconsistent (i.e. poisoned).

3.3 Security Definition

In this subsection, we will propose a security formulation for client-side deduplication. Our formulation will address the protection of *useful* partial information (particularly any physical bit $F[i]$ in the sensitive file F) from the malicious outside adversary or the honest-but-curious cloud server: Roughly speaking, PPT outside/inside adversary cannot learn any *new* information on any physical bit $F[i]$ of file F from client-side deduplication process beyond the side channel leakage.

The $\mathcal{CSD}$ security game $\mathsf{G}_{\mathcal{A}}^{\mathcal{CSD}}(\xi_0, \xi_1)$ between a PPT adversary $\mathcal{A}$ and a challenger w.r.t. $\mathcal{CSD}$ scheme $(\mathcal{E}, \mathcal{D}, \langle \mathcal{P}, \mathcal{V} \rangle)$ is defined as below, where $\xi_0 > \xi_1 \geq \lambda$. Here ξ_0 is the lower bound of min-entropy of the challenged file F at the beginning of the game, and the adversary is allowed to learn at most $(\xi_0 - \xi_1)$ bits information of file F from the challenger.

Setup. The description of $(\mathcal{E}, \mathcal{D}, \langle \mathcal{P}, \mathcal{V} \rangle)$ is made public. Let F be sampled from any distribution over $\{0,1\}^M$ with min-entropy $\geq \xi_0$, where the public integer parameter $M \geq \xi_0$ is polynomially bounded in λ. The challenger sends $\mathsf{hash}(F)$ to the adversary $\mathcal{A}$.

Learning-I. The adversary $\mathcal{A}$ can make a LEAK-QUERY in the following form to the challenger:

- LEAK-QUERY(Func): This query consists of a PPT-computable function Func. The challenger responses this query by computing $y := \mathsf{Func}(F)$ and sending y to the adversary, where the bit-length of output y is required to be smaller than $(\xi_0 - \xi_1)$.

Commit. The adversary $\mathcal{A}$ chooses a subset of v indices $i_1, \ldots, i_v$ from $[1, |F|]$, where $v \geq 1$ and $v + |y| \leq \xi_0 - \xi_1$. The challenger finds the subsequence $\alpha \in \{0,1\}^v$ of F, such that, for each $j \in [1, v], \alpha[j] = F[i_j]$. The challenger chooses a random bit $b \in \{0,1\}$ and sets $\alpha_b := \alpha$ and $\alpha_{1-b} \xleftarrow{\$} \{0,1\}^v$. The challenger sends (α_0, α_1) to the adversary $\mathcal{A}$.

Guess-I. Let $\mathsf{View}_{\mathcal{A}}^{\mathtt{Commit}}$ denote the view of the adversary $\mathcal{A}$ at this moment. Given $\mathsf{View}_{\mathcal{A}}^{\mathtt{Commit}}$ as input, another PPT algorithm (called "extractor") $\mathcal{A}^*$ outputs a guess $b_{\mathcal{A}^*} \in \{0,1\}$ of value b.

Learning-II. The adversary $\mathcal{A}$ can adaptively make queries to the challenger, where concurrent queries are not allowed[10] and each query is in one of the following forms:

- ENCODE-QUERY: The challenger responses the ENCODE-QUERY by running the probabilistic encoding algorithm on F to generate $(\tau, C_0, C_1) := \mathcal{E}(F, 1^\lambda)$ and sending (C_0, C_1) to the adversary. The adversary can make exactly one query in this type.
- VERIFY-QUERY: The challenger, running the prover algorithm $\mathcal{P}$ with input F, interacts with adversary $\mathcal{A}$ which replaces the verifier algorithm $\mathcal{V}$, to obtain $(y_0; y_1, y_2) := \langle \mathcal{P}(F), \mathcal{A} \rangle$. The adversary knows the values of y_1 and y_2, and can make polynomially many queries in this type.
- PROVE-QUERY: The challenger, running the verifier algorithm $\mathcal{V}$ with input C_0, interacts with the adversary $\mathcal{A}$ which replaces the prover algorithm $\mathcal{P}$, to obtain $(y_0; y_1, y_2) := \langle \mathcal{A}, \mathcal{V}(C_0) \rangle$. The adversary $\mathcal{A}$ knows the value of y_0, and can make polynomially many queries in this type.

Guess-II. The adversary $\mathcal{A}$ outputs a guess $b_{\mathcal{A}} \in \{0,1\}$ of value b.

DEFINITION 2 (SECURE $\mathcal{CSD}$). *Let integer λ be the security parameter and $\xi_0 > \xi_1 \geq \lambda$. We say a $\mathcal{CSD}$ scheme $(\mathcal{E}, \mathcal{D}, \langle \mathcal{P}, \mathcal{V} \rangle)$ is (ξ_0, ξ_1)-secure, if for any PPT (inside or*

[10]Similar to Halevi *et al.* [20], concurrent PROVE-QUERY and VERIFY-QUERY will allow the adversary to replay messages back and forth between these two queries, and eliminate the possibility of any secure and efficient solution to client-side deduplication. Therefore, both this work and Halevi *et al.* [20] do not allow concurrent queries of different types in the security formulation. We clarify that, concurrent queries of the same type can be supported. Thus, in the real application, the cloud storage server (verifier) can safely interact with multiple cloud users (prover) w.r.t. the same file concurrently.

outside) adversary $\mathcal{A}$, there exists some PPT extractor algorithm $\mathcal{A}^*$, such that in the security game $\mathsf{G}_{\mathcal{A}}^{\mathcal{CSD}}(\xi_0, \xi_1)$,

$$\begin{aligned} &\Pr[\mathcal{A} \textit{ finds } b \textit{ in } \textbf{\textit{Guess-II}} \textit{ phase }] \\ \leq &\Pr[\mathcal{A}^* \textit{ finds } b \textit{ in } \textbf{\textit{Guess-I}} \textit{ phase }] + negl(\lambda). \end{aligned} \quad (1)$$

Equivalently, the above Equation (1) *can be written as*

$$\Pr[b_{\mathcal{A}} = b] \leq \Pr[b_{\mathcal{A}^*} = b] + negl(\lambda). \quad (2)$$

Remarks on the security formulation.

- Our formulation (particularly, Equation (1) and (2) in Definition 2) requires that $\Pr[b_{\mathcal{A}} = b] \leq \Pr[b_{\mathcal{A}^*} = b] + negl(\lambda)$, which means the adversary $\mathcal{A}$ essentially cannot learn any *new* information on physical bits $F[i_1] \ldots F[i_v]$ in file F during **Learning-II** phase. We emphasize that it is important to ask *some* extractor $\mathcal{A}^*$ instead of the adversary $\mathcal{A}$ to make a guess $b_{\mathcal{A}^*}$ before **Learning-II**, to exclude a trivial plausible attack: Adversary $\mathcal{A}$ intentionally outputs a random guess of b before **Learning-II**, and outputs its maximum-likelihood of b after **Learning-II**, in order to increase the difference between success probability in **Guess-I** and **Guess-II**. Note that this requirement follows the style of original definition of semantic security (Definition 5.2.1 in Goldreich [18]).

- The adversary is allowed to obtain the long encoding C_1 of users' data file F in the above security game, since in real applications, C_1 is typically stored in the large but potentially insecure secondary storage, as mentioned in Section 3.1.1.

- A $\mathcal{CSD}$ scheme does not have any master secret key. Therefore, the adversary $\mathcal{A}$ himself/herself can find answers to any queries (i.e. ENCODE-QUERY, VERIFY-QUERY, PROVE-QUERY, and LEAK-QUERY, etc) w.r.t any input file F' that is owned by $\mathcal{A}$, without help of the challenger.

- If the long encoding C_1 is obtained by encrypting file F using the convergent encryption [14, 15], i.e. encrypting the file F under AES method with some hash value $\mathsf{hash}'(F)$ as encryption key, then the adversary (i.e. the curious cloud server) will have both ciphertext C_1 and decryption key $\mathsf{hash}'(F)$, and thus obtain the file F, where
  - the ciphertext C_1 is given by the challenger in the security game;
  - the convergent encryption key $\mathsf{hash}'(F)$ can be obtained by making a LEAK-QUERY in **Learning-I** phase.

  Therefore, convergent encryption is insecure in our security game due to the bounded leakage setting. One may argue that this was not a fair setting for convergent encryption scheme, since leakage of encryption key will render any encryption scheme insecure. However, if the encryption key is chosen independent on the plaintext (i.e. the file F), then LEAK-QUERY in our setting will be unable to reveal the encryption key. Thus the distinctive feature of convergent encryption—deriving encryption key from plaintext—becomes a two-bladed-sword: on one side, it makes deduplication of encrypted data possible; on the other side, it becomes the security vulnerability in the higher level of security formulation. We will give a more detailed comparison between convergent encryption and our approach later in Section 4.3.

- For similar reasons as above, the "hash-as-a-proof" method is also insecure in our leakage setting. Therefore, when the owner of file F does not understand well what information of F has been leaked by other owners of the same file, our scheme will be much more preferable than the efficient "hash-as-a-proof" method.

3.4 Comparison of two formulations: Proofs of Ownership and Client-side Deduplication

In this subsection, we compare the formulation of PoW [20] and our formulation of $\mathcal{CSD}$. Recall that more details on PoW is given in Section 2.

Both PoW [20] and $\mathcal{CSD}$ formulation aim to secure deduplication mechanism in cloud storage service and protect the confidentiality of users' data in the setting that a bounded amount of information of users' data has been leaked. However, the two formulations differ in both breadth and depth of protection of confidentiality of users' data and also differ in the leakage setting:

- **Breadth**: PoW only formulates the protection of confidentiality of users' data from outside adversaries; $\mathcal{CSD}$ formulation protects the confidentiality of users' data from both outside adversaries and inside adversaries (i.e. the honest-but-curious cloud storage server). In other words, PoW formulation trusts the cloud storage server in data confidentiality, but $\mathcal{CSD}$ formulation doesn't.

- **Depth**: PoW formulation only prevents attackers from recovering the whole user file F and *potentially* allows attackers to recover some partial unknown information of F; $\mathcal{CSD}$ gives a stronger formulation and prevents attackers from recovering any unknown bits $F[i]$, which implies that the adversary cannot recover F. Here, we make two clarifications: (1) The nature of deduplication problem inevitably provides adversaries a way to do brute search for unknown information by observing whether client-side deduplication occurs[11], and thus make semantic security impossible. Therefore, any solution to client-side deduplication cannot achieve semantic security and has to reveal some partial information. Our formulation will protect an important type of partial information: the physical bits $F[i]$'s in file F. (2) Although PoW formulation does not address protection of partial information, the constructions in Halevi *et al.* [20] indeed protects the privacy of physical bits $F[i]$'s against outside adversaries (Of course, this is not true for inside adversaries).

- **Leakage**: PoW allows a stronger leakage setting than $\mathcal{CSD}$. PoW allows *multi-time* leakage at any time

[11]Probabilistic deduplication that saves the duplicated copy only with a certain probability (say 0.5) still allows brute force attack, since the success probability of deduplication can be amplified close to 1 via repetition.

except during the course of the interactive proof between the verifier and prover, while $\mathcal{CSD}$ allows only *one-time* leakage before $\mathcal{CSD}$ scheme starts to execute. Putting the PoW formulation in our context, the security game of PoW consists of multiple interleaved executions of **Learning-I** and **Learning-II** phases, i.e. (**Learning-I**, **Learning-II**, **Learning-I**, **Learning-II**, ...), subject to an additional constraint that the adversary cannot make the LEAK-QUERY and PROVE-QUERY concurrently. On the other side, our $\mathcal{CSD}$ allows larger amount of leakage than the PoW formulation (precisely Definition 2 and 3 in PoW [20]).

4. SECURE CLIENT-SIDE DEDUPLICATION

4.1 Construction

We present the construction of a $\mathcal{CSD}$ scheme ($\mathcal{E}$, $\mathcal{D}$, $\langle \mathcal{P}, \mathcal{V} \rangle$) in Figure 2 on page . Since this construction relies on universal hash function to achieve leakage-resilience, we call the constructed $\mathcal{CSD}$ scheme as UH-CSD. Suppose Alice is the first user who uploads file F. She will execute algorithm $\mathcal{E}$ with file F and security parameter 1^λ as input and obtain a short secret encryption key τ, a short encoding $C_\tau \in \{0,1\}^{3\lambda}$ and a long encoding C_F. Alice will send both C_τ and C_F to the cloud storage server Bob. Bob will compute the hash value $\mathsf{hash}(C_F)$, put C_τ in secure and small primary storage, and put C_F in the potentially insecure but large secondary storage. At the last, Bob will add ($\mathtt{key} = \mathsf{hash}(F), \mathtt{value} = (\mathsf{hash}(C_F), C_\tau)$ into his lookup database. Suppose Carol is another user who tries to upload the same file F after Alice. Carol will send $\mathsf{hash}(F)$ to the cloud storage server Bob. Bob finds that $\mathsf{hash}(F)$ is already in his lookup database. Then Bob, who is running algorithm $\mathcal{V}$ with C_τ as input, interacts with Carol, who is running algorithm $\mathcal{P}$ with F as input. At the end of interaction, Carol will learn τ and Bob will compare the hash value $\mathsf{hash}(C_F)$ provided by Carol with the one computed by himself. Later, Carol is allowed to download C_F from Bob at any time and decrypt it to obtain the file F by running algorithm $\mathcal{D}(\tau, C_F)$.

4.2 Security Analysis

THEOREM 1. *Let* $\xi_0 > \xi_1 = 2\lambda + \Omega(\lambda)$. *Let* hash *be a collision-resistant full domain hash function. Suppose the encryption scheme* E *is private-key semantic secure (Definition 5.2.1 in Goldreich [18]) and the hash function family* $\{\mathsf{h}_k\}$ *is a universal hash. Then the* UH-CSD *scheme in Figure 2 is* (ξ_0, ξ_1)*-secure. (Proof is in Appendix A)*

The proof of correctness of UH-CSD scheme under Definition 1 is straightforward. We save the details. Notice that the *leakage rate* (i.e. the ratio of the amount of leakage to the entropy of the sensitive file) of our scheme UH-CSD is $1 - \xi_1/\xi_0$, which could be close to 1 for suitable ξ_0 and ξ_1.

4.3 Comparison with Convergent Encryption

4.3.1 *Our custom encryption method is derived from convergent encryption*

Convergent encryption encrypts a file F as $\mathsf{Enc}_{\mathsf{hash}(F)}(F)$. In our scheme UH-CSD given in Figure 2, the encryption of file F is $(s, \mathsf{h}_s(F) \oplus \tau, \mathsf{Enc}_\tau(F))$. This encryption method can be treated as a natural extension of convergent encryption which overcomes the below shortcomings of convergent encryption.

Figure 2: The construction of a $\mathcal{CSD}$ scheme, denoted as UH-CSD. Let $\mathsf{E} = (\mathsf{KeyGen}, \mathsf{Enc}, \mathsf{Dec})$ be a symmetric encryption scheme with λ $(= \rho)$ bits long key length and $\mathsf{h}_k : \{0,1\}^* \to \{0,1\}^\rho$ be a key-ed hash function. Notice that the random coin of Enc will be put in the generated ciphertext.

$\mathcal{E}(F, 1^\lambda)$	$\mathcal{D}(\tau, C_F)$
1. $\tau := \mathsf{KeyGen}(1^\lambda) \in \{0,1\}^\lambda$. 2. $s \xleftarrow{\$} \{0,1\}^\lambda$. 3. $C_F := \mathsf{Enc}_\tau(F)$. 4. $C_\tau := (s, \mathsf{h}_s(F) \oplus \tau, \mathsf{hash}(F))$. 5. Output (τ, C_τ, C_F).	1. $F' := \mathsf{Dec}_\tau(C_F)$ 2. Output F'.

$\langle \mathcal{P}(F'), \mathcal{V}(C_\tau) \rangle$

V1: Parse C_τ as $(s, \mathsf{h}_s(F) \oplus \tau, \mathsf{hash}(F))$. Send $(s, \mathsf{h}_s(F) \oplus \tau)$ to the prover.

P1: Compute the secret key y_0 as below

$y_0 := \mathsf{h}_s(F') \oplus (\mathsf{h}_s(F) \oplus \tau)$, where $\oplus$ refers to XOR.

Encrypt[a] F' with key y_0 to generate ciphertext $C_{F'}$ and compute the hash value $y_2 := \mathsf{hash}(C_{F'})$ of the ciphertext. Send y_2 to verifier.

V2: Let $H_{C_F} := \mathsf{hash}(C_F)$ be computed previously when receiving ciphertext C_F. If $y_2 = H_{C_F}$, set $y_1 := \mathtt{accept}$, otherwise $y_1 := \mathtt{reject}$.

[a] As mentioned in the overview in Section 1, this encryption step is required to compute the hash value $\mathsf{hash}(C_{F'})$, which will help the verifier (i.e. cloud storage server) to detect poison attack.

Revocation of encryption key. It is very difficult, if not impossible, to revoke the encryption key of convergent encryption, when the current encryption key is compromised. Suppose a user tries to encrypt file F using $\mathsf{hash}(F)$ as AES encryption key, and finds that the value of $\mathsf{hash}(F)$ has already been revealed to Internet by some other owner of file F. He may switch to use $\mathsf{hash}'(F)$ as encryption key where $\mathsf{hash}'(\cdot)$ is another secure hash function. Meanwhile, the user has to broadcast this switch of hash function to all future users. This approach will face two issues: (1) The number of different secure hash function is very limited. (2) Users may abuse the above hash-revoking functionality. A natural fixes to the above two issues are: (1) Use a secure key-ed hash function and revoke the hash key if necessary. (2) It is not necessary that every user adopts the same hash function (i.e. the same keyed-hash function and hash key) to generate the AES encryption key. Every user can independently choose a new hash key without notifying others. As a result, a user can encrypt a file F in this way: Randomly choose a hash key s and generate the ciphertext $(s, \mathsf{AES}_{\mathsf{h}_s(F)}(F))$.
Can any hash value be a valid encryption key? It is a coincidence that the range of hash function (e.g. $\mathtt{SHA256}$)

is consistent with the key space of encryption method (e.g. AES). Many other encryption schemes have special key generating algorithm and the generated key should have a particular structure, for example, some public key encryption schemes. Therefore, convergent encryption cannot generalize to generic encryption scheme. Our proposed encryption method overcomes this weakness, by invoking the key generating algorithm of the underlying encryption method to generate an encryption key and protect this generated encryption key using a one-time pad. Let $(\mathsf{KeyGen}, \mathsf{Enc}, \mathsf{Dec})$ be the underlying encryption method. The ciphertext of F will be $(s, \mathsf{h}_s(F) \oplus \tau, \mathsf{Enc}_\tau(F))$, where the hash key s is randomly generated and the underlying encryption key τ is generated by algorithm KeyGen.

Leakage Resilient. More importantly, convergent encryption is insecure if a bounded amount of efficiently-extractable information of the plaintext F is leaked *before* encryption (i.e. the leakage setting of $\mathcal{CSD}$ in this paper). Our encryption method is resilient to such bounded leakage of the plaintext F, in the random oracle (assuming h is a random oracle) or in the standard model (assuming h is pairwise-independent hash function). It is worth pointing out that no encryption schemes could be secure in the stronger leakage setting model (e.g. POW [20]) where leakage may occur *before* and *after* the encryption process.

4.3.2 Advantage of Convergent Encryption

Convergent encryption can be used for both client-side and server-side deduplication. In contrast, our encryption method can be used only for client-side deduplication, since the one round interaction in the client-side deduplication is essential for our solution to synchronize the hash key. Unsurprisingly, both convergent encryption and our custom encryption method are not semantically secure [18].

5. PERFORMANCE

We have implemented a prototype of the proposed scheme UH-CSD with SHA256 as the collision-resistant full domain hash function $\mathsf{hash}(\cdot)$, and with $\mathtt{SHA256}(k\|x)$ as the keyed-hash[12] $\mathsf{h}_k(x)$, and AES encryption[13] as the semantic-secure symmetric cipher E. The hash function SHA256 [29] and the symmetric cipher AES [11] are provided in OpenSSL [31] library (version 1.0.0g). The whole program is written in C language and compiled with GCC 4.4.5. It runs in a single process. Our implementation is not optimized and further performance improvements can be expected.

The test machine is a laptop computer, which is equipped with a 2.5GHz Intel Core 2 Duo mobile CPU (model T9300), a 3GB PC2700-800MHZ RAM and a 7200RPM hard disk. The test machine runs 32 bits version of Gentoo Linux OS with kernel 3.1.10. The file system is EXT4 with 4KB page size.

We run the proposed client-side deduplication scheme UH-CSD over files[14] of size 128MB, 256MB, 512MB, and 1024MB, respectively. The running time of the proof protocol (i.e. interactive algorithm $\langle \mathcal{P}, \mathcal{V} \rangle$) in UH-CSD is reported in Figure 3, compared with network transfer time of test files without encryption or deduplication. The running time of encoding algorithm $\mathcal{E}$ is very close to (and smaller than) the interactive algorithm $\langle \mathcal{P}, \mathcal{V} \rangle$. The decoding algorithm $\mathcal{D}$ is just the AES decryption algorithm, which is more efficient than $\mathcal{E}$. Here we save the actual running time for $\mathcal{E}$ and $\mathcal{D}$. All measurement represents the mean of 5 trails. Since the variants are very small, we do not report it.

We observe that, for small files, the saving in uploading time is small if the network *upload* speed is as fast as 5Mbps or even 20Mbps, but saving in server storage still matters to the cloud storage server and is independent on network speed. We remark that, leakage resilient server-side deduplication over encrypted files remains an open problem.

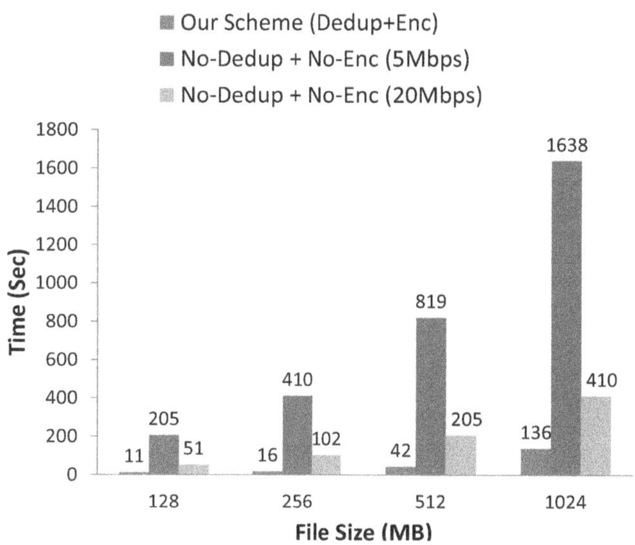

Figure 3: Comparison between the running time of the proof protocol (i.e. interactive algorithm $\langle \mathcal{P}, \mathcal{V} \rangle$) of our client-side deduplication scheme UH-CSD and the network transfer time of files without encryption.

6. CONCLUSION AND FUTURE WORK

In this paper, we addressed an important security concern in cross-user client-side deduplication of encrypted files in the cloud storage: confidentiality of users' sensitive files against both outside adversaries and the honest-but-curious cloud storage server in the bounded leakage model. On technique aspect, we enhanced and generalized the convergent encryption method, and the resulting encryption scheme could support client-side deduplication of encrypted file in the bounded leakage model.

We clarify that this paper adopts a weaker leakage setting than Halevi *et al.* [20]. Our unpublished full paper [42] with stronger result adopts the same leakage setting as Halevi *et al.* w.r.t. outside adversaries. Furthermore, construction of secure client-side deduplication scheme in the strong leakage setting w.r.t. both outside adversaries and honest-but-curious cloud storage server is in our future work.

[12] We treat SHA256 as a random oracle and choose $\mathtt{SHA256}(k\|x)$ to simulate the universal hash function $\mathsf{h}_k(x)$ in order to achieve high performance. Software performance of universal hash is reported by Nevelsteen and Preneel [27].

[13] AES encryption in CBC mode with fresh random IV, where IV will be a part of the ciphertext.

[14] Our test files are generated by encrypting a large file using AES method with a random encryption key, so could be considered as random files.

Acknowledgment

We thank anonymous reviewers for their helpful suggestions.

References

[1] Joël Alwen, Yevgeniy Dodis, and Daniel Wichs. Leakage-Resilient Public-Key Cryptography in the Bounded-Retrieval Model. In *CRYPTO '09: Annual International Cryptology Conference on Advances in Cryptology*, pages 36–54, 2009.

[2] Amazon. AWS Customer Agreement. `http://aws.amazon.com/agreement/`.

[3] Paul Anderson and Le Zhang. Fast and secure laptop backups with encrypted de-duplication. In *Proceedings of the 24th international conference on Large installation system administration*, LISA'10, pages 1–8, 2010.

[4] Giuseppe Ateniese, Randal Burns, Reza Curtmola, Joseph Herring, Lea Kissner, Zachary Peterson, and Dawn Song. Provable data possession at untrusted stores. In *CCS '07: ACM conference on Computer and communications security*, pages 598–609, 2007.

[5] Boaz Barak, Yevgeniy Dodis, Hugo Krawczyk, Olivier Pereira, Krzysztof Pietrzak, François-Xavier Standaert, and Yu Yu. Leftover Hash Lemma, Revisited. In *CRYPTO*, pages 1–20, 2011.

[6] Boaz Barak, Oded Goldreich, Russell Impagliazzo, Steven Rudich, Amit Sahai, Salil Vadhan, and Ke Yang. On the (im)possibility of obfuscating programs. *J. ACM*, 59(2):6:1–6:48, 2012.

[7] Mihir Bellare, Sriram Keelveedhi, and Thomas Ristenpart. Message-locked encryption and secure deduplication. Cryptology ePrint Archive, Report 2012/631, 2012. `http://eprint.iacr.org/`.

[8] Lawrence Carter and Mark Wegman. Universal classes of hash functions (Extended Abstract). In *STOC '77: ACM symposium on Theory of computing*, pages 106–112, 1977.

[9] Ee-Chien Chang and Jia Xu. Remote Integrity Check with Dishonest Storage Server. In *ESORICS '08: European Symposium on Research in Computer Security: Computer Security*, pages 223–237, 2008. `http://eprint.iacr.org/2008/346`.

[10] CNET. Who owns your files on Google Drive? `http://news.cnet.com/8301-1023_3-57420551-93/who-owns-your-files-on-google-drive/`.

[11] Joan Daemen and Vincent Rijmen. *The Design of Rijndael: AES - The Advanced Encryption Standard.* 2002.

[12] Ernesto Damiani, S. De Capitani di Vimercati, Sara Foresti, Sushil Jajodia, Stefano Paraboschi, and Pierangela Samarati. Key management for multi-user encrypted databases. In *StorageSS '05: Proceedings of ACM workshop on Storage security and survivability*, pages 74–83, 2005.

[13] Yevgeniy Dodis, Rafail Ostrovsky, Leonid Reyzin, and Adam Smith. Fuzzy Extractors: How to Generate Strong Keys from Biometrics and Other Noisy Data. *SIAM J. Comput.*, 38(1):97–139, 2008.

[14] John Douceur, Atul Adya, William Bolosky, Dan Simon, , and Marvin Theimer. Reclaiming space from duplicate files in a serverless distributed file system. In *ICDCS '02: International Conference on Distributed Computing Systems*, 2002.

[15] John Douceur, William Bolosky, and Marvin Theimer. US Patent 7266689: Encryption systems and methods for identifying and coalescing identical objects encrypted with different keys, 2007.

[16] Dropbox. Dropbox Privacy Policy. `https://www.dropbox.com/privacy`.

[17] Dropship. Dropbox api utilities, April 2011. `https://github.com/driverdan/dropship`.

[18] Oded Goldreich. *Foundations of Cryptography: Volume 2, Basic Applications.* 2004.

[19] Google. Google Terms of Service. `http://www.google.com/policies/terms/`.

[20] Shai Halevi, Danny Harnik, Benny Pinkas, and Alexandra Shulman-Peleg. Proofs of ownership in remote storage systems. In *CCS '11: ACM conference on Computer and communications security*, pages 491–500, 2011. `http://eprint.iacr.org/2011/207`.

[21] Shulman-Peleg A. Harnik D., Pinkas B. Side Channels in Cloud Services: Deduplication in Cloud Storage. *IEEE Security and Privacy Magazine, special issue of Cloud Security*, 8(6), 2010.

[22] Apple Inc. Apple Privacy Policy (Covering iCloud). `http://www.apple.com/privacy/`.

[23] Ari Juels and Burton S. Kaliski, Jr. Pors: proofs of retrievability for large files. In *CCS '07: ACM conference on Computer and communications security*, pages 584–597, 2007.

[24] Luis Marques and Carlos Costa. Secure deduplication on mobile devices. In *OSDOC '11: Workshop on Open Source and Design of Communication*, pages 19–26, 2011.

[25] Microsoft. Microsoft Services Agreement. `http://windows.microsoft.com/en-US/windows-live/microsoft-service-agreement`.

[26] Ilya Mironov, Omkant Pandey, Omer Reingold, and Gil Segev. Incremental deterministic public-key encryption. In *EUROCRYPT '12: Proceedings of the 31st Annual international conference on Theory and Applications of Cryptographic Techniques*, pages 628–644, 2012.

[27] Wim Nevelsteen and Bart Preneel. Software Performance of Universal Hash Functions. In *EUROCRYPT '99, International Conference on the Theory and Application of Cryptographic Techniques*, pages 24–41.

[28] Wee Keong Ng, Yonggang Wen, and Huafei Zhu. Private data deduplication protocols in cloud storage. In *SAC '12: Proceedings of the 27th Annual ACM Symposium on Applied Computing*, pages 441–446, 2012.

[29] NIST. National Institute of Standards and Technology. Secure hash standard (SHS). FIPS 180-2, August 2002.

[30] Tatsuaki Okamoto. Provably Secure and Practical Identification Schemes and Corresponding Signature Schemes. In *CRYPTO '92: Annual International Cryptology Conference on Advances in Cryptology*, pages 31–53, 1993.

[31] OpenSSL. OpenSSL Project. `http://www.openssl.org/`.

[32] Roberto Di Pietro and Alessandro Sorniotti. Boosting EfñĄciency and Security in Proof of Ownership for Deduplication. In *ASIACCS '12: ACM Symposium on Information, Computer and Communications Security (Full Paper)*, 2012.

[33] Hovav Shacham and Brent Waters. Compact Proofs of Retrievability. In *ASIACRYPT '08*, pages 90–107, 2008.

[34] SNIA. Understanding Data De-duplication Ratios. white paper.

[35] SpiderOak-Blog. Why SpiderOak doesn't de-duplicate data across users.

[36] Mark Storer, Kevin Greenan, Darrell Long, and Ethan Miller. Secure Data Deduplication. In *StorageSS '08: ACM international workshop on Storage security and survivability*, pages 1–10, 2008.

[37] Twitter. Tweetdeck. http://money.cnn.com/2012/03/30/technology/tweetdeck-bug-twitter/.

[38] Mark Wegman and Larry Carter. New Hash Functions and Their Use in Authentication and Set Equality. *J. Comput. Syst. Sci.*, pages 265–279, 1981.

[39] Wikipedia. Comparison of online backup services. http://en.wikipedia.org/wiki/Comparison_of_online_backup_services.

[40] wired.com. Dropbox Left User Accounts Unlocked for 4 Hours Sunday. http://www.wired.com/threatlevel/2011/06/dropbox/; http://blog.dropbox.com/?p=821.

[41] Jia Xu and Ee-Chien Chang. Towards efficient proof of retrievability in cloud storage. In *ASIACCS '12: Proceedings of the 7th ACM Symposium on Information, Computer and Communications Security (Full Paper)*, 2011. http://eprint.iacr.org/2011/362.

[42] Jia Xu, Ee-Chien Chang, and Jianying Zhou. Leakage-resilient client-side deduplication of encrypted data in cloud storage. Cryptology ePrint Archive, Report 2011/538, 2011. http://eprint.iacr.org/.

[43] Qingji Zheng and Shouhuai Xu. Secure and efficient proof of storage with deduplication. In *CODASPY '12: ACM conference on Data and Application Security and Privacy*, pages 1–12, 2012.

APPENDIX

A. PROOF OF THEOREM 1

PROOF. For any PPT adversary $\mathcal{A}_{\mathcal{CSD}}$ against the UH-CSD scheme in Figure 2, we construct a PPT adversary $\mathcal{A}_{\mathsf{E}}$ against the underlying private-key semantic secure encryption scheme E. We emphasize that we adopt the equivalent alternative definition of "private-key semantic secure encryption" given by Goldreich [18].

Construction of $\mathcal{A}_{\mathsf{E}}$: The adversary $\mathcal{A}_{\mathsf{E}}$ is given a ciphertext $C_F = \mathsf{E.Enc}_\tau(F)$ where the encryption key τ and the input file F are unknown and F has at least ξ_0 bits min-entropy. $\mathcal{A}_{\mathsf{E}}$ is allowed to learn any output of $\mathsf{Func}(F)$ from the oracle $\mathcal{O}^F$, where the PPT-computable function Func is chosen by $\mathcal{A}_{\mathsf{E}}$.

$\mathcal{A}_{\mathsf{E}}$ can simulate a security game $\mathsf{G}^{\mathtt{Sim}}$ as below, where $\mathcal{A}_{\mathsf{E}}$ plays the role of challenger and $\mathcal{A}_{\mathcal{CSD}}$ plays the role of adversary:

Setup. $\mathcal{A}_{\mathsf{E}}$ learns the hash value $\mathsf{hash}(F)$ from the oracle $\mathcal{O}^F$ and sends $\mathsf{hash}(F)$ to $\mathcal{A}_{\mathcal{CSD}}$.

Learning-I. $\mathcal{A}_{\mathsf{E}}$ simply forwards LEAKQUERY made by $\mathcal{A}_{\mathcal{CSD}}$ to the oracle $\mathcal{O}^F$ and forwards the response given by the oracle to $\mathcal{A}_{\mathcal{CSD}}$.

Commit. $\mathcal{A}_{\mathsf{E}}$ learns the value of the challenged subsequence $\alpha = F[i_1]\|\ldots\|F[i_v]$ from the oracle $\mathcal{O}^F$ and then exactly follows the rest part of **Commit** phase in the real game $\mathsf{G}_{\mathcal{A}}^{\mathcal{CSD}}$.

Guess-I. Denote the output of the extractor $\mathcal{A}^*_{\mathcal{CSD}}$ as $b^{\mathtt{Sim}}_{\mathcal{A}^*_{\mathcal{CSD}}} \in \{0,1\}$.

Learning-II. Challenger $\mathcal{A}_{\mathsf{E}}$ answers the following queries made by $\mathcal{A}_{\mathcal{CSD}}$:

- ENCODE-QUERY: In response to the encode query, the challenger $\mathcal{A}_{\mathsf{E}}$ independently and randomly chooses $\hat{\tau} \xleftarrow{\$} \mathsf{KeyGen}(1^\lambda)$ and $s_1, s_2 \xleftarrow{\$} \{0,1\}^\lambda$, and set $C_{\hat{\tau}} := (s_1, s_2, \mathsf{hash}(F))$. Let $(C_0, C_1) = (C_{\hat{\tau}}, C_F)$ and sends (C_0, C_1) to $\mathcal{A}_{\mathcal{CSD}}$. Recall that $\mathcal{A}_{\mathsf{E}}$ is given the ciphertext C_F, and $\mathsf{hash}(F)$ is obtained from the oracle $\mathcal{O}^F$ in the **Setup** phase.
- VERIFY-QUERY: $\mathcal{A}_{\mathsf{E}}$ runs the prover algorithm and $\mathcal{A}_{\mathcal{CSD}}$ replaces the verifier algorithm. Denote with (u_1, u_2) the message received from $\mathcal{A}_{\mathcal{CSD}}$. If $(u_1, u_2) = (s_1, s_2)$, then send $\mathsf{hash}(C_F)$ to $\mathcal{A}_{\mathcal{CSD}}$; otherwise, send a random value $H \xleftarrow{\$} \{0,1\}^\lambda$ to $\mathcal{A}_{\mathcal{CSD}}$.
- PROVE-QUERY: $\mathcal{A}_{\mathsf{E}}$ runs $\mathcal{V}(C_0)$ to interact with adversary $\mathcal{A}_{\mathcal{CSD}}$ which replaces the prover algorithm, following the description in game $\mathsf{G}_{\mathcal{A}}^{\mathcal{CSD}}$ exactly.

Guess-II. The adversary $\mathcal{A}_{\mathcal{CSD}}$ outputs a guess $b^{\mathtt{Sim}}_{\mathcal{A}_{\mathcal{CSD}}} \in \{0,1\}$ of b. The game $\mathsf{G}^{\mathtt{Sim}}$ simulated by $\mathcal{A}_{\mathsf{E}}$ completes.

At the end, $\mathcal{A}_{\mathsf{E}}$ outputs $\alpha_{b^{\mathtt{Sim}}_{\mathcal{A}_{\mathcal{CSD}}}} \in \{\alpha_0, \alpha_1\}$ and wins the semantic-security game w.r.t. encryption scheme E if $\alpha_{b^{\mathtt{Sim}}_{\mathcal{A}_{\mathcal{CSD}}}} = \alpha = F[i_1]\|\ldots\|F[i_v]$.

So far, $\mathcal{A}_{\mathsf{E}}$ has received at most $(\lambda + \xi_0 - \xi_1)$ bits (in term of length) message about the unknown file F from the oracle $\mathcal{O}^F$. Thus, after leakage from the oracle, the unknown file F should have at least $(\xi_1 - \lambda) = \lambda + \Omega(\lambda)$ bits min-entropy, according to Lemma 2.2 in Dodis *et al.* [13].

CLAIM 1. *Suppose E is private key ciphertext-indistinguishable and $\{\mathsf{h}_k(\cdot)\}$ be a universal hash family. The simulated game $\mathsf{G}^{\mathtt{Sim}}$ is computationally indistinguishable with the real game $\mathsf{G}^{\mathtt{Real}} = \mathsf{G}^{\mathcal{CSD}}_{\mathcal{A}_{\mathcal{CSD}}}(\xi_0, \xi_1)$, to the view of adversary $\mathcal{A}_{\mathcal{CSD}}$.*

PROOF PROOF OF CLAIM 1. In game $\mathsf{G}^{\mathtt{Real}}$ all messages that the adversary $\mathcal{A}_{\mathcal{CSD}}$ obtain from the challenger are (derived from) $(\mathfrak{S}^{\mathtt{Real}}, \mathsf{hash}(F), s, \mathsf{h}_s(F) \oplus \tau, C_F)$, where $\mathfrak{S}^{\mathtt{Real}} = (y, \alpha)$ and y is the output of LEAK-QUERY and α is computed in **Commit** phase. Similarly, in game $\mathsf{G}^{\mathtt{Sim}}$, the counterpart is $(\mathfrak{S}^{\mathtt{Sim}}, \mathsf{hash}(F), s_1, s_2, C_F)$. For the same adversary $\mathcal{A}_{\mathcal{CSD}}$ with the same random coin, $\mathfrak{S}^{\mathtt{Real}} = \mathfrak{S}^{\mathtt{Sim}}$. So we just write them as $\mathfrak{S}$ for simplicity.

Sample a file F' from $\{0,1\}^{|F|}$ following the same distribution from which F is sampled. Generate a key $\tau' = \mathsf{E.KeyGen}(1^\lambda)$ and encrypt F' with key τ' to obtain ciphertext $C_{F'} = \mathsf{E.Enc}_{\tau'}(F')$. Let $X \approx_c Y$ denote that random variable X and Y are computationally-indistinguishable. We have

$$(\mathfrak{S}, \mathsf{hash}(F), s, \mathsf{h}_s(F) \oplus \tau, C_F) \tag{3}$$
$$\approx_c (\mathfrak{S}, \mathsf{hash}(F), s, \mathsf{h}_s(F) \oplus \tau, C_{F'}) \tag{4}$$
$$\approx_c (\mathfrak{S}, \mathsf{hash}(F), s_1, s_2, C_{F'}) \tag{5}$$
$$\approx_c (\mathfrak{S}, \mathsf{hash}(F), s_1, s_2, C_F). \tag{6}$$

We explain the above equations as below. Eq (3) $\approx_c$ Eq (4) is because E is private key ciphertext-indistinguishable: Given information $(\mathfrak{S}, \mathsf{hash}(F), s, \mathsf{h}_s(F) \oplus \tau)$ about F, the unknown file F still has at least $\Omega(\lambda)$ entropy, hence its encryption C_F is computationally indistinguishable from an encryption $C_{F'}$ of a random file $F' \in \{0,1\}^{|F|}$ under a random encryption key τ', where F' is sampled following the same distribution from which F is sampled.

Eq (4) $\approx_c$ Eq (5) is followed directly from the leftover hash lemma [5] which applies to the universal hash $\{\mathsf{h}_k\}$. Note that $C_{F'}$ is independent on other terms in these two equations.

Eq (5) $\approx_c$ Eq (6) is again implied by the ciphertext-indistinguishability property of the encryption scheme E. Note that s_1, s_2 are independent on the other terms in these two equations.

Therefore, Claim 1 is proved.

□

CLAIM 2. *There exists some PPT extractor $\mathcal{A}^*_{\mathcal{CSD}}$, such that* $\Pr[b^{\mathtt{Sim}}_{\mathcal{A}_{\mathcal{CSD}}} = b^{\mathtt{Sim}}] \leq \Pr[b^{\mathtt{Sim}}_{\mathcal{A}^*_{\mathcal{CSD}}} = b^{\mathtt{Sim}}] + negl(\lambda)$.

PROOF. In **Learning-II** phase of $\mathsf{G}^{\mathtt{Sim}}$, the challenger $\mathcal{A}_\mathsf{E}$ does not make any new queries to $\mathcal{O}^F$, and all responses that $\mathcal{A}_\mathsf{E}$ provided to $\mathcal{A}_{\mathcal{CSD}}$ are computed from randomly sampled values and information that $\mathcal{A}_{\mathcal{CSD}}$ has already known before **Learning-II** (i.e. the hash value $\mathsf{hash}(F)$), except the ciphertext C_F.

It is straightforward that

$$\Pr[\mathcal{A}_\mathsf{E}^{\mathcal{O}^F}(C_F, |F|) = \alpha] = \Pr[b^{\mathtt{Sim}}_{\mathcal{A}_{\mathcal{CSD}}} = b^{\mathtt{Sim}}].$$

Since the underlying encryption scheme E is semantic secure (See the definition in Exercise 18 of Chapter 5 in Goldreich [18]), there exists a PPT algorithm $\mathcal{B}$, such that

$$\Pr[\mathcal{A}_\mathsf{E}^{\mathcal{O}^F}(C_F, |F|) = \alpha] \leq \Pr[\mathcal{B}^{\mathcal{O}^F}(|F|) = \alpha] + negl(\lambda).$$

We construct the extractor $\mathcal{A}^*_{\mathcal{CSD}}$ based on algorithm $\mathcal{B}$: Let $\alpha_\mathcal{B}$ be the output of $\mathcal{B}$. If $\alpha_\mathcal{B} = \alpha_{\hat{b}} \in \{\alpha_0, \alpha_1\}$ for some $\hat{b} \in \{0,1\}$, then $\mathcal{A}^*_{\mathcal{CSD}}$ outputs $b^{\mathtt{Sim}}_{\mathcal{A}^*_{\mathcal{CSD}}} := \hat{b}$; otherwise $\mathcal{A}^*_{\mathcal{CSD}}$ outputs a random bit $b^{\mathtt{Sim}}_{\mathcal{A}^*_{\mathcal{CSD}}} \xleftarrow{\$} \{0,1\}$. We have

$$\begin{aligned} \Pr[b^{\mathtt{Sim}}_{\mathcal{A}_{\mathcal{CSD}}} = b^{\mathtt{Sim}}] =& \Pr[\mathcal{A}_\mathsf{E}^{\mathcal{O}^F}(C_F, |F|) = \alpha] \\ \leq& \Pr[\mathcal{B}^{\mathcal{O}^F}(|F|) = \alpha] + negl(\lambda) \\ \leq& \Pr[b^{\mathtt{Sim}}_{\mathcal{A}^*_{\mathcal{CSD}}} = b^{\mathtt{Sim}}] + negl(\lambda). \end{aligned} \quad (7)$$

Therefore, there exists some PPT extractor $\mathcal{A}^*_{\mathcal{CSD}}$, such that $\Pr[b^{\mathtt{Sim}}_{\mathcal{A}_{\mathcal{CSD}}} = b^{\mathtt{Sim}}] \leq \Pr[b^{\mathtt{Sim}}_{\mathcal{A}^*_{\mathcal{CSD}}} = b^{\mathtt{Sim}}] + negl(\lambda)$.

Furthermore, Claim 1 implies that

$$|\Pr[b^{\mathtt{Sim}}_{\mathcal{A}_{\mathcal{CSD}}} = b^{\mathtt{Sim}}] - \Pr[b^{\mathtt{Real}}_{\mathcal{A}_{\mathcal{CSD}}} = b^{\mathtt{Real}}]| \leq negl(\lambda) \quad (8)$$

$$|\Pr[b^{\mathtt{Sim}}_{\mathcal{A}^*_{\mathcal{CSD}}} = b^{\mathtt{Sim}}] - \Pr[b^{\mathtt{Real}}_{\mathcal{A}^*_{\mathcal{CSD}}} = b^{\mathtt{Real}}]| \leq negl(\lambda). \quad (9)$$

Combine Eq (7), Eq (8) and Eq (9), we have

$$\Pr[b^{\mathtt{Real}}_{\mathcal{A}_{\mathcal{CSD}}} = b^{\mathtt{Real}}] \leq \Pr[b^{\mathtt{Real}}_{\mathcal{A}^*_{\mathcal{CSD}}} = b^{\mathtt{Real}}] + negl(\lambda).$$

□

Therefore, the client-side deduplication scheme UH-CSD is (ξ_0, ξ_1)-secure according to Definition 2.

Data-Oblivious Graph Algorithms for Secure Computation and Outsourcing

Marina Blanton, Aaron Steele, and Mehrdad Aliasgari
Department of Computer Science and Engineering
University of Notre Dame
{mblanton,asteele2,maliasga}@nd.edu

ABSTRACT

This work treats the problem of designing data-oblivious algorithms for classical and widely used graph problems. A data-oblivious algorithm is defined as having the same sequence of operations regardless of the input data and data-independent memory accesses. Such algorithms are suitable for secure processing in outsourced and similar environments, which serves as the main motivation for this work. We provide data-oblivious algorithms for breadth-first search, single-source single-destination shortest path, minimum spanning tree, and maximum flow, the asymptotic complexities of which are optimal, or close to optimal, for dense graphs.

Categories and Subject Descriptors

G.2.2 [**Discrete Mathematics**]: Graph Theory—*graph algorithms*; K.6 [**Management of Computing and Information Systems**]: Security and Protection

Keywords

Graph algorithms; oblivious execution; secure computation

1. INTRODUCTION

Cloud computing has become prevalent today and allows for convenient on-demand access to computing resources which enable clients to meet their unique needs. Such services are used for as diverse a range of applications as management of personal photos, carrying out computationally intensive scientific tasks, or outsourcing building and maintenance of an organization's computing infrastructure. Placing large volumes of one's data and computation outside one's immediate control, however, raises serious security and privacy concerns, especially when the data contains personal, proprietary, or otherwise sensitive information. To protect such information while being able to utilize external computing resources, secure processing of data used for specialized tasks has become an active area of research. Examples of such results include secure and verifiable storage outsourcing (e.g., [19]) and secure outsourcing of common tasks such as linear algebra operations (e.g., [4]) or sequence comparisons (e.g., [6]). There is, however, a lack of efficient techniques for computation on protected data in outsourced environments for most commonly used algorithms and data structures. To help eliminate this void, we develop data-oblivious algorithms for fundamental graph problems. Data-oblivious, or just oblivious, execution is defined as having the same sequence of operations regardless of the input data and data-independent memory accesses, which makes it suitable for use in outsourced tasks. Note that the great majority of data structures and algorithms commonly used in practice are not data-oblivious and thus reveal information about data, while naive approaches for achieving data-obliviousness incur a substantial increase in computation time over best-known solutions (consider, e.g., search, binary trees, etc.). Therefore, a careful design of data structures and algorithms is essential for bringing the complexity of oblivious execution as close to the complexity of its non-oblivious counterpart as possible.

While secure computation on sensitive data in outsourced and similar environments serves as the main motivation for our data-oblivious techniques, we would like to abstract the presentation of the developed techniques from a concrete setting or underlying mechanisms for securing the data. Nevertheless, to guarantee privacy of the data used in the computation, throughout this work we assume that the computation proceeds on protected data (e.g., using suitable secure multi-party computation techniques) and the only values that can be observed are memory accesses and results of the computation that lead to accessing specific memory locations. Data privacy is then guaranteed if the memory accesses are data-independent or oblivious.

In this work we focus on classical graph problems such as breadth-first search (BFS), shortest path, minimum spanning tree, and maximum flow and construct data-oblivious algorithms of practical performance. Several of them are fundamental graph problems with many uses both by themselves (e.g., for traversing a social network graph) and as building blocks for more complex problems on graphs. It therefore would be desirable to obtain data-oblivious solutions suitable for use in secure computation and outsourcing for the above problems.

1.1 Related work

One topic closely related to secure and oblivious computation is Oblivious RAM (ORAM) [23, 33, 24]. ORAM was introduced as the means of secure program execution in an untrusted environment. Only the CPU with a small amount

ASIA CCS'13, May 8–10, 2013, Hangzhou, China.

of internal memory ($O(1)$ or $O(n^a)$ for $0 < a < 1$, where n is the size of used external memory) is assumed to be trusted, while the program itself and the data are stored encrypted on untrusted storage. The goal is to ensure that no information about the program is revealed by observing its memory accesses and therefore the accesses must appear random and independent of the actual sequence of accesses made by the program. There is a renewed interest in ORAM due to the emergence of cloud computing and storage services, and recent results include [39, 40, 34, 1, 17, 27, 7, 36, 37, 29]. ORAM techniques then can be used to make a non-oblivious algorithm oblivious at the cost of polylogarithmic (in the size of the data) amortized cost per access. Current solutions achieve $O((\log n)^2)$ overhead per access (if random functions are used or server's storage is superlinear in n; otherwise, the overhead is $O((\log n)^3)$ per access). A comparison of several ORAM schemes and their complexities can be found in [1].

Another related topic is private information retrieval (PIR) (see, e.g., [14, 15, 30, 11, 22, 31, 38] among others), where a server holds a database and a clients wants to a retrieve a record at a specific position with the goal that the server should not learn what record was accessed. Symmetric PIR (SPIR) solutions also require that the user should not learn anything about any other records in the database except the record of interest. Current PIR solutions exist in both the information-theoretic and computational settings.

Privacy-preserving graph algorithms have become a recent area of interest with the prevalent adoption of location based services (LBS) for cloud services and mobile devices. With the widespread use of LBS there is an increased need for maintaining users' privacy. The approach in [32] utilizes PIR techniques to query a data-set while keeping the user's position, path, and desired destination private. Also motivated by LBS is the work proposed in [18]. It introduced data-oblivious algorithms for calculating classical geometric problems, such as convex hull and all nearest neighbors, using secure multi-party computation.

Other solutions that focus on privacy-preserving graph algorithms include algorithms designed for a public or joint graph. In particular, [8] presents a two-party solution for computing all pairs shortest distance and single source shortest path when each party holds a portion of the graph. The solution uses Yao's protocol [41] as the primary building block and achieves better efficiency than naively applying Yao's protocol to existing graph algorithms. Similarly, [20] presents a two-party method for privately performing an $A^\star$ search over a public graph where different parties hold the edge weights and the heuristics. Additionally, there are several techniques for constructing a graph privately. One such solution is [21], which shows how many clients, each possessing a portion of the distributed graph, can privately construct a graph so that an adversary is unable to de-anonymize any part of it after the construction process.

There are also recent data-oblivious graph algorithms designed for the external-memory model. This model assumes that the client has access to some amount of working-memory that is inaccessible to any adversary, but the remaining storage is outsourced. The solution in [26] introduces data-oblivious algorithms for compaction, selection, and sorting. A final technique of interest is [28], which introduces data-oblivious algorithms for graph drawing problems (e.g., Euler Tours, Treemap Drawings) and develops efficient compressed-scanning techniques, which in turn allow for private data to be searched in an efficient manner.

Similar to the external-memory model is the cache-oblivious model, where algorithms are designed to perform optimally using a CPU-cache without the size of the cache as an explicit-parameter. Several cache-oblivious algorithms and data structures are described in [2, 10, 9, 3]. Of particular interest is [10], where the authors describe cache-oblivious solutions for breadth-first search and shortest path.

1.2 Our contributions

In this work, we present data-oblivious algorithms for several fundamental graph algorithms, namely, breadth-first search, single-source single-destination (SSSD) shortest path, minimum spanning tree, and maximum flow. Given graph $G = (V, E)$ as the input, our solutions assume adjacency matrix representation of the graph, which has size $\Theta(|V|^2)$ and is asymptotically optimal for dense graphs with $|E| = \Theta(|V|^2)$. Our oblivious solutions achieve the following complexities: $O(|V|^2)$ for BFS, $O(|V|^2)$ for SSSD shortest path, $O(|V|^2)$ for the minimum spanning tree, and $O(|V|^3|E|\log(|V|))$ for the maximum flow. This performance is optimal, or close to optimal, for dense graphs and outperforms applying ORAM techniques to the fastest conventional non-oblivious algorithms for these problems.

Because our algorithms use adjacency matrix representation, they are not well suited for sparse graphs. For such graphs, combining ORAM with the best non-oblivious algorithm is likely to result in superior performance, even though ORAM techniques are often not particularly fast in practice. We leave the design of efficient oblivious algorithms for sparse graphs as a direction for future work.

Our solutions assume that numeric values can be compared, multiplied, and added in a protected form; the cost of such operations is considered to be constant (see section 1.3 for a specific instantiation of these operations). We also rely on random permutation of a vector as our building block. This functionality can be accomplished by assigning random values to each element of the vector and sorting them according to the assigned values. Oblivious sorting can be accomplished, e.g., using the techniques in [25] at cost $O(n \log n)$ for a set of size n. We therefore assume that a vector can be randomly permuted at the same asymptotic cost.

All of the graph algorithms considered in this work proceed in iterations. At the core of achieving the above complexities obliviously is the idea that we do not need to touch all locations of the adjacency matrix at each iteration to hide information about the access pattern, but are able to access a single row of the matrix per iteration. Because we access each row of the matrix exactly once and the accesses are performed in random order, we show that the accesses are data oblivious. We first develop our solution for the BFS problem and then extend it with path reconstruction to obtain a solution to the SSSD shortest path problem. Both BFS and SSSD shortest path solutions are consecutively used to build our data-oblivious algorithm for the maximum flow problem. Lastly, we also present an oblivious algorithm for the minimum spanning tree problem that utilizes similar techniques.

1.3 Applications to secure computation and outsourcing

As mentioned earlier, secure outsourcing serves as the main motivation for our data-oblivious graph algorithms.

We therefore sketch how our algorithms can be used in outsourced environments for securely computing on protected data. We utilize the setting in which the computation is carried out by multiple computational nodes, which allows us to formulate the problem as secure multi-party computation.

To allow for as general problem formulation as possible, we place all participants into the following three categories: (i) the party or parties who hold private inputs; (ii) the party or parties who learn the outcome of the computation, and (iii) the parties who conduct the computation. There are no constraints on how these three groups are formed, and a single entity can be involved in a solution taking on one or more of the above roles. This framework formulation naturally fits several broad categories of collaborative and individual computing needs. For example, a number of parties with private inputs can engage in secure function evaluation among themselves and learn the result (or their respective results). They can also choose a subset among themselves, a number of outside parties, or a combination of the above to carry out the computation. Note that this includes the important use case of a single entity outsourcing its computation to computational servers, in which case the data owner is the only input and output party.

The algorithms that we present can be realized in this setting using a number of underlying techniques such as linear secret sharing, threshold homomorphic encryption, or Boolean garbled circuits. Then if we, for instance, utilize an information-theoretically secure linear secret sharing scheme (such as [35]), any linear combination of secret-shared values is computed locally and multiplication is the very basic interactive building block. Comparisons such as less-than and equality tests can be performed efficiently using, for instance, techniques of [13]. In this setting, the arithmetic is efficient and complexity is measured in the number of interactive operations (while in settings that rely on public-key cryptography, the complexity will additionally need to be measured in the number of modular exponentiations). This gives us a secure implementation of our data-oblivious graph algorithms which are suitable for outsourced environments as the sequence of operations they execute does not reveal any information about the data.

2. PROBLEM DEFINITION

In this work we consider graph problems which take graph $G = (V, E)$ as part of their input. We assume that the graph is specified in the form of the adjacency matrix M. To avoid ambiguities, we explicitly specify the adjacency matrix representation used in our description. The adjacency matrix M is a $|V| \times |V|$ matrix containing Boolean values (if the edges do not have weights), where the row corresponding to node $v \in V$ contains information about the edges leaving v. If $(v, u) \in E$, then the cell at row v and column u is set to 1; otherwise, it is set to 0. For undirected graphs, both $M_{v,u}$ and $M_{u,v}$ are set to 1, where $M_{i,j}$ refers to the cell at row i and column j. $M_{v,v}$ is set to 0. Without loss of generality, we assume that the nodes are numbered 1 through $|V|$. For problems that use weighted graph G, we will specify the differences to the adjacency matrix representation at the time of specifying the respective graph problem.

Because secure outsourced computation is used to motivate this work, we assume that the computation proceeds on properly protected data. This means that all inputs and intermediate results are not known to the party or parties carrying out the computation unless we explicitly open their values for the purposes of accessing data at specific locations. For concreteness of exposition, we use notation $[x]$ to indicate that the value of x is protected from the entities performing the computation.

To maintain data privacy, we must guarantee that the computational parties do not learn anything about the data during the execution of the algorithm. Because each private value is assumed to be adequately protected, the only way for the computational parties to deduce information about the private data is when the sequence of instructions or algorithm's memory access pattern depends on the data. Thus, to guarantee data privacy, we formally formulate data-oblivious execution of a deterministic algorithm as follows:

DEFINITION 1. *Let d denote input to a graph algorithm. Also, let $A(d)$ denote the sequence of memory accesses that the algorithm makes. The algorithm is considered data-oblivious if for two inputs d and d' of equal length, the algorithm executes the same sequence of instructions and access patterns $A(d)$ and $A(d')$ are indistinguishable to each party carrying out the computation.*

Without loss of generality, in the description that follows, we use arithmetic operations to implement Boolean operations. In particular, we write $a \cdot b$ to implement conjunction $a \wedge b$ and we write $(1-a)$ to implement complement $\overline{a}$ for Boolean a. We also use notation $(a \stackrel{?}{=} b)$ and $(a \stackrel{?}{<} b)$ to denote conditions comparing two values, which produce a bit.

3. BREADTH-FIRST SEARCH

Breadth-first search is one of the most basic algorithms for searching a graph, but it is applicable to a large number of problems and is the basis for many important graph algorithms. Given a graph $G = (V, E)$ and a source vertex $s \in V$, BFS systematically explores the edges of G to compute the distance (measured in the number of edges) from s to each reachable vertex in G, where nodes at distance i from s are discovered before nodes at distance $i + 1$.

The conventional algorithm for BFS starts from the source node s and colors the nodes white, gray, and black, where the nodes start white and may later become gray and black. The intuition is that white nodes have not been discovered yet, gray nodes have been discovered, but nodes adjacent to them may have not been discovered yet, and black nodes have been discovered themselves as well as their adjacent nodes. The conventional algorithm works by maintaining a queue with gray nodes which initially contains only the source s. When a gray node is being removed from the queue, it becomes black and all white nodes adjacent to it are colored gray and added to the queue.

Now notice that directly applying this algorithm to obtain a data-oblivious version presents problems: as we must protect the graph information, we cannot reveal how many adjacent nodes a vertex has, which means that when removing a node from the queue $O(|V|)$ nodes must be added to the queue (many of which will be just empty records not corresponding to actual nodes). This results in an exponentially growing queue and is clearly unacceptable.

To protect the structure of the graph, we design our algorithm to work on adjacency matrix representation which allows us to achieve $O(|V|^2)$ runtime. This is optimal for dense graphs with $|E| = \Theta(|V|^2)$ and any time an adjacency

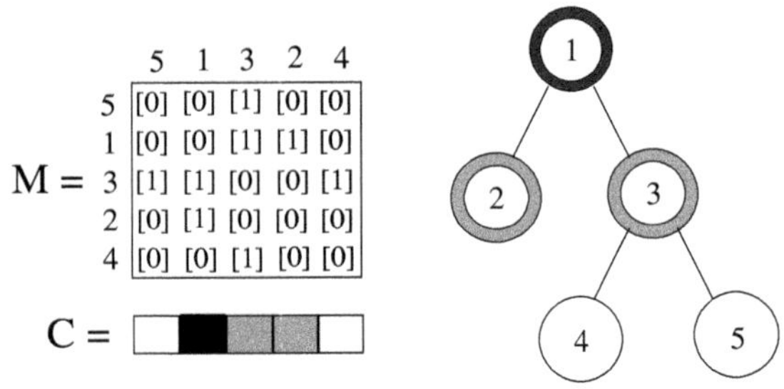

Figure 1: Illustration of the BFS algorithm.

matrix is used. We leave the problem of an efficient data-oblivious algorithm for sparse graphs open.

The intuition behind our algorithms is as follows: instead of maintaining a queue of gray nodes, node coloring is maintained in a (protected) vector C of size $|V|$. Initially, only the source is colored gray. Using the row of the adjacency matrix M corresponding to the source, we obliviously update the node coloring in vector C using the adjacency information of the source s. Once this is accomplished, we obliviously choose one of the gray nodes with the smallest distance from the source from the vector, reveal its location in the adjacency matrix, and use its adjacency information to update the vector C. This process is then repeated until all rows of the matrix have been processed.

We illustrate the idea on an example graph in Figure 1. In the figure, the (protected) matrix M contains node adjacency information where the nodes (i.e., both rows and columns) are randomly permuted. For example, row and column 1 correspond to node 5 in the graph. In the figure, the source node 1 has already been processed and colored black, and the nodes adjacent to it (nodes 2 and 3) has been colored gray. This information is stored in vector C, where the location of (black) node 1 after the permutation is equal to 2 and the locations of (gray) nodes 2 and 3 after the permutation are equal to 4 and 3, respectively.

There are subtleties in implementing the logic described above. In general, revealing what row (and thus the corresponding node) is being processed next can leak information about the structure of the graph. A naive way of hiding this information is to touch each element of matrix M when only one row is being used. We, however, can ensure that no information about the graph is revealed even if only a single row is accessed. This is achieved using two crucial ideas: (i) the order of the nodes of M is random and (ii) the next gray node to be processed is chosen from all candidates at random. The former means that the nodes are either randomly numbered in the adjacency matrix or they are randomly permuted (i.e., both the rows and columns of the matrix are randomly and consistently permuted). The latter means that rather than choosing the first node that meets the condition for the next iteration of the algorithm, it should be chosen from all candidates at random in order to hide the number of candidate nodes (and thus information about graph connectivity).

Also, the graph must be connected in order for the above algorithm to work. That is, to ensure that there is always a way to proceed with another row of the matrix, all nodes must be reachable from the source. For clarity of exposition, we first provide a solution that assumes that the graph is connected and later extend it to work with arbitrary graphs.

3.1 Basic algorithm for connected graphs

On input adjacency matrix $[M]$ of graph $G = (V, E)$ with randomly ordered nodes and source node $s \in V$:

Algorithm 1:

1. Create vector C of size $|V|$, each element of which corresponds to a node of G using the same node ordering as in M. Each element of C contains three fields: color, distance from s, and parent node in the BFS tree.
2. Initialize C to set the color of all nodes except s to white and the color of s to gray. The distance of s is set to 0 and the distance of all other nodes is set to ∞ (which for practical purposes can be set to $|V|$, i.e., larger than any actual distance). The parent of each node is set to a special character $\perp$ indicating that the parent is undefined. The values of the elements of C are protected throughout the execution.
3. Set working node v to s and retrieve row M_v of M.
4. Update C using M_v as follows:
   (a) For $i = 1$ to $|V|$, if $[M_{v,i}]$ is 1 and the color of C_i is white, set the color of C_i to gray, the parent of C_i to v, and the distance of C_i to the distance of C_v incremented by 1.
   (b) Set the color of C_v to black.

   Oblivious execution of conditional statements can be realized by executing both branches of the if-statement (in cases when only one branch is present, the second branch corresponds to keeping the original values). For instance, executing statement
   if $([cond])$ then $[a] = [b]$ else $[a] = [c]$
   becomes rewritten as
   $[a] = [cond] \cdot [b] + (1 - [cond]) \cdot [c]$
   In our case, the executed code is:
   1.for $i = 1$ to $|V|$ do
   2. $[cond] = ([M_{v,i}] \stackrel{?}{=} 1) \cdot ([C_i.color] \stackrel{?}{=} white)$
   3. $[C_i.color] = [cond] \cdot gray + (1 - [cond]) \cdot [C_i.color]$
   4. $[C_i.parent] = [cond] \cdot v + (1 - [cond]) \cdot [C_i.parent]$
   5. $[C_i.dist] = [cond]([C_v.dist] + 1) + (1 - [cond])[C_i.dist]$
5. Obliviously choose one of the gray nodes in C with the smallest distance from the source at random. Data-oblivious execution in this case means that the algorithm must be independent of the k number of nodes in C to choose from, yet random selection requires $1/k$ probability of any qualifying node to be selected.
   To achieve this, we use C to create another vector C' in which the elements that contain gray nodes with the smallest distance from s in C are set to contain their index values and all other elements are set to 0. We also choose a random permutation π of node indices and assign the key $\pi(i)$ to each element i in C' with a non-zero value. All other elements have the key of 0. We select the element with the maximum key and retrieve the index of the node stored with that element as the next node to process. In more detail, we execute:
   1. $[min] = |V|$
   2. for $i = 1$ to $|V|$ do
   3. $[cond_i] = ([C_i.color] \stackrel{?}{=} gray)$
   4. $[cond'_i] = ([C_i.dist] \stackrel{?}{<} [min])$
   5. $[min] = [cond_i][cond'_i][C_i.dist] + (1 - [cond_i] \cdot [cond'_i])[min]$
   6. for $i = 1$ to $|V|$ do
   7. $[cond''_i] = ([C_i.dist] \stackrel{?}{=} [min])$

8. $[C'_i.value] = [cond_i][cond''_i] \cdot i$
9. $[C'_i.key] = [cond_i][cond''_i] \cdot [\pi(i)]$
10. $[max] = 0$
11. $[i_{max}] = 0$
12. for $i = 1$ to $|V|$ do
13. $[cond] = ([C'_i.key] \overset{?}{>} [max])$
14. $[max] = [cond][C'_i.key] + (1 - [cond])[max]$
15. $[i_{max}] = [cond][C'_i.value] + (1 - [cond])[i_{max}]$
16. $i_{max} = open([i_{max}])$
17. $v = open([C'_{i_{max}}.value])$

In the above, $\pi : [1, |V|] \to [1, |V|]$ is a random permutation of node indices and $open(\cdot)$ corresponds to revealing (or opening) its argument. For efficiency reasons, we can skip all previously processed (i.e., black) nodes as we know that they no longer will be selected.

6. Use chosen node v to retrieve row M_v of M, and repeat steps 4–6 of the algorithm $|V| - 1$ times.

Because our algorithm always selects a gray node with the shortest distance from the source for the next iteration, it correctly implements the BFS queue.

3.2 Supporting general graphs

Algorithm 1 described in the previous section works as long as the graph G is connected, i.e., there is at least a single gray node to choose from at each iteration of the algorithm. In general, however, the input graph is not guaranteed to be connected and we next show how to modify the algorithm to ensure that it is suitable for arbitrary graphs.

The main idea behind the change is that we introduce fake nodes that will be chosen by the algorithm once there are no gray nodes to process in one of the algorithm iterations. In general, the algorithm may terminate after as few as only a single iteration, which means that we need extra $|V| - 1$ nodes to simulate the remaining algorithm steps. We therefore modify the adjacency matrix M to include $|V| - 1$ extra rows and columns corresponding to fake nodes which are placed at random indices within the matrix. Every fake node is made adjacent to all other fake nodes, but none of them are adjacent to the original nodes of the graph.

The modified matrix M can be formed by appending $|V| - 1$ rows and columns to the original matrix and randomly and consistently permuting its rows and columns. The new location of the source node s then needs to be revealed. For the purposes of our algorithm we also store a bit vector F of size $2|V| - 1$ with the nodes ordered in the same way as in the matrix in which the element at location v is set iff node v is fake. This vector is formed by initializing its first $|V|$ elements to 0 and the remaining elements to 1 and permuting it using the same permutation as in the matrix.

With this modified setup information, we are ready to proceed with the BFS algorithm. The basic structure of Algorithm 1 and most of its steps remain unchanged, and the only modification that we introduce is the way a node is chosen for the next algorithm iteration. That is, the algorithm must account for the fact that there might be no gray nodes to choose from at any given iteration and it should proceed with choosing a fake node.

5. Create vector C' as before, but now before choosing one element from it, we check whether it contains at least one element to choose from. If it does, we leave it unmodified; otherwise, we mark all fake nodes as gray using vector F. One of the qualifying nodes of C' is chosen as before as the node to be processed in the next iteration. To test whether there is a least one node to choose from, we can simply test whether the value of $[min]$ has not been updated (i.e., it is still $|V|$). More precisely, we execute the following code between lines of 9 and 10 of the original code of step 5:

1. $[cond] = ([min] \overset{?}{=} |V|)$
2. for $i = 1$ to $2|V| - 1$ do
3. $[C'_i.value] = [C'_i.value] + [cond][F_i] \cdot i$
4. $[C'_i.key] = [C'_i.key] + [cond][F_i] \cdot [\pi(i)]$

The remaining code remains unchanged with the exception that all for-loops now range from 1 to $2|V| - 1$.

3.3 SSSD shortest path

The above algorithm already computes the distance from the source node to all other nodes in the graph. In certain problems, however, the knowledge of the shortest path itself is required. In this section we therefore show how to obliviously reconstruct the shortest path from the source s to a given destination node t. Both BFS and shortest path computation are used as the building blocks in our solution to the maximum flow problem.

For the ease of exposition, we divide our description in two parts: we first present a solution that reconstructs the path, where the information about the path itself is protected, but the length of the path is revealed. We then show how to modify this solution to also hide the length of the path.

3.3.1 Basic solution

On input $G = (V, E)$ described by its adjacency matrix M, source node $s \in V$, and destination node $t \in V$, our solution first uses BFS to compute the distances from s to all other nodes and then starts from node t and retrieves parent node information from vector C. A simple oblivious solution to the path reconstruction problem can be implemented in $O(|V|^2)$ time. In particular, we can scan the vector C up to $|V|$ times, each time retrieving and storing the parent of the current node on the path. Then because in each round we touch all $|V|$ nodes (or all $2|V| - 1$ nodes when fake nodes are used), no information about the node added to the path is revealed. Also note that because the BFS algorithm requires $\Omega(|V|^2)$ time, this solution would not increase the asymptotic complexity of the overall solution.

An asymptotically more efficient solution would be to retrieve one element of C at a time, where each consecutive element is set to the parent node of the previously retrieved node, and use ORAM techniques to hide information about the elements of C that have been accessed. This increases both the storage necessary to maintain ORAM for C as well as adds polylogarithmic computational overhead for each memory access. Because of the complexity of ORAM techniques and their simulation in a privacy-preserving framework in particular, this approach would provide computational advantage only for very large graphs.

Also note that logic similar to what we have previously used for BFS does not work here. That is, suppose we randomly shuffle vector C and directly access its element to retrieve the next node on the path. This shuffling guarantees that there is no correlation between the row accessed during BFS and vertices accessed during path reconstruction. Unfortunately, translating parent information of the accessed location to its shuffled version cannot be performed obliviously in constant time and we are back to either scanning the entire vector or applying ORAM techniques to access the mapping information. For these reasons, we next describe a simple solution of $O(|V|^2)$ complexity.

In the description that follows, we conservatively assume that the location of destination node t in the adjacency matrix is not known (i.e., it is protected). This algorithm can then be easily modified for the case when the location of t in M and C is known. The initial solution, i.e., the one that does not hide the size of the path, is as follows:

Algorithm 2:

1. Execute BFS to compute shortest distances from source s to all other nodes.
2. Initialize the path P to $\langle [t] \rangle$. Set the current working node $[v]$ to $[t]$.
3. Scan vector C to find node $[w]$ such that $[w] = [C_v.parent]$. In more detail, we execute:
   1. $[w] = 0$
   2. for $i = 1$ to $2|V| - 1$ do
   3. $\quad [cond] = ([v] \stackrel{?}{=} i)$
   4. $\quad [w] = [cond] \cdot [C_i.parent] + (1 - [cond]) \cdot [w]$
4. Update the path as $P = \langle [w], P \rangle$ and set $[v]$ to $[w]$.
5. Repeat steps 3–5 until $[v] = s$.

When the location of t in the adjacency matrix is known, we can skip step 3 in the first iteration and directly add $[C_t.parent]$ to P and set $[v]$ to value $[C_t.parent]$ in step 4.

3.3.2 Hiding the length of the path

To protect information about the length of the path, we need to ensure that the algorithm always performs $|V| - 1$ iterations and produces a path of length $|V|$. To achieve this, we instruct the algorithm to continue adding nodes to the path if the source s is reached, but the path is shorter than $|V|$ vertices long.

To be able to hide the length of the path from s to t, we must hide the fact that $v = s$ and proceed with adding nodes to the path. For that reason, we add s to the path, but set one of the fake nodes as the current working node $[v]$. The algorithm will then keep adding fake nodes to the path until its length becomes $|V| - 1$.

To ensure that a sufficient number of fake nodes is added without repetitions, we set their parent information to form a cycle of length $|V| - 1$. Because the algorithm is oblivious, the parents of the fake nodes can be assigned in an arbitrary manner. We choose to assign them sequentially to form one cycle. More precisely, the parent of fake node v is set to fake node w with the highest index less than v (and the parent of the fake node with the lowest index is set to the fake node with the highest index). For that purpose, we utilize vector F to update parent information of fake nodes in vector C.

In more detail, the computation that we need to perform after step 1 of Algorithm 2 is as follows. We first scan F to find the index of the fake node with the highest number. It will be used to initialize the parent value (to be used for the fake node with the lowest index). We then scan the nodes in C updating parent values of fake nodes and the current parent value.

1. $[parent] = 0$
2. for $i = 1$ to $2|V| - 1$ do
3. $\quad [parent] = [F_i] \cdot i + (1 - [F_i])[parent]$
4. for $i = 1$ to $2|V| - 1$ do
5. $\quad [C_i.parent] = [F_i] \cdot [parent] + (1 - [F_i])[C_i.parent]$
6. $\quad [parent] = [F_i] \cdot i + (1 - [F_i])[parent]$

The only other change that we need to make to Algorithm 2 is to hide the fact that source s has been reached by selecting a fake node. To achieve this, we randomly select one of the fake nodes and set the current working node to that fake node if the source has been reached. Because we do not know at which step of the computation the source might be reached, we have to perform such testing at each iteration of the algorithm. Fortunately, because the source node can be reached only once, we can pre-select one of the fake nodes at the beginning of the algorithm (instead of doing it at each iteration), but test each time whether it should be chosen. This means that after assigning parent node information to the fake nodes as described above, we select one of the nodes at random as follows:

1. choose (protected) random permutation $\pi(\cdot)$
2. for $i = 1$ to $2|V| - 1$ do $[v_i] = [F_i] \cdot [\pi(i)]$
3. $[u] = 0$
4. for $i = 1$ to $2|V| - 1$ do
5. $\quad [cond_i] = ([v_i] \stackrel{?}{>} [u])$
6. $\quad [u] = [cond_i] \cdot i + (1 - [cond_i])[u]$

Lastly, we modify steps 4–5 of Algorithm 2 to the following:

4. Prepend $[w]$ to the path, i.e., $P = \langle [w], P \rangle$. If $[w]$ is different from s, set $[v]$ to $[w]$; otherwise, set $[v]$ to a randomly chosen fake node. That is, compute:
   1. $[cond] = ([w] \stackrel{?}{=} s)$
   2. $[v] = [cond] \cdot [u] + (1 - [cond])[w]$
5. Repeat steps 3–5 $|V| - 2$ times.

3.3.3 Handling unreachable destination

The algorithm that we described so far for shortest path computation works if there is a path from s to t. If, however, node t cannot be reached from the source node, the algorithm must still proceed with the computation protecting the fact that there is no path. According to our BFS algorithm, node t will have its parent set to $\perp$ if there is no path from s to t. We can use this information to slightly modify the algorithm and proceed with one of the fake nodes if $\perp$ has been reached. This introduces a very minor change to the algorithm described in section 3.3.2 because we can reuse the selected fake node for our purposes. This is a safe modification to the algorithm because a path will never simultaneously contain s and $\perp$. We therefore update step 4 of Algorithm 2 to the following:

4. Prepend $[w]$ to the path, i.e., $P = \langle [w], P \rangle$. If $[w]$ is different from s or $\perp$, set $[v]$ to $[w]$; otherwise, set $[v]$ to a randomly chosen fake node. That is, compute:
   1. $[cond_1] = ([w] \stackrel{?}{=} s)$
   2. $[cond_2] = ([w] \stackrel{?}{=} \perp)$
   3. $[cond] = [cond_1] + [cond_2] - [cond_1][cond_2]$
   4. $[v] = [cond] \cdot [u] + (1 - [cond])[w]$

In the above, we use $a + b - a \cdot b$ to implement $a \vee b$.

3.4 Analysis

For all of our oblivious graph algorithms, we first analyze their time complexity followed by their security analysis.

3.4.1 Complexity analysis

In this section, we analyze the complexities of our BFS algorithm and shortest path reconstruction.

It is easy to see that the complexity of steps 1–2 and 4–5 of the BFS algorithm (both the basic and general versions) is

$O(|V|)$, while the complexity of steps 3 and 6 is $O(1)$. Then because steps 4–6 are executed $|V|$ times, the overall runtime of $O(|V|^2)$, which is optimal for the adjacency matrix representation and for graphs with $|E| = \Theta(|V|^2)$. It also outperforms any solution that combines the conventional algorithm with ORAM when $|E| = \Omega(|V|^2/\log(|V|)^2)$.

If the labeling of the graph nodes is not guaranteed to be random (i.e., the labeling can reveal information about the structure of the graph) or the graph is not guaranteed to be connected and the fake nodes need to be placed at random locations in the graph, the nodes of the graph will need to be randomly permuted. When this step is implemented using oblivious sorting, its complexity is $O(|V|^2 \log(|V|))$, which dominates the complexity of the algorithm.

Our algorithm for SSSD shortest path computation also has $O(|V|^2)$ time. In particular, after executing BFS in step 1 of the algorithm, the only steps that have non-constant time is one-time pre-processing of fake nodes and step 3, both with complexity $O(|V|)$. Because steps 3–4 are performed $|V|-1$ times, we obtain the overall runtime of $O(|V|^2)$.

3.4.2 *Security analysis*

To show security of our BFS and shortest path algorithms, we show that they are oblivious with respect to Definition 1.

THEOREM 1. *The BFS algorithm is data-oblivious.*

PROOF. To prove this theorem, we analyze each major operation in the algorithm with respect to Definition 1. We show that for any given input graph $G = (V, E)$ and source s, (i) the sequence of executed instructions is the same as for all other input graphs with the same number of nodes $|V|$ and (ii) the memory accesses are indistinguishable from the memory accesses when the input G is a randomly generated graph with $|V|$ nodes.

The first three steps of the algorithm are independent of the input graph G, and step 1 is also independent of the source node s. This means that step 1 is exactly the same for all possible inputs. Step 2 performs the same operations for all inputs, but accesses and updates node s in C with different information than other nodes. Because according to the solution, s has a random location in the graph, its position is indistinguishable for real and randomly generated input graphs G. The same applies to step 3 of the algorithm.

Steps 4 and 5 execute the same sequence of instructions for all input graphs and access all memory locations of C and M_v in exactly the same manner, therefore they are identical for all input graphs. The only part that remains is to show that revealing the locations of $|V|$ nodes which are being processed by the algorithm cannot be used to extract information about the input $\langle G, s\rangle$. In particular, because of randomized order of the nodes, the revealed locations are random and cannot be used to extract information about the structure of the graph. Furthermore, when selecting the next candidate node, one of them is selected at random, which also protects information about the number of candidate nodes. We obtain that the revealed locations are random and the memory accesses are indistinguishable from those of randomly generated graphs. □

THEOREM 2. *The SSSD shortest path algorithm is data-oblivious.*

PROOF. Similar to the proof of BFS algorithm, we consider all steps of the SSSD shortest path algorithm and show that they are data-oblivious according to our definition.

Step 1 executes BFS and is data-oblivious according to Theorem 1. All remaining steps (including fake node selection) execute exactly the same sequence of operations for all inputs graphs and access exactly the same memory locations for all input graphs. This means that the execution and memory accesses are identical for all inputs and the algorithm is data-oblivious. □

4. MAXIMUM FLOW

In this section we provide an oblivious solution to another graph problem, namely, maximum flow. Before we can proceed with its description, we need to provide background information.

4.1 Background

A flow network is a directed graph $G = (V, E)$, where each edge $(v, u) \in E$ has a non-negative capacity $c(v, u) \geq 0$ (and if $(v, u) \notin E$, $c(v, u) = 0$). Given a source vertex $s \in V$ and a sink vertex $t \in V$, a flow f in G is a function $f : V \times V \to \mathbb{R}$ that must satisfy the properties of capacity constraint (i.e., for all $v, u \in V$, $f(v, u) \leq c(v, u)$), skew symmetry (i.e., for all $v, u \in V$, $f(v, u) = -f(u, v)$), and flow conservation (i.e., for all $v \in V \setminus \{s, t\}$, $\sum_{u \in V} f(v, u) = 0$). The value of a flow in the flow network is defined as the total flow out of the source s $|f| = \sum_{u \in V} f(s, u)$, and the maximum-flow problem is to find a flow of maximum value.

A standard way of computing maximum flow relies on the Ford-Fulkerson method, which proceeds as follows: We initialize the flow to 0, and while there exists an augmenting path p, we augment the flow f along the path p. Here an augmenting path is defined as a path from the source s to the sink t which can admit more flow and therefore can be used to increase the overall flow of the network.

In implementing the above high-level logic, existing algorithms rely on the notion of residual network, which intuitively consists of edges that can admit more flow. In more detail, given a flow network $G = (V, E)$ and a flow f, the residual network of G induced by f is $G_f = (V, E_f)$ with edges of positive capacity $E_f = \{(v, u) \in V \times V \mid c_f(v, u) > 0\}$, where $c_f(v, u) = c(v, u) - f(v, u)$ is the residual capacity of (v, u). An augmenting path p is then a simple path from s to t in the residual network G_f. This means that each edge on the path admits additional positive flow, and the (residual) capacity of p is defined as $c_f(p) = \min\{c_f(v, u) \mid (v, u)$ is on $p\}$.

The basic structure of the approach, which corresponds to the Ford-Fulkerson algorithm, is then as follows: On input $G = (V, E)$, $s \in V$, and $t \in V$,

1. for each $(v, u) \in E$ do
2. $\quad f(v, u) = 0$
3. $\quad f(u, v) = 0$
4. while there exists path p from s to t in residual network G_f do
5. $\quad c_f(p) = \min\{c_f(v, u) \mid (v, u)$ is on $p\}$
6. $\quad$ for each (v, u) in p do
7. $\quad\quad f(v, u) = f(v, u) + c_f(p)$
8. $\quad\quad f(u, v) = -f(v, u)$

This algorithm has complexity $O(|E| \cdot |f_{max}|)$, where f_{max} is the maximum flow returned by the algorithm, as finding a path in each iteration of the algorithm involves $O(|E|)$ time. For networks with integral capacities and small f_{max}, the runtime of this algorithm is good; however, in the general case a variant known as the Edmonds-Karp algorithm is preferred. If we use a shortest path from s to t on line 4 of the algorithm, we obtain the Edmonds-Karp algorithm with complexity $O(|V| \cdot |E|^2)$. It is guaranteed to find the maximum flow in $O(|V| \cdot |E|)$ iterations, each of which involves $O(|E|)$ time (see, e.g., [16]). Finding a shortest path can be accomplished using BFS.

4.2 Oblivious algorithm

In our oblivious solution, we follow the overall structure of the algorithm and use the implementation of BFS and shortest path computation from Section 3. Because our oblivious BFS algorithm processes one node at a time and reveals its location in the adjacency matrix, we need to shuffle the rows and columns of the matrix between each iteration of the maximum flow solution to hide all access patterns. We also now must maintain the residual network and obliviously update it after computing the augmenting path p.

We obtain the solution, which takes as the input a flow network $G = (V, E)$ with the capacity function stored in the (protected) adjacency matrix M, source node s, and sink node t. We assume that positive capacity of edge (v, u) is stored in $[M_{v,u}]$, and $M_{v,u} = 0$ indicates that there is no edge from v to u. The algorithm proceeds as follows:

Algorithm 3:

1. Expand matrix M with $|V|-1$ fake nodes inserted into M consistently as rows and columns. The capacity of all edges (v, u) and (u, v) connecting two fake nodes v and u is set to c_{max}, where c_{max} refers to the maximum possible capacity of an edge. The capacity of all edges (v, u) and (u, v), where v is a node of the original graph G and u is a fake node is set to 0. As before, the information about which nodes of M are fake is maintained in bit vector F. The location of fake nodes does not need to be randomized at this step.
2. Create (protected) matrix M' for storing residual network G_f and initialize each element of it $M'_{i,j}$ to $M_{i,j}$. Also create (protected) matrix L for storing the flow function and initialize each element of it $L_{i,j}$ to 0.
3. Repeat the computation that follows $|V| \cdot |E|$ times.
4. Apply a random permutation to the rows and columns of M' and L as well as to the elements of F consistently. Reveal the location of s after the permutation, but keep the location of t protected.
5. Execute SSSD shortest path algorithm on the graph defined by M' using source s and destination $[t]$. Because our algorithms in Section 3 were defined for unweighted graphs, we need to introduce changes to work with weighted graphs and also preserve information about the capacity of each edge on the path. For that reason, for the purposes of the BFS algorithm, node u is considered to be adjacent to node v if the capacity of the edge (v, u) is positive, i.e., $M'_{v,u} > 0$. In addition, when updating vector C in step 4 of Algorithm 1 with parent and distance from the source information, we also preserve information about the capacity of the edge from the parent to the current node using matrix M'. This information is stored in field $C_i.capacity$ and is updated together with $C_i.parent$ and $C_i.dist$. That is, we execute:

   1. for $i = 1$ to $|V|$ do
   2. $\quad [cond] = ([M'_{v,i}] \stackrel{?}{>} 0) \cdot ([C_i.color] \stackrel{?}{=} white)$
   3. $\quad [C_i.color] = [cond] \cdot gray + (1 - [cond]) \cdot [C_i.color]$
   4. $\quad [C_i.parent] = [cond] \cdot v + (1 - [cond]) \cdot [C_i.parent]$
   5. $\quad [C_i.dist] = [cond]([C_v.dist] + 1) + (1 - [cond]) \cdot [C_i.dist]$
   6. $\quad [C_i.capacity] = [cond][M'_{v,i}] + (1 - [cond]) \cdot [C_i.capacity]$

To preserve edge capacity in the reconstructed path P, we also introduce changes to Algorithm 2. In particular, instead of storing a single node $[w]$ in P, we now store a tuple of the form $\langle [v_1], [v_2], [v_3], [c] \rangle$. Here $[v_1]$ is the same as previously stored, i.e., the first elements of the stored tuples form a path in the graph padded in the beginning with fake nodes to be of length $|V|$. Nodes v_2 and v_3 represent an edge, and c is capacity. In almost all cases, $v_1 = v_2$ and (v_3, v_2) is the edge on the path of capacity c, i.e., v_3 is stored as the first element of the preceding tuple. The only time when this does not hold is during special cases when the source or no path symbol $\perp$ has been reached during path computation. In those cases, we set v_1 to s or $\perp$, but the edge (v_3, v_2) is between two fake nodes, and c is the capacity of that edge in M' when $v_1 = s$ and $c = 0$ when $v_1 = \perp$. Setting the capacity in such a way will allow us to ensure that the capacity of the path is computed correctly. For reasons that will soon become apparent, we also set $[v_2] = [v_1]$, $[v_3] = [t]$, and $[c] = 0$ in the last iteration of path reconstruction (i.e., the first tuple on the path). The path reconstruction algorithm becomes:

(a) Initialize the path P to empty and set $[v] = [v'] = [t]$. Randomly choose a fake node as before and store it in $[u]$.

(b) Repeat $|V| - 1$ times:

   i. Scan vector C to retrieve parent and edge capacity information and compute data to be added to the path by executing the following:

      1. $[w] = 0$
      2. $[c] = 0$
      3. for $i = 1$ to $2|V| - 1$ do
      4. $\quad [cond] = ([v] \stackrel{?}{=} i)$
      5. $\quad [w] = [cond][C_i.parent] + (1 - [cond])[w]$
      6. $\quad [c] = [cond][C_i.capacity] + (1 - [cond])[c]$
      7. $[cond] = (v' \stackrel{?}{=} \perp)$
      8. $[v_1] = [v']$
      9. $[v_2] = [v]$
      10. $[v_3] = [w]$
      11. $[c] = (1 - [cond])[c]$
      12. $[cond_1] = ([w] \stackrel{?}{=} s)$
      13. $[cond_2] = ([w] \stackrel{?}{=} \perp)$
      14. $[cond] = [cond_1] + [cond_2] - [cond_1][cond_2]$
      15. $[v'] = [w]$
      16. $[v] = [cond] \cdot [u] + (1 - [cond])[w]$

      where, as before u, is a randomly selected fake node. The purpose of v' here is simply to store node information from the previous round (in case v is set to u, while we need to remember w as well).

   ii. Prepend $\langle [v_1], [v_2], [v_3], [c] \rangle$ to P.

(c) Set $[v_1] = [v']$, $[v_2] = [v']$, $[v_3] = [t]$, and $[c] = [0]$ and prepend $\langle [v_1], [v_2], [v_3], [c] \rangle$ to P.

6. Compute the residual capacity $[c_f]$ of the path in P. Oblivious execution of this step is not difficult:

   1. $[c_f] = [c_{max}]$
   2. for $i = 2$ to $|V|$ do
   3. $\quad [cond] = ([P_i.capacity] \stackrel{?}{<} [c_f])$
   4. $\quad [c_f] = [cond] \cdot [P_i.capacity] + (1 - [cond])[c_f]$

   Here P_i is the ith element of the path P. Note that the first tuple on the path is ignored as its capacity is always 0, i.e., only $|V| - 1$ edges are used in computing the path's capacity.

7. To be able to update the residual network M' and flow function L, we obliviously rotate the entries in the path P to have the first record contain an edge that leaves the source s. This will allow us to proceed with one edge of P at a time updating cells of M' and L in the next step.

   When there is no path between s and t, vertex s does not appear in P. To hide this information, we replace $\perp$ with s in tuple $\langle [t], [t], [\perp], [0] \rangle$.[1] This does not affect correctness of the algorithm. We also ignore the preceding record (of the form $\langle [s], [v_2], [v_3], [c] \rangle$ or $\langle [\perp], [v_2], [v_3], [c] \rangle$), i.e., process only $|V| - 1$ edges.

   A simple way of performing oblivious rotation of the tuples in P is as follows:

   1. for $i = 1$ to $|V| - 1$ do
   2. $\quad [cond] = ([P_i.v_3] \stackrel{?}{=} s)$
   3. $\quad [temp] = [P_1]$
   4. $\quad$ for $j = 1$ to $|V| - 1$ do
   5. $\quad\quad [P_j] = [cond][P_j] + (1 - [cond])[P_{j+1}]$
   6. $\quad [P_{|V|}] = [cond][P_{|V|}] + (1 - [cond])[temp]$
   7. remove P_1 from P

   For simplicity of exposition, we use assignment $[x] = [P_i]$ to indicate that the entire tuple at the ith position of P is stored in variable x (where the value of i is known, while the content of values at P_i is not).

   After rotating the elements of P in this manner, we have that P stores a path from s to t padded at the end with fake nodes and edges between them (and if no path exists, there will be a path in P, but its capacity is 0, which means that the residual network will not be modified). Note that there is an edge leaving t of capacity 0 (i.e., the first element of P prior to its rotation) which correctly forms the path and transitions to a fake node (after which all nodes on the path are fake). Figure 2 demonstrates how P is formed before and after path rotation for two cases: (i) when a path from s to t is present and (ii) when there is no path from s to t. Notation v_i refers to an original node of the graph, and notation f_i is used for fake graph nodes.

8. Update residual network M' and flow function L. Now because we use random labeling of vertices in the graph, we can update M' and L by revealing the path information (i.e., open edges (v_3, v_2) in each element of P) and update the corresponding cells of M' and L with (still protected) path capacity information. We, however, already revealed a certain sequence of nodes during BFS computation, and to ensure that there is no correlation between memory accesses during BFS and during residual network update, we need to randomly shuffle the nodes again. This time, in addition to permuting the rows and columns of M' and L, as well as vector F, we also need to map each node v contained in path P to its permuted index $\pi(v)$. A simple way of accomplishing this is to scan the $\langle i, [\pi(i)] \rangle$ pairs for each v in P and replace v with $\pi(i)$ when $v = i$. That is, for each node v included in P, we execute:

   1. for $i = 1$ to $2|V| - 1$ do
   2. $\quad [cond] = ([v] \stackrel{?}{=} i)$
   3. $\quad [\pi(v)] = [cond][\pi(i)] + (1 - [cond])[\pi(v)]$

   It is important that the location of t is not known after the random permutation in each iteration and the nodes on any given path do not repeat. The cells of M' and L are then updated as follows:

   1. for $i = 1$ to $|V| - 1$ do
   2. $\quad$ open v_3 and v_2 in P_i
   3. $\quad [L_{v_3,v_2}] = [L_{v_3,v_2}] + [c_f]$
   4. $\quad [L_{v_2,v_3}] = -[L_{v_3,v_2}]$
   5. $\quad [M'_{v_3,v_2}] = [M'_{v_3,v_2}] - [c_f]$
   6. $\quad [M'_{v_2,v_3}] = [M'_{v_2,v_3}] + [c_f]$

[1]This step can be performed at the time of forming the tuples in P, but for clarity of presentation we choose to form the path correctly in step 5.

Before		After
$P_1 = \langle f_3, f_3, t, 0 \rangle$	$\Longrightarrow$	$P_1 = \langle v_2, v_2, s, c_3 \rangle$
$P_2 = \langle f_2, f_2, f_3, c_5 \rangle$		$P_2 = \langle v_1, v_1, v_2, c_2 \rangle$
$P_3 = \langle s, f_1, f_2, c_4 \rangle$		$P_3 = \langle t, t, v_1, c_1 \rangle$
$P_4 = \langle v_2, v_2, s, c_3 \rangle$		$P_4 = \langle f_3, f_3, t, 0 \rangle$
$P_5 = \langle v_1, v_1, v_2, c_2 \rangle$		$P_5 = \langle f_2, f_2, f_3, c_5 \rangle$
$P_6 = \langle t, t, v_1, c_1 \rangle$		

(a) Existing path before and after path rotation

Before		After
$P_1 = \langle f_5, f_5, t, 0 \rangle$	$\Longrightarrow$	$P_1 = \langle t, t, s, 0 \rangle$
$P_2 = \langle f_4, f_4, f_5, c_4 \rangle$		$P_2 = \langle f_5, f_5, t, 0 \rangle$
$P_3 = \langle f_3, f_3, f_4, c_3 \rangle$		$P_3 = \langle f_4, f_4, f_5, c_4 \rangle$
$P_4 = \langle f_2, f_2, f_3, c_2 \rangle$		$P_4 = \langle f_3, f_3, f_4, c_3 \rangle$
$P_5 = \langle \perp, f_1, f_2, c_1 \rangle$		$P_5 = \langle f_2, f_2, f_3, c_2 \rangle$
$P_6 = \langle t, t, \perp, 0 \rangle$		

(b) Unreachable destination before and after path rotation

Figure 2: Example of path computation for $|V| = 6$.

At the end of the computation we output the maximum flow as the total flow $\sum_{i=1}^{|V|} [L_{s,i}]$ leaving the source node s. If the flow information is desired with respect to the original node labeling, the algorithm can maintain the composition of random permutations used at each iteration.

4.3 Analysis

4.3.1 Complexity analysis

To show that our algorithm runs in $O(|V|^3|E|\log(|V|))$ time, we analyze the complexities of each step of it. In the algorithm, steps 1, 2, 5, and 7 take $O(|V|^2)$ time, while the complexity of step 6 is $O(|V|)$ and that of steps 4 and 8 is $O(|V|^2 \log(|V|))$. Because steps 4–8 are repeated $O(|V||E|)$ times, we obtain the overall complexity as claimed. Our complexity is higher by a factor $\log(|V|)$ than that of Edmonds-Karp's algorithm when the adjacency matrix is used or the graph is dense with $|E| = O(|V|^2)$. It also outperforms any solution that combines the Edmonds-Karp algorithm with ORAM when $|E| = \Omega(|V|^2/\log(|V|))$.

4.3.2 *Security analysis*

We show the security of the maximum-flow algorithm as before using Definition 1.

THEOREM 3. *The maximum-flow algorithm is data-oblivious.*

PROOF. As before, we analyze each major operation in the algorithm with input $G = (V, E)$, s, and t. In steps 1 and 2, we perform identical operations and touch exactly the same locations for all inputs with $|V|$ nodes and therefore the steps are data-oblivious. Step 3 is merely an iterator with the same number of iterations for all graphs with $|V|$ nodes and $|E|$ edges, so we proceed directly to step 4. Step 4 only executes a data-oblivious shuffle operation and is therefore oblivious. Step 5 calls BFS and SSSD shortest path algorithms with small changes to preserve edge capacity information. These algorithms have been shown to be data-oblivious in Theorems 1 and 2, respectively. Step 6 performs identical operations and accesses the same memory locations for all graphs with $|V|$ nodes and is therefore data-oblivious. In step 7, the algorithm rotates the path, once again executing identical operations for all inputs and accessing the same memory locations. Lastly, in step 8, we execute the same operations for all graphs, but accessed memory locations differ. Our algorithm crucially relies on two facts to achieve data-obliviousness: (i) the vertices are randomly permuted prior to any access is made and (ii) all updates follow a path of size $|V|$, which always starts with the source s and has no repeated nodes on it. This means that, besides for the first accessed node s, all other memory accesses are random and cannot be distinguished from accesses used for a randomly generated input graph. Lastly, because the algorithm proceeds in iterations, we ensure that every time a subset of memory locations is accessed (i.e., during BFS and when updating the residual network), the nodes are randomly permuted right before that operations. This ensures that all accesses cannot be distinguished from a random sequence of vertex accesses. □

5. MINIMUM SPANNING TREE

Given the already developed techniques, it is not difficult to design an oblivious algorithm for computing the minimum spanning tree that runs in $O(|V|^2)$ time (unless the node labeling reveals information about the structure of the graph). Our algorithm uses the structure of Prim's algorithm, and we start by describing background information.

5.1 Background

On input connected undirected weighted graph $G = (V, E)$, Prim's algorithm initializes the spanning tree $T = (V', E')$ to a single arbitrary vertex from V and no edges. Then until $V' \neq V$, the algorithm selects an edge (v, u) with the minimum weight such that $v \in V'$ and $u \notin V'$ and sets $V' = V' \cup \{u\}$ and $E' = E' \cup \{(v, u)\}$. The complexity of this algorithm depends on the graph representation and data structures used. Using the adjacency matrix, the algorithm runs in $O(|V|^2)$ time; using the adjacency list, the algorithm's performance is $O(|E| \log(|V|))$ or $O(|E| + |V| \log(|V|))$ if a binary heap or Fibonacci heap is used, respectively.

5.2 Oblivious algorithm

In order to compute a minimum spanning tree obliviously without incurring a large amount of additional overhead, we maintain a vector with candidate nodes to potentially be added to the tree T and select one of them which has an edge with the minimum weight connecting it to T. The vector is updated each time a node is added to V'. To efficiently implement this step, we crucially rely on the ability to process a single row of the matrix, which can be safely performed only when the node labeling is random and does not reveal information about the structure of the graph. We obtain the algorithm given below, which takes as input adjacency matrix M representing the graph G. Because edge weights in G can be arbitrary, we assume that each cell $M_{v,u}$ of M contains two fields: (i) Boolean field *adjacent*, which is set if $(v, u) \in E$ and (ii) numeric value *weight* indicating the weight of the edge (v, u).

Algorithm 4:

0. If the numbering of the nodes in G may convey information about the structure of the graph, randomly and consistently permute the rows and columns of adjacency matrix M.
1. Choose a single node v from G, e.g., at position 1 in M. Set $V' = \{v\}$ and $E' = \emptyset$. Create (protected) vector C of size $|V|$, where each element of C, C_i, contains two fields: *weight* and *parent*. Initialize each $[C_i.weight]$ to ∞ (which for practical purposes can be any constant larger than any weight in M) and each $[C_i.parent]$ to $\perp$.
2. Retrieve row at position v from M and update C using M_v. That is, for each i we store in C_i the minimum of C_i and the weight at $M_{v,i}$ if v is adjacent to i. Then if $C_i.weight$ is updated to the weight stored at $M_{v,i}$, we also store v in $C_i.parent$. The pseudo-code for oblivious implementation of this step is:
   1. for $i = 1$ to $|V| - 1$ do
   2. $\quad [cond] = ([M_{v,i}.weight] \stackrel{?}{<} [C_i])[M_{v,i}.adjacent]$
   3. $\quad [C_i.weight] = [cond][M_{v,i}.weight] + (1 - [cond]) \cdot [C_i.weight]$
   4. $\quad [C_i.parent] = [cond] \cdot v + (1 - [cond])[C_i.parent]$
3. Locate the minimum element of C, $C_{i_{min}}$, considering only locations i such that $i \notin V'$. That is, execute:
   1. $[min] = \infty$
   2. $[i_{min}] = 0$
   3. for $i = 1$ to $|V| - 1$ do
   4. $\quad$ if $i \notin V'$ then
   5. $\quad\quad [cond] = ([C_i.weight] \stackrel{?}{<} [min])$
   6. $\quad\quad [min] = [cond][C_i.weight] + (1 - [cond])[min]$
   7. $\quad\quad [i_{min}] = [cond] \cdot i + (1 - [cond])[i_{min}]$
   8. $i_{min} = open(i_{min})$
4. Update the minimum spanning tree by setting $V' = V' \cup \{i_{min}\}$, $E' = E' \cup \{([C_{i_{min}}.parent], [i_{min}])\}$, and $v = i_{min}$. Repeat steps 2–4 $|V| - 2$ times and output the edges E'.

Recall that the nodes of the graph are randomly permuted and the order in which they are added to V' does not reveal information about the graph. This means that the nodes that form the set V' are not protected, where each node is added to V' exactly once. This means that in step 3 we need to consider only the nodes that have not yet been added to the minimum spanning tree.

Correctness of this algorithm follows from the correctness of Prim's algorithm.

5.3 Analysis

5.3.1 Complexity analysis

The performance of our algorithm is not difficult to assess: steps 1–3 take $O(|V|)$ time, while step 4 takes $O(1)$ time. Because these steps are repeated $O(|V|)$ times, we arrive at $O(|V|^2)$ overall time. It is optimal with matrix adjacency representation and when $|E| = \Theta(|V|^2)$. If step 0 is necessary, however, the overall time raises to $O(|V|^2 \log(|V|))$.

5.3.2 Security analysis

To show that the minimum spanning tree algorithm is oblivious, we as before follow Definition 1 and show that the revealed information is indistinguishable from random.

THEOREM 4. *The minimum spanning tree algorithm is data-oblivious.*

PROOF. We analyze each major operation in the algorithm with input $G = (V, E)$. Step 0 calls random shuffling, which can be performed obliviously. Step 1 does not use the input, only the input's size. In step 2, after retrieving a row of the adjacency matrix, the execution that follows uses the same instructions and accesses the same memory for all input graphs. Then, because this step is executed multiple times, each iteration accesses a row corresponding to a unique node. Since we know that the node ordering is random and each node is used exactly once, the access pattern induces a random permutation on the set of nodes and is therefore indistinguishable from access patterns of randomly chosen graphs. Steps 3 and 4 execute the same instructions and access the same memory locations for all input graphs and are therefore data-oblivious. □

6. BUILDING SECURE PROTOCOLS

In this section we briefly illustrate how our oblivious algorithms can be used to build protocols for graph problems suitable for secure computation and outsourcing. We use the BFS algorithm for illustration purposes.

Using the setting of section 1.3 with respect to computation participants, we denote the computational parties as $P_1, \ldots, P_n$ and define security in presence of semi-honest participants (who follow the prescribed computation, but might attempt to learn additional information about private data from the messages that they receive) as follows:

DEFINITION 2. *Let parties $P_1, \ldots, P_n$ engage in a protocol Π that computes function $f(\mathsf{in}_1, \ldots, \mathsf{in}_n) = (\mathsf{out}_1, \ldots, \mathsf{out}_n)$, where in_i and out_i denote the input and output of party P_i, respectively. Let $\mathrm{VIEW}_\Pi(P_i) = (\mathsf{in}_i, r_i, m_1, \ldots, m_k)$ denote the view of participant P_i during the execution of protocol Π, which is formed by its input, internal random coin tosses r_i, and messages $m_1, \ldots, m_k$ passed between the parties during protocol execution. Let $I = \{P_{i_1}, P_{i_2}, \ldots, P_{i_t}\}$ denote a subset of the participants for $t < n$ and $\mathrm{VIEW}_\Pi(I)$ denote the combined view of participants in I during the execution of protocol Π (i.e., the union of the views of the participants in I). We say that protocol Π is t-private in presence of semi-honest adversaries if for each coalition of size at most t there exists a probabilistic polynomial time simulator S_I such that $\{S_I(\mathsf{in}_I, f(\mathsf{in}_1, \ldots, \mathsf{in}_n)\} \equiv \{\mathrm{VIEW}_\Pi(I), \mathsf{out}_I\}$, where $\mathsf{in}_I = \bigcup_{P_i \in I}\{\mathsf{in}_i\}$, $\mathsf{out}_I = \bigcup_{P_i \in I}\{\mathsf{out}_i\}$, and $\equiv$ denotes computational or statistical indistinguishability.*

To obtain a solution that complies with the above security definition, we employ a (n, t) threshold linear secret sharing scheme (such as [35]), using which a secret s is split into n shares. Then any $t + 1$ shares can be used to reconstruct s exactly, while possession of t or fewer shares information-theoretically reveals no information about s. Using such techniques for data protection, we obtain the following secure protocol for BFS computation:

1. The input party/parties distribute shares of the graph (in the form of adjacency matrix M) to the computational parties and indicate what node is the source.
2. If the node labels are not random, the computational parties use a random permutation to randomize the graph representation.
3. The parties execute Algorithm 1 on shares using the following building blocks for computing with shares:
   (a) addition and subtraction of shares, multiplication of a shared value by a known or shared value constitute elementary operations;
   (b) comparisons and equality tests are implemented using protocols LT and Eq, respectively, from [13];
   (c) random permutation is achieved by first choosing random labels for all elements to be permuted, e.g., by calling $\mathsf{PRandFld}$ from [13], and obliviously sorting the elements using the chosen random labels as the sorting key as, e.g., shown in [5].
4. The parties send their shares of the result to the output party/parties.

Security of the above solution can be shown based on the facts that (1) the building blocks have been previously proven secure in the same security model and (2) composition of secure building blocks leads to the security of the overall solution using the composition theorem from [12]. In more detail, to build a simulator as specified in Definition 2, we can invoke the corresponding simulators of the building blocks to simulate the view of a coalition of computational parties which is indistinguishable from the real protocol execution.

7. CONCLUSIONS

In this work we design data-oblivious algorithms for several classical graph problems, namely, breadth-first search, single-source single-destination shortest path, minimum spanning tree, and maximum flow. The algorithms are designed to work on protected data and have applications to secure computation and outsourcing. The algorithms have optimal, or close to optimal, performance for dense graphs or when adjacency matrix is used to represent the input graphs. It is an open problem to design efficient data-oblivious algorithms for sparse graphs.

8. ACKNOWLEDGMENTS

This work was supported in part by grants CNS-1223699 from the National Science Foundation, FA9550-13-1-0066 from the Air Force Office of Scientific Research, and FA8750-11-2-0218 from Air Force Research Laboratory. Any opinions, findings, and conclusions or recommendations expressed in this publication are those of the authors and do not necessarily reflect the views of the funding agencies.

9. REFERENCES

[1] M. Ajtai. Oblivious RAMs without cryptographic assumptions. In *STOC*, pages 181–190, 2010.

[2] L. Arge, M. Bender, E. Demaine, B. Holland-Minkley, and J. Munro. Cache-oblivious priority queue and graph algorithm applications. In *STOC*, pages 268–276, 2002.

[3] L. Arge, M. Bender, E. Demaine, B. Holland-Minkley, and J. Munro. An optimal cache-oblivious priority queue and its application to graph algorithms. *SIAM Journal on Computing*, pages 1672–1695, 2007.

[4] M. Atallah, K. Frikken, and S. Wang. Private outsourcing of matrix multiplication over closed semi-rings. In *SECRYPT*, pages 136–144, 2012.

[5] M. Blanton and E. Aguiar. Private and oblivious set and multiset operations. In *ASIACCS*, 2012.

[6] M. Blanton, M. Atallah, K. Frikken, and Q. Malluhi. Secure and efficient outsourcing of sequence comparisons. In *ESORICS*, pages 505–522, 2012.

[7] D. Boneh, D. Mazieres, and R. Popa. Remote oblivious storage: Making oblivious RAM practical. Technical Report MIT-CSAIL-TR-2011-018, MIT, 2011.

[8] J. Brickell and V. Shmatikov. Privacy-preserving graph algorithms in the semi-honest model. In *ASIACRYPT*, pages 236–252, 2005.

[9] G. Brodal. Cache-oblivious algorithms and data structures. In *SWAT*, pages 3–13, 2004.

[10] G. Brodal, R. Fagerberg, U. Meyer, and N. Zeh. Cache-oblivious data structures and algorithms for undirected breadth-first search and shortest paths. In *SWAT*, pages 480–492, 2004.

[11] C. Cachin, S. Micali, and M. Stadler. Computationally private information retrieval with polylog communication. In *EUROCRYPT*, pages 402–414, 1999.

[12] R. Canetti. Security and composition of multiparty cryptographic protocols. *Journal of Cryptology*, 13(1):143–202, 2000.

[13] O. Catrina and S. de Hoogh. Improved primitives for secure multiparty integer computation. In *SCN*, pages 182–199, 2010.

[14] B. Chor, O. Goldreich, E. Kushilevitz, and M. Sudan. Private information retrieval. In *FOCS*, pages 41–50, 1995.

[15] B. Chor, E. Kushilevitz, O. Goldreich, and M. Sudan. Private information retrieval. *Journal of the ACM*, pages 965–981, 1998.

[16] T. Cormen, C. Leiserson, R. Rivest, and C. Stein. *Introduction to algorithms*. The MIT Press, 2009.

[17] I. Damgård, S. Meldgaard, and J. Nielsen. Perfectly secure oblivious RAM without random oracles. In *TCC*, pages 144–163, 2011.

[18] D. Eppstein, M. Goodrich, and R. Tamassia. Privacy-preserving data-oblivious geometric algorithms for geographic data. In *GIS*, pages 13–22, 2010.

[19] C. Erway, A. Kupcu, C. Papamanthou, and R. Tamassia. Dynamic provable data possession. In *CCS*, pages 213–222, 2009.

[20] P. Failla. Heuristic search in encrypted graphs. In *SECURWARE*, pages 82–87, 2010.

[21] K.B. Frikken and P. Golle. Private social network analysis: how to assemble pieces of a graph privately. In *ACM Workshop on Privacy in the Electronic Society (WPES)*, pages 89–98, 2006.

[22] C. Gentry and Z. Ramzan. Single-database private information retrieval with constant communication rate. In *ICALP*, pages 803–815, 2005.

[23] O. Goldreich. Towards a theory of software protection and simulation by oblivious RAMs. In *STOC*, pages 182–194, 1987.

[24] O. Goldreich and R. Ostrovsky. Software protection and simulation on oblivious RAMs. *Journal of the ACM*, 43(3):431–473, 1996.

[25] M. Goodrich. Randomized Shellsort: A simple oblivious sorting algorithm. In *SODA*, pages 1262–1277, 2010.

[26] M. Goodrich. Data-oblivious external-memory algorithms for the compaction, selection, and sorting of outsourced data. In *SPAA*, pages 379–388, 2011.

[27] M. Goodrich and M. Mitzenmacher. Privacy-preserving access of outsourced data via oblivious RAM simulation. In *ICALP*, pages 576–587, 2011.

[28] M. Goodrich, O. Ohrimenko, and R. Tamassia. Data-oblivious graph drawing model and algorithms. Arxiv preprint arXiv:1209:0756, 2012.

[29] E. Kushilevitz, S. Lu, and R. Ostrovsky. On the (in) security of hash-based oblivious RAM and a new balancing scheme. In *SODA*, pages 143–156, 2012.

[30] E. Kushilevitz and R. Ostrovsky. Replication is not needed: Single database, computationally-private information retrieval. In *FOCS*, pages 364–373, 1997.

[31] H. Lipmaa. An oblivious transfer protocol with log-squared total communication. In *Information Security Conference (ISC)*, pages 314–328, 2005.

[32] K. Mouratidis and M. Yiu. Shortest path computation with no information leakage. *VLDB Endowment*, 5(8):692–703, 2012.

[33] R. Ostrovsky. Efficient computation on oblivious RAMs. In *STOC*, pages 514–523, 1990.

[34] B. Pinkas and T. Reinman. Oblivious RAM revisited. In *CRYPTO*, pages 502–519, 2010.

[35] A. Shamir. How to share a secret. *Communications of the ACM*, 22(11):612–613, 1979.

[36] E. Shi, E. Stefanov, and M. Li. Oblivious RAM with $O((\log N)^3)$ worst-case cost. In *ASIACRYPT*, pages 197–214, 2011.

[37] E. Stefanov, E. Shi, and D. Song. Towards practical oblivious RAM. In *Network and Distributed System Security Symposium (NDSS)*, 2012.

[38] S. Wang, X. Ding, R. Deng, and F. Bao. Private information retrieval using trusted hardware. In *ESORICS*, pages 49–64, 2006.

[39] P. Williams and R. Sion. Usable PIR. In *Network and Distributed System Security Symposium (NDSS)*, 2008.

[40] P. Williams, R. Sion, and B. Carbunar. Building castles out of mud: Practical access pattern privacy and correctness on untrusted storage. In *CCS*, pages 139–148, 2008.

[41] A. Yao. How to generate and exchange secrets. In *FOCS*, pages 162–167, 1986.

SecLaaS: Secure Logging-as-a-Service for Cloud Forensics

Shams Zawoad
University of Alabama at Birmingham
Birmingham, Alabama
35294-1170
zawoad@cis.uab.edu

Amit Kumar Dutta
University of Alabama at Birmingham
Birmingham, Alabama
35294-1170
adutta@cis.uab.edu

Ragib Hasan
University of Alabama at Birmingham
Birmingham, Alabama
35294-1170
ragib@cis.uab.edu

ABSTRACT

Cloud computing has emerged as a popular computing paradigm in recent years. However, today's cloud computing architectures often lack support for computer forensic investigations. Analyzing various logs (e.g., process logs, network logs) plays a vital role in computer forensics. Unfortunately, collecting logs from a cloud is very hard given the black-box nature of clouds and the multi-tenant cloud models, where many users share the same processing and network resources. Researchers have proposed using log API or cloud management console to mitigate the challenges of collecting logs from cloud infrastructure. However, there has been no concrete work, which shows how to provide cloud logs to investigator while preserving users' privacy and integrity of the logs. In this paper, we introduce Secure-Logging-as-a-Service (*SecLaaS*), which stores virtual machines' logs and provides access to forensic investigators ensuring the confidentiality of the cloud users. Additionally, *SeclaaS* preserves proofs of past log and thus protects the integrity of the logs from dishonest investigators or cloud providers. Finally, we evaluate the feasibility of the scheme by implementing *SecLaaS* for network access logs in OpenStack – a popular open source cloud platform.

Categories and Subject Descriptors

C.2.4 [**Computer Communication Networks**]: Distributed Systems—*Cloud Computing*; K.6.m [**Management of Computing and Information Systems**]: Miscellaneous

General Terms

Security

Keywords

Cloud Forensics, Forensic Investigation, Cloud Security, Logging-as-a-Service

1. INTRODUCTION

Cloud computing offers infinite infrastructure resources, very convenient pay-as-you-go service, and low cost computing. As a result, cloud computing has become one of the most dominant computing paradigms in recent years. Today, small and high level companies are attracted to cloud computing because it does not require any kind of local infrastructure setup, and, at the same time, it is highly cost effective. According to Khajeh-hossainei, an organization can save 37% of its cost just by moving their IT infrastructure from an outsourced data center to Amazon's cloud [17]. Market Research Media stated in one of their recent reports that the cloud computing market is expected to grow at a 30% compound annual growth rate and will reach $270 billion in 2020 [20]. Garner Inc. states that the strong growth of cloud computing will bring $148.8 billion revenue by 2014 [12]. From the research work of INPUT, it is clear that Cloud computing is equally popular in both Government and private industry; their report identifies that the federal cloud market is expected to expand to $800 million by 2013 [15].

Cloud computing opens a new horizon of computing for business and IT organizations. However, at the same time, malicious individuals can easily exploit the power of cloud computing. Attackers can attack applications running inside the cloud. Alternatively, they can launch attacks from machines inside the cloud. These issues are the primary concerns of *Cloud Forensics*. An annual report of the Federal Bureau of Investigation (FBI) states that, the size of the average digital forensic case is growing 35% per year in the United States. From 2003 to 2007, it increased from 83GB to 277 GB [11]. As a result, forensic experts are devising new techniques for digital forensics. There are several forensics analysis schemes and tools available in market. Unfortunately, none of them are suitable for the dynamic nature of cloud computing. Many of the implicit assumptions made in regular forensics analysis (e.g., physical access to hardware) are not valid for cloud computing. Hence, for cloud infrastructure, a special branch of digital forensics has been brought up by researchers - *Cloud Forensics*. Cloud forensics offers new challenges and has opened new research problems for security and forensics experts, which are important from both technical and legal point of view.

The process of digital forensics starts with acquiring the digital evidence. In a cloud, the evidence could be the image of virtual machines, files stored in cloud storage, and logs provided by cloud service providers (CSP). However, collecting these evidences, specially logs from cloud infrastructure, is extremely difficult because cloud users or investigators have very little control over the infrastructure. Currently, to collect logs from cloud, investigators are dependent on the CSP. Investigators need to issue a subpoena to the

CSP to acquire the logs of a particular user. However, they need to believe the CSPs blindly, as there is no way to verify whether the CSPs are providing valid logs or not. Moreover, if an adversary shuts down the virtual machine (VM) she is using, there is no way to collect logs from the terminated VM.

To overcome the challenges of acquiring logs from cloud infrastructure, Bark et al. proposed that the CSPs can provide network, process, and access logs to customers by a read-only API [5]. Dykstra et al. recommended a cloud management plane for using in Infrastructure-as-a-Service model [10]. However, they did not show how we can practically implement those schemes. Additionally, log information is highly sensitive and user's privacy issues are directly related to it. Previous studies do not provide a secure way of revealing the logs while maintaining user privacy. Moreover, it is vital to ensure that logs are not tampered with before exposing to investigators. For a successful forensic scheme based on logs, these issues must be resolved in a secure and trustworthy manner.

In this paper, we take the first step towards exposing a publicly available secure log service. This service can be used by forensic investigators to identify malicious activities that took place in virtual machines of a cloud system.

To illustrate the specific problem we look at, we present the following hypothetical scenario:

Alice is a successful businesswoman who runs a shopping website in cloud. The site serves a number of customers every day and her organization generates a significant amount of profit from it. Therefore, if the site is down even for a few minutes, it will seriously hamper not only their profit but also the goodwill. Mallory, a malicious attacker decided to attack Alice's shopping website. She rented some machines in cloud and launched a Distributed Denial of Service (DDoS) attack to the shopping website using those rented machines. As a result, the site was down for an hour, which had quite negative impact on Alice's business. Consequently, Alice asked a forensic investigator to investigate the case. The investigator found that Alice's website records the visiting customer's IP address. Analyzing the visiting customers records, the investigator found that Alice's website was flooded by some IP addresses which are owned by a cloud service provider. Eventually, the investigator issued a subpoena to the corresponding cloud provider to provide him the network logs for those particular IP addresses. On the other hand, Mallory managed to collude with the cloud provider after the attack. Therefore, while providing the logs to the investigator, the cloud provider supplied tampered log to the investigator, who had no way to verify the correctness of the logs. Under this circumstance, Mallory will remain undetected. Even if the cloud provider was honest, Mallory could terminate her rented machines and left no traces of the attack. Hence, the cloud provider could not give any useful logs to the investigator.

To mitigate the challenges discussed in the above scenario, we propose the notion of Secure-Logging-as-a-Service (SecLaaS) in this paper.

Contributions: The contributions of this paper are as follows:

1. We propose a scheme of revealing cloud users' logs for forensics investigation while preserving the confidentiality of users' logs from malicious cloud employee or external entity;
2. We introduce Proof of Past Log (PPL) – a tamper evident scheme to prevent the cloud service provider or investigators from manipulating the logs after-the-fact.
3. We evaluate the proposed scheme using a open source cloud computing platform.

Organization: The rest of this paper is organized as follows. Section 2 provides some background information and challenges of cloud forensics in terms of logging. Section 3 describes the adversary's capabilities and possible attacks on logging-as-a-service. Section 4 presents our SecLaaS scheme and Section 5 provides security analysis of the scheme. In Section 6, we provide the implementation and performance evaluation of our scheme on an open source cloud software, OpenStack. Section 7 discusses the usability of our proposed schemes. In Section 8, we provide an overview of related research about logging in cloud forensics, and finally, we conclude in Section 9.

2. BACKGROUND AND CHALLENGES

With the increasing popularity of cloud computing, there is a significant interest in the law-enforcement community to extend digital forensics techniques in the context of a cloud. In this section, we present the definitions of digital forensics and cloud forensics, motivation behind our work, and discuss the challenges of logging-as-a-service for cloud forensics.

2.1 Digital Forensics

Digital forensics is the process of preserving, collecting, confirming, identifying, analyzing, recording, and presenting crime scene information. Wolfe defines digital forensics as *"A methodical series of techniques and procedures for gathering evidence, from computing equipment and various storage devices and digital media, that can be presented in a court of law in a coherent and meaningful format"* [30]. According to a definition by NIST [16], computer forensics is an applied science to identify an incident, collection, examination, and analysis of evidence data. While doing so, maintaining the integrity of the information and strict chain of custody for the data is mandatory. Several other researchers define computer forensics as the procedure of examining computer system to determine potential legal evidence [18, 24].

2.2 Cloud Forensics

Cloud forensics can be defined as applying computer forensics procedures in a cloud computing environment. As cloud computing is based on extensive network access, and as network forensics handles forensic investigation in private and public network, Ruan et al. defined cloud forensics as a subset of network forensics [25]. They also identified three dimensions in cloud forensics – technical, organizational, and legal. Cloud forensics procedures will vary according to the service and deployment model of cloud computing. For Software-as-a-Service (SaaS) and Platform-as-a-Service (PaaS), we have very limited control over process or network monitoring. Whereas, we can gain more control in Infrastructure-as-a-Service (IaaS) and can deploy some forensic friendly logging mechanism. The first three steps of computer forensics, identification, collection, and organization of evidence will vary for different service and deployment model. For example, the evidence collection procedure of SaaS and IaaS will not be same. For SaaS, we solely depend on the CSP to get the application log, while in IaaS, we can acquire the virtual machine image from the customer and can enter into examination and analysis phase. On the other hand, in the private deployment model, we have physical access to the digital evidence, but we merely can get physical access to public deployment model.

2.3 Motivation

Though cloud computing offers numerous opportunities to different level of consumers, many security issues of cloud environment have not been resolved yet. According to a recent IDCI survey, 74% of IT executives and CIO's referred to security as the main reason to prevent their migration to the cloud services model [8]. Some recent and well-publicized attacks on cloud computing platform justify the concern with security. For example, a botnet attack on Amazon's cloud infrastructure was reported in 2009 [2].

Besides attacking cloud infrastructure, adversaries can use the cloud to launch attack on other systems. For example, an adversary can rent hundreds of virtual machines to launch a Distributed Denial of Service (DDoS) attack. After a successful attack, she can erase all the traces of the attack by turning off the virtual machines. A criminal can also keep her secret files (e.g., child pornography, terrorist documents) in cloud storage and can destroy all her local evidence to remain clean. When law enforcement investigates such a suspect, the suspect can deny launching a DDoS attack. At present, there is no way to claim that an adversary access a certain network at a given time.

Researchers are working to protect the cloud environment from different types of attacks. However, in case of an attack, we also need to investigate the incident, i.e., we need to carry out a digital forensic investigation in the cloud. Besides protecting the cloud, we need to focus on this issue. Unfortunately, there has been little research on adapting digital forensics for use in cloud environments. In this paper, we address this problem, which has significant real-life implications in law enforcement investigating cybercrime and terrorism.

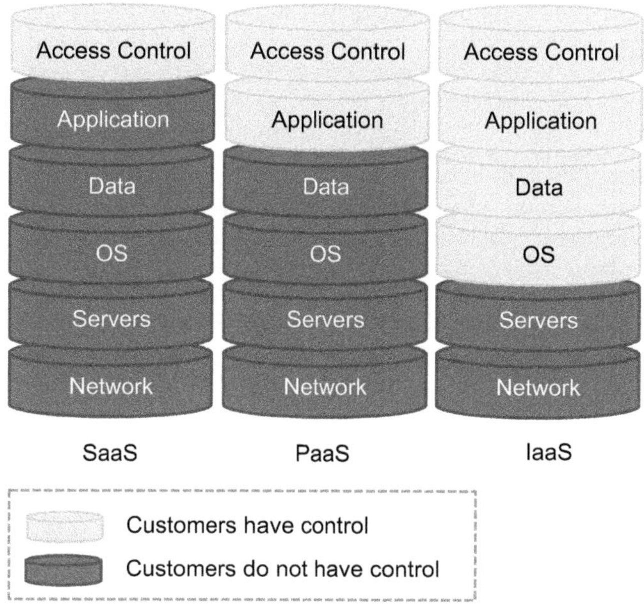

Figure 1: Customers' control over different layers in different service model

2.4 Challenges

Analyzing logs from different processes plays a vital role in digital forensic investigation. Process logs, network logs, and application logs are really useful to identify a malicious user. However, gathering this crucial information in cloud environment is not as simple as it is in privately owned computer system, sometimes even impossible. The inherent characteristics of cloud have made the forensic log-analysis a nightmare for the forensic investigators. It is very difficult to collect and prove the validity of the logs to the court authority. For example, how can an investigator collect network logs of malicious VMs, which have been already terminated by the attacker after launching a DDoS attack last month? We must find secure techniques for storing and providing logs to investigators, which also need to be admissible in a court of law as valid evidence. Many things can complicate the log collection process. A malicious CSP can change the logs while providing the logs to investigators. Clients may question the integrity of any such logs, claiming that the forensic investigators or the prosecution and the CSP have colluded to plant evidence in the cloud. The following reasons also make the log collection and providing the proof of the logs challenging in cloud.

Reduced Level of Control, and Dependence on the CSP: One of the challenges of collecting logs securely from cloud is the users' or investigators' reduced level of control over the cloud environment. In traditional computer forensics, the investigators have full control over the evidence (e.g., router logs, process logs, hard disk). Currently, to acquire the logs, we extensively depend on the CSPs. The availability of the logs varies depending on the service model. Figure 1 shows the control of customers in different layers for the three different service models – IaaS, PaaS, and SaaS. From the figure, we can observe that cloud users have highest control in IaaS and least control in SaaS. This physical inaccessibility of the evidence and lack of control over the system make evidence acquisition a challenging task in cloud forensics. In SaaS, customers do no get any log of their system, unless the CSP provides the logs. In PaaS, it is only possible to get the application log from the customers. To get the network log, database log, or operating system log we need to depend on the CSP. For example, Amazon does not provide load balancer log to the customers [3]. In a recent research work, Marty mentioned that he was unable to get MySql log data from Amazon's Relational Database Service [21]. In IaaS, customers do not have the network or process logs. Several other problems come along with the less control issue. For example, we need to depend on the cloud service providers for evidence acquisition, which in turn brings the honesty issue of the CSP's employee, who is not a certified forensic investigator. CSPs can always tamper the logs as they have the full control over the generated logs. Additionally, CSPs are not always obligated to provide all the necessary logs.

Decentralization. In cloud infrastructure, log information is not located at any single centralized log server; rather logs are decentralized among several servers. Multiple users' log information may be co-located or spread across multiple servers. Moreover, there are several layers and tiers in cloud architecture. Logs are generated in each tier. For example, application, network, operating system, and database – all of these layers produce valuable logs for forensic investigation. Collecting logs from these multiple servers and layers and providing it to investigators in a secure way is extremely challenging.

Accessibility of Logs. The logs generated in different layers are required to be accessible to different stakeholders of the system, e.g., system administrator, forensic investigator, and developer. System administrators need relevant log to troubleshoot the system; developers need the required log to fix the bug of the application; forensic investigators need logs, which can help in their investigation. Hence, there should be some access control mechanism, so that everybody will get what they need exactly – nothing more, nothing less and obviously, in a secure way. We should not expect that a malicious cloud employee, who can violate the privacy of the users gets access to users' log information.

Multi-tenancy: In cloud computing, multiple virtual machines

(VM) can share the same physical infrastructure, i.e., log for multiple customers may be co-located. The nature of this infrastructure is different from the traditional single owner computer system. Hence, while collecting logs for one user, other users' data can be mingled with the log evidence. An alleged user can claim that the log contains information of other users, not her. The investigator then needs to prove it to the court that the provided logs indeed belongs to the malicious user. Moreover, we need to preserve the privacy of the other tenants. For both of these issues, collecting and providing logs to the investigator is challenging in cloud paradigm.

Chain of custody: The chain of custody is one of the most vital issues in traditional digital forensic investigation. The chain of custody should clearly depict how the evidence was collected, analyzed, and preserved in order to be presented as admissible evidence in court [29]. In traditional forensic procedure, it starts with gaining the physical control of the evidence, e.g., computer, hard disk. However, in cloud forensics, this step is not possible due to the multi jurisdictional laws, procedures, and proprietary technology in cloud environment [27, 14]. Collecting logs from cloud infrastructure, analyzing, and presenting the proof of logs need to follow this chain of custody. We must clarify certain things to maintain the chain of custody, e.g., how the logs were generated and stored, and who had the access to the logs.

Presentation: The final step of digital forensic investigation is presentation, where an investigator accumulates his findings and presents to the court as the evidence of a case. Challenges also lie in this step of cloud forensics [22]. Proving the integrity of network and process logs in front of jury for traditional computer forensics is relatively easy, compared to the complex structure of cloud computing. Presenting the logs and the proofs of integrity of the logs to the court in an admissible way is challenging for cloud computing.

3. THREAT MODEL

In this section, we describe the attacker's capability, possible attacks on logs, and properties of our proposed system. Before describing the threat model, we first define the important terms to clarify the threat model.

3.1 Definition of terms

- *User:* A user is a customer of the cloud service provider (CSP), who uses the CSP's storage service. A user can be malicious or honest.
- *Log:* A log can be the network log, process log, operating system logs, or any other logs generated in cloud for a VM.
- *Proof of Past Logs (PPL):* The *PPL* contains the proof of logs to verify whether some logs belong to a particular user or not.
- *Investigator:* An investigator is a professional forensic investigator, who needs to collect necessary logs from cloud infrastructure in case of any malicious incident.
- *CSP:* The Cloud Service Provider (CSP) will generate the *PPL* and give access to the logs to users and investigators through an API or management console.
- *Log Chain (LC):* The *LC* maintains the correct order of the logs. From the *LC*, it can be verified that the CSP or the investigators provide logs in the actual order of log generation.
- *Auditor:* Most likely, the auditor will be the court authority that will verify the correctness of the logs from the *PPL* and *LC*.
- *Intruder:* An intruder can be any malicious person including a employee of the CSP, who wants to reveal user's activity from the *PPL* or from the log storage.

3.2 Attacker's Capability

In our threat model, we assume that the users and the investigators do not trust the CSPs, and both of them can be malicious. We assume that a CSP is honest at the time of publishing the *PPL* and *LC*. However, during the investigation, CSP can collude with a user or an investigator and provide tampered logs, for which the *PPL* and *LC* have already been published. User, investigator, and CSP can collude with each other to provide fake logs to the auditor. A user cannot modify the logs by him, but he can collude with CSP to alter the logs. An investigator can present false log information to the court to frame an honest user or can collude with a malicious user to save her from accusation. The CSP can also repudiate any published *PPL*. An intruder can acquire the *PPL* of a user to learn the user's activity. A malicious cloud employee can also be an intruder.

3.3 Possible Attacks

There can be different types of attacks on providing log API. CSP can remove some crucial logs or can reorder the logs. A user can deny the ownership of any logs. Even an investigator can present invalid logs to the court. Below we mention some of the possible attacks:

- *Privacy violation:* If the CSP published the PPL publicly on the web, any malicious person can acquire the published PPL and try to learn about the logs from the proof. Even if logs are kept unpublished, an otherwise honest employee of the CSP who has access to the log storage can identify the activity of the user from the stored logs.
- *Log modification:* A dishonest CSP, while colluding with user or investigator can modify the logs, either to save a malicious user or to frame a honest user. If an investigator is not trustworthy, he can also tamper with the logs before presenting the logs to the court. There can be three types of contamination of logs:
    1. Removal of crucial logs
    2. Planting of false logs
    3. Modification of the order of the logs
- *Repudiation by CSP:* An otherwise honest CSP can deny a published PPL/LC after-the-fact.
- *Repudiation by User:* As data are co-mingled in the cloud, a malicious user can claim that the logs contain another cloud user's data.

3.4 System Property

In designing SecLaaS, our goal is to ensure the secure preservation of cloud users' logs in a persistent storage. Our mechanism should prevent any malicious party to produce a false proof of past logs *PPL*. A false *PPL* attests the presence of a log record for a user, which the user does not actually own. Once the proof has been published, the CSP can neither modify the proof nor repudiate any published proof. Additionally, we must prevent false implications by malicious forensic investigators. Based on our analysis, a secure log service for clouds should possess the following integrity and confidentiality properties:

- **I1:** The CSP cannot remove a log entry from the storage after publishing the PPL.
- **I2:** The CSP cannot change the order of a log from its actual order of generation.
- **I3:** The CSP cannot plant false log after-the-fact.
- **I4:** An investigator cannot hide or remove a log entry at the time of presenting logs to court.
- **I5:** An investigator cannot change the actual order of a log entry at the time of presenting evidences to court.
- **I6:** An Investigator cannot present phony log to the court.
- **I7:** The CSP cannot repudiate any previously published proof of logs.
- **C1:** From the published proof of log, no adversaries can recover any log.
- **C2:** A malicious cloud employee will not be able to recover logs from the log storage.

4. THE SecLaaS SCHEME

In this section, we present SecLaaS – our system for secure retrieval of logs and storage of the proof of past logs. Initially, we provide an overview of the mechanism, followed by the schematic and the protocol specific description of the system.

4.1 Overview

A VM in the cloud can attack other VMs inside the cloud or can attack a computing device outside the VM. The attacker VM can also attack the *Node Controller* (NC) to launch a side channel attack [23]. Figure 2 presents an overview of storing the logs in a secured way and making it available to forensic investigators in case of such attacks. Malicious activity of a VM can be found from various logs generated in the NC, on which the VM is running. For each running VM, our system first extracts various kinds of logs from the NC and will store in a persistent log database. Hence, terminating the VM will not prevent SecLaaS to provide useful logs during investigation. While saving logs in log database, SecLaaS ensures the integrity and confidentiality of the logs. After saving a log entry in the log database, the system will additionally store the proof of this entry in the proof database. When an investigator wants logs of a particular IP to investigate an incident, he can get the necessary logs by an API call or from the cloud management plane. In order to prove the logs as admissible evidence, the investigator can provide the proof of the logs along with the logs.

4.2 Schematic Description

In this section, we present the schematic description of our system. SecLaaS extracts log information from different log sources and generates a Log Entry *LE*. A Log Entry *LE* for network log is defined as follows:

$$LE = < FromIP, ToIP, T_{\mathrm{L}}, Port, UserId > \quad (1)$$

To ensure the confidentiality of users' log, some information of the *LE* can be encrypted using a common public key of the security agencies. The Encrypted Log Entry *ELE* is prepared as follows:

$$ELE = < E_{P_{K_a}}(ToIP, Port, UserId), FromIP, T_{\mathrm{L}} > \quad (2)$$

where P_{K_a} is the common public key of all the agencies. We cannot encrypt all the fields of the *LE* as the CSP needs to search the storage by some fields. To preserve the correct order of the log entry, we will use a hash-chain scheme. We refer the hash-chain as Log Chain (*LC*), which will be generated as follows:

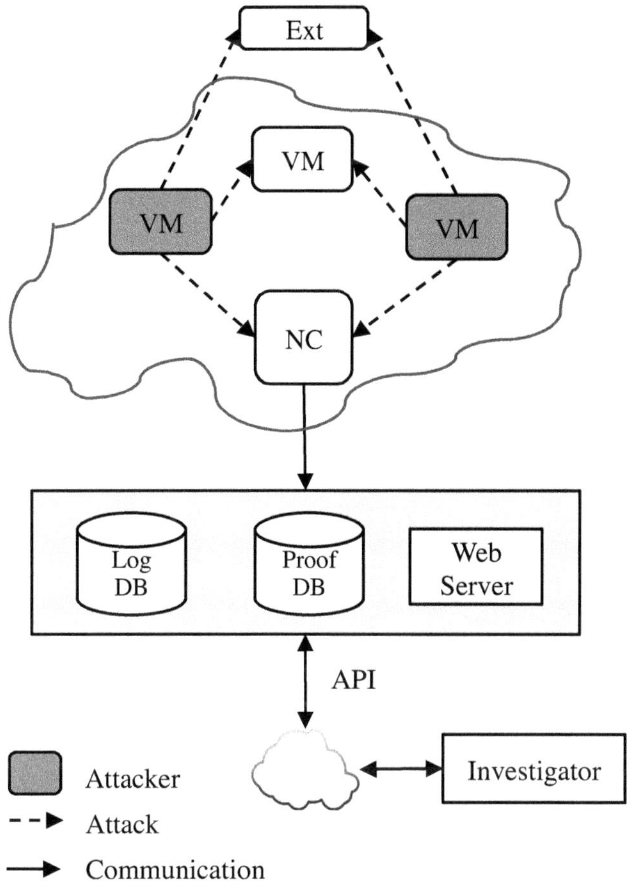

Figure 2: Overview of SecLaaS

$$LC = < H(ELE, LC_{\mathrm{Prev}}) > \quad (3)$$

where LC_{Prev} is the Log Chain *LC* of the previous entry of the persistent storage. Each entry for the persistent log database *DBLE* is constituted of ELE and LC,

$$DBLE = < ELE, LC > \quad (4)$$

The proof of this *DBLE* will be inserted into a accumulator. We denote this as Accumulator Entry *AE*. At the end of each day, CSP retrieves the AE_D of that day and generates the Proof of Past Log *PPL* as follows:

$$PPL = < H(AE_{\mathrm{D}}), S_{\mathrm{PKc}}(AE_{\mathrm{D}}), t > \quad (5)$$

where *H(AE)* is the hash of *AE*, t represents the proof generation time, and $S_{PKc}(AE)$ is the signature over *AE* using the private key of the CSP, PKc.

4.3 System Details

In this section, we present how the log insertion, proof generation, and verification of SecLaaS work. We consider the network log to describe the entire system. After generating the Log Entry *LE* the system will work for any type of logs.

4.3.1 Log and Proof Insertion

Figure 3 illustrates the detail flow of log retrieval, secured log insertion, and PPL generation and below is the details of the system.

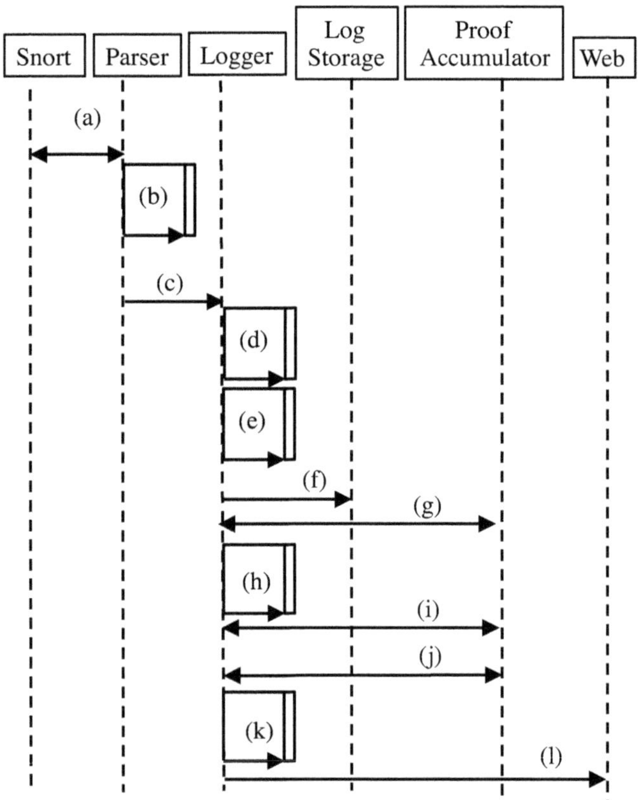

Figure 3: Process Flow of Retrieving Log and Storing the PPL

(a) The parser module first communicates with the log sources to collects different types of logs. For example, to store network log the parser listens the Snort [1].

(b) After acquiring logs from different sources, the parser then parses the collected log and generates the Log Entry *LE*.

(c) The parser module sends the Log Entry *LE* to the logger module to further process the *LE*.

(d) The logger module, upon receiving the *LE* from the parser, encrypts some confidential information using the public key of the security agencies and generates the Encrypted Log Entry *ELE*. The private key to decrypt the log can be shared among the security agencies. For the network log, some crucial information that we can encrypt includes: destination IP, port, and user information.

(e) After generating the *ELE*, the logger module then creates the Log Chain *LC* by using equation 3. In section 5, we will discuss how this can prevent reordering and deletion of logs.

(f) The logger module then prepares the entry for the log storage *DBLE* using the Encrypted Log Entry *ELE* and the Log Chain *LC*, and sends the *DBLE* to the log storage to add the new entry.

(g) After creating the database entry *DBLE*, the logger module communicates with the proof storage to retrieve the latest accumulator entry.

(h) In this step, the logger generates the proof of the database entry *DBLE*, i.e., the logger creates a new entry for the accumulator *AE* and updates the last retrieved accumulator entry with the newly generated *AE*.

(i) The logger module sends the updated accumulator entry to the accumulator storage to store the proof.

(j) At the end of each day, the logger retrieves the last accumulator entry of each static IP, which we denote as AE_D.

(k) According to equation 5, the logger then creates the Proof of Past Log *PPL* using the AE_D.

(l) After computing the Proof of Past Log *PPL*, the logger will publish the *PPL* and the public key of CSP on the web. These information can also be available by RSS feed to protect it from manipulation by the CSP after publishing the proof. We can also build a trust model by engaging other CSPs in the proof publication process. Whenever one CSP publishes a PPL, that PPL will also be shared between other CSPs. Therefore, we can get a valid proof as long as one CSP is honest.

4.3.2 Verification

When an investigator wants to investigate an incident, he will first gather the required log either by calling log API or from the cloud management console. While presenting the evidence to the court, he needs to provide the collected logs and also the proof of the logs. There will be two steps to verify the provided logs. In the first step, the auditor will verify the integrity of the proof and the individual log entry. In the next step, he will verify the order of the log.

Integrity Verification: Figure 4 shows the the process flow of individual log entry verification.

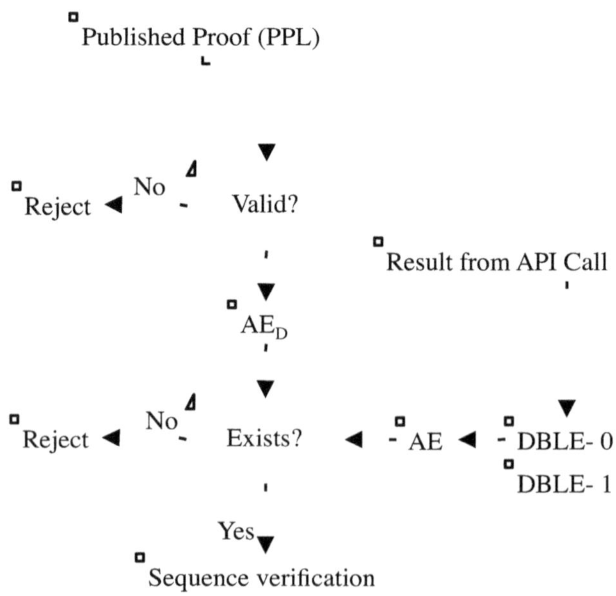

Figure 4: Log Verification Process Flow

9 [1] http://www.snort.org

The verification process starts from checking the validity of the published Proof of Past Log *PPL*. To do so, first, the auditor decrypts the $S_{PKc}(AE_D)$ using the public key of the CSP and he will get the AE_D. Then the auditor generates the hash value from the dycrypted AE_D. If the generated hash and the $H(AE_D)$ of the *PPL* matches, then the auditor accepts the *PPL* as a valid proof of log, otherwise he rejects the verification process.

In the next step, the auditor generates the Accumulator Entry *AE* for each *DBLE*. Then, he will check whether the calculated *AE* exists in the AE_D. If exists, then the auditor proceeds towards log order verification process, otherwise he rejects the provided log information.

Sequence Verification: Figure 5 illustrates the log order verification process, where we verify whether the current log (*DBLE1*) is actually after the previous log (*DBLE0*) in the original sequence of log generation. In the figure 5, *ELE0* denotes the Encrypted Log Entry *ELE* of the first log and *ELE1* represents the same for the second log. To verify the correct order, the auditor calculates the Log Chain *LCa* from the first Log Chain *LC0* and the second Encrypted Log *ELE1* according to the following equation.

$$LCa = < H(ELE1, LC0) > \qquad (6)$$

If *LCa* matches with the 2nd Log Chain *LC1* then the auditor accepts the logs, otherwise he rejects it.

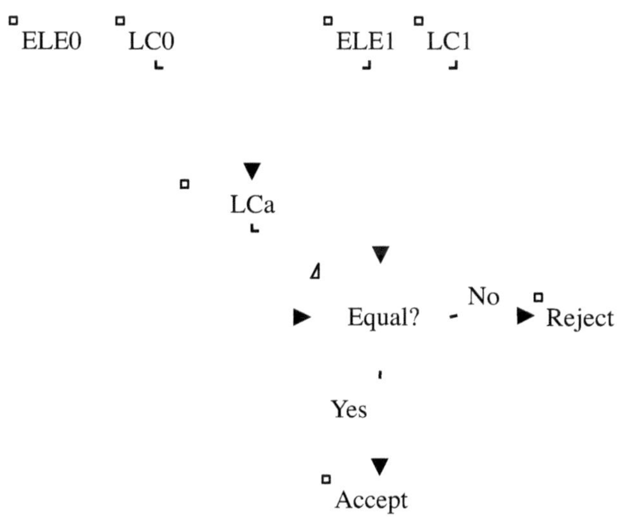

Figure 5: Log Order Verification Process Flow

5. SECURITY ANALYSIS

As CSPs have control over generating the logs and the proofs, they can always tamper with the logs. After acquiring logs through API or management console, investigators can also alter the logs before presenting it to court. Therefore, here we propose a tamper evident scheme. Any violation of the integrity and confidentiality properties, as mentioned in Section 3 can be detected during the verification process.

In our collusion model, there are three entities involved – CSP, user, and investigator. All of them can be malicious individually or can collude with each other. We denote an honest CSP as C, a dishonest CSP as $\bar{C}$, an honest user as U, a dishonest user as $\bar{U}$, an honest investigator as I, and a dishonest investigator as $\bar{I}$. Hence, there can be total eight possible combinations of collusion. Table 1 presents all the combinations of collusion, possible attacks for each collusion, and required security properties to defend that collusion. Here, we discuss how our proposed system can ensure all the security properties, which are required to protect collusion between CSP, user, and investigator.

- *I1, I2, I4, I5*: A CSP can collude with the cloud user or the investigator and can remove crucial log information. Also, while providing logs through the API or the management console, the CSP can simply hide some crucial log entries. An Investigator can also hide logs before at the time of presenting evidence to court, though he have received correct logs through the log API. However, at the verification stage, our system can detect any such removal of log entries. Let us assume that there are three log entries DBLE0, DBLE1, and DBLE2 and their proof has already been published. Now, if CSP removes DBLE1 and provides only DBLE0 and DBLE2 to the investigator, then this removal can be easily detected at the sequence verification stage. In this case, the hash of LC0 and ELE2 will not match with the LC2 because the original LC2 was calculated by hashing LC1 and ELE2. In the same way, an auditor can detect the re-ordering of logs. For example, while providing the logs to an auditor, if the CSP or investigator provides the log in DBLE0, DBLE2, DBLE1 order, then using the same technique, the auditor can identify that DBLE2 does not come after DBLE0 in actual generation order. A CSP can further try to change the DBLE2 by replacing the original LC2 with a new Log Chain value so that, in the sequence verification process, the order breaking will not be detected. However, an attempt of changing the DBLE2 will be detected during the individual log entry verification phase. The accumulator entry of the fake DBLE2 will not exist in the published Proof of Past Log PPL.

- *I3, I6*: A colluding CSP can plant false log information while providing the log to the investigator. However, if the CSP does this after publishing the proof, our system can detect these phony logs. A dishonest investigator can also try to frame an honest user by presenting fake logs to the court. Suppose, $DBLE_F$ is the fake log and the auditor generates the Accumulator Entry AE_F for this log. If it's fake, then AE_F will not be present in the AE_D of the Proof of Past Log *PPL* and the auditor can reject that incorrect log.

- *I7*: After publishing the proof of past log *PPL*, CSP cannot repudiate the published proof as the accumulator entry AE_D is signed by CSP's private key. Nobody other than the CSP can use that private key to sign the AE_D. Hence, after decrypting the signed value and generating hash on the decrypted value, if it matches with the hashed AE_D value, the CSP cannot repudiate the published value. Additionally, if the CSP comes up with a false PPL_f in place of a published *PPL*, then it will be easily detected. In that case, the $H(AE_D)$ of the published *PPL* and the $H(AE_{Df})$ of the false PPL_f will not be same. As the CSP has already signed the AE_D of the published *PPL* using its private key, it cannot deny the published value.

- *C1, C2*: To store the proof of the logs, we propose to use an accumulator function, which will ensure the C1 property, i.e., from the proof of logs, adversaries cannot recover any log. We implement our scheme using Bloom filter and One-Way Accumulator, which can ensure this property. While storing the log data in persistent storage, we propose to encrypt some crucial information e.g., user id, destination IP, etc by using a common public key of all the investigator agencies. Hence, a

Is Honest?			Notation	Attack	Required Security Properties
CSP	User	Investigator			
✓	✓	✓	C U I	No attack	None
×	✓	✓	$\bar{C}$ U I	Reveal user activity from logs	C2
✓	×	✓	C $\bar{U}$ I	Recover other cloud users' log from published proof	C1
✓	✓	×	C U $\bar{I}$	Remove, reorder, and plant fake logs	I4, I5, I6
✓	×	×	C $\bar{U}\bar{I}$	Remove, reorder, and plant fake logs	I4, I5, I6
×	✓	×	$\bar{C}$ U $\bar{I}$	Remove, reorder, plant fake logs, and repudiate published PPL	I1, I2, I3, I4, I5, I6, I7
×	×	✓	$\bar{C}\bar{U}$ I	Remove, reorder, plant fake logs, and repudiate published PPL	I1, I2, I3, I7
×	×	×	$\bar{C}\bar{U}\bar{I}$	Remove, reorder, plant fake logs, and repudiate published PPL	I1, I2, I3, I4, I5, I6, I7

Table 1: Collusion model, possible attacks and required security properties

malicious cloud employee cannot retrieve plain log information from the persistent storage; e.g., identifying the visiting IPs of a particular user will not be possible by the malicious cloud employee. In this way, our scheme can ensure the C2 property.

6. IMPLEMENTATION AND EVALUATION

In this section, we present the implementation of SecLaaS on OpenStack and performance analysis of the scheme using different types of accumulators.

6.1 Implementation

System Setup: We used Openstack[1] and Snort for testing and implementation of our project. OpenStack is an open source cloud computing software and Snort is a free lightweight network intrusion detection system. We created the virtual environments with VirtualBox (a free virtualization software)[2] running on a single Ubuntu machine. Figure 6 illustrates the system setup and below is the description of the system:

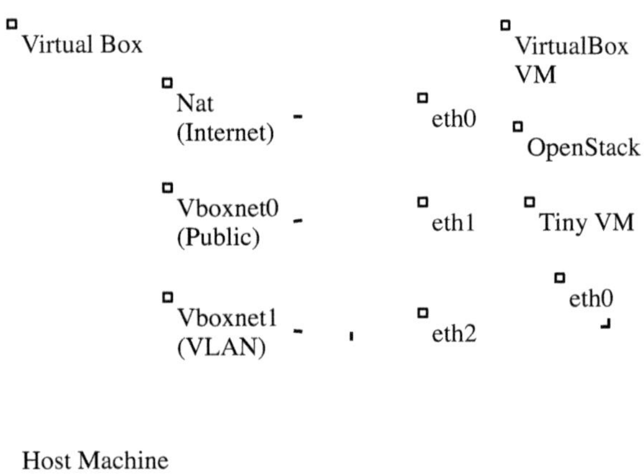

Figure 6: Prototype Environment Configuration [28]

- Host machine's hardware configuration: Intel Core I7 quad core CPU, 16 GB ram and 750 GB hard drive. Ubuntu 12.04 LTS 64-bit is used as Host Operating System.

9 [1]http://www.openstack.org
9 [2]https://www.virtualbox.org

- VirtualBox 4.1.22 r80657 for Ubuntu 12.04 LTS
- Openstack (Essex release, came out by the end of April 2012) installation in VirtualBox; for simplicity, the system had one node controller. Configuration of vitualized cloud controller: Intel 2.5Ghz Dual Core cpu, 8 GB ram and 20 GB hard drive. Ubuntu 12.04 64-bit Sever edition is used as the Operating system for Openstack setup.
- In the virtualized environment, the Cloud Controller required following network adapter configuration in VirtualBox to work properly:
  - Adapter 1: Attached to NAT- eth0 of the Cloud controller is connected here.
  - Adapter 2: Host-only network for Public interface- connected with eth1 (IP was set to 172.16.0.254, mask 255. 255.0.0, dhcp disbaled)
  - Adapter 3: Host-only network for Private (VLAN) interface connected with eth2 (IP to 11.0.0.1, mask 255. 0.0.0, dhcp disbaled)
- We used RSA (2048 bit) for signature generation and SHA-2(SHA-256) hash function for hashing.

We set up Snort in node controller to track the network activity of the virtual machines. We added two virtual machines: the first one had private IP: 11.1.0.3 and public IP: 172.16.1.1; while the other had private IP: 11.1.0.5 and public IP: 172.16.1.3. Here is a sample Snort log:

"11/19-13:43:43.222391 11.1.0.5:51215 -> 74.125.130.106:80 TCP TTL:64 TOS:0x0 ID:22101 IpLen:20 DgmLen:40 DF ***A***F Seq: 0x3EA405D9 Ack: 0x89DE7D Win: 0x7210 TcpLen: 20"

This log tells that the virtual machine with private IP 11.1.0.5 performed a http request to machine 74.125.130.160. By reverse engineering Openstack's "nova" mysql database, it is also possible to find out the static private IP and user information from a public IP. We used the references among FloatingIps, FixedIps and Instances tables to resolve the user id for a particular log entry. Figure 7 shows the relation between these three tables.

We implemented the Proof of Past Log *PPL* scheme of the SecLaaS using two accumulators. One is BloomFilter [6] and another

Figure 7: Resolving User ID from Public IP

is One-Way Accumulator [4]. The steps from (g) to (k) will work differently for the two accumulators.

BloomFilter: A Bloom filter is a probabilistic data structure with no false negatives rate, which is used to check whether an element is a member of a set or not [6]. Bloom filter stores the membership information in a bit array. Bloom filters decrease the element insertion time and membership checking time. The only drawback of the Bloom filter is the probability of finding false positives. However, we can decrease the false positive probability by using a large bit array.

To use the Bloom filter as a proof, we use one bloom filter for one static IP for each day. That means, one Bloom filter stores the proof of all the logs of one static IP for a particular day. In step (g), the logger retrieves the bloom filter from the proof storage, which holds the bit positions for the previously inserted logs of the day. In step (h), while creating the accumulator entry *AE*, the logger will generate the k number of bit positions for the database entry *DBLE* by hashing the log for k times. Then, the logger updates the previously retrieved Bloom filter with the newly generated *AE* and sends the updated Bloom filter to the proof storage. At the end of each day, the CSP will retrieve the Bloom filter entry of each static IP AE_D and create the proof of past log PPL for that day using equation 5.

In the verification phase, after verifying the validity of the published proof, the auditor will hash the log entry that he has received from the API call and calculate the bit positions of the Bloom filter. Then he will compare these bit positions with the published AE_D. If all the calculated bit positions are set in the published Bloom filter AE_D, then the verifier will be sure about the validity of the log. One single false bit position means the log entry is not valid.

One-Way Accumulator: A One-Way accumulator is a cryptographic accumulator, which is based on RSA assumption and provides the functionality of checking the membership of an element in a set [4]. This scheme works with zero false negative and false positive probability. Initially, we create the public and private values for the accumulator. The private values are two large prime numbers P and Q. The first public value is N, where N = P*Q and the second public value is a large random number which is the initial seed X.

In step (g), the logger retrieves the accumulator entry *AE*. If there is no proof entry, i.e., the *AE* is empty for an IP on a day, then the *AE* of the first *DBLE* of the day is generated using the following equation

$$AE = X^{\mathrm{H(DBLE)}} modN \quad (7)$$

where *H(DBLE)* is a numeric hash value of *DBLE*. If the retrieved *AE* is not empty, then the new *AE* will be generated using the following equation

$$AE = AE^{\mathrm{H(DBLE)}} modN \quad (8)$$

The logger module then sends the calculate *AE* to the proof storage.

At the end of the day, the logger retrieves the last accumulator entry AE_D and creates the proof of past log *PPL* for the day using equation 5. The logger needs to do some additional computation here comparing with the Bloom filter. It will generate an identity for each *DBLE* and tagged it with the *DBLE*. If there are k number of *DBLE* on a day then the identity *ID* of the ith DBLE will be calculated using the following equation

$$ID = X^{\mathrm{H(DBLE_1)\,H(DBLE_2)...\,H(DBLE_{i-1})\,H(DBLE_{i+1})....H(DBLE_k)}} \quad (9)$$

While verifying the validity of the $DBLE_i$, the verifier computes $ID^{H(DBLE_i)} mod\ N$ and compares it with AE_D. If $AE_D = ID^{H(DBLE_i)} mod\ N$, then the verifier will be sure about the validity of the log.

6.2 Evaluation

To evaluate the performance of our scheme we ran our experiment using multiple accumulators. For Bloom filter, we used two configurations: 1% false positive (FP) probability with 5000 items, and 2% FP with 10000 items. Then, for One-Way accumulator, we choose 32-bit and 64-bit P,Q, and X.

Figure 8 shows the performance analysis of log insertion including generating and updating the proof of log, i.e., the time required to complete the steps from (b) to (i) of the Figure 3. We found that for all the accumulators, time increases linearly with the increase of log size. For the two Bloom filter configurations, we noticed nearly similar time. However, we found a significant amount of increase in time when changed the one-way accumulator from 32-bit to 64-bit.

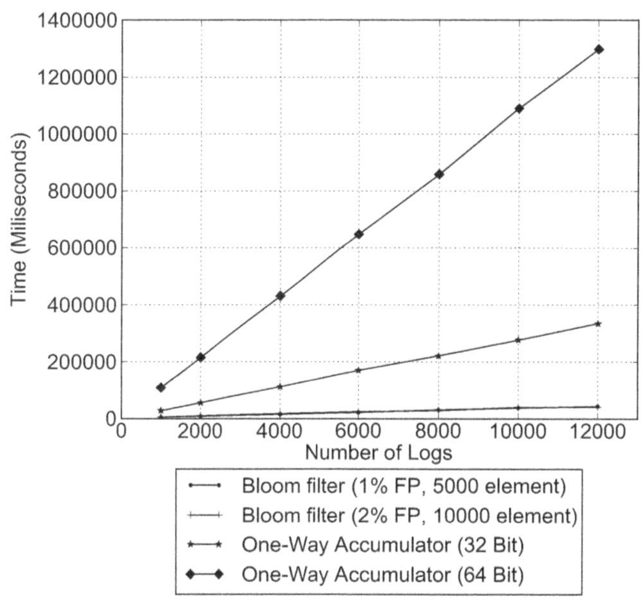

Figure 8: Performance Analysis of Log Insertion Using Different Accumulators

Figure 9 illustrates the performance analysis of generating the Proof of Past Log of a day for different accumulators, i.e., the time required to complete the steps from (j) to (l) of the figure 3. For the two different Bloom filters, we found nearly constant amount of time. On the contrary, for the One-Way accumulators, we found a linear increase in time for different sizes of logs. It is obvious, because in the later case, we need to compute the identity of each log entry using the equation 9 which has *O(n)* time complexity. [13].

Figure 10 presents the time for verifying the validity of each log. For all of the accumulators, we found nearly constant amount of time with the increase in log size. However, the time required for

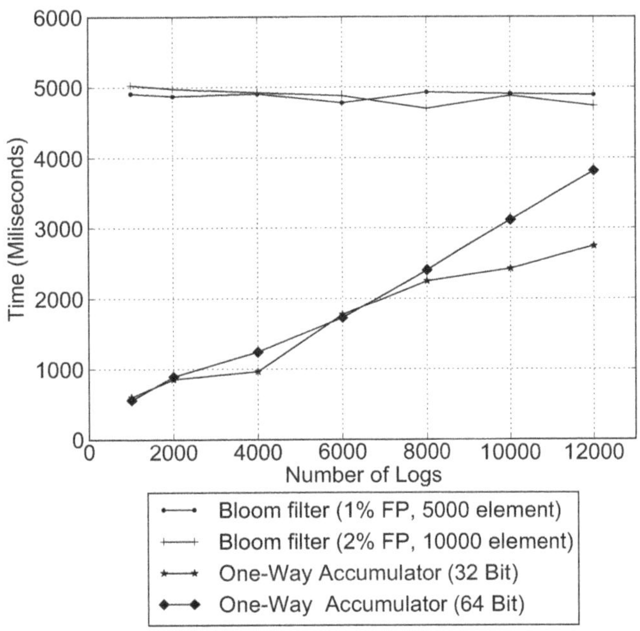

Figure 9: Performance Analysis of PPL Generation Using Different Accumulators

32-bit accumulator is higher than the Bloom filters and for 64-bit accumulator the time is significantly higher than its counterparts.

To identify the performance degradation of NC for storing log, we first run a RSA encryption on a 16 MB file for several times and measure the average execution time without running the snort logger in NC. At that time, two VMs were running on that NC. Then we start the snort service and again measure the average execution time of encryption on the same data. From these two execution times we measured the performance overhead, which is only 1.6%.

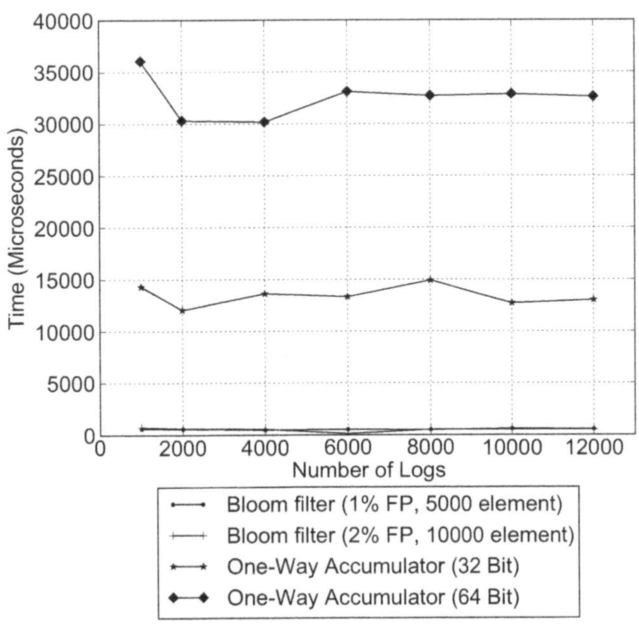

Figure 10: Performance Analysis of Log Verification Using Different Accumulators

7. DISCUSSION

Our experimental result shows that the Bloom filter outperforms the One-Way accumulator for all the tasks: log insertion, *PPL* generation, and log verification. However, Bloom filter is a probabilistic accumulator, which can state about the existence of a log in the *PPL* with certain probability. It works with zero false negative probability though. On the other hand, One-Way accumulator works with zero false positive probability. This means, in Bloom filter there is still some chance of planting false log information by the CSP or the investigator, which is not possible in One-Way accumulator. The later one always finds a valid log entry with zero false positive probability. However, we can decrease the false positive probability of the Bloom filter by allocating more space to the bit array. For example, to ensure 1% FP for 10,000 elements we need 91133 bits or 11.12 KBytes storage and to ensure 0.1% FP for the same number of elements we need 111945 bits or 13.67 KBytes storage. In our scheme we use one Bloom filter for one static IP for each day. If we have n number of static IP then for 0.1% FP and 10,000 logs we will require n * 13.67 KBytes storage each day and n*4989.55 KBytes in one year.

For the One-Way accumulator, proof requires a very small amount of storage. The 32-bit accumulator requires 19 Bytes for the final accumulator entry, whereas, the 64-bit requires 39 Bytes. However, to complete the verification in O(1) time, we need to pre-compute the identity of each log record and store it along with the logs. For a 32-bit accumulator, we require 10 Bytes for each identity and for the 64-bit the requirement is 20 Bytes. That means, for 10,000 records, we need 97.67 KBytes storage with the 32-bit accumulator and 195.35 KBytes for the 64-bit accumulator. Hence, we get a 600% increase in storage in one year for the 32-bit accumulator comparing with the 0.1% FP for 10,000 records. However, this extra storage can provide us zero false positive probability. Therefore, we need to choose whether we will go for accommodating higher storage with the One-Way accumulator or tolerating a little false positive probability with the Bloom filter. Moreover, the Bloom filter will give us better performance in all the required tasks.

8. RELATED WORKS

As logging information is one of the prime needs in forensic investigation, several researchers have explored this problem across multiple dimensions. Marty proposed a log management solution, which can solve several challenges of logging, discussed in Section 2 [21]. In his solution, after enabling logging on all infrastructure components to collect logs, he proposed to establish a synchronized, reliable, bandwidth efficient, and encrypted transport layer to transfer log from the source to a central log collector. Final step deals with ensuring the presence of the desired information in the logs. The proposed guideline tells us to focus on when to log, what to log, and how to log. The answer of when to log depends on the use-cases, such that business relevant logging, operations based logging, security (forensics) related logging, and regulatory and standards mandates. At minimum, he suggested to log the time-stamps record, application, user, session ID, severity, reason, and categorization, so that we can get the answer of what, when, who, and why (4 W). However, this work does not provide any solution for logging network usage, file metadata, process usage, and many other important sources of evidence.

As a solution of forensic investigation, Zafarullah et al. proposed logging provided by OS and the security logs [32]. In order to investigate the digital forensics in cloud, they set up cloud computing environment using Eucalyptus. Using Snort, Syslog, Log Analyzer (e.g., Sawmill), they were able to monitor the Eucalyptus behaviour and log all internal and external interaction of Eucalyptus components. For their experiment, they launched a DDoS attack from two virtual machine and analyzed bandwidth usage log and processor usage log to detect the DDoS attack. From the logs in */var/eucalyptus/jetty-request-05-09-xx* file on Cloud Controller (CC) machine, it is possible to identify the attacking machine IP, browser type, and content requested. From these logs, it is also possible to determine the total number of VMs controlled by a single Eucalyptus user and the VMs communication patterns. Their experiment shows that if the CSPs come forward to provide better logging mechanism, cloud forensics will get a better advancement.

To make the network, process and access logs available to the customer, Bark et al. proposed exposing read-only API by the CSP [5]. By using these APIs, customer can provide valuable information to investigator. In PaaS, customers have full control on their application and can log variety of access information in a configurable way. So for PaaS, they proposed a central log server, where customer can store the log information. In order to protect log data from possible eavesdropping and altering action, customers can encrypt and sign the log data before sending it to the central server. In the same context, Dykstra et al. recommended a cloud management plane, for using in IaaS model [10]. From the console panel, customers, as well as investigators can collect VM image, network, process, database logs, and other digital evidence, which cannot be collected in other ways.

Secure logging has been discussed in several research works [19, 1, 26]. However, none of these works focus on secure logging in cloud environment, specially providing secure logging as a service. Moreover, they did not consider the logger as dishonest. In the threat model of current secure logging works, researchers consider attacks on privacy and integrity from external entity. These works do not consider collusion between different entities. The closest work that we can relate to our work is a secure logging scheme proposed by Yavuz et al., which provides public verifiability of audit logs for distributed system [31]. Using their proposed scheme, time required for logging and verification increase with the number of logs. On the other hand, in our system, time required for log verification is almost constant with number of logs using various types of accumulators.

The solution proposed by Marty provided a guideline for logging criteria and answered some importation questions, e.g., what are the information we need to log, how to log and when to log. Zafarullah et al. showed that it is possible to collect necessary logs from cloud infrastructure, while Bark et al. and Dykstra et al. proposed for public API or management console to mitigate the challenges of log acquisition. However, none of them proposed any scheme of storing the logs in Cloud and making it available publicly in a secure way. Dyskstra et al. mentioned that the management console requires an extra level of trust and the same should hold for APIs. In this paper, we took the first step towards providing a solution to mitigate these challenges. Combining all the previous solutions and our scheme will drive towards making the Cloud more forensics friendly.

9. CONCLUSION AND FUTURE WORK

Logs from different sources, e.g., network, process, database are a crucial source of evidence for forensics investigation. However, collecting logs from cloud is challenging as we have very little control over clouds compared to traditional computing systems. Till now, investigators need to depend on the CSP to collect logs of different sources. To make the situation even worse, there is no way to verify whether the CSP is providing correct logs to the investigators or the investigators presenting valid logs to the court. Moreover, while providing the logs to the investigators, the CSPs need to preserve the privacy of the cloud users. Unfortunately, there has been no solution which can make the logs available to the investigators and at the same time, can preserve the confidentiality and integrity of the logs. In this paper, we proposed *SecLaaS*, which can be the solution to store and provide logs for forensics purpose securely. This scheme will allow the CSP to store the logs while preserving the confidentiality of the cloud users. Additionally, an auditor can check the integrity of the logs using the Proof of Past Log *PPL* and the Log Chain *LC*. We ran our proposed solution on OpenStack and found it practically feasible to integrate with the cloud infrastructure.

Preserving the logs and the proofs of the logs will also increase the auditability of cloud environment. Using our scheme, it is possible to store and provide any types of logs from which we can get all the activities of cloud users. Auditability is a vital issue to make the cloud compliant with the regulatory acts, e.g., Sarbanes-Oxley (SOX) [9] or The Health Insurance Portability and Accountability Act (HIPAA) [7]. Hence, implementing SecLaaS will make the cloud more compliant with such regulations, leading to widespread adoption of clouds by major businesses and healthcare organizations.

In future, we will integrate other logs besides the snort log with our proof-of-concept application. Moreover, we will continue experiment on different accumulators to find the best fitted accumulator algorithm with SecLaaS. And finally, we will implement SecLaaS as a module of OpenStack.

Acknowledgment

This research was supported by a Google Faculty Research Award, the Office of Naval Research Grant #N000141210217, the Department of Homeland Security Grant #FA8750-12-2- 0254, and by the National Science Foundation under Grant #0937060 to the Computing Research Association for the CIFellows Project.

10. REFERENCES

[1] R. Accorsi. On the relationship of privacy and secure remote logging in dynamic systems. In *Security and Privacy in Dynamic Environments*, volume 201, pages 329–339. Springer US, 2006.

[2] Amazon. Zeus botnet controller. `http://aws.amazon.com/security/security-bulletins/zeus-botnet-controller/`. [Accessed July 5th, 2012].

[3] AWS. Amazon web services. `http://aws.amazon.com`. [Accessed July 5th, 2012].

[4] J. Benaloh and M. De Mare. One-way accumulators: A decentralized alternative to digital signatures. In *Advances in CryptologyâĂŤEUROCRYPTâĂŹ93*, pages 274–285. Springer, 1994.

[5] D. Birk and C. Wegener. Technical issues of forensic investigatinos in cloud computing environments. *Systematic Approaches to Digital Forensic Engineering*, 2011.

[6] B. Bloom. Space/time trade-offs in hash coding with allowable errors. *Communications of the ACM*, 13(7):422–426, 1970.

[7] Centers for Medicare and Medicaid Services. The health insurance portability and accountability act of 1996 (hipaa). `http://www.cms.hhs.gov/hipaa/`, 1996. [Accessed July 5th, 2012].

[8] Clavister. Security in the cloud. `http://www.clavister.com/documents/resources/white-papers/clavister-whp-security-in-the-cloud-gb.pdf`. [Accessed July 5th, 2012].

[9] Congress of the United States. Sarbanes-Oxley Act. http://thomas.loc.gov, 2002. [Accessed July 5th, 2012].

[10] J. Dykstra and A. Sherman. Acquiring forensic evidence from infrastructure-as-a-service cloud computing: Exploring and evaluating tools, trust, and techniques. *DoD Cyber Crime Conference*, January 2012.

[11] FBI. Annualreport for fiscal year 2007. 2008 Regional Computer Forensics Laboratory Program, 2008. [Accessed July 5th, 2012].

[12] Gartner. Worldwide cloud services market to surpass $68 billion in 2010. http://www.gartner.com/it/page.jsp?id=1389313, 2010. [Accessed July 5th, 2012].

[13] M. Goodrich, R. Tamassia, and J. Hasić. An efficient dynamic and distributed cryptographic accumulator. *Information Security*, pages 372–388, 2002.

[14] G. Grispos, T. Storer, and W. Glisson. Calm before the storm: The challenges of cloud computing in digital forensics. *International Journal of Digital Crime and Forensics (IJDCF)*, 2012.

[15] INPUT. Evolution of the cloud: The future of cloud computing in government. http://iq.govwin.com/corp/library/detail.cfm?ItemID=8448&cmp=OTC-cloudcomputingma042009, 2009. [Accessed July 5th, 2012].

[16] K. Kent, S. Chevalier, T. Grance, and H. Dang. Guide to integrating forensic techniques into incident response. *NIST Special Publication*, pages 800–86, 2006.

[17] A. Khajeh-Hosseini, D. Greenwood, and I. Sommerville. Cloud migration: A case study of migrating an enterprise it system to iaas. In *proceedings of the 3rd International Conference on Cloud Computing (CLOUD)*, pages 450–457. IEEE, 2010.

[18] D. Lunn. Computer forensics–an overview. *SANS Institute*, 2002, 2000.

[19] D. Ma and G. Tsudik. A new approach to secure logging. *Trans. Storage*, 5(1):2:1–2:21, Mar. 2009.

[20] Market Research Media. Global cloud computing market forecast 2015-2020. http://www.marketresearchmedia.com/2012/01/08/global-cloud-computing-market/. [Accessed July 5th, 2012].

[21] R. Marty. Cloud application logging for forensics. In *In proceedings of the 2011 ACM Symposium on Applied Computing*, pages 178–184. ACM, 2011.

[22] D. Reilly, C. Wren, and T. Berry. Cloud computing: Pros and cons for computer forensic investigations. 2011.

[23] T. Ristenpart, E. Tromer, H. Shacham, and S. Savage. Hey, you, get off of my cloud: exploring information leakage in third-party compute clouds. In *Proceedings of the 16th ACM conference on Computer and communications security*, pages 199–212. ACM, 2009.

[24] J. Robbins. An explanation of computer forensics. *National Forensics Center*, 774:10–143, 2008.

[25] K. Ruan, J. Carthy, T. Kechadi, and M. Crosbie. Cloud forensics: An overview. In *proceedings of the 7th IFIP International Conference on Digital Forensics*, 2011.

[26] B. Schneier and J. Kelsey. Secure audit logs to support computer forensics. *ACM Trans. Inf. Syst. Secur.*, 2(2):159–176, May 1999.

[27] M. Taylor, J. Haggerty, D. Gresty, and R. Hegarty. Digital evidence in cloud computing systems. *Computer Law & Security Review*, 26(3):304–308, 2010.

[28] Tikal. Experimenting with OpenStack Essex on Ubuntu 12.04 LTS under VirtualBox. http://bit.ly/LFsVUY, 2012. [Accessed November 30th, 2012].

[29] J. Vacca. *Computer forensics: computer crime scene investigation*, volume 1. Delmar Thomson Learning, 2005.

[30] J. Wiles, K. Cardwell, and A. Reyes. *The best damn cybercrime and digital forensics book period.* Syngress Media Inc, 2007.

[31] A. Yavuz and P. Ning. Baf: An efficient publicly verifiable secure audit logging scheme for distributed systems. In *Computer Security Applications Conference, 2009. ACSAC '09. Annual*, pages 219 –228, dec. 2009.

[32] Z. Zafarullah, F. Anwar, and Z. Anwar. Digital forensics for eucalyptus. In *Frontiers of Information Technology (FIT)*, pages 110–116. IEEE, 2011.

An Empirical Study on the Software Integrity of Virtual Appliances: Are You Really Getting What You Paid For?

Jun Ho Huh
University of Illinois at Urbana-Champaign, USA
jhhuh@illinois.edu

Mirko Montanari
University of Illinois at Urbana-Champaign, USA
mmontan2@illinois.edu

Derek Dagit
University of Illinois at Urbana-Champaign, USA
dagit@illinois.edu

Rakesh B. Bobba
University of Illinois at Urbana-Champaign, USA
rbobba@illinois.edu

Dong Wook Kim
University of Illinois at Urbana-Champaign, USA
kim628@illinois.edu

Yoonjoo Choi
Dartmouth College, USA
yoonjoo@cs.dartmouth.edu

Roy Campbell
University of Illinois at Urbana-Champaign, USA
rhc@illinois.edu

ABSTRACT

Virtual appliances (VAs) are ready-to-use virtual machine images that are configured for specific purposes. For example, a virtual machine image that contains all the software necessary to develop and host a JSP-based website is typically available as a "Java Web Starter" VA. Currently there are many VA repositories from which users can download VAs and instantiate them on Infrastructure-as-a-Service (IaaS) clouds, allowing them to quickly launch their services. This marketplace, however, lacks adequate mechanisms that allow users to a priori assess whether a specific VA is really configured with the software that it is expected to be configured with. This paper evaluates the integrity of software packages installed on real-world VAs, through the use of a software whitelist-based framework, and finds that indeed there is a lot of variance in the software integrity of packages across VAs. Analysis of 151 Amazon VAs using this framework shows that about 9% of real-world VAs have significant numbers of software packages that contain unknown files, making them potentially untrusted. Virus scanners flagged just half of the VAs in that 9% as malicious, demonstrating that virus scanning alone is not sufficient to help users select a trustable VA and that a priori software integrity assessment has a role to play.

Categories and Subject Descriptors

D.4.6 [**Software**]: OPERATING SYSTEMS—*Security and Protection*

ASIA CCS'13, May 8–10, 2013, Hangzhou, China.

General Terms

Security, Experimentation, Measurement, Verification

Keywords

IaaS, Virtual appliances, Whitelists, Software integrity

1. INTRODUCTION

Cloud computing has become popular over the years as organizations have tried to reduce the complexities and costs associated with deploying and maintaining an internal IT infrastructure. Infrastructure-as-a-Service (IaaS) cloud model [24], where users can build and configure virtual computing infrastructure by renting computing platform resources (e.g., [1, 6]) in the form of virtual machines, is a popular cloud computing service model. On these rented virtual machines, users have administrative privileges to install any piece of software they want and can configure them to meet their needs. Many services (e.g., [2, 10, 4, 5, 8]) that either provide preconfigured virtual machine disk images or enable sharing of such virtual machine disks have emerged to support the IaaS paradigm.

Virtual machine (VM) disk images that are preconfigured with the necessary software to support specific workflows of functions are known as *virtual appliances* (VAs), and services providing repositories of VAs are referred to as *appliance stores*. A VM image with a Tomcat web server fully configured to host Java web services and use the underlying MySQL database is an example of a VA. In the rest of the paper, VA and VM image will be used interchangeably. The benefits of this publisher-consumer model (see Figure 1) are clear: consumers enjoy the convenience of downloading a VA that suits their needs and launching a service quickly; and publishers can get paid for providing the VA[1]. This market-

[1]Other models exist in which publishers provide the virtual machine images to promote their software, or promote the use of their IaaS infrastructure, or simply share their images for the common good. All of these settings provide benefits, tangible or otherwise, to both consumers and publishers.

place, however, lacks adequate mechanisms that allow users to a priori assess whether a specific VA is correctly equipped with the software packages that it claims to contain. This paper sheds light on the integrity of software packages in real-world VAs, and demonstrates a clear need for a mechanism to assess their software integrity.

There are many well-known security challenges associated with the cloud computing paradigm (e.g., [19, 11, 27, 14, 29, 28]), and the security implications of the publisher-consumer model associated with VA stores (e.g., [32, 17]) are an important sub-class. A malicious publisher could install malware in a VA and attempt to read private information while a consumer uses the VA. A careless publisher might unintentionally publish a VA that is infected with a virus, is configured insecurely (e.g., running an unpatched version of an operating system, or with unintentional backdoors), or contains sensitive private information. Further, an irresponsible publisher could publish an incomplete VA that is missing integral software packages that it claims to have. Some software packages might be partially installed (i.e., missing critical files) or have files that are modified (i.e., different from the version provided by the software vendor). VAs with such partially installed or modified packages will not meet the expectations of VA consumers, who would expect the VAs to be configured with all the packages claimed by publishers, and that those packages are unmodified (unless otherwise noted). Currently, checking the integrity and identify of the installed software packages is a time-consuming process that must be performed manually by the consumer after the instantiation of the VA. Such information, if available during the selection process, could provide early insight into whether a VA is suitable and can be trusted. Some VAs come with build-scripts and logs, but without any guarantees that these logs are correct and accurate. The same is true for VA publishers and store providers: they lack mechanisms to verify the integrity of VA software before allowing them to be published, leading to security and quality assurance problems in appliance stores.

This paper presents an empirical study on the integrity of software packages in real-world VAs, assessed using a whitelist-based framework, showing that there is a significant variance in the software integrity of software packages across VAs. Our findings show that about 9% of the evaluated real-world VAs have significant software integrity issues and would probably not meet consumers' expectations. Surely, consumers should be informed about such VAs when selecting a VA. We would like to point out that while a well-configured VA with authentic software packages can still be susceptible to software vulnerabilities that are present in legitimate packages, the whitelisting approach can be used to avoid VAs with partially installed or modified packages[2] thus mitigating risk for consumers.

Whitelist-based approaches are hard to realize in general computing environments, given the variety and vast amount of existing software [20]. Our hypothesis is that in the appliance store model, a majority of consumers use fairly well-known, popular pieces of software, and thus the whitelist sizes will be more manageable. To test this hypothesis, we constructed a whitelist for our sample VAs and indeed found that the whitelist size growth decelerates.

[2]Note that there is a legitimate need for modified or customized software packages. Our focus is on VAs meant for well established workflows with standard software stacks. None of the VAs that we found to have "modified" software in our analysis claimed to be custom or modified VAs in their description.

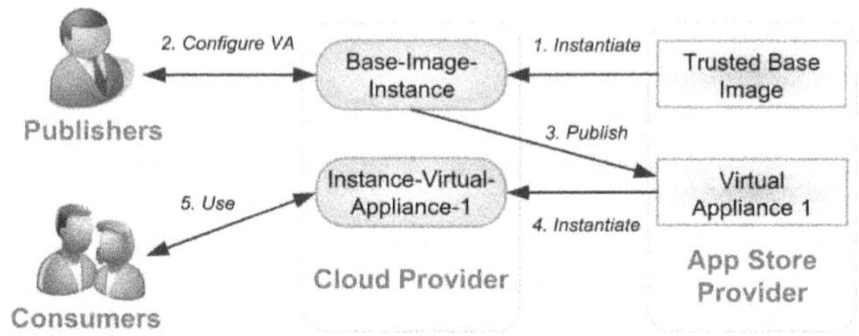

Figure 1: A typical virtual appliance store

Contributions: The key contributions of this paper are (1) an empirical evaluation of the software integrity in real-world VAs, showing that there is a *significant variation*, (2) demonstration of the *usefulness* of assessing the software integrity in VAs to help customers choose correctly configured VAs, and (3) demonstration of the *feasibility* and *scalability* of using whitelists for integrity assessment.

Organization: Section 2 covers related work. Section 3 provides an overview of the assessment framework used for analyzing real-world VAs in Section 4. Section 5 discusses the security implications. Our conclusions and future work are in Section 6.

2. RELATED WORK

Over the last few years, researchers and industry practitioners have been looking at challenges associated with cloud computing, including security and reliability challenges [19, 11, 26, 15, 27, 32, 14, 28, 29, 34, 16, 18, 33]. Here, we focus our discussion on past work [26, 32, 34, 17, 21] that addresses security risks and challenges associated with VA stores and with the management of VAs in general, which is most relevant to our work.

Reimer et al. [26] address the challenges posed by the sprawl of virtual machine images that need to be mounted for maintenance tasks. Specifically they propose Mirage Image Format (MIF), which simplifies the maintenance and administration process. To our knowledge, Wei et al. [32] are the first work to identify the security challenges and risks associated with the VA store model and with managing VAs in general. They proposed an image management framework with access control for VAs, filtering to remove unwanted information (e.g., browsing history), provenance tracking of images for accountability, and image maintenance (e.g., virus scanning), and argues for the benefits of such a system. In a later work [34], a scalable offline patching tool called *Nüwa*, which leverages MIF to significantly reduce the overhead of patching, was proposed. Accountability through provenance tracking, virus and malware scanning, and software patching are good security hygiene practices and do provide some assurances to the consumer regarding the security and trustworthiness of the VA. However, they alone are not sufficient. Accountability is after the fact and may come too late for some consumers, and the proportion of new or near zero-day malware that is detected through scanning is not very high [12]. In contrast, our approach provides more explicit information regarding a VA's software configuration to the consumer and is complementary to the aforementioned work.

Bugiel et al. [17] analyzed over a thousand Amazon Machine Images (VAs for the Amazon Elastic Compute Cloud) and found sensitive information (e.g., keys and credentials) and SSH backdoors inadvertently left in the VAs. To mitigate the risks to publishers and consumers they suggest countermeasures that include running filters and scanners on VAs and use of ratings based on certain verifiable properties and reputation, but they do not elaborate on the properties of interest or the rating system. There are in fact appliance stores that provide ratings for VAs [8]. However, the primary basis for the ratings seems to be user opinion, which, while quite useful, may not be objective, and may be based on many factors, of which software integrity and security are only a part. Furthermore, in ratings-based systems that are based on reputation and user feedback, it takes time for VAs to gather ratings and publishers to earn reputation, and it will be easy for brand-name providers to build their reputation even though many of the smaller players may be providing VAs that are as good or better than their well-known counterparts. In contrast, our software integrity assessment approach provides a means to rate VAs before publishing them, and it could be used either on its own or as part of a larger ratings ecosystem that also considers reputation of provider and user feedback.

Jayaram et al. [21] undertook an empirical study of similarity between VAs to enable efficient design of VA management systems. Their study considered block-level similarity in VAs to enable efficient design of de-duplication schemes, reduce storage, and improve image distribution, among other things. Our work focuses on use of similarity in VMs at the software or file level to gauge the size of whitelists that need to be maintained by a VA store provider. Moreover, to the best of our knowledge, we are the first group to analyze comprehensively the software integrity of VAs used in the real world.

Techniques based on file analysis similar to the one we used have been used previously in the context of operating systems. Seminal work from Kim et al. [23] and Vincenzetti et al. [31] introduced mechanisms for detecting changes in critical files in the system that might indicate the presence of rootkits or of other illicit modifications. These techniques are based on comparing files in the disk with a previously-generated hash of the file to detect changes. More recent work [25] extended such approaches to perform the analysis in a virtual machine environment, so that modifications to the files cannot be hidden even when the kernel is compromised. The goal of such techniques is to identify at *runtime* malware or malicious modifications to critical files by detecting changes from a trusted configuration (i.e., the initial configuration of the system). While we use similar file integrity based techniques, our focus is on analyzing the software integrity of a VA to assess the "initial" configuration. Further, our analysis discusses ratings for assessed software in contrast to earlier efforts that stop at detecting changes.

3. SOFTWARE INTEGRITY ASSESSMENT FRAMEWORK

To study the integrity of software packages in real-world VAs, we designed an integrity assessment framework based on software whitelisting techniques. At the heart of the

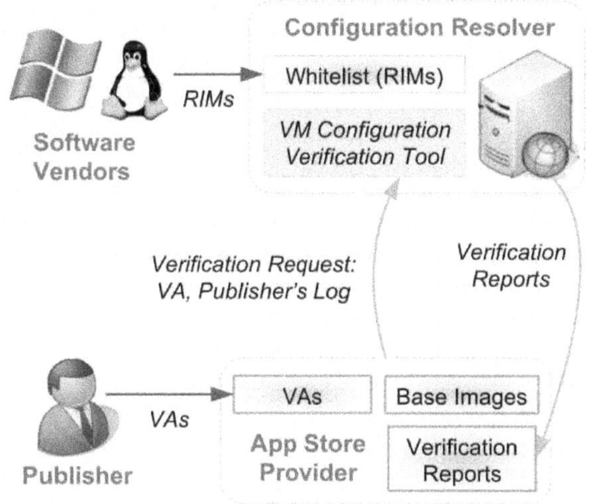

Figure 2: Software integrity assessment framework overview

framework is a Configuration Resolver (described in Section 3.1) that uses a VM Configuration Verification Tool (VM-CVT) to generate a "verification report" on a given VA (see Figure 2). Those tools use information provided by publishers and by software producers to verify the integrity of the software installed on the system, and to identify any modified or missing files. Based on this information, we introduce a simple rating system to score the installed software. The verification reports provide consumers with explicit information about the configuration of the system that they can use to gauge its suitability without having to download and instantiate the VA. Providers can use the verification reports to ensure that published VAs respect basic requirements by specifying policies. For example, a cloud provider could create a more trustable store by introducing a policy stating that all of the installed software and updates need to be fully verified. Such policies are discussed further in Section 5.3 as means to enhance VA security.

Our integrity validation is based on reference integrity measurements (RIMs) [30] published by the vendors for their software. The RIMs contain the signed hash values of all the files in the software and metadata, including a description of each of them. This is a common software practice, and many software vendors already provide signed checksums for files (e.g., `rpm` packages contain MD5 checksums for the files), allowing users to verify the file integrity. In fact, the U.S. National Institute of Standards and Technology (NIST) maintains a software registry, called the National Software Reference Library (NSRL) [7], that provides RIMs for software. Bit9, a private security company, also maintains a massive software registry [3].

The framework can take advantage of additional information made available from publishers and from software vendors. In particular, we take advantage of the fact that VAs are often generated from trusted base images made available from the repository provider (e.g., Amazon Linux AMIs [1]) to speed up the analysis. This technique also allows a provider to whitelist implicitly proprietary software without having to register with third party Resolver. Moreover, a *log* describing the list of all software (and updates) installed on a VA, as well as the version information at the time of

VA publication, can be generated by the publishers and submitted with the VA to simplify the analysis. If publishers submit fabricated logs (e.g., lying about the installed packages), their VAs will get low integrity scores as the installed files will not match the RIMs and will appear unreliable.

Software vendors can provide additional metadata associated with the RIMs for identifying the file types (e.g., source code or image file) and their *criticality* for the software package (e.g., critical or non-critical). The criticality of a file may be determined by looking at how much of the expected software behavior can be affected when that file gets modified, replaced or deleted. Based on this guideline, a vendor, for instance, could treat executables and source files as critical, and compressed files and data files as non-critical. We use a similar rule to design the prototype (see Section 4.1.3).

3.1 Configuration Resolver

The Configuration Resolver communicates with the software vendors or third-party RIMs aggregators (e.g., NSRL or Bit9) to maintain an up-to-date database of the RIMs (the whitelist), also keeping track of the history of the RIMs, since not every VA will be equipped with the latest versions of software. It also maintains a list of RIMs that have expired or been revoked. Existing security services companies (e.g., anti-virus companies) are potential candidates for providing Configuration Resolver services to the cloud providers. The VMCVT is also managed by the Resolver and is explained next.

3.2 VM Configuration Verification Tool (VMCVT)

The VMCVT verifies the integrity of a VA by computing hashes for all of its files and comparing them with trusted hashes. Specifically, when a VA is created from a trusted base image, the tool mounts the filesystems of both the VA and base VM, hashes all of the files in both, and compares the absolute paths and hashes of the files to compile a list of files that have not changed from the base as well as the files that have been *added*, *modified*, or *deleted*. Files that are the same will be marked as "verified."

Files that have been added or modified are verified against the RIMs. The tool uses the publisher's log (which contains the exact software versions) to determine which sets of RIMs should be used to verify the software and updates installed on the VA. Added or modified files for which a corresponding RIM could not be identified are marked as "unverified." From the unverified files, the tool separates out the configuration files and performs a "diff" against the base image version (when available), or against the default version included in the software. In cases where a base VM is not available for the initial comparison, the VMCVT simply verifies all of the files in the VA against the RIMs. Although this would increase the number of files that need to be checked against the RIMs (which can be expensive if a third party service is utilized), the contents of the verification reports should not be affected.

The RIM's metadata is used to identify which files need to be present for a certain version of software to be considered complete and unmodified. For deleted files, the tool identifies the files that should not have been deleted based on that metadata. The tool also marks deleted base image files that belong to partially removed/uninstalled packages. Those files are all marked as "missing."; files that are part of completely uninstalled packages are not marked.

3.2.1 *Verification report*

Based on the verification results, the tool generates a digitally signed "verification report" that states the following:

- the list of installed software (and updates), including the version information;
- the list of verified, unverified, and missing files for each installed software; the file types, indicating whether a file is an executable, source file, web page, configuration file, image file, compressed file, or data file;
- the list of configuration files that have changed for each installed software and the content differences;
- the "integrity score" (explained in Section 3.2.2) for each installed software;
- the list of unverified files that are not part of the installed software;
- and the hash of the entire VA image.

The report provides consumers with an explicit list of the differences between the VA and the trusted software. Such a list provides a guide for inspecting changed files and configurations in the installed VA. For modified configuration files, in addition to the list, we provide consumers with the list of differences between the VA file and a trusted configuration file. There may also be unverified files that are not part of any installed software/update packages. All unverified files are grouped by their file types (see above), and a summary of the number of unverified files for each file type is shown in the report. The last element of the report is the hash of the entire VA image. This hash value allows the consumer to check the integrity of the VA (as it is described in the report) before launching it. The generated report is signed and published together with the VA.

The VMCVT can be made available to publishers so they may test and evaluate their VAs prior to publication. If the publishers believe that their legitimate software is failing the integrity checks, they may submit a report to the cloud provider (or a third party managing the Configuration Resolver), asking for a review and inclusion of their software. Such a practice will also help publishers get their logs correct. If their logs are listing wrong versions of software or are missing certain directory paths for a software package, their VAs will get low integrity scores; the verification reports will show which files have failed the integrity checks or which files are missing. It is the responsibility of publishers to use that information to correct their logs and add any missing directory paths before publishing their VAs.

3.2.2 *Software integrity score and expected behavior*

To summarize the analysis, the VMCVT tool considers each piece of software installed in a VA and assigns it a score on a scale of 1 to 3 using the rules described below. The score represents the *integrity* and *completeness* of the software as perceived by the VMCVT, indicating the extent to which a software will operate and *behave as expected*.

- integrity score ③—the software has *no* unverified or missing files, with the exception of configuration files;

- integrity score ②—the software has no unverified or missing critical files, but may have unverified or missing *non critical* files;

- integrity score ①—the software has unverified or missing *critical* files.

Installed software that has an integrity score of ③ is considered fully verified and integrity-protected, falling under the "clean or high-integrity installation" category. Clean installation indicates that a particular software package will behave as expected. A complete verification of such software requires only an inspection of the configuration files. Software that has a score of ② has only files deemed non critical missing or unverified; hence, the impact on expected behavior is considered to be limited. Such software is categorized as "partially clean or medium-integrity". Score of ① represents the "modified or low-integrity" category as such software may have critical files missing or unverified. It is highly likely that such low-integrity packages, which may have unverified or missing critical files, will not operate in an expected manner. These integrity scores allow the VA consumer to identify easily software packages that have a standard unmodified installation and those that have been modified or have a nonstandard installation.

4. ANALYSIS OF REAL-WORLD VIRTUAL APPLIANCES

In this section we shed light on the integrity of software packages in real-world VAs using the framework described in the previous section. By looking at the variance in the results, we gauge how useful the verification reports would be for consumers in selecting well installed VAs, and for providers in removing suspicious VAs from their stores. We generated verification reports for 151 randomly picked Amazon VAs and analyzed them.

4.1 Methodology and assumptions

4.1.1 Sampling method

We sampled the content of the Amazon market [2] by randomly selecting publicly available Amazon Machine Images (AMIs). We instantiated the selected images and obtained their disk content using a set of commands equivalent to the AMI tools provided by Amazon. As our prototype takes advantage of the Red Hat package manager (`rpm` database) in validating image content (in order to ease the verification process as explained further in Section 4.1.3), we focus on `rpm`-based distributions, which represent a significant portion of the Linux images available in the market.

Our random sample is composed of 151 images from an estimated pool of 2,300 valid `rpm`-based AMIs available in the Amazon US-east zone. We created the sample pool by acquiring the list of available AMIs through EC2 API calls. The call returned a list of 8,798 images that are available for instantiation. This list was reduced to 4,513 AMIs after filtering out Windows and `dpkg`-based Linux images. We randomly selected 300 images from this filtered list and tried instantiating them. About 50% of the selected images failed at boot due to errors, lack of user credentials or product codes, leaving us with a sample of 151 successfully instantiated VAs.

Table 1: Three randomly selected sub groups. The numbers shown here are average values.

	Group 1	*Group 2*	*Group 3*
Total number of files	52,886	51,909	52,786
Number of verified files	52,137	51,183	52,036
Number of software packages	427	427	426

4.1.2 Representativeness of the samples

We evaluated whether our sample size is sufficiently large to represent the entire sample pool (instantiable `rpm`-based AMIs estimated at 50% of the total 4,513). If our samples are a representative set, the randomly shuffled subgroups (partial sets) of the samples should also represent our samples (i.e., show similar characteristics) with respect to the properties of interest, which are the number of files, number of verified (and unverified) files and number of software packages (see Table 1). To verify whether they do, we randomly shuffled the 151 VAs and divided them into three subgroups, repeating this process 30 times and computing the average values. The three groups are hypothesized to share similar characteristics and closely represent each other. To test this hypothesis, we compare the properties of interest between the three subgroups. The results are shown in Table 1. There is no significant difference among the three randomly generated groups. Moreover, our statistical analysis did not reject the null hypothesis, showing $\chi^2 = 0.11$, $p = 0.99$, and degrees of freedom of 4.

4.1.3 The VMCVT prototype

Using a combination of shell and python scripts, we constructed a prototype implementation of the VMCVT tailored to the experiment at hand. Given a base image and a derived VA, it first generates the checksums by hashing all of the files in both images. It compares the hash values between the two images and creates a list of *added* files, *modified* files, and *deleted* files. The `rpm verify` command is then used to verify the integrity of the files that were added, modified, or deleted through `rpm`; `rpm verify` checks every file installed through `rpm` against its database of MD5 checksums and checks cryptographic signatures. The VMCVT keeps track of all the files that failed `rpm verify`, including the deleted files that should not have been deleted (these are what we refer to as the *missing* files). This reduces the effort needed to manually create the whitelist, since we are satisfied with the files that are verified through `rpm`. In a production implementation, though, each and every file in a VA should be checked against a common whitelist.

VMCVT then checks the remaining added, modified and deleted files (i.e., those that are unknown to `rpm`) against the whitelist that we created (see Section 4.1.4) and marks the files that are unverified or missing. To figure out whether an entire software package was removed from the base image (without using `rpm`), we inspected the differences in the file directories between the two images. If an entire directory for a software package was removed from the base, we assumed that this software package was completely uninstalled. Deleted files that were part of this directory were ignored and did not affect the software integrity scores.

The above generated unverified/missing list, together with the list that failed `rpm verify`, contains the complete list of unverified added and modified files as well as the missing files

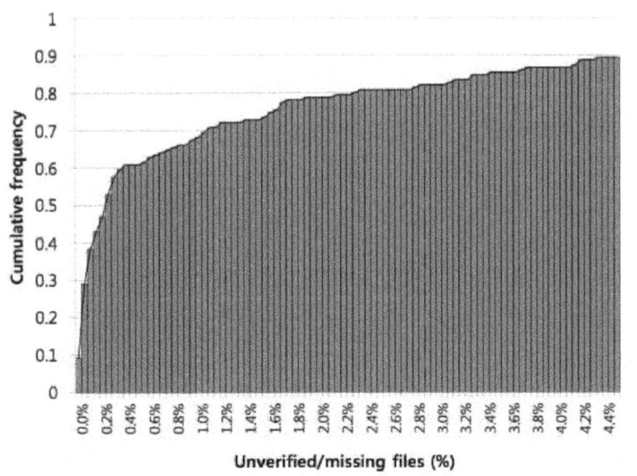

Figure 3: Cumulative frequency of VAs vs. percentage of unverified/missing files

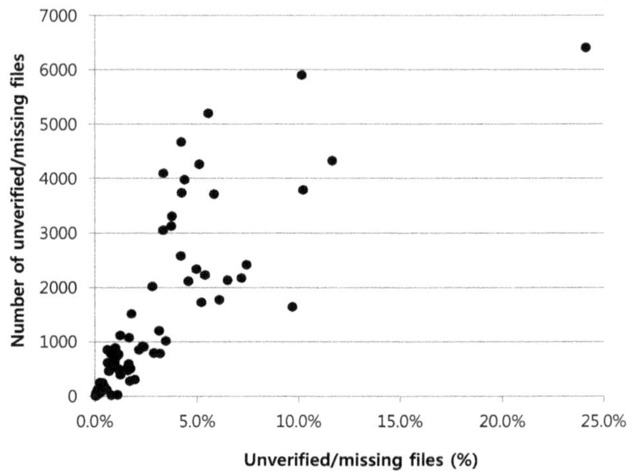

Figure 4: Number of unverified/missing files vs. percentage of unverified/missing files

that should not have been deleted. The VMCVT classifies these files into the file types shown in Section 3.2.1 using the `file` command. Then, by figuring out which software package each unverified/missing file belongs to, the VMCVT computes the integrity score of the installed software using the rules described in Section 3.2.2. Because the metadata regarding the criticality of files are not available, we treat all executables, source files, web pages, and image files as "critical", and compressed or other data files as "non critical." As the final step, the VMCVT generates a verification report (see Section 3.2.1).

4.1.4 Constructing the whitelist

A software whitelist was constructed manually based on the list of added, modified, and deleted files that `rpm verify` did not know about. This "rpm-unknown list" was generated for all the VAs, keeping track of the absolute paths and file names. By examining the absolute paths and the file contents, we figured out the exact software versions that were installed (without using `rpm`) on each VA; we then downloaded the source code and binary packages (e.g., tar.gz or zip files) for them from their respective vendor websites. After extracting the packages, we created MD5 checksums (the RIMs) for every file that was contained in the packages and added them to the whitelist. Here, we assumed that the downloaded packages represented trusted sources.

4.2 VA classification based on the percentage of unverified/missing files

4.2.1 Classification method

Figure 3 shows the cumulative frequency of unverified/missing file percentages. 90% of the VAs have less than 4.5% of unverified or missing files. Our intuition was that the number of unverified/missing files would indicate, to some degree, the integrity level of software in a VA. As the first step to study their characteristics, we first looked for a correlation between the number of unverified/missing files and the total number of files in a VA, but found none. We did find, however, a correlation between the number of unverified/missing files and the percentage (Figure 4, Pearson's: 0.84). Using that correlation, we classify the VAs into the following three "Integrity Level Groups" (ILG) to help demonstrate common VA characteristics:

- "ILG A"—44 VAs are in this group, and they have less than 0.1% of unverified/missing files;
- "ILG B"—59 VAs are part of this group, and they have 0.1-1% of unverified/missing files;
- "ILG C"—48 VAs in this group have > 1% of unverified/missing files.

4.2.2 Characteristics of the ILGs

The characteristics of the three groups are shown in Table 2. All average values shown in the table are "truncated average" values that exclude the top 20% and bottom 20% of the results; truncated average [22] was used to accommodate for the high variances we saw in the results. First, the average number of files in the VAs is similar across the three ILGs, averaging around 52,300. The average number of unverified/missing files, on the other hand, is significantly higher in ILG C (1,915) than in ILGs A (34) and B (172). That was expected, as the ILGs are classified based on the percentage of unverified/missing files and the average number of files across the three groups is the same.

Table 3 contains the number of software packages with each integrity score as well as the proportion of unverified/missing critical and non critical files, again showing the truncated average values for the three ILGs. The table indicates that the VAs in ILG C have a relatively larger number of software packages with low integrity scores (② and ①), while most of the software packages installed on VAs in ILG A have an integrity score of ③. Note, score ② represents a partially clean or medium-integrity package and ① a modified or low-integrity package (see Section 3.2.2).

The total number of packages averages around 420 in all three groups. About 99% of the software packages in the 44 VAs in ILG A are high-integrity, and the numbers are not too different for the 59 VAs in ILG B. Both groups have a small number of low-integrity and medium integrity packages. There is a big jump, however, as we reach ILG C,

Table 2: **Integrity Level Groups for VAs and their file characteristics**

	No. VAs	Avg. % of unverified/missing files	Avg. # files		Avg. # unverified/missing file types				
			All files	Unverified/missing	Exe	Src/web	Img	Comp	Data
ILG A	44	0.06%	53,737	29	3	2	1	0	23
ILG B	59	0.34%	50,292	165	16	24	3	1	121
ILG C	48	3.67%	52,198	1,871	291	635	28	6	911

Table 3: Average number of software packages installed and average number of software packages to which the three integrity scores have been assigned

	Avg. # software	Avg. # unverified/missing		Avg. % of unverified/missing		Avg. # integrity scores			Avg. time
		non critical files	critical files	non critical files	critical files	③	②	①	
ILG A	428	23	6	68%	18%	426	1	0	37s
ILG B	424	122	43	71%	25%	416	2	1	35s
ILG C	415	917	954	48%	50%	346	20	45	36s

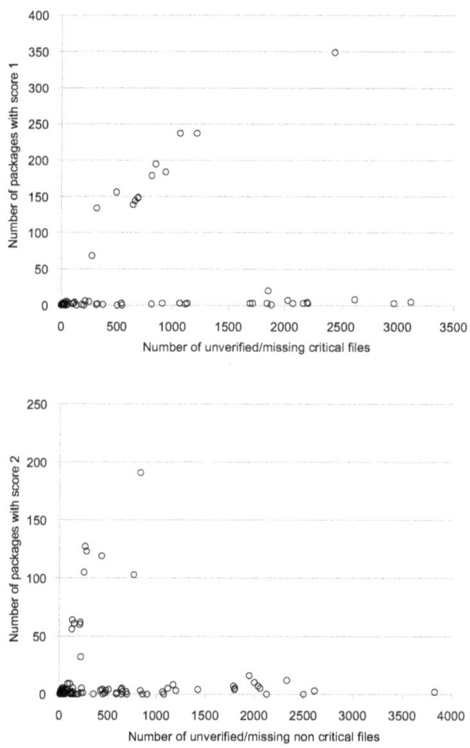

(A)

(B)

Figure 5: (A) # unverified/missing critical files vs. # score 1; (B) # unverified/missing non critical files vs. # score 2

whose VAs have, on average, 45 low-integrity packages and 20 medium-integrity packages. Those are about 11% and 4.8% of the total number of packages, respectively. We note this as our first key finding.

Finding 1: across the VAs, there is high variance in the number of unverified/missing files and the number of low-integrity and medium-integrity software packages.

While the majority of the unverified/missing files belong to what we deemed the non critical file category (data and compressed files) for ILGs A and B (68% and 71%, respectively), this proportion decreases significantly in ILG C (48%). In contrast, the proportion of unverified/missing files that belong to what we deemed the critical file category (executables, source/web files, image files) starts small in ILG A (18%), but increases significantly in ILG C (50%). This can be explained by the huge jump in the number of unverified/missing executables and src/web files from ILG A to ILG C.

Graphs (A) and (B) in Figure 5 plot the number of unverified/missing critical files against the number of packages with score ①, and the number of unverified/missing non critical files against the number of packages with score ②, respectively. Interestingly, in most VAs, both the unverified/missing critical files and non critical files are *concentrated* in a small number of packages. The graphs indicate that there is no strong correlation between the number of unverified/missing files and the number of packages they influence. Hence, the absolute number of unverified/missing files does not give a good estimation of the number of partially clean or modified packages. Our second key finding is as follows.

Finding 2: The number of unverified/missing files is not a good indicator of the number of medium-integrity or low-integrity packages installed.

There are some VAs, however, for which the unverified or missing files do spread out across a large number of packages, and these are the 12 or so outliers that we see in the two graphs. These outliers seem to show some correlation between the number of packages and the number of unverified/missing files, and so we investigate these further in Section 4.3.

4.2.3 *Verification time*

Table 3 shows that the entire verification process takes about 36 seconds on average across all three groups. This is how long publishers evaluating the configuration of their VA prior to publishing it will have to wait to see the results. The performance overhead could be significantly amortized by using the Mirage image library [13], which uses the Mirage Image Format [26] (see Appendix A).

4.3 VAs with large number of low-integrity packages

4.3.1 *Classification method*

In this second part of the analysis, we further investigate the outliers identified in Figure 5, which have noticeably larger number of partially clean and modified packages.

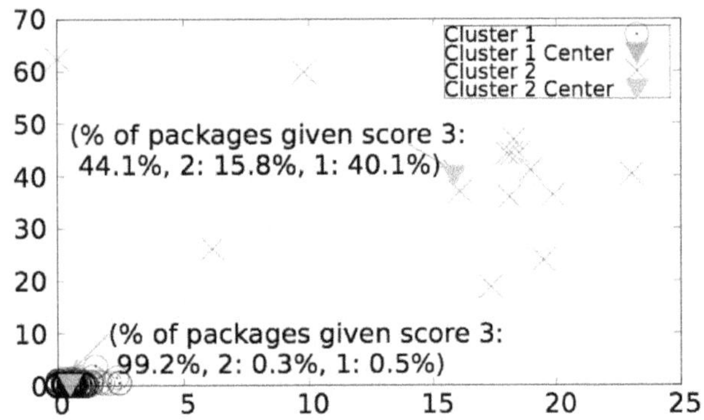

Figure 6: VA clusters based on the percentage of the software packages given each of the three integrity scores; k-means clustering with $k = 2$ is used.

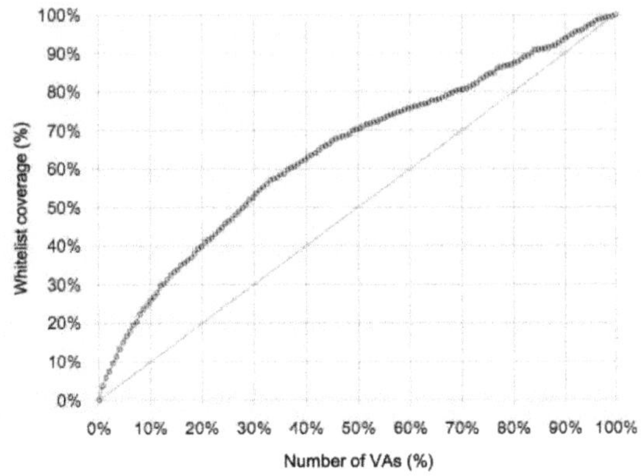

Figure 7: Percentage of the VAs used vs. percentage of the whitelist coverage

We do so by forming VA clusters based on the percentages of packages given integrity scores ① and ②. The k-means clustering method—an unsupervised learning algorithm—is used to identify the clusters ($k = 2$). Figure 6 shows the two distinct VA clusters: Cluster 1 represents VAs with low percentages of packages given scores ① and ②, comprising 137 VAs; Cluster 2 represents VAs with high percentages of packages given those two scores, consisting of 14 VAs.

4.3.2 *Characteristics of the VA clusters*

Table 4 summarizes the characteristics of the two VA clusters, showing that the 14 VAs in Cluster 2 have significantly high percentage of packages with scores ① and ②; the absolute numbers are also very high, averaging 171 and 77 for scores ① and ②, respectively. Only 43% of the packages are cleanly installed (high-integrity). The percentage of the unverified/missing critical files is high too, averaging 73%. In contrast, for the VAs in Cluster 1, 99% of the packages are cleanly installed, and the percentage of unverified/missing critical files averages only 22%. Considering that our samples are a representative set (see Section 4.1.2), here is our third key finding:

Finding 3: About 9% of the VAs have a significant portion of low-integrity and medium-integrity packages installed.

To cross-validate the results, we look at which ILGs the 14 VAs in Cluster 2 belong to: all belong to ILG C (the group with the highest percentage of unverified/missing files), except for one that belongs to ILG B. We took a closer look at the 14 potentially untrusted VAs and the types of unverified files they contain. Most of the files are ones that failed the `rpm verify` checks. A significant portion of the files are common system files like `/bin/cut` and `/bin/grep` which `rpm verify` flagged as not matching its database of trusted hashes for the package versions that are supposed to be installed. Although it is hard to say why those files ended up being unverified (and this is out of the scope), our virus scan results (see Section 5.2) show that 41 of those files are potentially malicious, infecting 7 of the 14 VAs. The VAs are from different publishers, were built to support different functions, and do not share common base images. What is most worrying about those 14 VAs is that none of them, in their name or VA description, mention anything about software customization or modification efforts. Just looking at the VA descriptions, they all appear to have only the cleanly installed, standard packages.

Interestingly, we found a correlation between the number of packages given score ① and the number of packages given score ② (see Figure 8, Appendix B, Pearson's: 0.65). This implies that VAs with a high number of medium-integrity packages, such as the VAs we see in Cluster 2, tend also to have a high number of low-integrity packages.

4.4 Scalability evaluation: whitelist size

This section analyzes the size of the whitelist that we created (including the `rpm` database used) to verify our sample of 151 VAs and the 47 base images, showing how the size changes as the number of VAs grows. Figure 7 plots the percentage of the VAs used against the percentage of the whitelist coverage (i.e., how much of the total whitelist was created). To get the values, we randomly shuffled all the VAs and incrementally generated the whitelist, adding the per-VA whitelist entries (the hashes used to verify a specific VA) to a global whitelist while removing any duplicate entries; the size of the global whitelist was recorded after each step. We repeated this process 10 times to get the average size for each incremental step, and used these average values to calculate the percentage values.

Intuitively, we see that the whitelist coverage percentage grows as the number of VAs used to construct the whitelist goes up. It does so more rapidly at first, but then slows down progressively. We can see that after 20% of the VAs are used, about 40% of the total global whitelist is already constructed. The size of the final global whitelist was 1,697,469 hash entries. The growth rate slows down because some software packages are installed on multiple VAs, and thus do not add any extra entries to the whitelist. Table 5 shows the top 10 most commonly installed software packages as well as their file counts (not counting the packages installed on the base images). `perl5.8.8`, which has 3,115 files, was installed commonly across 27 VAs. `python2.6` is another popular package that appeared on 22 VAs, containing 3,915 files. Such packages are responsible for the decreasing growth rate that we see on the graph.

4.5 Base image analysis

We found the base images for the VAs from the Amazon appliance store by looking at their Linux distribution

Table 4: Characteristics of the VA clusters

	# VAs	Avg. % of unverified/missing		Avg. % of integrity scores			Avg. no. of integrity scores		
		non critical files	critical files	③	②	①	③	②	①
Cluster 1	137	68%	22%	99%	0.5%	0.3%	418	2	1
Cluster 2	14	26%	73%	43%	17%	40%	205	77	171

Table 5: Most common software packages installed

Software package	*# VAs*	*# Files*
perl5.8.8	27	3,115
selinux-policy-2.4.6	22	23
python2.6	22	3,915
kernel-2.6.21.7-2.fc8xen	15	251
kernel-2.6.18-xenU-ec2-v1.0	15	325
python2.4	12	3,038
kernel-2.6.18-238.9.1.el5xen	12	99
ec2-ami-tools-1.3-56066	11	165
rubygems-1.3.3	11	307

Table 6: Base image characteristics: truncated average values for all base images

All files	*Unverified/ missing files*	*# software*	③	②	①
38,928	5	386	385	0	0

and version information. In most cases, the base images were published by the distribution providers (e.g., Redhat, CentOS) or directly by Amazon (with the Amazon Linux AMI). It would be reasonable to expect that these base images would be clean and contain only verified files. Here we analyze their characteristics and find that it is indeed the case.

To verify 151 VAs, we downloaded 47 base images, each of which we used to verify about 3 derived images. Table 6 shows the average values for the base images. The average number of unverified/missing files is 5, which is much smaller than the number we saw for ILG A (see Table 4). Based on that characteristic and the fact that the number of low-integrity packages is sufficiently small, we claim that the base images are trusted and our assumption is valid.

4.6 On the usefulness of VA integrity assessment

Findings 1 and 3 clearly demonstrate the need for a priori software integrity assessment of VAs. When a consumer pays for a VA to use, she has the right to know that a VA is configured properly and meets her software integrity expectations.

To illustrate the usefulness of software integrity evaluation, in Table 7 we show example choices (among the 151 VAs studied) that a consumer would see when selecting a VA that provides software package `php-5.3`. The table is sorted by increasing number of unverified/missing files. The variances in the integrity scores are quite significant. VAs 5 and 7 both have more than 100 low-integrity and medium-integrity packages (scores ② and ①). Such VAs are part of Cluster 2 (see Table 4). The first two VAs have some number of unverified/missing files but the counts for the scores ② and ① are all 0. This is because all of their unverified/missing files are of type configuration files, not affecting the integrity scores.

High variances in the number of low-integrity and medium-

Table 7: Example VAs that provide php 5.3

	All files	*Unverified/ missing files*	*# software*	③	②	①
VA 1	34,410	21	351	351	0	0
VA 2	28,606	27	352	352	0	0
VA 3	63,719	48	472	470	2	0
VA 4	90,977	94	711	707	3	1
VA 5	101,582	617	711	454	123	134
VA 6	91,873	886	715	711	4	0
VA 7	74,039	2,018	641	301	103	237

integrity packages among VAs that provide similar functions reinforces the need for a priori software integrity assessment. Verification reports would help a consumer avoid VAs that are configured badly and choose ones with high-integrity packages. Providers could also make use of the verification reports to admit VAs for standard workflows only when they have high-integrity software packages. Compliance policies can be defined in terms of the integrity scores, allowing, for example, only the VAs with 1 or fewer low-integrity packages and 2 or fewer medium integrity packages to be published (using the ILG B average as a guideline). A more flexible policy might state that a VA for well-known workflow with more than 2 medium integrity packages can still be published but must always be monitored with runtime monitoring mechanisms when instantiated. Such policies would allow the repository providers to be more permissive (or open) and still maintain a reliable infrastructure.

4.7 Implementation challenges

We faced several challenges when implementing the framework and verifying Amazon VAs. This section highlights some of these challenges and suggests workarounds.

4.7.1 Origin of unverified/missing files

Identifying the software package to which an unverified or missing file belongs is not always trivial. Some packages get installed manually (i.e., without using `rpm`), and the files get located in different directories. For such files (unknown to `rpm`) that are unverified or missing, we manually identified keywords from their absolute path and matched them with software packages. The VMCVT prototype, while computing the integrity scores, used these keywords to figure out which package's score an unverified/missing file should influence. As a workaround for a production VMCVT, the publisher's log (see Section 3) should list all the directories under which a package's files are installed. The VMCVT would use that information to determine the origin of the unverified/missing files when rating the packages.

4.7.2 Creating a whitelist for proprietary software

A small number of VAs had proprietary/commercial software packages installed (e.g., Sugarcrm, Ideamax2 Gold, IBMSoftware/ITMAgent), and we could not find the source or binary files or signed MD5 checksums for them. Hence, we selected one VA with a particular commercial software package installed and created a whitelist for the package by hashing the files from this VA instance, under the assumption that it was a trusted source. Other VA instances that also had the package installed were verified using this whitelist. We suggest that in a real system publishers submit a software review request to the Configuration Resolver provider, asking for their software to be reviewed and added to the whitelist.

4.7.3 Dealing with `tmp` *and* `var` *files*

We note that VAs typically have a lot of temporary files such as those in the `tmp` or `var` directory, and that there is no way to meaningfully verify their integrity. `tmp` files are less worrying, since they do not affect VA behavior, and we ignored them. We were less certain about the contents of `var` folders, though, and examined the types of files they contained before deciding to ignore them. Table 9 (see Appendix C) shows the top 10 most common `var` files across all the VAs we looked at. Most of the `var` files are log files and `rpm` database files, which, we argue, can also be deleted without affecting the VA behavior.

5. SECURITY IMPLICATIONS

So far, we have studied the integrity levels of real-world VAs in terms of their software identity, integrity, and completeness. Those software properties indicate the extent to which a VA will behave and operate as expected, and are also integral to figuring out how secure a VA is. This section discusses the security implications of measuring the VA integrity levels and suggests a few ways to add useful security metrics into the framework.

5.1 Expanding the integrity scores with blacklisting

There is room for improving the integrity scores to say more about the trustworthiness of the software packages. We suggest complementing our whitelist-based approach with "blacklisting" via scanning of all unverified files for viruses: score ⓪ is given to a package that fails an anti-virus check. Packages with score ⓪ fall under the "malicious" category.

A VA might be planted with a small number of malicious files (that are *known* to virus scanners), but overall has a high percentage of cleanly installed packages (score ③). Packages that contain those files will get score ⓪ though, and that VA will be flagged immediately as malicious. If those malicious files are new, clever and unknown to virus scanners, the maximum score it can get through our system is ②, which indicates a package that might not behave in an expected manner. Even though the integrity scores are not a silver bullet solution, it will provide some indication as to how trustworthy a package is whereas a virus scanner (or any other malware detection solution) will likely miss it.

5.2 Checking the unverified files against virus scanners

To demonstrate the usefulness of the combined approach and expanded integrity scores (see above), we checked all the unverified files against VirusTotal[3] [9], which is an online tool that runs multiple anti-virus engines and returns aggregated reports on the files. For a total of 111,981 unverified files (from all 151 VAs), VirustTotal flagged 45 of them as potentially malicious, which were spread out over 10 VAs (about 7% of the samples). Table 8 shows a summary of the viruses flagged by VirusTotal.

[3]We thank VirusTotal for allowing us to scan a huge volume of files.

Table 8: Viruses found from 111,981 unverified files

Report	*# Found*	*# VAs*	*Files*
Heuristic.BehavesLike. Exploit.CodeExec.F	34	8	setkeycodes; pt_chown; word-list-compress; ...
TROJ_GEN.F47V0802	7	3	mksock; setcap
Win32.Trojan	2	1	0000; 0001 (postgres 9.1)
Win32.TRCrypt.Upkm	1	1	batik-svg-dom-1.7.jar (tomcat 6.0)
Heuristic.LooksLike. HTML.Suspicious-URL.H	1	1	BookmarkFile.pod (perl 5)

"Heuristic.BehavesLike.Exploit" is the most common report, being flagged 34 times across 8 VAs. To ensure that those are not false alarms, we looked at the number of times that a suspicious file (e.g., `setkeycodes`, `ctrlaltdel`) appeared in the unverified files list across all the VAs, and the number of times it was flagged. For most of those files, there was a difference between the two numbers, indicating that the flagged ones are, indeed, potentially malicious. For instance, `ctrlaltdel` appeared in 13 of the VAs as unverified, but only 6 of them were flagged by VirusTotal.

All of the 10 VAs infected with a potentially malicious file belong to ILG 3 (see Table 4). This is somewhat expected as the number of unverified files is much greater in ILG 3. Interestingly though just 7 of those VAs belong to Cluster 2 (see Table 4) which is characterized by the large number of low-integrity packages. We note that from the 14 VAs that are part of Cluster 2, virus scanners flagged only about 50% of them as potentially untrustworthy. This further reinforces the need for our framework as it can also flag the other 50% that are equally suspicious and would likely not meet a customers expectations when considering a VA for a standard (or well-known) workflow (e.g., an Apache web server).

Packages that contain these malicious files will get a score of ⓪, discouraging consumers from picking VAs that contain them. The providers could define a policy that disallows such VAs from being published in the first place. The results clearly indicate the need to check VAs against virus scanners. To that end, our approach will give a great performance boost by minimizing the number of files that need to be checked. Only the unverified files (i.e., files that fail whitelist checks) need to be checked, which represent only a small portion of the total number of files; 90% of the VAs contain less than 4.5% of unverified files (see Figure 3). For VAs that fall under ILG C (see Table 3), this would imply checking only about 2,012 unverified files through a virus scanner, as opposed to checking all 52,198 files.

5.3 Combining integrity scores with security vulnerability assessment results

The expanded integrity scores would still provide only a partial view of how secure a VA might be. A VA that is equipped with 100% cleanly installed packages is *likely to be* more secure than a VA that has some modified packages, but not always. If some of the clean packages have known security vulnerabilities, then the first VA might be equally insecure or even more insecure than the second VA. Hence, to make strong claims about VA security, we also need to consider whether packages have known security vulnerabilities, and combine this information with their integrity scores.

Numerous security companies today assess software vulnerabilities and report them. This information could be used to build a more complete verification report, also indicating which packages are known to be secure or have known vulnerabilities. The corresponding integrity scores would then provide a stronger indication of how secure that VA is. We imagine that the providers could make use of this information and define a whitelist of software packages or combinations of packages that are known to be secure; it would also be possible to create a blacklist of packages that are known to have security vulnerabilities or bugs. Compliance policies could then be defined based on the whitelist and blacklist, specifying which software combinations are allowed to be published and which should not be admitted.

6. CONCLUSIONS AND FUTURE WORK

We studied the integrity of software packages in real-world virtual appliances (VAs) through a software whitelist-based framework, and found high variance in the software integrity across VAs. Our analysis of 151 Amazon VAs showed that about 9% of those real-world VAs were configured with a significant portion of modified (low-integrity) packages without any indication of the publishers' efforts to customize them—demonstrating the need for a priori assessment of software integrity to help consumers pick correctly installed VAs. Virus scanners were able to flag only about half of the VAs in that 9%, showing that the whitelisting-based integrity assessment has a role to play, and may complement blacklisting techniques like virus scanners in helping consumers pick more reliable VAs. While whitelist-based techniques are hard to realize in practice, we find that because a large number of software packages are installed on multiple VAs in the case of appliance stores, the rate at which the whitelist size grows slows down over time.

In the future, we plan to further expand the integrity assessment framework to compute an overall trust score for a VA based on the integrity scores assigned to individual software packages. In doing so, we will consider other factors such as the criticality of a software package and consumer's preference order for packages. Future work may also look at VAs that are built on Windows or other Linux distributions (e.g., `dpkg`-based distributions) to further validate the key findings presented in this paper.

Acknowledgments

This material is based on research sponsored in part by the Air Force Research Laboratory and the Air Force Office of Scientific Research, under agreement number FA8750-11-2-0084. The authors would like to thank Jenny Applequist for her editorial assistance, and the anonymous reviewers for their careful attention and insightful comments.

7. REFERENCES

[1] Amazon Elastic Compute Cloud. http://aws.amazon.com/ec2/.

[2] Amazon Machine Images (AMIs). http://aws.amazon.com/amis.

[3] Bit9 Global Software Registry. http://www.bit9.com/products/bit9-global-software-registry.php.

[4] BitNami Virtual Images. http://bitnami.org/learn_more/virtual_machines.

[5] CUBRID Virtual Images. http://www.cubrid.org/virtual_machine_images.

[6] IBM SmartCloud.

[7] NIST National Software Reference Library. http://www.nsrl.nist.gov/.

[8] Thunderflash Pre-Built Virtual Images. http://thunderflash.com/.

[9] Virustotal. https://www.virustotal.com/.

[10] VMWare Solution Exchange. https://solutionexchange.vmware.com/.

[11] Security Guidance for Critical Areas of Focus in Cloud Computing. http://www.cloudsecurityalliance.org/guidance/csaguide.pdf, April 2009.

[12] AntiVirus Performance Statistics. http://winnow.oitc.com/malewarestats.php, August 2012.

[13] G. Ammons, V. Bala, T. Mummert, D. Reimer, and X. Zhang. Virtual machine images as structured data: the mirage image library. In *Proceedings of the 3rd USENIX conference on Hot topics in cloud computing*, HotCloud'11, pages 22–22, Berkeley, CA, USA, 2011. USENIX Association.

[14] M. Armbrust, A. Fox, R. Griffith, A. D. Joseph, R. Katz, A. Konwinski, G. Lee, D. Patterson, A. Rabkin, I. Stoica, and M. Zaharia. A view of cloud computing. *Commun. ACM*, 53(4):50–58, Apr. 2010.

[15] K. D. Bowers, A. Juels, and A. Oprea. Hail: a high-availability and integrity layer for cloud storage. In E. Al-Shaer, S. Jha, and A. D. Keromytis, editors, *ACM Conference on Computer and Communications Security*, pages 187–198. ACM, 2009.

[16] K. D. Bowers, M. van Dijk, A. Juels, A. Oprea, and R. L. Rivest. How to tell if your cloud files are vulnerable to drive crashes. In Y. Chen, G. Danezis, and V. Shmatikov, editors, *ACM Conference on Computer and Communications Security*, pages 501–514. ACM, 2011.

[17] Bugiel, Sven and Nürnberger, Stefan and Pöppelmann, Thomas and Sadeghi, Ahmad-Reza and Schneider, Thomas. AmazonIA: when elasticity snaps back. In *Proceedings of the 18th ACM conference on Computer and communications security*, pages 389–400, New York, NY, USA, 2011. ACM.

[18] B. Danev, R. J. Masti, G. O. Karame, and S. Capkun. Enabling secure VM-vTPM migration in private clouds. In *Proceedings of the 27th Annual Computer Security Applications Conference*, pages 187–196, New York, NY, USA, 2011. ACM.

[19] T. Garfinkel and M. Rosenblum. When virtual is harder than real: security challenges in virtual machine based computing environments. In *Proceedings of the 10th conference on Hot Topics in Operating Systems - Volume 10*, HOTOS'05, pages 20–20, Berkeley, CA, USA, 2005. USENIX Association.

[20] J. H. Huh, H. Kim, J. Lyle, and A. Martin. Achieving attestation with less effort: an indirect and configurable approach to integrity reporting. In *Proceedings of the sixth ACM workshop on Scalable trusted computing*, pages 31–36, New York, NY, USA, 2011. ACM.

[21] K. R. Jayaram, C. Peng, Z. Zhang, M. Kim, H. Chen, and H. Lei. An empirical analysis of similarity in virtual machine images. In *Proceedings of the Middleware 2011 Industry Track Workshop*, Middleware '11, pages 6:1–6:6, New York, NY, USA, 2011. ACM.

[22] John A. Rice. *Mathematical Statistics and Data Analysis*, chapter 10, page 365. Duxbury Press, 2 edition, 1994.

[23] G. Kim and E. Spafford. The design and implementation of tripwire: A file system integrity checker. In *Proceedings of the 2nd ACM Conference on Computer and Communications Security*, pages 18–29. ACM, 1994.

[24] N. Leavitt. Is cloud computing really ready for prime time? *Computer*, 42(1):15 –20, jan. 2009.

[25] N. Quynh and Y. Takefuji. A novel approach for a file-system integrity monitor tool of xen virtual machine. In *Proceedings of the 2nd ACM symposium on Information, computer and communications security*, pages 194–202. ACM, 2007.

[26] D. Reimer, A. Thomas, G. Ammons, T. Mummert, B. Alpern, and V. Bala. Opening black boxes: using semantic information to combat virtual machine image sprawl. In *Proceedings of the fourth ACM SIGPLAN/SIGOPS international conference on Virtual execution environments*, VEE '08, pages 111–120, New York, NY, USA, 2008. ACM.

[27] T. Ristenpart, E. Tromer, H. Shacham, and S. Savage. Hey, you, get off of my cloud: exploring information leakage in third-party compute clouds. In *Proceedings of the 16th ACM conference on Computer and communications security*, CCS '09, pages 199–212, New York, NY, USA, 2009. ACM.

[28] T. Ristenpart and S. Yilek. When good randomness goes bad: Virtual machine reset vulnerabilities and hedging deployed cryptography. In *NDSS*. The Internet Society, 2010.

[29] H. Takabi, J. Joshi, and G. Ahn. Security and privacy challenges in cloud computing environments. *Security Privacy, IEEE*, 8(6):24 –31, nov.-dec. 2010.

[30] TCG. TCG Infrastructure Working Group Architecture Part II - Integrity Management. `http://www.trustedcomputinggroup.org/resources/infrastructure_work_group_architecture_part_ii__integrity_management_version_10`, November 2006.

[31] D. Vincenzetti and M. Cotrozzi. Anti tampering program. In *Proceedings of the Fourth {USENIX} Security Symposium, Santa Clara, CA*. USENIX, 1993.

[32] J. Wei, X. Zhang, G. Ammons, V. Bala, and P. Ning. Managing security of virtual machine images in a cloud environment. In *Proceedings of the 2009 ACM workshop on Cloud computing security*, CCSW '09, pages 91–96, New York, NY, USA, 2009. ACM.

[33] Y. Zhang, A. Juels, A. Oprea, and M. K. Reiter. Homealone: Co-residency detection in the cloud via side-channel analysis. In *IEEE Symposium on Security and Privacy*, pages 313–328. IEEE Computer Society, 2011.

[34] W. Zhou, P. Ning, X. Zhang, G. Ammons, R. Wang, and V. Bala. Always up-to-date: scalable offline patching of vm images in a compute cloud. In *Proceedings of the 26th Annual Computer Security Applications Conference*, ACSAC '10, pages 377–386, New York, NY, USA, 2010. ACM.

APPENDIX

A. AMORTIZING PERFORMANCE OVERHEADS

Mirage [13, 26] generates an index of a VA using the VA's filesystem structure in order to simplify maintenance and management of a large collection of images. During index generation, Mirage already computes hashes of all the files in the VA to detect duplicate files in the system. Integration of our framework with the Mirage library would involve only the addition of one step to look up the hash of the files in a whitelist when a file is first encountered. Thus, in addition to the other benefits of using Mirage, including reduced storage cost and query capabilities, the cost of producing the verification reports can be significantly amortized.

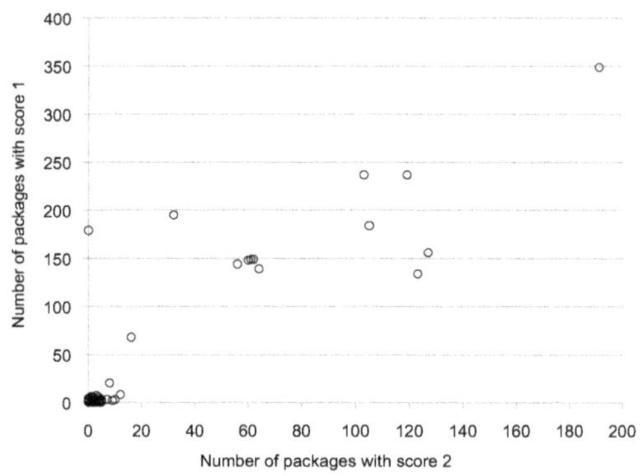

Figure 8: Number of software packages with score 1 vs. number of software packages with score 2

B. NUMBER OF MEDIUM-INTEGRITY PACKAGES VS NUMBER OF LOW-INTEGRITY PACKAGES

There is a correlation between the number of medium-integrity (partially clean) packages and the number of low-integrity (modified) packages (see Figure 8). As the number of packages with score ① increases, so does the number of packages with score ②, and vice versa.

C. TMP AND VAR FILES

Table 9: Most common files in "var" folders

# VAs	*Absolute path*
151	var/log/wtmp
151	var/log/spooler
151	var/log/secure
151	var/log/messages
151	var/log/maillog
151	var/lib/rpm/Triggername
151	var/lib/rpm/Sigmd5
151	var/lib/rpm/Sha1header
151	var/lib/rpm/Requireversion
151	var/lib/rpm/Requirename

As Figure 9 shows, most of the `var` files are log files and `rpm` database files, which, we argue, can be deleted without affecting the VA behavior. The `rpm` database, if needed, can be reconstructed after a VA is instantiated. We propose that temporary files be scrubbed from the VA before publication. Similarly, files in the `var` directory should also be scrubbed before publication, unless the software vendor metadata or the publisher's log explicitly mention the file as necessary.

Expressive Search on Encrypted Data

Junzuo Lai
School of Information Systems
Singapore Management University
Singapore 178902
Dept. of Computer Science
Jinan University
Guangzhou 510632, China
junzuolai@smu.edu.sg

Xuhua Zhou*
Dept. of Computer Science
Shanghai Jiao Tong University
Shanghai, China 200240
xuhuazhou2010@gmail.com

Robert H. Deng
School of Information Systems
Singapore Management University
Singapore 178902
robertdeng@smu.edu.sg

Yingjiu Li
School of Information Systems
Singapore Management University
Singapore 178902
yjli@smu.edu.sg

Kefei Chen
School of Science
Hangzhou Normal University
Zhejiang, China 310036
kfchen2004@gmail.com

ABSTRACT

Different from the traditional public key encryption, searchable public key encryption allows a data owner to encrypt his data under a user's public key in such a way that the user can generate search token keys using her secret key and then query an encryption storage server. On receiving such a search token key, the server filters all or related stored encryptions and returns matched ones as response.

Searchable pubic key encryption has many promising applications. Unfortunately, existing schemes either only support simple query predicates, such as equality queries and conjunctive queries, or have a superpolynomial blowup in ciphertext size and search token key size.

In this paper, based on the key-policy attribute-based encryption scheme proposed by Lewko *et al.* recently, we present a new construction of searchable public key encryption. Compared to previous works in this field, our construction is much more expressive and efficient and is proven secure in the standard model.

Categories and Subject Descriptors

E.3 [**Data Encryption**]: Public Key Cryptosystems; H.3 [**Information Storage and Retrieval**]: Information Search and Retrieval

*Corresponding author

General Terms

Design, Security

Keywords

Searchable Public Key Encryption, Public Key Encryption with Keyword Search, Anonymous Key-Policy Attribute-Based Encryption

1. INTRODUCTION

Consider a user Bob who sends email to another user Alice encrypted under Alice's public key. An email gateway is configured to check whether emails contain the keyword "urgent" so that it could route emails accordingly. On the other hand, Alice does not wish to give the gateway the ability to decrypt all her messages. Boneh *et al.* in [7] define and construct a mechanism – public key encryption with keyword search (PEKS) that allows Alice to provide a token key to the gateway so that the latter can test whether the word "urgent" is a keyword in the email while without learning anything else about the email.

In a PEKS scheme [7], to send an encrypted email message msg with keywords $s_1, \ldots, s_n$ to Alice, Bob computes

$$[E_{A_{pub}}(msg)||\mathsf{PEKS}(A_{pub}, s_1)|| \cdots ||\mathsf{PEKS}(A_{pub}, s_n)],$$

where A_{pub} is Alice's public key. After Alice submitting a token key $\mathrm{TK}_{s'}$ for keyword s', the email gateway can decide whether this encrypted message matches with $\mathrm{TK}_{s'}$ by testing each keyword encryption $\mathrm{PEKS}(A_{pub}, s_i)$ against $\mathrm{TK}_{s'}$ iteratively. If there exists s_i satisfying $s_i = s'$, the gateway deems that the entire encrypted message is matched and then takes appropriate actions on the email accordingly. The first part of the ciphertext above, i.e., $E_{A_{pub}}(msg)$, is normally ignored when studying a PEKS scheme, since its security can be achieved via the traditional public key encryption. Except stated explicitly, ciphertexts of a public key encryption with keyword search scheme consist of only the keyword encryptions. Therefore, informally, the basic security requirement for a public key encryption with keyword search is that the email gateway or the storage server does not learn any information about encrypted keywords unless it has the knowledge of a matched token key.

The PEKS scheme in [7] only supports equality queries. However, more expressive search filters, e. g., conjunctive keywords or even boolean formulas, are required in many applications. For comparability, we use the email gateway application scenario in [7] to illustrate this. Two typical expressive search filters at the gateway may look like:

Eg. 1: *sender: Bob* AND *priority: urgent*

Eg. 2: (*sender: Bob* AND *priority: urgent*) OR *subject: recruitment*

where the first example means that Alice wants the email gateway to return all urgent emails sent by Bob, and the second one means that all emails either sent by Bob with urgent priority or having a subject of recruitment should be forwarded to Alice.

With regard to the conjunctive keyword search problem presented by Eg. 1, security concerns and high storage overhead (i.e., storage overhead is exponential in the number of keyword fields) overrule two straightforward solutions, namely set intersection and meta keywords [26, 16]. The set intersection is built upon simple PEKS scheme, such as the one in [7]. Given a conjunction of keywords, the gateway is provided with a search capability for every individual keyword in the conjunction. For every keyword, the gateway finds the set of ciphertexts that match that keyword, then returns the intersection of all the sets. This approach allows the gateway to learn a lot of extra information in addition to the results of the conjunctive query. The meta keyword approach is to define a meta keyword for every possible conjunction of keywords. These meta keywords are then associated with messages like regular keywords. The obvious drawback of this approach is that m keywords require 2^m meta keywords in order to accommodate all possible conjunctive queries. We refer interested readers to [26, 16] for more details.

The schemes proposed in [26, 19, 10, 20, 33] solve the conjunctive keyword search problem in the public-key setting. Especially, Boneh and Waters [10] present a general framework for analyzing and constructing searchable public key encryption (S-PKE)[1] schemes for various families of predicates. Boneh and Waters [10] then construct public key encryption schemes that support comparison queries (such as greater-than) and general subset queries. They also support arbitrary conjunctions. However, to the best of our knowledge, except S-PKE schemes constructed from inner-product predicate encryption [20], existing schemes mostly focus on conjunctive keyword search and do not work in the situations involving keyword disjunction, such as Eg. 2.

The notion of inner-product predicate encryption (IPE) is first introduced by Katz *et al.* [20]. In a predicate encryption scheme, secret keys correspond to predicates and ciphertexts are associated with a set of attributes; a secret key SK_f corresponding to a predicate f can be used to decrypt a ciphertext associated with an attribute set I if and only if $f(I) = 1$. The special case of inner product predicates is obtained by having each attributes set correspond to a vector $\vec{x}$ and each predicate $f_{\vec{v}}$ correspond to a vector $\vec{v}$, where $f_{\vec{v}}(\vec{x}) = 1$ iff $\vec{x} \cdot \vec{v} = 0$ ($\vec{x} \cdot \vec{v}$ denotes the standard inner-product). IPE can be extended to construct a solution to disjunctive keyword search. However, as shown in [20], the solution has a *superpolynomial* blowup in ciphertext size and search token key size, thus is not efficient. As described in [20], the conjunction predicate AND_{s_1,s_2} where $\mathsf{AND}_{s_1,s_2}(x_1, x_2) = 1$ iff both $x_1 = s_1$ and $x_2 = s_2$, and the disjunction predicate OR_{s_1,s_2} where $\mathsf{OR}_{s_1,s_2} = 1$ iff either $x_1 = s_1$ or $x_2 = s_2$, can be encoded as the following bi-variate polynomials

$$\mathsf{AND}_{s_1,s_2}\text{:}\ p'(x_1, x_2) = r \cdot (x_1 - s_1) + (x_2 - s_2),$$

$$\mathsf{OR}_{s_1,s_2}\text{:}\ p''(x_1, x_2) = (x_1 - s_1) \cdot (x_2 - s_2),$$

where r in the first polynomial is a random element chosen from a proper domain. Following this principle, we could easily encode Eg. 2 as a tri-variate polynomial

$$\begin{aligned}
&\text{Eg.}2_{s_1,s_2,s_3}(x_1, x_2, x_3) \\
&= (r \cdot (x_1 - s_1) + (x_2 - s_2)) \cdot (x_3 - s_3) \\
&= 0 \cdot x_1x_2x_3 + 0 \cdot x_1x_2 + r \cdot x_1x_3 + 1 \cdot x_2x_3 \\
&\quad +(-rs_3) \cdot x_1 + (-s_3) \cdot x_2 + (-rs_1 - s_2) \cdot x_3 \\
&\quad +s_3(rs_1 + s_2),
\end{aligned}$$

where s_1, s_2, s_3 denote "*sender: Bob*", "*priority: urgent*" and "*subject: recruitment*", respectively. To support the search type in Eg. 2, the sender Bob could take advantage of IPE to encrypt his email with the vector $\vec{x} = (x_1x_2x_3, x_1x_2, x_1x_3, x_2x_3, x_1, x_2, x_3, 1)$, where x_1, x_2, x_3 are keywords in the email; the receiver Alice generates her token key using the predicate vector $\vec{v} = (0, 0, r, 1, -rs_3, -s_3, -rs_1 - s_2, s_3(rs_1 + s_2))$ (i.e., the coefficients of the above equation). Obviously, the gateway can successfully match Alice's query to the encrypted email as long as the search filter Eg. 2 is satisfied. As pointed out in [20], the complexity of the resulting scheme is proportional to d^t, where t is the number of variables and d is the maximum degree (of the resulting polynomial) in each variable.

These facts motivate us to construct an efficient public key encryption supporting expressive search.

1.1 Our Contribution

In this paper, based on the key-policy attribute-based encryption (KP-ABE) scheme proposed by Lewko *et al.* [22] recently, we present an efficient construction of a S-PKE scheme which supports arbitrary monotone boolean predicate, such as Eg. 1 and Eg. 2. We prove that our scheme is secure in the standard model. Inheriting from [22], our scheme has the restriction that each keyword field can only be used once in a predicate.

In a KP-ABE scheme [30, 17], every ciphertext is associated with a set of attributes (i.e., keywords in S-PKE), and every user's secret key is associated with an access structure on attributes (i.e. search predicate in S-PKE). A user is able to decrypt a ciphertext only if the set of attributes associated with the ciphertext satisfies the access structure associated with the user's private key. However, the attributes (i.e., keywords) associated with ciphertexts in existing KP-ABE scheme, including Lewko *et al.* 's KP-ABE scheme [22], is not anonymous, thus a KP-ABE scheme can not be used

[1] S-PKE is a generalization of PEKS. PEKS refers to public key encryption supporting simple encrypted keyword search predicates such as equality queries while S-PKE refers to public key encryption supporting more expressive keyword search predicates.

as S-PKE. In our S-PKE scheme, the ciphertext does not reveal any information about the keywords. In fact, our S-PKE scheme can be easily extended to obtain the first efficient anonymous KP-ABE scheme, in which given a ciphertext a probabilistic polynomial time adversary cannot learn any information about the associated attribute set.

1.2 Related Work

In this section, we briefly review literature on searchable public key encryption.

Boneh *et al.* [7] initiate the research on PEKS and give a specific construction, known as the BDOP-PEKS scheme, which only supports equality queries. Abdalla *et al.* [1] formally define the property of consistency for PEKS, and state the relation between PEKS and anonymous identity based encryption (IBE). Based on different techniques or conditions, several PEKS constructions, also supporting equality queries, are presented in [13, 21, 5]. Noticing that the encryption in BDOP-PEKS and its followings is not invertible, Fuhr and Paillier [15] introduce the ability of decryption to searchable encryption (DSE) in which the receiver is allowed to decrypt the keyword ciphertext using an additional decryption algorithm.

Park *et al.* [26] propose the notion of public key encryption with conjunctive keyword search (PECK), then Hwan and Lee [19] make improvement on the sizes of ciphertext and private key, and extend the technique to multi-user setting. Zhang and Zhang [33] study a similar problem, namely conjunctive with subset keywords search (PECSK). Bringer *et al.* [11] take advantage of Bloom Filter [6] to construct an error-tolerant searchable encryption, permitting to search on encrypted data with only an approximation of some keywords. Boneh and Waters [10] present a general framework for analyzing and constructing S-PKE for various families of predicates and construct several schemes that support arbitrary conjunctions. Katz *et al.* [20] propose the notion of inner-product predicate encryption (IPE), which can be extended to construct S-PKE with disjunctive keyword search. However, as shown in [20], the resulting solution suffers from a *superpolynomial* blowup in ciphertext size and search token key size.

In addition to designing schemes with more expressive search criteria, there are also efforts [2, 34] studying the combination of a public key encryption (PKE) scheme and a PEKS scheme. Baek *et al.* [2] call their proposed scheme "PKE/PEKS" and define its security against chosen ciphertext attack (IND-PKE/PEKS-CCA). The resulting construction is based on a variation of ElGamal encryption and BDOP-PEKS, and is proved secure in the random oracle model. Zhang and Imai [34] give a generic construction which is based on secure PEKS schemes and tag-KEM/DEM schemes, and is proved secure without random oracles.

Some literature aim at enhancing the original security definition of PEKS in [7]. Byun *et al.* [12] define off-line keyword guessing (KG) attacks and show that BDOP-PEKS is insecure against KG attacks unless there exists a secure channel between the receiver and the server (i.e., the storage server or the email gateway). PEKS schemes which are immune to KG attacks (namely PEKS with a designated server, dPEKS for short [29] and its enhancement [27], or secure channel free PEKS, SCF-PEKS for short [3, 18, 14]) have also been reported in the literature. Tang and Chen [32] provide another approach to resist KG attacks, but their construction seems more like a searchable encryption scheme in the private-key setting. The security of the search token key (trapdoor), which ensures an adversary cannot learn the search criteria from a token key generated by the receiver, is discussed in [9, 28] and it is known that such a security notion can only be achieved in the private-key setting [31].

We omit the literature on searchable encryption in the private-key setting since they are outside the scope of this paper.

1.3 Organization

The rest of this paper is organized as follows. Section 2 gives some preliminaries and formally defines security of our S-PKE. Section 3 describes the proposed construction and its security proof. Section 4 concludes the paper.

2. PRELIMINARIES

If S is a set, then $s \xleftarrow{\$} S$ denotes the operation of picking an element s uniformly at random from S. Let $\mathbb{N}$ denote the set of natural numbers. If $\lambda \in \mathbb{N}$ then 1^λ denotes the string of λ ones. Let $z \leftarrow \mathsf{A}(x, y, \dots)$ denote the operation of running an algorithm A with inputs $(x, y, \dots)$ and output z. A function $f(\lambda)$ is *negligible* if for every $c > 0$ there exists a λ_c such that $f(\lambda) < 1/\lambda^c$ for all $\lambda > \lambda_c$.

2.1 Access Structures

DEFINITION 1 (ACCESS STRUCTURE [4]). *Let $\{P_1, \dots, P_n\}$ be a set of parties. A collection $\mathbb{A} \subseteq 2^{\{P_1,\dots,P_n\}}$ is monotone if $\forall B, C$: if $B \in \mathbb{A}$ and $B \subseteq C$, then $C \in \mathbb{A}$. An access structure (respectively, monotone access structure) is a collection (respectively, monotone collection) $\mathbb{A}$ of non-empty subsets of $\{P_1, \dots, P_n\}$, i.e., $\mathbb{A} \subseteq 2^{\{P_1,\dots,P_n\}} \backslash \{\emptyset\}$. The sets in $\mathbb{A}$ are called authorized sets, and the sets not in $\mathbb{A}$ are called unauthorized sets.*

In our context, keywords play the role of parties and we restrict our attention to monotone access structures (i.e., search predicates). It is possible to (inefficiently) realize general access structures using our techniques by treating the negation of a keyword as a separate keyword.

2.2 Linear Secret Sharing Schemes

Our construction will employ linear secret-sharing schemes (LSSS). We use the definition adapted from [4]:

DEFINITION 2 (LINEAR SECRET-SHARING SCHEMES). *A secret sharing scheme Π over a set of parties $\mathcal{P}$ is called linear (over $\mathbb{Z}_p$) if*

1. *The shares for each party form a vector over $\mathbb{Z}_p$.*
2. *There exists a matrix $\mathbf{A}$ with ℓ rows and n columns called the share-generating matrix for Π. For all $i = 1, \dots, \ell$, the ith row of $\mathbf{A}$ is labeled by a party $\rho(i)$ (ρ is a function from $\{1, \dots, \ell\}$ to $\mathcal{P}$). When we consider the column vector $v = (s, r_2, \dots, r_n)$, where $s \in \mathbb{Z}_p$ is the secret to be shared, and $r_2, \dots, r_n \in \mathbb{Z}_p$ are randomly chosen, then $\mathbf{A}v$ is the vector of ℓ shares of the secret s according to Π. The share $(\mathbf{A}v)_i$ belongs to party $\rho(i)$.*

It is shown in [4] that every linear secret-sharing scheme according to the above definition also enjoys the linear reconstruction property, defined as follows. Suppose that Π

is an LSSS for the access structure $\mathbb{A}$. Let $S \in \mathbb{A}$ be any authorized set, and let $I \subset \{1, \ldots, \ell\}$ be defined as $I = \{i|\rho(i) \in S\}$. Then there exist constants $\{\omega_i \in \mathbb{Z}_p\}_{i \in I}$ such that, if $\{\lambda_i\}$ are valid shares of any secret s according to Π, then $\sum_{i \in I} \omega_i \lambda_i = s$. Let A_i denotes the ith row of $\mathbf{A}$, we have $\sum_{i \in I} \omega_i A_i = (1, 0, \ldots, 0)$. These constants $\{\omega_i\}$ can be found in time polynomial in the size of the share-generation matrix $\mathbf{A}$ [4]. Note that, for unauthorized sets, no such constants $\{\omega_i\}$ exist.

Boolean Formulas Access structures (i.e., search predicates) might also be described in terms of monotonic boolean formulas. Using standard techniques [4] one can convert any monotonic boolean formula into an LSSS representation. We can represent the boolean formula as an access tree. An access tree of ℓ nodes will result in an LSSS matrix of ℓ rows. We refer the reader to the appendix of [24] for a discussion on how to perform this conversion.

2.3 Searchable Public Key Encryption

During the rest of the paper, we follow the general framework and notation for S-PKE as defined in [10]. Suppose a user Bob is about to send an encrypted document to Alice with keywords $s_1, \ldots, s_n$, where n is the number of keyword fields. If documents were emails for example, we could define 4 keyword fields, such as "From", "To", "Date" and "Subject". Bob sends the following message:

$$[E_{A_{pub}}[msg], \ \mathsf{S\text{-}PKE}(A_{pub}, (s_1, \ldots, s_n))],$$

where A_{pub} is Alice's public key, msg is the email body, and S-PKE is an algorithm with properties discussed below. To simplify the description, we ignore $E_{A_{pub}}[M]$ that can be encrypted with any secure public key encryption.

Without loss of generality, we assume that all keyword fields are defined for every document, as employed in [16]. From here onwards, we identify a document with the vector of n keywords. We denote a document by $\mathrm{D} = (s_1, \ldots, s_n)$, where s_i is the keyword of document D in the ith keyword field. A S-PKE scheme consists of the following four algorithms:

$\mathsf{Setup}(1^\lambda)$ takes as input a security parameter λ and outputs a public key PK and secret key SK.

$\mathsf{Encrypt}(\mathsf{PK}, \mathrm{D} = (s_1, \ldots, s_n))$ takes as input the public key PK and a document $\mathrm{D} = (s_1, \ldots, s_n)$ and outputs a ciphertext C_{D}.

$\mathsf{GenToken}(\mathsf{PK}, \mathsf{SK}, \mathcal{P})$ takes as input the public key PK, the secret key SK and a predicate $\mathcal{P}$ and outputs a token key $\mathsf{TK}_{\mathcal{P}}$.

$\mathsf{Test}(\mathsf{PK}, \mathsf{TK}_{\mathcal{P}}, C_{\mathrm{D}})$ takes as input the public key PK, a token key $\mathsf{TK}_{\mathcal{P}} \leftarrow \mathsf{GenToken}(\mathsf{PK}, \mathsf{SK}, \mathcal{P})$ and a ciphertext $C_{\mathrm{D}} \leftarrow \mathsf{Encrypt}(\mathsf{PK}, \mathrm{D} = (s_1, \ldots, s_n))$. It outputs "yes" if the keywords $(s_1, \ldots, s_n)$ satisfies the predicate $\mathcal{P}$ (i.e., $\mathcal{P}(\mathrm{D}) = 1$) and "no" otherwise.

We now give the security model for S-PKE in the sense of semantic-security. We need to ensure that an $\mathsf{S-PKE.Encrypt}$ $(A_{pub}, \mathrm{D} = (s_1, \ldots, s_n))$ does not reveal any information about D unless $\mathsf{TK}_{\mathcal{P}}$ with $\mathcal{P}(\mathrm{D}) = 1$ is available. We define security against an active attacker who is able to obtain token keys $\mathsf{TK}_{\mathcal{P}}$ for any $\mathcal{P}$ of his choice. Even under such attack the attacker should not be able to distinguish an encryption of a document D_0 from an encryption of a document D_1 for which he did not obtain the token key $\mathsf{TK}_{\mathcal{P}}$ such that $\mathcal{P}(\mathrm{D}_0) = 1$ or $\mathcal{P}(\mathrm{D}_1) = 1$. Formally, we define security against an active attacker $\mathcal{A}$ using the following game between a challenger and the attacker:

Setup: The challenger runs $\mathsf{Setup}(1^\lambda)$ to obtain a public key PK and secret key SK. It gives the public key PK to the adversary $\mathcal{A}$ and keeps SK to itself.

Query phase 1: The adversary $\mathcal{A}$ adaptively queries the challenger for token keys corresponding to predicates $\mathcal{P}_1, \ldots, \mathcal{P}_q$. In response, the challenger runs $\mathsf{TK}_{\mathcal{P}_i} \leftarrow \mathsf{GenToken}(\mathsf{PK}, \mathsf{SK}, \ \mathcal{P}_i)$ and gives the token key $\mathsf{TK}_{\mathcal{P}_i}$ to $\mathcal{A}$, for $1 \le i \le q$.

Challenge: The adversary $\mathcal{A}$ submits two documents $\mathrm{D}_0, \mathrm{D}_1$, subject to the restriction that, D_0 and D_1 cannot satisfy any of queried predicates. The challenger selects a random bit $\beta \in \{0, 1\}$, sets $C_{\mathrm{D}_\beta} \leftarrow \mathsf{Encrypt}(\mathsf{PK}, \mathrm{D}_\beta)$ and sends C_{D_β} to the adversary as its challenge ciphertext.

Query phase 2: The adversary continues to adaptively query the challenger for token keys corresponding to predicates with the added restriction that none of these can be satisfied by D_0 or D_1.

Guess: The adversary $\mathcal{A}$ outputs its guess $\beta' \in \{0, 1\}$ for β and wins the game if $\beta = \beta'$.

The advantage of the adversary in this game is defined as $|\Pr[\beta = \beta'] - 1/2|$ where the probability is taken over the random bits used by the challenger and the adversary.

DEFINITION 3. *A S-PKE scheme is secure if all polynomial time adversaries have at most a negligible advantage in this security game.*

2.4 Composite Order Bilinear Groups

We will construct our scheme in composite order bilinear groups whose order is the product of four distinct primes. Composite order bilinear groups were first introduced in [8].

Let $\mathcal{G}$ be an algorithm that takes as input a security parameter 1^λ and outputs a tuple $(p_1, p_2, p_3, p_4, \mathbb{G}, \mathbb{G}_T, e)$, where p_1, p_2, p_3, p_4 are distinct primes, $\mathbb{G}$ and $\mathbb{G}_T$ are cyclic groups of order $N = p_1 p_2 p_3 p_4$, and $e : \mathbb{G} \times \mathbb{G} \to \mathbb{G}_T$ is a map such that

1. (Bilinear) $\forall g, h \in \mathbb{G}, a, b \in \mathbb{Z}_N, e(g^a, h^b) = e(g, h)^{ab}$;

2. (Non-degenerate) $\exists g \in \mathbb{G}$ such that $e(g, g)$ has order N in $\mathbb{G}_T$.

We further require that multiplication in $\mathbb{G}$ and $\mathbb{G}_T$, as well as the bilinear map e, are computable in time polynomial in λ. We use $\mathbb{G}_{p_1}, \mathbb{G}_{p_2}, \mathbb{G}_{p_3}, \mathbb{G}_{p_4}$ to denote the subgroups of $\mathbb{G}$ having order p_1, p_2, p_3, p_4, respectively. Observe that $\mathbb{G} = \mathbb{G}_{p_1} \times \mathbb{G}_{p_2} \times \mathbb{G}_{p_3} \times \mathbb{G}_{p_4}$. Note also that if $g_1 \in \mathbb{G}_{p_1}$ and $g_2 \in \mathbb{G}_{p_2}$ then $e(g_1, g_2) = 1$. A similar rule holds whenever e is applied to elements in distinct subgroups.

We now state the complexity assumptions we use. Utilizing the theorems proposed in [20], one can easily to prove that the assumptions hold in the generic group model.

ASSUMPTION 1. *Let $\mathcal{G}$ be as above. We define the following distribution:*

$$(p_1, p_2, p_3, p_4, \mathbb{G}, \mathbb{G}_T, e) \leftarrow \mathcal{G}(1^\lambda),\ N = p_1p_2p_3p_4,$$
$$g \xleftarrow{\$} \mathbb{G}_{p_1},\ X_3 \xleftarrow{\$} \mathbb{G}_{p_3},\ X_4 \xleftarrow{\$} \mathbb{G}_{p_4},$$
$$D = (\mathbb{G}, \mathbb{G}_T, N, e, g, X_3, X_4),$$
$$T_1 \xleftarrow{\$} \mathbb{G}_{p_1} \times \mathbb{G}_{p_2},\ T_2 \xleftarrow{\$} \mathbb{G}_{p_1}.$$

The advantage of an algorithm $\mathcal{A}$ in breaking Assumption 1 is defined as

$$Adv^1_{\mathcal{A}} = |\Pr[\mathcal{A}(D, T_1) = 1] - \Pr[\mathcal{A}(D, T_2) = 1]|.$$

DEFINITION 4. *we say $\mathcal{G}$ satisfies Assumption 1 if for any polynomial time algorithm $\mathcal{A}$, $Adv^1_{\mathcal{A}}$ is negligible.*

ASSUMPTION 2. *Let $\mathcal{G}$ be as above. We define the following distribution:*

$$(p_1, p_2, p_3, p_4, \mathbb{G}, \mathbb{G}_T, e) \leftarrow \mathcal{G}(1^\lambda),\ N = p_1p_2p_3p_4,$$
$$g, X_1 \xleftarrow{\$} \mathbb{G}_{p_1},\ X_2, Y_2 \xleftarrow{\$} \mathbb{G}_{p_2},\ X_3, Y_3 \xleftarrow{\$} \mathbb{G}_{p_3},\ X_4 \xleftarrow{\$} \mathbb{G}_{p_4},$$
$$D = (\mathbb{G}, \mathbb{G}_T, N, e, g, X_1X_2, Y_2Y_3, X_3, X_4),$$
$$T_1 \xleftarrow{\$} \mathbb{G}_{p_1} \times \mathbb{G}_{p_2} \times \mathbb{G}_{p_3},\ T_2 \xleftarrow{\$} \mathbb{G}_{p_1} \times \mathbb{G}_{p_3}.$$

The advantage of an algorithm $\mathcal{A}$ in breaking Assumption 2 is defined as

$$Adv^2_{\mathcal{A}} = |\Pr[\mathcal{A}(D, T_1) = 1] - \Pr[\mathcal{A}(D, T_2) = 1]|.$$

DEFINITION 5. *we say $\mathcal{G}$ satisfies Assumption 2 if for any polynomial time algorithm $\mathcal{A}$, $Adv^2_{\mathcal{A}}$ is negligible.*

ASSUMPTION 3. *Let $\mathcal{G}$ be as above. We define the following distribution:*

$$(p_1, p_2, p_3, p_4, \mathbb{G}, \mathbb{G}_T, e) \leftarrow \mathcal{G}(1^\lambda),\ N = p_1p_2p_3p_4,$$
$$s \xleftarrow{\$} \mathbb{Z}_N,\ g, h \xleftarrow{\$} \mathbb{G}_{p_1}, g_2, X_2 \xleftarrow{\$} \mathbb{G}_{p_2},$$
$$X_3 \xleftarrow{\$} \mathbb{G}_{p_3},\ X_4, Z' \xleftarrow{\$} \mathbb{G}_{p_4},\ B_{24}, D_{24} \xleftarrow{\$} \mathbb{G}_{p_2} \times \mathbb{G}_{p_4},$$
$$D = (\mathbb{G}, \mathbb{G}_T, N, e, g, g_2, hX_2, hZ', g^sB_{24}, X_3, X_4, T),$$
$$T_1 = h^sD_{24},\ T_2 \xleftarrow{\$} \mathbb{G}_{p_1} \times \mathbb{G}_{p_2} \times \mathbb{G}_{p_4}.$$

The advantage of an algorithm $\mathcal{A}$ in breaking Assumption 3 is defined as

$$Adv^3_{\mathcal{A}} = |\Pr[\mathcal{A}(D, T_1) = 1] - \Pr[\mathcal{A}(D, T_2) = 1]|.$$

DEFINITION 6. *we say $\mathcal{G}$ satisfies Assumption 3 if for any polynomial time algorithm $\mathcal{A}$, $Adv^3_{\mathcal{A}}$ is negligible.*

3. OUR PROPOSED CONSTRUCTION

Recall that a document D is identified by a vector of n keywords $(s_1, \ldots, s_n)$, where s_i is the keyword of document D in the ith keyword field. For notational purposes, let i denote the ith keyword field. Our S-PKE scheme supports arbitrary monotone boolean predicate. We express an monotone boolean predicate by an LSSS $(\mathbf{A}, \rho, \mathcal{T})$, where $\mathbf{A}$ is a $\ell \times m$ share-generating matrix, ρ is a map from each row of $\mathbf{A}$ to a keyword field (i.e., ρ is a function from $\{1, \ldots, \ell\}$ to $\{1, \ldots, n\}$), $\mathcal{T}$ can be parsed as $(t_{\rho(1)}, \ldots, t_{\rho(\ell)})$ and $t_{\rho(i)}$ is the value (i.e., keyword) of keyword field $\rho(i)$ specified by the predicate.

Using our notations, a document $\mathrm{D} = (s_1, \ldots, s_n)$ satisfies a predicate $(\mathbf{A}, \rho, \mathcal{T})$ if and only if there exist $\mathcal{I} \subseteq \{1, \ldots, \ell\}$ and constants $\{\omega_i\}_{i\in\mathcal{I}}$ such that

$$\sum_{i\in\mathcal{I}} \omega_i A_i = (1, 0, \ldots, 0) \text{ and } s_{\rho(i)} = t_{\rho(i)} \text{ for } \forall i \in \mathcal{I},$$

where A_i denotes the ith row of $\mathbf{A}$. We also say that $\mathcal{I} \subseteq \{1, \ldots, \ell\}$ satisfies $(\mathbf{A}, \rho)$ if there exist constants $\{\omega_i\}_{i\in\mathcal{I}}$ such that $\sum_{i\in\mathcal{I}} \omega_i A_i = (1, 0, \ldots, 0)$. We define $\mathrm{I}_{\mathbf{A},\rho}$ as the set of minimum subsets of $\{1, \ldots, \ell\}$ that satisfies $(\mathbf{A}, \rho)$. By "minimum", we mean the subset cannot become smaller while still satisfying $(\mathbf{A}, \rho)$.

The proposed S-PKE scheme consists of the following algorithms:

Setup(1^λ): The setup algorithm first runs $\mathcal{G}(1^\lambda)$ to obtain $(p_1, p_2, p_3, p_4, \mathbb{G}, \mathbb{G}_T, e)$ with $\mathbb{G} = \mathbb{G}_{p_1} \times \mathbb{G}_{p_2} \times \mathbb{G}_{p_3} \times \mathbb{G}_{p_4}$, where $\mathbb{G}$ and $\mathbb{G}_T$ are cyclic groups of order $N = p_1p_2p_3p_4$. Next it chooses $g, u, h_1, \ldots, h_n \in \mathbb{G}_{p_1}$, $X_3 \in \mathbb{G}_{p_3}$, $X_4, Z, Z_0, Z_1, \ldots, Z_n \in \mathbb{G}_{p_4}$ and $\alpha \in \mathbb{Z}_N$ uniformly at random. The public key is published as $\mathsf{PK} = (N, gZ, e(g, g)^\alpha, U = uZ_0, \{H_i = h_i \cdot Z_i\}_{1\le i\le n}, X_4)$. The secret key is $\mathsf{SK} = (g, u, h_1, \ldots, h_n, X_3, \alpha)$.

Encrypt$(\mathsf{PK}, \mathrm{D} = (s_1, \ldots, s_n) \in \mathbb{Z}_N^n)$: This encryption algorithm chooses $s \in \mathbb{Z}_N$ and $Z_{1,0}, \{Z_{1,i}\}_{1\le i\le n} \in \mathbb{G}_{p_4}$ uniformly at random. The corresponding ciphertext $C_\mathrm{D} = (\widetilde{C}, C_0, \{C_i\}_{1\le i\le n})$ is computed as

$$\widetilde{C} = e(g, g)^{\alpha s},\ C_0 = (gZ)^s \cdot Z_{1,0},\ C_i = (U^{s_i} H_i)^s \cdot Z_{1,i}.$$

GenToken$(\mathsf{PK}, \mathsf{SK}, \mathcal{P} = (\mathbf{A}, \rho, \mathcal{T}))$: Suppose $\mathbf{A}$ is an $\ell \times m$ matrix, ρ is a map from each row A_j of $\mathbf{A}$ to $\{1, \ldots, n\}$ and $\mathcal{T} = (t_{\rho(1)}, \ldots, t_{\rho(\ell)}) \in \mathbb{Z}_N^\ell$. The key generation algorithm chooses a random vectors $v \in \mathbb{Z}_N^m$ such that $\mathbf{1} \cdot v = \alpha$. (Here, $\mathbf{1}$ denotes the vector with the first entry equal to 1 and the rest equal to 0). For each row A_j of $\mathbf{A}$, it chooses a random $r_j \in \mathbb{Z}_N$ and random elements $R_{1,j}, R_{2,j} \in \mathbb{G}_{p_3}$. The token key $\mathsf{TK}_\mathcal{P} = ((\mathbf{A}, \rho, \mathcal{T}),\ \{K_{1,j}, K_{2,j}\}_{1\le j\le \ell})$ is computed as

$$K_{1,j} = g^{A_j \cdot v}(u^{t_{\rho(j)}} h_{\rho(j)})^{r_j} R_{1,j},\ K_{2,j} = g^{r_j} R_{2,j}.$$

Test$(\mathsf{PK}, \mathsf{TK}_\mathcal{P}, C_\mathrm{D})$: Let $C_\mathrm{D} = (\widetilde{C}, C_0, \{C_i\}_{1\le i\le n})$ and $\mathsf{TK}_\mathcal{P} = ((\mathbf{A}, \rho, \mathcal{T}),\ \{K_{1,j}, K_{2,j}\}_{1\le j\le \ell})$. The test algorithm first calculates $\mathrm{I}_{\mathbf{A},\rho}$ from $(\mathbf{A}, \rho)$, where $\mathrm{I}_{\mathbf{A},\rho}$ denotes the set of minimum subsets of $\{1, \ldots, \ell\}$ that satisfies $(\mathbf{A}, \rho)$. It then checks if there exists an $\mathcal{I} \in \mathrm{I}_{\mathbf{A},\rho}$ that satisfies

$$\widetilde{C} = \prod_{i\in\mathcal{I}} (e(C_0, K_{1,i})/e(C_{\rho(i)}, K_{2,i}))^{\omega_i},$$

where $\sum_{i\in\mathcal{I}} \omega_i A_i = (1, 0, \ldots, 0)$. If no element in $\mathrm{I}_{\mathbf{A},\rho}$ satisfies the above equation, it outputs "no". Otherwise, it outputs "yes".

3.1 Efficiency

The size of the public key, a token key and a ciphertext are $(n + 3)|\mathbb{G}| + |\mathbb{G}_T|$, $2\ell|\mathbb{G}|$ and $(n + 1)|\mathbb{G}| + |\mathbb{G}_T|$, respectively, where $|\mathbb{G}|$ and $|\mathbb{G}_T|$ are the lengths of the bit-representation of a group element in $\mathbb{G}$ and $\mathbb{G}_T$ respectively. Compared with expressive searchable public key encryption scheme constructed from IPE [20], the size of a ciphertext (resp. a key) in our scheme is *linear* with the number of keyword fields n (resp. the size of the search predicate), not *superpolynomial.*

For a search predicate $(\mathbf{A}, \rho, \mathcal{T})$, let $\iota_1 = |\mathrm{I}_{\mathbf{A},\rho}|$, $\mathrm{I}_{\mathbf{A},\rho} = \{\mathrm{I}_1, \ldots, \mathrm{I}_{\iota_1}\}$ and $\iota_2 = |\mathrm{I}_1| + \cdots + |\mathrm{I}_{\iota_1}|$. The computational costs of an encryption, and a test under $(\mathbf{A}, \rho, \mathcal{T})$ are $(n+1)t_{\mathbb{G}\_m\_e} + t_{\mathbb{G}_T\_e}$ and $\leq 2\iota_2 t_p + \iota_1 t_{\mathbb{G}_T\_m\_e}$, respectively, where t_p, $t_{\mathbb{G}_T\_e}$, $t_{\mathbb{G}\_m\_e}$ and $t_{\mathbb{G}_T\_m\_e}$ are the computational costs of bilinear map, exponentiation in $\mathbb{G}_T$, multi exponentiation in $\mathbb{G}$ and $\mathbb{G}_T$, respectively.

3.2 Discussion

Our S-PKE scheme is based on the KP-ABE scheme [22] proposed by Lewko *et al.* . Since the attributes (i.e., keywords) in Lewko *et al.* 's KP-ABE scheme [22] is not anonymous, it is not enough to obtain a secure S-PKE. On the contrary, our S-PKE scheme can be easily extended to obtain the first efficient anonymous KP-ABE scheme. We remark that the KP-ABE scheme in [22] works in a small universe of attributes, while the keywords in our S-PKE scheme have a large universe (i.e., $\mathbb{Z}_N$).

Similar to the KP-ABE scheme in [22], our proposed S-PKE scheme has the restriction that each keyword field can only be used once in a predicate, which is called one-use S-PKE. We can obtain a secure S-PKE scheme where keyword fields are used multiple times (up to a constant number of uses fixed at setup) from a one-use scheme by applying the generic transformation given in Lewko *et al.* [22]. While the transformation does incur some cost in public key and ciphertext size, it does not increase the size of the token key. Utilizing the techniques proposed by Lewko and Waters [25] recently, it is possible to eliminate the above mentioned efficiency loss and allow unrestricted use of keyword fields while still proving security in the standard model.

3.3 Security

Note that in our construction, all components of the public key (except $e(g,g)^\alpha$) and the ciphertext (except $\widetilde{C} = e(g,g)^{\alpha s}$) have a $\mathbb{G}_{p_4}$ part. This formation allows us to prove that the ciphertext C_{D} does not reveal any information about D unless $\mathsf{TK}_{\mathcal{P}}$ with $\mathcal{P}(\mathrm{D}) = 1$ is available; it does not affect the test algorithm, since no component in a token key has a $\mathbb{G}_{p_4}$ part. We now state the security theorem of our S-PKE scheme.

THEOREM 1. *If Assumptions 1, 2 and 3 hold, then the proposed S-PKE scheme is secure.*

PROOF. Following the approach by Lewko and Waters [23], we define two additional structures: *semi-functional* ciphertexts and *semi-functional* keys. These will not be used in the real system, but will be used in our proof.

Semi-functional Ciphertext Let g_2 denote a generator of the subgroup $\mathbb{G}_{p_2}$. A semi-functional ciphertext is created as follows. We first use the encryption algorithm to form a normal ciphertext $C_{\mathrm{D}} = (\widetilde{C}, C_0, \{C_i\}_{1\leq i\leq n})$. Then, we choose a random exponent $c \in \mathbb{Z}_N$. We also choose random values $\eta_i \in \mathbb{Z}_N$ associated to keywords. The semi-functional ciphertext C is set to be

$$\left(\widetilde{C},\ C_0 \cdot g_2^c,\ \{C_i \cdot g_2^{c\eta_i}\}_{1\leq i\leq n}\right).$$

It should be noted that the values η_i are chosen randomly once and then fixed - these same values will also be involved in semi-functional keys which we will define below.

Semi-functional Key A semi-functional key will take on one of two forms. To create a semi-functional key, we first use the key generation algorithm to form a normal token key $\mathsf{TK}_{\mathcal{P}} = ((\mathbf{A}, \rho, \mathcal{T}),\ \{K_{1,j}, K_{2,j}\}_{1\leq j\leq \ell})$. Then, we choose random values $\gamma_j \in \mathbb{Z}_N$ associated with row j of the $\ell \times m$ matrix $\mathbf{A}$. We also choose a random vector $w \in \mathbb{Z}_N^m$ and set $\delta_j = A_j \cdot w$. The semi-functional key of type 1 is set as

$$((\mathbf{A}, \rho, \mathcal{T}),\ \{K_{1,j} \cdot g_2^{\delta_j + \gamma_j \eta_{\rho(j)}},\ K_{2,j} \cdot g_2^{\gamma_j}\}_{1\leq j\leq \ell}).$$

A semi-functional key of type 2 is formed without the terms $g_2^{\gamma_j \eta_{\rho(j)}}$ and $g_2^{\gamma_j}$ (one could also interpret this as setting $\gamma_j = 0$):

$$((\mathbf{A}, \rho, \mathcal{T}),\ \{K_{1,j} \cdot g_2^{\delta_j},\ K_{2,j}\}_{1\leq j\leq \ell}).$$

We will prove the security of our scheme based on Assumptions 1, 2 and 3 using a hybrid argument over a sequence of games. The first game, Game_{real} is the real security game (the ciphertext and all token keys are normal). In the next game, Game_0, all of token keys will be normal, but the challenge ciphertext will be semi-functional. We let q denote the number of token key queries made by the attacker. For k from 1 to q and ϱ from 1 to n, we define

$\mathsf{Game}_{k,1}$: In this game, the challenge ciphertext is semi-functional, the first $k-1$ token keys are semi-functional of type 2, the kth token key is semi-functional of type 1, and the remaining token keys are normal.

$\mathsf{Game}_{k,2}$: In this game, the challenge ciphertext is semi-functional, the first k token keys are semi-functional of type 2, and the remaining token keys are normal.

$\mathsf{Game}_{\mathsf{Final}_\varrho}$: In this game, all token keys are semi-functional of type 2, and the challenge ciphertext $C_{\mathrm{D}_\beta} = (\widetilde{C}, C_0, \{C_1, \ldots, C_n\})$ is a semi-functional encryption of D_β with $C_1, \ldots, C_\varrho$, each of which is randomly chosen from $\mathbb{G}_{p_1} \times \mathbb{G}_{p_2} \times \mathbb{G}_{p_4}$.

For notational purposes, we think of $\mathsf{Game}_{0,2}$ as another way of denoting Game_0 and $\mathsf{Game}_{\mathsf{Final}_0}$ as another way of denoting $\mathsf{Game}_{q,2}$. In the final game, $\mathsf{Game}_{\mathsf{Final}_n}$, all token keys are semi-functional, and the ciphertext $C_{\mathrm{D}_\beta} = (\widetilde{C}, C_0, \{C_1, \ldots, C_n\})$ is a semi-functional encryption with $C_1, \ldots, C_n$ randomly chosen from $\mathbb{G}_{p_1} \times \mathbb{G}_{p_2} \times \mathbb{G}_{p_4}$ (thus the ciphertext is independent of D_0 and D_1 provided by the adversary). It is clear that in the final game, no adversary can have advantage greater than 0.

We prove that these games are indistinguishable in the following four lemmas. Therefore, we conclude that the advantage of the adversary in Game_{real} (i.e., the real security game) is negligible. This completes the proof of Theorem 1. □

LEMMA 1. *Suppose that $\mathcal{G}$ satisfies Assumption 1. Then* Game_{real} *and* Game_0 *are computationally indistinguishable.*

PROOF. Suppose there exists an algorithm $\mathcal{A}$ that distinguishes Game_{real} and Game_0. Then we can build an algorithm $\mathcal{B}$ with non-negligible advantage in breaking Assumption 1. $\mathcal{B}$ is given g, X_3, X_4, T and will simulate Game_{real} or Game_0 with $\mathcal{A}$. $\mathcal{B}$ chooses $Z, Z_0, Z_1, \ldots, Z_n \in \mathbb{G}_{p_4}$ and $\alpha, a_0, a_1, \ldots, a_n \in \mathbb{Z}_N$ uniformly at random. It then sets $u = g^{a_0}, h_1 = g^{a_1}, \ldots, h_n = g^{a_n}, U = uZ_0, H_1 = h_1 Z_1, \ldots, H_n = h_n Z_n$, and sends $\mathcal{A}$ the public key:

$$\mathsf{PK} = (N, gZ, e(g,g)^\alpha, U, \{H_i\}_{1\leq i\leq n}, X_4).$$

It can generate normal token keys in response to $\mathcal{A}$'s token key requests by using the key generation algorithm, since it knows the $\mathsf{SK} = (g, u, h_1, \ldots, h_n, X_3, \alpha)$.

At some point, $\mathcal{A}$ sends $\mathcal{B}$ two documents $\mathrm{D}_0, \mathrm{D}_1$. $\mathcal{B}$ chooses $\beta \in \{0,1\}$ randomly and does the following:

1. $\mathcal{B}$ chooses random values $\widetilde{Z}_{1,0}, \{\widetilde{Z}_{1,i}\}_{1\le i\le n} \in \mathbb{G}_{p_4}$.
2. Let $\mathrm{D}_\beta = (s_{\beta,1}, \ldots, s_{\beta,n})$. $\mathcal{B}$ computes
$$\widetilde{C} = e(g^\alpha, T),\ C_0 = T \cdot \widetilde{Z}_{1,0},\ C_i = T^{a_0 s_{\beta,i} + a_i} \cdot \widetilde{Z}_{1,i}.$$
3. $\mathcal{B}$ sets the challenge ciphertext as $C_{\mathrm{D}_\beta} = (\widetilde{C}, C_0, \{C_i\}_{1\le i\le n})$ and sends it to $\mathcal{A}$.

If $T \xleftarrow{\$} \mathbb{G}_{p_1} \times \mathbb{G}_{p_2}$, let $T = g^s g_2^c$, then
$$\widetilde{C} = e(g,g)^{\alpha s},\ C_0 = (gZ)^s Z_{1,0} \cdot g_2^c,$$
$$C_i = (U^{s_{\beta,i}} H_i)^s Z_{1,i} \cdot g_2^{c\eta_i},$$
where $Z_{1,0} = \widetilde{Z}_{1,0} Z^{-s}$, $Z_{1,i} = \widetilde{Z}_{1,i}(Z_0^{s_{\beta,i}} Z_i)^{-s}$, $\eta_i = a_0 s_{\beta,i} + a_i$. This is a semi-functional ciphertext and $\mathcal{B}$ simulates Game_0. We note that the values of $a_0, s_{\beta,i}, a_i$ modulo p_1 are uncorrelated from their values modulo p_2, so this is properly distributed. If $T \xleftarrow{\$} \mathbb{G}_{p_1}$, it is easy to observe that this is a normal ciphertext and $\mathcal{B}$ simulates Game_{real}. Hence, $\mathcal{B}$ can use the output of $\mathcal{A}$ to distinguish between these possibilities for T. □

LEMMA 2. *Suppose that $\mathcal{G}$ satisfies Assumption 2. Then $\mathsf{Game}_{k-1,2}$ and $\mathsf{Game}_{k,1}$ are computationally indistinguishable.*

PROOF. Suppose there exists an algorithm $\mathcal{A}$ that distinguishes $\mathsf{Game}_{k-1,2}$ and $\mathsf{Game}_{k,1}$. Then we can build an algorithm $\mathcal{B}$ with non-negligible advantage in breaking Assumption 2. $\mathcal{B}$ is given $g, X_1X_2, Y_2Y_3, X_3, X_4, T$ and will simulate $\mathsf{Game}_{k-1,2}$ or $\mathsf{Game}_{k,1}$ with $\mathcal{A}$. $\mathcal{B}$ chooses $Z, Z_0, Z_1, \ldots, Z_n \in \mathbb{G}_{p_4}$ and $\alpha, a_0, a_1, \ldots, a_n \in \mathbb{Z}_N$ uniformly at random. It then sets $u = g^{a_0}, h_1 = g^{a_1}, \ldots, h_n = g^{a_n}, U = uZ_0, H_1 = h_1 Z_1, \ldots, H_n = h_n Z_n$, and sends $\mathcal{A}$ the public key:
$$\mathsf{PK} = (N, gZ, e(g,g)^\alpha, U, \{H_i\}_{1\le i\le n}, X_4).$$
Note that $\mathcal{B}$ knows the secret key $\mathsf{SK} = (g, u, h_1, \ldots, h_n, X_3, \alpha)$ associated with PK. Let us now explain how $\mathcal{B}$ answers the jth key query for a predicate $(\mathbf{A}, \rho, \mathcal{T} = (t_{\rho(1)}, \ldots, t_{\rho(\ell)}))$.

For $j < k$, $\mathcal{B}$ creates a semi-functional key of type 2 by choosing a random vector v such that $\mathbf{1} \cdot v = \alpha$, a random vector w', random exponents $r_i \in \mathbb{Z}_N$, random elements $R_{1,i}, R_{2,i} \in \mathbb{G}_{p_3}$, and setting:
$$K_{1,i} = g^{A_i \cdot v}(u^{t_{\rho(i)}} h_{\rho(i)})^{r_i} R_{1,i}(Y_2Y_3)^{A_i \cdot w'},\ K_{2,i} = g^{r_i} R_{2,i}.$$
We note that this is a properly distributed semi-functional key of type 2 because the value of $A_i \cdot w'$ modulo p_2 is uncorrelated to its value modulo p_3.

For $j > k$, $\mathcal{B}$ creates a normal token key by running the key generation algorithm since it knows SK.

To answer the kth key quest for $(\mathbf{A}, \rho, \mathcal{T} = (t_{\rho(1)}, \ldots, t_{\rho(\ell)}))$, $\mathcal{B}$ chooses a random vector v' such that $v' \cdot \mathbf{1} = \alpha$, a random vector w such that $w \cdot \mathbf{1} = 0$, random exponents $\widetilde{\gamma}_i \in \mathbb{Z}_N$, random elements $\widetilde{R}_{1,i}, \widetilde{R}_{2,i} \in \mathbb{G}_{p_3}$ and sets:
$$K_{1,i} = g^{A_i \cdot v'} T^{A_i \cdot w} T^{\widetilde{\gamma}_i(a_0 t_{\rho(i)} + a_{\rho(i)})} \widetilde{R}_{1,i},\ K_{2,i} = T^{\widetilde{\gamma}_i} \widetilde{R}_{2,i}.$$

We have the following observations. If $T \xleftarrow{\$} \mathbb{G}_{p_1} \times \mathbb{G}_{p_2} \times \mathbb{G}_{p_3}$, then T can be written as $g^r g_2^d R$, and
$$K_{1,i} = g^{A_i \cdot v}(u^{t_{\rho(i)}} h_{\rho(i)})^{r_i} R_{1,i} \cdot g_2^{\delta_i + \gamma_i \eta_{\rho(i)}},$$
$$K_{2,i} = g^{r_i} R_{2,i} \cdot g_2^{\gamma_i},$$
where $v = v' + rw$, $r_i = r\widetilde{\gamma}_i$, $\delta_i = dA_i w$, $\gamma_i = d\widetilde{\gamma}_i$, $R_{1,i} = R^{A_i w + \widetilde{\gamma}_i(a_0 t_{\rho(i)} + a_{\rho(i)})} \widetilde{R}_{1,i}$, $R_{2,i} = R^{\widetilde{\gamma}_i} \widetilde{R}_{2,i}$, $\eta_{\rho(i)} = a_0 t_{\rho(i)} + a_{\rho(i)}$. This is a semi-function key of type 1. Note that the values of $\widetilde{\gamma}_i, a_0, t_{\rho(i)}, a_{\rho(i)}$ modulo p_1 are uncorrelated from their values modulo p_2. If $T \xleftarrow{\$} \mathbb{G}_{p_1} \times \mathbb{G}_{p_3}$, this is a properly distributed normal token key.

At some point, $\mathcal{A}$ sends $\mathcal{B}$ two documents $\mathrm{D}_0, \mathrm{D}_1$. $\mathcal{B}$ chooses $\beta \in \{0,1\}$ randomly and does the following:

1. $\mathcal{B}$ chooses random values $\widetilde{Z}_{1,0}, \{\widetilde{Z}_{1,i}\}_{1\le i\le n} \in \mathbb{G}_{p_4}$.
2. Let $\mathrm{D}_\beta = (s_{\beta,1}, \ldots, s_{\beta,n})$. $\mathcal{B}$ computes
$$\widetilde{C} = e(g^\alpha, X_1X_2),\ C_0 = (X_1X_2) \cdot \widetilde{Z}_{1,0},$$
$$C_i = (X_1X_2)^{a_0 s_{\beta,i} + a_i} \cdot \widetilde{Z}_{1,i}.$$
3. $\mathcal{B}$ sets the challenge ciphertext as $C_{\mathrm{D}_\beta} = (\widetilde{C}, C_0, \{C_i\}_{1\le i\le n})$ and sends it to $\mathcal{A}$.

If we let $X_1X_2 = g^s g_2^c$, then
$$\widetilde{C} = e(g,g)^{\alpha s},\ C_0 = (gZ)^s Z_{1,0} \cdot g_2^c,$$
$$C_i = (U^{s_{\beta,i}} H_i)^s Z_{1,i} \cdot g_2^{c\eta_i},$$
where $Z_{1,0} = \widetilde{Z}_{1,0} Z^{-s}$, $Z_{1,i} = \widetilde{Z}_{1,i}(Z_0^{s_{\beta,i}} Z_i)^{-s}$, $\eta_i = a_0 s_{\beta,i} + a_i$. This is a semi-functional ciphertext. Note that the values of $a_0, s_{\beta,i}, a_i$ modulo p_1 are uncorrelated from their values modulo p_2.

Similar to the analysis in the proof of Lewko *et al.* 's KP-ABE scheme [22], the kth key and the challenge ciphertext are properly distributed. We can thus conclude that, if $T \xleftarrow{\$} \mathbb{G}_{p_1} \times \mathbb{G}_{p_2} \times \mathbb{G}_{p_3}$, then $\mathcal{B}$ has properly simulated $\mathsf{Game}_{k,1}$. If $T \xleftarrow{\$} \mathbb{G}_{p_1} \times \mathbb{G}_{p_3}$, then $\mathcal{B}$ has properly simulated $\mathsf{Game}_{k-1,2}$. Hence, $\mathcal{B}$ can use the output of $\mathcal{A}$ to distinguish between these possibilities for T. □

LEMMA 3. *Suppose that $\mathcal{G}$ satisfies Assumption 2. Then $\mathsf{Game}_{k,1}$ and $\mathsf{Game}_{k,2}$ are computationally indistinguishable.*

PROOF. Suppose there exists an algorithm $\mathcal{A}$ that distinguishes $\mathsf{Game}_{k,1}$ and $\mathsf{Game}_{k,2}$. Then we can build an algorithm $\mathcal{B}$ with non-negligible advantage in breaking Assumption 2. $\mathcal{B}$ is given $g, X_1X_2, Y_2Y_3, X_3, X_4, T$ and will simulate $\mathsf{Game}_{k,1}$ or $\mathsf{Game}_{k,2}$ with $\mathcal{A}$. $\mathcal{B}$ chooses $Z, Z_0, Z_1, \ldots, Z_n \in \mathbb{G}_{p_4}$ and $\alpha, a_0, a_1, \ldots, a_n \in \mathbb{Z}_N$ uniformly at random. It then sets $u = g^{a_0}, h_1 = g^{a_1}, \ldots, h_n = g^{a_n}, U = uZ_0, H_1 = h_1 Z_1, \ldots, H_n = h_n Z_n$, and sends $\mathcal{A}$ the public key:
$$\mathsf{PK} = (N, gZ, e(g,g)^\alpha, U, \{H_i\}_{1\le i\le n}, X_4).$$
The responses to all key queries and challenge ciphertexts are the same as in Lemma 2, except that the kth query which is given below.

To answer the kth key quest for $(\mathbf{A}, \rho, \mathcal{T} = (t_{\rho(1)}, \ldots, t_{\rho(\ell)}))$, $\mathcal{B}$ chooses a random vector v such that $v \cdot \mathbf{1} = \alpha$, a random vector w, random exponents $\widetilde{\gamma}_i \in \mathbb{Z}_N$, random elements $\widetilde{R}_{1,i}, \widetilde{R}_{2,i} \in \mathbb{G}_{p_3}$ and sets:
$$K_{1,i} = g^{A_i \cdot v}(Y_2Y_3)^{A_i \cdot w} T^{\widetilde{\gamma}_i(a_0 t_{\rho(i)} + a_{\rho(i)})} \widetilde{R}_{1,i},\ K_{2,i} = T^{\widetilde{\gamma}_i} \widetilde{R}_{2,i}.$$

We have the following observations. If $T \xleftarrow{\$} \mathbb{G}_{p_1} \times \mathbb{G}_{p_2} \times \mathbb{G}_{p_3}$, then T can be written as $g^r g_2^d R$, and

$$K_{1,i} = g^{A_i \cdot v}(u^{t_{\rho(i)}} h_{\rho(i)})^{r_i} R_{1,i} \cdot g_2^{\delta_i + \gamma_i \eta_{\rho(i)}},$$
$$K_{2,i} = g^{r_i} R_{2,i} \cdot g_2^{\gamma_i},$$

where $r_i = r\tilde{\gamma}_i$, $\delta_i = \log_{g_2} Y_2 \cdot A_i w$, $\gamma_i = d\tilde{\gamma}_i$, $R_{1,i} = Y_3^{A_i \cdot w} R^{A_i w + \tilde{\gamma}_i(a_0 t_{\rho(i)} + a_{\rho(i)})} \tilde{R}_{1,i}$, $R_{2,i} = R^{\tilde{\gamma}_i} \tilde{R}_{2,i}$, $\eta_{\rho(i)} = a_0 \cdot t_{\rho(i)} + a_{\rho(i)}$. This is a semi-function key of type 1. Note that the values of $\tilde{\gamma}_i, a_0, t_{\rho(i)}, a_{\rho(i)}$ modulo p_1 are uncorrelated from their values modulo p_2. If $T \xleftarrow{\$} \mathbb{G}_{p_1} \times \mathbb{G}_{p_3}$, this is a properly distributed semi-functional key of type 2.

We can conclude that, if $T \xleftarrow{\$} \mathbb{G}_{p_1} \times \mathbb{G}_{p_2} \times \mathbb{G}_{p_3}$, then $\mathcal{B}$ has properly simulated $\mathsf{Game}_{k,1}$. If $T \xleftarrow{\$} \mathbb{G}_{p_1} \times \mathbb{G}_{p_3}$, then $\mathcal{B}$ has properly simulated $\mathsf{Game}_{k,2}$. Hence, $\mathcal{B}$ can use the output of $\mathcal{A}$ to distinguish between these possibilities for T. □

LEMMA 4. *Suppose that $\mathcal{G}$ satisfies Assumption 3. Then $\mathsf{Game}_{\mathsf{Final}_{\varrho-1}}$ and $\mathsf{Game}_{\mathsf{Final}_{\varrho}}$ are computationally indistinguishable.*

PROOF. Suppose there exists an algorithm $\mathcal{A}$ that distinguishes $\mathsf{Game}_{\mathsf{Final}_{\varrho-1}}$ and $\mathsf{Game}_{\mathsf{Final}_{\varrho}}$. Then we can build an algorithm $\mathcal{B}$ with non-negligible advantage in breaking Assumption 3. $\mathcal{B}$ is given $(g, g_2, hX_2, hZ', g^s B_{24}, X_3, X_4, T)$ and will simulate $\mathsf{Game}_{\mathsf{Final}_{\varrho-1}}$ or $\mathsf{Game}_{\mathsf{Final}_{\varrho}}$ with $\mathcal{A}$. $\mathcal{B}$ chooses $Z, Z_0, Z_1, \ldots, Z_{\max\{1,\varrho-1\}}, Z_{\varrho+1} \ldots, Z_n \in \mathbb{G}_{p_4}$ and $\alpha, a_0, a_1, \ldots, a_{\max\{1,\varrho-1\}}, a_{\varrho+1} \ldots, a_n \in \mathbb{Z}_N$ uniformly at random. It then sets $u = g^{a_0}, h_1 = g^{a_1}, \ldots, h_{\max\{1,\varrho-1\}} = g^{\max\{1,\varrho-1\}}$, $h_{\varrho+1} = g^{\varrho+1} \ldots, h_n = g^{a_n}, U = uZ_0$, $H_1 = h_1 Z_1$, $\ldots$, $H_{\max\{1,\varrho-1\}} = h_{\max\{1,\varrho-1\}} \cdot Z_{\max\{1,\varrho-1\}}$, $H_\varrho = hZ'$ (implicitly setting $h_\varrho = h$ and $Z_\varrho = Z'$), $H_{\varrho+1} = h_{\varrho+1} Z_{\varrho+1}, \ldots$, $H_n = h_n Z_n$, and sends $\mathcal{A}$ the public key:

$$\mathsf{PK} = (N, gZ, e(g,g)^\alpha, U, \{H_i\}_{1 \le i \le n}, X_4).$$

Each time $\mathcal{B}$ is asked to provide a key for $(\mathbf{A}, \rho, \mathcal{T} = (t_{\rho(1)}, \ldots, t_{\rho(\ell)}))$ (where $\mathbf{A}$ is an $\ell \times m$ matrix), $\mathcal{B}$ creates a semi-functional key of type 2 by choosing a random vector v such that $\mathbf{1} \cdot v = \alpha$, a random vector w, random exponents $r_i \in \mathbb{Z}_N$, random elements $R_{1,i}, R_{2,i} \in \mathbb{G}_{p_3}$, and setting:

$$K_{1,i} = \begin{cases} g^{A_i \cdot v}(g^{a_0 t_{\rho(i)}}(hX_2))^{r_i} R_{1,i} g_2^{A_i \cdot w}, & \text{if } \rho(i) = \varrho; \\ g^{A_i \cdot v}(g^{a_0 t_{\rho(i)} + a_{\rho(i)}})^{r_i} R_{1,i} g_2^{A_i \cdot w}, & \text{otherwise.} \end{cases}$$
$$K_{2,i} = g^{r_i} R_{2,i}.$$

We note that $K_{1,i}$ can be written as $g^{A_i \cdot v}(u^{t_{\rho(i)}} h_{\rho(i)})^{r_i} R_{1,i} \cdot g_2^{\delta_i}$, where $\delta_i = A_i \cdot w + r_i \log_{g_2} X_2$ if $\rho(i) = \varrho$ or $\delta_i = A_i \cdot w$, so this is a properly distributed semi-functional key of type 2.

At some point, $\mathcal{A}$ sends $\mathcal{B}$ two documents $\mathrm{D}_0, \mathrm{D}_1$. $\mathcal{B}$ chooses $\beta \in \{0,1\}$ randomly and does the following:

1. $\mathcal{B}$ chooses random values $\tilde{Z}_{1,0}, \{\tilde{Z}_{1,i}\}_{\varrho \le i \le n} \in \mathbb{G}_{p_4}$.
2. Let $\mathrm{D}_\beta = (s_{\beta,1}, \ldots, s_{\beta,n})$. $\mathcal{B}$ chooses random elements $C_1, \ldots, C_{\max\{1,\varrho-1\}} \in \mathbb{G}_{p_1} \times \mathbb{G}_{p_2} \times \mathbb{G}_{p_4}$ and computes
$$\tilde{C} = e(g, g^s B_{24})^\alpha, \quad C_0 = g^s B_{24} \cdot \tilde{Z}_{1,0},$$
$$C_\varrho = (g^s B_{24})^{a_0 s_{\beta,\varrho}} \cdot T \cdot \tilde{Z}_{1,\varrho},$$
$$\{C_i = (g^s B_{24})^{a_0 s_{\beta,i} + a_i} \cdot \tilde{Z}_{1,i}\}_{\varrho < i \le n}.$$
3. $\mathcal{B}$ sets the challenge ciphertext as $C_{\mathrm{D}_\beta} = (\tilde{C}, C_0, \{C_i\}_{1 \le i \le n})$ and sends it to $\mathcal{A}$.

Let B_2, B_4 be the $\mathbb{G}_{p_2}, \mathbb{G}_{p_4}$ parts of B_{24} respectively. If $T = h^s D_{24}$, then

$$\tilde{C} = e(g,g)^{\alpha s}, \quad C_0 = (gZ)^s Z_{1,0} \cdot g_2^c,$$
$$\{C_i = (U^{s_{\beta,i}} H_i)^s Z_{1,i} \cdot g_2^{c\eta_i}\}_{\varrho \le i \le n},$$

where $c = \log_{g_2} B_2$, $Z_{1,0} = B_4 \tilde{Z}_{1,0} Z^{-s}$, $Z_{1,\varrho} = B_4^{a_0 s_{\beta,\varrho}} D_4 \tilde{Z}_{1,\varrho}(Z_0^{s_{\beta,\varrho}} Z')^{-s}$, $\{Z_{1,i} = B_4^{a_0 s_{\beta,i} + a_i} \tilde{Z}_{1,i}(Z_0^{s_{\beta,i}} Z_i)^{-s}\}_{\varrho < i \le n}$, $\eta_\varrho = (\log_{g_2}(B_2^{a_0 s_{\beta,\varrho}} D_2))/c$, $\{\eta_i = a_0 s_{\beta,i} + a_i\}_{\varrho < i \le n}$, D_2, D_4 are the $\mathbb{G}_{p_2}, \mathbb{G}_{p_4}$ parts of D_{24} respectively. Note that the values of $a_0, s_{\beta,i}, a_i$ modulo p_1 are uncorrelated to their values modulo p_2. This is a properly distributed semi-functional ciphertext with $C_1, \ldots, C_{\max\{1,\varrho-1\}}$ random in $\mathbb{G}_{p_1} \times \mathbb{G}_{p_2} \times \mathbb{G}_{p_4}$. If $T \xleftarrow{\$} \mathbb{G}_{p_1} \times \mathbb{G}_{p_2} \times \mathbb{G}_{p_4}$, this is a properly distributed semi-functional ciphertext with $C_1, \ldots, C_\varrho$ randomly chosen from $\mathbb{G}_{p_1} \times \mathbb{G}_{p_2} \times \mathbb{G}_{p_4}$.

We can conclude that, if $T = h^s D_{24}$, then $\mathcal{B}$ has properly simulated $\mathsf{Game}_{\mathsf{Final}_{\varrho-1}}$. If $T \xleftarrow{\$} \mathbb{G}_{p_1} \times \mathbb{G}_{p_2} \times \mathbb{G}_{p_4}$, then $\mathcal{B}$ has properly simulated $\mathsf{Game}_{\mathsf{Final}_{\varrho}}$. Hence, $\mathcal{B}$ can use the output of $\mathcal{A}$ to distinguish between these possibilities for T. □

4. CONCLUSIONS

Based on the KP-ABE scheme proposed by Lewko *et al.* [22] recently, we presented a new construction of searchable public key encryption. Our scheme supports arbitrary monotone boolean predicates. Compared to previous works in this field, our construction is more expressive and efficient and is proven secure in the standard model.

5. ACKNOWLEDGEMENTS

The authors thank the anonymous reviewers for their helpful comments. This work is in part supported by the Office of Research, Singapore Management University. The first author is partially supported by Natural Science Foundation of China (No. 61272534). The second and fifth authors are supported by National Science Foundation of China (No. 61133014).

6. REFERENCES

[1] M. Abdalla, M. Bellare, D. Catalano, E. Kiltz, T. Kohno, T. Lange, J. Malone-Lee, G. Neven, P. Paillier, and H. Shi. Searchable encryption revisited: Consistency properties, relation to anonymous ibe, and extensions. In *CRYPTO*, pages 205–222, 2005.

[2] J. Baek, R. Safavi-Naini, and W. Susilo. On the integration of public key data encryption and public key encryption with keyword search. In *ISC*, pages 217–232, 2006.

[3] J. Baek, R. Safavi-Naini, and W. Susilo. Public key encryption with keyword search revisited. In *ICCSA (1)*, pages 1249–1259, 2008.

[4] A. Beimel. *Secure Schemes for Secret Sharing and Key Distribution.* PhD thesis, Israel Institute of Technology, 1996.

[5] M. Bellare, A. Boldyreva, and A. O'Neill. Deterministic and efficiently searchable encryption. In *CRYPTO*, pages 535–552, 2007.

[6] B. H. Bloom. Space/time trade-offs in hash coding with allowable errors. *Commun. ACM*, 13(7):422–426, 1970.

[7] D. Boneh, G. D. Crescenzo, R. Ostrovsky, and G. Persiano. Public key encryption with keyword search. In *EUROCRYPT*, pages 506–522, 2004.

[8] D. Boneh, E.-J. Goh, and K. Nissim. Evaluating 2-dnf formulas on ciphertexts. In *TCC*, pages 325–341, 2005.

[9] D. Boneh, E. Kushilevitz, R. Ostrovsky, and W. E. S. III. Public key encryption that allows pir queries. In *CRYPTO*, pages 50–67, 2007.

[10] D. Boneh and B. Waters. Conjunctive, subset, and range queries on encrypted data. In *TCC*, pages 535–554, 2007.

[11] J. Bringer, H. Chabanne, and B. Kindarji. Error-tolerant searchable encryption. In *ICC*, pages 1–6, 2009.

[12] J. W. Byun, H. S. Rhee, H.-A. Park, and D. H. Lee. Off-line keyword guessing attacks on recent keyword search schemes over encrypted data. In *Secure Data Management*, pages 75–83, 2006.

[13] G. D. Crescenzo and V. Saraswat. Public key encryption with searchable keywords based on jacobi symbols. In *INDOCRYPT*, pages 282–296, 2007.

[14] L. Fang, W. Susilo, C. Ge, and J. Wang. A secure channel free public key encryption with keyword search scheme without random oracle. In *CANS*, pages 248–258, 2009.

[15] T. Fuhr and P. Paillier. Decryptable searchable encryption. In *ProvSec*, pages 228–236, 2007.

[16] P. Golle, J. Staddon, and B. R. Waters. Secure conjunctive keyword search over encrypted data. In *ACNS*, pages 31–45, 2004.

[17] V. Goyal, O. Pandey, A. Sahai, and B. Waters. Attribute-based encryption for fine-grained access control of encrypted data. In *ACM Conference on Computer and Communications Security*, pages 89–98, 2006.

[18] C. Gu, Y. Zhu, and H. Pan. Efficient public key encryption with keyword search schemes from pairings. In *Inscrypt*, pages 372–383, 2007.

[19] Y. H. Hwang and P. J. Lee. Public key encryption with conjunctive keyword search and its extension to a multi-user system. In *Pairing*, pages 2–22, 2007.

[20] J. Katz, A. Sahai, and B. Waters. Predicate encryption supporting disjunctions, polynomial equations, and inner products. *IACR Cryptology ePrint Archive*, 2007:404, 2007.

[21] D. Khader. Public key encryption with keyword search based on k-resilient ibe. In *ICCSA (3)*, pages 298–308, 2006.

[22] A. B. Lewko, T. Okamoto, A. Sahai, K. Takashima, and B. Waters. Fully secure functional encryption: Attribute-based encryption and (hierarchical) inner product encryption. In *EUROCRYPT*, pages 62–91, 2010.

[23] A. B. Lewko and B. Waters. New techniques for dual system encryption and fully secure hibe with short ciphertexts. In *TCC*, pages 455–479, 2010.

[24] A. B. Lewko and B. Waters. Decentralizing attribute-based encryption. In *EUROCRYPT*, pages 568–588, 2011.

[25] A. B. Lewko and B. Waters. New proof methods for attribute-based encryption: Achieving full security through selective techniques. In *CRYPTO*, pages 180–198, 2012.

[26] D. J. Park, K. Kim, and P. J. Lee. Public key encryption with conjunctive field keyword search. In *WISA*, pages 73–86, 2004.

[27] H. S. Rhee, J. H. Park, W. Susilo, and D. H. Lee. Improved searchable public key encryption with designated tester. In *ASIACCS*, pages 376–379, 2009.

[28] H. S. Rhee, J. H. Park, W. Susilo, and D. H. Lee. Trapdoor security in a searchable public-key encryption scheme with a designated tester. *Journal of Systems and Software*, 83(5):763–771, 2010.

[29] H. S. Rhee, W. Susilo, and H.-J. Kim. Secure searchable public key encryption scheme against keyword guessing attacks. *IEICE Electronics Express*, 6(5):237–243, 2009.

[30] A. Sahai and B. Waters. Fuzzy identity-based encryption. In *EUROCRYPT*, pages 457–473, 2005.

[31] E. Shen, E. Shi, and B. Waters. Predicate privacy in encryption systems. In *TCC*, pages 457–473, 2009.

[32] Q. Tang and L. Chen. Public-key encryption with registered keyword search. In *EuroPKI*, pages 163–178, 2009.

[33] B. Zhang and F. Zhang. An efficient public key encryption with conjunctive-subset keywords search. *J. Network and Computer Applications*, 34(1):262–267, 2011.

[34] R. Zhang and H. Imai. Generic combination of public key encryption with keyword search and public key encryption. In *CANS*, pages 159–174, 2007.

Towards Asymmetric Searchable Encryption with Message Recovery and Flexible Search Authorization

Qiang Tang
APSIA group, SnT, University of Luxembourg
6, rue Richard Coudenhove-Kalergi, L-1359
Luxembourg
qiang.tang@uni.lu

Xiaofeng Chen
State Key Laboratory of Integrated Service
Networks (ISN)
Xidian University, Xian 710071, P.R. China
xfchen@xidian.edu.cn

ABSTRACT

When outsourcing data to third-party servers, searchable encryption is an important enabling technique which simultaneously allows the data owner to keep his data in encrypted form and the third-party servers to search in the ciphertexts. Motivated by an *encrypted email retrieval and archive* scenario, we investigate asymmetric searchable encryption (ASE) schemes which support two special features, namely message recovery and flexible search authorization. With this new primitive, a data owner can keep his data encrypted under his public key and assign different search privileges to third-party servers. In the security model, we define the standard IND-CCA security against any outside attacker and define adapted ciphertext indistinguishability properties against inside attackers according to their functionalities. Moreover, we take into account the potential information leakage from trapdoors, and define two trapdoor security properties. Employing the bilinear property of pairings and a deliberately-designed double encryption technique, we present a provably secure instantiation of the primitive based on the DLIN and BDH assumptions in the random oracle model.

Categories and Subject Descriptors

E.3 [**Data Encryption**]: Public Key Cryptosystems

Keywords

Cloud Computing; Data Outsourcing; Searchable Encryption; Privacy

1. INTRODUCTION

To protect outsourced data and services in the cloud computing environment, cryptographic researchers have devoted a lot of efforts to searchable encryption techniques. Such techniques are particularly interesting because they allow a data owner to encrypt his data and outsource the cipheretxts while still being able to let third-party service providers search on his behalf without leaking any unnecessary information. Roughly speaking, searchable encryption schemes fall into two categories. One category is symmetric searchable encryption (SSE) schemes, represented by the work of Song, Wagner, and Perrig [12]. In this category, only the data owner can contribute searchable contents. The other category is asymmetric searchable encryption (ASE) schemes, represented by the work of Boneh et al. [4]. In this category, the concept of public key encryption is employed so that every entity can contribute searchable contents. A detailed survey and analysis of existing searchable encryption (both SSE and ASE) schemes can be found in [13].

Motivated by an *encrypted email retrieval and archive* scenario, as described below, we investigate ASE schemes which support two special features: *message recovery* and *flexible search authorization*. The message recovery feature requires that a ciphertext not only allows the data owner to recover the plaintext but also allows third-party servers to search in it. The flexible searchable authorization feature requires that the data owner can authorize a third-party server in three different ways: (1) authorize the server to search any message at the data owner's interest by assigning a message-dependent trapdoor (i.e. the server can only determine whether the message encoded in the trapdoor is equal to the plaintext inside a ciphertext); (2) authorize the server to search any message at the server's interests by assigning a master trapdoor (i.e. the server can choose a message at its will and see whether it is equal to the plaintext inside any ciphertext); (3) authorize the server to perform both types of searches. Throughout the paper, we refer to this new type of ASE schemes as $\text{ASE}^{\dagger\dagger}$.

1.1 Encrypted Email Retrieval and Archive

Suppose that Bob is an employee of the company COM and emails sent to him are required to be encrypted and stored in an email server managed by COM. Suppose that Alice wants to send an email to Bob, then she can encrypt the email using Bob's public key and send the ciphertext to COM's email server. Note that, here, Bob is the owner of his emails. In practice, the underlying encryption scheme should satisfy the following requirements.

1. When Bob is traveling around, he may want to selectively retrieve and read his emails from COM's email server. Thus, the encryption scheme should allow the email server to search on Bob's behalf to identify those at his interests.
2. A malicious user can send Bob encrypted emails, which contain malwares or viruses. Thus, the encryption

scheme should allow the email server to scan the encrypted emails to identify malicious contents.

3. Bob may change his job over the time, so that he may want to archive his emails during different jobs in a cloud server, such as Gmail. Bob can simply forward all his encrypted emails to the archive server. Later on, Bob may need to selectively retrieve some of the emails, therefore, the encryption scheme should allow the cloud server to search on Bob's behalf as in the first requirement.

With an $ASE^{\dagger\dagger}$ scheme, the message recovery feature guarantees that only Bob can decrypt the encrypted emails while he can still authorize the servers to search on his behalf. The flexible search authorization feature allows Bob to assign a master trapdoor to COM's email server so that the latter can scan malicious contents inside the encrypted emails, and this feature also allows Bob to assign message-dependent trapdoors to COM's email server and the cloud server to search emails at his interests.

1.2 Related Work

As surveyed in [13], the majority of existing ASE schemes are index-based, which means that they only aim at supporting search over scrambled keywords and typically do not allow the data owner to recover the keywords. By definition, these schemes do not allow the servers to search directly over the contents, therefore their functionality is far from what an $ASE^{\dagger\dagger}$ scheme is aimed for. On the other hand, an $ASE^{\dagger\dagger}$ scheme fulfills the purpose of these index-based ASE schemes.

Fuhr and Paillier [7] and Hofheinz and Weinreb [8] investigated the concept of ASE with message recovery. Their formulations only allow the data owner to assign message-dependent trapdoors to third-party servers, thus provide less functionality than $ASE^{\dagger\dagger}$. As to the security models, the authors only consider information leakage from ciphertexts and allow the servers to easily recover the information encoded in the trapdoors. In [7], if a match is found then the server immediately knows the plaintext in the ciphertext, while, in [8], the to-be-searched message is sent to the server in plaintext. In practice, this may be regarded as a serious security weakness.

Ibraimi et al. [9] pushed forward the concept of ASE with message recovery and proposed a new primitive PKEDS, namely public key encryption with delegated search. With a PKEDS scheme, a data owner can authorize third-party servers in two ways: (1) authorize a server to search any message at the server's interests by assigning a master trapdoor; (2) authorize a server to search messages at the data owner's interests by assigning message-dependent trapdoors. In their formulation, authorization (2) implies authorization (1), because search based on message-dependent trapdoors also requires a master trapdoor as input. In other words, the data owner must assign a master trapdoor to a server in order to ask the latter to perform any search. This fact conflicts with the *least privilege* principle in information security and is undesirable. For instance, in the aforementioned application scenario, Bob may not want to assign a master trapdoor to the cloud server to let the latter probe all his emails.

Recently, Tang et al. in [14] refined PKEDS and proposed a primitive similar to $ASE^{\dagger\dagger}$. However, in their security model, only IND-CPA security is considered and the notion of soundness is also weaker than that in this paper.

1.3 Contribution

In this paper, we formulate a new primitive, namely $ASE^{\dagger\dagger}$. With an $ASE^{\dagger\dagger}$ scheme, the data owner can keep data in encrypted form while still be able to recover the plaintext and authorize third-party servers to search on his behalf. The authorization to a server is through assigning the appropriate trapdoors: message-dependent trapdoors, master trapdoors, or both types of trapdoors. In contrast to [9], a search based on a message-dependent trapdoor does not require a master trapdoor as input, and this implies a significant security improvement. In practice, the data owner can authorize different servers based on their trustworthiness.

With respect to the functionality of $ASE^{\dagger\dagger}$, we provide a comprehensive definition for the soundness property. The property guarantees that not only the encryption/decryption algorithms work well but also the decryption and the test algorithms are bilaterally consistent with each other. We present a fine-grained security model by considering four categories of attackers, including an outside attacker and three types of curious servers based on the trapdoors they receive. We define the standard IND-CCA security against an outside attacker, and define adapted ciphertext indistinguishability security properties against the curious servers. Moreover, we define two trapdoor security properties to model the possible information leakages from message-dependent trapdoors. This security model is stronger than that in [14].

The soundness property turns out to be very difficult to be satisfied. Hybrid constructions (e.g. [1, 16]) and other constructions (e.g. [14]) do not satisfy this property. Based on bilinear pairing techniques and a deliberately-designed double encryption technique, we propose a new $ASE^{\dagger\dagger}$ scheme and prove its security based on DLIN and BDH assumptions in the random oracle model.

1.4 Organization

In Section 2, we formulate the concept of $ASE^{\dagger\dagger}$ and define the soundness property. In Section 3, we present a fine-grained security model for $ASE^{\dagger\dagger}$. In Section 4, we present an IND-CCA secure scheme and analyse its security. In Section 5, we conclude the paper.

2. DEFINITION OF $ASE^{\dagger\dagger}$

Throughout the paper, we use the following notation. $x||y$ means the concatenation of x and y, P.P.T. means probabilistic polynomial time, $x \in_R X$ means that the element x is chosen from the set X uniformly at random, and $x \xleftarrow{\$} \mathcal{A}(m_1, m_2, \cdots; \mathcal{O}_1, \mathcal{O}_2, \cdots)$ means that x is the output of the algorithm $\mathcal{A}$ which runs with the input $m_1, m_2, \cdots$ and access to the oracles $\mathcal{O}_1, \mathcal{O}_2, \cdots$.

2.1 Primitive Formulation

In general, an $ASE^{\dagger\dagger}$ scheme involves the following entities. In each server category, there can be multiple servers, but we only consider one for the simplicity of description.

- A data owner, who is supposed to receive encrypted messages. This entity is also referred to as *receiver* in encryption schemes.
- Type-I server $\mathcal{S}_1$, which receives message-dependent trapdoors from the data owner. It can test whether the

message encoded in a trapdoor is equal to the plaintext inside any given ciphertext.

- Type-II server $\mathcal{S}_2$, which receives a master trapdoor from the data owner. It can choose a message at its will and test whether it is equal to the plaintext inside any given ciphertext.
- Hybrid server $\mathcal{S}_h$, which is both Type-I and Type-II.
- Senders, who may send messages to the data owner.

Let λ be the security parameter. Formally, an ASE†† scheme consists of the following algorithms.

- $\mathsf{rKeyGen}(\lambda)$: Run by the data owner, it outputs a public/private key pair (PK_r, SK_r). Let the message space be denoted as $\mathcal{W}$.
- $\mathsf{Encrypt}(w, PK_r)$ Run by a message sender, it outputs a ciphertext c_w for a message $w \in \mathcal{W}$.
- $\mathsf{Decrypt}(c_w, SK_r)$: Run by the data owner, it outputs the plaintext w or an error message $\perp$.
- $\mathsf{sKeyGen}(\lambda)$: Run by a server ($\mathcal{S}_1$, $\mathcal{S}_2$, or $\mathcal{S}_h$), it outputs a public/private key pair (PK_s, SK_s)
- $\mathsf{TrapGen}_1(w, PK_s, SK_r)$: Run by the data owner, it generates a message-dependent trapdoor $t_{w,s}$ for the server with public key PK_s.
- $\mathsf{Test}_1(c_w, t_{w',s}, SK_s)$: Run by the server with message-dependent trapdoor $t_{w',s}$ and private key SK_s, it returns 1 if $w' = w$ and 0 otherwise.
- $\mathsf{TrapGen}_2(PK_s, SK_r)$: Run by the data owner, it outputs a master trapdoor $t_{*,s}$ for the server with public key PK_s.
- $\mathsf{Test}_2(c_w, w', t_{*,s}, SK_s)$: Run by the server with the master trapdoor $t_{*,s}$ and private key SK_s, it returns 1 if $w' = w$ and 0 otherwise.

Note that ($\mathsf{rKeyGen}, \mathsf{Encrypt}, \mathsf{Decrypt}$) define a standard PKE scheme. As indicated in [4], the trapdoors should be transmitted to the servers through a secure channel (with confidentiality), otherwise any attacker will be able to obtain the trapdoors and search over the ciphertexts. However, this issue has not been formally addressed in the primitive formulation and security model in [4]. Here, we explicitly provide the $\mathsf{sKeyGen}$ algorithm so that potential servers can run this algorithm to generate a key pair, with which the data owner can generate trapdoors in an encrypted form by running the $\mathsf{TrapGen}_1$ and $\mathsf{TrapGen}_2$ algorithms.

2.2 Soundness Property

Similar to the case for other primitives, the first property we want is soundness, defined as follows.

DEFINITION 1. *An ASE†† scheme is sound if, for any* $(PK_r, SK_r) = \mathsf{rKeyGen}(\lambda)$ *and* $(PK_s, SK_s) = \mathsf{sKeyGen}(\lambda)$, *the following conditions are satisfied.*

1. *For any* $w \in \mathcal{W}$, $\mathsf{Decrypt}(\mathsf{Encrypt}(w, PK_r), SK_r) = w$ *always holds.*
2. *For any ciphertext* c, $\mathsf{Decrypt}(c, SK_r) = w$ *if and only if* $\mathsf{Test}_1(c, t_{w,s}, SK_s) = 1$.
3. *For any ciphertext* c, $\mathsf{Decrypt}(c, SK_r) = w$ *if and only if* $\mathsf{Test}_2(c, w, t_{*,s}, SK_s) = 1$.

In the above definition, the first condition means that the encryption/decryption functionality works well. The second and the third conditions define the bilateral consistency property between the decryption and test algorithms.

The "if" condition guarantees that any matched ciphertext by the test algorithms can be successfully decrypted and the resulted plaintext will be equal to that assumed in the test algorithms. For instance, if a test algorithm indicates c is an encryption of w, then the decryption algorithm will not output $w' \neq w$ or an error $\perp$. Basically, this property guarantees that there is no "false positive" in the search process.

The "only if" condition guarantees that if a ciphertext can be successfully decrypted then the test algorithms should be able to properly match it. This property guarantees that no targeted ciphertext will be missed by the test algorithms. Take the *encrypted email retrieval and archive* scenario as example, this property is crucial for the email server not to miss any malicious contents in the encrypted emails.

3. THE SECURITY MODEL

We assume that the message senders possess a valid copy of the receiver's public key and the receiver possesses valid copies of the public keys of the servers. How to securely distribute these public keys should follow some standard practice, and we skip the discussion in this paper.

As to the security of an ASE†† scheme, there are two main privacy concerns.

- One concern is the leakage of plaintext information from ciphertexts, which is a standard concern for all encryption schemes. Given ciphertexts, an attacker can try to deduce information about the encrypted plaintexts. Particularly, for an ASE†† scheme, knowledge about the (un-encrypted) master trapdoors or message-dependent trapdoors will provide additional advantage to the attacker. Therefore, we will consider the following types of attackers.
  - Outside attacker: This type of attacker is not assigned with any type of (unencrypted) trapdoors.
  - Curious Type-I server $\mathcal{S}_1$: This type of attacker has been assigned with only message-dependent trapdoors generated under its public key.
  - Curious Type-II server $\mathcal{S}_2$: This type of attacker has only been assigned with a master trapdoor generated under its public key.
  - Curious hybrid server $\mathcal{S}_h$: This type of attacker has been assigned with a master trapdoor and message-dependent trapdoors generated under its public key.

  It is clear that a hybrid server is more powerful than others. However, due to the fact that the data owner may employ all three types of servers, it is necessary to consider the maximal level of security against each of them independently.
- The other concern is information leakage from message-dependent trapdoors. For example, the Type-I server $\mathcal{S}_1$ receives message-dependent trapdoors so that it

knows which ciphertext matches a received trapdoor. However, $\mathcal{S}_1$ should not know the message encoded in the trapdoor, or equivalently, $\mathcal{S}_1$ should not know what is the plaintext of the matched ciphertext. Furthermore, any entity other than $\mathcal{S}_1$ should not learn anything about the trapdoor. The same security requirement applies to a hybrid server $\mathcal{S}_h$ because it also gets message-dependent trapdoors from the receiver.

To facilitate our security definitions, we first detail all the potential oracles accessible to an attacker. Based on our assumption that there is only one server of every type, so that each key generation oracle ($\mathsf{sKeyGen}_1$, $\mathsf{sKeyGen}_2$, $\mathsf{sKeyGen}_h$) can only be queried once. Trivially, the key request oracles can only be queried after the corresponding key generation oracles have been queried.

- $\mathsf{sKeyGen}_1$: The challenger runs the $\mathsf{sKeyGen}$ algorithm to generate (PK_{s_1}, SK_{s_1}) for the Type-I server $\mathcal{S}_1$, and returns PK_{s_1}.
- $\mathsf{sKeyReq}_1$: The challenger returns SK_{s_1}.
- $\mathsf{sKeyGen}_2$: The challenger runs the $\mathsf{sKeyGen}$ algorithm to generate (PK_{s_2}, SK_{s_2}) for the Type-II server $\mathcal{S}_2$, and returns PK_{s_2}.
- $\mathsf{sKeyReq}_2$: The challenger returns SK_{s_2}.
- $\mathsf{sKeyGen}_h$: The challenger runs the $\mathsf{sKeyGen}$ algorithm to generate (PK_{s_h}, SK_{s_h}) for the hybrid server $\mathcal{S}_h$, and returns PK_{s_h}.
- $\mathsf{sKeyReq}_h$: The challenger returns SK_{s_h}.
- $\mathsf{TrapGen}_1$ query with a message w and PK_s as input: The challenger returns $\mathsf{TrapGen}_1(w, PK_s, SK_r)$. In this case s can be either s_1 or s_h
- $\mathsf{TrapGen}_2$ query with a public key PK_s: The challenger returns $\mathsf{TrapGen}_2(PK_s, SK_r)$. In this case s can be either s_2 or s_h
- $\mathsf{rKeyReq}_r$: The challenger returns the receiver's private key SK_r.
- $\mathsf{Decrypt}$ query with input c: The challenger returns $\mathsf{Decrypt}(c, SK_r)$.

In the following, we present security definitions with respect to the aforementioned two types of privacy concerns.

3.1 Ciphertext Security Properties

Against an outside attacker, in Definition 2, we define an IND-CCA property similar to the IND-CCA security for PKE. Informally, the property says that, given some ciphertexts and various oracle accesses (including decryption oracle), an attacker cannot learn anything about the plaintexts.

DEFINITION 2. *An ASE†† scheme is IND-CCA secure against an outside attacker if the attacker's advantage (i.e. $|Pr[b' = b] - \frac{1}{2}|$) is negligible in the game shown in Figure 1.*

In the attack game, $w_0, w_1 \in \mathcal{W}$ and *state* is some state information generated by the attacker. In Phase 4, the attacker is not allowed to query the $\mathsf{Decrypt}$ oracle with c_{w_b}. In this game, an outside attacker is modeled since it is not allowed to query any server's private key through $\mathsf{sKeyReq}_1$, $\mathsf{sKeyReq}_2$, or $\mathsf{sKeyReq}_h$. In practice, an outside attacker can eavesdrop on the transmission of (encrypted) trapdoors, so that we offer it access to both trapdoor generation oracles with its own inputs.

1. $(PK_r, SK_r) \xleftarrow{\$} \mathsf{rKeyGen}(\lambda)$
2. $(w_0, w_1, state) \xleftarrow{\$} \mathcal{A}(PK_r; \mathsf{sKeyGen}_1, \mathsf{sKeyGen}_2, \mathsf{sKeyGen}_h, \mathsf{TrapGen}_1, \mathsf{TrapGen}_2, \mathsf{Decrypt})$
3. $b \in_R \{0,1\},\ c_{w_b} = \mathsf{Encrypt}(w_b, PK_r)$
4. $b' \xleftarrow{\$} \mathcal{A}(PK_r, state, c_{w_b}; \mathsf{sKeyGen}_1, \mathsf{sKeyGen}_2, \mathsf{sKeyGen}_h, \mathsf{TrapGen}_1, \mathsf{TrapGen}_2, \mathsf{Decrypt})$

Figure 1: IND-CCA Security

REMARK 1. *To define the IND-CPA security against an outside attacker, we only need to disable the* $\mathsf{Decrypt}$ *oracle in the game shown in Figure 1. In fact, if an ASE†† scheme is secure under Definition 3 and Definition 7 (given below), then it is automatically IND-CPA secure. Due to the limit of space, we put the proof of this result in the full version of this paper.*

Against a curious Type-I server $\mathcal{S}_1$, in Definition 3, we define a ciphertext indistinguishability (CI) security property similar to the security definition of PEKS [4]. Informally, this property says that, given some ciphertexts *for which $\mathcal{S}_1$ has not obtained the corresponding message-dependent trapdoors*, then it cannot learn anything about the plaintexts.

DEFINITION 3. *An ASE†† scheme achieves CI security against a curious Type-I server $\mathcal{S}_1$, if the attacker's advantage (i.e. $|\Pr[b' = b] - \frac{1}{2}|$) is negligible in the game shown in Figure 2.*

1. $(PK_r, SK_r) \xleftarrow{\$} \mathsf{rKeyGen}(\lambda)$
2. $(w_0, w_1, state) \xleftarrow{\$} \mathcal{A}(PK_r; \mathsf{sKeyGen}_1, \mathsf{sKeyGen}_2, \mathsf{sKeyGen}_h, \mathsf{TrapGen}_1, \mathsf{TrapGen}_2, \mathsf{sKeyReq}_1)$
3. $b \in_R \{0,1\},\ c_{w_b} = \mathsf{Encrypt}(w_b, PK_r)$
4. $b' \xleftarrow{\$} \mathcal{A}(PK_r, state, c_{w_b}; \mathsf{sKeyGen}_1, \mathsf{sKeyGen}_2, \mathsf{sKeyGen}_h, \mathsf{TrapGen}_1, \mathsf{TrapGen}_2, \mathsf{sKeyReq}_1)$

Figure 2: CI Security against a Type-I Server

In the attack game, $w_0, w_1 \in \mathcal{W}$ and *state* is some state information generated by the attacker. In step 2 and 4, the $\mathsf{TrapGen}_1$ oracle should have not been queried with (w_0, PK_{s_1}) or (w_1, PK_{s_1}). In this game, a Type-I server is modeled since the attacker is allowed to obtain SK_{s_1} through a $\mathsf{sKeyReq}_1$ query, but has no access to SK_{s_2} or SK_{s_h}. The oracle access to message-dependent trapdoors for any messages except for w_0 and w_1 is indeed a big privilege for the attacker. Most existing IND-CPA secure PKE schemes, such as ElGamal [6], normally allow some sort of homomorphism, so that they will not be secure if directly used in constructing an ASE†† scheme. As an example, it is easy to verify that the PKEDS scheme in [9] is not secure under this definition.

Against a curious hybrid server $\mathcal{S}_h$, we define a special CI security property. Compared with Definition 2 and Definition 3, the speciality lies in that the challenger randomly chooses the challenge messages w_0, w_1 instead of letting the attacker do. The rationale is that, for a curious hybrid server $\mathcal{S}_h$, if it can choose the messages then it can easily tell which message is encrypted by running the Test_2 algorithm.

DEFINITION 4. *An ASE†† scheme achieves CI security against a curious hybrid server $\mathcal{S}_h$, if the attacker's advantage (i.e. $|\Pr[b' = b] - \frac{1}{2}|$) is negligible in the game shown in Figure 3.*

1. $(PK_r, SK_r) \xleftarrow{\$} \mathsf{rKeyGen}(\lambda)$
2. $state \xleftarrow{\$} \mathcal{A}(PK_r; \mathsf{sKeyGen}_1, \mathsf{sKeyGen}_2, \mathsf{sKeyGen}_h, \mathsf{TrapGen}_1, \mathsf{TrapGen}_2, \mathsf{sKeyReq}_h)$
3. $b \in_R \{0, 1\}$, $w_0, w_1 \in_R \mathcal{W}$, $c_b = (\mathsf{Encrypt}(w_0, PK_r), \mathsf{Encrypt}(w_b, PK_r))$
4. $b' \xleftarrow{\$} \mathcal{A}(PK_r, state, c_b; \mathsf{sKeyGen}_1, \mathsf{sKeyGen}_2, \mathsf{sKeyGen}_h, \mathsf{TrapGen}_1, \mathsf{TrapGen}_2, \mathsf{sKeyReq}_h)$

Figure 3: CI Security against a Hybrid Server

In the attack game, *state* is some state information generated by the attacker. In this game, a hybrid server is modeled since the attacker is allowed to obtain SK_{s_h} through a $\mathsf{sKeyReq}_h$ query, but has no access to SK_{s_1} or SK_{s_2}. This property guarantees that, given some ciphertexts, the attacker can neither recover the plaintexts nor find the equality relationship of the plaintexts. Therefore, it provides much stronger security guarantee than the standard one-wayness property used in [9] and the enhanced one-wayness property by Bellare, Boldyreva and O'Neill [2].

Similarly, we can define a special CI security property against a curious Type-II server, as shown in Definition 5. It is clear that if an ASE†† scheme is secure under Definition 4 then it is also secure under Definition 5.

DEFINITION 5. *An ASE†† scheme achieves CI security against a curious Type-II server $\mathcal{S}_2$, if the attacker's advantage (i.e. $|\Pr[b' = b] - \frac{1}{2}|$) is negligible in the game shown in Figure 3, where the $\mathsf{sKeyReq}_h$ oracle is replaced with the $\mathsf{sKeyReq}_2$ oracle.*

REMARK 2. *It is worth noting that the security properties defined in Definition 4 and 5 only make sense when the message space of the encryption scheme is not polynomial size. when the message space is polynomial size, the attacker can perform brute-force attacks. This fact is also true for Definition 6.*

3.2 Trapdoor Security Properties

In Definition 6, we define the universal trapdoor one-wayness property. The property guarantees that no entity can recover the message encoded in a message-dependent trapdoor. Note that this will require the message space $\mathcal{W}$ not to be polynomial size. In the attack game, an outside attacker and various servers are modeled because the attacker can obtain the private keys $(SK_{s_1}, SK_{s_2}, SK_{s_h})$ through the key request oracles $(\mathsf{sKeyReq}_1, \mathsf{sKeyReq}_2, \mathsf{sKeyReq}_h)$.

DEFINITION 6. *An ASE†† scheme achieves universal trapdoor one-wayness if the attacker's advantage (i.e. $\Pr[w = w']$) is negligible in the game shown in Figure 4. Note that the output value t in Phase 2 can be either 1, or 2, or h.*

1. $(PK_r, SK_r) \xleftarrow{\$} \mathsf{rKeyGen}(\lambda)$
2. $(t, state) \xleftarrow{\$} \mathcal{A}(PK_r; \mathsf{sKeyGen}_1, \mathsf{sKeyGen}_2, \mathsf{sKeyGen}_h, \mathsf{TrapGen}_1, \mathsf{TrapGen}_2, \mathsf{sKeyReq}_1, \mathsf{sKeyReq}_2, \mathsf{sKeyReq}_h)$
3. $w \in_R \mathcal{W}$, $t_{w,s_t} = \mathsf{TrapGen}_1(w, PK_{s_t}, SK_r)$
4. $w' \xleftarrow{\$} \mathcal{A}(PK_r, state, t_{w,s_t}; \mathsf{sKeyGen}_1, \mathsf{sKeyGen}_2, \mathsf{sKeyGen}_h, \mathsf{TrapGen}_1, \mathsf{TrapGen}_2, \mathsf{sKeyReq}_1, \mathsf{sKeyReq}_2, \mathsf{sKeyReq}_h, \mathsf{rKeyReq}_r)$

Figure 4: Universal Trapdoor One-wayness

To further strengthen the universal trapdoor one-wayness property, in Definition 7, we define the *message-dependent trapdoor indistinguishability* property. Informally, the property guarantees that, for the Type-I server $\mathcal{S}_1$ or the hybrid server $\mathcal{S}_h$, no other entity can learn anything about the messages encoded in message-dependent trapdoors generated under its public key.

DEFINITION 7. *An ASE†† scheme achieves trapdoor indistinguishability if the attacker's advantage (i.e. $|\Pr[b' = b] - \frac{1}{2}|$) is negligible in the game shown in Figure 5.*

1. $(PK_r, SK_r) \xleftarrow{\$} \mathsf{rKeyGen}(\lambda)$
2. $(w_0, w_1, t, state) \xleftarrow{\$} \mathcal{A}(PK_r; \mathsf{sKeyGen}_1, \mathsf{sKeyGen}_2, \mathsf{sKeyGen}_h, \mathsf{TrapGen}_1, \mathsf{TrapGen}_2, \mathsf{sKeyReq}_{t'}, \mathsf{sKeyReq}_2)$
3. $b \in_R \{0, 1\}$, $t_{w_b} = \mathsf{TrapGen}_1(w_b, PK_{s_t}, SK_r)$
4. $b' \xleftarrow{\$} \mathcal{A}(PK_r, state, t_{w_b}; \mathsf{sKeyGen}_1, \mathsf{sKeyGen}_2, \mathsf{sKeyGen}_h, \mathsf{TrapGen}_1, \mathsf{TrapGen}_2, \mathsf{sKeyReq}_{t'}, \mathsf{sKeyReq}_2, \mathsf{rKeyReq}_r)$

Figure 5: Trapdoor Indistinguishability

In the attack game, $w_0, w_1 \in \mathcal{W}$, the values t, t' are chosen (by the attacker) from $\{1, h\}$ and $t \neq t'$, and *state* is some state information generated by the attacker. This guarantees that $\mathsf{sKeyReq}_t$ is not queried in the game. We allow the attacker to request the receiver's private key in Phase 2 and all servers' private keys except for SK_{s_t}. It captures our intention that, even if the receiver's private key is leaked or compromised, then the attacker still cannot figure out what the receiver has searched for. This is similar to the forward secrecy property in key establishment protocols.

4. IND-CCA SECURE ASE††

In this section, we first briefly mention some attempts, and then describe a novel ASE†† scheme and analyse its security properties in our security model.

4.1 Some Attempts

To construct an ASE†† scheme, a very natural direction is to follow the hybrid method used in [16]. Generically, the data owner can have two key pairs (pk_1, sk_1) and (pk_2, sk_2), and the ciphertext of a message m is in the form of $c = (c_1, c_2) = (\mathsf{Encrypt}_1(m, pk_1), \mathsf{Encrypt}_2(m, pk_2))$, where c_1 can be used for decryption and c_2 can be used for search. With this methodology, the difficulty lies in that we need to provide a knowledge proof that c_1 and c_2 contain the same message in order to achieve the soundness property. Moreover, how to generate master trapdoors and message-dependent trapdoors is not a straightforward task either. It remains as an open problem to have a generic construction for ASE[††].

Instead of a generic construction, we may follow the semi-hybrid method in [14]: use a standard PKE scheme as a main component and employ a key-private IBE scheme to facilitate trapdoor constructions. In order to achieve the soundness property, we still need to provide a knowledge proof to guarantee that the decryption and the test functions are consistent with each other (note that the scheme in [14] does not satisfy the third requirement in Definition 1). Moreover, it is also an interesting future work to improve this scheme to achieve IND-CCA security.

4.2 Proposed Scheme and its Soundness

In the proposed scheme, the intuition behind the Encrypt algorithm is that a message is protected by two layers of encryption (see the element c_2). The first layer of encryption makes use of a hash value of the message and can only be removed by the receiver, while the second layer makes use of a hash value of the message and a bilinear pairing technique and it can only be removed with an appropriate message-dependent trapdoor or a master trapdoor. The search algorithms (Test_1 and Test_2) are made possible by employing the bilinear property of pairings.

- $\mathsf{sKeyGen}(\lambda)$: It generates a key pair (PK_s, SK_s) for a standard PKE scheme $(\mathsf{KeyGen}', \mathsf{Encrypt}', \mathsf{Decrypt}')$, which has the message space $\mathcal{M}$.

- $\mathsf{rKeyGen}(\lambda)$: It generates (PK_r, SK_r) as follows.

  1. Generate the pairing parameters: group $\mathbb{G}$ of prime order p, a bilinear map $\hat{e}$: $\mathbb{G} \times \mathbb{G} \to \mathbb{G}_T$, randomly chosen generators g_1 and g_2 of $\mathbb{G}$.
  2. Select a symmetric key encryption scheme $(\mathsf{Enc}, \mathsf{Dec})$ whose key space is $\mathcal{M}$, and select three hash functions: $\mathsf{H}_1 : \mathbb{G}_T \to \mathbb{G}$, $\mathsf{H}_2 : \mathbb{G} \to \mathbb{G}$, and $\mathsf{H}_3 : \{0,1\}^* \to \{0,1\}^\lambda$.
  3. Select $x_1, x_2 \in_R \mathbb{Z}_p$ and generate the receiver's key pair (PK_r, SK_r) where $PK_r = (g_1^{x_1},\ g_2^{x_2})$ and $SK_r = (x_1,\ x_2)$.

  Besides PK_r, the pairing parameters $(\mathbb{G}, \mathbb{G}_T, \hat{e}, p, g_1, g_2)$, symmetric key encryption scheme $(\mathsf{Enc}, \mathsf{Dec})$, and hash functions H_1, H_2, and H_3 should be made public. The receiver's message space $\mathcal{W}$ is $\mathbb{G}$.

- $\mathsf{Encrypt}(w, PK_r)$: It selects $r_1, r_2 \in_R \mathbb{Z}_p$ and generates a ciphertext $c_w = (c_1, c_2, c_3, c_4, c_5)$ as follows.

$$c_1 = g_1^{r_1},\ c_2 = \mathsf{H}_1(\hat{e}(g_1^{x_1}, \mathsf{H}_2(w)^{r_1})) \cdot \mathsf{H}_2(w)^{r_1} \cdot w,$$

$$c_3 = g_2^{r_2},\ c_4 = g_2^{x_2 r_2} \cdot \mathsf{H}_2(w)^{r_1},$$

$$c_5 = \mathsf{H}_3(c_1||c_2||c_3||c_4||\mathsf{H}_2(w)||\hat{e}(g_1^{x_1}, \mathsf{H}_2(w)^{r_1})).$$

- $\mathsf{Decrypt}(c_w, SK_r)$: It parses c_w as $(c_1, c_2, c_3, c_4, c_5)$ and computes $w = \frac{c_2}{\mathsf{H}_1(\hat{e}(g_1^{x_1}, \frac{c_4}{c_3^{x_2}})) \cdot \frac{c_4}{c_3^{x_2}}}$, and outputs w if the following equalities hold.

$$\hat{e}(c_1, \mathsf{H}_2(w)) = \hat{e}(g_1, \frac{c_4}{c_3^{x_2}})$$

$$c_5 = \mathsf{H}_3(c_1||c_2||c_3||c_4||\mathsf{H}_2(w)||\hat{e}(g_1^{x_1}, \frac{c_4}{c_3^{x_2}}))$$

  If either equality does not hold, the algorithm outputs an error symbol $\perp$.

- $\mathsf{TrapGen}_1(w, PK_s, SK_r)$: It selects $y \in_R \mathbb{Z}_p$ and computes the message-dependent trapdoor $t_{w,s} = (v_7, v_8)$ as follows.

$$v_0 = \mathsf{H}_2(w)^{x_1 y},\ v_1 = \mathsf{H}_2(w)^{x_1},\ v_2 = \hat{e}(g_1^{x_1 y}, w^{-1}),$$

$$v_3 = g_1^{x_1 y},\ v_4 = g_1^{x_1 x_2 y},\ v_5 = \mathsf{H}_2(w),$$

$$k_1 \in_R \mathcal{M},\ v_7 = \mathsf{Enc}(v_0||v_1||v_2||v_3||v_4||v_5, k_1),$$

$$v_8 = \mathsf{Encrypt}'(k_1, PK_s).$$

- $\mathsf{Test}_1(c, t_{w,s}, SK_s)$: It performs as follows.

  1. Parse c as $(c_1', c_2', c_3', c_4', c_5')$ and $t_{w,s}$ as (v_7, v_8).
  2. Decrypt v_8 to obtain k_1 and decrypt v_7 to recover $v_0, v_1, v_2, v_3, v_4, v_5$.
  3. Output 1 if the following equalities hold, and output 0 otherwise.

$$\frac{\hat{e}(v_3, c_4')}{\hat{e}(v_4, c_3')} = v_2 \cdot \hat{e}(v_3, \frac{c_2'}{\mathsf{H}_1(\hat{e}(c_1', v_1))}) = \hat{e}(c_1', v_0) \quad (1)$$

$$c_5' = \mathsf{H}_3(c_1'||c_2'||c_3'||c_4'||v_5||\hat{e}(c_1', v_1)) \quad (2)$$

- $\mathsf{TrapGen}_2(PK_s, SK_r)$: It generates a master trapdoor $t_{*,s} = (u_1, u_2)$, where $z \in_R \mathbb{Z}_p$, $k_2 \in_R \mathcal{M}$,

$$u_1 = \mathsf{Enc}(x_1||g_1^z||g_1^{x_2 \cdot z}, k_2),\ u_2 = \mathsf{Encrypt}'(k_2, PK_s).$$

- $\mathsf{Test}_2(c, w, t_{*,s}, SK_s)$: It performs as follows.

  1. Parse c as $(c_1', c_2', c_3', c_4', c_5')$ and $t_{*,s}$ as (u_1, u_2).
  2. Decrypt u_2 to obtain k_2 and decrypt u_1 to recover $(x_1, g_1^z, g_1^{x_2 \cdot z})$.
  3. Output 1 if the following equalities hold, and output 0 otherwise.

$$\hat{e}(g_1^z, \frac{c_4'}{\frac{c_2'}{\mathsf{H}_1(\hat{e}(c_1', \mathsf{H}_2(w)^{x_1})) \cdot w}}) = \hat{e}(g_1^{x_2 \cdot z}, c_3') \quad (3)$$

$$\hat{e}(g_1, \frac{c_2'}{\mathsf{H}_1(\hat{e}(c_1', \mathsf{H}_2(w)^{x_1})) \cdot w}) = \hat{e}(c_1', \mathsf{H}_2(w)) \quad (4)$$

$$c_5' = \mathsf{H}_3(\alpha||\mathsf{H}_2(w)||\hat{e}(c_1', \mathsf{H}_2(w)^{x_1})), \quad (5)$$

  where $\alpha = c_1'||c_2'||c_3'||c_4'$.

REMARK 3. *The element c_5 in the ciphertext is solely for the purpose of achieving IND-CCA security against an outside attacker. To achieve IND-CPA security against an outside attacker and preserve all other security properties, we can eliminate c_5 in* Encrypt*, the verification of c_5 in* Decrypt*, the verifications of c_5' in* Test_1 *and* Test_2 *in the above scheme. Moreover, the element v_5 in* $\mathsf{TrapGen}_1$ *can also be eliminated. Due to the limit of space, we put the proof of this simplified scheme in the full version of this paper.*

THEOREM 1. *The proposed ASE†† scheme is sound according to Definition 1.*

Proof. To prove the theorem, we need to show that the requirements in Definition 1 are satisfied.

Referring to the Encrypt algorithm, given a ciphertext $c_w = (c_1, c_2, c_3, c_4, c_5)$, it is straightforward to verify that the following equalities hold with probability 1.

$$\frac{c_2}{\mathsf{H}_1(\hat{e}(g_1^{x_1}, \frac{c_4}{c_3^{x_2}})) \cdot \frac{c_4}{c_3^{x_2}}} = w, \;\; \hat{e}(c_1, \mathsf{H}_2(w)) = \hat{e}(g_1, \frac{c_4}{c_3^{x_2}})),$$

$$c_5 = \mathsf{H}_3(c_1||c_2||c_3||c_4||\mathsf{H}_2(w)||\hat{e}(g_1^{x_1}, \frac{c_4}{c_3^{x_2}})).$$

Referring to the definition of the Decrypt algorithm, the above equalities means that $\mathsf{Decrypt}(c_w, SK_r) = w$. As a result, the first requirement in Definition 1 is satisfied.

As to the second requirement, we prove two things.

- If $\mathsf{Decrypt}(c, SK_r) = w$ holds, we prove that the equality $\mathsf{Test}_1(c, t_{w,s}, SK_s) = 1$ holds. Let $c = (c_1', c_2', c_3', c_4', c_5')$. Referring to the definition of the Decrypt algorithm, $\mathsf{Decrypt}(c, SK_r) = w$ implies the following equalities.

$$\frac{c_2'}{\mathsf{H}_1(\hat{e}(g_1^{x_1}, \frac{c_4'}{(c_3')^{x_2}})) \cdot \frac{c_4'}{(c_3')^{x_2}}} = w \qquad (6)$$

$$\hat{e}(c_1', \mathsf{H}_2(w)) = \hat{e}(g_1, \frac{c_4'}{(c_3')^{x_2}})) \qquad (7)$$

$$c_5' = \mathsf{H}_3(c_1'||c_2'||c_3'||c_4'||\mathsf{H}_2(w)||\hat{e}(g_1^{x_1}, \frac{c_4'}{(c_3')^{x_2}})). \qquad (8)$$

Based on these equalities and the definition of $t_{w,s}$, we have the following.

$$\begin{aligned}\frac{\hat{e}(v_3, c_4')}{\hat{e}(v_4, c_3')} &= \frac{\hat{e}(g_1^{x_1 y}, c_4')}{\hat{e}(g_1^{x_1 y}, (c_3')^{x_2})} \\ &= \hat{e}(g_1^{x_1 y}, \frac{c_4'}{(c_3')^{x_2}}) \\ &= \hat{e}(c_1', v_0)\end{aligned}$$

$$\begin{aligned}& v_2 \cdot \hat{e}(v_3, \frac{c_2'}{\mathsf{H}_1(\hat{e}(c_1', v_1))}) \\ =\; & \hat{e}(g_1^{x_1 y}, w^{-1}) \cdot \hat{e}(g_1^{x_1 y}, \frac{c_2'}{\mathsf{H}_1(\hat{e}(c_1', \mathsf{H}_2(w)^{x_1}))}) \\ =\; & \hat{e}(g_1^{x_1 y}, w^{-1}) \cdot \hat{e}(g_1^{x_1 y}, \frac{c_2'}{\mathsf{H}_1(\hat{e}(g_1^{x_1}, \frac{c_4'}{(c_3')^{x_2}})) \cdot w}) \\ =\; & \hat{e}(g_1^{x_1 y}, \frac{c_4'}{(c_3')^{x_2}}) \\ =\; & \hat{e}(c_1', v_0)\end{aligned}$$

$$c_5' = \mathsf{H}_3(c_1'||c_2'||c_3'||c_4'||v_5||\hat{e}(c_1', v_1))$$

Based on the definition of Test_1, we can conclude that $\mathsf{Test}_1(c, t_{w,s}, SK_s) = 1$ holds.

- If $\mathsf{Test}_1(c, t_{w,s}, SK_s) = 1$ holds, we prove that the equality $\mathsf{Decrypt}(c, SK_r) = w$ holds. Let $c = (c_1', c_2', c_3', c_4', c_5')$, we can re-write it in the following form.

$$c_1' = g_1^{r_1'},\; c_2' = \mathsf{H}_1(\hat{e}(g_1^{x_1}, \mathsf{H}_2(w)^{r_1'})) \cdot \mathsf{H}_2(w)^{r_1'} \cdot w^*,$$

$$c_3' = g_2^{r_2'},\; c_4' = g_2^{x_2 r_2'} \cdot \mathsf{H}_2(w)^{r_1'} \cdot w^\dagger,\; c_5' = c_5',$$

for some $r_1', r_2', w^*, w^\dagger$. The equalities associated with labels (1) and (2) in the definition of Test_1 lead to the following equalities.

$$\frac{\hat{e}(g_1^{x_1 y}, g_2^{x_2 r_2'} \cdot \mathsf{H}_2(w)^{r_1'} \cdot w^\dagger)}{\hat{e}(g_1^{x_1 x_2 y}, g_2^{r_2'})} = \hat{e}(g_1^{r_1'}, \mathsf{H}_2(w)^{x_1 y}),$$

$$\begin{aligned}& \hat{e}(g_1^{x_1 y}, w^{-1}) \cdot \hat{e}(g_1^{x_1 y}, \frac{\mathsf{H}_1(\hat{e}(g_1^{x_1}, \mathsf{H}_2(w)^{r_1'})) \cdot \mathsf{H}_2(w)^{r_1'} \cdot w^*}{\mathsf{H}_1(\hat{e}(g_1^{r_1'}, \mathsf{H}_2(w)^{x_1}))}) \\ =\; & \hat{e}(g_1^{r_1'}, \mathsf{H}_2(w)^{x_1 y}),\end{aligned}$$

$$c_5' = \mathsf{H}_3(c_1'||c_2'||c_3'||c_4'||\mathsf{H}_2(w)||\hat{e}(c_1', \mathsf{H}_2(w)^{x_1})).$$

The first equality implies $w^\dagger = 1$, and the second equality implies that $w^* = w$. Clearly, $\mathsf{Decrypt}(c, SK_r) = w$ holds.

As to the third requirement, we prove two things.

- If $\mathsf{Decrypt}(c, SK_r) = w$ holds, we prove that the equality $\mathsf{Test}_2(c, w, t_{*,s}, SK_s) = 1$ holds. Let $c = (c_1', c_2', c_3', c_4', c_5')$. Based on these equalities associated with labels (6),(7),(8) and the definition of $t_{*,s}$, we have the following.

$$\begin{aligned}\hat{e}(g_1^z, \frac{c_4'}{\frac{c_2'}{\mathsf{H}_1(\hat{e}(c_1', \mathsf{H}_2(w)^{x_1}))\cdot w}}) &= \hat{e}(g_1^z, \frac{c_4'}{\frac{c_2'}{\mathsf{H}_1(\hat{e}(g_1^{x_1}, \frac{c_4'}{(c_3')^{x_2}}))\cdot w}}) \\ &= \hat{e}(g_1^z, \frac{c_4'}{\frac{c_4'}{(c_3')^{x_2}}}) \\ &= \hat{e}(g_1^z, (c_3')^{x_2}) \\ &= \hat{e}(g_1^{x_2 \cdot z}, c_3')\end{aligned}$$

$$\begin{aligned}&\hat{e}(g_1, \frac{c_2'}{\mathsf{H}_1(\hat{e}(c_1', \mathsf{H}_2(w)^{x_1}))\cdot w}) \\ = \; &\hat{e}(g_1, \frac{c_2'}{\mathsf{H}_1(\hat{e}(g_1^{x_1}, \frac{c_4'}{(c_3')^{x_2}}))\cdot w}) \\ = \; &\hat{e}(g_1, \frac{c_4'}{(c_3')^{x_2}}) \\ = \; &\hat{e}(c_1', \mathsf{H}_2(w))\end{aligned}$$

$$c_5' = \mathsf{H}_3(c_1'||c_2'||c_3'||c_4'||\mathsf{H}_2(w)||\hat{e}(c_1', \mathsf{H}_2(w)^{x_1})).$$

As a result, the equalities associated with labels (3), (4), and (5) in the definition of Test_2 hold, we can conclude that $\mathsf{Test}_2(c, w, t_{*,s}, SK_s) = 1$.

- If $\mathsf{Test}_2(c, w, t_{*,s}, SK_s) = 1$, we prove that $\mathsf{Decrypt}(c, SK_r) = w$. Let $c = (c_1', c_2', c_3', c_4', c_5')$, we can re-write it in the following form.

$$c_1' = g_1^{r_1'},\ c_2' = \mathsf{H}_1(\hat{e}(g_1^{x_1}, \mathsf{H}_2(w)^{r_1'})) \cdot \mathsf{H}_2(w)^{r_1'} \cdot w^*,$$

$$c_3' = g_2^{r_2'},\ c_4' = g_2^{x_2 r_2'} \cdot \mathsf{H}_2(w)^{r_1'} \cdot w^\dagger,\ c_5' = c_5,$$

for some $r_1', r_2', w^*, w^\dagger$. The equalities associated with labels (3), (4), and (5) in the definition of Test_2 lead to the following equalities.

$$\hat{e}(g_1^z, \frac{g_2^{x_2 r_2'} \cdot \mathsf{H}_2(w)^{r_1'} \cdot w^\dagger}{\frac{\mathsf{H}_1(\hat{e}(g_1^{x_1}, \mathsf{H}_2(w)^{r_1'}))\cdot \mathsf{H}_2(w)^{r_1'} \cdot w^*}{\mathsf{H}_1(\hat{e}(g_1^{r_1'}, \mathsf{H}_2(w)^{x_1}))\cdot w}}) = \hat{e}(g_1^{x_2 \cdot z}, g_2^{r_2'}),$$

$$\hat{e}(g_1, \frac{\mathsf{H}_1(\hat{e}(g_1^{x_1}, \mathsf{H}_2(w)^{r_1'})) \cdot \mathsf{H}_2(w)^{r_1'} \cdot w^*}{\mathsf{H}_1(\hat{e}(g_1^{r_1'}, \mathsf{H}_2(w)^{x_1})) \cdot w}) = \hat{e}(g_1^{r_1'}, \mathsf{H}_2(w)),$$

$$c_5' = \mathsf{H}_3(c_1'||c_2'||c_3'||c_4'||\mathsf{H}_2(w)||\hat{e}(c_1', \mathsf{H}_2(w)^{x_1})).$$

The second equality implies that $w^* = w$. Based on this fact, it is straightforward to verify that the first equality implies $w^\dagger = 1$. Based on the proof of the first requirement, it is clear that $\mathsf{Decrypt}(c, SK_r) = w$ holds.

All three conditions required by Definition 1 hold, the theorem follows. □

4.3 Security Analysis

Let the pairing parameters $param = (\mathbb{G}, \mathbb{G}_T, \hat{e}, p)$ be defined in the same way as in the $\mathsf{rKeyGen}$ algorithm. We briefly review the Decision Linear (DLIN) assumption [3] and Bilinear Diffie-Hellman (BDH) assumption [5].

The DLIN assumption is as follows: any P.P.T. attacker adv can only distinguish T_0 and T_1 with a negligible advantage $|\Pr[Adv(param, T_0) = 1] - \Pr[Adv(param, T_1) = 1]|$, where $x, y \in_R \mathbb{Z}_p$, $g_a, g_b, g_c, \theta \in_R \mathbb{G}$, and

$$T_0 = (g_a,\ g_b,\ g_c,\ g_a^x,\ g_b^y,\ g_c^{x+y}),$$

$$T_1 = (g_a,\ g_b,\ g_c,\ g_a^x,\ g_b^y,\ \theta).$$

The BDH assumption is as follows: given $(param, g, g^x, g^y, g^z)$ where $x, y, z \in_R \mathbb{Z}_p$ and g is a generator of $\mathbb{G}$, any P.P.T. attacker adv can only compute $\hat{e}(g, g)^{x \cdot y \cdot z}$ with a negligible probability.

In proving one of the theorems, we also use the assumption that inverting the bilinear map $\hat{e}$ is hard, and this problem has been shown to be equivalent to the discrete logarithm problem in $\mathbb{G}$ and $\mathbb{G}_T$ [10, 15]. Certainly, this assumption is weaker than both DLIN and BDH assumptions.

THEOREM 2. *The proposed scheme achieves CI security against a curious Type-I server $\mathcal{S}_1$ in the random oracle model given: (1)* (KeyGen', $\mathsf{Encrypt}'$, $\mathsf{Decrypt}'$) *and* (Enc, Dec) *are IND-CPA secure; (2) the DLIN and BDH assumptions.*

Proof. Under Definition 3, let Game0 be the attack game defined in Figure 2. Let the attacker's advantage be ϵ_0.

Next, consider a game Game1, where the challenger performs in the same way as in Game0, except for the following. For any $\mathsf{TrapGen}_2$ query with the input PK_s, the challenger returns $t_{*,s} = (u_1, u_2)$, generated as follows: select $k_2', k_2'' \in_R \mathcal{M}$, $\gamma \in_R \mathbb{Z}_p$ and $\alpha, \beta \in_R \mathbb{G}$, compute $u_1 = \mathsf{Enc}(\gamma||\alpha||\beta, k_1')$ and $u_2 = \mathsf{Encrypt}'(k_2'', PK_s)$. In this game, let the attacker's advantage be ϵ_1. Based on the IND-CPA definition for encryption schemes, $|\epsilon_1 - \epsilon_0|$ is negligible if both (KeyGen', $\mathsf{Encrypt}'$, $\mathsf{Decrypt}'$) and (Enc, Dec) are IND-CPA secure.

Next, consider a game Game2, where the challenger performs in the same way as in Game1, except for the following.

- At the beginning of the game, the challenger chooses $y^* \in_R \mathbb{Z}_p$ and constructs a list for the random oracle H_2. If H_2 is queried with input w, the challenger first checks the list. If there is already a hash value for w, the challenger returns this value; otherwise, the challenger chooses $r_w \in_R \mathbb{Z}_p$ and returns $g_1^{r_w}$ as the hash value, and then adds $(w, g_1^{r_w}, r_w)$ to the list.

- For a $\mathsf{TrapGen}_1$ query with input w and PK_s, the challenger generates the message-dependent trapdoor $t_{w,s} = (v_7, v_8)$ as follows.

$$y_w \in_R \mathbb{Z}_p,\ v_0 = g_1^{x_1 \cdot y^* \cdot r_w \cdot y_w},\ v_1 = g_1^{x_1 \cdot r_w},$$

$$v_2 = \hat{e}(g_1^{x_1 \cdot y^* \cdot y_w}, w^{-1}), v_3 = g_1^{x_1 \cdot y^* \cdot y_w}, v_4 = g_1^{x_1 \cdot x_2 \cdot y^* \cdot y_w},$$

$$v_5 = g_1^{r_w},\ k_1 \in_R \mathcal{M},\ v_8 = \mathsf{Encrypt}'(k_1, PK_s),$$

$$v_7 = \mathsf{Enc}(v_0||v_1||v_2||v_3||v_4||v_5, k_1).$$

In this game, let the attacker's advantage be ϵ_2. Given that H_2 is modeled as a random oracle, this game is identical to Game1 so that $\epsilon_2 = \epsilon_1$.

Next, consider a game Game3, where the challenger performs in the same way as in Game2, except that the challenge $c_{w_b} = (c_1, c_2, c_3, c_4, c_5)$ is generated as follows.

$$r_1, r_2 \in_R \mathbb{Z}_p,\ R \in_R \mathbb{G}_2,\ R' \in_R \{0,1\}^\lambda,\ c_1 = g_1^{r_1},$$

$$c_2 = R{\cdot}\mathsf{H}_2(w_b)^{r_1}{\cdot}w_b, c_3 = g_2^{r_2}, c_4 = g_2^{x_2 \cdot r_2}{\cdot}\mathsf{H}_2(w_b)^{r_1}, c_5 = R'.$$

In Game3, let the attacker's advantage be ϵ_3. If H_1 and H_3 are modeled as random oracles, then Game3 differs from Game2 only if $\hat{e}(g_1^{x_1}, \mathsf{H}_2(w_b)^{r_1})$ has been queried to H_1 or $*||\hat{e}(g_1^{x_1}, \mathsf{H}_2(w)^{r_1})||*$ has been queried to H_3 where the $*$ can be anything. Based on the Difference lemma [11] and CLAIM 1 (stated and proven below), $|\epsilon_3 - \epsilon_2|$ is negligible based on the BDH assumption.

Next, consider a new game Game4, where the challenger performs in the same way as in Game3, except that the challenge $c_{w_b} = (c_1, c_2, c_3, c_4, c_5)$ is generated as follows.

$$r_1, r_2 \in_R \mathbb{Z}_p,\ R \in_R \mathbb{G}_2,\ R' \in_R \{0,1\}^\lambda,\ c_1 = g_1^{r_1},\ c_2 = R,$$

$$c_3 = g_2^{r_2},\ c_4 = g_2^{x_2 r_2} \cdot \mathsf{H}_2(w_b)^{r_1},\ c_5 = R'.$$

Clearly, Game4 is identical to Game3. Let the attacker's advantage be ϵ_4, so that we have $\epsilon_4 = \epsilon_3$. Based on its definition, ϵ_4 can also be regarded as the attacker's advantage in distinguishing X_0 and X_1, where $x_1, x_2, y^*, r_1, r_2 \in_R \mathbb{Z}_p$. Other public parameters (e.g. pairing parameters and w_0, w_1) are described in the game.

- $X_0 = (g_1, g_2, g_1^{x_1 \cdot y^*}, g_1^{x_1}, g_2^{x_2}, g_1^{x_1 \cdot x_2 \cdot y^*}, g_1^{r_1}, g_2^{r_2}, g_2^{x_2 r_2} \cdot \mathsf{H}_2(w_0)^{r_1})$
- $X_1 = (g_1, g_2, g_1^{x_1 \cdot y^*}, g_1^{x_1}, g_2^{x_2}, g_1^{x_1 \cdot x_2 \cdot y^*}, g_1^{r_1}, g_2^{r_2}, g_2^{x_2 r_2} \cdot \mathsf{H}_2(w_1)^{r_1})$

Let this distinguishing problem be referred to as D1.

Reduction step 2a. D1 is equivalent to distinguish Y_0 and Y_1, where $x_2, r_1, r_2 \in_R \mathbb{Z}_p$ and $h_1 \in_R \mathbb{G}$. Note that $g_1^{x_1 \cdot y^*}$ is set to be h_1 and $g_1^{x_1}$ is removed.

- $Y_0 = (g_1,\ g_2,\ h_1,\ g_2^{x_2},\ h_1^{x_2},\ g_1^{r_1}, g_2^{r_2},\ g_2^{x_2 r_2} \cdot \mathsf{H}_2(w_0)^{r_1})$
- $Y_1 = (g_1,\ g_2,\ h_1,\ g_2^{x_2},\ h_1^{x_2},\ g_1^{r_1}, g_2^{r_2},\ g_2^{x_2 r_2} \cdot \mathsf{H}_2(w_1)^{r_1})$

Let this distinguishing problem be referred to as D2.

Reduction step 2b. D2 can be reduced to distinguish Z_0 and Z_1, where $x_2, r_1, r_2 \in_R \mathbb{Z}_p$, $h_1, h_2, \delta \in_R \mathbb{G}$.

- $Z_0 = (g_1,\ g_2,\ h_1,\ h_2,\ g_2^{x_2},\ h_1^{x_2},\ g_1^{r_1},\ g_2^{r_2},\ g_2^{x_2 r_2} \cdot h_2^{r_1})$
- $Z_1 = (g_1,\ g_2,\ h_1,\ h_2,\ g_2^{x_2},\ h_1^{x_2},\ g_1^{r_1},\ g_2^{r_2},\ \delta)$

Let this distinguishing problem be referred to as D3.

Reduction step 2c. D3 is equivalent to distinguish U_0 and U_1, where $x_2, x_3, r_1, r_2 \in_R \mathbb{Z}_p$, $h_1, \delta \in_R \mathbb{G}$. Note that h_2 is replaced by $g_2^{x_3}$.

- $U_0 = (g_1,\ g_2,\ h_1,\ g_2^{x_3},\ g_2^{x_2},\ h_1^{x_2},\ g_1^{r_1},\ g_2^{r_2},\ g_2^{x_2 r_2} \cdot g_2^{x_3 r_1})$
- $U_1 = (g_1,\ g_2,\ h_1,\ g_2^{x_3},\ g_2^{x_2},\ h_1^{x_2},\ g_1^{r_1},\ g_2^{r_2},\ \delta)$

Let this distinguishing problem be referred to as D4.

Reduction step 2d. D4 is equivalent to distinguish V_0 and V_1, where $\alpha, x_3, r_1, r_2 \in_R \mathbb{Z}_p$, $h_1, h_2, \delta \in_R \mathbb{G}$. Note that x_2 is set to be $\alpha + x_3$.

- $V_0 = (g_1, g_2, h_1, g_2^{x_3}, g_2^{\alpha + x_3}, h_1^{\alpha + x_3}, g_1^{r_1}, g_2^{r_2}, g_2^{\alpha r_2 + x_3(r_1 + r_2)})$
- $V_1 = (g_1,\ g_2,\ h_1,\ g_2^{x_3},\ g_2^{\alpha + x_3},\ h_1^{\alpha + x_3},\ g_1^{r_1},\ g_2^{r_2},\ \delta)$

Let this distinguishing problem be referred to as D5.

Reduction step 2e. D5 can be reduced to distinguish W_0 and W_1, where $x_3, r_1, r_2 \in_R \mathbb{Z}_p$, $h_1, h_2, \delta \in_R \mathbb{G}$. The reduction is based on the fact that it is straightforward to construct a D5's instance from W_0 and W_1.

- $W_0 = (g_1,\ g_2,\ h_1,\ g_2^{x_3},\ h_1^{x_3},\ g_1^{r_1},\ g_2^{r_2},\ g_2^{x_3(r_1 + r_2)})$
- $W_1 = (g_1,\ g_2,\ h_1,\ g_2^{x_3},\ h_1^{x_3},\ g_1^{r_1},\ g_2^{r_2},\ \delta)$

Let this distinguishing problem be referred to as D6.

Reduction step 2f. D6 can be reduced to the DLIN problem in $\mathbb{G}$. The reduction is because from a DLIN instance we can construct a D6 instance as follows, where $\gamma \in_R \mathbb{Z}_p$.

- $T'_0 = (g_a,\ g_b,\ g_b^{\gamma},\ g_c,\ g_c^{\gamma},\ g_a^{x},\ g_b^{y},\ g_c^{x+y})$
- $T'_1 = (g_a,\ g_b,\ g_b^{\gamma},\ g_c,\ g_c^{\gamma},\ g_a^{x},\ g_b^{y},\ \theta)$

Based on all above reductions, in Game4, the attacker's advantage ϵ_4 is negligible based on the DLIN assumption. As a result, ϵ_0 is negligible based on all the assumptions mentioned in the theorem. □

CLAIM 1. *In Game3 of the proof of Theorem 2, the attacker can only succeed in querying H_1 with $\hat{e}(g_1^{x_1}, \mathsf{H}_2(w_b)^{r_1})$ for a negligible probability based on the BDH assumption.*

Proof. In Game3, besides the public parameters, the challenger only needs $(g_1^{x_1 \cdot y^*}, g_1^{x_1 \cdot x_2 \cdot y^*})$ in order to answer the attacker's oracle queries. The computational problem (referred to as P1) is defined as follows.

- Input: $(g_1, g_2, g_1^{x_1 \cdot y^*}, \mathsf{H}_2(w_b), g_1^{x_1}, g_2^{x_2}, g_1^{x_1 \cdot x_2 \cdot y^*}, g_1^{r_1}, g_2^{r_2}, g_2^{x_2 r_2} \cdot \mathsf{H}_2(w_b)^{r_1})$, where $x_1, x_2, y^*, r_1, r_2 \in_R \mathbb{Z}_p$. Other public parameters are described in the game.
- Output: $\hat{e}(g_1^{x_1}, \mathsf{H}_2(w_b)^{r_1})$.

The problem P1 can be equivalently rephrased as follows, referred to as P2. Note that $g_1^{x_1 \cdot y^*}$ is set to be h_1 and $\mathsf{H}_2(w_b)$ is set to be h_2.

- Input: $(g_1, g_2, h_1, h_2, g_1^{x_1}, g_2^{x_2}, h_1^{x_2}, g_1^{r_1}, g_2^{r_2}, g_2^{x_2 r_2} {\cdot} h_2^{r_1})$, where $x_1, x_2, r_1, r_2 \in_R \mathbb{Z}_p$, $h_1 \in_R \mathbb{G}$ and $h_2 \in_R \mathbb{G}$.
- Output: $\hat{e}(g_1^{x_1}, h_2^{r_1})$ or $\hat{e}(g_1^{x_1}, g_2^{x_2 r_2})$.

The problem P2 can be reduced to the following problem (referred to as P3).

- Input: $(g_1, g_2, g_1^{x_1}, g_2^{x_2}, g_1^{x_2}, g_2^{r_2})$, where $x_1, x_2, r_1, r_2 \in_R \mathbb{Z}_p$.

- Output: $\hat{e}(g_1^{x_1}, g_2^{x_2 r_2})$.

The reduction is based on the fact that, from a P3's instance $(g_1, g_2, g_1^{x_1}, g_2^{x_2}, g_1^{x_2} g_2^{r_2})$, we can construct a P2's instance $(g_1, g_2, g_1^{\alpha}, g_2^{\beta - r_2}, g_1^{x_1}, g_2^{x_2}, (g_1^{x_2})^{\alpha}, g_1^{x_2+\gamma}, g_2^{r_2}, g_2^{x_2\beta+\beta\gamma-r_2\gamma})$ for $\alpha, \beta, \gamma \in_R \mathbb{Z}_p$. Furthermore, the problem P3 can be reduced to the following problem (referred to as P4).

- Input: $(g_2,\ g_2^{x_1},\ g_2^{x_2},\ g_2^{r_2})$, where $x_1, x_2, r_2 \in_R \mathbb{Z}_p$.
- Output: $\hat{e}(g_2, g_2)^{x_1 \cdot x_2 \cdot r_2}$.

The reduction is based on the fact that, from a P4's instance $(g_2,\ g_2^{x_1},\ g_2^{x_2},\ g_2^{r_2})$, we can construct a P2's instance $(g_2^{\tau},\ g_2,\ g_2^{x_1 \cdot \tau},\ g_2^{x_2},\ g_2^{x_2 \cdot \tau}\ g_2^{r_2})$ for $\alpha, \beta, \gamma \in_R \mathbb{Z}_p$. P4 is indeed the BDH problem. The claim now follows. □

THEOREM 3. *The proposed scheme achieves IND-CCA security against an outside attacker in the random oracle model given: (1)* $(\mathsf{KeyGen}', \mathsf{Encrypt}', \mathsf{Decrypt}')$ *and* $(\mathsf{Enc}, \mathsf{Dec})$ *are IND-CPA secure; (2) the DLIN and BDH assumptions.*

Proof. Under Definition 2, let Game0 be the attack game defined in Figure 1. Let the attacker's advantage be ϵ_0.

Next, consider a game Game1, where the challenger performs in the same way as in Game0, except for the following. For any $\mathsf{TrapGen}_1$ query with the input w and PK_s, the challenger returns $t_{w,s} = (v_7, v_8)$, generated as follows: select $v_0, v_1, v_2, v_3, v_4, v_5, k_1$ uniformly at random from their corresponding domains, compute

$$v_7 = \mathsf{Enc}(v_0||v_1||v_2||v_3||v_4||v_5, k_1)$$

and $v_8 = \mathsf{Encrypt}'(k_1, PK_s)$. In this game, let the attacker's advantage be ϵ_1. Based on the IND-CPA definition for encryption schemes, $|\epsilon_1 - \epsilon_0|$ is negligible if both $(\mathsf{KeyGen}', \mathsf{Encrypt}', \mathsf{Decrypt}')$ and $(\mathsf{Enc}, \mathsf{Dec})$ are IND-CPA secure.

Next, consider a game Game2, where the challenger performs in the same way as in Game1, except for the following. For any $\mathsf{TrapGen}_2$ query with the input PK_s, the challenger returns $t_{*,s} = (u_1, u_2)$ which is generated as follows: select $k_2', k_2'' \in_R \mathcal{M}$, $\gamma \in_R \mathbb{Z}_p$ and $\alpha, \beta \in_R \mathbb{G}_1$, compute $u_1 = \mathsf{Enc}(\gamma||\alpha||\beta, k_2')$ and $u_2 = \mathsf{Encrypt}'(k_2'', PK_s)$. In this game, let the attacker's advantage be ϵ_2. Based on the IND-CPA definition for encryption schemes, $|\epsilon_2 - \epsilon_1|$ is negligible if both $(\mathsf{KeyGen}', \mathsf{Encrypt}', \mathsf{Decrypt}')$ and $(\mathsf{Enc}, \mathsf{Dec})$ are IND-CPA secure.

Next, consider a game Game3, where the challenger performs in the same way as in Game2, except that it answers the $\mathsf{Decrypt}$ oracle as follows. Given a ciphertext c, the challenger parses it as $(c_1, c_2, c_3, c_4, c_5)$ and performs as follows.

1. Compute $w = \frac{c_2}{\mathsf{H}_1(\hat{e}(g_1^{x_1}, \frac{c_4}{c_3^{x_2}})) \cdot \frac{c_4}{c_3^{x_2}}}$.
2. If $c_1||c_2||c_3||c_4||\mathsf{H}_2(w)||\hat{e}(g_1^{x_1}, \frac{c_4}{c_3^{x_2}})$ has not been queried to H_3 or w has not been queried to H_2, abort by outputting an error symbol $\perp$. In addition, if there are different inputs to either oracle and result in the same output, abort by outputting an error symbol $\perp$.
3. If $c_5 = \mathsf{H}_3(c_1||c_2||c_3||c_4||\mathsf{H}_2(w)||\hat{e}(g_1^{x_1}, \frac{c_4}{c_3^{x_2}}))$ and $\hat{e}(c_1, \mathsf{H}_2(w)) = \hat{e}(g_1, \frac{c_4}{c_3^{x_2}}))$ output w. Otherwise, abort by outputting an error symbol $\perp$.

This game is identical to Game2 unless the following event occurs: there is a ciphertext c, the $\mathsf{Decrypt}$ algorithm outputs w but the challenger returns $\perp$. If H_2 and H_3 are modeled as random oracles, the event occurs with a negligible probability. In this game, let the attacker's advantage be ϵ_3. Based on the Difference lemma [11], $|\epsilon_3 - \epsilon_2|$ is negligible in the random oracle model.

Next, consider a game Game4, where the challenger performs in the same way as in Game3, except that it answers the $\mathsf{Decrypt}$ oracle as follows. Given a ciphertext c, the challenger parses it as $(c_1, c_2, c_3, c_4, c_5)$ and performs as follows.

1. Check whether there is an input $c_1||c_2||c_3||c_4||\alpha_1||\alpha_2$ to H_3, where $\alpha_1 \in \mathbb{G}$ and $\alpha_2 \in \mathbb{G}_T$ and there is an input $w' \in \mathbb{G}$ to H_2 such that
$$c_5 = \mathsf{H}_3(c_1||c_2||c_3||c_4||\alpha_1||\alpha_2), \mathsf{H}_2(w') = \alpha_1.$$
If there is no such inputs or there are more than one inputs satisfying the equalities, aborts by outputting an error symbol $\perp$.
2. Compute $\beta = \frac{c_2}{\mathsf{H}_1(\alpha_2) \cdot w'}$.
3. Check whether the following equalities hold.
$$\hat{e}(g_1^{x_1}, \beta) = \alpha_2,\ \hat{e}(\frac{c_4}{\beta}, g_2) = \hat{e}(g_2^{x_2}, c_3),$$
$$\hat{e}(c_1, \mathsf{H}_2(w')) = \hat{e}(g_1, \beta).$$
If so, output w', otherwise aborts by outputting an error symbol $\perp$.

This game is identical to Game3, but the receiver's private keys are not required to answer the oracle queries. In this game, let the attacker's advantage be ϵ_4, so that $\epsilon_3 = \epsilon_4$.

Next, consider a game Game5, where the challenger performs in the same way as in Game4, except that the challenge $c_{w_b} = (c_1, c_2, c_3, c_4, c_5)$ is generated as follows.

$$r_1, r_2 \in_R \mathbb{Z}_p,\ R \in_R \mathbb{G}_2,\ R' \in_R \{0,1\}^{\lambda},\ c_1 = g_1^{r_1},$$

$$c_2 = R \cdot \mathsf{H}_2(w_b)^{r_1} \cdot w_b, c_3 = g_2^{r_2}, c_4 = g_2^{x_2 \cdot r_2} \cdot \mathsf{H}_2(w_b)^{r_1}, c_5 = R'.$$

In this game, let the attacker's advantage be ϵ_5. Note the fact that the challenger answers $\mathsf{Decrypt}$ oracle access without the knowledge of the receiver's private keys (i.e. $\mathsf{Decrypt}$ oracle access gives the attacker no actual privilege), therefore an outside attacker infact has less privilege than a curious Type-I server. Based on the analysis from Game3 and the following games in the proof of Theorem 2, we can conclude that ϵ_5 is negligible based on the the DLIN assumption and the BDH assumption. The theorem now follows. □

THEOREM 4. *The proposed scheme achieves CI security against a hybrid server* $\mathcal{S}_h$ *in the random oracle model given: (1)* $(\mathsf{KeyGen}', \mathsf{Encrypt}', \mathsf{Decrypt}')$ *and* $(\mathsf{Enc}, \mathsf{Dec})$ *are IND-CPA secure; (2) the DLIN and BDH assumptions.*

Proof. Under Definition 4, suppose the attacker's advantage is ϵ_0 in the attack game defined in Figure 3. With respect to the proposed scheme, it is clear that, given a master trapdoor $t_{*,s}$, an attacker can simulate the $\mathsf{TrapGen}_1$ oracle by itself. As a result, ϵ_0 can also be regarded as the attacker's advantage in distinguishing X_0 and X_1, where $x_1, x_2, r_1, r_2, r_1', r_2', z \in_R \mathbb{Z}_p$, and $w_0, w_1 \in_R \mathbb{G}$, and any $c_1, c_2, c_3, c_4, c_1', c_2', c_3', c_4'$. Other public parameters are described in the game.

- $X_0 = (x_1, g_1, g_2, g_1^{x_1}, g_2^{x_2}, g_1^{z}, g_1^{x_2 \cdot z}, g_1^{r_1}, \mathsf{H}_1(\hat{e}(g_1^{x_1}, \mathsf{H}_2(w_0)^{r_1}))\cdot \mathsf{H}_2(w_0)^{r_1}\cdot w_0, g_2^{r_2}, g_2^{x_2 r_2}\cdot \mathsf{H}_2(w_0)^{r_1}, \mathsf{H}_3(c_1||c_2||c_3||c_4||\mathsf{H}_2(w_0)||\hat{e}(g_1^{x_1}, \mathsf{H}_2(w_0)^{r_1})), g_1^{r_1'}, \mathsf{H}_1(\hat{e}(g_1^{x_1}, \mathsf{H}_2(w_0)^{r_1'}))\cdot \mathsf{H}_2(w_0)^{r_1'}\cdot w_0, g_2^{r_2'}, g_2^{x_2 r_2'}\cdot \mathsf{H}_2(w_0)^{r_1'}, \mathsf{H}_3(c_1'||c_2'||c_3'||c_4'||\mathsf{H}_2(w_0)||\hat{e}(g_1^{x_1}, \mathsf{H}_2(w_0)^{r_1'})))$

- $X_1 = (x_1, g_1, g_2, g_1^{x_1}, g_2^{x_2}, g_1^{z}, g_1^{x_2 \cdot z}, g_1^{r_1}, \mathsf{H}_1(\hat{e}(g_1^{x_1}, \mathsf{H}_2(w_0)^{r_1}))\cdot \mathsf{H}_2(w_0)^{r_1}\cdot w_0, g_2^{r_2}, g_2^{x_2 r_2}\cdot \mathsf{H}_2(w_0)^{r_1}, \mathsf{H}_3(c_1||c_2||c_3||c_4||\mathsf{H}_2(w_0)||\hat{e}(g_1^{x_1}, \mathsf{H}_2(w_0)^{r_1})), g_1^{r_1'}, \mathsf{H}_1(\hat{e}(g_1^{x_1}, \mathsf{H}_2(w_1)^{r_1'}))\cdot \mathsf{H}_2(w_1)^{r_1'}\cdot w_1, g_2^{r_2'}, g_2^{x_2 r_2'}\cdot \mathsf{H}_2(w_1)^{r_1'}, \mathsf{H}_3(c_1'||c_2'||c_3'||c_4'||\mathsf{H}_2(w_1)||\hat{e}(g_1^{x_1}, \mathsf{H}_2(w_1)^{r_1'})))$

Let the above distinguishing problem be denoted as I1.

Reduction step 4a. I1 is equivalent to distinguish Y_0 and Y_1, where $x_1, x_2, r_1, r_2, r_1', r_2', z \in_R \mathbb{Z}_p$ and $w_0, w_1, h_0, h_1 \in_R \mathbb{G}$, and any $c_1, c_2, c_3, c_4, c_1', c_2', c_3', c_4'$. Note that $\mathsf{H}_2(w_0)$ is set to be h_0 and $\mathsf{H}_2(w_1)$ is set to be h_1.

- $Y_0 = (x_1, g_1, g_2, g_1^{x_1}, g_2^{x_2}, g_1^{z}, g_1^{x_2 \cdot z}, g_1^{r_1}, \mathsf{H}_1(\hat{e}(g_1^{x_1}, h_0^{r_1}))\cdot h_0^{r_1}\cdot w_0, g_2^{r_2}, g_2^{x_2 r_2}\cdot h_0^{r_1}, \mathsf{H}_3(c_1||c_2||c_3||c_4||h_0||\hat{e}(g_1^{x_1}, h_0^{r_1})), g_1^{r_1'}, \mathsf{H}_1(\hat{e}(g_1^{x_1}, h_0^{r_1'}))\cdot h_0^{r_1'}\cdot w_0, g_2^{r_2'}, g_2^{x_2 r_2'}\cdot h_0^{r_1'}, \mathsf{H}_3(c_1'||c_2'||c_3'||c_4'||h_0||\hat{e}(g_1^{x_1}, h_0^{r_1'})))$

- $Y_1 = (x_1, g_1, g_2, g_1^{x_1}, g_2^{x_2}, g_1^{z}, g_1^{x_2 \cdot z}, g_1^{r_1}, \mathsf{H}_1(\hat{e}(g_1^{x_1}, h_0^{r_1}))\cdot h_0^{r_1}\cdot w_0, g_2^{r_2}, g_2^{x_2 r_2}\cdot h_0^{r_1}, \mathsf{H}_3(c_1||c_2||c_3||c_4||h_0||\hat{e}(g_1^{x_1}, h_0^{r_1})), g_1^{r_1'}, \mathsf{H}_1(\hat{e}(g_1^{x_1}, h_1^{r_1'}))\cdot h_1^{r_1'}\cdot w_1, g_2^{r_2'}, g_2^{x_2 r_2'}\cdot h_1^{r_1'}, \mathsf{H}_3(c_1'||c_2'||c_3'||c_4'||h_1||\hat{e}(g_1^{x_1}, h_1^{r_1'})))$

Let this distinguishing problem be denoted as I2.

Reduction step 4b. Let Y_0' and Y_1' be the following. Straightforwardly, CLAIM 2 (stated and proven below) also implies that, for any P.P.T. attacker, given Y_1', it can only compute $\hat{e}(g_1^{x_1}, h_0^{r_1})$ and $\hat{e}(g_1^{x_1}, h_1^{r_1'})$ with a negligible probability based on the BDH assumption.

- $Y_0' = (x_1, g_1, g_2, g_1^{x_1}, g_2^{x_2}, g_1^{z}, g_1^{x_2 \cdot z}, g_1^{r_1}, g_2^{r_2}, g_2^{x_2 r_2}\cdot h_0^{r_1}, g_1^{r_1'}, g_2^{r_2'}, g_2^{x_2 r_2'}\cdot h_0^{r_1'})$

- $Y_1' = (x_1, g_1, g_2, g_1^{x_1}, g_2^{x_2}, g_1^{z}, g_1^{x_2 \cdot z}, g_1^{r_1}, g_2^{r_2}, g_2^{x_2 r_2}\cdot h_0^{r_1}, g_1^{r_1'}, g_2^{r_2'}, g_2^{x_2 r_2'}\cdot h_1^{r_1'})$

As a result, I2 is equivalent to distinguish Y_0' and Y_1' based on the BDH assumption in the random oracle model. Let this distinguishing problem be denoted as I3.

Reduction step 4c. I3 is equivalent to distinguish Z_0 and Z_1, where $x_2, r_1, r_2, r_1', r_2', z \in_R \mathbb{Z}_p$ and $h_0, h_1 \in_R \mathbb{G}$. Note that x_1, g^{x_1} are removed.

- $Z_0 = (g_1, g_2, g_2^{x_2}, g_1^{z}, g_1^{x_2 \cdot z}, g_1^{r_1}, g_2^{r_2}, g_2^{x_2 r_2}\cdot h_0^{r_1}, g_1^{r_1'}, g_2^{r_2'}, g_2^{x_2 r_2'}\cdot h_0^{r_1'})$

- $Z_1 = (g_1, g_2, g_2^{x_2}, g_1^{z}, g_1^{x_2 \cdot z}, g_1^{r_1}, g_2^{r_2}, g_2^{x_2 r_2}\cdot h_0^{r_1}, g_1^{r_1'}, g_2^{r_2'}, g_2^{x_2 r_2'}\cdot h_1^{r_1'})$

Let this distinguishing problem be denoted as I4.

Reduction step 4d. I4 can be reduced to distinguish U_0 and U_1, where $x_2, r_1', r_2', z \in_R \mathbb{Z}_p$ and $h_0, h_1 \in_R \mathbb{G}$. The reduction is based on the fact that we can construct $g_1^{r_1}, g_2^{r_2}, g_2^{x_2 r_2}\cdot h_0^{r_1}$ based on h_0.

- $U_0 = (g_1, g_2, g_2^{x_2}, g_1^{z}, g_1^{x_2 \cdot z}, h_0, g_1^{r_1'}, g_2^{r_2'}, g_2^{x_2 r_2'}\cdot h_0^{r_1'})$

- $U_1 = (g_1, g_2, g_2^{x_2}, g_1^{z}, g_1^{x_2 \cdot z}, h_0, g_1^{r_1'}, g_2^{r_2'}, g_2^{x_2 r_2'}\cdot h_1^{r_1'})$

Let this distinguishing problem be denoted as I5.

It is easy to see that I5 can be reduced to the distinguishing problem D2 defined at *Reduction step 2a.* in the proof of Theorem 2 (based on a D2's instance, by adding either $\mathsf{H}_2(w_0)$ or $\mathsf{H}_2(w_1)$, we can obtain an I5's instance). Therefore, from the proof of Theorem 2, a P.P.T. attacker can only have a negligible advantage in solving I5 based on the DLIN assumption. As a result, ϵ_0 is negligible based on all the assumptions mentioned in the theorem. □

CLAIM 2. *In the proof of Theorem 4, for any P.P.T. attacker, given either Y_0', it can only compute $\hat{e}(g_1^{x_1}, h_0^{r_1})$ and $\hat{e}(g_1^{x_1}, h_0^{r_1'})$ with a negligible probability based on the BDH assumption.*

Proof. Due to the symmetry, we only need to prove that, for any P.P.T. attacker, given either Y_0', it can only compute $\hat{e}(g_1^{x_1}, h_0^{r_1})$ with a negligible probability.

Reduction step a1. Given $(x_1, g_1, g_2, g_1^{x_1}, g_2^{x_2}, g_1^{z}, g_1^{x_2\cdot z}, g_1^{r_1}, g_2^{r_2}, g_2^{x_2 r_2}\cdot h_0^{r_1})$, we can construct Y_0^* as follows, where $\alpha, \beta \in_R \mathbb{Z}_p$. $Y_0^* = (x_1, g_1, g_2, g_1^{x_1}, g_2^{x_2}, g_1^{z}, g_1^{x_2\cdot z}, g_1^{r_1}, g_2^{r_2}, g_2^{x_2 r_2}\cdot h_0^{r_1}, g_1^{r_1 \alpha}, g_2^{r_2 \alpha + \beta}, (g_2^{x_2 r_2}\cdot h_0^{r_1})^{\alpha}\cdot g_2^{x_2 \beta})$. It is easy to verify that Y_0^* has an identical distribution to that of Y_0'. So, the problem can be reduced to: given $(x_1, g_1, g_2, g_1^{x_1}, g_2^{x_2}, g_1^{z}, g_1^{x_2\cdot z}, g_1^{r_1}, g_2^{r_2}, g_2^{x_2 r_2}\cdot h_0^{r_1})$, the attacker computes $\hat{e}(g_1^{x_1}, g_2^{x_2 r_2})$.

Reduction step a2. The above problem can be reduced to: given $(x_1, g_1, g_2, g_1^{x_1}, g_2^{x_2}, g_1^{z}, g_1^{x_2\cdot z}, g_2^{r_2})$, the attacker computes $\hat{e}(g_1^{x_1}, g_2^{x_2 r_2})$.

Reduction step a3. The above problem can be reduced to: given $(g_1, g_2, g_2^{x_2}, g_1^{z}, g_1^{x_2\cdot z}, g_2^{r_2})$, the attacker computes $\hat{e}(g_1, g_2^{x_2 r_2})$.

Reduction step a4. This above reduced problem is equivalent to: given $(f_1^{x_3}, g_2, g_2^{x_2}, f_1, f_1^{x_2}, g_2^{r_2})$, where $f_1 \in_R \mathbb{G}$ and $x_3 \in_R \mathbb{Z}_p$, the attacker can only compute $\hat{e}(f_1^{x_3}, g_2^{x_2 r_2})$ with a negligible probability. The equivalence is based on the fact that $g_1, g_1^{z}, g_1^{x_2\cdot z}$ are replaced with $f_1^{x_3}, f_1, f_1^{x_2}$ respectively, which is simply a change of notation. Referring to the reduction from P3 to P4 in the proof of CLAIM 1, this can be reduced to the BDH problem. The claim follows. □

Note that this theorem implies that the proposed scheme is also secure under Definition 5.

The following theorem is straightforward, so that we skip a formal proof.

THEOREM 5. *The proposed scheme achieves universal trapdoor one-wayness under Definition 6, given: (1) the hash function H_2 is one-way; (2) the bilinear map $\hat{e}$ is one-way.*

THEOREM 6. *The proposed scheme achieves trapdoor indistinguishability property if* $(\mathsf{KeyGen}', \mathsf{Encrypt}', \mathsf{Decrypt}')$ *and* $(\mathsf{Enc}, \mathsf{Dec})$ *are IND-CPA secure.*

Proof. Under Definition 7, according to the attack game definition in Figure 5, the attacker can query $\mathsf{sKeyReq}_{t'}$ for $t' = 1$ or $t' = h$. Without loss of generality, we assume that the attacker queries $\mathsf{sKeyReq}_1$. This means $t = h$, and the challenge $t_{w_b} = (v_7, v_8)$ are defined as follows.

$$y \in_R \mathbb{Z}_p, v_0 = \mathsf{H}_2(w_b)^{x_1 y}, v_1 = \mathsf{H}_2(w_b)^{x_1}, v_2 = \hat{e}(g_1^{x_1 y}, w_b^{-1}),$$

$$v_3 = g_1^{x_1 y},\ v_4 = g_1^{x_1 x_2 y},\ v_5 = \mathsf{H}_2(w_b),\ k_1 \in_R \mathcal{M},$$

$$v_7 = \mathsf{Enc}(v_0||v_1||v_2||v_3||v_4||v_5, k_1), v_8 = \mathsf{Encrypt}'(k_1, PK_{s_h}).$$

Note the fact that revealing the receiver's private key x_1, x_2 will not affect the security of the encryption schemes, namely $(\mathsf{KeyGen}', \mathsf{Encrypt}', \mathsf{Decrypt}')$ and $(\mathsf{Enc}, \mathsf{Dec})$. Without oracle access to SK_{s_h}, the attacker can learn nothing about w_b from t_{w_b}. Then, the theorem follows. □

5. CONCLUSION

In this paper, we have formulated the concept of ASE[††] and attempted to provide a comprehensive security model for the primitive. As to the relationships of various security properties, the IND-CPA security against an outside attacker is implied by the ciphertext indistinguishability property against a Type-I server and the message-dependent trapdoor indistinguishability property (to be presented in the full paper due to space limit). Compared with previous works, such as those from [7, 8, 9], our formulation defines more flexible functionalities and our security model reflects a higher level of security guarantees. The security of the proposed instantiation relies on the standard DLIN and BDH assumptions in the random oracle model, which plays an important role in the double encryption structure. It is an interesting future work to investigate an instantiation in the standard model, namely without using random oracle. With regard to trapdoor security, we have designed universal one-wayness property against all attackers. There is a possibility of replacing it with the augmented notion proposed in [2] or an even stronger notion similar to the ciphertext indistinguishability property against Type-II attacker in our security model. We consider this to be another line of future work.

6. REFERENCES

[1] J. Baek, R. Safavi-Naini, and W. Susilo. On the integration of public key data encryption and public key encryption with keyword search. In S. K. Katsikas et al., editor, *Proceedings of the 9th international conference on Information Security*, volume 4176 of *LNCS*, pages 217–232. Springer, 2006.

[2] M. Bellare, A. Boldyreva, and A. O'Neill. Deterministic and efficiently searchable encryption. In *Advances in cryptology — CRYPTO 2007*, pages 535–552. Springer, 2007.

[3] D. Boneh, X. Boyen, and H. Shacham. Short group signatures. In M. K. Franklin, editor, *Advances in Cryptology — CRYPTO 2004*, volume 3152 of *LNCS*, pages 41–55. Springer, 2004.

[4] D. Boneh, G. Di Crescenzo, R. Ostrovsky, and G. Persiano. Public Key Encryption with Keyword Search. In C. Cachin and J. Camenisch, editors, *Advances in Cryptology — EUROCRYPT 2004*, volume 3027 of *LNCS*, pages 506–522. Springer, 2004.

[5] D. Boneh and M. K. Franklin. Identity-based encryption from the weil pairing. In J. Kilian, editor, *Advances in Cryptology — CRYPTO 2001*, volume 2139 of *LNCS*, pages 213–229. Springer, 2001.

[6] T. ElGamal. A public key cryptosystem and a signature scheme based on discrete logarithms. In G. R. Blakley and D. Chaum, editors, *Advances in Cryptology - CRYPTO 1984*, volume 196 of *LNCS*, pages 10–18. Springer, 1985.

[7] T. Fuhr and P. Paillier. Decryptable searchable encryption. In *Proceedings of the 1st international conference on Provable security*, pages 228–236, 2007.

[8] D. Hofheinz and E. Weinreb. Correlation-resistant storage via keyword-searchable encryption. Cryptology ePrint Archive: Report 2008/423, 2008.

[9] L. Ibraimi, S. Nikova, P. H. Hartel, and W. Jonker. Public-key encryption with delegated search. In J. Lopez and G. Tsudik, editors, *Applied Cryptography and Network Security - 9th International Conference*, volume 6715 of *LNCS*, pages 532–549, 2011.

[10] K. Karabina, E. Knapp, and A. Menezes. Generalizations of verheul's theorem to asymmetric pairings. Technical Report CACR 2012-03, University of Waterloo, 2012.

[11] V. Shoup. Sequences of games: a tool for taming complexity in security proofs. http://shoup.net/papers/, 2006.

[12] D. X. Song, D. Wagner, and A. Perrig. Practical Techniques for Searches on Encrypted Data. In *Proceedings of the 2000 IEEE Symposium on Security and Privacy*, pages 44–55, 2000.

[13] Q. Tang. Search in encrypted data: Theoretical models and practical applications. Cryptology ePrint Archive: Report 2012/648, 2012.

[14] Q. Tang, Y. Zhao, X. Chen, and H. Ma. Refine the concept of public key encryption with delegated search. Cryptology ePrint Archive: Report 2012/654, 2012.

[15] E. R. Verheul. Evidence that xtr is more secure than supersingular elliptic curve cryptosystems. *J. Cryptol.*, 17(4):277–296, 2004.

[16] R. Zhang and H. Imai. Generic combination of public key encryption with keyword search and public key encryption. In *Proceedings of the 6th international conference on Cryptology and network security*, pages 159–174. Springer, 2007.

Boolean Symmetric Searchable Encryption

Tarik Moataz
Telecom Bretagne, Institut Mines-Telecom
Rennes, France
tarik.moataz@telecom-bretagne.eu

Abdullatif Shikfa
Bell Labs Research, Alcatel-Lucent
Nozay, France
abdullatif.shikfa@alcatel-lucent.com

ABSTRACT

In this article we tackle the issue of searchable encryption with a generalized query model. Departing from many previous works that focused on queries consisting of a single keyword, we consider the the case of queries consisting of arbitrary boolean expressions on keywords, that is to say conjunctions and disjunctions of keywords and their complement. Our construction of boolean symmetric searchable encryption BSSE is mainly based on the orthogonalization of the keyword field according to the Gram-Schmidt process. Each document stored in an outsourced server is associated with a label which contains all the keywords corresponding to the document, and searches are performed by way of a simple inner product. Furthermore, the queries in the BSSE scheme are randomized. This randomization hides the search pattern of the user since the search results cannot be associated deterministically to queries. We formally define an adaptive security model for the BSSE scheme. In addition, the search complexity is in $O(n)$ where n is the number of documents stored in the outsourced server.

Categories and Subject Descriptors

[**Security and privacy**]: Cryptography; [**Information systems**]: Data management systems—*Data encryption*; [**Theory of computation**]: Computational complexity and cryptography

General Terms

Security

Keywords

Searchable encryption, Boolean queries, Randomized queries, Outsourced storage

1. INTRODUCTION

Outsourcing is a major trend in computing and it reached its peak with the advent of cloud computing. The issue is that the outsourcing service cannot always be fully trusted. A reasonable assumption is to consider such a server as semi-honest in the sense that it performs operations well but wants to access the data of the user. However, the need to access client's data (e.g., files, email, location) is often at odds with clients' data privacy needs, and in particular the requirement to protect their data (usually through encryption). Reconciling these contradictory requirements, and achieving computation on encrypted data, is an important research and engineering problem.

We focus here on the specific case of indexing and searching on documents. Typically to perform a search on a set of documents, one creates an index of keywords included in the documents and later on performs the search on the index. This does not raise major issues if the querier (the one who requests the search) and the storage provider (the one who stores the documents and the index and who actually performs the search and returns the result) fully trust each other. The difficulties arise when the querier wants to protect the privacy of its data and stores the documents as well as the indexes encrypted, and sends encrypted queries also.

1.1 Problem statement

The problem to solve is therefore to allow performing encrypted searches on a collection of encrypted documents. Solution already exist in the literature and the field of searchable encryption is already mature. However most solutions focus on simple keyword match.

In this article we target a more general model, boolean searchable symmetric encryption, which achieves all basic boolean searches over encrypted data. The basic boolean operations are : the disjunction, the conjunction and the negation. The disjunctive search allows the user to search for encrypted documents containing $word_1$ *or* $word_2$... *or* $word_n$. The conjunctive search allows the user to search for the encrypted documents containing: "$word_1$ *and ... and* $word_n$" and finally the negative search allows the user to search for all encrypted documents which do not contain particular words.

1.2 State of the art

Searchable encryption has been an active research area and many quality works have been published in the beginning of the 2000's, both concerning the symmetric search setting [21, 13, 11, 12, 8, 2] as well as the asymmetric one [5, 4, 10, 1]. Those early schemes focused however only on single keyword query and thus single keyword searches.

Since 2004, several articles focus on extending the search options to enable especially conjunctive search [15, 6, 18,

ASIA CCS'13, May 8–10, 2013, Hangzhou, China.

19, 22, 23, 16, 3, 7]. A natural question arises then: why do we need special schemes for performing conjunctive queries? One trivial solution is indeed to perform several queries for one keyword each and then performing set intersection on all the results. However, this simple solution allows the servers to distinguish which documents contain which -encrypted-keyword, and hence leaks information which cannot be tolerated in some cases as mentioned in [15]. Furthermore, this solution comes at the price of higher communication overhead: consider the case where the first query corresponds is matched by 1000 files and the second query by 2000 files while the intersection of the two keywords is only 10 files. In that case the number of transmitted files is several order of magnitudes more than what is required and the situation could be even worse if the query consists on multiple conjunctions. Another solution would be to define a meta-keyword for each possible keywords conjunction; however this solution remains infeasible in practice. Indeed, for each document containing m keywords, the user has to store 2^m other meta-keywords. This solution has thus an exponential storage complexity associated with an increased complexity of search computation on the server.

Golle et al.'s scheme [15] presents the first solution for symmetric conjunctive searchable encryption search. In fact, this scheme is based on bilinear map functions and the Decisional Diffie-Hellman assumption as a security model. We note that there is a similar scheme [18] tackling asymmetric search with similar features as [15]. The main difference between these two schemes besides the fact that they deal with different search setting (symmetric/asymmetric), is that Golle et al. introduce an off-line query. In fact for performance reasons, Golle et al. decided to divide the search query into two parts: an off-line part stored in the outsourced server while storing the whole data, and a common on-line query similar to other schemes. However both these schemes have such a complexity that they cannot be implemented in realistic scenarios due to the exponential storage space that they require. On the other hand, Boneh and Waters [6] present a real improvement in searchable encryption search options. Indeed, they introduce a feasible solution for conjunctive, subset and range queries. They shows that the basic idea is very expensive and needs an exponential data storage, both for the ciphertext or for the trapdoors. Thereby, they present an idea based on Hidden Vector System (HVE) that sharply reduces the ciphetext storage complexity for equality, comparison and subset conjunctions. However, their solution focuses only on conjunctive search queries and did not consider other boolean search options. In addition, Katz et al. [17] present a predicate encryption scheme enabling disjunctive search over encrypted data. Finally, Shi & al. [20] presented an interesting solution to deal with multi-dimensional range query over encrypted data, but as in [6], they only consider conjunctive operations.

To sum up, to the best of our knowledge, there are no previous solutions in the literature supporting boolean searches on encrypted data, i.e. conjunctive, disjunctive and negation search operations at the same time.

1.3 Our contribution

Our construction BSSE is the first scheme that supports generic boolean search over encrypted data. It is based on an original idea of considering keywords as vectors and using the Gram-Schmidt process to orthogonalize and then orthonormalize them. It further makes use of a very efficient operation, the inner product, to perform searches at the server side. The inner product indeed leverages the orthonormalized keywords to efficiently test if a boolean expression query matches the label corresponding to an encrypted document or not. It is an application of mathematical and computing principles to practical security, especially in the searchable encryption scope.

In addition to that, our BSSE scheme presents new features for encrypted documents stored in the server. Indeed, the query sent for retrieving encrypted documents is randomized, which means that the server sees different queries even if the same boolean query is sent several times. Consequently, this feature prevents deterministic association between a query and the corresponding set of documents and makes statistical analysis harder. We introduce the new security notion of *randomized queries* verified by the BSSE scheme. On the other hand, the label that we associate with each encrypted document can be seen as the secure index defined by Goh [13] or the secure scrambled array defined by Curtmola et al. [11]. However, the secure index leaks some meta-information about the number of keywords stored in the Bloom filter, while our secure labels do not leak such information about its content or about the number of keywords that are encrypted before being sent to the server. Our contribution is based on an adversarial model similar to the one defined by Curtmola et al. [11], the differences arise first in the nature of the query sent to the server, and second in the fact that the search pattern of our construction is hidden since we introduce the notion of *randomized query*. Our adversarial model lies in the adaptive setting: this model considers adversaries able to adapt their query in function of previous received queries and search outcomes.

1.4 Outline

The rest of the article is as follows: in section 2 we introduce our notations and some cryptographic reminders while section 3 defines the algorithms of BSSE and the associated security definitions. In section 4 we present the details of our original construction of BSSE and in section 5 we evaluate our construction both from a security and performance perspective. Finally we conclude in section 6.

2. PRELIMINARIES

In this section we start by introducing the notations that will be used in the rest of the article.

2.1 Notations

Let $\mathcal{X}$ be a distribution, we write $x \leftarrow \mathcal{X}$ to denote that x is a sample from the distribution $\mathcal{X}$. $x \stackrel{R}{\leftarrow} \mathcal{S}$ represents a string sampled uniformly at random from the set $\mathcal{S}$. $|x|$ denotes the string length. If x and y are two strings then $x\|y$ designates the string concatenation. If $\mathcal{A}$ is an algorithm then $x \leftarrow \mathcal{A}(...)$ represents the result of applying the algorithm $\mathcal{A}$ to given arguments. k denotes a security parameter. $\delta_{i,j}$ represents the Kronecker symbol such that: if $i = j$ $\delta_{i,j} = 1$ else $\delta_{i,j} = 0$.

Let $D = \{D_1, ..., D_n\}$ be a set of documents, and $W = \{w_1, ..., w_l\}$ be the set of all unique keywords in all documents in D, where $\forall i \in [\![1, l]\!]$ $w_i \in \{0,1\}^*$ (the method for extracting keywords from documents is out of the scope of this work) . We denote the set of positive integers smaller than l by $I = [\![1, l]\!]$. $M_i \subseteq I$ represents the indexes in W of the keywords corresponding to the document D_i. We consider $M = \{M_i\}_{i \in [\![1,n]\!]}$ the collection of all these index sets.

Boolean expressions: Let $\mathcal{W} = \{w_1, ..., w_r\}$ be a set of keywords. A boolean expression on $\mathcal{W}$ has the form: $\mathcal{B}(\mathcal{W}) = \diamond w_1 \star \diamond w_2 \star ... \star \diamond w_r$, where $\star$ is a binary logical operator $\star \in \{\wedge, \vee\}$ and $\diamond$ is a unitary operator to indicate that the keyword can be complemented ($\diamond \in \{\varnothing, \neg\}$)

All boolean expressions can be expressed in disjunctive normal form and in the following we will only consider this form:

$$\mathcal{B}(\mathcal{W}) = \bigvee_{q=1}^{t} \bigwedge_{j \in I_q} \diamond w_j \ ,$$

where $I_q \subseteq I$ for $1 \le q \le t$.

2.2 Cryptographic reminders

In this section we recall some useful and classical definitions in cryptography.

Advantage: Let $\mathcal{A} : \{0,1\}^n \rightarrow \{0,1\}$ be an efficient probabilistic algorithm, and R_1, R_2 two random variables. We define the **distinguishing probability** of $\mathcal{A}$, or the **advantage** of $\mathcal{A}$, for R_1 and R_2 as the ability to distinguish between these two random variables :

$$Adv(\mathcal{A}) = |P_r[\mathcal{A}(R_1) = 1] - P_r[\mathcal{A}(R_2) = 1]|$$

DEFINITION 2.1. ***Symmetric encryption:*** *a symmetric encryption scheme consists of three algorithms $(\mathcal{K}, \mathcal{E}, \mathcal{D})$. The key generation $\mathcal{K}$ takes a security parameter k as input and returns the secret key k_{secret}, we write $k_{secret} \leftarrow \mathcal{K}(k)$. The encryption algorithm $\mathcal{E}$ takes the secret key, the plain text* m $\in \{0,1\}^*$ *and a random string* r *as inputs and returns a ciphertext $C \in \{0,1\}^*$, we write $C \leftarrow \mathcal{E}(k_{secret}, m, r)$ or $C \xleftarrow{R} \mathcal{E}(k_{secret}, m)$. The decryption algorithm $\mathcal{D}$ takes the secret key and the ciphertext as inputs and returns the plaintext* m *or $\perp$ (in case of error) and we write respectively : $m \leftarrow \mathcal{D}(k_{secret}, \mathcal{E}(k_{secret}, m, r))$ and $\perp \leftarrow \mathcal{D}(k_{secret}, C)$, the symmetric encryption chosen for the following will be semantically secure.*

DEFINITION 2.2. ***Semantic security[14]:*** *a symmetric encryption scheme $(\mathcal{K}, \mathcal{E}, \mathcal{D})$ is semantically secure if for every probabilistic polynomial time algorithm $\mathcal{A}$ there exists a probabilistic polynomial time algorithm $\mathcal{A}'$ such that for every probability ensemble $\{X_k\}_{k \in \mathbb{N}}$ with $|X_k| \le poly(k)$ where k is the security parameter; every pair of polynomial bounded functions $f, h : \{0,1\}^* \rightarrow \{0,1\}^*$ we have:*

$$\Pr[\mathcal{A}(1^k, \mathcal{E}_K(X_k), h(1^k, X_k) = f(1^k, X_k)] < \Pr[\mathcal{A}'(1^k, 1^{|X_k|}, h(1^k, X_k) = f(1^k, X_k)] + \epsilon \ ,$$

where $K \leftarrow \mathcal{K}(k)$

The probability in these terms is taken over X_k as well as over the internal coin tosses of either algorithms $\mathcal{K}, \mathcal{E}$ and $\mathcal{A}$ or $\mathcal{A}'$. $\{X_k\}_{k \in \mathbb{N}}$ represents the distribution of plaintexts, f represents the information that the adversary tries to obtain and h represents a priori partial information about the plaintext.

DEFINITION 2.3. ***Pseudo-random functions:*** *Let us consider a function $f : \{0,1\}^s \times \{0,1\}^n \rightarrow \{0,1\}^l$ and the family F of all maps $\{0,1\}^n \rightarrow \{0,1\}^l$. We say that f is a (t, ϵ, q)-pseudo-random function if:*
-f is efficiently computable in polynomial time.
-For any t-time probabilistic polynomial algorithm $\mathcal{A}$:

$$Adv(\mathcal{A}) = |\Pr[\mathcal{A}^{f(k,.)} = 1] - P_r[\mathcal{A}^{r(.)} = 1]| < \epsilon \ ,$$

with $k \xleftarrow{R} \{0,1\}^s$ and $r \xleftarrow{R} F$. When f is bijective (in this case $n = l$) then it is a pseudo-random permutation.

3. BSSE DEFINITIONS

We now focus on definitions specific to our problem. Recall that we are targeting the following scenario: a user wants to store his data D in an outsourced server while protecting data's confidentiality : the documents are thus encrypted. Later on, the user may query the server for some documents. Contrary to previous works we do not target only simple queries consisting of only one keyword, we rather consider an enhanced query model consisting of general boolean expressions on keywords expressed in disjunctive normal form. In the following subsections, we define formally the 4-steps algorithm for encrypting and querying for encrypted data, and then define its security.

3.1 BSSE algorithm definitions

DEFINITION 3.1. *Boolean Symmetric Searchable Encryption (BSSE) over a set of documents D consists of four polynomial-time algorithms BSSE=(Gen,Enc,Query,Test) such that:*

- *$(K_1, K_2) \leftarrow Gen(1^k)$: a key generation algorithm, which takes a security parameter k as input and outputs two secret keys K_1 and K_2.*
- *$(\mathcal{X}, C) \leftarrow Enc(K_1, K_2, W, M, D)$: a probabilistic algorithm, which takes as input the collection of documents D, the keyword set W, the set of indexes M and the secret keys K_1 and K_2, and outputs the collection of encrypted documents $C = \{C_1, ..., C_n\}$ and a set of labels $\mathcal{X} = \{\mathcal{X}_1, ..., \mathcal{X}_n\}$ where each label $\mathcal{X}_i$ is associated with the corresponding encrypted document C_i.*
- *$\mathcal{Q} \leftarrow Query(K_1, \mathcal{B}(\mathcal{W}))$: a probabilistic algorithm, which takes as input a boolean expression $\mathcal{B}(\mathcal{W})$ and the key K_1, and outputs an encoded query.*
- *$\mathcal{L} \leftarrow Test(\mathcal{Q}, \mathcal{X})$: a deterministic algorithm, which takes as input a query $\mathcal{Q}$ and the set of labels $\mathcal{X}$ and outputs the list of ciphertexts $\mathcal{L} \subseteq C$ matching the query.*

3.2 BSSE security definitions

In the following, we define the security requirements of BSSE. We consider the requirements proposed by previous works on symmetric searchable encryption and add two new concepts corresponding to the extended features offered by BSSE.

Before tackling security, our scheme like any cryptographic scheme should be consistent and in our context this means the following:

DEFINITION 3.2. ***Consistency:*** *Let $\mathcal{B}(\mathcal{W})$ be a boolean expression and $\mathcal{L}$ the set of encrypted documents that match*

$\mathcal{B}(\mathcal{W})$. *If $\mathcal{Q}$ is the result of the probabilistic algorithm* ***Query*** *on $\mathcal{B}(\mathcal{W})$, we say that the scheme is consistent if:*

$$Test(\mathcal{Q}, \mathcal{X}) = \mathcal{L}$$

Since the model of adversary is adaptive, we consider security definitions similar to those introduced first by Curtmola et al. in [11] and which take into account the adaptive adversarial model. Some but not all of these definitions are adapted in order to set well for the scheme (we take the same notations and naming as [11]).

DEFINITION 3.3. ***History:*** *Let D be a set of documents and $\mathsf{W} = \{\mathcal{W}_1, \mathcal{W}_2, ..., \mathcal{W}_q\}$ be a vector of q sets of keywords. A History is a tuple $H = (D, \mathsf{W})$.*

DEFINITION 3.4. ***Access Pattern:*** *Let D be a set of documents and $\mathsf{W} = \{\mathcal{W}_1, \mathcal{W}_2, ..., \mathcal{W}_q\}$ be a vector of q sets of keywords, and their q corresponding boolean queries $\{\mathcal{B}(\mathcal{W}_1), \mathcal{B}(\mathcal{W}_2), ..., \mathcal{B}(\mathcal{W}_q)\}$. The access pattern of the History $H = (D, \mathsf{W})$ is the tuple*

$$\alpha(H) = (D(\mathcal{B}(\mathcal{W}_1)), D(\mathcal{B}(\mathcal{W}_2)), ..., D(\mathcal{B}(\mathcal{W}_q)))$$

where $D(\mathcal{B}(\mathcal{W}_i))$ represents the set of documents' identifiers which match the boolean query $\mathcal{B}(\mathcal{W}_i)$.

DEFINITION 3.5. ***Search Pattern:*** *Let D be a set of documents and $\mathsf{W} = \{\mathcal{W}_1, \mathcal{W}_2, ..., \mathcal{W}_q\}$ be a vector of q sets of keywords, and their q corresponding boolean queries $\{\mathcal{B}(\mathcal{W}_1), \mathcal{B}(\mathcal{W}_2), ..., \mathcal{B}(\mathcal{W}_q)\}$. The search pattern of the History $H = (D, \mathsf{W})$ is a binary symmetric matrix $(\sigma(H))_{1 \le i,j \le q}$ such that: $\sigma(H)_{i,j} = 1$ if $\mathcal{B}(\mathcal{W}_i) = \mathcal{B}(\mathcal{W}_j)$, and 0 otherwise.*

The *access pattern* represents the results of queries sent to the server, and more specifically the document identifiers corresponding to each query, on the other hand the *search pattern* represents a transcript of queries searched for, the matrix σ records the iteration of the search, a special case of this matrix is the identity matrix if all keyword' sets are different from each others. In addition, the server knows the documents' size. Hence, we will capture all of these information leakage in the *trace* notion.

DEFINITION 3.6. ***Trace:*** *Let D be a set of documents and $\mathsf{W} = \{\mathcal{W}_1, \mathcal{W}_2, ..., \mathcal{W}_q\}$ be a vector of q sets of keywords, and their q corresponding boolean queries $\{\mathcal{B}(\mathcal{W}_1), \mathcal{B}(\mathcal{W}_2), ..., \mathcal{B}(\mathcal{W}_q)\}$. The trace induced by the History $H = (D, \mathsf{W})$ is the tuple $\gamma(H) = (|D_1|, |D_2|, ..., |D_n|, \alpha(H), \sigma(H))$ where $\alpha(H)$ and $\sigma(H))$ are respictively the access and the search pattern induced by the history H.*

Our adversarial model is based on a simulation-based definition (adversary and simulator), this notion as illustrated in [11], compared to *oracle* access, does not weaken our adversarial definition since we deal with a set of documents instead of individual ones. Adaptive security means that the adversary generates the history adaptively, that means that he takes into account the results of previous queries (search and access pattern) each time before sending the next query from a chosen set of documents generated initially. The only additional requirement is that the view of the adversary (labels, cyphertexts, generated queries) generated from an adversarial and adaptively chosen history should be simulatable given only the trace. We are going to detail the simulation-based experiments in the following.

DEFINITION 3.7. ***Adaptive semantic security*** *: Let $BSSE = (Gen, Enc, Query, Test)$ be the boolean symmetric searchable scheme, k be the security parameter, $\mathcal{A} = (\mathcal{A}_0, ..., \mathcal{A}_q)$ be an adversary such that $q \in \mathbb{N}$ and $\mathcal{S} = (\mathcal{S}_0, ..., \mathcal{S}_q)$ be a simulator, we consider the following probabilistic experiments $\boldsymbol{Real}_{BSSE,\mathcal{A}}(k)$ and $\boldsymbol{Sim}_{BSSE,\mathcal{A},\mathcal{S}}(k)$:*

$\boldsymbol{Real}_{BSSE,\mathcal{A}}(k)$
- $(K_1, K_2) \leftarrow Gen(1^k)$
- $(D, st_{\mathcal{A}}) \leftarrow \mathcal{A}_0(1^k)$
- $(\mathcal{X}, C) \leftarrow Enc(K_1, K_2, W, M, D)$
- $(\mathcal{B}(\mathcal{W}_1), st_{\mathcal{A}}) \leftarrow \mathcal{A}_1(st_{\mathcal{A}}, \mathcal{X}, C)$
- $\mathcal{Q}_1 \leftarrow Query(K_1, \mathcal{B}(\mathcal{W}_1))$
- *for* $2 \le i \le q$:
  - $(\mathcal{B}(\mathcal{W}_i), st_{\mathcal{A}}) \leftarrow \mathcal{A}_i(st_{\mathcal{A}}, \mathcal{X}, C)$
  - $\mathcal{Q}_i \leftarrow Query(K_1, \mathcal{B}(\mathcal{W}_i))$
- *let* $\mathcal{Q} = \{\mathcal{Q}_1, ..., \mathcal{Q}_q\}$
- *output* $o = (\mathcal{X}, C, \mathcal{Q})$ *and* $st_{\mathcal{A}}$

$\boldsymbol{Sim}_{BSSE,\mathcal{A},\mathcal{S}}(k)$
- $(D, st_{\mathcal{A}}) \leftarrow \mathcal{A}_0(1^k)$
- $(\mathcal{X}, C, st_{\mathcal{S}}) \leftarrow \mathcal{S}_0(\gamma(D))$
- $(\mathcal{B}(\mathcal{W}_1), st_{\mathcal{A}}) \leftarrow \mathcal{A}_1(st_{\mathcal{A}}, \mathcal{X}, C)$
- $(\mathcal{Q}_1, st_{\mathcal{S}}) \leftarrow \mathcal{S}_1(st_{\mathcal{S}}, \gamma(D, \mathcal{B}(\mathcal{W}_1)))$
- *for* $2 \le i \le q$:
  - $(\mathcal{B}(\mathcal{W}_i), st_{\mathcal{A}}) \leftarrow \mathcal{A}_i(st_{\mathcal{A}}, \mathcal{X}, C, \mathcal{Q}_1, ..., \mathcal{Q}_{i-1})$
  - $(\mathcal{Q}_i, st_{\mathcal{S}}) \leftarrow \mathcal{S}_i(st_{\mathcal{S}}, \gamma(D, \mathcal{B}(\mathcal{W}_1), ..., \mathcal{B}(\mathcal{W}_i)))$
- *output* $o = (\mathcal{X}, C, \mathcal{Q})$ *and* $st_{\mathcal{A}}$

where $st_{\mathcal{A}}$ and $st_{\mathcal{S}}$ are respectively the state of the adversary and the simulator.
We say that BSSE is adaptively semantically secure if for all polynomial time adversaries $\mathcal{A} = (\mathcal{A}_0, ..., \mathcal{A}_q)$ such that q is polynomial in k, there exists a non-uniform polynomial-size simulator $\mathcal{S} = (\mathcal{S}_0, ..., \mathcal{S}_q)$ such that for all polynomial size $\mathcal{H}$ we have:

$$Adv(\mathcal{H}) = \Pr[\mathcal{H}(o, st_{\mathcal{A}}) = 1] - \Pr[\mathcal{H}(o', st'_{\mathcal{A}}) = 1] \le \epsilon$$

where $(o, st_{\mathcal{A}}) \leftarrow Real_{BSSE,\mathcal{A}}(k)$, $(o', st'_{\mathcal{A}}) \leftarrow Sim_{BSSE,\mathcal{A},\mathcal{S}}(k)$ and the probabilities are over the coins of Gen, Enc and Query.

As defined previously, our scheme introduces a randomized query that enables the user to hide his search pattern. In fact, queries are different even if they are generated for the same boolean query. We formalize this notion as follows:

DEFINITION 3.8. ***Randomized query*** *Let $(\mathcal{Q})_{1 \le i \le l}$ be l queries generated with $Query(K_1, .)$ with the same key K_1 and possibly for the same boolean expression $\mathcal{B}(\mathcal{W})$. We say that the scheme has* (l, ϵ)***-randomized query*** *if:*

$$\forall\ i, j \in \{1, l\}\ \Pr(\mathcal{Q}_i = \mathcal{Q}_j) < \epsilon$$

This completes our presentation of BSSE security requirement, and we now present our BSSE construction.

4. BSSE CONSTRUCTION

Our overall approach is to consider the keywords as vectors and to pad them so that they form a family of independent vectors. From such a family we construct an orthonormal basis of the resulting vector space, which means that each keyword is associated with one vector of this orthonormal base. Then to each document we associate the vector resulting of the addition of all vectors. The queries are then also expressed as a vector in the same vector space using the same basis and we roughly use scalar product between the vector of the query and the vector of a document to decide whether a document satisfies a query or not. We thus first present the setup phase of constructing the orthonormal basis and then present the construction of the four algorithm of BSSE.

4.1 BSSE pre-construction

The BSSE algorithm is based on a pre-construction over the unique keywords set $W = \{w_1, ..., w_l\}$ (see section 2 for notations). The aim of this construction is to create from W an orthonormal keyword family. This last is obtained by applying the Gram-Schmidt algorithm to a free family derived from W.

To be more precise, we consider all keywords to be of equal length m (this can be achieved by padding all keywords so that they have the same length).

Let consider a pseudo-random permutation $\pi : \{0,1\}^s \times \{0,1\}^m \to \{0,1\}^m$ and a key K_1 generated from the $Gen(1^k)$ where k is security parameter, for each keyword w_i in W, we apply π_{K_1} in order to create a permuted set of keywords such that $W = \{\pi_{K_1}(w_1), ..., \pi_{K_1}(w_l)\}$.

Let us consider the canonical base of size l $\{e_1, ..., e_l\}$ such that $\forall i \in \{1, l\}$ $|e_i| = l$, and $e_i = (\delta_{i,1}, \ldots, \delta_{i,j}, \ldots, \delta_{i,l})$. In order to create a free family $(f_i)_{1\leq i\leq l}$ from W, we concatenate each permuted keyword $\pi_{K_1}(w_i)$ with e_i such that $f_i = \pi_{K_1}(w_i)\|e_i$ and $|f_i| = m + l$.

We consider the inner scalar product ϕ :

$$\begin{array}{cccc} \phi & : & \{0,1\}^{m+l} \times \{0,1\}^{m+l} & \to & [\![0, m+l]\!] \\ & & x, y & \mapsto & \sum_{k=1}^{m+l} x_k y_k \end{array}$$

THEOREM GRAM-SCHMIDT [9] 4.1. *If $(f_i)_{i\in\{0,l\}}$ is a free family of a pre-Hilbert space, there exists an orthogonal family $(o_i)_{i\in\{0,l\}}$ such that:*

- $\phi(o_i, o_j) = \phi(o_i, o_i)\delta_{i,j}$,
- $Vect(f_1, ..., f_l) = Vect(o_1, ..., o_l)$,
- *The scalar product $\phi(f_i, o_i)$ is strictly positive.*

DEFINITION 4.1. *The orthogonal projection p on a vector line led by u is defined as:*

$$p_u(v) = \frac{\phi(v,u)}{\phi(u,u)} u$$

Orthogonal Keyword Construction: Let us consider $(f_i)_{1\leq i\leq l}$ a free family. According to the Theorem 4.1, there is an orthogonal family $(o_i)_{i\in\{0,l\}}$. We can construct this family following the Gram-Schmidt process:

$$o_i = f_i - \sum_{k=1}^{i-1} p_{o_k}(f_i) \qquad for\ 1 \leq i \leq l$$

$(o_i)_{i\in\{0,l\}}$ forms the new keyword orthogonal family that verifies:

$$\forall i, j \in I,\ \phi(o_i, o_j) = \phi(o_i, o_i).\ \delta_{i,j} \tag{1}$$

In addition, we can normalize the vectors to obtain an orthonormal family:

$$\forall k \in I\ u_k = \frac{1}{\sqrt{\phi(o_k, o_k)}}.o_k$$

At the end of this phase, we should point out that each orthonormal keyword u_i corresponds to a keyword w_i. In addition, if we take into account the order of the free family $(f_i)_{1\leq i\leq l}$ then the construction according to the Gram Schmidt algorithm produces a unique orthonormal family. If we change the order of keywords, the algorithm outputs a different orthonormal family. The number of possible families generated by this construction is $l!$, and the order can hence be considered as a parameter that can enhance security if kept secret.

We should point out that the keyword set W will be stored on the client side in a table $\mathcal{T}$, and that optionally this table could also contain the orthonormal keywords corresponding to them $\{u_k\}_{1\leq k\leq l}$. Here ends the setup phase, in the following, we will detail the 4-steps BSSE algorithm.

4.2 Construction of BSSE algorithms

We first detail the two steps of the BSSE algorithm namely *Gen* and *Enc* steps generically, and then for the *Query* and *Test* steps, we present them gradually depending on the operations supported.

4.2.1 Key generation step

Gen(k): The user generates randomly two secret keys of size s where s depends on the security parameter k that is to say $(K_1, K_2) \xleftarrow{R} \{0,1\}^{2s}$.

4.2.2 Encryption step

$Enc(K_1, K_2, W, M, D)$: From the entire set of unique keywords $(w_i)_{i\in\{0,l\}}$, according to the BSSE pre-construction we produce the orthonormal keyword family $(u_i)_{i\in\{0,l\}}$. These orthonormal families can either be retrieved or recomputed depending on whether they were stored in table $\mathcal{T}$ or not.

Each document D_i in the set D contains a set M_i of unique keywords $(u_k)_{k\in M_i}$ where M_i is the set of keywords indexes of D_i. The output of this step is the encrypted document $\mathcal{E}_{K_2}(D_i)$, and the associated label $\mathcal{X}_i = \sum_{k\in M_i} u_k$ for each $i \in \{1, n\}$, where $\mathcal{E}$ is the encryption algorithm of a semantically secure symmetric encryption scheme.

Once all documents are encrypted, they will be stored in the outsourced server with their associated labels, see table 1.

$\mathcal{E}_{K_2}(D_1)$	$\mathcal{X}_1$
...	...
$\mathcal{E}_{K_2}(D_i)$	$\mathcal{X}_i$
...	...
$\mathcal{E}_{K_2}(D_n)$	$\mathcal{X}_n$

Table 1: Encrypted documents in the outsourced server

4.2.3 Search and Test phases

We now explain the search phase. This phase is generic as well but we will explain it in three steps for the sake of clarity: first how to proceed with conjunctive queries, then how to take into account disjunctions as well and finally how to deal with negations and hence any boolean expression in disjunctive normal form defined earlier in section 2.

Conjunctive Search: We consider a user who wants to search for all documents that match the query: "w_{i_1} *and* w_{i_2} *...and* w_{i_r}" in the outsourced server. First of all, these keywords are converted in the orthonormal base either by regenerating this base or by retrieving the corresponding vectors in table $\mathcal{T}$ if available. The user only selects the orthonormal keywords $(u_i)_{i\in I_q}$ corresponding to keywords existing in the boolean expression where I_q is the set of keyword indexes in the boolean expression.

Afterwards the user picks at random $|I_q|$ strictly positive integers $\{a_k\}_{1\leq k\leq |I_q|}$ such that $a_k \xleftarrow{R} \{1,..,2^r\}$ where r is a sufficiently large positive number. The query is then constructed as follows:

$$\mathcal{Q} = Query(K_1, \bigwedge_{k\in I_q} w_k) = \frac{1}{\sum_{k\in I_q} a_k} . \sum_{k\in I_q} a_k u_k$$

The user sends then to the server the query $\mathcal{Q}$.

On the server side, for each document D_i, the server will perform the following verification:

$$Test(\mathcal{Q}, \mathcal{X}_i) = \phi(\mathcal{Q}, \mathcal{X}_i) = \mathcal{Q}.^t\mathcal{X}_i$$

$$\begin{aligned} Test(\mathcal{Q}, \mathcal{X}_i) &= \frac{1}{\sum_{k\in I_q} a_k} \sum_{k\in I_q} \sum_{l\in M_i} a_k \phi(u_k, u_l) \\ &= \frac{1}{\sum_{k\in I_q} a_k} \sum_{k\in I_q \bigcap M_i} a_k \end{aligned}$$

- If all keywords exist in the document C_i i.e. $I_q \bigcap M_i = I_q$ then: $Test(\mathcal{Q}, \mathcal{X}_i) = 1$ and we add C_i to $\mathcal{L}$.
- Else the document C_i does not contain all keywords i.e. $I_q \bigcap M_i \subsetneq I_q$ and then : $Test(\mathcal{Q}, \mathcal{X}_i) \neq 1$.

Conjunctive and disjunctive search: In this paragraph, we consider a more general case where the user wants to search for the following query:

$$\bigvee_{k=1}^{t} \bigwedge_{j\in I_{q_k}} w_j$$

Again the user translates the keywords in the corresponding orthonormal keywords.

For each $k \in [\![1, t]\!]$, the user picks at random $|I_{q_k}|$ integers $\{a_{k,i}\}_{1\leq i\leq |I_{q_k}|}$ such that $a_{k,i} \xleftarrow{R} \{1,..,2^r\}$ where r is a sufficiently large positive number. We then define:

$$\mathcal{S}_k = \frac{1}{\sum_{j\in I_{q_k}} a_{k,j}} \sum_{j\in I_{q_k}} a_{k,j} u_j$$

The user sends to the server the query in the form of the following matrix:

$$\mathcal{Q} = Query(K_1, \bigvee_{k=1}^{t} \bigwedge_{j\in I_{q_k}} w_j) = \begin{pmatrix} \mathcal{S}_1 \\ \vdots \\ \mathcal{S}_t \end{pmatrix}$$

For each document C_i, the server performs the following verification:

$$\begin{aligned} Test(\mathcal{Q}, \mathcal{X}_i) &= \phi(\mathcal{Q}, \mathcal{X}_i) = \mathcal{Q}.^t\mathcal{X}_i = \begin{pmatrix} \phi(\mathcal{S}_1, \mathcal{X}_i) \\ \vdots \\ \phi(\mathcal{S}_t, \mathcal{X}_i) \end{pmatrix} \\ &= \begin{pmatrix} \frac{1}{\sum_{j\in I_{q_1}} a_{1,j}} \sum_{j\in I_{q_1} \bigcap M_i} a_{1,j} \\ \vdots \\ \frac{1}{\sum_{j\in I_{q_t}} a_{t,j}} \sum_{j\in I_{q_t} \bigcap M_i} a_{t,j} \end{pmatrix} \end{aligned}$$

The result of this operation is interpreted as follows:

- If $\exists k \in [\![1, t]\!]$ such that $(Test(\mathcal{Q}, \mathcal{X}_i))_k = 1$ then the document C_i exactly match the query and we add C_i to $\mathcal{L}$, as it means that at least one conjunction was satisfied, hence the disjunction of these disjunctions is satisfied.
- Else $\forall k \in [\![1, t]\!]$, $(Test(\mathcal{Q}, \mathcal{X}_i))_k \neq 1$ then the document does not match the query, because no conjunction was satisfied so the whole disjunction is not satisfied either.

So far, we have seen general queries dealing with any number of conjunctions or disjunctions in the same query. However, to complete our scheme, we need to take into account the negation as well. In the following paragraph, we detail our full solution which is the first solution that supports general boolean searches over encrypted data.

Boolean search: We now consider the case of a general boolean expression search taken in the disjunctive normal form as follows:

$$\bigvee_{k=1}^{t} \bigwedge_{j\in I_{q_k}} \diamond w_j$$

For this construction, we introduce a new keyword which we consider as belonging to all documents by default. To this extent, during the u_i construction, we add to all documents an arbitrary keyword (a word which does not exist in reality). After the Gram-Schmidt process, we thus have one orthonormal keyword J existing in all documents D_i and at the same time $((u_i)_{i\in I}, J)$ represents an orthonormal family.

Let us now consider each conjunction $L_k = \bigwedge_{j\in I_{q_k}} \diamond w_j$ separately. I_{q_k} represents the indexes of keywords in each disjunction L_k. We split I_{q_k} into two partitions such that: $I_{q_k} = \{P_{q_k}, N_{q_k}\}$ where P_{q_k} and N_{q_k} represent respectively the indexes of affirmative and negative keywords in the boolean expression. For each $k \in [\![1, t]\!]$, we pick at random $|P_{q_k}|$ integers $\{a_{k,i}\}_{1\leq i\leq |P_{q_k}|}$, $|N_{q_k}|$ negative numbers $\{c_{k,i}\}_{1\leq i\leq |N_{q_k}|}$ and one random integer b_k such that $a_{k,i} \xleftarrow{R} \{1,..,2^r\}$, $c_{k,i} \xleftarrow{R} \{-2^r, ..., -1\}$ and $b_k \xleftarrow{R} \{1,..,2^r\}$ where r is a sufficiently large positive number. We then define:

$$\mathcal{S}_k = \frac{1}{b_k + \sum_{j\in P_{q_k}} a_{k,j}} (\sum_{j\in N_{q_k}} c_{k,j} u_j + \sum_{j\in P_{q_k}} a_{k,j} u_j + b_k J)$$

The user sends to the server the following query:

$$\mathcal{Q} = Query(K_1, \bigvee_{k=1}^{t} \bigwedge_{j\in I_{q_k}} \diamond w_j) = \begin{pmatrix} \mathcal{S}_1 \\ \vdots \\ \mathcal{S}_t \end{pmatrix}$$

For each document C_i, the server performs the following verification:

$$Test(\mathcal{Q}, \mathcal{X}_i) = \phi(\mathcal{Q}, \mathcal{X}_i) = \mathcal{Q}.^t\mathcal{X}_i = \begin{pmatrix} \phi(\mathcal{S}_1, \mathcal{X}_i) \\ \vdots \\ \phi(\mathcal{S}_t, \mathcal{X}_i) \end{pmatrix}$$

$$= \begin{pmatrix} \frac{1}{b_1 + \sum_{j\in P_{q_1}} a_{1,j}} (\sum_{j\in N_{q_1} \bigcap M_i} c_{1,j} + \sum_{j\in P_{q_1} \bigcap M_i} a_{1,j} + b_1) \\ \vdots \\ \frac{1}{b_t + \sum_{j\in P_{q_t}} a_{t,j}} (\sum_{j\in N_{q_t} \bigcap M_i} c_{t,j} + \sum_{j\in P_{q_t} \bigcap M_i} a_{t,j} + b_t) \end{pmatrix}$$

The result of this operation is interpreted as follows:

- If $\exists k \in [\![1,t]\!]$ *such that* $(Test(\mathcal{Q},\mathcal{X}_i))_k = 1$ then the document C_i exactly match the query and we add C_i to $\mathcal{L}$. Indeed it means that in the corresponding disjunctions all affirmative keywords are in the document and non negative keyword is present (otherwise the result would be smaller than one).

- Else $\forall k \in [\![1,t]\!]$, $(Test(\mathcal{Q},\mathcal{X}_i))_k \neq 1$ then the document doesn't match the query.

After performing the $Test$ algorithm, the server sends to the user the set of all encrypted documents $\mathcal{L}$ that match the query.

Note: The reason why we have included a joker J in all documents is to deal with some particular cases. For instance, the case where we have only negative keywords in the disjunction L_k. Then $|P_{q_k}| = 0$. Without a joker, the query $\mathcal{S}_k$ would have to be divided by 0. By adding the joker, we prevent this case from occurring and the division is at least by a.

5. EVALUATION

In this section we evaluate the security and performance of our solution.

We first note that our scheme is consistent in the sense of definition 3.2. We already hinted to that during the explanation of our construction. Basically we consider each conjunction separately and if at least one of them is satisfied then the whole disjunction is satisfied as well. The test operation at each conjunction level is basically a scalar product between the query and the labels which are both vectors of a vector space expressed with an orthonormal base. Hence if a keyword is present in both the query and the label the result is one otherwise it is zero.The overall result is divided by the weight of positive keywords in the conjunction. If a negative keyword is both in the query and the keyword it will produce a negative component in the sum, hence the sum in the inner product will be strictly less than the divisor.

5.1 Security

THEOREM 5.1. *BSSE probabilistic Query algorithm has randomized queries.*

PROOF. We are going to consider two situations, the first case is when we have the same boolean query $\mathcal{B}(\mathcal{W})$, the second one will be more general since we are going to take queries generated for different boolean queries. We are aiming to prove that:

$$\forall\, i,j \in \{1,l\}\ \Pr(\mathcal{Q}_i = \mathcal{Q}_j) < \epsilon\ ,$$

where ϵ tends to zero.

The first case: let us consider the following boolean query $\mathcal{B}(\mathcal{W})$, we apply the probabilistic algorithm Query$(K_1,.)$ (with the same key K_1) q times such that $\{\mathcal{Q}_1, \mathcal{Q}_2, ..., \mathcal{Q}_q\}$ represents the resulting queries. Having the same boolean query implies that, for all queries, we will have the same affirmative and negative keywords in all disjunctions. Let us consider the ith query:

$$\mathcal{Q}_i = \begin{pmatrix} \mathcal{S}_{i,1} \\ \vdots \\ \mathcal{S}_{i,t} \end{pmatrix}$$

where

$$\mathcal{S}_{i,k} = \frac{1}{b_k^{(i)} + \sum_{j\in P_{q_k}^{(i)}} a_{k,j}^{(i)}}\Big(\sum_{j\in N_{q_k}^{(i)}} c_{k,j}^{(i)} u_j + \sum_{j\in P_{q_k}^{(i)}} a_{k,j}^{(i)} u_j + b_k^{(i)} J\Big)$$

We note $\alpha_k^{(i)} = \frac{1}{b_k^{(i)} + \sum_{j\in P_{q_k}} a_{k,j}^{(i)}}$.

Let us take the lth query, since we have the same boolean query we simplify the notation in this case to $N_{q_k}^{(i)} = N_{q_k}^{(l)} = N_{q_k}$ and $P_{q_k}^{(i)} = P_{q_k}^{(l)} = P_{q_k}$. We have then for the first row of the queries:

$$\Pr(\mathcal{S}_{i,1} - \mathcal{S}_{l,1} = 0) = \Pr\big(\textstyle\sum_{j\in N_{q_1}} (\alpha_1^{(i)} c_{1,j}^{(i)} - \alpha_1^{(l)} c_{1,j}^{(l)}) u_j + $$
$$+ \textstyle\sum_{j\in P_{q_1}} (\alpha_1^{(i)} a_{1,j}^{(i)} - \alpha_1^{(l)} a_{1,j}^{(l)}) u_j + (\alpha_1^{(i)} b_1^{(i)} - \alpha_1^{(l)} b_1^{(l)}) J = 0\big)$$

Since we have that $P_{q_1} \bigcap N_{q_1} \bigcap \{J\} = \emptyset$ (each keyword in the combination is unique) and orthonormalized keywords which represent a free family, we then have for a fixed i and l:

$$\Pr(\mathcal{S}_{i,1} - \mathcal{S}_{l,1} = 0) = \prod_{j\in P_{q_1}} \Pr\Big(a_{1,j}^{(i)} = \frac{\alpha_1^{(l)}}{\alpha_1^{(i)}} a_{1,j}^{(l)}\Big) \times$$
$$\prod_{j\in N_{q_1}} \Pr\Big(c_{1,j}^{(i)} = \frac{\alpha_1^{(l)}}{\alpha_1^{(i)}} c_{1,j}^{(l)}\Big) \times \Pr\Big(b_1^{(i)} = \frac{\alpha_1^{(l)}}{\alpha_1^{(i)}} b_1^{(l)}\Big)$$

For all $j \in N_{q_1} \bigcup P_{q_1}$ we have $c_{1,j}^{(l)}$, $c_{1,j}^{(i)}$ and $b_1^{(l)}$, $b_1^{(i)}$, $a_{1,j}^{(l)}$, $a_{1,j}^{(i)}$ are respectively sampled uniformly at random from $\{-2^r, ..., -1\}$ and $\{1, .., 2^r\}$ where r is a sufficiently large positive number, we then have:

$$\begin{aligned}\Pr(\mathcal{S}_{i,1} - \mathcal{S}_{l,1} = 0) &= \tfrac{1}{2^n}^{|P_{q_1}|} \times \tfrac{1}{2^n}^{|N_{q_1}|} \times \tfrac{1}{2^n} \\ &= (\tfrac{1}{2^n})^{|P_{q_1}| + |N_{q_1}| + 1}\end{aligned}$$

We should point out that all rows are independent. Consequently for all rows we will obtain:

$$\begin{aligned}\Pr(\mathcal{Q}_i - \mathcal{Q}_l = 0) &= \textstyle\prod_{k=1}^{t} Pr(\mathcal{S}_{i,k} - \mathcal{S}_{l,k} = 0) \\ &= (\tfrac{1}{2^n})^{\sum_{k=1}^{t}(|P_{q_k}| + |N_{q_k}| + 1)}\end{aligned}$$

We say that the scheme has $(l, (\frac{1}{2^n})^{\sum_{k=1}^{t}(|P_{q_k}|+|N_{q_k}|+1)})$-randomized query.

The second case, we consider different boolean query for each time the user sends a query to the server. Consequently, we will have different affirmative and negative keywords for each query $\mathcal{Q}_i$.

We have then for the first row of queries $\mathcal{Q}_i$ and $\mathcal{Q}_l$:

$$\Pr(\mathcal{S}_{i,1} - \mathcal{S}_{l,1} = 0) = \Pr\big(\textstyle\sum_{j\in N_{q_1}^{(i)}} \alpha_1^{(i)} c_{1,j}^{(i)} u_j + \sum_{j\in P_{q_1}^{(i)}} \alpha_1^{(i)} a_{1,j}^{(i)} u_j + \alpha_1^{(i)} b_1^{(i)} J - \sum_{j\in N_{q_1}^{(l)}} \alpha_1^{(l)} c_{1,j}^{(l)} u_j - \sum_{j\in P_{q_1}^{(l)}} \alpha_1^{(l)} a_{1,j}^{(l)} u_j - \alpha_1^{(l)} b_1^{(l)} J = 0\big)$$

Since we can have negative keywords in the first query existing in the affirmative keywords sets of the second query and vice versa, we should take into account all the possible combinations such that:

- $\Gamma_1^{(1)} = N_{q1}^{(i)} \bigcap (N_{q1}^{(l)} \bigcup P_{q1}^{(l)})$
- $\Gamma_2^{(1)} = P_{q1}^{(i)} \bigcap (N_{q1}^{(l)} \bigcup P_{q1}^{(l)})$
- $\Gamma_3^{(1)} = N_{q1}^{(i)} \setminus (N_{q1}^{(l)} \bigcup P_{q1}^{(l)})$
- $\Gamma_4^{(1)} = P_{q1}^{(i)} \setminus (N_{q1}^{(l)} \bigcup P_{q1}^{(l)})$
- $\Gamma_5^{(1)} = N_{q1}^{(l)} \setminus (N_{q1}^{(i)} \bigcup P_{q1}^{(i)})$
- $\Gamma_6^{(1)} = P_{q1}^{(l)} \setminus (N_{q1}^{(i)} \bigcup P_{q1}^{(i)})$

The feature owned by this repartition is the creation of six partitions such that:

$$\Gamma_1^{(1)} \bigcap \Gamma_2^{(1)} \bigcap \Gamma_3^{(1)} \bigcap \Gamma_4^{(1)} \bigcap \Gamma_5^{(1)} \bigcap \Gamma_6^{(1)} \bigcap \{J\} = \emptyset$$

Consequently, the probability of distinguishing two rows of different queries is:

$\Pr(\mathcal{S}_{i,1}-\mathcal{S}_{l,1}=0)=\Pr(\sum_{j\in\Gamma_1^{(1)}}(\alpha_1^{(i)}c_{1,j}^{(i)}-\alpha_1^{(l)}(a_{1,j}^{(l)}+c_{1,j}^{(l)}))u_j+\sum_{j\in\Gamma_2^{(1)}}(\alpha_1^{(i)}a_{1,j}^{(i)}-\alpha_1^{(l)}(a_{1,j}^{(l)}+c_{1,j}^{(l)}))u_j+\sum_{j\in\Gamma_3^{(1)}}\alpha_1^{(i)}c_{1,j}^{(i)}u_j+\sum_{j\in\Gamma_4^{(1)}}\alpha_1^{(i)}a_{1,j}^{(i)}u_j-\sum_{j\in\Gamma_5^{(1)}}\alpha_1^{(l)}c_{1,j}^{(l)}u_j-\sum_{j\in\Gamma_6^{(1)}}\alpha_1^{(l)}a_{1,j}^{(l)}u_j+(\alpha_1^{(i)}b_1^{(i)}-\alpha_1^{(l)}b_1^{(l)})J=0)$

Since we are dealing with orthonormal keywords we have than:

$\Pr(\mathcal{S}_{i,1}-\mathcal{S}_{l,1}=0)=\prod_{j\in\Gamma_1^{(1)}}\Pr(c_{1,j}^{(i)}=\frac{\alpha_1^{(l)}}{\alpha_1^{(i)}}(a_{1,j}^{(l)}+c_{1,j}^{(l)}))\times\prod_{j\in\Gamma_2^{(1)}}\Pr(a_{1,j}^{(i)}=\frac{\alpha_1^{(l)}}{\alpha_1^{(i)}}(a_{1,j}^{(l)}+c_{1,j}^{(l)}))\times\prod_{j\in\Gamma_3^{(1)}}\Pr(c_{1,j}^{(i)}=0)\times\prod_{j\in\Gamma_4^{(1)}}\Pr(a_{1,j}^{(i)}=0)\times\prod_{j\in\Gamma_5^{(1)}}\Pr(c_{1,j}^{(l)}=0)\times\prod_{j\in\Gamma_6^{(1)}}\Pr(a_{1,j}^{(l)}=0)\times\Pr(b_1^{(i)}=\frac{\alpha_1^{(l)}}{\alpha_1^{(i)}}b_1^{(l)})$

All sample here are independent from each others, this will imply that:

$$\Pr(c_{1,j}^{(i)}=\frac{\alpha_1^{(l)}}{\alpha_1^{(i)}}(a_{1,j}^{(l)}+c_{1,j}^{(l)}))=\Pr(c_{1,j}^{(l)})+\Pr(a_{1,j}^{(l)})-\Pr(c_{1,j}^{(l)}).\Pr(a_{1,j}^{(l)})$$

$$\Pr(c_{1,j}^{(i)}=\frac{\alpha_1^{(l)}}{\alpha_1^{(i)}}(a_{1,j}^{(l)}+c_{1,j}^{(l)}))=\frac{1}{2^{n-1}}-\frac{1}{2^{2n}}\leq\frac{1}{2^{n-1}}$$

Consequently:

$\Pr(\mathcal{S}_{i,1}-\mathcal{S}_{l,1}=0)\leq\frac{1}{2^{n-1}}^{|\Gamma_1^{(1)}|+|\Gamma_2^{(1)}|}\times\frac{1}{2^n}^{|\Gamma_3^{(1)}|+|\Gamma_4^{(1)}|+|\Gamma_5^{(1)}|+|\Gamma_6^{(1)}|+1}$

Knowing that all rows are independent. Consequently for all rows we will obtain:

$$\Pr(\mathcal{Q}_i-\mathcal{Q}_l=0)=\prod_{k=1}^{t}Pr(\mathcal{S}_{i,k}-\mathcal{S}_{l,k}=0)$$

$\Pr(\mathcal{Q}_i-\mathcal{Q}_l=0)\leq\frac{1}{2^{n-1}}^{\sum_{k=1}^{t}(|\Gamma_1^{(k)}|+|\Gamma_2^{(k)}|)}\times\frac{1}{2^n}^{\sum_{k=1}^{t}(|\Gamma_3^{(k)}|+|\Gamma_4^{(k)}+|\Gamma_5^{(k)}|+|\Gamma_6^{(k)}|+1)}$ □

We have shown that the queries generated are randomized which makes it harder for the server to differentiate them. However the server can know if an orthonormalized keyword exists or not in the query, it can be done by performing an inner product between a randomized generated sequence and the query, the probability to find it out is equal to $\frac{|P_{q_K}|+|N_{q_K}|}{2^{m+l}}$ where m+l is the size of the orthonormalized encrypted keywords, we fix the lower bound of keyword size to m=100 , consequently when $W=w_1$ (which is the most favorable case for the attacker, i.e. there is only one keyword stored in labels), the probability to find it is equal to $\frac{|P_{q_k}|+|N_{q_K}|}{2^{100}}$, on the other hand if we are storing for example 1000 keywords in labels, the size of keywords will then be equal to m+l=1100 and the probability will be equal then to $\frac{|P_{q_k}|+|N_{q_K}|}{2^{1100}}$. Hence it makes it difficult to check if a given query contains a particular keyword, further more, even if the server knows that a query contains a particular keyword, he can not find out if it is an affirmative or negative one with a probability equal to $\frac{1}{2}$. That means that the server can not distinguish if the user search for documents containing this keyword or not. Even if keywords are already secured by using a pseudo-random permutation which make them indistinguishable from a random sample of strings, the aim of our randomized queries is to hide the search pattern of the user, making queries not deterministic.

THEOREM 5.2. *If $\mathcal{E}$ is a semantically secure symmetric encryption scheme and π is a pseudo-random permutation, then BSSE scheme is adaptively semantically secure.*

PROOF. We are going to describe a polynomial-size simulator $\mathcal{S}=\{\mathcal{S}_0,...,\mathcal{S}_q\}$ such that for all polynomial-size adversaries $\mathcal{A}=\{\mathcal{A}_0,...,\mathcal{A}_q\}$ the outputs $(o,st_{\mathcal{A}})$ of $Real_{BSSE,\mathcal{A}}(k)$ and $(o',st'_{\mathcal{A}})$ of $Sim_{BSSE,\mathcal{A},\mathcal{S}}(k)$ are computationally indistinguishable. We are going to detail each steps of the simulator in the experiment $Sim_{BSSE,\mathcal{A},\mathcal{S}}(k)$.

$\mathcal{S}_0(1^k,\gamma(D))$: in this phase the simulator based on the history that he knows at this step (i.e. $\gamma(D)$)(which means only the number and the size of documents from the trace), $\mathcal{S}_0$ will generate at random for each $i\in\{1,...,n\}$ $C'_i\xleftarrow{R}\{0,1\}^{|D_i|}$ and $\mathcal{X}'_i\xleftarrow{R}\{0,1\}^{m+l}$. $\mathcal{S}_0$ then outputs $o=(\mathcal{X}',\mathcal{C}',st_{\mathcal{S}})$. Since $st_{\mathcal{A}}$ does not include K_1, $\mathcal{X}'$ is indistinguishable from a real label, otherwise we can distinguish between the output of π and random values of size m+l which is not true in accordance to the pseudo-random permutation definition. Likewise, since $st_{\mathcal{A}}$ does not include K_2, the output C' of the simulator $\mathcal{S}_0$ is indistinguishable from the real cyphertexts, otherwise we can distinguish between the output of a semantically secure symmetric encryption scheme from a random values of size $|D_i|$.

$\mathcal{S}_1(st_{\mathcal{S}},\gamma(D,\mathcal{B}(\mathcal{W}_1))$: At this phase the simulator has the knowledge of all document' identifiers corresponding to the boolean query, but the simulator $\mathcal{S}_1$ has no information about the structure of the boolean query (i.e. the number of keywords in each conjunction, how many negative or affirmative keywords exist in the boolean query and also the number of disjunctions in the structure). Let consider $D(\mathcal{B}(\mathcal{W}_1))$ the set of all these identifiers. For all $1\leq j\leq|D(\mathcal{B}(\mathcal{W}_1))|$, $\mathcal{S}_1$ chooses at first step an association between each document identifier and a generated label at the previous phase such that $(D(\mathcal{B}(\mathcal{W}_1))_i,\mathcal{X}'_i)$ are pairwise distinct (because each document has one and only one associated label). $\mathcal{S}_1$ then creates $\mathcal{Q}'_1\xleftarrow{R}\{0,1\}^{m+l}\times\{0,1\}^a$ (where a representes the simulator guess about the number of disjunctions) such that for all $1\leq j\leq|D(\mathcal{B}(\mathcal{W}_1))|$, there exists at least one $1\leq k\leq a$ such that the inner product $\phi(\mathcal{Q}'_{1,k},\mathcal{X}'_j)=1$. $\mathcal{S}_1$ stores the association between $\mathcal{Q}'_1$ and $\mathcal{B}(\mathcal{W}_1)$ in $st_{\mathcal{S}}$, then outputs $(\mathcal{Q}'_1,st_{\mathcal{S}})$. Since $st_{\mathcal{A}}$ does not include K_1, the output $\mathcal{Q}'_1$ is indistinguishable from a real generated query $\mathcal{Q}_1$, otherwise we can distinguish between the output of a pseudo-random permutation from a random sequence of bits of size $\{0,1\}^{m+l}\times\{0,1\}^a$.

$\mathcal{S}_i(st_{\mathcal{S}},\gamma(D,\mathcal{B}(\mathcal{W}_1),...,\mathcal{B}(\mathcal{W}_i)))$ for $2\leq i\leq q$: $\mathcal{S}_i$ check first if the boolean query $\mathcal{B}(\mathcal{W}_i)$ was queried before. The simulator has only to verify in the trace if there exists a $1\leq j\leq i-1$ such that $\sigma_{i,j}=1$. If $\sigma_{i,j}=0$, the boolean query has not previously queried for, the simulator $\mathcal{S}_i$ then follows the same steps as $\mathcal{S}_0$, however he has to make sure that any label associated to document' identifiers are pairwise distinct. On the other hand, if the boolean query has appeared before, the simulator $\mathcal{S}_i$ takes the same generated query. Finally it outputs $(\mathcal{Q}'_i,st_{\mathcal{S}})$ where the output $\mathcal{Q}'_i$ is indistinguishable from a real generated query $\mathcal{Q}_i$, otherwise we can distinguish between the output of a pseudorandom permutation from a random sequence of bits of size $\{0,1\}^{m+l} \times \{0,1\}^a$.

At the end the simulution the output $\mathcal{Q}' = (\mathcal{Q}'_1, ..., \mathcal{Q}'_q)$ is indistinguishable from a real generated query $\mathcal{Q} = (\mathcal{Q}_1, ..., \mathcal{Q}_q)$ and consequently the output of the experiments $Sim_{BSSE,\mathcal{A},\mathcal{S}}(k)$ and $Real_{BSSE,\mathcal{A}}(k)$ are indistinguishable. □

5.2 Performance

From a performance perspective, the most critical step of searchable encryption is the search phase as it is performed by the server for potentially many users. The search complexity of our construction is $O(n)$ inner products where n is the number of documents stored in the server. This is as good as the solution provided by Goh in [13], except that our scheme can handle much more complex queries. The search phase is illustrated in figure 1, in this example, we show that the search is linear in the number of documents, we take also three different cases where documents stored in the server are associated to labels containing 1000, 2000 or 4000 keywords (for all searches we have taken the same boolean query, we did not take into account the network transmission delay or the query construction time). We emphasize that the increase of labels size implies a longer computation phase for each document.

The main novelty in our scheme is the pre-computation phase with the Gram-Schmidt process. As we mentioned there are two options: either performing this computation once and storing the result in a table resulting in $O(l)$ storage complexity or trading storage for computation and performing the Gram-Schmidt process for each query. Intuitively, since storage is cheap the first solution would be more advantageous. However we implemented the keyword orthonormalization construction which include the Gram-Schmidt and the permutation processes and found out that this computation phase is still reasonable although quadratic in the number of keywords as can be seen in figure 2.

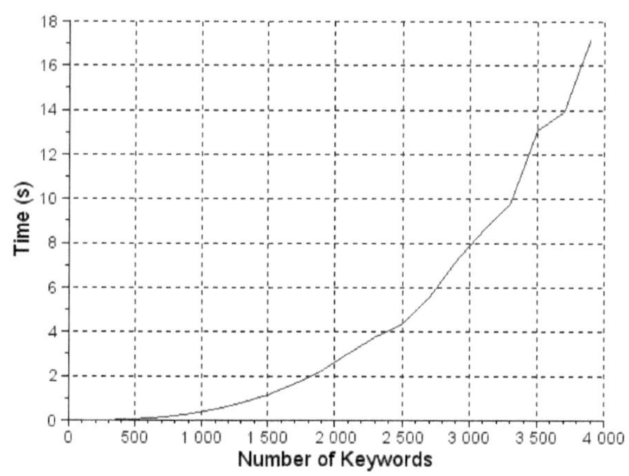

Figure 2: Orthogonal keywords construction

In this figure, the *number of keywords* denotes the size of the free family that we have constructed in the first section. As can be seen on the graph, the construction of one thousand orthogonal keywords takes less 0.5 second which is acceptable given that these computations are performed on a client side. The algorithm used was not optimized in any way, and it simply proves that the solution of recomputing the base at each time is feasible (although we believe that the storage alternative is more efficient). In addition to this implementation, we have tested the performance of the off-line construction of labels in the client side, see figure 3 (we do not take into account the extraction of keywords from documents).

6. CONCLUSION

Searchable encryption is a mechanism which enables to perform secure searches over encrypted data stored on outsourced servers. Our work focused on the enhancement of search options over encrypted data. In particular we presented BSSE, the first efficient scheme which supports any boolean expression query, that is to say that our scheme to perform queries composed of conjunction, disjunction and negation of keywords. Our scheme is based on an original approach making use of the Gram-Schmidt orthogonalization process to encode keywords that are further used in labels and queries and inner products to perform the search and leveraging on the orthonormal basis properties. We have proven our scheme secure in an adaptive adversarial model by means of a simulation based formal proof. The efficiency of the scheme is also supported by the fact that the search complexity is in $O(n)$ where n is the number of documents stored in the outsourced server, while the orthogonalization process complexity is quadratic but can be performed offline and can benefit from pre-computations.

As future work, we aim to further enhance the search option as boolean operations only represent a first step towards supporting queries in the form of generic regular expressions over encrypted data. On the other hand, it would be also interesting to create a similar boolean scheme in an asymmetric setting.

7. REFERENCES

[1] M. Abdalla, M. Bellare, D. Catalano, E. Kiltz, T. Kohno, T. Lange, J. Malone-Lee, G. Neven, P. Paillier, and H. Shi. Searchable encryption revisited: Consistency properties, relation to anonymous ibe, and extensions. *J. Cryptology*, 21(3):350–391, 2008.

[2] G. Amanatidis, A. Boldyreva, and A. O'Neill. Provably-secure schemes for basic query support in outsourced databases. *DBSec*, pages 14–30, 2007.

[3] L. Ballard, S. Kamara, and F. Monrose. Achieving efficient conjunctive keyword searches over encrypted data. *ICICS*, pages 414–426, 2005.

[4] M. Bellare, A. Boldyreva, and A. O'Neill. Deterministic and efficiently searchable encryption. *CRYPTO*, pages 535–552, 2007.

[5] D. Boneh, G. D. Crescenzo, R. Ostrovsky, and G. Persiano. Public key encryption with keyword search. *EUROCRYPT*, pages 506–522, 2004.

[6] D. Boneh and B. Waters. Conjunctive, subset, and range queries on encrypted data. *TCC*, pages 535–554, 2007.

[7] J. W. Byun, D. H. Lee, and J. Lim. Efficient conjunctive keyword search on encrypted data storage system. *EuroPKI*, pages 184–196, 2006.

[8] Y.-C. Chang and M. Mitzenmacher. Privacy

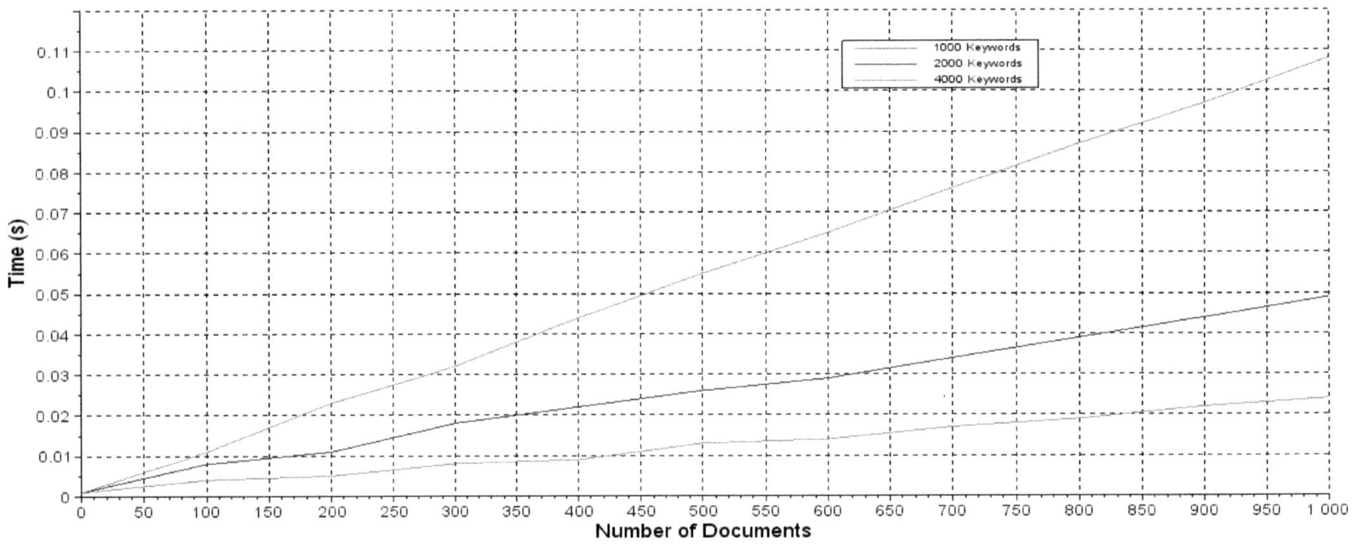

Figure 1: Search phase

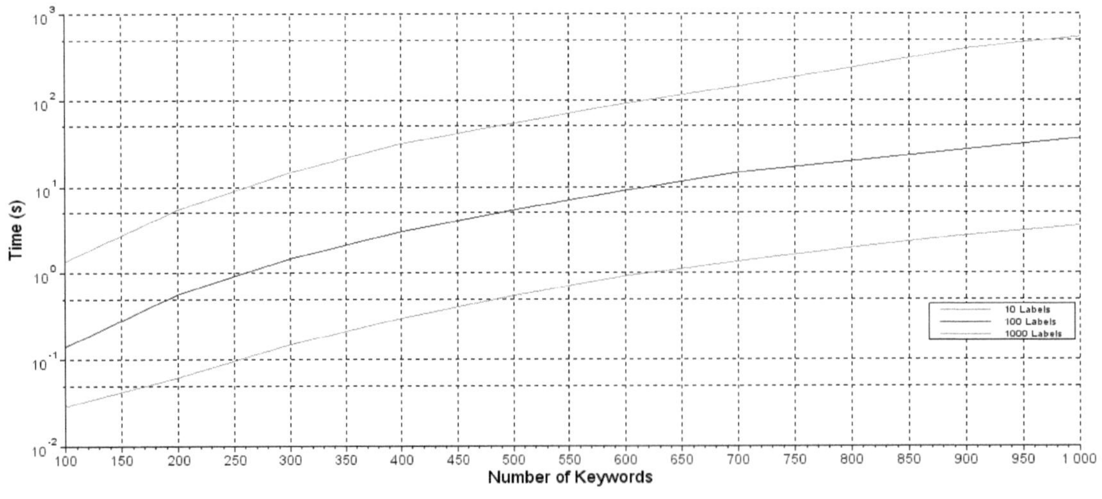

Figure 3: Labels creation

preserving keyword searches on remote encrypted data. *ACNS*, pages 442–455, 2005.

[9] W. Cheney and D. R. Kincaid. Linear algebra: Theory and applications. pages 544–558.

[10] G. D. Crescenzo and V. Saraswat. Public key encryption with searchable keywords based on jacobi symbols. *Progress in Cryptology - INDOCRYPT 2007, 8th International Conference on Cryptology in India, Chennai, India, December 9-13, 2007, Proceedings*, 4859:282–296, 2007.

[11] R. Curtmola, J. A. Garay, S. Kamara, and R. Ostrovsky. Searchable symmetric encryption: improved definitions and efficient constructions. *ACM Conference on Computer and Communications Security*, pages 79–88, 2006.

[12] R. Curtmola, J. A. Garay, S. Kamara, and R. Ostrovsky. Searchable symmetric encryption: Improved definitions and efficient constructions. *Journal of Computer Security*, 19(5):895–934, 2011.

[13] E.-J. Goh. Secure indexes. *IACR Cryptology ePrint Archive*, 2003:216, 2003.

[14] O. Goldreich. Foundations of cryptography volume 2 basic applications. 2004.

[15] P. Golle, J. Staddon, and B. R. Waters. Secure conjunctive keyword search over encrypted data. *ACNS*, pages 31–45, 2004.

[16] Y. H. Hwang and P. J. Lee. Public key encryption with conjunctive keyword search and its extension to a multi-user system. *Pairing*, pages 2–22, 2007.

[17] J. Katz, A. Sahai, and B. Waters. Predicate encryption supporting disjunctions, polynomial equations, and inner products. pages 146–162, 2008.

[18] D. J. Park, K. Kim, and P. J. Lee. Public key encryption with conjunctive field keyword search. *WISA*, pages 73–86, 2004.

[19] E.-K. Ryu and T. Takagi. Efficient conjunctive keyword-searchable encryption. *AINA Workshops (1)*, pages 409–414, 2007.

[20] E. Shi, J. Bethencourt, H. T.-H. Chan, D. X. Song, and A. Perrig. Multi-dimensional range query over encrypted data. *IEEE Symposium on Security and Privacy*, pages 350–364, 2007.

[21] D. X. Song, D. Wagner, and A. Perrig. Practical techniques for searches on encrypted data. *IEEE Symposium on Security and Privacy*, pages 44–55, 2000.

[22] P. Wang, H. Wang, and J. Pieprzyk. An efficient scheme of common secure indices for conjunctive keyword-based retrieval on encrypted data. *WISA*, pages 145–159, 2008.

[23] P. Wang, H. Wang, and J. Pieprzyk. Keyword field-free conjunctive keyword searches on encrypted data and extension for dynamic groups. *CANS*, pages 178–195, 2008.

Multi-Channel Broadcast Encryption

Duong Hieu Phan
LAGA, University of Paris 8
Paris, France

David Pointcheval
DI/ENS, CNRS – ENS – INRIA
Paris, France

Viet Cuong Trinh
LAGA, University of Paris 8
Paris, France

ABSTRACT

Broadcast encryption aims at sending a content to a large arbitrary group of users at once. Currently, the most efficient schemes provide constant-size headers, that encapsulate ephemeral session keys under which the payload is encrypted. However, in practice, and namely for pay-TV, providers have to send various contents to different groups of users. Headers are thus specific to each group, one for each channel: as a consequence, the global overhead is linear in the number of channels. Furthermore, when one wants to zap to and watch another channel, one has to get the new header and decrypt it to learn the new session key: either the headers are sent quite frequently or one has to store all the headers, even if one watches one channel only. Otherwise, the zapping time becomes unacceptably long.

This paper deals with encapsulation of several ephemeral keys, for various groups and thus various channels, in one header only, and we call this new primitive *Multi-Channel Broadcast Encryption* – MCBE: one can hope for a much shorter global overhead and a much shorter zapping time since the decoder already has the information to decrypt any available channel at once. Our candidates are private variants of the Boneh-Gentry-Waters scheme, with a constant-size global header, independently of the number of channels.

Categories and Subject Descriptors

D.4.6 [**Security and Protection**]: Cryptographic controls

Keywords

Broadcast encryption, pay-TV

1. INTRODUCTION

Broadcast encryption has been widely and deeply studied as it is a core primitive for many concrete applications. In the following, we focus on the pay-TV scenario, in which users own decoders to decode only the channels they subscribed to. In this context, the broadcaster sends several channels at the same time, to different groups of users or target sets.

Unfortunately, previous broadcast encryption models only dealt with one single content and one single target set at a time. This was a first reasonable goal to get such an efficient broadcast encryption scheme, but not quite relevant to practice. In fact, TV systems contain many channels, with different sets of privileged users. One could argue that this scenario is covered by the usual systems, applying independent broadcast encryption schemes for each channel. However, even with a constant-size ciphertext (header) broadcast encryption, this results in a very inefficient scheme: the bandwidth, or header size, linearly grows in the number of channels, which could be very large. Of course, one header is enough to decrypt one channel, but in case of zapping from one channel to another channel, one has to start from scratch, and namely to wait for the reception of the new appropriate header, which can take some time, unless the decoder stores all the headers all the time.

These bandwidth and zapping-time problems lead to new efficiency criteria, with a common solution: a broadcast encryption with a short *global header* for multiple channels. The problem of optimizing the bandwidth already appeared in the context of classical (one-channel) broadcast encryption: a broadcast encryption can trivially be constructed from any encryption scheme, by encrypting the session key under each user's key. But this induces a cost that is linear in the number of users. It took more than a decade from the introduction of the primitive [5] to come up with an optimal solution: without considering the description of the target set, the header is of constant size [2]. This BGW solution [2] is particularly interesting even if it is still not practical, due to the high decryption complexity: the latter is indeed linear in the size of the target set.

Our new primitive MCBE, for Multi-Channel Broadcast Encryption, addresses the bandwidth and zapping-time problems. In the following, we show that it is possible to solve these problems in an optimal way: a constant-size global header, independently of the number of channels (and of the users too). Unfortunately, this is still an asymptotic result, with room for improvement in practice. Actually, our solutions suffer from the same weakness as the above BGW scheme: the decryption has to take into account all the public keys of the users involved in all the target sets. It is thus not quite efficient. However, it seems that this problem is unavoidable when one compacts the information for all the targeted users into one constant-size ciphertext. We also notice that there is a simple and similar trade-off between

ASIA CCS'13, May 8–10, 2013, Hangzhou, China.

the ciphertext size and the decryption time as in [2] by partionning the set of channels into different subsets and then encrypting to each of these subsets. The union of the target sets in a ciphertext is smaller, but there are more ciphertexts. Our objective is therefore to show that the bandwidth and zapping-time problems in the multi-channel setting can be improved from trivial techniques, as BGW did in the one-channel setting.

Finally, we emphasize that the solution requires some new techniques that we will develop in the section 1.2. In particular, we have to deal with the problem of encapsulating different and independently-looking session keys for the different channels into one constant-size element only. We will then prove the security in the new multi-channel setting.

1.1 Broadcast Encryption Schemes

Broadcast encryption was first described by Fiat and Naor in [5] but receives much attention since the work of Naor, Naor, and Lotspiech [8] in which they presented a symmetric-key subset-cover framework along with a security model and a security analysis. Dodis and Fazio [4] presented the first public-key CCA-secure scheme. Boneh, Gentry, and Waters [2] designed a fully collusion-resistant scheme and proposed a security model where the adversary can corrupt any user, except the users in the challenge target set. With their scheme, the adversary had to precise this challenge target set before knowing the parameters of the system, hence the so-called *selective model*. Delerablée constructed a selectively secure ID-based BE [3] in the random oracle model. Thereafter, Gentry and Waters [6] defined the *adaptive model*, where the adversary can corrupt users and then adaptively choose the challenge target set, and provided adaptively secure schemes in the standard and the random oracle models. Waters [11] and Lewko *et. al.* [7] used dual system encryption to achieve adaptive security. Recently, a scheme that achieves all desired properties (constant-size ciphertexts, adaptive and CCA security) has been presented in [9] but it relies on rather non-standard assumptions.

Phan, Pointcheval and Strefler [10] recently gave a global picture of the relations between the security notions for broadcast encryption. However, our setting of multi-channel broadcast encryption goes beyond their considerations, because the adversary could corrupt some users of one channel to break the security of the other channels. The sessions keys of all channels should indeed be compacted into one ciphertext only, there are thus some relations between these keys inside one session and the security model has to take these relations into account.

1.2 Contributions

We first propose a formalization of the problem, with the so-called *Multi-Channel Broadcast Encryption* – MCBE. Because of some constraints between the various target sets, we introduce the *dummy-helper technique* that helps to prove the security. We eventually propose two constructions, derived from the Boneh-Gentry-Waters (BGW) [2] scheme. They are private broadcast encryption schemes, with the following properties:

- The first construction is, asymptotically, very competitive with the BGW scheme. In fact, it achieves the constant-size header, independently of the number of channels, while the private decryption key size remains linear in the number of the channels that a user has subscribed to. In addition, it is fully-collusion resistant against basic selective adversaries, *i.e.* adversaries who can only ask corruption queries to get the decryption keys of users in the selective security model (the challenge target set is announced before having seen the global parameters). This is also the security level that the original BGW scheme achieves and our security proof holds under the standard assumption $\mathsf{n-BDHE}$, as in the original BGW scheme [2].
- The second construction improves on the previous one, to resist to strong selective adversaries who have the power of basic selective adversaries plus unlimited access to encryption and decryption queries, while keeping the parameter sizes and computational assumptions unchanged. To this aim, we introduce the *dummy-helper technique* and make use of a *random oracle* [1]. Our scheme is more efficient than the CCA version of the BGW scheme [2] but our dummy-helper technique actually requires the random oracle model.

Dummy-helper technique. In the multi-channel setting, since the session keys of all channels are compacted in only one ciphertext, even if they have to look independent for adversaries, there exists an implicit relation between them, which could be known by the simulator without the whole knowledge of the master key. The *dummy-helper technique* consists in adding a new channel for one additional dummy user. We then get the following interesting properties:

1. For the security analysis: it gives our simulator the possibility to decrypt this channel and get the corresponding session key. This is then sufficient for the simulator to derive the other session keys and successfully answer any decryption query, if the simulator knows the above implicit relation between the encapsulated keys;
2. In practice: by eventually publishing the decryption key of the dummy user, it introduces a channel that can be decoded by all the users registered in the system. It can then be used to send them the program or ads.

We implement this dummy-helper technique in the random oracle model. It is worth noting that, though working in a more complex setting of multi-channel broadcast encryption, the security is achieved under the standard assumption $\mathsf{n-BDHE}$ as in the BGW scheme.

2. MULTI-CHANNEL BROADCAST ENCRYPTION

2.1 Syntax

In this section we describe the model for a multi-channel broadcast encryption system. Formally, such a system consists of four probabilistic algorithms:

Setup(λ)**:** Takes as input the security parameter λ, it generates the global parameters param of the system, including n the maximal number of users (receivers are implicitly represented by integers in $\{1, \ldots, n\}$), and returns a master key MSK and an encryption key EK. If the scheme allows public encryption, EK is public, otherwise EK is kept private, and can be seen as part of MSK.

Extract(i, MSK)**:** Takes as input the user's index i, together with the master key, and outputs the user's private key d_i.

Encrypt$(S_1, S_2, \ldots, S_m, \mathsf{EK})$**:** Takes as input m subsets (or target sets) $S_1, S_2, \ldots, S_m$ where, for $i = 1, \ldots, m$, $S_i \subseteq \{1, \ldots, n\}$, and the encryption key EK. It outputs $(\mathsf{Hdr}, K_1, K_2, \ldots, K_m)$ where Hdr encapsulates the ephemeral keys $(K_i)_{i=1,\ldots,m} \in \mathcal{K}$. The key K_i will be associated to the subset S_i. We will refer to Hdr as the broadcast ciphertext, or *header*, whereas this header together with the description of all the target sets is called the *full header*.

Decrypt$(S_1, S_2, \ldots, S_m, \mathsf{Hdr}, j, d_j, i)$: Takes as input a full header $(S_1, S_2, \ldots, S_m, \mathsf{Hdr})$, a user $j \in \{1, \ldots, n\}$ and its private key d_j, together with a subgroup index $i \in \{1, \ldots, m\}$. If $j \in S_i$, then the algorithm outputs the ephemeral key $K_i \in \mathcal{K}$.

For correctness, we require that for all $S_i \subseteq \{1, \ldots, n\}$ and $j \in S_i$, if $(\mathsf{EK}, \mathsf{MSK}) \leftarrow \mathbf{Setup}(\lambda)$, $d_j \leftarrow \mathbf{Extract}(j, \mathsf{MSK})$ and $(\mathsf{Hdr}, K_1, \ldots, K_m) \leftarrow \mathbf{Encrypt}(S_1, S_2, \ldots, S_m, \mathsf{EK})$, one then should get $K_i = \mathbf{Decrypt}(S_1, S_2, \ldots, S_m, \mathsf{Hdr}, j, d_j, i)$.

In practice, the goal of such ephemeral keys is to encrypt the payload, which consists of m messages $M_1, \ldots, M_m$ to be broadcast to the sets $S_1, \ldots, S_m$ respectively. They will thus be encrypted under the symmetric keys $K_1, \ldots, K_m$ into the ciphertexts $\mathsf{CM}_1, \ldots, \mathsf{CM}_m$ respectively. The overall data the broadcaster sends consists of $(S_1, S_2, \ldots, S_m, \mathsf{Hdr}, \mathsf{CM}_1, \mathsf{CM}_2, \ldots, \mathsf{CM}_m)$ where $(S_1, S_2, \ldots, S_m, \mathsf{Hdr})$ is the *full header* and $(\mathsf{CM}_1, \mathsf{CM}_2, \ldots, \mathsf{CM}_m)$ is often called the *encrypted payload*.

2.2 Security Model

We define the security of a multi-channel broadcast encryption system by the following game between an attacker $\mathcal{A}$ and a challenger, in the Real-or-Random setting:

Setup. The challenger runs the **Setup** algorithm to generate the global parameters param of the system, and returns a master key MSK and an encryption key EK. If the scheme is asymmetric, EK is given to $\mathcal{A}$, otherwise it is part of the MSK, and thus kept secret. Corruption and decryption lists Λ_C, Λ_D are set to empty lists.

Query phase 1. The adversary $\mathcal{A}$ adaptively asks queries:

1. Corruption query for the i-th user: the challenger runs $\mathbf{Extract}(i, \mathsf{MSK})$ and forwards the resulting private key to the adversary. The user i is appended to the corruption list Λ_C;
2. Decryption query on the full header $(S_1, \ldots, S_m, \mathsf{Hdr})$ with $u \in \{1, \ldots, n\}$ and $j \in \{1, \ldots, m\}$. The challenger answers with $\mathbf{Decrypt}(S_1, \ldots, S_m, \mathsf{Hdr}, u, d_u, j)$. The pair (Hdr, S_j) is appended to the decryption list Λ_D;
3. Encryption query (if EK is private) for the target sets $(S_1, S_2, \ldots, S_m)$. The challenger answers with $\mathbf{Encrypt}(S_1, S_2, \ldots, S_m, \mathsf{EK})$.

Challenge. The adversary $\mathcal{A}$ outputs t target sets $S_1^*, \ldots, S_t^* \subseteq \{1, \ldots, n\}$ and an index j, which specifies the attacked target set S_j^*.
The challenger runs $\mathbf{Encrypt}(S_1^*, S_2^*, \ldots, S_t^*, \mathsf{EK})$ and gets $(\mathsf{Hdr}^*, K_1^*, K_2^*, \ldots, K_t^*)$. Next, the challenger picks a random $b \xleftarrow{\$} \{0,1\}$. If $b = 1$, it picks a random $K_j^* \xleftarrow{\$} \mathcal{K}$. It outputs $(\mathsf{Hdr}^*, K_1^*, \ldots, K_t^*)$ to $\mathcal{A}$.
Note that if $b = 0$, K_j^* is the real key, encapsulated in Hdr^*, and if $b = 1$, K_j^* is random, independent of the header.

Query phase 2. The adversary $\mathcal{A}$ continues to adaptively ask queries as in the first phase.

Guess. The adversary $\mathcal{A}$ eventually outputs its guess $b' \in \{0, 1\}$ for b.

We say the adversary wins the game if $b' = b$, but only if $S_j^* \cap \Lambda_C = \emptyset$ and $(\mathsf{Hdr}^*, S_j^*) \notin \Lambda_D$. We then denote by $\mathbf{Succ}^{\mathsf{ind}}(\mathcal{A}) = \Pr[b' = b]$ the probability that $\mathcal{A}$ wins the game, and its advantage is

$$\begin{aligned}\mathbf{Adv}^{\mathsf{ind}}(\mathcal{A}) &= 2 \times \mathbf{Succ}^{\mathsf{ind}}(\mathcal{A}) - 1 \\ &= \Pr[1 \leftarrow \mathcal{A}|b = 1] - \Pr[1 \leftarrow \mathcal{A}|b = 0].\end{aligned}$$

DEFINITION: ***Full Security.*** A multi-channel broadcast encryption scheme is said $(t, \varepsilon, q_C, q_D, q_E)$-secure if for any t-time algorithm $\mathcal{A}$ that makes at most q_C corruption queries, q_D decryption queries, and q_E encryption queries, one has $\mathbf{Adv}^{\mathsf{ind}}(\mathcal{A}) \leq \varepsilon$. We denote by $\mathbf{Adv}^{\mathsf{ind}}(t, q_C, q_D, q_E)$ the advantage of the best t-time adversary.

There are two classical restricted scenarios: a *selective* attacker provides the target sets $S_1^*, S_2^*, \ldots, S_t^* \subseteq \{1, \ldots, n\}$ and index j, which specifies the attacked target set S_j^*, at the beginning of the security game, and one can also restrict the adversary not to ask some queries.

DEFINITION: ***Basic Selective Security.*** A multi-channel broadcast encryption scheme is said to be (t, ε, q_C)-selectively secure if it is $(t, \varepsilon, q_C, 0, 0)$-secure against a selective adversary. We denote by $\mathbf{Adv}^{\mathsf{b-ind}}(t, q_C)$ the advantage of the best t-time basic selective adversary.

Note that in the public broadcast setting (where encryption is public), this just excludes decryption queries: we allow CPA adversaries.

DEFINITION: ***Strong Selective Security.*** A multi-channel broadcast encryption scheme is said to be $(t, \varepsilon, q_C, q_D, q_E)$-selectively secure if it is $(t, \varepsilon, q_C, q_D, q_E)$-secure against a selective adversary. We denote by $\mathbf{Adv}^{\mathsf{s-ind}}(t, q_C, q_D, q_E)$ the advantages of the best t-time strong selective adversaries.

This definition is much stronger since it not only allows decryption queries in the public setting, but also encryption queries in the private setting.

2.3 Disjoint Target Sets

Before presenting our construction in details, we want to stress that our solution requires that all the target sets of the distinct channels are disjoint. Fortunately, this is compatible with our target application of Pay-TV: whenever a user subscribes for a new channel, he is given a new key for decrypting that channel, and it is resonable to consider that the two keys are independent. More formally, in our systems, we assume there are several channels, which are encrypted to independent target sets of users. The users in the appropriate target sets own decryption keys specific to each channel:

- When a user u registers to the system, he receives a smart card with decryption keys (d_u^i) for every channel i. But at the broadcast time, channel i is encrypted for the target set with the subscribers to this channel only (a subset of the decryption keys);

- Another possibility is to first define U_i the set of all the possible decryption keys for the channel i. When a user u subscribes to a channel i, he receives a key $d_u^i \in U_i$.

In both above cases, the target sets are subsets of predetermined and disjoint sets of keys. As a consequence, the target sets S_i are disjoint too. The drawback is that we have to define many keys in the system.

In a more general setting, in order to limit this number of keys, one could think about sharing keys for several channels. Then, it would make the setting incompatible with our solutions which require disjoint target sets. On the other hand, will reducing the number of keys, it would reduce privacy protection too since would be able to know which channels are registered by similar users, and derive some profiles. Alternatively, in order to limit the global number of keys, one could reassign keys when a user unsubscribes from a channel to another channel.

Anyway, in the following, at a time t, when the broadcaster encapsulates keys for several target sets S_i, we assume them to be disjoint.

3. PRELIMINARIES

3.1 Computational Assumptions

We first recall the definition of the classical Computational Diffie-Hellman (CDH) assumption:

DEFINITION: CDH ***Assumption.*** The $(t, \varepsilon)-$CDH assumption says that for any t-time adversary $\mathcal{A}$ that is given $(g, g^r, h) \in \mathbb{G}$, its probability to output h^r is bounded by ε:

$$\mathbf{Succ}^{\mathsf{cdh}}(\mathcal{A}) = \Pr[\mathcal{A}(g, g^r, h) = h^r] \leq \varepsilon.$$

Stronger assumptions have been introduced by Boneh-Gentry-Waters [2]. They both imply the above CDH assumption.

DEFINITION: BDHE ***Assumption.*** The $(t, n, \varepsilon)-$BDHE assumption says that for any t-time adversary $\mathcal{A}$ that is given $(g, h, g^{\alpha^1}, \ldots, g^{\alpha^n}, g^{\alpha^{n+2}}, \ldots, g^{\alpha^{2n}}) \in \mathbb{G}^{2n+1}$, its probability to output $e(g, h)^{\alpha^{n+1}} \in \mathbb{G}$ is bounded by ε:

$$\mathbf{Succ}^{\mathsf{bdhe}}(\mathcal{A}) =$$

$$= \Pr[\mathcal{A}(g, h, g_1, \ldots, g_n, g_{n+2}, \ldots, g_{2n}) = e(g_{n+1}, h)] \leq \epsilon.$$

DEFINITION: DBDHE ***Assumption.*** The $(t, n, \varepsilon)-$DBDHE assumption says that for any t-time adversary $\mathcal{A}$ that is given $(g, h, g^{\alpha^1}, \ldots, g^{\alpha^n}, g^{\alpha^{n+2}}, \ldots, g^{\alpha^{2n}}) \in \mathbb{G}^{2n+1}$, and a candidate to the BDHE problem, that is either $e(g, h)^{\alpha^{n+1}} \in \mathbb{G}$ or a random value T, cannot distinguish the two cases with advantage greater than ε:

$$\mathbf{Adv}^{\mathsf{dbdhe}}(\mathcal{A}) =$$

$$\left| \begin{array}{l} \Pr[\mathcal{A}(g, h, g_1, \ldots, g_n, g_{n+2}, \ldots, g_{2n}, e(g_{n+1}, h)) = 1] \\ - \Pr[\mathcal{A}(g, h, g_1, \ldots, g_n, g_{n+2}, \ldots, g_{2n}, T)) = 1] \end{array} \right| \leq \epsilon.$$

3.2 BGW Overview

As a warm up, we first recall the BGW scheme [2], on which our constructions will rely.

Setup(λ)**:** Let $\mathbb{G}$ be a bilinear group of prime order p. The algorithm first picks a random generator $g \in \mathbb{G}$ and a random scalar $\alpha \in \mathbb{Z}_p$. It computes $g_i = g^{\alpha^i} \in \mathbb{G}$ for $i = 1, 2, \ldots, n, n+2, \ldots, 2n$. Next, it picks a random scalar $\gamma \in \mathbb{Z}_p$ and sets $v = g^\gamma \in \mathbb{G}$.
The public key is $\mathsf{EK} = (g_1, \ldots, g_n, g_{n+2}, \ldots, g_{2n}, v)$, whereas the private decryption key of user $i \in \{1, \ldots, n\}$ is $d_i = v^{\alpha^i}$. These decryption keys are sent by the **Extract** algorithm.

Encrypt(S, EK)**:** Pick a random scalar $r \in \mathbb{Z}_p$, and set $K = e(g_{n+1}, g)^r$, where $e(g_{n+1}, g)$ can be computed as $e(g_n, g_1)$ from EK.
Next, set: $\mathsf{Hdr} = (g^r, (v \cdot \prod_{j \in S} g_{n+1-j})^r)$, and output (Hdr, K).

Decrypt$(S, \mathsf{Hdr}, i, d_i, \mathsf{EK})$**:** Parse $\mathsf{Hdr} = (C_1, C_2)$, output $K = e(g_i, C_2)/e(d_i \cdot \prod_{j \in S, j \neq i} g_{n+1-j+i}, C_1)$.

Trivially, when one wants to broadcast m different messages to m different sets $S_1, S_2, \ldots, S_m$, one can combine m independent BGW schemes:

Setup(λ)**:** As in the BGW scheme.

Encrypt$(S_1, S_2, \ldots, S_m, \mathsf{EK})$**:** Pick random scalars $r_1, \ldots, r_m \in \mathbb{Z}_p$, and set

$$K_1 = e(g_{n+1}, g)^{r_1}, \ldots, K_m = e(g_{n+1}, g)^{r_m}$$

$$\mathsf{Hdr} = \left((g^{r_1}, (v \cdot \prod_{j \in S_1} g_{n+1-j})^{r_1}), \ldots, (g^{r_m}, (v \cdot \prod_{j \in S_m} g_{n+1-j})^{r_m}) \right)$$

Decrypt$(S_1, \ldots, S_m, \mathsf{Hdr}, i, (\mathsf{EK}, d_i), j)$**:** Extract $C_1 = g^{r_j}$, $C_2 = (v \cdot \prod_{j \in S_j} g_{n+1-j})^{r_j}$ from Hdr and decrypt as in BGW.

3.3 Intuition

One can note that, in the above "trivial" construction, the number of elements in the header is $2m$, and we want to reduce it. A first attempt is by reusing the same random scalar in all the ciphertexts, which leads to a header of size $m + 1$:

$$\mathsf{Hdr} = \left(g^r, (v \cdot \prod_{j \in S_1} g_{n+1-j})^r, \ldots, (v \cdot \prod_{j \in S_m} g_{n+1-j})^r \right).$$

However, this reuse of random coins suffers from a simple attack: the same random coins result in the same session keys for all channels and a subscriber of a channel can decrypt all channels, since the session key is $e(g_{n+1}, g)^r$. Different r's are thus required in each session keys, but not necessarily totally independent. Our idea is to add an element $X_i \in \mathbb{G}$ corresponding to users $i = 1, \ldots, n$, and to adapt the session key and Hdr using scalars x_i, where $X_i = g^{x_i}$, for $i = 1, \ldots, n$,

$$K_1 = e(g_{n+1}, g)^{r + \sum_{j \in S_1} x_j}, \ldots, K_m = e(g_{n+1}, g)^{r + \sum_{j \in S_m} x_j},$$

$$\mathsf{Hdr} = \Bigg(g^r, (v \cdot \prod_{j \in S_1} g_{n+1-j})^{r+\sum_{j\in S_1} x_j}, \ldots,$$

$$(v \cdot \prod_{j \in S_m} g_{n+1-j})^{r+\sum_{j\in S_m} x_j}\Bigg)$$

The above step shorten the header to $m+1$ elements, with no more easy attack. But our goal is to have a constant number of elements:

$$\mathsf{Hdr} = \Bigg(g^r, (v \cdot \prod_{j \in S_1} g_{n+1-j})^{r+\sum_{j\in S_1} x_j} \times \cdots \times$$

$$(v \cdot \prod_{j \in S_m} g_{n+1-j})^{r+\sum_{j\in S_m} x_j}\Bigg)$$

where we essentially multiply all the ciphertexts together. And, magically, it works because a user in a set S_i can cancel out all the terms $(v \cdot \prod_{j\in S_k} g_{n+1-j})^{r+\sum_{j\in S_k} x_j}$ for $k \neq i$ in this product and transform it into his corresponding ciphertext in S_i.

Of course, security has to be proven, this is the goal of the next section to prove the basic selective security. Limitation not to ask decryption nor encryption queries is quite strong, and is the main drawback of the first scheme MCBE_1. And thus, we provide a second construction MCBE_2 that covers strong selective adversaries. For that, we replace $\prod_{j\in S_k} X_j$ by a value outputted by a random oracle on the set S_k and the value g^r at the time of encryption. It will prevent malleability. The *dummy-helper technique* will make the rest.

4. MULTI-CHANNEL BROADCAST ENCRYPTION I – MCBE_1

4.1 Description

Let us now describe formally our first construction MCBE_1. We will then prove its basic selective security.

Setup(λ)**:** The algorithm takes as input the security parameter λ, it generates the global parameters param of the system as follows: Let $\mathbb{G}$ be a bilinear group of prime order p. The algorithm first picks a random generator $g \in \mathbb{G}$ and a random $\alpha \in \mathbb{Z}_p$. It computes $g_i = g^{\alpha^i} \in \mathbb{G}$ for $i = 1, 2, \ldots, n, n+2, \ldots, 2n$. Next, it picks a random $\gamma \in \mathbb{Z}_p$ and sets $v = g^\gamma \in \mathbb{G}$. It also picks additional random scalars $x_1, x_2, \ldots, x_n \in \mathbb{Z}_p$ and sets $X_1 = g^{x_1}, X_2 = g^{x_2}, \ldots, X_n = g^{x_n}$. The master secret key is $\mathsf{MSK} = (g, v, \alpha, \gamma, x_1, x_2, \ldots, x_n)$, while the encryption key (that is private to the broadcaster) is $\mathsf{EK} = (g, v, g_{n+1}, x_1, x_2, \ldots, x_n)$. The public global parameters are $(g_1, \ldots, g_n, g_{n+2}, \ldots, g_{2n}, X_1, X_2, \ldots, X_n)$, whereas the private decryption key is $d_i = v^{\alpha^i}$, for $i \in \{1, \ldots, n\}$. These decryption keys are sent by the **Extract** algorithm.
We note that if a user registers to t different channels, he will possess t different private decryption keys: n will be the product of the number of users and the number of channels.

Encrypt($S_1, S_2, \ldots, S_m, \mathsf{EK}$)**:** Pick a random scalar $r \overset{\$}{\leftarrow} \mathbb{Z}_p$, set $K_k = e(g_{n+1}, g)^{r+\sum_{j\in S_k} x_j}$ for $k = 1, \ldots, m$. Next, set

$$\mathsf{Hdr} = \left(g^r, \prod_{k=1}^{k=m} (v \cdot \prod_{j\in S_k} g_{n+1-j})^{r+\sum_{j\in S_k} x_j}\right).$$

The broadcaster knows $g_{n+1}, x_1, \ldots, x_n$ from EK. It eventually outputs $(\mathsf{Hdr}, K_1, K_2, \ldots, K_m)$.

Decrypt($S_1, \ldots, S_m, \mathsf{Hdr}, i, d_i, k$)**:** Parse $\mathsf{Hdr} = (C_1, C_2)$. If $i \in S_k$ then one computes

$$\begin{aligned}
K_k &= \frac{e(g_i, C_2)}{e(d_i \cdot \prod_{j\in S_k, j\neq i} g_{n+1-j+i}, C_1 \cdot \prod_{j\in S_k} X_j)} \cdot \\
&\quad \frac{1}{\prod_{\ell=1, \ell\neq k}^{\ell=m} e(d_i \cdot \prod_{j\in S_\ell} g_{n+1-j+i}, C_1 \cdot \prod_{j\in S_\ell} X_j)} \\
&= \frac{e(g_i, C_2)}{e(d_i \cdot \prod_{j\in S_k, j\neq i} g_{n+1-j+i}, g^{r+\sum_{j\in S_k} x_j})} \cdot \\
&\quad \frac{1}{\prod_{\ell=1, \ell\neq k}^{\ell=m} e(d_i \cdot \prod_{j\in S_\ell} g_{n+1-j+i}, g^{r+\sum_{j\in S_\ell} x_j})} \\
&= \frac{e(g^{\alpha^i}, \prod_{\ell=1}^{\ell=m} (v \cdot \prod_{j\in S_\ell} g_{n+1-j})^{r+\sum_{j\in S_\ell} x_j})}{e(v^{\alpha^i} \cdot (\prod_{j\in S_k, j\neq i} g_{n+1-j})^{\alpha^i}, g^{r+\sum_{j\in S_k} x_j})} \cdot \\
&\quad \frac{1}{\prod_{\ell=1, \ell\neq k}^{\ell=m} e(v^{\alpha^i} \cdot (\prod_{j\in S_\ell} g_{n+1-j})^{\alpha^i}, g^{r+\sum_{j\in S_\ell} x_j})} \\
&= \frac{e(g^{\alpha^i}, (v \cdot \prod_{j\in S_k} g_{n+1-j})^{r+\sum_{j\in S_k} x_j})}{e(v^{\alpha^i} \cdot (\prod_{j\in S_k, j\neq i} g_{n+1-j})^{\alpha^i}, g^{r+\sum_{j\in S_k} x_j})} \cdot \\
&\quad \prod_{\ell=1, \ell\neq k}^{\ell=m} \frac{e(g^{\alpha^i}, (v \cdot \prod_{j\in S_\ell} g_{n+1-j})^{r+\sum_{j\in S_\ell} x_j})}{e(v^{\alpha^i} \cdot (\prod_{j\in S_\ell} g_{n+1-j})^{\alpha^i}, g^{r+\sum_{j\in S_\ell} x_j})} \\
&= \frac{e((v \cdot \prod_{j\in S_k} g_{n+1-j})^{\alpha^i}, g^{r+\sum_{j\in S_k} x_j})}{e((v \cdot \prod_{j\in S_k, j\neq i} g_{n+1-j})^{\alpha^i}, g^{r+\sum_{j\in S_k} x_j})} \cdot \\
&\quad \prod_{\ell=1, \ell\neq k}^{\ell=m} \frac{e((v \cdot \prod_{j\in S_\ell} g_{n+1-j})^{\alpha^i}, g^{r+\sum_{j\in S_\ell} x_j})}{e((v \cdot \prod_{j\in S_\ell} g_{n+1-j})^{\alpha^i}, g^{r+\sum_{j\in S_\ell} x_j})} \\
&= e(g_{n+1-i}^{\alpha^i}, g^{r+\sum_{j\in S_k} x_j}) = e(g_{n+1}, g^{r+\sum_{j\in S_k} x_j}) \\
&= e(g_{n+1}, g)^{r+\sum_{j\in S_k} x_j}
\end{aligned}$$

We used the relations $d_i = v^{\alpha^i}, g_{n+1-j+i} = g_{n+1-j}^{\alpha^i}$, and $g_{n+1-i}^{\alpha^i} = g_{n+1}$.

Remark 1. *In MCBE_1, the encryption key EK contains $g_{n+1}, x_1, x_2, \ldots, x_n$ and thus cannot be public: this is a private variant of BGW scheme. Actually, g_{n+1} is not really required, but the x_i's would be enough to break the semantic security, and thus cannot be public either. However, the broadcaster does not need to know α, γ to encrypt, and without them it cannot generate decryption keys for users. We can separate the role of group manager (who generates the decryption keys) and broadcaster (who encrypts and broadcasts the content).*

4.2 Security Result

We now prove the semantic security of the first scheme.

THEOREM 1. *The* MCBE_1 *system achieves the basic selective security under the* DBDHE *assumption in* $\mathbb{G}$. *More precisely, if there are* n *users,*

$$\mathbf{Adv}^{\mathsf{b-ind}}(t, q_C) \leq 2 \times \mathbf{Adv}^{\mathsf{dbdhe}}(t', n),$$

for $t' \leq t + (mn + nq_C)T_e$ *where* T_e *is the time complexity for computing an exponentiation and* m *is the maximum number of channels in the system.*

PROOF. Let us assume there exists an adversary $\mathcal{A}$ which breaks the semantic security of our first scheme, we build an algorithm $\mathcal{B}$ that has the same advantage in deciding the DBDHE problem in $\mathbb{G}$. This algorithm $\mathcal{B}$ proceeds as follows:

Init. Algorithm $\mathcal{B}$ first takes as input a DBDHE instance $(g, G, g_1, \ldots, g_n, g_{n+2}, \ldots, g_{2n}, T)$, where T is either $e(g_{n+1}, G)$ or a random element of $\mathbb{G}$. It implicitly defines α: $g_i = g^{\alpha^i}$. $\mathcal{B}$ then runs $\mathcal{A}$, and since we are in the selective model, it receives m sets $S_1, \ldots, S_m$ and an index k that $\mathcal{A}$ wishes to be challenged on.

Setup. $\mathcal{B}$ now generates the public global parameters and private keys d_i, for $i \notin S_k$: it first chooses a random scalar $r \in \mathbb{Z}_p$ and sets $h = g^r$, and $h_i = g_i^r$, for $i = 1, \ldots, n$. One chooses a random index η in S_k, and for $i \in \{1, \ldots, n\} \backslash \{\eta\}$, one chooses a random scalar $x_i \in \mathbb{Z}_p$, and computes $X_i = g^{x_i}$. One eventually sets $X_\eta \stackrel{\text{def}}{=} G / \prod_{i \in S_k \backslash \{\eta\}} X_i = g^{x_\eta}$: All the scalars x_i are known, excepted x_η. $\mathcal{B}$ gives $\mathcal{A}$ the public global parameters:

$$(g_1, \ldots, g_n, g_{n+2}, \ldots, g_{2n}, X_1, X_2, \ldots, X_n)$$

$\mathcal{B}$ has to compute all the private decryption keys d_i except for $i \in S_k$: It chooses a random $u \in \mathbb{Z}_p$ and sets

$$v \stackrel{\text{def}}{=} g^u \cdot (\prod_{j \in S_k} g_{n+1-j})^{-1}$$

$$d_i \stackrel{\text{def}}{=} g_i^u / (\prod_{j \in S_k} g_{n+1-j+i}) =$$

$$g^{u \cdot \alpha^i} \cdot (\prod_{j \in S_k} g_{n+1-j})^{-\alpha^i} = v^{\alpha^i}$$

On can remark that $\mathcal{B}$ can compute, without explicitly knowing α, $\prod_{j \in S_k} g_{n+1-j+i}$ for any $i \notin S_k$, and cannot when $i \in S_k$. Moreover, since $d_i = v^{\alpha^i}$, it satisfies the specifications of the schemes.

Challenge. To generate the challenge for $\mathcal{A}$, $\mathcal{B}$ first computes $\mathsf{Hdr} = (C_1, C_2)$ by setting $C_1 = h$, and

$$\begin{aligned}
C_2 &= (h^u \cdot G^u) \cdot \prod_{\substack{\ell=1 \\ \ell \neq k}}^{\ell=m} \left(h^u \cdot \left(\frac{\prod_{j \in S_\ell} h_{n+1-j}}{\prod_{j \in S_k} h_{n+1-j}} \right) \cdot \right. \\
&\qquad \left. (v \cdot \prod_{j \in S_\ell} g_{n+1-j})^{\sum_{j \in S_\ell} x_j} \right) \\
&= (g^u)^{r + \sum_{i \in S_k} x_i} \cdot \prod_{\substack{\ell=1 \\ \ell \neq k}}^{\ell=m} \left(g^{ur} \cdot \left(\frac{\prod_{j \in S_\ell} g_{n+1-j}}{\prod_{j \in S_k} g_{n+1-j}} \right)^r \right. \\
&\qquad \left. \cdot \left(v \prod_{j \in S_\ell} g_{n+1-j} \right)^{\sum_{j \in S_\ell} x_j} \right) \\
&= \left(v \prod_{j \in S_k} g_{n+1-j} \right)^{r + \sum_{i \in S_k} x_i} \prod_{\substack{\ell=1 \\ \ell \neq k}}^{\ell=m} \left(\frac{g^u}{\prod_{j \in S_k} g_{n+1-j}} \right)^r \\
&\qquad \left(\prod_{j \in S_\ell} g_{n+1-j} \right)^r \left(v \prod_{j \in S_\ell} g_{n+1-j} \right)^{\sum_{j \in S_\ell} x_j} \\
&= \left(v \prod_{j \in S_k} g_{n+1-j} \right)^{r + \sum_{i \in S_k} x_i} \prod_{\substack{\ell=1 \\ \ell \neq k}}^{\ell=m} \left(v \prod_{j \in S_\ell} g_{n+1-j} \right)^r \\
&\qquad \left(v \prod_{j \in S_\ell} g_{n+1-j} \right)^{\sum_{j \in S_\ell} x_j} \\
&= \left(v \prod_{j \in S_k} g_{n+1-j} \right)^{r + \sum_{i \in S_k} x_i} \\
&\qquad \prod_{\substack{\ell=1 \\ \ell \neq k}}^{\ell=m} \left(v \prod_{j \in S_\ell} g_{n+1-j} \right)^{r + \sum_{j \in S_\ell} x_j} \\
&= \prod_{\ell=1}^{\ell=m} \left(v \prod_{j \in S_\ell} g_{n+1-j} \right)^{r + \sum_{j \in S_\ell} x_j}
\end{aligned}$$

We used the following notations and relations $h = g^r$ and $g_{n+1-j}^r = h_{n+1-j}$. Note that $\mathcal{B}$ knows all the values x_i, excepted $x_{i_{k,t}}$, that appears in $h^u \cdot G^u = (v \cdot \prod_{j \in S_k} g_{n+1-j})^{r + \sum_{j \in S_k} x_j}$. To generate session keys, $\mathcal{B}$ first computes, for all $i \neq k$, $K_i = e(g_n, g_1)^{\sum_{j \in S_i} x_j} \cdot e(g_n, h_1)$, and sets $K_k = T \cdot e(g_n, h_1)$. It outputs $(\mathsf{Hdr}, K_1, \ldots, K_m)$ as the challenge to $\mathcal{A}$.

Note that, for $i \neq k$, $K_i = e(g_{n+1}, g)^{r + \sum_{j \in S_i} x_j}$, and, if T is the correct value, $K_k = e(g_{n+1}, G) \cdot e(g_n, h_1) = e(g_{n+1}, g^{\sum_{j \in S_k} x_j}) \cdot e(g_{n+1}, g^r) = e(g_{n+1}, g)^{r + \sum_{j \in S_k} x_j}$. If T is random, the latter is also random.

Guess. $\mathcal{A}$ outputs its guess b' for b. If $b' = b$ the algorithm $\mathcal{B}$ outputs 0 (indicating that $T = e(g_{n+1}, G)$). Otherwise, it outputs 1 (indicating that T is random in $\mathbb{G}_1$). From the above remark, if T is the correct value, $\Pr[\mathcal{B} = 1] = \Pr[b' = b] = (\mathbf{Adv}^{\mathsf{ind}}(\mathcal{A}) + 1)/2$. However, if T is a random value, $\Pr[\mathcal{B} = 1] = 1/2$: $\mathbf{Adv}^{\mathsf{dbdhe}}(\mathcal{B}) = \mathbf{Adv}^{\mathsf{ind}}(\mathcal{A})/2$.

□

5. MULTI-CHANNEL BROADCAST ENCRYPTION II – MCBE_2

We now improve the previous scheme to allow encryption and decryption queries. To this aim, we will need a random oracle.

5.1 Dummy-Helper Technique

First, in order to achieve semantic security, we still have to embed the critical element from the n − BDHE instance in the challenge header related to the specific target set S_k. In the previous scheme, it was implicitly embedded in the X_η, or at least in one of them. But then, if this element is involved in a decryption query, the simulator cannot answer, hence the limitation for the adversary not to ask decryption queries. For the same reason, it was not possible to simulate encryption queries with this critical value.

Using a random oracle, it is possible to embed this element at the challenge time only, and then, instead of a deterministic $\sum_{i \in S_j} x_i$ one can use a random y_j implicitly defined by Y_j given by a random oracle. With the knowledge of the discrete logarithm y_j (excepted in the challenge ciphertext), the simulator is able to answer all encryption queries, but this is still not enough to answer decryption queries: the simulator has no idea about the random scalar r involved in the ciphertext, whereas it has to compute $e(g_{n+1}, g)^r$. But this can be done by adding a dummy set for which the session key can be computed by the simulator. In this case, we apply the *dummy-helper technique* to prove the security.

5.2 Description

Setup(λ)**:** it takes as input the security parameter λ, and generates the global parameters param of the system as follows: Let $\mathbb{G}$ be a bilinear group of prime order p; pick a random generator $g \in \mathbb{G}$ and a random scalar $\alpha \in \mathbb{Z}_p$; compute $g_i = g^{\alpha^i} \in \mathbb{G}$ for $i = 1, 2, \ldots, 2n$; pick a random scalar $\gamma \in \mathbb{Z}_p$ and set $v = g^\gamma \in \mathbb{G}$ and $d_n = v^{\alpha^n}$. The algorithm also uses a random oracle $\mathcal{H}$ onto $\mathbb{G}$.

The master key is $\mathsf{MSK} = (g, v, \alpha, \gamma)$, the private encryption key is $\mathsf{EK} = \mathsf{MSK}$ and the public global parameters are $(g_1, \ldots, g_n, g_{n+2}, \ldots, g_{2n}, d_n)$, whereas the private decryption key is $d_i = v^{\alpha^i}$, for $i \in \{1, \ldots, n\}$. These decryption keys are sent by the **Extract** algorithm.

Encrypt($S_1, \ldots, S_m, \mathsf{EK}$)**:** Pick a random scalar $r \in \mathbb{Z}_p$; set $S_{m+1} = \{n\}$, for each set S_i, for $i = 1, \ldots, m+1$ compute $Y_i = \mathcal{H}(i, g^r)$ ($Y_i = g^{y_i}$, for some unknown scalar y_i), and for $i = 1, \ldots, m+1$

$$K_i = e(g_{n+1}, Y_i) \cdot e(g_{n+1}, g)^r = e(g_{n+1}, g)^{r+y_i}$$

Eventually compute $\mathsf{Hdr} = (C_1, C_2, C_3)$ as follows:

$$\begin{aligned} C_1 &= g^r \\ C_2 &= \prod_{i=1}^{i=m+1} \left(Y_i^{\gamma + \sum_{j \in S_i} \alpha^{n+1-j}} \cdot \left(v \cdot \prod_{j \in S_i} g_{n+1-j} \right)^r \right) \\ &= \prod_{i=1}^{i=m+1} \left(v \cdot \prod_{j \in S_i} g_{n+1-j} \right)^{r+y_i} \\ C_3 &= \mathcal{H}(C_1, C_2)^r \end{aligned}$$

Note that the broadcaster knows both α and γ to compute C_2. It outputs $(\mathsf{Hdr}, K_1, \ldots, K_{m+1})$.

Decrypt($S_1, \ldots, S_m, \mathsf{Hdr}, i, d_i, k$)**:** Set $S_{m+1} = \{n\}$, parse $\mathsf{Hdr} = (C_1, C_2, C_3)$. If $i \in S_k$ then one first checks whether $e(C_1, \mathcal{H}(C_1, C_2)) = e(g, C_3)$, computes $Y_i = \mathcal{H}(i, g^r)$, for $i = 1, \ldots, m+1$, and computes (as in the previous scheme, where the Y_j's replace some products of the X_i's)

$$\begin{aligned} K_k &= \frac{e(g_i, C_2)}{e(d_i \cdot \prod_{j \in S_k, j \neq i} g_{n+1-j+i}, C_1 \cdot Y_k)} \cdot \\ & \quad \frac{1}{\prod_{\ell=1, \ell \neq k}^{\ell=m+1} e(d_i \cdot \prod_{j \in S_\ell} g_{n+1-j+i}, C_1 \cdot Y_\ell)} \\ &= e(g_{n+1}, g)^{r+y_k} \end{aligned}$$

Note that $d_i = v^{\alpha^i}, g_{n+1-j+i} = g_{n+1-j}^{\alpha^i}$, and $g_{n+1-i}^{\alpha^i} = g_{n+1}$.

5.3 Security

THEOREM 2. *The MCBE_2 system achieves the strong selective security under the DBDHE assumption in $\mathbb{G}$. More precisely, if there are n users,*

$$\mathbf{Adv}^{\mathsf{s\text{-}ind}}(t, q_C, q_D, q_E) \leq 2 \times \mathbf{Adv}^{\mathsf{dbdhe}}(t', n) +$$

$$+2 \times \mathbf{Succ}^{\mathsf{cdh}}(t'') + 2/p,$$

for $t' \leq t + (nq_C + nmq_D + nmq_E)T_e + (mq_D + mq_E)T_p + mq_D T_{lu}$ and $t'' \leq t + (q_C + q_D + nmq_E)T_e + (q_D + mq_E)T_p + q_D T_{lu}$, where T_e, T_p are the time complexity for computing an exponentiation, a pairings, T_{lu} is the time complexity for a look up in a list, and m is the maximum number of channels in the system.

PROOF. We organize our proof in three games:

1. **Game 0:** The real strong selective security game between an adversary and a challenger.

2. **Game 1:** This is similar to Game 0 with a following exception: if we denote $\mathsf{Hdr} = (C_1, C_2, C_3)$ the challenge header, then any decryption query on a different header $\mathsf{Hdr}' = (C_1, C_2', C_3')$, but with the same C_1, we answer $\perp$ (*i.e.* invalid ciphertext). We can show that this exception happens with negligible probability under the CDH assumption.

3. **Game 2:** We can now safely answer all decryption queries $\mathsf{Hdr}' = (C_1, C_2', C_3')$ by $\perp$ and the others using either a valid decryption key or d_n. Using the programmability of the random oracle, and thus the knowledge of the y_i, one can easily simulate the encryption queries. Eventually, the semantic security then relies on the DBDHE assumption.

Game 1: In this game, we know all the secret keys, but answer $\perp$ to a decryption query $\mathsf{Hdr}' = (C_1, C_2', C_3')$, with the same first C_1 as in the challenge header. Our algorithm $\mathcal{B}$ is given a CDH instance $g, A = g^{r^*}, B$, and should answer $C = B^{r^*}$. It runs the adversary $\mathcal{A}$:

- since we consider selective attacks only, the target sets are known from the beginning, and $\mathcal{B}$ can thus first generate the challenge header using r^* as random scalar, without knowing it: $C_1 = A$. Since $\mathcal{B}$ knows MSK, and namely α and γ, it can compute the appropriate C_2: $v^{r^*} = A^\gamma$ and $g_i^r = A^{\alpha^i}$. It then programs $\mathcal{H}(C_1, C_2) = g^u$ for a random scalar u and sets $C_3 = A^u$. The triple (C_1, C_2, C_3) is a perfect header;

- answers all the hash queries $\mathcal{H}(A,X)$, for any X, by B^t for some randomly chosen scalar t;

- answers all the other queries with MSK.

Let us now assume that $\mathcal{A}$ asks for a valid decryption query $(S'_1,\ldots,S'_{m'+1},k',\mathsf{Hdr}')$ in which $C'_1 = A$. Since $C'_3 = \mathcal{H}(C_1,C'_2)^{r^*} = B^{r^*\cdot t}$ for a known value t, one can extract $C = B^{r^*} = (C'_3)^{1/t}$, which breaks the CDH assumption. $\mathbf{Succ}^{\mathsf{ind}}(\mathcal{A}) - \mathbf{Succ}_1(\mathcal{A}) \le \mathbf{Succ}^{\mathsf{cdh}}(\mathcal{B})$.

Game 2: We now assume there exists a selective adversary $\mathcal{A}$ that breaks the semantic security of our scheme while decryption queries with the same C_1 as in the challenge are answered by $\perp$. We build an algorithm $\mathcal{B}$ that has twice the advantage in deciding the DBDHE in $\mathbb{G}$. As said above, the programmability of the random oracle will help simulating the encryption queries, and the dummy set will help answering the decryption queries. In game 2.1, the algorithm $\mathcal{B}$ is defined as follows:

Init. Algorithm $\mathcal{B}$ first takes as input a DBDHE instance $(g,G,g_1,\ldots,g_n,g_{n+2},\ldots,g_{2n},T)$ where $T = e(g_{n+1},G)$. It implicitly defines α: $g_i = g^{\alpha^i}$. $\mathcal{B}$ then runs $\mathcal{A}$ to receive m^* sets $S^*_1,\ldots,S^*_{m^*}$ and an index k^* that $\mathcal{A}$ wishes to be challenged on. Note that $n \notin S^*_{k^*}$ because the decryption key d_n is public. $\mathcal{B}$ makes use of a random oracle $\mathcal{H}$ which output is a random element in $\mathbb{G}$, and a hash List is initially set empty list, to store all the query-answer, with additional information, when possible. Namely, for a query q, with answer $Y = g^y$, the tuple (q,Y,y) is stored. Sometimes, y will not be known, and thus replaced by $\perp$.

Setup. $\mathcal{B}$ needs to generate the public global parameters and decryption keys d_i, $i \notin S^*_{k^*}$: it chooses a random $u \in \mathbb{Z}_p$ and sets $v \stackrel{\mathsf{def}}{=} g^u/\prod_{j\in S^*_{k^*}} g_{n+1-j}$. It then computes

$$d_i \stackrel{\mathsf{def}}{=} g_i^u / \prod_{j\in S^*_{k^*}} g_{n+1-j+i} =$$

$$= g^{u\cdot\alpha^i}\cdot\left(\prod_{j\in S^*_{k^*}} g_{n+1-j}\right)^{-\alpha^i} = v^{\alpha^i}$$

Eventually, $\mathcal{B}$ gives $\mathcal{A}$ the public global parameters $(g_1,\ldots,g_n,g_{n+2},\ldots,g_{2n},d_n)$.

Phase 1. Since we now allow encryption and decryption queries, let show how they can be answered. We first start by the hash queries:

1. There are two kinds of useful hash queries, $(j,u) \in \mathbb{Z}_p \times \mathbb{G}$ or $(u_1,u_2) \in \mathbb{G}^2$. But for any query q, if it has already been asked, the same answer is sent back. Otherwise, $\mathcal{B}$ chooses a random scalar $y \stackrel{\$}{\leftarrow} \mathbb{Z}_p$ and sets $\mathcal{H}(q) = g^y$. It appends the appropriate tuple (q,g^y,y) to the hash List.

2. For an encryption query $(S_1,S_2,\ldots,S_m)$, $\mathcal{B}$ makes the ciphertext as follows: it first chooses a random scalar $r \in \mathbb{Z}_p$ and sets $S_{m+1} = \{n\}$, and $Y_i = \mathcal{H}(i,g^r) = g^{y_i}$ for $i = 1,\ldots,m+1$: y_i is obtained from the hash List. To generate $\mathsf{Hdr} = (C_1,C_2,C_3)$, $\mathcal{B}$ sets $C_1 = g^r$, and computes

$$C_2 = \prod_{i=1}^{m+1}\left(v\cdot\prod_{j\in S_i} g_{n+1-j}\right)^{r+y_i}, C_3 = \mathcal{H}(C_1,C_2)^r$$

and $K_i = e(g_n,g_1)^{r+y_i}$, for $i = 1,\ldots,m+1$.

3. For a decryption query $(S_1,\ldots,S_{m+1},\mathsf{Hdr},i,k)$ in the name of user $i \in S_k$, $\mathcal{B}$ decrypts as follows: it first checks whether $S_k \subseteq S^*_{k^*}$ or not. In the negative case, it finds $j \in S_k\backslash S^*_{k^*}$, and using d_j it can decrypt as the decryption oracle would do; in the positive case
   - $\mathcal{B}$ uses d_n to decrypt, using the decryption oracle, and obtain $K_{m+1} = e(g_{n+1},g)^{r+y_{m+1}}$;
   - $\mathcal{B}$ extracts, from the hash List for $\mathcal{H}(m+1,C_1)$, the value y_{m+1}, and computes
   $$L = \frac{K_{m+1}}{e(g_{n+1},g)^{y_{m+1}}} = e(g_n,g_1)^r$$
   - $\mathcal{B}$ extracts, from the hash List for $\mathcal{H}(k,C_1)$, the value y_k, and computes the session key
   $$K_k = L\times e(g_n,g_1)^{y_k} = e(g_{n+1},g)^{r+y_k}$$

Challenge. The challenge has to be generated on the target sets $S^*_1,\ldots,S^*_{m^*}$, with the index k^* for the indistinguishability of the key:

- $\mathcal{B}$ first chooses a random scalar $r^* \in \mathbb{Z}_p$ and sets $h = g^{r^*}$, and $h_i = g_i^{r^*}$ for $i = 1,\ldots,n$;
- it chooses a random scalar $z^* \in \mathbb{Z}_p$ and sets $\mathcal{H}(k^*,h) = Y^*_{k^*} = G/g^{z^*}$, which is the value $Y^*_{k^*} = g^{y^*_{k^*}}$ for an unknown $y^*_{k^*}$. The tuple $((k^*,h),Y^*_{k^*},\perp)$ is appended to the hash List;
- $\mathcal{B}$ asks for the other values $Y^*_i = \mathcal{H}(i,h) = g^{y^*_i}$, for $i = 1,\ldots,k^*-1,k^*+1,\ldots,m^*+1$

Note that $S^*_{m^*+1} = \{n\}$, then $\mathcal{B}$ generates $\mathsf{Hdr}^* = (C^*_1,C^*_2,C^*_3)$ by setting $C^*_1 = h$ and $C^*_3 = \mathcal{H}(C^*_1,C^*_2)^{r^*}$, where (as in the previous proof)

$$C^*_2 = \left(h^u\cdot(Y^*_{k^*})^u\right)\prod_{\substack{\ell=1\\ \ell\neq k^*}}^{\ell=m^*}\left(h^u\cdot\left(\frac{\prod_{j\in S^*_\ell} h_{n+1-j}}{\prod_{j\in S^*_{k^*}} h_{n+1-j}}\right.\right.$$

$$\left.\left.\left(v\prod_{j\in S^*_\ell} g_{n+1-j}\right)^{y^*_\ell}\right)\right)$$

$$= \prod_{\ell=1}^{\ell=m^*}\left(v\prod_{j\in S^*_\ell} g_{n+1-j}\right)^{r^*+y^*_\ell}$$

To generate the session keys, $\mathcal{B}$ first computes

$$K^*_i = e(g_n,g_1)^{y^*_i}\cdot e(g_n,h_1) = e(g_{n+1},g)^{r^*+y^*_i}, \quad i \neq k^*$$

It then sets

$$K^*_{k^*} = \frac{T\cdot e(g_n,h_1)}{e(g_{n+1},g^{z^*})}$$

It gives $(\mathsf{Hdr}^*,K^*_1,\ldots,K^*_{m^*+1})$ as the challenge to $\mathcal{A}$.

Note that when $T = e(g_{n+1}, G)$, with $G = Y^*_{k^*} g^{z^*}$,

$$K^*_{k^*} = \frac{e(g_{n+1}, Y^*_{k^*} g^{z^*}) \cdot e(g_n, h_1)}{e(g_{n+1}, g^{z^*})}$$

$$= e(g_{n+1}, g)^{y^*_{k^*}} \cdot e(g_{n+1}, g)^{r^*} = e(g_{n+1}, g)^{r^* + y^*_{k^*}}$$

Phase 2. $\mathcal{B}$ responds as in the first phase. Note that, if $\mathcal{A}$ asks a decryption query with $C_1 = C_1^*$, $\mathcal{B}$ simply answers $\perp$.

In this game 2.1, the advantage of $\mathcal{A}$ is unchanged, except in case of problem during the programmation of $\mathcal{H}$, which is required once only, and the query has already been asked with probability $1/p$: $\mathbf{Succ}_1(\mathcal{A}) - \mathbf{Succ}_{2.1}(\mathcal{A}) \le 1/p$. In a game 2.2, we replace T by a random element in $\mathbb{G}$: $\mathbf{Succ}_{2.2}(\mathcal{A}) = 1/2$, whereas $\mathbf{Succ}_{2.1}(\mathcal{A}) - \mathbf{Succ}_{2.2}(\mathcal{A}) \le \mathbf{Adv}^{\mathsf{dbdhe}}(\mathcal{B})$.

As a consequence,

$$\mathbf{Succ}^{\mathsf{s-ind}}(\mathcal{A}) \le \mathbf{Succ}^{\mathsf{cdh}}(\mathcal{B}_1) + \mathbf{Adv}^{\mathsf{dbdhe}}(\mathcal{B}_2) + 1/p + 1/2,$$

where $\mathcal{B}_i$ denotes the simulator $\mathcal{B}$ in Game i. □

6. CONCLUSION

We initiate the new research line on multi-channel broadcast encryption. Our objective is to optimize the ciphertext size while maintaining the polynomial-time complexity of all the algorithms. We propose two efficient schemes with constant-size ciphertexts and leave some challenging open problems:

- As already mentioned in the introduction, our schemes share the same weakness as with BGW scheme: the decryption takes into account of all the corresponding public keys of the users in all the target sets. It is thus not quite efficient. A trade-off between the ciphertext size and the decryption time can be done by partitioning the sets for each channel into subsets and then encrypting to each of these subsets. A better solution than this trade-off would definitely be very interesting.

- While privacy concerns imply independent keys for all the channels a user subscribed to, this however also leads to large decryption keys for users (linear in the number of channels). One could prefer to have shorter or even constant-size keys, sacrificing on privacy. This problem is quite related to the above one.

- Our first scheme achieves the basic selective security level in the standard model while our second scheme achieves the strong selective security level, which resists to both CPA and CCA, but in the random oracle model. Ruling out the random oracle seems quite challenging because of the implicit relations between session keys.

Acknowledgments

This work was partially supported by the French ANR-09-VERSO-016 BEST Project, the European Commission through the ICT Program under Contract ICT-2007-216676 ECRYPT II and partially conducted within the context of the International Associated Laboratory Formath Vietnam (LIAFV).

7. REFERENCES

[1] M. Bellare and P. Rogaway. Random oracles are practical: A paradigm for designing efficient protocols. In V. Ashby, editor, *ACM CCS 93: 1st Conference on Computer and Communications Security*, pages 62–73. ACM Press, Nov. 1993.

[2] D. Boneh, C. Gentry, and B. Waters. Collusion resistant broadcast encryption with short ciphertexts and private keys. In V. Shoup, editor, *Advances in Cryptology – CRYPTO 2005*, volume 3621 of *Lecture Notes in Computer Science*, pages 258–275. Springer, Aug. 2005.

[3] C. Delerablée. Identity-based broadcast encryption with constant size ciphertexts and private keys. In K. Kurosawa, editor, *Advances in Cryptology – ASIACRYPT 2007*, volume 4833 of *Lecture Notes in Computer Science*, pages 200–215. Springer, Dec. 2007.

[4] Y. Dodis and N. Fazio. Public key trace and revoke scheme secure against adaptive chosen ciphertext attack. In Y. Desmedt, editor, *PKC 2003: 6th International Workshop on Theory and Practice in Public Key Cryptography*, volume 2567 of *Lecture Notes in Computer Science*, pages 100–115. Springer, Jan. 2003.

[5] A. Fiat and M. Naor. Broadcast encryption. In D. R. Stinson, editor, *Advances in Cryptology – CRYPTO'93*, volume 773 of *Lecture Notes in Computer Science*, pages 480–491. Springer, Aug. 1994.

[6] C. Gentry and B. Waters. Adaptive security in broadcast encryption systems (with short ciphertexts). In A. Joux, editor, *Advances in Cryptology – EUROCRYPT 2009*, volume 5479 of *Lecture Notes in Computer Science*, pages 171–188. Springer, Apr. 2009.

[7] A. B. Lewko, A. Sahai, and B. Waters. Revocation systems with very small private keys. In *2010 IEEE Symposium on Security and Privacy*, pages 273–285. IEEE Computer Society Press, May 2010.

[8] D. Naor, M. Naor, and J. Lotspiech. Revocation and tracing schemes for stateless receivers. In J. Kilian, editor, *Advances in Cryptology – CRYPTO 2001*, volume 2139 of *Lecture Notes in Computer Science*, pages 41–62. Springer, Aug. 2001.

[9] D. H. Phan, D. Pointcheval, S. F. Shahandashti, and M. Strefler. Adaptive CCA broadcast encryption with constant-size secret keys and ciphertexts. In *ACISP 2012*, LNCS 7372, pages 308–321. Springer, 2012.

[10] D. H. Phan, D. Pointcheval, and M. Strefler. Security notions for broadcast encryption. In J. Lopez and G. Tsudik, editors, *ACNS 11: 9th International Conference on Applied Cryptography and Network Security*, volume 6715 of *Lecture Notes in Computer Science*, pages 377–394. Springer, June 2011.

[11] B. Waters. Dual system encryption: Realizing fully secure IBE and HIBE under simple assumptions. In S. Halevi, editor, *Advances in Cryptology – CRYPTO 2009*, volume 5677 of *Lecture Notes in Computer Science*, pages 619–636. Springer, Aug. 2009.

Comparative Study of Multicast Authentication Schemes with Application to Wide-Area Measurement System

Yee Wei Law
Department of EEE
The University of Melbourne
Australia
ywlaw@unimelb.edu.au

Zheng Gong*
School of Computer Science
South China Normal University
Guangzhou 510631, China
cis.gong@gmail.com

Tie Luo
Institute for Infocomm Research
A*STAR, Singapore
luot@i2r.a-star.edu.sg

Slaven Marusic
Department of EEE
The University of Melbourne
Australia
slaven@unimelb.edu.au

Marimuthu Palaniswami
Department of EEE
The University of Melbourne
Australia
palani@unimelb.edu.au

ABSTRACT

Multicasting refers to the transmission of a message to multiple receivers at the same time. To enable authentication of sporadic multicast messages, a conventional digital signature scheme is appropriate. To enable authentication of a multicast data stream, however, an authenticated multicast or multicast authentication (MA) scheme is necessary. An MA scheme can be constructed from a conventional digital signature scheme or a multiple-time signature (MTS) scheme. A number of MTS-based MA schemes have been proposed over the years. Here, we formally analyze four MA schemes, namely BiBa, TV-HORS, SCU+ and TSV+. Among these MA schemes, SCU+ is an MA scheme we constructed from an MTS scheme designed for secure code update, and TSV+ is our patched version of TSV, an MA scheme which we show to be vulnerable. Based on our simulation-validated analysis, which complements and at places rectifies or improves existing analyses, we compare the schemes' computational and communication efficiencies relative to their security levels. For numerical comparison of the schemes, we use parameters relevant for a smart (power) grid component called wide-area measurement system. Our comparison shows that TV-HORS, while algorithmically unsophisticated and not the best performer in all categories, is the most balanced performer. SCU+, TSV+ and by implication the schemes from which they are extended do not offer clear advantages over BiBa, the oldest among the schemes.

*Zheng Gong is also affiliated with State Key Laboratory of Information Security, Institute of Information Engineering, Chinese Academy of Sciences, Beijing 100093, China.

ASIA CCS'13, May 8–10, 2013, Hangzhou, China.

Categories and Subject Descriptors

C.2.0 [**Computer-Communication Networks**]: Security and protection; K.6.5 [**Management of Computing and Information Systems**]: Security and Protection—*Authentication*

General Terms

Algorithms; Security

Keywords

Multicast authentication; multiple-time signature scheme; smart grid; wide-area measurement system

1. INTRODUCTION

A **multicast authentication (MA)** scheme enables receivers of a multicast packet to authenticate the sender, and ensures no entities besides the sender can send authenticated packets to the multicast group. At the core of every MA scheme, lies a signature scheme. Conventional digital signature algorithms such as DSA and ECDSA (see FIPS 186-3) can sign a practically unlimited number of distinct messages with a private key, but they have high computation and memory requirements. The simplest digital signature-based MA scheme appends a digital signature to every multicast message, which is fine for sporadic multicast messages, but computationally prohibitive for multicast data streams. The latest research on digital signature-based MA scheme focuses on *signature amortization* [27], i.e., spreading a signature across many packets such that signature verification cost is amortized and the loss of a small number of packets does not forestall the verification of successfully received packets. However, all signature amortization schemes require a number of packets to be assembled before their collective signature can be verified, incurring a delay that can be problematic for real-time applications.

Real-time requirements have been motivating investigation of **multiple-time signature (MTS)** schemes [22] for multicast authentication. An MTS scheme can sign a fixed number of distinct messages using a public/private key pair. Although they generally produce longer signatures, they have much lower computation and memory requirements than conventional digital signature schemes. MTS-based MA schemes are thus suitable for real-time systems, and one such system is a smart grid component called **wide-area measurement system (WAMS)**, which in fact motivates this work. A number of MTS-based MA schemes have been proposed over the years, yet due to inadequate analysis, a systematic comparison of these schemes is lacking, preventing a scientific approach to the selection of MA schemes for real-time systems, including WAMS.

Our contributions include a rigorous, simulation-validated analysis and a methodical comparison of four MTS-based MA schemes, namely BiBa [20], TV-HORS [31], SCU+ and TSV+ [13]. Among these MA schemes, SCU+ is an MA scheme we constructed from an MTS scheme designed for secure code update [29], and TSV+ [13] is a patched version of Tunable Signing and Verification (TSV) [15]. These MA and MTS schemes are chosen because they are either highly cited or tailored to smart grids. Our analysis fills the gaps of, and at places rectifies or improves existing analyses. Our comparison shows that TV-HORS has the most balanced computational and communication costs; and that contrary to common perception, recent sophisticated designs do not necessarily trump older over-criticized designs. Our work is motivated by the need to find an efficient MA scheme for the WAMS, and hence the parameters of our comparison are tailored to the WAMS. However, our analysis is system-independent, and the parameter values used for comparison can be adapted to any other system. This work is meant to serve as the first step of an ongoing series of comparative studies of MTS-based MA schemes.

This article is organized as follows. Section 2 discusses related work. Section 3 lists the mathematical symbols and notation used in this work. Section 4 presents an overview of MA schemes. We present our analysis of BiBa, TV-HORS, SCU+ and TSV+ in Section 5. In Sections 6 and 7, we introduce the WAMS and compare the four schemes using parameter values relevant for the WAMS. Section 8 concludes the paper.

2. RELATED WORK

Active research on wireless sensor networks for the past decade can be said to have spurred interest in broadcast/ multicast authentication and therefore multiple-time signature (MTS) schemes. This is largely due to the resource constraints on a typical sensor node that favor low computational complexity and small code size. Recently, smart grid researchers are also turning to MTS schemes for real-time multicast authentication.

It is well known that one-time signatures (a subset of MTS schemes) were first considered by Lamport [12]. Among subsequent schemes that improve upon Lamport's impractical construction, BiBa [20] counts as a benchmark for its simple ingenuity. TESLA and its variant μTESLA [21] are more lightweight than BiBa but the use of delayed signature verification in these schemes precludes them from real-time multicast authentication. HORS [23] improves upon BiBa by generating shorter signatures for the same security level, and has inspired many variants (e.g., [14,19]) and extensions. TV-HORS [31] is an extension of HORS to an MA scheme. TSV [15] is both a variant and an extension of HORS, because it is both an MTS scheme and an MA scheme. TV-HORS and TSV were both motivated by smart grid applications, making them ideal candidates for comparison here. SCU [29] was designed for wireless sensor networks, and has an interesting design, so including it in our comparison introduces diversity. Using Katz's taxonomy [10], all schemes studied here are chain-based stateful schemes. As Steinwadt et al. [25] noted, naming a single superior MTS scheme (and accordingly, MA scheme) is nontrivial.

In the following, we discuss related work in the context of the wide-area measurement system (WAMS). In the North American SynchroPhasor Initiative (NASPI), data multicasting from a phasor measurement unit (a WAMS node) to multiple control centers is seen as a necessity. Bobba et al. [3] proposed a Policy-Based Encryption System (PBES) to secure sharing of data, including WAMS data, between utilities. PBES was not designed for securing WAMS traffic itself. NASPI has yet to standardize on an MA scheme for the purpose.

IEC 61850 is a series of standards on *substation automation*, i.e., the automation of data acquisition, control, protection, diagnostics and monitoring functions within substations (where most WAMS nodes are located). As part of the series, IEC 61850-90-5 governs the IEC 61850-compliant transmission of IEEE C37.118-formatted WAMS data. The standard specifies Group Domain of Interpretation (GDOI, see RFC 6407) for securing the distribution of group keys, and IPsec (see RFC 4301) for securing IP multicast using the group keys. However, GDOI does not support mutual authentication among group members [17, Section 4.3]. Furthermore, IPsec relies on a shared group key for encryption, which can be abused by a rogue member to forge messages to the whole group (see RFC 5374). Zhang and Gunter [33] proposed using IPsec for securing multicast WAMS data, but did not point out the pitfalls as we do here.

Researchers at the Future Renewable Electric Energy Delivery and Management (FREEDM) Systems Center implemented TV-HORS on their testbed [32], but did not provide the elaborate justification we provide here.

Recently, Law et al. [13] proposed a key management scheme for the WAMS that specifies TV-HORS for securing multicast data streams. Their choice is based on a simulation-based comparison between TV-HORS and TSV+. Our comparison here covers more schemes, and is both analytical and empirical.

Within the wireless sensor network community, several studies have been performed to evaluate the efficiency of various signature schemes. For example, Seys and Preneel [24] compared the *energy-efficiency* of ECDSA, Lamport-Diffie and HORS one-time signature schemes. Their results show that for less than 15000 signatures, HORS is the most energy-efficient, whereas for more than 15000 signatures, Lamport-Diffie is better.

3. NOTATION AND DEFINITIONS

For the ensuing discussion, the following mathematical notation is used:

H	one-way hash function;
M, c	message and counter respectively;
t	number of elements of a private key tuple;
λ	last index of a one-way chain;
$\mathscr{S}$	see Definition 1;
$\mathscr{C}_\sigma, \mathscr{C}_v$	expected number of hash operations required for signing and verification respectively;
$\mathscr{L}_\sigma$	expected number of bits of a signature;

r	number of signatures generated per epoch;
$\|x\|$	bit-length of x when x is a bit-string;
l_H	bit-length of a *truncated* hash value;
$\mathsf{PRF}(K, M)$	pseudorandom function (PRF) with key K and plaintext M;
$\text{Split}_k()$	function that splits a bit-string into k sub-strings.

Note that a one-way hash function is a hash function that is preimage-resistant and second preimage-resistant. Furthermore, the following definitions are used:

DEFINITION 1. *Assuming a polynomial-time adversary $\mathcal{A}$ can successfully execute an existential forgery on a scheme S with probability p, then the security level of the scheme is $\mathscr{S}(\mathcal{A}, S) \triangleq -\log_2(p)$.*

DEFINITION 2. *[26] Let f be a function from X to Y, and $x_1, \ldots, x_k \in X$. Suppose that values $y_i = f(x_i)$ have been determined for $i = 1, \ldots, k$. Then, f is k-wise independent, if for all $x \in X \setminus \{x_1, \ldots, x_k\}$ and all $y \in Y$,*

$$\Pr[H(x) = y | y_1 \leftarrow H(x_1) \wedge \cdots \wedge y_k \leftarrow H(x_k)] = 1/|Y|.$$

4. MULTICAST AUTHENTICATION USING MULTIPLE-TIME SIGNATURE SCHEMES

Most MTS schemes can be divided into the following parts:

- a private key tuple $(s_1, s_2, \ldots, s_t)$ consisting of t fixed-length random strings;
- a key generation algorithm for generating a public key tuple from a private key tuple;
- a signature generation algorithm that, based on the message to be signed, selects elements of a private key, and generate signature elements from the selected private key elements;
- a signature verification algorithm that checks if the public key elements can be derived from the received signature elements.

Constructing an MA scheme from an MTS scheme requires two key "ingredients". **The first ingredient** is *one-way chains.* Since a key pair can generate only a fixed number of signatures, to sign a message stream of unlimited length, the key pair must be refreshed once its usage limit is reached. The de facto standard technique is to use the private key of an expired key pair as the public key of a new key pair. In the case of BiBa, this means generating the one-way chain as $s_{i,\lambda}, s_{i,\lambda-1}, \ldots, s_{i,0}$, where $s_{i,j-1} = H(s_{i,j}), \forall i = 1, \ldots, t$ and $j = 1, \ldots, \lambda$. As such, $s_{i,j}$ is *both* the private key element corresponding to public key element $s_{i,j-1}$, *and* the public key element corresponding to the private key element $s_{i,j+1}$ (see Fig. 1). Since this technique requires a private key to have the same number of elements as the corresponding public key, this technique does not apply to tree-based MTS schemes (e.g., [11]). **The second ingredient** is clock/time synchronization, which is essential for the security of MTS-based MA schemes, as explained in Fig. 1. In a smart grid, this requirement is satisfiable by the draft standard IEEE PC37.238, which specifies a common profile for the use of IEEE 1588-2008 Precision Time Protocol in power system protection, control, automation and data communication applications utilizing an Ethernet communications architecture.

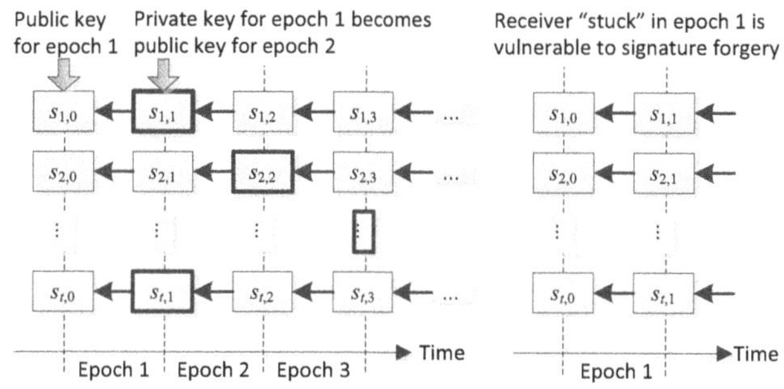

Figure 1: Some (not all) private key elements are disclosed for each signature generated. However, as time passes, an attacker can capture enough signatures (in this example, the "thick boxes") to reconstruct the whole $s_{1,1}, s_{2,1}, \ldots, s_{t,1}$. It is therefore necessary to deprecate private keys by (i) dividing time into epochs, (ii) keeping track of the active private key corresponding to the current epoch, and (iii) synchronizing the clocks of the sender and receivers.

In the approach depicted in Fig. 1, during epoch j, the active private key elements are $s_{1,j}, \ldots, s_{t,j}$. We call this approach **uniform chain traversal**. Using uniform chain traversal, in the first epoch, when a signature containing $s_{i_1,1}, s_{i_2,1}, \ldots, s_{i_k,1}$ is received, H is invoked k times to check $H(s_{i_1,1}) \stackrel{?}{=} s_{i_1,0}$, $H(s_{i_2,1}) \stackrel{?}{=} s_{i_2,0}$, and so on. Assume every private key is used to generate only one signature. In the second epoch, suppose a signature containing $s_{j_1,2}, s_{j_2,2}, \ldots, s_{j_k,2}$ is received, and $\{i_1, \ldots, i_k\} \cap \{j_1, \ldots, j_k\} = \emptyset$. This time, H would have to be invoked $2k$ times to check $H^2(s_{j_1,2}) \stackrel{?}{=} s_{j_1,0}$, $H^2(s_{j_2,2}) \stackrel{?}{=} s_{j_2,0}$, and so on. Therefore, in uniform chain traversal, the verification cost, i.e., expected number of hash operations for verifying a signature, varies from signature to signature.

An alternative approach, which we call **nonuniform chain traversal**, is meant to ensure the verification cost stays at a minimum from signature to signature. Using nonuniform chain traversal, the first active private key is $s_{1,1}, s_{2,1}, \ldots, s_{t,1}$. Without loss of generality, suppose $s_{1,1}, s_{2,1}, \ldots, s_{k,1}$ have been used for a signature, the active private key now becomes $s_{1,2}, s_{2,2}, \ldots, s_{k,2}, s_{k+1,1}, \ldots, s_{t,1}$. As explained in Fig. 1, it is essential that receivers keep track of the active private key, but the loss of synchrony between the epoch counter and the key-chain indices of the active private key means time can no longer be used as a reference. The sender can disclose the key-chain indices of the active private key with every signature, but by blocking packets to a receiver, an attacker can cause a receiver to lose track of the current active private key. Once the attacker has collected enough signatures, it will be able to forge packets to the receiver. In order to keep the verification cost at a minimum for every signature, nonuniform chain traversal inadvertently compromises the receivers' ability to track the active private key and exposes them to signature forgery. Therefore in comparison, uniform chain traversal is more robust and is adopted for all MA schemes in this paper, at the expense of higher verification cost.

The discussion above glosses over a particular caveat of one-way chains. If H is to be k-wise independent (see Definition 2), following Bradford et al.'s analysis [4], the size of the domain of H must be $\Omega((\lambda+1)^{\lfloor k/2 \rfloor})$, where $\lambda+1$ is the length of the one-way chain. There are many ways to expand the domain of H (below, $s_{i,\lambda}$ is randomized, $i = 1, \dots, t$, $j = 1, \dots, \lambda$):

1. Use a separate *salt chain* $K_{j-1} = \mathsf{PRF}(K_j, 0)$, and set $s_{i,j-1} = \mathsf{PRF}(s_{i,j}, K_{i,j})$, $\forall j = 2, \dots, \lambda$ [20].
2. Use a separate salt chain $K_{j-1} = H(K_j)$, and set $s_{i,j-1} = H(s_{i,j} \| K_j)$ [31].
3. Use a synchronized counter c_j, e.g., the epoch counter, and set $s_{i,j-1} = H(s_{i,j} \| c_j)$ [5].
4. Set $s_{i,j-1} = H(s_{i,j} \| s_{i,j+1})$, $s_{i,\lambda-1} = H(s_{i,\lambda})$ [5].

Above, note that replacing $\mathsf{PRF}(K, M)$ with $H(K\|M)$ is valid provided H can be modelled as a random oracle [2].

The performance of an MA scheme is evaluated in terms of the computational complexity of signature generation and signature verification ($\mathscr{C}_\sigma$ and $\mathscr{C}_v$), and in terms of communication overhead ($\mathscr{L}_\sigma$). However, these performance metrics are only meaningful with respect to the achievable security level ($\mathscr{S}$). In other words, we are interested in how MA schemes compare with each other in terms of (i) $\mathscr{L}_\sigma / \mathscr{S}$, (ii) $\mathscr{C}_\sigma / \mathscr{S}$, and (iii) $\mathscr{C}_v / \mathscr{S}$. The *lower* an MA scheme scores in all these metrics, the better the MA scheme. The next section presents our description and analysis of BiBa, TV-HORS, SCU+ and TSV+.

5. ANALYSIS OF MULTICAST AUTHENTICATION SCHEMES

The MA schemes BiBa, TV-HORS, SCU+ and TSV+ are analysed in terms of signing cost, verification cost, signature length, and security level.

For the assessment of security levels, the following models are used: (i) **random oracle model** [2]: hash function outputs are uniformly distributed at random; (ii) **Dolev-Yao model** [8]: an attacker can intercept, modify, fabricate messages. We add the condition that is implicit in the literature (e.g., in [20]): an attacker cannot completely disrupt clock synchronization; for example, an attacker can block all messages to a receiver, but cannot prevent the receiver from advancing its clock from one epoch to the next.

For the evaluation of computational cost for signing, it is assumed that a sender caches all *non-intermediate* one-way chain elements–for BiBa and TV-HORS, this means *all* one-way chain elements ("keys" hereafter). *Intermediate* keys are only applicable to SCU+ and TSV+, and are defined in Section 5.3 and 5.4 respectively. In practice, a sender would employ algorithms like Coppersmith and Jakobsson's [6] to reduce the number of cached keys at the expense of higher signing cost, but our assumption is meant to put all schemes on an equal footing for comparison.

We emphasize that all formulas for $\mathscr{C}_\sigma$, $\mathscr{C}_v$, $\mathscr{L}_\sigma$ below have been validated with simulations.

In Algorithms 1 to 4 below, we denote a private key tuple by $(s_1, \dots, s_t)$, a public key tuple by $(v_1, \dots, v_t)$, a message by M, a counter by c, and a state tuple by $(S_1, \dots, S_t)$.

5.1 BiBa

Algorithm 1 shows the BiBa MTS scheme. Our strategy is to determine $\mathscr{S}$, $\mathscr{C}_\sigma$, $\mathscr{C}_v$ and $\mathscr{L}_\sigma$ in turn.

Algorithm 1: The BiBa MTS scheme

$k \triangleq$ number of elements of a signature tuple
Key generation$(s_1, s_2, \dots, s_t)$:
  $(v_1, v_2, \dots, v_t) \leftarrow (\mathsf{PRF}(s_1, 0), \mathsf{PRF}(s_2, 0), \dots, \mathsf{PRF}(s_t, 0))$
Signing$(M, s_1, s_2, \dots, s_t)$:
  $c \leftarrow 0$
  repeat
    if $\exists I \subseteq \{1, \dots, t\}$ s.t. $|I| = k$, $\mathsf{PRF}(H(M\|c), s_i)$ is the same $\forall i \in I$ **then**
      $\{i_1, i_2, \dots, i_k\} \leftarrow I$
      return $(c, s_{i_1}, s_{i_2}, \dots, s_{i_k})$
    end if
    $c \leftarrow c + 1$
  end repeat
Verification$(M, c, \sigma_1, \sigma_2, \dots, \sigma_k)$:
  if $\sigma_i \neq \sigma_j$, $\forall i \neq j$, $1 \leq i, j \leq k$ **and** $\exists i \in \{1, \dots, t\}$ s.t. $\mathsf{PRF}(\sigma_j, 0) = v_i$, $\forall j \in \{1, \dots, k\}$ **and** $\mathsf{PRF}(H(M\|c), \sigma_j)$ is the same $\forall j \in \{1, \dots, k\}$ **then**
    return "accept"
  else
    return "reject"
  end if

$\mathscr{S}$ and $\mathscr{C}_\sigma$ are related to P_S, an essential parameter of BiBa denoting "the probability that the sender can find a signature in one trial" [20], but this definition is imprecise. There are 4 ways by which the sender can find a signature:

1. increment c until there is exactly one bin with exactly k balls;
2. increment c until there is exactly one bin with at least k balls;
3. increment c until there is at least one bin with exactly k balls;
4. increment c until there is at least one bin with at least k balls.

Our validation of [20, Figure 5] suggests BiBa uses the 4th method above, but [20, Appendix A] indicates the 3rd method is used instead. To simplify our evaluation, we assume the 3rd method is used, i.e., $P_S \triangleq$ the probability of finding at least one bin with exactly k balls. P_S is related to $\mathscr{B}$ (defined in Lemma 1) as

$$P_S = \mathscr{B}(n, t, k)/n^t, \tag{1}$$

where n is the cardinality of the range of PRF. Consistent with our definition of P_S, the original security analysis of BiBa remains valid [20, p. 31], i.e.,

$$\mathscr{S} = \log_2 \frac{n^{rk-1}}{\binom{rk}{k}(n-1)^{rk-k}}. \tag{2}$$

LEMMA 1. *The number of ways to distribute t balls in n bins with at least one bin having exactly k balls is*

$$\mathscr{B}(n, t, k) \triangleq \sum_{i=1}^{\lfloor t/k \rfloor} \left[(-1)^{i-1} \binom{n}{i} (n-i)^{t-ik} \prod_{j=0}^{i-1} \binom{t-jk}{k} \right].$$

PROOF. Let A_i be the event that bin i has exactly k balls. $\bigcup_{i=1}^{n} A_i$ is the event that at least one bin has exactly k balls. Using the inclusion-exclusion principle, we have

$$|\bigcup_{i=1}^{n} A_i| = \sum_{i=1}^{n} |A_i| - \sum_{i,j} |A_i \cap A_j| + \cdots =$$

$$\binom{n}{1}\binom{t}{k}(n-1)^{t-k} - \binom{n}{2}\binom{t}{k}\binom{t-k}{k}(n-2)^{t-2k} + \cdots$$

There are at most $\lfloor t/k \rfloor$ bins with exactly k balls, so there are only $\lfloor t/k \rfloor$ terms in the expression above. □

To compute $\mathscr{C}_\sigma$ and $\mathscr{C}_v$, we replace all invocations of $\mathsf{PRF}(K, M)$ with $H(K \| M)$, which is valid provided H can be modelled as a random oracle [2]. Using (1), and [20, Table 2],

$$\mathscr{C}_\sigma = (1+t)n^t / \mathscr{B}(n,t,k). \tag{3}$$

During signature verification, a message is (i) hashed once together with a counter; (ii) the resultant hash is hashed with each signature element; and (iii) each signature element is verified against the last received signature element on the same one-way chain. The number of hash operations for item (iii) is given by Lemma 2. Therefore,

$$\mathscr{C}_v = 1 + k + \frac{k}{r[1-(1-1/t)^k]}. \tag{4}$$

LEMMA 2. *Denote by r the number of signatures generated per epoch. In uniform chain traversal, to verify a single signature element (of k in total), the expected number of hash operations required is $\left[r\left(1-(1-1/t)^k\right)\right]^{-1}$.*

PROOF. Without loss of generality, suppose a signature contains $s_{1,j}$. The corresponding public key element is one of $s_{1,0}, s_{1,1}, \ldots, s_{1,j-1}$, with "distances" from $s_{1,j}$ being $j, j-1, \ldots, 1$ respectively. Let A_d be the event that the distance is d ($d = 1, \ldots, j$). First, consider A_1, which only occurs if

1. $s_{1,j}$ has *not* been used in any signature in the current epoch yet: this has a probability of $(\frac{t-1}{t})^{xk}$, where x is the number of signatures that have already been generated in the current epoch; and
2. $s_{1,j-1}$ has been used in a signature last epoch: this has a probability of $\left[1-(\frac{t-1}{t})^{rk}\right]$.

So $\Pr[A_1] = (\frac{t-1}{t})^{xk}\left[1-(\frac{t-1}{t})^{rk}\right]$. Similarly, we have $\Pr[A_d] = (\frac{t-1}{t})^{[x+(d-1)r]k}\left[1-(\frac{t-1}{t})^{rk}\right]$. Now, the expected distance (conditioned on x) can be computed as

$$E[d|x] = \lim_{j\to\infty}\sum_{d=1}^{j} d\Pr[A_d]$$
$$= \left(\frac{t-1}{t}\right)^{(x-r)k}\left[1-\left(\frac{t-1}{t}\right)^{rk}\right]\lim_{j\to\infty}\sum_{d=1}^{j} d\left(\frac{t-1}{t}\right)^{drk}.$$

Substituting $q \triangleq \frac{t-1}{t}$ and summing the infinite series in the expression above, we have

$$E[d|x] = q^{(x-r)k}(1-q^{rk})\frac{q^{rk}}{(1-q^{rk})^2} = \frac{q^{xk}}{1-q^{rk}},$$

and finally

$$E[d] = \sum_{x'=0}^{r-1} E[d|x=x']\Pr[x=x'] = \frac{1}{r(1-q^k)}.$$

□

A BiBa signature consists of a counter and k signature elements. Let the maximum value of a counter be c_{max}, then c_{max} happens at a probability of $\epsilon \triangleq (1-P_S)^{c_{max}-1}P_S$. Note that $c_{max} \geq 1 \iff \epsilon \leq P_S$. If we want to use a short string to represent c_{max}, then we should keep ϵ low, e.g., 10^{-4}, and make sure $P_S \geq \epsilon$. With this consideration,

$$\mathscr{L}_\sigma = \lceil \log_2(c_{max}+1)\rceil + k\,l_H$$
$$= \left\lceil \log_2\left(\log_{1-P_S}\frac{\epsilon}{P_S}+2\right)\right\rceil + k\,l_H. \tag{5}$$

5.2 TV-HORS

Algorithm 2 shows the HORS/TV-HORS MTS scheme. Our strategy is to determine $\mathscr{C}_\sigma$, $\mathscr{L}_\sigma$, $\mathscr{C}_v$ and $\mathscr{S}$ in turn.

Algorithm 2: The HORS/TV-HORS MTS scheme

$k \triangleq$ intended number of elements of a signature tuple
Key generation$(s_1, s_2, \ldots, s_t)$:
  $(v_1, v_2, \ldots, v_t) \leftarrow (H(s_1), H(s_2), \ldots, H(s_t))$
Signing$(M, s_1, s_2, \ldots, s_t)$:
  $(i_1, i_2, \ldots, i_k) \leftarrow \mathrm{Split}_k(H(M))$
  $\Sigma \leftarrow (s_{i_1}, s_{i_2}, \ldots, s_{i_k})$ with redundant elements removed
  return Σ
Verification(M, Σ):
  $(i_1, i_2, \ldots, i_k) \leftarrow \mathrm{Split}_k(H(M))$
  if $\exists \sigma \in \Sigma$ s.t. $H(\sigma) = v_i, \forall i \in \{i_1, i_2, \ldots, i_k\}$ **then**
    return "accept"
  else
    return "reject"
  end if

TV-HORS' signing cost is the same as HORS', i.e.,

$$\mathscr{C}_\sigma = 1.$$

Unlike BiBa, a HORS/TV-HORS signature may not always contain k distinct signature elements, because the signing function may produce redundant elements. According to Lemma 3,

$$\mathscr{L}_\sigma = \frac{l_H}{t^k}\sum_{i=1}^{k} i i! \binom{t}{i}\left\{ {k \atop i} \right\}. \tag{6}$$

LEMMA 3. *The expected number of occupied bins if k balls are randomly thrown into t empty bins is $\frac{1}{t^k}\sum_{i=1}^{k} i i!\binom{t}{i}\left\{ {k \atop i} \right\}$, where $\left\{ {\cdot \atop \cdot} \right\}$ denotes a Stirling number of the second kind.*

PROOF. Let A_i be the event that i bins are occupied. There are $\binom{t}{i}$ ways to choose i from t empty bins, and $\left\{ {k \atop i} \right\}$ ways to divide k balls into i bins. Furthermore, there are $i!$ ways to arrange the i chosen bins. Therefore, $\Pr[A_i] = \binom{t}{i}\left\{ {k \atop i} \right\} i!/t^k$. $\sum_{i=1}^{k} i\Pr[A_i]$ gives us the expectation we need. □

Nevertheless, as $t/k \to \infty$, $\mathscr{L}_\sigma \to k$. Therefore to compute $\mathscr{C}_v$, we can re-use Lemma 2, i.e.,

$$\mathscr{C}_v = 1 + \frac{k}{r[1-(1-1/t)^k]}. \tag{7}$$

For estimating $\mathscr{S}$, suppose M_{att} is the message whose signature is to be forged. Let A_i denote the event that the attacker has captured i signature elements from r signatures; and B_j denote the event that $H(M_{att})$ requires j signature elements. The expected probability of forgery is

$$\sum_{j=1}^{k}\sum_{i=1}^{rk}\left(\frac{i}{t}\right)^j \Pr[A_i]\Pr[B_j],$$

where $\Pr[A_i]$ and $\Pr[B_j]$ are given by Lemma 3. Therefore,

$$\mathscr{S} = (rk + k)\log_2 t - \log_2 \sum_{i=1}^{rk} \sum_{j=1}^{k} \frac{i^j i! j! \binom{t}{i} \binom{t}{j} \left\{ {rk \atop i} \right\} \left\{ {k \atop j} \right\}}{t^j}. \quad (8)$$

When $t = 1024$, (8) requires $k \geq 14$ for at least 80 bits of security; whereas the widely used approximation [23] $\mathscr{S} = k \log_2 t - k \log_2(rk)$ requires $k \geq 13$.

5.3 SCU+

Algorithm 3 shows the SCU/SCU+ MTS scheme. Due to SCU's design, nonuniform chain traversal seems like a natural fit for SCU+, but as discussed in Section 5.1, uniform chain traversal is more robust and is used in SCU+ instead. Fig. 2 shows the epoch-j private key as $(s_{1,rj}, \ldots, s_{t,rj})$. All keys between $s_{i,r(j-1)}$ and $s_{i,rj}$, where $i = 1, \ldots, t$ and $j \geq 1$, are called *intermediate* keys. Our strategy is to determine $\mathscr{L}_\sigma$, $\mathscr{C}_\sigma$, $\mathscr{C}_v$ and $\mathscr{S}$ in turn.

Algorithm 3: The SCU/SCU+ MTS scheme

$(S_1, S_2, \ldots, S_t) \triangleq$ state tuple
Key generation$(s_1, s_2, \ldots, s_t)$:
  $S_i \leftarrow r$, $v_i \leftarrow H^r(s_i)$, $\forall i \in \{1, \ldots, t\}$
Signing$(M, s_1, s_2, \ldots, s_t)$:
  $c_z \leftarrow$ number of 0's in $H(M)$
  $I \leftarrow$ set of bit positions in $H(M)\|c_z$ where bit value is 1
  $S_i \leftarrow S_i - 1$, $\forall i \in I$
  return $(H^{S_i}(s_i) : i \in I)$
Verification$(M, \sigma_1, \sigma_2, \ldots, \sigma_k)$:
  $c_z \leftarrow$ number of 0's in $H(M)$
  $I \leftarrow$ set of bit positions in $H(M)\|c_z$ where bit value is 1
  if $k = |I|$ **and**
    $\exists i \in I$, $x_{ij} \in \mathbb{N}^+$, s.t. $H^{x_{ij}}(\sigma_j) = v_i$, $\forall j \in \{1, \ldots, k\}$ **then**
    $v_i \leftarrow \sigma_j$, $\forall H^{x_{ij}}(\sigma_j) = v_i$
    return "accept"
  else
    return "reject"
  end if

$\mathscr{L}_\sigma$ is proportional to the expected number of 1-bits in $H(M)\|c_z$. In $H(M)$, the expected number of 1-bits is $l_\mathcal{H}/2$ (we distinguish $l_\mathcal{H}$ from the normal hash length l_H because typically $l_\mathcal{H} \geq l_H$ by design). In c_z, the expected number of 1-bits varies with $l_\mathcal{H}$, because c_z may be longer than necessary to represent the number of 0's in $H(M)$. In fact, c_z is of length

$$l_c \triangleq |c_z| = \lceil \log_2 (l_\mathcal{H} + 1) \rceil, \quad (9)$$

and by t's definition, $t = l_\mathcal{H} + l_c$. Fig. 3 shows the probability of having 1 at the ith bit of c_z for $l_\mathcal{H} = 128, \ldots, 248$. For the case $l_\mathcal{H} = 160$ (the length of a SHA-1 or RIPEMD-160 hash value), bits 4-8 are 1 at a probability of 1/2, bit 2 is almost always 1, bit 3 is almost always 0, and bit 1 is always 0; in other words,

$$\mathscr{L}_\sigma = \left[l_\mathcal{H} \frac{1}{2} + (l_c - 3)\frac{1}{2} + 1 \right] l_H = \frac{t-1}{2} l_H.$$

For the general case, it is simpler to use the approximation

$$\mathscr{L}_\sigma = t\, l_H / 2. \quad (10)$$

Signing cost varies with the state variables corresponding to the 1-bits of $H(M)\|c_z$. Without loss of generality, let us consider the first bit of $H(M)\|c_z$. Within an epoch, if the first bit is 1 once, which occurs at a probability of $\binom{r}{1}(\frac{1}{2})(\frac{1}{2})^{r-1}$, then the accumulative signing cost is $r-1$. If the first bit is 1 twice, which occurs at a probability of $\binom{r}{2}(\frac{1}{2})^2(\frac{1}{2})^{r-2}$, then the accumulative signing cost is $(r-1)+(r-2)$. By extension, the expected signing cost for the first bit is given by

$$\frac{1}{2^r} \sum_{i=1}^{r} \left[\binom{r}{i} \sum_{j=1}^{i} (r-j) \right] = \frac{3r(r-1)}{8}.$$

The expected signing cost for all bits is then

$$\mathscr{C}_\sigma = 3tr(r-1)/16. \quad (11)$$

When a signature is received, each of the $t/2$ (on average) signature elements needs to be verified. Without loss of generality, let us consider a signature element corresponding to the first bit of $H(M)\|c_z$. Let A_d be the event that this signature element requires d hash operations to verify, which occurs when the past $d-1$ signatures do not contain a signature element corresponding to the first bit of $H(M)\|c_z$, *but* the dth signature in the past does, i.e., $\Pr[A_d] = \frac{1}{2}(1-\frac{1}{2})^{d-1}$, and

$$\mathscr{C}_v = 1 + \frac{t}{2} \sum_{d=1}^{\infty} d \Pr[A_d] = 1 + t. \quad (12)$$

To determine $\mathscr{S}$, we estimate the success probability of signature forgery during epoch j. To forge a signature for M_{att}, an attacker needs to supply $s_{i_1,j}, \ldots, s_{i_k,j}$, where i_1,

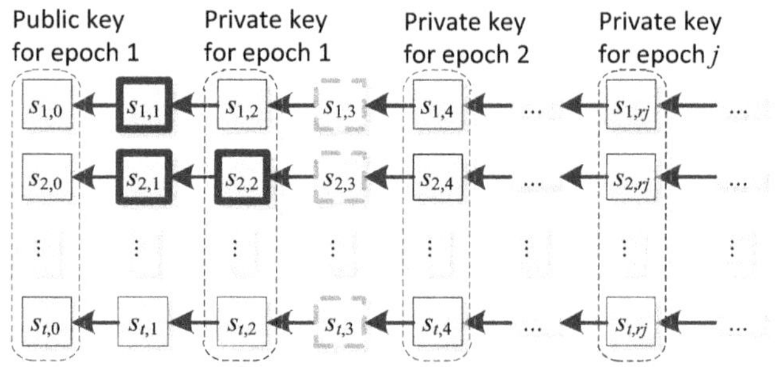

Figure 2: SCU+ with uniform chain traversal and $r = 2$. Suppose in epoch 1, two signatures are received: $(s_{1,1}, s_{2,1})$ and $(s_{2,2})$ (note "thick boxes"). In epoch 2, to verify signature $(s_{1,3}, s_{2,3}, s_{t,3})$ (note "orange dashed boxes"), a receiver checks $H^2(s_{1,3}) \stackrel{?}{=} s_{1,1}$, $H(s_{2,3}) \stackrel{?}{=} s_{2,2}$, and $H^3(s_{t,3}) \stackrel{?}{=} s_{t,0}$.

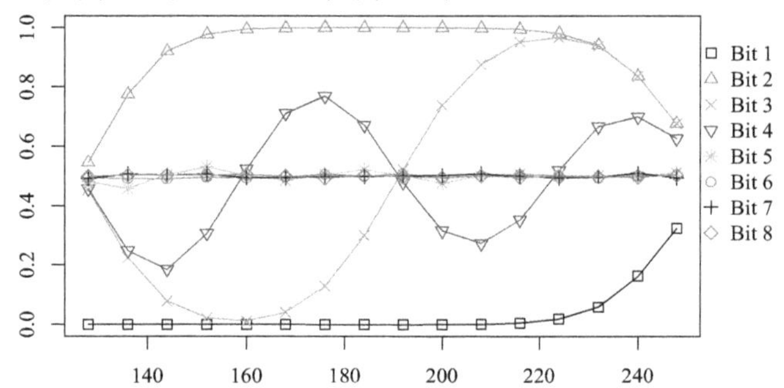

Figure 3: Y-axis: probability of having bit value 1 at the ith bit of c_z. X-asix: $l_\mathcal{H}$.

$\ldots, i_k$ correspond to the positions of 1-bits in $H(M_{att})\|c_z$. Suppose the attacker has already captured r signatures for epoch j. The success probability of signature forgery, Pr[forgery$|k$], is the probability that bit positions $i_1, \ldots, i_k$ are covered by a subset of the r captured signatures. For the case of $l_{\mathcal{H}} = 160$, there is almost always a bit position among $i_1, \ldots, i_k$ that corresponds to bit 2 in c_z (see Fig. 3), so the attacker only has to match $k-1$ bits to the bit positions that are already compromised, i.e., $\Pr[\text{forgery}] = (1 - 1/2^r)^{k-1}$. For the general case, it is simpler to use the approximation $\Pr[\text{forgery}] = (1 - 1/2^r)^k$. Next, let us consider the probability of having k 1-bits in $H(M_{att})\|c_z$, denoted by $\Pr[k]$. If we denote by A_i the event that $H(M_{att})$ has i 1-bits, and by B_i the event that c_z has i 1-bits, then $\Pr[k] = \sum_{i=0}^{\min(k,l_{\mathcal{H}})} \Pr[A_i]\Pr[B_{k-i}|A_i]$. Therefore,

$$\begin{aligned}
\Pr[\text{forgery}] &= \sum_{k'=1}^{t} \Pr[\text{forgery}|k = k'] \Pr[k = k'] \\
&= \sum_{k'=1}^{t} \left[(1 - 1/2^r)^{k'} \sum_{i=0}^{\min(k',l_{\mathcal{H}})} \Pr[A_i]\Pr[B_{k'-i}|A_i] \right] \\
&= \sum_{k'=1}^{t} \left[(1 - 1/2^r)^{k'} \sum_{i=0}^{\min(k',l_{\mathcal{H}})} \frac{\binom{l_{\mathcal{H}}}{i}}{2^{l_{\mathcal{H}}}} I(l_c, l_{\mathcal{H}} - i, k' - i) \right],
\end{aligned}$$

and

$$\begin{aligned}
\mathscr{S} = l_{\mathcal{H}} - \log_2 \Bigg\{ \sum_{k'=1}^{t} \Bigg[&(1 - 1/2^r)^{k'} \\
&\sum_{i=0}^{\min(k',l_{\mathcal{H}})} \binom{l_{\mathcal{H}}}{i} I(l_c, l_{\mathcal{H}} - i, k' - i) \Bigg] \Bigg\},
\end{aligned} \tag{13}$$

where $I(l_c, l_{\mathcal{H}} - i, k' - i)$ is defined by Definition 3.

DEFINITION 3. *$I(l_c, b_1, b_2)$ is 1 if the following has a solution, and 0 if otherwise:*

$$\begin{bmatrix} 2^{l_c-1} & 2^{l_c-2} & \cdots & 2^0 \\ 1 & 1 & \cdots & 1 \end{bmatrix} \begin{bmatrix} x_{l_c-1} \\ \vdots \\ x_0 \end{bmatrix} = \begin{bmatrix} b_1 \\ b_2 \end{bmatrix},$$

$$x_0, \ldots, x_{l_c-1} \in \{0, 1\}.$$

Note the system of equations above does not always have a solution, e.g., when $l_c \geq b_1 = b_2 = 2$. A closed-form expression for $I(l_c, b_1, b_2)$ is unknown.

5.4 TSV+

TSV+ inherits TSV's most notable features:

- For the same k, TSV becomes more secure than HORS by imposing an order/sequence on the signature elements.
- For efficiency, the order is imposed on individual groups, and not across all signature elements.
- So that signature elements are not interchangeable between groups, TSV releases keys at different levels of the one-way chains depending on the group.

However, TSV+ introduces two main enhancements: firstly, so that it is comparable to other MA schemes, we enable TSV+ to support multiple signatures within an epoch; secondly, TSV+ uses uniform chain traversal because it is more robust than nonuniform chain traversal (which is used in TSV), as we explained in Section 4. As shown in Algorithm 4, TSV+ uses a state tuple (like SCU/SCU+ does) to keep track of intermediate keys (between a public/private key pair or a pair of adjacent private keys). The number of intermediate keys is $(wg - 1)$, as shown in Fig. 4, where w is by design the smallest integer such that the probability of a one-way chain being used for more than w out of r signatures in an epoch is less than 10^{-4}. The probability of a one-way chain being used is $\binom{t-1}{k-1}/\binom{t}{k} = \frac{k}{t}$, so w is the smallest integer such that

$$\sum_{i=w+1}^{r} \binom{r}{i} \left(\frac{k}{t}\right)^i \left(1 - \frac{k}{t}\right)^{r-i} < 10^{-4}. \tag{14}$$

Fig. 4 shows the epoch-j private key as $(s_{1,wgj}, \ldots, s_{t,wgj})$, and in that example, $w = 2$. For analysis, our strategy is to first determine $\mathscr{L}_\sigma$ and $\mathscr{C}_\sigma$, which are closely related; then $\mathscr{C}_v$ and $\mathscr{S}$.

Algorithm 4: The TSV+ MTS scheme

$k \triangleq$ number of elements of a signature tuple
$g \triangleq$ number of groups
$n_i \triangleq$ number of $\log_2 t$-bit strings in group i, $\forall i = 1, \ldots, g$, under the constraint $\sum_{i=1}^{g} n_i = k$
$w \triangleq$ smallest integer that satisfies (14)
$(S_1, \ldots, S_t) \triangleq$ state
Key generation$(s_1, s_2, \ldots, s_t)$:
  $S_i \leftarrow wg$, $v_i \leftarrow H^{wg}(s_i)$, $\forall i \in \{1, \ldots, t\}$
Signing$(M, s_1, s_2, \ldots, s_t)$:
  $c \leftarrow 0$
  repeat
    $(h_1, h_2, \ldots, h_g) \leftarrow \text{Split}_g(H(M\|c))$
    $(i_1, \ldots, i_{n_1}) \leftarrow \text{Split}_{n_1}(h_1)$
    $(i_{n_1+1}, \ldots, i_{n_1+n_2}) \leftarrow \text{Split}_{n_2}(h_2)$
    $\cdots$
    $(i_{k-n_g+1}, \ldots, i_k) \leftarrow \text{Split}_{n_g}(h_g)$
    if each of $h_1, \ldots, h_g$ consists of decreasing elements **and** $i_1, \ldots, i_k$ are distinct **then**
      $S_{i_j} \leftarrow S_{i_j} - 1$, $\forall j \in \{1, \ldots, n_1\}$
      $S_{i_j} \leftarrow S_{i_j} - 2$, $\forall j \in \{n_1 + 1, \ldots, n_1 + n_2\}$
      $\cdots$
      $S_{i_j} \leftarrow S_{i_j} - g$, $\forall j \in \{k - n_g + 1, \ldots, k\}$
      return $(c, H^{S_{i_1}}(s_{i_1}), \ldots, H^{S_{i_k}}(s_{i_k}))$
    end if
    $c \leftarrow c + 1$
  end repeat
Verification$(M, c, \sigma_1, \sigma_2, \ldots, \sigma_k)$:
  $(h_1, h_2, \ldots, h_g) \leftarrow \text{Split}_g(H(M\|c))$
  $(i_1, \ldots, i_{n_1}) \leftarrow \text{Split}_{n_1}(h_1)$
  $(i_{n_1+1}, \ldots, i_{n_1+n_2}) \leftarrow \text{Split}_{n_2}(h_2)$
  $\cdots$
  $(i_{k-n_g+1}, \ldots, i_k) \leftarrow \text{Split}_{n_g}(h_g)$
  if each of $h_1, \ldots, h_g$ consists of decreasing elements **and** $i_1, \ldots, i_k$ are distinct **and** $\exists i \in \{i_1, \ldots, i_k\}$, $x_{ij} \in \mathbb{N}^+$, s.t. $H^{x_{ij}}(\sigma_j) = v_i$, $\forall j \in \{1, \ldots, k\}$ **then**
    $v_i \leftarrow \sigma_j$, $\forall H^{x_{ij}}(\sigma_j) = v_i$
    return "accept"
  else
    return "reject"
  end if

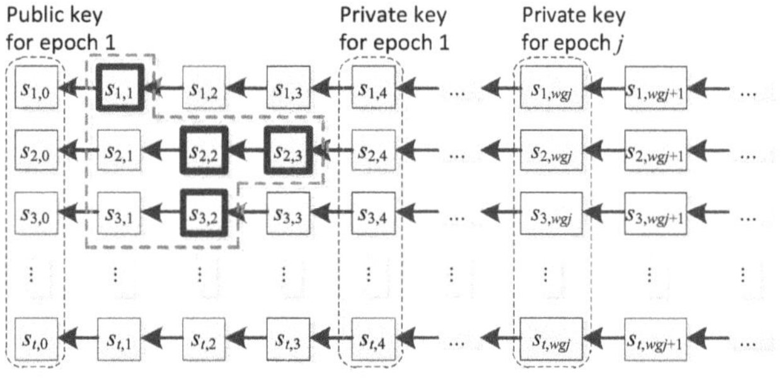

Figure 4: TSV+ with uniform chain traversal, and $k = g = w = 2$, $n_1 = n_2 = 1$, so $wg = 4$ (see Algorithm 4 for definition of symbols). In this example, suppose corresponding to the first message M_1, $H(M_1 \| c_1) = 1\|2$, so the first signature is $(H^{4-1}(s_{1,4}), H^{4-2}(s_{2,4})) = (s_{1,1}, s_{2,2})$. Suppose corresponding to the second message M_2, $H(M_2 \| c_2) = 2\|3$, so the second signature is $(H^{2-1}(s_{2,4}), H^{4-2}(s_{3,4})) = (s_{2,3}, s_{3,2})$.

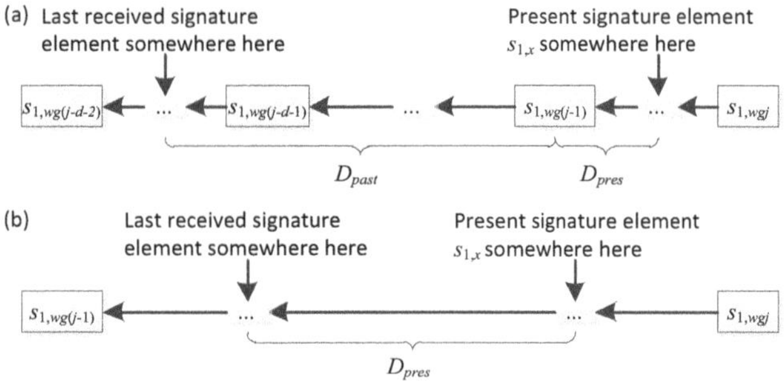

Figure 5: Two cases to be considered for the derivation of $\mathscr{C}_v$ for TSV+.

The probability of finding an acceptable c is correctly given by $P_S \triangleq k!\binom{t}{k}/(t^k \prod_{i=1}^{g} n_i!)$ [15]. Using the same reasoning for (5),

$$\mathscr{L}_\sigma = \left\lceil \log_2 \left(\log_{1-P_S} \frac{\epsilon}{P_S} + 2 \right) \right\rceil + k\, l_H,$$

where ϵ is a small user-defined constant, e.g., 10^{-4}. After finding c, the signer invokes H at a multiplicity that depends on the state tuple $(S_1, \ldots, S_t)$ at the last line of Algorithm 4 – we need to calculate the expected value of this multiplicity. Let $A_{g_1, g_2, \ldots, g_i}$ denote the event that a one-way chain is used for i out of r signatures in an epoch, and for the i signatures, the one-way chain belongs to groups $g_1, g_2, \ldots, g_i$ respectively. Event $A_{g_1, g_2, \ldots, g_i}$ occurs at a probability of

$$\binom{r}{i} \left(\frac{k}{t}\right)^i \left(1 - \frac{k}{t}\right)^{r-i} \frac{n_{g_1} n_{g_2} \cdots n_{g_i}}{k^i},$$

during which the *expected* number of hash invocations is

$$\frac{(wg - g_1) + (wg - g_1 - g_2) + \cdots + (wg - g_1 - g_2 - \cdots - g_i)}{r}.$$

Since there are t one-way chains,

$$\mathscr{C}_\sigma = \frac{1}{P_S} + t \sum_{i=1}^{r} \binom{r}{i} \left(\frac{k}{t}\right)^i \left(1 - \frac{k}{t}\right)^{r-i} \sum_{g_1=1}^{g} \cdots \sum_{g_i=1}^{g} \frac{n_{g_1} \cdots n_{g_i}}{k^i} \frac{(wg - g_1) + \cdots + (wg - g_1 - g_2 - \cdots - g_i)}{r}. \tag{15}$$

For the case $n_1 = n_2 = \cdots = n_g = k/g$, (15) can be simplified as

$$\begin{aligned}\mathscr{C}_\sigma &= \frac{1}{P_S} + \frac{t}{r} \sum_{i=1}^{r} i \binom{r}{i} \left(\frac{k}{t}\right)^i \left(1 - \frac{k}{t}\right)^{r-i} \cdots \\ &\qquad \left[\left(w - \frac{i+1}{4} \right) g - \frac{i+1}{4} \right] \\ &= \frac{1}{P_S} + kwg + k(g+1) \left(\frac{k - kr}{4t} - \frac{1}{2} \right). \end{aligned} \tag{16}$$

(15) can be further simplified when k/t and r are small. In this case, the probability that a one-way chain is used more than once in an epoch is negligible. In an epoch, a one-way chain is chosen at a probability of k/t, and when chosen, it belongs to one of g groups. The probability of the one-way chain belonging to group i is given by n_i/k ($i = 1, \ldots, g$). Therefore,

$$\begin{aligned}\mathscr{C}_\sigma &= \frac{1}{P_S} + t \left[\left(1 - \frac{k}{t}\right) \cdot 0 + \sum_{i=1}^{g} \frac{k}{t} \frac{n_i}{k} (wg - i) \right] \\ &= \frac{1}{P_S} + \sum_{i=1}^{g} n_i (wg - i). \end{aligned} \tag{17}$$

Deriving $\mathscr{C}_v$ is more involved. Suppose the present signature element $s_{1,x}$, which without loss of generality, falls on the first one-way chain. If $s_{1,x}$ belongs to the *first* signature in an arbitrary epoch j, then by definition, the last received signature element on the same one-way chain must be from a past epoch. The expected "distance" between $s_{1,x}$ and the last received signature element (see Fig. 5(a)) is the sum of

- $D_{past} \triangleq$ the expected distance between the last received signature element and $s_{1,wg(j-1)}$; and
- $D_{pres} \triangleq$ the expected distance between $s_{1,wg(j-1)}$ and $s_{1,x}$.

So when $s_{1,x}$ belongs to a first signature, which occurs at a probability of $1/r$, the distance is $D_{past} + D_{pres}$.

Now, suppose $s_{1,x}$ belongs to the *second* signature in epoch j. The last received signature element on the same one-way chain can either be

1. from a past epoch, at a probability of $q \triangleq 1 - k/t$; or
2. from the current epoch, at a probability of $1 - q$.

For the first case, we showed that the expected distance is $D_{past} + D_{pres}$, whereas for the second case, the expected distance is D_{pres} (see Fig. 5(b)). So when $s_{1,x}$ belongs to a second signature, which occurs at a probability of $1/r$, the expected distance is $(D_{past} + D_{pres})q + D_{pres}(1 - q)$.

Applying the reasoning above to the cases that $s_{1,x}$ belongs to the third signature, the fourth signature and so on, we can write the expected distance between $s_{1,x}$ and the last received signature element as

$$\frac{1}{r}\sum_{i=1}^{r}\{(D_{past}+D_{pres})q^{i-1}+D_{pres}(1-q^{i-1})\}$$
$$=\frac{(1-q^r)}{r(1-q)}D_{past}+D_{pres}. \tag{18}$$

The estimation of $\mathscr{C}_v$ is now reduced to the estimation of D_{past} and D_{pres} in (18). D_{pres} is simply

$$D_{pres}=\frac{1}{g}\sum_{i=1}^{g} i=\frac{g+1}{2}. \tag{19}$$

To find D_{past}, let $A_{d,g_1,g_2,\ldots,g_i}$ denote the event that

- the past epoch and the present epoch in Fig. 5(a) are separated by d epochs of no signature, where $d \geq 0$;
- and in the past epoch, i signature elements have been received, which belong to group $g_1, g_2, \ldots, g_i$ respectively.

Event $A_{d,g_1,g_2,\ldots,g_i}$ occurs at a probability of

$$\binom{r}{i}\left(\frac{k}{t}\right)^i\left(1-\frac{k}{t}\right)^{(d+1)r-i}\frac{n_{g_1}n_{g_2}\cdots n_{g_i}}{k^i},$$

during which the distance is

$$dwg+(wg-g_1-g_2-\cdots-g_i).$$

Therefore,

$$D_{past}=\sum_{d=0}^{\infty}\sum_{i=1}^{r}\sum_{g_1=1}^{g}\cdots\sum_{g_i=1}^{g}$$
$$[(d+1)wg-g_1-\cdots-g_i]\Pr[A_{d,g_1,g_2,\ldots,g_i}]. \tag{20}$$

When $n_1=n_2=\cdots=n_g=k/g$, (20) can be simplified as

$$D_{past}=\sum_{d=0}^{\infty}\sum_{i=1}^{r}\binom{r}{i}\left(\frac{k}{t}\right)^i\left(1-\frac{k}{t}\right)^{(d+1)r-i}$$
$$\left[(d+1)wg-\frac{i}{2}g-\frac{i}{2}\right]$$
$$=\frac{2twg-kr(g+1)}{2t[1-(1-k/t)^r]}. \tag{21}$$

Substituting (19) and (21) back into (18), the expected distance between $s_{1,x}$ and the last received signature element becomes simply $\frac{twg}{kr}$. Since a signature has k signature elements,

$$\mathscr{C}_v=twg/r, \tag{22}$$

for the special case of uniform group sizes.

Now, we look at $\mathscr{S}$. In Fig. 4 where $k=2$, we can see that if an attacker (i) manages to capture the signature elements marked by thick frames, namely $s_{1,1}$, $s_{2,2}$, $s_{2,3}$ and $s_{3,2}$, and (ii) block these signature elements from the recipients, then the attacker can forge a signature using *any* two of the elements surrounded by the red dashed contour. In reality, r captured signatures use at most rk distinct one-way chains, but for small k/t and r, it is approximately true that r captured signatures use exactly rk distinct one-way chains (and we used the same approximation for (17)). Let us denote by $\mathcal{G}_i$ the set of one-way chains corresponding to the captured signature elements of groups $i, i+1, \ldots, g$ ($i=1,\ldots,g$). Hence, $\mathcal{G}_g$ has rn_g elements, $\mathcal{G}_{g-1}$ has $r(n_{g-1}+n_g)$ elements, and so on. An attacker successfully forges a signature if he/she is able to ensure the forged group-g signature elements lie on any n_g one-way chains from the set $\mathcal{G}_g$; the forged group-$(g-1)$ signature elements lie on n_{g-1} one-way chains from the set $\mathcal{G}_{g-1}$ that are distinct from previously chosen one-way chains; and so on. In other words, the number of ways to forge a signature is

$$\binom{rn_g}{n_g}\binom{rn_{r-1}+(r-1)n_g}{n_{g-1}}\cdots$$
$$\binom{rn_1+(r-1)(n_g+n_{g-1}+\cdots+n_2)}{n_1}$$
$$=\binom{rn_g}{n_g}\prod_{i=1}^{g-1}\binom{rn_i+(r-1)\sum_{j=i+1}^{g}n_j}{n_i}. \tag{23}$$

When $r=1$, (23) reduces to 1, consistent with intuition. Therefore,

$$\mathscr{S}=\log_2\frac{\binom{t}{k}}{\binom{rn_g}{n_g}\prod_{i=1}^{g-1}\binom{rn_i+(r-1)\sum_{j=i+1}^{g}n_j}{n_i}}. \tag{24}$$

When $r=1$, (24) becomes $\log_2\binom{t}{k}$, which is different from Li and Cao's $k\log_2 t$, because they consider $H(M_{att})$ instead of $H(M_{att}\|c_{att})$.

6. APPLICATION TO WIDE-AREA MEASUREMENT SYSTEMS

The analysis in the previous section forms the basis for comparison of BiBa, TV-HORS, SCU+ and TSV+. For the comparison to be performed in the context of the wide-area measurement system (WAMS), an introduction to the WAMS is given here. A WAMS is essentially a high-speed network of **phasor measurement units (PMUs)**, whose sole objective is to report voltage and current phasor measurements (amplitude, frequency and phase). Given enough real-time phasors, the state of the grid (voltage and phase angle of each bus) can be tracked, giving the utility enhanced "situational awareness" about its system. This enhanced situational awareness provides many advantages: improved operation planning, optimized transmission assets utilization, system stabilization, disturbances containment, etc. In fact, the lack of this level of situational awareness is one of the factors that contributed to the infamous 2003 North America and 2003 Italy blackouts [28,30].

The WAMS consists of four components: (i) synchronized PMUs (also called *synchrophasors*), (ii) phasor data concentrators (PDCs), (iii) wide area network (WAN), and (iv) real-time database and data archiver [16]. Fig. 6 shows the four-layer generic architecture of the WAMS [16]. The PMUs in Layer 1 report voltage and current phasors that are time-stamped with high-precision internal clocks and the Global Positioning System at 10-30 frames per second, enabling the correlation of phasor measurements across a wide grid area. The PMUs transmit the data in the IEEE C37.118 format to the PDCs in Layer 2 via the WAN. The PDCs correlate the time-tagged data, and forward the data to the Applications Data Buffer in Layer 3. The Applications Data Buffer monitors the data for losses, errors and synchronization, in addition to supplying the data in the required format to the applications in Layer 4. Layer 4 consists of the Real Time Database and Data Archiver, which is responsible for collecting and archiving data for post-incident analysis and assessment. Layer 4 also contains applications for monitoring, control and protection functions.

PMUs are required to multicast phasor data to multiple consumers including PDCs for communication redundancy, whereas PDCs at the same hierarchical level are required to share data with each other through multicast [1].

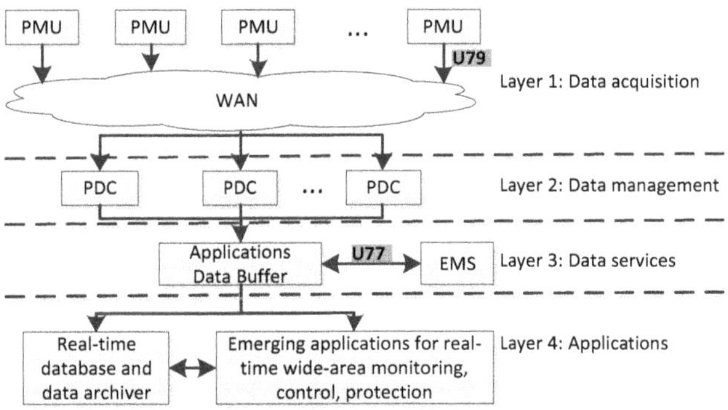

Figure 6: Generic architecture of the WAMS. U77 and U79 are logical interfaces defined in NISTIR 7628 [18].

7. COMPARING MULTICAST AUTHENTICATION SCHEMES

With application to the WAMS in mind, the MA schemes BiBa, TV-HORS, SCU+ and TSV+ are evaluated in terms of the metrics (i) $\mathscr{L}_\sigma \,/\, \mathscr{S}$, (ii) $\mathscr{C}_\sigma \,/\, \mathscr{S}$, and (ii) $\mathscr{C}_v \,/\, \mathscr{S}$; and compared with each other. The parameters of each scheme are set under the following constraints:

- **Security level**: Each scheme must provide a security level of *at least* 80 bits.
- **Signature length**: A recent simulation study [9] suggests that a signature should be *at most* 300 bytes, because a C37.118 frame can be as much as 1200 bytes long while a WAN typically supports a maximum transmission unit (MTU) of 1500 bytes.
- **Hash length**: Since second preimage resistance is weaker than preimage resistance, we are primarily concerned with the former. For any truncated hash of SHA-1, SHA-224, SHA-256 and SHA-512, the actual second preimage resistance is influenced by the preimage length, but an 80-bit truncated SHA-384 hash has a second preimage resistance of exactly 80 bits [7]. So, we set $l_H = 80$ assuming SHA-384 hashes are truncated to 80 bits.
- **Number of one-way chains**: With the exception of SCU+, we fix $t = 1024$ following standard practice [15,20,23,31].

By default, we configure the parameters according to Table 1 to satisfy the constraints above, as well as to minimize the signature length due to the large data volume in a WAMS. For BiBa, two configurations are provided: BiBa_0 is the default configuration, whereas BiBa_1 satisfies the additional constraint $\mathscr{C}_\sigma \leq 10\mathscr{C}_v$. Compared to BiBa_0, BiBa_1 trades off communication efficiency for signing efficiency, but cannot support $r \geq 4$. For SCU+ and TSV+, there are no suitable parameter values that satisfy the above constraints for $r \geq 3$. Two TSV+ configurations are provided: TSV+_0 is the default configuration, whereas TSV+_1 satisfies the additional constraints $\mathscr{C}_\sigma \leq 10\mathscr{C}_v$ and $\mathscr{C}_v \leq 10\mathscr{C}_\sigma$ (i.e., $\mathscr{C}_\sigma$ and $\mathscr{C}_v$ are at most one order of magnitude different from each other). Compared to TSV+_0, TSV+_1 is meant to provide more balanced signing and verification costs.

Table 1: Configurations used for comparison, found through exhaustive search.

Configuration	$r = 1$	2	3	4	5
BiBa_0 (k,n)	9, 1414	12, 618	15, 358	19, 215	24, 137
BiBa_1 (k,n)	11, 256	17, 123	24, 72		
TV-HORS (k)	14	18	21	25	29
SCU+ ($l_{\mathcal{H}}$)	185	405			
TSV+_0 (k,g,n,w)	11, 11, $\{1,1,1,1,1,1,1,1,1,1,1\}$, 1	18, 9, $\{4,4,3,2,1,1,1,1,1\}$, 1			
TSV+_1 (k,g,n,w)	11, 3, $\{2,3,6\}$, 1	18, 9, $\{4,4,3,2,1,1,1,1,1\}$, 1			

Fig. 7 to 9 show the bar charts for $\mathscr{L}_\sigma \,/\, \mathscr{S}$, $\mathscr{C}_\sigma \,/\, \mathscr{S}$, and $\mathscr{C}_v \,/\, \mathscr{S}$. The charts show only $r = 1, 2$ since not all configurations support $r \geq 3$. BiBa_0 is the best performer in signature length but has far poorer efficiency in signing than the others. BiBa_1 produces slightly longer signatures but has significantly better signing efficiency than BiBa_0, so BiBa_1 is more practical. SCU+ is efficient in signing and verification but requires far longer signatures than the others for the same security level. TSV+ (both TSV+_0 and TSV+_1) is more efficient than TV-HORS in signature length when $r = 1$, but not when $r = 2$; moreover, TSV+ is several orders of magnitude slower than TV-HORS in signing and verification. Compared to TSV+_0, TSV+_1 has more balanced signing and verification costs, and is hence better for senders and receivers with similar capabilities. Despite its algorithmic simplicity, TV-HORS is a good performer in all categories. Although proposed after BiBa, SCU and TSV do not offer clear advantages over BiBa.

8. CONCLUSION AND FUTURE WORK

This work is motivated by the need for an efficient multicast authentication (MA) scheme to secure the required real-time multicast traffic within a wide-area measurement system (WAMS). For real-time systems like the WAMS, an MA scheme is best constructed from a multiple-time signature (MTS) scheme rather than a conventional digital signature scheme [15, 31]. Instead of designing yet another MTS-based MA scheme from scratch, this work executes the common sense of first attempting to find suitable candidates among the many MTS-based MA schemes already proposed to date. To this end, we first identify four representative MA schemes, namely BiBa, TV-HORS, SCU+ and TSV+. Among these MA schemes, SCU+ is an MA scheme we constructed from an MTS scheme designed for secure code update [29], and TSV+ is our patched version of TSV [15], an MA scheme which we show to be vulner-

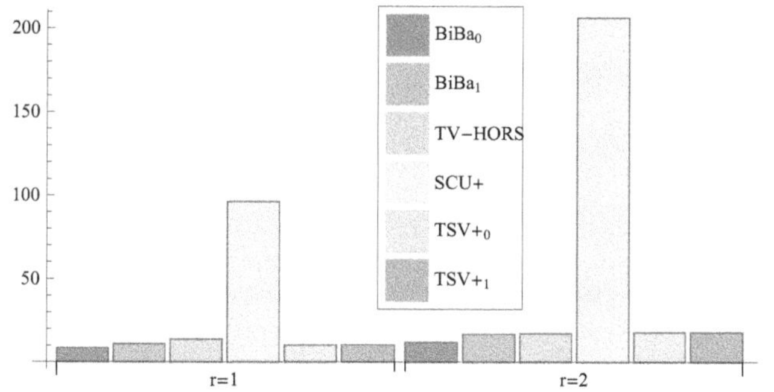

Figure 7: A plot of $\mathscr{L}_\sigma$ / $\mathscr{S}$ against r. Lower is better.

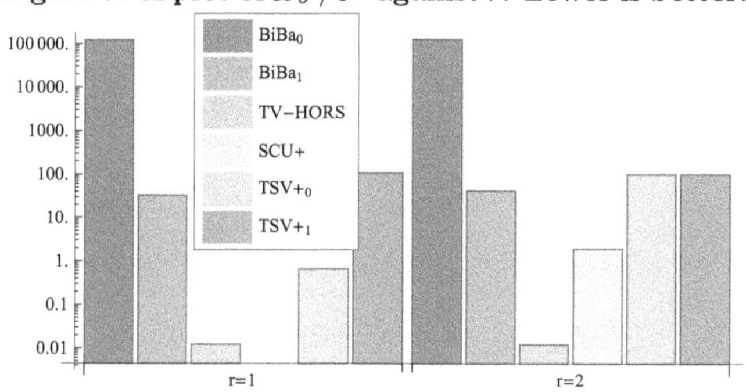

Figure 8: A plot of $\mathscr{C}_\sigma$ / $\mathscr{S}$ against r. Lower is better.

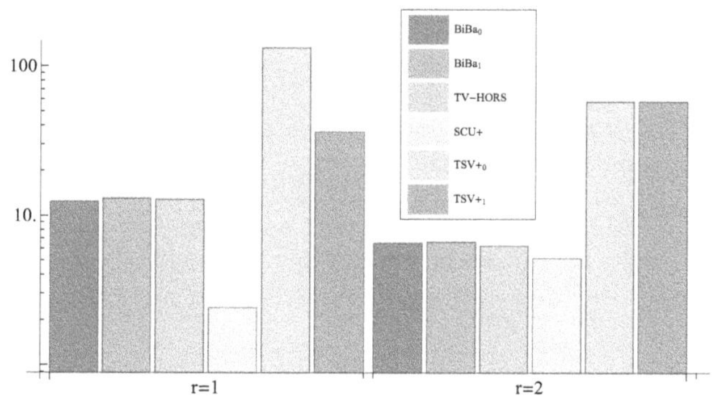

Figure 9: A plot of $\mathscr{C}_v$ / $\mathscr{S}$ against r. Lower is better.

able. We then provide rigorous mathematical analysis of these schemes. Our simulation-validated analysis fills the gaps of, and at places rectifies or improves existing analyses. Based on our analysis, our comparison shows that TV-HORS, while algorithmically unsophisticated, has the most balanced computational and communication efficiencies relative to security levels. SCU+, TSV+ and by extension SCU and TSV do not offer clear advantages over BiBa, the oldest among the studied schemes. As a follow-up to this preliminary study, we aim to expand our analysis and comparison to cover more schemes. Theoretical accounting of memory costs is nontrivial and will be attempted in future work. Just as naming a single superior MTS scheme is nontrivial [25], naming a single superior MA scheme is equally nontrivial. This preliminary work serves as a first step, and already we know that TV-HORS has set a benchmark.

Acknowledgment

The authors would like to thank Dr Gina Kounga and Dr Anthony Lo for reviewing an early draft of this paper, and Prof Ahmad-Reza Sadeghi for shepherding this paper. Yee Wei Law is partly supported by the Institute for a Broadband-Enabled Society, the ARC under the Discovery Project grant DP1095452, and the EC under contract FP7-ICT-2009-257992 "SmartSantander". GONG Zheng is supported by NSFC (61100201, 61070217), Foundation for Distinguished Young Talents in Higher Education of Guangdong (LYM11053), and Guangzhou Science and Technology Plan Project (11C42090777).

9. REFERENCES

[1] M. Adamiak, B. Kasztenny, and W. Premerlani. Synchrophasors: definition, measurement, and application. In *59th Annual Georgia Tech Protective Relaying*, 2005.

[2] M. Bellare and P. Rogaway. Random oracles are practical: a paradigm for designing efficient protocols. In *CCS '93: Proceedings of the 1st ACM conference on Computer and communications security*, pages 62–73. ACM, 1993.

[3] R. Bobba, H. Khurana, M. AlTurki, and F. Ashraf. PBES: a policy based encryption system with application to data sharing in the power grid. In *Proceedings of the 4th International Symposium on Information, Computer, and Communications Security*, ASIACCS '09, pages 262–275, New York, NY, USA, 2009. ACM.

[4] P. G. Bradford and O. V. Gavrylyako. Foundations of security for hash chains in ad hoc networks. *Cluster Computing*, 8(2):189–195, 2005.

[5] P. G. Bradford and O. V. Gavrylyako. Hash chains with diminishing ranges for sensors. *Int. J. High Performance Computing and Networking*, 4(1/2):31–38, 2006.

[6] D. Coppersmith and M. Jakobsson. Almost optimal hash sequence traversal. In *Financial Cryptography*, volume 2357 of *Lecture Notes in Computer Science*, pages 102–119. Springer Berlin / Heidelberg, 2003.

[7] Q. Dang. Recommendation for applications using approved hash algorithms. NIST Special Publication 800-107, Computer Security Division, Information Technology Laboratory, NIST, Feb. 2009.

[8] D. Dolev and A. Yao. On the security of public key protocols. *IEEE Trans. Inf. Theory*, 29(2):198–208, Mar. 1983.

[9] P. Kansal and A. Bose. Smart grid communication requirements for the high voltage power system. In *2011 IEEE Power and Energy Society General Meeting*, pages 1–6, July 2011.

[10] J. Katz. *Digital Signatures*. Springer, 2010.

[11] I. Krontiris and T. Dimitriou. Authenticated in-network programming for wireless sensor networks. *Ad-Hoc, Mobile, and Wireless Network*, 4104:390–403, 2006.

[12] L. Lamport. Constructing digital signatures from a one-way function. Technical Report SRI-CSL-98, SRI Intl. Computer Science Laboratory, Oct. 1979.

[13] Y. W. Law, M. Palaniswami, G. Kounga, and A. Lo. WAKE: Key Management Scheme for Wide-Area

Measurement Systems in Smart Grid. *IEEE Communications Magazine*, Jan. 2013, in press.

[14] J. Lee, S. Kim, Y. Cho, Y. Chung, and Y. Park. HORSIC: An efficient one-time signature scheme for wireless sensor networks. *Information Processing Letters*, 112(20):783–787, 2012.

[15] Q. Li and G. Cao. Multicast authentication in the smart grid with one-time signature. *IEEE Transactions on Smart Grid*, 2(4):686–696, 2011.

[16] C. Martinez, M. Parashar, J. Dyer, and J. Coroas. Phasor Data Requirements for Real Time Wide-Area Monitoring, Control and Protection Applications. White paper, EIPP – Real Time Task Team, Jan. 2005.

[17] C. Meadows and P. Syverson. Formalizing GDOI group key management requirements in NPATRL. In *Proceedings of the 8th ACM conference on Computer and Communications Security*, CCS '01, pages 235–244. ACM, 2001.

[18] NIST. Guidelines for smart grid cyber security. IR 7628, Aug. 2010.

[19] Y. Park and Y. Cho. Efficient one-time signature schemes for stream authentication. *Journal of Information Science and Engineering*, 22(3):611–624, 2006.

[20] A. Perrig. The BiBa one-time signature and broadcast authentication protocol. In *CCS '01: Proceedings of the 8th ACM conference on Computer and Communications Security*, pages 28–37. ACM, 2001.

[21] A. Perrig, R. Szewczyk, V. Wen, D. Culler, and J. Tygar. SPINS: Security Protocols for Sensor Networks. In *Proceedings of the 7th Ann. Int. Conf. on Mobile Computing and Networking*, pages 189–199. ACM Press, 2001.

[22] J. Pieprzyk, H. Wang, and C. Xing. Multiple-time signature schemes against adaptive chosen message attacks. In *Selected Areas in Cryptography*, volume 3006 of *Lecture Notes in Computer Science*, pages 88–100. Springer Berlin / Heidelberg, 2004.

[23] L. Reyzin and N. Reyzin. Better than BiBa: Short One-Time Signatures with Fast Signing and Verifying. In *Information Security and Privacy*, volume 2384 of *LNCS*, pages 144–153. Springer-Verlag, 2002.

[24] S. Seys and B. Preneel. Power consumption evaluation of efficient digital signature schemes for low power devices. In *IEEE International Conference on Wireless And Mobile Computing, Networking And Communications (WiMob'2005)*, pages 79–86, Aug. 2005.

[25] R. Steinwandt and V. I. Villányi. A one-time signature using run-length encoding. *Information Processing Letters*, 108(4):179 – 185, 2008.

[26] D. R. Stinson. Some observations on the theory of cryptographic hash functions. *Designs, Codes and Cryptography*, 38(2):259–277, 2006.

[27] C. Tartary, H. Wang, and S. Ling. Authentication of digital streams. *IEEE Transactions on Information Theory*, 57(9):6285–6303, Sept. 2011.

[28] UCTE. *Final Report of the Investigation Committee on the 28 September 2003 Blackout in Italy*, Apr. 2004.

[29] O. Ugus, D. Westhoff, and J.-M. Bohli. A ROM-friendly secure code update mechanism for WSNs using a stateful-verifier τ-time signature scheme. In *WiSec '09: Proceedings of the second ACM conference on Wireless network security*, pages 29–40. ACM, 2009.

[30] U.S.-Canada Power System Outage Task Force. *Final Report on the August 14, 2003 Blackout in the United States and Canada: Causes and Recommendations*, Apr. 2004.

[31] Q. Wang, H. Khurana, Y. Huang, and K. Nahrstedt. Time valid one-time signature for time-critical multicast data authentication. In *IEEE INFOCOM 2009*, pages 1233 –1241, Apr. 2009.

[32] W. Wang, Y. Xu, and M. Khanna. A survey on the communication architectures in smart grid. *Computer Networks*, 55(15):3604–3629, 2011.

[33] J. Zhang and C. A. Gunter. Application-aware secure multicast for power grid communications. *International Journal of Security and Networks*, 6(1):40–52, 2011.

Gadge Me If You Can

Secure and Efficient Ad-hoc Instruction-Level Randomization for x86 and ARM

Lucas Davi[1,2], Alexandra Dmitrienko[3], Stefan Nürnberger[2], Ahmad-Reza Sadeghi[1,2,3]

[1]Intel Collaborative Research Institute for Secure Computing (ICRI-SC) at TU-Darmstadt, Germany

[2]CASED/System Security Lab Technische Universität Darmstadt Darmstadt, Germany

[3]Fraunhofer Institute for Secure Information Technology Darmstadt, Germany

ABSTRACT

Code reuse attacks such as return-oriented programming are one of the most powerful threats to contemporary software. ASLR was introduced to impede these attacks by dispersing shared libraries and the executable in memory. However, in practice its entropy is rather low and, more importantly, the leakage of a single address reveals the position of a whole library in memory. The recent mitigation literature followed the route of randomization, applied it at different stages such as source code or the executable binary. However, the code segments still stay in one block. In contrast to previous work, our randomization solution, called *XIFER*, (1) disperses all code (executable and libraries) across the whole address space, (2) re-randomizes the address space for each run, (3) is compatible to code signing, and (4) does neither require offline static analysis nor source-code. Our prototype implementation supports the Linux ELF file format and covers both mainstream processor architectures x86 and ARM. Our evaluation demonstrates that *XIFER* performs efficiently at load- and during run-time (1.2% overhead).

Categories and Subject Descriptors

D.4.6 [**Operating Systems**]: Security and Protection

General Terms

Security

Keywords

software diversity; randomization; ASLR; return-oriented programming, return-into-libc

1. INTRODUCTION

Security-critical operations such as online banking are increasingly performed by widespread everyday-software. This makes them an appealing target for various attacks, in particular runtime attacks which subject a process to an adversary's control. Albeit control-flow attacks on software are known for about two decades, they are still one of the major threats to software on desktop PCs and mobile devices. The NIST vulnerability database reported 663 buffer errors in 2011, and 724 for 2012 [26]. The broad introduction of non-executable memory, write-xor-execute ($W \oplus X$) for short, successfully mitigates code injection attacks but gave rise to a form of attacks that re-use existing code by intelligently stitching small code fractions, so-called *gadgets*, together in order to execute arbitrary code. These gadgets are well selected so that they end in an instruction that transfers control to the next gadget, e.g. a `ret` (return) instruction which pops its target off the stack. Hence, the name *return-oriented programming* (ROP [31]).

Since these attacks rely on exact addresses of the instructions they want to abuse, Address Space Layout Randomization (ASLR, e.g. [29]) debuted as the next move in that cat-and-mouse game. By randomizing the base address of loaded code and data in memory, ASLR in theory makes it infeasible for an attacker to predict the location of gadgets in memory. However, low entropy [32] and the fact that a single leaked pointer makes it possible to calculate relative addresses called for yet another step of defense. Such a means of defense has come to light in the form of finer and finer code randomization that, in contrast to ASLR, also shuffles the code itself, not just its base address. Despite the fact that such randomization is a simple idea, its implementation is highly involved and several approaches exist in the literature, ranging from compiler-based solutions [6, 11, 18] to run-time solutions [16] that randomize the program either once or even constantly during its lifetime [13].

As we elaborate in Section 3, most of the existing works have at least one of the following drawbacks: they (i) need access to source code, (ii) do not cover the whole address space, e.g. no loaded libraries and the code segment stays in one block, (iii) do not re-randomize at each process start, or (iv) touch the executable file rendering them incompatible to code signing which is prevalent for commercial software and mandatory in modern app stores.

In order to compare and measure existing software diversity methods, we establish a set of properties that make a randomization solution *ideal*, i.e., featuring the best trade-off among these properties. These properties are: (1) mitigation of code reuse attacks (ROP and return-into-libc),

ASIA CCS'13, May 8–10, 2013, Hangzhou, China.

(2) high diversification entropy, (3) efficient re-randomization, (4) required input information (e.g., source code or other side information), (5) code coverage (code parts that cannot be diversified), (6) compliance to code signing, (7) performance, (8) space consumption (memory and disk) (9) shared library support, and (10) generality vs. specificity, i.e. applicability or limit to certain hardware architectures.

To our surprise, recent software diversity and randomization approaches do *not* fulfil these criteria (as shown in the comparison in Table 1). In particular, they randomize the code only within its segment so that it stays as one block, which means a leaked pointer is always surrounded by the remainder of the code. Moreover, most of the existing work avoids re-randomization for each run, probably due to efficiency concerns. Hence, all binaries of a system remain unchanged leading to an increasing advantage for an adversary over time. Further, many existing solutions touch the main executable file which courrupts a potential digital signature that could be in place.

Our Contribution. In this paper, we present a novel tool, called *XIFER* that adequately addresses the aforementioned requirements for an ideal randomization tool:

- It achieves an instruction granularity randomization.
- It randomizes all sections of an executable and library (not just `.text`) and disperses fractions of the executable segments so that they do not stay in one block, changing *all* relative relations of code and data so that leaked pointers cannot be used to calculate relative addresses.
- The randomization takes place on-the-fly – not requiring an offline static analysis – leading to a different address space layout for each process.
- The high randomization performance needed for an on-the-fly solution stems from a technique we call *partial dis- and reassembly.* This technique leverages the fact that most of the instructions do not reference code or data and hence do not need to be dis- or re-assembled.

We evaluated our prototype for x86 and ARM by using the benchmark suite SPEC CPU2006. Our evaluation results (see Section 6) demonstrate that *XIFER* efficiently performs randomization so that the resulting runtime overhead is only 1.2%, and the linear load-time overhead achieves 5500 kBit/s.

2. BACKGROUND

A simplified view of a code reuse attack is shown in Figure 1. It shows an abstract memory layout of a vulnerable application, where an adversary hijacks the intended execution flow of the application by exploiting a memory-related vulnerability on the stack (e.g., a buffer overflow). In step 1, the adversary exploits the vulnerability to inject a number of pointers (Return Address 1 to 3) on the stack (step 1). Each of these addresses point to a certain code sequence residing in the linked libraries or the executable itself. The executable/library code segment contains a number of functions, where each function consists of several so-called basic blocks (BBLs). A BBL is a sequence of machine instructions with a single entry and exit instruction, where the latter one can be any branch instruction the processor supports (e.g., return, indirect/direct jump or call).

After the adversary has subverted the execution flow by, e.g. overwriting the program's return address, the execution

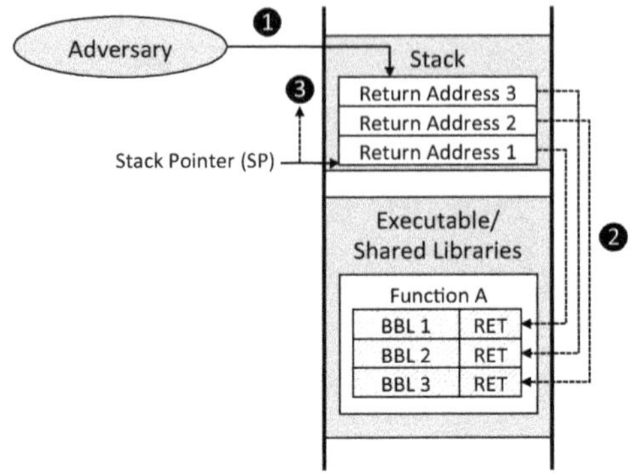

Figure 1: Memory view of a stack-based code reuse attack.

is redirected to BBL 1 (step 2). After BBL 1 has executed, the terminating return instruction (RET) pops the next return address off the stack (Return Address 2), thereby increasing the stack pointer by one memory word, and redirects the execution to BBL 3 and so forth.

In practice, an application includes a large number of BBLs. While return-into-libc attacks would target execution of known *functions*, return-oriented programming enables the combination of arbitrary code sequences. Both attack techniques form a Turing-complete basis for building arbitrary (malicious) programs [31, 34]. Since code-reuse attacks have become the standard runtime attack vector against desktop and mobile computing platforms [35, 14, 30, 17, 25, 20, 7, 19], our goal in this paper is to present an efficient mitigation technique that entirely prevents these attacks.

3. RELATED WORK

To mitigate the threat of code reuse attacks, recent proposals apply various randomization techniques. We define a set of properties that we think are vital for secure, effective and efficient randomization and use them to motivate the design and implementation of our approach. In Table 1, these properties are also compared to existing randomization proposals realized as compiler extensions [6, 11, 18, 4], system-wide base address randomization (ASLR) [29], and binary rewriting tools [21, 28, 16, 36].

P1 – Effectiveness Against Code Reuse Attacks. The solution shall mitigate return-into-libc and ROP, since both have been shown to be Turing-complete [31, 34].

P2 – Entropy. A strong randomization scheme must provide enough entropy to render brute-force attacks infeasible and should change relative distances between code and data as well in order to render a leaked pointer futile.

P3 – Randomization Frequency. A randomization tool should be able to re-diversify a program for each execution. Otherwise, an adversary could acquire knowledge about the memory layout of the diversified program by running the program various times and launching brute-force attacks.

P4 – Input Information. The fact that a randomization solutions requires access to source code or to additional information like debug symbols or relocation information.

P5 – Code Coverage. The recent past has shown that only a piece of un-randomized code is in many cases sufficient to launch a code-reuse attack [12]. Hence, randomizing the whole address space is a necessity.

P6 – Compliance to Code Signing. Code signing is a mandatory feature in nearly all modern app stores [24, 15, 9]. A randomization tool should not touch the executable file in order to keep the signature intact.
P7 – Performance. Performance is influenced by the fact that (1) randomized code pieces must be connected to retain their original order and (2) the reduced locality of code is more prone to cache misses.
P8 – Memory and Disk Space. The randomization solution might need a static analysis or caches to be built and maintained in an offline phase or at run-time. A large size of either of these is undesirable.
P9 – Library Support. Libraries exist to ease application development and to save space on disk. As code-reuse attacks typically leverage code residing in libraries, it is crucial that a randomization tool can also be applied to libraries.
P10 – Target Hardware Architectures. A randomization tool should be general enough to be easily ported to another processor architecture.

Compiler-Based Randomization.
The original randomization approach targets and proposes a compiler-based solution [6]. Recently, Franz et al. [11, 18] have explored the feasibility of a compiler-based approach for large-scale software diversity in the mobile market. The authors suggest that app store providers integrate a multi-compiler (diversifier) in the code production process. However, this approach has two shortcomings: App store providers have no access to the app source code. This requires the multicompiler to be deployed on the developer side, who has to deliver thousands of different app instances to the app store. Secondly, an app instance gets only randomized once. A former randomization work by Bhatkar et al. [4] does not suffer from this shortcoming, because it deploys a source code transformer that enables re-randomization for each run.

In general, compiler based solutions have the potential to provide among all randomization approaches the highest degree of entropy due to the access to source code. However, as argued above, source code is rarely available in practice and current app store models are not compatible to a multicompiler approach.

Binary Instrumentation Based Randomization.
These techniques directly operate on the application binary to perform code randomization. In particular, Kil et al. [21] introduce address space layout permutation (ASLP). The proposed scheme statically rewrites ELF executables to permute all functions and data objects of an executable. Moreover, the Linux kernel has been modified to increase the entropy for the base address randomization of shared libraries. Although, the presented scheme is efficient and supports re-randomization, it is not directly compatible to code signing, provides a lower randomization entropy compared to instruction or basic block (BBL) randomization, and does not apply code randomization to shared libraries.

Recently, Pappas et al. [28] introduced ORP, a static rewriting tool which randomizes instructions and registers within a BBL to mitigate ROP attacks. However, ORP cannot prevent return-into-libc attacks (which have shown to be Turing-complete [34]), since all functions remain at their original position. In contrast, ILR (instruction location randomization) [16] translates each address to a randomized version while executing in a process virtual machine. For this, a program needs to be analyzed and re-assembled during a static analysis phase which induces significant run-time performance and space overhead (the rewriting rules reserve on average 104 MB for only one benchmark of SPEC CPU). ILR also suffers from code coverage deficiencies due to imprecision of the static analysis phase. This concerns in particular the destination addresses of indirect jumps, indirect calls, and function returns. Specifically, ILR does not attempt to resolve destination addresses of indirect calls [16, 6] allowing an adversary to launch return-into-libc like attacks. Both ORP and ILR cannot re-randomize a binary for each program.

In contrast, STIR [36] randomizes and permutes BBLs for each execution and hence provides a higher randomization entropy compared to ORP and ILR. However, STIR still requires a static analysis and rewriting phase which is not compatible to code signing. Moreover, it suffers from a high space overhead, because the file size of a stirred program increases by 73 % on average.

Finally, Guiffrida et al. [13] propose a fine-grained randomization proposal for operating system kernels. Besides stack, heap and code randomization, it allows re-randomization of a module at runtime but is limited to microkernels.

4. DESIGN

In this section, we first present the design of our randomization tool *XIFER*, after which we elaborate on several technical challenges and present our solutions thereof.

4.1 High-Level Design of XIFER

As already alluded to, the goal of our tool is to fulfill all of the aforementioned properties and criteria by randomizing the *complete* address space for every start of a process. This randomization deliberately tears code apart subjecting relative distances within code to change. This is achieved by randomizing the position of each executable and library segment (such as `.text`, `.init`, `.ctors`, `.data`, `.bss`) in memory and additionally twirling the code so that leaked pointers do not reveal anything about the remainder of the code or data.

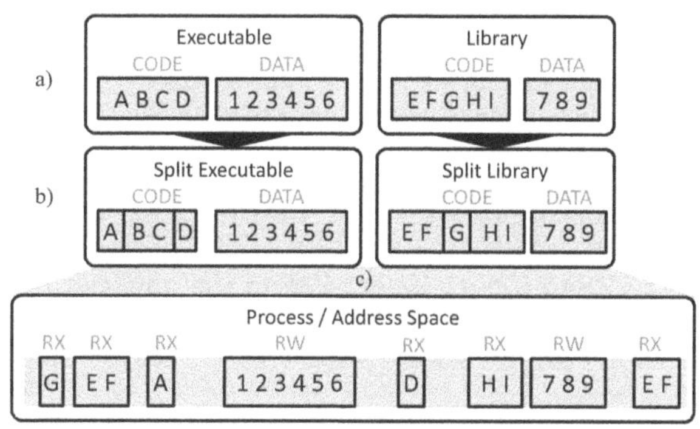

Figure 2: High-level process of the randomization

As depicted in Figure 2, the idea of our randomization approach is to cover the complete executable code of a process, which consists of the executable file itself and all loaded libraries. For that purpose, we apply the necessary randomization steps right before execution of a process starts, but after all the necessary code has been loaded into the address

Criteria and Properties	Multicompiler [6, 11, 18]	Source Transformer [4]	ASLR e.g.,[29]	ASLP [21]	ORP [28]	ILR [16]	STIR [36]	*XIFER*
P1 - Effectiveness								
a) ROP	yes	yes	partially	yes	yes	yes	yes	yes
b) Return-into-Libc	yes	yes	partially	yes	no	no	partially	yes
P2 - Entropy	very high	medium	low	medium	medium	high	high	high
P3 - Frequency	one time	many	many	many	one time	many	many	many
P4 - Input Information	source code	source code	reloc	reloc	none	none	none	reloc
P5 - Code Coverage	high	high	high	high	medium	medium	high	high
P6 - Code Signing Comp.	partially	partially	yes	yes	yes	no	no	yes
P7 - Performance	no impl.*	≈ 11%	0%	0%	≈ 1%	≈ 13%	≈ 6.6%	≈ 1.2%
P8 - Memory Consumption								
a) Memory Footprint	no impl.*	low**	0%	low**	0%	high	≈ 37%	≈ 5%
b) Disk Space	no impl.*	low**	0	low**	0	≈ 104MB	73%	0/7%
P9 - Library Support								
a) Static Libraries	yes	yes	yes	yes	yes	yes	yes	yes
b) Shared Libraries	yes	partially	yes	partially	yes	no	no	yes
P10 - Target Architectures	no impl.*	x86 ELF	x86,ARM, ...	x86 ELF	x86 PE	x86 ELF	x86 ELF/PE	x86/ARM ELF

Table 1: Comparison of Existing Randomization Methods to *XIFER*

* In [6], only an experimental setup has been used. The author mentions that the approach induces negligible performance and space overhead. In contrast, [11, 18] provide no implementation and evaluation.
** No precise numbers provided in [4, 21].

space by the linker (step *a* in Figure 2). In order to randomize the individual code segments and to intermix them (all library code and executable code), the code is cut into arbitrarily small pieces (step *b*). In the last step *c*, all code pieces are spread across the whole address space.

In order to make the individual steps work, we have to overcome several challenges. The most obvious one is the fact, that code cannot simply be re-arranged without breaking its semantics since all control flow information (e.g., branch addresses) is outdated. This challenge is addressed by building upon binary rewriting techniques, that is disassembling code, understanding its semantics and subjecting it to the desired changes and re-assembling it. The binary rewriter's duty is to make sure that all changes are reflected in the output code but also its original semantics are still preserved. With a rewriter in place, we can intelligently choose the points where we cut the code into pieces and the rewriting process preserves the original control flow despite its shuffled layout. A dynamic translation approach – mimicking the changes in a virtual machine – is out of the question because of its poor performance [23, 5, 16] due to its piecemeal and constant translation process.

The existing binary rewriting approaches do not feature load-time static rewriting and are not customizable to our needs. Further, a full-blown rewriter is over the top for our needs and hence does not deliver the performance for an ad hoc translation at process start-up. Hence, we built a rewriter from scratch as detailed in the implementation (Section 5). For the time being, we take the binary rewriter for granted and first explain our randomization solution.

4.2 Randomization

The randomization's goal is that no instruction remains at its original relative distance to any other instruction. This ensures that leaked pointers do not reveal any information about their surrounding code. The locality of code is further kept minimal, i.e. it is split apart, so that an attacker cannot guess anything about the surrounding of any byte of code. Other randomization solutions that keep the code segment as one block always reveal that for any leaked pointer at position x the remainder of the code must be in the interval $]x-s, x+s[$ while s is the size of the code. Our deliberate low locality on the other hand is achieved by splitting the code at certain positions right between two subsequent instructions. These cuts result in code that is broken in pieces whose order can be shuffled. The binary rewriter automatically takes care of keeping formerly subsequent instructions that have been severed in a sequential control flow (see Figure 3).

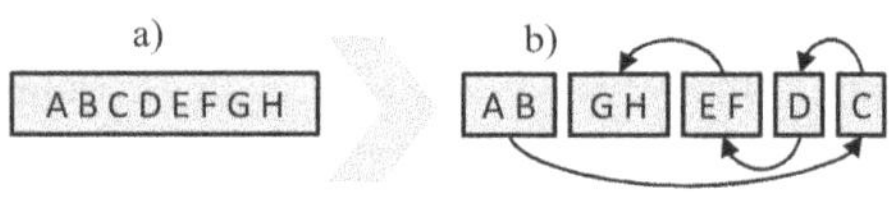

Figure 3: Splitting of code into several interconnected pieces.

4.3 Piece Size

Another challenge is the fact that splitting code in too many pieces imposes a lot of pressure on the processor's instruction cache (as can be seen in the evaluation section 6) since the locality of code has been destroyed. This is why we constrain the amount and position of the cuts:
Positions: When possible, we leverage already existing control transfer instructions (e.g. `jump`, `call`) as a splitting boundary. This has two advantages: First, there is no need to connect the severed control flow later, when the pieces have been moved away from each other, because there is already a control flow instruction that can be adjusted. Second, when using an already existing control flow instruction, the cache miss penalty is most likely to be identical to the original program.
Amount: The maximum possible entropy of a 32-bit user mode process however is limited to 2^{31} (2 GB). The entropy of 13 permutations is already larger than that ($13! = 6,227,020,800 \approx 2^{32.5}$). Hence, it actually makes no sense to split more than 12 times[1] for 32-bit system or more than 16 times for a 64-bit system (with 48 bit address user space).

[1] 12 times splitting makes 13 pieces, $log_2 13! \approx 32.54$ bits of entropy

4.4 Compliance

As described in Section 3, an ideal randomization tool needs to comply with several criteria. While not all of the criteria can be fulfilled at the same time, in this section, we explain why we think the best trade-off of them is met by *XIFER*.

P1 – Effectiveness. The most important property (P1) concerns the effectiveness of our solution against code reuse attacks, which is the main objective of this paper. We achieve complete mitigation of code-reuse attacks by diversifying the location of each instruction, so that the addresses needed to mount a successful code-reuse attack remain unknown. In addition, *XIFER* is not vulnerable to *disclosure attacks*, i.e. the address of a known function is leaked to the adversary allowing him to revert the memory structure of the executable, a library or even the whole application. This is due to the fact that all offsets between functions, basic blocks of code and even instructions have been randomly changed. Even if the permutation and the memory layout of one specific instance is known, the adversary cannot assume that the target device is using this instance, since our diversification is re-applied for each application run.

P2 – Entropy. Our diversification techniques also yield very high diversification entropy, i.e. every instruction is moved from its original location, and their relative distance to each other is completely random. With the permutation of arbitrarily small code pieces, we achieve an entropy of $n!$, while n denotes the number of pieces the code was divided into. While n can be arbitrarily large, a suitable position to cut is a boundary of a basic block. Our tests have shown that on average 15.5% of all instructions are such boundaries. This means that for a sample 1000-instruction-binary we end up with 156 code pieces to shuffle. This entropy of 156! is already higher than the ultimate ASLR solution that would provide an entropy of 2^{48} on a 64-bit system. Moreover, besides code transformation, we randomize the location of each data section to achieve a fully-randomized memory layout.

P3 – Frequency and P6 – Signing. Since *XIFER* performs its operation entirely at runtime, we are able to automatically randomize a program for each run (*P3*) while keeping compliance to code signing (*P6*).

P4 – Input Information. Instructions may reference other data or other code (control flow). These references need to be detected reliably in order to adjust them accordingly to the new randomized position. In most cases, references are encoded in the instruction (e.g. a branch) and can be detected automatically by disassembling the code. However, the value of a register at run-time could represent an address which is used to direct control flow (indirect jump) or to dereference data (pointer). Calculating the value of that register beforehand is a highly involved task. For instance, `mov $0x8067ab, %eax` might represent an arbitrary number being moved to `%eax` or it might be an address to which `eax` should point. Relocation information solves this problem by pointing to positions in code and data that represent addresses. Fortunately, all address references, including C++ vTables, and indirect jumps benefit from that identification and can be rewritten reliably. There exists literature about how to reliably disassemble and rewrite code when no relocation information is available [33]. However, these solutions suffer from significant space problems, and cannot accurately determine indirect jump targets. Hence, for the time being, our solution was implemented using relocation information.

Note that *P7*-*P10* are implementation-related and we will show how *XIFER* addresses these criteria in the implementation section (Section 5) and the performance evaluation (Section 6). To meet the code coverage property (*P5*) our solution faces and tackles several technical challenges that are addressed in the following subsection.

4.5 Technical Challenges

Our goal is to achieve all properties mentioned in Section 3. This poses several technical challenges that are mainly related to the code coverage criterion (*P5*). Without loss of generality, examples herein are given in x86 assembler.

CH1 – Function Returns.

The position of a split at the end of one and the beginning of the respective next basic block is not always a trivial case. Simple cases are instructions that unconditionally transfer control to another point in the program, e.g., a jump instruction (`jmp 0x1234`) for x86. These instructions can simply be rewritten to their new address in memory to which the original target has been moved. However, this does not hold for function calls or conditional branches as they feature an *implicit fall-through* control flow, i.e. the control flow will continue at the next instruction after returning from the call. The same is true for conditional branches which *might* continue at the next instruction depending on the outcome of a comparison. In either case, the subsequent instruction to which the control flow would implicitly fall through can potentially be moved away because it is now being part of a different code piece that has been shuffled away.

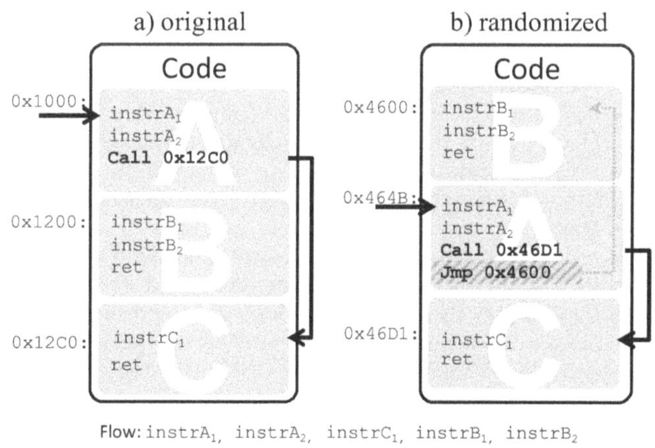

Figure 4: Implicit fall-through control flow requires the injection of jumps in order to retain the original sequence.

Consider the example shown in Figure 4. After the execution returns from the call to `0x12C0` (original), it will continue execution after the `call` instruction by falling through to code piece B (`instrB1, instrB2`). The randomized version would also return to the position right after the `call` where it left off. In contrast to the original, the code has been moved away and the control flow would fall through to code piece C, which is wrong. Consequently, we need to insert a jump in order to connect control flow with the original code piece that now resides at a different position.

CH2 – Position-Independent Code (PIC).
PIC can start execution immediately without the need for certain instructions to be adjusted. It became widely adopted with the introduction of ASLR in modern operating systems. The avoidance of addresses is achieved by using only relative addressing for code branches, function calls and even data. Consequently, the absolute addresses in memory may shift, i.e. the base address of code and data may change, but the relative distances within code and data must stay intact in order for the relative calculations not to become stale. Typically, the code calculates its own address in memory and references code and data relative to its current position. Data is referenced using the Global Offset Table (GOT) by knowing the relative distance to the GOT and indexing the GOT to read and write data.

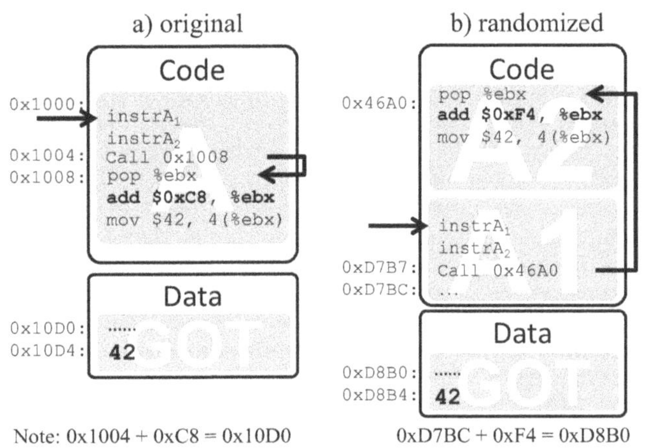

Figure 5: Position-independent code (PIC) needs to be adjusted as it assumes relative relations inside the code.

As we move around code pieces to deliberately change code layout and hence their relative distances to each other, the calculation of the GOT is no longer correct. This necessitates a detection of PIC and its correction in order for the data references to be still valid after the shuffling of code pieces. Consider the example in Figure 5 which depicts a common write operation to data residing in the GOT. In this example, a `call` instruction was issued by the compiler that targets the next instruction. This next instruction (`pop %ebx`) in turn pops the return address off the stack that has been placed there by the `call`. With this trick, register `ebx` now holds the absolute address in memory of where the call should return, in this case the absolute position of the `pop` instruction.

In the randomized version (Figure 5b), the instructions have not only been split apart but also the `pop %ebx` instruction no longer pops its *own absolute address* off the stack. To compensate for this effect, we correct the offset added to the register `ebx` in order to restore the reference to the GOT. To apply this correction in a general fashion, we detect call targets that do not return but rather save a position of the stack that holds the return address. We then follow the register to find the instruction that adds the offset to the found position. When the original offset plus its calculated own position equals the GOT, we have found such a case and correct that offset by rewriting the instruction (e.g. `add $0xF4, %ebx`).

CH3 – C++ Exceptions.
When the *GCC* compiler compiles C++ exceptions, a so-called *unwind frame* is created for every function in a special section called `.eh_frame`. It contains information about how to restore the stack and registers when returning to the point in code that actually catches the exception. This list works similar to a hash map which is indexed with the position in memory that has thrown the exception and provides information about which function catches that particular exception. The information about how to unwind the stack is stored as a stack machine bytecode that is interpreted at run-time while the exception is thrown. This *GCC*-specific machine language incorporates code load addresses and assumes relative offsets, e.g. how to get to the beginning of a function and where the frame pointer is stored. However, since we randomized the code this information is no longer valid. The `.eh_frame` would need to be rewritten according to the changed layout, which we actually did not implement, and leave as future work. Note that with the exception of ILR [16], *GCC* C++ exceptions are also not supported by other randomization approaches.

CH4 – Intermixed Code and Data.
Compilers often optimize code by aligning functions in memory so that they start at the beginning of a cache line. The inevitable gap before aligned functions is sometimes filled with data in the middle of code. A typical fall-through disassembly[2] is thus not possible, as it would interpret alignment zeros (garbage) or intentional data as instructions. To prevent this, code must not be disassembled in a linear fashion but should rather recursively follow the control-flow with respect to indicators for the start of an instruction. There are plenty of such indicators, e.g. targets of control-flow such as function calls, conditional and unconditional branches. We leverage this information to re-align the disassembly process based on discovered control-flow targets and sanity checks that ensure all references from and to code and data stay in their respective segment and target only the beginning of an instruction. We are aware of the fact that from a theoretical point of view, there might be corner cases in which such a disassembly might not be complete. However, we could not find any such cases and a reliable disassembly is not the main topic of this paper but a mere building block. However, it should be noted that an incorrect disassembly could lead to program crashes.

CH5 – Shared Libraries.
Typically, applications link by default to a number of shared libraries (e.g., the Unix standard library libc). Due to space and efficiency reasons, the code pages of these libraries are shared among the various simultaneously running processes. Hence, when every process gets a randomized version of the same shared library, the operating system can no longer benefit from code sharing across multiple processes because their content has changed. For flexibility reasons, we offer the user or system administrator to decide himself whether to favor security (i.e., allowing *XIFER* to generate multiple randomized library versions on-demand for each application) or benefit from shared library memory savings (i.e., randomizing a library once at boot or in a static offline phase). In particu-

[2]The process of disassembling instructions from the beginning of the code in a linear fashion to its end.

lar, for the latter case, we have implemented the possibility to write the randomized libraries back to disk as ELF files. Since shared system libraries are typically not included in digital signatures for signed applications, randomizing them (e.g. every boot or in an offline phase) does not pose a threat to the signatures as long as the main executable remains untouched. Due to our support of shared libraries, we accurately address *P9*.

5. IMPLEMENTATION

The very heart of *XIFER* is the binary rewriter that is responsible for keeping the program semantics despite all the changes we subject the program to. We implemented the rewriter for both mainstream processor architectures Intel x86 and ARM for Linux and its Android derivative. Our prototype can be applied to all binaries that comply with the Linux ELF format which is the default file format for all Linux distributions and includes executable files as well as shared libraries. The code is entirely written in `C++` and contains 7183 source code lines (SLOC).

The binary rewriter builds the foundation for disassembling instructions, injecting instructions, relocating code and adjusting references. This is enabled by keeping an additional layer of references above the original x86/ARM instructions that keep track of references to/from code and to/from data. Transformation to achieve the desired randomization are applied directly on selected instructions. There code can teared apart and new addresses can be assigned to the individual newly generated code pieces. Lastly, the code pieces are transformed back to x86 or ARM instructions either on-the-fly in the address space of a process or can be written as an ELF file on disk, if need be. Most of the code is architecture-agnostic, only special cases of ARM and x86 are handled in small fractions of the total code. It is, however, not possible to transform a program from x86 to ARM due to the unique design that treats most of the code as a black-box because it does not operate on an intermediate language. Any other processor architecture can be easily supported when a disassembler for the architecture is available. Thus, *XIFER* accurately addresses property *P10* (see Section 3).

5.1 Internal Workflow

We explain the implementation of our on-the-fly rewriter with the running example of ARM code shown in Figure 6. The example also applies to Intel x86 instructions. This example assumes a very simplified program that consists of a branch (`beq`) whose target is the `mov` instruction located at address `837c`. This example explains why not all of the code needs to be disassembled while still guaranteeing an instruction-granularity randomization. In the following, the main steps involved in the rewriting process are explained:

1. *Loading* the executable.
2. *Disassembling* the bytecode on-the-fly.
3. Building a *reference graph* of the executable.
4. Applying *code transformation.*
5. Writing the executable back to memory (*fixation*) so that it can start executing

Step 1 — Executable Loading.

In general, a program is composed of different loadable (code and data) segments. These segments originate from different files like the executable file itself and shared libraries the program depends on. Therefore, we intercept the loading of the ELF executable file after all its dependent libraries have been loaded in the address space but before execution begins. To achieve that, we use the `LD_PRELOAD` environment variable that defines shared libraries that are forced to be present in an address space by the operating system. Originally intended to fix bugs of legacy software by allowing the override the symbol resolution of shared libraries, it is a perfect anchor point for our randomization solution. *XIFER* is compiled as a library (`librewrite.so`) to be automatically injected using `LD_PRELOAD` and features a special section called `.init` that is guaranteed to get executed before any other code. At that moment in time, the program that is to be randomized and all libraries the program depends on (including our injected `librewrite.so`) are now marked as loaded by the Linux kernel. This does not mean that they are actually loaded. Luckily, Linux uses a mechanism referred to as *lazy binding* for the resolution of symbols in shared libraries and on-demand paging for the loading of code in the main executable as well as all shared libraries. We use the second advantage of the `LD_PRELOAD` mechanism, i.e. overriding symbols, to replace the on-demand resolution of symbols with calls to our `librewrite.so`. This way, the rewriter and the standard Linux linker do not get in each other's way and do not do double the work.

By iterating over the mapped shared libraries and the executable file, we load file handlers to the particular ELF files representing the code and data of all loaded libraries. If available, we read the relocation information from the ELF file as well. However, relocation information provide *no* information about *local* data/code references within a segment, but *XIFER* still requires this information in order to accurately perform code transformations within a code segment, which makes a disassembly component a necessity.

Step 2 — Disassembly.

The main goal of the dis- and reassembly process is to make it partial as to avoid putting labor in instructions that are not modified in the rewriting process anyway. This is true for the majority of instructions as they do not reference memory addresses and hence their byte representation is not affected by the randomization.

ARM and x86 instructions are composed of a mandatory opcode, potentially followed by other information, most importantly an immediate value such as a memory address. This eases the processes of disassembly as we can simply use a look-up table for the opcodes to know whether a particular instruction uses an immediate value that encodes an address. If it is not, it is simply treated as a black-box – a blob of bytes. This feature does not only increase the performance[3] but due to inspecting only instructions of interest, it also comes in handy because *'uninteresting'* black-box instructions are treated no different from data. This is an advantage compared to classical disassemblers such as *IDA Pro* or *objdump* which need to decide whether some position is either code or data.

[3]Our experiments showed that our disassembler is faster by factor 10 compared to binutil's disassembler that objdump internally uses.

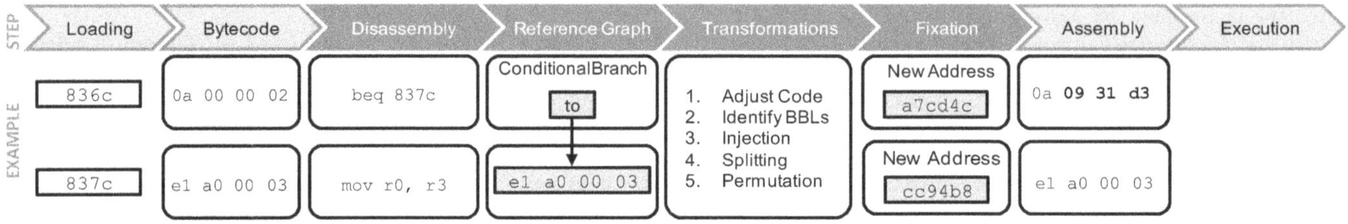

Figure 6: Processing Steps of the Rewriter

In contrast to ARM, which features a fixed-length instruction set, that of x86 is variable. That means, going through the code to find instructions is not a matter of indexing them by a multiple of 4 – which would be the instruction length for ARM. Hence, for x86 it is impossible to predict where an instruction begins without knowing where the preceding one ended. Again, we use the same look-up process to identify an opcode and look-up its length. The opcode also includes information about whether optional immediate values make the instruction longer. Once the length has been determined, the way is paved to start over at the next instruction and so forth.

All instructions that reference data or code are candidates that potentially need to be rewritten later and act as input to build the reference graph that provides information as to which other part of code or data an instruction refers to. The reference graph is similar to relocation information in the sense that it is architecture-agnostic and only saves which part of an instruction encodes an absolute or relative address. As an optimizing step, all instructions still have the original bytecode attached, so that they can be written back to memory and are either left untouched in case of a black-box instruction or are rewritten with aid of the reference graph that tells which part of the instruction encodes an immediate value that needs to be adjusted. Further, x86 and ARM are both generic enough, so that changing address-dependent information in a bytecode is limited to changing two's complement bits in a masked representation of the bytecode. This enables an efficient rewriting process as all of the original bytes are copied to their new memory position and are adjusted with the aid of the reference graph.

Step 3 — Building The Reference Graph.

To resolve code and data references, we build a reference graph by decoding only those instructions that are known to potentially refer to addresses. As an example decoding, we explain this process using the `beq` instruction in Figure 6. This instruction's bytecode is `0a 00 00 02`, while `0a` represents the opcode (`beq`) and `00 00 02` encodes the two's complement representation of a relative addressing. This two's complement representation of the decimal number 2 must be multiplied by 4 (as ARM instructions can only target addresses that are aligned by a multiple of 4). The `beq` instruction is stored at address `836c` which leads to an absolute target of `836c` $+2 \cdot 4 =$ `8374`. Moreover, due to the pipelined architecture of ARM, the program counter `PC` always points two pipeline stages (or 8 bytes) ahead, which leads to `837c` as absolute target for the `beq` instruction. This decoding is stored as additional *FastDecode* (see Table 2) information and is attached to the instruction, so that it can be used to generate appropriate bytecodes later. Particularly, *FastDecode* information includes whether it is signed or unsigned, a bit mask, a bit shift and a summand. This coding is generic enough to enable the rewriter to later write back addresses to an instruction without understanding the instruction itself, and is faster than assembling an instruction. In this example, the attached information would store the values depicted in Table 2 for the `beq` instruction.

Info	Value
Opcode	0a
Signedness	Signed
Bit Mask	0x00ffffff
Bit Shift	2 (left)
Summand	8

Table 2: *FastDecode* information codes how to write back a part of an instruction in an assembler-agnostic way

Using this information, we build the reference graph. The reference graph (see Figure 7) is built by introducing a layer of indirection. Each instruction is routed through the layer of indirection to the instruction it refers to. This is the most important step as it keeps references to the original instruction, even through they might be moved in memory.

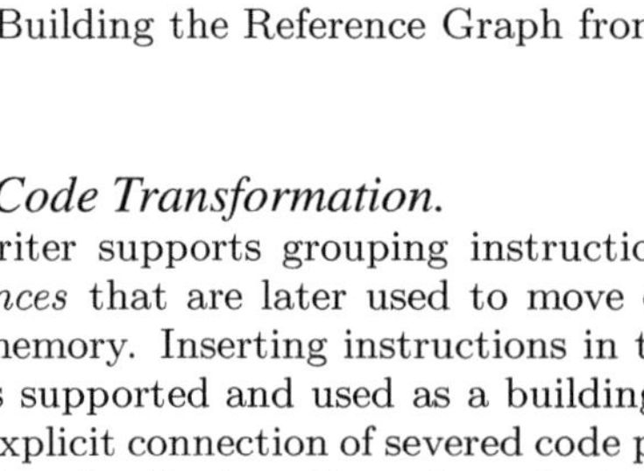

Figure 7: Building the Reference Graph from Instructions

Step 4 — Code Transformation.

The rewriter supports grouping instructions to *Instruction Sequences* that are later used to move code pieces together in memory. Inserting instructions in the *Instruction Sequence* is supported and used as a building block for the necessary explicit connection of severed code pieces (see challenge CH1) or for the insertion of `nop` instructions that do not change the program behavior.

The actual idea of randomization, or code piece permutation respectively, is implemented as operations that can be applied to *Instruction Sequences* or individual instructions. Prior to these operations, the challenge *CH1* of implicit control-flow needs to be tackled. We do this by introducing explicit branches (e.g., `jump 0x1234`) at the end of a code piece in case the original control-flow exhibited an implicit fall-through as described in Section 4. Using this trick, we can then move the code to different memory positions and the reference graph (see Step 3) will later ensure that this injected jump points corresponding code piece that is ought to be connected to it. The number of those artificial code pieces can be specified by the security parameter.

Step 5 — Fixation & Assembly.

This step assigns a randomly chosen address to each code piece. To keep the number of wasted memory pages low, code pieces are grouped to sections that resemble the size of a memory page (e.g. 4 kB). Should an instruction or code piece not fit in (a multiple of) a memory page, it must go into the next one. The induced overhead of this procedure is rather low, as the aligned instructions on ARM allow exactly 1024 instructions to take place in a memory page without the need for padding. For x86, there is no standard instruction length or alignment, hence how many bytes need to be padded at the end of a memory page varies depending on the instruction. Our empirical studies have shown that the padding in the range of 2 to 4 bytes, albeit the theoretical maximum of one complete instruction that does not fit is 14 bytes[4].

We then assign randomly chosen addresses to the created sections (being a multiple of a page size in length). The number of sections, again, depends on the security parameter, as it is not necessary to introduce a higher entropy than the theoretical limit of either 32 bits of 64 bit address space.

Now every section has an address assigned. We then write back the instructions to their respective new addresses in memory while adjusting all code and data references with aid of the reference graph. This can be efficiently done, as the original bytecode of an instruction is still stored behind the layer of indirection by the reference graph. If we encounter an instruction that references code or data, a new bytecode with the adjusted address has to be emitted to the corresponding new location of that instruction with aid of the attached *FastDecode* information. If the instruction does not need any memory address corrections, the old bytecode is simply copied without the need to assemble a new instruction (see Figure 8).

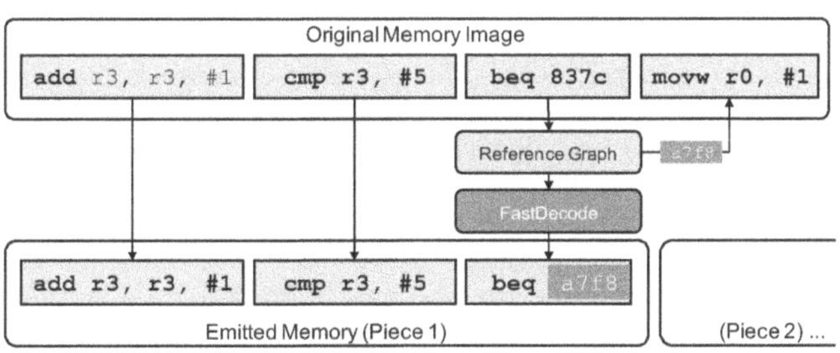

Figure 8: Emitting the Final Opcodes to Memory

Finishing.

Once the rewriting process is done and execution could start, we unload the rewriter by releasing the mapped memory that `librewrite.so` occupies. This is easily doable as the entire code of the rewriter resides in the `.init` section which is never executed again after it has been finished. This procedure is similar to how the Linux kernel frees memory after the initialization of modules has finished.

Debugging.

Debugging a software-diversified process is rather difficult in comparison to the normal process for two reasons: First, the debugger expects the instructions in memory to be in the same order as the program that is stored on disk. However, *XIFER* completely permuted the layout and addresses of the process and thereby renders the debug symbols obsolete. Since our randomization also shuffles the memory pages at process load time, this has to be propagated back to the debugger. In order to solve this problem, debug symbols according to the permutation of our tool are emitted when requested by the user. We chose to rewrite the debug information of an ELF file to a different file according to the randomization. Currently we support the common DWARF [1] file format which can be read e.g. by the 'gdb' debugger. The gdb debugger can then step through the code, inspect variables etc. as if the program were unmodified.

The second debugging issue is that two subsequent executions of a program result in completely different memory layouts, which makes it harder for humans to understand memory-related faults in the program in question. Further, symbolic debuggers usually allow to visually track data structures based on their addresses. If the addresses would change between every program run, finding the new addresses is labor-intensive. We avoided this issue by adding a debug flag to `librewrite.so` that indicates that the same random seed should be used for every execution of the program, resulting in the same memory layout for each executing with the debug flag enabled. The consequence is that every run of a particular program ends up in exactly the same address space layout of a process with every single instruction being at the exact same address across multiple runs. For all intends and purposes, this is against common sense of randomizing a process in the first place but greatly helps debugging a randomized process because all the variables reside at the very same address across different process runs or even reboots.

6. EVALUATION

In this section we evaluate the effectiveness of our randomization solution empirically as well as theoretically. In order to demonstrate the efficiency, we used industry standard performance benchmarks (SPEC CPU2006) as well as micro benchmarks for the important control flow instructions.

6.1 Practical Security Evaluation

In order to test the effectiveness, two experiments were performed: (1) Calculating the *gadget elimination*. A comparison of found gadgets before and after the randomization. (2) **Mitigation of an exploit** targeted to a vulnerable program.

Gadget Elimination. We used the 12 benchmark programs from the SPEC CPU2006 suite (see Table 3) to find ROP gadgets using the program *ROPgadget* [2]. After randomizing the code and writing it to an ELF file, ROPgadget was run again to check whether and how many gadgets have stayed at the original position.

[4]The longest x86 instruction consumes 15 bytes

Benchmark	ROP gadgets	Remaining gadgets
400.perlbench	67	0
401.bzip2	51	0
403.gcc	194	0
429.mcf	45	0
445.gobmk	105	0
456.hmmer	58	0
458.sjeng	57	0
462.libquantum	45	0
464.h264ref	79	0
471.omnetpp	168	0
473.astar	91	0
483.xalancbmk	460	0

Table 3: Overview of the SPEC CPU2006 integer benchmark suite. The C++ benchmarks are listed in gray.

Exploit. To demonstrate the effectiveness against an exploit, we constructed a sample program that is vulnerable to code-reuse attacks. The code is shown in Appendix A. The function `foo()` opens a file of which the path and length are provided as parameters. Then, `fgets()` reads as many characters as specified by the `file_length` parameter and copies them into the local buffer `buf` without checking its bounds. This in turn allows an adversary to divert the control-flow by overflowing the buffer and eventually overwriting the return address of `foo()`, and inject a ROP payload on the stack. We used ROPgadget to find gadgets in the executable and then successfully mounted a shellcode exploit.

After the randomized program has been started the exploit failed. Even intentionally disclosing addresses with `printf ("%x", &foo)` no longer works as the relative offsets in the code segment have been changed.

6.2 Performance Evaluation

We evaluated the performance of *XIFER* on the Intel x86 platform and conducted micro benchmarks on the ARM platform. To evaluate the efficiency, we used the SPEC CPU2006 integer benchmark suite for the x86 version.

Runtime Overhead on Intel x86.

All benchmarks were performed on an Intel Core i7-2600 CPU running at 3.4 GHz with 8 GB of DDR3-SDRAM. We excluded two of the total twelve benchmarks because they use C++ exceptions (see section 4.5 CH3). We compiled all benchmarks using gcc-4.5.3 and the uClibc C library. All measurements include a *complete* randomization of the entire address space including the executable and all shared libraries. We examined three different randomization configurations: (Config-1) Maximum entropy of 52 bits, (Config-2) Forcing a split of code after exactly 15 instructions (Config-3) Strict BBL permutation: all found BBls are split into code pieces.

The results of our evaluation are summarized in Figure 9 and demonstrate that *XIFER* is highly efficient and hence addresses property *P7* (see Section 3). For *Config-1*, which already achieves an entropy of 52 bits, the overhead is only 5%.

Runtime Overhead on ARM.

In contrast to our evaluation on x86, we conducted micro benchmarks for ARM. In particular, we used an Android Nexus S device running Android version 4.0.3. To perform precise measurements, we leveraged the ARM hardware clock cycle counter (CCNT) which is part of the system co-processor (CP15). To measure the runtime overhead of our prototype, we developed an application that calculates 10,000 times the SHA-1 hash of a 1K buffer with padding. The second micro benchmark is a standard bubble sort algorithm being run on an array of 1024 reverse-ordered elements, so all elements in the array need to be touched (worst-case scenario). Again, measured 10,000 times and averaged. In average, the runtime overhead for the diversified executable is only 1.52% for the SHA-1 benchmark, and 1.92% for the bubble sort algorithm.

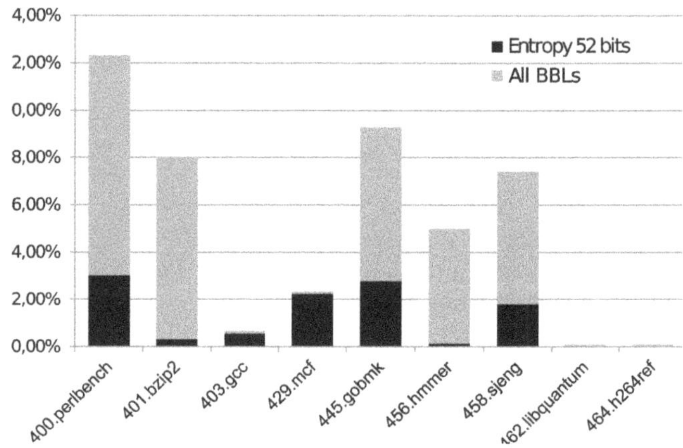

Figure 9: Runtime Measurements with SPEC CPUint2006

Cache Miss Penalty. We also evaluated the cache effects of *XIFER*. Since wild jumping in the code due to the randomization thwarts the locality of code that a processor cache assumes, it has negative impact. For this impact to be measured, we handcrafted code that consists of `add`-instructions whose input depends on the prior output. These instructions are aligned in memory so that they start at the beginning of a cache line and re-occur in memory so that every cache set and every cache line is filled after execution. The total number of instructions exactly fit the entire L1 cache of the Intel Core i7 CPU.

We then split the instruction sequences by inserting jumps between them while keeping the original number of interdependent `add` instructions. This effectively decreases the number of instructions that are executed per cache line before jumping to the next location. As the `jmp` instructions are inserted in equal distance to split the sequences, the total number of instructions to execute grows larger than the L1 cache, which leads to cache misses and lines being evicted from the cache in order to load new lines. The total runtime of all instructions in the cache was measured 100,000 times.

For our benchmark system, equipped with an Intel Core i7-2600 (32 KB L1 cache, 64 bytes per line), we found an acceptable minimum length of 6 `add` instructions (12 bytes) before a jump. This yields a negligible overhead of 0.4%. On the other hand, smaller sizes induce a significant number of cache misses. For instance, when we lower the maximum number of instructions between a jump to 2 instructions, we notice 90.3% overhead due to cache misses as every 3rd instruction is a jump instruction. Hence, we suggest not to set the granularity of randomization to ≤ 6 since this still achieves a very high entropy.

6.3 Rewriting Time

Based on the SPEC CPUint2006 benchmarks we also evaluated the time *XIFER* requires to rewrite and randomize a program. In average, the throughput of the rewriting is 5500 kBit/s which demonstrates the efficiency of our approach.

6.4 Memory Overhead

The possibility to write out ELF executable or shared library files might increase their file size compared to the original, because the code is more bloated and additional segment have been introduced to cope with the different load addresses.

File Size. Encapsulating each memory page in a separate segment in the ELF file requires the allocation of one section header and one program header per page. A section header is 40 bytes and the ELF program header is 32 bytes which leads to an overhead of 72 bytes per 4096 byte memory page, or $\approx 1.76\%$. Figure 10 depicts both, the increase of instructions due the static translation as well as the increase of the ELF section and program headers. `librewrite.so` itself occupies 72 kB when loaded.

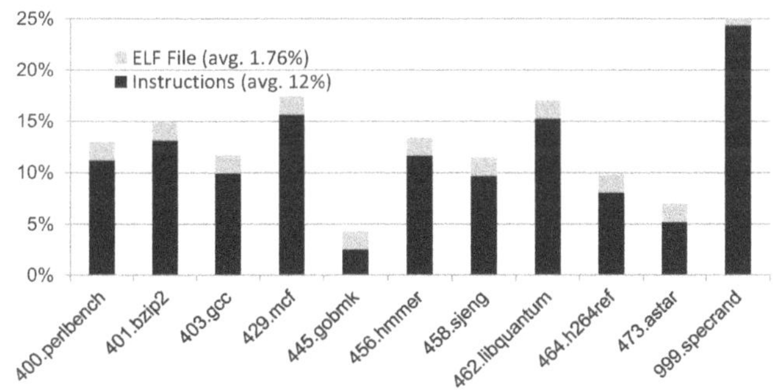

Figure 10: Memory overhead after static translation.

Run-Time. During run-time, the `librewrite.so` has to loaded once into the address space of a process. It is, however, unloaded after the `.init` section has finished and the target program has been randomized. The code size of `librewrite.so` that is temporarily mapped into an address space is $\approx 90kB$. The overhead due to the inserted instruction varies. On average, it increases the code by $\approx 5\%$.

In summary, our approach has a negligible impact on filesize and memory. Hence, we accurately address criterion *P8*.

7. OTHER MITIGATION TECHNIQUES

In this paper we focused on randomization-based countermeasures against code reuse attacks. The main advantage of these defenses resides in the fact that they typically require no access to source code, perform efficiently, and are already deployed in its basic form (ASLR) on today's commodity systems. Nevertheless, and for the sake of completeness, we briefly elaborate on the most well-known countermeasures against return-oriented programming like attacks in the following.

One of the first defense techniques against runtime attacks (that are based on corrupting return addresses) is StackGuard [8], a compiler extension which inserts random stack canaries before return addresses on the stack. A more comprehensive defense is provided by StackGhost [10] which encrypts return addresses on the stack and proposes the concept of return address stacks (i.e., shadow stacks) to keep valid copies of return addresses in a protected memory area. However, these defenses only focus on specific code reuse attacks, and can typically be circumvented by a sophisticated adversary. Another compiler-based solution against ROP attacks was proposed by Onarlioglu et al. [27]. The authors propose a compiler extension for Intel x86 to eliminate the so-called unintended instruction sequences of a program.

A very well-known binary based solution against code-reuse attacks is monitoring of the program flow which has been originally proposed by Kiriansky et al. [22]. In particular, control-flow integrity (CFI) [3] ensures that a program only follows legitimate execution paths. However, CFI induces more performance overhead than randomization-based proposals [3].

8. CONCLUSION

Runtime attacks that reuse existing code pieces (e.g., return-oriented programming and return-into-libc) are a prevalent attack vector against today's applications. In this paper, we tackle these attacks and present the design and implementation of an efficient mitigation technique that is inspired by the principle of *software diversity*. Our software diversity tool *XIFER* accurately mitigates code-reuse attacks by diversifying the structure of an application for each run by means of binary rewriting at the load-time of the application. At the heart of *XIFER* is our binary rewriter which disassembles application binaries on-the-fly, performs code transformations and assembles new application instances with new memory layouts, while still covering the entire semantics of the initial program. *XIFER* is fully dynamic, highly effective (provides a high randomization entropy based on a security parameter), and efficient (induces only 1.2% of runtime overhead in average). Moreover, it requires no access to source codes (which are rarely available in practice), and is compatible to application signatures. Our reference implementation targets both ARM and Intel x86 processors. In order to achieve a highly efficient and effective dynamic binary rewriter, we had to overcome a number of challenges which we highlighted in this paper.

9. REFERENCES

[1] Dwarf 2.0 debugging format standard. `http://www.dwarfstd.org/doc/dwarf-2.0.0.pdf`.

[2] ROPgadget. `http://shell-storm.org/project/ROPgadget/`.

[3] M. Abadi, M. Budiu, U. Erlingsson, and J. Ligatti. Control-Flow Integrity: Principles, Implementations, and Applications. In *ACM Conference on Computer and Communications Security (CCS)*, 2005.

[4] S. Bhatkar, R. Sekar, and D. C. DuVarney. Efficient techniques for comprehensive protection from memory error exploits. In *USENIX Security Symposium*. USENIX Association, 2005.

[5] D. Bruenning. *Efficient, Transparent and Comprehensive Run-time Code Manipulation*. PhD thesis, Massachusetts Institute of Technology, 2004.

[6] F. B. Cohen. Operating system protection through program evolution. *Computer & Security*, 12(6):565–584, Oct. 1993.

[7] comex. `http://www.jailbreakme.com//#`.

[8] C. Cowan, C. Pu, D. Maier, H. Hintony, J. Walpole, P. Bakke, S. Beattie, A. Grier, P. Wagle, and Q. Zhang. StackGuard: Automatic Adaptive Detection and Prevention of Buffer-Overflow Attacks. In *USENIX Security Symposium*, 1998.

[9] R. Enderle. Windows 8: The app store to rule them all? `http://www.conceivablytech.com/9973/products/windows-8-the-app-store-to-rule-them-all`.

[10] M. Frantzen and M. Shuey. StackGhost: Hardware Facilitated Stack Protection. In *USENIX Security Symposium*, 2001.

[11] M. Franz. E unibus pluram: massive-scale software diversity as a defense mechanism. In *Proceedings of the 2010 workshop on New security paradigms*, pages 7–16. ACM, 2010.

[12] G. Fresi Roglia, L. Martignoni, R. Paleari, and D. Bruschi. Surgically Returning to Randomized lib(c). In *ACSAC*, 2009.

[13] C. Giuffrida, A. Kuijsten, and A. S. Tanenbaum. Enhanced operating system security through efficient and fine-grained address space randomization. In *USENIX Security Symposium*, 2012.

[14] D. Goodin. Apple QuickTime backdoor creates code-execution peril. `http://www.theregister.co.uk/2010/08/30/apple_quicktime_critical_vuln/`, 2010.

[15] Google Play. `https://play.google.com/store`.

[16] J. D. Hiser, A. Nguyen-Tuong, M. Co, M. Hall, and J. W. Davidson. ILR: Where'd My Gadgets Go? In *IEEE Symposium on Security and Privacy*, 2012.

[17] V. Iozzo and R. Weinmann. PWN2OWN contest. `http://blog.zynamics.com/2010/03/24/ralf-philipp-weinmann-vincenzo-iozzo-own-the-iphone-at-pwn2own/`, 2010.

[18] T. Jackson, B. Salamat, A. Homescu, K. Manivannan, G. Wagner, A. Gal, S. Brunthaler, C. Wimmer, and M. Franz. Compiler-generated software diversity. In *Moving Target Defense*. 2011.

[19] X. Jiang. GingerMaster: First android malware utilizing a root exploit on Android 2.3 (Gingerbread). `http://www.csc.ncsu.edu/faculty/jiang/GingerMaster/`, 2011.

[20] M. Keith. Android 2.0-2.1 Reverse Shell Exploit, 2010. `http://www.exploit-db.com/exploits/15423/`.

[21] C. Kil, J. Jun, C. Bookholt, J. Xu, and P. Ning. Address space layout permutation (ASLP): Towards fine-grained randomization of commodity software. In *ACSAC*, 2006.

[22] V. Kiriansky, D. Bruening, and S. P. Amarasinghe. Secure Execution via Program Shepherding. In *USENIX Security Symposium*, 2002.

[23] C. Luk, R. Cohn, R. Muth, H. Patil, A. Klauser, G. Lowney, S. Wallace, V. Reddi, and K. Hazelwood. Pin: building customized program analysis tools with dynamic instrumentation. In *ACM SIGPLAN Notices*, volume 40, pages 190–200. ACM, 2005.

[24] Mac App Store. `http://itunes.apple.com/us/app/apple-store/id375380948?mt=8`.

[25] C. Miller and D. Blazakis. Pwn2Own contest. `http://www.ditii.com/2011/03/10/pwn2own-iphone-4-running-ios-4-2-1-successfully-hacked/`, 2011.

[26] National Institute of Standards and Technology. National vulnerability database statistics. `http://web.nvd.nist.gov/view/vuln/search`.

[27] K. Onarlioglu, L. Bilge, A. Lanzi, D. Balzarotti, and E. Kirda. G-Free: defeating return-oriented programming through gadget-less binaries. In *ACSAC'10, Annual Computer Security Applications Conference*, Dec. 2010.

[28] V. Pappas, M. Polychronakis, and A. D. Keromytis. Smashing the gadgets: Hindering return-oriented programming using in-place code randomization. In *IEEE Symposium on Security and Privacy*, 2012.

[29] PaX Team. PaX Address Space Layout Randomization (ASLR). `http://pax.grsecurity.net/docs/aslr.txt`.

[30] M. J. Schwartz. Adobe Acrobat, Reader under attack from zero-day exploit. `http://www.informationweek.com/news/security/vulnerabilities/227400016`, 2010.

[31] H. Shacham. The Geometry of Innocent Flesh on the Bone: Return-into-libc Without Function Calls (on the x86). In *ACM Conference on Computer and Communications Security (CCS)*, 2007.

[32] H. Shacham, E. jin Goh, N. Modadugu, B. Pfaff, and D. Boneh. On the Effectiveness of Address-space Randomization. In *ACM Conference on Computer and Communications Security (CCS)*, 2004.

[33] M. Smithson, K. Anand, A. Kotha, K. Elwazeer, N. Giles, and R. Barua. Binary rewriting without relocation information. Technical report, University of Maryland, 2010.

[34] M. Tran, M. Etheridge, T. Bletsch, X. Jiang, V. Freeh, and P. Ning. On the expressiveness of return-into-libc attacks. In *Proceedings of the 14th international conference on Recent Advances in Intrusion Detection*. Springer-Verlag, 2011.

[35] P. Vreugdenhil. Pwn2Own 2010 Windows 7 Internet Explorer 8 exploit. `http://vreugdenhilresearch.nl/Pwn2Own-2010-Windows7-InternetExplorer8.pdf`, 2010.

[36] R. Wartell, V. Mohan, K. W. Hamlen, and Z. Lin. Binary stirring: Self-randomizing instruction addresses of legacy x86 binary code. In *ACM Conference on Computer and Communications Security (CCS)*, 2012.

APPENDIX

A. VULNERABLE PROGRAM

```
FILE *sFile;
void foo(char *path, file_length) {
    char buf[8];
    sFile = fopen(path, ''r'');
    fgets(buf, file_length, sFile);
    fclose(sFile);
}
```

Enforcing System-Wide Control Flow Integrity for Exploit Detection and Diagnosis

Aravind Prakash
Department of EECS
Syracuse University
arprakas@syr.edu

Heng Yin
Department of EECS
Syracuse University
heyin@syr.edu

Zhenkai Liang
School of Computing
National Univ of Singapore
liangzk@comp.nus.edu.sg

ABSTRACT

Modern malware like Stuxnet is complex and exploits multiple vulnerabilites in not only the user level processes but also the OS kernel to compromise a system. A main trait of such exploits is manipulation of control flow. There is a pressing need to diagnose such exploits. Existing solutions that monitor control flow either have large overhead or high false positives and false negatives, hence making their deployment impractical. In this paper, we present Total-CFI, an efficient and practical tool built on a software emulator, capable of exploit detection by enforcing system-wide Control Flow Integrity (CFI). Total-CFI performs *punctual guest OS view reconstruction* to identify key guest kernel semantics like processes, code modules and threads. It incorporates a novel thread stack identification algorithm that identifies the stack boundaries for different threads in the system. Furthermore, Total-CFI enforces a CFI policy - a combination of whitelist based and shadow call stack based approaches to monitor indirect control flows and detect exploits. We provide a proof-of-concept implementation of Total-CFI on DECAF, built on top of Qemu. We tested 25 commonly used programs and 7 recent real world exploits on Windows OS and found 0 false positives and 0 false negatives respectively. The boot time overhead was found to be no more than 64.1% and the average memory overhead was found to be 7.46KB per loaded module, making it feasible for hardware integration.

Categories and Subject Descriptors

D.4.7 [**Operating Systems**]: Organization and Design; D.4.6 [**Operating Systems**]: Security and Protection, Invasive software

Keywords

Exploit Diagnosis, Exploit Detection, Virtual Machine Introspection, Vulnerability Detection, Software Security

1. INTRODUCTION

Software exploiting attacks have become increasingly sophisticated in the past few years, due to the wide deployment of various defense mechanisms (e.g., Non-Executable bit and Address Space Layout Randomization). As an infamous example, Stuxnet exploits multiple zero-day vulnerabilities to successfully circumvent the existing defense mechanisms and infiltrate into the victim computers [28]. Therefore, there is a pressing need to effectively detect and analyze newly emerging exploit attacks.

Unfortunately, previous research efforts on exploit detection and diagnosis either fall short in precision or incur prohibitive performance overhead. Many prior exploit detection efforts only detect single-stage exploit attacks, thus fall short in analyzing sophisticated multi-stage attacks (e.g., Stuxnet [28]). To detect and analyze multi-stage exploit attacks, PointerScope [40] takes a type inference approach. It detects pointer misuses as key attack steps by inferring pointer and non-pointer types through dynamic binary execution. Although PointerScope has demonstrated to be a feasible approach, it has two major limitations: 1) since it performs type inference on an instruction trace, its analysis overhead is prohibitively high and traces can be very large; and 2) as type conflicts often appear in benign program execution, PointerScope has considerable number of special cases to handle.

In this paper, we aim to provide an accurate and efficient solution to the problem of exploit detection and analysis. Such a solution should be accurate and have close to zero false positive and false negative. It should also be efficient to keep up with highly interactive and computation intensive program execution. To achieve these design goals, we leverage control flow integrity (CFI [2]). The notion of CFI dictates that normal program execution should follow its control flow graph, which can be statically determined at compilation time. It is widely accepted that the majority of software exploit attacks (e.g., buffer overflow, return-to-libc, ROP, use-after-free, etc.) violate this inherent program property.

However, there are several key challenges to enforce CFI in practice. (1) First of all, CFI requires a *complete* control-flow graph (CFG), which is computed from source code of the whole program (including the main executable and the shared libraries). In practice, we do not have access to the source code of all these program components. (2) Moreover, prior CFI enforcement efforts primarily focus on monitoring a single user-level process, however we often do not know the vulnerable process or worse, processes that are exploited in a target system. System wide enforcement of CFI in the operating system kernel has not received sufficient investigation. As a result, kernel exploit detection and diagnosis is still missing. Also, practical CFI enforcement needs to properly handle various special cases, such as dynamically generated code, setjmp/longjmp, user-level thread (or Fiber), etc. Additionally, to be considered as a candidate for CFI enforcement in the hardware, the system must not only be robust but also provide strong performance guarantees.

In this paper, we present the design and implementation of *Total-CFI*, which detects and provides a strong basis for analyzing exploits by enforcing CFI for system-wide binary execution. Total-CFI overcomes the first challenge by computing a conservative but complete CFG directly from binary code modules. Built on top of a CPU emulator, Total-CFI monitors and validates every indirect control transition of the whole-system execution, thus addressing the second challenge. Total-CFI also carefully copes with various extraordinary control transfers to minimize false positives. Furthermore, it leverages the hardware `CR3` register to identify processes. It introduces a novel algorithm to identify thread stack layout by directly referring to the `ESP` register in the CPU. Total-CFI relies directly on the hardware to infer several key pieces of information required in CFI enforcement therefore making it viable to integrate Total-CFI into the hardware.

To evaluate the efficacy and performance of Total-CFI, we have implemented it as a proof-of-concept plug-in on top of DECAF [1, 37] in 3.7K lines of C code. We evaluated Total-CFI in terms of accuracy in identifying the exploits and in terms of performance overhead introduced on the system. We performed experiments on 25 popular and widely used programs on Windows XP and Windows 7 and found 0 false positives, while we were able to successfully identify 8 real world exploits including 1 kernel exploit, with 0 false negatives. In terms of performance, Total-CFI introduced boot time overhead of up to 64.1%. Moreover, we found that on an average, Total-CFI requires 7.46KB of space per module that is loaded in the guest OS. Such an acceptable overhead allows for integration of Total-CFI into the hardware. Furthermore, we ran Total-CFI on Pass Mark CPU and memory benchmark. We report a CPU benchmark overhead of 4.4% and a memory benchmark overhead of 19.8% over Qemu [6], further supporting the possibility of integration into the hardware.

In summary, this paper makes the following contributions:

- We propose to enforce system-wide CFI to detect and analyze exploit attacks.
- We manage to overcome several key challenges that hinder CFI enforcement in practice.
- We propose *Punctual OS View Extraction* to extract guest OS semantics in a timely manner.
- We propose a novel hardware based thread identification algorithm to identify the threads in the guest OS.
- We design and implement a prototype Total-CFI as a plugin for DECAF, and demonstrate its effectiveness and efficiency.

In Section 2 we provide some insights into Stuxnet, a complex malware and derive the design goals for Total-CFI. In Section 2 we also provide the high level architecture of Total-CFI. In Section 3 and Section 4 we discuss the internals of the PVE component and the CFI component of Total-CFI. We present the evaluation results in Section 5 followed by related work in Section 7. Finally, we conclude with Section 8.

2. PROBLEM MOTIVATION AND SOLUTION OVERVIEW

Modern attacks are complex [28] and involve multiple exploits. To be able to successfully diagnose the multiple exploits involved, it is necessary to have robust and reliable CFI enforcement which can capture cross-process control flow violations. In this section, we motivate the problem by first providing some insights into a complex malware - Stuxnet[1] then, we highlight some concerns in existing CFI techniques that make them unsuitable to detect state-of-the-art malware. Finally, we derive some design goals to address the challenges in system-wide CFI.

Stuxnet is a malware that targets SCADA systems running Siemens SIMATIC WinCC or Siemens SIMATIC Step 7, where SCADA in general refers to computer systems that monitor and control industrial processes. In a typical infection scenario, the malware enters a network through an infected usb drive and then spreads to different systems in the network via open network shares. The malware includes a rootkit component that not only hides itself but also enables the infected system to be remotely controlled by an attacker. In summary Stuxnet:

1. Exploits the Microsoft Windows Shortcut LNK/PIF Files Automatic File Execution Vulnerability (MS10-046) to load a DLL into memory.
2. Exploits the Microsoft Windows Print Spooler Service Remote Code Execution Vulnerability (MS10-061) to spread to other vulnerable systems on the LAN.
3. Exploits the vulnerability in Microsoft Windows Share Service (MS08-067) to spread through the Share Service to all the shared drives on the network.
4. Exploits the undocumented Keyboard Layout File related vulnerability (or an undocumented vulnerability in Task Scheduler) to escalate privilege and install a rootkit.

It is worth noting that Stuxnet is *system centric* as opposed to *application centric*. Modern malware exploit multiple vulnerabilities in one or more services/processes and they often involve a rootkit component that is installed in the kernel. Most previous efforts are incapable of detecting exploits involving multiple processes/services. Existing solutions either focus on identifying if a given process has been exploited or not [40], or they focus on hardening any given process to prevent it from being exploited [2]. Knowing what processes to monitor is a requirement for current exploit diagnosis solutions. Scaling such solutions to monitor the entire system imposes severe performance overhead hence making them impractical. We can derive the following requirements for

[1]Here, we consider Stuxnet only as a representative. Malware such as Flamer, Gauss, etc. that followed Stuxnet are equally complex and involve exploiting multiple vulnerabilities spanning across multiple processes.

an exploit diagnosis engine to be able to cope with the state-of-the-art and emerging malware. Exploit diagnosis should:

1. Be system wide - to capture exploits involving multiple processes and the OS kernel.
2. Be practical for deployment with reasonable overhead.
3. Have close to zero false positives and false negatives.

Total-CFI is built with the above design goals in mind. As a proof-of-concept, we leverage the environment provided by software emulation to perform whole system monitoring. In the following section, we provide the high level details of Total-CFI. Then, we present the technical details and design choices behind the core components of Total-CFI.

2.1 Solution Overview

Total-CFI leverages full system emulation to monitor the guest operating system from its inception. The overview of Total-CFI can be found in Figure 1. At a high level, Total-CFI consists of 2 components, *CFI-enforCement* (CFIC) and *Punctual OS View Extraction* (PVE).

OS view extraction. The PVE component extracts semantic information such as kernel objects from the guest, captures events in the guest OS which are pertinent to CFI enforcement, and supports opcode-specific interception of guest OS execution. Here, we refer to events such as process start, process exit, thread start and module load as pertinent events, since such information is required in CFI enforcement. Precise and timely identification of such events is crucial in order to ensure that there is no difference between the time the event actually occurs and the time the semantic values are used by Total-CFI. Timing differences lead to inaccurate assessment of exploit location and context. Though it is possible to identify pertinent events by installing hooks in the guest OS, such an approach is unsuitable for system-wide CFI. Firstly, hooks are not scalable across different OSs since an expert will have to manually identify the precise hooking points for each supported OS. Secondly, hooks installed in the guest OS are prone to modification by a compromised kernel. Finally, capturing the pertinent events directly at the hardware is closer to the actual time of the event occurring as opposed to hooking.

At a high level, PVE component first identifies the guest kernel load address and the global data structures. This is important since most data structures in the kernel are traversable from global data structures [26]. Then, PVE component installs hooks in the VMM to intercept execution whenever a new entry is added into the code cache of the TLB in the CPU. By keeping track of the new entries to code cache of the TLB, PVE component infers if a new module has started.

Similarly, by checking for new entries in the `CR3` register, PVE component detects new processes, and by checking the process list at regular intervals, it detects if a process is active or has exited. PVE component also incorporates a novel technique that instruments the VMM to intercept the guest OS execution when instructions with specific opcodes are encountered. This feature of PVE component is used by CFIC to enforce control flow integrity. Furthermore, PVE component retrieves the running thread information by directly accessing the thread related data structures in the guest OS. In fact, CFIC maintains a shadow call stack and requires the currently executing thread information during every indirect control transfer instruction to verify the control flow integrity. However, performing such accesses frequently can seriously impair the performance of the system. PVE component implements a novel *Stack Layout Identification* algorithm to identify the stack layout of the currently executing thread with minimized access to guest OS's memory. More details about the algorithm and PVE component are provided in Section 3.

CFI enforcement. CFIC acts on the semantic information and event related information gathered by PVE component and enforces the CFI model on the executing guest OS. In a nutshell, CFIC via PVE component intercepts every indirect control transfer instruction and every call instruction executed by the guest OS. CFIC maintains shadow memory to reflect states of processes and threads in the guest OS, both at the user level and the kernel level. When a `call` or an indirect `jmp` instruction is encountered in a process address space, CFIC ensures that the target address belongs to a pre-determined whitelist. A whitelist for a specific module consists of all the statically determinable target addresses for indirect control flow in the module, such as the elements of relocation table and the export table. A whitelist for a process is a union of all the whitelists of all the modules loaded in the process address space. When a `call` instruction is encountered, the address of the instruction succeeding the `call` instruction becomes a valid target for a `ret` instruction and is therefore pushed onto the currently executing thread's shadow stack. Information regarding the currently executing thread is obtained from PVE component.

When a `ret` instruction is encountered, CFIC pops the target address from the thread's shadow stack. However, if the target address is not found on the shadow stack, CFIC treats the access as a potential exploit and stops execution. Section 4 details the internals of CFIC.

3. PUNCTUAL OS VIEW EXTRACTION (PVE)

PVE component builds a bottom-up semantic view of the guest OS from the Virtual Machine Monitor (VMM) and extracts the kernel data structures and deciphers process, module and thread related information, which are necessary to accurately pinpoint the exploit context during CFI enforcement. PVE also instruments the VMM to intercept the guest OS when executing `call`, `ret` and indirect `jmp` instructions.

3.1 System Load and Global Data Structure Identification

Identifying the global data structures is a first step towards reconstruction of guest view because most relevant data structures like process structure, modules list, etc., are traversable from the globals. PVE refers to certain CPU registers to identify guest kernel load address and global data. When the guest OS loads, one of the first tasks performed by the OS is to set up the System Call Table and the Interrupt Descriptor Table. We leverage the fact that system call handlers are located inside the OS kernel to compute the address where the kernel is loaded. For example in Windows, at the start of the system, we monitor the CPU till we find a valid entry in the `sysenter_eip` register. Then, we scan backwards for all the page-aligned addresses till we

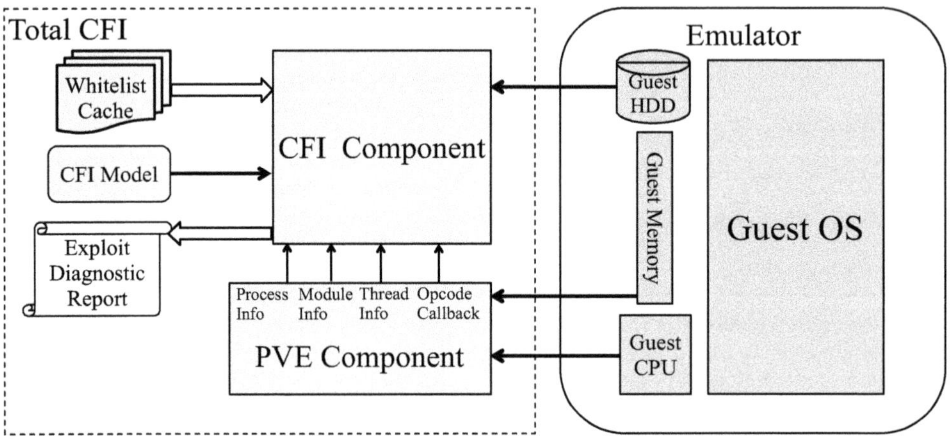

Figure 1: Architecture Overview of Total-CFI

find the MSDOS header - 0x5a4d and the NT signature - 0x00004550.

Using a set of pre-determined addresses to locate the globals may work if ASLR is disabled however, if ASLR is enabled, which is the default case in Windows 7, such an approach does not work. Once the kernel is located, PVE component parses the kernel binary to extract the exported kernel symbols. In flavors of Windows OS, when the execution is in the kernel, the base of the FS register in the CPU contains the address of a global data structure called KPCR. Once the KPCR structure is identified, other kernel data structures like the process list, thread list, module list, etc., can be reached by traversing from the KPCR structure.

3.2 Process Identification

The value of the `CR3` register in the CPU is unique for every process running in the system. PVE component leverages this fact to identify new processes in the system. With respect to the CFIC, process information is required for two reasons. Firstly, it is needed to associate the identified exploits to the actual process that is exploited. Secondly, process information is needed to associate whitelists with processes. When a new module is loaded in a process address space, the process whitelist is updated with the whitelist corresponding to the module that was loaded. In modern x86 CPUs the `CR3` control register is unique per process and is used by the CPU to translate virtual address to physical address. PVE component monitors the `CR3` register for new entries which have not been previously encountered. A new value in the `CR3` register implies that a new process has started. When a new entry is found in the `CR3` register, PVE component traverses the list of processes starting from the previously identified global data structure (KPCR in Windows) to identify the new process that was started.

Furthermore, detecting process exits is necessary to release the process related shadow memory used by Total-CFI so as not to result in memory leaks. Both Linux and Windows OSs contain the process exit time as members in the process data structure. At regular pre-configured intervals, PVE component scans the list of process structures for non-zero process exit time. If found, PVE component deletes the corresponding process from its records. Whenever a process starts or a process exits, PVE component appropriately notifies the CFI component of Total-CFI.

3.3 Module Identification

A new module executed in the process address space results in a new entry in code cache of the TLB. PVE component maintains a list of already encountered entries in the code cache of TLB and when a new entry is encountered, it traverses the process' module list to retrieve a new module, if present. In Total-CFI, we instrument the VMM's TLB cache handling code to insert callbacks such that, whenever a new entry is made into the code cache of the TLB, PVE component callback handler traverses the process' module list[2] and identifies new modules, if any.

If no new module is found in the process module list (which is possible if the new entry corresponds to a new code page in an already existing module), PVE component adds the entry to the list of already encountered TLB code cache entries. When a new module is identified, PVE component extracts the full path of the module (Eg: LDR_DATA_TABLE_ENTRY. FullDllName in Windows), base address and size of the module.

Module unload. It is necessary to identify the module unloads to ensure that there are no redundant entries in the process whitelist hence reducing the possibilities of dangling pointer type of attacks. The code cache of TLB can only indicate if a module is loaded. Capturing the module unload event is less straightforward. Total-CFI considers a module unloaded/modified only if the code pages of the module are overwritten. Therefore, if an entry is created in the write cache of the TLB that also exists in the code cache of the TLB, a module is said to be unloaded. The list of modules is updated during such an event to precisely identify the unloaded modules.

3.4 Thread Stack Layout Identification

We use a novel technique described in Algorithm 1 to identify the thread stack layout in the executing guest OS. Identifying threads is critical in identifying the appropriate execution stack, which in-turn is correlated to the shadow stack maintained by the CFIC.

[2]The process list can be retrieved by traversing `PsActiveProcessHead` in Windows and `task_struct.task_list` in Linux

Algorithm 1 Algorithm to identify the Thread Stack Layout. All actions except † either act upon Total-CFI's shadow memory or retrieve information from the guest CPU. † retrieves information from guest memory.

```
procedure GetStack(GuestCPU CPU)
    Process ← ProcFromCR3(CPU.CR3)
    Thread ← Process.GetCurrThread()
    new_thread ← FALSE
    if PAGE(CPU.ESP) = PAGE(Thread.StackEnd) then
        stack ← Thread.GetShadowStack()
    else                                          ▷ Slow lookup
        Thread ← GetCurrThreadFromGuest()†
        if Thread ∉ Process then
            Process.Add(Thread)
            Thread.InitShadowStack()
            newThread ← TRUE
        end if
        Thread.StackEnd ← CPU.ESP& PAGE_MASK
        stack ← Thread.Stack        ▷ If cpu in kernel, kernel stack, else user stack.
        if new_thread then
            Thread.StackStart ← PAGE(CPU.ESP)
        end if
        if Thread.StackEnd overlaps with any
    thread.StackStart where thread ∈ Process then
            DeleteThread(thread)
            Process.Remove(thread)
        end if
    end if
    return stack
end procedure
```

Based on the context of execution, threads can be classified into:

(a) User level thread - These threads operate in user privilege level. They are assigned a stack in user space.

(b) Kernel level thread - These threads operate in kernel privilege level. They are assigned a stack in the kernel region.

(c) User managed thread - These threads are optionally created and managed by the application code. They operate within the context of a user level thread and share the thread's stack.

Identifying each of the above types of threads is challenging. Specifically, CFIC is interested in efficiently identifying the precise shadow stack to act upon. In a given thread of execution, normally the `ESP` register changes through `push, pop, add, sub` instructions to allocate and reclaim stack space. More importantly, sizes of such allocations and deallocations are often small. Moreover, the OS allocates stack space for threads at page granularity. Therefore, we devise an algorithm to identify stack space partition by examining the `ESP` movement. Algorithm 1 leverages from the above observations to identify the current thread context. It minimizes the performance overhead imposed due to accessing the guest memory by only accessing the guest memory when the execution context switches to a different thread within the same process. All other accesses are performed by either looking up the shadow memory or by directly accessing the guest CPU, which are both considerably faster. Furthermore, `ESP` register in the CPU contains the base address of the current stack frame. As long as the `ESP` register value remains in the same page in a given address space, the execution must be in the same thread context.

By keeping track of a thread's stack boundaries both in the kernel and the userland[3], PVE component can identify the thread that is running by referring to the `ESP` register at any given point. However, if it encounters a value in the `ESP` register that does not belong to any of the known threads, it is possible that a new thread has started or that a thread's stack frame has grown. Every user level thread structure, irrespective of the OS contains information regarding user managed threads (if any). PVE component determines if the current context of execution is in a user managed thread by referring to the thread structure. In such a case, it obtains the thread ID and the user managed thread ID of the currently executing thread (for example in Windows, if the execution is in the kernel, the thread id can be obtained by traversing through the path, `KPCR→Prcb→CurrentThread→Cid→UniqueThread` and if the execution is in the user level, thread ID can be obtained by directly accessing the `CurrentThreadId` member of the Thread Information Block (TIB)).

In flavors of Windows OS, when the execution is in user mode, the base address of the `FS` register contains the TIB. Other than the thread ID, TIB also contains the user managed thread (Fiber) information. Since a majority of applications do not implement Fibers, in the interest of performance Total-CFI does not support Fibers in the default setting, however it is configurable to do so. A thread's stack can increase and overlap another thread's stack base only if the latter thread has terminated. Therefore, if PVE component finds an overlap, it deletes the thread whose stack base gets overlapped. If Total-CFI is configured to use Fibers, it uses the combination of thread ID and fiber ID as the shadow stack identifier.

3.5 Dynamically Generated Code

We further PVE to identify dynamically generated code. Execution of dynamically generated code portray the following characteristics:

(i) Firstly, the page containing the dynamically generated code must be written to memory and made executable (specially on systems with DEP enabled) before it is executed.

(ii) Secondly, control transitions from non-dynamic to dynamic code follow a finite pre-set path.

PVE component tracks the entries in the code and the write caches of TLB to identify dynamically generated code. If an entry in the write cache of the TLB were to appear in the code cache of the TLB, PVE component identifies it as dynamically generated code. Details on CFI enforcement for dynamic code can be found in Section 4.3.

4. CONTROL FLOW INTEGRITY ENFORCEMENT (CFIC)

CFI enforcement component closely interacts with the PVE component and enforces a system-wide instruction level CFI model and diagnoses exploits by detecting violations to the CFI model. Exploits alter the normal control flow by manipulating the derived code addresses (e.g., function pointers).

[3] *Requested Privilege Level* of the CS register on the CPU to determine the *Current Privilege Level*(CPL). A CPL value of 0 indicates kernel.

Total-CFI's CFI model is based on the following two observations:

(a) Most of the control flow is restricted by a pre-determined subset of code that forms the entry point of branch targets. For example, targets of `call, jmp` instructions must adhere to the statically determined call graph.

(b) A `ret` instruction must return to an address succeeding a `call` instruction that was previously encountered.

CFIC requests PVE component to dispatch callbacks whenever the guest OS executes instructions with opcodes that correspond to variants of `call, ret` instructions or with opcodes that correspond to variants of indirect `jmp` instruction. As per observation (a), when indirect `jmp, call` instructions are encountered, CFIC checks if the target address is in the whitelist. If not found, the CFI model is violated and CFIC reports that the instruction is a part of a possible exploit. Based on observation (b), CFIC maintains two shadow call stacks per executing thread in the system. One stack shadows the user level stack of the thread and the other shadows the kernel level stack. Whenever a `call` instruction is encountered, CFIC pushes the return address to the corresponding shadow stack (kernel level shadow stack if operating in kernel mode and user level shadow stack if operating in user mode) of the currently executing thread.

When a `ret` instruction is encountered, CFIC pops the target address of the return instruction from the appropriate stack of the currently executing thread. If the target address is not found on the shadow stack and the target address does not belong to dynamically generated code, CFIC reports the `ret` instruction to be a part of a potential exploit.

In the remainder of this section, we present the internal details and challenges addressed by CFIC.

4.1 Target Whitelist

The addresses within the relocation table and the export table of the binary constitute the module whitelist. With compatibility in mind, most modern binaries are compiled to be relocatable. When the loader cannot load a binary at its default location, it performs relocation. The loader refers to the relocation table and fixes the addresses of the entries in the relocation table. Indirectly addressable code must be relocatable. Similarly, export table contains the functions that a given module exposes for use by the other modules. Addresses of such functions are resolved at runtime based on the actual load address of the dependent modules. Therefore entries of the relocation table and the export table of a module together form valid branch targets for a module.

Irrespective of the guest OS being executed, the binary loader first needs to load the entire module binary to memory before performing relocation fix-ups (if any) and transferring control to the module. However, when the control reaches the module entry, it is possible that the guest OS memory manager has paged out the relocation table from the memory. To optimize the whitelist extraction, CFIC first tries to retrieve the relocation and export tables corresponding to a module directly from the guest memory. If the pages corresponding to relocation and export table are paged out, CFIC accesses the binary file corresponding to the module on the guest file system and extracts the relocation table and export table from the binary.

For each process in the system, CFIC maintains a sorted array of loaded modules in the process address space. It also maintains a hashtable that maps the base address of a module to the whitelist corresponding to the module. When a module is loaded, CFIC first checks the whitelist cache for the whitelist corresponding to the module, only if not present, it constructs the whitelist for the module and adds the whitelist to the whitelist cache.

When CFIC encounters an indirect `call, jmp` instruction, it performs a binary search for the target address in the loaded modules array of the process. Here, the binary search returns a negative insertion point if the search failed. If the returned search value is an even negative index, then the target address does not belong to any of the modules and is treated as a violation of CFI model. However, if the return value is an odd negative index, the target address belongs to the module with base address equal to the address at index - 1 in the loaded module array. Note that it is not possible for the return value to be positive since the target address cannot be equal to the start address or the end address of a module. Therefore, if the return value is non-negative, the address is considered not to be present.

The whitelist lookup is performed with a time complexity of $\lg(n)$, where n is the number of modules in the process address space. Although maintaining a single hash-table for each process with all the whitelists corresponding to all the modules in its address space will suffice, such an approach leads to severe memory overhead due to redundancies, because several common modules (like NTDLL.DLL, KERNEL32.DLL, etc.) will be present in every process whitelist.

Whitelist caching. The whitelist for a binary is statically determined and therefore remains the same across different execution instances. As an optimization, the whitelist is generated only once per binary file and the generated whitelist is stored in the whitelist cache as a [*file's md5 checksum, whitelist*] pair. When a new file is loaded, as an optimization, Total-CFI first checks the whitelist cache to determine if the whitelist has already been extracted, only if the file is being encountered for the first time, Total-CFI extracts the whitelist from the file and adds the whitelist to the whitelist cache.

4.2 Shadow Call Stack

To keep track of the `call-ret` pairs during the execution of a thread in the guest OS, CFIC maintains two shadow call stacks per thread - one for user mode execution and one for kernel mode execution. It relies on PVE component to notify when a `call, ret` instruction is executed on the guest CPU at which time, CFIC first identifies the context in which the instruction was executed. The context constitutes the process, thread and the user/kernel mode the instruction was executed in. From the context, CFIC identifies the appropriate shadow call stack using Algorithm 1. Then, if the instruction were a `call` instruction, CFIC pushes the address of the succeeding instruction on to the identified shadow stack. Conversely, if the instruction were a `ret` instruction, CFIC pops the target address off the shadow stack. If the address is not present in the shadow stack, CFIC reports an exploit.

Special control flows. Though strict pairing between `call-ret` pairs account for a majority of control transfers, there are certain special control transfers that make CFI enforcement via shadow call stack monitoring challenging. Below, we consider such special control flow scenarios.

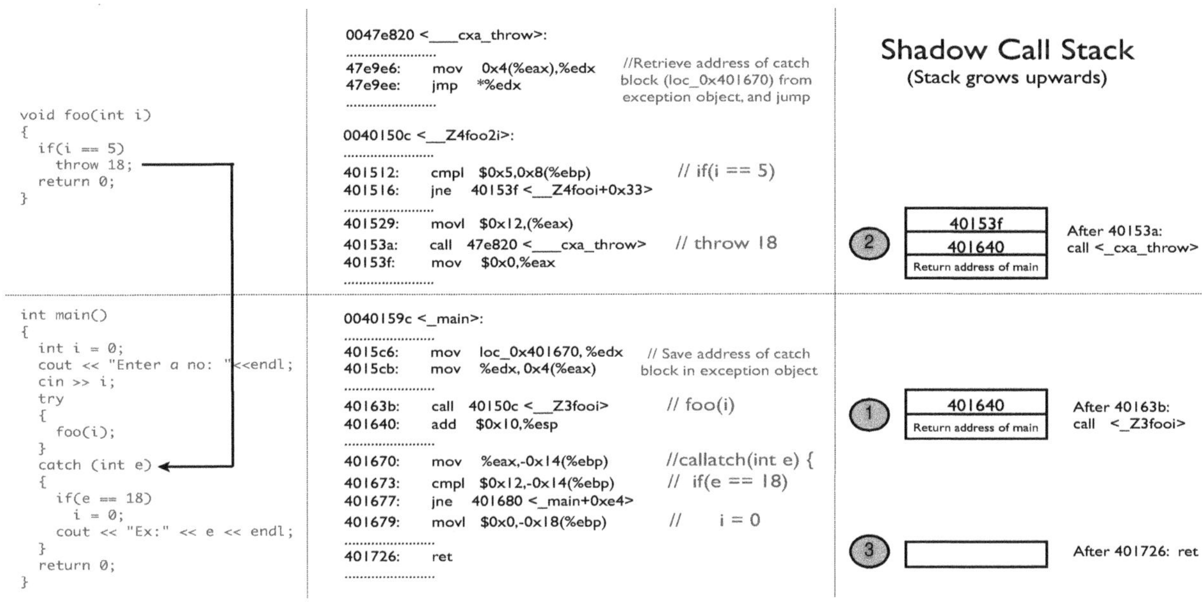

Figure 2: Shadow Call Stack Behavior During a C++ Exception. Input value is 5.

Handling of Exceptions: Exception handling is a mechanism to handle anomalous events that often change the normal control flow of a program. Figure 2 describes the handling of such exceptions by Total-CFI. Column 1 lists the source code of a program that raises and handles an exception. Column 2 lists the simplified version of the corresponding code in assembly, obtained when the code is compiled using the MinGW-g++ cross compiler. The exception handler or the `catch` block is relocatable and hence appears as an entry in the relocation table. During compile time, the compiler stores the address of such a block in the exception object. At runtime, when an exception is thrown, the throw statement translates to a call to *__cxa_throw*, which in turn retrieves the address of the `catch` block from the exception object, rewinds the stack and transfers control to the `catch` block via an indirect jump. Column 3 of Figure 2 shows the contents of shadow stack at different stages during the program execution. On the one hand, during the `jmp` instruction, Total-CFI verifies that the branch address is a part of the program's whitelist and lets the instruction pass, but on the other hand, when the *main* function returns (stage 4 in Column 3 of Figure 2), the CFIC recognizes that the return address is not at the top of the shadow stack and therefore pops all the items up to and including the return address of *main* function from the top of the stack.

Handling of `setjmp/longjmp`: In C and its flavors, `setjmp` and `longjmp` are used to save and restore the CPU environment respectively to transfer control to a predetermined location. During `setjmp`, the environment including the contents of the CPU registers are cached in a user provided buffer, and during `longjmp`, the CPU register contents are restored from the buffer. Upon encountering a `setjmp`, Total-CFI records the value of the program counter where the control will be transferred to during `longjmp`. When a `longjmp` is encountered, Total-CFI verifies the target address to be the same as the value of the program counter as recorded during the previous `setjmp` in the current execution context. Mismatch in the target address is flagged as a potential exploit.

Kernel mode to User mode call backs: Typically, the control transfers from user mode to kernel mode happen through the `sysenter, int` instructions and back from kernel mode to user mode via `sysexit, iret` instructions respectively. However, in Windows, NTDLL maintains a set of entry points that are used by the kernel to invoke certain functionality on behalf of the user mode [31]. Some such NTDLL APIs are: *KiUserExceptionDispatcher, KiUserApcDispatcher, KiRaiseUserExceptionDispatcher* and *KiUserCallbackDispatcher*. They are used by the kernel as a trampoline to invoke functionality in the user mode. Kernel saves the processor state and alters the thread stack to accomplish such a call. When the kernel alters the execution of a thread and transitions to user mode, the return address may not coincide with the expected return address at the top of the stack. To address this problem, for every stack in the system, CFIC maintains a hash-table wherein, for every `ESP` register value encountered during a call instruction as key, it stores the position of the entry in the stack as value. When a return instruction is encountered, it first checks the `ESP` register's value in the hash-table to find the position on the shadow stack and then, pops all the elements up to the position off the stack. Such an approach is reasonable since the stack is dictated primarily by the `ESP` register and a rewind of the ESP register would imply a clean-up of the stack. If the `ESP` register value is not found in the hash-table, the instruction is flagged as a potential exploit.

4.3 Dynamically generated code

CFIC leverages observation 2 in Section 3.5. Most control transfers to dynamically generated code happen through a pre-determined path. Initially, CFIC is trained to accumulate the possible control paths that lead to dynamically generated code in a particular application. This is done by recording the shadow stacks for the valid control flows that lead to dynamically generated code. An intersection of such

```
00000a44 <foo>:
..................
a69: call  9a7<__i686.get_pc_thunk>
a6e: add   $0x1586,%ebx //Offset of f()
a74: call  *%ebx
..................
000009a7 <__i686.get_pc_thunk>:
9a7: mov   (%esp),%ebx
9aa: ret
```

Figure 3: Position Independent Code. *__i686.get_pc_thunk()* retrieves the value of the IP from the stack. *foo()* uses the return address to calculate address of *f()*

paths is used as a signature that is enforced during execution. Here, it is possible that the dynamic code generation library is loaded at different locations on each instance it is loaded. Therefore, CFIC maintains the signature as a [*module:offset*] pair to validate across load instances.

During normal execution, when CFIC encounters a branch target that is not in the whitelist, it first checks if the target belongs to dynamically generated code, next it checks the shadow call stack to check if the shadow call stack satisfies the dynamic code signature for the application.

4.4 Non-relocatable binaries

Though most binaries are relocatable, some legacy code can be non-relocatable. In such cases, PVE component statically analyzes the binaries to extract all the statically identifiable addresses - the ones that either occur as constant address operands in the disassembly or the ones that have a function prologue. Though this approach includes addresses which may not be valid targets, such a conservative approach will reduce false positives.

4.5 Branch Tables or Jump Tables

A jump table is an array of function pointers or an array of machine code jump instructions. Calls to the functions (or code blocks) in the array are made through indirect addressing using the base address of the jump table and the offset of the desired code block in the table. We make two key observations about jump tables:

1. The base address of a jump table must be relocatable and therefore contains an entry in the relocation table.

2. Every entry in the jump table must point to a valid code block.

Total-CFI takes a liberal approach to handle jump tables. For every entry in the relocation table, Total-CFI checks if the content at that address points to code, if so, it treats it as a potential base address of a jump table. It traverses the table for consecutive entries that point to code and adds them to the whitelist.

4.6 Position Independent Code (PIC)

The addressing in PIC does not rely on any particular position in the program address space. Conceptually, PIC identifies the current value of Program Counter (PC) and addresses different code blocks as offsets from the PC. Figure 3 shows a typical example of PIC. The current version of Total-CFI does not support PIC, however it is possible to parse the binary to scan for target address generation patterns. For example, Wartell et al. [34] scan the binary to identify call instructions and perform simple data-flow analysis to identify instructions that use the generated address in an arithmetic computation.

Name	Version	`.reloc` present?	*Fiber* present?	*Dyn Code* present?
Calculator	6.1	✗	✗	✗
Notepad	6.1	✗	✗	✗
Internet Explorer	8.0	✓	✓	✓
Firefox	3.5	✓	✗	✓
Adobe Reader	8.1.1	✗	✗	✓
Google Talk	1.0.72	✓	✗	✓
Microsoft Paint	6.1	✗	✗	✗
Windows Media Player	12.0	✓	✗	✓
XPS viewer	6.1	✓	✓	✓
Yahoo Messenger	8.1.0.29	✗	✗	✗
Apple Quicktime	7.69.80.9	✓	✗	✓
Apple iTunes	10.2	✓	✗	✗
Process Explorer	15.05	✓	✗	✗
Filezilla	0.9.40.0	✗	✗	✗
Google chrome	18.0.1025	✓	✗	✓
Windows Messenger	4.7	✗	✗	✗
RealPlayer	11	✗	✗	✓
DivX Player	6.2.5	✗	✗	✓
Winamp	5.2	✗	✗	✓
VLC Media Player	1.1.11	✗	✗	✓
Skype	5.10.0.116	✓	✗	✓
Registry Editor	6.1	✗	✗	✗

Table 1: List of applications that were tested for false positives on Windows OS. "`.reloc` present?" column indicates if relocation table was present, and "*Fiber* present?" indicates if the application had user level threads or Fibers.

Guest OS	Qemu	PVE only	PVE + CFIC	PVE + CFIC + WL_Cache
WinXPSP3	48s	1m 5s	1m 27s	1m 15s
Win7SP1	1m 57s	2m 36s	3m 26s	3m 12s

Table 3: Times taken to boot Windows 7 and XP till the login screeen is reached. PVE only represents the time taken with only the PVE component enabled. Similarly, PVE +CFIC and PVE +CFIC+WL_Cache correspond to both components enabled and both components enabled with whitelist caching respectively.

5. EVALUATION

Total-CFI was implemented as a plugin for DECAF [1, 37], which is a modification of Qemu [6] version 1.0.1, a full system emulator. DECAF modifies dynamic translation code of Qemu to incorporate opcode specific callbacks into the translation blocks. It also modifies the TLB cache manipulation code to dispatch a callback whenever an entry is made to the TLB cache. In all, Total-CFI consists of 3.8K lines of C code. In this section, we present the evaluation of Total-CFI. All the experiments were performed on a system with Intel core i7, 2.93GHz Quad core processor and 8GB of RAM, running Ubuntu 10.04 with Linux kernel

CVE	Application (Version)	Attack Technique	Exploit EIP	Target EIP	Vulnerable Module
CVE-2010-0249	Internet Explorer (6.0)	Uninitialized memory. Heap spray	0x7dc98c85	0x0c0d0c0d	mshtml.dll
CVE-2010-3962	Internet Explorer (6.0)	Incorrect variable initialization. Heap spray	0x71a51440	0x71a52c66	mswsock.dll
CVE-2011-0073	FireFox 3.5.0	Dangling pointer abuse	0x00346e54	0x01730ee5	js3250.dll
CVE-2011-0257	QuickTime 7.6	Buffer overflow	0x0044888d	0x00194ab0	QuickTime-Player.exe
CVE-2006-1016	Internet Explorer (6.0)	Stack Overflow	0x773f67a8	0x0c112402	ws2_32.dll
CVE-2009-3672	Internet Explorer (6.0)	Incorrect variable initialization. Heap-spray	0x74913ff2	0x0013e0d4	mshtml.dll
CVE-2006-1359	Internet Explorer (6.0)	Incorrect variable initialization. Heap-spray	0x7c8097f3	0x77c3210d	mshtml.dll
CVE-2010-4398*	Windows 7 kernel	Improper driver interaction. Buffer overflow	0x95dca042	0xb8cb8694	win32k.sys

Table 2: Summary of Exploits. Exploit EIP indicates the EIP address where the indirect control transfer was reported. Target EIP corresponds to the branch address and the vulnerable module is the module containing the EIP. * is a kernel exploit

System state	# files in cache	Total Size (B)	Avg. size per file (KB)	# files without .reloc
Login screen	263	1725024	6.405	0
Desktop UI visible	385	2853392	7.237	7
Boot completed	454	3496120	7.52	9
3 programs running	504	3900312	7.557	9
5 programs running	645	5672224	8.588	13

Table 4: Memory Overhead for whitelist cache with respect to number of files in the cache at different system states on Windows 7. "#files without `.reloc`" refers to the number of files in the file cache that did not have reloc section in the binary. For such files, the binaries were parsed to extract statically determinable addresses.

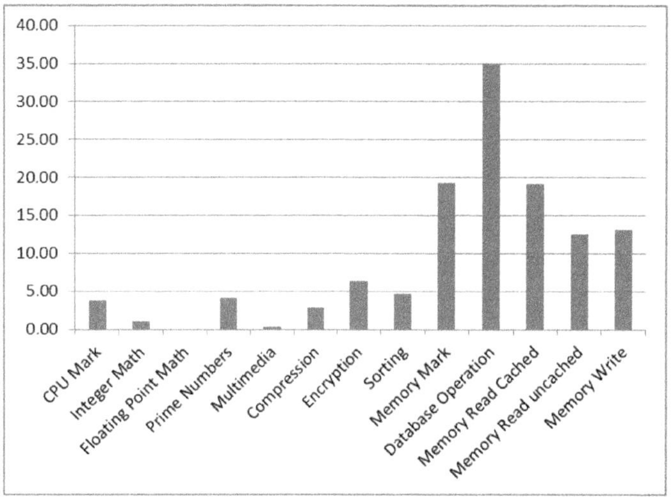

Figure 4: Percentage overhead of Total-CFI with respect to Qemu 1.0.1 for Pass Mark CPU and Memory benchmark on Windows 7 SP1.

2.6.32-44-generic-pae. We evaluate Total-CFI on two factors - accuracy and performance.

False-Positive evaluation. To measure its accuracy, we tested benign applications and exploits on Total-CFI to check for false positives and false negatives. We ran Total-CFI on 25 common applications that are listed in Table 1, on Windows XP and Windows 7. We observed that several preloaded application executables in Windows do not contain relocation table in them. For such executables, we parsed the PE file and extracted statically determinable addresses into the whitelist. 0 exploits were reported in all 25 applications.

False-Negative evaluation. To check the effectiveness of Total-CFI on detecting exploits, we ran Total-CFI on 7 recent real-world exploits that have exploits available in the MetaSploit framework, and one Windows 7 kernel exploit. All the exploits were detected. The summary of exploits and their detection are listed in Table 2. It is worth noting that the kernel exploit, CVE-2010-4398, which starts as a user mode program, exploits a vulnerability in Win32k.sys and eventually escalates privilege. A crafted REG_BINARY value for `SystemDefaultEUDCFont` registry key is inserted to cause a stack-based buffer overflow in the *RtlQueryRegistryValues* function in Win32k.sys. Monitoring a user program (or a set of user programs) alone is insufficient in identifying such an attack. It is essential to monitor both the kernel code and the user level code to diagnose such attacks. Detailed results are tabulated in Table 2.

Performance evaluation. We capture the performance overhead introduced by Total-CFI under two categories. (1) Execution overhead and (2) Memory overhead. We conducted experiments to measure the boot time execution overhead introduced by Total-CFI on Windows 7 and XP. The results are listed in Table 3. We consider boot time execution overhead under performance evaluation because variety of activities occur during the system boot that span across system level, user level and IO. Moreover, most module loads happen during the system boot. Therefore, the boot process is perhaps the worst case scenario with respect to performance overhead imposed by Total-CFI. Optionally, Total-CFI can be turned on after the boot process and it can monitor the execution of all the newly created processes from that point forward.

In Table 3, Total-CFI is configured in 3 modes. PVE only mode performed no CFI enforcement, but generated the process, module and thread related information from the guest. Whereas PVE +CFIC is the setup with both PVE and CFI components turned on and lastly, PVE +CFIC+WL_Cache corresponds to the configuration with whitelist cache containing all the binaries loaded during the boot process. With whitelist caching enabled, the overhead on Windows XP and 7 were found to be 56.2% and 64.1% respectively. Keeping integration with hardware in mind, we also captured the memory overhead introduced by Total-CFI to maintain the shadow whitelist during the boot process on Windows 7.

The memory overhead indicates the amount of memory required to store the whitelists. The results are tabulated in Table 4. We found the average overhead per file to be 7.46KB. We observed that the whitelist for files without `.reloc` section tend to be larger in size since Total-CFI takes a conservative approach to extract all the statically determinable addresses from the binary. Furthermore, from our experiments, we found that even with large number of programs in the memory (such as Microsoft Office applications, Adobe flash, IE, Google Chrome and so on), no more than 1000 files were present in the whitelist cache. At the rate of 7.46KB per file, one would need to set aside approximately 8MB in the hardware for the whitelists, which is conceivable. In combination with a carefully designed cache flush policy to accommodate for even larger number of files in the memory, we believe that integrating whitelist management into the hardware is not far fetched.

Furthermore, we ran the Pass Mark[4] CPU and Memory benchmark on Total-CFI. The results are shown in Figure 4. The CPU benchmark on Total-CFI showed an average overhead of 4.4% over Qemu and the memory benchmark showed an average overhead of 19.8%.

6. DISCUSSION AND FUTURE WORK

6.1 Security Analysis

Evading Total-CFI. In this paper, we address the attacks that arise due to control flow violations. Most attacks in the wild are control hijacking attacks, where attacker executes malicious payload by diverting control flow. However, there exist data only attacks [12] that do not hijack control flow (e.g., bad system configuration resulting in unintended privilege escalation). Such attacks are out of our scope. That said, works in the past [9, 15, 8] have focussed on addressing data integrity concerns. There also exist techniques based on dynamic taint analysis [24, 39, 13] wherein, input data is marked as tainted and tracked through memory to ensure that they do not end up in security critical data structures. PVE component relies on integrity of kernel data to guarantee the correctness of perceived events in the guest kernel. Since PVE component retrieves the data directly from the guest OS kernel data structures, attacks that tamper with the kernel data will mislead Total-CFI.

Furthermore, non-control flow side channel attacks, physical attacks and attacks that target the VMM are also out of Total-CFI's scope. Attacks against the VMM have been demonstrated in the past [4].

Exploits within whitelist. Total-CFI treats the entries in the whitelist as legal entries for indirect branch operations. Therefore, all the function entry points (such as libc functions) belong to the whitelist. This gives rise to a possibility for an attacker to craft an attack such that the jump/call target is an entry within the whitelist. Currently, Total-CFI is vulnerable to such `jump-or-call-to-libc` type of attacks. Note that return-to-libc will be captured by Total-CFI due to the violation in the shadow call stack.

[4]Passmark Software: http://www.passmark.com

6.2 Integration with Hardware

While enforcing CFI at the hardware provides improved performance and stricter security, low performance overhead is a primary requirement for a functionality to be implemented at the hardware. An average overhead of 7.46KB per loaded binary in the memory allows a hardware engineer to set aside a designated amount of memory for whitelist cache (say 8MB or 16MB). This makes it feasible to move system-wide CFI enforcement to the hardware. CFIMon [36] has made efforts in this direction. Though CFIMon uses performance counters in the CPU to aid in CFI enforcement, the actual CFI enforcement happens in software modules.

While the core functionality can be accomplished by directly accessing the guest CPU and memory, guest OS data structures are needed to reconstruct guest semantic view. It is possible to design a system wherein, the PVE component and CFIC are embedded in the guest CPU, and the module whitelists for CFI enforcement are retrieved and shared to the CPU by a privileged kernel module. We intend to pursue this direction as future research.

7. RELATED WORK

Vulneratbility Detection and Testing. There have been several works [21, 7, 32, 20] that discover vulnerabilities in binaries either through static analysis or through dynamic analysis or both. Fundametally, we differ from such works since our goal is to dynamically identify when a particular vulnerability is exploited and not to detect the vulnerability itself. However, once an exploit has been detected, Total-CFI does help in analyzing the exploit and identifying the vulnerability that was exploited.

Exploit Diagnosis. Exploit diagnosis and mitigation has attracted several research efforts [40, 27, 30, 11, 14] in the past. PointerScope [40] defines control and data types and performs instruction level type propagation to identify type misuses, which is usually the case during an exploit. However, PointerScope suffers from severe performance overhead. On the other hand, tainting based approaces [27, 30, 39] are prone to under and overtainting problems that cause false positives.

Program Integrity Models. Control Flow Integrity is just one of many Program Integrity Models such as Data Flow Integrity, Software Fault Isolation, etc. Abadi et al. [2] highlighted and formalized Control Flow Integrity in binary programs. They enforce CFI during a program execution by embedding CFI checks in the form of inlined reference monitors into the binary during compilation.

Along similar lines, WIT [3] uses points analysis to identify the objects that a program can write to at compile time and prevents the program from writing to other objects. Both [2, 3] approaches rely on the program source code, which is not always available. Kiriansky et al. [25] proposed Program Shepherding, where the control flows are restricted based on the origin and the target of the control transfer. Their solution does not address whole system CFI enforcement.

Davi et al. [16] introduce CFI on ARM. They combine static analysis with dynamic binary rewriting to moderate control flow. CFIMon [36], on the other hand, uses hardware performance counters to identify control flow violations. In an offline mode, it gathers the set of legitimate target addresses for each branch instruction and in the online mode, it gathers and analyzes the traces from branch counters, and uses heuristics to identify control flow violations. Though it relies on the hardware for trace gathering, actual enforcement happens in the the software. Some prior efforts also made use of relocation tables and export address tables to validate the legitimacy of indirect jump targets. HookScout [38] tracks function pointers in the kernel and automatically generates hook detection policy. More recently, FPGate [35] retrieves a list of indirect code targets from relocation table and export table, and performs binary rewriting to validate indirect jump/call targets and prevent control-flow hijacking attacks.

To address data attacks, DFI [8] first performs static analysis to capture legitimate data flows in the form of data flow graph and then ensures that the data flow at runtime satisfies the computed graph.

Software Fault Isolation [33] focusses on software based isolation of untrusted modules within an address space. They modify an untrusted binary to prevent it from branching out of or changing memory outside the memory region allocated to it. This way, untrusted modules cannot exploit other modules. Though such a solution is feasible for individual modules, it is not feasible for system-wide monitoring.

Virtual Machine Introspection. Introspecting a virtual machine often requires interpreting the low level bits and bytes of guest OS kernels to high level semantic state. This is a non-trivial task, because of the semantic-gap [10]. Garfinkel and Rosenblum [19] introduced VMI in intrusion detection and Jiang et al. used VMI in to detect malware [23]. Early approaches (e.g., [29, 23, 5, 39]) use manual efforts in combination with installing hooks in the guest OS to locate the kernel objects in the guest OS. Recent advances largely automate this process [17, 18]. In our work, we identify OS entities like process, thread and modules directly from the CPU. We also minimize the access to guest memory in the better interest of performance and devise a new thread stack identification algorithm to cope with performance requirements.

OS-Sommelier [22] takes a memory only approach to fingerprint the guest OS in the cloud. It identifies the kernel code and computes a hash to fingerprint the OS. Though its approach is scalable to different OSs, such an approach is more suitable for memory forensics than for punctual OS view extraction.

8. CONCLUSION

In this paper, we presented Total-CFI, a proof-of-concept implementation of system-wide CFI enforcement. To accomplish system-wide CFI, we performed Punctual Guest OS View Extraction and introduced a novel Thread Stack Layout Identification algorithm to gather semantic information from the guest in a timely manner. We evaluated Total-CFI and found 0 false positives and false negatives. We found a memory overhead of 7.46KB per loaded module and an execution overhead of 64.1% hence making it feasible for integration with hardware.

9. ACKNOWLEDGEMENTS

We would like to thank anonymous reviewers for their comments, and in particular our shepherd, Lenx Wei, for his help in addressing the concerns of those reviewers. We would also like to thank Masters students at Syracuse University, Shengming Xu and Haoru Zhao for their help in evaluating Total-CFI. This research was supported in part by McAfee Inc., NSF grant #1018217, NSF grant #1054605 and Singapore MoE grant R-252-000-460-112. Any opinion, findings, conclusions, or recommendations expressed are those of the authors and not necessarily of the funding agencies.

10. REFERENCES

[1] DECAF: Binary Analysis Platform. Sycurelab, Syracuse University. `http://code.google.com/p/decaf-platform/`.

[2] Abadi, M., Budiu, M., Erlingsson, U., and Ligatti, J. Control-flow integrity principles, implementations, and applications. *ACM Trans. Inf. Syst. Secur. 13*, 1 (Nov. 2009), 4:1–4:40.

[3] Akritidis, P., Cadar, C., Raiciu, C., Costa, M., and Castro, M. Preventing memory error exploits with wit. In *Proceedings of the 2008 IEEE Symposium on Security and Privacy* (2008), SP '08.

[4] Bahram, S., Jiang, X., Wang, Z., Grace, M., Li, J., Srinivasan, D., Rhee, J., and Xu, D. Dksm: Subverting virtual machine introspection for fun and profit. In *Proceedings of the 29th IEEE International Symposium on Reliable Distributed Systems (SRDS'10)* (2010).

[5] Baiardi, F., and Sgandurra, D. Building trustworthy intrusion detection through vm introspection. In *Proceedings of the Third International Symposium on Information Assurance and Security* (2007), IEEE Computer Society.

[6] Bellard, F. Qemu, a fast and portable dynamic translator. In *USENIX Annual Technical Conference, FREENIX Track* (April 2005).

[7] Cadar, C., Dunbar, D., and Engler, D. KLEE: unassisted and automatic generation of high-coverage tests for complex systems programs. In *Proceedings of the 8th USENIX conference on Operating systems design and implementation (OSDI'08)*.

[8] Castro, M., Costa, M., and Harris, T. Securing software by enforcing data-flow integrity. In *Proceedings of the 7th symposium on Operating systems design and implementation*, OSDI '06.

[9] Chaudhuri, A., Naldurg, P., and Rajamani, S. A type system for data-flow integrity on windows vista. *SIGPLAN Not. 43*, 12 (Feb. 2009), 9–20.

[10] Chen, P. M., and Noble, B. D. When virtual is better than real. In *Proceedings of the Eighth Workshop on Hot Topics in Operating Systems* (2001).

[11] Chen, S., Xu, J., Nakka, N., Kalbarczyk, Z., and Iyer, R. Defeating memory corruption attacks via pointer taintedness detection. In *Dependable Systems and Networks, 2005. DSN 2005. Proceedings. International Conference on* (2005).

[12] Chen, S., Xu, J., Sezer, E. C., Gauriar, P., and Iyer, R. K. Non-control-data attacks are realistic threats. In *Proceedings of the 14th conference on USENIX Security Symposium* (2005), SSYM'05.

[13] Clause, J., Li, W., and Orso, A. Dytan: a generic dynamic taint analysis framework. In *Proceedings of the 2007 International Symposium on Software Testing and Analysis (ISSTA'07)*.

[14] Costa, M. Vigilante: End-to-end containment of internet worms. In *Proceedings of the 20th ACM*

Symposium on Operating Systems Principles (SOSP'05).

[15] COSTA, M., CASTRO, M., ZHOU, L., ZHANG, L., AND PEINADO, M. Bouncer: securing software by blocking bad input. In *Proceedings of 21st ACM SIGOPS Symposium on Operating Systems Principles (SOSP'07).*

[16] DAVI, L., DMITRIENKO, R., EGELE, M., FISCHER, T., HOLZ, T., HUND, R., NÃIJRNBERGER, S., AND REZA SADEGHI, A. Mocfi: A framework to mitigate control-flow attacks on smartphones. In *In Proceedings of the Network and Distributed System Security Symposium (NDSS'12).*

[17] DOLAN-GAVITT, B., LEEK, T., ZHIVICH, M., GIFFIN, J., AND LEE, W. Virtuoso: Narrowing the semantic gap in virtual machine introspection. In *Proceedings of the IEEE Symposium on Security and Privacy (Oakland)* (May 2011).

[18] FU, Y., AND LIN, Z. Space traveling across vm: Automatically bridging the semantic-gap in virtual machine introspection via online kernel data redirection. In *Proceedings of the 2012 IEEE Symposium on Security and Privacy* (San Francisco, CA, May 2012).

[19] GARFINKEL, T., AND ROSENBLUM, M. A virtual machine introspection based architecture for intrusion detection. In *Proceedings of Network and Distributed Systems Security Symposium (NDSS'03)* (February 2003).

[20] GODEFROID, P., KLARLUND, N., AND SEN, K. Dart: directed automated random testing. In *Proceedings of the 2005 ACM SIGPLAN conference on Programming language design and implementation* (2005), PLDI '05, ACM.

[21] GODFROID, P., LEVIN, M. Y., AND MOLNAR, D. Automated whitebox fuzz testing. In *Proceedings of the 15th Annual Network and Distributed System Security Symposium (NDSS'08)* (February 2008).

[22] GU, Y., FU, Y., PRAKASH, A., LIN, Z., AND YIN, H. Os-sommelier: memory-only operating system fingerprinting in the cloud. In *Proceedings of the Third ACM Symposium on Cloud Computing*, SoCC '12.

[23] JIANG, X., WANG, X., AND XU, D. Stealthy malware detection through VMM-based "out-of-the-box" semantic view reconstruction. In *Proceedings of the 14th ACM conference on Computer and Communications Security (CCS'07)* (October 2007).

[24] KANG, M. G., MCCAMANT, S., POOSANKAM, P., AND SONG, D. Dta++: Dynamic taint analysis with targeted control-flow propagation. In *Proceedings of the 18th Annual Network and Distributed System Security Symposium (NDSS'11).*

[25] KIRIANSKY, V., BRUENING, D., AND AMARASINGHE, S. P. Secure execution via program shepherding. In *Proceedings of the 11th USENIX Security Symposium* (2002), USENIX Association.

[26] LIN, Z., RHEE, J., ZHANG, X., XU, D., AND JIANG, X. Siggraph: Brute force scanning of kernel data structure instances using graph-based signatures. In *Proceedings of the 18th Annual Network and Distributed System Security Symposium (NDSS'11).*

[27] NEWSOME, J., AND SONG, D. Dynamic taint analysis for automatic detection, analysis, and signature generation of exploits on commodity software. In *Proceedings of the 12th Annual Network and Distributed System Security Symposium (NDSS'05).*

[28] NICOLAS FALLIERE, LIAM O MURCHU, E. C. Symantec stuxnet dossier. `http://www.symantec.com/content/en/us/enterprise/media/security_response/whitepapers/w32_stuxnet_dossier.pdf`.

[29] PETRONI, N. L., JR., FRASER, T., MOLINA, J., AND ARBAUGH, W. A. Copilot - a coprocessor-based kernel runtime integrity monitor. In *Proceedings of the 13th USENIX Security Symposium* (2004).

[30] PORTOKALIDIS, G., SLOWINSKA, A., AND BOS, H. Argos: an emulator for fingerprinting zero-day attacks. In *EuroSys 2006* (April 2006).

[31] RUSSINOVICH, M., SOLOMON, D. A., AND IONESCU, A. *Windows Internals. 5th Ed.* Microsoft press, 2009.

[32] SEN, K., MARINOV, D., AND AGHA, G. Cute: a concolic unit testing engine for c. In *Proceedings of the 10th European software engineering conference held jointly with 13th ACM SIGSOFT international symposium on Foundations of software engineering* (2005).

[33] WAHBE, R., LUCCO, S., ANDERSON, T. E., AND GRAHAM, S. L. Efficient software-based fault isolation. In *Proceedings of the 14th SymPosium on Operating System Principles* (1993).

[34] WARTELL, R., MOHAN, V., HAMLEN, K. W., AND LIN, Z. Binary stirring: self-randomizing instruction addresses of legacy x86 binary code. In *Proceedings of the 2012 ACM conference on Computer and communications security*, CCS '12.

[35] WEI, T., ZHANG, C., CHEN, Z., DUAN, L., SZEKERES, L., MCCAMANT, S., AND SONG, D. Fpgate: The last building block for a practical cfi solution, technical report for microsoft bluehat prize contest. Tech. rep., Apr 2012.

[36] XIA, Y., LIU, Y., CHEN, H., AND ZANG, B. Cfimon: Detecting violation of control flow integrity using performance counters. In *Dependable Systems and Networks (DSN) 2012.*

[37] YAN, L. K., AND YIN, H. Droidscope: seamlessly reconstructing the os and dalvik semantic views for dynamic android malware analysis. In *Proceedings of the 21st USENIX conference on Security symposium - 2012*, USENIX Association.

[38] YIN, H., POOSANKAM, P., HANNA, S., AND SONG, D. HookScout: Proactive binary-centric hook detection. In *Proceedings of Seventh Conference on Detection of Intrusions and Malware & Vulnerability Assessment (DIMVA'10)* (July 2010).

[39] YIN, H., SONG, D., MANUEL, E., KRUEGEL, C., AND KIRDA, E. Panorama: Capturing system-wide information flow for malware detection and analysis. In *Proceedings of the 14th ACM Conference on Computer and Communication Security (CCS'07).*

[40] ZHANG, M., PRAKASH, A., LI, X., LIANG, Z., AND YIN, H. Identifying and analyzing pointer misuses for sophisticated memory-corruption exploit diagnosis. In *Proceedings of 19th Annual Network & Distributed System Security Symposium* (2012).

Secure Cloud-assisted Location Based Reminder

Xinxin Zhao
Arizona State University
zhao.xinxin@asu.edu

Lingjun Li
Arizona State University
li.lingjun@asu.edu

Guoliang Xue
Arizona State University
xue@asu.edu

ABSTRACT

In this paper, we propose a secure cloud-assisted location based reminder system. The proposed system is secure and responsive. Our system outsources the location testing task — testing whether the current location is near a reminder location — to the cloud server such that the device synchronization is not necessary in the system. This feature makes the proposed system more responsive, especially when the reminder message is of large size, e.g., audio, images. Above all, the proposed system protects a user's location privacy and the confidentiality of the reminder message. The system is designed in a way that the cloud server can perform location testing for a user but cannot learn about her current location, reminder locations, and reminder messages. We prove the security of the proposed system and demonstrate its efficiency using simulations on a Motorola Droid smartphone.

Categories and Subject Descriptors

Computer Systems Organization [**subcategory**]: COMPUTER-COMMUNICATION NETWORKS

Keywords

location based reminder, cloud, location privacy, mobile security

1. INTRODUCTION

Reminder applications are becoming the must-have applications on mobile phones. The reminder application helps users to remind themselves of specific things at a specific time. To add a reminder, users need to enter a short reminder message associated with a reminder time in mobile phones. The reminder time could be a specific time or a repeated time in the near future. The mobile phone will alert (e.g., ringing or popping out the reminder message) users at the reminder time. The past few years have witnessed an exponential growth of smart devices. The most distinctive feature of smart devices is that they are equipped with various sensors. According to [16], most smart devices are equipped with high-precision localization sensors, usually based on GPS receivers, access points or triangulation with nearby base stations.

Thanks to the localization sensors, more and more reminder applications on smartphones now have an additional functionality, which allows users to mark a location during the setting of a reminder. We call this location a *reminder location*. The smartphone will alert a user when her current location is close to the reminder location. The extent of closeness is pre-set by the user and is called the *reminder distance*. For example, a person can add a location based reminder such that the smart device can remind her to buy some grocery when she is near a Walmart, or to refill the medicine when she passes by a CVS or a Walgreens.

The goal of this paper is to construct a secure cloud-assisted location based reminder system, which protects the user's location privacy and the confidentiality of reminder messages. The system hides a user's reminder messages, reminder locations, and current locations from the cloud server. However, the system still enables the cloud server to decide whether a user's current location is within any reminder area. To this end, we propose a novel area representation method that reduces an inequality testing, i.e. distance comparison, to an equality testing, i.e. finding set intersections. Specifically, the system divides the surface of the Earth into small squares (forming a grid), represents a reminder area by a set of squares, and associates them with the encrypted reminder message using searchable symmetric encryption. The smart device represents the current area in the same way and lets the cloud server search the nearby reminder locations without exposing any location information and the reminder message.

The contributions of our paper are as follows.

- We propose a secure cloud-assisted location based reminder system, which protects the user's location information and reminder message from an untrusted cloud server. We prove the security of the system under the decisional Diffie-Hellman (DDH) assumption [18] in the random oracle model (ROM) and demon-

Zhao, Li and Xue are all affiliated with Arizona State University, Tempe, AZ 85287. Email: {zhao.xinxin, li.lingjun, xue}@asu.edu. This research was supported in part by ARO grant W911NF-09-1-0467, and NSF grants 1217611 and 0901451. The information reported here does not reflect the position or the policy of the federal government.

ASIA CCS'13, May 8–10, 2013, Hangzhou, China.

strate its efficiency using simulations on a Motorola Droid phone.

- We propose a novel method to represent the reminder area, which reduces the time complexity for checking whether a user is in a particular reminder location area. The representation is easy to construct and lightweight in space overhead.

- Based on our novel representation approach, we propose to use searchable symmetric encryption to construct a secure cloud-assisted location based reminder system. Considering the computational limitations, we propose to use a lightweight pseudo random function [11], which is proved to be secure in the random oracle model.

The rest of our paper is organized as follows. We formulate the problem of the secure cloud-assisted location based reminder in Section 2. We demonstrate the system construction in Section 3. We represent our simulation results in Section 4. We present the related work in Section 5, and concludes our paper in Section 6.

2. PROBLEM FORMULATION

In this section, we present the system model, threat model, and design goals.

2.1 System Model and Threat Model

Our system consists of a cloud server s and multiple users. Since all the interactions are between a user and the cloud server, we use u to denote the user in discussion. The user u may have multiple smart devices (e.g., tablets or smartphones) with the same account. Since our system does not need data synchronization, the system interactions are independent of devices and only specific to a user account. Therefore, we use a smart device to denote any one used by the user. When user u adds a reminder to her smart device, u can enter a reminder time, a reminder location, and a reminder message, which could be text, audio, or image.

When a user u is enrolled in the system for the first time, she sets a reminder distance d_u and a parameter a_u. In order to reduce the inequality testing to equality testing, our system uses an approximate distance to estimate the distance between the reminder location and the user's current location Given a reminder location, a reminder area is a disc centered at the location with radius of d_u. When u enters a reminder area, the smart device retrieves the reminder message from the cloud server and alerts the user. The user may have multiple reminders on the cloud server. Thus, the cloud server should test the current location with each reminder area and send all the valid reminder messages. The user may have multiple smart devices. The cloud sever allows a user to combine any device with her account and reminds the user on any device that is in a reminder area.

We consider an honest but curious cloud server. A cloud server will follow the protocol specification and not intentionally destroy users' reminder service in order to maintain its business reputation. However, the server may be curious to learn the user's reminder contents and location privacy. Our system does not rely on a trusted third party so as to simplify the system architecture and enhance the system performance.

2.2 Design Goals

In the following, we summarize the goals that a secure cloud-assisted location based reminder system should achieve.

- Location privacy: The cloud server should not know a user's reminder locations and her current location;

- Confidentiality: The server should not be able to learn about a user's reminder content.

- Workability: The cloud server should be able to check whether a user's current location is in a reminder area and send the associated reminder message to the user.

- Efficiency: The system should be efficient in computation, data usage, and server space consumption.

- Responsiveness: A user's device should not need to synchronize to catch up reminders from the cloud server, even when the device has been offline for a long time.

3. SYSTEM CONSTRUCTION

In this section, we describe our system construction in details. First we introduce the tessellation on the surface of the Earth. Based on the tessellation, our system constructs a novel reminder area representation. We then present our system constructions.

3.1 Tessellation on the Surface of the Earth

To represent a proximity area, Narayanan *et al.* [14] proposed to tessellate the earth surface by hexagons. To deal with the curved Earth surface, they divided the Earth into strips by latitude, one strip per degree of latitude, and tessellated each strip separately. Since the curvature of the Earth within a single strip can be ignored, the small cells can thus keep approximately the same everywhere. Following the same idea, we tessellate the earth surface by squares, instead of hexagons. The tessellation forms a grid on the surface of the Earth. We note that the reminder area is usually small and the curvature of the Earth passing through the area is too small to ignore. Let G_u denote the tessellation on the plane, which is defined by user u and consists of squares with the side length a_u. Each square has a unique identifier and we use $id_{(x,y)}$ to denote the square identifier that the location point (x, y) is located.

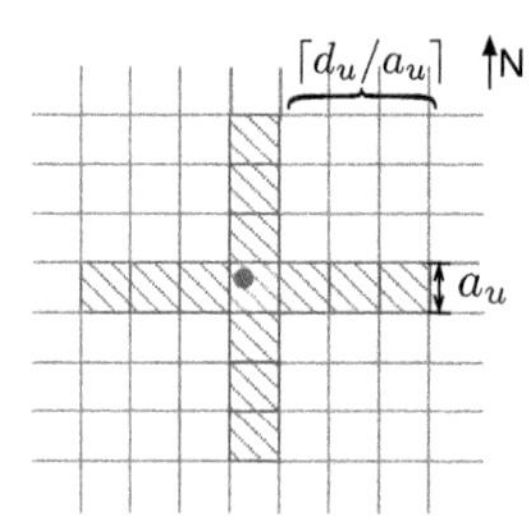

Figure 1: Using cross to represent an area

3.2 A Novel Area Representation

We try to find an area representation that reduces the space consumption while still keeping the inaccuracy small.

Our idea is to use an equal-armed cross to represent a reminder area, as shown in Figure 1. The cross consists of multiple small squares in the tessellation and is centered at the square where the interested location (a reminder location or a current location) is located. Each arm of the cross contains $\lceil d_u/a_u \rceil$ squares. Suppose there is a point (x, y), e.g., the blue point in Figure 1. Let $id^{i,*}_{(x,y)}$, $i = \pm1, \pm2, \cdots, \pm(\lceil d_u/a_u \rceil)$ denote the i-th square identifier to the east/west (negative numbers denote the west of the center square) of square $id_{(x,y)}$. Let $id^{*,i}_{(x,y)}$ $i = \pm1, \pm2, \cdots, \pm(\lceil d_u/a_u \rceil)$ denote the i-th square identifier to the north/south (negative numbers denote the south of the center square) of square $id_{(x,y)}$. All the squares in the cross, i.e., shaded squares in Figure 1, are called *valid* squares with respect to (x, y).

Procedure 1: CreateCross(x, y, d_u, a_u)

Data: location (x, y), reminder distance d_u, square side length a_u
Result: valid square identifier set $\mathcal{ID}_{(x,y)}$
$\mathcal{ID}_{(x,y)} \leftarrow \emptyset$;
$\mathcal{ID}_{(x,y)} \leftarrow \mathcal{ID} \cup \{id_{(x,y)}\}$;
for $i \leftarrow 1$ *to* $\lceil d_u/a_u \rceil$ **do**
$\quad \mathcal{ID}_{(x,y)} \leftarrow \mathcal{ID}_{(x,y)} \cup \{id^{*,i}_{(x,y)}, id^{*,-i}_{(x,y)}, id^{i,*}_{(x,y)}, id^{-i,*}_{(x,y)}\}$;
end

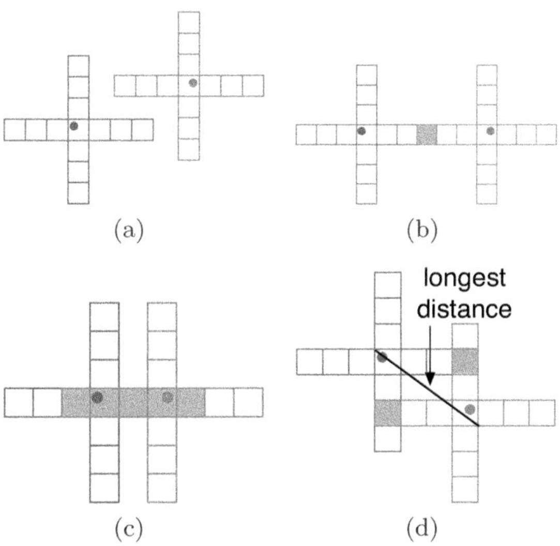

Figure 2: Four cases of cross intersection

Given a location (x, y), a distance d_u, and square side length a_u, Procedure 1 shows how to construct a valid square identifier set $\mathcal{ID}_{(x,y)}$. Using valid square identifier set, we can test whether the distance between the two points (x', y') and (x, y) is smaller than d_u or not. We calculate two valid square identifier sets $\mathcal{ID}_{(x',y')}$ and $\mathcal{ID}_{(x,y)}$ with respect to (x', y') and (x, y), respectively, and test whether they have intersections. If they do not have any intersection, it indicates that their distance is longer than d_u, as shown in Figure 2(a). In Figure 2, we use the blue point and the red point to respectively denote (x, y) and (x', y'). If they have intersections, we can further estimate the distance between the two locations by using the intersected squares. If there are two interactions on two different directions, as shown in Figure 2(d), the user knows that the distance between the two locations is no longer than the longest distance between any two points in the two squares. Therefore, this distance can be used to bound the distance between the two locations. If the intersection squares are on the same line, the user can also estimate the distance. An example is shown in Figure 2(b), there is only one intersection square and the user can know that the distance between the two locations is longer than $2d_u - a_u$. If the number of interaction squares is more than $\lceil d_u/a_u \rceil + 1$, e.g. Figure. 2(c), the distance between the two locations is shorter than d_u.

The original location testing is an inequality testing, in which the user needs to compute the distance between the two locations and compare it with the reminder distance. Our novel representation reduces this inequality testing to an equality testing, in which users test if there are intersections between the valid square identifier sets of the two locations. This reduction allows us to do location testing even when the reminder location and current location is transformed by a one-way trapdoor function.

3.3 System Construction

Our system consists of three protocols: user enrollment protocol, location marking protocol, and location search protocol.

Function: A user gets enrolled in the system.

1. The User sends request to the cloud server s.
2. s sets up a new user account and assigns a user ID, e.g. u, to the user.
3. User u sets up her reminder distance d_u and square side length a_u. d_u and a_u are securely stored on the user's local storage. u also generates two pseudo random function keys k_1, k_2 and keeps them secret in the local storage.

Figure 3: User enrollment

The user enrollment protocol is shown in Figure 3. When a user is enrolled, she selects her reminder distance d_u and square side length a_u. We note that a user may have multiple smart devices. If the user wants to combine one more device to her account, she needs to have the keys k_1 and k_2 stored in the device. The user can transfer the keys from a smart device which is already combined with her account to the newly added one through a secure out-of-band channel, such as near field communication (NFC) channel or blue tooth.

Figure 4 presents the location marking protocol. Our system needs a semantic secure symmetric encryption scheme, in which $Enc_k(\cdot)$ and $Dec_k(\cdot)$ are respectively the encrypting and decrypting algorithm under the key k. In the protocol, a user constructs the valid square identifier set with respect to the reminder location. The user encrypts the reminder message with a random key and then generates an index set for the reminder location. In each index tuple $t_{id} =< f(k_1, id), Enc_{g(k_2,id)}(k) >$, the first element is a pseudo random number generated by square identifier id, and the second one is an encryption using another pseudo random number derived from id as key. The first one is used for search. If a pseudo random number $f(k_1, id')$ matches

Function: User u marks location (x, y) as a reminder location with a reminder message m.

1 u randomly selects a symmetric encryption key k and encrypts the reminder message as $C = Enc_k(m)$.

2 u gets the valid square identifier set $\mathcal{ID}_{(x,y)}$ ←CreateCross $(x,\ y,\ d_u,\ a_u)$. For each $id \in \mathcal{ID}_{(x,y)}$, u calculates an index tuple $t_{id} =< f(k_1, id), Enc_{g(k_2,id)}(k) >$ and puts t_{id} into an index set $\mathcal{T}$. Finally, user u shuffles the elements in $\mathcal{T}$ and submits $< \mathcal{T}, C >$ to the cloud server s

3 Cloud server s maintains a sorted index set $\mathcal{IND}_u$ for each user u. The index set is sorted by the first trapdoor value $f(k_1, id)$ in the index tuple t_{id}. Upon receiving $< \mathcal{T}, C >$ from u, s stores C, obtains a reference r to C, and updates $\mathcal{IND}_u$ as $\mathcal{IND}_u = \mathcal{IND}_u \cup \{< b, c, r > \mid < b, c > \in \mathcal{T}\}$

Figure 4: Location marking

$f(k_1, id)$, it is the case that $id' = id$ because f is a secure pseudo random function. The user can use the second pseudo random function to derive the decryption key for $Enc_{g(k_2,id)}(k) >$ and obtain the message encryption key k. The user submits the index set and the ciphertext of the reminder message to the cloud server, which merges the submission to the user's $\mathcal{IND}_u$

After marking the locations in her reminder, a user periodically checks whether her current location is within any reminder area. This process is specified by the location search protocol in our system, which is presented in Figure 5.

Generally speaking, in the location search protocol, user u checks whether the cross centered at the current location intersects with any cross stored on the cloud server and retrieves the corresponding reminder messages if the condition in Step **3** is satisfied. Specifically, if the two crosses intersect on a square id, the trapdoor $ind_{id} = f(k_1, id)$ calculated in the location marking protocol and the location search protocol are identical. Therefore, if any non-empty $\mathcal{INT}$ is returned by the s in Step **2**, u knows not only that the cross centered at the current location intersects with the cross centered at a certain reminder location but also the exact intersection squares by looking up the trapdoors in the table T. In Step **3**, the user filters out the out-of-range reminder message references, because it is possible that the distance between the two locations is longer than d_u even if their crosses are intersected, e.g. Figures 2(b) and 2(d). For the remaining references, the user retrieves and decrypts them to obtain the reminder messages. The key point here is that the symmetric key for encrypting the reminder message is stored in the ciphertexts encrypted by $g(k_2, id)$, where k_2 is known to the user and id is any square identifier in the cross, including the identifier of the intersection square. Since the user has the intersection square identifier id, she can derive the key $g(k_2, id)$ and then obtain the symmetric key to decrypt the ciphertext of the reminder message.

4. SIMULATION

We evaluate the performance of our system through simulations. The simulations were carried out on an Motorola

Function: Check whether user u's current location (x, y) is within any reminder area.

1 u gets the valid square identifier set $\mathcal{ID}_{(x,y)}$ ←CreateCross $(x,\ y,\ d_u,\ a_u)$. For each $id \in \mathcal{ID}_{(x,y)}$, u calculates trapdoor $ind_{id} = f(k_1, id)$ and puts it in set $\mathcal{I}$. u shuffles set $\mathcal{I}$ and submits it to the cloud server s. u also keeps all calculated $< ind_{id}, id >$'s in a table T.

2 Upon receiving $\mathcal{I}$ from u, cloud server s searches the user's index set $\mathcal{IND}_u$ against each element in $\mathcal{I}$ and returns $\mathcal{INT} = \{< b, c, r > \mid b \in \mathcal{I} \wedge < b, c, r > \in \mathcal{IND}_u\}$ to the user.

3 Upon receiving $\mathcal{INT}$, the user u constructs a request set $\mathcal{R}$ as follows:

- If a reference r appears for no less than $\lceil d_u/a_u \rceil + 1$ times in $\mathcal{INT}$, add it in $\mathcal{R}$;
- If a reference r appears for two times in $\mathcal{INT}$, find the two corresponding intersection square. If they are on the same line, drop r and continue to check the next available reference. Otherwise, assume the two square identifiers are $id^{i,*}_{(x,y)}$ and $id^{*,j}_{(x,y)}$ $i, j = \pm 1, \cdots, \pm(\lceil d_u/a_u \rceil)$. If ${a_u}^2 \cdot (|i| + 1)^2 + {a_u}^2 \cdot (|j| + 1)^2 \le {d_u}^2$, then u adds r in $\mathcal{R}$.

u sends $\mathcal{R}$ to s.

4 Upon receiving a request set $\mathcal{R}$ from user u, s sends a ciphertext set $\mathcal{C} = \{< C, r > \mid r \in \mathcal{R}\}$ to the user.

5 Upon receiving $\mathcal{C}$ from the server, the user decrypts the ciphertexts. For each $< C_i, r_i > \in \mathcal{C}$, u finds a square identifier id from table T such that $< b, c, r >_j \in \mathcal{INT}$, $r_i = r_j$, and $ind_{id} = b_j$. u decrypts C_i using the decryption key $k_i = Dec_{g(k_2,id)}(c_j)$ to obtain the reminder message M_i.

6 After obtaining message M_i, user u may want to remove it from the cloud server. The user then sends a delete request and the corresponding reference r_i to the cloud server s. s removes C_i and all elements containing r_i in $\mathcal{IND}_u$.

Figure 5: Location search

Droid phone and a laptop, which simulates a cloud server. The smartphone runs an Android 2.2 OS and has a 550MHz ARM A8 processor, 256MB memory, and a 16GB SD card. The laptop has a 2.53GHz Intel Dual Core CPU and a 4GB memory. The cryptographic library used in our simulation is Crypto++5.6.1 [7] and we ported it to the Android platform. We use SHA-1 and SHA512 as the hash functions in our simulations and AES encryption in CFB mode as the semantic secure symmetric encryption in the system.

Our pseudo random function uses a hash-into-group hash function, which hashes an input into a group element. We first tested the success probability of the construction of the hash function, which took up to 10 attempts before it outputs *fail*. For test purpose, we hashed 1000 random numbers into four groups of different orders, 128, 160, 512, and 1024

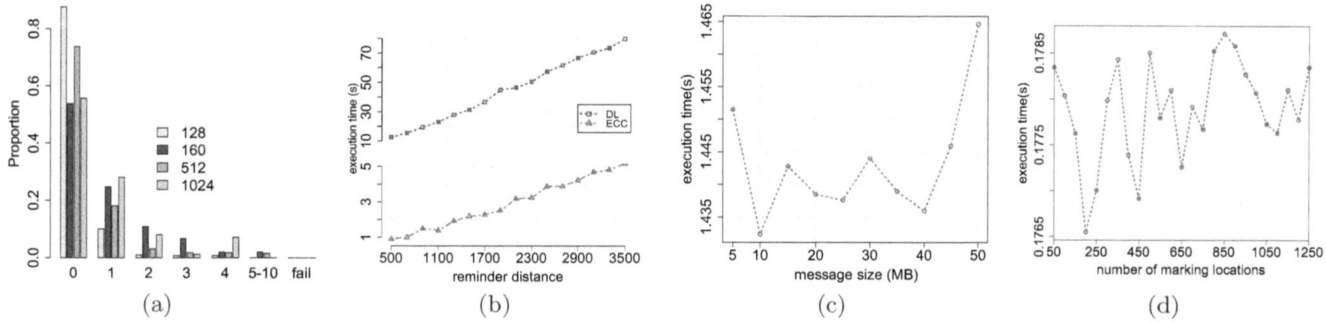

Figure 6: Simulation result

bit groups. The first two group orders are popular selections for the elliptic curve groups (ECC) and the last two are for the quadratic residue groups over the Galois field mod a safe prime number, which is called DL groups in this section. We obtained 160-bit and 512-bit outputs by using SHA-1 and SHA-512. In order to obtain a 128-bit output, we used SHA-1 and took the 128 least significant bits as the output. To obtain 1024-bit output, the binary string of the input number was divided into two halves, which are hashed into a 512-bit string, respectively. The two resulting hash values were then concatenated to form a 1024 bit output. The number of attempts needed to hash each number into the desired group was recorded. The hash is called *fail* if the attempt number is beyond 10. Figure 6(a) shows the result. We can see that, in our experiments, to hash a number into 128-bit and 1024-bit groups, we needed to try no more than 4 times. For the 160-bit and 512-bit groups, we needed to re-hash a small portion of numbers for more than 5 times. We note that, in our experiments, all 1000 random numbers were successfully hashed into the groups in 10 attempts.

When a user marks a location, most of the time is spent on the calculation of the trapdoors of the cross squares, which is dependent on $\lceil d_u/a_u \rceil$. We recorded the time needed for a user to compute $< \mathcal{T}, C >$ for a location while the square side length a_u is fixed at 20 meters and the reminder distance is from 500 meters to 3500 meters, with 200-meter increment. Our system works on an underlying cyclic group, on which the DDH problem is hard. There are two types of groups, a quadratic residue subgroup of a Galois field mode a safe prime (DL group) and an elliptic group (ECC group). To compare their performances, we implemented our system over a 1024-bit DL group and a 160-bit ECC group, which has an equivalent security level according to NIST's guidance [15]. The recorded computation time needed for different reminder distance over two groups are shown in Figure 6(b). First, it can be seen from the figure that the computation time increases linearly with the increase of the reminder distance. Second, it is shown that the implementation based on DL group costs much more time than the one based on ECC group does. This is because the DL group size is much larger than the ECC group size and a group operation costs more time in a larger group.

Our system does not encrypt the reminder message using the second trapdoor $g(k_2, id)$. We use a random key to encrypt the reminder message and use the second trapdoor to encrypt the random key. This saves time when the message size goes large. We tested the computation time of a user's marking location for different sizes of reminder messages. We fixed $a_u = 20$ meters and $d_u = 1000$ meters and increase the reminder message size from 5M bytes to 50M bytes with 5M-byte increment. The result is shown in Figure 6(c). As shown in the figure, the execution time did not change much, from 1.435 seconds to 1.465 seconds, because the extra time is just spent on the symmetric encryption, which is fast.

On the cloud server side, our system lets the server keep a sorted set of (by trapdoor value) user index tuples to speed up the search for a trapdoor. We used a laptop to simulate the server and used a red-black tree [6] to maintain the sorted index set. We recorded the time for searching a random location when different numbers of the locations have been marked. The number of marking locations is from 50 to 1250, with an increment of 50. We note that marking a location or searching a location incurs $4\lceil d_u/a_u \rceil + 1$ searches in the set. The result is shown in Figure 6(d). It can be seen from the figure that with the increase of the number of marking location, the search time increases but not linearly with the number, because the red-black tree's leaf nodes are not on the same level and the searching may end at some internal node. It is also shown in the figure that the time variance gap is small, from 0.1765 second to 0.1787 second.

5. RELATED WORK

In this section, we review two security areas that are related to our work: location privacy and searchable encryption. We also differentiate our work with prior works.

5.1 Location Privacy

Prior studies on location privacy mainly focused on privacy issues in the traditional location based service (LBS), in which a user sends the LBS server a location query and the server returns the local information around the query location. An extensive literature on location privacy utilizes the idea of K-anonymity or location cloaking [1, 10, 13]. In this approach, a user's query is relayed by a trusted third party to the LBS server. The trusted third party does not query the LBS server until K different user queries are collected from the same area. In this way, the LBS server cannot tell which user queries a specific location. The assumption that the third party is fully trusted by users may become a security weakness of this approach. Kalnis *et al.* [12] pointed out that there exists certain scenarios in which K-anonymity approach leaks the private location information to malicious entities. In addition, these approaches employ complex server query processing techniques and entail the transmission of large quantities of intermediate results between users and the trusted third party. Recently, Zhao *et al.* [20] proposed a novel approach utilizing data structures to protect the locaton privacy in location based social networks.

Different from the above works, we study the location privacy issue in a cloud-assisted location based reminder system, in which a user not only provides a location query, i.e. the current location, but also sets a reminder location and both locations are oblivious to the server. In addition, the server is enabled to test whether the current location appears in a reminder area even if it does not know the current locations and the reminder locations.

5.2 Searchable Encryption

In order to enable the square IDs to be searchable, we propose to use the idea of searchable symmetric encryption. To the best of our knowledge, searchable encryption was first proposed by Song *et al.* [17] and has received extensive attentions [2] . The earlier progress has been made on its efficiency and security definition formalization. The searchable index schemes proposed by Goh [9], and Chang and Mitzenmacher [5] reduce the search cost to one proportional to the number of files. In [8], Curtmola *et al.* summarized previous studies and gave a new formal security definition on searchable symmetric encryption. Recently, a conjunctive similarity keyword search over encrypted data is proposed by Wang *et al.* [19]. Their scheme is proved to be non-adaptive semantic security defined in [8]. Regarding the public key setting, Boneh *et al.* gave the first construction on searchable asymmetric encryption, where anyone with the public key can write to the data stored in the server but only the ones with the private key can search on the data [3]. Conjunctive keyword search, subset query, and range query over encrypted data [4] were also proposed in the public key setting.

6. CONCLUSIONS

In this paper, we have proposed a secure location based reminder system. This system enables the cloud server to securely search over the reminders stored in it. In addition, the user can receive the reminder message from the cloud server with whichever smart device she brought with her as long as she is in the reminder area. Utilizing the searchable symmetric encryption, we have proposed a novel area representation approach, which is easy to construct and achieves lightweight storage overhead. We have theoretically proved the security of our system. The evaluation of our system has shown that it is efficient to implement.

7. REFERENCES

[1] B. Bamba, L. Liu, P. Pesti, and T. Wang, "Supporting anonymous location queries in mobile environments with privacygrid," in *WWW*, 2008, pp. 237–246.

[2] M. Bellare, A. Boldyreva, and A. O'Neill, "Deterministic and efficiently searchable encryption," in *CRYPTO*, 2007, pp. 535–552.

[3] D. Boneh, G. D. Crescenzo, R. Ostrovsky, and G. Persiano, "Public key encryption with keyword search," in *EUROCRYPT*, 2004, pp. 506–522.

[4] D. Boneh and B. Waters, "Conjunctive, subset, and range queries on encrypted data," in *TCC*, 2007, pp. 535–554.

[5] Y.-C. Chang and M. Mitzenmacher, "Privacy preserving keyword searches on remote encrypted data," in *ACNS*, 2005, pp. 442–455.

[6] T. Cormen, C. Leiserson, R. Rivest, and C. Stein, *Introduction to algorithms.* MIT press, 2001.

[7] Crypto++, http://www.cryptopp.com/, 2010.

[8] R. Curtmola, J. A. Garay, S. Kamara, and R. Ostrovsky, "Searchable symmetric encryption: improved definitions and efficient constructions," in *ACM CCS*, 2006, pp. 79–88.

[9] E.-J. Goh, "Secure indexes," *IACR Cryptology ePrint Archive*, vol. 2003, p. 216, 2003.

[10] M. Gruteser and D. Grunwald, "Anonymous usage of location-based services through spatial and temporal cloaking," in *MobiSys*, 2003.

[11] S. Jarecki and X. Liu, "Fast secure computation of set intersection," *Security and Cryptography for Networks*, pp. 418–435, 2010.

[12] P. Kalnis, G. Ghinita, K. Mouratidis, and D. Papadias, "Preserving anonymity in location based services," National University of Sigapore, Tech. Rep., 2006.

[13] M. F. Mokbel, C.-Y. Chow, and W. G. Aref, "The new casper: Query processing for location services without compromising privacy," in *VLDB*, 2006, pp. 763–774.

[14] A. Narayanan, N. Thiagarajan, M. Lakhani, M. Hamburg, and D. Boneh, "Location privacy via private proximity testing," in *Proc. of NDSS*, vol. 2011, 2011.

[15] *Implementation Guidance for FIPS PUB 140-2 and the Cryptographic Module Validation Program*, NIST, CSE, Mar. 2011, available at http://csrc.nist.gov/groups/STM/cmvp/documents/fips140-2/FIPS1402IG.pdf.

[16] R. Shokri, G. Theodorakopoulos, J.-Y. L. Boudec, and J.-P. Hubaux, "Quantifying location privacy," in *IEEE Symposium on Security and Privacy*, 2011, pp. 247–262.

[17] D. X. Song, D. Wagner, and A. Perrig, "Practical techniques for searches on encrypted data," in *IEEE Symposium on Security and Privacy*, 2000, pp. 44–55.

[18] D. Stinson, *Cryptography: theory and practice.* CRC press, 2006.

[19] C. Wang, K. Ren, S. Yu, and K. M. R. Urs, "Achieving usable and privacy-assured similarity search over outsourced cloud data," in *INFOCOM*, 2012, pp. 451–459.

[20] X. Zhao, L. Li, and G. Xue, "Checking in without worries: Location privacy in location based social networks," in *INFOCOM*, 2013.

DroidChameleon: Evaluating Android Anti-malware against Transformation Attacks

Vaibhav Rastogi, Yan Chen, and Xuxian Jiang†
Northwestern University, †North Carolina State University
vrastogi@u.northwestern.edu, ychen@northwestern.edu, jiang@cs.ncsu.edu

ABSTRACT

Mobile malware threats (e.g., on Android) have recently become a real concern. In this paper, we evaluate the state-of-the-art commercial mobile anti-malware products for Android and test how resistant they are against various common obfuscation techniques (even with known malware). Such an evaluation is important for not only measuring the available defense against mobile malware threats but also proposing effective, next-generation solutions. We developed DroidChameleon, a systematic framework with various transformation techniques, and used it for our study. Our results on ten popular commercial anti-malware applications for Android are worrisome: none of these tools is resistant against common malware transformation techniques. Moreover, the transformations are simple in most cases and anti-malware tools make little effort to provide transformation-resilient detection. Finally, in the light of our results, we propose possible remedies for improving the current state of malware detection on mobile devices.

Categories and Subject Descriptors

D.4.6 [**Operating Systems**]: Security and Protection—*Invasive software (e.g., viruses, worms, Trojan horses)*

General Terms

Security

Keywords

Mobile; malware; anti-malware; Android

1. INTRODUCTION

Mobile computing devices such as smartphones and tablets are becoming increasingly popular. Unfortunately, this popularity attracts malware authors too. It has been reported that on Android, the most popular smartphone platform [11], malware has constantly been on the rise [13]. With the growth of malware, the platform has also seen an evolution of anti-malware tools, with a range of free and paid offerings now available in the official Android app market, Google Play.

In this paper, we aim to evaluate the efficacy of anti-malware tools on Android in the face of various evasion techniques. For example, polymorphism is used to evade detection tools by transforming a malware in different forms ("morphs") but with the same code. Metamorphism is another common technique that can mutate code so that it no longer remains the same but still has the same behavior. For ease of presentation, we use the term polymorphism in this paper to represent both obfuscation techniques. In addition, we use the term 'transformation' broadly, to refer to various polymorphic or metamorphic changes.

Polymorphic attacks have long been a plague for traditional desktop and server systems. While there exist earlier studies on the effectiveness of anti-malware tools on PCs [8], our domain of study is different in that we exclusively focus on mobile devices like smartphones, which require different ways for anti-malware design. Also, malware on mobile devices have recently escalated their evolution but the capabilities of existing anti-malware tools are largely not yet understood. In the meantime, simple forms of polymorphic attacks have already been seen in the wild [26].

To evaluate existing anti-malware software, we develop a systematic framework called *DroidChameleon* with several common transformation techniques that may be used to transform Android applications automatically. Some of these transformations are highly specific to the Android platform only. Based on the framework, we pass known malware samples (from different families) through these transformations to generate new variants of malware, which are verified to possess the originals' malicious functionality. We use these variants to evaluate the effectiveness and robustness of popular anti-malware tools.

Our results on ten popular anti-malware products, some of which even claim resistance against malware transformations, show that all the anti-malware products used in our study have little protection against common transformation techniques. Our results also give insights about detection models used in existing anti-malware and their capabilities, thus shedding light on possible ways for their improvements. We hope that our findings work as a wake-up call and motivation for the community to improve the current state of mobile malware detection.

To summarize, this paper makes the following contributions.

- We systematically evaluate anti-malware products for Android regarding their resistance against various transformation techniques in known malware. For this purpose, we developed DroidChameleon, a systematic framework with various transformation techniques to facilitate anti-malware evaluation.
- We have implemented a prototype of DroidChameleon and used it to evaluate ten popular anti-malware products for Android. Our findings show that all of them are vulnerable to common evasion techniques. Moreover, we find that 90% of the signatures studied do not require static analysis of bytecode.
- Based on our evaluation results, we also explore possible ways to improve current anti-malware solutions. Specifically, we point

ASIA CCS'13, May 8–10, 2013, Hangzhou, China.

out that Android eases developing advanced detection techniques because much code is high-level bytecodes rather than native codes. Furthermore, certain platform support can be enlisted to cope with advanced transformations.

2. BACKGROUND

Android is an operating system for mobile devices such as smartphones and tablets. It is based on the Linux kernel and provides a middleware implementing subsystems such as telephony, window management, management of communication with and between applications, managing application lifecycle, and so on.

Applications are programmed primarily in Java though the programmers are allowed to do native programming via JNI (Java native interface). Instead of running Java bytecode, Android runs Dalvik bytecode, which is produced from Java bytecode. In Dalvik, instead of having multiple `.class` files as in the case of Java, all the classes are packed together in a single `.dex` file.

Android applications are made of four types of components, namely activities, services, broadcast receivers, and content providers. These application components are implemented as classes in application code and are declared in the AndroidManifest (see next paragraph). The Android middleware interacts with the application through these components.

Android application packages are jar files containing the application bytecode as a `classes.dex` file, any native code libraries, application resources such as images, config files and so on, and a manifest, called AndroidManifest. It is a binary XML file, which declares the application package name, a string that is supposed to be unique to an application, and the different components in the application. It also declares other things (such as application permissions) which are not so relevant to the present work. The AndroidManifest is written in human readable XML and is transformed to binary XML during application build.

Only digitally signed applications may be installed on an Android device. Signing keys are usually owned by individual developers and not by a central authority, and there is no chain of trust. All third party applications run unprivileged on Android.

3. FRAMEWORK DESIGN

In this work, we focus on the evaluation of anti-malware products for Android. Specifically, we attempt to deduce the kind of signatures that these products use to detect malware and how resistant these signatures are against changes in the malware binaries. In this paper, we generally use the term transformation to denote semantics preserving changes to a program. Since we are dealing with malware, we only care about the interested semantics such as sending SMS message to a premium number and not things like change of application name in the system logs.

In this work, we develop several different kinds of transformations that may be applied to malware samples while preserving their malicious behavior. Each malware sample undergoes one or more transformations and then passes through the anti-malware tools. The detection results are then collected and used to make deductions about the detection strengths of these anti-malware tools.

We classify our transformations as trivial (which do not require code level changes or changes to meta-data stored in AndroidManifest), those which result in variants that can still be detected by advanced static analyses involving data-flow and control-flow analysis (DSA), and those which can render malware undetectable by static analysis (NSA). In the rest of this section, we describe the different kinds of transformations that we have in the DroidChameleon framework. Where appropriate we give examples, using original and transformed code. Transformations for Dalvik bytecode are given in Smali (as in Listing 1), an intuitive assembly language for Dalvik bytecode.

```
const-string v10, "profile"
const-string v11, "mount -o remount rw system\nexit\n"
invoke-static {v10, v11}, Lcom/android/root/Setting;->
    runRootCommand(Ljava/lang/String;Ljava/lang/String;)
    Ljava/lang/String;
move-result-object v7
```

Listing 1: A code fragment from DroidDream malware

3.1 Trivial Transformations

Trivial transformations do not require code-level changes or changes to meta-data stored in AndroidManifest. These transformations are meant to defeat signatures based on whole files (or a part of file that changes simply by reorganizing file sections) or the key used to sign an application package. We have the following two transformations for this purpose.

Repacking. Recall that Android packages are signed jar files. These may be unzipped with the regular zip utilities and then repacked again with tools offered in the Android SDK. Once repacked, applications are signed with custom keys (the original developer keys are not available). Detection signatures that match the developer keys or a checksum of the entire application package are rendered ineffective by this transformation. Note that this transformation applies to Android applications only; there is no counterpart in general on traditional Desktop operating systems although the malware in the latter are known to use sophisticated packers for the purpose of evading anti-malware tools.

Disassembling and Reassembling. The compiled Dalvik bytecode in `classes.dex` of the application package may be disassembled and then reassembled back again. The various items in a dex file may be arranged or represented in different ways and thus a compiled program may be represented in more than one form. Signatures that match the whole `classes.dex` are beaten by this transformation. Signatures that depend on the order of different items in the dex file will also likely break with this transformation. Similar assembling/disassembling also applies to the resources in an Android package and to the conversion of AndroidManifest between binary and human readable formats.

3.2 Transformation Attacks Detectable by Static Analysis (DSA)

The application of DSA transformations does not break all types of static analysis. Specifically, forms of analysis that describe the semantics, such as data flows are still possible. Only simpler checks such as string matching or matching API calls may be thwarted.

Changing Package Name. Every application is identified by a package name unique to the application. This name is defined in the package's AndroidManifest. We change the package name in a given malicious application to another name.

Identifier Renaming. Most class, method, and field identifiers in bytecode can be renamed. We note that several free obfuscation tools such as ProGuard [5] provide identifier renaming. Listing 2 presents an example transformation for code in Listing 1.

```
const-string v10, "profile"
const-string v11, "mount -o remount rw system\nexit\n"
invoke-static {v10, v11}, Lcom/hxbvgH/IWNcZs/jFAbKo;->
    axDnBL(Ljava/lang/String;Ljava/lang/String;)Ljava/
    lang/String;
move-result-object v7
```

Listing 2: Code in Listing 1 after identifier renaming

Data Encryption. The dex files contain all the strings and array data that have been used in the code. These strings and arrays may be used to develop signatures against malware. To beat such signatures we can keep these in encrypted form. Listing 3 shows code in Listing 1, transformed by string encryption.

```
const-string v10, "qspgjmf"
invoke-static {v10}, Lcom/EncryptString;->applyCaesar(
     Ljava/lang/String;)Ljava/lang/String;
move-result-object v10
const-string v11, "npvou!.p!sfnpvou!sx!tztufn]ofyju]o"
invoke-static {v11}, Lcom/EncryptString;->applyCaesar(
     Ljava/lang/String;)Ljava/lang/String;
move-result-object v11
invoke-static {v10, v11}, Lcom/android/root/Setting;->
     runRootCommand(Ljava/lang/String;Ljava/lang/String;)
     Ljava/lang/String;
move-result-object v7
```

Listing 3: Code in Listing 1 after string encryption. Strings are encoded with a Caesar cipher of shift +1.

Call Indirections. This transformation can be seen as a simple way to manipulate call graph of the application to defeat automatic matching. Given a method call, the call is converted to a call to a previously non-existing method that then calls the method in the original call. This can be done for all calls, those going out into framework libraries as well as those within the application code. This transformation may be seen as trivial function outlining (see function outlining below).

Code Reordering. Code reordering reorders the instructions in the methods of a program. This transformation is accomplished by reordering the instructions and inserting `goto` instructions to preserve the runtime execution sequence of the instructions. Listing 4 shows an example reordering.

```
goto :i_1
:i_3
invoke-static {v10, v11}, Lcom/android/root/Setting;->
     runRootCommand(Ljava/lang/String;Ljava/lang/String;)
     Ljava/lang/String;
move-result-object v7
goto :i_4  # next instruction
:i_2
const-string v11, "mount -o remount rw system\nexit\n"
goto :i_3
:i_1
const-string v10, "profile"
goto :i_2
```

Listing 4: Code in Listing 1 reverse ordered

Junk Code Insertion. These transformations introduce code sequences that are executed but do not affect rest of the program. Detection based on analyzing instruction (or opcode) sequences may be defeated by junk code insertion. Junk code may constitute simple `nop` sequences or more sophisticated sequences and branches that actually have no effect on the semantics.

Encrypting Payloads and Native Exploits. In Android, native code is usually made available as libraries accessed via JNI. However, some malware such as DroidDream also pack native code exploits meant to run from a command line in non-standard locations in the application package. All such files may be stored encrypted in the application package and be decrypted at runtime. Certain malware such as DroidDream also carry payload applications that are installed once the system has been compromised. These payloads may also be stored encrypted. We categorize payload and exploit encryption as DSA because signature based static detection is still possible based on the main application's bytecode. These are easily implemented and have been seen in practice as well (e.g., DroidKungFu malware uses encrypted exploit).

Function Outlining and Inlining. In function outlining, a function is broken down into several smaller functions. Function inlining involves replacing a function call with the entire function body. These are typical compiler optimization techniques. However, outlining and inlining can be used for call graph obfuscation also.

Other Simple Transformations. There are a few other transformations as well, specific to Android. Debug information, such as source file names, local and parameter variable names, and source line numbers may be stripped off. Moreover, non-code files and resources contained in Android packages may be renamed or modified.

Composite Transformations. Any of the above transformations may be combined with one another to generate stronger obfuscations. While compositions are not commutative, anti-malware detection results should be agnostic to the order of application of transformations for the cases discussed above.

3.3 Transformation Attacks Non-Detectable by Static Analysis (NSA)

These transformations can break all kinds of static analysis. Some encoding or encryption is typically required so that no static analysis scheme can infer parts of the code. Parts of the encryption keys may even be fetched remotely. In this scenario, interpreting or emulating the code (i.e., dynamic analysis) is still possible but static analysis becomes infeasible.

Reflection. The Java reflection API allows a program to invoke a method by using the name of the methods. We can convert any method call into a call to that method via reflection. This makes it difficult to analyze statically which method is being called. A subsequent encryption of the method name can make it impossible for any static analysis to recover the call.

Bytecode encryption. Code encryption tries to make the code unavailable for static analysis. The relevant piece of the application code is stored in an encrypted form and is decrypted at runtime via a decryption routine. Code encryption has long been used in polymorphic viruses; the only code available to signature based antivirus applications remains the decryption routine, which is typically obfuscated. To accomplish this the majority of the malware code may be stored in an encrypted dex file that is decrypted and loaded dynamically through a user-defined class loader.

4. IMPLEMENTATION

Apart from function outlining and inlining, we applied all other DroidChameleon transformations to the malware samples. We have implemented most of the transformations so that they may be applied automatically to the application. Automation implies that the malware authors can generate polymorphic malware at a very fast pace.

We utilize the Smali/Baksmali [6] and its companion tool Apktool [1] for our implementation. Our code-level transformations are implemented over Smali. Moreover, disassembling and assembling transformation uses Apktool. This has the effect of repacking, changing the order and representation of items in the `classes.dex` file, and changing the AndroidManifest (while preserving the semantics of it). All other transformations in our implementation (apart from repacking) make use of Apktool to unpack/repack application packages.

5. RESULTS

We begin by describing our anti-malware and malware dataset, followed by our methodology, and then discuss our findings. We

Table 1: Anti-malware products evaluated. All tools collected in February 2013.

Vendor	Product	Package name	Version	# downloads
AVG	Antivirus Free	com.antivirus	3.1	50M-100M
Symantec	Norton Mobile Security	com.symantec.mobilesecurity	3.3.0.892	5M-10M
Lookout	Lookout Mobile Security	com.lookout	8.7.1-EDC6DFS	10M-50M
ESET	ESET Mobile Security	com.eset.ems	1.1.995.1221	500K-1M
Dr. Web	Dr. Web anti-virus Light	com.drweb	7.00.3	10M-50M
Kaspersky	Kaspersky Mobile Security	com.kms	9.36.28	1M-5M
Trend micro	Mobile Security Personal Ed.	com.trendmicro.tmmspersonal	2.6.2	100K-500K
ESTSoft	ALYac Android	com.estsoft.alyac	1.3.5.2	5M-10M
Zoner	Zoner Antivirus Free	com.zoner.android.antivirus	1.7.2	1M-5M
Webroot	Webroot Security & Antivirus	com.webroot.security	3.1.0.4547	500K-1M

Table 2: Malware samples used for testing anti-malware tools

Family	Package name	SHA-1 code	Date found	Remarks
DroidDream	com.droiddream. bowling-time	72adcf43e5f945ca9f72 064b81dc0062007f0fbf	03/2011	Root exploit
Geinimi	com.sgg.spp	1317d996682f4ae4cce6 0d90c43fe3e674f60c22	10/2011	Information exfiltration; bot-like capabilities
Fakeplayer	org.me.androidappli-cation1	1e993b0632d5bc6f0741 0ee31e41dd316435d997	08/2010	SMS trojan
Bgserv	com.android.vending. sec-tool.v1	bc2dedad0507a916604f 86167a9fa306939e2080	03/2011	Information exfiltration; bot-like capabilities; SMS trojan
BaseBridge	com.keji.unclear	508353d18cb9f5544b1e d1cf7ef8a0b6a5552414	05/2011	Root exploit; SMS trojan packed as payload
Plankton	com.crazyapps.angry. birds.rio.unlocker	bee2661a4e4b347b5cd2 a58f7c4b17bcc3efd550	06/2011	Dynamic code loading

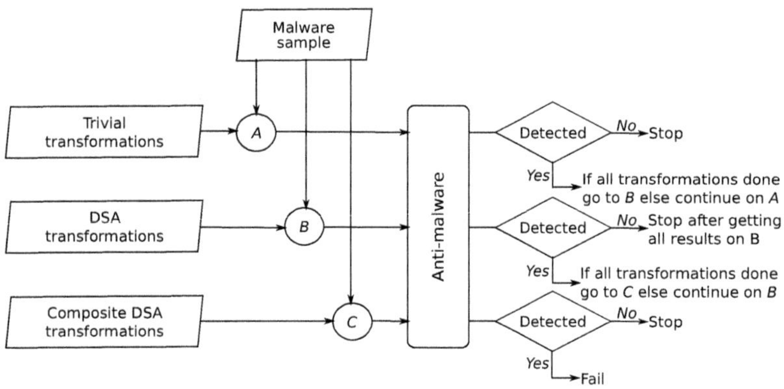

Figure 1: Evaluating anti-malware

Table 3: Key to Table 4. Transformations coded with single letters are trivial transformations. All others are DSA. We did not need NSA transformations to thwart anti-malware tools.

Code	Technique
P	Repack
A	Dissassemble & assemble
RP	Rename package
EE	Encrypt native exploit or payload
RI	Rename identifiers
RF	Rename files
ED	Encrypt strings and array data
CR	Reorder code
CI	Call indirection
JN	Insert junk code
All transformations contain P	
All transformations except P contain A	

evaluated ten anti-malware tools, which are listed in Table 1. We selected the most popular products; in addition, we included Kaspersky, ESET, and Trend Micro, which were then not very popular but are well established vendors in the security industry. We had to omit a couple of products in the most popular list because they would fail to identify many original, unmodified malware samples we tested. All the products were downloaded directly from the official Android app market, Google Play, in February 2013.

Our malware set is summarized in Table 2. We used a few criteria for choosing malware samples. First, all the anti-malware tools being evaluated should detect the original samples. Second, the malware samples should be sufficiently old so that signatures against them are well stabilized. All the samples in our set were discovered in or before October 2011. All the samples are publicly available on Contagio Minidump [22]. Finally, as seen in the table, the set spans over multiple malware kinds, from root exploits to information stealing.

As has already been discussed, we transform malware samples using various techniques discussed in Section 3 and pass them through anti-malware tools we evaluate. Our methodology is depicted in Figure 1. For every malware-antimalware pair, we begin testing with trivial transformations and then proceed with transformations that are more complex. Each transformation is applied to a malware sample (of course, some like exploit encryption apply only in certain cases) and the transformed sample is passed through anti-malware. If detection breaks with trivial transformations, we stop (all DSA and NSA transformations also result in trivial transformations). Next, we apply all the DSA transformations. If detection still does not break, we apply combinations of DSA transformations. In general there is no well-defined order in which transformations should be applied (in some cases a heuristic works; for example, malware that include native exploits are likely to be detected based on those exploits). Fortunately, in our study, we did not need to apply combinations of more than two transformations to break detection. When applying combinations of transformations, we stopped when detection broke.

Our results with all the malware samples are summarized in Table 4. This table gives the minimal transformations necessary to evade detection for malware-anti-malware pairs. For example, DroidDream requires both exploit encryption and call indirection to evade Dr. Web's detection. These minimal transformations also give insight into what kind of detection signatures are being used. We next describe our key findings in the light of the detection results.

Table 4: Evaluation summary. Please see Table 3 for key. '+' indicates the composition of two transformations.

	DroidDream	Geinimi	Fakeplayer	Bgserv	BaseBridge	Plankton
AVG	RP	RI	RP + RI	RI	RI	RP + RI
Symantec	RI	RI	RP + RI	RI + ED	ED	P
Lookout	P	RI + ED	RP + RI	RI + ED	EE + ED	RI
ESET	RI + EE	ED	RI	RI	EE + ED	RI + ED
Dr. Web	EE + CI	CI	CI	CI	EE + CI	CI
Kaspersky	EE + ED	RI	RI	RI + ED	EE + ED	A
Trend M.	EE + RF	RI	A	A	EE + RF	A
ESTSoft	RP	RP	RP	RP	RP	RP
Zoner	A	RI	A	A	A	RI
Webroot	RI	RI	RP	RI	RP	RI

Finding 1 *All the studied anti-malware products are vulnerable to common transformations.* All the transformations appearing in Table 4 are easy to develop and apply, redefine only certain syntactic properties of the malware, and are common ways to transform malware. Transformations like identifier renaming and data encryption are easily available using free and commercial tools [4, 5]. Exploit and payload encryption is also easy to achieve. Such transformations may already be seen in the wild in current malware, e.g., Geinimi variants have encrypted strings [19] and DroidKungFu malware uses encrypted exploit code [3].

We found that only Dr. Web uses a somewhat more sophisticated algorithm for detection. Our findings indicate that the general detection scheme of Dr. Web is as follows. The set of method calls from every method is obtained. These sets are then used as signatures and the detection phase consists of matching these sets against sets obtained from the sample under test.

Finding 2 *At least 43% signatures are not based on code-level artifacts.* That is, these are based on file names, checksums (or binary sequences) or information easily obtained by the PackageManager API. We also found all AVG signatures to be derived from the content of AndroidManifest only (and hence that of the PackageManager API). Changing component names in AndroidManifest while keeping the code same was sufficient to break AVG's detection.

Finding 3 *90% of signatures do not require static analysis of bytecode. Only one of ten anti-malware tools was found to be using static analysis.* Names of classes, methods, and fields, and all the strings and array data are stored in the `classes.dex` file as they are and hence can be obtained by content matching. The only signatures requiring static analysis of bytecode are those of Dr. Web because it extracts API calls made in various methods.

6. DEFENSES

6.1 Semantics-based Malware Detection

We point out that owing to the use of bytecodes, which contain high-level structural information, analyses of Android applications becomes much simpler than those of native binaries. Hence, semantics-based detection schemes could prove especially helpful in the case of Android. For example, Christodorescu et al. [9] describe a technique for semantics based detection. Their algorithms are based on unifying nodes in a given program with nodes in a signature template (nodes may be understood as abstract instructions), while preserving def-use paths described in the template. Since this technique is based on data flows rather than a superficial property of the program such as certain strings or names of methods being defined or called, it is not vulnerable to any of the transformations (all of which are trivial or DSA) that show up in Table 4.

Semantics-based detection is quite challenging for native codes; their analyses frequently encounters issues such as missing information on function boundaries, pointer aliasing, and so on [16, 25]. Bytecodes, on the other hand, preserve much of the source-level information, thus easing analysis. We therefore believe that anti-malware tools have greater incentive to implement semantic analysis techniques on Android bytecodes than they had for developing these for native code.

6.2 Support from Platform

Note that the use of code encryption and reflection (NSA transformations) can still defeat the above scheme. Code encryption does not leave much visible code on which signatures can be developed. The use of reflection simply hides away the edges in the call graph. If the method names used for reflective invocations are encrypted, these edges are rendered completely opaque to static analysis. Furthermore, it is possible to use function outlining to thwart any forms of intra-procedural analysis as well. Owing to these limitations, the use of dynamic monitoring is essential.

Recall that anti-malware tools in Android are unprivileged third party applications. This impedes many different kinds of dynamic monitoring that may enhance malware detection. We believe special platform support for anti-malware applications is essential to detect malware amongst stock Android applications. This can help malware detection in several ways. For example, a common way to break evasion by code encryption is to scan the memory at runtime. The Android runtime could provide all the classes loaded using user-defined class loaders to the anti-malware application. Once the classes are loaded, they are already decrypted and anti-malware tools can analyze them easily.

Google Bouncer performs offline dynamic analysis for malware detection [18]. Such scanning however has its own problems, ranging from detection of the dynamic environment to the malicious activity not getting triggered in the limited time for which the analysis runs; Bouncer is no exception to this [21, 27]. We therefore believe offline emulation must be supplemented by strong static analysis or real-time dynamic monitoring.

7. RELATED WORK

Evaluating Anti-malware Tools. Zheng et al. [28] also studied the robustness of anti-malware against Android malware recently. They implement a subset of our transformations, use them to generate several malware variants, and test these on VirusTotal, a webservice that tests submitted samples against over 40 anti-virus products. Their results however only show the change in overall detection percentages as the transformations are applied. Our results are much stronger in that we can show that all anti-malware tools actually succumb for all malware samples tested. Moreover, we also deduce the weaknesses and strengths of some of the products.

Christodorescu and Jha [8] conducted a study similar to ours on desktop anti-malware applications nine years ago. They also arrived at the conclusion that these applications have low resilience against malware obfuscation. Our study is based on Android anti-malware, and we include several aspects in our study that are unique to Android. Furthermore, our study comes after much research on obfuscation resilient detection, and we would expect the proposed techniques to be readily integrated into new commercial products. Finally, our study is orthogonal to studies about completeness of detection of anti-malware tools such as those by AV-Test.org [2].

Obfuscation Techniques. Collberg et al. [12] review and propose different types of obfuscations. DroidChameleon provides only a few of the transformations proposed by them. Nonetheless, the set of transformations provided in DroidChameleon is comprehensive in the sense that they can break typical static detection techniques used by anti-malware. Off-the-shelf tools like Proguard [5] and Klassmaster [4] provide renaming of classes and class members, flow obfuscation, and string encryption. While the goal of these tools is to evade manual reverse engineering, we aim at thwarting analysis by automatic tools.

Obfuscated Malware Detection. Obfuscation resilient detection is based on semantics rather than syntac. As discussed earlier, Christodorescu et al. [9] present one such technique. Christodorescu et al. [10] and Fredrikson et al. [14] attempt to generate semantics based signatures by mining malicious behavior automatically. Kolbitsch et al. [17] also propose similar techniques. The last three works are for behavior-based detection and use different behavior representations such as data dependence graphs and information flows between system calls. Due to lower privileges for anti-malware tools on Android, these approaches cannot directly apply to these tools presently.

Smartphone Malware Research. Many works have been done towards discovery and characterization of smartphone malware [7, 15, 20, 23, 24, 29, 30]. Our work is distinct from these as we try to evaluate the efficacy of existing tools against transformed malware.

8. CONCLUSION

We evaluated ten anti-malware products on Android for their resilience against malware transformations. To facilitate this, we developed DroidChameleon, a systematic framework with various transformation techniques. Our findings show that all the anti-malware products evaluated are susceptible to common evasion techniques. Finally, we explored possible ways to improve the current situation and develop next-generation solutions.

We refer the interested readers to http://list.cs.northwestern.edu/mobile for further information about this work, such as more detailed technical reports.

References

[1] Android-apktool: A tool for reengineering Android apk files. http://code.google.com/p/android-apktool/.

[2] Test: Malware Protection for Android, March 2012. http://www.av-test.org/en/tests/android/.

[3] DroidKungFu. http://www.csc.ncsu.edu/faculty/jiang/DroidKungFu.html.

[4] Zelix Klassmaster. http://www.zelix.com/klassmaster/.

[5] ProGuard. http://proguard.sourceforge.net/.

[6] Smali: An assembler/disassembler for Android's dex format . http://code.google.com/p/smali/.

[7] I. Burguera, U. Zurutuza, and S. Nadjm-Tehrani. Crowdroid: behavior-based malware detection system for android. In *Proceedings of the 1st ACM workshop on Security and privacy in smartphones and mobile devices*, 2011.

[8] M. Christodorescu and S. Jha. Testing malware detectors. In *Proceedings of the ACM SIGSOFT international symposium on Software testing and analysis*, 2004.

[9] M. Christodorescu, S. Jha, S. Seshia, D. Song, and R. Bryant. Semantics-aware malware detection. In *Security and Privacy, 2005 IEEE Symposium on*, 2005.

[10] M. Christodorescu, S. Jha, and C. Kruegel. Mining specifications of malicious behavior. In *Proceedings of the the 6th ACM ESEC-FSE*, 2007.

[11] CNET, February 2013. http://news.cnet.com/8301-1035_3-57569402-94/android-ios-combine-for-91-percent-of-market/.

[12] C. Collberg, C. Thomborson, and D. Low. A taxonomy of obfuscating transformations. Technical report, Department of Computer Science, The University of Auckland, New Zealand, 1997.

[13] F-Secure. Mobile Threat Report Q3 2012. http://www.f-secure.com/static/doc/labs_global/Research/Mobile%20Threat%20Report%20Q3%202012.pdf.

[14] M. Fredrikson, S. Jha, M. Christodorescu, R. Sailer, and X. Yan. Synthesizing near-optimal malware specifications from suspicious behaviors. In *Security and Privacy (SP), 2010 IEEE Symposium on*, 2010.

[15] M. Grace, Y. Zhou, Q. Zhang, S. Zou, and X. Jiang. Riskranker: scalable and accurate zero-day android malware detection. In *Proceedings of the 10th international conference on Mobile systems, applications, and services*, MobiSys '12, 2012.

[16] L. Harris and B. Miller. Practical analysis of stripped binary code. *ACM SIGARCH Computer Architecture News*, 33(5): 63–68, 2005.

[17] C. Kolbitsch, P. Comparetti, C. Kruegel, E. Kirda, X. Zhou, and X. Wang. Effective and efficient malware detection at the end host. In *Proceedings of the 18th conference on USENIX security symposium*, 2009.

[18] H. Lockheimer. Android and security, February 2012. http://googlemobile.blogspot.com/2012/02/android-and-security.html.

[19] Lookout. Geinimi Trojan Technical Analysis. http://blog.mylookout.com/blog/2011/01/07/geinimi-trojan-technical-analysis/.

[20] Y. Nadji, J. Giffin, and P. Traynor. Automated remote repair for mobile malware. In *Proceedings of the 27th Annual Computer Security Applications Conference*, 2011.

[21] J. Oberheide. Dissecting android's bouncer, June 2012. https://blog.duosecurity.com/2012/06/dissecting-androids-bouncer/.

[22] M. Parkour. Contagio Mobile. Mobile Malware Mini Dump. http://contagiominidump.blogspot.com/.

[23] H. Peng, C. Gates, B. Sarma, N. Li, Y. Qi, R. Potharaju, C. Nita-Rotaru, and I. Molloy. Using probabilistic generative models for ranking risks of android apps. In *Proceedings of the 2012 ACM conference on Computer and communications security*, 2012.

[24] V. Rastogi, Y. Chen, and W. Enck. AppsPlayground: Automatic Security Analysis of Smartphone Applications. In *Proceedings of ACM CODASPY 2013*, February 2013.

[25] P. Saxena, R. Sekar, and V. Puranik. Efficient fine-grained binary instrumentationwith applications to taint-tracking. In *Proceedings of the 6th annual IEEE/ACM international symposium on Code generation and optimization*, 2008.

[26] Symantec. Server-side Polymorphic Android Applications. http://www.symantec.com/connect/blogs/server-side-polymorphic-android-applications.

[27] R. Whitwam. Circumventing Google's Bouncer, Android's anti-malware system, June 2012. http://www.extremetech.com/computing/130424-circumventing-googles-bouncer-androids-anti-malware-system.

[28] M. Zheng, P. Lee, and J. Lui. Adam: An automatic and extensible platform to stress test android anti-virus systems. *DIMVA*, July 2012.

[29] Y. Zhou and X. Jiang. Dissecting android malware: Characterization and evolution. *Security and Privacy, IEEE Symposium on*, 2012.

[30] Y. Zhou, Z. Wang, W. Zhou, and X. Jiang. Hey, you, get off of my market: Detecting malicious apps in official and alternative android markets. In *Proceedings of the 19th Network and Distributed System Security Symposium*, 2012.

Bind Your Phone Number with Caution: Automated User Profiling Through Address Book Matching on Smartphone

Yao Cheng
Institute of Software
Chinese Academy of Sciences
Beijing 100190, China
chengyao@is.iscas.ac.cn

Lingyun Ying
Institute of Software
Chinese Academy of Sciences
Beijing 100190, China
yly@is.iscas.ac.cn

Sibei Jiao
Institute of Software
Chinese Academy of Sciences
Beijing 100190, China
jiaosibei@is.iscas.ac.cn

Purui Su
Institute of Software
Chinese Academy of Sciences
Beijing 100190, China
supurui@is.iscas.ac.cn

Dengguo Feng
Institute of Software
Chinese Academy of Sciences
Beijing 100190, China
feng@is.iscas.ac.cn

ABSTRACT

Due to the cost-efficient communicating manner and attractive user experience, messenger applications have dominated every smartphone in recent years. Nowadays, Address Book Matching, a new feature that helps people keep in touch with real world contacts, has been loaded in many popular messenger applications, which unfortunately as well brings severe privacy issues to users. In this paper, we propose a novel method to abuse such feature to automatically collect user profiles. This method can be applied to any application equipped with Address Book Matching independent of mobile platforms. We also build a prototype on Android to verify the effectiveness of our method. Moreover, we integrate profiles gathered from different messenger applications and provide insights by performing a consistency and authenticity analysis on user profile fields. As our experiments show, the abuse of Address Book Matching can cause severe user privacy leakage. Finally, we provide some countermeasures for developers to avoid this issue when designing messenger applications.

Categories and Subject Descriptors

D.4.6 [**Security and Protection**]: Invasive software; K.4.1 [**Public Policy Issues**]: Privacy

General Terms

Experimentation, Security

Keywords

Smartphone Applications, Privacy, User Profiling, Address Book Matching

1. INTRODUCTION

Messenger applications (e.g., WeChat [2] and Kik Messenger), which provide instant messaging service for mobile users, have become killer applications in mobile application markets (e.g., App Store and Google Play). It is reported that WeChat, a wildly popular instant messaging application for mobile devices, has doubled its users base in only 6 months, from 100 million in March 2012 to 200 million in September 2012. Messenger applications are popular for their flexible multimedia communication experience and economic boon. Before enjoying such convenience, users are asked to register an application account by uploading their personal information. For easily being discovered by friends, users prone to register with genuine personal information which is supposed to be visible to friends, however, in the real world this situation is not exactly as expected. Besides, messenger applications often recommend a list of other users by exploring contact lists in smartphones, which highs up the probability of exposing users' privacy to the wild.

Our work is motivated by the fact that an increasing number of messenger applications start to add Address Book Matching to their features. This new feature allows user to upload his contact list to the application server from which returns basic personal information of a list of other users who exist in the contact list, for further identification.

Our study shows that the abuse of Address Book Matching, equipped by most popular messenger applications, will bring about severe leak of user privacy in a large scale. Specifically, we make the following contributions: 1) We propose a novel approach to automatically collect user profiles via messenger applications on mobile platform for the first time in this area. Our method discovers a connection explicitly set up between user's virtual account and his real-world phone number, both of which would be revealed to strangers; 2) We implement a prototype to verify our approach. Through single-application analysis and cross-application integration, the experiments show the effectiveness of our method. Moreover, we conduct a consistency and authenticity analysis, which discloses wider genuine information of victimized users; 3) After analyzing main factors that make this attack possible, we provide some suggestions for mobile application developers to evade user privacy leakage caused by abusing Address Book Matching feature.

2. PROBLEM FORMULATION

The basic idea of our method is presented in this section. Address Book Matching will first be introduced, along with the discussion of its potential security risk.

2.1 Abusing Address Book Matching

From a mobile application designer view, Address Book Matching is essentially introduced to set up connections between user's application account and his phone number which can be regarded as user's unique ID. It allows user to upload address books (mainly including names and phone numbers) to the application server end from which returns a list of user accounts that are bound with phone numbers in those address books. It is mandatory for a user to bind his own phone number with his application account before using Address Book Matching. It assumes that each user behaves regularly and the contacts truly reflect his real social relations. However, in reality attackers are not fabled. The exposure of the aforementioned binding exhibits high potential risk if is abused by attackers.

We propose an approach to reveal the danger of abusing Address Book Matching in terms of user privacy leakage. Our method is applicable to any messenger application featured with Address Book Matching. Specifically, we achieve this goal by uploading a forged address book which consists of contact information of a large number of recommended individuals and randomly chosen phone number owners. Thus we can obtain all the phone numbers along with corresponding account profiles in the response from the application server end. Practically, some applications store users' phone numbers via hashing. However, such protection does little impact on our method, since Address Book Matching completes the phone number match process in the view of regular users; phone numbers are stored as plain text in address book. Our approach totally abuses Address Book Matching, rather than reading hash values by our effort.

2.2 Information Integration and Analysis

The content and scheme of user profiles vary from applications to applications. The information returned from one application can give us a set of user profiles, while integration across different applications would give us a more comprehensive user profiles.

2.2.1 *Horizontal Collaboration*

Horizontal collaboration contains two parts: broad union and deep intersection. Broad union integrates all distinct accounts corresponding to different phone numbers, producing a large set of user profiles. Deep intersection aims at digging more comprehensive information of one single real identity. The key point behind deep intersection is that phone numbers, as globally unique identifiers for real identities, might connect with multiple accounts in different applications.

2.2.2 *Vertical Penetration*

Many large online service providers like Facebook and Tencent who also produce mobile applications open ID login services to other entities. As a result, some applications (e.g., instant messaging and social networking) share the same set of login IDs and many of them are also supported by the same vendor. Penetrating applications using the same login ID is called vertical penetration which can provides us with more detailed personal information of that ID than horizontal collaboration, yet at the expense of more manual work. A case of this will be analyzed and demonstrated later in Section 4.5.

2.2.3 *Consistency and Authenticity Analysis*

Applications aim at meeting users' different social needs, such as meeting new friends or contacting old friends. Different needs determine users' diverse profile filling patterns. Our analysis of the consistency and authenticity targets at multiple user profiles collected from different applications simultaneously. In terms of consistency, we trust the field value as long as it is consistent across all applications. For inconsistent ones, a winner-take-all scheme is employed. For authenticity analysis, particularly in terms of user name, we use *Bai Jia Xing*, a database containing hundreds of most commonly used Chinese surnames with coverage over 90%, and regular patterns to search for users' possible official names.

2.3 Malicious Usage

In online environment, people tend to trust accounts with detailed and reasonable-looking profiles. However, this common sense will lead to potential misjudgment if attackers collect a larger number of user profiles in an automated way. Generally, attackers may perform the following operations by abusing Address Book Matching:

- Attackers may obtain users' profiles and the associated phone numbers. In other words, a link between *real* identity and *virtual* account has been established. Deeper understanding about the phone number owners can assist greatly to performing further attacks, such as cracking user accounts and social engineering.
- Attackers, especially spammers, may determine whether particular phone numbers are in use or not by querying the server with a list of unidentified phone numbers. Meanwhile, with collected information (e.g., gender, region and hobby), they may also perform contextually targeted advertising.
- Through information integration, comprehensive personal data can be retrieved, such as school, department, working company and etc., with which virtual network fraud can be carried out readily. In this case, the more applications users register, the more personal information will be given away.
- Attackers may clone identities leveraging the profiles they obtained. Clone means duplicating identities on particular applications. The more comprehensive profiles gathered by attackers, the closer the clones are to real identities.

3. SYSTEM OVERVIEW

In this section, we present the overall system architecture of abusing Address Book Matching as shown in Figure 1, then we elaborate the abusing method and implementation details.

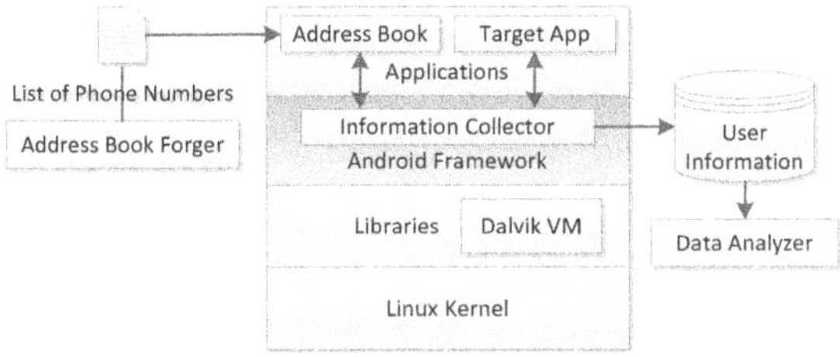

Figure 1: System Architecture

3.1 Method

Our method performs in three phases, i.e., address generation, information collection and data analysis which respectively achieve for forging address book, collecting information from the return list and correlating data for analysis.

Address Generation. This phase is to automatically generate lists of target phone numbers. Attackers typically enumerate all the interested phone numbers. The output of this phase will be used as input for querying target application servers; thus it should be formatted as a standard contact for compatibility.

Information Collection. We feed above mentioned contact lists to target applications via Address Book Matching. Applications then return a list of users within the contact lists, each of which has already registered an account and bound phone number with this account. Information Collector handles all user data returned from the applications. The implementation details are discussed in Section 3.2.

Data Analysis. This phase focuses on understanding the data obtained by previous phases more deeply. Our analysis includes two parts, i.e., single-application analysis and cross-application integration (see Section 4).

3.2 Implementation

This section elaborates the implementation of our system, particularly Information Collector. Different from crawling webpages, collecting user profiles from mobile devices is more complicated. The access to application data is often protected by mobile systems from other applications or processes for security purpose. We choose Android as the platform to build our prototype. At kernel level, each application has a low-privilege user ID and runs in its own process. At Android system level, when runs, each Android application is isolated by the Dalvik VM. The permission policies provided by Android system prevent applications from abusing each other's components, which makes our collection more complicated as the components of target application are not directly accessible.

To tackle this issue, we finally decide to monitor the Android APIs because: 1) local application database is often encrypted as well as the network traffic, however, decryption itself is an already hard problem with high cost; 2) hijacking the Intent is unworkable as the huge Intent messages are hard to target. Also, using Intent to deliver user information is not always the best choice for application developers; 3) modifying target applications to log down user data can avoid the decryption. However, this approach is subject to scalable problem since we need to rewrite specific modules for different applications, which is tedious and pleonastic. In fact, the Android framework wraps Linux system calls as specialized APIs for easily creating rich and novel applications, meaning that we do not need to make modifications for various applications. Instead, we just need to monitor each Android API the application invokes. Therefore, we monitor the APIs called by target applications which are responsible for handling user profiles. The monitor logs down user profile related data in real-time and in raw format, which will be post-processed for further analysis. Besides, for scalability purpose, we leverage ADB shell commands to automatically process the activity change by recording the motion and replaying it till the collection is done.

4. EXPERIMENT AND DATA ANALYSIS

In this section, we show our experiments and data analysis results. We first make some ethical considerations to prevent our method from being misused as best as we can.

4.1 Ethical Considerations

Our experiments randomly forge an address book and collect corresponding user profiles from messenger applications, which may bring about ethical and legal issues, similar with the cases in [9, 3]. However, this is the most efficient and effective way to conduct our experiments since 1) other methods like survey always suffer from inflating user concerns about privacy [6], by directly asking questions about privacy; 2) Comparatively, empirical experiments are more reliable. The collected data could reflect situations in the real world which are more convincing; 3) We carefully design experiments to avoid privacy issues, insuring the following principles fulfilled: a) The collected user data will never flow out of our experiment devices or be provided to any other irrelevant individual or organization; b) We will never perform any further penetration attempt to any account, except for proving our idea in this paper.

4.2 Single Application Analysis

We choose WeChat (Version 4.2 for Android), an aforementioned popular mobile messenger application with Address Book Matching, as our object to conduct the experiments.

Phone numbers in mainland China consist of 11 digits, not including two digits Mainland China Country Code (+86). We forge a list of 100,000 contacts of two separate continuous phone number segments, i.e., +861521063 - +861521070 and +861521098 - +861521099. After binding our test phone number to a newly applied WeChat account, Address Book

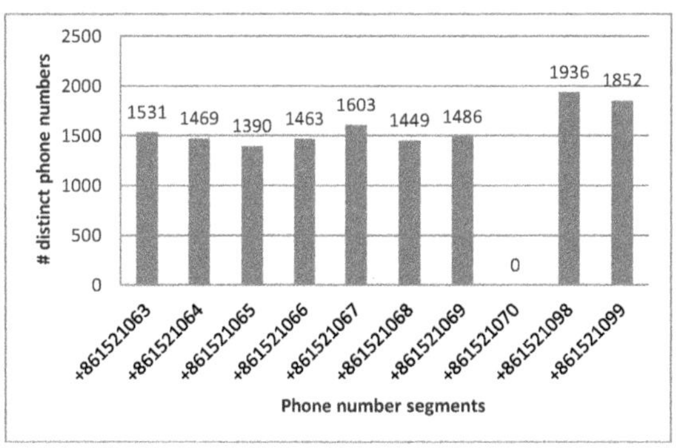

Figure 2: The Distribution of Active Phone Numbers Across all the Segments

Matching component of WeChat is activated. Upon uploading the forged address book to the server, WeChat begins to recommend all the accounts that already have been bound with phone numbers in the address book. As mentioned in Section 3.2, we modified Android system to monitor and record all the APIs specified to text and image display, thus we send a sequence of ADB shell "sendevent" commands to the simulator to start an ADB motion simulation which imitates a sequence of user behaviors of browsing the contacts profile. Finally, we succeed to obtain all fields of WeChat user profiles via Address Book Matching, they are WeChat ID, display name, phone number, gender, region, and "what's up". Generally, display name is close to the user's real name, as the display name is meant for identification by real world friends.

In the following step, we conduct a statistic analysis. Among 100,000 phone numbers, overall 14,179 registered users are matched (denoted as W), thus the user penetration rate is 14.18%. This number is consistent with the official released data 13.73%, which verifies the representative of our sampling. The distribution of the phone numbers in each segment is shown in Figure 2. Note that no user profiles correspond to the phone number segment +861521070, which is unallocated yet according to our further investigation. Among those 14179 users, not all of them upload their information completely. Over 55.13% of them have their profile fully completed as shown in Figure 3, indicating that most of the users indeed rely on social applications to maintain their social connections, thus they eager to put more details of their personal information online to facilitate the interaction with other people. Only 12.43% fill nothing but the required fields of WeChat ID and name. Even so, their phone numbers already connect with the WeChat accounts and may be leaked to attackers.

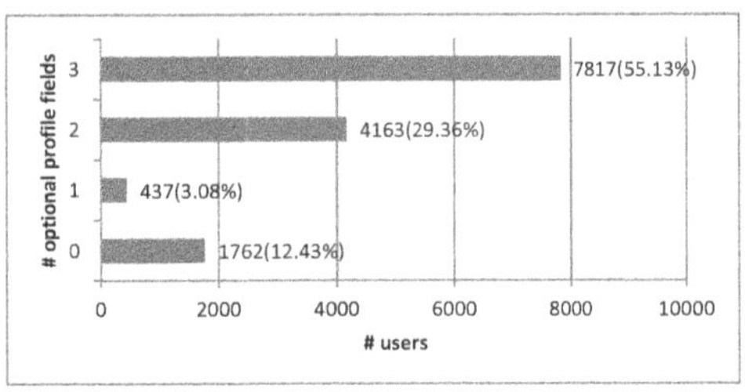

Figure 3: The Completeness of Optional Profile Fields

4.3 Horizontal Collaboration

Horizontal collaboration performs across multiple applications. We choose MiTalk Messenger [1] as our target which is a free and cross-platform messenger application whose users has already reached 17 million until August 2012. MiTalk provides a seamless connection between users' accounts and their address books to facilitate users' social connections.

In this phase, we use segment +861521098 which has uncovered maximum number of user profiles in previous experiment on WeChat. In the experiment, among those 10,000 phone numbers, 360 MiTalk user accounts have been returned (denoted as M) from MiTalk server and been added to our MiTalk friend list. MiTalk provides a more comprehensive profile scheme, including user name, gender, birthday, photo links, location he currently stays, and the names of schools and companies. Almost all the users have more than one photo link which can be visited directly via browser, and 30% of them maintain school and company name fields which are regarded as privacy for strangers.

As illustrated in Section 2.2, horizontal collaboration includes two parts: broad union and deep intersection. 116 new users are found in MiTalk. After broad union we totally have 2052 phone numbers and corresponding user accounts. In terms of deep intersection, we integrate user profiles obtained from WeChat and MiTalk and totally 244 users are found to have registered both applications. We then perform consistency and authenticity analysis based on these 244 users profiles (denoted as D).

4.4 Consistency and Authenticity Analysis

We find that 34.02% of users in D share similar display names in both applications, which inspires us to further find out in what degree their names are the official names. We leverage the Chinese *Bai Jia Xing* (aka. Hundred Family Surnames) data set to match official names. Typical Chinese names begin with one of the surnames in *Bai Jia Xing* as family name and followed by a given name of one or two Chinese characters length. We believe that display names in formal Chinese full name format are probably users' official names. Through regular pattern matching, we investigate the display names from W, M and D to find out the ratio of formal Chinese full name in respective dataset. The results are shown in Table 1 that in M 48.33% of users' display names are probably official names, which is quite higher than that of W, 31.24%, indicating that MiTalk users tend to register accounts using official names. This may be ascribed to the fact that MiTalk promotionally recommends users to provide personal information as truthful as possible when registering, so as to facilitate the contact with their friends. This "seamless" strategy makes users lean to trust other MiTalk users. They naturally believe that the uploaded profiles are only prepared for their friends by default. The case in D outperforms those in W and M that it reaches a ratio of 59.43%, which implies users, registered in both applications, are more adhered to and trust messenger applications; thus are more willing to use real personal information for registration. It also makes us realize that through deep intersection, more comprehensive and truthful user profiles can be aggregated not only because of the correlation but also the user habit.

We also spot another odd phenomenon - the gender inconsistency. Presented in Table 2, 54.1% of the users in D have their gender fields contradicted in both applications. This phenomenon is called inconsistency which has also been found in different types of social networks [3]. Moreover, among all "gender inconsistency" users, those who appear "male in MiTalk and female in WeChat" account for 68.9% (90/132). To further determine the likely-to-be-true gender of each user, the winner-take-all strategy which favors majority cases if more than two choices are available could be employed. Yet, it is still hard to confirm a user's gender just by comparing results from two applications. However, we have noticed that the proportion of male and female gender is more unbalanced in WeChat profiles. In addition, based on the name authenticity analysis presented above, we deem user profiles in MiTalk to be more trustworthy which is convincing after we manually check the display name field of inconsistency users in D.

Dataset	Total # Users	# Users with Real Name	Rate
WeChat (W)	14179	4430	31.24%
MiTalk (M)	360	174	48.33%
Deep intersection set (D)	244	145	59.43%

Table 1: Comparison of Name Authenticity Analysis Results from Three Dataset - WeChat, MiTalk and Deep Intersection Set

MiTalk / WeChat	Male	Female
Male	37	42
Female	90	57

Table 2: Gender Consistency of Deep Intersection Set

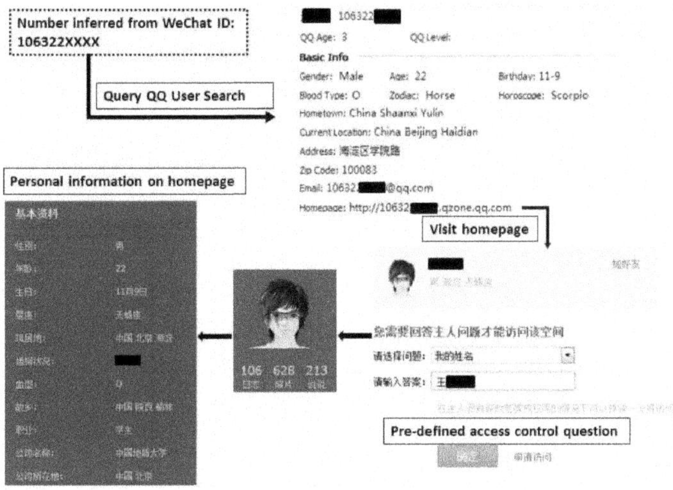

Figure 4: Vertical Penetration Process

4.5 Vertical Penetration

In this section, we manually conduct a case study using deep intersection dataset D to concisely demonstrate vertical penetration.

Created by the same company, WeChat is designed to share the same account database as QQ which is the most popular IM (Instant Messaging) software in China. We choose a user whose WeChat ID is "t106322XXXX" with last four digits replaced by letter 'X' for privacy concern. As shown in Figure 4, we query "user search function" of QQ using the last ten digits of the WeChat ID as QQ account ID by our speculation, and as expected a single match finally returns. We then collect additional profile fields except those from WeChat, from which we know that this QQ account belongs to a young man from Shannxi, China, born on November 9th 1990, and currently stays in Beijing, China. In addition, we also find his personal home page address. However, until now we are not sure the owner of this QQ account is exactly the one has WeChat ID of "t106322XXXX". We then go one step further - visiting his personal home page, where a pre-defined question asking for the answer of "real name" shows up for access control. By inputting the display name of this WeChat account, we successfully log in! Until now, we can safely conclude that the aforementioned QQ account and WeChat account are indeed belong to the same person. Finally, the home page reveals more information about the young man as shown in Table 3 which also illustrates incremental profile fields we collect along each step.

Step	Field	Value
0	WeChat ID	t106322XXXX
	WeChat name	WangXX
	Phone number	+861521098XXXX
1	QQ name	TianXX
	Date of birth	1990.11.09
	Blood type	O
	Hometown	Yulin, Shannxi Province
	Address	Xueyuan Rd, Haidian Distinct
	Zip code	100083
	E-mail address	106322XXXX@qq.com
	Home page	http:// X.qzone.qq.com
2	Real name	WangXX
	Marital status	X
	Blogs	106 blogs
	Photos	628 photos
	Status	213 status
	Occupation	Student
	Name of school	China University of Geosciences

Table 3: Profile Fields Seized in Each Penetration Step

We can also infer from the pre-defined access control question that the home page is supposed to be visited by the friends who really know this young man. However, we, as strangers, have successfully obtained his personal information, even private secretes.

We argue that for attackers who intend to crack someone, manual work should be a price worth paying. As it can be seen that vertical penetration can provide a wealth of information which is crucial for earning trust when camouflaging, boosting the implementation of online fraud.

5. COUNTERMEASURE

By abusing Address Book Matching, we have successfully gathered users' phone numbers and corresponding user accounts in large scale, however, attackers will be no exception. It seems like throwing away the apple because of the core to disable this promising function. Therefore, in this section, we propose several feasible alternatives for developers to avoid this issue by tackling the key factor - the establishment of connection between real phone numbers and virtual user accounts in flawed messenger applications: 1) Make phone numbers be queried with a one-at-a-time constraint, which meets users' needs and meanwhile removes the chance for attackers to launch large scale attacks. 2) Only return the display names of application accounts, rather than detailed profiles nor corresponding phone numbers. This could be applied to applications with real-name system in which users can identify their real world friends by the official names. 3) Return the phone numbers and the names which should be exactly identical to the ones stored in the address book. No account information should be accessed, making sure no virtual and real identity connections are acquired. 4) Recommend a user U those who also have U in their address books. This is the way particularly suitable for phone number based relationship management which is generally bilateral.

6. RELATED WORK

Privacy Analysis of Applications. Privacy issues of applications are widely studied. W. Enck et al. [8] implements TanitDroid, a dynamic taint analysis tool, that modifies Android's Dalvik VM and determines the violation of privacy when there is tainted sensitive data flowing out of the phone. PiOS [7] is another privacy leakage analyzing tool performing static taint analysis on iOS application binaries. In terms of preventing privacy leakage, works such as MockDroid [4] and TISSA [11] allow users to give resource access permission to an application, whereas the system reports empty or unavailable when application requests access. Differently, we leverage existing Address Book Matching function to gather private user profiles without considering the information flow internally.

Automated User Profiling. Leyla Bilge et al. [5] achieves automated identity theft via profiling users on SNS (Social Networking Services) by employing friendship relation. M. Balduzzi et al. [3] abuses registered e-mail address querying which is designed for users to search friends on SNS. They collected a large number of users' personal information by crawling and connecting their profiles. In our paper, we try to get user information from applications and the techniques are quite different from crawling webpages.

Problems with Misusing Phone Numbers in Communication Applications. Sebastian Schrittwieser et al. [10] analyze several popular message and VoIP applications, finding that most such applications treat users' phone numbers as their account identifiers. They show some security flaws in terms of authentication mechanisms, such as account hijack or ID spoof. Differing from their work, our experiments show that contact matching technology, adopted by many messenger applications today, has exhibited severe privacy leakage problem because phone numbers are used as external identifiers of user accounts which can be exploited by attackers to seize other users' accounts.

7. CONCLUSION

We exploit Address Book Matching, a novel feature of smartphone messenger applications, to collect user profiles. To our best knowledge, this work is the first to collect user profiles via smartphone applications. The consistency and authenticity of collected user profiles are further analyzed. Results show that smartphone users prone to leave truthful personal information in mobile messenger applications, however, for different applications, users' confidence may vary due to different registration rules and promotion strategies. Finally, we analyze main factors that make this abusing possible and provide countermeasures for application developers. Developers should be cautious and do their best to prevent the exposure of the mapping information when users' phone numbers are linked with their application accounts.

8. ACKNOWLEDGMENTS

This work was supported by National Program on Key Basic Research Project (Grant No. 2012CB315804), National Natural Science Foundation of China (Grant No. 61073179 and No. 91118006), National Science and Technology Major Project (Grant No. 2011ZX03002-005-02) and Beijing Municipal Natural Science Foundation (Grant No. 4122086).

9. REFERENCES

[1] Mitalk messenger. `http://www.miliao.com/`.

[2] Wechat. `http://weixin.qq.com`.

[3] M. Balduzzi, C. Platzer, T. Holz, E. Kirda, D. Balzarotti, and C. Kruegel. Abusing social networks for automated user profiling. In S. Jha, R. Sommer, and C. Kreibich, editors, *Recent Advances in Intrusion Detection*, volume 6307 of *Lecture Notes in Computer Science*, pages 422–441. 2010.

[4] A. R. Beresford, A. Rice, N. Skehin, and R. Sohan. Mockdroid: trading privacy for application functionality on smartphones. In *Proceedings of the 12th Workshop on Mobile Computing Systems and Applications*, HotMobile '11, pages 49–54, New York, NY, USA, 2011.

[5] L. Bilge, T. Strufe, D. Balzarotti, and E. Kirda. All your contacts are belong to us: automated identity theft attacks on social networks. In *Proceedings of the 18th international conference on World wide web*, WWW '09, pages 551–560, New York, NY, USA, 2009.

[6] A. Braunstein, L. Granka, and J. Staddon. Indirect content privacy surveys: measuring privacy without asking about it. In *Proceedings of the Seventh Symposium on Usable Privacy and Security*, SOUPS '11, pages 15:1–15:14, New York, NY, USA, 2011.

[7] M. Egele, C. Kruegel, E. Kirda, and G. Vigna. PiOS: Detecting privacy leaks in iOS applications. In *Proceedings of the 18th Annual Network & Distributed System Security Symposium (NDSS)*, Feb. 2011.

[8] W. Enck, P. Gilbert, B.-G. Chun, L. P. Cox, J. Jung, P. McDaniel, and A. N. Sheth. Taintdroid: an information-flow tracking system for realtime privacy monitoring on smartphones. In *Proceedings of the 9th USENIX conference on Operating systems design and implementation*, OSDI'10, pages 1–6, Berkeley, CA, USA, 2010.

[9] M. Jakobsson, N. Johnson, and P. Finn. Why and how to perform fraud experiments. *IEEE Security and Privacy*, 6(2):66–68, Mar. 2008.

[10] S. Schrittwieser, P. Fruehwirt, P. Kieseberg, M. Leithner, M. Mulazzani, M. Huber, and E. Weippl. Guess who is texting you? evaluating the security of smartphone messaging applications. In *Network and Distributed System Security Symposium (NDSS 2012)*, 2012.

[11] Y. Zhou, X. Zhang, X. Jiang, and V. W. Freeh. Taming information-stealing smartphone applications (on android). In *Proceedings of the 4th international conference on Trust and trustworthy computing*, TRUST'11, pages 93–107, Berlin, Heidelberg, 2011.

Towards Preventing QR Code Based Attacks on Android Phone using Security Warnings

Huiping Yao
Computer Science and Engineering
New Mexico Tech
Socorro, NM 87801, USA
hyao@nmt.edu

Dongwan Shin
Computer Science and Engineering
New Mexico Tech
Socorro, NM 87801, USA
doshin@nmt.edu

ABSTRACT
QR (Quick Response) code has become quite popular in recent years due to its large storage capacity, ease of generation and distribution, and fast readability. However, it is not likely that users will be able to find out easily the content encoded, typically URLs, until after they scan QR codes. This makes QR codes a perfect medium for attackers to conceal and launch their attacks based on malicious URLs. We believe that security hardening on QR code scanners is the most effective way to detect and prevent the potential attacks exploiting QR codes. However, little attention has been paid to the security features of QR code scanners so far in literature. In this paper, we investigated the current status of existing QR code scanners in terms of their detection of malicious URLs exploited for two well-known attacks: phishing and malware. Our study results show the existing scanners either cannot detect or can very poorly detect those two attacks. Hence, we propose a QR code solution called *SafeQR* that enhances the detection rate of malicious URLs by leveraging two existing security APIs to detect phishing and malware attacks: *Google Safe Browsing API* and *Phishtank API.* Additionally, a visual warning scheme was carefully designed and implemented to enable users to better heed warnings. A user study was designed and conducted to investigate the effectiveness of our scheme compared with the methods adopted by existing QR code scanners.

Categories and Subject Descriptors
K.6.5 [**Management of Computing and Information Systems**]: Security and Protection – *Invasive software, unauthorized access.*

Keywords
QR code security; phishing; malware; visual warning; user study

1. INTRODUCTION
As smartphones have become very popular recently, more and more services based on smartphones have been being developed and adopted. Quick response codes (QR codes), as one of such services, have been widely used due to their easy generation and distribution, large storage capacity, and fast readability [1]. They are often found on public signage, usually to direct users to a website with more information about a product, company, or location. According to a recent survey, 75% of retailers now offer customers this 2D technology to interact with and track potential buyers [2]. Customers can scan them with QR code scanners on their smart phone or tablet to receive coupons, discounts, or other information related to products or services.

However, due to the fact that their encoding scheme is more machine-readable than human-readable, the only way for users to figure out what information is encoded in a QR code is to scan it. This makes QR codes a perfect medium for attackers to conceal and launch their attacks based on malicious contents embedded such as phishing URLs [3]. Besides, people often cannot resist their own curiosity to scan the codes they have come across. This innate obfuscation of QR codes and the curiosity make QR codes one of the biggest hidden security threats [4]. So far, little attention has been paid to the security features of QR code scanners in literature. In this paper, we investigated the current status of existing QR code scanners in terms of their detection of malicious URLs exploited for two well-known attacks: phishing and malware. Our study results show that the existing scanners either cannot detect or can very poorly detect those two attacks. Hence, we propose a QR code solution called *SafeQR* that enhances the detection rate of malicious URLs by leveraging two existing security APIs for effectively dealing with phishing and malware attacks: *Google Safe Browsing API* and *Phishtank API.* Additionally, our solution offers an effective visual warning scheme to help users better informed of imminent threats. Finally, an experiment involving a user study was designed and conducted to investigate the effectiveness of our scheme compared with the methods adopted by existing QR code scanners.

The rest of the paper is organized as follows: Section 2 discusses the potential threats in QR codes, and reveals the problems of the existing QR code applications on Android phone. Section 3 presents our approach to tackle the existing problems. Section 4 discusses the methodology and design of a user study developed to test the effectiveness of our approach as well as some results from the user study. Section 5 concludes the paper with our future research work.

2. BACKGROUND
In this section, we first discuss the main potential threats brought about by QR codes, and then we present our study results which show that the vast majority of Android-based QR code scanners did not provide users any security information for the scanned URLs.

ASIACCS'13, May 8–10, 2013, Hangzhou, China.

2.1 Potential Threats of QR Codes

Usually, a QR code directs users to the website of their interest, which can provide further information or services. By scanning a carefully manipulated QR code, however, they may be taken to a malicious website. Attackers can use QR codes for various types of attacks. However, two critical attacks we are interested in for this study are phishing and malware attacks.

- **Phishing**: phishing is an activity that tricks people to divulge their sensitive information by masquerading as a trustworthy entity [5, 6]. A QR code can redirect users to a fake bank that looks exactly like the real bank. A normal user cannot see the differences and type in his or her information and hand it to the attackers [7-9].
- **Malware**: A malicious QR code can be used to redirect users to a URL containing malware[3]. An early example of malware attacks through QR code was that people were fooled into scanning a QR code and downloading a malicious application, which sent off multiple text messages to a number that charged users $5 per SMS message [10].

There are other types of attacks such as social engineering [7, 11] and cross-site attacks [12]. Though interesting, a complete study to analyze these rather rare cases was out of purview of our study.

2.2 Defects of Existing QR Code Readers

QR code scanners are the only direct way to detect QR code based attacks. In this study, we first investigated the security related features of existing QR code scanners, and then focused on the detection rates of existing QR code scanners with security indicators.

2.2.1 Security Related Features of QR Code Apps

Google Play, the official Android Market, is used as a reliable source for users to download the apps. On Jun 6, 2012, we searched for QR code scanning apps from Google Play market using keywords "QR code", "QR code reader", and "QR code scanner," and we found 31 apps. We installed them in a HTC Nexus One with Android 2.3.6 and used them to scan QR codes with benign URLs randomly chosen from DMOZ [13], a manually edited directory, and QR codes with malicious URLs selected from PhishTank [14], a blacklist of phishing URLs, to find out the features of these scanners.

Table 1 shows the results of our study on the security related features of those scanners. Due to the space limitation, we only listed the software names. Though the names (Barcode Scanner) of No. 2 and No. 3 are the same, No. 2 scanner was developed by ZXing Team, while No. 3 was developed by George Android.

As we can see from Table 1, 74.19% (23/31) of the apps had a feature called user confirmation. The user confirmation feature provides users with a confirmation page that displays the URL encoded in the QR code, along with an option to continue/discontinue to visit the website. This allows users to find out where they will visit, and thus offers a way to involve users to potentially prevent attacks by inspecting the decoded URL. Interestingly, two readers, *QR Droid* and *QR Droid Private*, had an additional feature called preview. The preview feature provides the preview of the scanned website so that users will be able to figure out roughly if the site is actually the one that they are initially interested in (this is especially true in case that they actually visited the site previously.). On the other hand, an app called *QuickMark* had an interesting utility function based on the history of QR code usage. The function displays two usage data: *Scan* and *Click Rate*. The first is the total number of times that a QR code has been scanned using this app, and the second pertains to the number of times a QR code has been clicked divided by *Scan*.

Table 1. Security related features of QR code scanners

No.	Application	User Confirmation	Preview	Security Warning
1	AT&T Code Scanner	✓		
2	Barcode Scanner	✓		
3	Barcode Scanner	✓		
4	BeeTagg QR Reader	✓		
5	RedLaser Barcode& QR Scanner	✓		
6	Codee QR Code Reader	✓		
7	DTEScanner			
8	Google Goggles	✓		
9	HandyShopping Barcode Scanner	✓		
10	HP CodeScan			
11	i-nigma Barcode Scanner	✓		
12	Mobiletag QR Code Scanner			
13	NeoReader	✓		
14	Norton Snap QR Code Reader	✓		✓
15	Scan			
16	ScanLife Barcode & QR Reader			
17	Scanner Pro	✓		
18	SHARP QR Code Reader	✓		
19	ShopSavvy Barcode Scanner			
20	QuickMark Barcode	✓		
21	QR Barcode Scanner	✓		
22	QR Barcode Scanner – Lite	✓		
23	QR barcode scanner			
24	QR Code Reader / Scanner			
25	QR Droid	✓	✓	
26	QR Droid Private	✓	✓	
27	QR Pal – QR & Barcode Scanner	✓		✓
28	QR Pro	✓		
29	QR Reader for Android	✓		
30	QR Rewords	✓		
31	UberScanner	✓		

However, none of the features seems to provide users with information that will help them make a better security decision. As to the user confirmation feature, users see only a URL and they often click to continue to visit the site without much thinking about the security consequences. Unsophisticated users may even not be able to recognize malicious URLs. The preview feature can be easily abused, incapable of providing users with needed security information either; a well designed fraudulent website can only give a false sense of security to users.

Among the 31 scanners, only two (0.06%), "*Norton Snap QR Code Reader*" and "*QR Pal – QR & Barcode Scanner*", had a feature called security warning. Upon scanning a QR code containing a malicious URL, *Norton Snap* and *QR Pal* equipped with the feature display the URL along with a warning message before loading the website of the URL, as shown in Figure 1.

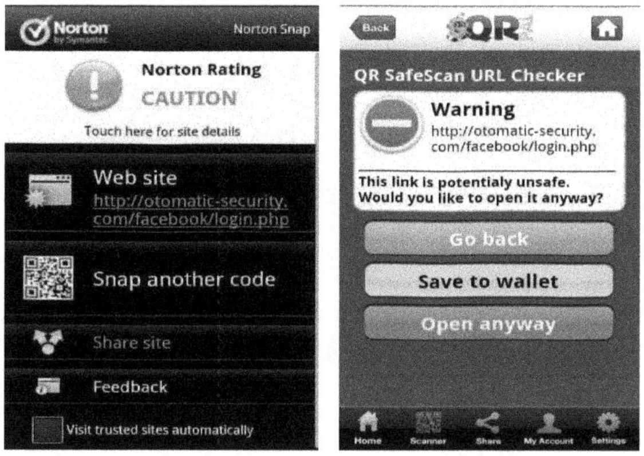

(a) (b)

Figure 1. Warning messages displayed in two existing QR code scanners: (a) Norton Snap; (b) QR Pal

2.2.2 *Evaluation of Two Scanners*

Since *Norton Snap* and *QR Pal* provide various levels of security warnings for malicious URLs, it would be interesting to study the accuracy of those warnings. During the testing for measuring their detection rates, we applied the default settings of those two QR code scanners. We investigated two types of errors, false positive and false negative errors. Since we believe false negative errors are more serious that false positive ones, we conducted a more extensive testing on false negative errors.

2.2.2.1 *Benign URLs*

To study whether the two QR scanners had false positive errors or not, we used DMOZ Open Directory Project as data set, of which all entities are vetted by editors who also go through a vetting process [15], and randomly chose one URL from each of the 10 random subcategories under 15 categories as follows: Arts, Business, Computers, Games, Health, Home, Kids and Teens, News, Recreation, Reference, Regional, Science, Shopping, Society, and Sports. Out of 150 URLs, *Norton Snap* took 16 URLs as untested rating, 2 as malicious, and the remaining 132 as trusted ones. So the false positive rate of *Norton Snap* was 1.33% (2/150). *QR Pal* took all the 150 URLs as benign, thus we concluded that *QR Pal* did not have false positive errors.

2.2.2.2 *Phishing URLs*

Testing phishing URLs, drawn from PhishTank, was conducted based on the following plan: first, we conducted a preliminary testing using a small sample size to see if the two scanners had a desirable detection rate for phising URLs, by setting 90% detection rate as a baseline that is usually achieved by email-based phishing detection tools [16] (1st phase). Since their performance was way below than the baseline, we decided to conduct a more comprehensive testing with a larger sample size (2nd phase).

The First Phase

When analyzing a site, *Norton Snap* goes through the Norton Safe Web [17] to identify trusted and malicious sites, and gives the corresponding ratings. Specifically, Norton Safe Web will evaluate a site by computing the three factors, *computer threats*, *identity threats*, and *annoyance factors*. A site is identified malicious if Norton detects one or more threats from any of the above three factors. If none of the three threats was found, then the site is identified as a trusted one. *Norton Snap* has five security ratings as follows: WARNING_MALICIOUS, CAUTION, UNTESTED, SAFE, and SECURED. WARNING_MALICIOUS or CAUTION rating indicates that the URL is malicious. UNTESTED means the site has not been tested yet. SAFE or SECURED implies that the URL is trusted. The difference between WARNING_MALICIOUS and CAUTION is that: if a site is identified to have one or more *computer threats* regardless of the other two factors, it is rated as WARNING_MALICIOUS, otherwise CAUTION. If a site passes all the above three factors, Norton Safe Web then checks if it satisfies at least one item of Ecommerce Safety [18], if yes, the site is rated as SECURED, otherwise, rated as SAFE.

In the first phase, we randomly picked and scanned 15 phishing URLs on Jun 28, 2012 using the two apps. Our analysis result showed that, for *Norton Snap*, only 26.67% (4/15) URLs were predicted correct, 60% (9/15) scans did not provide any security information by displaying "UNTESTED", which actually does not provide helpful warnings for users to decide to open a site, 13.33% (2/15) cases gave the wrong security information. *QR Pal* either provides warning information for malicious URLs or opens the URL for the benign URLs. Our analysis result on *QR Pal* showed that 66.67% (10/15) were predicted correct, and it opened the remaining 5 malicious URLs directly without any user confirmation.

Through the preliminary testing, we discovered that neither *Norton Snap* nor *QR Pal* had a desirable detection rate for phishing URLs. Their performance was way below the threshold of 90%.

The Second Phase

In the second phase, we randomly chose 400 out of 6131 entries (6.52%) from the PhishTank dataset downloaded on Jul 7, 2012.

Table 2. Detection of phishing attacks

	Correct		Neutral		Incorrect	
	#	%	#	%	#	%
Norton Snap	112	28.00	150	37.50	137	34.25
QR Pal	111	27.75	0	0.00	289	72.25

The results in Table 2 show that both *Norton Snap* and *QR Pal* were quite disappointing in terms of their detection of phishing attacks. *Norton Snap* only detected 28% (112/400, 95% Confidence, CI-4.25%) of the URLs correctly. For 37.5% of (150/400) the scans, *Norton Snap* did not provide any security warning for users. What is worse, it provided wrong security information for the remaining 34.25% (137/400) cases. For *QR Pal*, it did not contain the neutral case. The detection rate was only 27.75% (111/400, 95% Confidence, CI-4.24%).

2.2.2.3 *Malware URLs*

Malware attack is another popular attack on Android. We tested the two QR code scanners against the attacks using malware obtained from http://malgenomeproject.org [3]. Specifically, we used malware under the drive-by download categories (*GGTracker*, *Jifake*, *Spitmo*, and *Zitmo*) and repackaging malware categories (*AnserverBot* and *DroidKungFu*). Each of the four drive-by downloaded categories contains only one malware. We chose all these four malware for testing. For *AnserverBot* and *DroidKungFu*, we randomly picked one malware from each category. We uploaded these six malware to a personal homepage http://infohost.nmt.edu/~hyao, and tested the corresponding URLs, thus simulating a zero-day attack.

We found that both *Norton Snap* and *QR Pal* detected very poorly for all the six situations. *Norton Snap* only checked if the domain of the URL was secure or not. Therefore, since our test domain was not classified as malicious, it gave the security rating "Safe", so that users could easily download the malware. On the other hand, *QR Pal* did not download any of these malware with its default setting. However, it could download all these six malware if its default *Built-in browser* setting in the general settings was turned off, which could be easily done by users who just followed the instructions displayed by the app, upon receiving the download request from the users.

3. OUR SOLUTION

Our approach called *SafeQR* aimed at two goals in order to address the poor performance problem of the existing QR code scanners. The first was to enhance the effectiveness of detecting malicious URLs used for phishing and malware attacks. The second was to improve user perception of security when a QR code is scanned and used so that users can make a better security decision. For this purpose, we focused on how to improve the effectiveness of the security warnings by providing a better mobile UI design.

3.1 Malicious URLs Detection

Two well-known security APIs, *Google Safe Browsing API* [19] and *PhishTank API* [20], were adopted for our solution to improve the effectiveness of detecting malicious URLs.

Safe Browsing, developed by Google, is a service that enables applications to check URLs against Google's constantly updated lists of suspected phishing and malware websites. For simplicity, we chose *Safe Browsing Lookup API* [21], queried the URLs through HTTP GET request, and got the state of the URL(s) directly. PhishTank contains a blacklist of phishing URLs consisting of manually verified websites. PhishTank provides API for developers to lookup a URL's status in their database. We used the API to query a URL's status, thus further enhancing the capability for detecting phishing scams.

In addition, if the URL string ends with ***.apk***, this means that a non-official Android market application will be downloaded to a user's mobile phone. Specifically, if "Unknown Sources" setting is checked in Android, the app will be automatically downloaded to the user's Android device. If the setting is not checked, a dialog will pop up to ask the user if she wants to check the option to download the app. Users can easily tick "Unknown Sources" option just by following the instructions in the dialog, and then download and install the application. Our solution checked if the URL ended with "***.apk***", and if yes, then we provided potential warnings to users, as shown in Figure 2.

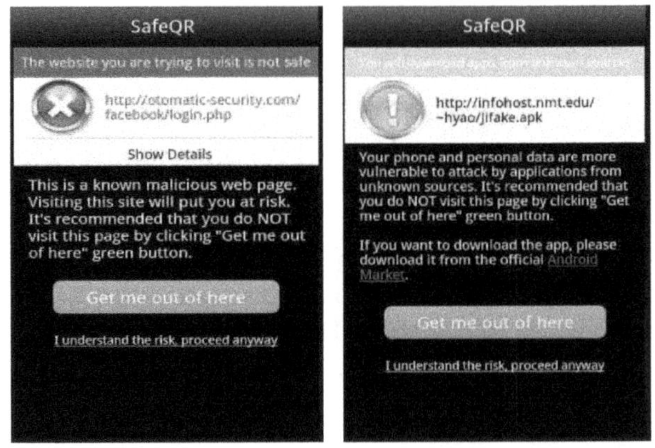

(a) (b)

Figure 2. Our UI designs: (a) UI for phishing or malware attacks; (b) UI for the URL ending with ".apk"

The algorithm for our solution is outlined in the following pseudo code.

```
Algorithm detection(url)
status = 0;
if google-safebrowsing-code(url)==200
    if contains-phishing && contains-malware
        status = 3;
    else if contains-malware
        status = 2;
    else if contains-phishing
        status =1;
if in-phishtank-list(url)
    if status == 0 || status==2
        status++;
if status == 0 && end-with-apk(url)
    status = 4;
return status;
```

In the code, the *status* variable representing the security status of a URL has 5 different values. The *status* variable is initialized to 0, implying a safe URL. Its value becomes 1 when a URL is detected as a phishing one, 2 when a URL is detected as a malware website, 3 when a URL is a both phishing and malware site, 4 when a URL is checked safe through the Safe Browsing and PhishTank but ends with "***.apk***".

3.2 Security Warning Design

Studies have shown that warning designs affect user's decision to obey or ignore the warnings [22, 23], thus designing an effective security warning by providing better risk perception is a significant part of our solution. We applied existing warning design recommendations [22, 24] and Microsoft's NEAT (Neat, Explained, Actionable, Tested) & SPRUCE (Source, Process, Risk, Unique, Choices and Evidence) [25] into our warning message design as follows: we first set the warnings mainly with black and red colors, since this color combination is quite effective to prevent users from attacks [22]; we made the safest button "Get me out of here", as a recommended action, most visible by setting its background green, which also helps users think of safe actions; since users are likely to ignore lengthy text, we only include necessary words in the warning, giving users a chance to click "Show Details" button for viewing details. In addition, only simple words are used, for users will not understand or will misinterpret technical jargons.

Two screenshots of our solutions are captured in Figure 2 (a) (b) respectively. Figure 2 (a) shows the display of our solution when a user scans a QR code with a phishing or malware URL. Figure 2 (b) shows the screenshot when a user scans a QR code whose encoded URL ends with "***.apk***".

If the website addressed by the URL is a phishing website or contains malware, we immediately return a negative evaluation, which is shown in Figure 2(a). If the URL ends with "***.apk***", we return an uncertain evaluation, shown in Figure 2(b), for ***.apk*** file may be not a malware. Compared to *Norton Snap* and *QR Pal*, our security warning design is better understood by providing the sufficient risk details and the recommended action without using technical jargons, and this is confirmed by the results of our user study in the next section.

4. USER STUDY

To investigate the effectiveness of our solution, we designed a user study to explore the effectiveness of the security warning in our solution. Our user study also was to compare our solution against the existing two solutions we discussed in the previous section. We also wanted to study how all of these solutions compare against the absence of any visual security warning.

4.1 Design and Recruitment

We defined four separate user groups, each of which was exposed to a different warning provided by a different QR code scanner.

- Group 1: Exposed to the attack with no warning, using *QR-code Scanner*
- Group 2: Exposed to the attack with *Norton Snap*'s warning
- Group 3: Exposed to the attack with *QR Pal*'s warning
- Group 4: Exposed to the attack with our designed warning

In the design of our experiment, we performed a power analysis to determine the minimum sample size that we would require to test our hypotheses. We chose an error of 0.05 and a power of 0.8, common among such experiments, and determined a minimum sample size of 19 subjects across the four user groups. Based on this analysis, we chose to recruit 80 participants with which 20 subjects in each of the four groups. It was necessary for participants to have a Facebook account, so that we can test their reactions after scanning a QR code containing the URL of the Facebook authentication page.

Our participants, 20 females and 60 males, have a high education rate, with all having completed at least high school degree, and 83.75% (67/80) having or currently pursuing undergraduate college degrees. Age groups of our participants include 38 from 10-20 years old, 32 from 21-30 years old, 5 from 31-40 years old, 2 from 41-50 years old, 2 from 51-60 years old, 1 from 61-70 years old. Relative to security knowledge, the participants in our sample are very sophisticated, with only 7 claiming they have poor security knowledge. 54% (43/80) of the participants replied that they had good security knowledge or above.

4.2 Experiment

During the experiment, we randomly assign 20 participants to each of four test groups. Each group was exposed to the phishing attack; three groups were given a specific warning respectively by *Norton Snap*, *QR Pal*, and *SafeQR*, and the fourth group was not warned at all by using *QR-code Scanner*. To avoid the framing effect, we did not want the users to be aware that we were testing their reaction to a security warning. In addition, we wanted the users to be exposed to the warnings as an abnormality or exceptional condition. To achieve both goals, we told the participants that we were investigating whether they made full use of smart phone apps, and evaluating the usability of using QR code scanners to access websites.

The hypotheses we wanted to test were on (1) the user unawareness of malicious QR codes, (2) the effectiveness of QR code based phishing attacks, and (3) the helpfulness of security warnings. Finally, we wanted to verify that our proposed warning design would be more effective than the existing solutions.

The participants were given an Android smart phone where the QR code apps were installed and were told that we would be interested in improving the usability of the apps, so they were encouraged to use the app with their real credentials to help us achieve the goal. Besides, users were asked to act as if they were using their own phone, in that all decisions they made should be the same as if they were being made on their own private phone. Security was never explicitly mentioned. Although we initially thought the "make all the decisions as if this was your phone" statement could bring focus on security issues, the results showed us that this was not the case.

4.3 User Study Results

Each user was asked to take an exit survey, which was also used to test our hypotheses. Our first hypothesis was tested by asking the subjects whether they had ever thought of any security problems caused by QR codes. Out of 80 subjects, 67.5% (54/80) were not aware that malicious QR codes existed. Hence, we concluded that most of the users were unaware of malicious QR codes. From the results of our second hypothesis testing, we learned that without any added security mechanisms, the phishing attack is highly effective, for 100% (20/20) of the participants opened the link and 75% (15/20) of them submitted their Facebook username and password. We also found that the security warnings provided by the QR code scanners, i.e. *Norton Snap*, *QR Pal*, and *SafeQR*, helped users perceive potential dangers and avoid phishing attacks, as shown in Table 3. Hence, our third hypothesis was confirmed. The results obtained for our proposed solution *SafeQR* proved more efficient than the existing solution. In terms of opening the link, *Norton Snap* only led 5 out of 20 users to not open the link, *QR Pal* led 11 out of 20 users to not open the link, while our solution *SafeQR* led 17 out of 20 users to not open the link. In terms of submission, both *Norton Snap* and *QR Pal* led 13 out of 20 users to not submit their credentials. Our solution *SafeQR* led 18 out of 20 users to not submit their credentials.

Table 3. For different study groups, # of open and submit

	Open	Submit
QR-code Scanner	20	15
Norton Snap	15	7
QR Pal	9	7
SafeQR	3	2

5. CONCLUSION AND FUTURE WORK

In this paper, we presented an approach to preventing QR code-based phishing and malware attacks. Specifically, we first studied the current status of existing QR code scanners in terms of their detection rate for malicious URLs. Then we proposed our solution to detect malicious URLs more effectively by using two well-know security APIs along with visual security warning design. Lastly we discussed our user study design to evaluate the effectiveness of the proposed solution.

This research is by no means complete. Firstly, our immediate future work is to analyze the data from user study to provide some insightful guidelines about designing effective security warnings on mobile phones. Secondly, in this paper, we mainly focused on security hardening on QR code scanners. Our future research direction will include an extensive study of whether it is possible to enhance the security of QR code itself. This could be achieved by considering some cryptographic methods applied to QR codes for the purpose of certification and identification. Lastly, the phishing URLs that we used for this study came from PhishTank. In our immediate future work, we are going to use sample URLs from other sources such as Spamscatter to evaluate our proposed solution.

6. ACKNOWLEDGMENTS

This work was partially supported at the Secure Computing Laboratory at New Mexico Tech by the grant from the National Science Foundation (NSF-IIS-0916875). The authors would like to thank Chen Sun from NM Tech and three anonymous reviewers for their valuable comments and suggestions.

7. REFERENCES

[1] C. Woo Bong, H. Keon il, L. Won Gyu, P. Won Hyung, and C. Tai Myoung, "The New Vulnerability of Service Set Identifier (SSID) Using QR Code in Android Phone," in International Conference on Information Science and Applications (ICISA), Washington, DC, USA, 2011.

[2] CNET. (2012). *The Dark Side of QR Codes.* Available: http://news.cnet.com/8301-1009_3-57464276-83/the-dark-side-of-qr-codes/.

[3] Z. Yajin and J. Xuxian, "Dissecting Android Malware: Characterization and Evolution," Security and Privacy (SP), 2012 IEEE Symposium on, 2012.

[4] D. Winder. (2012). *Five Hidden Security Threats.* Available: http://www.pcpro.co.uk/features/374896/five-hidden-security-threats.

[5] R. Dhamija, J. D. Tygar, and M. Hearst, "Why phishing works," Proceedings of Conference on Human Factors in Computing Systems, Montréal, Québec, Canada, 2006.

[6] P. Soni, S. Firake, and B. B. Meshram, "A phishing analysis of web based systems," Proceedings of the 2011 International Conference on Communication, Computing; Security, Rourkela, Odisha, India, 2011.

[7] P. Kieseberg, M. Leithner, M. Mulazzani, L. Munroe, S. Schrittwieser, M. Sinha, and E. Weippl, "QR code security," Proceedings of the 8th International Conference on Advances in Mobile Computing, Paris, France, 2010.

[8] V. Sharma, "A Study of Malicious QR Codes," International Journal of Computational Intelligence and Information Security (IJCIIS), vol. 3, May 2012.

[9] A. P. Felt and D. Wagner, "Phishing on Mobile Devices," the WEB 2.0 Security and Privacy (W2SP), Oakland, California, USA, 2011.

[10] *Monthly Malware Statistics: September 2011.* Available: www.securelist.com/en/analysis/204792195/Monthly_Malware_Statistics_September_2011

[11] L. Borrett. (2011). *Beware of Malicious QR Codes.* Available: http://www.abc.net.au/technology/articles/2011/06/08/3238443.htm

[12] A. Kieyzun, P. J. Guo, K. Jayaraman, and M. D. Ernst, "Automatic creation of SQL Injection and cross-site scripting attacks," Proceedings of the 31st International Conference on Software Engineering, 2009.

[13] Netscape. *DMOZ Open Directory Project.* Available: http://www.dmoz.org

[14] *PhishTank.* Available: http://www.phishtank.com

[15] J. Ma, L. K. Saul, S. Savage, and G. M. Voelker, "Beyond blacklists: learning to detect malicious web sites from suspicious URLs," Proceedings of the 15th ACM SIGKDD, Paris, France, 2009.

[16] S. Abu-Nimeh, D. Nappa, W. Xinlei, and S. Nair, "Distributed Phishing Detection by Applying Variable Selection Using Bayesian Additive Regression Trees," in Communications, 2009. ICC '09.

[17] *Norton Safe Web.* Available: http://safeweb.norton.com/

[18] Norton. *Ecommerce Safety.* Available: https://safeweb.norton.com/help/ecommerce_safety

[19] *Safe Browsing API.* Available: https://developers.google.com/safe-browsing

[20] OpenDNS. *PhishTank API Information.* Available: http://www.phishtank.com/api_info.php

[21] *Safe Browsing Lookup API Developer's Guide.* Available: https://developers.google.com/safe-browsing/lookup_guide

[22] J. Sunshine, S. Egelman, H. Almuhimedi, N. Atri, and L. F. Cranor, "Crying wolf: an empirical study of SSL warning effectiveness," Proceedings of the 18th conference on USENIX security symposium, Montreal, Canada, 2009.

[23] S. Egelman, L. F. Cranor, and J. Hong, "You've been warned: an empirical study of the effectiveness of web browser phishing warnings," Proceedings of Conference on Human Factors in Computing Systems, Italy, 2008.

[24] L. Zeltser. (2011). *How to Design Security Warnings to Protect Users.* Available: http://blog.zeltser.com/post/3638747689/designing-security-warnings

[25] R. Reeder, E. C. Kowalczyk, and A. Shostack, "Helping Engineers Design NEAT Security Warnings," presented at the Symposium On Usable Privacy and Security (SOUPS) Pittsburgh, PA, USA, 2011.

Time Evolving Graphical Password for Securing Mobile Devices

Zhan Wang* zwang@lois.cn

Jiwu Jing* jing@lois.cn

Liang Li† lli@jdl.ac.cn

*State Key Laboratory of Information Security, Institute of Information Engineering, Chinese Academy of Sciences, Beijing 100093, China

†Key Laboratory of Intelligent Information Processing, Institute of Computing Technology, Chinese Academy of Sciences, Beijing 100190, China

ABSTRACT

Increasingly widespread use of mobile devices for processing monetary transactions and accessing business secrets has created a great demand on securing mobile devices. Poorly designed authentication mechanisms (e.g., screen lock and SIM card lock) on mobile devices either make users feel a hassle to lock the devices, or are vulnerable to attacks, such as shoulder surfing and smudge attack.

In this paper, we propose a new login option for unlocking mobile devices called *Time-Evolving Graphical Password* (TEGP), which improves the strength of the password gradually over time by evolving the distortion degree of the images in the challenge portfolio without changing the pass images. By taking advantage of the extraordinary human ability to recall images, TEGP authenticates users by asking them to recognize the pass images which are transformed from the images uploaded by the user at registration. To achieve desired security and remain the usability, we present two metrics, *Information Retention Rate* (IRR) and *Password Diversity Score* (PDS), to advise the selection and distortion of the pass images and decoy images. Our experimental results show the memorability from the perspective of users, and the ability of TEGP to defend against various attacks.

Categories and Subject Descriptors

K.6.5 [**Management of Computing and Information Systems**]: Security and Protection—*Authentication*

General Terms

Security

Keywords

Mobile Device Security, Graphical Password, Edge Detection

ASIA CCS'13, May 8–10, 2013, Hangzhou, China.

1. INTRODUCTION

Increasingly widespread use of mobile devices for processing monetary transactions and accessing business secrets has created a great demand on securing mobile devices. An unlocked mobile device is open for everyone to see and manipulate. This can pose a security risk if you store valuable or confidential information on the device, or install certain applications (e.g. mobile cloud) that contain private data. For example, emails on the device alone can reveal large amount of information about the person, including where you bank, the names of family members, and so on. If enterprise emails are accessible on the device, there can be competitive information, salaries, system passwords, etc. If any of those emails contain links, often clicking on them will sign you into the websites, such as online banking, social network page, or a corporate portal.

Poorly designed authentication mechanisms (e.g., screen lock and SIM card lock) on mobile devices either make users feel a hassle to lock the devices, or are vulnerable to attacks. In current market, there are two common practices for mobile device authentication. The first one is based on the passcode adopted by Apple products. That is, the user to be authenticated is asked to input the correct 4-digital passcode to unlock the screen. The second one is based on the dot selection and sliding. The user has to slide across multiple dots on the screen with correct sequence. Both authentication mechanisms suffer from the shoulder-surfing attack especially when users choose a simple pattern for quick accessing the device. They are also prone to the smudge attacks where the oily residues on the touch screen leak the passcode or sliding pattern to the attackers.

In this paper, we propose a new login option for unlocking mobile devices called *Time-Evolving Graphical Password* (TEGP) to defend against various attacks. By taking advantage of the extraordinary human ability to recall images, TEGP authenticates users by asking them to recognize the pass images which are transformed from the images uploaded by the user at registration. With time passed, TEGP can be configured to increase the distortion degree of the images in the challenge portfolio, so as to improve the strength of the password gradually over time without changing the pass images. To our best knowledge, this is the first work which explores the time-evolving feature of graphical password, which is a unique ability for improving the strength of the password without changing the password.

Fig. 1 illustrates an instance of TEGP. The user chooses and uploads images he or she would like to set as pass im-

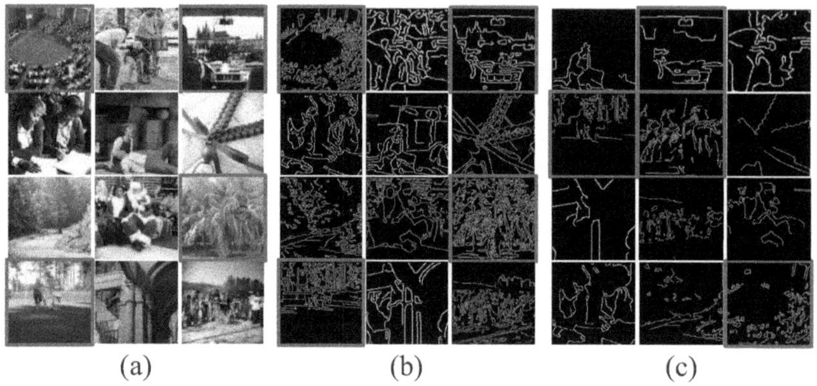

Figure 1: Time Evolving Graphical Password

ages (highlighted by the red frames in Fig. 1.a). To generate the challenge portfolio (Fig. 1.b), we extract the edge information of all the pass images and decoy images in Fig. 1.a. We design a training phase and allow the user to adjust the amount of edge information presented in the images and be comfortable with memorizing them. If you are experiencing difficulties to find the relationship between the images in Fig. 1.a and Fig. 1.b at the first look, the reason is that you do not have the adequate knowledge of the original pass images. That is, it evidently increases the difficulty to devise the shoulder-surfing attack. In addition, the adoption of distorted images also mitigates the guessing attack from acquaintances. Furthermore, TEGP allows the user to adopt the challenge images with reduced edge information as shown in Fig. 1.c for improving security, while the user still can recognize the pass images based on the previous visual memory. This will leave an unexpected display to the attackers who try to figure out the password by shoulder-surfing multiple times. Additionally, we rearrange the display sequence of the images in the challenge portfolio for each authentication to resist the smudge attack.

To bring TEGP into practice, we present two metrics, *Information Retention Rate* (IRR) and *Password Diversity Score* (PDS), to advise the selection and distortion of the pass images and decoy images. Users may upload very diverse pictures, either with clear shapes or with complex lines. Consequently, if we apply a fixed distortion parameter, it might either expose some images meanwhile or cause some images unrecognisable. IRR is used to rate the loss of information entropy of images after distortion and helps to determine the moderate distortion parameter. The decoy images are selected by the system for covering the pass images. If the decoy images are very different with all pass images, which will sell out the pass images. PDS is used to evaluate the variance of all the images in the challenge portfolio and guides the selection of decoy images.

We conducted experiments to apply IRR and PDS for generating optimal challenge portfolio, and then evaluate the memorability of TEGP by users and the ability of TEGP to defend against shoulder-surfing attack. The experimental results demonstrate that the users quickly become skilled at being authenticated with a good recall rate and acceptable login time with TEGP.

This paper is organized as follows. Section 2 describes the related works. Section 3 presents our time evolving graphical password mechanisms. Section 4 describes the prototype of TEGP. Section 5 discusses the experimental results. Section 6 concludes this paper.

2. TEGP

2.1 Overview

In this paper, we propose a new login option for unlocking mobile devices called *Time-Evolving Graphical Password* (TEGP) to defend against various attacks. By taking advantage of the extraordinary human ability to recall images, TEGP authenticates users by asking them to recognize the pass images which are transformed from the images uploaded by the user at registration. The user chooses and uploads images he or she would like to set as pass images. The decoy images are chosen by the system. To generate the challenge portfolio, we extract the edge information of all the pass images and decoy images. We design a training phase and allow the user to adjust the amount of edge information presented in the images and be comfortable with memorizing them. With time passed, when the users are more familiar with their distorted pass images, TEGP can be configured to increase the distortion degree of the images in the challenge portfolio, so as to improve the strength of the password gradually over time without changing the pass images. This increase the difficulty of shoulder-surfing attack, meanwhile users still are able to identify them based on the knowledge of original and previous versions of the pass images. Therefore, TEGP includes three functional parts: *Registration*, *Authentication* and *Time-Evolving*.

In *Registration*, the following steps are executed:

- Private images uploading: when a user requests to register, he/she is asked to provide some images as the materials of the graphical password from his/her mobile devices.

- Validating Private Images: we check if users have uploaded insecure set of images. For example, a user uploads three similar images that could allow attackers to identify pass images among the decoy images immediately.

- Generating Challenge Portfolio: the system that applies TEGP chooses a certain number of decoy images from its database. These images together with private images uploaded by users are distorted by a specific distortion algorithm. All distorted images construct the challenge portfolio.

- User training: training is a chance for users to practice the authenticating process and get familiar with the new password system. Users also can examine if the pass images can be identified easily by themselves. If users dislike the images in the challenging set that are displayed during authentication, they can apply to refresh the challenging set by restarting the step of private images uploading.

In *Authentication*, all images in the challenge portfolio are displayed to the user in random order. To be authenticated, the user must identify the correct pass images. If users fail too many times, their accounts will be locked as many other password systems do.

In *Time-Evolving*, TEGP can be configured to improve the strength of password manually or automatically over time. The *time-evolving* operation reduces the details of the challenge images by applying different distortion parameters. The users can refuse to evolve the password when they find trouble to recognize the pass images with less information. The users can also request to roll-back to a limited number of previous passwords. In order to prevent such requests issued by an attacker, the system can validate the issuer by using other authentication approach, such as PIN number.

2.2 Metrics

To bring TEGP into practice, we present two metrics, *Information Retention Rate* (IRR) and *Password Diversity Score* (PDS), to advise the selection and distortion of the pass images and decoy images. Users may upload very diverse pictures, either with clear shapes or with complex lines. Consequently, if we apply a fixed distortion parameter, it might either expose some images meanwhile or cause some images unrecognisable. IRR is used to rate the loss of information entropy of images after distortion and helps to determine the moderate distortion parameter. The decoy images are selected by the system for covering the pass images. If the decoy images are very different with all pass images, which will sell out the pass images. PDS is used to evaluate the variance of all the images in the challenge portfolio and guides the selection of decoy images.

2.2.1 Information Retention Ratio (IRR)

Information Retention Ratio (IRR) indicates that how much information of the image is retained after the distortion. IRR is obtained by calculating the ratio of the entropies between the original private images and their distorted versions. In the Shannon's information theory, entropy is a measurement of the uncertainty associated with a random variable. Here, it is a statistical measure of randomness that can be used to characterize the texture of the image. IRR is defined as follows:

$$IRR = \frac{\mathsf{H}(\mathsf{F}(\mathsf{I}))}{\mathsf{H}(I)}, \quad \mathsf{H}(I) = -\sum p_i \times \log(p_i) \tag{1}$$

where $\mathsf{H}(I)$ represent the entropy of an image I and p_i is the probability of grayscale i in the image I.

We can obtain an average IRR related to different distortion algorithms, which indicates how much information retention is needed to help users to recall the images, while still making it difficult to recognize a distorted image by attackers. The experiments section will show how to select the distortion parameter with the help of IRR.

2.2.2 Password Diversity Score (PDS)

The selection of decoy images in the challenge portfolio will affect both security and usability. On one hand, if all images in the challenging set seem very similar, that will achieve more security, but increase the difficulty of recognizing the pass images. On the other hand, if all images in the challenging set seem very diverse, that will achieve good usability, but increase the risk of exposing the pass images. Therefore, the ideal status is that decoy images can well conceal the pass images. Password Diversity Score (PDS) is proposed to evaluate the diversity of images in the challenging set and help to choose the decoy images automatically.

At first, gray statistical histogram is introduced to represent an image. For image p, it is described as follows:

$$Histo_p = (g_{p1}, g_{p2} \dots g_{pt}) \tag{2}$$

where g_{pi} is the total number of gray value i in picture p. t is the number of dimensionality, and it is set as 256 in our experiment empirically. We calculate the Euclidean distance of image p and q by:

$$Dis_{pq} = \sum_{l=1}^{t} (g_{pl} - g_{ql})^2 \tag{3}$$

Then we can obtain the Euclidean distances between each pair of images. Our goal is to hide the pass images among the decoy images. It can be described that the distances should be very uniform. Here, the variance of these distances is used to imply if those pass images can be well covered. We calculate:

$$PDS = Var(Dis_{12}, Dis_{13} \dots Dis_{xy}), \quad where\ 1 \leq x, y \leq k + m\ and\ x \neq y. \tag{4}$$

where k and m are the number of pass images and decoy images, respectively. If the value of PDS is very large, it poses the risk that attackers can distinguish the pass images easily. If the value of PDS is small, users still can identify the pass images easily since they have the prior knowledge of the original images.

3. EVALUATIONS

We conducted experiments to apply IRR and PDS for generating optimal challenge portfolio, then evaluate the memorability of TEGP by users and the ability of TEGP to defend against shoulder-surfing attack.

3.1 Experiment Settings

We recruited 102 volunteers including university faculties, diverse major students and staffs from IT companies. They are all active users of mobile devices. Table 1 shows the distributions of participants on age and careers. All 103 volunteers participated in the distortion parameters experiment.

Table 1: The distributions of participants on age and careers.

	<=25	26∼35	36∼45	46∼55
Student	18	3	0	0
Faculty	1	9	9	18
IT Staff	8	15	18	3

3.2 Distortion Parameters

In TEGP, private images are distorted by the edge detection algorithm which is called Canny Operator [2]. After the distortion, we obtain the pass images which are the profile versions of private images. In a pass image, some edges are left according to different parameters. Leaving too many edges and too few edges both bring difficulties to identify the pass images for users. All 103 participants attended this experiment to help us to determine the usable distortion parameters which would have a direct bearing on the security and convenience of the mechanism. All the volunteers install our application on their own mobile phones with Android OS systems.

Canny parameter is the level of edge retrieval (0.9∼0.1) using Canny Operator. In the experiment, we will give each participant one identical original image and ask them to pick up its profiled version from a nine images distortion set. This distortion set is comprises of the profile version of the original image together with other 8 decoy images. We will display the distortion set from the most distortion (0.9 profiled levels, with majority edges) to the least distortion (0.1 profiled level) with the step of 0.05. We investigate that at which level participants can recognize the profile version of the original image. Above experiment will repeat for 20 times. Finally, we set the default parameter as 0.55.

Then we calculate the average IRR value for each parameter (0.9∼0.1) of Canny Operator. Ten thousands images are involved in the calculation. We observe that 2.32%-3.246% is the acceptable range related to 0.4-0.55 level.

When a new images uploaded by a user, we first use the default 0.55 level to distort the images and obtain the related IRR value. If the value falls into the range of 2.32%-3.246%, this new images are accepted by system. Otherwise, we use lower distortion level at least 0.4 to distort the images and repeat the above process until the IRR value falls into the accepted range. If we cannot obtain a reasonable IRR value, users are required to re-upload the image.

3.3 Decoy Images

Aforementioned metric PDS is used to evaluate the quality of the selection of decoy images. All candidate decoy images in the database of the system are distorted according to above method. Here, we will show how to select decoy images in the scheme of TEGP with the help of PDS.

When pass images are settled, we will select an optimal decoy image by calculating the PDS value of a specific decoy image together with existing pass images in the challenging set. In this fashion, we will select and add new optimal decoy image into the challenging set one by one. Eventually, we obtain an optimal challenging set. Fig. 2 compares two challenging set. Fig. 2 (Left) has a PDS value of 3.9204×10^{-4}. Fig. 2 (Right) has a PDS value of 1.5682×10^{-4}. In Fig. 2 (Right), all images in the challenging set seem very uniform. However, in Fig. 2 (Left), some images with clear shape, others with complex background. In this situation, attackers tend to try the clear one, which might be the pass images with a high probability. By calculating the PDS value and selecting the challenging set with smaller PDS value, we can avoid the above problem.

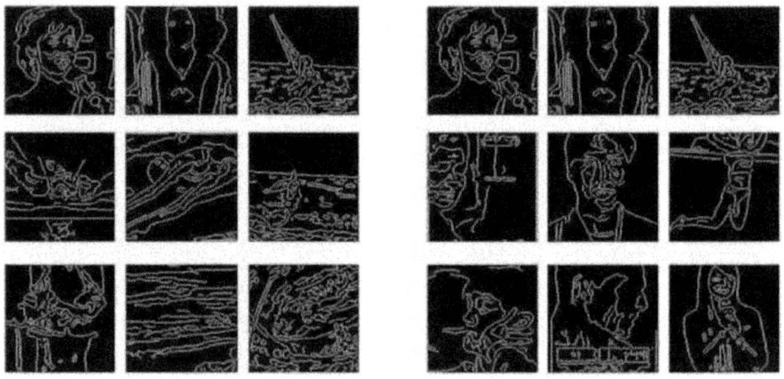

Figure 2: Challenging sets with different PDS values.

3.4 Memorability and Efficiency

Participants are required to register their pass images by using our system for lock/unlock cell phones at the first day, and pass the authentications in five periods during two weeks. In the fourth (one week later) and fifth period (two weeks later), after E-mail invitation, 98 and 93 participants respectively returned to our experiments. We divide all the participants into three groups evenly. Each group contains participant with different ages and careers. The participants in first group are authenticated without time-evolving feature. The rest of two groups use the time-evolving graphical password with different evolving interval. Group "with time-evolving I" reduces the distortion details every two days. Group "with time-evolving II" reduces the distortion details every day. Fig. 1.b and Fig. 1.c illustrate the appearance of the challenge portfolio between the first day and two weeks later for Group "with time-evolving II".

In a recent research [14] on the habit of people using smart phones, the study subjects checked their phones 34 times a day on average. We ask each participant to try the authentication every 6 minutes within one hour in each period. We record the attempts for passing each authentication. We allow roll-back to the previous password once after 3 attempts for the two groups with time-involving feature. The success rates for each period are shown in Table 2.

In the first two periods, all the participants can pass the authentication within three attempts. We find the first roll-back request on the third day from Group "with time-evolving II". In the fifth period, we observe the first authentication failure after 3 attempts and one roll-back. Therefore, the users of TEGP quickly become skilled at being authenticated with a good recall rate.

We recorded the login time for all the successful authentications within three attempts. The average login time is 5.1 seconds, which is acceptable for unlocking a mobile device.

3.5 Shoulder-Surfing Attack

Shoulder-surfing is an attack where the login is observed at the real-time, and reused in a later replay attack. It is extremely difficult to quantify the threat of the shoulder-surfing. Here, we assume that each login of mobile device

Table 2: Success rate for each authentication period (The number of successful authentications through 1 attempt/2 attempts/3 attempts/roll-back once/total number of authentications)

Groups	First Day	Second Day	Third Day	One Week Later	Two Weeks Later
w/o time-evolving	317/18/5/-/340	324/14/2/-/340	324/15/1/-/340	285/23/2/-/310	256/29/5/-/290
with time-evolving I	320/19/1/0/340	322/18/0/0/340	316/23/1/0/340	263/59/8/1/330	290/39/9/2/340
with time-evolving II	313/25/2/0/340	327/12/0/0/340	326/20/3/1/340	291/40/7/2/340	247/45/6/1/300

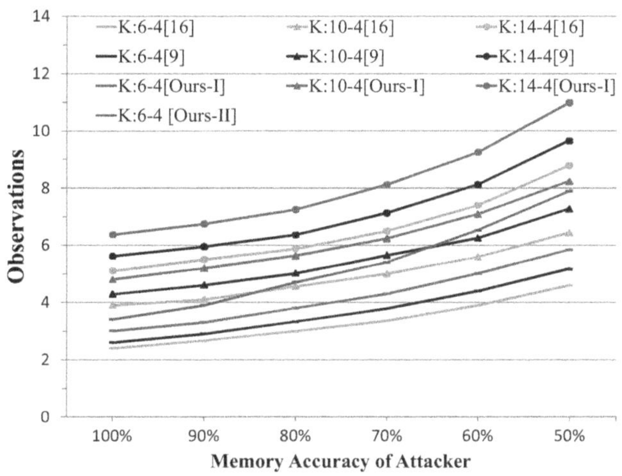

Figure 3: Number of observations required by an attacker to learn one pass image.

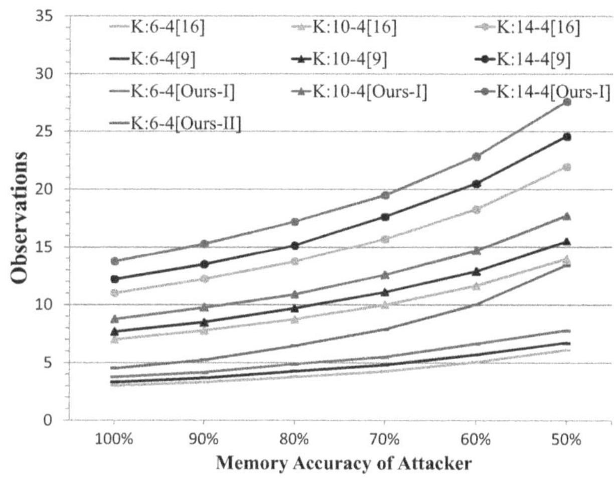

Figure 4: Number of observations required by an attacker to learn all the pass images.

will be observed, particularly by friends and colleagues, with whom users might neglect to shield their input process.

Our time evolving mechanism aims to increase the number of observations required for an attacker to mount reply attack in the future. In this section, we quantify the impact of shoulder-surfing attack on our mechanism. In order to investigate the capabilities of an attacker, the follow steps are executed: (1) The attacker is invited to observe legitimate user entry of each pass image at the login phase; (2) The attacker then has one attempt to authenticate, if the attacker can identify 4 images that he has previously observed being selected, the attack is successful, otherwise unsuccessful; (3) Repeat above two steps, with the knowledge of the attacker increasing each time until an attack is successful.

Fig. 3 and Fig. 4 report the number of observations required by an attacker to learn only one and all pass images respectively for unlocking mobile devices. We ask the attackers to estimate their memory accuracy range from 100% to 50%. In the parameter "K:a-b", "a" indicates the total number of images in the challenge portfolio; "b" indicates the number of pass images required for one successful login, and it is set to 4 in this paper. We compare our method with two popular graphical password schemes [16] and [9]. All the results are collected through 10,000 simulated logins.

There are four observations: 1) More observations are needed for successfully devising shoulder-surfing attack against the distortion-based graphical passwords([9] and ours), which provide higher security than the recognition based graphical passwords [16] without distortion; 2) With the scale of challenge portfolio increasing, the required observations for successful login are also increased; (3) When attackers are more confident with their memory accuracy, they use less number of observations; (4) The performance of our scheme with time evolving (K:6-4 pattern) is closed to that of the method [16](K:10-4 pattern), which means that our method requires a small size of challenge portfolio for achieving the same security.

4. DISCUSSION

4.1 Theoretical Password Spaces

Authentication systems with small theoretical password spaces in user choice of passwords are especially vulnerable to guessing attacks. The theoretical password space of TEGP depends on the number of pass images and the total number of images in challenge portfolio. The widely used 4-digital passcode has a 10^4 space. TEGP can achieve the same security by authenticating user with 4 pass images among 12 images in the challenge portfolio.

4.2 Password Reset

With the automated time-evolving feature, the users may experience the difficulty to recognized newly evolved password which has less information than previous one. The users can also request to roll-back to a limited number of previous passwords. However, the attackers cannot recognize the previous passwords without the knowledge of the original images. Another countermeasure is that the system validates the issuer by using other authentication approaches, such as PIN number, since our TEGP is one of the options for mobile device authentication.

5. RELATED WORK

Graphical password systems [1, 17, 18, 10, 19, 13] gain more attention recently as a promising alternative of text-based password in the both academia and industry. Most graphical password mechanisms are based on recognizing system provided images, such as faces [4], identifying user uploaded images [9, 16, 15], recognizing the category of pass images [12] or recalling a sequence of actions such as, clicking [7, 3, 6, 5] and drawing [8].

Mobile devices with touch screen, such as smart phones and PDA, have become an ideal application environment [9, 11] for graphical password. [9] is the most related work to ours, which utilizes oil-painting effect to filter user uploaded images from their devices. However, it inadequately solves the following problems. First, it allows users to upload their own images without further checking. For example, if all the pass images are very similar with each other, pass images may compromise themselves. Therefore, some validation or a guideline should be given for choosing secure and optimal graphical password. Second, they determine a fixed distortion parameter to process the images, which cannot fit all images with different textual. Finally, they claim the login time as 18 seconds which is not acceptable for unlocking a mobile device.

6. CONCLUSIONS

In this paper, we propose a new authentication mechanism called time evolving graphical password (TEGP) for securing mobile devices. The salient feature of TEGP is time-evolving, which improves the strength of the password gradually over time by evolving the distortion degree of the images in the challenge portfolio without changing the pass images. TEGP authenticates users by asking them to recognize the pass images distorted by the edge retrieval algorithm. We conduct experiments with 102 volunteers to show the memorability of the time evolving passwords from the perspective of users, and to evaluate the ability of TEGP to defend against shoulder-surfing attack.

7. ACKNOWLEDGEMENTS

This material is based upon work supported by the National Natural Science Foundation of China under grant 70890084/G021102, by the Knowledge Innovation Program of CAS (No. YYYJ-1013), by the Strategy Pilot Project of Chinese Academy of Sciences (Sub-project XDA06010702), and by the National High Technology Research and Development Program of China (863 Program) under grant 2013AA01A214.

8. REFERENCES

[1] A. D. Angeli, L. Coventry, G. Johnson, and K. Renaud. Is a picture really worth a thousand words? exploringthe feasibility of graphical authentication systems. *Int. J. Hum.-Comput. Stud.*, 63(1-2):128–152, 2005.

[2] J. Canny. A computationnal approach to edge detection. *IEEE Trans. Pattern Anal. Mach. Intell.*, 8:679–698, 1986.

[3] S. Chiasson, A. Forget, R. Biddle, and P. van Oorschot. User interface design affects security: patterns in click-based graphical passwords. *Int. J. Inf. Secur*, 8(6):387–398, 2009.

[4] Passfaces corporation. In *The Science Behind Passfaces. Company white paper.*

[5] S. Chiasson. Usable authentication and click-based graphical passwords. In *Ph.D. Thesis, School of Computer Science, Carleton University, Ottawa, Canada*, 2008.

[6] S. Chiasson, A. Forget, E. Stobert, P. van Oorschot, and R. Biddle. Multiple password interference in text passwords and click-based graphical passwords. In *ACM Conference on Computer and Communications Security*, pages 500–511, 2009.

[7] A. E. Dirik, N. Memon, and J.-C. Birget. Modeling user choice in the passpoints graphical password scheme. In *Proceedings of the 3rd symposium on Usable privacy and security*, SOUPS '07, pages 20–28, New York, NY, USA, 2007. ACM.

[8] P. Dunphy and J. Yan. Do background images improve "draw a secret" graphical passwords? In *Proceedings of the 14th ACM conference on Computer and communications security*, CCS '07, pages 36–47, New York, NY, USA, 2007. ACM.

[9] E. Hayashi, R. Dhamija, N. Christin, and A. Perrig. Use your illusion: secure authentication usable anywhere. In *Symposium on Usable Privacy and Security*, pages 35–45, 2008.

[10] D. Hong, S. Man, B. Hawes, and M. Mathews. A password scheme strongly resistant to spyware. In *Proc. International conference on security and management*, 2004.

[11] W. Z. Khan, M. Y. Aalsalem, and Y. Xiang. A graphical password based system for small mobile devices. *CoRR*, abs/1110.3844, 2011.

[12] R. A. Khot, K. Srinathan, and P. Kumaraguru. Marasim: a novel jigsaw based authentication scheme using tagging. In *Proceedings of the SIGCHI Conference on Human Factors in Computing Systems*, CHI '11, pages 2605–2614, New York, NY, USA, 2011. ACM.

[13] Microsoft, http://windows.microsoft.com.

[14] A. Oulasvirta, T. Rattenbury, L. Ma, and E. Raita. Habits make smartphone use more pervasive. *Personal and Ubiquitous Computing*, 16(1):105–114, 2012.

[15] T. Pering, M. Sundar, J. Light, and R. Want. Photographic authentication through untrusted terminals. *IEEE Pervasive Computing*, 2(1):30–36, 2003.

[16] T. Takada, T. Onuki, and H. Koike. Awase-e: Recognition-based image authentication scheme using users' personal photographs. In *Innovations in Information Technology*, pages 1–5, 2006.

[17] X. Suo, Y. Zhu, and G. Owen. Graphical passwords: A survey. In *Proc. Annual Computer Security Applications Conference*, pages 463–472, 2005.

[18] D. Weinshall and S. Kirkpatrick. Passwords you'll never forget, but can't recall. In *Proc. Conference on Human Factors in Computing Systems*, pages 1399–1402, 2004.

[19] S. Wiedenbeck, J. Waters, J. Birget, A. Brodskiy, and N. Memon. Authentication using graphical passwords: effects of tolerance and image choice. In *Proc. Symposium on Usable Privacy and Security*, pages 1–12, 2005.

DroidAlarm: An All-sided Static Analysis Tool for Android Privilege-escalation Malware

Yibing Zhongyang, Zhi Xin, Bing Mao and Li Xie
State Key Laboratory for Novel Software Technology
Department of Computer Science and Technology
Nanjing University
Sophie.xuer@gmail.com, {zxin,maobing,lixie}@nju.edu.cn

ABSTRACT

Since smartphones have stored diverse sensitive privacy information, including credit card and so on, a great deal of malware are desired to tamper them. As one of the most prevalent platforms, Android contains sensitive resources that can only be accessed via corresponding APIs, and the APIs can be invoked only when user has authorized permissions in the Android permission model. However, a novel threat called privilege escalation attack may bypass this watchdog. It's presented as that an application with less permissions can access sensitive resources through public interfaces of a more privileged application, which is especially useful for malware to hide sensitive functions by dispersing them into multiple programs. We explore privilege-escalation malware evolution techniques on samples from Android Malware Genome Project. And they have showed great effectiveness against a set of powerful antivirus tools provided by VirusTotal. The detection ratios present different and distinguished reduction, compared to an average 61% detection ratio before transformation. In order to conquer this threat model, we have developed a tool called DroidAlarm to conduct a full-spectrum analysis for identifying potential capability leaks and present concrete capability leak paths by static analysis on Android applications. And we can still alarm all these cases by exposing capability leak paths in them.

Categories and Subject Descriptors

D.2.0 [**Software Engineering**]: General—*Protection mechanisms*; D.4.6 [**Operating Systems**]: Security and protection—*Invasive software*

General Terms

Security

Keywords

Android; Capability Leaks; Privilege Escalation Attack; Malware Transformation; Static Analysis

1. INTRODUCTION

It is generally agreed that users are shifting their primary Internet accessing terminals from PCs to mobile devices. And they can download plenty of applications from diverse markets, such as Google Play, an official Android Market that contains over 600,000 applications. However, these applications are also carrying a huge number of unprecedented threats. According to F-Secure's Mobile Threat Report Q3 2012 [10], mobile malware cases on Android platform almost increased 16 times in the first three quarters of 2012, and the total number in the third quarter of 2012 is as high as 51,447. Many malware have shown great interests in user's privacy information for profit, including SMS messages [17], credit card information [9], audio records [18] and so on. In order to restrict potential privacy information abuse, Android raises a permission-based security model [4]. Every application must explicitly propose a list of required permissions to users before the installation procedure, which may expose its malicious intention. However, a new emerging threat model called *privilege escalation attack* [7] can bypass this permission authorization mechanism. It aims at the event that a authorization list contains suspicious sensitive permissions, which may lead to malicious usage. By utilizing the threat model, sensitive permissions and relevant malicious usage will be dispersed into different applications.

According to existing references [19, 22, 3, 13], antivirus scanners may first calculate hash values of the tested file to decide whether it is a known malware. If it is an unknown sample, scanners then filter sensitive permissions and then use CFG analysis, heuristics-based methods and so on to verify whether it is malicious. So in the above threat model, it is hard for scanners to identify potential malicious intentions with multiple individual permission lists and applications.

There are some analysis and detection methods against *privilege escalation attack*. Dynamic methods focus on call-graph analysis [11], the call-chain of IPC [8] or a customized monitoring framework [5] to address part or all of the attack. However, these methods have problems with path coverage and input effectiveness. There are also some static techniques to analyze the attack in the stock image [12] or applications [6, 16]. However, they only care about public interfaces on middleware layer and do not cover other forms of public interfaces on kernel layer, including file system and network sockets, which may miss some capability leak cases. For example, NickyBot [14] puts its audio files on SDcard, which may be obtained by its potential cooperator. And Chan et.al. [6] use decompiled applications to analyze, which suffers unsatisfactory decompiled results and potential obfuscation interruption.

In this paper, we show the seriousness of privilege-escalation evolution on Android malware and also have developed an automatic static analyzer to alarm these cases. We build three privilege-escalation transformation models to explain the former issue based on different functionalities. Different functionalities are mapped to implement in different cooperation ways. The communication between applications makes full use of different forms of public interfaces.

To alarm the attack, we have developed DroidAlarm to perform the automatic static analysis process. Relevant files (i.e. the configuration file and the executable file) are parsed for sensitive permissions and public interfaces. If an application has both sensitive permissions and public interfaces, relationships between sensitive permissions and public interfaces will be built to reveal its capability leaks. API calls corresponding with these permissions will be located. Then, based on the cooperation ways concerning these APIs, existing public interfaces will be checked and the data flow will be followed to get concrete capability leak paths, which will raise users' vigilance on potential *privilege escalation attacks*. Since DroidAlarm is off-line, there is a limited overhead.

We analyze 49 well-known malware from Android Malware Genome Project [20] with DroidAlarm. These 49 covers each malware family infected by the similar payload, so they already involve all malware patterns in this project. They have not widely adopted privilege escalation to advance themselves yet. Only 6% have unintentional capability leaks. However, we explore the construction models to transform them into privilege-escalation malware based on different functionalities. After transformation, the detection ratios have significantly reduced, compared to an average 61% before transformation. So these privilege-escalation malware bypass most antivirus tools on VirusTotal [1]. However, we can still alarm all these cases.

In particular, our contributions are as follows:

- We explore three styles of privilege-escalation malware transformation techniques based on their different functionalities.
- We build the first bytecode-based static capability leak analyzer, DroidAlarm, on all kinds of communication channels, including ICC, file system and network sockets.

The rest of this paper is arranged as follows: Section 2 gives a whole background of Android and *privilege escalation attack*; Section 3 presents the design and implementation of our malware transformation techniques and DroidAlarm in detail; Section 4 provides the evaluation results; Section 5 summarizes related work and the distinguishing features of our work; Section 6 illustrates the limitations and future work; Section 7 concludes the paper.

2. BACKGROUND

Before presenting our work, the relevant background of Android and the threat model will be briefly highlighted.

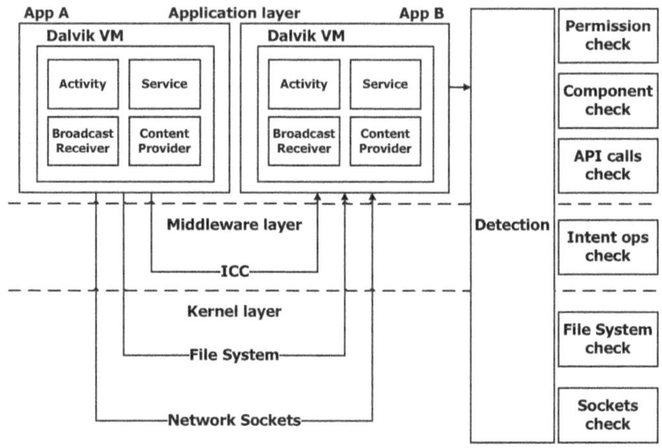

Figure 1: System Overview

2.1 Android

Fig. 1 shows Android internals from the viewpoints of possible communication channels and our check spots. Android is a free, open source, Linux-based OS designed for smartphones. Possible communication channels are built at multiple layers, including middleware layer and kernel layer. Communication channels at kernel layer contain Linux file system and network sockets. At middleware layer, the communication channel is usuallly built at the granularity of components, called *inter component communication* (ICC). An Android application consists of four types of components, which are Activities, Services, Broadcast Receivers and Content Providers. The ties between different components are Intents, which contain data and actions to transmit. At application layer, each application has a unique uid and runs on an independent Dalvik Virtual Machine (DVM). In Android, APIs related with sensitive resources, like sending SMS messages, are protected by different permissions.

2.2 Privilege Escalation Attack

Privilege escalation attack needs to be distinguished from *privilege escalation* or root exploit [21]. For root exploit, malware intentionally utilize Android platform-level vulnerabilities to get root privileges. However, *privilege escalation attack* mostly focuses on borrowing others' permissions to reach its malicious goal instead of actually getting them.

We divide privilege escalation attacks into code cooperation and data cooperation. In code cooperation, app A with less permissions can invoke components of a more privileged app B, like Activity launching function startActivity, to conduct sensitive behaviors. And in data cooperation app A can get sensitive data from app B through various communication channels, such as intent.getExtras().

There may exist a dispute that the attacker needs to install at least two malicious apps on a single user's device to complete the procedure. However, it can be easily handled in two aspects. On one hand, an attacker can explore capability leaks in other apps [12, 6, 16] and utilize them to reach their malicious goals. On the other hand, if one attack app has been installed, it can use the function of app recommendation to wheedle the user to install other ones. Then they can cooperate to achieve their malicious intention.

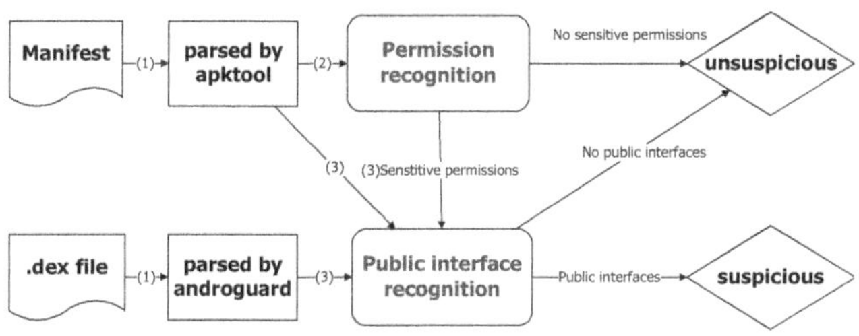

Figure 2: overview of suspicious analysis

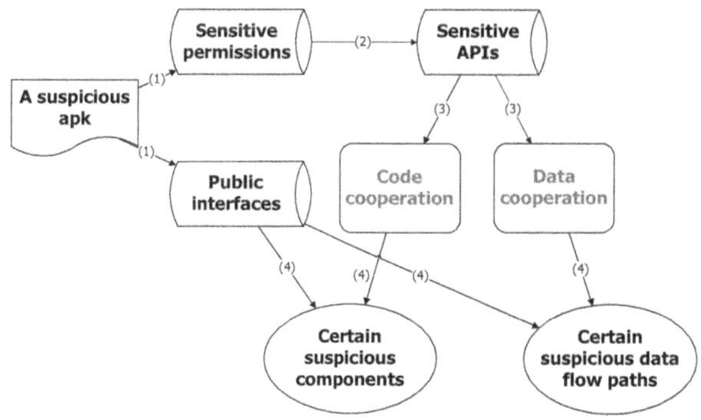

Figure 3: overview of leak path analysis

3. DESIGN

After presenting the threat model, the design of DroidAlarm and our malware transformation techniques will be introduced.

3.1 DroidAlarm

Fundamentally, in order to reduce privilege escalation attack, DroidAlarm aims at alarming capability leaks of an application that potentially give rise to the attack. The whole solution is divided into two modules and four submodules.

3.1.1 permission recognition

Pemission recognition is a submodule in Fig. 2. In our threat model, sensitive resources is the purpose of capability leaks. Since sensitive permission is related with accessing sensitive resource, it is an important target in this module.

Android categorizes the protection levels of permissions into four levels: normal, dangerous, signature and signatureOrSystem. The threat level increases one by one. In our solution, as we do not think normal permissions can be actual risks, we define all other permissions as sensitive permissions. The threat level of permissions in a third-party application is generally no higher than dangerous, and malware are no exceptional. With permissions extracted from AndroidManifest by apktool [2], we check every item with permission protection levels file and maintain those whose protection levels are not normal. If sensitive permissions are found, analysis procedure jumps to the next step.

3.1.2 public interface recognition

Based on our threat model, public interfaces are vital targets in the following submodule (Fig. 2). They have various forms in AndroidManifest and the dex file. In AndroidManifest they are mostly related with components. Firstly, Activities, Services and BroadcastReceivers containing a user-defined intent-filter can be triggered by other applications. So this kind of intent-filter is a striking feature. Secondly, Content Providers have a "true" exported value unless they are designed for Android 4.2 or higher versions. Finally, as multiple applications that have a same uid can share their resources to make privilege escalation attack, another conspicuous attribute is sharedUserId. However, more public interfaces at three communication channel levels are extracted from the dex file by androguard [3]. For ICC, Intent is the key. For file system, all the four types of file, SharedPreference, traditional Java file, SQLite and SDcard file, can be considered as public interfaces. Ultimately, all the network socket classes also need to be recognized.

Combined with the results of permission recognition, if a malware has such interfaces, it is suspicious. However, suspicious analysis module is not sufficient to settle this issue. When a malware has public interfaces that do not reveal sensitive resources protected by corresponding permissions, it is a false positive case. To reduce the rate of false positive, we need to filter the preliminary result set.

3.1.3 code cooperation

Based on suspicious analysis module, we construct a concrete path from a sensitive permission to a public interface. We locate sensitive APIs corresponding to the above permissions, and divide them on the cooperation way.

Code cooperation that focuses on APIs with some sensitive behaviors is a submodule in Fig. 3. Regarding public interfaces obtained, we firstly check whether components containing the above APIs have such an interface or not. If so, a leak path will be built. Besides, we also need to check the above components to see if other components that invoke one of them have public interfaces. Then we repeat the process iteratively. Eventually, we may build a leak path, a sensitive API with relevant permissions releasing resources during a series of component invoking, where the component at the highest invoking level contains a public interface.

3.1.4 data cooperation

Data cooperation is a more complex submodule in Fig. 3, where we concentrate on APIs concerning sensitive data and APIs that provide sensitive behaviors to generate sensitive data. A simple algorithm is presented as Algorithm 1.

As variables are stored in registers on Dalvik bytecode, we extract every method that initially calls one of the above APIs, and gets a sensitive register set with related variables. Then we analyze each instruction as a sequence of intraprocedural CFG of these methods. Based on different types of instructions, we make four decisions. Firstly, if any register in the set is changed, we remove it. Secondly, if data of any sensitive register is passed to another one that is not in the set, we record the influenced register in the set. Thirdly, if there is a public interface, which exposes all or part of the sensitive register set, we can then build a certain capability leak path. Fourthly, if a function call involves any sensitive registers, we track it iteratively. In the end, if the return value of the method is in one register of the set, we refer to interprocedural CFG and check its invokers similarly. Consequently, there will be a conclusion that whether a malware has a capability leak path where sensitive data obtained from an API with sensitive permissions are transmitted to a public interface during a series of function calls.

Algorithm 1 algorithm of data flow tracking

```
[method, SOURCE] = extract sensitive API()
for every instruction I as a sequence of intraprocedural
  CFG of method do
  if i changes va in SOURCE then
    SOURCE.remove(va)
  end if
  if i assigns vb in SOURCE to vc then
    SOURCE.merge(vc)
  end if
  if i provides a public interface and its parameters in
   SOURCE then
    build a certain suspicious data flow path
  end if
  if i is a function call and its parameters in SOURCE
   then
    track the function and its parameters in SOURCE
  end if
end for
if method has return value then
  expand CFG to find its invokers and check them with
  the return value
end if
```

3.2 Malware transformation techniques

Since existing malware have not widely adopted privilege-escalation construction techniques as discussed in Section 4, we are inspired to explore the techniques based on their different functionalities. According to the survey [21], there are mainly four payload functionalities, namely root exploit, remote control, financial charges, and personal information stealing. However, root-exploit malware aims at getting root privilege by invoking particular native codes, without requiring any sensitive permissions. So we focus on other three kinds of functionalities. As malware may have multiple functionalities, they may need to be split into different parts that achieve different functionalities.

Remote control is usually the prerequisite of financial charges and personal information stealing. This kind of malware receives network or SMS bot commands. Since this behavior may need INTERNET or RECEIVE_SMS permission, we separate this procedure from malware. We implement it as a way of code cooperation. When an application receives bot commands, it will start the exported components with corresponding behaviors in other applications.

Financial charges is a typical functionality for profit. In this scenario, malware send SMS messages or make background phone calls to premium-rate numbers. However, some malware hard-code premium-rate numbers and message contents, so we disperse content preparation and SMS sending into different applications. Background phone calls making can be transformed similarly. This kind of transformation can be also implemented as code cooperation. Intent containing these relevant information in one application starts the exported components with sensitive behaviors in other applications.

Personal information stealing is another functionality that is mostly related with malware motivations. In personal information stealing, malware are interested in collecting privacy information. Since information collecting and sending are the two main aspects in these malware, we can put these two procedures into two applications to reduce requested permissions in each one and crack the complete semantic. It is a typical scenario of data cooperation.

4. EVALUATION

In this section, we demonstrate the effect of our malware transfromation techniques against traditional methods on VirusTotal [1] and the evaluation results of DroidAlarm. We present our experiment results in Section 4.1. Furthermore, we put forward case studies for how to utilize unintentional leak in malware and how to construct privilege-escalation malware in Section 4.2.

Our test set picks up 49 malware from Android Malware Genome Project [20]. All experiments of DroidAlarm are performed on Ubuntu 10.04, which has a 512MB of RAM.

4.1 Result Overview

Firstly, we check each existing malware on VirusTotal [1] and their detection ratio is 61% on average. Secondly, we run DroidAlarm on these samples and verified the correctness manually. Thirdly, we recheck transformed malware.

Table 1 shows our progressive analysis results. Suspicious analysis module reports 33 samples are suspicious, which means they have both sensitive permissions and public interfaces. But sensitive permissions can be used independently from public interfaces in many scenarios. So in order to reduce false positive, we filter the results by building concrete paths from sensitive resources protected by permissions to public interfaces.

According to the results, we found the leak paths for three malware, namely AnserverBot, Bgserv and NickyBot. On one hand, most malware use sensitive resources internally. On the other hand, some malware send sensitive data to a fixed remote destination, and do not provide an open interface for other local applications. Compared with 67% alarm rate before leak path analysis to 6% after it, the number of false positive cases is largely reduced. For all leak paths in these three malware, we confirm them. Since we do not think they are actually designed that way, we define them as unintentional paths, which may also lead to *privilege escalation attack*. NickyBot will be discussed in detail in Section 4.2 to explore the potential exploitation.

In our experiments, we construct privilege-escalation malware by dispersing sensitive operations into multiple applications with three models introduced in Section 3.2. This transformation evade part of detections on VirusTotal [1], and Table 1 has showed detection ratios after our transformation. Because most of these methods only target a single application and cannot get a complete malicious pattern.

Based on transformable functionalities, we choose FakePlayer, GPSSMSSpy, DroidDream and GamblerSMS to perform all the three transformation models. From Table 1, we find the detection ratios have really declined a lot, which shows the techniques are very effective. We will describe the leak paths of DroidDream in detail in the following case studies. In order to compare different specific implementation techniques, we do further experiments, such as reducing the number of APIs, and removing auxiliary classes or its host app. And we find the detection ratios have different changes as shown in Table 1.

Some previous research works [12, 6] do detect part of transformed malware we build, by analyzing public interface forms of Intent and component. However, through analysis, we find that the forms of network sockets and file system are also used frequently. So these two forms cannot be ignored. As we recognize public interfaces on all the communication channels, we can easily find these public interfaces, and build concrete paths to show the capability leaks.

Table 1: Results on Existing Malware

Malware	O.r.	S.p.	P.i.	C.p.	F.m.	T.r.	Detected
ADRD	59%	✓	✓	✗	2	-	-
AnserverBot	70%	✓	✓	✓	2,3	-	-
Asroot	45%	✓	✗	✗	1	-	-
BaseBridge	86%	✓	✓	✗	1,2,3	-	-
BeanBot	39%	✓	✓	✗	2,3,4	-	-
Bgserv	70%	✓	✓	✓	2,3,4	-	-
CoinPirate	61%	✓	✓	✗	2,3,4	-	-
CruseWin	70%	✓	✗	✗	2,3,4	-	-
DogWars	75%	✓	✓	✗	3	-	-
DroidC.	34%	✓	✓	✗	1,2	-	-
DroidDeluxe	73%	✓	✗	✗	1	-	-
DroidDream	79%	✓	✗	✗	1,4	0%-77%	✓
DDlight	59%	✓	✓	✗	2,4	-	-
DroidK.F.1	64%	✓	✓	✗	1,2,4	-	-
DroidK.F.2	66%	✓	✓	✗	1,2,4	-	-
DroidK.F.3	61%	✓	✓	✗	1,2,4	-	-
DroidK.F.4	64%	✓	✓	✗	2	-	-
DroidK.F.S.	32%	✓	✓	✗	1,2,4	-	-
DroidK.F.U.	33%	✓	✓	✗	-	-	-
Endofday	68%	✓	✓	✗	2,3,4	-	-
FakeNetflix	68%	✓	✗	✗	4	-	-
FakePlayer	70%	✓	✗	✗	3	25%-38%	✓
GamSMS	46%	✓	✗	✗	3	20%	✓
Geinimi	61%	✓	✓	✗	2,3,4	-	-
GGTracker	73%	✓	✗	✗	3,4	-	-
GinMaster	80%	✓	✓	✗	1,2,4	-	-
GoldDream	61%	✓	✗	✗	2,3,4	-	-
Gone60	48%	✓	✓	✗	4	-	-
GPSSMSSpy	55%	✓	✓	✗	2,3,4	5%	✓
HippoSMS	50%	✓	✓	✗	3	-	-
Jifake	70%	✓	✗	✗	3	-	-
jSMSHider	66%	✓	✓	✗	2,3,4	-	-
Kmin	73%	✓	✗	✗	2,3	-	-
LoveTrap	66%	✓	✓	✗	3	-	-
NickyBot	45%	✓	✓	✓	2,3,4	-	-
NickySpy	61%	✓	✓	✗	2,3,4	-	-
Pjapps	58%	✓	✓	✗	2,3,4	-	-
Plankton	59%	✓	✓	✗	2	-	-
RogueLemon	33%	✓	✓	✗	2,3,4	-	-
RogueSPP.	71%	✓	✓	✗	3	-	-
SMSRep.	58%	✓	✗	✗	3,4	-	-
SndApps	59%	✓	✓	✗	4	-	-
Spitmo	69%	✓	✗	✗	2,3,4	-	-
Tapsnake	50%	✓	✗	✗	4	-	-
Walkinwat	72%	✓	✗	✗	3	-	-
YZHC	59%	✓	✓	✗	2,3,4	-	-
zHash	56%	✓	✓	✗	1	-	-
Zitmo	73%	✓	✗	✗	4	-	-
Zsone	64%	✓	✓	✗	3	-	-

S.p., P.i., C.p., F.m., O.r., T.r. and Detected mean sensitive permissions, public interfaces, certain path and functionality model, original ratio, transformed ratio and detected by DroidAlarm respectively.

1, 2, 3 and 4 mean root exploit, remote control, financial charges and personal information stealing respectively.

DroidK.F.U. is not malicious itself, but wheedles the user to install a malicious update, which is not included in our models.

4.2 Case Studies

To have a profound understanding of our work, we discuss two cases in depth. NickyBot is used to show unintentional capability leak in existing malware. And DroidDream variants demonstrate the practical pattern of personal information stealing.

4.2.1 NickyBot

NickyBot [14] records the surrounding sounds and stores audio files on SDcard. The lack of protection against SDcard files leads to the capability leak. The left part of the broken line in Fig. 4 included in the malware illustrates the capability leak path from RECORD_AUDIO permission to storing files on SDcard. An unprivileged attack app without the RECORD_AUDIO permission may obtain the audio files on SDcard and sends them out. It is a special case that an API with sensitive behavior generates sensitive data. The file type can be changed to one of other three forms mentioned before with an operation mode as MODE_WORLD_READABLE.

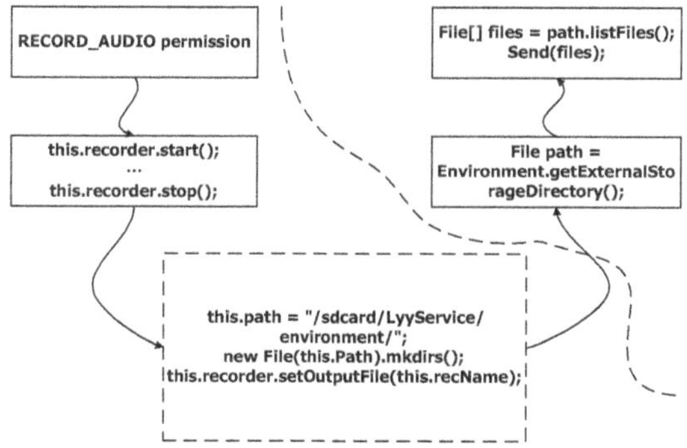

Figure 4: Capability Leak Path of NickyBot

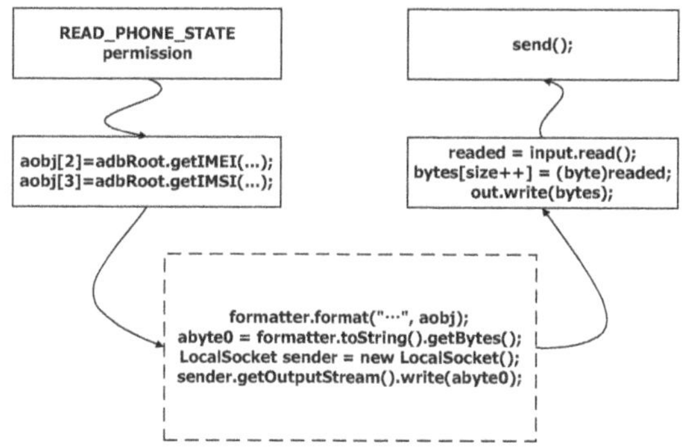

Figure 5: Capability Leak Path of DroidDream Variant

4.2.2 DroidDream Variant

DroidDream [15] collects users' IMEI and IMSI, and sends them through URL network socket connection. It is a behavior of personal information stealing. As using URL network socket connection needs INTERNET permission, which is too loud and may be easily caught by traditional methods, we change it to LocalSocket that does not need such a permission. The leak path is demonstrated in Fig. 5. When the attack app obtains IMEI and IMSI, it codes and sends them through a LocalSocket. The unprivileged app without READ_PHONE_STATE permission receives the data from a same SOCKET_ADDRESS LocalSocket and sends them to our ftp server. The sensitive data can be also transmitted through other public interfaces.

5. RELATED WORK

There are several detection methods against *privilege escalation attack*. To discover the attack, IPC Inspection [11] combines call-graph analysis with communication tracking between applications, while QUIRE [8] tracks the call-chain of IPC and creates lightweight signatures for applications. A system-centric solution [5] monitors applications by extending relevant components in the OS system and decide whether there is a *privilege escalation attack* or not by customized permission policies. Compared to the above three methods, DroidAlarm does not require modifying the OS system or applications. And we use static analysis methods to avoid the low rate of path coverage and input effectiveness in dynamic methods. Woodpecker [12], CHEX [16] and DroidChecker [6] use static analysis techniques to detect capability leaks only at ICC level but not file system or network sockets. Woodpecker [12] allows for stock phone images but not applications. And DroidChecker [6] aims at source codes of applications, facing the difficulty of obtaining source codes and unsatisfactory decompiled results. However, DroidAlarm analyzes Dalvik bytecode directly. Meanwhile, we concentrate on all the communication channels to achieve an all-sided analysis.

6. LIMITATIONS AND FUTURE WORK

Our work has two aspects to improve. Firstly, since most auxiliary resources have a certain version, DroidAlarm is currently limited to Android 2.2. However, we can add a module to recognize the version and choose proper resources. Secondly, DroidAlarm ignores using data through file system or network sockets from other apps to do sensitive behaviors. However, we can settle it by building leak paths from pulic interfaces to sensitive permissions with existing methods.

7. CONCLUSIONS

In this paper, we demonstrate DroidAlarm to alarm *privilege escalation attack* by analyzing capability leaks in Android malware statically. Three malware out of 49 are considered to expose capability leaks. We then transform some malware with *privilege escalation attack* technique in three models to bypass traditional detections on VirusTotal [1]. And the detection ratio decreases in different levels on each variant. However, these cases can still trigger our alarm.

8. ACKNOWLEDGMENTS

We would like to thank our shepherd, Xuxian Jiang, and the anonymous reviewers for their comments. This work was supported in part by grants from the Chinese National Natural Science Foundation (61073027, 61272078, 90818022, and 61021062), and the Chinese National 863 High-Tech Program (2011AA01A202).

9. REFERENCES

[1] Virustotal. *https://www.virustotal.com/*.

[2] android-apktool. *http://code.google.com/p/android-apktool/*, 2011.

[3] androguard. *http://code.google.com/p/androguard/*, 2012.

[4] Permissions. *http://developer.android.com/guide/-topics/security/permissions.html#arch*, 2012.

[5] S. Bugiel, L. Davi, A. Dmitrienko, T. Fischer, A. Sadeghi, and B. Shastry. Towards taming privilege-escalation attacks on android. In *Proceedings of the 19th Network and Distributed System Security Symposium*, 2012.

[6] P. P. Chan, L. C. Hui, and S. Yiu. Droidchecker: Analyzing android applications for capability leak. In *ACM Conference on Security and Privacy in Wireless and Mobile Networks*, 2012.

[7] L. Davi, A. Dmitrienko, A.-R. Sadeghi, and M. Winandy. Privilege escalation attacks on android. In *Proceedings of the 13th Information Security Conference*, 2010.

[8] M. Dietz, S. Shekhar, Y. Pisetsky, A. Shu, and D. S.Wallach. Quire: Lightweight provenance for smart phone operating systems. In *Proceedings of the 20th USENIX Security Symposium*, 2011.

[9] F-Secure. Warning on possible android mobile trojans. *http://www.f-secure.com/weblog/archives/00001852.-html*, 2010.

[10] F-Secure. Mobile threat report q3 2012. *http://www.f-secure.com/static/doc/labs_global/Research/Mobile%20 Threat%20Report%20Q3%202012.pdf*, 2012.

[11] A. P. Felt, H. Wang, A. Moschuk, S. Hanna, and E. Chin. Permission re-delegation: Attacks and defenses. In *Proceedings of the 20th USENIX Security Symposium*, 2011.

[12] M. Grace, Y. Zhou, Z. Wang, and X. Jiang. Systematic detection of capability leaks in stock android smartphones. In *Proceedings of the 19th Network and Distributed System Security Symposium*, 2012.

[13] M. Grace, Y. Zhou, Q. Zhang, S. Zou, and X. Jiang. Riskranker: Scalable and accurate zero-day android malware detection. In *Proceedings of the 10th International Conference on Mobile Systems, Applications and Services*, 2012.

[14] X. Jiang. Security alert: New nickibot spyware found in alternative android markets. *http://www.csc.ncsu.edu/faculty/jiang/NickiBot/*, 2011.

[15] Lookout. Update: Security alert: Droiddream malware found in official android market. *https://blog.mylookout.com/2011/03/security-alert-malware-found-in-official-android-market-droiddream/*, 2011.

[16] L. Lu, Z. Li, Z. Wu, W. Lee, and G. Jiang. Chex: statically vetting android apps for component hijacking vulnerabilities. In *Proceedings of the 2012 ACM conference on Computer and communications security*, 2012.

[17] D. Maslennikov. First sms trojan for android. *http://www.securelist.com/en/blog/2254/First_SMS_-Trojan_for_Android*, 2010.

[18] R. Schlegel, K. Zhang, X. Zhou, M. Intwala, A. Kapadia, and X. Wang. Soundcomber: A stealthy and context-aware sound trojan for smartphones. In *Proceedings of the 18th Network and Distributed System Security Symposium*, 2011.

[19] V. Svajcer. Deceiving permissions - rules for android malware detection. In *RSA Conference*, 2012.

[20] Y. Zhou and X. Jiang. Android malware genome project. *http://www.malgenomeproject.org/*, 2012.

[21] Y. Zhou and X. Jiang. Dissecting android malware: Characterization and evolution. In *Proceedings of the 33rd IEEE Symposium on Security and Privacy*, 2012.

[22] Y. Zhou, Z. Wang, W. Zhou, and X. Jiang. Hey, you, get off of my market: Detecting malicious apps in official and alternative android markets. In *Proceedings of the 19th Network and Distributed System Security Symposium*, 2012.

k-Anonymous Reputation

Sebastian Clauß
Technische Universität Dresden
Institute of Systems Architecture
01062 Dresden, Germany
Sebastian.Clauss@tu-dresden.de

Stefan Schiffner
Center for Advanced Security Research Darmstadt
Mornewegstrasse 32
64293 Darmstadt, Germany
Stefan.Schiffner@cased.de

Florian Kerschbaum
SAP Applied Research
Vincenz-Prießnitz-Strasse 1
76131 Karlsruhe, Germany
Florian.Kerschbaum@sap.com

ABSTRACT

While performing pure e-business transactions such as purchasing software or music, customers can act anonymously supported by, e.g., anonymous communication protocols and anonymous payment protocols. However, it is hard to establish trust relations among anonymously acting business partners. Anonymous reputation systems have been proposed to mitigate this problem. Schiffner et al. recently proved that there is a conflict between anonymity and reputation and they established the non-existence of certain privacy-preserving reputation functions. In this paper we argue that this relationship is even more intricate. First, we present a reputation function that deanonymizes the user, yet provides *strong anonymity* (SA) according to their definitions. However, this reputation function has no *utility*, i.e., the submitted ratings have no influence on the resulting reputation values. Second, we show that a reputation function having utility requires the system to choose new independently at random selected pseudonyms (for all users it has utility for) on every new rating as a necessary condition to provide strong anonymity according to the aforementioned definition. Since some persistence of pseudonyms is favorable, we present a more secure, but also more usable definition for anonymous reputation systems that allows persistency yet guaranties k-anonymity. We further present a definition for *rating secrecy* based on a threshold. Finally, we propose a practical reputation function, for which we prove that it satisfies these definitions.

Categories and Subject Descriptors

K.4.1 [**Computers and Society**]: Public Policy Issues—*Privacy*; C.2.4 [**Computer-Communication Networks**]: Distributed Systems—*Distributed applications*

General Terms

Algorithms, Design, Security

Keywords

Reputation, E-Commerce, Privacy Enhancing Technology, k-Anonymity

1. INTRODUCTION

Many Internet applications integrate reputation systems [22]. Thus, before interacting with others, users investigate the reputation profiles of their potential interaction partners to evaluate their trustworthiness. This is even more important if business partners protect their privacy by the means of anonymous communication and payment. Such a scenario is realistic at least in cases where electronic goods like software, music or movies are traded. However, the use of reputation systems themselves raises privacy concerns. For example, as noted by Bygrave [3], ratings and reputation values may be seen as personal data. After various reputation systems providing different types of privacy guarantees (e.g., rater anonymity or ratee anonymity) have been proposed [1, 32, 25, 31, 11, 29], Schiffner et al. [27] recently formalized this notion of privacy within a reputation system, based on a formalization of anonymity and unlinkability aspects in communication systems by Bohli et al. [13, 2]. Further, Schiffner et al. formally showed a conflict between privacy and reputation and established the non-existence of certain privacy-preserving reputation functions, i.e., reputation functions with certain combinations of properties cannot fulfill strong anonymity definitions.

In this paper we study this relationship more in depth. After restating the main aspects of the privacy definition of [27] in Section 3.1, in Section 4 we present a reputation function that provides *strong anonymity* (SA) and has no utility according to the definitions of [27], but completely violates the privacy of the user, i.e., enables to learn the users long term identifier. Second, we show that a reputation function *having utility* requires the system to choose new independently at random selected pseudonyms (for all users it has utility for) on every new rating as a necessary condition for providing SA according to the [27] definition. This also includes *rating secrecy*, that is, individual ratings are kept secret in order to avoid retaliation. In Section 5, we give a new definition for user privacy based on k-anonymity [24], which overcomes these identification issues and at the same time is less restrictive with regards to utility of the reputation function. For the same reason, we give a definition for weaker rating secrecy based on a threshold, which, when set appropriately, enables a reasonable degree of rating secrecy, while not restricting utility of the reputation function significantly. We propose a practical reputation function, for which we prove that it satisfies these definitions, and conclude in Section 6.

2. RELATED WORK

An overview on reputation system architectures is, e.g., provided by Voss in [31] while possible reputation functions are, e.g., outlined in [20, 14]. Next to electronic marketplaces reputation systems are often used in peer-to-peer networks [4, 15] where also the resistance to forged ratings is important [17]. For an economic introduction, we refer to Dellarocas' work [6]. In [31, 11, 29] the design of reputation systems is investigated from the perspective of privacy and security requirements, which is our focus. More concrete proposals for privacy-preserving reputation systems are provided in [1, 32, 25, 31, 11, 29, 21, 7]. While all these proposals lack a common formal privacy model, such a model was proposed by Schiffner et al. [27] based on a formalization of anonymity and unlinkability aspects in communication systems by Bohli et al. [13, 2]. In our work we base on this formalization.

In order to obtain anonymity of raters and ratees, it needs to be ensured that many users are indistinguishable by an adversary, so that they are in large anonymity sets. The possibility of recognizing users by reputation is limited if the set of possible reputations is limited [29] or if the reputation is only published as an estimated reputation, as proposed by Dellarocas in [5]. Androulaki et al. and Steinbrecher use transaction pseudonyms to avoid linkability between transactions [1, 28]. Further, convertible credentials [29], electronic cash [1, 25] and a modified DC-network [26] have been proposed for the protection of rater privacy. Finally, Kerschbaum proposes a secure computation of the reputation system [16]. In our work, we also utilize transaction pseudonyms, and we do not place restrictions on the way communication is shaped. Hence, all these proposals for the way how ratings are communicated as well as how reputation is queried remain possible.

Another issue having negative effects on the practical utility of reputation systems is the possibility of so called revenge ratings. This is an issue especially in systems where mutual ratings between transaction partners are possible. In [23] Resnick and Zeckhauser systematically evaluate eBay feedback data. They observe that only 0.3% of all transactions were rated negative, but in case one partner rated negative the probability of the other partner rating negative was over 37%. Countermeasures based on payment methods are proposed in [19], which require an additional accounting mechanism. Other countermeasures based on statistical correction are proposed in [30]. The actual effect in case users know about this correction is rather unclear. In this work we keep individual ratings confidential towards the rater to prevent retaliation. We utilize Goodrich's and Kerschbaum's [12] method for aggregation of ratings, which with a high probability hides the actual ratings while preserving high accuracy of the calculated reputation values.

The work of this paper is also related to approaches used in the context of differential privacy (e.g., see [10]) for hiding the contribution of a single value to the whole through the introduction of random noise or assumed randomness. Unfortunately there are two primary reasons that such approaches are not a good fit for protecting the privacy of reputation scores in online auctions. Firstly, up to our knowledge, existing schemes for protecting the privacy of individual values that contribute to a sum require that at least the number of items included in the sum is known, c.f. Dinur and Nissim [8]. However, since the total number of transactions at a certain point in time is unknown and unbounded, it appears difficult making such previous differential-privacy methods work. Second, a reputation score evolves over time. So we cannot assume that such scores are drawn uniformly at random from some distribution, say, as is needed in the differential-privacy scheme of Duan [9]. Likewise, a participant that queries his or her own reputation after every transaction can be modeled as issuing correlated summation queries. Obfuscation techniques for such queries have been presented, c.f. Li et al. [18]; however, it appears that theses techniques introduce too much noise to be of practical use for reputation aggregation.

3. REPUTATION SYSTEM PROPERTIES

3.1 Privacy Definition

In this section, we introduce our privacy definitions. Note that privacy is a complex term, which is not easily translated into properties of technical systems. In this work we present techniques to retain information confidential that could be used to attack users privacy, namely we aim to hide who has been rated and how someone has been rated, i.e., ratee anonymity and rating secrecy.

We first briefly restate the system model of [27]. A reputation system, denoted by $\mathcal{S}$ in the sequel, enables its users to rate each other, and provides access to a reputation query interface over which users may query the current reputation value of the users. A reputation system consists of an enumerable but not necessarily finite user identifier space U such that $|U| \geq 2$, a pseudonym space P such that $|P| \geq |U|$, a space of ratings M, a space of reputation values V, and a *probabilistic algorithm* ϕ that maps all sequences of the form $((u, u', m)_1, (u, u', m)_2 \ldots, (u, u', m)_n) \in (U^2 \times M)^n$ to a function that maps users to reputation values. We call ϕ the 'reputation function' and use $\phi(\mathcal{H})(u)$ to denote the reputation value in V that the algorithm produces for a given user u on input the sequence $\mathcal{H}$. A reputation system also provides the following four algorithms.

- The *initialization protocol* Init takes as input a security parameter $k \in \mathbb{N}$ and produces any required cryptographic material. It also sets the initial history $\mathcal{H}$ to be the empty sequence.
- The *rating protocol* Rate takes as input a triple $(u, u', m) \in U^2 \times M$ and appends it to $\mathcal{H}$. The rating protocol may also produce some output. Invoking the rating protocol means that user u rates user u' with the rating m.
- The *pseudonym generation protocol* $\mathsf{NewPseudo}$ takes as input a user $u \in U$ and outputs a pseudonym $p \in P$.
- The *reputation query protocol* $\mathsf{GetReps}$ outputs, for each $u \in U$, the pair $(\mathsf{NewPseudo}(u), v) \in P \times V$, where $\mathsf{NewPseudo}(u)$ is used to generate a fresh pseudonym for user u and where $v = \phi(\mathcal{H})(u)$ represents u's current reputation. The order in which the pairs are output depends on $\mathcal{S}$.

Experiment $\mathbf{Exp}^{\mathsf{SA}-b}_{\mathcal{S},\mathcal{A}}(k)$ $g \leftarrow \mathcal{A}^{\mathtt{rate}((\cdot,\cdot,\cdot),(\cdot,\cdot,\cdot)),\mathtt{getreps}()}$ return $g == b$

Figure 1: Experiment $\mathbf{Exp}^{\mathsf{SA}-b}_{\mathcal{S},\mathcal{A}}(k)$ where $\mathcal{S}$ is the system under attack, $\mathcal{A}$ the adversary, $b \in \{0,1\}$ the challengers coin toss, g the adversary's guess about the coin toss, k is $\mathcal{S}$'s security parameter.

Based on this system model [27] defines privacy properties in the following way. We only recapitulate the strongest anonymity variant SA (strong anonymity), which is sufficient for our work.

Let $n = |\mathcal{H}|$ denote the number of times Rate is invoked in a given time period. The correspondence between Rate invocations and the set of raters and ratees is modeled as two functions $\sigma, \rho \in \mathfrak{F}$, where $\mathfrak{F} = \{f : \{1, \dots, n\} \mapsto U\}$ is the space of functions that map (the serial number of) each Rate invocation to the user identifier space. That is, if $(u, u', \cdot)$ is the parameter triple of the ith Rate invocation, then, for all $i \in \{1, \dots, n\}$, $\sigma(i)$ returns u and $\rho(i)$ returns u'.

The adversary's goal is to identify σ, ρ, or some 'interesting property' of these functions, possibly with respect to only some subset of Rate invocations, through interaction with, or observation of, the system $\mathcal{S}$. As SA does not require any restrictions to σ and ρ, we only restate the definition of function distinguishability with respect to SA:

Definition 1. With respect to a subset of invocations $I \subseteq \{1, 2, \dots, n\}$, two functions $f_0, f_1 \in \mathfrak{F}$, $f_0 \neq f_1$, are said to be SA-*distinguishable* in any case.

The adversary, denoted by $\mathcal{A}$, adaptively controls the usage of $\mathcal{S}$. Its interaction with $\mathcal{S}$ is modeled via queries in an experiment that a challenger arranges for $\mathcal{A}$ (c.f. Fig. 1). At the beginning of this game, the challenger sets up $\mathcal{S}$ by executing the Init algorithm. The challenger then selects a bit $b \in \{0,1\}$, uniformly at random, and then offers the following interfaces to $\mathcal{A}$, through which the system can be controlled.

- $\mathtt{rate}((\cdot,\cdot,\cdot),(\cdot,\cdot,\cdot))$:
  on input $(u_0, u'_0, m_0), (u_1, u'_1, m_1) \in (U^2 \times M)^2$, the challenger executes $\mathsf{Rate}(u_b, u'_b, m_b)$ and forwards $\mathcal{S}$'s output, if any, to $\mathcal{A}$.
- $\mathtt{getreps}()$:
  the challenger invokes GetReps and forwards $\mathcal{S}$'s output to $\mathcal{A}$.

$\mathcal{A}$ may issue a number of queries over these interfaces and, at some point in time, outputs a guess bit $g \in \{0,1\}$. We say that $\mathcal{A}$ wins the experiment if and only if $g = b$, and its advantage is given by:

$$\mathbf{Adv}_{\mathcal{S},\mathcal{A}}(k) = \left|\Pr(\mathbf{Exp}^{\mathsf{SA}-0}_{\mathcal{S},\mathcal{A}}(k) = 0) - \Pr(\mathbf{Exp}^{\mathsf{SA}-1}_{\mathcal{S},\mathcal{A}}(k) = 0)\right|$$

In order to describe the course of an experiment we introduce the following notation. Let n and λ denote the number of $\mathtt{rate}$ and $\mathtt{getreps}$ queries respectively, that $\mathcal{A}$ has issued up to the point in time it outputs g in an $\mathbf{Exp}^{\mathsf{SA}-b}_{\mathcal{S},\mathcal{A}}(k)$ experiment. We define the functions $\sigma_0, \sigma_1, \rho_0, \rho_1$ such that, for all $i \in \{1, 2, \dots, n\}$, $\sigma_0(i) = u_{0,i}$, $\sigma_1(i) = u_{1,i}$, $\rho_0(i) = u'_{0,i}$, $\rho_1(i) = u'_{1,i}$, where $((u_{0,i}, u'_{0,i}, m_{0,i}), (u_{1,i}, u'_{1,i}, m_{1,i}))$ is the parameter tuple of $\mathcal{A}$'s ith $\mathtt{rate}$ query. We further define the subsets of $\mathtt{rate}$ query invocation serial numbers

$$I_1 = \{1, \dots, c_1\},$$
$$I_2 = \{1, \dots, c_2\},$$
$$\vdots$$
$$I_\lambda = \{1, \dots, c_\lambda\}.$$

where, for all $j \in \{1, \dots, \lambda\}$, c_j denotes the number of $\mathtt{rate}$ queries $\mathcal{A}$ has issued up to the point in time it issues the jth $\mathtt{getrep}$ query. Finally, the main definition of [27] for SA is as follows:

Definition 2. A reputation system $\mathcal{S}$ is said to computationally provide 'rater-SA', denoted S/SA (resp. 'ratee-SA', denoted R/SA) if and only if

- $\mathcal{A}$ is restricted to $\mathtt{rate}$ sequences $(u_0, u'_0, \cdot)$, $(u_1, u'_1, \cdot), \dots$ such that σ_0 and σ_1 (resp. ρ_0 and ρ_1) are SA-distinguishable with respect to all $I \in 2^{\{I_1, \dots, I_\lambda\}}$,
- $\mathbf{Adv}_{\mathcal{S},\mathcal{A}}(k) \leq \epsilon(k)$ for a negligible function $\epsilon(k)$, and
- $\mathcal{A}$'s running time is polynomial in k.

$\mathcal{S}$ is said to statistically provide notion $\cdot$/SA if and only if the first condition applies. Finally, $\mathcal{S}$ is said to unconditionally provide $\cdot$/SA if and only if $\mathbf{Adv}_{\mathcal{S},\mathcal{A}}(k) = 0$.

In this paper, we deal with privacy issues of the rater and ratee. We consider the raters privacy fairly well protected by the means of anonymous communication as well as rating secrecy presented in this paper in Section 5.2.

Ratee privacy is formally modeled as strong ratee anonymity (R/SA). Further, ratee privacy deserves more protection, since while a rater has full control over his rating, a ratee has only limited control over her reputation. Hence, we consider ratee privacy as more challenging. In terms of the model presented in the beginning of this section we can say, that the output of GetReps usually reveals some information about ratees by means of the reputation values. For the rest of the paper, we therefore only consider issues with respect to providing R/SA. Also note that this definition provides ratee anonymity under arbitrary rating sequences, i.e., a successful attack on anonymity using any sequence of ratings is sufficient to render the system insecure. This also means that if the system provides anonymity, this anonymity is preserved regardless of the future ratings that may occur.

3.2 Utility of a Reputation Function

The core assumption of a reputation system is, that information about the past behavior of a reputation item allows to predict its future behavior. A reputation function aggregates ratings, that is feedback of past users, of reputation items to a compact reputation value, which can be used to make predictions.

Aggregation might cause information loss. The utility of a reputation function reflects how much a user can learn from an item's reputation value about the item's past. Since it is not clear, what exactly a user needs to assess a reputation item's past, we model utility as mutual information between input and output of the reputation function. Obviously, there are many reputation functions that have a non-zero utility according to our definition that are useless for users, however every function having value for users has a non-zero utility.

More formally speaking, a reputation function ϕ has non-negligible utility "if, for some $l \geq 1$, there exist two distinct histories $\mathcal{H}, \mathcal{H}'$ of length l and a function γ that maps users to reputation values, such that $\Pr(\phi(\mathcal{H}) = \gamma) > \Pr(\phi(\mathcal{H}') = \gamma) + \epsilon(k)$, where ϵ denotes a negligible function." [27]

Two distinct histories $\mathcal{H}, \mathcal{H}'$ of length $l \geq 1$ are said to be distinguishable, if there exists a function γ that maps users to reputation values, such that $\Pr(\phi(\mathcal{H}) = \gamma) > \Pr(\phi(\mathcal{H}') = \gamma) + \epsilon(k)$, where ϵ denotes a negligible function.

This means, that a reputation function with non-negligible utility always states a non-negligible influence on the mapping of reputation values to users.

3.3 Two Examples

In order to shed some light on the rationale behind the aforementioned system specification and privacy definition, we describe two reputation systems in terms of the model above and argue about their privacy issues.

Example 1. The set of ratings is $M = \{m\}$, and the set of reputation values is $V = \{1, \ldots, |U|\}$. The reputation function ϕ proceeds as follows: First, it calculates, for each $u \in U$, the sum of ratings received by u. Then it ranks all users according to this sum; the user with the highest sum is mapped to reputation value 1, the user with second highest sum to value 2, and so on. Ties are broken randomly, such that, in the end, every user is assigned a unique reputation value from V. The protocols of this example system proceed as follows:

- The *initialization protocol* Init takes as input a security parameter $k \in \mathbb{N}$ and produces any required cryptographic material. It also sets the initial history $\mathcal{H}$ to be the empty sequence.
- The *rating protocol* Rate takes as input a triple $(u, u', m) \in U^2 \times M$ and appends it to $\mathcal{H}$. It does not produce any output.
- The pseudonym generation protocol NewPseudo(u) returns for every user a new pseudonym chosen uniformly at random from the set of all possible pseudonyms (i.e., on every new protocol execution the user pseudonym is changed uniformly at random.)
- The reputation query protocol GetReps outputs the (pseudonym, reputation value)-pairs (NewPseudo(u), v) in an order that is chosen independently and uniformly at random for every new protocol execution. Note that NewPseudo(u) chooses for every user independently and uniformly at random a pseudonym from the entire pseudonym space.

It is easy to see that this reputation system has utility, as in any case users with more ratings are ranked higher than users with less ratings. Further, it can be proven that this reputation function provides strong ratee anonymity R/SA (for details we refer to [27].) In short, this results from the fact that the output of GetReps is independent from Rate-invocations. User pseudonyms and the output order are chosen independently at random for every GetReps call, while the list of reputations remains constant.

However for many applications more persistent pseudonyms are desirable or even necessary. The second example relaxes the restrictions on how users chose their pseudonyms.

Example 2. The system specification is identical to the one of example 1, with one difference: NewPseudo(.) returns for every user a new pseudonym chosen uniformly at random from the set of all possible pseudonyms, with the exception of one user, u_i. For u_i, the pseudonym stays always the same, i.e., the output of NewPseudo(u_i) is constant. (This models the property that there is a user u_i who does not want to change his pseudonym or even just knows his randomly chosen pseudonym.) ◇

Similar to example 1, it is easy to see that this reputation system has utility. However, different from example 1, this reputation function does *not* provide ratee anonymity R/SA, as shown by the following attack:

Assume that the adversary knows which user u_i does not change its pseudonym. Let $i' \neq i$. The adversary $\mathcal{A}$ first invokes $\texttt{rate}((\cdot, u_i, m), (\cdot, u_i, m))$ and then $\texttt{getreps}()$, by that obtaining a (pseudonym, reputation value)-list L_1. By this, $\mathcal{A}$ knows that user u_i is rated at top, independent of the bit b chosen by the challenger. Now $\mathcal{A}$ invokes twice $\texttt{rate}((\cdot, u_i, m), (\cdot, u_i', m)), u_i' \neq u_i$, and then $\texttt{getreps}()$, obtaining a (pseudonym, reputation value)-list L_2. If the challenge bit is $b = 0$ in list L_2 the top rank is assigned to the same pseudonym as in list L_1. If $b = 1$ this is not the case, as now u_i' has more ratings than u_i. So, the adversary $\mathcal{A}$ is able to correctly guess b.

Further, we can relax the assumption that the adversary knows which user changes its pseudonym, by repeating the attack for every user i, assuming that this user is the one that never changes its pseudonym.

4. UTILITY VERSUS ANONYMITY

In this section we first give an example of a reputation function that provides R/SA according to Definition 2, but nevertheless the adversary gets to know the user identities behind all pseudonyms. Secondly, we prove that in case of using persistent pseudonyms strong anonymity is only possible for reputation functions without utility.

4.1 Identification While Providing R/SA

Example 3. The system specification is identical to the one of example 1, with one difference:

For every user $u_i \in U$ the reputation function ϕ_c statically outputs its user id u_i as her reputation value. ◇

As the output of ϕ_c is static and not influenced by ratings, an adversary under Definition 2 cannot get any advantage in guessing b. This is because the attack is based on `rate` invocations, which are in fact ignored by ϕ_c. On the other hand, by invoking `getreps` the adversary gets the user ids of every user, by that identifying them.

This issue arises, because Definition 2 effectively addresses how the difference of reputation values from one `getreps` invocation to another `getreps` invocation (induced by `rate` invocations) can be used to attack various aspects of anonymity. On the other hand, Definition 2 does not take into account information about the mapping of reputation values to users the adversary receives *directly* from the reputation values output by getreps.

In order to overcome this issue, in Section 5.1 we propose a privacy definition, which also addresses direct information gain from the reputation values.

4.2 Utility and Persistent Pseudonyms

In some applications persistent pseudonyms are desirable. Within the system model specified in Section 3.1 pseudonym persistency can be implemented within an instantiation of NewPseudo. In this case NewPseudo is a stateful algorithm in which the output at a certain time depends statistically on earlier outputs. Furthermore, a user's pseudonyms are persistent if and only if the probability, that a certain user is using a given pseudonym, is higher than pure chance would suggest. We now formulate the following theorem:

THEOREM 1. *Within a reputation system $\mathcal{S}$ under attack according to Definition 2 a reputation function with utility can only provide* R/SA *if no user, for which the reputation function has utility, uses persistent pseudonyms.*

In order to prove this theorem, we first define a reputation function to have utility for a user in case the reputation value of this user can change dependent on ratings submitted for arbitrary users:

Definition 3. Let ϕ be a reputation function with utility which is used in a reputation system $\mathcal{S}$. Let u be a user in that system. Let $\mathcal{H}, \mathcal{H}'$ be two distinct histories of length $l \geq 1$. Let γ, γ' be two functions mapping users to reputation values, where γ maps user u to a reputation value other than γ' maps user u to. ϕ is said to have *utility for user u* if there exist two such histories $\mathcal{H}, \mathcal{H}'$ and two such mappings γ, γ' so that $\Pr(\phi(\mathcal{H}) = \gamma) > \Pr(\phi(\mathcal{H}') = \gamma) + \epsilon(k)$, while $\Pr(\phi(\mathcal{H}') = \gamma') > \Pr(\phi(\mathcal{H}) = \gamma') + \epsilon(k)$, where ϵ denotes a negligible function. If this holds for a pair of histories $\mathcal{H}, \mathcal{H}'$, these histories are said to be *distinguishable for user u*.

We now state the following lemma:

LEMMA 2. *If a reputation function ϕ has utility, it has utility for at least one user u within system $\mathcal{S}$.*

We can now proof Theorem 1 by contradiction, i.e., we need to show that if users use persistence pseudonyms there exists a general attack on R/SA.

Assume a system $\mathcal{S}$ under attack according to Definition 2, which has a reputation function with utility and a provides R/SA in case all users get independent at random chosen pseudonyms for each getreps call. Further assume a user u that uses persistence pseudonyms, for which the reputation function has utility, that is, there exists a pair of histories $(\mathcal{H}, \mathcal{H}')$, such that $\Pr(\phi(\mathcal{H}') = \gamma') > \Pr(\phi(\mathcal{H}) = \gamma') + \epsilon(k)$ holds.

We now need to construct two sequences $(b = 0, b = 1)$ of rate and getreps the adversary can distinguish if one user with utility uses persistent pseudonyms. First, the adversary invokes in both sequences rate in such a way that the user u with persistent pseudonyms has the same reputation when calling getreps. Then the adversary invokes rate in $b = 0$ in such a way that the history of the system is $\mathcal{H}$. For $b = 1$ the adversary invokes rate in such a way that the history of the system is $\mathcal{H}'$, so that $\mathcal{H}$ is distinguishable from $\mathcal{H}'$ for user u. After finally calling getreps in $b = 0$ and in $b = 1$, the adversary can distinguish the two outputs with the help of Definition 3. This is possible as the adversary can detect u by his persistent pseudonym and, because $\mathcal{H}$ and $\mathcal{H}'$ are distinguishable for user u, u is mapped to different reputation values.

This is the case for every user the reputation function has utility for. If all users, for which the reputation function has utility, get new independent at random chosen pseudonyms, the adversary cannot recognize them with a non-negligible advantage, and therefore cannot recognize the difference in u's mapping to its reputation value with a non-negligible advantage, and cannot successfully attack the system. Note that the realistic assumption that a user might always know his own pseudonym is a problem in this scenario. If the adversary is a user, the reputation function providing R/SA cannot have utility for this user, and therefore for all users which always know their pseudonyms.

5. TOWARDS PRACTICAL PRIVACY-FRIENDLY REPUTATION

In the sequel, we re-use the system specification from Section 3.1, but introduce the following change:

We add an additional algorithm setPseudonymPolicy in order to govern the selection and possible re-usage of user pseudonyms. It takes as input a policy description and does not produce output.

In order to be usable in practice, we identify two important privacy requirements of a reputation function and the reputation system that utilizes this function. The system should provide:

1. k-Anonymity: If a user chooses to get a new pseudonym, he should stay anonymous within a relevant group of users (i.e., k-anonymous).
2. Rating Secrecy: If a user is rated, he should not get to know the rating value. This is important to prevent revenge ratings, as already highlighted by Goodrich and Kerschbaum [12]. If a system does not restrict revenge ratings, the actual reputation values the system aggregates are to a large extent devaluated, as ratings are not only based on the subject/action to rate, but to a great extent on how other users have rated or how other users are expected to rate.

We further aim at providing a system that is flexible with regards to users' anonymity. So, users should be allowed to use their pseudonym for a longer period of time, i.e., for a certain amount of transactions, i.e., ratings, if they want to.

Definition 2 does not fit these requirements, as reputation functions with utility are only allowed in case users get a new pseudonym on every new rating. In the following we state definitions for these goals.

5.1 k-Anonymity

In order to overcome the issues discussed in Section 4 we propose an alternative privacy definition, which considers both, the direct information gained by an invocation of GetReps as well as the information gained by an adaptive active attack.

As mentioned before, we aim at specifying a reputation system, where users, when getting a new independent at random chosen pseudonym, remain k-anonymous, while other users might choose to keep their pseudonym. So, we define k-anonymity on a per-user basis.

We first specify an experiment using a reputation system $\mathcal{S}$ provided by a challenger to an adversary $\mathcal{A}$. The adversary is allowed to invoke the GetReps and the Rate algorithm of $\mathcal{S}$.

1. The adversary $\mathcal{A}$ can do any number of Rate and/or GetReps invocations on $\mathcal{S}$.
2. $\mathcal{A}$ does one GetReps invocation, by that getting a list of pairs (reputation value, pseudonym) (V, P).
3. $\mathcal{A}$ does *one* additional Rate invocation.
4. the challenger selects a set $P_A \subset P$ containing the pseudonyms of all users, which shall be k-anonymous after this Rate invocation. The challenger uses setPseudonymPolicy in order to input this policy to $\mathcal{S}$.
5. The adversary does a final GetReps invocation, by that getting another list of pairs (V', P').

We now define k-anonymity with respect to this experiment:

Definition 4. A system $\mathcal{S}$ is said to provide k-anonymity, if and only if after the final step, for any pair $(v, p) \in (V, P)$ where $p \in P_A$, the adversary is not able to decide with a probability greater than $\frac{1}{k}$ whether a pair $(v', p') \in (V', P'), p' \neq p$ belongs to the same user.

This definition essentially describes that if a user changes his pseudonym, he shall be k-anonymous (for the next output of GetReps.)[1]

In order for k-anonymity to have the desired anonymizing effect within a real-world implementation, it must further be ensured, that an adversary cannot easily impersonate many users. Our system definition does not state any restrictions with regards to enrolment of users. So, real-world implementations may require users to enroll non-anonymous (e.g. by proving their civil identity etc.), or—more general—require enrolment to be reasonably costly.

5.2 Rating Secrecy

Rating secrecy is already covered by Definition 2. We define an attack on rating secrecy according to Definition 2 as.

Definition 5. Assume a system $\mathcal{S}$ under attack by an adversary $\mathcal{A}$. $\mathcal{A}$ selects two users $u_i, u_j \in U$, and he selects two different rating values $m_0, m_1 \in M$, $m_0 \neq m_1$. After doing any number of **getreps** and **rate** invocations (which do not already form a successful attack), $\mathcal{A}$ invokes **rate**$((u_i, u_j, m_0), (u_i, u_j, m_1))$, i.e., he inputs two ratings, which only differ in their value. Finally, the adversary invokes **getreps**.

Using this attack, we define strong rating secrecy.

[1]If a user is assumed to always know his own pseudonym, he cannot be k-anonymous regarding his view on the system. However, we can cover this issue by a pseudonym policy for the attacking user, which states that this user keeps his pseudonym, and therefore does not want to be k-anonymous in this scenario.

Definition 6. Assume a system $\mathcal{S}$ under attack as specified in Definition 5. The system $\mathcal{S}$ is said to provide *strong rating secrecy*, if and only if in any such attack the adversary does not get an advantage according to Definition 2. For that, rating secrecy can be provided computationally, statistically or unconditionally depending on which of the specific conditions stated in Definition 2 are fulfilled.

As strong rating secrecy is covered by the Definition 2, the proof of Section 4.2 is also valid for this definition, i.e., a system $\mathcal{S}$ can only keep strong rating secrecy in case all users, for which the reputation function has utility, get new independent at random chosen pseudonyms after any new rating.

In order to allow various pseudonym selection policies, we state a definition for weak rating secrecy by modifying the success criterion for an attack on rating secrecy, so that a system provides weak rating secrecy if the advantage of the adversary is below a given probability threshold.

Definition 7. Assume a system $\mathcal{S}$ under attack as specified in Definition 5. The system $\mathcal{S}$ is said to provide *weak rating secrecy*, if and only if $\mathbf{Adv}_{\mathcal{S},\mathcal{A}}(k) < t$, $0 < t < 1$, where t is a probability threshold.

5.3 A Practical Privacy-friendly Reputation System

We consider a system $\mathcal{S}$ as specified in Section 3.1, with the additional algorithm setPseudonymPolicy as specified in the beginning of Section 5. The system has $n > 0$ users U and $1 < l \ll n$ reputation levels. The system keeps three subsets of users, $U_{id} \subset U$ containing the users who want to keep their pseudonym, $U_{anon} \subset U$ who want to get a new pseudonym and want to be k-anonymous, and a set $U_{dc} \subset U$ of user who don't care about their pseudonyms. Every user $u \in U$ is in exactly one of these subsets. $0 < k \leq n$ is the anonymity set size.

Further, the system keeps a mapping of users to pseudonyms (U, P_S) consisting of the current pseudonyms of all users, a mapping of users to reputation values (U, V_S) consisting of the current reputation values of all users, and the history $\mathcal{H}$ of ratings.

The protocols of the reputation system are specified as follows:

- The *initialization protocol* Init takes as input a security parameter $k \in \mathbb{N}$, which is the anonymity set size. Furthermore, it takes a security parameter o, $0 < o < 1$ for the rating secrecy, which is the fraction of ratings that are considered for reputation calculation—see the rating protocol for details. Lastly, the initial history $\mathcal{H}$ is set to be the empty sequence.
- The *rating protocol* Rate takes as input a triple $(u, u', m) \in U^2 \times M$ and appends it to $\mathcal{H}$. This protocol does not produce output.
- the *pseudonym policy protocol* setPseudonymPolicy takes as input a set of users, who want to keep their pseudonyms, and another distinct set of users, who want to stay k-anonymous. It does not produce output. The protocol updates the sets of users U_{id}, U_{anon} and U_{dc} accordingly.

- The *pseudonym generation protocol* NewPseudo takes as input a mapping $(U, V'_{\mathcal{S}})$ of reputation values of all users and outputs for each $u \in U$ a mapping $(U, P'_{\mathcal{S}})$ consisting of a pseudonym $p \in P$ for all users, or $\perp$ in case the current system state does not allow creating a valid set of pseudonyms. The output is generated as follows:

  1. Based on the current mapping of pseudonyms $(U, P_{\mathcal{S}})$ a new mapping of pseudonyms $(U, P'_{\mathcal{S}})$ is created. Within the new mapping, all users in set U_{anon} get a new independent at random chosen pseudonym, while the other users keep their old pseudonym.
  2. For all i reputation levels, $0 < i \leq l$, the following algorithm is executed:
     (a) For reputation level i three sets $U_{anon,i} \subset U_{anon}$, $U_{id,i} \subset U_{id}$ and $U_{dc,i} \subset U_{dc}$ are built, containing the users of the respective sets having reputation level i within the input mapping of reputation levels $(U, V'_{\mathcal{S}})$.
     (b) If $U_{anon,i} = \emptyset$ continue with the next reputation level.
     (c) If $|U_{anon,i} \cup U_{dc,i}| < k$ output $\perp$.
     (d) Within $(U, P'_{\mathcal{S}})$, set new independent at random chosen pseudonyms for $k - |U_{anon,i}|$ users, which are selected at random from $U_{dc,i}$.
  3. Output the new mapping of pseudonyms for all users $(U, P'_{\mathcal{S}})$.

- The *reputation query protocol* GetReps outputs, for each $u \in U$, the pair $(p, v) \in P \times V$, or it outputs $\perp$ in case the current state does not allow to choose pseudonyms according to the pseudonym policy.[2] The output is generated as follows:

  1. In case no new rating has been added to the history $\mathcal{H}$ since the last call of GetReps, output mappings (p, v) for each user from the mappings $(U, P_{\mathcal{S}})$ and $(U, V_{\mathcal{S}})$, i.e., the same output as in the previous call of GetReps.
  2. In case at least one rating has been added to the history $\mathcal{H}$ since the last call of GetReps, a new mapping of reputation values $(U, V'_{\mathcal{S}})$ is generated by performing the following algorithm for every user $u \in U$.
     (a) From the history $\mathcal{H}$, the set of all ratings for user u, $M_u \subset \mathcal{H}$ is built.
     (b) A set $M_{u,o} \subset M_u$ is built by selecting $o \cdot |M_u|$ ratings at random from M_u.
     (c) The reputation of user u is calculated as $v'_u = \lceil l \cdot o \cdot \sum_{i=1}^{|M_{u,o}|} m_{u,o,i} \rceil$ where $m_{u,o,i} \in M_{u,o}$.
     (d) The new reputation value v'_u is set in the mapping $(U, V'_{\mathcal{S}})$.
  3. NewPseudo($(U, V'_{\mathcal{S}})$) is invoked. If the output of NewPseudo is $\perp$, this is also the output of this invocation of GetReps.
  4. If the output of NewPseudo is a mapping of pseudonyms $(U, P'_{\mathcal{S}})$, the current mappings of pseudonyms and reputation values are updated: $(U, P_{\mathcal{S}}) := (U, P'_{\mathcal{S}})$ and $(U, V_{\mathcal{S}}) := (U, V'_{\mathcal{S}})$.
  5. Output mappings (p, v) for each user from the mappings $(U, P_{\mathcal{S}})$ and $(U, V_{\mathcal{S}})$.

[2]This is the case when k-anonymity cannot be guaranteed for all users $u \in U_{anon}$.

5.4 Discussion

The system's reputation function, i.e., the algorithm for calculating reputation values from the rating history, is specified in step 2 of the reputation query protocol. For a given user u, it derives reputation values from a randomly selected subset of ratings for this user u. The size of the subset is governed by the security parameter o.

In [12] it has been proved, that the expected error probability of an aggregated reputation value does not depend on the overall number of ratings, i.e., aggregating a fixed number of ratings is sufficient. So, e.g., an expected error probability of 10% requires 16 ratings. They conclude that it is sufficient and efficient to sample a (small) fixed number of ratings.

From these finding we can conclude, that our reputation function has utility according to the specification of utility in Section 3.2, i.e., ratings have a non-negligible influence on the reputation values of users.

We now prove k-anonymity[3] and weak rating secrecy for this protocol.

THEOREM 3. *The system $\mathcal{S}$ either provides k-anonymity according to Definition 4, or it outputs $\perp$.*

PROOF. There are two cases. Either GetReps outputs $\perp$. If it does not output $\perp$, after at least one new rating has been added to the history $\mathcal{H}$ (which is done in Step 3 of the k-anonymity experiment) for every user $u \in U_{anon}$ the protocol NewPseudo selects new independent at random chosen pseudonyms for u and at least $k - 1$ other users within the same reputation level. For every reputation level, this ensures that from the output of GetReps the adversary cannot decide which of these at least k users is u. So the probability for deciding on the correct user is less or equal $\frac{1}{k}$ for every user $u \in U_{anon}$.

We further need to consider the issue, that the fact that GetReps switches from outputting reputation values to outputting $\perp$, or vice versa, may be a valuable information for an adversary. With respect to the k-anonymity Definition 4 the adversary cannot benefit from this information for the following reasons. Different outputs of *getreps* are independent from each other with respect to the users' k-anonymity, as always new independently chosen pseudonyms are used. So, even if the adversary knows the reputation level of a user u wishing to stay k-anonymous, this user u stays anonymous within k users for any new output of GetReps. If GetReps switches to outputting $\perp$ after some new rating by the adversary, this only informs the attacker on the fact, that some reputation level (or even a reputation level known to the adversary) now contains less than k

[3]Note, we abstract from anonymity issues, which might be introduced by information an adversary could gain from the communication system utilized. As stated in Section 2 various approaches for using anonymous communication in reputation systems have been proposed. However, a specific design for using anonymous communication within the system proposed here is out of scope of this work.

users wishing to stay anonymous. However, this information does not help the adversary in getting additional information about the pseudonym-user relation within any of the sets of k-anonymous users from the previous GetReps output. □

THEOREM 4. *The system $\mathcal{S}$ provides weak rating secrecy according to Definition 7.*

PROOF. There are two cases. Either GetReps outputs $\perp$. In this case, the adversary does not get any information about reputation values, and thus also no information about a rating. For the other case, where GetReps outputs pairs (p, v) for every user, we need to evaluate the algorithm for reputation calculation within the GetReps algorithm. The reputation value for every user u is calculated from only $o \cdot |M_u|$ ratings for this user, where $0 < o < 1$. The ratings actually used for calculation of the user's reputation value are selected at random. Therefore, the probability for any rating to be not used for calculation of the reputation values is $1 - o$.

Further, if a rating is not used for calculation of a reputation value, the adversary cannot decide on the rating's value. Therefore, for system $\mathcal{S}$, the adversary cannot get an advantage better than the threshold $\frac{1+o}{2}$. □

While providing the desired properties, the system has the drawback that, depending on the current state, GetReps might produce no output. This depends on (a) the relation between the total number of users $|U|$ and the numbers of users, who want to be k-anonymous respective who want to keep their pseudonym, and the value of k. Further, it depends on (b) the numbers of users within the different reputation levels, and by that on the ratings.

While the parameters of (a) appear to be manageable for a given system depending on its usage scenario, the parameters of (b) solely depend on the users' rating behavior. On the other hand, the rating behavior should not be influenced by technical system parameters, as a rating should solely base on the subject or action to be rated.

5.5 Enhanced Reputation Query Protocol

In order to get rid of the parameters (b), we propose another reputation query protocol, which varies the number of reputation levels in order to create groups of users in each reputation level, which are big enough for the requested anonymity parameter k. Therefore l is interpreted as the upper bound of the number of reputation levels[4].

In order to do this, within an updated GetReps protocol Step 2.(d) is replaced by the following two steps, which check for each reputation level, whether the requested k-anonymity is possible. If not, neighboring reputation levels are merged until k-anonymity is satisfied. Only in case k-anonymity is not possible within the whole set of users, the protocol outputs $\perp$:

2. (d) For all i reputation levels, $0 < i \leq l$, the following algorithm is executed:
   i. For reputation level i three sets $U_{anon,i} \subset U_{anon}$, $U_{id,i} \subset U_{id}$ and $U_{dc,i} \subset U_{dc}$ are built, containing the users of the respective sets having reputation level i within the input mapping of reputation levels $(U, V'_{\mathcal{S}})$.
   ii. If $U_{anon,i} = \emptyset$ or $|U_{anon,i} \cup U_{dc,i}| \geq k$ continue with the next reputation level.
   iii. If $i = l$ output $\perp$.
   iv. For all users having $v' = i$ set $v' = i + 1$ (i.e., merge reputation levels i and $i + 1$.)

   (e) The new reputation value v'_u is set in the mapping $(U, V'_{\mathcal{S}})$.

[4]For simplicity, we use 1 as a lower bound for the number of reputation levels. We note, that the protocol could easily be modified for higher lower bounds, however this increases the risk for empty GetReps output.

It is easy to see, that this modification as well preserves k-anonymity (or it outputs $\perp$). Further, it preserves weak rating secrecy, as the rating selection mechanism (Step 2.(b)) remains unchanged.

Different to the protocol in Section 5.3 this protocol produces output independent from the rating behavior of the users. The reputation system also preserves utility in case more than one reputation level remains.

6. CONCLUSION

This paper contributes to the discussion on limits of reputation systems with respect to privacy aspects. Based on a previous formalization of privacy and utility aspects of reputation systems [27], we show that a reputation system with utility can only provide strong recipient anonymity in case all users of the system (for which it has utility) are unlinkable by their pseudonyms for every new request for reputation values. Further, we present a reputation function that provides strong recipient anonymity according to the definition of [27], but nevertheless violates the privacy of the user by revealing the user's long term identifier.

Based on these findings, we state two new privacy definitions for reputation systems, namely a k-anonymity definition and a definition for weak rating secrecy. These definitions allow the specification of practical reputation systems while providing certain privacy guarantees, which are especially important even for the utility a reputation system provides in the context it is applied in. We further give two examples of reputation systems satisfying these definitions.

While the proposed systems in Sections 5.3 and 5.5 are rather proof of concepts, in future work we will examine such reputation systems more thoroughly aiming at better utility under similar conditions.

7. REFERENCES

[1] E. Androulaki, S. G. Choi, S. M. Bellovin, and T. Malkin. Reputation systems for anonymous networks. In *PETS '08: Proceedings of the 8th international symposium on Privacy Enhancing Technologies*, pages 202–218, Berlin, Heidelberg, 2008. Springer-Verlag.

[2] J.-M. Bohli and A. Pashalidis. *Relations Among Privacy Notions*, pages 362–380. Springer-Verlag, Berlin, Heidelberg, 2009.

[3] L. Bygrave. *Data Protection Law, Approaching Its Rationale, Logic and Limits.* Kluwer Law International, The Hague, London, New York, 2002.

[4] E. Damiani, S. De Capitani di Vimercati, S. Paraboschi, and P. Samarati. Managing and sharing servents' reputation in p2p systsms. *IEEE Transactions on Knowledge and Data Engineering*, 15(4):840–854, 2003.

[5] C. Dellarocas. Immunizing online reputation reporting systems against unfair ratings and discriminatory behavior. In *EC '00: Proceedings of the 2nd ACM conference on Electronic commerce*, pages 150–157, New York, NY, USA, 2000. ACM Press.

[6] C. Dellarocas. The digitization of word-of-mouth: Promise and challenges of online feedback mechanisms. *Management Science*, pages 1407–1424, October 2003.

[7] C. Dellarocas. Research note – how often should reputation mechanisms update a trader's reputation profile? *Information Systems Research*, 17(3):271–285, 2006.

[8] I. Dinur and K. Nissim. Revealing information while preserving privacy. In *PODS '03: Proceedings of the twenty-second ACM SIGMOD-SIGACT-SIGART Symposium on Principles of Database Systems*, pages 202–210, New York, USA, 2003. ACM.

[9] Y. Duan. Privacy without noise. In *Proc. of 18th ACM Conf. on Information and Knowledge Management.*

[10] C. Dwork. Differential privacy: a survey of results. In *Proc. of the 5th Int. Conf. on Theory and Applications of Models of Computation (TAMC)*, pages 1–19. Springer, 2008.

[11] ENISA. Position paper. reputation-based systems: a security analysis. available from `http://www.enisa.europa.eu/act/it/privacy-and-trust/reputation-systems/reputation-based-systems-a-security-analysis` (last visit 24/08/12), 2007.

[12] M. T. Goodrich and F. Kerschbaum. Privacy-enhanced reputation-feedback methods to reduce feedback extortion in online auctions. In *1st ACM Conference on Data and Application Security and Privacy (CODASPY)*, 2011.

[13] A. Hevia and D. Micciancio. An indistinguishability-based characterization of anonymous channels. In N. Borisov and I. Goldberg, editors, *Privacy Enhancing Technologies*, volume 5134 of *Lecture Notes in Computer Science*, pages 24–43. Springer Berlin / Heidelberg, 2008. 10.1007/978-3-540-70630-4 3.

[14] A. Jøsang, R. Ismail, and C. Boyd. A survey of trust and reputation systems for online service provision. *Decision Support Systems*, 43(2):618 – 644, 2007. Emerging Issues in Collaborative Commerce.

[15] S. Kamvar, M. Schlosser, and H. Garcia-Molina. The eigentrust algorithm for reputation management in p2p networks. In *Proceedings of the 12th International Conference on World Wide Web*, pages 640–651, 2003.

[16] F. Kerschbaum. A verifiable, centralized, coercion-free reputation system. In *Proceedings of the 8th ACM workshop on Privacy in the electronic society*, WPES '09, pages 61–70, New York, NY, USA, 2009. ACM.

[17] F. Kerschbaum, J. Haller, Y. Karabulut, and P. Robinson. Pathtrust: A trust-based reputation service for virtual organization formation. In *Proceedings of the 4th International Conference on Trust Management*, pages 193–205, 2006.

[18] C. Li, M. Hay, V. Rastogi, G. Miklau, and A. McGregor. Optimizing linear counting queries under differential privacy. In *Proc. of 29th ACM Symp. on Principles of Database Systems of Data (PODS)*, pages 123–134, 2010.

[19] N. Miller, P. Resnick, and R. Zeckhauser. Eliciting informative feedback: The peer prediction method. *Management Science*, 51(9):1359–1373, 2005.

[20] L. Mui. *Computational Models of Trust and Reputation: Agents, Evolutionary Games, and Social Networks.* PhD Thesis, Massachusetts Institute of Technology, 2003.

[21] E. Pavlov, J. S. Rosenschein, and Z. Topol. Supporting privacy in decentralized additive reputation systems. In *The Second International Conference on Trust Management*, pages 108–119, Oxford, United Kingdom, March 2004.

[22] P. Resnick, K. Kuwabara, R. Zeckhauser, and E. Friedman. Reputation systems. *Communications of the ACM*, 43(12):45–48, 2000.

[23] P. Resnick and R. Zeckhauser. Trust among strangers in internet transactions: Empirical analysis of ebay's reputation system. *Advances in Applied Microeconomics*, 11:127–157, 2002.

[24] P. Samarati. Protecting respondents' identities in microdata release. *IEEE Transactions on Knowledge and Data Engineering*, 13(6):1010–1027, 2001.

[25] S. Schiffner, S. Clauß, and S. Steinbrecher. Privacy and liveliness for reputation systems. In *Proceedings of 2009 European PKI Workshop (EuroPKI'09).* Springer, 2010. (to appear).

[26] S. Schiffner, S. Clauß, and S. Steinbrecher. Privacy, liveliness and fairness for reputation. In I. Cerná, T. Gyimóthy, J. Hromkovic, K. G. Jeffery, R. Královic, M. Vukolic, and S. Wolf, editors, *SOFSEM 2011: Theory and Practice of Computer Science - 37th Conference on Current Trends in Theory and Practice of Computer Science*, volume 6543 of *Lecture Notes in Computer Science*, page 16, Nový Smokovec,SK, 2011. Springer.

[27] S. Schiffner, A. Pashalidis, and E. Tischhauser. On the limits of privacy in reputation systems. In *Proceedings of WPES'11.*

[28] S. Steinbrecher. Design options for privacy-respecting reputation systems within centralised internet communities. In *Proceedings of IFIP Sec 2006, 21st IFIP International Information Security Conference: Security and Privacy in Dynamic Environments*, volume 201 of *IFIP*, pages 123–134. Springer, May 2006.

[29] S. Steinbrecher. Enhancing multilateral security in and by reputation systems. In *Proceedings of the IFIP/FIDIS Internet Security and Privacy Summer School, Masaryk University Brno, 1-7 September 2008, to be published by Springer*, volume 298 of *IFIP AICT*, pages 135–150. Springer, 2009.

[30] J. Traupman and R. Wilensky. Robust reputations for peer-to-peer marketplaces. In *Proceedings of the 4th International Conference on Trust Management*, pages 382–396, 2006.

[31] M. Voss. Privacy preserving online reputation systems. In *International Information Security Workshops*, pages 245–260. Kluwer, 2004.

[32] M. Voss, A. Heinemann, and M. Mühlhäuser. A Privacy Preserving Reputation System for Mobile Information Dissemination Networks. In *First International Conference on Security and Privacy for Emerging Areas in Communications Networks (SECURECOMM'05)*, pages 171–181. IEEE, 2005.

Privacy-preserving Smart Metering with Regional Statistics and Personal Enquiry Services*

Cheng-Kang Chu
Institute for Infocomm Research
1 Fusionopolis Way
Singapore, 138632
ckchu@i2r.a-star.edu.sg

Joseph K. Liu†
Institute for Infocomm Research
1 Fusionopolis Way
Singapore, 138632
ksliu@i2r.a-star.edu.sg

Jun Wen Wong
Institute for Infocomm Research
1 Fusionopolis Way
Singapore, 138632
jwwong@i2r.a-star.edu.sg

Yunlei Zhao
Fudan University
Shanghai 201203, China
Chinese Academy of Sciences
Beijing 100093, China
yunleizhao@gmail.com

Jianying Zhou
Institute for Infocomm Research
1 Fusionopolis Way
Singapore, 138632
jyzhou@i2r.a-star.edu.sg

ABSTRACT

In smart grid, households may send the readings of their energy usage to the utility and a third-party service provider which provides analyzed statistics data to users. User privacy becomes an important issue in this application. In this paper, we propose a new cryptographic-based solution for the privacy issue in smart grid systems. The advantages of our system are twofold: Households can send authenticated energy consumption readings to a third-party service provider anonymously. The service provider learns only the region where the readings come from but not their respective identities. On the other hand, users with personal secret information can enquiry their usage history records or regional statistics.

Formal security analysis is provided to show that our scheme is secure. We further analyze the performance of our system by giving simulation results.

Categories and Subject Descriptors

K.4.1 [**Public Policy Issues**]: Privacy; K.6.5 [**Security and Protection**]: Authentication

Keywords

Smart metering; privacy; authentication

*Funded by EMA project SecSG-EPD090005RFP(D).

†Corresponding author.

1. INTRODUCTION

Smart grid is a network of computers and power infrastructures that monitors and manages energy usage. Each energy supplier, for example, an electrical company, maintains operational centers that receive usage information from collector devices placed throughout the served area. The collector device reports usage readings to the operational centers using a long-haul communication media such as the Internet. The energy suppliers manage transmission and perform billing based on these readings.

The smart grid relies on smart meters, which can report energy consumption back to the energy suppliers. These new smart meters will replace the electromechanical meters that are attached to most households across the world today. Smart meters can track usage as a function of the time of day, disconnect a customer via software, or send out alarms in case of problems.

Smart appliances, which are being developed and sold by some of the world's largest manufacturers, feed smart meters with real-time information about energy consumption down to the appliance level, and thus enhance the intelligent grid. This precision allows energy companies to analyze peak power usage times and set energy rates accordingly. In turn, households will gain a tool for more efficient management of their energy consumption, which they could use to lower costs and conserve energy.

When the grid, the meter, and the appliances are implemented and integrated, consumers will be able to fine-tune their energy consumption to get the best rates while the energy suppliers will be able to manage energy distribution more effectively and identify or resolve problems remotely. For example, the smart meter could start a washing machine in a specific time interval when the electricity price is low. On the other hand, the washing machine could inform the smart meter of its power consumption for the wash cycle so that the smart meter can forward this information to the electrical company to optimize the resource planning.

Moreover, in US the White House has issued a call-to-

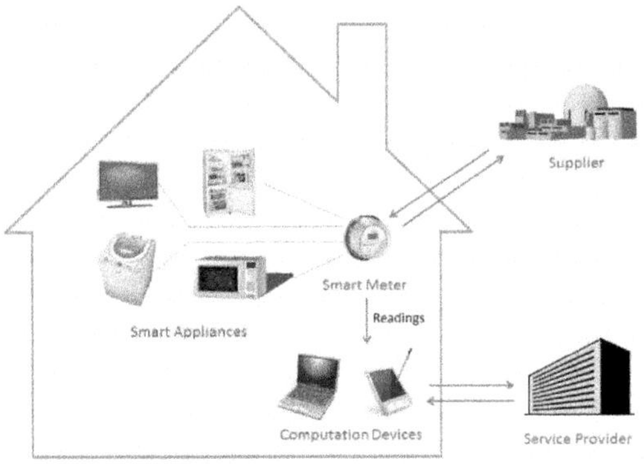

Figure 1: Our Smart Metering Architecture

action for industry to provide electricity customers with easy access to their energy usage data in a consumer-friendly and computer-friendly format via Internet. Green Button [17] is then setup by some utility companies in US in response to this call. It provides a web interface and smartphone apps for consumers to choose the most economical rate plan for their use patterns. It may also deliver customized energy-efficiency tips or provide easy-to-use tools to size and finance rooftop solar panels. Most importantly, it can conduct virtual energy audits that can cut costs for building owners and speed the initiation of retrofits.

Figure 1 shows the smart metering system architecture. Smart meter is the key device for energy consumption measurement. It certifies the readings and provides the measurement data to the supplier or other in-home computational devices. Note that the smart meter still connects to the supplier and is assumed tamper-resistant. Therefore the supplier is convinced that the received readings are valid. Some smart appliances can also be connected to the meter for further control. On input the certified readings, the computational devices can communicate with the service provider and enjoy data analysis or other services. A similar architecture can also be found in [38]. Note that we assume the service provider collects the readings from the users and provides the statistics services for free. It may be an academic organization, a government agency, or the supplier itself. So user's identity is not required for the service provider. Moreover, as user's computation devices are not tamper-resistant, the readings may still need to be authenticated by the smart meter. So we can assume the readings sent to the computational devices are authenticated by the authentication process in the smart meter using a common key. Then the service provider can be assured that the readings are generated by the smart meters without learning where they are from. The authentication process could be implemented by a strongly unforgeable signature scheme. The meter signs each reading with its timestamp. This avoids collaborative-replay attacks because the service provider can verify that every signature is unique at a time.

PRIVACY CONCERN. One of the major concerns of users in this architecture is privacy. Although the smart metering brings with it many new data collection and information sharing capabilities related to energy usage, these technologies in turn introduce concerns about privacy. Since the fine-grained measurements sent by smart meters will form a library of personal information, the mishandling of which could be highly invasive of consumer privacy. These meter readings may leak personal information of households because it is easy to identify the usage of specific electrical appliances [19]. Then the household's lifestyle can be inferred. For example, with the use of nonintrusive appliance load monitors (NALM) [28] which analyze energy signatures to track appliance usage patterns, we can learn when someone is at home by recognizing that lights are switched on; what the cooking times are by recognizing that an oven or microwave is used; or even how many times the family go to the restroom at night by recognizing that the restroom light is switched on. Thus it provides means to identify appliance usage even when multiple household load signatures are aggregated [27].

In our system, there are two parties that may receive households' readings. The energy supplier needs the readings for billing and settlement or profiling processes. On the other side, the service provider may need these data for providing analyzed information. Therefore, there are different privacy issues against these two parties respectively:

- Energy Supplier: For time-of-use billing, the supplier can send tariffs to the meter and let the meter compute the monthly bills according to the readings. So the supplier only learns the final bill instead of individual consumption readings at some time point. For settlement or profiling process, the supplier can determine or estimate the fraction of grid production by aggregation and sampling techniques. It hides the individual consumption of a set of users at some time point.

- Service Provider: In order to provide a fine-grained analysis service and allow users to query their personal history records, individual readings for every user at every time point should be sent to the service provider. Therefore we cannot hide the individual readings from the service provider. The aggregation or sampling techniques above are not applicable. Instead, we should hide the user's identity directly. At the same time, the readings still need to be authenticated.

In the case that the service provider and the supplier are the same party, privacy concern will be more important.

SECURITY REQUIREMENTS. As there are many protocols designed for the energy supplier side (see Section 2) already, we focus on the service provider side only. We consider a privacy-preserving protocol which provides the service provider with all individual *authenticated* readings. The protocol must satisfy the following requirements:

1. **Privacy.** The service provider cannot learn the identification information of users.

2. **Reading Authentication.** The service provider can be assured that the readings it received are from authenticated users.

3. **Regional Statistics.** The service provider knows that a reading is from a particular group or region, yet it does not know who the sender is. This is to let the service provider generate a regional statistics report or document.

4. **Personal Enquiry.** The service provider allows users to enquiry their usage history records. That means the readings which come from the same user should be linked even though the identification information is unknown to the service provider. We also require that no user can put any data and reading into other users' accounts. That is, no user can affect the correctness and accuracy of other users' readings.

5. **No Trusted Third Party.** The protocol involves users and the service provider only. The privacy does not rely on any trusted third party.

At first glance the above properties seem contradictory. For example, how can the service provider know that the anonymous readings are sent by an authenticated user? Or how can users enquiry their previous records if they send the readings anonymously? In this paper, we propose a solution which accommodates all properties defined above.

1.1 Our Contribution

In this paper, we address the privacy issue against the service provider in smart metering. We propose a solution equipped with all security requirements mentioned above. It is specifically designed to protect user privacy on the service provider side while enabling personal enquiry and regional statistic services simultaneously. Our system is constructed by some cryptographic techniques. We also give an analysis to prove that it fulfills all security requirements. Simulation result of our prototype shows that our system is practical to be used.

We note that since our solution is designed for the service provider side only, other billing or settlement profiling protocols (described in Section 2) may be used on the energy supplier side to provide additional privacy and security protection.

1.2 Organization

The rest of the paper is organized as follows. In Section 2, we explain why existing privacy techniques cannot be used in our application. The main scheme is presented in Section 3 and its security is analyzed in Section 4. Then we implement our scheme and show the simulated results in Section 5. Section 6 concludes this paper.

2. RELATED WORKS

There are many cryptographic primitives existed in the literature to preserve privacy in different situations. We review some of those commonly used techniques here, and explain why they are not suitable to be used in our scenario. We also explain an engineering infrastructure that can be used to provide privacy in smart grid systems. A summary of these related works is shown in Table 1.

SMART METERING PROTOCOLS: There are many privacy-preserving protocols for smart metering. These protocols are mainly designed against the energy supplier. Billing protocols [32, 3, 12, 38, 11, 21] allow energy supplier to charge users according to their consumption. The energy supplier can learn the final consumption or bill of a user for a period of time (e.g. every week or every month), but not the individual readings over any sub-period (e.g. every hour). Therefore "regional statistics" and "personal enquiry" are not provided.

On the other hand, settlement or profiling protocols [15, 26, 42] provide information to the supplier to learn the energy consumption of a set of users at some time without revealing any information about their individual consumption. These aggregated data can be used to estimate the fraction of grid production. Again, the "personal enquiry" cannot be provided in these protocols. Figure 2 shows the two directions to hide readings for billing and settlement or profiling.

In addition to those protocols for supplier, Efthymiou and Kalogridis [14] proposed a method to anonymize the meter readings targeted for service provider. In their method, the readings can be authenticated and tied with a particular distribution sub-station. However, it needs a third party escrow mechanism, which is not widely acceptable in practice.

ANONYMOUS CREDENTIALS: An anonymous credential system, first introduced by Chaum [8], is a system in which a user obtains a credential from an issuer and demonstrates the possession of the credential to a verifier who only has the public information of the issuer. Apart from the fact of the user's ownership of a credential granted by an issuer, the verifier cannot get the identity of the user, even if the verifier colludes with the issuer.

Camenisch and Lysyanskaya proposed the first practical multi-show credential system [5]. Their system allows a user to unlinkably and anonymously demonstrate possession of a credential as many times as necessary without involving the issuer. Nevertheless, unlinkability is not our goal. The credentials come from the same user should be linked in our system. Liu and Wong [31] proposed a *restricted* anonymous multi-show credential system, which only allows a user to demonstrate his possession of a credential once in a given event. On the other hand, the same credential can still be shown anonymously in another event without being linked.

In terms of functionality, the restricted credential system is similar to our solution. However, there is still one major difference. It only allows the verifier to know whether a particular user has demonstrated his credential once or more. It does not show the group that the user belongs to, i.e., it is not able to provide the regional statistics information.

GROUP SIGNATURE: Group signatures, first introduced by Chaum and van Heyst [9], allow a user to sign messages on behalf of the whole group. The user obtains a user secret key at the beginning, from a group manager. The signature generated from this user's secret key does not tell who the actual signer is, yet the verifier knows that the signer belongs to a particular group. However, the group manager can revoke the anonymity at any time. Besides, group signatures are also unlinkable. One cannot detect whether two signatures are generated by the same signer. It is easy to see that a group signature scheme does not fulfill our requirements described in the previous section.

RING SIGNATURE: Ring signatures, first introduced by Rivest et al. [39], allow a user to sign messages on behalf of a set of users without the need to reveal his/her identity. Different from a group signature scheme, the group formation is spontaneous and there exists no group manager to revoke the identity of the signer. That is, under the assumption that each user is already associated with a public key of some standard signature scheme, a user can form a group

	Regional Statistics	Personal Enquiry	No Trusted Third Party
Billing Protocols (e.g. [32, 3, 12, 38, 11, 21])	×	×	✓
Settlement/profiling Protocols (e.g. [15, 26, 42])	✓	×	✓
Service Provider Protocols (e.g. [14])	✓	✓	×
Anonymous Credentials (e.g. [5, 31])	×	×	✓
Group Signatures (e.g. [9])	✓	×	×
Ring Signatures (e.g. [39, 30])	×	×	✓
Battery-Assisted Power Transformation [23, 22]	×	×	✓
Our Solution	✓	✓	✓

Table 1: A comparison between our solution and related works regarding the properties we need. We omit the "Privacy" and "Reading Authentication" requirements because they are satisfied in all solutions here.

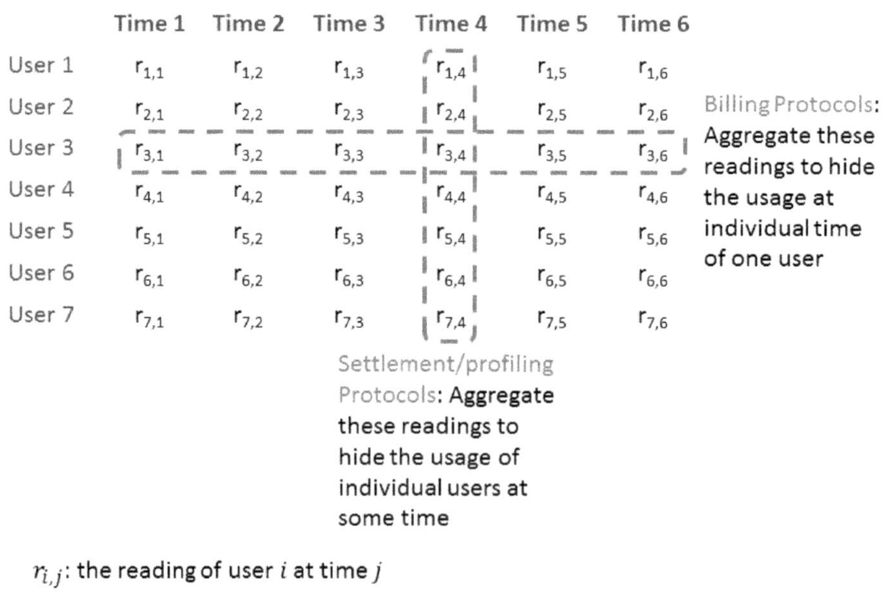

Figure 2: Privacy protection in existing systems

by simply collecting the public keys of all the group members including his own. These diversion group members can be totally unaware of being conscripted into the group.

Once again the ring signature is unlinkable. That is, given two signatures, no one can tell whether they are generated by the same signer or not. A linkable variant was introduced in [30]. In this new notion, users are assumed to sign only once, in which case they enjoy anonymity in full. However, if a user signs twice (or k times, in general), anyone can tell if two signatures are produced by the same user or not. The anonymity is still preserved, though the verifier knows that the two signatures are generated by the same signer.

There is also another limitation of ring signature. For every ring signature, the actual signer needs to include (and know) the identities or public keys of other users within the group, in order to preserve his anonymity. In the smart metering system, it is impossible for a user to know all members within a building or a region. Thus linkable ring signature is not suitable in our case.

BATTERY-ASSISTED POWER TRANSFORMATION: All the above mentioned protocols use cryptographic algorithms to protect privacy. Besides that, another approach is to use engineering infrastructure. As energy storage is one of the key underpinnings in a smart grid system [43, 44], it can be also used to provide privacy to individual user. A rechargeable battery and a bidirectional inverter can be used to provide selective control and power mixing of a number of electricity sources to cover consumption demands. [23, 22] provides a comprehensive description in this direction.

While it is not a complete substitute for smart metering protocols, different measures should co-exist to offer a range of data privacy services.

3. THE PROPOSED SYSTEM

There are five phases in our system. In **Setup phase**, the service provider first chooses a master secret key and a list of system parameter. In **Registration phase**, the service provider (with the master secret key) carries an interactive protocol with each registered user. The result of this protocol allows the user to have a pseudonym, a query key and a signing key which contain some secret information that is unknown to the service provider. The user should keep his query key and signing key in a secret place. In **Metering phase**, the user signs a reading with that signing key and sends the authenticated reading to the service provider through an anonymous channel [7, 35, 36, 34, 16, 18, 45, 20]. No one (including the service provider) can tell who authenticates this reading. In **Analysis phase**, the service provider can collect, link, and analyze the readings. Finally, in **Query phase**, users can check the energy consumption of different regions, or look up their personal usage records using his query key and pseudonym.

3.1 The Construction

Our scheme integrates the cryptographic techniques from [6, 10, 29]. It is mainly based on an identity-committable signature scheme. We transform it to a strongly unfrogeable signature to avoid the replay attacks. The description of our protocol is as follows. We define $\mathcal{SP}$ as the service provider and $\mathcal{U}_{ID}$ as the user with an l-level identity $ID = (id_1, id_2, \ldots, id_l)$ in our system. Each identity is given according to the location of the user. We set t out of l levels of the identity to be revealed for regional statistics, where t is the exposing level defined by $\mathcal{SP}$. For example, an identity may consist of (*country*, *state*, *city*, *area*, *building*, *unit*) and expose (*country*, *state*, *city*, *area*) in the signature where $l = 6$ and $t = 4$. More details can be found in the proof-of-concept prototype (Section 5.2). Our scheme is described as follow and sketched in Figure 3.

Setup phase.

$\mathcal{SP}$ chooses groups $\mathbb{G}$ and $\mathbb{G}_T$ with order p, a bilinear map $e : \mathbb{G} \times \mathbb{G} \to \mathbb{G}_T$ and a generator g of $\mathbb{G}$ as defined in Appendix A.1. It computes $y_1 \leftarrow g^{x_1}$ and $y_2 \leftarrow g^{x_2}$, where $x_1, x_2 \in_R \mathbb{Z}_p^*$. Let $H_1 : \{0,1\}^* \to \mathbb{G}_1$, $H_2 : \{0,1\}^* \to \mathbb{Z}_p^*$ and $H_2' : \{0,1\}^* \to \mathbb{Z}_p^*$ be three collision-resistant hash functions. It also chooses an asymmetric encryption scheme $\mathcal{E}$ with encryption key K_e and decryption key K_d [1]. Furthermore, $\mathcal{SP}$ defines an exposing level t so that it can learn t out of l levels of a user's identity for regional statistics. $\mathcal{SP}$ announces the system parameter as $PK = (\mathbb{G}_1, \mathbb{G}_2, \mathbb{G}_T, e, g, y_1, y_2, H_1, H_2, H_2', t, K_e)$ and keeps the master secret key $MSK = (x_1, x_2, K_d)$ secret.

Registration phase.

For each user $\mathcal{U}_{ID}$ with identity $ID = (id_1, id_2, \ldots, id_l)$, $\mathcal{SP}$ uses the master secret key MSK to compute

$$h_{id_i} \leftarrow H_1(id_i), \quad h'_{id_i} \leftarrow (h_{id_i})^{x_1} \quad \text{and} \quad \hat{h}_{id_i} \leftarrow (h_{id_i})^{x_1 x_2}$$

for all $i = 1, 2, \ldots, l$, and sends $sk'_{ID} = (\{h_{id_i}, h'_{id_i}, \hat{h}_{id_i}\}_{i \in [1,\ldots,l]})$ to $\mathcal{U}_{ID}$.

Upon receiving sk'_{ID}, $\mathcal{U}_{ID}$ randomly chooses $w_{ID}, s_{ID} \in \mathbb{Z}_p^* \backslash \{1\}$ and computes

$$\begin{aligned} P_{ID} &\leftarrow y_2^{w_{ID}}, \\ h' &\leftarrow h'_{id_1} \cdots h'_{id_t} (h'_{id_{t+1}} \cdots h'_{id_l})^{s_{ID}}, \\ h &\leftarrow (h_{id_{t+1}} \cdots h_{id_l})^{s_{ID}}, \\ \hat{h} &\leftarrow \hat{h}_{id_1} \cdots \hat{h}_{id_t} (\hat{h}_{id_{t+1}} \cdots \hat{h}_{id_l})^{s_{ID}}. \end{aligned}$$

The value $sk_{ID} = (P_{ID}, h, h', \hat{h}, w_{ID})$ is defined as the signing key. The value P_{ID} is called the *pseudonym* of $\mathcal{U}_{ID}$ and w_{ID} is defined as the query key.

Metering phase.

For a reading m[2], $\mathcal{U}_{ID}$ carries out the following steps:

1. He first randomly generates a chameleon hash function (defined in Appendix A.2) $\mathcal{H} : \mathcal{M} \times \mathcal{R} \to \mathcal{S}$ with a trapdoor key tk, where $\mathcal{M}, \mathcal{R}$ and $\mathcal{S}$ are the spaces defined in $\mathcal{H}$.

2. Then he picks random $m' \in \mathcal{M}, r' \in \mathcal{R}$ and $s \in \mathbb{Z}_p^*$ and uses his signing key to compute

$$\begin{aligned} h_1 &\leftarrow H_1(id_1)H_1(id_2) \cdots H_1(id_t), \\ c &\leftarrow \mathcal{H}(m'; r'), \\ u &\leftarrow (h')^s, \\ v &\leftarrow \hat{h}^{w_{ID}s + H_2(c,h,h_1,h',u,P_{ID})}. \end{aligned}$$

[1] In practice, it may be a hybrid encryption such as KEM-DEM encryption mechanism [25, 33, 1, 2, 13] for efficient implementation.

[2] As described in Section 1, the reading may be authenticated by the meter.

Setup phase

- $\mathcal{SP}$: generate system parameter $PK = (\mathbb{G}, \mathbb{G}_T, e, g, y_1, y_2, H_1, H_2, H_2', t, K_e)$ and master secret key $MSK = (x_1, x_2, K_d)$

Registration phase

- $\mathcal{SP}$ interacts with $\mathcal{U}_{ID}$ using MSK to generate $sk_{ID} = (P_{ID}, h, h', \hat{h}, w_{ID})$ with pseudonym P_{ID} and query key w_{ID} which are *unknown* to $\mathcal{SP}$.

Metering phase

- $\mathcal{U}_{ID} \longrightarrow \mathcal{SP}$: ctxt which is the ciphertext for the reading m and the corresponding signature $\sigma = (P_{ID}, id_1, \ldots, id_t, \sigma', r, \mathcal{H}, H_3, \tilde{R}, \tilde{t})$

Analysis phase

- $\mathcal{SP}$: decrypt ctxt using K_d and verify the validity of σ and record (m, σ) in the database

Query phase

- Regional enquiry queries by everyone for any region $\mathcal{R} = \{id_1, \ldots, id_{t'}\}$ where $t' \leq t$: identifying signatures with $\mathcal{R}$
- Personal enquiry by $\mathcal{U}_{ID}$: identifying signatures with P_{ID} using knowledge of w_{ID}

Figure 3: The sketch of our scheme

3. Let $\sigma' = (h, h', u, v)$. $\mathcal{U}_{ID}$ chooses a collision-resistant hash function $H_3 : \{0,1\}^* \to \mathcal{M}$ and uses the trapdoor key tk to find r such that $\mathcal{H}(H_3(m||\sigma'); r) = \mathcal{H}(m'; r')$.

4. Let $\sigma'' = (P_{ID}, id_1, \ldots, id_t, \sigma', r, \mathcal{H}, H_3)$.[3] $\mathcal{U}_{ID}$ computes a non-interactive proof of knowledge of w_{ID} such that $P_{ID} = y_2^{w_{ID}}$ as follow:
   - He first randomly generates $\tilde{r} \in \mathbb{Z}_p$ and computes $\tilde{R} \leftarrow y_2^{\tilde{r}}$.
   - He also computes $\tilde{c} \leftarrow H_2'(\tilde{R}, \sigma'', m)$ and $\tilde{t} \leftarrow \tilde{r} + \tilde{c}w_{ID} \bmod p$.

5. Let $\sigma = (\sigma'', \tilde{R}, \tilde{t})$ which is regarded as the signature generated by $\mathcal{U}_{ID}$. Finally, $\mathcal{U}_{ID}$ uses the encryption key K_e to encrypt $\{m, \sigma\}$ into ciphertext ctxt and sends ctxt to $\mathcal{SP}$ through an anonymous channel (one may refer to **Remark** below for further explanation on anonymous channel).

Analysis phase.
Upon receiving ctxt, $\mathcal{SP}$ decrypts it using its decryption key K_d to get m and $\sigma = (\sigma'', \tilde{R}, \tilde{t})$ where $\sigma'' = (P_{ID}, id_1, \ldots, id_t, \sigma', r, \mathcal{H}, H_3)$. $\mathcal{SP}$ then computes

$$\begin{aligned} h_1 &\leftarrow H_1(id_1)H_1(id_2)\cdots H_1(id_t), \\ z &\leftarrow H_2\big(\mathcal{H}(H_3(m||\sigma'); r), h, h_1, h', u, P_{ID}\big), \\ \tilde{c} &\leftarrow H_2'(\tilde{R}, \sigma'', m), \end{aligned}$$

and check whether

$$e(hh_1, y_1) \stackrel{?}{=} e(h', g), \quad (1)$$

$$e(v, g) \stackrel{?}{=} e(u, P_{ID})e(h', y_2)^z, \quad (2)$$

$$y_2^{\tilde{t}} \stackrel{?}{=} \tilde{R}{P_{ID}}^{\tilde{c}}$$

If *all* equations hold, $\mathcal{SP}$ inserts m and σ into the database. Otherwise, it discards m and σ. Note that all readings with the same pseudonym P_{ID} (from the same user) or the same $\{id_1, id_2, \ldots, id_t\}$ (from the same region) can be linked so that $\mathcal{SP}$ can do some analysis and regional statistics.

[3]In practice, the standard name of the hash function H_3 (e.g. SHA-1) can be put inside σ'' instead of the complete description of the hashing algorithm which may save a lot of space.

Query phase.
There are two kinds of queries:

- Regional statistics enquiry: This query may be opened to public. Everyone can check the energy consumption in some region and learn the comparison or other statistics between different regions. Upon receiving a query for a region $\mathcal{R} = \{id_1, \ldots, id_{t'}\}$ where $t' \leq t$, $\mathcal{SP}$ retrieves all signatures with the column $\mathcal{R}$ and returns these signatures and the corresponding messages. One can check for the validity of these signatures. If they are valid, one can be convinced that these readings are correct.

- Personal enquiry: User $\mathcal{U}_{ID}$ sends the pseudonym P_{ID} and a proof of knowledge of w_{ID} such that $P_{ID} = y_2^{w_{ID}}$ to $\mathcal{SP}$. The proof of knowledge can be constructed by the following interactive protocol:
  - User $\mathcal{U}_{ID}$ randomly generates $\tilde{r} \in \mathbb{Z}_p$ and computes $\tilde{R} \leftarrow y_2^{\tilde{r}}$. He sends $\tilde{R}$ to $\mathcal{SP}$.
  - $\mathcal{SP}$ randomly selects $\tilde{c} \in \mathbb{Z}_p$ and sends it to user $\mathcal{U}_{ID}$.
  - User $\mathcal{U}_{ID}$ computes $\tilde{t} \leftarrow \tilde{r} + \tilde{c}w_{ID} \bmod p$ and sends it back to $\mathcal{SP}$.
  - $\mathcal{SP}$ checks whether $y_2^{\tilde{t}} \stackrel{?}{=} \tilde{R}{P_{ID}}^{\tilde{c}}$.

  If the equation hold, $\mathcal{SP}$ retrieves all the readings and signatures corresponding to P_{ID} and sends to $\mathcal{U}_{ID}$. If these signatures are valid, $\mathcal{U}_{ID}$ can be convinced that these readings are from him.

Remark: Our solution hides users' identities on the application layer. However on the physical layer, it is still possible to find out the origin of an authenticated reading by recognizing the origin of the network address. For this issue, we use an anonymous channel [7, 35, 36, 34, 16, 18, 45, 20] to hide the source of the reading. Note that although anonymous channel allows a sender to send any data in an anonymous way, authentication will be lost. The receiver does not know whether the information is sent from an authenticated user. Besides, regional statistics information cannot be provided if we use anonymous channel solely. Thus we need to combine our solution on the application level together with an anonymous channel.

4. SECURITY ANALYSIS

In this section, we show that our system satisfies the five security requirements of privacy, reading authentication, regional statistics, personal enquiry and no trusted third party as described in Section 1. Before starting to present the security analysis, we first give a description of the computational assumptions on which the security of our system relies on.

4.1 Computational Assumptions

To prove the security of our scheme, we assume the following mathematical problems are hard: the discrete logarithm (DL) problem and the Diffie-Hellman problem with chosen bases (CB-CDH). They are described as follow.

DEFINITION 1 (DL ASSUMPTION). *For any probabilistic polynomial time algorithm $\mathcal{A}$, $\Pr[\mathcal{A}(g, g^a) = a]$ is negligible, where $g, g^a \in_R \mathbb{G}$.*

DEFINITION 2 (CB-CDH ASSUMPTION). *For any probabilistic polynomial time algorithm $\mathcal{A}$, $\Pr[\mathcal{A}(g, g^a, g^b) = (h, h^{ab})]$ is negligible, where $g, g^a, g^b \in_R \mathbb{G}, h \in \mathbb{G}_1\backslash\{e_{\mathbb{G}}\}$, $e_{\mathbb{G}}$ is the identity of $\mathbb{G}$.*

4.2 Analysis

4.2.1 Privacy

We only analyze the application layer since the anonymity of the physical layer is guaranteed by anonymous channel. For a signature $\sigma = (P_{ID}, h, h', u, v, r, \tilde{R}, \tilde{t})$, we assume $l = t + 1$ without loss of generality. $P_{ID}, \tilde{R}, \tilde{t}$ contain nothing related to ID_l. For h, h', since for every h_{id^*} there exists a s_{ID^*} such that $h = (h_{id^*})^{s_{ID^*}}$, $h' = h'_{id_1} \cdots h'_{id_t}(h_{id^*}^{x_1})^{s_{ID^*}}$ and $\hat{h} = \hat{h}_{id_1} \cdots \hat{h}_{id_t}(h_{id^*}^{x_1 x_2})^{s_{ID^*}}$, there are n different possibilities where n is the number of id_l that $\mathcal{SP}$ has issued. That is, the probability of guessing correctly for the identity of the signer is no better than wild guess. For u, v, the same logic follows.

4.2.2 Reading Authentication

We prove by reduction that the signature of our scheme is unforgeable. That is, our scheme satisfies reading authentication.

We state that if there is an algorithm $\mathcal{A}$ that forges a signature σ^* under adaptively chosen message attack, then there exists another algorithm $\mathcal{B}$ that can solve CB-CDH problem with non-negligible probability. Here we also provide $\mathcal{A}$ a metering oracle. When $\mathcal{A}$ inputs an identity and a message, the oracle outputs a valid signature. This is a more powerful attacking tool provided to the adversary. The hash function H_1 and H_2 are modeled as random oracles [4] which can be simulated by $\mathcal{B}$.

The proof begins by given a CB-CDH problem instance $(g, g^a, g^b) \in \mathbb{G}$ for some $a, b \in \mathbb{Z}_p$, $\mathcal{B}$ is asked to output $(z, z^{ab}) \in \mathbb{G}$. $\mathcal{B}$ processes as follow:

Setup Let $y_1 \leftarrow g^a$ and $y_2 \leftarrow g^b$. We model H_1 and H_2 as random oracles. In addition, $\mathcal{B}$ also provides a *metering oracle* to $\mathcal{A}$ which simulates the metering phase. Let q_{H_1} be the maximum number of queries to H_1. We assume the adversary has queried H_1 for an identity before making any metering query or forging any signature with respect to that identity.

Query For the i-th H_1 query, upon received id_i, $\mathcal{B}$ then randomly selects $z_i \in_R \mathbb{Z}_p$ and backbatches random oracle $H_1(id_i) \leftarrow g^{z_i}$ for $1 \leq i \leq q_{H_1}$.

For the j-th metering query, upon received an identity $ID = (id_1, \ldots, id_t, id_{t+1}, \ldots, id_l)$ and a message m_j, it first looks up the table to find out the pseudonym P_{ID} and the corresponding query key w_{ID}. If it is not existed, it randomly selects $w_{ID} \in \mathbb{Z}_p$, sets $P_{ID} \leftarrow y_2^{w_{ID}}$ and puts (ID, P_{ID}, w_{ID}) in the table. Otherwise, it retrieves from the table for later use. $\mathcal{B}$ then performs the following steps:

1. Choose a chameleon hash function $\mathcal{H} : \mathcal{M} \times \mathcal{R} \rightarrow \mathcal{S}$ with a trapdoor tk. Choose another hash function $H_3 : \{0,1\}^* \rightarrow \mathcal{M}$. Compute
$$h_1 \leftarrow H_1(id_1) \cdots H_1(id_t),$$
$$c \leftarrow \mathcal{H}(m'; r')$$
for some randomly generated $m' \in \mathcal{M}$ and $r' \in \mathcal{R}$.
2. Choose a random $w \in_R \mathbb{Z}_p$ and set
$$h \leftarrow g^{w(z_{t+1}+\ldots+z_l)},$$
$$h' \leftarrow (g^a)^{(z_1+\ldots+z_t)+w(z_{t+1}+\ldots+z_l)}$$
3. Choose a random $z', k \in_R \mathbb{Z}_p$ and set
$$v \leftarrow (g^b)^k,$$
$$u \leftarrow g^{\frac{k}{w_{ID}}}(g^a)^{\frac{-z'\left((z_1+\ldots+z_t)+w(z_{t+1}+\ldots+z_l)\right)}{w_{ID}}}$$
and the random oracle $H_2(c, h, h_1, h', u, P_{ID}) \leftarrow z'$.
4. Let $\sigma' = (h, h', u, v)$. Use the trapdoor tk to find r such that $\mathcal{H}(H_3(m_j || \sigma'); r) = \mathcal{H}(m'; r')$.
5. Generate $\tilde{R}$ and $\tilde{t}$ according to the algorithm (Metering phase step 4) as it has the knowledge of w_{ID}.
6. Output the signature
$$\sigma = \left(\{P_{ID}, id_1, \ldots, id_t, \sigma', r, \mathcal{H}, H_3\}\tilde{R}, \tilde{t}\right).$$

This is a valid signature as can be seen from the following facts:

- The formation of h and h' can be seen trivially for the same distribution as the real signature.
- We let $Z = (z_1 + \ldots + z_t) + w(z_{t+1} + \ldots + z_l)$. Then we have $h' = g^{aZ}$. For u, we have

$$\begin{aligned} u &= g^{\frac{k}{w_{ID}}}(g^a)^{\frac{-z'\left((z_1+\ldots+z_t)+w(z_{t+1}+\ldots+z_l)\right)}{w_{ID}}} \\ &= g^{\frac{k}{w_{ID}}}(g^{aZ})^{-\frac{z'}{w_{ID}}} \\ &= (g^{aZ})^{\frac{k}{aZw_{ID}} - \frac{z'}{w_{ID}}} \\ &= (h')^s \end{aligned}$$

where

$$s = \frac{k}{aZw_{ID}} - \frac{z'}{w_{ID}} \quad (3)$$

– From equation (3), we have $k = \left(s + \frac{z'}{w_{ID}}\right) aZw_{ID}$.

– For v, we then have

$$\begin{aligned} v &= (g^b)^k \\ &= g^{baZw_{ID}\left(s+\frac{z'}{w_{ID}}\right)} \\ &= (g^{abZ})^{sw_{ID}+z'} \end{aligned}$$

which has the same distribution as the one produced according to the algorithm.

Output When $\mathcal{A}$ outputs a valid signature $\sigma^* = (P^*_{ID}, id^*_1, \ldots, id^*_t, \sigma'^*, r^*, \mathcal{H}^*, H^*_3, \tilde{R}^*, \tilde{t}^*)$ for the message m^* where $\sigma'^* = (h^*, h'^*, u^*, v^*)$, using rewind technique [37] to rewind $\mathcal{A}$ with the same random tape but with a different value of H_2 random oracle. It results another valid signature $\tilde{\sigma} = (P^*_{ID}, id^*_1, \ldots, id^*_t, \tilde{\sigma}', \tilde{r}, \mathcal{H}^*, H^*_3, \tilde{R}^*, \tilde{t}^*)$ where $\tilde{\sigma}' = (h^*, h'^*, u^*, \tilde{v})$ with a different value of $\tilde{v} \neq v^*$ while other elements remain the same value [4] except $r^* \neq \tilde{r}$. Note that although $r^* \neq \tilde{r}$, we still have $\mathcal{H}(H_3(m^*||\sigma'^*); r^*) = \mathcal{H}(H_3(m^*||\tilde{\sigma}'); \tilde{r})$ which is equal to $\mathcal{H}(m'; r')$ for some randomness m', r', which should be the same in both signatures generation.

Let $c^* = \mathcal{H}(H_3(m^*||\sigma'^*); r^*)$, z'^* as the output of the first time $H_2(c^*, h^*, h_1{}^*, u^*, P^*_{ID})$ query and $\tilde{z}'$ as the output of the second time $H_2(c^*, h^*, h_1{}^*, u^*, P^*_{ID})$ query. Let $h^*_1 = H_1(id^*_1) \cdots H_1(id^*_t)$. Output $\left(h^* h^*_1, (\frac{v^*}{\tilde{v}})^{\frac{1}{z'^* - \tilde{z}'}}\right)$ as the solution of the problem instance. We can see that it is the required solution, from the following facts:

– Since both signatures σ'^* and $\tilde{\sigma}'$ pass through verification equations, from equation (1), we have

$$\begin{aligned} e(h^* h^*_1, y_1) &= e(h'^*, g) \\ e(h^* h^*_1, g^a) &= e(h'^*, g) \\ e((h^* h^*_1)^a, g) &= e(h'^*, g) \end{aligned}$$

That is, we have

$$h'^* = (h^* h^*_1)^a \quad (4)$$

– From equation (2), we have

$$\begin{aligned} e(v^*, g) &= e(u^*, P^*_{ID}) e(h'^*, y_2)^{z'^*} \\ &= e((h'^*)^{s^*}, y_2^{w^*_{ID}}) e(h'^*, y_2)^{z'^*} \\ &\quad \text{for some } s^*, w^*_{ID} \in \mathbb{Z}_p \\ &= e((h'^*)^{s^* w^*_{ID} + z'^*}, y_2) \\ &= e((h'^*)^{b(s^* w^*_{ID} + z'^*)}, g) \\ &= e((h^* h^*_1)^{ab(s^* w^*_{ID} + z'^*)}, g) \\ &\quad \text{from equation (4)} \\ \therefore v^* &= (h^* h^*_1)^{ab(s^* w^*_{ID} + z'^*)} \end{aligned}$$

[4]Note that although $\tilde{R}^*, \tilde{t}^*$ maybe different for the second signature, it does not affect the rest of the proof. Thus we simply assume that they are the same.

– Finally we have

$$\begin{aligned} \left(\frac{v^*}{\tilde{v}}\right)^{\frac{1}{z'^* - \tilde{z}'}} &= \left(\frac{(h^* h^*_1)^{ab(s^* w^*_{ID} + z'^*)}}{(h^* h^*_1)^{ab(s^* w^*_{ID} + \tilde{z}')}}\right)^{\frac{1}{z'^* - \tilde{z}'}} \\ &= \left((h^* h^*_1)^{ab(z'^* - \tilde{z}')}\right)^{\frac{1}{z'^* - \tilde{z}'}} \\ &= (h^* h^*_1)^{ab} \end{aligned}$$

which is our required solution.

4.2.3 *Regional Statistics*

As an identity is divided into l levels while the first t levels (regarding the area and region) are known (which are also part of the signature), $\mathcal{SP}$ knows that this signature and the corresponding reading come from a user from a particular area or region. It can generate statistics report easily by grouping all users from the same region.

4.2.4 *Personal Enquiry*

The user carries an interactive protocol with $\mathcal{SP}$ to authenticate himself for the personal enquiry. If $\mathcal{SP}$ accepts the authentication, it means the user is the owner of P_{ID}. This is a standard Schnorr authentication [40] and security can be guaranteed under the Discrete Logarithm (DL) Assumption. That is, without the knowledge of the query key w_{ID}, no one can succeed in the authentication.

On the other side, a user cannot generate a signature with the pseudonym belonged to an existing user. Again we prove this property by reduction. We assume that if there is an algorithm $\mathcal{A}$ that can produce a valid signature with the pseudonym $P_{ID'}$ belonged to an existing user $\mathcal{U}_{ID'}$, there exists another algorithm $\mathcal{B}$ that can solve DL problem with non-negligible probability.

The proof begins by given a DL problem instance $(g, g^a) \in \mathbb{G}$ for some $a \in \mathbb{Z}_p$, $\mathcal{B}$ is asked to output $a \in \mathbb{Z}_p$. $\mathcal{B}$ processes as follow:

Setup $\mathcal{B}$ generates everything (master secret key and public parameter) according to the algorithm, except that it models H'_2 as random oralce. It generates signatures for all users, except for a user $\mathcal{U}_{ID'}$. For $\mathcal{U}_{ID'}$, it generates the pseudonym as $P_{ID'} \leftarrow (g^a)^{x_2}$. By modeling H'_2 as random oracle, it can also generate valid $\tilde{R}$ and $\tilde{t}$ even without the knowledge of $w_{ID'}$ such that $P_{ID'} = y_2^{w_{ID'}}$.

Output After querying the n signatures for a list of users $\mathcal{L}_\mathcal{U} = \{\mathcal{U}_{ID_1}, \ldots, \mathcal{U}_{ID_n}\}$ such that $\mathcal{U}_{ID'} \in \mathcal{L}_\mathcal{U}$, $\mathcal{A}$ outputs a signature and the corresponding pseudonym. With probability $1/n$, $\mathcal{A}$ outputs the signature for $\mathcal{U}_{ID'}$ which contains a proof of knowledge of $w_{ID'}$ where $P_{ID'} = y_2^{w_{ID'}}$. Note that $\mathcal{A}$ is not allowed to have the signing key and query key for the user with respect to this pseudonym. $\mathcal{B}$ then uses standard rewind technique [37] to rewind $\mathcal{A}$ to obtain the value of $w_{ID'}$, which is equal to the solution of the problem instance.

4.2.5 *No Trusted Third Party*

In our scheme, $\mathcal{SP}$ knows nothing related to the *real* identity of a signature. The pseudonym contains the query key which is unknown to $\mathcal{SP}$ (user only needs to prove the knowledge of query key in a zero knowledge way such that $\mathcal{SP}$ gets no information about it). On the other side, $\mathcal{SP}$ cannot modify or forge a signature from any user as the verification equations in analysis phase guarantee the unforgeability of the signature.

5. SIMULATION RESULTS

5.1 Computation Cost

For the service provider side, the main computation cost is on signature verification in **Analysis phase**, which needs some pairing computation. However, the service provider usually owns very powerful back-end servers for computation, so we do not need to worry the computation cost on this side. For the user side, there are three kinds of operations: modular exponentiation (analogous to scalar multiplication on an elliptic curve), chameleon hash function and collision-resistant hash function. Collision-resistant hash function is a negligibly light computation when compared to others. Since chameleon hash function can be implemented by exponentiation operations, the only actual operation the user needs to compute is modular exponentiation. In particular, we do not need any pairing computation, which is a much more expensive algorithm compared to modular exponentiation. Therefore, as long as a device can do modular exponentiation efficiently, it can be used by the user in our system.

We are going to present the simulation of our application scenario in the rest of this section. The results include the proof-of-concept prototype and the performance evaluation.

5.2 Proof-of-Concept Prototype

In our simulation, the identity of a user is defined as a tuple of (*country*, *state*, *city*, *area*, *building*, *unit*). The exposing level is 4, that is, the *country*, *state*, *city* and *area* of origin of a signature can be recognized. The signed message consists of a timestamp and the reading at that time. So the reading table in the $\mathcal{SP}$ database contains the following columns:

P_{ID}	T	reading	country	state	city	area

where P_{ID} and T stand for pseudonym and timestamp respectively.

The system is shown in Figure 4 and 5. Through the web pages, the user can query the total power consumption of some region or his personal power consumption on a specific day/week/month. We assume the meters send the readings to the service provider every hour. The system also shows the usage comparing to the average usage. This allows people to understand their consumption and encourages them to save more energy.

For the regional statistics (Figure 4), the service provider first lists the readings from the queried region on the queried date in the database. Then it sums up the readings to get the total consumption from that region and compares it with the average readings. For the personal enquiry (Figure 5), the user can compute the proof of knowledge of his query key and upload his pseudonym (P_{ID}) together with the proof. Then the service provider retrieves the readings from the database according to the uploaded pseudonym. The user can also compare his personal consumption with the average consumption.

5.3 Performance Evaluation

In our simulation, we use a desktop to play the role of the service provider. The desktop is equipped with an Intel Core 2 Duo CPU @ 3.00GHz and Microsoft Windows XP 32-bit operating system. Since **Setup phase** is performed in the beginning only, we consider the performance of **Analysis phase** here, which mainly verifies the signature of every signed reading. The function is implemented by Java language with MIRACL library. With the help of precomputation, the desktop can finish a verification in 81.36 milliseconds.

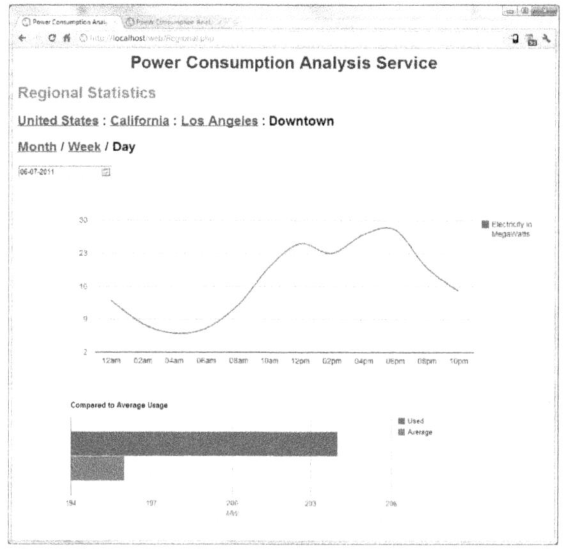

Figure 4: Regional statistics: The consumption of a region on a specific day/week/month can be publicly queried.

For the user side, since **Setup phase** and **Query phase** can be performed efficiently, we consider **Metering phase** on a laptop and a smart phone, which are two common devices for home use. The laptop is equipped with an Intel Core 2 Duo CPU @ 2.53GHz and Microsoft Windows XP 32-bit operating system. The smart phone is equipped with a Qualcomm Snapdragon processor @ 1GHz and Microsoft Windows Mobile 6.5. The function is implemented by C# with bouncy castle library. The signature generation time is 90.27 milliseconds on the laptop and 879 milliseconds on the smart phone.

In summary, we can see that the computation time on the service provider side is acceptable, not to mention that the server used by the service provider will be much more powerful in the reality. For the users, even they sign the readings on the smart phone, the signing time is still less than one second. This shows the practicability of our scheme.

6. CONCLUSIONS

In this paper, we first figure out the privacy concern of the third-party services in smart grid. Then we propose a scheme to protect users' privacy. With this scheme, the user can anonymously send individual authenticated meter readings to the service provider so that the service provider can provide fine-grained analyses of the readings for personal enquiry and regional statistic. We give a formal security analysis and also provide the implementation result of our scheme. It shows that the scheme is practical and can be performed on common computation devices for home use.

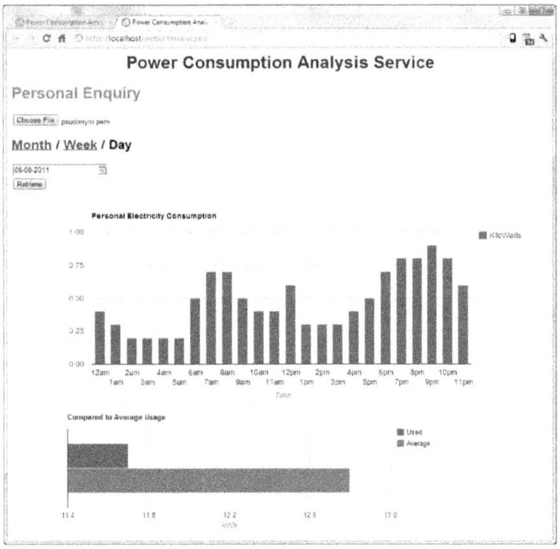

Figure 5: Personal Enquiry: Users can upload their pseudonyms with the corresponding proof and query their consumption on a specific day/week/month.

7. REFERENCES

[1] M. Abe, R. Gennaro, K. Kurosawa, and V. Shoup. Tag-KEM/DEM: A New Framework for Hybrid Encryption and A New Analysis of Kurosawa-Desmedt KEM. In *EUROCRYPT*, volume 3494 of *Lecture Notes in Computer Science*, pages 128–146. Springer, 2005.

[2] J. Baek, D. Galindo, W. Susilo, and J. Zhou. Constructing Strong KEM from Weak KEM (or How to Revive the KEM/DEM Framework). In *SCN*, volume 5229 of *Lecture Notes in Computer Science*, pages 358–374. Springer, 2008.

[3] J. Balasch, A. Rial, C. Troncoso, B. Preneel, I. Verbauwhede, and C. Geuens. Pretp: Privacy-preserving electronic toll pricing. In *Proceedings of the 19th USENIX Security Symposium (USENIX '10)*, pages 63–78. USENIX Association, 2010.

[4] M. Bellare and P. Rogaway. Random oracles are practical: A paradigm for designing efficient protocols. In *ACM Conference on Computer and Communications Security*, pages 62–73. ACM, 1993.

[5] J. Camenisch and A. Lysyanskaya. An Efficient System for Non-Transferable Anonymous Credentials with Optional Anonymity Revocations. In *EUROCRYPT 2001*, volume 2045 of *Lecture Notes in Computer Science*, pages 93–118. Springer, 2001.

[6] J. Cha and J. Cheon. An Identity-Based Signature from Gap Diffie-Hellman Groups. In *PKC 2003*, volume 2567 of *Lecture Notes in Computer Science*, pages 18–30. Springer, 2003.

[7] D. Chaum. Untraceable Electronic Mail, Return Addresses, and Digital Pseudonyms. *Communications of the ACM*, 24(2):84–88, Feb. 1981.

[8] D. Chaum. Security without Identification: Transaciton Systems to Make Big Brother Obsolete. *Communications of the ACM*, 28(10):1030–1044, Oct. 2001.

[9] D. Chaum and E. van Heyst. Group Signatures. In *EUROCRYPT 1991*, volume 547 of *Lecture Notes in Computer Science*, pages 257–265. Springer, 1991.

[10] C.-K. Chu and W.-G. Tzeng. Identity-Committable Signatures and Their Extension to Group-Oriented Ring Signatures. In *ACISP 2007*, volume 4586 of *Lecture Notes in Computer Science*, pages 323–337. Springer, 2007.

[11] G. Danezis, M. Kohlweiss, and A. Rial. Differentially private billing with rebates. In *Proceedings of the 13th International Conference on Information Hiding (IH '11)*, volume 6958 of *Lecture Notes in Computer Science*, pages 148–162. Springer, 2011.

[12] W. de Jonge and B. Jacobs. Privacy-friendly electronic traffic pricing via commits. In *Proceedings of the 5th International Workshop on Formal Aspects in Security and Trust (FAST '08)*, volume 5491 of *Lecture Notes in Computer Science*, pages 143–161. Springer, 2008.

[13] Y. Desmedt, R. Gennaro, K. Kurosawa, and V. Shoup. A new and improved paradigm for hybrid encryption secure against chosen-ciphertext attack. *J. Cryptology*, 23(1):91–120, 2010.

[14] C. Efthymiou and G. Kalogridis. Smart grid privacy via anonymization of smart metering data. In *Proceedings of the 1st IEEE International Conference on Smart Grid Communications (SmartGridComm 2010)*, pages 238–243. IEEE, 2010.

[15] F. D. Garcia and B. Jacobs. Privacy-friendly Energy-metering via Homomorphic Encryption. In *Proceedings of the 6th Workshop on Security and Trust Management (STM '10)*, volume 6710 of *Lecture Notes in Computer Science*, pages 226–238. Springer, 2010.

[16] P. Golle and M. Jakobsson. Reusable anonymous return channels. In *WPES*, pages 94–100. ACM, 2003.

[17] Green Button. Pacific Gas and Electric, San Diego Gas and Electric . http://www.greenbuttondata.org/.

[18] J. Groth. Efficient Maximal Privacy in Boardroom Voting and Anonymous Broadcast. In *Financial Cryptography 2004*, volume 3110 of *Lecture Notes in Computer Science*, pages 90–104. Springer, 2004.

[19] G. W. Hart. Nonintrusive appliance load monitoring. In *Proceedings of the IEEE*, pages 1870–1891. IEEE, 1992.

[20] A. Hevia and D. Micciancio. An indistinguishability-based characterization of anonymous channels. In *Privacy Enhancing Technologies*, volume 5134 of *Lecture Notes in Computer Science*, pages 24–43. Springer, 2008.

[21] M. Jawurek, M. Johns, and F. Kerschbaum. Plug-in privacy for smart metering billing. In *Proceedings of the 11th Privacy Enhancing Technologies Symposium (PETS '11)*, volume 6794 of *Lecture Notes in Computer Science*, pages 192–210. Springer, 2011.

[22] G. Kalogridis, R. Cepeda, S. Z. Denic, T. A. Lewis, and C. Efthymiou. Elecprivacy: Evaluating the privacy protection of electricity management algorithms. *IEEE Trans. Smart Grid*, 2(4):750–758, 2011.

[23] G. Kalogridis, C. Efthymiou, T. Lewis, S. Denic, and R. Cepeda. Privacy for smart meters: Towards undetectable appliance load signatures. In *Proceedings of the 1st IEEE International Conference on Smart

Grid Communications (SmartGridComm 2010), pages 232–237. IEEE, 2010.
[24] H. Krawczyk and T. Rabin. Chameleon Signatures. In *NDSS*. The Internet Society, 2000.
[25] K. Kurosawa and Y. Desmedt. A New Paradigm of Hybrid Encryption Scheme. In *CRYPTO*, volume 3152 of *Lecture Notes in Computer Science*, pages 426–442. Springer, 2004.
[26] K. Kursawe, G. Danezis, and M. Kohlweiss. Privacy-friendly aggregation for the smart-grid. In *Proceedings of the 11th Privacy Enhancing Technologies Symposium (PETS '11)*, volume 6794 of *Lecture Notes in Computer Science*, pages 175–191. Springer, 2011.
[27] H. Y. Lam, G. S. K. Fung, and W. K. Lee. A novel method to construct taxonomy electrical appliances based on load signatures. *IEEE Trans. Consum. Electron.*, 53(2):653–660, 2007.
[28] C. Laughman, K. Lee, R. Cox, S. Shaw, S. Leeb, L. Norford, and P. Armstrong. Power signature analysis. *IEEE Power and Energy Magazine*, 1(2):56–63, 2003.
[29] J. K. Liu, M. H. Au, W. Susilo, and J. Zhou. Short Generic Transformation to Strongly Unforgeable Signature in the Standard Model. In *ESORICS 2010*, volume 6345 of *Lecture Notes in Computer Science*, pages 168–181. Springer, 2010.
[30] J. K. Liu, V. K. Wei, and D. S. Wong. Linkable Spontaneous Anonymous Group Signature for Ad Hoc Groups (Extended Abstract). In *ACISP 2004*, volume 3108 of *LNCS*, pages 325–335. Springer, 2004.
[31] J. K. Liu and D. S. Wong. A Restricted Multi-show Credential System and Its Application on E-Voting. In *ISPEC*, volume 3439 of *Lecture Notes in Computer Science*, pages 268–279. Springer, 2005.
[32] A. Molina-Markham, P. Shenoy, K. Fu, E. Cecchet, and D. Irwin. Private memoirs of a smart meter. In *Proceedings of the 2nd ACM Workshop on Embedded Sensing Systems for Energy-Efficiency in Building (BuildSys '10)*, pages 61–66. ACM, 2010.
[33] W. Nagao, Y. Manabe, and T. Okamoto. A Universally Composable Secure Channel Based on the KEM-DEM Framework. In *TCC*, volume 3378 of *Lecture Notes in Computer Science*, pages 426–444. Springer, 2005.
[34] W. Ogata, K. Kurosawa, K. Sako, and K. Takatani. Fault tolerant anonymous channel. In *ICICS*, volume 1334 of *Lecture Notes in Computer Science*, pages 440–444. Springer, 1997.
[35] C. Park, K. Itoh, and K. Kurosawa. Efficient anonymous channel and all/nothing election scheme. In *EUROCRYPT*, volume 765 of *Lecture Notes in Computer Science*, pages 248–259. Springer, 1993.
[36] C. Park, K. Itoh, and K. Kurosawa. Efficient Anonymous Channel and All/Nothing Election Scheme. In *EUROCRYPT' 93*, volume 765 of *Lecture Notes in Computer Science*, pages 248–259. Springer, 1994.
[37] D. Pointcheval and J. Stern. Security arguments for digital signatures and blind signatures. *J. Cryptology*, 13(3):361–396, 2000.
[38] A. Rial and G. Danezis. Privacy-friendly smart metering. Technical Report MSR-TR-2010-150, Microsoft Research, November 2010.
[39] R. L. Rivest, A. Shamir, and Y. Tauman. How to Leak a Secret. In *ASIACRYPT 2001*, volume 2248 of *Lecture Notes in Computer Science*, pages 552–565. Springer, 2001.
[40] C.-P. Schnorr. Efficient identification and signatures for smart cards. In *CRYPTO*, volume 435 of *Lecture Notes in Computer Science*, pages 239–252. Springer, 1989.
[41] A. Shamir and Y. Tauman. Improved Online/Offline Signature Schemes. In *CRYPTO 2001*, volume 2139 of *Lecture Notes in Computer Science*, pages 355–367. Springer, 2001.
[42] E. Shi, T.-H. H. Chan, E. G. Rieffel, R. Chow, and D. Song. Privacy-preserving aggregation of time-series data. In *Proceedings of the 18th Network and Distributed System Security Symposium (NDSS 2011)*. The Internet Society, 2011.
[43] P. Vytelingum, T. Voice, S. D. Ramchurn, A. Rogers, and N. R. Jennings. Agent-based micro-storage management for the smart grid. In *AAMAS*, pages 39–46. IFAAMAS, 2010.
[44] P. Vytelingum, T. Voice, S. D. Ramchurn, A. Rogers, and N. R. Jennings. Theoretical and practical foundations of large-scale agent-based micro-storage in the smart grid. *J. Artif. Intell. Res. (JAIR)*, 42:765–813, 2011.
[45] Z. Wang, J. Deng, and R. B. Lee. Mutual anonymous communications: A new covert channel based on splitting tree mac. In *INFOCOM*, pages 2531–2535. IEEE, 2007.

APPENDIX

A. CRYPTOGRAPHIC PRELIMINARIES

A.1 Bilinear Groups

Let $\mathbb{G}_1, \mathbb{G}_2$ and $\mathbb{G}_T$ be multiplicative cyclic groups of prime order p and $e : \mathbb{G}_1 \times \mathbb{G}_2 \rightarrow \mathbb{G}_T$ be a map with the following properties:

- Bilinear: $\forall g_1 \in \mathbb{G}_1, g_2 \in \mathbb{G}_2$ and $a, b \in \mathbb{Z}$, $e(g_1^a, g_2^b) = e(g_1, g_2)^{ab}$.
- Non-degenerate: for some $g_1 \in \mathbb{G}_1$ and $g_2 \in \mathbb{G}_2$, $e(g_1, g_2) \neq 1$.

We say that $(\mathbb{G}_1, \mathbb{G}_2)$ is a bilinear group pair if all the above operations are efficiently computable.

A.2 Chameleon Hash Function

A chameleon hash function [24, 41] is a trapdoor one-way hash function whose collisions are efficiently computable with knowledge of the trapdoor. That is, for a chameleon hash function $\mathcal{H}$ and its trapdoor tk, there is a deterministic polynomial-time algorithm that on input tk, a pair (m, r) and an additional message m', outputs a value r' such that $\mathcal{H}(m; r) = \mathcal{H}(m'; r')$. One can find some concrete constructions in [24, 41].

Protecting Privacy by Sanitizing Personal Data: a New Approach to Anonymous Credentials

Sébastien Canard
Orange Labs, Caen, France
Applied Crypto Group
sebastien.canard@orange.com

Roch Lescuyer*
Normandie Université, UNICAEN,
GREYC, ENSICAEN, CNRS, Caen, France
roch.lescuyer@ensicaen.fr

ABSTRACT

Anonymous credential systems allow users to obtain certified credentials from organizations and use them later without being traced. For instance, a student will be able to prove, using his student card certified by the University, that he is a student living *e.g.* in Hangzhou without revealing other information given by the student card, such as his name or studies. Besides, sanitizable signatures enable a designated person, called the sanitizer, to modify some parts of a signed message in a controlled way, such that the message can still be verified *w.r.t.* the original signer.

We propose in this paper to formalize the following new idea. A user gets from the organization a signed document certifying personal data (*e.g.* name, address, studies, etc.) and plays the role of the sanitizer. When showing his credential, he uses sanitization techniques to hide the information he does not want to reveal (*e.g.* name, studies or complete address), and shows the resulting document, which is still seen as a document certified by the organization. Unfortunately, existing sanitizable signatures can not directly be used for this purpose. We thus seek for generic conditions on them to be used as anonymous credentials. We also provide a concrete construction based on standard assumptions and secure in the random oracle model.

Categories and Subject Descriptors

E.3 [**Data Encryption**]: Public Key Cryptosystems; K.4.1 [**Computers and Society**]: Public Policy Issues—*Privacy*; K.6.5 [**Management of Computing and Information Systems**]: Security and Protection—*Authentication*

General Terms

Algorithms, Design, Security.

Keywords

Cryptographic protocols, Anonymous credentials, Privacy-enhancing systems, Sanitizable signatures

1. INTRODUCTION

ANONYMOUS CREDENTIALS. Digital signatures are a widely used cryptographic tool in our everyday life. We here focus on the use of digital signatures in credential systems where a user gets some certified credentials issued by some organization. These certified credentials enable the user to prove access rights, get some advantages, or anything else which requires authentication. For instance a student is given an electronic student card by the University and later uses this credential to prove that he actually is a student, if necessary. Anonymous credential systems [16] allow users to prove possession of credentials in an anonymous way. When a credential is shown, a proof of possession is done without leaking (even for the issuing organization) any knowledge about neither the owner of the credential, nor the credential itself.

Several ways exist to design anonymous credential systems and some industrial companies seem to be interested. As an evidence, IBM's Idemix system [24] is based on the use of group signatures (or some close variants), as well as several existing papers [9, 10, 2, 8, 20, 15]. Microsoft prefers the Brands' blind signature approach [4] for their UProve technology [25], as in [23]. Finally, the Orange operator seems to be interested in the use of aggregate signatures [14].

SANITIZABLE SIGNATURES. Sanitizable signatures [1] enable a designated entity, called the *sanitizer*, to modify some parts of a signed message in a controlled way defined by the initial signer at the creation of the signature.

In fact, there are multiple ways to handle such a problem and a lot of works on this subject can be found in the cryptographic literature. In a nutshell, (i) the sanitization procedure may be public [26, 5], or restricted to a designated sanitizer [1, 13, 6, 7, 11], (ii) the signed document can be a bit string [1, 13, 6, 11] or have a particular data structure [5], (iii) the sanitizer may be allowed to modify some parts of the message, or simply remove them [26, 5], and (iv) the capacity of sanitization can be given to a single sanitizer [1, 13, 6, 11], or to a group of sanitizers [7, 12].

Let us consider the case where only a designated sanitizer is able to modify a given message. The signer divides a message $m \in \{0,1\}^*$ into N blocks $m_1, \ldots, m_N$, defines the set $\textsc{adm} \subseteq [1, N]$ of blocks that are said *admissible* (*i.e.* allowed to be subsequently modified) and then signs the whole message, using some key related to the sanitizer. Using this key, the sanitizer is next able to modify the admissible parts of the given message so that the resulting signature is still valid under the signer's public key.

* This work was done while the second author was at Orange Labs, Caen, France and ENS, Paris, France.

FROM SANITIZABLE SIGNATURES TO ANONYMOUS CREDENTIALS. Our idea in this paper is to use sanitizable signatures to design anonymous credentials, which is, to the best of our knowledge, a new approach. The underlying idea is very simple: the organization in the anonymous credential system is a signer and the user who obtain certified credentials plays the role of a sanitizer.

For example, a student (acting as a sanitizer) obtains a signature (the certified credential) by the University (which plays the role of the signer) on the message: $m :=$ 'Name: Wang Li; City: Hangzhou; Studies: Computer Science'. This message is split by the signer into six sub-messages $m_1 :=$ 'Name:', $m_2 :=$ 'Wang Li', $m_3 :=$ '; City:', $m_4 :=$ 'Hangzhou', $m_5 :=$ '; Studies:', and $m_6 :=$ 'Computer Science', who also sets $\text{ADM} = \{2, 4, 6\}$. Finally, the whole message m is signed to obtain σ, using some key related to the student / sanitizer, and sends the whole result (m, σ) to the student / sanitizer. The latter can next prove *e.g.* that he is a student living in Hangzhou without revealing his name, nor his studies. For this purpose, he can use the appropriate key to replace m_2 and m_6 by some predefined non relevant symbol (like "#") of adequate length, to obtain the message $m :=$ 'Name: ######; City: Hangzhou; Studies: ######' and modify the University's signature accordingly. This student is hence anonymous inside the group of students living in Hangzhou but proves such a belonging since the University's signature is valid. This simple approach seems to work but unfortunately, "the devil is on the details" and, as we will see, additional work is necessary.

ADDITIONAL NEEDS FOR SANITIZABLE SIGNATURES. An anonymous credential system should be anonymous and unforgeable (see *e.g.* [14, 9]). Regarding sanitizable signature schemes, different formal security models can be found in the literature. An initial one has been designed by Brzuska *et al.* [6] and gives main procedures (for signing, sanitizing, verifying, proving accountability of a signature and judging these proofs) and basic security properties (immutability, accountability and transparency). Based on this initial model, Brzuska *et al.* [7] have next added the unlinkability property, Canard and Jambert [11] have integrated some extensions and Canard *et al.* [12] have formalized the concept for multiple signers and sanitizers. These models do not entirely match our needs for the design of a secure anonymous credential system. More precisely, the following questions need to be addressed.

Verification without sanitizing key. The verification of a sanitizable signature generally needs *the sanitizer's public key*. In our case, since we want the user / sanitizer to be anonymous, a signature should be checkable without the sanitizer's public key. We thus modify the verification algorithm accordingly.

Traceability of signatures. The above feature obviously gives the anonymity of the sanitizer. This implies, for security purpose, the necessity for an algorithm to recover the actual designated sanitizer of a given message-signature pair. However, in order to prevent organizations to trace users, this algorithm should be carried out by a separate authority. This algorithm has no equivalent in the literature of sanitizable signatures. We will call this procedure FINDSAN, since it is the counterpart of the FINDORI algorithm in the model of [12]. The reader will notice that the above tracing procedure should not be confused with the opening procedure in anonymous primitives, like group signatures. In a group signature scheme, the opening aims at finding the actual producer of a given signature, whereas we just want here to recover the actual *designated* sanitizer of a given message/signature pair. The latter could have not modified or even seen the signature.

Dealing with the proof algorithm. The above tracing procedure is not designed to decide whether a given signature has been sanitized or not, which is the role of the PROVE algorithm in [6] and the ALGOPEN procedure in [12]. Such a procedure is no more useful in our case. In front of a student proving that he is a student, the verifier sees that the document has been sanitized, and some fields hidden, but this is not a matter of concern here. In fact, we will see that we do not need the accountability properties of sanitizable signatures, at least as long as there is no opening algorithm in the anonymous credential system. Such a procedure would imply a notion of non-frameability, and, consequently, an accountability property for the signer.

Restrictions on admissible values. We want the true sanitizer to be able to hide admissible parts of the message. However, we do not want him to modify as he wants these admissible parts so as to be able to prove false statements about his attributes (*e.g.* proving to be a student in Hangzhou while being a student in Shanghai). We can use for this purpose the `LimitSet` extension of sanitizable signatures given by Canard and Jambert [11], which limits the set of possible modifications on a single block of the message.

Replaying sessions. Another problem we have to deal with is that an eavesdropper could intercept sanitized documents and replay them as often as she wants. To fix this problem, we let the verifier send at each authentication a random value and the sanitizer modify a specific admissible part of the message with this value. The latter may depend on the session data, like the verifier's identity and the current time.

COMPARISON WITH RELATED WORKS. The solution we propose in this paper is a new way to design an anonymous credential system. As said before, previous designs make use of group [24], blind [25] or aggregate signature schemes [14]. We here propose to use sanitizable signatures. The efficiency and security of the practical scheme we propose in this paper are close to the one of related works, as we will see at the end of the paper (see Section 5). Our main aim is clearly to propose a new approach. In fact, we provide two main contributions in this paper. The first one is a generic analysis of the properties for sanitizable signatures in order to use them in a privacy-preserving context. The second one is a concrete construction to show that those ideas can be efficiently implemented. Our proposal also introduces a new use case to sanitizable signatures, which has not yet been proposed in the related literature. Furthermore, we may expect that any improvement in sanitizable signatures will improve the related anonymous credential system.

One of the most significant interest of our solution *w.r.t.* related work is that there is no need, in our solution, for an interaction between the user and the issuer during the issuing protocol, contrary to other existing systems. The issuer produces a credential only with the sanitizer's public key. In both group and blind signature based anonymous credential systems, it is necessary for the user to add a random part of the secret key which will be certified by the organization with the attributes, and which is new for each credential. This is mainly due to the unforgeability of the used signatures. In our case, this is not necessary and only the user public key is useful. Moreover, in pseudonym systems, an interactive registration is necessary in order to obtain a non-interactive anonymous credential issuing [2]. This might be preferable in practice when the issuer needs to issue credentials without being on-line with the sanitizer. The price to cost is that there is no anonymity during the registration process. Nevertheless, if one needs some anonymity during the registration process, one can add this property at the cost of an interaction.

Another advantage of our sanitizable signature based solution is that the size of a credential is lower than what exists today [24, 25, 14].

Independently to our work, [17] addresses the problematic of anonymous authentication in cloud storage, which implies some kind of "attributes hiding". However, their techniques for attributes hiding are based on group signatures techniques rather than sanitization ones (in a strict sense). More precisely, the hiding of an attribute in [17] consists in not revealing the attribute by only proving its knowledge, using for this purpose Groth-Sahai based non-interactive zero-knowledge proofs. In our case, the attribute hiding is done by using the technique of sanitizable signatures, and thus *e.g.* chameleon hash functions [1], accumulators [11] or the split into two related signatures [7]. In our construction, we use a mix of the last two ideas, and thus not a zero-knowledge proof of a hidden attribute.

One may argue that our proposal does not permit to prove some well chosen statements on non-revealed attributes, such as proving that one has to prove that he has more than 65 years, which is the case for some existing proposals. In fact, in some particular cases, such as the one in the last sentence, we can use the power of the `LimitSet` extension of sanitizable signatures by permitting the user to replace his date of birth by *e.g.* the sub message "more that 65".

ORGANIZATION OF THE PAPER. First of all, anonymous credentials are formalized in Section 2. Then Section 3 formally defines sanitizable signatures according to our needs. Section 4 shows how to build anonymous credentials from this new type of sanitizable signatures. Finally, a concrete sanitizable signature scheme is presented in Section 5.

2. ANONYMOUS CREDENTIALS

2.1 Notations

All along the paper, we distinguish two types of security properties. First of all, for any adversary $\mathcal{A}$ against a property prop and any security parameter $\lambda \in \mathbb{N}$, the success probability of $\mathcal{A}$ is the probability that the related experiment outputs 1. We say that the whole scheme verifies prop if this success is negligible (as a function of λ) for any polynomial-time $\mathcal{A}$. Then, for a decisional experiment, a challenge bit $b \in \{0,1\}$ is set and for any adversary $\mathcal{A}$ against a property prop and any security parameter λ, the advantage of $\mathcal{A}$ is $\Pr\left[\mathbf{Exp}_{\mathcal{A}}^{\mathsf{prop}\text{-}1}(\lambda) = 1\right] - \Pr\left[\mathbf{Exp}_{\mathcal{A}}^{\mathsf{prop}\text{-}0}(\lambda) = 1\right]$. We next say that the whole scheme verifies prop if this advantage is negligible for any polynomial-time $\mathcal{A}$.

2.2 Definition

We base the security model below on the work done in [9, 14], for the case of one single organization (see [14] for multiple organizations). An *anonymous credential system* $\mathcal{AC}$ involves users, an organization[1] and verifiers, and is given by the following algorithms.

SETUP. This algorithm takes as input a security parameter λ and outputs global parameters gpk, which are given to all algorithms as auxiliary input, and the key pair $(\mathsf{opk}, \mathsf{osk})$ for the organization.

USERKG. This algorithm, on input $j \in \mathbb{N}$, produces a key pair $(\mathsf{upk}[j], \mathsf{usk}[j])$ for the user j.

OBTAIN $\leftrightarrow$ ISSUE. The OBTAIN algorithm takes a user secret key $\mathsf{usk}[j]$ and the organization public key opk; the ISSUE algorithm takes the organization secret key osk, a user public key $\mathsf{upk}[j]$ and a list of messages / attributes $\{m_n\}_{n=1}^{N}$. At the end of the protocol, the user U obtained a credential C.

SHOW $\leftrightarrow$ VERIFY. The SHOW algorithm takes a user secret key $\mathsf{usk}[j]$, the organization public key opk, a list of messages $\{m_n\}_{n=1}^{N'}$ and a credential C; the VERIFY public algorithm takes the organization public key opk and a list of messages $\{m_n\}_{n=1}^{N'}$. At the end of the protocol, the VERIFY algorithm outputs a bit $b \in \{0,1\}$.

2.3 Security properties

We now focus on the security properties related to anonymous credential systems. Experiments are played between a challenger and an adversary $\mathcal{A}$, which may in addition call the following oracles. The tables cred and reg are global variables which maintain the issued credentials and the certified attributes.

ADDU(j). It adds j to the set HU of honest users, executes USERKG and updates the j-th entry of $\mathsf{upk}[j]$ and $\mathsf{usk}[j]$. It finally returns $\mathsf{upk}[j]$.

CRPTU(j, pk_U). It adds j to the set CU of corrupted users and sets $\mathsf{upk}[j] \leftarrow \mathsf{pk}_U$.

USK(j). It returns $(\mathsf{usk}[j], \mathsf{cred}[j])$ and put j in the set KU of users for whom the secrets are known[2].

SNDTOU$_{algo}(j, m)$ (resp. SNDTOO(osk, m)) , with $algo \in$ {ISSUE, VERIFY}. It plays the role of the user j (resp. the organization) which receives the message m from the corrupted organization – the ISSUE case – or a corrupted verifier – the VERIFY case – (resp. the user j). This may modify the entry $\mathsf{cred}[j]$ (resp. $\mathsf{reg}[j]$).

GETCRED($\mathsf{osk}, j, \{m_n\}_{n=1}^{N}$). It permits the honest user j to obtain a credential on $\{m_n\}_{n=1}^{N}$ from the organization. The oracle plays the role of both the user and the organization. The credentials are added to $\mathsf{cred}[j]$, the attributes to $\mathsf{reg}[j]$ and the view is returned.

[1]For sake of the exposition, we have only one signer in our model, even if the case of multiple signers can easily be studied, using [6, 11, 7].

[2]The reason for the distinction between corruption and secret key leakage concerns the anonymity property. A user must remain anonymous even if his secret keys leak.

LoR$(b, j_0, j_1, \{m'_{n'}\}_{n'=1}^{N'})$. It takes as input a pair (j_0, j_1) of honest users and a set $\{m'_{n'}\}_{n'=1}^{N'}$ of N' attributes certified to both j_0 and j_1. It plays the role of the user j_b with the set $\{m'_{n'}\}_{n'=1}^{N'}$ with the adversary during a verification protocol.

Correctness. First of all, such a system has to be *correct.* This means that honestly obtained credentials must be accepted by an honest verifier.

Unforgeability. Regarding the unforgeability property, the aim of the adversary is to prove that she is in possession of some credentials issued by the organization while this is not the case. The adversary may (i) interact with the organization as corrupted user with the help of the SndToO oracle, (ii) ask for an honest user to obtain credentials on her choice with the help of the GetCred oracle and (iii) interact with an honest user as a corrupted verifier with the help of the SndToU$_{\text{Verify}}$ oracle. A forgery may appear by different ways. For instance, a new credential has been produced. Another possibility is that several corrupted users could have colluded and shared their credentials.

In order to properly address this property, we need to introduce the notion of *identity extractor*, which formalizes the idea that there exists a well-defined identity underlying an accepted authentication protocol, even if the user is anonymous. We require that there exists a pair of algorithms {ESetup, Extract} such that (i) parameters returned by ESetup are indistinguishable from those returned by Setup and (ii) the Extract algorithm correctly identify the underlying identity. More formally, we require that:

- For all adversary $\mathcal{A}$ there exists a negligible function $\nu(\cdot)$ s.t. $\Pr\left[\mathsf{gpk}_0 \leftarrow \text{Setup}(1^\lambda);\ \mathsf{gpk}_1 \leftarrow \text{ESetup}(1^\lambda);\ b \xleftarrow{\$} \{0,1\};\ b' \leftarrow \mathcal{A}(\mathsf{gpk}_b) : b' = b\right] < \frac{1}{2} + \nu(\lambda)$
- For all $(\mathsf{gpk}, \mathsf{ek}) \leftarrow \text{ESetup}(1^\lambda)$, all j, all $(\mathsf{pk}_{U_j}, \mathsf{sk}_{U_j})$, all $\{m_{n'}\}_{n'=1}^{N'}$ and all C, if $\text{Show}(\mathsf{osk}, \mathsf{sk}_{U_j}, \{m_{n'}\}_{n'=1}^{N'}, C \leftrightarrow \text{Verify}(\mathsf{osk}, \{m_{n'}\}_{n'=1}^{N'})$ outputs 1, then $\text{Extract}(\mathsf{ek}, \mathit{view}) = j$, where *view* is the view of the protocol.

With this notion of extractor, the unforgeability experiment Unforgeability(λ) is given as follows.

- $(\mathsf{gpk}, (\mathsf{opk}, \mathsf{osk}), \mathsf{ek}) \leftarrow \text{ESetup}(1^\lambda)$
- $\mathcal{O} \leftarrow \{\text{AddU}(\cdot), \text{USK}(\cdot), \text{CrptU}(\cdot,\cdot), \text{SndToO}(\mathsf{osk},\cdot), \text{SndToU}_{\text{Verify}}(\cdot,\cdot), \text{GetCred}(\mathsf{osk},\cdot,\cdot)\}$
- $(\{m'_n\}_{n'=1}^{N'}, \mathit{state}) \leftarrow \mathcal{A}^{\mathcal{O}}(\mathsf{gpk}, \mathsf{opk})$
- $\mathcal{A}(\mathit{state})$ interacts with $\text{Verify}(\mathsf{opk}, \{m'_n\}_{n'=1}^{N'})$. The view of the protocol is denoted *view* and the response of the Verify algorithm is denoted d
- $j \leftarrow \text{Extract}(\mathsf{ek}, \mathit{view})$
- Return 1 if $d = 1$ and $\exists n' \in [1, N']$ s.t. $\forall(\{m_n\}_{n=1}^{N}) \in \mathsf{reg}[j]$, $m_{n'} \notin \{m_n\}_{n=1}^{N}$.

A scheme $\mathcal{AC}$ is $(t, \ell_{\max}, q_S, \epsilon)$-unforgeable if the probability for an adversary $\mathcal{A}$ to win the Unforgeability(λ) game in time t, with no more than $\ell_{\max}$ messages in each list of messages, after q_S queries to the SndToO and GetCred oracles is at most ϵ. A scheme $\mathcal{AC}$ is said *unforgeable* if ϵ is negligible (as a function of λ) for all polynomial-time $\mathcal{A}$.

Anonymity. Regarding the anonymity, the aim of the adversary is to distinguish between two chosen honest users which one is showing his credentials. For this purpose, the adversary has access to a *left-or-right* LoR challenge oracle. Moreover the adversary can interact with honest users as a corrupted organization in order to issue them some credentials or as a corrupted verifier with the help of the SndToU oracles. The formal anonymity experiment Anonymity$_{\mathcal{A}}(\lambda)$ is given as follows.

- $(\mathsf{gpk}, (\mathsf{opk}, \mathsf{osk})) \leftarrow \text{Setup}(1^\lambda)$
- $b \xleftarrow{\$} \{0,1\}$
- $\mathcal{O} \leftarrow \{\text{AddU}(\cdot), \text{USK}(\cdot), \text{CrptU}(\cdot), \text{SndToU}_{\text{Issue}}(\cdot,\cdot), \text{SndToU}_{\text{Verify}}(\cdot,\cdot), \text{LoR}(b,\cdot,\cdot,\cdot)\}$
- $b' \leftarrow \mathcal{A}^{\mathcal{O}}(\mathsf{gpk}, (\mathsf{opk}, \mathsf{osk}))$
- Return 1 if $b = b'$.

A scheme $\mathcal{AC}$ is $(t, \ell_{\max}, q_C, \epsilon)$-*anonymous* if the probability for an adversary $\mathcal{A}$ to win the Anonymity(λ) game in time t, with no more than $\ell_{\max}$ messages in each list of messages, after q_C queries to the LoR oracle is at most ϵ. A scheme $\mathcal{AC}$ is said *anonymous* if ϵ is negligible (as a function of λ) for all polynomial-time $\mathcal{A}$.

3. OUR NEW PRIVACY-AWARE MODEL FOR SANITIZABLE SIGNATURES

We now formalize sanitizable signature schemes and their security according to the considerations we discussed in introduction. Our goal is to model the security requirements we need in order to use sanitizable signatures as anonymous credentials. Our work is based on [6, 11, 7, 12]. A *sanitizable signature scheme* $\mathcal{SAN}$ is given by the following seven algorithms involving one signer and several sanitizers. The verification and judge procedures only involve public data.

Admissible modifications. As in [6, 11, 7], admissible modifications are modelled by two functions adm and mod. adm contains a description of (i) a division of a message $m \in \{0,1\}^*$ of length t into N blocks of respective lengths $\{t_n\}_{n=1}^{N}$ such that $\sum_{n=1}^{N} t_n = t$, (ii) a subset $A \subseteq [1, N] \subseteq \mathbb{N}$ indicating the indexes of the modifiable blocks. adm being fixed, the function fixadm maps a message m to the its fixed part m_{FIX}, *i.e.* the concatenation of all non-admissible blocks. By misuse of notation, A is also denoted adm.

mod maps a message m to a modified message $m' = \text{mod}(m)$ and is essentially a set of pairs (n, m'_n) specifying that the block n has to be replaced by the message m'_n. adm(mod) $\in \{0,1\}$ indicates whether the modification instructions mod matches the admissible modifications adm.

Setup. This algorithm takes as input a security parameter λ and outputs global parameters gpk, implicitly given to all algorithms as auxiliary input. A secret key tsk for tracing is produced as well. Finally, the key pair $(\mathsf{spk}, \mathsf{ssk})$ for the signer is also produced.

SanKG. This algorithm, on input $j \in \mathbb{N}$, produces the key pair (denoted $(\mathsf{pk}_{\mathsf{san}}[j], \mathsf{sk}_{\mathsf{san}}[j])$) for the j-th sanitizer.

Sign. This algorithm takes as input the secret signing key ssk, a sanitizer public key $\mathsf{pk}_{\mathsf{san}}[j]$, a message $m \in \{0,1\}^*$ and a description adm *w.r.t.* m. It outputs a signature σ (or an error message $\perp$). We assume that adm can always be recovered from a signature σ.

Sanitize. This algorithm takes as input a sanitizer secret key $\mathsf{sk}_{\mathsf{san}}[j]$, the signer public key spk, a message $m \in$

$\{0,1\}^*$, a signature σ and the modification instructions MOD the sanitizers wants to carry out (according to ADM). It outputs a message m' and a signature σ' (or an error message $\perp$).

VERIFY. This algorithm takes as input the signer public key spk, a message m and a signature σ. It outputs a bit $b \in \{0,1\}$.

FINDSAN. This algorithm takes as input the tracing key tsk, the signer public key spk, a message m and a signature σ. It outputs the index $j \in \mathbb{N}$ of a sanitizer and a proof τ ($j = 0$ means that the algorithm can not trace the signature).

JUDGE. This algorithm takes as input the signing public key spk, a sanitizer public key $\mathsf{pk_{san}}[j]$, a message m, a signature σ and a proof τ. It outputs a bit $b \in \{0,1\}$, indicating whether the tracing is correct.

Set of Admissible Values. As we need the `LimitSet` extension [11], we introduce the set $\mathcal{V} := \{V_i \subseteq \{0,1\}^{t_i} \mid i \in \text{ADM}\}$. Each V_i defines the set for the modifications that can be done by the sanitizer for the block m_i. Following [11], the SIGN and SANITIZE procedures are not modified, but the definition ADM is modified in order to contain the public set $\mathcal{V}$.

SECURITY PROPERTIES. We now focus on the security properties related to our notion of sanitizable signatures. It uses several oracles which mostly corresponds to the execution of the above procedures (we do not detail them since they exactly correspond to those given above). The sole exception is the LOR oracle which is formally described below.

$\text{LOR}(b, \mathsf{spk}, (j_0, m_0, \sigma_0, \text{MOD}_0), (j_1, m_1, \sigma_1, \text{MOD}_1))$. This oracle first stops if ADM_0 and ADM_1 (contained in the signatures) are different, if the signatures are not valid, or if $(m_0, \text{MOD}_0, \text{ADM}_0) \not\equiv (m_1, \text{MOD}_1, \text{ADM}_1)$ (which means that the resulting message, after modification, are different). It next sanitizes the message m_b, according to $\mathsf{sk_{san}}[j_b]$ to obtain (m'_b, σ'_b). It finally returns (m'_b, σ'_b).

Correctness. The correctness property states that honestly generated (*signing correctness*) and sanitized (*sanitizing correctness*) signatures will be accepted by the verifier, and that honestly generated proofs on valid signatures (*proof correctness*) will be accepted by the judge. The latter implies that honestly generated signatures will trace to the actual designated sanitizer.

Traceability. The traceability property states that all valid signatures should trace to a particular sanitizer. An adversary against this game aims at forging a valid signature such that the tracing procedure will not identify the designated sanitizer or will not be able to prove its charge.

- $(\mathsf{gpk}, \mathsf{tsk}, (\mathsf{ssk}, \mathsf{spk})) \leftarrow \text{SETUP}(1^\lambda)$
- $(m^*, \sigma^*) \leftarrow \mathcal{A}^{\text{SIGN}(\mathsf{ssk},\cdot,\cdot,\cdot)}(\mathsf{gpk}, \mathsf{tsk}, \mathsf{spk})$
- $(j^*, \tau^*) \leftarrow \text{FINDSAN}(\mathsf{tsk}, \mathsf{spk}, m^*, \sigma^*)$
- Return 1 if
  - $\text{VERIFY}(\mathsf{spk}, m^*, \sigma^*) = 1$ and
  - $j^* = 0$ or $\text{JUDGE}(\mathsf{spk}, \mathsf{pk_{san}}[j^*], m^*, \sigma^*), \tau^*) = 0$.

Immutability. The immutability property states that a sanitizer will not sanitize a part of the message he is not allowed to sanitize. In our case, this can happen either if a modification on a fixed part has been done or if a modification on a modifiable part has been done with a disallowed value.

- $(\mathsf{gpk}, \mathsf{tsk}, (\mathsf{ssk}, \mathsf{spk})) \leftarrow \text{SETUP}(1^\lambda)$
- $(m^*, \sigma^*) \leftarrow \mathcal{A}^{\text{SIGN}(\mathsf{ssk},\cdot,\cdot,\cdot)}(\mathsf{gpk}, \mathsf{tsk}, \mathsf{spk})$
- $(j^*, \tau^*) \leftarrow \text{FINDSAN}(\mathsf{tsk}, \mathsf{spk}, m^*, \sigma^*)$
- Return 1 if
  - $\text{VERIFY}(\mathsf{spk}, m^*, \sigma^*) = 1$ and
  - $\text{JUDGE}(\mathsf{spk}, \mathsf{pk_{san}}[j^*], m^*, \sigma^*, \tau^*) = 1$ and
  - $\forall (\mathsf{pk_{san}}[j], m_i, \text{ADM}_i)$ queried to the SIGN oracle s.t. $\mathsf{pk_{san}}[j^*] = \mathsf{pk_{san}}[j]$, $m^* \notin \{\text{MOD}(m_i) | \text{MOD} \subseteq \text{ADM}_i\}$.

Unforgeability. The unforgeability property states that nobody is able to forge a valid signature without the signing secret key, nor any sanitizer's secret key. An adversary against this property is given two challenge keys, a signer public key and a sanitizer public key, together with oracles for signing and sanitizing.

- $(\mathsf{gpk}, \mathsf{tsk}, (\mathsf{ssk}, \mathsf{spk})) \leftarrow \text{SETUP}(1^\lambda)$
- $(\mathsf{pk_{san}}[j^*], \mathsf{sk_{san}}[j^*]) \leftarrow \text{SANKG}()$
- $\mathcal{O} \leftarrow \{\text{SIGN}(\mathsf{ssk}, \cdot, \cdot, \cdot), \text{SANITIZE}(\mathsf{sk_{san}}[j^*], \cdot, \cdot, \cdot, \cdot)\}$
- $(m^*, \sigma^*) \leftarrow \mathcal{A}^{\mathcal{O}}(\mathsf{gpk}, \mathsf{tsk}, \mathsf{spk}, \mathsf{pk_{san}}[j^*])$
- $(j, \tau) \leftarrow \text{FINDSAN}(\mathsf{tsk}, \mathsf{spk}, m^*, \sigma^*)$
- Return 1 if
  - $j = j^*$ and
  - $\text{VERIFY}(\mathsf{spk}, m^*, \sigma^*) = 1$ and
  - $\forall (\mathsf{pk_{san}}[j], m_j)$ queried to the SIGN oracle, we have $(\mathsf{pk_{san}}[j^*], m^*) \neq (\mathsf{pk_{san}}[j], m_j)$ and
  - m^* has never been queried to the SANITIZE oracle.

Remark 1. The accountability properties [6, 11, 7] state that signers and sanitizers are responsible for the signatures they produced but we do not need them for the security of our main construction. As long as we do not have an opening algorithm in the anonymous credential system, we do not need the accountability properties.

Unlinkability. The unlinkability property states that it is infeasible to distinguish sanitized signatures that have been produced from the same message and / or on the same sanitizer. We only have a trace-restricted unlinkability notion, since an adversary could easily win the game with a trace oracle on her challenge. The adversary is given a *left-or-right* oracle described above. Moreover, the adversary may add new sanitizers in the system thanks to the SANKG oracle.

- $(\mathsf{gpk}, \mathsf{tsk}, (\mathsf{ssk}, \mathsf{spk})) \leftarrow \text{SETUP}(1^\lambda)$
- $b \xleftarrow{\$} \{0,1\}$
- $\mathcal{O} \leftarrow \{\text{SANKG}(), \text{SIGN}(\mathsf{ssk}, \cdot, \cdot, \cdot), \text{SANITIZE}(\cdot, \mathsf{spk}, \cdot, \cdot, \cdot), \text{LOR}(b, \mathsf{spk}, \cdot, \cdot), \text{FINDSAN}(\mathsf{tsk}, \mathsf{spk}, \cdot, \cdot)\}$
- $b' \leftarrow \mathcal{A}^{\mathcal{O}}(\mathsf{gpk}, \mathsf{spk})$
- Return 1 if
  - $b = b'$ and
  - $\mathcal{A}$ has not queried any (m', σ') output by the LOR oracle to the FINDSAN oracle.

Remark 2. We could ask for a stronger notion than the existing one in the state of the art (cf. [7]), the notion of *strong unlinkability*. According to this notion, the sanitizable signatures must be unlinkable even if the adversary is given the secret signing key. Formally, the game is the same as the standard unlinkability game, except that the adversary is given the secret signing key (and, in this case, a SIGN oracle is no more useful).

Remark 3. The transparency property [6, 11, 7] states that it is infeasible to distinguish a sanitized signature from a freshly generated signature. In our case, this is not a matter of concern if a verifier notices that a value has been hidden.

4. FROM SANITIZABLE SIGNATURES TO ANONYMOUS CREDENTIALS

We are now in possession of all the elements we need to build anonymous credentials from sanitizable signatures.

4.1 A Black-box Reduction

Given a sanitizable signature scheme $\mathcal{SAN}$ as described in the above section, we define the following anonymous credential system $\mathcal{AC}$. We consider in the following that the character # stands for predefined non relevant symbol that will be used by sanitizers to hide an information.

SETUP AND KEY GENERATION. The $\mathcal{AC}$.SETUP procedure runs the $\mathcal{SAN}$.SETUP procedure. We thus set the global parameter gpk as that from the sanitizable signature scheme, the organization key pair $(\mathsf{opk}, \mathsf{osk})$ corresponds to the one of the signer (output by $\mathcal{SAN}$.SETUP). Next, each user can execute the $\mathcal{AC}$.USERKG by using exactly the $\mathcal{SAN}$.SANKG of the sanitizable signature scheme, becoming a sanitizer.

CREDENTIAL ISSUING. On input the public key $\mathsf{upk}[j]$ of a requesting user and a list of messages / attributes $\{m_n\}_{n=1}^{N} \in \{0,1\}^*$ (m_n of length t_n) related to this user, the organization O specifies $\text{ADM} \subseteq [1, N]$ at his convenience[3]. A specific field with index n is next chosen. It will be used by the verifier to verify the freshness of the SHOW / VERIFY protocol.

Afterwards, for all admissible blocks with an index $n \neq \mathsf{n}$, the signer defines a set of admissible values $\mathcal{V}_n$ which contains 2 elements: a first element corresponding to the true sub-message m_n he is signing (*i.e.* the user attribute) and a second element corresponding to some predefined non relevant symbol (like several # characters) of adequate length (a black value for an admissible block is always assumed to be allowed). Then, O produces the signature

$$\sigma := \mathcal{SAN}.\text{SIGN}(\mathsf{osk}, \mathsf{upk}[j], m, \text{ADM})$$

and sends σ to U. User U records σ as his credential C.

CREDENTIAL SHOWING. We consider a verifier who aims at checking that user U_j has previously obtained credentials on some attributes $\mathbb{M} = \{m'_n\}_{n=1}^{N'}$. Let us assume that user U_j obtained a credential C (a sanitizable signature) on some attributes $\{m_n\}_{n=1}^{N}$ such that $\mathbb{M} \subseteq \{m_n\}_{n=1}^{N}$.

The verifier first sends a random value r to user U_j, depending of the context of the verification, like the concatenation of some verifier's data and the current time. Then the user U_j sanitizes his credential according to

$$\text{MOD} := \{(\mathsf{n}, \mathsf{r})\} \cup \left(\bigcup_{n' | m'_n \notin \mathbb{M}} \{(n', \#)\}\right),$$

and computes the pair

$$(m', \sigma') \leftarrow \mathcal{SAN}.\text{SANITIZE}(\mathsf{usk}[j], \mathsf{opk}, m, \sigma, \text{MOD}).$$

In a nutshell, the user sanitizes the attributes he wants to hide (replacing them by some # characters) and set the submessage with index n to the random challenge r. The user finally sends (m', σ') to the verifier who checks the resulting signature by testing whether $\mathcal{SAN}.\text{VERIFY}(\mathsf{opk}, m', \sigma') = 1$ (and verifying the use of value r).

[3]In practice, static fields are stated to non admissible while the true user attributes are stated to admissible (see our example in the introduction).

4.2 Security Analysis

The security of the resulting construction relies on the underlying building block: we prove the security of the $\mathcal{AC}$ scheme under the assumption that $\mathcal{SAN}$ is a secure sanitizable signature scheme in the sense of Section 3.

4.2.1 *From $\mathcal{SAN}$.{ Traceability, Immutability, Unforgeability } to $\mathcal{AC}$.Unforgeability*

First of all, we define the extractor of the unforgeability experiment to be the execution of the tracing algorithm FINDSAN. In our scheme, a showing protocol corresponds to the showing of a modified document (m', σ'). Given such a pair, the FINDSAN algorithm is able to return the identity of the designated sanitizer, *i.e.* the identity of the user.

Let now $\mathcal{A}$ be an adversary against the unforgeability of the $\mathcal{AC}$ scheme. The winning condition for the experiment is: for j returned by ESETUP, for all $\{m_n\}_{n=1}^{N} \in \mathsf{reg}[j]$, there exists $\mathsf{n}' \in [1, N']$ s.t. $m_{\mathsf{n}'} \notin \{m_n\}_{n=1}^{N}$. Informally, no user (honest or not) has obtained a certification on the whole specified list. In other words, for each delivered credential, there is at least an attribute which is not in the list.

Let us consider the following cases, in excluding order.

- Type I. The (m', σ') pair is not traceable. We construct an adversary against the traceability property of the sanitizable signature scheme.
- Type II. Otherwise, let assume that the tracing of σ' returned an index j^*. If $j^* \in HU$ is honest, then we construct an adversary against the unforgeability property of the sanitizable signature scheme.
- Type III. Otherwise, $j^* \in CU \cup KU$ is dishonest or his keys have been compromise. We then construct an adversary against the immutability property of the sanitizable signature scheme.

Let us describe each adversary in more details. The challenger maintains the HU, KU and CU lists of users.

Type I. $\mathcal{A}$ successfully pass a showing protocol, but the signature (m', σ') is not traceable, meaning that the extractor is not able to identify a user underlying the view of the protocol. This means that the tracing algorithm is not able to trace the signature (m', σ'). We construct an adversary $\mathcal{B}$ against the traceability of $\mathcal{SAN}$. The ADDU oracle is simulated by generating fresh key pairs. The SNDTOO oracle, together with the GETCRED oracle, on list $\{m_n\}_{n=1}^{N}$ are simulated thanks to $\mathcal{B}$'s own SIGN oracle on a message m constructed following the issuing protocol. The SNDTOU$_{\text{VERIFY}}$ oracle is simulated by computing the authentication ($\mathcal{B}$ owns users' keys). When $\mathcal{A}$ successfully pass the protocol, (m', σ') is not traceable, and the game is won.

Type II. $\mathcal{A}$ successfully pass the showing protocol, and the FINDSAN algorithm returns j^* such that $j^* \in HU$. We construct an adversary $\mathcal{B}$ against the unforgeability property of the $\mathcal{SAN}$ scheme. $\mathcal{B}$ gets two challenge keys (a signer one and a sanitizer one). $\mathcal{B}$ generates other keys and guesses that the forgery will concern his challenge key. The SNDTOO and GETCRED oracles on lists $\{m_n\}_{n=1}^{N}$ are simulated thanks to the own SIGN challenger's oracle on a message m built as in the issuing protocol. The $\textsc{SndToU}_{\textsc{Verify}}$ oracle is simulated by $\mathcal{B}$ with his own sanitization oracle (or computes the response himself if he knows the keys). Let now assume that $\mathcal{A}$ successfully pass the showing protocol. Since j^* has not got the whole specified credential, then (from the winning condition) the $\textsc{Sign}$ oracle has not been queried on the list. Since the sanitization oracle is queried only on previously generated signatures, the list did not appear in the sanitization oracle queries. We then got a valid and non-trivial forgery (m', σ') and the game is won.

Type III. $\mathcal{A}$ successfully pass the showing protocol, and the $\textsc{FindSan}$ algorithm returns j^* such that $j^* \in CU \cup KU$. We construct an adversary $\mathcal{B}$ against the immutability property of the $\mathcal{SAN}$ scheme. The $\textsc{AddU}$, $\textsc{SndToO}$, $\textsc{GetCred}$ oracles on attributes lists $\{m_n\}_{n=1}^N$ and the $\textsc{SndToU}_{\textsc{Verify}}$ oracle are simulated as in the two games above. When $\mathcal{A}$ successfully pass the showing protocol, the winning condition says that for all $\{m_n\}_{n=1}^N \in \mathsf{reg}[j^*]$, there exists $\mathsf{n}' \in [1, N']$ such that $m_{\mathsf{n}'} \notin \{m_n\}_{n=1}^N$. We want to show that the immutability winning condition is satisfied. The last condition says that for all $(j_k, m_k, \textsc{adm}_k)$ queried during the attack, we have $\mathsf{pk}_{\mathsf{san}}[j^*] \neq \mathsf{pk}_{\mathsf{san}}[j_k]$ or $m^* \notin \{\textsc{mod}(m_k) \mid \textsc{mod} \text{ s.t. } \textsc{adm}_k(\textsc{mod}) = 1\}$. This means that, when the $\mathsf{pk}_{\mathsf{san}}[j^*]$ key appeared in a query to the $\textsc{Sign}$ oracle, the message m^* did not come from the modification of one of the m_k queried to the oracle. One can easily be convinced of that, since for all obtained credential (*i.e.* call to the $\textsc{Sign}$ oracle), there exists m^* which was not involved. These attributes can not be the masking value, since the latter is always allowed. The game against the immutability is then won.

In the beginning of the game, the challenger chooses one of the three game above with uniform probability. Therefore, the security loss is $1/3|\mathcal{U}|$, which is still negligible if the underlying probabilities are negligible. □

4.2.2 From $\mathcal{SAN}$.Unlinkability to $\mathcal{AC}$.Anonymity

The anonymity property of the anonymous credential system directly relies on the unlinkability fo the sanitizable signature scheme. Let $\mathcal{A}$ be an adversary against the anonymity of the $\mathcal{AC}$ scheme. We construct an adversary $\mathcal{B}$ against the unlinkability of the $\mathcal{SAN}$ scheme. Regarding the *left-or-right* challenge, $\mathcal{A}$ asks for a showing protocol from the user j_0 or j_1 on a list $\{m'_{n'}\}_{n'=1}^{N'}$ such that both users already got a credential on them. Since both credentials have already been issued, $\mathcal{B}$ retrieves them and computes a query to his own oracle with the corresponding modification instructions $\textsc{mod}$. Finally, $\mathcal{A}$ returns a bit b that $\mathcal{B}$ returns to her own challenge. The probability that $\mathcal{B}$ wins is that of $\mathcal{A}$.

We note that, if $\mathcal{SAN}$ is a strongly unlinkable sanitizable scheme, then the credential system is anonymous against a corrupted organization. If $\mathcal{SAN}$ is an unlinkable sanitizable scheme, then the credential system is proven anonymous only with respect to an honest organization, and the adversary *is not given* the organization secret key. In other words, the anonymity level of the credential system depends on the unlinkability level of the underlying signature scheme.

5. A CONCRETE CONSTRUCTION

We now give a concrete sanitizable signature scheme that meets the features described in Section 3 and that can thus directly be used to design a secure anonymous credential system as shown just above.

5.1 Useful tools

Let us consider an asymmetric bilinear environment $(p, \mathbb{G}_1, \mathbb{G}_2, \mathbb{G}_T, e)$ where p is prime, $\mathbb{G}_1$, $\mathbb{G}_2$ and $\mathbb{G}_T$ three groups or order p, and e a map $e : \mathbb{G}_1 \times \mathbb{G}_2 \to \mathbb{G}_T$ such that (bilinearity) for all $g_1, g_2 \in \mathbb{G}_1 \times \mathbb{G}_2$ and $a, b \in \mathbb{Z}_p$, we have $e(g_1^a, g_2^b) = e(g_1, g_2)^{ab} = e(g_1^b, g_2^a)$ and (non-degenerate) for all $g_1, g_2 \in \mathbb{G}_1 \setminus \{\mathbf{1}_{\mathbb{G}_1}\} \times \mathbb{G}_2 \setminus \{\mathbf{1}_{\mathbb{G}_2}\}$, we have $e(g_1, g_2) \neq \mathbf{1}_{\mathbb{G}_T}$.

Proofs of knowledge. Our scheme makes use of zero-knowledge proofs of knowledge (ZKPK) which correspond to interactive protocols during which a prover convinces a verifier that he knows a set $(\alpha_1, \ldots, \alpha_q)$ of secret values verifying a given relation R without revealing any information about the known secrets. We denote by $\mathrm{PoK}\{\langle \alpha_1, \ldots, \alpha_q \rangle : \mathsf{R}(\alpha_1, \ldots, \alpha_q)\}$ such a proof of knowledge. A signature of knowledge, denoted SoK in the following and taking on input a message m, is a ZKPK that has been transformed into a signature scheme, using the Fiat-Shamir heuristic [19], and secure in the random oracle model[4].

Pseudo-random functions. We will also need pseudo-random functions (PRF). A PRF is given by a pair of algorithms $\mathcal{PRF} := \{\textsc{KeyGen}, \mathrm{PRF}\}$ (*cf.* Appendix A.2).

Bilinear signatures. Finally, the signer is in the following able to produce a particular signature related to the Boneh-Boyen one [3]. Let $\mathcal{PRF} := \{\textsc{KeyGen}, \mathrm{PRF}\}$ be a pseudo-random function and $(p, \mathbb{G}_1, \mathbb{G}_2, \mathbb{G}_T, e)$ be an asymmetric bilinear environment. During the $\textsc{KeyGen}$ procedure, the signer picks $g_1, u_1, \ldots, u_\ell, v \xleftarrow{\$} \mathbb{G}_1$, $g_2 \xleftarrow{\$} \mathbb{G}_2$ and $\gamma \xleftarrow{\$} \mathbb{Z}_p$, and next computes $X := {g_2}^\gamma$ and $Y := {h_2}^\gamma$. Next, $\mathsf{pk} := (X, Y)$, $\mathsf{sk} := \gamma$ and parameters are

$$(p, \mathbb{G}_1, \mathbb{G}_2, \mathbb{G}_T, e, g_1, u_1, \ldots, u_\ell, v, g_2, h_2)$$

During the $\textsc{Sign}$ procedure, the signer generates two scalars $r \leftarrow \mathrm{PRF}(k, 0\|m_1\|m_2\| \ldots \|m_\ell)$ and $s \leftarrow \mathrm{PRF}(k, 1\|m_1\|m_2 \| \ldots \|m_\ell)$ thanks to the pseudo-random function, sets $S := (g_1 {u_1}^{m_1} \cdots {u_\ell}^{m_\ell} v^s)^{\frac{1}{\gamma + r}}$ and returns $\sigma := (S, r, s)$.
The $\textsc{Verify}$ algorithm consists in checking that

$$e(S, {g_2}^r X) \stackrel{?}{=} e(g_1 {u_1}^{m_1} \cdots {u_\ell}^{m_\ell} v^s, g_2).$$

This scheme is studied in details in Appendix A.3.

5.2 Algorithms

5.2.1 Setup

Global parameters. An asymmetric bilinear group environment $(p, \mathbb{G}_1, \mathbb{G}_2, \mathbb{G}_T, e)$ is generated. Random generators g_1, u_1, u_2, $v \xleftarrow{\$} \mathbb{G}_1$ and $g_2 \xleftarrow{\$} \mathbb{G}_2$ are chosen, together with $\alpha \xleftarrow{\$} \mathbb{Z}_p$. $h_1 := {g_1}^\alpha$ and $h_2 := {g_2}^\alpha$ are computed. Let $\mathcal{H}_1 : \{0,1\}^* \longrightarrow \mathbb{Z}_p$ be a cryptographic hash function. The global parameters of our system are

$$\mathsf{gpk} := (p, \mathbb{G}_1, \mathbb{G}_2, \mathbb{G}_T, e, g_1, u_1, u_2, v, h_1, g_2, h_2, \mathcal{H}_1)$$

and the tracing key is $\mathsf{tsk} := \alpha$.

Signer's keys. The signer picks $\gamma \xleftarrow{\$} \mathbb{Z}_p$ and computes $X := {g_2}^\gamma$ and $Y := {h_2}^\gamma$ as the keys for the signature scheme. It also picks a key for the pseudo-random function $k \leftarrow \mathcal{PRF}.\textsc{KeyGen}(1^\lambda)$. The signer's public key is $\mathsf{spk} := (X, Y)$ and the corresponding secret key is $\mathsf{ssk} := (\gamma, k)$.

[4]For example, $\mathrm{SoK}\{\langle \alpha \rangle : h = g^\alpha\}(m)$ denotes a Schnorr proof of knowledge of a discrete logarithm transformed into a signature scheme by the Fiat-Shamir heuristic.

Sanitizers' keys. Finally, each sanitizer j picks $y_j \xleftarrow{\$} \mathbb{Z}_p$, sets $\mathsf{pk}_{\mathsf{san}}[j] := z_j = u_1^{y_j}$ and $\mathsf{sk}_{\mathsf{san}}[j] := y_j$. Each value (j, z_j) is registered in the table $\mathsf{pk}_{\mathsf{san}}$.

5.2.2 Structure of a signature

Let us now explain the idea behind our construction. Let us assume that a signer wants to sign a message $m \in \{0,1\}^*$ of length t, split into N blocks of length $\{t_n\}_{n=1}^{N}$ with admissible blocks $M \subseteq \{1, \dots, N\}$ and the specific field $\mathsf{n} \in \mathbb{N} \setminus M$. According to our needs, we have $\text{ADM} := \{(\mathsf{n}, \top)\} \cup \{(n, \bot) \mid n \in M\}$, which means that (i) m_n could be modified with any value $m'_\mathsf{n} \in \{0,1\}^{t_\mathsf{n}}$ while (ii) a value m_n for $n \in M$ should only be replaced, if needed, by a specific black block of message. The designated sanitizer with index j is represented by his value z_j. A sanitizable signature is next divided into three parts.

Handling the fixed part. The fixed part m_{FIX} of the message is handled by giving a signer's signature on both y_j (the sanitizer's secret) and h_{FIX} (a hash value of the fixed part), using the key generated by the signer during the set-up phase. The resulting signature corresponds to three elements (S, r, s) where S is a group element and where the randomness (r, s) are derived via the $\mathcal{PRF}$.

However, to reach unlinkability, we should not give directly the signature and we use for this purpose encryption techniques. We thus compute an ElGamal encryption [21] $(B_{0:1}, B_{0:2})$ of the S part of the signature. During the SIGN procedure, the signer picks at random $t_0 \in \mathbb{Z}_p$ and generates $c_1 \leftarrow \mathcal{H}_1(m_{\text{FIX}} \| \text{ADM})$. It next generates the randomness of the signature scheme as $r \leftarrow \text{PRF}(k, 0\|0\|m_{\text{FIX}}\|\text{ADM})$ and $s \leftarrow \text{PRF}(k, 0\|1\|m_{\text{FIX}}\|\text{ADM})$, and computes the signature $S := (g_1 z_j u_2^{c_1} v^s)^{\frac{1}{\gamma+r}}$. The ElGamal ciphertext of S is finally $(B_{0:1} := g_1{}^{-t_0}, B_{0:2} := S h_1{}^{t_0})$ and is sent to the sanitizer (while the initial signature is not). During the SANITIZE procedure, as the ElGamal encryption scheme is randomizable, the sanitizer can easily randomize such a couple $(B_{0:1}, B_{0:2})$, which provides us the unlinkability property. For this purpose, the sanitizer chooses $t'_0 \in \mathbb{Z}_p$ and computes $(B'_{0:1} := B_{0:1} g_1{}^{-t'_0}, B'_{0:2} := B_{0:2} h_1{}^{t'_0})$.

Handling the admissible part. As an admissible sub-message is necessarily an accepted sub-message, the sanitizer may not modify it but the verifier should be convinced that it has been validated by the signer. For this purpose, we use the same technique as for the fixed part, but for each admissible sub-messages m_n, independently of each other. Next each admissible block is signed and next encrypted, as above (again to reach the unlinkability property). During the SIGN procedure, the signer picks, for all $n \in M$, a value $t_i \xleftarrow{\$} \mathbb{Z}_p$, computes $h_n \leftarrow \mathcal{H}_1(m_{\text{FIX}}\|\text{ADM}\|n\|m_n)$ thanks to the hash function and $r_n \leftarrow \text{PRF}(k, n\|0\|m_{\text{FIX}}\|\text{ADM}\|m_n)$ and $s_n \leftarrow \text{PRF}(k, n\|1\|m_{\text{FIX}}\|\text{ADM}\|m_n)$ thanks to the pseudo-random function. The signature is $G_n := (g_1 z_j u_2{}^{h_n} v^{s_n})^{\frac{1}{(\gamma+r_n)}}$ and the resulting ciphertext is $(B_{n:1} := g_1{}^{-t_n}, B_{n:2} := G_n h_1{}^{t_n})$. All these steps are repeated for all admissible parts of the message. During the SANITIZE procedure, the modification instructions MOD are given by the set $\{(\mathsf{n}, m'_\mathsf{n})\} \cup \{(n, \bot)\}_{n \in J}$, meaning that the sanitizer wants to replace m_n by m'_n and hide a subset $J \subseteq M$. There are consequently two cases for admissible parts, as explained above.

In the first case, the sanitizer is able to modify as desired an admissible part of the message. As we do not need the transparency property (contrary to existing sanitizable signature constructions), the sanitized message-signature pair can still include the ElGamal ciphertext of the signer's signature with the initially signed sub-message m_i. The new sub-message m'_n should however be signed by the authorized sanitizer, which will be done in the proof of validity of the whole signature below. Moreover, as we need unlinkability, the ElGamal ciphertext should be randomized, as for the fixed part above. Next, for all $n \in M \setminus J$, the sanitizer again randomizes each ElGamal ciphertext by choosing at random $t'_n \in \mathbb{Z}_p$ and computing $(B'_{n:1} := B_{n:1} g_1{}^{-t'_n}, B'_{n:2} := B'_{n:2} h_1{}^{t'_n})$.

In the second case, the sanitizer can only replace the admissible sub-message by a sequence of $|m_n|$ times the '#' character. As this black value is, by convention, always allowed, we do not need to supply a proof that the value is indeed authorized (contrary to [11]). However, as the initially signed sub-message should not be retrieved, we ask the sanitizer to remove the corresponding ElGamal ciphertext. Thus, for all $n \in J$, $(B_{n:1} \| B_{n:2})$ is removed from the sanitized signature. Any other modification of such part of the message will be obviously refused by the verifier who has access to the variable ADM.

Proof of validity of the whole signature. It finally remains to prove the validity of the signature. In fact, from the sanitizer's point of view, we only need to prove the knowledge of y_j in the above relations to prove that this signature comes from an authorized entity (since y_j is signed S). The verification procedure of a signature whether coming from SIGN or SANITIZE should be unique. However, in our case, the signer does not know the secret value y_j. Our solution consists in using a proof of the "or" statement [18, 27].

5.2.3 Signing, sanitizing, verifying and tracing

Putting all these elements together, a signature on a message m *w.r.t.* admissible modifications ADM is a signature of knowledge, on the message

$$\mathsf{M} := B_{0:1} \| B_{0:2} \| r \| s \| \bigcup_{n \in M} (B_{n:1} \| B_{n:2} \| r_n \| s_n) \| m \| \text{ADM} \| X \| Y$$

that should be produced as:

$$U := \text{SoK}\Big\{ \langle \rho \rangle \ : \ e(g_1, g_2)^\rho = e(g_1, X) \ \vee$$
$$\Big(e(u_1, g_2)^\rho = \frac{e(B_{0:2}, g_2{}^r X) e(B_{0:1}, h_2{}^r Y)}{e(g_1 u_2{}^{c_1} v^s, g_2)} \wedge \forall n \in M,$$
$$e(u_1, g_2)^\rho = \frac{e(B_{n:2}, g_2{}^{r_n} X) e(B_{n:1}, h_2{}^{r_n} Y)}{e(g_1 u_2{}^{h_n} v^{s_n}, g_2)} \Big) \Big\} (\mathsf{M})$$

SIGNING. During the SIGN procedure, the signer, knowing γ such that $X = g_2{}^\gamma$ can perform the first part of the proof.

SANITIZING. During the SANITIZE procedure, the sanitizer, knowing y_j can perform the second part of the "or" proof, but this time with the message $\mathsf{M}' :=$

$$B'_{0:1} \| B'_{0:2} \| r \| s \| \| \bigcup_{n \in M \setminus J} (B'_{n:1} B'_{n:2} \| r_n \| s_n) \| m' \| \text{ADM} \| X \| Y.$$

VERIFICATION. A sanitizable signature is next composed of the ElGamal ciphertexts together with the randomness $r, r_1, \dots, s, s_1, \dots$ and signature of knowledge U. The verification procedure simply consists in verifying the signature U.

TRACING. The authority in possession of $\mathsf{tsk} := \alpha$ can decrypt the ElGamal ciphertext $(B_{0:1}, B_{0:2})$ to retrieve $S^* = B_{0:2}{B_{0:1}}^{\alpha}$, and use the table $\mathsf{pk}_{\mathsf{san}}$ to retrieve an entry (j, z_j) such that $e(S^*, {g_2}^r X) = e(g_1 z_j {u_2}^{c_1} v^s, g_2)$. The proof of validity of such a tracing is done by proving that the decryption has been correctly done, using the knowledge of α.

SECURITY . The following theorem is proved in Appendix A.

Theorem 1 *The $\mathcal{SAN}$ scheme is a traceable, immutable, unforgeable and unlinkable signature scheme in the sense of the Section 3 if the* DDH *problem is hard in $\mathbb{G}_1$ and the* SDH *problem is hard in $(\mathbb{G}_1, \mathbb{G}_2)$, in the random oracle model.*

5.3 Comparisons

We have now introduced all the useful elements to make a quantitative comparison between related works and an anonymous credential system based on the sanitizable signature of this section. We then give in Figure 1 several elements of comparison between our solutions and [25, 24, 14]. "This1" is the scheme described above, while "This2" is a variant with some optimizations described in Appendix B.

Underlying algebraic structures may be RSA environment [24], $\mathbb{Z}_n^*$ with a subgroup of prime order q [25], or bilinear groups $(\mathbb{G}_1, \mathbb{G}_2)$ over a base field $\mathbb{F}_p$ [14]. L denotes the maximum number of attributes issued by an authority into a single credential. Bandwidth columns denote the exchanged quantity of data during the protocols runnings, whereas the other columns denote a cost either in memory (size of a public key or a certificate) or in computation (number of operations in the underlying algebraic structure). We give asymptotic complexity as well as some concrete sizes in bits.

As said before, Figure 1 explicitly shows that our solutions are direct signatures, and thus, the issuing process is for free for the user. Moreover, in our variant This2, our multiple-use credential is rather short, compared to existing schemes. UProve's technology is based on blind signatures, so brings very efficient but single-use credentials. The other solutions bring multiple-use credentials.

6. CONCLUSION

In this paper, we proposed a new way to design anonymous credential systems. As an interesting remark, one can notice that an existing way to design anonymous credential (as stated before) is to use group signatures. The question now is the exact link between these primitives. How do articulate each other group signatures, (strongly) unlinkable sanitizable signatures and anonymous credentials ?

Acknowledgements

This work has been supported by the ANR-11-INS-0013 LYRICS Project. We are also grateful to Amandine Jambert for helpful discussions on sanitizable signatures, to Sherman Chow for his help on this final version and to anonymous referees for their valuable comments.

7. REFERENCES

[1] Giuseppe Ateniese, Daniel H. Chou, Breno de Medeiros, and Gene Tsudik. Sanitizable signatures. In *ESORICS '05*, pages 159–177, 2005.

[2] Mira Belenkiy, Melissa Chase, Markulf Kohlweiss, and Anna Lysyanskaya. P-signatures and noninteractive anonymous credentials. In *TCC*, pages 356–374, 2008.

[3] Dan Boneh and Xavier Boyen. Short signatures without random oracles and the sdh assumption in bilinear groups. *Journal of Cryptology*, 21(2):149–177, 2008.

[4] Stefan Brands. *Rethinking PKI and digital certificates - building in privacy.* PhD thesis, Eindhoven Institute of Technology, 1999.

[5] Christina Brzuska, Heike Busch, Özgür Dagdelen, Marc Fischlin, Martin Franz, Stefan Katzenbeisser, Mark Manulis, Cristina Onete, Andreas Peter, Ber tram Poettering, and Dominique Schröder. Redactable signatures for tree-structured data: Definitions and constructions. In *ACNS '10*, pages 87–104, 2010.

[6] Christina Brzuska, Marc Fischlin, Tobias Freudenreich, Anja Lehmann, Marcus Page, Jakob Schelbert, Dominique Schröder, and Florian Volk. Security of sanitizable signatures revisited. In *PKC '09*, pages 317–336, 2009.

[7] Christina Brzuska, Marc Fischlin, Anja Lehmann, and Dominique Schröder. Unlinkability of sanitizable signatures. In *PKC '10*, pages 444–461, 2010.

[8] Jan Camenisch, Markulf Kohlweiss, and Claudio Soriente. An accumulator based on bilinear maps and efficient revocation for anonymous credentials. In *PKC '09*, pages 481–500. Springer, 2009.

[9] Jan Camenisch and Anna Lysyanskaya. An efficient system for non-transferable anonymous credentials with optional anonymity revocation. In *EUROCRYPT '01*, pages 93–118, 2001.

[10] Jan Camenisch and Anna Lysyanskaya. Signature schemes and anonymous credentials from bilinear maps. In *CRYPTO '04*, pages 56–72, 2004.

[11] Sébastien Canard and Amandine Jambert. On extended sanitizable signature schemes. In *CT-RSA '10*, pages 179–194. Springer, 2010.

[12] Sébastien Canard, Amandine Jambert, and Roch Lescuyer. Sanitizable signatures with several signers and sanitizers. In *AFRICACRYPT '12*, pages 35–52. Springer, 2012.

[13] Sébastien Canard, Fabien Laguillaumie, and Michel Milhau. Trapdoor sanitizable signatures and their application to content protection. In *ACNS '08*, pages 258–276. Springer, 2008.

[14] Sébastien Canard and Roch Lescuyer. Anonymous credentials from (indexed) aggregate signatures. In *Digital Identity Management '11*, pages 53–62. ACM, 2011.

[15] Sébastien Canard, Roch Lescuyer, and Jacques Traoré. Multi-show anonymous credentials with encrypted attributes in the standard model. In *CANS '11*, pages 194–213. Springer, 2011.

[16] David Chaum and Jan-Hendrik Evertse. A secure and privacy-protecting protocol for transmitting personal information between organizations. In *CRYPTO '86*, pages 118–167. Springer, 1986.

[17] Sherman S. M. Chow, Yi Jun He, Lucas Chi Kwong Hui, and Siu-Ming Yiu. Spice – simple privacy-preserving identity-management for cloud environment. In *ACNS '12*, pages 526–543. Springer, 2012.

[18] Ronald Cramer, Ivan Damgard, and Berry Schoenmakers. Proofs of partial knowledge and simplified design of witness hiding protocols. In *CRYPTO '94*, pages 174–187. Springer, 1994.

	Keys (L attributes per credential)						Issuing Protocol		
	Group Structures		PK Size	Credential Size			User	Issuer	Bw.
[25]	$p = 1024, q = 128$		$O(L)$	$O(1)$	$\approx 3 \cdot \lvert G_q \rvert + 128$		$O(L) : G_q$	$O(L) : G_q$	$O(1)$
[24]	$\mathbb{Z}_n$: RSA 1024		$O(L)$	$O(1)$	$\approx 2600 + 1024$		$O(L) : \mathbb{Z}_n$	$O(L) : \mathbb{Z}_n$	$O(L)$
[14]	$\mathbb{F}_p : \lvert \mathbb{G}_1 \rvert \approx 170$		$O(1)$	$O(L)$	$170 \cdot L + \approx 510 + \lvert \mathbb{G}_2 \rvert$		$O(L) : \mathbb{G}_2$	$O(L) : \mathbb{G}_1$	$O(L)$
This1	$\mathbb{F}_p : \lvert \mathbb{G}_1 \rvert \approx 170$		$O(1)$	$O(L)$	$(2L + 2) \cdot (\lvert \mathbb{G}_1 \rvert + \lvert \mathbb{Z}_p \rvert)$		$\mathbf{O(1)}$	$O(L) : \mathbb{G}_1$	$O(L)$
This2	$\mathbb{F}_p : \lvert \mathbb{G}_1 \rvert \approx 170$		$O(L)$	$O(1)$	$\mathbf{2} \cdot \lvert \mathbb{G}_\mathbf{1} \rvert + \mathbf{3} \cdot \lvert \mathbb{Z}_\mathbf{p} \rvert$		$\mathbf{O(1)}$	$O(L) : {\mathbb{G}_1}^a$	$O(1)$
	Showing Protocol								
	User	Verifier	Bw.	User	Verifier	Bw.	User	Verifier	Bw.
	Single-use			*Single-use k-out-of-L attributes*			*K-use L-attributes*		
[25]	$O(L)$		$O(L)$	$O(k) : G_q$		$O(L) : \mathbb{Z}_q$	$O(KL) : G_q$		$O(KL)$
[24]	$O(L) : \mathbb{Z}_n$		$O(1)$	$O(L) : \mathbb{Z}_n$		$O(L-k)$	$O(L) : \mathbb{Z}_n$		$O(1)$
[14]	$O(1)$ pair.; $O(L) : \mathbb{G}_2$		$O(1)$	$O(1)$ pair.; $O(L) : \mathbb{G}_2$		$O(1)$	$O(1)$ pair.; $O(L) : \mathbb{G}_2$		$O(1)$
This1	$O(1)$	$O(L)$ pair.	$O(L)$	$O(1)$	$O(L)$ pair.	$O(k)$	$O(L) : \mathbb{G}_1$	$O(L)$ pair.	$O(L)$
This2	$O(1)$	$O(L) : \mathbb{G}_1$	$O(1)$	$O(L-k) : {\mathbb{G}_T}^b$	$O(L-k) : {\mathbb{G}_T}^b$	$O(L-k)$	$O(1)$	$O(L) : \mathbb{G}_1$	$O(1)$

[a] May be accelerated as a single multi-exponentiation (contrary to [14] and This1). [b] May be accelerated as a single multi-exponentiation.

Figure 1: Comparisons between our solution and the related works

[19] Amos Fiat and Adi Shamir. How to prove yourself: Practical solutions to identification and signature problems. In *CRYPTO'86*, pages 186–194. Springer, 1987.

[20] Georg Fuchsbauer. Commuting signatures and verifiable encryption. In *EUROCRYPT '11*, pages 224–245. Springer, 2011.

[21] Taher El Gamal. A public key cryptosystem and a signature scheme based on discrete logarithms. In *CRYPTO '84*, pages 10–18. Springer, 1984.

[22] Shafi Goldwasser, Silvio Micali, and Ronald L. Rivest. A digital signature scheme secure against adaptive chosen-message attacks. *SIAM Journal of Computing*, 17(2):281–308, 1988.

[23] Jorge Guajardo, Bart Mennink, and Berry Schoenmakers. Anonymous credential schemes with encrypted attributes. In *CANS '10*, pages 314–333. Springer, 2010.

[24] IBM. Identity mixer - Idemix, 2010.

[25] Microsoft. U-Prove community technology, 2010.

[26] Kunihiko Miyazaki, Goichiro Hanaoka, and Hideki Imai. Digitally signed document sanitizing scheme based on bilinear maps. In *ASIACCS '06*, pages 343–354. ACM, 2006.

[27] Alfredo De Santis, Giovanni Di Crescenzo, Giuseppe Persiano, and Moti Yung. On monotone formula closure of SZK. In *FOCS '94*, pages 454–465. IEEE, 1994.

APPENDIX

A. PROOF OF THEOREM 1

We now analyze the security of our concrete scheme. First of all, we state the hard problems under which the security relies. Then we give a proof of security of the underlying signature scheme and, at last, we prove the main properties.

A.1 Hard Problems

Let us fix a group $\mathbb{G}$ of prime order p and a bilinear environment $(p, \mathbb{G}_1, \mathbb{G}_2, \mathbb{G}_T, e)$. For a problem X, the X assumption in $\mathbb{G}$ says that X is hard to solve in $\mathbb{G}$.

Problem 2 (DL) *Given $(g, h) \in \mathbb{G}^2$, find α s.t. $h = g^\alpha$.*

Problem 3 (DDH) *Given $(g, g^a, g^b, g^c) \in \mathbb{G}^4$, for uniformly picked $g \xleftarrow{\$} \mathbb{G}$, and $a, b \xleftarrow{\$} \mathbb{Z}_p$, decide whether $c = ad \bmod p$ or if c was also uniformly picked in $\mathbb{Z}_p$.*

Problem 4 (XDH) *Given $\mathbb{G}_1$, $\mathbb{G}_2$ and $\mathbb{G}_T$, while the* DDH *problem could be easy in $\mathbb{G}_2$, the* XDH *assumption states that the* DDH *problem is still hard in $\mathbb{G}_1$.*

Problem 5 (q-SDH) *Given $(g_1, {g_1}^\theta, {g_1}^{\theta^2}, \ldots, {g_1}^{\theta^q}, g_2, {g_2}^\theta) \in {\mathbb{G}_1}^{q+1} \times \mathbb{G}_2^2$ for a uniformly picked $\theta \xleftarrow{\$} \mathbb{Z}_p^*$, find a pair $(c, {g_1}^{1/(\theta+c)}) \in \mathbb{Z}_p \times \mathbb{G}_1$.*

A.2 Dealing with the PRF

The security of a PRF states that for any efficient algorithm $\mathcal{B}$ the value $\left| \Pr[\mathcal{B}^{\mathrm{PRF}(k,\cdot)}(1^\lambda) = 1] - \Pr[\mathcal{B}^{\mathcal{R}}(1^\lambda) = 1] \right|$ is negligible, where the first probability is taken over $k \leftarrow \textsc{KeyGen}(1^\lambda)$ and $\mathcal{B}$'s internal coin tosses, and the second probability is over the choice of the random function $\mathcal{R} : \{0,1\}^* \to \{0,1\}^\lambda$ and $\mathcal{B}$'s randomness. We replace in the security proofs all invocation to the PRF by a truly random value in $\mathbb{Z}_p$ and maintain a list of these values in order to be consistent. Then, for any successful adversary against each property we construct a successful distinguisher against the PRF. In each case, we can decide whether an adversary has won or not. In the case of the immutability, this decidability follows from the fact that we can decide if a message m' belongs to the set of the possible resulting of a sanitization of a message m with respect to admissibility $\textsc{adm}$. Indeed, it suffices to parse the message block per block and to check whether the modification has been allowed or not.

A.3 Signing Several Messages

We study here the signature scheme, denoted $\mathcal{S}^{(\ell)}$, used in our sanitizable signature scheme and prove its unforgeability.

Theorem 6 *If the* SDH *problem is (q, t', ϵ')-hard in $(\mathbb{G}_1, \mathbb{G}_2)$, then the $\mathcal{S}^{(1)}$ signature scheme is (q_S, t, ϵ)-unforgeable as long as $t \le t' - O(q^2)$, $q_S \le q$ and $\epsilon \le \epsilon'/q$.*

Proof. Let $\mathcal{A}$ be an adversary that (q_S, N, t, ϵ)-win the EUF-CMA existential unforgeability game against chosen message attacks for signature schemes (cf. [22]). We set $q = q_S$. Let $(g_1, {g_1}^\theta, {g_1}^{\theta^2}, \ldots, {g_1}^{\theta^q}, g_2, {g_2}^\theta)$ be a q-SDH challenge, for an unknown and uniformly distributed $\theta \in \mathbb{Z}_p^*$. We build an algorithm $\mathcal{B}$ which outputs $(c, {g_1}^{1/(\theta+c)})$, for a $c \in \mathbb{Z}_p \backslash \{-\theta\}$.

Parameters. Pick $k \xleftarrow{\$} \{1, \ldots, q\}$ and scalars $r_1, s_1 \ldots, r_q$, s_q in order to simulate the signatures. For $\{r_1, \ldots, r_q\} \in \mathbb{F}_p$, define polynomials P, P_-, P_{m-} on $\mathbb{F}_p[Z]$ respectively by

$$\prod_{n=1}^{q} (Z + r_n - r_k), \prod_{\substack{n=1 \\ n \neq k}}^{q} (Z + r_n - r_k), \prod_{\substack{n=1 \\ n \neq m \\ n \neq k}}^{q} (Z + r_n - r_k).$$

Expanding P on θ, we get $P(\theta) = \sum_{n=0}^{q} a_n \theta^n$ for some $\{a_n\}_{n=0}^{q}$ depending on the r_n. Since we know ${g_1}^{\theta^n}$ by the q-SDH challenge, we are able to compute ${g_1}^{P(\theta)}$, without knowledge of θ. Set $X \leftarrow {g_2}^{\theta} {g_2}^{-r_k}$ as a challenge key. Pick $s_k, \alpha, \beta, \delta, \rho \xleftarrow{\$} \mathbb{Z}_p^*$, and compute ${g_1}' \leftarrow {g_1}^{\beta(\rho P(\theta) - s_k P_-(\theta))}$, $u \leftarrow {g_1}^{\beta \delta P(\theta)}$, $v \leftarrow {g_1}^{\beta P_-(\theta)}$, $h_2 \leftarrow {g_2}^{\alpha}$, $Y \leftarrow ({g_2}^{\theta} {g_2}^{-r_k})^{\alpha}$. Thanks to $\alpha, \beta, \delta, \rho$, the public key is correctly distributed. Finally $\mathcal{A}$ is given $(p, \mathbb{G}_1, \mathbb{G}_2, \mathbb{G}_T, e, {g_1}', u, v, g_2, h_2, X, Y)$.

Simulating $\text{SIGN}(m_n)$. We keep a counter n for the queries. If $n = k$, we set $A_k \leftarrow {g_1}^{\beta P_-(\theta)(\rho + \delta m_k)}$ and if $n \neq k$, $A_n \leftarrow {g_1}^{\beta P_{n-}(\theta)(\rho\theta + \delta\theta m_n + s_n - s_k)}$.

Forgery. $\mathcal{A}$ eventually outputs a valid and non-trivial forgery $(m_*, (A_*, r_*, s_*))$ such that

$$A_* = ({g_1}' u^{m_*} v^{s_*})^{\frac{1}{\gamma + r_*}} = {g_1}^{\beta P_-(\theta)(\rho\theta + \delta\theta m_* + s_* - s_k)\frac{1}{\theta + r_* - r_k}}$$

Solving the SDH challenge. We distinguish two cases.

I. $r_* \notin \{r_1, \ldots, r_q\}$. Let us consider $\beta P_-(\theta)(\rho\theta + \delta\theta m_* + s_* - s_k)$ as a polynomial A in θ. If we carry out the euclidean division of A by $(\theta + r_* - r_k)$, we get Q and R such that $A(\theta) = (\theta + r_* - r_k)Q(\theta) + R(\theta)$. As $(\theta + r_* - r_k)$ is a first degree polynomial $X - (r_k - r_*)$, we know that $R(\theta) = A(r_k - r_*)$, so we are able to compute: $C := R(\theta) = A(r_k - r_*) = \left[\prod_{n=1, n\neq k}^{q} (r_n - r_*)\right] ((\rho + \delta m_*)(r_k - r_*) + s_* - s_k)$. As $A_* = {g_1}^{\frac{C}{\theta + r_* - r_k} + Q(\theta)}$, we are able to compute C and ${g_1}^{Q(\theta)}$ from the SDH challenge. We again have two cases.

1. $(s_* - s_k) \neq (\rho + \delta m_*)(r_* - r_k)$. In this case, $C \neq 0$. Compute ${g_1}^{\frac{1}{\theta + r_* - r_k}} = (A_* {g_1}^{-Q(\theta)})^{\frac{1}{C}}$, set $c = r_* - r_k$, and $(c, {g_1}^{1/(\theta+c)})$ as a solution to the SDH challenge.
2. $(s_* - s_k) = (\rho + \delta m_*)(r_* - r_k)$. $\mathcal{B}$ returns $\perp$.

II. $r_* \in \{r_1, \ldots, r_q\}$. We distinguish two cases.

1. $r_* = r_k$. In this case, we have $({A_*}^{s_k} {A_k}^{-s_*})^{\frac{1}{s_k - s_*}} = {g_1}^{\beta(P(\theta)(\rho(s_k - s_*) + \delta(m_* s_k - m_k s_*))\frac{1}{s_k - s_*} - P_-(\theta) s_k)\frac{1}{\theta}}$. Since P vanishes in 0, but not P_-, then when we divide $\beta(P(\theta)(\rho(s_k - s_*) + \delta(m_* s_k - m_k s_*))\frac{1}{s_k - s_*} - P_-(\theta) s_k)$ by θ and we get R and Q such that $C := R(0) = -\beta s_k \left[\prod_{n=1, n\neq k}^{q} (r_n - r_*)\right]$ and $({A_*}^{s_k} {A_k}^{-s_*})^{\frac{1}{s_k - s_*}} = {g_1}^{\frac{C}{\theta} + Q(\theta)}$ with $C \neq 0$. $\mathcal{B}$ computes
$$ {g_1}^{1/\theta} = (({A_*}^{s_k} {A_k}^{-s_*})^{\frac{1}{s_k - s_*}} {g_1}^{-Q(\theta)})^{1/C}, $$
sets $c = 0$ and obtains a SDH solution $(0, {g_1}^{1/\theta})$.
2. $r_* \neq r_k$. $\mathcal{B}$ returns $\perp$.

We now have to estimate the probability for $\mathcal{B}$ to fail. No information is available about k from $\mathcal{A}$'s point of view. So if $r_* \in \{r_1, \ldots, r_q\}$, $r_* = r_k$ with probability $1/q$, and if $r_* \notin \{r_1, \ldots, r_q\}$, $(s_* - s_k) = (\rho + \delta m_*)(r_* - r_k)$ with probability no more than $1/q$. Therefore if $\mathcal{A}$ outputs a valid forgery with probability ϵ', then the probability ϵ that $\mathcal{B}$ solve the challenge is ϵ'/q. We need $O(q)$ multiplications in order to compute ${g_1}^{P(\theta)}$, and we need to carry out an division of a polynomial of degree q by a polynomial of degree 1, which we can do in $O(q^2)$. □

Theorem 7 *If the $\mathcal{S}^{(1)}$ scheme is (q_S, t, ϵ)-EUF-CMA-secure, then the $\mathcal{S}^{(\ell)}$ scheme is $(q_S, t - O(\ell), \epsilon/2\ell)$-EUF-CMA-secure.*

Proof. Let $\mathcal{A}$ an adversary that (q_S, t, ϵ)-breaks the $\mathcal{S}^{(\ell)}$ scheme. $\mathcal{A}$ outputs valid and non-trivial $(m_{*1}, \ldots, m_{*\ell})$, (S_*, r_*, s_*) with probability greater than ϵ. Let $u_1, \ldots, u_\ell$ be parameters for the $\mathcal{S}^{(\ell)}$ scheme. We distinguish two types of adversary.

Type I. With probability greater than $\epsilon/2$, for all i and $(m_{i1}, \ldots, m_{i\ell})$ queried during the attack, we have $\prod_{j=1}^{\ell} u_j^{m_{*j}} \neq \prod_{j=1}^{\ell} u_j^{m_{ij}}$.

Type II. With probability greater than $\epsilon/2$, $\exists i \in [1, q_S]$ s.t. $(m_{i1}, \ldots, m_{i\ell})$ has been queried during the attack and we have $\prod_{j=1}^{\ell} u_j^{m_{*j}} = \prod_{j=1}^{\ell} u_j^{m_{ij}}$.

From type I adversaries we construct an EUF-CMA adversary $\mathcal{B}$ against the $\mathcal{S}^{(1)}$ scheme. From type II adversary, we construct an adversary against the DL problem.

Type I. Given $(p, \mathbb{G}_1, \mathbb{G}_2, \mathbb{G}_T, e, g_1, u, v, g_2, h_2)$, together with public key (X, Y) for the $\mathcal{S}^{(1)}$ scheme, for all $j \in \{1, \ldots, \ell\}$, $\mathcal{B}$ picks $\alpha_j \xleftarrow{\$} \mathbb{Z}_p^*$ and computes $u_j := u^{\alpha_j}$. $\mathcal{A}$ is next given the parameters. When $\mathcal{A}$ asks for a signature on $(m_{i1}, \ldots, m_{i\ell})$, $\mathcal{B}$ asks for a signature on $m_i = \sum_{j=1}^{\ell} \alpha_j m_{ij}$. The response (A_i, r_i, s_i) is transferred to $\mathcal{A}$. Adversary $\mathcal{A}$ eventually outputs $(m_{*1}, \ldots, m_{*\ell}), (A_*, r_*, s_*)$. $\mathcal{B}$ sets $m_* = \sum_{j=1}^{\ell} \alpha_j m_{*j}$ and returns $(m_*, (A_*, r_*, s_*))$.

Type II. Given a discrete logarithm challenge $(g, h) \in \mathbb{G}_1$, we pick $k \xleftarrow{\$} [1, l]$, $v, h' \xleftarrow{\$} \mathbb{G}_1$, $g_2 \xleftarrow{\$} \mathbb{G}_2$ and $\gamma, \alpha_1, \ldots, \alpha_\ell \xleftarrow{\$} \mathbb{Z}_p^*$. Set $(u_1, \ldots, u_{k-1}, u_k, u_{k+1}, \ldots, u_\ell) := (g^{\alpha_1}, \ldots, g^{\alpha_{k-1}}, h^{\alpha_k}, g^{\alpha_{k+1}}, \ldots, g^{\alpha_\ell})$. $\mathcal{A}$ is given the parameters and the public key $(X, Y) = (g^{\gamma}, {g_2}^{\gamma})$. For each request $(m_{i1}, \ldots, m_{i\ell})$, a $\mathcal{S}^{(\ell)}$ signature is produced with γ. When $\mathcal{A}$ returns a forgery $(m_{*1}, \ldots, m_{*\ell}, (A_*, r_*, s_*))$, with probability greater than $\epsilon/2$, $\exists i \in [1, q_S]$ s.t. $\prod_{j=1}^{\ell} u_j^{m_{*j}} = \prod_{j=1}^{\ell} u_j^{m_{ij}}$, and

$$\log_g h = \frac{\sum_{j=1, j\neq k}^{\ell} \alpha_j (m_{*j} - m_{ij})}{\alpha_k (m_{*k} - m_{ik})}$$

if $\alpha_k(m_{ik} - m_{*k}) \neq 0$. It remains to bound the probability that $\alpha_k(m_{ik} - m_{*k}) = 0$. Since $(m_{i1}, \ldots, m_{i\ell}) \neq (m_{*1}, \ldots, m_{*\ell})$, and since k has been uniformly picked and is independent of the adversary's view, then the probability that $m_{ik} \neq m_{*k}$ is at least $1/\ell$.

In front of an adversary $\mathcal{A}$, we pick a Type I or II adversary with probability $1/2$, so the global security loss is $\epsilon/2\ell$. □

A.4 Security of the Main Scheme

Theorem 8 (Traceability) *The $\mathcal{SAN}$ scheme is traceable in the sense of definition 3.*

(Sketch of proof). Let $\mathcal{A}$ be a successful attacker against the traceability property. If the tracing is not able to conclude, then a valid signature has been produced during the attack on a new y^*, which contradict the unforgeability of the underlying signature scheme. More precisely, we extract the underlying y^* via the forking lemma and, since we know the extraction key, we extract a signature on y^* and win the unforgeability game against the $\mathcal{S}^{(2)}$ scheme. □

Theorem 9 (Immutability) *The $\mathcal{SAN}$ scheme is immutable in the sense of definition 3.*

Proof. The immutability property refers to the fact that an adversary with oracle access to the signer algorithm cannot create a valid message-signature for a sanitizer with key y^* s.t. for all previous queries i we either have $y^* \neq y_i$ or $m^* \notin \{\text{MOD}(m) \mid \text{MOD} \text{ with } \text{ADM}_i(\text{MOD}) = 1\}$. Moreover, recall that this condition takes into account the case where a non-allowed modification has been done on a modifiable block.

Let $\mathcal{A}$ be an efficient successful attacker against immutability. We construct a forger $\mathcal{B}$ against the $\mathcal{S}^{(2)}$ signature scheme. $\mathcal{B}$ receives as input a public key $\mathsf{pk}_s := (X, Y)$ of the signature scheme. $\mathcal{B}$ generates parameters for our scheme, including $h_1 := g_1{}^{\alpha}$ and $h_2 := g_2{}^{\alpha}$, knowing α. When $\mathcal{A}$ ask $\mathcal{B}$ for a signature on m for sanitizer y with respect to admissible modifications ADM, $\mathcal{B}$ ask her own oracle for $|M|+1$ signatures on $(y, \mathcal{H}(m_{\text{FIX}}\|\text{ADM}))$ and $(y, \mathcal{H}(m_{\text{FIX}}\|\text{ADM}\|n\|m_n))$ for all $n \in M$ and generates a sanitizable signature thanks to these individual signatures. $\mathcal{A}$ eventually outputs y^*, m^* and $\sigma^* = (B^*_{0:1}, B^*_{0:2}, r^*, s^*, \{(B^*_{n:1}, B^*_{n:2}, r^*_n, s^*_n)\}_{n\in M^*}, R^*, s^*_y, \text{ADM}^*)$. Let m^*_{FIX} be the fixed part of message m^* *w.r.t* ADM^*, *i.e.* $m^*_{\text{FIX}} := \text{FIX}_{\text{ADM}^*}(m^*)$. If $\mathcal{A}$ succeeds, then

$y^* \neq y_i$ for all query i.
Then our adversary $\mathcal{B}$ computes $S^* := B_{0:2}B_{0:1}{}^{\alpha}$ and returns (S^*, r^*, s^*) as forgery on $(y^*, m^*_{\text{FIX}}\|\text{ADM}^*)$.

Or $(y^*, m^*_{\text{FIX}}\|\text{ADM}^*)$ has not been submitted by $\mathcal{B}$ to the signature oracle before.
We then have $\text{FIX}_{\text{ADM}^*}(m^*) \neq \text{FIX}_{\text{ADM}_i}(m_i)$, *i.e.* $m^*_{\text{FIX}} \neq m_{\text{FIX},i}$. Then $\mathcal{B}$ computes $S^* := B_{0:2}B_{0:1}{}^{\alpha}$ and returns (S^*, r^*, s^*) as forgery on $(y^*, m^*_{\text{FIX}}\|\text{ADM}^*)$.

Or $\text{FIX}_{\text{ADM}^*}(m^*) = \text{FIX}_{\text{ADM}_i}(m_i)$, but since $m^* \notin \{\text{MOD}(m) \mid \text{MOD} \text{ s.t. } \text{ADM}_i(\text{MOD}) = 1\}$, there exists $n \in M^*$ such that m_n is not an admissible value.
Set $G^* := B_{0:2}B_{0:1}{}^{\alpha}$. (G^*, r^*_n, s^*_n) is a valid and non trivial forgery on $(y^*, m^*_{\text{FIX}}\|\text{ADM}^*\|n\|m_n)$.

In either case we have a forgery on $\mathcal{S}^{(2)}$, which concludes the proof. □

Theorem 10 (Unforgeability) *The $\mathcal{SAN}$ scheme is unforgeable in the sense of definition 3.*

(Sketch of proof). Our scheme is unforgeable since both (1) the signature scheme $\mathcal{S}^{(2)}$ for the fixed part and (2) the signature of the modifiable part are unforgeable (which is an adaptation of the proof of the Schnorr signature). □

Theorem 11 (Unlinkability) *The $\mathcal{SAN}$ scheme is (trace-restricted-)unlinkable in the sense of definition 3.*

Proof. Given access to a signing oracle, a sanitizing oracle and a trace oracle, an adversary cannot essentially distinguish left or right outputs of a *left-or-right* type LoR oracle better than a flipping coin. In addition, the adversary has access to an oracle for generating sanitizers' keys. The LoR oracle is initialized with a random bit b, takes two pairs $(j_0, m_0, \text{MOD}_0, \sigma_0, \text{ADM}_0)$ and $(j_1, m_1, \text{MOD}_1, \sigma_1, \text{ADM}_1)$ where $\text{ADM}_0 = \text{ADM}_1$, both signatures are valid, the modified message $\text{MOD}_0(m_0) = \text{MOD}_1(m_1)$ coincide, and both sanitizers j_0 and j_1 are able to sanitize the respective messages. It outputs $\text{SANITIZE}(\mathsf{sk}_{\mathsf{san},j_b}, m_b, \sigma_b, \text{MOD}_b)$.

First note that for each query to LoR, we have $\text{ADM} := \text{ADM}_0 = \text{ADM}_1$ and since admissible modifications do not change the fixed part of messages and the modified messages of a query coincide, it also holds $m_{\text{FIX},0} = m_{\text{FIX},1}$. So the r, s signature component for the fixed parts must be the same for both messages. If an adversary submit distinct r, s, both signatures $S_{\text{FIX},0} \neq S_{\text{FIX},1}$ contains in the $B_{0:1}, B_{0:2}$ part are valid, and it contradicts the strong unforgeability of $\mathcal{S}$: since an honest signer produces those signatures deterministically, it has output at most one signature.

Hence, such a query with identical messages and distinct signatures for the fixed part must thus contain a forgery, and the signature part r, s for the sanitized messages must be identical. By a similar argument, the r_n, s_n part for each restricted modification $n \in M$ must be identical.

Since a new signature for the same message $m'_0 = m'_1$ is computed from scratch, both signatures are identically distributed in both cases. It follows that the probability of predicting b is $\frac{1}{2}$, unless the adversary creates a forgery against $\mathcal{S}^{(2)}$. We notice that the $B_{n:1}, B_{n:2}$ parts, including the case $n = 0$, are randomized. They are then identically distributed for each sanitization. For a successful distinguisher against the unlinkability game, we are able to construct a distinguisher against the semantic security of those ElGamal encryptions, namely the DDH assumption in $\mathbb{G}_1$. □

Remark 4. Unfortunately, our scheme is not *strongly* unlinkable. The reason is that the signer is supposed to be honest in the generation of the randomness. If not, he could use different randomness to link several sanitizing processes.

B. A MORE EFFICIENT SCHEME

We now give a slightly more efficient variant of our scheme. Instead of producing a signature per attributes, the issuer certifies them together in a single signature. We use the $\mathcal{S}^{(\ell)}$ signature scheme, for $\ell = |M|+2$. For $(S, r, s) \in \mathbb{G}_1 \times \mathbb{Z}_p \times \mathbb{Z}_p$, we have :

$$S := \left(g_1 u_0{}^{h_0} u_1{}^{y_j} \prod_{n=2}^{|M|+1} u_n{}^{h_{\phi(n-1)}} v^s\right)^{\frac{1}{\gamma+r}}$$

where ϕ maps the modifiable blocs to their position in the message. A signature σ on a message m is

$$\sigma := \Big((B_1, B_2, r, s), (\vec{R}, \vec{c}, \vec{s_\rho}), \text{ADM}\Big)$$

The signature of knowledge becomes :

$$(\vec{R}, \vec{c}, \vec{s_\rho}) := \text{SoK}\Big\{\Big\langle \rho, \{h_n\}_{\substack{n\in M \\ n\notin K}} \Big\rangle \ : \ e(g_1, g_2)^{\rho} = e(g_1, X) \vee$$

$$\left. e(u_1, g_2)^{\rho} = \frac{e(B_2, g_2{}^r X) e(B_1, h_2{}^r Y)}{e(g_1 u_0{}^{h_{\text{FIX}}} \prod_{n=2}^{|M|+1} u_n{}^{h_{\phi(n-1)}} v^s, g_2)} \right\}(\mathsf{M})$$

on the message $\mathsf{M} := B_1\|B_2\|r\|s\|m\|\text{ADM}\|X\|Y$.

An Information-Flow Type-System for Mixed Protocol Secure Computation

Florian Kerschbaum
SAP Applied Research
Karlsruhe, Germany
florian.kerschbaum@sap.com

ABSTRACT

There are a number of domain-specific programming languages for secure computation. Out of those, the ones that are based on generic programming languages support mixing different protocol primitives and enable implementing a wider, possibly more efficient range of protocols. On the one hand, this may result in better protocol performance. On the other hand, this may lead to insecure protocols. In this paper we present a security type system that enables mixing protocol primitives in a generic programming language, but also ensures that well-typed programs are secure in the semi-honest model. Consequently, a compiled protocol must be secure. We show an extension of the L1 language with our security type system and evaluate the implementation of two protocols from the literature. This shows that our type system supports the provably secure implementation even of complex protocols.

Categories and Subject Descriptors

D.4.6 [**Operating Systems**]: Security and Protection—*Cryptographic controls*; D.3.4 [**Programming Languages**]: Processors—*Compilers*

General Terms

Security, Programming Languages

Keywords

Secure Two-Party Computation, Type System, Programming, Domain-Specific Language, Information Flow

1. INTRODUCTION

Secure (two-party) computation [37] allows two parties to compute a function f over their joint, private inputs x and y, respectively. No party can infer anything about the other party's input (e.g. y) except what can be inferred from one's own input (e.g. x) and output (e.g. $f(x, y)$). Secure computation has many applications, e.g. in the financial sector, and has been successfully deployed in commercial and industrial settings [5, 6, 7, 23].

Secure computation protocols are notoriously difficult to implement. First, they can encompass arbitrary functionality in the joint computation. Second, they need to follow a rigorous approach to security. Often special protocols – mixing several different primitives – are developed for important problems. This is expected to provide more efficient protocols due to insight into the problem domain [16]. These protocols then require a manual verification and security proof.

Current domain-specific programming languages for secure computation [2, 4, 9, 18, 19, 28, 35] do not adequately address these problems. Either they are tied to a specific protocol [2, 4, 18, 28] or they enable implementing insecure protocols [9, 19, 35]. On the one hand, if they are tied to a specific protocol, then this protocol may be proven secure manually independent of the functionality. Such a proof extends to all protocols implemented in this language, but the language prevents implementing many special, possibly more efficient protocols. On the other hand, if the programming language is built upon a generic programming language, such as Python [9] or Java [19, 35], all special protocols can be implemented. Yet, this allows the programmer to also implement insecure protocols that do not withstand security verification.

In this paper we address this problem in a novel way. We build upon the L1 language [35] which is a domain-specific language extension for secure computation based on Java that allows mixing several different protocol primitives. Although we use this specific language, our approach is generic and can be adapted for any domain-specific language that allows mixing protocols and is not tied to a protocol, e.g. [9, 19]. We augment the L1 language with a novel security type system. This type system provably ensures that well-typed programs are secure in the semi-honest model of secure computation. Loosely speaking, in the semi-honest model – as defined by Goldreich [15] – all parties follow the protocol description, but may keep a record of the interaction and try to infer additional information about the other party's input. Protocols secure in the semi-honest model provably prevent any such inference which includes many real-world attacks, such as insider attacks.

Through the use of the type system the compiler statically verifies the security of the protocol during compilation. Only secure protocols are compiled. The programmer is immediately notified about any (possible) security violation.

ASIA CCS'13, May 8–10, 2013, Hangzhou, China.

We evaluate our security type system for the L1 language by implementing two examples. The first one is a simple multiplication protocol based on [14]. The second one is a complex protocol for privacy-preserving string processing from [21]. This example shows that even such protocols – using a wide variety of protocol primitives in intricate ways – can be effectively implemented in our type system. Furthermore, the security proof of our type system also extends to this new implementation. This underpins the protocol's manual security proof from [21] by formal verification.

In summary, this paper contributes

- a *security type system* for mixed protocol secure computation.
- an *integration* of this type system into the L1 language.
- a *proof* that any well-typed program is secure in the semi-honest model.
- an *evaluation* using two exemplary protocols for multiplication and substring creation. This shows that implementing any functionality and even complex protocols using our type system is feasible.

The remainder of this paper is structured as follows. In Section 2 we review the related work in secure computation and security type systems. Section 3 reviews the protocol primitives implemented for secure computation and Section 4 reviews the L1 language. Sections 5 and 6 present our main results. Section 5 describes how re-randomization is implemented and Section 6 describes the security type system. We show our examples in Section 7 and conclude the paper in Section 8.

2. RELATED WORK

Our work is related to domain-specific languages (and extensions) for secure computation [2, 4, 9, 18, 19, 28, 32, 35] and type systems for information flow security [12, 29, 34, 36].

Domain-specific languages (DSL) for secure computations can be coarsely classified into those based on a specific secure computation protocol, such as [2, 4, 18, 28], and those that extend a generic programming language [9, 19, 35]. DSLs based on specific secure computation protocols implement functionality solely based on those protocols. This may not limit implementable functionality, since these protocols are generic and support any functionality, but it may limit performance, since problem-specific protocol primitives are usually expected to yield better efficiency. Nevertheless, once the specific protocols is proven secure, any implemented functionality is also secure.

The first such domain-specific language for secure computation was FairPlay [28]. It implemented Yao's two-party, garbled circuit protocol [37]. It was later extended to multi-party computations in [2] based on the multi-party version of Yao's protocol [1]. It introduced the SFDL programming language which provided an abstraction of the ideal functionality, i.e. the function computed by the secure protocol. Programs in SFDL are translated into a binary circuit which is interpreted using Yao's protocol. Secure multi-party computations were also implemented by ShareMind [4] based on the information-theoretically secure protocol of [3]. Its programming language is rather simplistic and resembles machine language. This makes programming harder, but also makes it easier to ensure security. All these single primitive DSLs generally suffer from poor performance of the compiled protocols. Mixing primitives in a DSL, such as Yao's protocol and homomorphic encryption, was introduced in TASTY [18]. Nevertheless, they restricted themselves to one provably secure protocol of [25] and enabled annotations in the language naming the primitive to use.

DSLs as extensions of existing, generic programming languages can implement any protocol based on mixing (almost) arbitrary primitives. This allows implementing the most efficient protocols, but also does not prevent the programmer from making mistakes and implementing insecure protocols. The first such language is VIFF [9] based on the Python language. Although it is catered for the multiparty protocol using linear secret sharing of [8] which is also considered as the basis for the related SCML programming language of [32], in its implementation it supports the full power of Python. Similar extensions have been presented for the Java language [19, 35]. The framework of [19] is based on Yao's protocol [37] and currently produces the most efficient protocols, but provides very little language support. A significant portion of its efficiency gain is based on optimizations by the programmer, which also invites mistakes likely at the expense of security. Implementing some such optimizations in the compiler was proposed in [22].

Our security type system is based on the L1 language [35] for mixed protocol secure computation, although it can be adapted for other mixed-protocol languages. This language also currently provides little support to the programmer for protocol security.

Our idea of supporting security in a programming language by a type system has been applied to information flow security. Information flow is concerned with confidentiality breaches between principals in computer systems [26]. Its lattice policy model of security levels has been introduced in [10]. Later, it was shown that compliance with such a policy can be statically verified given the program source code [11]. Volpano et al. were the first to recognize that these policies can also be implemented using a type system [36]. Their type system also influenced our construction of our secure computation security type system.

Since type systems can be effectively handled by programmers as well as compilers, this sparked recent interest in language-based information flow security. A very good survey of current research can be found in [34]. There is also an implementation of a type system for information flow security in Java called JIF [29].

Recently information flow types have been applied to homomorphic encryption [12]. These do not yet cover secure computations using homomorphic encryption, since all secure computations involve admissible information flows. Therefore the basic typing assumption of non-interference does not hold in secure computation. Our security type system must therefore cater for a more complicated use case.

3. SECURE COMPUTATION

We implement secure computations using the primitives of homomorphic encryption, secret sharing, garbled circuits and oblivious transfer. These primitives can be combined in many ways and even potentially result in an insecure protocol. For simplicity we only consider secure two-party computation.

3.1 Homomorphic Encryption

Secure computation can be implemented based on additively homomorphic encryption. On the one hand, opposed to fully homomorphic encryption [13] additively homomorphic encryption only implements addition (modulo a key-dependent constant) as the homomorphic operation. On the other hand, additively homomorphic encryption is as fast as standard public-key cryptography.

Let $E_X(x)$ denote the encryption of plaintext x encrypted under X's (Alice's or Bob's) public key and $D_X(c)$ the corresponding decryption of ciphertext c. Then the homomorphism can be expressed as

$$D_X(E_X(x) \cdot E_X(y)) = x + y$$

The following property can be easily derived

$$D_X(E_X(x)^y) = xy$$

In our implementation we mainly use Paillier's encryption system [33], although our language also supports different encryption systems such as Naccache-Stern [31] and allows the programmer to extend the available encryption systems. Paillier's and Naccache-Stern's encryption systems are not only public-key, but also secure against chosen plaintext attacks (IND-CPA). This semantic security implies that two different ciphertexts – even of the same plaintext – cannot be distinguished (without the private key).

In secure computation based on homomorphic encryption it is usually sufficient to use one key per party. Our type system - differently than [12] – therefore does not protect against mixing ciphertexts from different keys.

3.2 Secret Sharing

In order to implement the full functionality of secure computation we augment homomorphic encryption with secret sharing. Each variable is secretly shared between Alice and Bob. Let x be the variable secretly shared and p be the modulus of the homomorphic encryption. Then Alice has x_A and Bob has x_B, such that

$$x = x_A + x_B \bmod p$$

In the following we show how secret sharing and homomorphic encryption can be used to implement any functionality in secure computation. Many other combinations of homomorphic encryption and secret sharing may already be insecure, e.g. by revealing dependent shares. In order to implement any ideal functionality it suffices to implement addition and multiplication. Addition of $x = x_A + x_B$ and $y = y_A + y_B$ (of the same bit-length l) can be implemented locally by addition of each party's shares. Multiplication $z = x \cdot y$ needs to be implemented as a protocol. Let r be a uniformly random number in $\mathbb{Z}_p$. We use the commonly used protocol in Figure 1 from [14].

$A \longrightarrow B$	$E_A(x_A), E_A(y_A)$
$B \longrightarrow A$	$E_A(c) = E_A(x_A)^{y_B} E_A(y_A)^{x_B} E_A(r)$
A	$z_A = x_A y_A + c \bmod p$
B	$z_B = x_B y_B - r \bmod p$

Figure 1: Multiplication Protocol

It is easy to verify that $z_A + z_B = (x_A + x_B)(y_A + y_B)$.

3.3 Oblivious Transfer

Oblivious transfer (OT) is a protocol between a sender and a receiver. As input the sender has n messages $x_0, \ldots, x_{n-1}$ and the receiver has an index i ($0 \leq i < n$). After the protocol the receiver obtains x_i as output. A secure OT protocol ensures that the sender has not learnt i and the receiver has learnt nothing about the other messages x_j ($j \neq i$).

OT can be used as a primitive in mixed secure computations. It facilitates a distributed "if" statement, where one party has the results of the branches and the other party the condition. Furthermore, although we do not implement this protocol in our language, OT can also be used as the sole primitive to implement generic secure computation protocols [15].

There are a number of efficient implementations of OT including [30]. Nevertheless, all OT involve some public key operations. In [20] the authors describe a technique to extend a constant number of OTs to polynomially many OTs using symmetric cryptography.

3.4 Garbled Circuits

Yao introduced secure two-party computation in [37]. He uses a technique called garbled circuits. Garbled circuits allow the computation of any function, but in our language we combine garbled circuits with secret shares and somewhat restrict its functionality. Our restriction is that each input and output (function result) must be a secret share. Let Alice's input be a and Bob's input b which both may be a secret share. Let f be the function to be computed using garbled circuits. We then automatically augment the circuit using an additional input r by the circuit encryptor. The circuit then computes the function f':

$$f'(a, b, r) = f(a, b) - r$$

The circuit encryptor obtains no output from the secure computation, but uses its locally chosen r as return value. The circuit decryptor obtains the new function result f'. These two return values are secret shares of the function result $f(a, b)$ which can be subsequently used in computations using homomorphic encryptions or garbled circuits.

Secrets as inputs are automatically reconstructed at the beginning of the function. For example, let x_A, y_A and x_B, y_B be shares of x and y, respectively. Then the function $f(x, y) = x < y$ to compare x and y is implemented as follows:

$$f''(x_A, y_A, x_B, y_B, r) = ((x_A + x_B) < (y_A + y_B)) - r$$

4. THE L1 LANGUAGE

We extend the L1 language [35] for mixed-protocol secure computation with our security type system. L1 is a simplification and extension of Java that incorporates features to better support secure computation. The two L1 features important for this paper are messaging support and player-specific code.

Messaging support enables the parties to send each other messages. It is implemented using a framework for asynchronous communication available to all parties. In its basic form the **send** command is invoked as in Listing 1. Here, we send the contents of the variable *_var* to the player with identifier 2 under the message name *var_id*. The variable *_var* is cleared after the **send** command, such that it cannot be reused. This ensures that each variable is sent at most once (which is also always sufficient).

```
send(2, _var, "var_id");
```

Listing 1: Send Command

Player identifiers are assigned globally in a configuration file at the start of the protocol execution. An identifier is associated with an IP address and port number where the player's code receives messages. Message names must be unique for a receiving party. Otherwise, messages may be lost due to overlapping communication.

The other party can receive a previously sent value using the **read** command. We show in Listing 2 an example where the player receives content under the message name *var_id*.

```
var = read("var_id");
```

Listing 2: Receive Command

A particular design principle of the L1 language is that the code for all players is integrated into one program. This one (and the same) program is then executed by all players. This corresponds to secure computation where the functionality (and the protocol) is known to all players in advance. Now in order to implement the sending and receiving of messages between the parties – using the **send** and **read** commands – we use player-specific code.

Player-specific code is only executed at a player with a specific identifier. This player's identifier is placed at the start of the statement. If no such identifier is present, the statement is executed by all parties. We can therefore implement the send and receive commands within the same program. The following Listing 3 combines the code of the previous two listings.

```
1: send(2, _var, "var_id");
2: var = read("var_id");
```

Listing 3: Message Sending and Receiving

4.1 Data Types

The L1 language has its own type system for basic, non-security relevant types. This type system is orthogonal to our security type system. L1 provides data types for multi-precision integers (**int**), public (**pubk**) and private (**privk**) keys. We extend this for the purpose of our type system with data types for secret shares and ciphertexts.

L1 implements copy semantics for variable assignments. Even if a complex variable (such as a share, ciphertext or a multi-precision integer) is assigned, then the assigned value is copied onto the memory of the assigned variable. This differs from the pointer semantics of Java where only a reference is assigned.

A secret share type is parameterized by the modulus of the share. We create only additive shares, i.e., let p be the modulus and s ($s < p$) be the secret. Then there are n ($n \geq 2$) shares s_i, such that

$$\sum_{i=0}^{n-1} s_i (\text{mod } p) = s$$

A secret share can be internally represented as an object containing a multi-precision integer – the value of the secret share – and a reference to a multi-precision integer – the modulus. An integer can be implicitly converted into a secret share, but converting a secret share to an integer requires a type cast. When an expression is assigned to a secret share, a modulo operation is performed automatically. We declare a secret share at each party as in Listing 4.

```
int p = getModulus(pubKey);
share<p> s;
```

Listing 4: Secret Share Declaration

There are several ciphertext data types. It not only depends on the encryption system used, but also for each public key encryption system there are two types of ciphertexts: one where the player has the private key and one where the player has only the public key. Our assumption is that for ciphertexts where the player has the private key no other player can access the plaintext, i.e. the private key is indeed confidential. Similar to secret shares we parameterize the ciphertext type with the respective (public or private) key.

A ciphertext contains by definition a secret share as payload. This is not explicitly declared in our language extension. Furthermore, this secret share has the same modulus as the homomorphic operation in the encryption system using the specified key. A ciphertext can be internally represented as an object containing multi-precision integers – the ciphertext value – and a reference to a key. Again, an integer can be implicitly converted into a ciphertext, but converting a ciphertext to an integer requires a type cast. We declare ciphertexts as in Listing 5.

```
pubk pubKey;
cipher<pubKey> c;

prvk privKey;
cipher<privKey> d;
```

Listing 5: Ciphertext Declaration

4.2 Cryptographic Protocols

As cryptographic protocol primitives we need to support garbled circuits and oblivious transfer. We implement garbled circuits using two commands. Yao's protocol has a sender who encrypts (and garbles) the circuit and a receiver who – after obtaining the decryption keys via oblivious transfer – decrypts the circuit. The ideal functionality to be computed in the circuit is specified as a function (in C like syntax). Each input variable is shared between both parties and reconstructed at the beginning of the circuit. Each party then inputs its shares as parameters to the command. The output is also automatically shared. The random variable r (using the modulus of the result) is automatically appended as input by the encryptor to the circuit. It is filled with a random input chosen by the command and returned as a result from the command. The output (the result of the circuit) is returned as a function result to the decryptor (subtracted by the random value of the encryptor).

In the ideal functionality we currently only support arithmetic operations (addition, subtraction, multiplication, division) and comparison (equal, lower, lower or equal). In Listing 6 we implement the circuit for the above function $f''(x_A, y_A, x_B, y_B, r) = ((x_A + x_B) < (y_A + y_B)) - r$.

Oblivious transfer is also implemented via two commands: one for sending and one for receiving. The sending command takes an array as parameter. Its length is the number

```
share<(1 << 32)> x, y;
share<2> r;

1: r = gcencrypt(2,
        "bool compare(int x, int y) {
                compare = x < y;
        }",
        x, y);
2: r = gcdecrypt(1, "compare", x, y);
```

Listing 6: Garbled Circuit Protocol

of possible choices in the oblivious transfer. The input array is cleared after sending. The receiving command takes the index as a parameter and outputs the selected message. We need to specify the communicating party in the receive command as well, since it is an interactive protocol. In Listing 7 we show an example for an 1-out-of-2 oblivious transfer.[1]

```
1 1: share<p> _x*[2];
2 2: share<p> y;
3 2: int b;
4
5 1: otsend(2, "id", _x);
6 2: y = otread(1, "id", b);
```

Listing 7: Oblivious Transfer

5. REVERSE TAINT CHECKING

Our type system ensure that only freshly and independently chosen random variables are revealed to the other party. We distinguish these variables by typing them as *untainted* (versus tainted for variables during processing). Each basic data type of secret share, ciphertext or array can be tainted or untainted. We perform the operation of re-randomization for reverse taint checking in order to create the untainted type of variables.

Taint checking[2] is a programming language operation that allows to track whether inputs are processed safely. It helps to prevent certain common programming errors, such as SQL injections or buffer overflows. In taint checking mode each variable set by an input is tainted. Then after certain checks have been performed, e.g. in PERL using regular expressions, the variable can be cleared as untainted. Every time a tainted variable is used in a dangerous operation, such as accessing a database, the program is aborted with a fatal error.

We call our operation reverse taint checking, because on the one hand we use a similar variable tracking mechanism (although statically verifiable), but on the other hand the taint flag is set by local operations and checked on sending data. In detail our algorithm works as follows. Each variable is either tainted or untainted. All variables assigned by expressions are tainted, since they are result of a local computation. It does not matter whether the assignment expression is a **read** command, some arithmetic operation or even a constant. Every regularly assigned variable is tainted.

The fundamental check is that the sending commands do not accept tainted variables as parameters. If the parameter is tainted, then the program does not type check. Only untainted variables may be sent to the other parties.

[1]The star in line 1 is explained after introducing our security type system.

[2]http://perldoc.perl.org/perlsec.html#Taint-mode

We define the semantics of an untainted variable as we use it in our security type system: An untainted variable is independently *random*, *immutable*, and *use-once*. Independently random implies that it contains indepedently randomly distributed data. It is immutable, as it can only be set by special commands and the randomness is chosen by the system. An untainted variable can only be used once and is cleared after use in a send or re-randomization command. We summarize these properties in Definition 1. Next, we need to describe how to create an untainted variable.

DEFINITION 1. *An untainted variable is an immutable object that contains random data independent of all other* untainted *variables. It can be used in send and re-randomization command and is cleared after use.*

5.1 Re-Randomization

We perform a type-dependent re-randomization operation in order to create untainted variables. Re-randomization renews the randomness in a tainted variable using fresh random input. Depending on the type we perform different re-randomization operations. A re-randomize operation is written as in Listing 8.

```
share<p> s;
cipher<pubKey> c;
cipher<privKey> d;
share<p> x[2];
share<p> r;

_s = rerandomize(s, r);
_c = rerandomize(c, r);
_d = rerandomize(d, null);
_x = rerandomize(x, r);
```

Listing 8: Re-Randomization

Our different re-randomization operations are as follows:

- *Secret share*: A secret share is associated with a modulus p. We choose a fresh random number r uniformly in $\mathbb{Z}_p$. Let s be the value of the secret share. We return the updated

  $$\_s \leftarrow s - r \pmod{p}$$

  and copy unto the randomization parameter (given to the function) the new secret share with the value r.

- *Ciphertext with public key only*: The ciphertext and the contained secret share are associated with the same modulus p. We again choose a fresh random number r uniformly in $\mathbb{Z}_p$. We encrypt the negation of this random number r using the associated public key. Let $E(-r)$ denote this encryption and c denote the ciphertext of the re-randomized variable. We return the following, updated ciphertext

  $$\_c \leftarrow c \cdot E(-r)$$

  and copy unto the randomization parameter the new secret share with the value r.

- *Ciphertext with private key*: We assume that the possessor of the private key is its sole possessor. Therefore the contained secret share cannot be accessed by any other party and there is no need to re-randomize it. Nevertheless, we need to re-randomize the ciphertext. We use the usual re-randomization of IND-CPA secure, homomorphic encryption. Let d denote this ciphertext. We return it as

$$\_d \leftarrow d \cdot E(0)$$

The randomization parameter is not used.

- *Array*: For arrays we choose the randomization parameter (i.e. r, $E(r)$ or $E(0)$) only once. The same randomization is applied to each element of the array and the randomization parameter is copied as usual. The return array contains the re-randomized elements. Note that only the array is untainted by this operation. Each element in the array remains tainted. This enables using the array in an oblivious transfer command, but prevents the elements from being sent individually using the send command.

5.2 Re-Randomization with Untainted Variables

Sometimes it is necessary to choose random shares before performing an operation and sending the result. An instance of this requirement occurs in our second example of substring creation in Section 7.2. Usually an untainted variable is created during a re-randomization command and then sent over the network, but instead of directly sending it, the untainted variable can be used to re-randomize another variable. The semantics of this re-randomization operation is that the untainted input parameter is cleared after the re-randomization command (just as it is cleared after a send command), such that it cannot be reused. Yet, the return variable is then subsequently untainted and can be used as such. Listing 9 shows an example where an untainted variable is used to re-randomize another variable. The variable $\_s$ is null after the execution of this code.

The complete re-randomization function is depicted in Algorithm 1. It is easy to verify from Algorithm 1 that after re-randomization one variable – the untainted one – is independently randomly distributed as required by Definition 1. This holds even in case of re-randomization with untainted variables.

```
share<p> r;

// _s = s - r
// r is set and chosen fresh
_s = rerandomize(s, r);
// _t = t - _s
// _s is used and then cleared
_t = rerandomize(t, _s);
```

Listing 9: Re-Randomization with Untainted Variable

5.3 Output

If all variables were only to be sent as re-randomized variables, no useful result could ever be obtained, i.e. every (useful) secure computation has an admissible information flow. Therefore we need to allow reconstruction of the output values from the secret shares. Let s be a secret share designated as output. We use an output statement in L1 in order to declare output, but before we can output the result the parties need to exchange result secret shares. The output command is restricted to secret shares as parameter. They can only send the secret share if the variable is untainted. The programmer needs to declare that this variable is designated for output before the output is actually performed. We perform a mutual exchange of (result) shares between two parties as in Listing 10.

Algorithm 1 Rerandomization Algorithm

```
function RERANDOMIZEUNIT(v, r)
    if INSTANCEOF(v, SecretShare) then
        return v − r
    end if
    if INSTANCEOF(v, PublicKeyCipherText) then
        return v· ENCRYPT(−r)
    end if
    if INSTANCEOF(v, PrivateKeyCipherText) then
        return v· ENCRYPT(0)
    end if
end function
function RERANDOMIZE(v, r)
    if ISTAINTED(r) then
        if INSTANCEOF(v, Array) then
            m ← GETMODULUS(v_1)
            r ← RANDOM(m)
            for all v_i ∈ v do
                v_i ← RERANDOMIZEUNIT(v_i, r)
            end for
            return v
        else
            m ← GETMODULUS(v)
            r ← RANDOM(m)
            return RERANDOMIZEUNIT(v, r)
        end if
    else
        if INSTANCEOF(v, Array) then
            for all v_i ∈ v do
                v_i ← RERANDOMIZEUNIT(v_i, r)
            end for
            r ← null
            return v
        else
            v' ← RERANDOMIZEUNIT(v, r)
            r ← null
            return v'
        end if
    end if
end function
```

```
share<p> s_prime;

_s = output(s);
send(id() % 2 + 1, _s, "result_share");
s_prime = read("result_share");
s = s + s_prime;
output("result = ", s);
```

Listing 10: Share Reconstruction and Output

6. SECURITY TYPE SYSTEM

In our L1 language we augment the regular type system with a security type system where every expression carries both a type (such as secret share, ciphertext or array) and a security type (tainted or untainted). Following the definition of [34] our security type system is a collection of typing rules that describe what security type is assigned to a program (or expression), based on the types of sub-programs (sub-expressions). We write $\vdash exp : t$ to mean that the expression exp has security type $t \in \{tainted, untainted\}$ according to the typing rules. This assertion is known as a typing judgment. Similar to the construction of the security type system for information flow [36] we use a security context $[sc]$ associated with a label of the program counter. We use the security context to prevent the programmer from sending messages depending on the truth value of expressions with tainted variables. Untainted variables are random and therefore not useful for branching conditions. This limitation is important, since not only the content of a message may reveal information, but just sending the message itself. The typing judgment $[sc] \vdash C$ means that the program C is typable in the security context sc.

Figure 2 presents the typing rules for a simplification of the L1 language. We omit loops, since they can only have a constant number of iterations in secure computation and can therefore be unrolled. We also omit the typing rules for the regular type system. Expression security types and security contexts can be either tainted or untainted. A typing rule is an inference rule that describes how types are assigned. The statements above the line must be fulfilled for the rule to be applied, yielding the statement below the line.

According to the rules [E1-3], some variables (u) have security type untainted. All other expressions and variables have type tainted. This includes expression which include (but are not limited to) occurrences of untainted variables. Rules [R1-2] ensure that untainted variables can only be assigned by re-randomization and output statements, but are otherwise static.

Consider the rules [M1-6]. The message sending commands can only be executed in an untainted security context. Of those commands the ones that transmit plaintext messages as payload (**send**, **otsend**) can only transmit untainted variables. Received variables are always immediately tainted.

The typing rules [C1-5] control the security context in a composite program. The commands **skip** and variable assignment are typable in any context. Branches must be typable in a tainted context. This is justified by the above requirement that message sending commands may not be executed depending on tainted variables. The rules [C4-5] enable composing programs similar to [36] including the subsumption rule. It enables sending messages before or after a tainted context (branching).

In L1 we declare untainted variables using an asterisk after the type declaration. We show examples for a share and an untainted array in Listing 11 (and already in Listing 7).

$$\vdash u : untainted \quad (E-1)$$

$$\frac{var \neq u}{\vdash var : tainted} \quad (E-2)$$

$$\frac{exp \neq u}{\vdash exp : tainted} \quad (E-3)$$

$$\frac{\vdash var_1 : untainted}{[sc] \vdash var_1 = \textbf{rerandomize}(exp, var_2)} \quad (R-1)$$

$$\frac{\vdash var : untainted}{[sc] \vdash var = \textbf{output}(exp)} \quad (R-2)$$

$$\frac{\vdash var : untainted}{[untainted] \vdash \textbf{send}(var)} \quad (M-1)$$

$$\frac{\vdash var : tainted}{[untainted] \vdash var = \textbf{read}()} \quad (M-2)$$

$$\frac{\vdash var : untainted}{[untainted] \vdash \textbf{otsend}(var)} \quad (M-3)$$

$$\frac{\vdash var : tainted}{[untainted] \vdash var = \textbf{otread}(exp)} \quad (M-4)$$

$$\frac{\vdash var : tainted}{[untainted] \vdash var = \textbf{gcencrypt}(exp, \dots)} \quad (M-5)$$

$$\frac{\vdash var : tainted}{[untainted] \vdash var = \textbf{gcdecrypt}(exp, \dots)} \quad (M-6)$$

$$[sc] \vdash \textbf{skip} \quad (C-1)$$

$$\frac{\vdash var : tainted}{[sc] \vdash var = exp} \quad (C-2)$$

$$\frac{[tainted] \vdash C_1 \quad [tainted] \vdash C_2}{[sc] \vdash \textbf{if } exp \textbf{ then } C_1 \textbf{ else } C_2} \quad (C-3)$$

$$\frac{[sc] \vdash C_1 \quad [sc] \vdash C_2}{[sc] \vdash C_1; C_2} \quad (C-4)$$

$$\frac{[tainted] \vdash C}{[untainted] \vdash C} \quad (C-5)$$

Figure 2: Security Typing Rules

```
share<p>* _s;
share<p> _a*[2];
```

Listing 11: Untainted Variable Declaration

6.1 Proofs

Goldreich [15] defines security in the semi-honest model. The view $VIEW^{\Pi}(x, y)$ of a party during protocol Π on this party's input x and the other party's input y is its input x, the outcome of its coin tosses and the messages received during the execution of the protocol.

DEFINITION 2. *We say a protocol Π computing $f(x,y)$ is secure in the* semi-honest *model, if for each party there exist a polynomial-time simulator S given the party's input and output that is computationally indistinguishable from the party's view $VIEW^{\Pi}(x,y)$:*

$$S(x, f(x,y)) \stackrel{c}{=} VIEW^{\Pi}(x,y)$$

Our main theorem states that well-typed program is compiled into a protocol secure in the semi-honest model.

THEOREM 3. *Let program C implement function $\mathcal{F}$. Then if C is well-typed ($\vdash C$), it implements $\mathcal{F}$ securely in the semi-honest model.*

Proof Outline: We need to show that we can construct a simulator for each party's view. The problem is that the view is determined by the messages received and not sent (where type safety applies). We therefore assume that the entire protocol is written in L1 using player-specific code. Then we simulate the messages received from the other party. We give only the construction for one party's simulator – due to the symmetry of the language it also applies to the other party.

Furthermore, differently from information flow type system there is always an admissible information flow in secure computation. This is due to the nature of the joint computation, i.e. some result is revealed. It is therefore not possible to strictly distinguish between the program and the information conveyed. We capture this using the output variables.

PROOF. We now construct the simulator for one party. The simulator consists of input, coin tosses and messages received. Input can be simulated by the real input (which is given to the simulator) and coin tosses are chosen by the Java random number generator. This leaves the received messages to be simulated.

Let $m_1, \ldots, m_n$ be the messages received. We then construct the simulator step-by-step, i.e. for each message received we add a simulated message. This means, we start with the empty simulator S_0 which does nothing. Then we construct S_i from S_{i-1} by appending a simulation of message m_i to S_{i-1}. Therefore it is first necessary to note that the number of messages is constant.

LEMMA 4. *In a well-typed program ($\vdash C$) the number of sending statements (**send**, **otsend**, **gcencrypt**) is constant.*

Note that no sending statement may appear inside a branching statement. We can complete the proof by induction over the structure of C.

We now construct a simulated message for each message m_i received. We refer to Goldreich's composition theorem [15] for our cryptographic protocols. Oblivious transfer and Yao's protocol can be simulated as an oracle transmitting only the resulting messages, as long as the implementations have been proven secure in the semi-honest model. The security proof for our implementation of oblivious transfer can be found in [30]. The security proof for generic Yao's protocol can be found in [27]. We can therefore simulate only the resulting messages.

There are two types of messages received: output and intermediate.

Output messages: These are messages marked as output by the other party. Let m_i be such a message received. In a correctly implemented program these messages are indeed output at the local party (after combination with a local share). Therefore the local party can simulate the message as follows: Let o be the output which is given to the simulator and l the local share for reconstructing this value. The local share l can be computed from the local state of the party up to that point in the program. This must be the case, since otherwise the party could not produce the output. Then the message is simulated as

$$m_i = o - l \pmod{p}$$

Intermediate messages: These are messages re-randomized by the other party. Let m_i be such a message received. In a well-typed program these message are all randomly independently distributed.

LEMMA 5. *In a well-typed program all re-randomized variables are independently randomly distributed.*

This follows directly from our construction of re-randomization. After the algorithm of re-randomization (see Algorithm 1) only one untainted variable is assigned. This variable is always independently randomly distributed. Therefore the set of re-randomized variables is always a set of independently randomly distributed variables. Note that local and output variables may be dependent on the re-randomized variables.

The received message m is then simulated as an independent random variable depending on the type of the message (secret share or public-key ciphertext)

$$m_i \leftarrow_R \mathbb{Z}_p / E(\mathbb{Z}_p)$$

This completes the simulator – all messages can be simulated – and consequently the proof of semi-honest security. □

7. EXAMPLES

7.1 Multiplication

We first show a simple example. We implement the multiplication protocol using homomorphic encryption and secret shares from Figure 1 based on [14]. This protocol shows that using our typed language any functionality can be implemented securely (Listing 12).

The encrypt command (lines 33-34) returns a tainted variable. Subsequently (lines 35-36) we re-randomize the ciphertext (with known private key). This may seem superfluous, since the randomness is fresh during encryption, but the contained share may be tainted in ciphertexts with public key only. We chose this re-randomization approach over a polymorphic command based on return type. We see in line 47 that using the **rerandomize** command we can omit explicitly adding a random variable. Instead we use the randomness introduced by the re-randomization command.

7.2 Substring Creation

As a second example we show the substring creation protocol of Jensen and Kerschbaum [21]. This protocol is one of a set of three for privacy-preserving string processing. It is particularly suited to show the power of our language and type system. First, it uses all of our primitives: Yao's garbled circuits, oblivious transfer, secret shares and homomorphic encryption. Second, it's security proof is not obvious, such that a well-typed implementation underpins its provable security.

The protocol operates on strings. A string consists of a length field and a fixed-length array of characters. Each string is secretly shared between Alice and Bob, such that neither Alice nor Bob know its contents or even length.

The purpose of the protocol is to create a substring of length t of a string x starting at position t. The resulting substring will also be secretly shared – just as the inputs, i.e. there is no output reconstruction. The protocol starts with a rotation of the shared string to the left by s characters. Then the trailing characters beyond the length field are masked, such that they are all set to 0 again. Each share of the string is masked separately, such that the masking protocol is executed twice with the roles reversed. The composed main protocol is shown in Listing 13.

The types for strings (containing length field and array) are defined in a separate header (not shown for brevity) This header of the protocol also loads the cryptographic keys for the homomorphic encryption. Both a private, public keypair and a public key (of the other party) are read from disk. In this protocol we use Naccache-Stern encryption [31], because the modulus of the homomorphic operation must be the same at both parties. This cannot be achieved using Paillier's encryption system [33]. We use a fixed modulus of 256 in order to simplify integration with Yao's protocol. In [21] the authors use Goldwasser-Micali encryption [17],

```
int multiply (int s, int t) {
    // key variables
    1: prvk privKey;
    pubk pubKey;

    // read keys from hard disk
    1: {
        privKey = readKey("priv.key");
        pubKey = getPublicKey(privKey);
    }
    2: pubKey = readKey("pub.key");

    int n = getModulus(pubKey);
    share<n> x, y, z;

    // initialize shares
    x = s;
    y = t;

    // declare ciphertexts
    1: {
        cipher<privKey> e, f, g;
        cipher<privKey>* _e, _f;
    }
    2: {
        cipher<pubKey> e, f, g;
        cipher<pubKey>* _g;
        share<n> r;
    }

    // send 1st message
    1: {
        e = encrypt(privKey, x);
        f = encrypt(privKey, y);
        _e = rerandomize(e, null);
        _f = rerandomize(f, null);
        send(2, _e, "E_A(x_A)");
        send(2, _f, "E_A(y_A)");
    }
    // receive 1st message, send 2nd
    2: {
        e = read("E_A(x_A)");
        f = read("E_A(y_A)");

        g = modPow(e, y, n);
        g *= modPow(f, x, n);
        _g = rerandomize(g, r);
        send(1, _g, "E_A(c)");
    }
    // receive 2nd message, compute result
    1: {
        g = read("E_A(c)");

        z = decrypt(g) + x * y;
    }
    // also compute result at Bob
    2: z = r + x * y;

    return (int) z;
}
```

Listing 12: Secure Multiplication

because they need its properties in a subsequent protocol. In this paper we do not implement this subsequent protocol and can therefore use the more efficient Naccache-Stern encryption system. There are also some helper functions for adding (secret shares), rotating strings, re-randomization functions for the different string types, en-/decryption functions and message sending functions – all not shown for brevity.

Next we consider the sub-protocols of the substring protocol. The complete rotation protocol is presented in Listing 14. The rotation protocol rotates the string, such that the initial character is at position 0. This is achieved us-

```
string substring (string x,
                  share<n> s,
                  share<n> t) {
    string a, b, sub;

    sub = rotate(x, s);
    sub.len = t;
    1: a = mask_send(sub);
    2: a = mask_recv(sub);
    1: b = mask_recv(sub);
    2: b = mask_send(sub);

    return add(a, b);
}
```

Listing 13: Substring: Composed Protocol

```
string rotate (share<n> s, string in) {
    string x, r;
    _string _x;
    string_priv e;
    _string_priv _e;
    string_pub y;
    _string_pub _y;

    x = in;
    1:{
        _x = rerandomize_shift(x, s);
        _e = encrypt_string(x);
        send_string("1", _e);
    }
    2:{
        y = recv_string_pub("1");
        rotate_left(y, s);
        _y = rerandomize_string(y, r);
        rotate_right(r, s);
        x = add(x, r);
        send_string("2", _y);
        _e = encrypt_string(x);
        send_string("3", _e);
    }
    1:{
        e = recv_string_priv ("2");
        x = decrypt_string(e);
        rotate_left(x, s);
        y = recv_string_pub("3");
        rotate_left(y, s);
        _y = rerandomize_string(y, _x);
        send_string("4", _y);
    }
    2:{
        e = recv_string_priv ("4");
        x = decrypt_string(e);
        rotate_left(x, s);
    }

    return x;
}
```

Listing 14: Substring: Rotation

```
_string rerandomize_shift (string x,
                           share<n> s) {
    _string _x;
    string r;

    x = rotate_right(x, s);
    _x = rerandomize_string(x, r);
    x = r;
    x = rotate_left(x, s);
    return _x;
}
```

Listing 15: Substring: Initial Re-randomization

ing the following technique: Each share is encrypted and sent to the other parties which then rotates it by its share of the initial position. It then re-randomizes it and returns it before it is rotated by the local share of the initial position. This requires a particular interlocking technique. Before Alice can send her share to Bob she needs to re-randomize it with the re-randomization values for Bob's share (these will cancel out). She cannot do this after she receives the returned ciphertexts from Bob, since they are then already rotated. Therefore, in order to implement this protocol Alice needs to choose the re-randomization parameters for Bob's share before she sends her share. This is achieved by re-randomization using untainted variables created in the **rerandomize_shift** function (Listing 15). This function creates an untainted variable containing a random share that is already rotated by the local share of the initial position.

We have implemented the entire protocol in one function where each party's code is implemented using player-specific code. We chose this type of implementation, since the two sides of the protocol are quite asymmetric. In line 11 (Listing 14) there is the call of the initial re-randomization creating the untainted variable. This untainted variable is then used for re-randomization in line 31 (Listing 14) completing the interlock technique. Bob can simply add his re-randomization to his share in line 20 (Listing 14).

```
string mask_send (string x) {
    share<2> b;
    string m, y, r;
    _string _y;
    string_priv e;
    _string_priv _e;

    m.len = n;
    for (int i = 0; i < n; i++)
        if (i < x.len)
            m.chars[i] = 1;
        else
            m.chars[i] = 0;
    _e = encrypt_string(m);
    send_string("5", _e);
    b = gcdecrypt("compare",
                  (int) x.len, 0);
    e = otread("6", (int) b);
    y = decrypt_string(e);
    _y = rerandomize_string(y, r);
    send_string("7", _y);
    return r;
}
```

Listing 16: Substring: Masking Sender

The masking protocol creates a 0, 1 encrypted string which is used to mask the local share. Care has to be taken, if the shares of the length of the substring wrap around the modulus. Two cases need to be prepared: one in which they do and one in which they do not wrap around. The correct one is chosen by oblivious transfer. The condition is computed by a comparison implemented as a Yao's protocol. Therefore we see also these primitives in this protocol.

The masking protocol is executed twice with roles reversed. Therefore we implemented each side of the protocol as a separate function. Listing 16 shows one side and Listing 17 the other. Then each side is called with the local input in the composed protocol (Listing 13 lines 6–9).

```
string mask_recv (string x) {
    share<2> b;
    string r, q;
    string_pub y, e, f[2];
    string_pub _f*[2];

    y = recv_string("5");
    f[0].len = x.len;
    f[1].len = x.len;
    for (int i = 0; i < x.len; i++)
        f[0].chars[i] =
            encrypt(pubKeyExt, 1);
    for (int i = x.len; i < n; i++)
        f[0].chars[i] =
            y.chars[i - x.len];
    for (int i = 0; i < n - x.len; i++)
        f[1].chars[i] =
            y.chars[i + n - x.len];
    for (int i = n - x.len;
            i < n - x.len; i++)
        f[1].chars[i] =
            encrypt(pubKeyExt, 0);
    for (int i = 0; i < n; i++)
        e.chars[i] =
            f[0].chars[i] * f[0].chars[i];
    for (int i = 0; i < n; i++) {
        f[0].chars[i] =
            modPow(e.chars[i],
                   (int) x.chars[i]);
        f[1].chars[i] =
            e.chars[i] *
            encrypt(pubKeyExt, n-1);
        f[1].chars[i] =
            modPow(f[1].chars[i],
                   (int) x.chars[i]);
    }
    b = gcencrypt(
        "bool compare(int x, int y) {
            compare = x < y;
        }",
        (int) x.len, n);
    if (b == 1) {
        e = f[0];
        f[0] = f[1];
        f[1] = e;
    }
    _f = rerandomize(f, r);
    otsend(_f);
    q = recv_string("7");
    q = add(q, r);
    return q;
}
```

Listing 17: Substring: Masking Receiver

8. CONCLUSIONS

In this paper we consider the problem of ensuring semi-honest security of secure computations implemented in domain-specific languages. Currently, either the language is based on a specific protocol which has been proven secure or on a generic language that enables also implementing insecure protocols. We are the first to propose a type system to limit the implementable protocols to only secure ones, but also enable the programmer to freely choose the protocol primitives. We prove that any well-typed program is secure in the semi-honest model of secure computation. This presents a new trade-off between security and performance for the implementation of secure computations. Our type system enables implementing complex protocols, such as one of our examples for privacy-preserving string processing.

Future work is to extend the security guarantees to stronger security models, such as the malicious model. In the malicious model the parties may behave arbitrarily and still privacy of inputs and correctness of outputs are preserved. Such a security guarantee can be of practical relevance in highly sensitive secure computations such as e-voting.

9. REFERENCES

[1] D. Beaver, S. Micali, and P. Rogaway. The Round Complexity of Secure Protocols. *Proceedings of the 22nd ACM Symposium on Theory of Computing*, 1990.

[2] A. Ben-David, N. Nisan, and B. Pinkas. FairplayMP: A System for Secure Multi-Party Computation. *Proceedings of the 15th ACM Conference on Computer and Communications Security*, 2008.

[3] M. Ben-Or, S. Goldwasser, and A. Wigderson. Completeness Theorems for Non-Cryptographic Fault-Tolerant Distributed Computation. *Proceedings of the 20th ACM Symposium on Theory of Computing*, 1988.

[4] D. Bogdanov, S. Laur, and J. Willemson. Sharemind: A Framework for Fast Privacy-Preserving Computations. *Proceedings of the 13th European Symposium on Research in Computer Security*, 2008.

[5] D. Bogdanov, R. Talviste, and J. Willemson. Deploying Secure Multi-Party Computation for Financial Data Analysis. *Proceedings of the 16th International Conference on Financial Cryptography and Data Security*, 2012.

[6] P. Bogetoft, D. L. Christensen, I. Damgård, M. Geisler, T. P. Jakobsen, M. Krøigaard, J. D. Nielsen, J. B. Nielsen, K. Nielsen, J. Pagter, M. I. Schwartzbach, and T. Toft. Secure Multiparty Computation Goes Live. *Proceedings of the 13th International Conference on Financial Cryptography and Data Security*, 2009.

[7] O. Catrina, and F. Kerschbaum. Fostering the Uptake of Secure Multiparty Computation in E-Commerce. *Proceedings of the International Workshop on Frontiers in Availability, Reliability and Security*, 2008.

[8] R. Cramer, I. Damgård and U. Maurer. Efficient General Secure Multi-Party Computation from any Linear Secret-Sharing Scheme. *Proceedings of EUROCRYPT*, 2000.

[9] I. Damgård, M. Geisler, M. Krøigaard, and J. B. Nielsen. Asynchronous Multiparty Computation: Theory and Implementation. *Proceedings of the 12th International Conference on Practice and Theory in Public Key Cryptography*, 2009.

[10] D. Denning. A Lattice Model of Secure Information Flow. *Communications of the ACM 19(5)*, 1976.

[11] D. Denning, and P. Denning. Certification of Programs for Secure Information Flow. *Communications of the ACM 20(7)*, 1977.

[12] C. Fournet, J. Planul, and T. Rezk. Information-Flow Types for Homomorphic Encryptions. *Proceedings of the 18th ACM Conference on Computer and Communications Security*, 2011.

[13] C. Gentry. Fully Homomorphic Encryption using Ideal Lattices. *Proceedings of the 41st ACM Symposium on Theory of Computing*, 2009.

[14] B. Goethals, S. Laur, H. Lipmaa, and T. Mielikäinen. On Private Scalar Product Computation for Privacy-Preserving Data Mining. *Proceedings of the 7th International Conference on Information Security and Cryptology*, 2004.

[15] O. Goldreich. Secure Multi-party Computation. Available at *www.wisdom.weizmann.ac.il/~oded/pp.html*, 2002.

[16] S. Goldwasser. Multi-Party Computations: Past and Present. *Proceedings of the 16th ACM Symposium on Principles of Distributed Computing*, 1997.

[17] S. Goldwasser, and S. Micali. Probabilistic Encryption. *Journal of Computer and Systems Science 28(2)*, 1984.

[18] W. Henecka, S. Kögl, A.-R. Sadeghi, T. Schneider, and I. Wehrenberg. TASTY: Tool for Automating Secure Two-Party Computations. *Proceedings of the 17th ACM Conference on Computer and Communications Security*, 2010.

[19] Y. Huang, D. Evans, J. Katz, and L. Malka. Faster Secure Two-Party Computation Using Garbled Circuits. *Proceedings of the USENIX Security Symposium*, 2011.

[20] Y. Ishai, J. Kilian, K. Nissim, and E. Petrank. Extending Oblivious Transfers Efficiently. *Proceedings of CRYPTO*, 2003.

[21] M. Jensen, and F. Kerschbaum. Towards Privacy-Preserving XML Transformation. *Proceedings of the 9th IEEE International Conference on Web Services*, 2011.

[22] F. Kerschbaum. Automatically Optimizing Secure Computation. *Proceedings of the 18th ACM Conference on Computer and Communications Security*, 2011.

[23] F. Kerschbaum, A. Schröpfer, A. Zilli, R. Pibernik, O. Catrina, S. de Hoogh, B. Schoenmakers, S. Cimato, and E. Damiani. Secure Collaborative Supply Chain Management. *IEEE Computer 44 (9)*, 2011.

[24] F. Kerschbaum, and A. Sorniotti. RFID-based Supply Chain Partner Authentication and Key Agreement. In *Proceedings of the 2nd ACM Conference on Wireless Network Security (WISEC)*, 2009.

[25] V. Kolesnikov, A.-R. Sadeghi, and T. Schneider. Modular Design of Efficient Secure Function Evaluation Protocols. *Cryptology ePrint Archive Report 2010/079*, 2010.

[26] B. Lampson. A Note on the Confinement Problem. *Communications of the ACM 16(10)*, 1973.

[27] Y. Lindell, and B. Pinkas. A Proof of Yao's Protocol for Secure Two-Party Computation. *Journal of Cryptology 22(2)*, 2009.

[28] D. Malkhi, N. Nisan, B. Pinkas, and Y. Sella. Fairplay - A Secure Two-party Computation System. *Proceedings of the USENIX Security Symposium*, 2004.

[29] A. Myers. JFlow: Practical Mostly-Static Information Flow Control. *Proceedings of the 26th ACM Symposium on Principles of Programming Languages*, 1999.

[30] M. Naor, and B. Pinkas. Efficient Oblivious Transfer Protocols. *Proceedings of the Symposium on Data Structures and Algorithms*, 2001.

[31] D. Naccache, and J. Stern. A New Public-Key

Cryptosystem Based on Higher Residues. *Proceedings of the ACM Conference on Computer and Communications Security*, 1998.

[32] J. D. Nielsen and M. I. Schwartzbach. A Domain-Specific Programming Language for Secure Multiparty Computation. *Proceedings of the ACM Workshop on Programming Languages and Analysis for Security*, 2007.

[33] P. Paillier. Public-Key Cryptosystems Based on Composite Degree Residuosity Classes. *Proceedings of EUROCRYPT*, 1999.

[34] A. Sabelfeld, and A. Myers. Language-Based Information-Flow Security. *IEEE Journal on Selected Areas in Communications 21(1)*, 2003.

[35] A. Schröpfer, F. Kerschbaum, and G. Müller. L1 - An Intermediate Language for Mixed-Protocol Secure Computation. *Proceedings of the IEEE Computer Software and Applications Conference*, 2011.

[36] D. Volpano, G. Smith, and C. Irvine. A Sound Type System for Secure Flow Analysis. *Journal of Computer Security 4(3)*, 1996.

[37] A. Yao. Protocols for Secure Computations. *Proceedings of the IEEE Symposium on Foundations of Computer Science*, 1982.

Robust Network Traffic Identification with Unknown Applications

Jun Zhang
School of Information Technology
Deakin University, Australia
jun.zhang@deakin.edu.au

Chao Chen
School of Information Technology
Deakin University, Australia
zvm@deakin.edu.au

Yang Xiang
School of Information Technology
Deakin University, Australia
yang.xiang@deakin.edu.au

Wanlei Zhou
School of Information Technology
Deakin University, Australia
wanlei@deakin.edu.au

ABSTRACT

Traffic classification is a fundamental component in advanced network management and security. Recent research has achieved certain success in the application of machine learning techniques into flow statistical feature based approach. However, most of flow statistical feature based methods classify traffic based on the assumption that all traffic flows are generated by the known applications. Considering the pervasive unknown applications in the real world environment, this assumption does not hold. In this paper, we cast unknown applications as a specific classification problem with insufficient negative training data and address it by proposing a binary classifier based framework. An iterative method is proposed to extract unknown information from a set of unlabelled traffic flows, which combines asymmetric bagging and flow correlation to guarantee the purity of extracted negatives. A binary classifier is used as an application signature which can operate on a bag of correlated flows instead of individual flows to further improve its effectiveness. We carry out a series of experiments in a real-world network traffic dataset to evaluate the proposed methods. The results show that the proposed method significantly outperforms the-state-of-art traffic classification methods under the situation of unknown applications present.

Categories and Subject Descriptors

C.2.3 [**COMPUTER-COMMUNICATION NETWORKS**]: Network Operations

Keywords

Network Security; Traffic Classification; Unknown Traffic

1. INTRODUCTION

As a basic technique for securing modern information infrastructure, traffic classification aims to identify the generating application/protocol of any network traffic in the complex environment [18, 4, 11]. For instance, identifying network traffic is an essential requirement in full deployment of QoS control and intrusion detection [21, 10]. Traditional traffic classification methods rely on checking IP port numbers since early well-known applications always use own unique ports for transmitting IP packets. In the last decade, many applications employ dynamic ports to evade port-based traffic identification. Taking this problem into account, current industry products apply deep inspection to traffic classification that checks the applications' signature strings in the payload of IP packets [14]. While the payload-based methods are accurate, they fail to handle encrypted applications and protect user privacy. Recent research in the area tends to investigate flow (i.e., successive IP packets) statistical features which can be extracted from IP headers without deep inspection. Substantial attention has been paid on the application of machine learning techniques into statistical feature based traffic classification [18].

Traffic classification using flow statistical features is conventionally considered as a multi-class classification problem and addressed by using supervised and unsupervised machine learning algorithms. In supervised traffic classification [16, 1, 2, 5, 9], a classification model is learned by using the labelled training samples from each predefined traffic class. By contrast, the unsupervised (clustering) methods [23, 7, 3] automatically group a set of unlabelled training samples and apply the clustering results to construct a traffic classifier with the assistance of other tools such as payload-based software. In the multi-class framework, it is assumed that any testing flow comes from a predefined traffic class. While these multi-class classification methods have reported good performance, they cannot effectively handle the emerging applications which are unknown to the traffic classification system.

In another point of view, statistical feature based traffic classification can break down to a series of detection problems by revisiting the concept of application signature. In payload based traffic classification, any testing flow is inspected to determine whether it contains the signature of a known application. If an application signature is found, the flow is categorized to the application-based traffic class. Otherwise, the flow does not belong to the class. When the flow does not contain the signature of any known application, it is labelled as unknown. Based on the signatures of known applications, we can see that any traffic flows of either known or unknown applications can be identified. In the same way, once any known application has a statistical feature based signature, all flows generated by known and unknown applications can be dealt with straightforward. However, few work has contributed to propose a statistical feature based signature including one-class support vector machine (SVM) [9] and normalized threshold [6]. Moreover, one-class SVM classifier usually suffers from poor decision boundary without the negative information and the normalized threshold method is heuristic.

Naturally, we can cast signature construction using statistical feature as a binary classification problem. For a target known application, the positive class consists of the flows generated by the application and the negative class is created by all other flows. To obtain an accurate classifier, we need the sufficient positive and negative training samples. It is generally accepted that the sufficient training samples for known applications can be easily obtained. Considering the specific binary classification problem, we can have a sufficient positive training set, but the negative training set is always insufficient since unknown applications present. Note that it will lead to a biased classifier by simply using the training samples of other known applications as the negatives. The problem of insufficient negative training set has not been addressed in previous work [9, 6] on statistical feature based signature.

This paper is aimed to achieve robust network traffic classification with unknown applications, which relaxes the unrealistic assumption that all classes are known to the classifier. The following lists the major contributions.

- Considering unknown applications, we formulate statistical feature based signature construction as a specific binary classification problem with insufficient negative samples.
- To solve this problem, we develop a generic approach to extract unknown application information from a set of unlabelled traffic flows and propose to incorporate asymmetric bagging and flow correlation to guarantee the purity.
- We propose to use a binary classifier as the application signature which can operate on a bag of correlated flows instead of individual flows.

For performance evaluation, a series of experiments are carried out in two real-world network traffic datasets. The results show that the binary classifier based signature is superior to the-state-of-art traffic classification methods when unknown applications present.

The rest of this paper is organized as follows. Section 2 states the research problem by analysing the existing methods using flow statistical features. In Section 3, a new method for unknown information extraction is proposed, which is followed by a binary classifier based application signature in Section 4. Section 5 reports the experiments and results. Finally, Section 6 concludes this paper.

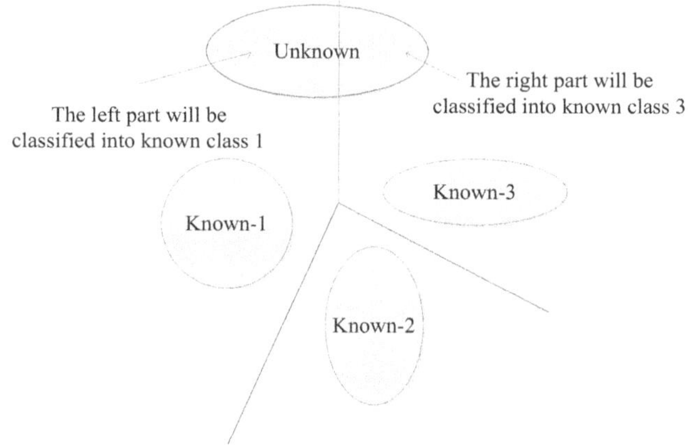

Figure 1: Problem of unknown application

2. PROBLEM STATEMENT AND ANALYSIS

What is the problem of traffic classification with unknown applications and why is it difficult to solve? In this section, we answer this question by providing a critical analysis on several typical methods using flow statistical features.

Considering a real-world network scenario, the traffic dataset, Ω, consists of K known classes and U unknown classes, $\Omega = \{\omega_1, ..., \omega_K, \overline{\omega}_1, ..., \overline{\omega}_U\}$. A known class ω_k is corresponding to an application which is well known by the traffic classification system. In this paper, it means that a set of labelled flow samples, ψ_k, is available for a known class, ω_k. By contrast, an unknown class is related to an unknown application of the system and no labelled flow samples are available. This scenario is common since a lot of new applications are emerging every day on Internet. Given the labelled flow samples $\{\psi_1, ..., \psi_K\}$, the problem is how to identify the class of the flows in Ω. In this paper, a flow consists of successive IP packets with the same 5-tuple: *source_ip*, *source_port*, *destination_ip*, *destination_port*, *transport_protocol*. A number of statistical features, such as number of IP packets, are used to represent a flow $\mathbf{x}$ for traffic classification.

Conventional traffic classification methods address a K-class classification problem, which is under the multi-class framework and cannot deal with unknown classes [16, 1, 2, 5, 9, 23, 7, 3, 12, 15, 17]. These methods use the labelled flow samples to form a training set straightforward, $T = \{\psi_1, ..., \psi_K\}$, and employ a learning algorithm to seek a classification model. The classifier trained by using T will classify any flows into a known class. Thus, the flows in the unknown classes, $\{\overline{\omega}_1, ..., \overline{\omega}_U\}$, will be inaccurately classified into K known classes. Traditional pattern classification algorithms assume that all testing flows come from the known (predefined) classes, so they all suffer from this issue. Fig. 1 illustrates the problem of unknown application class under the multi-class framework. We take the classification algorithm using decision boundary as an example. One can see that the optimized decision boundary in the feature space can effectively separate the known classes, but it will inaccurately classify the flows of unknown class into known class 1 and known class 3.

A semi-supervised method [8] was proposed to address

a $(K+1)$-class classification problem, which extended the multi-class framework by taking unknown applications into account. Firstly, some mixture of labelled and unlabelled training samples are grouped into k clusters by using traditional clustering algorithms such as k-means. Then, k clusters are mapped to $\omega_1, ..., \omega_K$, or unknown according to the locations of the labelled (supervised) training samples. For traffic classification, a flow will be predicted to the class of its nearest cluster. Although this method demonstrates the capability of dealing with unknown applications, it is heuristic in nature, e.g., the number of clusters, k, is manually adjusted. Moreover, a large k is necessary for generating high-purity clusters, while it will lead to many false unknown clusters.

In contrast to the perspective of multi-class classification, the signature-based approach offers another view for traffic classification. By using statistical feature based signature, any testing flow can be determined whether it belongs to a known class. If the flow does not belong to any known class, it is classified as unknown. The early work on normalized threshold [6] for statistical feature based signature is heuristic, which does not investigate the information of unknown applications. One-class SVM [9] can be considered as another type of signature. For a known class ω_k, the training samples in ψ_k are used to learn a one-class SVM and other training samples in $\bigcup_{i=1,i\neq k}^{i=K} \psi_i$ are used to adjust the decision boundary. There are two issues: firstly, one-class SVM normally needs a large number of training samples and the modified method cannot outperform a traditional two-class SVM; secondly, the decision boundary is still poor due to the lake of the information about unknown classes, $\{\overline{\omega}_1, ..., \overline{\omega}_U\}$.

The above analysis motives us to create statistical feature based signatures by taking unknown applications into account. Since the signature based decision is binary, 'Yes' or 'No', we cast signature construction as a binary classification problem. However, this binary classification problem is special because only a partial negative training set is available in a real world scenario. In detail, for class ω_k, the positive training set is $T_P = \psi_k$ and the partial negative training set is $T_K = \bigcup_{i=1,i\neq k}^{i=K} \psi_i$. If we use T_P and T_K to train a classifier C_k, the decision boundary of C_k will be biased towards the negative class in the feature space. In other words, some negative samples from $\{\overline{\omega}_1, ..., \overline{\omega}_U\}$ will be inaccurately classified to the positive class. The reason is that a sufficient negative training set should include the information of unknown classes, while T_K does not. Therefore, the key problem in our work is how to obtain the information of unknown classes, $\{\overline{\omega}_1, ..., \overline{\omega}_U\}$.

Inspired by the semi-supervised methods [8], we realize that the samples randomly collected from the target network are unlabelled, but they must contain traffic flows generated by unknown applications. Formally, we can obtain a set of unlabelled samples, $\Omega_r \subset \Omega$. Then, the key problem becomes how to extract the examples of unknown classes from Ω_r. Once some flows from unknown classes are identified, they can be combined with T_K to form a better negative training set, T_N. Then, T_P and T_N can be used to learn an accurate binary classifier which is considered as the signature of ω_k

3. UNKNOWN INFORMATION EXTRACTION

This section presents our generic method of unknown information extraction. The proposed method is able to extract the negative samples from a large number of unlabelled network traffic during an iterative process. Moreover, two new techniques, asymmetric bagging and flow correlation, are applied to improve the effectiveness of the iterative extraction.

input : positive trainng set T_P; partial negative training set T_K; unlabelled data set Ω_r; a binary classification algorithm Φ
output: sufficient negative training set T_N

$T_N \leftarrow T_K$;
`// create output flow set`
Use Φ to create a classifier C_0 from T_P and T_K;
`// the biased classifier` C_0 `will produce many false positives`
Classify Ω_r by C_0;
Put positive samples classified by C_0 into P_1;
Put negative samples classified by C_0 into N_1;
$i = 1$
while $N_i \neq \emptyset$ **do**
    $T_N = T_N \cup N_i$;
    Use Φ to create a classifier C_i from T_P and T_N;
    Classify P_i by C_i;
    Put positive samples classified by C_i into P_{i+1};
    Put negative samples classified by C_i into N_{i+1};
    $i = i + 1$;
end
Return T_N;

Algorithm 1: Unknown sample extraction

3.1 An Iterative Extraction Method

This method aims to identify the negative samples as much as possible during an iterative process from the unlabelled data which contains universal traffic flows of known and unknown classes. The detailed process is listed in Algorithm 1. We take the class ω_k as an example to illustrate the proposed method. For the binary classification, the positive class is ω_k and the negative class includes other known classes $\Omega'_K = \{\omega_2, ..., \omega_K\}$ and all unknown classes $\overline{\Omega}_U = \{\overline{\omega}_1, ..., \overline{\omega}_U\}$.

To start this process, we need an initial binary classifier given $T_P = \psi_k$ and $T_N = T_K = \bigcup_{i=1,i\neq k}^{i=K} \psi_i$. As mentioned before, if we use the classification algorithm Φ to create a classifier from T_P and T_K, the biased classifier will inaccurately classify some negative samples as positive. However, in the other hand, it definitely can produce some new negative samples which may be from some known or unknown classes. Based on this observation, we can construct an initial binary classifier C_0 from T_P and T_K.

$$C_0 = \Phi(T_P, T_K), \quad (1)$$

Then, C_0 is used to classify the unlabelled data set Ω_r, which produces a positive sample set P_1 and a negative sample set N_1.

$$C_0 : \Omega_r = P_1 \cup N_1. \quad (2)$$

In the training data, T_P is sufficient, while T_K is insufficient without the samples for $\{\overline{\omega}_1, ..., \overline{\omega}_U\}$. Therefore, the decision boundary of C_0 will be biased towards the negative class. Ideally, P_1 will consist of the flows of ω_k and many flows of unknown classes $\overline{\Omega}_U$. N_1 will contain the flows of Ω'_K and some flows of unknown classes $\overline{\Omega}_U$. We can see that the negative samples N_1 classified by C_0 could be high-purity and contains the information of unknown classes. Thus, N_1 can be complementary to the negative training set T_N. In addition, P_1 includes many flows of unknown classes, which motivates us to further extract the unknown information from P_1.

Iteratively, we can update the negative training set T_N by adding a high-purity negative set N_i and train a more accurate classifier C_i from T_P and T_N. In fact, a series of binary classifiers push the decision boundary back to the real borderline of the positive class step by step. The stop criteria is $N_i = \emptyset$, then T_N becomes steady. Finally, we have

$$T_N = T_K \cup N_1 ... \cup N_i, \ (where\ N_i = \emptyset). \tag{3}$$

This iterative process has the capability of effectively extract unknown information from a large number of unlabelled network traffic and significantly improve the negative training set.

The proposed method can employ most traditional classification algorithms such as SVM and random forest for implementation. A similar method [22] has been proposed for web page classification without negative examples, which combines two algorithms: 1-DNF for constructing the initial classifier and SVM for the iterative process. By considering the traffic classification situation, 1-DNF is difficult to incorporate partial negative training samples. We need a new classifier to ensure the high-purity of classified negative flows and thus to extract unknown information accurately. Therefore, we propose to apply two techniques, asymmetric bagging and flow correlation, to construct a new classifier for unknown information extraction.

3.2 Asymmetric Bagging

The purity of extracted negative samples is critical to the performance of the final classifier used for testing. However, we observed that simply applying a traditional classifier in the iterative method cannot guarantee the high purity of obtained new negative samples. On the one hand, the insufficient negative training set will let the decision boundary be biased towards the negative class. On the other hand, the size of the negative training set is much larger than that of the positive training set. This imbalance will push the decision boundary towards the positive class. In short, the imperfect decision boundary cannot avoid false negative, i.e., classifying real positives as negatives, so the extracted negative set is not pure.

We propose to apply asymmetric bagging [19, 25] to help address the problem of low-purity negative samples. The idea is to train multiple classifiers using different negative samples and combine the results of these classifiers. Each time we randomly select a subset of negative samples with the same size to the positive training set. Then, we combine the selected negative training set with the positive training set to train a classifier.

$$C_i = \Phi(T_P, T_{Ni}), \ T_{Ni} \subset T_N \ \& \ \|T_{Ni}\| = \|T_P\|. \tag{4}$$

Once the classifier is ready, we use it to classify all testing flows. Therefore, we can train multiple classifiers to conduct classification and obtain multiple class labels for each flow. Finally, we combine the multiple predicted labels of each flow to make a final decision. For example, random forest can be used to train a classifier and the majority vote rule can be used to combine classification results.

$$assign\ \mathbf{x} \longrightarrow \omega_k\ if$$
$$\textstyle\sum_{i=1}^{i=A} C_i(\mathbf{x}) > \frac{A}{2}, \tag{5}$$

where $C_i(\mathbf{x}) = 1$ means flow $\mathbf{x}$ is classified to ω_k. $C_i(\mathbf{x}) = 0$ denotes that flow $\mathbf{x}$ is rejected by ω_k.

In the asymmetric bagging-based training process, the negative and positive training sets have the same size, which can avoid the unbalance problem. The decision boundary of a single classifier will be biased towards the negative class. Classifier combination can be helpful to improve the final classification accuracy. In this way, we can guarantee that the purity of the extracted negative samples is high.

3.3 Flow Correlation

Due to unknown applications, the decision boundary of a classifier trained by using insufficient negative samples could not be optimal, which will affect the quality of extracted unknown information. The previous work [24] showed that flow correlation can improve the traffic classification performance with an insufficient training set. Since flow correlation focuses on investigating the relationship among real world data, it can be applied to either known or unknown applications.

We propose to incorporate flow correlation into the asymmetric bagging-based classification process in order to further improve the effectiveness of unknown information extraction. We use the 3-tuple heuristic to determine correlated flows which are modelled by "bag of flows" (BoF).

- 3-tuple heuristic: in a certain period of time, the flows sharing the same 3-tuple {source_ip, destination_ip, transport_protocol}form a BoF.

For example, several flows initiated by different hosts are all connecting to a same host at TCP port 80 in a short period. These flows are very likely generated by the same application such as a web browser. In general, we can apply 3-tuple heuristic to determine BoFs, but it needs to answer which classes the BoFs belong to. With flow correlation, an aggregation classifier aims to classify BoFs instead of individual flows. Given a BoF $X = \{\mathbf{x}_1, ..., \mathbf{x}_B\}$, we aggregate the results of B flows produced by a binary classifier C to predict the class label of X. In this paper, we also employ the majority vote rule to perform flow prediction aggregation due to its good performance. A BoF X can be classified to class ω_k only if

$$\sum_{\mathbf{x} \in X} C(\mathbf{x}) > \|X\|/2. \tag{6}$$

Algorithm 2 shows the ensemble classification with asymmetric bagging and flow correlation.

4. APPLICATION SIGNATURE WITH FLOW CORRELATION

This section presents the new approach to construct application signatures with flow correlation (ASFC). Moreover, the technical justification is also provided to confirm the effectiveness of the proposed approach.

4.1 Proposed Approach: ASFC

Taking unknown applications into account, this work focus on developing a new application signature based traffic

input : positive trainng set T_P; negative training set T_N; testing set Ω_t; 3-tuple information set Ω_3; a binary classification algorithm Φ; number of asymmetric bagging classifier A
output: testing class label set L_t

```
Construct BoFs X = {X_j} by running 3-tuple
heuristic on Ω_3;
for i ← 1 to A do
    Obtain T_Ni by random sampling in T_N;
    // T_Ni has the same size to T_P
    Use Phi to create C_i from T_P and T_Ni
    for j ← 1 to ||X|| do
        for k ← 1 to ||X_j|| do
            Classify x_jk by C_i;
        end
        Use majority vote rule to aggregate the
        predictions of flows in BoF X_j.
    end
end
Obtain L_t by combining the predictions of BoFs
produced by A asymmetric bagging classifier
{C_1, ..., C_A};
Return L_t
```

Algorithm 2: Ensemble classification

classification. Section 3 has addressed the key problem of unknown information extraction. Another problem is how to construct the signature of each application using the available training data.

In this paper, we create a binary classifier and use it as the application signature in a fast way. Given the positive and negative training sets, we can train a binary classifier for an application. If a testing flow is classified to positive, it means this flow comes from the application. Otherwise, the testing flow does not belong to the application. With the consideration of flow correlation, we perform an ensemble classification on BoFs instead of individual flows. Then, the accuracy of the application signature can be improved by inspecting a bag of correlated flows simultaneously .

We summarize the proposed approach for signature based traffic classification as follows. For convenience, we take an application as example.

1. Collect a set of unlabelled traffic flows Ω_r from the target network traffic;
2. Extract a complete negative training set T_N from Ω_r using Algorithm 1;
3. Use T_P and T_N to train a binary classifier as the application signature;
4. Perform BoF-based classification according to Section 3.3.

Any traditional binary classification algorithm can be used in this approach. In our experiments, we select random forest as the basic classifier for our approach, but other algorithms are also tested.

4.2 Technique Justification

The proposed approach relies on the techniques of unknown information extraction. The critical problem is that the system does not have any training samples for unknown applications. The simple classifier by using the training samples of known applications cannot deal with the flows generated by unknown applications. The method of unknown information extraction was proposed to extract samples of unknown applications from a set of unlabelled network traffic. In unknown information extraction, asymmetric bagging balanced the positive and negative training sets which avoids decision boundary moving towards the positive class, so as to guarantee the purity of extracted negative samples. We further applied flow correlation in asymmetric bagging based classification, which can effectively increase the amount of extracted negative samples.

Moreover, the final ensemble classification process can improve the classification accuracy by investigating the relationship in real world data. Suppose $\varphi(\mathbf{x}, L)$ is a simple classifier (predictor) and L is the training set. The aggregation can be described as $\varphi_A(X, L) = E_{\mathbf{x}\in X}\varphi(\mathbf{x}, L)$. Let y be the class label of a flow $\mathbf{x}$ which belongs to a BoF X. Both y and $\mathbf{x}$ are random variables which are drawn from the distribution independent of the training set L. The average classification error on BoFs, estimated by the simple predictor $\varphi(\mathbf{x}, L)$, is $E_{y,\mathbf{x}\in X}(y - \varphi(\mathbf{x}, L))^2$. The corresponding classification error estimated by the aggregated predictor is $E_{y,\mathbf{x}\in X}(y - \varphi_A(X, L))^2$. Since

$$E_{y,\mathbf{x}\in X}\varphi^2(\mathbf{x}, L) \geq (E_{y,\mathbf{x}\in X}\varphi(\mathbf{x}, L))^2, \tag{7}$$

after some modification, we obtain

$$E_{y,\mathbf{x}\in X}(y - \varphi(\mathbf{x}, L))^2 \geq E_{y,\mathbf{x}\in X}(y - \varphi_A(X, L))^2. \tag{8}$$

This shows flow correlation can further improve the final classification performance.

Let us take the majority vote rule as an example for flow prediction aggregation. Suppose a basic classifier with an error rate of $p < 0.5$, it means the classifier is better than random guessing. Since the flows in a BoF are diverse, the classifier will make different errors in predicting the class of different flows. In the binary classification problem, for the majority vote to be incorrect, it requires that $\|X\|/2$ or more flows in X are classified incorrectly. The probability that r flows are classified incorrectly is

$$C_r^g \cdot p^r(1-p)^{g-r} = \frac{g!}{r!(g-r)!}p^r(1-p)^{g-r} \tag{9}$$

Therefore, the probability that the majority vote is incorrect is

$$P(error) = \sum_{r=\lfloor(\|X\|/2+1)/2\rfloor}^{\|X\|/2} \frac{g!}{r!(g-r)!}p^r(1-p)^{g-r}. \tag{10}$$

For example, given a BoF with $g = 30$ and the basic classifier with an error rate $p = 0.3$, the probability of the majority vote being incorrect is 0.006, which is much less than the individual classification error rate. In general, we have $P(error) < p$, so the classification accuracy can be improved by classifying BoFs with the majority vote rule.

5. PERFORMANCE EVALUATION

We carried out a series of network traffic identification experiments to evaluate the proposed ASFC approach. It was aimed to answer an important question, why the proposed ASFC approach is superior.

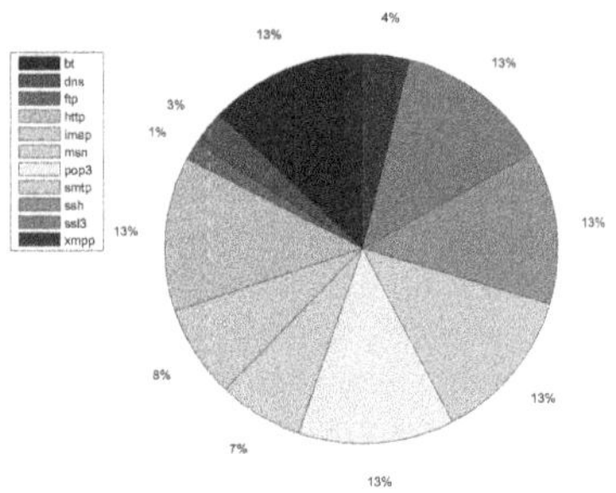

Figure 2: Dataset Distribution

Table 1: Unidirectional statistical features

Type	**Unidirection features**	**No.**
Packets	Number of packets transferred	2
Bytes	Volume of bytes transferred	2
Packet Size	Min., Max., Mean and Std Dev. of packet size	8
Inter-Packet Time	Min., Max., Mean and Std Dev. of Inter Packet Time	8
	Total	20

5.1 Datasets and Experiments

A real-world dataset was used to evaluate our new techniques in this work. The *isp* dataset is a full payload traffic dataset collected at a medium sized ISP in Australia [20]. In order to build ground truth, we have developed a deep payload inspection tool (DPI) which uses string signature of packets' payload to accurately identify the class of traffic flows. A number of application signatures were developed based on the previous experience and some open source tools like l7-filter (http://l7-filter.sourceforge.net) and Tstat (http://tstat.tlc.polito.it). For some encrypted or new applications, we inspected the flows manually to determine the class of flows. Finally, the *isp* dataset is constituted by over 80,000 traffic flows from 11 application-oriented classes. Figure 2 shows the distribution of traffic classes in the dataset. In the experiments, 20 unidirectional flow statistical features were extracted to represent traffic flows, which are listed in Table 1. We applied feature selection to further remove irrelevant and redundant features from the feature set [13]. The process of feature selection yielded 6 features for the isp dataset.

In the experiments, each dataset was separated into three parts, one for training, one for unlabelled, and the other for testing. The training, unlabelled, and testing parts have 25%, 25%, and 50% of traffic flows in the dataset, respectively. To simulate the problem unknown applications, the idea is to set several small classes to "unknown". In detail, the classes of DNS, FTP, IMAP, MSN, XMPP in the *isp* dataset are set as unknown. Therefore, the modified isp dataset consists of 6 known classes and 5 unknown classes. All unknown classes includes about 22% of traffic flows in the dataset. For each known class, 1,000 flow samples are randomly selected from the training part to form a supervised training set. It is important to note that no any samples of unknown classes are available for the classification system.

For the empirical study, we use different metrics to evaluate ASFC from multiple perspectives. All used metrics are derived from the four different possible outcomes of a single prediction for a two-class case, true positive (TP), true negative (TN), false positive (FP), and false negative (FN). A false positive is when the real "negative" is incorrectly classified as "positive". A false negative is when the real "positive" is incorrectly classified as "negative". True positives and true negatives are obviously correct classifications. The average performance over 100 runs are reported in this paper.

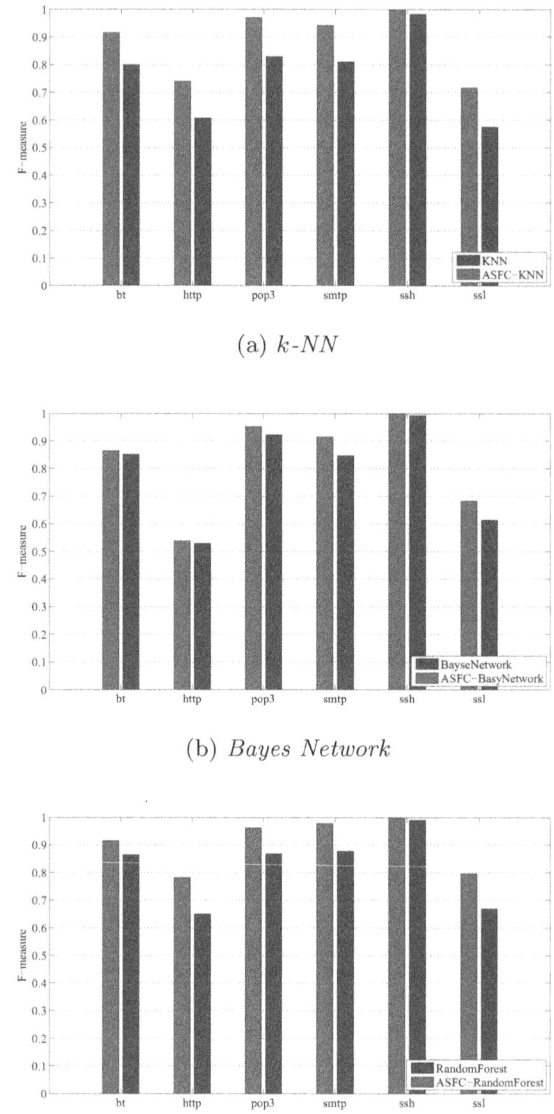

(a) *k-NN*

(b) *Bayes Network*

(c) *Random Forest*

Figure 3: F-measure on *isp*

5.2 Identification Performance of ASFC

We performed a set of experiments to test the identification performance of the proposed ASFC approach, in which three different learning algorithms, *k*-Nearest Neighbour (*k*-NN), Bayes Network (BayesNet) and RandomForest, are used to construct a basic binary classifier. Following the

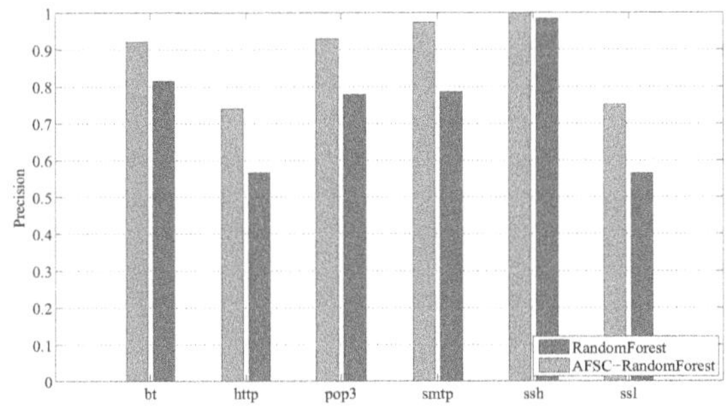

(a) Precision of each known class

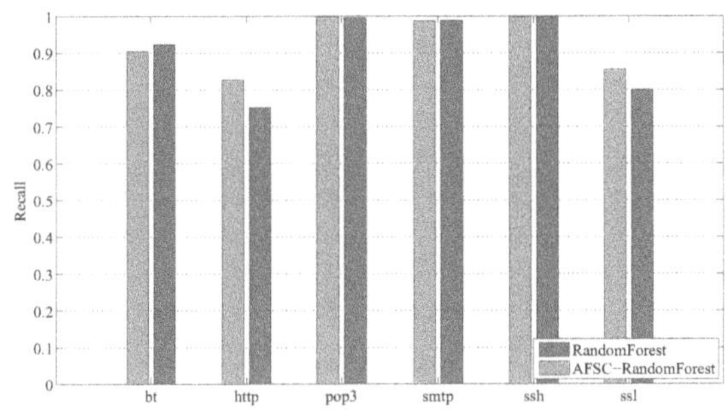

(b) Recall of each known class

Figure 4: Precision and recall of using random forest

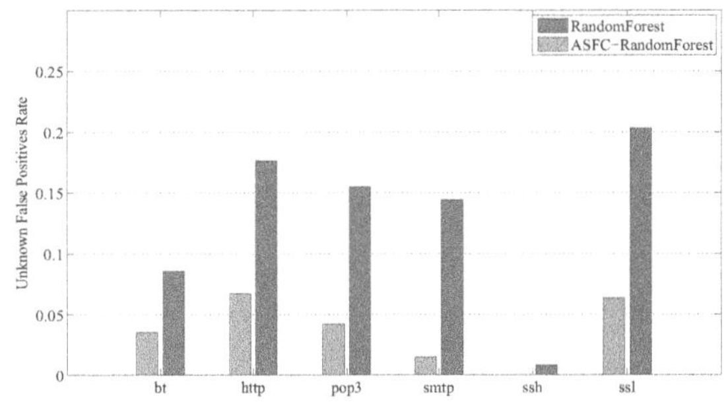

Figure 5: Unknown False Positives Rate

existing works in the area, we use *Precision*, *Recall* and *F-measure* to measure the final performance.

- Precision is defined as the ratio of correctly classified flows over all predicted flows in a class. It can be calculated by

$$Precision = \frac{TP}{TP + FP} \tag{11}$$

- Recall is defined as the ratio of correctly classified flows over all ground truth flows in a class. It can be calculated by

$$Recall = \frac{TP}{TP + FN} \tag{12}$$

- F-measure is a combination of precision and recall, it is a widely adopted metric to evaluate per-class performance. It can be calculated by

$$F - measure = \frac{2 * Precision * Recall}{Precision + Recall} \tag{13}$$

Fig. 3 shows the overall performance in terms of the F-measure. To demonstrate the advantage of ASFC, the results of solely using these learning algorithms for flow identification are also reported. One can see that ASFC can effectively improve the F-measure in each class which is independent to the basic learning algorithm. ASFC with random forest can achieve the best performance as shown in Fig. 3(c). For the classes, http, pop3, smtp, and ssl, the improvements can achieve 10%. Although ssh class is easy to identify, ASFC can further improve its F-measure. The F-measure of ASFC with random forest is higher than the method without ASFC about 5%.

We take ASFC with random forest as an example to study how ASFC can affect the precision and recall for each class. Fig. 4 reports the precision and recall of each class produced by ASFC with random forest, with comparison to the results of random forest. An important observation for this figure is that ASFC can significantly increase the precision without decreasing the recall of each class. It's why the F-measure got much improved. For example, the precision of smtp can be improved about 20% by using ASFC and its recall doesn't drop. In class ssl, ASFC can improve the precision over 15% and the recall by 5% as well.

5.3 Impact of Unknown Information Extraction

This section aims to investigate why unknown information extraction is helpful to deal with unknown applications. In this section, we introduce two new metrics, unknown flow extraction rate and unknown false positive rate. For convenience, the flows generated by unknown applications are called "unknown flows" in the rest of the paper.

- Unknown flow extraction rate is defined as the ratio of the number of extracted unknown flows to the amount of all unknown flows in the unlabelled dataset. A high rate means the proposed unknown extraction method can automatically and effectively collect the information of unknown applications from universal real world network traffic.

- Unknown false positive rate is defined as the ratio of the number of unknown flows that are inaccurately identified as known to the amount of flows that are classified to the known class. We use this rate to show the impact of unknown applications to flow identification. ASFC is developed to significantly reduce the unknown false positive rate.

Table 2 lists the unknown flow extraction rate of ASFC with random forest for each class. For example, the rate is 89.93% for bt that means our iterative method can automatically extract 89.93% from all unknown flows in the unlabelled dataset. In other words, we can successfully obtain the representative samples of unknown applications. Then, these extracted unknown flows are combined into the training set so as to effectively enhance the robustness of the final classifier. From Table 2, we can see that the proposed iter-

Table 2: Unknown flow extraction rate

Class	bt	http	pop3	smtp	ssh	ssl3
Extraction Rate	89.93%	58.79%	76.91%	82.44%	99.74%	63.91%

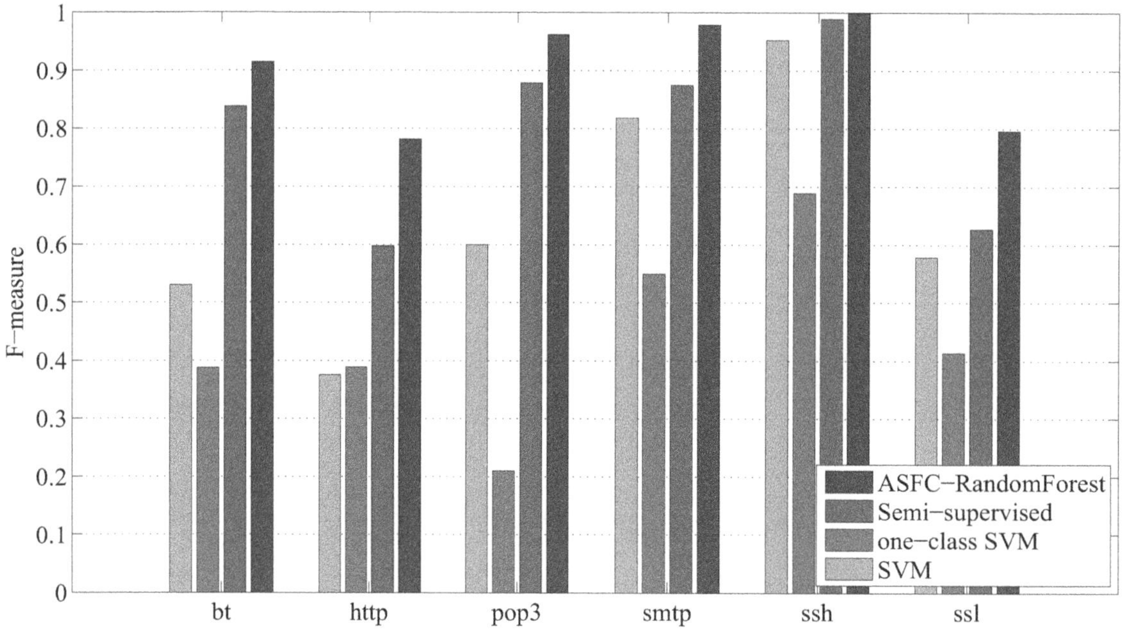

Figure 6: Methods comparison

ative method can extract most of unknown flows from the unlabelled dataset for robust classifier training of each class.

Moreover, we studied how the extracted unknown flows were helpful to improve the identification performance based on the unknown false positive rates as shown in Fig. 5. ASFC with random forest is compared to the method of solely using random forest. The results show that our ASFC with unknown information extraction can effectively reduce the amount of unknown flows that are inaccurately classified to known classes. For example, over 15% of unknown flows in the testing data are inaccurately identified as smtp, but this rate reduces to about 4% when using ASFC. In other words, the proposed approach can effectively detect the flows of unknown applications.

Based on the above results and analysis, we can draw an initial conclusion that ASFC can automatically extract sufficient information of unknown applications from unlabelled network traffic and utilize it to significantly reduce the unknown false positive rate. Therefore, ASFC demonstrates the strong capability of improving the robustness of flow identification.

5.4 Comparison with Other Methods

This section compares our proposed ASFC and two existing works, Erman's semi-supervised method [8] and Este's One-class SVM method [9]. In Erman's semi-supervised methods, we run k-means on the mixture data consisting of the training samples of known classes and the unlabelled data, where k was set to 400 based on the experiments. We found that the performance of flow identification using one-class SVM is very poor. Considering Este's idea, we used SVM to construct a binary classifier, thus the positive training samples and negative training samples can jointly optimize the decision boundary.

Fig. 6 reports the F-measures of these methods. Based on the results, the ranking list is ASFC-RandomForest, semi-supervised, SVM, and one-class SVM. In particular, our ASFC-RandomForest is much better than other competing methods. For example, ASFC-RandomForest outperforms semi-supervised over 15% in http class. The improvement is about 10% in other classes except ssh. For ssh, both methods achieve nearly 100% F-measure.

Fig. 7 shows the precision and recall of all competing methods. We can see that ASFC always significantly outperforms other methods in either precision or recall for any classes. For example, the precision of ASFC is higher than the second best, semi-supervised, over 15% in http class. The improvement on recall can achieve 20% in this class. In class ssl, the improvements on both precision and recall are over 15%. In general, ASFC has the superior capability to improve precision and recall when unknown applications present. The basic reason is ASFC can effectively extract unknown information and combine correlated flows to make more accurate decisions.

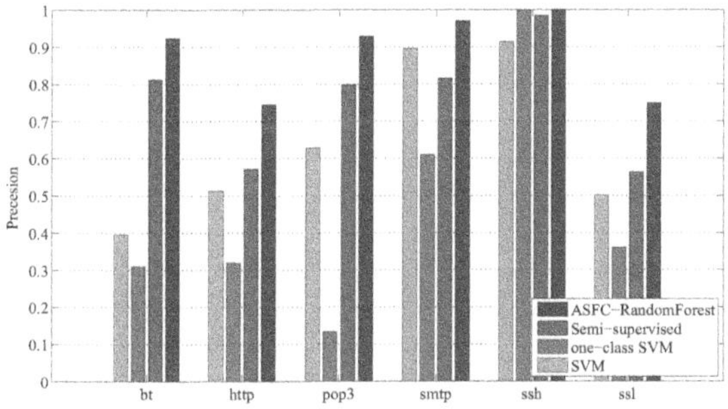

(a) Precision

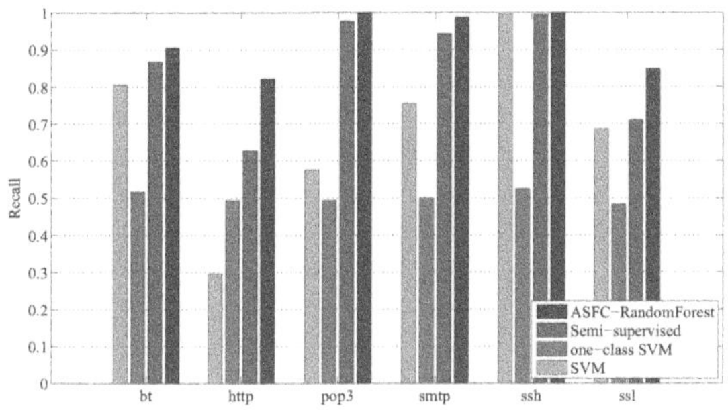

(b) Recall

Figure 7: Precision and recall of competing methods

6. CONCLUSION AND DISCUSSION

This paper solves a critical problem of traffic classification with unknown applications presented, which normally exists in real world networks. To address this problem, we proposed an iterative method to extract unknown information from a set of unlabelled traffic flows, which combines asymmetric bagging and flow correlation to guarantee the purity of extracted negatives. For traffic classification, a binary classifier was created as an application signature which can operate on a bag of correlated flows instead of individual flows to further improve classification accuracy. We carry out a series of experiments in a real-world network traffic dataset to evaluate the proposed methods. The results show that the proposed method outperforms the-state-of-art traffic classification methods under the extreme situation of unknown applications present. It is due to the high unknown extraction rate which leads to less unknown flows inaccurately classified to known classes.

Our empirical study suggests that the iterative extraction method only works when the asymmetric bagging and flow correlation techniques are used. Asymmetric bagging and flow correlation both have significant impact to the final performance. For example, with flow correlation, the F-measure can improve 5% for a class. By combining asymmetric bagging and flow correlation, the improvement can achieve 10%. In the future, we will develop practical software to facilitate unknown information extraction and apply it to real-time traffic classification.

7. REFERENCES

[1] T. Auld, A. W. Moore, and S. F. Gull. Bayesian neural networks for internet traffic classification. *IEEE Trans. Neural Netw.*, 18(1):223–239, January 2007.

[2] L. Bernaille and R. Teixeira. Early recognition of encrypted applications. In *Proceedings of the 8th international conference on Passive and active network measurement*, pages 165–175, Berlin, Heidelberg, 2007.

[3] L. Bernaille, R. Teixeira, I. Akodkenou, A. Soule, and K. Salamatian. Traffic classification on the fly. *SIGCOMM Comput. Commun. Rev.*, 36:23–26, April 2006.

[4] D. Bonfiglio, M. Mellia, M. Meo, and D. Rossi. Detailed analysis of skype traffic. *IEEE Trans. Multimedia*, 11(1):117 -127, Jan. 2009.

[5] D. Bonfiglio, M. Mellia, M. Meo, D. Rossi, and P. Tofanelli. Revealing skype traffic: when randomness plays with you. In *Proceedings of the conference on Applications, technologies, architectures, and protocols for computer communications*, pages 37–48, New York, NY, USA, 2007.

[6] M. Crotti, M. Dusi, F. Gringoli, and L. Salgarelli. Traffic classification through simple statistical fingerprinting. *SIGCOMM Comput. Commun. Rev.*, 37:5–16, January 2007.

[7] J. Erman, M. Arlitt, and A. Mahanti. Traffic classification using clustering algorithms. In *Proceedings of the SIGCOMM workshop on Mining network data*, pages 281–286, New York, NY, USA, 2006.

[8] J. Erman, A. Mahanti, M. Arlitt, I. Cohen, and C. Williamson. Offline/realtime traffic classification using semi-supervised learning. *Performance Evaluation*, 64(9-12):1194–1213, October 2007.

[9] A. Este, F. Gringoli, and L. Salgarelli. Support vector machines for tcp traffic classification. *Computer Networks*, 53(14):2476–2490, September 2009.

[10] Z. M. Fadlullah, T. Taleb, A. V. Vasilakos, M. Guizani, and N. Kato. DTRAB: combating against attacks on encrypted protocols through traffic-feature analysis. *IEEE/ACM Trans. Netw* 18(4):1234–1247, August 2010.

[11] A. Finamore, M. Mellia, M. Meo, and D. Rossi. KISS: Stochastic packet inspection classifier for UDP traffic. *IEEE/ACM Trans. Netw* 18(5):1505–1515, October 2010.

[12] E. Glatz and X. Dimitropoulos. Classifying internet one-way traffic. *SIGMETRICS Perform. Eval. Rev.*, 40(1):417–418, Jun. 2012.

[13] I. Guyon and A. Elisseeff. An introduction to variable and feature selection. *J. Mach. Learn. Res.*, 3:1157–1182, March 2003.

[14] P. Haffner, S. Sen, O. Spatscheck, and D. Wang. ACAS: automated construction of application signatures. In *Proceedings of the ACM SIGCOMM workshop on Mining network data*, pages 197–202, New York, NY, USA, 2005. ACM.

[15] Y. Jin, N. Duffield, J. Erman, P. Haffner, S. Sen, and Z.-L. Zhang. A modular machine learning system for flow-level traffic classification in large networks. *ACM Trans. Knowl. Discov. Data*, 6(1):4:1–4:34, Mar. 2012.

[16] H. Kim, K. Claffy, M. Fomenkov, D. Barman, M. Faloutsos, and K. Lee. Internet traffic classification demystified: myths, caveats, and the best practices. In

Proceedings of the ACM CoNEXT Conference, pages 1–12, New York, NY, USA, 2008.

[17] T. Nguyen, G. Armitage, P. Branch, and S. Zander. Timely and continuous machine-learning-based classification for interactive ip traffic. *IEEE/ACM Trans. Netw* 20(6):1880-1894, December 2012.

[18] T. T. Nguyen and G. Armitage. A survey of techniques for internet traffic classification using machine learning. *IEEE Commun. Surv. Tutor.* 10(4):56–76, Fourth Quarter 2008.

[19] D. Tao, X. Tang, X. Li, and X. Wu. Asymmetric bagging and random subspace for support vector machines-based relevance feedback in image retrieval. *IEEE Trans. Pattern Anal. Mach. Intell.* 28(7):1088–1099, July 2006.

[20] Y. Wang, Y. Xiang, J. Zhang, and S.-Z. Yu. A novel semi-supervised approach for network traffic clustering. In *International Conference on Network and System Security*, Milan, Italy, September 2011.

[21] Y. Xiang, W. Zhou, and M. Guo. Flexible deterministic packet marking: An ip traceback system to find the real source of attacks. *IEEE Trans. Parallel Distrib. Syst.* 20(4):567–580, April 2009.

[22] H. Yu, J. Han, and K. C.-C. Chang. PEBL: web page classification without negative examples. *IEEE Trans. Knowl. Data Eng.* 16(1):70–81, Jan. 2004.

[23] S. Zander, T. Nguyen, and G. Armitage. Automated traffic classification and application identification using machine learning. In *Annual IEEE Conference on Local Computer Networks*, pages 250–257, Los Alamitos, CA, USA, 2005.

[24] J. Zhang, Y. Xiang, Y. Wang, W. Zhou, Y. Xiang, and Y. Guan. Network traffic classification using correlation information. *IEEE Trans. Parallel Distrib. Syst.* 24(1):104–117, January 2013.

[25] J. Zhang and L. Ye. Content based image retrieval using unclean positive examples. IEEE Trans. Image Process. 18(10):2370–2375, Oct. 2009.

STRIDE: Sanctuary Trail – Refuge from Internet DDoS Entrapment

Hsu-Chun Hsiao* Tiffany Hyun-Jin Kim* Sangjae Yoo* Xin Zhang†
Soo Bum Lee* Virgil Gligor* Adrian Perrig*

*CyLab / Carnegie Mellon University †Google

{hchsiao, hyunjin, yoo, soobum, gligor, perrig}@cmu.edu xinzh@google.com

ABSTRACT

We propose STRIDE, a new DDoS-resilient Internet architecture that isolates attack traffic through viable bandwidth allocation, preventing a botnet from crowding out legitimate flows. This new architecture presents several novel concepts including tree-based bandwidth allocation and long-term static paths with guaranteed bandwidth. In concert, these mechanisms provide domain-based bandwidth guarantees within a trust domain – administrative domains grouped within a legal jurisdiction with enforceable accountability; each administrative domain in the trust domain can then internally split such guarantees among its endhosts to provide (1) connection establishment with high probability, and (2) precise bandwidth guarantees for established flows, regardless of the size or distribution of the botnet outside the source and the destination domains. Moreover, STRIDE maintains no per-flow state on backbone routers and requires no key establishment across administrative domains. We demonstrate that STRIDE achieves these DDoS defense properties through formal analysis and simulation. We also show that STRIDE mitigates emerging DDoS threats such as Denial-of-Capability (DoC) [6] and N^2 attacks [22] based on these properties that none of the existing DDoS defense mechanisms can achieve.

Categories and Subject Descriptors

C.2.0 [**Computer-Communication Networks**]: General; C.2.1 [**Computer-Communication Networks**]: Network Architecture and Design

Keywords

DDoS defense, DDoS-resilient Internet architecture, bandwidth allocation, bandwidth guarantees.

1. INTRODUCTION

DDoS attacks are still prevalent in the Internet today. In fact, a recent world-wide security survey [1] suggests that Botnet-driven DDoS attacks have become common as a low cost, high-profile form of cyber-protest. Both attack intensity and frequency have drastically accelerated: the largest reported attack size doubled every year, to more than 100 Gbps seen in 2010. The majority of network operators in the survey also ranked DDoS attacks as the biggest threat.

The recently proposed N^2 attack [22], also referred to as a Coremelt attack, poses a new threat and has not been effectively addressed by any system to date. In an N^2 attack, an adversary uses a large-scale botnet whose bots communicate only with each other to overload network links. Current DDoS defense mechanisms that attempt to eliminate undesired traffic are rendered ineffective, because all inter-bot traffic is desired by the bot endhosts. Other DDoS defense mechanisms that perform per-source or per-computation fair sharing at congested links may in fact give disproportionate advantage to sources with small uplink bandwidth or with high computational resources, respectively. Moreover, malicious domains can misuse per-source fair sharing by creating multiple bogus senders, and per-computation fair sharing may be too expensive to protect every data packet. Furthermore, global fair sharing implies global fate sharing – a source's share is affected by bots in distant domains over which the source has no influence.

Current DDoS countermeasures have encountered fundamental limitations to address the challenges we describe above to be compatible with the current Internet. Thus, an exciting research challenge is to study if a next-generation network infrastructure could be more effective against DDoS attacks – what architectural primitives can effectively defend against DDoS attacks?

In this paper, we formulate a new network architecture called STRIDE that provides *domain-based guarantees* for intrinsic DDoS protection within a Trust Domain (TD), which contains a set of contiguous Autonomous Domains (AD) with a common root of trust. Specifically, STRIDE provides precise bandwidth guarantees to AD-level paths, or the "sanctuary trails" that isolate attack traffic from legitimate communication. Each endpoint AD can then internally split the guarantee among its endhosts.

Our architecture is based on the following insights: (1) Bandwidth allocation is simple in a tree-based topology, as the available bandwidth can be split from the root down to each leaf; (2) with network capabilities encoded in packet-carried state and fixed bandwidth classes, routers can perform enforcement in a per-flow stateless fashion using probabilistic detection of the largest flows; (3) by combining a static long-term traffic class guaranteeing low bandwidth with a dynamically-allocated short-term traffic class guaranteeing high bandwidth, we can provide a variety of

ASIACCS'13, May 8–10, 2013, Hangzhou, China.

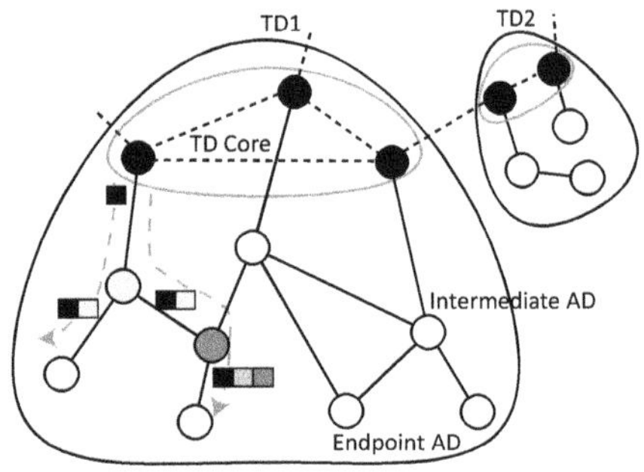

Figure 1: Trust domain (TD) example. Each node represents an AD, and the five black nodes represent the tier-1 ADs that constitute the TDCs. Each square corresponds to the node's path information.

guarantees to both privately communicating endhosts and public servers.

We leverage the SCION next-generation Internet architecture [31] to perform the tree-based bandwidth allocation: paths are created and available bandwidth is allocated as paths branch out like trees from the network core. The packet-carried forwarding state of SCION also provides us with a natural way to encode network capabilities [4, 26].

Note that existing schemes for bandwidth reservation (e.g., RSVP [8]) or Quality of Service are insufficient to guarantee timely end-to-end data delivery in the presence of DDoS adversaries for several reasons: (1) their reservation requests are unprotected against DDoS attacks, and (2) they lack lower-bound guarantees for reservable bandwidth. In contrast, STRIDE provides domain-based guarantees, achieving previously unachievable DDoS defense properties for communication within a TD; many of the properties also translate for communication between TDs.

STRIDE provably guarantees connection setup for private communication, bounds the waiting time for accessing public services, and provides precise bandwidth guarantees for established flows, all achieved regardless of the size or distribution of the botnet outside the source and destination ADs. These guarantees enable STRIDE to mitigate emerging threats such as the Denial-of-Capability (DoC) and N^2 attacks. Furthermore, STRIDE does not require backbone routers to keep per-flow state.

2. BACKGROUND: SCION

In this section, we review the SCION network architecture [31] that STRIDE leverages for DDoS resilience. Among new Internet architecture proposals [27, 31], we base our design on SCION because its notion of Trust Domain (TD) and secure top-down topology discovery enable tree-based bandwidth allocation within a uniform legal environment, whereby bandwidth guarantees and attack isolation can be enforced. Note that although SCION enables natural isolation of attack traffic from untrusted entities, it is still vulnerable to DDoS threats within a TD; in contrast, STRIDE seeks to provide domain-based bandwidth guarantees for intra-TD communication.

We consider an Internet topology at the Autonomous Domain (AD) level. In this topology, nodes represent ADs, each of which has several gateway routers (or interfaces) connecting it to neighboring ADs. Endpoint ADs provide Internet access to endhosts.

2.1 Trust Domains

SCION divides ADs on the Internet into several trust domains (TDs), where a TD is defined as "a set of ADs that agree on a coherent root of trust and have mutual accountability and enforceability for route computation under a common regulatory framework" [31]. Each TD contains a *TD Core* (TDC) consisting of the tier-1 ISPs that manage the TD. The primary advantage of such a trust domain division is that it avoids having a single root of trust for the entire Internet, which is difficult to unanimously agree on in practice. For ease of presentation, we will focus on operations within one TD unless we mention otherwise explicitly. Figure 1 illustrates these concepts.

2.2 Secure Top-Down Topology Discovery

For finding routing paths in SCION, the TDC periodically broadcasts *Path Construction Beacons* (PCB), which establish half-paths back to the TDC as they are disseminated throughout the network in a top-down manner (i.e., from the TDC to endpoint ADs). End-to-end communication paths are established by combining the source's and destination's half-paths.

PCBs are constructed as follows. The TDC initiates PCBs which contain *one-hop* paths starting from the core to its adjacent customer ADs with their expiration times. Upon receiving a PCB, an intermediate AD updates the PCB for each of its downstream ADs (e.g., customers and peers) with the authenticated local topology: the ith intermediate AD (AD_i) appends to the PCB the local path information, ingress and egress interfaces I_i, for a particular downstream AD (AD_{i+1}) followed by an opaque field O_i, which encodes the forwarding decision as ingress/egress points at AD_i.

$$I_i = ingress_i \| egress_i \| AD_{i+1},$$
$$O_i = \mathrm{MAC}_{K_i}(I_i \| O_{i-1}), \quad (1)$$

where $ingress_i$ and $egress_i$ stand for the ingress and egress interfaces of AD_i. O_i is computed using a secret key K_i known only to AD_i to protect the integrity of the routing information. Also, AD_i digitally signs PCBs to prevent fake route injection. Note that PCBs in SCION do not announce bandwidth availability.

AD_i propagates the updated PCB to the designated downstream AD (AD_{i+1}). AD_i repeats this process for other downstream ADs on different paths, and upon receiving PCBs, the downstream ADs learn the path to reach the TDC. PCBs travel along a special control channel which has isolated bandwidth from all data packets and hence is protected from data-plane DDoS attacks.

For each received PCB, an endpoint AD learns a series of interfaces and opaque fields as an unforgeable *path token* that represents the forwarding decisions of the corresponding path. To send packets on the path, the sender embeds in the packet header the path token, which reminds every intermediate AD of its own routing decision for carrying the packet based on its policy. Hence, no forwarding state at routers is needed.

Among all the half-paths that an endpoint AD learns from PCBs, the endpoint AD selects some as up-paths for reaching the TDC and some as down-paths for receiving packets from the TDC. To form end-to-end paths, the destination AD publishes its down-paths (i.e., the path tokens of these down-paths) to the Path Server, which is a DNS-like system, in the TDC. A source AD wishing to communicate with the destination can query the Path Server for the destination's down-paths. An end-to-end path is then constructed by splicing the source's and destination's half-paths.

3. THREAT MODEL

We consider massive Distributed Denial of Service (DDoS) attacks launched by a botnet, which consists of a large number of malware-infected bot endhost machines. In particular, we address two types of DDoS attacks: (1) disabling connection setup in capability-based protocols (DoC attack [6]), and (2) exhausting link bandwidth to crowd out established legitimate connections. In bandwidth exhaustion attacks, we especially focus on the N^2 attack [22], which aims to overload a target ISP's backbone network using a large number of legitimate-looking flows established among colluding bots (hence, any attempt to identify the attack based on the flow's bandwidth would fail). This attack is called an N^2 attack, because the N bots can open $O(N^2)$ connections among each other.

3.1 Desired Properties

We aim to achieve the following properties for a DDoS-resilient network architecture.

Domain-based guarantees within a TD. Precise bandwidth guarantees should be provided to communication between endpoint domains residing in the same TD, and each endpoint domain can internally split the guarantee among its endhosts based on its local policy. A domain that intends to achieve highly available communication could identify malicious bots and remove them to provide better guarantees to legitimate endhosts. Domain-based guarantees ensure that the effect of attacks is confined to infested domains, such that the endhosts can establish a bandwidth-guaranteed flow with high probability.

Robustness and efficiency. To be resilient to DDoS attacks, network elements require efficient protocols and network devices. indicates that the architecture should avoid per-flow or per-host state at backbone routers and should avoid expensive operations such as digital signature generation or verification in the fastpath.[1]

Flexible route control. Endpoint ADs should be able to control paths to avoid congestion: a source AD needs to have multiple paths to reach a destination; a destination AD should be able to hide/disclose paths for private/public communication, and change inbound paths to shift traffic.

3.2 Assumptions

In designing a new DDoS resilient Internet architecture, we only make two fundamental assumptions that can be justified using existing security mechanisms.

TDCs support congestion-free communication. Since the TDC topology is small and relatively fixed, ADs in the TDC can accurately assess and provision the capacity requirement of each link to ensure no packet loss for traffic below a certain rate. For example, given a topology and its link capacity, the TDC can adopt congestion-free routing [30] to determine how much congestion-free traffic it can support on each incoming link.

TDC detects and revokes malicious AD members. Since all ADs of a TD are within a uniform legal environment (as described in Section 2), the TDC can revoke the membership of misbehaving ADs. In Section 8, we discuss technical approaches to detect compromised or poorly-administered ADs that fail to correctly monitor traffic.

[1]Fastpath and slowpath refer to different processing elements that handle packet forwarding in a router. The fastpath refers to packet processing by dedicated hardware, for example, by the linecard in a router. The slowpath refers to packet processing by a general CPU, which often results in a packet processing latency that is usually about two orders of magnitude slower than the fastpath.

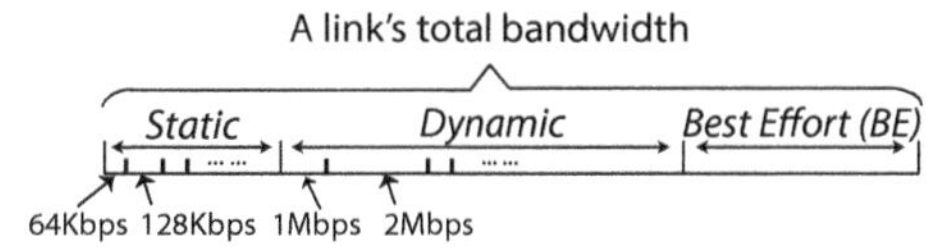

Figure 2: Three bandwidth classes of STRIDE.

4. STRIDE: DESIGN OVERVIEW

We first sketch how our new architecture, STRIDE, provides guaranteed end-to-end data delivery to legitimate flows even in the presence of DDoS attacks. In a nutshell, STRIDE protects end-to-end data delivery by establishing communication *channels*[2] that confine the effect of attacks to their originating domains, leaving other domains unaffected. Such communication channels also form the "sanctuary trails" that isolate attack traffic from legitimate communication. Moreover, since bots within a domain will compete among each other for a fixed amount of bandwidth, bandwidth guarantees can be provided to channels regardless of the size or distribution of the botnet outside the source and destination domains.

Hence, the primary challenge STRIDE faces are how to secure such channel establishment and adjust allocation in response to dynamic traffic patterns. STRIDE addresses these challenges by combining (1) a low bandwidth but long-lived and guaranteed traffic class for channel establishment with (2) a high bandwidth but short-term dynamically-allocated traffic class. Specifically, a channel is constructed using any of the 3 types of *bandwidth classes*: static, dynamic, and best effort (BE). Hybrid channels are also possible, as we explain later. For simplicity, we focus on presenting STRIDE using bidirectional links and elaborate how STRIDE can support directional links in Section 8.

Bandwidth classes. As shown in Figure 2, link bandwidth is split up into three bandwidth classes:

- *Static class* is for guaranteed, persistent long-term bandwidth that ADs allocate, for example to protect initial connection setup request packets between a source and a destination. Each AD allocates a small portion of its total bandwidth (e.g., 5–15%) to this class.
- *Dynamic class* is for guaranteed, short-term end-to-end bandwidth allocations, and supports high-capacity channels. The dynamic class may account for the majority of the link capacity (e.g., 60–65%).
- *Best-effort (BE) class* is allocated with the remainder of the bandwidth (e.g., 30%).

In case of congestion within a bandwidth class, it can take over the unused bandwidth from the other (uncongested) classes using statistical multiplexing.

[2]The bandwidth of a *path* is divided into separate *channels*, and multiple channels dynamically share the bandwidth within a path.

Within the static or dynamic bandwidth class, an AD can assign different bandwidth sub-classes (e.g., 500 Kbps, 1 Mbps, etc.) to individual paths/flows based on empirically measured flow size distribution. For example, the fraction of dynamic bandwidth allocated to the 1 Mbps sub-class can be derived based on the fraction of flows with 1 Mbps rate in the current Internet. We provide guidelines on how an AD can divide its total link capacity to the above three classes in Appendix A.

A half-path announced by a PCB is a path on the BE class, offering no guarantees. However, for each upstream AD, an endpoint AD can *activate* up to k BE half-paths to convert each into a static half-path, offering a guaranteed amount of bandwidth from that AD to the TDC. The parameter k is determined by contract and is enforced by provider ADs.

Static and BE channels. A communication channel is a conduit to carry traffic from a source to a destination. A single half-path can be used as a channel to reach the TDC, and two half-paths can be combined to form an end-to-end channel between a source and a destination. In this paper, we call the half-path of the source the up-path (as traffic on that channel traverses ADs upwards towards the TDC) and the half-path of the destination the down-path. If two BE half-paths (i.e., half-paths using the BE bandwidth class) are combined, the resulting channel is a BE channel, and similarly, two static half-paths create a static channel. Hybrid channels are also possible, and we will make use of a static up-path that is combined with a BE down-path. Packets flowing through different types of channels have different properties.

Dynamic channel. An endhost can send a request on static, BE, or hybrid channels to reserve a dynamic channel, which provides high bandwidth for a short amount of time (on the order of seconds to enable fast revocation). All ADs, including the destination, need to agree on the amount of bandwidth offered for the dynamic channel. This channel is similar to network capabilities [4, 26] with bandwidth guarantees [29].

One surprising aspect of STRIDE is that only the access routers in the endpoint ADs have to keep per-flow state for flow admission and policing. STRIDE does not require intermediate routers to maintain any per-flow or per-path state in the fastpath for forwarded traffic; only the initial channel establishment requests require slowpath operations for admission control. Performing per-flow management at edge routers is shown to be practical [21], and we can further relax this state requirement using probabilistic detection of the largest flows. With these ingredients, we construct a series of mechanisms that achieve the highly available communication we strive for.

4.1 Static Half-Path Setup

We now describe how an endpoint AD (AD_S) establishes a guaranteed static path (i.e., a path using static bandwidth class) to its TDC.

① **Bandwidth announcement:** The TDC assesses its link capacity and adds information regarding current reservable static bandwidth to periodic PCBs (TDC and PCBs are explained in Section 2). The TDC ensures no congestion on its internal links even if the announced static bandwidth becomes fully reserved. As a PCB travels from the TDC to endpoint ADs, each AD adds information regarding its own bandwidth availability. In particular, an intermediate AD splits the reservable bandwidth and announces the split amount for each of its children for the static path. Figure 3 depicts the bandwidth availability announcement (denoted by the numbers) of PCBs for the static path. This particular diagram shows that AD_S has three possible static paths:

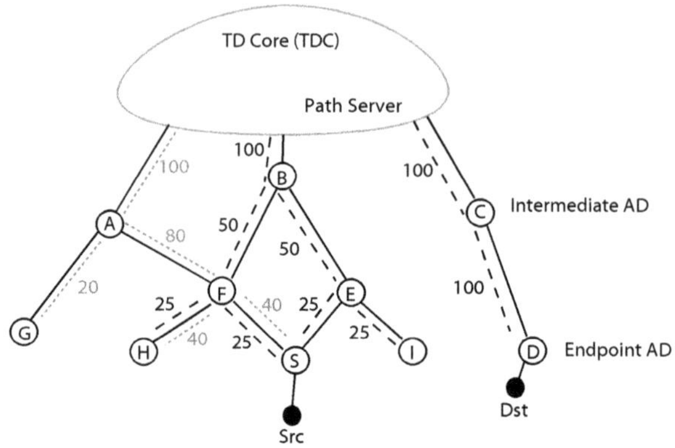

Figure 3: Illustration of STRIDE and how bandwidth is split and announced through PCBs.

❶ $TDC \stackrel{100}{\rightarrow} AD_A \stackrel{80}{\rightarrow} AD_F \stackrel{40}{\rightarrow} AD_S$
❷ $TDC \stackrel{100}{\rightarrow} AD_B \stackrel{50}{\rightarrow} AD_F \stackrel{25}{\rightarrow} AD_S$
❸ $TDC \stackrel{100}{\rightarrow} AD_B \stackrel{50}{\rightarrow} AD_E \stackrel{25}{\rightarrow} AD_S$

② **Activation:** Recall that each endpoint AD learns a set of BE half-paths to the TDC through PCB propagation. The AD can then activate k BE half-paths per provider AD to static class as follows. Upon receiving a PCB with the bandwidth availability information, AD_S sends an activation request to the TDC to reserve the static bandwidth as specified in the PCBs. While forwarding this request, all the intermediate ADs temporarily reserve the requested static bandwidth for AD_S if they can support the request. For example, AD_S may request 40 units of the static-class bandwidth through path ❶, 25 units through path ❷, etc.

③ **Confirmation:** The TDC sends a confirmation to AD_S for the guaranteed static path that AD_S (and all of its hosts) can use to reach the TDC. The new opaque fields constructed along the confirmation acts as a *static path token* that enables AD_S to communicate with the TDC on the static path. Note that STRIDE allows each endpoint AD to create up to k static half-paths per provider AD, and each endpoint AD has the freedom to keep a subset of static half-paths in private for privileged access, as we will explain in step ④ below, while registering others at the Path Server for public usage.

4.2 Static and BE Channel Setup

After the half-path setup, each endpoint AD obtains a set of static bandwidth-guaranteed half-paths in addition to a set of BE half-paths provided by the original PCBs. We now describe how two half-paths (i.e., an up-path and a down-path) can be combined to setup an end-to-end channel. Combining static and BE half-paths results in four types of channels (i.e., static, BE, static+BE, or BE+static) with different guarantees, as summarized in Table 1.

④ **End-to-end path selection:** When source *src* of AD_S wants to communicate with destination *dst* of AD_D, *src* queries the Path Server for the down-paths to reach *dst*. *Src* reaches the Path Server in the TDC using a BE or static half-path as a communication channel. The Path Server then returns unconcealed static down-paths and/or BE down-paths to *src*. Alternatively, *dst* can inform *src* of a private static path over an Out-Of-Band (OOB) channel. By combining one of its up-paths and one of the down-paths provided by the Path Server, *src* now establishes an end-to-end channel for sending a dynamic channel setup request.

4.3 Dynamic Channel Setup

⑤ **Dynamic channel setup request:** To acquire guaranteed bandwidth along a dynamic channel, *src* sends a dynamic channel setup request on this newly-established end-to-end (static, BE, or hybrid) channel. While this request travels toward *dst*, all ADs on the path specify the bandwidth that they can provide for the dynamic channel and forward it toward *dst*.

In case *src* sends packets beyond the allocated static bandwidth of the static up-path, AD_S sets an overuse bit on each extra packet to utilize unused static or BE bandwidth, thereby indicating that the extra packets are beyond permitted allocation for efficient traffic policing.

In case *src* cannot send the request on the static channel, possibly due to congestion on any of the announced down-paths, STRIDE flexibly allows the endpoint AD to send requests on the BE channel. We discuss several alternatives for channel composition and their priorities in Section 5.3.

⑥ **Dynamic-class bandwidth allocation:** When *dst* receives *src*'s dynamic channel setup request, *dst* can deduce sustainable dynamic-class bandwidth for *src* based on the reported dynamic-class bandwidth availability of all the intermediate ADs (e.g., minimum of the dynamic-class bandwidth allocations of all the intermediate ADs). Then *dst* sends this information to *src*, during which the *dynamic capability* is constructed on the return path. STRIDE provides flexible options for sending the reply. For example, *dst* may send the reply through the allocated dynamic channel or through the reverse channel that sent the request.

⑦ **Guaranteed data transmission:** When *src* receives the reply, it can enjoy sending data traffic using the dedicated dynamic-class bandwidth by embedding the dynamic capability in the data packets. Since this bandwidth is short-lived, *src* may renew this dynamic-class bandwidth using actual data packets. Similar to step ⑤, *src* can also send more than permitted dynamic bandwidth allocation, in which case AD_S sets the overuse bit for the extra data traffic.

5. STRIDE PROTOCOL DESCRIPTION

In this section, we elaborate on STRIDE's mechanisms.

5.1 Static Half-Path Setup

① **Bandwidth announcement:** Upon receiving a PCB, an intermediate AD (AD_i) forwards it to the downstream AD (AD_{i+1}) after appending the bandwidth information, which includes: (1) currently reservable static-class bandwidth for AD_{i+1}, (2) currently underutilized static-class bandwidth that AD_{i+1} can use, (3) currently available dynamic-class bandwidth, and (4) currently available BE-class bandwidth. This bandwidth information enables downstream ADs and endhosts to deduce congestion status and make informed decisions in selecting paths. When a PCB reaches an endpoint AD, it contains a path from the TDC to the endpoint AD with reservable bandwidth that can be provided once this path is activated (details in step ②).

On each PCB, AD_i adds an opaque field O_i, as described in Eq. (1). The resulting collection of opaque fields in the PCB access the BE channel on the route that the PCB traversed. Note that AD_i should use different MAC keys to construct opaque fields for different bandwidth classes and expiration time, so that an attacker cannot forge an opaque field for another traffic class or extend the expiration time. For example, the MAC key K_i can be derived from the master secret key $\hat{K}_i$: $K_i = F_{\hat{K}_i}(BE, timestamp)$, where $F(\cdot)$ is a pseudo-random function.

Path diversity vs. quality. The bandwidth announcement mechanism in STRIDE enables an AD to divide bandwidth among its customers. However, bandwidth allocation is still challenging because of the tradeoff between path diversity (i.e., the number of different paths) and path quality (i.e., the allocated bandwidth to each path). An intermediate AD can offer higher path diversity by propagating a PCB to more children (i.e., egress routers). Yet, increasing path diversity reduces the bandwidth that can be allocated to each child AD since the bandwidth contained in a PCB must be split for all children *recursively* as the PCB propagates to downstream ADs.

To address this issue, STRIDE uses a *bandwidth overbooking* technique to enhance path diversity and quality simultaneously as follows. An endpoint AD can choose up to k paths out of all announced paths. This indicates that not all announced paths will be activated. Hence, intermediate ADs can announce greater reservable bandwidth (i.e., overbook) to their downstream ADs than the actual link capacity, and defer the actual bandwidth reservation later during the activation step (②). However, if intermediate ADs overbook their bandwidth aggressively, path activation could be frequently denied. To address this issue, STRIDE allows each intermediate AD to overbook its bandwidth such that the probability of path-activation failure along its link is below a certain threshold, which we analyze in Appendix B.

From the bandwidth announcement in the latest PCB, an endpoint AD learns the amount of static bandwidth it may reserve on the corresponding route. Note that the actual allocation is not performed until the activation step (②), as the reservable bandwidth may be greater than what an intermediate AD can support because of overbooking or stale bandwidth information.

② **Activation:** An endpoint AD requests for a path activation along the reverse path to the TDC. Each request consists of desired (1) expiration time of the path, and (2) amount of static-class bandwidth, which does not exceed the announced reservable bandwidth in the latest PCB. STRIDE considers an activation request as a control message, like PCBs, that is protected from data-plane DDoS attacks. To avoid congestion on the control plane, each AD can rate-limit the activation requests on a per customer basis and advertise the limit with the reservable bandwidth during the announcement.

Upon receiving an activation request, an intermediate AD, which has sufficient unallocated bandwidth (i.e., spare capacity $\geq$ desired bandwidth), temporarily allocates the requested bandwidth for this path. Otherwise, the AD sends back an error message. Also, to minimize bandwidth waste, the AD recycles temporarily allocated bandwidth when the activation fails, which is indicated by the lack of a confirmation (step ③) or an error message, before the arrival of the next PCB.

For efficient bandwidth management (e.g., allocation and recycling), the expiration time and bandwidth is chosen from a pre-defined finite set of values. For example, the expiration time can be 6, 12, 18, or 24 hours; the sub-class bandwidth assigned to each activation request can be 64 Kbps, 128 Kbps, etc.

Each endpoint AD is allowed to activate up to k distinct paths per upstream AD (or provider). This policy is made in accordance with the observation that in the current Internet, large endpoint ADs often subscribe to multiple providers for increased capacity and path diversity. This k-path policy can be enforced either by the providers or the Path Server in the TDC.

③ **Confirmation:** The TDC informs the endpoint AD of a successful static path activation by sending a confirmation message along the activated path. The confirmation message contains the expiration time and the allocated bandwidth. The TDC also updates the Path Server to include this activated path. Upon receiving the valid confirmation from the TDC, each intermediate AD on the path converts the temporarily-allocated bandwidth to be long term (until the expiration time).

Before forwarding the confirmation to the next hop, the AD adds a new opaque field using a different MAC key K_i, derived from the master secret key $\hat{K}_i$ similarly as before: $K_i = F_{\hat{K}_i}(static, timestamp, BW)$, where BW is the amount of bandwidth allocated to the path. After receiving the confirmation, the endpoint AD can forward packets on the static channel of the path by including a *static path token*, which consists of the new opaque fields, in the header of packets.

5.2 Static and BE Channel Setup

After the half-path setup, endpoint ADs learn multiple long-term, bandwidth-guaranteed static half-paths (in addition to multiple BE half-paths) to communicate with the TDC.

④ **End-to-end path selection:** When an endhost *src* in AD_S attempts to make a connection to another endhost *dst* in AD_D, *src* contacts AD_S for path resolution. In turn, AD_S requests the Path Server in the TDC for a list of static paths to *dst*, and the server returns down-paths. As a result, the AD_S can select an up-path (from itself to its TDC) and a down-path (from the TDC to AD_D), and splice them to form an end-to-end path.

Path Server availability. A successful DDoS attack against Path Servers would disable end-to-end path establishment in STRIDE. Such an attack can be prevented in two ways: (1) the TDC can detect the origin of attack traffic against Path Servers and throttle their traffic, and (2) the bandwidth-guaranteed static paths can be used to contact a Path Server.

Path selection policy. If AD_S keeps on selecting paths based on the highest available bandwidth, it may end up selecting a single best path for all the source endhosts (besides *src*), eventually congesting this path. To resolve this issue, endpoint ADs in STRIDE perform probabilistic path selection as follows: the endpoint AD selects a path with probability proportional to the path bandwidth guarantees. With this policy, the endpoint ADs are more likely to select uncongested paths and reduce the average number of trials. We evaluate a specific instance of this policy in Section 7.

Private paths. We introduce the notion of private paths, which endpoint ADs can use to provide guaranteed down-paths to preferred endhosts. In a nutshell, an endpoint AD keeps a subset of half-paths as private and provides them to its destination endhosts such that they can selectively provide them to preferred sources. We define private services to be provided by those servers that can predict future customers (e.g., premium customers on Amazon). A private server providing access to a closed community can provide guaranteed connection setup to community members with private down-paths as follows: a destination can selectively disclose its private down-paths to preferred sources. The private paths can be distributed via OOB channels or by uploading encrypted private paths to a Path Server. As a result, a valued customer of Amazon, for example, can obtain a bandwidth-guaranteed static down-paths for sending dynamic channel setup requests to Amazon.

Table 1: Guarantees of dynamic channel setup delay and bandwidth for different types of end-to-end channels.

Up-path	Down-path	Delay	Bandwidth guaranteed?
Static	Static (private)	Constant	✓
Static	Static (public)	Linear	✓
Static	Best-effort	Linear	✗

5.3 Dynamic Channel Setup

Using a (BE, static, or hybrid) channel, *src* sends dynamic channel setup requests to establish an end-to-end dynamic channel. With such an end-to-end dynamic channel, STRIDE can provide bandwidth guarantees to short-term, high-bandwidth dynamic flows. Note that *src* can send any types of packets on the end-to-end channel, but we focus on the discussion of sending dynamic channel setup requests, as it is a part of our DDoS defense mechanism.

⑤ **Dynamic-channel setup request:** After selecting an end-to-end channel in step ④, a source endhost can send a dynamic channel setup request for guaranteed dynamic-bandwidth allocation. Table 1 describes guarantees of dynamic channel setup delay and bandwidth for different types of end-to-end channels.

A request header carries two additional indicators that enable congested intermediate ADs to efficiently control link bandwidth:

- *Overuse bit:* A source AD sets an overuse bit of a packet on a static up-path in case its endhost is sending packets more than the reserved static-class bandwidth of the up-path.
- *Congestion bit:* Any AD that experiences link congestion sets a congestion bit in BE packets.

Traffic priority of requests. When an AD receives more packets than what its outgoing links can afford, the AD has to discard some of them while maintaining the static-class bandwidth guarantee. Based on where the congestion occurs, we discuss different techniques to prioritize packets.

- *Host contention at source ADs:* Static bandwidth contention may occur on the source AD's outgoing links. Each AD can have a different way to resolve contention. For example, it can adopt a payment-based scheme: each client informs its host AD how much it is willing to pay for using a static up-path, and the endpoint AD can arrange based on some objectives (e.g., maximize the AD's revenue) [23]. For ease of analysis, we consider per-host fair share within an endpoint AD.

Table 2: Traffic priority of dynamic channel setup requests on the down-paths that experience link congestion.

Priority	Up-path	Bits set	How requests arrived
1	Static	-	Within allocated static BW
2	BE	-	On uncongested BE link
3	Static	Overuse	Beyond allocated static BW
4	BE	Congest	On congested BE link
5	Outside TD	-	From outside TD

- *Link congestion on up-paths:* During link congestion, each source domain obtains *a weighted fair share* of the available (unallocated, or allocated but unused) static bandwidth. A weighted fair share is proportional to the source domain's static allocation on the congested link. Hence, static packets with the overuse bit would be transmitted on the static channel up to the weighted fair share, and packets beyond the share are converted into BE packets to compete with the standard BE traffic.
- *Link congestion on down-paths:* STRIDE assigns priority levels to the packets such that low priority packets are dropped first in the case of congestion. Table 2 summarizes the priority levels ordered from the highest to the lowest.

This priority applies to both static and BE classes. If the congestion persists within the first-priority traffic, the congested link assigns to it a weighted fair share of the available static-class bandwidth proportional to each *destination domain*'s static allocation.

Determine reservable dynamic bandwidth. Recall that the dynamic bandwidth class has defined bandwidth sub-classes, such as 512 Kbps, 1 Mbps, 2 Mbps, etc., and each dynamic channel is associated with a given sub-class. Intuitively, to provide precise bandwidth guarantees for established flows, we have to ensure that every STRIDE-protected request packet can be offered sufficient dynamic bandwidth (e.g., at least 512 Kbps). Ideally, we would like to provide guaranteed flow bandwidth to every request packet traversing static channels, as Table 1 shows.

One key challenge here is how to flexibly determine the amount of reservable dynamic bandwidth for such requests. To address this challenge, STRIDE limits the rate of the dynamic channel setup requests (thus the dynamic allocation) within the static class to be *proportional to the static allocation.* For example, if each dynamic channel is guaranteed 10 units per second and expires in 2 seconds, and the rate limit is 3 requests per second, then the dynamic-class link bandwidth should be greater than $10 \cdot 2 \cdot 3$ units per second to accommodate the worst case where all requests arrive using the static class.

The AD assigns the smallest sub-class to the initial request and flexibly upgrades to a higher sub-class for the subsequent requests for the allocation renewal if the link is not congested. In the case of link congestion on the up-path (down-path), each source (destination) domain obtains a weighted fair share of the available dynamic-class bandwidth that is proportional to the source's (destination's) static allocation on the congested link.

Each AD on the path (including the destination AD) either approves the requested dynamic bandwidth, or indicates the maximum available bandwidth (which is at most the available bandwidth indicated by the previous AD).

⑥ **Dynamic-class bandwidth allocation:** Through a dynamic-channel setup request, a destination endhost can discover the bottleneck link(s) and the available dynamic-class bandwidth along the path. The destination constructs a reply packet, which carries (1) reserved dynamic-class bandwidth of this flow, (2) opaque fields (which include a *dynamic flow capability*), and (3) expiration time which indicates the lifetime of the guaranteed dynamic bandwidth. The destination also indicates which AD-to-AD link(s) is the bottleneck for determining the bandwidth reservation.

As the packet travels back to the source, ADs update their dynamic bandwidth allocation and opaque fields to accurately reflect the available bandwidth and reduce the potential waste of bandwidth. If the allocated end-to-end dynamic bandwidth does not meet the source's need (e.g., determined by an application service), the source may select an alternative path. Furthermore, the source can make an informed decision to avoid the bottleneck link when selecting an alternative path.

Capability update. A source can renew the short-term dynamic capability while communicating with the destination as follows: the sender sets a renewal bit in the header of the capability-protected dynamic-class packets. If the destination renews, the source AD invalidates the old dynamic capability (e.g., by keeping track of the latest capability for each flow and rejecting packets carrying old capabilities) to prevent misuse.

⑦ **Guaranteed data transmission:** Upon receiving a dynamic capability (step ⑥), *src* can use the end-to-end dynamic channel for guaranteed data transmission. *Src* can also flexibly choose other types of end-to-end channels for different guarantees.

Regulation. For per-flow bandwidth guarantees, endpoint ADs monitor per-flow data usage and regulate potential violation. For example, every endpoint AD ensures that the overuse bit is set in data packets whose flow rate exceeds the allocated value. ADs are responsible to drop some of the data packets with the overuse bit to resolve link congestion. For example, similar to the static channel regulation, the AD can drop packets that are beyond the weighted fair share of the source or the destination. In addition, intermediate ADs and the TDC can perform both real-time probabilistic monitoring and offline traffic analysis to identify misbehaving endpoint ADs that fail to regulate their clients. TDCs and ADs also monitor per-TD bandwidth usage of dynamic-class traffic at each interface at the TD boundary to isolate attack traffic from other TDs.

6. BANDWIDTH GUARANTEE ANALYSIS

We first show that STRIDE achieves domain-based guarantees for communication between the source (AD_{src}) and the destination (AD_{dst}) domains within a TD; specifically, we analyze what domain-based guarantees AD_{src} can obtain using different types of channels. We then discuss how an endpoint AD can divide such domain-based guarantees among its endhosts.

In Theorem 1, we show that by leveraging private down-paths, AD_{src} and AD_{dst} can establish bandwidth-guaranteed static channels for congestion-free communication. Let u_i and d_i be AD_i's total static up-path and down-path bandwidth allocations, respectively, where $1 \leq i \leq m$ and m is the number of ADs. Since each AD is expected to assess its bandwidth requirement of static half-paths based on its contractual agreements with human subscribers, u_i and d_i are constant irrespective of the number of ADs or the power of the botnet (which consists of compromised endhost machines). We denote $d^p(i,j)$ to be the total bandwidth of AD_j's private down-paths known only to AD_i.

THEOREM 1. *For private communication (using private static down-paths), AD_{src} can successfully send packets to AD_{dst} at rate $r^p = \min\{u_{src}, d^p(src, dst)\}$ without experiencing congestion on any intermediate links.*

Proof sketch: The first domain-based guarantee is straightforward. Since TDC is congestion-free (as described in Section 3.2), AD_{src} can establish end-to-end congestion-free channels by splicing its static up-paths and AD_{dst}'s private static down-paths. The sending rate of the resulting channels is dominated by the bottleneck bandwidth, which is the minimum of u_{src} and $d^p(src, dst)$. Both u_{src} and $d^p(src, dst)$ are independent of the botnet and other ADs' allocations. □

Note that r^p is a lower-bound guarantee of the sending rate, and the congestion-free property ensures that packets, such as connection setup requests, can be delivered at the first trial.

In Theorem 2, we show that AD_{src} can obtain a weaker guarantee (which depends on the static allocations of other ADs) when using public static down-paths. Let $U = \sum_{i=1}^{m} u_i$, and $U(i)$ be the total static up-path bandwidth activated by ADs that desire to communicate with AD_i. Let b_{dst} be the minimum cut of AD_{dst}'s BE-class bandwidth on the down-paths.

THEOREM 2. *For public communication (using unconcealed static down-paths), AD_{src} can successfully send packets to AD_{dst} at an average rate $r = u_{src}(\frac{d_{dst}}{U(dst)} + \frac{b_{dst}}{U})$.*

Proof sketch: Sources that desire to communicate with AD_{dst} compete for the limited bandwidth of static down-paths, d_{dst}. Sources do not need to compete with packets to other destinations on congested links because STRIDE performs weighted fair sharing on the static down-paths. To obtain the highest traffic priority and increase the chance of successful delivery, each source sends packets with no overuse bit on its static up-paths at full speed, resulting in $U(dst)$-amount of high-priority traffic to AD_{dst}. Hence, AD_{src} with u_{src} static allocation can send packets through static down-paths at rate $\frac{u_{src} \cdot d_{dst}}{U(dst)}$ (e.g., bit/s) on average. A similar result can be shown for the case where sources compete for the BE bandwidth on the congested links (b_{dst}). Since the BE class is shared, in the worst case AD_{src} has to compete with traffic between all ADs in the TD, resulting in a sending rate $\frac{u_{src} \cdot b_{dst}}{U}$. Hence, the average waiting time before successfully delivering a packet of size w is $\frac{w}{r}$. The waiting time for using the static channel and the static+BE channel is linear to $U(dst)$ and U, respectively. □

Theorem 3 shows that AD_{src} can obtain a lower-bound guarantee on the dynamic allocations to AD_{dst}. Let γ be the ratio of the dynamic-class bandwidth to the static-class bandwidth, and we assume γ is the same for every link for ease of description. $\Delta(i,j)$ is the guaranteed dynamic-class bandwidth that AD_i would like to allocate for communication with AD_j.

THEOREM 3. *For dynamic channels, STRIDE guarantees $\min\{\Delta(src, dst), \Delta(dst, src)\}$ amount of dynamic-class bandwidth for the flow aggregate between AD_{src} and AD_{dst}, where $\sum_{i=1}^{m} \Delta(src, i) \le \gamma \cdot u_{src}$ and $\sum_{i=1}^{m} \Delta(dst, i) \le \gamma \cdot d_{dst}$.*

Proof sketch: Because STRIDE performs weighted fair sharing within the dynamic class based on the static allocation, the flow aggregates from AD_{src} and to AD_{dst} are guaranteed to have $\gamma \cdot u_{src}$ and $\gamma \cdot d_{dst}$ bandwidth, respectively. Endpoint ADs can then freely divide the guaranteed bandwidth to flow aggregates going to/from different ADs. □

Splitting guaranteed bandwidth. Each endpoint AD decides how to divide its bandwidth guarantees among its endhosts based on its local policy. For example, a simple policy would be to split the static allocation based on per-host fair share. Suppose the sender and the receiver obtain a fair share u'_{src} and d'_{dst}, respectively, from their local domains. Similar to Theorems 1–3, we can show guarantees for the sender and receiver by replacing u_{src} with u'_{src} and d_{dst} with d'_{dst} in the proofs above. Moreover, because the number of dynamic channel setup requests within the static class is small and bounded, STRIDE can provide guaranteed high bandwidth to flows established through static channels. An interesting observation is that since compromised endhosts send traffic all the time whereas uncompromised endhosts do not, the available bandwidth is much higher than the fair share if the domain contains fewer bots.

Resilience against DoC and N^2 Attacks. Theorems 1 and 2 imply that a DoC attacker (1) cannot crowd out a capability request if the request is placed along a static channel with a private path, and (2) can delay a capability request at most by time linear to the static allocation of other domains if the request uses a public path. These bounds are independent of the size or distribution of the botnet. Theorem 3 implies that STRIDE can defend against the N^2 attack as the guaranteed flow bandwidth between the source and destination ADs is unaffected by bots outside those ADs.

7. EVALUATION

In this section, we evaluate STRIDE with respect to its effectiveness against DDoS attacks. We show the effectiveness of end-to-end bandwidth guarantees under large-scale attack scenarios. We also test the packet forwarding performance of STRIDE via real-field implementation.

Simulation setup. For realistic simulation, we use a CAIDA AS-relationship dataset to construct a TD; a tier-1 AD connecting to 2164 endpoint ADs is chosen as the TDC. Although the AS-relationship dataset does not include all interface-level paths, our analysis of the dataset reveals that AD-level path diversity is high enough to support STRIDE's path control and hence to evaluate STRIDE's path construction. Specifically, the endpoint ADs in the dataset have more than 40 different paths to the TDC on average. If interface-level paths are constructed, path diversity at endpoint ADs would become much higher since the number of paths grows exponentially as PCBs propagate downstream.

Bandwidth allocation. During PCB propagation, each AD allocates bandwidth to each child AD proportional to the child size. We assume that the size of an AD is proportional to its degree.

7.1 Resilience against DoC Attacks

We evaluate the resilience of STRIDE against DoC attacks, under the following simulation scenario. We randomly label one hundred ADs as *clean* (i.e., ADs containing no bots) and configure them to send traffic to a destination AD using 10 different down-paths (i.e., k=10), with a send rate equal to one tenth of the down-path capacity. Hence, in the absence of attacks, all down-paths are fully (but not overly) utilized. Then, we randomly label ADs as *contam-*

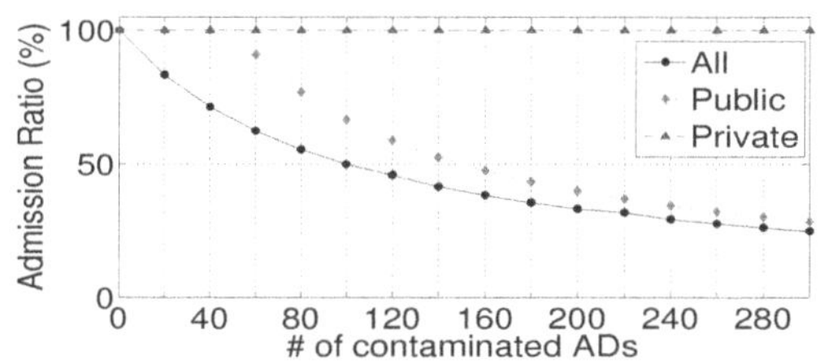

Figure 4: The admission ratio of legitimate packets for the number of contaminated ADs.

inated (ADs containing bots) and set their send rate equal to that of a clean AD so as to make individual contaminated ADs indistinguishable from clean ones. The number of contaminated ADs is increased from 0 to 300.

We evaluate the effectiveness of STRIDE against the above attacks in the following two scenarios.

1. *Public Paths:* All clean and contaminated ADs use the k activated down-paths to setup capabilities with the destination AD.
2. *Public and Private Paths:* Half of the clean ADs use a private down-path that was provided to the source ADs via a secret out-of-band channel. Meanwhile, the remaining half of the clean ADs and contaminated ADs use the public paths as before.

As an evaluation metric, we use the *admission ratio*, which is defined as the percentage of the legitimate packets (i.e., packets from the clean domains) that successfully traverse the bottleneck link/path.

Figure 4 shows that when all source ADs use the public paths ("All"), the admission ratio of the legitimate packets decreases as more contaminated ADs are added since the per-AD bandwidth decreases. When half of the clean ADs acquire a private path from the destination ("Private"), their packets are unaffected by the attack as the 100% admission ratio shows; and the packets of the remaining half of the clean ADs ("Public") obtained higher admission ratio along the public paths because the use of the private path reduced bandwidth contention along the public paths. This result illustrates how destination ADs can protect their valued customers' traffic from DDoS attacks in STRIDE.

While STRIDE enables private parties to use private paths to avoid congested static paths, it also protects clean ADs' traffic from large-scale DDoS attacks via packet prioritization: i.e., capability requests made through the static up-paths would have a higher priority than others through the best-effort up-paths. To examine this, we use the following simulation scenario: the attack strength is increased (by adding more attack sources within contaminated ADs) up to 10 times the bandwidth of the static up-paths; source ADs put the high priority marking on their outbound packets such that the bandwidth of high priority packets would not exceed that of the static up-paths (e.g., if the attack strength grows 10 times, 90% of attack packets would have a low priority marking). Legitimate source endhosts, on identifying congestion on static down-paths, use the best-effort down-paths; and attack source endhosts use the same path selection strategy as that of the legitimate sources to maximize their effects. The above attack scenario is the strongest attack scenario we consider for the given number of legitimate and attack sources since all packets from the 100 clean and 300 contaminated ADs compete for the bandwidth of public paths.

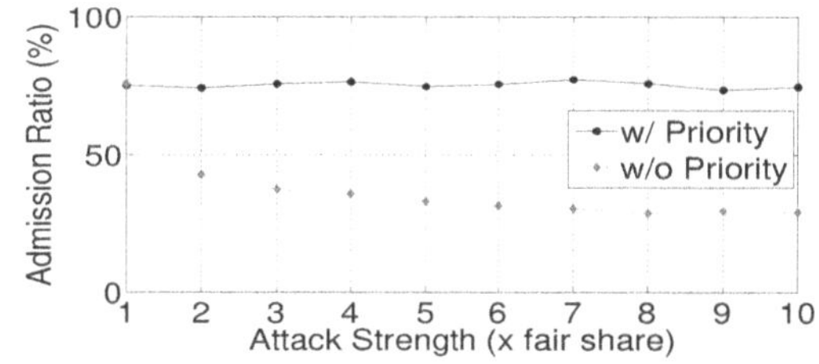

Figure 5: Effects of attack strength.

Figure 5 shows that even if the attack strength grows, the effects on legitimate traffic are marginal: attack sources, regardless of their strength, can only consume bandwidth proportional to their fair share both on the static and the best-effort channels. Meanwhile, 25% of legitimate packets sent through the static down-paths reach their destination without loss, and the other legitimate packets (i.e., 75% of them) sent through the best-effort channel reach the destination with a ratio close to 66.7%. Overall, 75% of the legitimate requests overcome the massive DDoS attack whose total send rate is 30 times higher than that of the legitimate sources, even if routers cannot distinguish between legitimate and attack packets. The figure also shows that without packet prioritization, the admission ratio of legitimate packets decreases as attack strength grows. This result shows the effectiveness of using static up-path and packet prioritization in STRIDE.

7.2 Flow Bandwidth Guarantees

STRIDE's bandwidth guarantees effectively isolate the bandwidth of attack traffic from that of legitimate traffic. As a consequence, in STRIDE, the effects of attacks are confined within the paths they follow regardless of whether attack sources flood a single path (or a link) or multiple paths simultaneously. We show this bandwidth isolation via large-scale simulations. For realistic simulations, we construct simulation topologies using a CAIDA SkitterMap [2], attach 10,000 legitimate sources to 200 ADs proportional to the AD size, and attach attack sources (hosts) to 100 ADs. Paths are probabilistically sampled from the SkitterMap to satisfy both the number of sources and the number of ADs. Legitimate sources control their packet sending rate based on the TCP congestion control mechanism, while attack sources send constant, high-rate traffic to flood a target link. We increase the attack size from 10K to 100K to compare STRIDE's bandwidth guarantees with those of a per-flow fair-sharing based mechanism. We consider a baseline case, labeled as "No Defense", where packets are randomly dropped during congestion.

Figure 6 shows the bandwidth used by the legitimate flows that originate from clean ADs. Under "No Defense", the legitimate flows obtain almost no bandwidth. DDoS attacks. When per-flow fair-sharing bandwidth control is em-

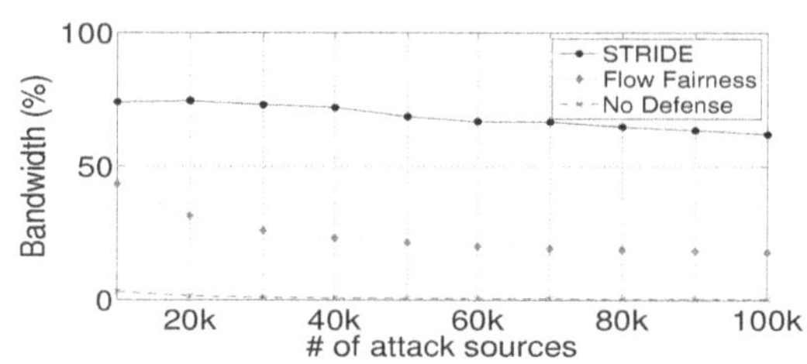

Figure 6: The effects of attack size.

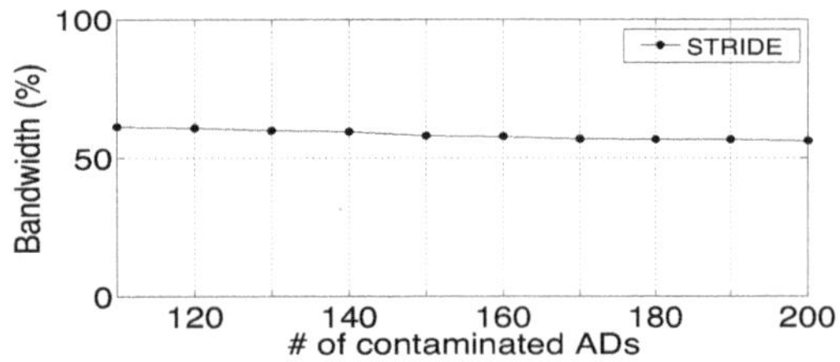

Figure 7: The effects of attack dispersion.

ployed, attack flows cannot completely exhaust the target's link bandwidth, yet the attack effects grow linearly with the attack size.

STRIDE provides consistent bandwidth guarantees to legitimate traffic under different attack sizes, which proves the effectiveness of path bandwidth isolation. The bandwidth of legitimate flows decreases slightly as the attack size grows, because (1) the extra bandwidth that is not fully used by some paths (due to the TCP congestion control) is shared by other flows, and (2) as the number of contaminated ADs increases, the number of clean ADs decreases (as the total number of ADs is fixed).

Next, we increase the number of contaminated ADs by 10 up to 200 ADs. As one can imagine, the bandwidth of legitimate flows decreases as Figure 7 shows. However, the effects of attack dispersion are marginal (i.e., proportional to the number of attack ADs) because the dynamic channel bandwidth is proportional to the static channel bandwidth and the static channel bandwidth that can be used by attack traffic is limited by the number of attack ADs in STRIDE.

7.3 Throughput

STRIDE introduces additional computational work for capability (or opaque field) verification. To gauge the computational overhead, we measure the throughput of a STRIDE router for various packet sizes and compare the result with that of the default IPv4 forwarding. We implement a STRIDE router as a user-space process using the Click Modular Router [10]. The capability generation and verification are implemented as CBC-MAC with AES-ni. We perform the measurement with a simple topology where a source and a destination are directly attached to a STRIDE router. NetPerf [3] is used for throughput measurement.

As described earlier, STRIDE forwards packets based on the interface identifier in the packet header; hence, unlike in today's routers, no overhead will be incurred for FIB (forwarding table) lookup. Meanwhile, the IPv4 forwarding in our experiments would produce the highest throughput that it can achieve since the FIB has only one entry in our network configuration.

Figure 8 shows that for small packets, both IPv4 and STRIDE routers under-utilize the link bandwidth while the IPv4 packet forwarding outperforms that of STRIDE; for

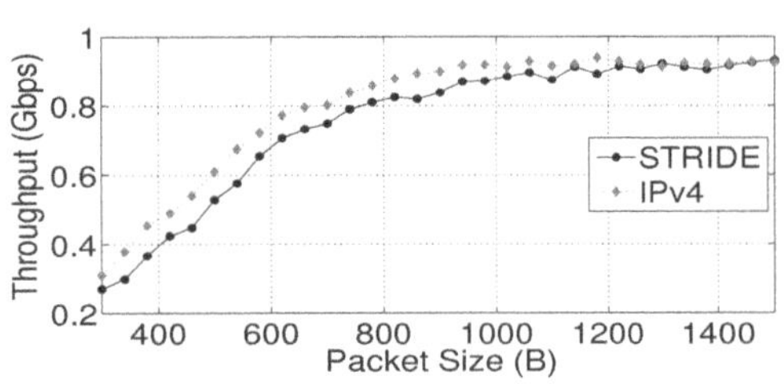

Figure 8: Throughput vs. packet size.

large packets, they both utilize more than 90% of the link bandwidth. In practice, the packet overhead becomes negligible since as small packets account for less than 10% of bandwidth and most of remaining packets are full sized [2].

8. DISCUSSION

Inter-TD traffic guarantees. While STRIDE provides domain-based guarantees for communication within a TD, many of the properties also translate for communication between TDs. For example, static channels on the up-paths still guarantee low-capacity throughput. Only BE channels are getting lower guarantees, as there is no explicit indication that the receiver desires the communication. Thus, establishing a connection to a public service that is under attack will be challenging for an external host. However, as soon as the service receives one initial packet and desires to serve that client, it can set up a dynamic channel with the same protected bandwidth guarantees within each TD assuming no congestion on high-capacity links between TDs.

Malicious ADs inside TD. ADs within a TD may get compromised, failing to regulate traffic. Although this is unlikely in well-administered ADs, attackers can nevertheless exploit software vulnerabilities in routers or administrative workstations. STRIDE can identify malicious ADs using neighborhood monitoring and existing fault detection protocols [32]. For example, if any AD sends more traffic than their allocated share, the AD must be malicious or misconfigured and the neighboring AD can block the offending traffic. As a technical defense, once the malicious AD is detected, hosts can avoid the malicious AD by selecting paths that avoid traversing that AD. Most importantly, since all ADs of a TD are within a uniform legal environment, the TDC can revoke the membership of misbehaving ADs.

Directional paths and asymmetric bandwidth requirements. In practice, network links may be directional or asymmetric with different bandwidths in the two directions. STRIDE can flexibly accommodate asymmetric paths with minimal modifications as follows. To request a packet on a directional path, the source puts in the header both the forward and backward paths. For example, dynamic flow capabilities can be requested on both the forward and return paths, and sent back to the other party through sufficient space allocated in the packet header. Bandwidth requests may be asymmetric, as in downloads the client-to-server bandwidth is one to two orders of magnitude smaller (acknowledgment packets are smaller than data packets). In this case, STRIDE supports asymmetric bandwidth allocations, where a uni-directional path token or flow capability is requested with different amounts of bandwidth depending on the direction.

9. RELATED WORK

Network-layer DDoS defense mechanisms can be largely classified into two categories: router-level bandwidth control and architectural extensions. One can generally combine approaches in these two categories for stronger properties.

Router-level bandwidth control. Typical router-level approaches to DDoS defense aim to filter or limit identified attack flows.

Filtering: Filtering approaches [5, 13] install filters against attack sources near their origins (i.e., source ADs) to prevent collateral damage of attack traffic. This would essentially require trust establishment between ADs and rely on source ADs' cooperation that would incur substantial overhead for managing flow state and packet inspection. In contrast, STRIDE facilitates natural trust relationships between ADs within the same trust domain. Network-layer capability schemes [26, 29] enable routers to perform stateless filtering without needing any trust on other routers, but are vulnerable to the DoC attack [6]. Though Portcullis [19] addresses the DoC attack, it requires high computational overhead even on benign source hosts.

Bandwidth throttling: Many bandwidth control mechanisms (especially the fair queueing mechanisms) proposed to date can be used to prevent some (malicious) flows from exhausting the network bandwidth [14, 15, 18, 21, 25]. However, per-flow or per-sender fair sharing does not provide any guarantees by design as the fair bandwidth becomes too low as more entities (e.g., flows) compete for a limited resource. Moreover, existing mechanisms require source authentication, which is difficult to provide efficiently.

Bandwidth guarantees: Existing approaches [7, 8] aiming to provide bandwidth guarantees to flows fail in cases where all available bandwidth is exhausted. FLoc [12] differentiates legitimate flows from attack flows to provide differential bandwidth guarantees. Low-rate attack flows, however, can often not be precisely distinguished from legitimate flows, thereby the lower bound of bandwidth may not be observed.

Architectural support. SCION inherently provides a default level of protection against DDoS attacks. For example, SCION's periodic topology discovery and resulting path-diversity by default enable agile path adjustment to avoid attacked areas. However, SCION itself does not provide any DDoS defense guarantee. Several other next-generation Internet architectures [17, 20, 27], instead of providing intrinsic DDoS resilience, aim to provide routing flexibility, path diversity [27], expressive routing policies [17, 20], etc. A line of multi-path routing protocols have also been proposed [9, 11, 16, 24, 28] to provide path diversity to the *source* nodes. Although the source nodes can utilize the path diversity to circumvent victim links/routers under a DDoS attack, the destinations are still left with little inbound traffic control. Furthermore, these protocols are built on top of the current Internet, thus still suffering from the underlying weaknesses of today's Internet. For example, local identification of attack sources can be imprecise or impossible to counter large-scale botnet attacks that do not directly target a specific service or endpoint.

10. CONCLUSION

A core goal of the STRIDE architecture is to achieve intrinsic DDoS defense with relatively simple routers. In particular, we avoid per-flow state in the fastpath, asymmetric cryptographic operations, reliance on untrustworthy domains, and key establishment across ADs. Even with our relatively simple operations, we can achieve protection against DDoS from large botnets. Reflecting on the STRIDE architecture, we observe that measured trust in ADs located within the same legal environment providing viable prosecution helps to simplify the architecture and results in higher efficiency, meanwhile the untrustworthy ADs outside the trust domain cannot inflict damage against local within-trust-domain communication. We anticipate that STRIDE provides a useful point in the design space to study holistic network architectures with strong DDoS defense properties.

11. ACKNOWLEDGMENTS

This research was supported by CyLab at Carnegie Mellon, and by support from NSF under awards CCF-0424422 and CNS-1040801. The views and conclusions contained here are those of the authors and should not be interpreted as necessarily representing the official policies or endorsements, either expressed or implied, of CMU, NSF or the U.S. Government or any of its agencies.

12. REFERENCES

[1] Arbor networks: Infrastructure security survey. `http://www.arbornetworks.com/sp_security_report.php`.

[2] CAIDA: The cooperative association for internet data analysis. `http://www.caida.org/`.

[3] Netperf benchmark. `http://www.netperf.org/netperf/`.

[4] T. Anderson, T. Roscoe, and D. Wetherall. Preventing Internet denial-of-service with capabilities. In *Proceedings of ACM HotNets*, 2003.

[5] K. Argyraki and D. R. Cheriton. Active internet traffic filtering: real-time response to denial-of-service attacks. In *Proceedings of the ATEC*, 2005.

[6] K. Argyraki and D. R. Cheriton. Network capabilities: The good, the bad and the ugly. In *Proceedings of ACM HotNets*, 2005.

[7] F. Bonomi and K. Fendick. The Rate-Based Flow Control Framework for the Available Bit Rate ATM Service. In *IEEE Network Magazine, vol. 9, no. 2*, 1995.

[8] R. Braden, L. Zhang, S. Berson, S. Herzog, and S. Jamin. Resource ReSerVation Protocol (RSVP) – Version 1 Functional Specification. RFC 2205 (Proposed Standard), Sept. 1997. Updated by RFCs 2750, 3936.

[9] P. B. Godfrey, I. Ganichev, S. Shenker, and I. Stoica. Pathlet routing. In *Proceedings of ACM SIGCOMM*, 2009.

[10] E. Kohler, R. Morris, B. Chen, J. Jannotti, and M. F. Kaashoek. The Click modular router. *ACM Transactions on Computer Systems*, August 2000.

[11] N. Kushman, S. Kandula, D. Katabi, and B. M. Maggs. R-BGP: Staying Connected In a Connected World. In *Proceedings of USENIX NSDI*, 2007.

[12] S. B. Lee and V. Gligor. FLoc: Dependable link access for legitimate traffic in flooding attacks. In *Proceedings of IEEE ICDCS*, 2010.

[13] X. Liu, X. Yang, and Y. Lu. To filter or to authorize: network-layer dos defense against multimillion-node botnets. In *Proceedings of ACM SIGCOMM*, 2008.

[14] X. Liu, X. Yang, and Y. Xia. NetFence: Preventing Internet Denial of Service from Inside Out. In *Proceedings of ACM SIGCOMM*, 2010.

[15] R. Mahajan, S. M. Bellovin, S. Floyd, J. Ioannidis, V. Paxson, and S. Shenker. Controlling high bandwidth aggregates in the network. *Comput. Commun. Rev.*, 2002.

[16] M. Motiwala, M. Elmore, N. Feamster, and S. Vempala. Path splicing. In *Proceedings of ACM SIGCOMM*, 2008.

[17] J. Naous, M. Walfish, A. Nicolosi, D. Mazieres, M. Miller, and A. Seehra. Verifying and enforcing

network paths with ICING. In *Proceedings of ACM CoNEXT*, 2011.

[18] R. Pan, B. Prabhakar, and K. Psounis. CHOKE, a stateless active queue management scheme for approximating fair bandwidth allocation. In *Proceedings of IEEE INFOCOM*, 2000.

[19] B. Parno, D. Wendlandt, E. Shi, A. Perrig, B. Maggs, and Y.-C. Hu. Portcullis: Protecting connection setup from denial-of-capability attacks. In *Proceedings of ACM SIGCOMM*, 2007.

[20] B. Raghavan and A. C. Snoeren. A system for authenticated policy-compliant routing. In *Proceedings of ACM SIGCOMM*, 2004.

[21] I. Stoica, S. Shenker, and H. Zhang. Core-stateless fair queueing: a scalable architecture to approximate fair bandwidth allocations in high-speed networks. *IEEE/ACM Trans. Netw.*, Feb. 2003.

[22] A. Studer and A. Perrig. The coremelt attack. In *Proceedings of ESORICS*, 2009.

[23] A. Vulimiri, G. A. Agha, P. B. Godfrey, and K. Lakshminarayanan. How well can congestion pricing neutralize denial of service attacks? In *Proceedings of ACM SIGMETRICS*, 2012.

[24] W. Xu and J. Rexford. MIRO: Multi-path Interdomain Routing. In *Proceedings of ACM SIGCOMM*, 2006.

[25] Y. Xu and R. Guérin. A double horizon defense design for robust regulation of malicious traffic. *Proceedings of SecureComm*, 2006.

[26] A. Yaar, A. Perrig, and D. Song. SIFF: A stateless internet flow filter to mitigate ddos flooding attacks. In *Proceedings of IEEE Symposium on Security and Privacy*, 2004.

[27] X. Yang, D. Clark, and A. W. Berger. NIRA: a new inter-domain routing architecture. *IEEE/ACM Trans. Netw.*, 2007.

[28] X. Yang and D. Wetherall. Source selectable path diversity via routing deflections. In *Proceedings of ACM SIGCOMM*, 2006.

[29] X. Yang, D. Wetherall, and T. Anderson. TVA: a dos-limiting network architecture. *IEEE/ACM Trans. Netw.*, Dec. 2008.

[30] B. Yener, Y. Ofek, and M. Yung. Combinatorial design of congestion-free networks. *IEEE/ACM Transactions on Networking (TON)*, 1997.

[31] X. Zhang, H.-C. Hsiao, G. Hasker, H. Chan, A. Perrig, and D. G. Andersen. SCION: Scalability, control, and isolation on next-generation networks. In *Proceedings of IEEE Symposium on Security and Privacy*, 2011.

[32] X. Zhang, Z. Zhou, H.-C. Hsiao, T. H.-J. Kim, A. Perrig, and P. Tague. ShortMAC: Efficient data-plane fault localization. In *Proceedings of NDSS*, 2012.

APPENDIX

A. LINK CAPACITY DIVISION

Using estimations, we show how STRIDE ADs can divide their link capacity to the three traffic classes, and how much static bandwidth can be allocated to a path.

Division of link capacity to three traffic classes. We provide guidelines that an AD can follow to divide its total link capacity to three traffic classes: static, dynamic, and BE. First, given that the current real-world link utilization is mostly below 30% based on the CAIDA dataset [2], allocating 30% of the link capacity to the BE class would satisfy legacy Internet traffic in most cases. Subsequently, assuming the static and dynamic classes are allocated s and d fractions of the link capacity, respectively, the following conditions should hold:

$$s + d = 1 - 30\% \tag{2}$$

$$40\text{Gbps} \times s > 500\text{Kbps} \times 10000 \tag{3}$$

The first condition ensures a link will not be overloaded when each bandwidth class is being fully utilized. The second condition assumes an OC-768 link capacity (40 Gbps) to be divided among around 10000 paths, such that each endpoint in a medium-size TD (e.g., a US TD with around 2200 ADs, according to the CAIDA dataset) can choose up to 10 paths in our experiment. The second condition requires the static bandwidth allocated to each path be no less than 500 Kbps.

Based on these guidelines, a reasonable example allocation is to divide $5 - 15\%$, $60 - 65\%$, and 30% link capacity to the static, dynamic, and BE traffic classes, respectively. In practice, an AD can adjust the numbers in the conditions based on its own link capacity, number of current paths that the AD supports, etc. Furthermore, when any bandwidth class is not fully utilized, other congested traffic class can take up all the bandwidth that is currently available.

B. BANDWIDTH OVERBOOKING

Section 5.1 introduces bandwidth overbooking for simultaneous enhancement of path quality and diversity, but with possible denial of path activation. To mitigate this issue, we suggest an appropriate overbooking ratio by analyzing the relationship between an overbooking ratio and the corresponding probability of path activation denial as follows.

In the following, we consider an intermediate AD AD_p wanting to determine its overbooking ratio. Let I_i and E_j represent the ith ingress interface from the providers ($0 \le i \le l$) and the jth egress interface to the customers ($0 \le j \le m$), respectively. Let each ingress interface connect to all m egress interfaces. We assume that m interfaces connect to n customer ADs (i.e., each customer AD has $\frac{m}{n}$ links to AD_p). Then, each customer AD has at least $\frac{l \cdot m}{n}$ distinct paths to the TDC through AD_p. In this setting, suppose each customer AD selects uniformly at random k out of the $\frac{l \cdot m}{n}$ paths to the TDC, then the probability that the customer ADs select I_i more than t times in total would be: $P_{I_i}(t) \approx 1 - \sum_{i=0}^{t} e^{-\lambda} \cdot \frac{\lambda^i}{i!}$, where $\lambda = \frac{n \cdot k}{m}$. This implies that I_i's bandwidth needs to be allocated to t egress interfaces (out of m interfaces), which would increase per-path bandwidth allocation by $\frac{t-\beta}{\beta}$, where $\beta = \frac{n \cdot k}{l}$ is the average number of activated paths through I_i. If $t \gg \frac{n \cdot k}{l}$, sufficient path diversity (as much as $\frac{t \cdot l}{n \cdot k}$) is provided to customer ADs.

As a result, AD_p may determine t such that the probability of the denial of path activation does not exceed some threshold P_{th} (i.e., $P_{I_i}(t) \le P_{th}$). For example, $P_{th} = 0.2$ means that 80% of path activation requests would be accepted on average; hence, the expected number of trials for successful path activation becomes 1.25. That is, P_{th} determines the number of requests that should be made by an endpoint AD until successful path activation.

Practical Verification of WPA-TKIP Vulnerabilities

Mathy Vanhoef
iMinds-DistriNet
KU Leuven
Mathy.Vanhoef@cs.kuleuven.be

Frank Piessens
iMinds-DistriNet
KU Leuven
Frank.Piessens@cs.kuleuven.be

ABSTRACT

We describe three attacks on the Wi-Fi Protected Access Temporal Key Integrity Protocol (WPA-TKIP). The first attack is a Denial of Service attack that can be executed by injecting only two frames every minute. The second attack demonstrates how fragmentation of 802.11 frames can be used to inject an arbitrary amount of packets, and we show that this can be used to perform a portscan on any client. The third attack enables an attacker to reset the internal state of the Michael algorithm. We show that this can be used to efficiently decrypt arbitrary packets sent towards a client. We also report on implementation vulnerabilities discovered in some wireless devices. Finally we demonstrate that our attacks can be executed in realistic environments.

Categories and Subject Descriptors

E.3 [**Data Encryption**]: Code breaking; C.2.0 [**Computer-Communication Networks**]: General—*Security and protection*; C.2.1 [**Computer-Communication Networks**]: Network Architecture and Design—*Wireless communication*

General Terms

Security, Experimentation, Verification

Keywords

802.11; WPA; TKIP; DoS; fragmentation; decryption; driver vulnerabilities

1. INTRODUCTION

Modern wireless networks are based on the IEEE 802.11 set of standards. These networks have gained popularity over the years and are nowadays widely used in different scenarios, ranging from personal use to high-profile commercial use. Because of the nature of wireless transmission special care must be taken to preserve the privacy and security of wireless networks. The original IEEE 802.11 standard supported a basic security algorithm called Wired Equivalent Privacy (WEP). Unfortunately WEP suffers from major design flaws and is considered completely broken [8, 19, 4].

An improvement of WEP is the Temporal Key Integrity Protocol (TKIP). Created as an intermediate protocol to the more secure CCMP, it was designed to run on existing WEP hardware [14, §11.4.1]. This affected many of the design decisions [14, 6]. Most notably it still uses WEP encapsulation and relies on a weak Message Integrity Check (MIC) algorithm called Michael [6]. Because the Michael algorithm provides inadequate security [20, 13, 22] countermeasures were added. TKIP and its countermeasures are explained in detail in Sect. 2.

Surprisingly, TKIP is still supported by a large number of networks. In Section 6.1 we report on an experiment where we collected information about wireless network usage in two Belgian municipalities and found that 71% of encrypted networks support TKIP. Furthermore, 19% of networks using encryption only allow TKIP.

In this paper we present a novel Denial of Service (DoS) attack on TKIP. Moreover, we take two ideas suggested in a paper by Beck [1] and significantly improve on them. In contrast with the paper of Beck, our improvements are also implemented and tested in practice. The first idea applies the known fragmentation attack on WEP [4] to TKIP. This allows an attacker to send an arbitrary amount of packets to a client. As a proof of concept we implemented a port scanner. The second idea is to construct a prefix that resets the internal state of the Michael algorithm. We will show that this enables us to efficiently decrypt arbitrary packets sent towards a client. We also report on several vulnerabilities found in the implementation of some wireless adapters and drivers. All attacks are designed against WiFi networks operating in infrastructure mode, and are tested when authentication is done using a passphrase and when using a personal username and password.

We hope that the publication of these novel attacks will motivate people to disable TKIP.

The remainder of this paper is organised as follows. Section 2 describes the details of the TKIP protocol. In Section 3 we explain the Denial of Service (DoS) attack. Section 4 discusses our fragmentation and portscan attack. In Section 5 the Michael state reset and decryption attack is explained. Section 6 investigates whether TKIP is still supported in practice and discusses experimental evaluation of our attacks. Finally, we summarise related work in Sect. 7 and conclude in Sect. 8.

ASIA CCS'13, May 8–10, 2013, Hangzhou, China.

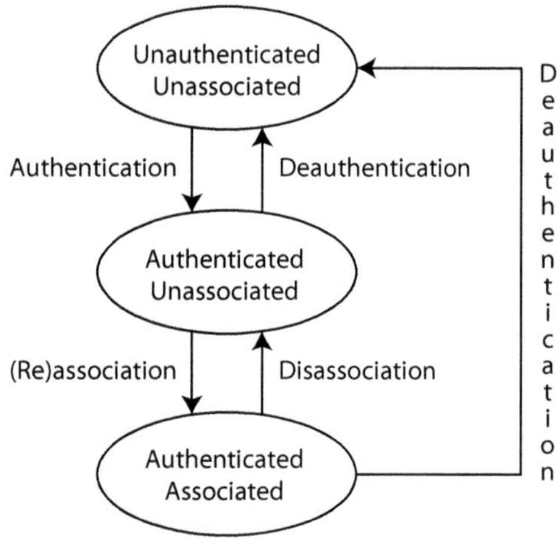

Figure 1: States a client can be in when connecting to a wireless network.

2. TEMPORAL KEY INTEGRITY PROTOCOL (TKIP)

This section describes the relevant parts of the IEEE 802.11 standard with a focus on the TKIP specification [14, §11.4.2]. We will also explain one of the first attacks on TKIP, called the Beck and Tews attack.

2.1 Connecting

A client connects to a wireless network by first authenticating and then associating with the Access Point. A state diagram of this process is shown in Fig. 1. There are two authentication methods. The first one is called Shared Key authentication and was based on WEP. Unfortunately this method is inherently insecure. Nowadays only the second method is used, called Open System authentication. As the name implies it imposes no real authentication. It is essentially just a formality, and if TKIP or CCMP is used actual authentication will happen at a later stage. Once authenticated, the client sends an association request to the AP. This request includes the secure authentication and encryption protocol it wants to use. If the AP supports the requested protocols the association is successful, and the AP informs the client that the association has completed.

Once authenticated and associated, a 4-way handshake is performed when using TKIP. The handshake negotiates the keys used by TKIP and is defined using IEEE 802.1X EAPOL-Key frames. This results in a 512-bit pairwise transient key (PTK) that is shared between the AP and client. From the PTK a 128-bit temporal encryption key (TK) is derived, as well as two 64-bit Message Integrity Check (MIC) keys: one for AP to client communication and one for client to AP communication. These keys are renewed after a user defined interval, commonly called the rekeying timeout. Most APs by default use a timeout of 1 hour. After a key has been negotiated, the client and AP can send encrypted data frames to each other.

The client or AP can end the connection at any time by sending a disassociation or deauthentication message. The older versions of the 802.11 standard left these messages unprotected. This means an attacker can forge them and forcibly close the connection between an AP and client. Continuously injecting such forged deauthentication packets causes a DoS attack [18]. This attack is well known and is called the deauthentication attack. It can be prevented by enabling protected management frames, a feature introduced in the IEEE 802.11w amendment [14, §4.5.4.9].

6	6	1	3	*var*
Destination	Source	Priority	Reserved	Payload

Figure 2: Input data given to the Michael algorithm. Destination and source represent MAC addresses. If the QoS extension is not used, priority is set to zero. Numbers denote the size of the field in bytes, where *var* defines a variable-length field.

	8	*var*	8	4
802.11 header	TSC	Data	MIC	ICV

⟵ unencrypted ⟶ ⟵ encrypted ⟶

Figure 3: Simplified format of an unfragmented TKIP frame.

2.2 Sender

When sending a TKIP frame first the MIC value of the MAC Service Data Unit (MSDU) is calculated. The purpose of the MIC is to protect both the integrity and authenticity of the message. Recall that the MSDU is essentially the complete data packet that needs to be transmitted, which is fragmented into smaller MAC Protocol Data Units (MPDU) fragments if it is too big to be sent at once. The MIC is calculated over the MSDU by the Michael algorithm. Michael is a keyed hash function taking two inputs: the input data to calculate the MIC of and a secret key. The input data given to the Michael algorithm is shown in Fig. 2. For calculating the MIC a different secret key is used for AP to client communication than is used for client to AP communication. Both MIC keys are derived from the PTK. The calculated MIC value is 8 bytes long. However, the Michael algorithm is not sufficiently secure. In particular it is possible to efficiently retrieve the MIC key given the data and calculated MIC value [20]. The designers realized this and included countermeasures in an attempt to mitigate potential attacks. These countermeasures are essential to our attacks and are discussed in detail in Sect. 2.3.

The MSDU concatenated with the MIC is fragmented into MAC Protocol Data Units (MPDUs) if necessary. At most 16 fragments (MPDUs) are supported. Each MPDU then undergoes WEP encapsulation. This is done so TKIP can be implemented on old WEP hardware. WEP encapsulation appends an Integrity Check Value (ICV) to the MPDU, which is simply a 32 bit CRC computed over the given data. Then it encrypts the packet using the RC4 stream cipher. The key used for encryption is called the WEP seed and is calculated by a mixing function that combines the temporal key (TK), transmitter MAC address, and the TKIP Sequence Counter (TSC). Figure 3 illustrates the final layout of an unfragmented TKIP frame.

Finally the resulting link-layer frame is constructed by adding the appropriate 802.11 headers. This includes the TKIP Sequence Counter (TSC), which is a replay counter that increases every time a MPDU frame is sent successfully. In Fig. 3 the TSC is given a size of 8 bytes, but in reality the TSC is only 6 bytes long. The remaining bits of the field are used for other purposes or are reserved. To prevent replay attacks the receiver drops frames that are not received in order. That is, the counter must always be increasing though gaps are allowed.

2.3 Receiver

When receiving a TKIP frame the client or AP first checks if the TSC is in order. If not, the frame is silently dropped. It then proceeds by checking if the ICV is correct. If not, the frame is also silently dropped. Once all MPDU's are received they are reassembled into the original MSDU and its MIC value is verified. If it is correct the frame is accepted and the receiver updates its TSC replay counter, otherwise the TKIP countermeasures kick in. These countermeasures were added to detect active attacks against the weak Michael algorithm [14, §11.4.2.4.1]. One class of such active attacks involves injecting multiple forged packets in the hope at least one of them has a valid MIC. In case such a packet is found the attacker learns the MIC key for the particular communication direction, as the MIC key can be derived when given the plaintext data and calculated MIC value.

The countermeasures are as follows [14, §11.4.2.4]:

- When a client receives a MSDU with an invalid MIC value it will send a MIC failure report to the AP.
- An AP receiving an invalid MIC value does not broadcast a MIC failure report but only logs the failure.
- If the AP detects two MIC failures within one minute all TKIP clients connected to the AP will be deauthenticated and the AP will not receive or transmit any TKIP-encrypted data frames for one minute. Once this minute is passed clients can reassociate with the AP and negotiate a new PTK.

2.4 Quality of Service Extension

Quality of Service (QoS) enhancements for wireless traffic were first defined in the IEEE 802.11e amendment. Most modern APs support this amendment [17]. It defines 8 different channels, each having their own QoS needs. A channel is defined by its Traffic Identifier (TID) and is internally represented by 4 bits, making some devices actually support 16 different channels. For the implementation of our attacks we assume only 2 QoS channels exist, though devices supporting more channels are also susceptible to our attacks. Tests showed that our assumption holds in practice. Additionally we note that by default most traffic is sent over the first QoS channel.

Essential for us is that each QoS channel has a separate TSC [14, §11.4.2.6]. This means that we can capture a packet transmitted on one QoS channel and replay it on another QoS channel having a lower TSC value. This enables an attacker to pass the TSC check, though he still has to pass the ICV and MIC checks in order to forge a message.

2.5 Beck and Tews Attack

One of the first known attacks on TKIP was discovered by Beck and Tews [20]. It was a variation of the chopchop attack on WEP [10] and works by decrypting a packet one byte at the time. Because this attack is used as the basis for the fragmentation and Michael reset attack it will be explained in detail.

First we will explain the chopchop attack when applied to a TKIP packet. It begins by taking an encrypted packet and removing the last byte. Let C denote the obtained shortened encrypted packet. With high probability the ICV of C is invalid. However, it can be corrected if one knows the plaintext value of the removed byte. Correcting the ICV of the unencrypted shortened message can be represented by $M' = M \oplus D$, where M is the unencrypted shortened message, D is the correction being applied, $\oplus$ denotes the XOR operator, and M' is the unencrypted shortened message with a valid CRC. It has been proven that D only depends on the plaintext value of the removed byte [10]. Interestingly we can apply this modification directly on C. Letting K denote the keystream used to encrypt the packet we get that $C = M \oplus K$. We can now make the following derivation:

$$\begin{aligned} C' = M' \oplus K &= (M \oplus D) \oplus K && (1)\\ &= (M \oplus K) \oplus D && (2)\\ &= C \oplus D && (3) \end{aligned}$$

The second equation follows from the associativity of the XOR operator. We see that C' is the encrypted shortened packet with a valid ICV, and that it can be obtained by directly applying the modification to the encrypted shortened packet C.

This technique can be used to decrypt TKIP packets sent towards the client as follows. An attacker tries all 2^8 possible values of the removed byte. For each guess the modification D is applied and the resulting packet is injected with a different priority y. Assuming that QoS channel y has a lower TSC it will pass the TSC check, and the client will decrypt the packet. Then the ICV is verified. If the guess of the attacker was wrong the ICV is invalid and packet will silently be dropped (without generating a MIC failure). On the other hand, if the guess was correct, the ICV will be valid. However, with high probability the MIC value of the shortened packet will be wrong. As a result the client sends a MIC failure report. Thus a correct guess can be detected by listening for the corresponding MIC failure. Note that an AP isn't vulnerable to the attack because it never sends MIC failure reports. To avoid triggering the TKIP countermeasures at most one byte can be decrypted each minute.

Because we must wait one minute after decrypting a byte, it is infeasible to decrypt all bytes using this method. Instead the Beck and Tews attack targets an ARP reply packet and decrypts only the ICV and the MIC value. This takes on average 12 to 15 minutes. The remaining content of the packet is guessed. A particular guess can be verified by checking if the calculated ICV of the predicted packet equals the decrypted ICV. If they match, the guess is very likely correct. Once the ARP reply has been decrypted we can use the inverse Michael algorithm to calculate the MIC key used for AP to client communication [20]. Combined with the keystream of the decrypted ARP reply, an attacker can now forge 3 to 7 packets having a length smaller or equal to the ARP reply. The precise number of packets that can be forged depends on the number of supported QoS channels.

3. DENIAL OF SERVICE

This section describes a novel attack we discovered. When closely inspecting the QoS extension to TKIP we notice that the keystream is independent of the priority (i.e., QoS channel) used to transmit the frame. On the other hand the calculated MIC value does depend on the priority of the MSDU. Say that we capture a packet sent with priority x and replay it with a different priority y. Assuming QoS channel y has a lower TSC it will pass the TSC check, and the receiver will use the correct keystream to decrypt the packet. As a result it will also pass the ICV check. However, the changed priority will cause the receiver to expect a different MIC value. Hence a MIC failure occurs. Replaying this packet a second time will trigger a second MIC failure. After these two MIC failures the AP will shut down all TKIP traffic for 1 minute. Repeating this process every minute will prevent any TKIP protected communication, effectively causing a DoS. If the network does not use QoS we can forge the QoS header when replaying the packet. As discovered by Morii and Todo, most clients will not check whether the network is actually using QoS and simply accept the packet [17]. Therefore the only requirement is that one or more clients *support* the QoS extension, which is true for most modern wireless adapters [17].

To implement the attack we had to patch the compat-wireless drivers of Linux. The original driver modified the QoS header when in monitor mode, which interfered with our attack. Monitor mode is a feature supported by some wireless adapters and drivers allowing one to capture all wireless traffic. It also enables injection of arbitrary 802.11 frames. Our tool monitors the traffic of a network and shows whether it supports TKIP or QoS. Additionally it shows a list of connected clients and basic statistics such as the cipher suite it is using, number of MIC failures, number of association failures, etc. When a vulnerable TKIP packet is captured its priority is changed and the packet is replayed. If the network isn't using QoS our tool will forge the QoS header. We show in Sect. 6 that our implementation was found to be very reliable.

Our attack appears to be one of the more effective DoS attacks that can be launched against a networking using TKIP. The simplicity of the attack not only makes it easy to implement and debug, it also assures it is applicable in many real world environments. In Section 7 a thorough comparison is given to currently known DoS attacks. Disabling the TKIP countermeasures prevents the attack. However, most APs do not provide this option, and with good reason. As mentioned in Sect. 2.3 the countermeasures were included to detect active attacks against the weak Michael algorithm [6].

Another option to prevent the attack is to make the keystream dependent on the priority. This entails a change in the protocol, meaning all devices implementing TKIP would have to be updated. Though preventing the attack, such a modification does not appear feasible in practice.

4. MORE AND BIGGER PACKETS

The Beck and Tews attack allows forging 3 to 7 packets of at most 28 bytes [20]. Our goal is twofold: we want to inject both more and bigger packets. In this section we assume the Beck and Tews attack has already been executed, meaning we have obtained the MIC key for AP to client communication. Our technique is similar to the fragmentation attack on WEP [4] and an improvement of a suggested attack by Beck [1]. To the best of our knowledge, this is the first time such an attack on TKIP has been implemented and proved to be possible.

4.1 Exploiting Fragmentation

Since we know the MIC key, sending additional packets only requires obtaining new keystreams. Recall that each TSC corresponds to a different keystream. Because RC4 is used, XORing an encrypted packet with its corresponding unencrypted packet unveils the keystream used during encryption. Hence, finding new keystreams reduces to predicting the content of encrypted packets.

All 802.11 packets start with a LLC and SNAP header of 8 bytes. If we know whether it is an ARP, IP, or EAPOL packet, this header can be predicted. Fortunately we can identify ARP and EAPOL packets based on their length, and can consider everything else to be IP. The first byte of an IP packet consists of the version and the header length, which is generally equal to 0x45. The second byte containing the Differentiated Services Field can be predicting based on the priority of the 802.11 frame [18]. Finally the next 2 bytes, representing the length of the IP packet, can be derived from the length of the 802.11 frame. As a result we can predict the first 4 bytes of an IP packet. Predicting the first 14 bytes of ARP packets is also possible. These bytes consist of the hardware type and size, protocol type and size, request type, and finally the MAC address of the sender. All these fields can be predicted in an IPv4 network. Because EAPOL packets are rarely transmitted we simply ignore them. Our predictions can then be XORed with the encrypted packets, revealing the first 12 bytes or more of the keystream.

These short keystreams are combined using fragmentation at the 802.11 layer. As mentioned in Sect. 2.2 the 802.11 protocol allows a frame to be fragmented into at most 16 MPDUs. Each MPDU must have a unique TSC value and is encrypted with the keystream corresponding to that TSC. All fragments must include an ICV, though the MIC value is calculated over the complete MSDU and can be spread over multiple MPDUs. Using fragmentation to combine the keystreams we can inject packets with 112 bytes of payload. Assuming there is enough traffic on the network, we can inject an arbitrary amount of packets. Compared to the Beck and Tews attack this is a significant improvement.

4.2 Implementation: Performing a Portscan

Implementing the fragmentation attack required patching the compat-wireless drivers of Linux. The original driver failed to correctly inject MPDUs, which caused the attack to fail. Using our patched driver we successfully tested the fragmentation attack against several devices, and found that our heuristics to predict the first bytes of packets were very accurate (see Sect. 6). To reduce packet loss we also detect whether the client acknowledged receiving the MPDU, otherwise the MPDU is retransmitted. As a proof of concept we implemented a port scanner. This requires injecting a large amount of packets and is thus ideal to test our fragmentation attack.

The port scanner is given a file containing the ports to scan and works by injecting TCP SYN request to each port. The SYN packets do not contain any options, assuring there is enough keystream to inject them. The encrypted TCP SYN-ACK reply is detected by its length. In practice most packets are larger than a SYN-ACK packet, meaning it has a distinctive short length. After scanning a port a TCP RST packet is always sent, even if no SYN-ACK was detected. This is done to prevent the client from retransmitting a potentially undetected SYN-ACK packet. Note that if the replies of the client can be sent to an IP under our control, we essentially have bidirectional communication and can connect to open ports.

The attack can be mitigated by preventing the Beck and Tews attack. This means disabling the client from sending MIC failure reports, or using a short rekeying time of 2 minutes or less [20].

5. DECRYPTING ARBITRARY PACKETS

In this section we describe a state reset attack on the Michael algorithm and we show how it can be used to decrypt arbitrary packets sent towards the client.

5.1 The Michael Algorithm

The state of the Michael algorithm is defined by two 32-bit words (L, R) called left and right. To calculate the MIC value of an 802.11 packet the state is first initialised to the MIC key. Then the data shown in Fig. 2 is processed, which is padded so that its length is a multiple of 4 bytes. Finally the calculation is finalised and resulting MIC value is outputted. All data is processed in 32-bit words by a block function.

The block function $B(L, R)$ is an unkeyed 4-round Feistel-type construction, taking as input a Michael state and returning a new state [14, §11.4.2.3.3]. When processing an input word M the next state is given by $B(L \oplus M, R)$. We will let L' stand for $L \oplus M$. The block function can be inverted [22], and its inverse is denoted by B^{-1}. Note that we can predictably influence L at the start of the block function using M. For convenience the notation $B((L, R), M)$ is used to represent $B(L \oplus M, R)$, denoting the new internal state after processing M.

5.2 Michael State Reset

If the Michael state ever returns to the initial state, all data processed so far has no influence on the MIC value. This idea can be used to construct a prefix packet which resets the state, allowing us to append a packet whose MIC value is calculated only over the appended data. As suggested by Beck [1] this could be done by appending two *magic words* to the prefix. These so-called magic words are ordinary 32-bits data words, but chosen in such a way so they reset the internal state of the Michael algorithm. Using them an attacker can append any encrypted packet to the prefix without invalidating the MIC value of the complete packet. Unfortunately Beck didn't provide a thorough theoretical analysis. We generalise the problem to finding a list of magic words that will transform a start state (L_s, R_s) to an end state (L_e, R_e).

We could try to use just one magic word M_1 to reset the state. In that case $B(L_s \oplus M_1, R_s)$ must return (L_e, R_e). Assuming that processing a random word M using the block function results in a random state, a guess for M_1 can be modelled as a Bernoulli trial with a success probability of 2^{-64}. Since we can try at most 2^{32} values for M, finding a solution reduces to having the first success after 2^{32} trails. This follows a geometric distribution. We get

$$\Pr[X \leq 2^{32}] = 1 - \left(1 - 2^{-64}\right)^{2^{32}} \approx 2{,}328 \cdot 10^{-10} \quad (4)$$

where X follows a geometric distribution with a success probability of 2^{-64}. Such a low chance of finding a solution is unusable.

A better option is to use two magic words, denoted by M_1 and M_2. This gives us one intermediate state (L_i, R_i) to work with. We begin by calculating $B^{-1}(L_e, R_e) = (L'_i, R_i)$. Note that we cannot calculate L_i because M_2 is still unknown. Nevertheless, this teaches us the required value for R_i. Then we brute force the first magic word M_1. For each possible value we apply the block function. If we obtain the required value for R_i we have found a valid intermediate state (L_i, R_i), since using the guessed value for M_1 and setting M_2 to $L_i \oplus L'_i$ results in a solution:

$$\begin{aligned} B(B((L_s, R_s), M_1), M_2) &= B(L_i \oplus M_2, R_i) && (5)\\ &= B(L'_i, R_i) && (6)\\ &= (L_e, R_e) && (7)\end{aligned}$$

The first equation is trivial. The second equation follows from our choice of M_2. Finally, the third equation follows from the calculation $B^{-1}(L_e, R_e) = (L'_i, R_i)$.

A solution is found if the guess for M_1 results in the required value for R_i, which has a probability of 2^{-32}. Similar to the previous case, the probability of finding a solution can be modelled by a geometric distribution:

$$\Pr[Y \leq 2^{32}] = 1 - \left(1 - 2^{-32}\right)^{2^{32}} = 0{,}6321\ldots \quad (8)$$

where Y follows a geometric distribution with a success probability of 2^{-32}. This implies that roughly 27% of the time 2^{32} calculations are performed yet no solution is found. As an experiment we ran 50000 runs where random Michael states had to be connecting using two magic words. In 31518 runs a solution was found, resulting in a success probability of 63,036%. This closely matches our analyses. On an 3,10 GHz Intel Core i5-2400 it took on average 11,14 seconds to find a solution, with a standard deviation of 6,33 seconds.

Another strategy is to use three magic words M_1, M_2, and M_3. This gives us two intermediate states (L_{i1}, R_{i1}) and (L_{i2}, R_{i2}) to work with. Again we start by calculating $B^{-1}(L_e, R_e) = (L'_{i2}, R_{i2})$. Next we take 2^{16} random values for the first magic word and compute the list of resulting intermediate states (L_{i1}, R_{i1}). We then try to brute-force the second magic word by applying the reverse Michael algorithm to the state (L'_{i2}, R_{i2}). If the resulting value for R_{i1} is in the list of earlier calculated states we have found a solution. Let L_{i1} be the value accompanying R_{i1}, then the magic words M_1, $M_2 = L_{i1} \oplus L'_{i1}$, and $M_3 = L_{i2} \oplus L'_{i2}$ provide a solution:

$$\begin{aligned} &B(B(B((L_s, R_s), M_1), M_2), M_3) \\ &\quad = B(B(L_{i1} \oplus M_2, R_{i1}), M_3) && (9)\\ &\quad = B(L_{i2} \oplus M_3, R_{i2}) && (10)\\ &\quad = (L_e, R_e) && (11)\end{aligned}$$

The first equation is trivial. The second and third follow from our choice of M_2 and M_3, and the usage of the reverse block function to calculate L'_{i1} and L'_{i2}.

The probability that the result of a guess for M_2 is in the list is 2^{-16}. The probability of finding a solution can again be modelled as a geometric distribution. We get

$$\Pr[Z \leq 2^{32}] = 1 - \left(1 - 2^{-16}\right)^{2^{32}} \approx 1 - 7{,}2 \cdot 10^{-28463} \quad (12)$$

where Z follows a geometric distribution with a success probability of $p = 2^{-16}$. Practically this shows that a solution will always be found. The average number of required guesses for the second word is $E[Z] = 1/p = 2^{16}$. As an experiment we ran 50000 runs where random states had to be connecting using three magic words. In all runs a solution was found. It took on average $65814 = 2^{16.017...}$ guesses for M_2 until a solution was found, with a standard deviation of 66407 guesses. On our 3,10 GHz Intel Core i5-2400 this corresponded to an average running time of 2,96 milliseconds. Though requiring more magic words, these results are significantly better than the previous two cases.

5.3 Decryption Attack

Our goal is to decrypt arbitrary packets sent towards the client. We will accomplish this by appending the targeted packet to a specially crafted prefix. The prefix will simulate the behaviour of a ping request, making the client echo back the appended data. We construct the prefix such that the reply is send to an IP under our control, meaning we will receive the plaintext content of the targeted packet, effectively decrypting it. To assure that the MIC value of the constructed packet is correct we will apply the Michael state reset attack to the ping-like prefix. This allows us to append the targeted packet without invalidating the MIC value. The resulting frame is sent to the client using the fragmentation attack.

Contrary to the suggestion by Beck [1] we cannot use an ICMP ping request as the prefix. This is because it includes a checksum calculated over the header and the data section. But since we do not know the plaintext data of the full packet, we cannot calculate a correct checksum. Instead we will construct a UDP prefix, where specifying a checksum is optional. Sending a UDP packet to a closed port results in an ICMP destination unreachable reply containing the first 8 bytes of the UDP packet. However, on Windows, Linux, and Android the ICMP unreachable reply contains a full copy of the original UDP packet. So when targeting these operating systems the client will reply with the complete content of our constructed packet. In particular this includes the plaintext content of the targeted packet. Even large packets can be quickly decrypted using this method.

Again we created a proof of concept tool in Linux. It listens for packets sent towards the client. Once a vulnerable packet has been captured, the UDP prefix is constructed and the Michael state reset attack is applied. The resulting UDP prefix is transmitted using the fragmentation attack, followed by the targeted packet (which is marked as the final fragment of the MSDU). In practice this means the final fragment will usually be bigger than all previous fragments. Though this is not allowed by the 802.11 specification, nearly all devices will accept the packet (see Sect. 6). The ICMP unreachable reply sent by the client will include the prefix, magic bytes, and the unencrypted data of the targeted packet. Among other things, this allows an attacker to decrypt a TCP packet, learn the sequence number, and hijack the TCP stream to inject arbitrary data [12]. As a consequence, malicious data could be injected when the client opens a website. Again the attack can be mitigated by preventing the Beck and Tews attack.

6. EXPERIMENTS

In this section we begin by investigating whether TKIP is still supported in practice. Then we evaluate how much devices adhere to the relevant aspects of the 802.11 standard, we report on implementation vulnerabilities discovered in some wireless devices, and we discuss how our findings impact the attacks. Finally our attacks are tested in realistic settings.

Table 1: Number of WiFi networks supporting the given encryption schemes for several regions. Note that one network can support multiple schemes.

Region	Open	WEP	TKIP	CCMP	#Networks
Leuven	381	618	3307	3143	5023
Heverlee	121	288	1212	1149	1886

6.1 Networks Supporting TKIP

Some new routers (for instance the Belkin N300 router) do not support TKIP anymore, in accordance with the security roadmap of the WiFi Alliance. The WiFi Alliance tests products and hands out certifications if they conform to certain standards. Their new roadmap specifies that, as of 2011, new APs are no longer allowed to support a TKIP only option [7]. Even mixed mode, which simultaneously allows TKIP and CCMP in the same network, is no longer a requirement. Finally, in 2014 TKIP is disallowed completely. Based on this one would think that TKIP is no longer widely supported. Surprisingly, we found the opposite to be true.

To investigate whether TKIP is still supported in practice we surveyed wireless networks in two Belgian municipalities (Leuven and Heverlee). Detecting networks was done using passive scanning, consisting of monitoring wireless traffic for beacon frames. These frames contain all information necessary to connect to a wireless network. In particular it includes the name of the network, the MAC address of the AP, and the encryption schemes supported by the network. During an initial tests we found that active scanning detected few additional networks (less than 6%), so in an attempt to prolong battery life only passive scanning was used.

Several trips, of around an hour long, were made on foot while scanning for networks. The raw capture was written to file and later analysed for beacon frames using a custom tool. This approach allowed us to make improvements to our tool after collecting the raw captures. We uniquely identified networks by their name. This is necessary because one network can be advertised by multiple APs. However we also encountered several APs advertising a network named after a vendor or product. These are default network names used by a particular device and do not represent the same network. Therefore we treated each of these APs as a unique network. Similarly there were several APs advertising a network with an empty name. These APs were also treated as an unique network.

In total we detected 6803 unique networks. The number of networks supporting a particular encryption scheme is shown in Table 1. Note that a handful of networks were present both in Leuven and Heverlee. We found that 93% of the networks used encryption, and that 66% supported TKIP. When considering only encrypted networks, 71% of them supported TKIP. Additionally, 19% of networks using encryption only allow TKIP. We believe the reason so many networks still support TKIP is because most routers, when configured to use WPA2, by default use mixed mode (allowing both TKIP and CCMP). We even observed that WEP is still used by 14% of encrypted networks.

Table 2: Results of various tests on different wireless adapters. For laptop and USB adapters, L and W denote it only works on Linux or Windows, respectively. Yes means it works on both, no means it works on neither. Open Source router firmware was tested on the Asus RT-N10.

		DoS	Fragmentation				Replay	Unenc.
			diff. size	eff. frag.	skip TSC	any MIC		
Laptop and USB:	Intel 4965AG	yes	yes	no	L	no	no	no
	Belkin F7D1102AZ	yes	yes	yes	yes	W	yes	L
	Belkin F5D7053	yes	L	L	L	L	yes	yes
	Alfa AWUS036h	yes	yes	yes	yes	W	W	W
	Ralink WA-U150BB	yes	yes	yes	yes	L	yes	W
Mobile Devices:	iPod MC086LL	yes	yes	no	yes	no	no	no
	iPad MC980NF	yes	yes	no	yes	no	no	no
Access Points:	Linksys WAG320N	yes	yes	yes	yes	no	no	no
	WRT54G 4.21.5	yes	yes	yes	yes	no	no	no
	Scarlet VDSL Box	yes	yes	no	no	yes	no	yes
	Cisco Aironet 1130 AG	yes	yes	yes	yes	no	no	no
	Asus RT-N10 1.0.2.4	yes	yes	yes	yes	no	no	no
	Tomato 1.28	yes	yes	yes	yes	no	yes	no
	DD-WRT v24-sp2	yes	yes	yes	yes	no	no	no

6.2 Adherence to the 802.11 Specification

The wireless adapters in Table 2 were tested for implementation details impacting our attacks. While doing this we encountered several vulnerabilities present in some wireless devices, and decided to test for their presence on all devices. Wireless routers and mobile devices were tested using their default configurations, with open source wireless router firmware was tested on the Asus RT-N10. Laptops and USB wireless adapters were tested on Linux using the compat-wireless 3.6.2-1-snp drivers, and on Windows using the default installed drivers.

For the DoS attack to work a client must send a MIC failure report when an invalid MIC has been detected, and the AP must shut down the network after two MIC failures. We found that all wireless adapters, in all configurations, implement this properly. As a result our DoS attack is applicable to all tested devices (see Table 2 column DoS). Additionally we tested the DoS on a network not using QoS, making our tool forge the QoS header. Again all adapters were vulnerable to the attack, confirming results by Todo et al. [21].

Fragmentation support has been tested for several properties. We found that the Belkin F5D7053 adapter on Windows incorrectly implemented fragmentation. Sending a fragment to this adapter always resulted in a MIC failure, though it worked fine under Linux. All other devices supported fragmentation.

An important property of fragmentation is whether the last fragment is allowed to be bigger than the previous fragments. Strictly speaking this is not allowed in the 802.11 standard [14, §9.5] yet we rely on this during the decryption attack. We found that all devices supporting fragmentation permitted this behaviour (see Table 2 column diff. size).

As mentioned in Sect. 2.3 the receiver should update its replay counter after reassembling the MSDU. This leaves open the possibility to send each fragment of an MSDU using the same TSC, keystream, and priority. If allowed, an attacker is then able to inject 16 fragments using only one keystream. In Table 2 column eff. frag. we see that this technique is indeed possible on several devices.

Finally we tested if fragments can be sent without using a sequential replay counter, i.e., whether we can skip TSC values. The 802.11 standard permits this behaviour for TKIP [14, §11.4.2.6], though some devices only accepted sequential TSC (see Table 2 column skip TSC).

Surprisingly we also discovered critical implementation flaws in some devices, particularly in wireless USB adapters. Several do not drop packets with an already used TSC, allowing an attacker to replay packets (see Table 2 column Replay). Another flaw is that certain devices do not verify the MIC value of a fragmented TKIP packet (see Table 2 column any MIC). This removes the requirement of first having to execute the Beck and Tews attack to obtain the MIC key. After all, if the MIC value is not verified we do not have to calculate it, meaning there is no reason to know the MIC key. On Windows the Ralink WA-U150BB drops fragmented packets with an incorrect MIC, but does not transmit a MIC failure. As mentioned in Sect. 3, dropping the packet but not sending a MIC failure can open the door for novel attacks on the MIC key.

More worrisome, we also encountered wireless adapters that accepted unencrypted packets while connected to an encrypted network (see Table 2 column Unenc.). In particular the vulnerability is present in the Windows drivers for the AWUS036h, a device popular for its wide reception and high transmission power. Another device susceptible to the attack was the Scarlet VDSL Box. This is a router is handed out by the Belgium ISP Scarlet when a customer buys a VDSL internet connection. The reason this case is interesting is because the vulnerability has a higher impact when present on APs. An attacker could abuse the flaw to easily inject arbitrary traffic into the network, but also to send packets to the internet using the public IP of the victim. It might be an interesting future research direction to test for additional implementation flaws on a wider array of devices.

6.3 Verifying Our Attacks

According to the 802.11 specification, the TKIP countermeasures must disable only TKIP traffic for one minute [14, §11.4.2.4]. However, most of the APs we tested disabled all

Table 3: Impact of our DoS attack on several Access Points. The term *Both* denotes that TKIP and CCMP traffic was disabled. Open source router firmware was tested on the Asus RT-N10.

Access Point	Protocols Disabled	
	WPA1	WPA2
Linksys WAG320N	Both	Both
WRT54G 4.21.5	TKIP	Both
Scarlet VDSL Box	Both	TKIP
Cisco Aironet 1130 AG	Both	Both
Asus RT-N10 1.0.2.4	Both	TKIP
Tomato 1.28	Both	Both
DD-WRT v24-sp2	Both	Both

wireless traffic, including CCMP protected traffic in mixed mode. When targeting these devices one single TKIP client allows an attacker to take down the complete wireless network. More precisely, the behaviour of an AP depends on whether WPA1 or WPA2 is being used (see Table 3). Practically the difference between WPA1 and WPA2 is mainly between supported encryption schemes: WPA1 mandates TKIP support and optionally allows CCMP, while the reverse is true for WPA2. For example, the WRT54G only supports TKIP when using WPA1, while the Belkin N300 only supports WPA2 with CCMP.

For the fragmentation, portscan, and decryption attack to work we need to be able to accurately predict the first 12 bytes of encrypted packets. To test our prediction algorithm we monitored TKIP traffic generated by visiting several websites for 20 minutes. In total 36643 data packets were captured, of whom 36642 were predicted correctly. The single miss was an EAPOL packet, which we purposely ignored. Further, we weren't able to capture all packets as an attacker, meaning the keystreams corresponding to certain TSCs remained unknown. In total 5680 packets were not captured, consisting of 13% of all traffic.

The fragmentation and portscan attack has been tested against Windows, Linux, iOS, and Android. When connected to a networking using TKIP, all were found to be vulnerable. Additionally we tested the attacks under two authentication mechanisms: the first being a shared passphrase and the second being a personal login and password verified using PEAP-MSCHAP v2. In both situations the attacks were successful. These tests clearly demonstrate the reliability of the attacks.

For the decryption attack we found that, when connected to a network using TKIP, Windows, Linux, and Android were vulnerable. Mac OS X and iOS only include the first 8 bytes of the UDP packet in the ICMP unreachable reply, meaning the targeted packet is never included. Nevertheless, we did receive ICMP unreachable replies from all operating systems, meaning the Michael reset attack worked in all cases. Installing a firewall which blocks the ICMP unreachable replies prevents the decryption attack, though the Michael reset attack itself will remain possible.

7. RELATED WORK

Several DoS attacks exist on wireless networks. Arguably the most well-known attack consists of forging deauthentication frames to either the client or AP [2]. This is possible because the deauthentication message is not protected using any keying material. Continuously sending them will result in a DoS attack. An advantage of our attack is that it requires replaying only two frames each minute to disconnect *every* client that is using TKIP. Hence our attack is stealthier and requires less power to execute. Furthermore, nowadays the deauthentication attack can be prevented by enabling protected management frames [14, §4.5.4.9]. Compared to the DoS attack of Glass and Muthukkumarasamy [9] our attack is easier to execute, since their attack requires a man-in-the-middle position. Another advantage is that our attack can be used to easily verify whether the client sends MIC failure reports, and thus to see if it is possible to perform the Beck and Tews attack [17]. The Beck and Tews attack could also be changed to a DoS attack. Once the first MIC failure has been detected, the corresponding packet could be injected a second time. However, on average 128 packets have to be injected before a MIC failure is triggered. This makes the attack easier to detect. It also isn't as easy to implement compared to our DoS attack, making it more difficult to execute in practice.

Könings et al. found DoS vulnerabilities at the physical and MAC layer of 802.11, some halting traffic for one minute with minimal packet injection [15]. Contrary to our DoS, their attacks do not simultaneously disconnect all clients connected to a network. Bicakci and Tavli give a survey on DoS attacks at the physical and MAC layer. The attacks they discuss require injecting a large amount of frames [3].

One of the first attacks on TKIP was found by Beck and Tews and decrypts an ARP reply packet. As a result the MIC key for AP to client communication can be obtained [20] and a few small packets can be injected. It only works if the QoS extension is enabled and takes 12–15 minutes to execute. Halvorson et al. improved the attack, allowing larger packets to be injected [11]. However their technique does not allow injection of more packets, and takes longer to execute compared to our fragmentation attack. Morii and Todo came up with a modification removing the requirement that the AP must have enabled QoS [17]. They found that even if QoS was not used, clients will still accept and process a QoS frame. This allows an attacker to forge the QoS header if not present. In another paper Todo et al. managed to reduce the execution time of the Beck and Tews attack to 8–9 minutes [21].

The Michael algorithm was designed by Ferguson [6]. It was quickly revealed to be invertible by Wool [22]. Based on this Wool suggested a related message attack on TKIP. Huang et al. showed that Michael is not collision free and suggested a packet forgery attack [13].

Beck suggested the use of two magic words to reset the Michael state [1]. Unfortunately, no thorough theoretical analysis was provided, and no implementation in a practical setting was given. Based on the fragmentation attack on WEP by Bittau et al. [4], Beck also suggested using the fragmentation attack on TKIP, though no implementation was made to verify his ideas.

Wireless drivers have been tested before for vulnerabilities [5]. Most of the techniques focus on fuzzing with the aim of finding code injection attacks. However, only a limited amount of research has been done to find logical implementation flaws. In particular we found no previous work testing for replay attacks, seeing if it is possible to inject plaintext packets into an encrypted network, or testing whether the MIC of fragmented TKIP packets is verified.

Moen et al. have analysed the key scheduling algorithm in TKIP [16]. They found that, given less than 10 RC4 packet keys, it is possible to recover the temporal key (TK) with a time complexity of $\mathcal{O}(2^{105})$ compared to a brute force attack with complexity $\mathcal{O}(2^{128})$.

8. CONCLUSION

TKIP was designed as an intermediary solution to mitigate the existing flaws of WEP. We found that TKIP is still supported by a large number of networks. Further, we showed that TKIP fails to provide sufficient security, by describing and implementing several new attacks. In addition, during our experiments, we identified several critical implementation vulnerabilities in several wireless devices. We conjecture that a more substantial test of these implementations will reveal more vulnerabilities.

Specifying a short rekeying interval prevents the fragmentation and decryption attack, unfortunately this does not prevent our DoS attack. To secure a wireless network it is strongly advised to only support the more secure CCMP.

9. ACKNOWLEDGEMENTS

The authors would like to thank the anonymous reviewers and the shepherd for their valuable feedback.

This research is partially funded by the Research Fund KU Leuven, and by the EU FP7 project NESSoS. With the financial support from the Prevention of and Fight against Crime Programme of the European Union (B-CCENTRE). Mathy Vanhoef holds a Ph. D. fellowship of the Research Foundation - Flanders (FWO).

10. REFERENCES

[1] M. Beck. Enhanced TKIP michael attacks. Retrieved 4 Februari, 2013, from `http://download.aircrack-ng.org/wiki-files/doc/enhanced_tkip_michael.pdf`.

[2] J. Bellardo and S. Savage. 802.11 denial-of-service attacks: real vulnerabilities and practical solutions. In *Proceedigns of the USENIX Security Symposium*, 2003.

[3] K. Bicakci and B. Tavli. Denial-of-service attacks and countermeasures in IEEE 802.11 wireless networks, 2009.

[4] A. Bittau, M. Handley, and J. Lackey. The final nail in WEP's coffin. In *IEEE Symposium on Security and Privacy*, pages 386–400, 2006.

[5] L. Butti and J. Tinnés. Discovering and exploiting 802.11 wireless driver vulnerabilities. *Journal in Computer Virology*, 4(1):25–37, 2008.

[6] N. Ferguson. Michael: an improved MIC for 802.11 WEP. *IEEE doc. 802.11-2/020r0*, Jan. 2002.

[7] G. Fleishman. Say goodbye to WEP and TKIP. Retrieved 26 November, 2012, from `http://bit.ly/cSFSvj`, 2010.

[8] S. R. Fluhrer, I. Mantin, and A. Shamir. Weaknesses in the key scheduling algorithm of RC4. In *Selected Areas in Cryptography*, pages 1–24, 2001.

[9] S. M. Glass and V. Muthukkumarasamy. A study of the TKIP cryptographic dos attack. In *15th International Conference on Networks*. IEEE, 2007.

[10] M. Guennoun, A. Lbekkouri, A. Benamrane, M. Ben-Tahir, and K. El-Khatib. Wireless networks security: Proof of chopchop attack. In *WOWMOM*, pages 1–4, 2008.

[11] F. M. Halvorsen, O. Haugen, M. Eian, and S. F. Mjølsnes. An improved attack on TKIP. In *14th Nordic Conference on Secure IT Systems*, NordSec '09, 2009.

[12] B. Harris and R. Hunt. Review: TCP/IP security threats and attack methods. *Computer Communications*, 22(10):885–897, 1999.

[13] J. Huang, J. Seberry, W. Susilo, and M. W. Bunder. Security analysis of michael: The IEEE 802.11i message integrity code. In *EUC Workshops*, pages 423–432, 2005.

[14] IEEE Std 802.11-2012 (Rev. of IEEE Std 802.11-2007). *Wireless LAN Medium Access Control (MAC) and Physical Layer (PHY) Specifications*, 2012.

[15] B. Könings, F. Schaub, F. Kargl, and S. Dietzel. Channel switch and quiet attack: New DoS attacks exploiting the 802.11 standard. In *LCN*, 2009.

[16] V. Moen, H. Raddum, and K. J. Hole. Weaknesses in the temporal key hash of WPA. *Mobile Computing and Communications Review*, 8(2):76–83, 2004.

[17] M. Morii and Y. Todo. Cryptanalysis for RC4 and breaking WEP/WPA-TKIP. *IEICE Transactions*, 94-D(11), 2011.

[18] S. Park, K. Kim, D. Kim, S. Choi, and S. Hong. Collaborative QoS architecture between DiffServ and 802.11e wireless LAN. In *Vehicular Technology Conference*, 2003.

[19] A. Stubblefield, J. Ioannidis, and A. D. Rubin. A key recovery attack on the 802.11b wired equivalent privacy protocol (wep). *ACM Trans. Inf. Syst. Secur.*, 7(2), 2004.

[20] E. Tews and M. Beck. Practical attacks against WEP and WPA. In *Proceedings of the second ACM conference on Wireless network security*, WiSec '09, 2009.

[21] Y. Todo, Y. Ozawa, T. Ohigashi, and M. Morii. Falsification attacks against WPA-TKIP in a realistic environment. *IEICE Transactions*, 95-D(2), 2012.

[22] A. Wool. A note on the fragility of the Michael message integrity code. *IEEE Transactions on Wireless Communications*, 3(5):1459–1462, 2004.

Faster Secure Two-Party Computation with Less Memory

Wilko Henecka
School of Mathematical Sciences
University of Adelaide
Australia
wilko.henecka@adelaide.edu.au

Thomas Schneider
European Center for Security and Privacy by Design (EC SPRIDE)
Technische Universität Darmstadt
Germany
thomas.schneider@ec-spride.de

ABSTRACT

Secure two-party computation is used as the basis for a large variety of privacy-preserving protocols, but often concerns about the low performance hinder the move away from non-private solutions.

In this paper we present an improved implementation of Yao's garbled circuit protocol in the semi-honest adversaries setting which is up to 10 times faster than previous implementations. Our improvements include (1) the first multi-threaded implementation of the base oblivious transfers resulting in a speedup of a factor of two, (2) techniques for minimizing the memory footprint during oblivious transfer extensions and processing of circuits, (3) compilation of sub-circuits into files, and (4) caching of circuit descriptions and network packets. We implement improved circuit building blocks from the literature and present for the first time performance results for secure evaluation of the ultra-lightweight block cipher PRESENT within 7 ms online time.

Categories and Subject Descriptors

D.4.6 [**Operating Systems**]: Security and Protection—*Cryptographic controls*; C.2.4 [**Computer-Communication Networks**]: Distributed Systems—*distributed applications*

General Terms

Security, Algorithms

Keywords

Secure Computation, Garbled Circuits, Efficiency, Privacy

1. INTRODUCTION

Secure two-party computation, often called secure function evaluation (SFE), allows two mutually mistrusting parties to compute an arbitrary function on their private inputs without revealing any information about their inputs beyond the function's output. Although the real-world deployment of SFE was believed to be very expensive for a relatively long time, the cost of SFE has been dramatically reduced in the recent years thanks to many algorithmic improvements and automatic tools, as well as faster computing platforms and communication networks. SFE enables a large variety of privacy-preserving applications such as electronic auctions [NPS99], data mining [LP09b], or biometric identification [BG11, HMEK11], to name a few.

Although other approaches exist, most practical applications of SFE, including the ones listed above, are based on Yao's garbled circuits technique [Yao86] for which many improvements have been proposed (we give a summary in §2.1). In this paper we focus on secure two-party computation based on garbled circuits in the semi-honest adversary model. In this model, the adversary is assumed to be "honest-but-curious", i.e., he honestly follows the protocol specification, but tries to learn additional information from the messages seen. Although this adversary model is very weak, it allows to construct highly efficient protocols for many application scenarios, e.g., for constructing privacy-preserving protocols that protect against attacks by insiders or future break-ins. We strongly believe that pushing the performance limits of such protocols is essential in order to promote secure computation as conceivable alternative to using protocols without provable privacy guarantees.

In order to make SFE usable in practice, several frameworks with different properties have been proposed as summarized in Table 1. These frameworks allow an application developer to describe the functionality that needs to be computed securely on a high level and abstract from the details of the underlying protocol. Fairplay [MNPS04, BNP08] allows to describe the functionality to be computed in a high-level language which is compiled into a boolean circuit in an offline pre-computation phase. This compilation allows to perform global optimizations such as eliminating dead code. Subsequently, TASTY [HKS+10] partitioned the garbled circuit protocol such that most expensive operations (w.r.t. both, communication and computation) are performed in the pre-computation phase. To reduce the memory footprint, VMCrypt [Mal11] introduced the concept of streaming, i.e., the garbled circuit is generated gate by gate and directly streamed into the network. To also reduce the memory footprint for the circuit, the programmer can compose the circuit by dynamically constructing and deconstructing sub-circuits. However, VMCrypt instantiates a new object for each gate such that its performance suffers from the additional overhead of garbage collection. Also in

ASIA CCS'13, May 8–10, 2013, Hangzhou, China.

Table 1: Frameworks for GC-based secure two-party computation in the semi-honest adversaries setting.

Framework	Compilation	Streaming	Memory and Additional Overhead During Online Phase
Fairplay [MNPS04, BNP08]	Yes	No	$\mathcal{O}$(circuit size)
TASTY [HKS+10]	Yes	No	$\mathcal{O}$(circuit size)
VMCrypt [Mal11]	No	Yes	depends on circuit, garbage collection
FastGC [HEKM11]	No	Yes	$\mathcal{O}$(max(size of sub-circuit)), garbage collection
GCParser [MZE12]	Yes	Yes	$\mathcal{O}$(max(size of sub-circuit)), garbage collection
[KSS12] (malicious, cluster)	Yes	Yes	$\mathcal{O}$(maximum working set(circuit)), usage counter
This Work	Yes	Yes	$\mathcal{O}$(max(max. working set(sub-circuit))), no online overhead

FastGC [HEKM11] the circuit is not compiled, but composed from sub-circuits and dynamically generated within a library. Also here, a new object is created for each gate of the sub-circuit which could be freed by the garbage collector when not used any more. GCParser [MZE12] extended the FastGC framework to read in a file which describes the way pre-defined sub-circuits should be put together; it also requires memory linear in the size of the sub-circuits. Most recently, the framework of [KSS12] implemented GC-based secure function evaluation in the malicious setting by exploiting the high degree of parallelism available in a cluster. In this framework, each gate carries a usage counter such that memory can be freed after the last use of the gate, but this requires additional overhead in the online phase.

1.1 Outline and Our Contributions

After giving related works in §1.2, an overview on Yao's garbled circuit protocol in §2.1.1, and a minor remark on the choice of security parameters in §2.2, we present the following contributions:

In §3 we present several optimizations that result in lower memory consumption and significantly better performance compared to previous frameworks. More specifically, we improve the implementation of base oblivious transfers using multi-threading (§3.1), reduce the memory consumption of oblivious transfer extensions (§3.2), compile sub-circuits into files (§3.4), and cache circuit descriptions and the communication (§3.5). In §3.3 we enhance streaming by considering both, the memory footprint of the circuit and the garbled circuit with no additional overhead in the online phase. As described in §1 and summarized in Table 1, most previous frameworks have a memory consumption which is linear in the size of the evaluated circuit. As shown in [JKSS10b] and implemented in the framework of [KSS12], it is sufficient to just hold the intermediate values in memory that are needed later on, called the working set. For example, the maximum size of the working set of a Karatsuba multiplication of two 128 bit values is 1,074 whereas the circuit has 57,000 gates. In VMCrypt [Mal11], the needed memory depends on the way a programmer creates and decorates circuit components, and the framework of [KSS12] requires additional overhead in the online phase to manage a usage counter and free unused memory (see §3.3 for details). The memory consumption of our engine for simultaneously streaming circuits and garbled circuits depends only on the maximum size of the working set with no additional overhead in the online phase.

In §4 we demonstrate that our implementation is substantially more efficient than previous frameworks. As applications we consider secure evaluation of the Hamming distance, fast multiplication, and computing the minimum. Moreover, we give performance results on securely computing the AES block cipher and for the first time on secure evaluation of the ultra-lightweight block cipher PRESENT.

As many previous frameworks do, we provide the source code of our implementation as open source software to foster future works and allow a fair performance comparison. The code is available for download at
`http://code.google.com/p/me-sfe/`.

1.2 Related Works

Faster secure two-party computation using garbled circuits (FastGC) [HEKM11] is the first software implementation of streamed garbled circuits. As optimization, FastGC implements the optimization of inputs that depend only on one party as described in [PSS09, Ker11]. The following applications are implemented within FastGC: AES, Hamming distance, and Levenstein / Smith-Waterman distance with applications in privacy-preserving genome and protein alignment. Subsequently, the FastGC framework was used for various other applications, including privacy-preserving biometric identification [HMEK11] and privacy-preserving set intersection [HEK12]. The FastGC framework was also extended to achieve stronger security guarantees [HKE12], and adapted for privacy-preserving applications on smartphones [HCE11a, HCE11b]. These and future applications of the FastGC framework (e.g., by using this framework for iris and fingerprint identification [BG11]) benefit from our improvements.

Many application scenarios require a low memory footprint, e.g., privacy-preserving applications on smartphones [HCE11a, HCE11b], generating garbled circuits in resource-restricted trusted hardware [JKSS10a], evaluating garbled circuits with a hardware accelerator [JKSS10b], or securely evaluating large functionalities in cloud computing scenarios [BNSS11].

Frameworks for secure two-party computation in the semi-honest adversaries setting can be classified into the traditional compilation paradigm, where the function to be computed is first compiled and the on-the-fly paradigm that generates circuits gate by gate from a library. The compilation paradigm is used in Fairplay [MNPS04, BNP08] and TASTY [HKS+10]. The on-the-fly paradigm is used in the FastGC framework [HMEK11] and VMCrypt [Mal11]. We provide the best of both worlds by compiling and optimizing sub-circuits once and dynamically composing these sub-circuits on-the-fly.

A compilation technique for memory-efficient on-the-fly generation of circuits from Fairplay's high-level description language was proposed in [MLB12]. Alternatively, circuits can also be compiled from ANSI C programs as shown in [HFKV12]. The FastGC framework [HEKM11] was recently extended to read in a description of how circuits are composed from hard-coded circuit building blocks [MZE12]. However, these techniques do not minimize the amount of memory needed during secure evaluation of the circuit.

A large-scale garbled circuits-based framework for secure computations with security against stronger active (malicious) adversaries was presented recently in [KSS12]. This framework uses the compilation paradigm and exploits the high level of parallelism available in grid computing infrastructures by running multiple instances of a garbled circuit protocol in parallel – one on each machine. We extend their ideas for memory-efficient secure evaluation of garbled circuits and use multi-threading within a single instance of the garbled circuit protocol. For completeness we note that an alternative approach to practical actively secure secure two-party computations is [NNOB12] which is based on OT extensions instead of garbled circuits.

2. PRELIMINARIES

2.1 Yao's Garbled Circuit Protocol

In the following we give a brief summary of Yao's garbled circuit protocol, its optimizations, and oblivious transfer. For a more detailed description we refer to [Sch12, Chapter 2] and for a proof of security to [LP09a].

2.1.1 Yao's Garbled Circuit Protocol

Yao's garbled circuit protocol [Yao86] allows two parties, a server and a client, to jointly compute a function f represented as boolean circuit on their respective private inputs x and y. On a very high level, the server (sometimes called creator) creates an encrypted, called garbled, version of f which is then sent to the client (sometimes called evaluator) who evaluates the function under encryption. To encrypt the function, for each wire of f, the server assigns two random-looking wire labels that correspond to the values 0 and 1, respectively. Afterwards, the server obliviously sends exactly those wire labels to the client that correspond to their inputs. For client's inputs this is done with a sub-protocol, called oblivious transfer, such that the server does not learn the client's inputs (see below for details). Additionally, for each gate G_i of f, the server creates and sends to the client a garbled table T_i with the following property: given the wire labels for G_i's inputs, T_i allows to recover *only* the wire label of the corresponding output of G_i, but nothing else. Now, the client can use the wire labels of the inputs together with the garbled tables T_i to evaluate the garbled circuit gate by gate and obtains the labels of the output wires. For these output labels (and only for them) the client obtains mappings to the plain values 0 and 1 from the server which allow to recover $f(x, y)$.

The following optimizations of garbled circuits and oblivious transfer are used in today's most efficient implementations of Yao's protocol, including [HKS+10,BG11,HEKM11, Mal11,KSS12] and our implementation.

2.1.2 Garbled Circuit Optimizations

The point-and-permute technique [NPS99] represents each wire label as a symmetric t-bit key and a permutation bit π, where t is a symmetric security parameter. The permutation bits of a gate's input wires are used as index to denote which table entry needs to be decrypted. The free XOR technique [KS08] allows to compute garbled linear gates, i.e., XOR and XNOR gates, without communication (no garbled table is needed) and only negligible computation. Thus, the dominating factor for the complexity of a circuit is the number of non-linear gates. Further, the garbled row reduction technique [NPS99, PSSW09] allows to reduce the garbled table by one entry.

2.1.3 Oblivious Transfer

In m-parallel Oblivious Transfer (OT) of ℓ-bit strings, denoted as OT_ℓ^m, the chooser inputs a vector of choice bits r_i, $i = 1 \dots m$ and the sender inputs a vector of pairs of ℓ-bit strings $(x_0, x_1)_i$, $i = 1 \dots n$. At the end of the protocol, the chooser learns the selected strings $x_{r_i,i}$, but nothing about the other strings $x_{1-r_i,i}$ whereas the sender learns nothing about the choices r_i.

In Yao's Garbled Circuit protocol $\ell = t+1$, whereas m corresponds to the number of input bits provided by the client which can be large. Using OT extensions of [IKNP03] it is possible to reduce a large number of OTs to a small number of only k OTs, where k is a security parameter. These remaining k base OTs are implemented with an efficient OT protocol which requires $\mathcal{O}(k)$ public-key operations, e.g., the OT protocol of [NP01].

In §3.1 we give implementation improvements for the base OTs of [NP01] and in §3.2 we show how the OT extension of [IKNP03] can be implemented with low memory footprint.

2.2 Minor Remark on the Choice of Security Parameters

As described in [HEKM11, Sect. 3.4], the implementation of FastGC uses SHA-1 as cryptographic hash function to encrypt the output wire labels of non-linear gates. Hence, the maximum security level that can be achieved is $t = 80$ bits. However, in the most recent implementation of FastGC [HEKM11] (version 0.1.1 released August 9, 2011), the last bit of the 80 bit wire labels is used as permutation bit for the point-and-permute technique of [NPS99] (cf. §2.1.2). As this permutation bit is known to the evaluator, the achieved symmetric security level is only 79 bits, but not 80 bits as stated in their paper. To correct this, we set Wire.labelBitLength=81 (actually we use 88 as internally a byte-oriented representation is used) to achieve a symmetric security level of exactly 80 bits. The longer wire labels result in a slight increase of communication. For the OT protocol of [NP01], implemented over $\mathbb{F}_P$ with a generator of order q, we use $|p| = 1024$ and $|q| = 160$ as these correspond to the current NIST recommendations for a symmetric security level of 80 bits.[1]

3. FASTER SECURE EVALUATION OF GARBLED CIRCUITS

We optimize several aspects of the FastGC framework [HEKM11] as described in the following.

We emphasize that our optimizations do not modify the underlying cryptographic protocols, but only the way they are implemented. Therefore, and because we work in the semi-honest setting, the proofs of security of the original protocols still hold for our optimizations.

[1]see `http://keylength.com`

3.1 Improved Base OTs

In order to improve the performance of the k base OTs with the OT protocol of [NP01], we split up the computationally intensive public key operations into multiple threads such that each of the N threads performs k/N base OTs (independent of each other). Furthermore, we tried to implement the base OTs over an elliptic curve instead of over $\mathbb{F}_P$ (the latter was used in the implementation of [HEKM11]). This resulted in a reduction of the communication complexity, but unfortunately no better runtimes. We assume that this is due to the additional overhead introduced by the Java VM. Our performance benchmarks for the improved base OT implementations are given in §4.1.

3.2 Extending OTs with Low Memory Footprint

A large number of m parallel OTs of ℓ-bit strings, OT_ℓ^m can be reduced to a small number of only k OTs of k-bit strings, OT_k^k, using OT extensions of [IKNP03] as implemented in [HEKM11]. The original protocol of [IKNP03] needs beyond the messages of the base OTs only two messages, but memory linear in m.

We reduce the memory requirement of this protocol by re-ordering its messages as follows: The OT extension construction of [IKNP03] proceeds in two steps: First, the large number of m parallel OTs of short ℓ-bit strings are reduced to k parallel OTs of long m-bit strings, cf. [IKNP03, Fig. 1]. These OTs are implemented using k parallel OTs of short k-bit keys that are then stretched into longer m-bit masks using a pseudo-random generator (PRG), cf. [IKNP03, Fig. 3].

A very efficient and standard way to implement a PRG is to successively apply a pseudo-random permutation (PRP) on a counter, i.e., $\mathrm{PRG}(k) = \mathrm{PRP}_k(0)||\mathrm{PRP}_k(1)||\ldots$. In practice, the PRP can be instantiated with a block cipher which operates on blocks of M bits, e.g., in our implementation we use AES-128 with $M = 128$. Now, in order to reduce the memory footprint, the OT extension construction of [IKNP03] can easily be split into smaller blocks where each block performs M parallel OTs. The overall protocol is shown in Fig. 1. W.l.o.g. and to simplify presentation we assume that m is a multiple of the block size, $m = BM$ (otherwise, the last messages are shorter) and that $\log B \leq M$. $\mathbf{T}$ denotes an $M \times k$ bit matrix, $\mathbf{T}^i$ its i-th column and $\mathbf{T}_j$ its j-th row. $G : \{0,1\}^M \times \{0,1\}^k \rightarrow \{0,1\}^M$ is a PRP; $H : \{0,1\}^{\lceil \log m \rceil} \times \{0,1\}^k \rightarrow \{0,1\}^\ell$ is a random oracle which can be implemented using a cryptographic hash function (in our implementation we use SHA-1).

Overall, our modified protocol can be seen as operating on a stream of data which is processed in chunks of size M.

Security and Correctness. The only modification that we applied to the original construction of [IKNP03] is that we re-order the messages sent into B rounds of communication. For semi-honest parties, this does not reveal any additional information. Therefore, the correctness and security of our optimized protocol in the semi-honest setting directly follows from the correctness and security of the original construction as proven in [IKNP03].

Performance. The computation and communication complexity of our protocol is exactly the same as that of the original construction of [IKNP03]. In contrast to the implementation in [HEKM11] which has a constant number of communication rounds but requires memory linear in the total number of OTs m, the memory consumption of our protocol is constant (for fixed block size M), but requires m/M rounds of communication. We give performance benchmarks for the improved OT extensions in §4.1.

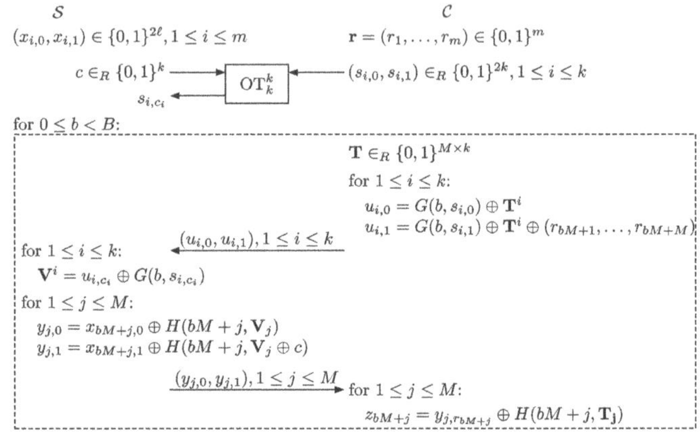

Figure 1: Extension of OT_ℓ^m with low memory footprint. W.l.o.g. $m = BM$ for B blocks of size M. G is a pseudo-random permutation and H is a random oracle.

3.3 Streaming Circuits and Garbled Circuits with Low Memory Footprint

As described in §1 and §1.2, previous frameworks for secure two-party computation in the semi-honest adversaries setting store the outgoing wire labels of each gate in memory and hence require memory linear in the size of the evaluated circuit. More recent frameworks [Mal11, HEKM11, MZE12] require memory only linear in the size of the sub-circuits, but these frameworks suffer from the low performance of memory management of many small objects (one object for each gate) and garbage collection.

During creation and evaluation of a (sub) circuit, only those wires need to be held in memory that are used in the future, i.e., are either an output wire of the circuit or used as input wires into a later gate. This set of wires is called the *working set*. Thus, the minimum size of the memory required to compute the circuit is defined by the maximum working set. As described and implemented in [KSS12], one strategy to keep track of the working set is to annotate to each wire a usage counter that is decremented on each use. When the counter reaches zero, the wire is deleted from the memory. This is similar to the way sub-circuits are dynamically constructed and deconstructed in VMCrypt [Mal11] and impedes additional overhead in the online phase.

We use a different approach where we shift the management of the working set from the online phase to the compilation phase in order to keep the online phase as lean as possible. In particular, the online phase of our approach neither requires a usage counter nor to allocate and later on free memory for each gate. During the compilation phase, the compiler determines the maximum size S of the working set and stores this along with the circuit description. The compiler allocates a slot ID (starting from 0) to each input wire of the circuit. Then, the circuit description is generated as an ordered list of gates where each gate is described as a tuple (output slot ID; input slot IDs; truth table). Whenever a slot ID is used for the last time, it is added to a list of available slot IDs and can be re-used as the output slot ID of a later gate. Note that all this is done in the offline (compilation) phase. In the online phase, an array with S slots is allocated where each slot can hold a wire label. Then, the gates are read one-by-one from the circuit description: the gate's input labels are taken from the slots given by its input slot IDs and the output label is stored to the slot specified by the gate's output slot ID. We internally keep a counter of the gate ID which is used to construct the garbled tables.

We emphasize that the compilation of a function has to be done just once, since the resulting circuit is independent from the inputs and can therefore be reused.

In future work, the compiler can be extended to re-arrange the order of the gates in order to reduce the size of the working set as described in [JKSS10b]. However, as noted in [KSS12], determining a topologic order of the circuit with the minimum size of the working set is known to be an NP-complete problem.

We further note that in principle generation of garbled circuits can be implemented such that the amount of memory is constant by pseudo-randomly deriving the wire labels from the gate ID as described in [JKSS10a]. However, this is essentially a time-memory tradeoff and cannot easily be combined with the highly efficient free XOR technique.

3.4 Sub-Circuit Compilation

A design goal of our framework was to keep the online phase of the circuit evaluation as lean as possible. Due to the topological ordering of the circuits the engine never has to hold more than one gate description in memory. Once a gate is processed, the information can be discarded (except for the intermediate wire labels). The circuit evaluation engine reads the circuit description from a file with a format similar to Fairplay. In order to re-use sub-circuits we use slot IDs to index the wires in the sub-circuit. A slot ID can be seen as a virtual register that can hold a wire label and can also be re-used. For the gate ID, which needs to be unique in the overall circuit as it is used for encrypting the non-linear gates, we use a counter which is incremented for each non-linear gate. A gate is described by its output slot ID, its input slot IDs, and its truth-table. An example is shown in Fig. 2 which describes the one-bit comparison circuit from [KSS09]. We also support a binary file format which is more efficient to read by our engine.

The circuit to be computed often consists of several calls to the same sub-function. For instance, AES consists of 160 S-Box, 10 AddRoundKey, and 9 MixColumns calls. Our framework allows for sub-circuits to be reused in a similar fashion like [HEKM11], except that we do not instantiate a new gate object in each invocation of the sub-circuit. Overall, for AES we do not create the entire circuit with 24,720 gates, but only 3 sub-circuits for the sub-functions listed above with a total of just 803 gates. As in our implementation the creation of the gate ID is decoupled from the circuit definition, we ensure that in every reuse of a sub-circuit all gates have a unique gate ID, and therefore the security of the underlying garbled circuit protocol, as proven in [LP09a, PSSW09], still holds.

We provide a compiler that converts circuits described in the format of [HEKM11] into our format.

Figure 2: A one-bit comparison circuit. Comments (from // on) are not part of the input.

```
inputsCreator: 0     // creator's input is r0
inputsEvaluator: 1   // evaluator's input is r1
outputsCreator:
outputsEvaluator: 0 // evaluator's output is r0
numberOfRegisters: 2
numberOfGates: 2
0;1,0;1              // r0 = r1 and r0
0;1,0;6              // r0 = r1 xor r0
```

3.5 Caching of Circuits and Communication

In order to improve the performance of garbled circuits evaluation we cache both, circuit descriptions and network packets during garbled circuit streaming resulting in a corresponding time-memory trade-off as described next.

In some circuits, the same sub-circuits are reused many times (cf. §3.4). Instead of reading the description of the sub-circuit from a file on every instantiation (or re-generating it as implemented in previous frameworks), we optionally cache its description once in memory. The memory consumption is 32 bytes per cached gate.

Sending the creator's input wire labels and the garbled tables straight after creation (as implemented in [HEKM11]) leads to an inefficient use of the network because of small packet sizes and an unnecessary large number of packets. By using fixed sized buffers on the communication channels we greatly improve the performance of the network usage.

In our benchmarks in §4 we use circuit caching and network buffers of size 9,000 byte.

4. PERFORMANCE BENCHMARKS AND APPLICATIONS

In the following we show that the implementation of our improvements described in the previous section results in substantially better performance than previous frameworks.

4.1 Oblivious Transfers

The following performance benchmarks were performed on two Apple computers with a dual core processor each (Intel Core i5 2.5GHz and Core i7 1.8GHz) running MacOS X 10.7.4 and Java 1.6.0_33, connected via 802.11n WIFI.

We observed that because of the JAVA just in time compiler, the runtime decreases in the first few runs due to compiler optimizations. Therefore, we executed each protocol 1,000 times and took the average. All benchmarks were executed with the default JAVA VM parameter.

The performance of our improved implementation for the base OTs (cf. §3.1) in comparison with the original implementation of [HEKM11] run in exactly the same setting is shown in Table 2. Due to the additional overhead for thread management, our multi-threaded implementation with one thread is slightly slower than the single-threaded one. However, when running with 4 threads on the dual core processors, our multi-threaded base OTs take 0.15 seconds, an improvement by factor 2 over the single-threaded version. We emphasize that we use multi-threading *only* for the base OTs and that for small circuit sizes this time is much longer than the online time (cf. Table 3).

Table 2: Comparison of base OT implementations.

single threaded	time
over $\mathbb{F}_P$ [HEKM11]	286 ms (100%)
over EC	560 ms (196%)
multi threaded (over $\mathbb{F}_P$)	**time**
1 thread	314 ms (110%)
2 threads	182 ms (64%)
4 threads	153 ms (53%)

For the OT extensions (independent of the performance of the base OTs), our improved implementation of the protocol of §3.2 can evaluate about 400,000 OTs per second, i.e., 2.5μs per OT, a factor 6 improvement over the 15μs reported in [HEKM11] (which used faster Intel Core Duos E8400 3GHz and a faster local area network).[2] We emphasize that this optimization is very beneficial for applications where the circuit has many inputs, e.g., for converting from homomorphic encryption to garbled circuits and subsequently finding the minimum, a very common building block in privacy-preserving protocols for biometric matching [HKS+10, HMEK11, BG11] (cf. the example we benchmark in §4.2.4).

4.2 Online Time

The following benchmarks were executed on a single iMac A1311 with an Intel Core i3 3GHz processor using the loopback network interface. We measured the *online* time, that is the time from after the connection is established and the base OTs are done until the end of the protocol. This time includes the time for the OT extensions and streaming, i.e., creating, transferring, and evaluating, the garbled circuit.

Our improved implementation evaluates about 500,000 non-linear gates per second (2μs per gate) on the same host (setting as described above) and about 350,000 non-linear gates per second (3μs per gate) over WLAN (setting as described in §4.1). In contrast, [HEKM11] reported 96,000 non-linear gates per second (10μs per gate) over a LAN.[3]

In the following we show that the online time of our framework when evaluated on the same host as described above (i.e., assuming an ideal network) is up to 10 times faster than that of previous frameworks; we give a comparison for small circuits with FastGC [HEKM11] in §4.2.1 and §4.2.2 (using optimized circuit constructions), for medium size circuits with FastGC [HEKM11] and TASTY [HKS+10] in §4.2.3, and for large circuits with FastGC [HEKM11] and VMCrypt in §4.2.4. The comparison between our implementation and FastGC [HEKM11], both executed on exactly the same machine, is summarized in Table 3.

4.2.1 Hamming Weight

Some applications, e.g., privacy-preserving face recognition [OPJM10], require to securely compute the Hamming distance $d_H(\vec{a}, \vec{b})$ between two ℓ-bit strings $\vec{a}$, $\vec{b}$. As shown in [HEKM11], this can be done by XORing $\vec{a}$ and $\vec{b}$ bitwise and computing its Hamming weight $h(\cdot)$, i.e., the number of "ones" in its binary representation: $d_H(\vec{a}, \vec{b}) = h(\vec{a} \oplus \vec{b})$.

Original Hamming Circuit. The authors of [HEKM11] propose to use a tree of addition circuits which requires approximately $\sum_{i=0}^{\lceil \log_2 \ell \rceil} \ell \frac{i}{2^i} = \ell(2 - \frac{\lceil \log_2 \ell \rceil + 2}{2^{\lceil \log_2 \ell \rceil}}) \approx 2\ell - \log_2 \ell$ non-linear gates. For the example of $\ell = 900$ given in their paper this yields approximately 1,790 non-linear gates.

Improved Hamming Circuit. We use the optimized Hamming weight circuit of [BP06] with size $\ell - h(\ell)$ non-linear gates, where $h(\ell)$ is the Hamming weight of ℓ. For $\ell = 900$ this yields $900 - h((1110000100)_2) = 896$ non-linear gates.

The resulting circuit sizes and performance results are shown in Table 3. Put together, the improved Hamming circuit together with our improved implementation is more than 10 times faster than the original Hamming circuit evaluated with FastGC [HEKM11] (6 w/o circuit optimization).

4.2.2 Block Ciphers

Oblivious evaluation of a block cipher where one party provides the key and the other party provides the message and obtains the ciphertext has many applications as summarized in [PSSW09]. These include oblivious pseudorandom functions (OPRFs) with applications to secure keyword searching [FIPR05] or secure set intersection [JL09], blind MACs, and blind encryption.

As noted in [HEKM11], the key schedule of the block cipher does not need to be computed securely within the garbled circuit. Instead, the party that knows the key can run the key schedule on the plain key data to expand it and provide the expanded key as input to the protocol.

AES. Secure evaluation of AES is commonly used as performance benchmark for secure computation frameworks, e.g., [PSSW09, HKS+10, HEKM11, KSS12].

Original AES Circuit. Excluding the key schedule, AES-128 consists of 10 rounds where in each round 16 S-boxes are evaluated. As shown in [HEKM11, Sect. 7], all other operations, e.g., MixColumns and AddRoundKey, can be performed using only free XOR gates. The S-box presented in [HEKM11] has 58 non-linear gates resulting in $58 \cdot 10 \cdot 16 =$ 9,280 non-linear gates for AES.

Improved AES Circuit. Instead, we implemented the S-box of [BP10] which consists of only 32 non-linear gates resulting in a total of $32 \cdot 10 \cdot 16 = 5{,}120$ non-linear gates for AES. We note that the AES circuit implemented in [KSS12] uses the same S-box and has 9,100 non-linear gates, but including the key schedule.

PRESENT. For the applications mentioned above, it might be sufficient to use a block cipher that does not provide the strong security guarantees of AES, but is more efficient to evaluate. An example for such an ultra-lightweight block cipher is PRESENT with a block length of 64 bit and an 80 bit key. PRESENT consists of 31 rounds where in each round a 4-bit S-box is applied 16 times in parallel. We implemented PRESENT using the S-box representation of [CHM11, CHM12] which has 4 non-linear gates. Overall, PRESENT requires $31 \cdot 16 \cdot 4 = 1{,}984$ non-linear gates.

[2] For completeness we note that [NNOB12] claim an actively-secure OT extension that can be implemented at about 500,000 OTs per second based on unpublished optimizations (cf. Appendix A and E in the full version of their paper).

[3] In the malicious setting, [KSS09] report 82,000 non-linear gates per second (12μs per gate) on a cluster and [NNOB12] 20,000 gates per second (50μs per gate) over an intranet.

Table 3: Comparison of circuit sizes and performance when run on the same machine.

	Circuit	non-linear gates		FastGC [HEKM11]		Our Implementation		
§4.2.1	Original Hamming	1,793	(100%)	64 ms	(100%)		8 ms	(13%)
	Improved Hamming	896	(44%)	39 ms	(61%)		6 ms	(9%)
§4.2.2	Original AES	9,280	(100%)	204 ms	(100%)		27 ms	(13%)
	Improved AES	5,120	(55%)	113 ms	(55%)		16 ms	(8%)
	PRESENT	1,984	(21%)	53 ms	(26%)		7 ms	(3%)
§4.2.3	Fast Multiplication	17,973		499 ms	(100%)		45 ms	(9%)
§4.2.4	Minimum	40,000,000		1,250 s	(100%)	a)	138 s	(11%)
						b)	272 s	(22%)
						c)	128 s	(10%)

Comparison. The size of the AES and PRESENT circuits and their performance comparison are shown in Table 3. The improved AES circuit evaluated with our improved implementation is about 12 times faster than the original AES circuit evaluated on the FastGC engine [HEKM11]. Due to its smaller gate count, PRESENT is almost 4 times faster than the original AES circuit and twice as fast as the improved AES circuit.

4.2.3 Fast Multiplication

To compare our implementation with FastGC [HEKM11] and TASTY [HKS+10] for circuits of medium size, we implemented secure multiplication using the fast multiplication method of Karatsuba and Ofman [KO62]. For multiplication of two 128 bit numbers, this circuit has 17,973 non-linear gates and took 45 ms online time to evaluate in our optimized framework.

The same circuit implemented in FastGC [HEKM11] and run on the same machine took 499 ms, i.e., more than 10 times longer. This result supports the fact that the improved online time of our implementation, in particular for medium size circuits, is substantially faster than that of [HEKM11], even without circuit-specific optimizations.

According to [HKS+10, Fig. 7], TASTY takes approximately 4,000 ms setup time and 700 ms online time to evaluate the same circuit on two desktop PCs with Intel Core 2 Duo CPU (E6850) running at 3GHz connected via Gigabit Ethernet. We emphasize that TASTY doesn't use streaming, but pre-computes the OT extensions and generates and transfers the garbled circuit already in the setup phase; hence, the online time in TASTY consists only of the very efficient online OTs of Beaver's construction [Bea95] and garbled circuit evaluation whereas our online time includes the computationally more expensive OT extensions and creating and transferring the garbled circuit which is minimized due to streaming. Overall, for this application, the online time of our improved framework is faster than TASTY by a factor of about 16 times, whereas our setup time is as low as 153 ms for the base OTs (cf. Table 2), i.e., 26 times faster.

4.2.4 Minimum

To compare our implementation with FastGC [HEKM11] and VMCrypt [Mal11] for circuits of large size, we implemented a circuit to compute the minimum of 10^6 20-bit numbers (half of the numbers are input by the server and the other half by the client) using a circuit similar to the one described in [Mal11, Fig. 2]. We use the OT extension with low memory footprint described in §3.2 such that the total memory consumption stays linear in the size of the subset and not in the order of the total number of inputs. The overall circuit has $2 \cdot (10^6 - 1) \cdot 20 \approx 40{,}000{,}000$ non-linear gates. There are different approaches to compute this functionality which demonstrate the flexibility of our system that allows to choose a trade-off between execution time (shown in Table 3) and memory consumption (shown in Table 4):

a) The first solution has the lowest OT overhead but requires the maximum amount of memory (800 MB) by doing the OTs for all inputs first and then evaluating a large circuit with a working set of size $2 * 10^7$. The runtime is 138 s.

b) As another extreme we can iteratively compare one input each at a time with the previously found minimum. This approach needs the minimal amount of memory (18.4 MB) but the maximal total runtime (272 s) as the OT protocol introduces a significant overhead.

c) Our framework allows to choose an intermediate approach where we iteratively compute the minimum of a subset of 500 inputs and the minimum of the previous iteration. This sub-circuit has 19,960 non-linear gates and a working set of 10,022 labels and is small enough to be cached in memory – the total memory requirement is 21.5 MB and the total runtime is 128 s.

When evaluating the same circuit on the same machine with FastGC [HEKM11], this took approximately 1,250 s and 189 MB memory, i.e., more than 9 times longer and 10 times more memory compared to our approach c).

According to [Mal11, Fig. 8], VMCrypt takes 44.5 min on a slower CPU (Thinkpad X301 laptop with 3 GB RAM and a 1.6 GHz Intel Core2 Duo processor running Ubuntu Linux over the loopback interface), i.e., about 10 times longer than our approach c), already considering the fact that our CPU is about twice as fast. We assume that our improved performance stems mainly from the fact that we do not allocate and free many small objects.

4.3 Memory Consumption

Measuring the memory consumption of a Java program is fuzzy, since released objects remain on the heap until the garbage collector deletes them and the garbage collector itself is managed by the Java virtual machine. Thus, the heap contains not only the currently used objects but also already released ones and therefore the size of the heap will be greater or equal to the size of the currently used objects. We measured the maximum heap consumption of every protocol, since this gives an indication of how much memory is needed for a runtime optimal execution. The protocols might run with smaller heap sizes, but then the virtual machine has to invoke the garbage collection more often which results in longer runtimes. Table 4 shows the maximum size of the heap during the execution for all protocols.

The memory consumption of FastGC is linear in the total number of gates whereas in our implementation it is linear in the size of the working set. Although for circuits that are divisible into many small sub-circuits like AES or PRESENT the memory consumption of both implementations is almost the same; the memory efficiency advantage of our implementation becomes obvious for larger circuits: Compared to FastGC we achieve a memory consumption reduction by factor 5 for Fast Multiplication and by factor 8 for Minimum for approach c) – by combining the repeated use of a sub-circuit and then just holding the working set in memory, we can evaluate the Minimum circuit with almost 140 million gates using only 21.5 MB of memory.

Acknowledgments. The authors would like to thank René Peralta for helpful discussions on efficient circuit constructions. The first author was supported by Australian Research Council grant DP0984063, by an Adelaide Scholarship International, and a Supplementary Scholarship by the Defence Systems Innovation Centre. The second author was supported by the German Federal Ministry of Education and Research (BMBF) within EC SPRIDE and by the Hessian LOEWE excellence initiative within CASED.

5. REFERENCES

[Bea95] D. Beaver. Precomputing oblivious transfer. In *Advances in Cryptology – CRYPTO'95*, volume 963 of *LNCS*, pages 97–109. Springer, 1995.

[BG11] M. Blanton and P. Gasti. Secure and efficient protocols for iris and fingerprint identification. In *European Symposium on Research in Computer Security (ESORICS'11)*, volume 6879 of *LNCS*, pages 190–209. Springer, 2011.

[BNP08] A. Ben-David, N. Nisan, and B. Pinkas. FairplayMP: a system for secure multi-party computation. In *ACM Conference on Computer and Communications Security (CCS'08)*, pages 257–266. ACM, 2008.

[BNSS11] S. Bugiel, S. Nürnberger, A.-R. Sadeghi, and T. Schneider. Twin Clouds: Secure cloud computing with low latency. In *Communications and Multimedia Security Conference (CMS'11)*, volume 7025 of *LNCS*, pages 32–44. Springer, 2011.

[BP06] J. Boyar and R. Peralta. Concrete multiplicative complexity of symmetric functions. In *Mathematical Foundations of Computer Science (MFCS'06)*, volume 4162 of *LNCS*, pages 179–189. Springer, 2006.

[BP10] J. Boyar and R. Peralta. A new combinational logic minimization technique with applications to cryptology. In *Symposium on Experimental Algorithms (SOA'10)*, volume 6049 of *LNCS*, pages 178–189. Springer, 2010.

[CHM11] N. T. Courtois, D. Hulme, and T. Mourouzis. Solving circuit optimisation problems in cryptography and cryptanalysis. In *2nd IMA Conference Mathematics in Defence*, 2011.

[CHM12] N. T. Courtois, D. Hulme, and T. Mourouzis. Solving circuit optimisation problems in cryptography and cryptanalysis. In *Special-purpose Hardware for Attacking Cryptographic Systems (SHARCS'12)*, pages 179–191, 2012.

[FIPR05] M. J. Freedman, Y. Ishai, B. Pinkas, and O. Reingold. Keyword search and oblivious pseudorandom functions. In *Theory of Cryptography Conference (TCC'05)*, volume 3378 of *LNCS*, pages 303–324. Springer, 2005.

[HCE11a] Y. Huang, P. Chapman, and D. Evans. Privacy-preserving applications on smartphones. In *USENIX Workshop on Hot Topics in Security (HotSec'11)*, 2011.

[HCE11b] Y. Huang, P. Chapman, and D. Evans. Secure computation on mobile devices, 2011. Poster at IEEE Symposium on Security and Privacy.

[HEK12] Y. Huang, D. Evans, and J. Katz. Private set intersection: Are garbled circuits better than custom protocols? In *Network and Distributed Security Symposium (NDSS'12)*. The Internet Society, 2012.

[HEKM11] Y. Huang, D. Evans, J. Katz, and L. Malka. Faster secure two-party computation using garbled circuits. In *USENIX Security Symposium (Security'11)*, pages 539–554. USENIX, 2011.

[HFKV12] A. Holzer, M. Franz, S. Katzenbeisser, and H. Veith. Secure two-party computations in ANSI C. In *ACM Conference on Computer and Communications Security (CCS'12)*, pages 772–783. ACM, 2012.

[HKE12] Y. Huang, J. Katz, and D. Evans. Quid-pro-quo-tocols: Strengthening semi-honest protocols with dual execution. In *IEEE Symposium on Security and Privacy*. IEEE, 2012.

[HKS+10] W. Henecka, S. Kögl, A.-R. Sadeghi, T. Schneider, and I. Wehrenberg. TASTY: Tool for Automating Secure Two-partY computations. In *ACM Conference on Computer and Communications Security (CCS'10)*, pages 451–462. ACM, 2010.

[HMEK11] Y. Huang, L. Malka, D. Evans, and J. Katz. Efficient privacy-preserving biometric identification. In *Network and Distributed System Security (NDSS'11)*. The Internet Society, 2011.

[IKNP03] Y. Ishai, J. Kilian, K. Nissim, and E. Petrank. Extending oblivious transfers efficiently. In *Advances in Cryptology – CRYPTO'03*, volume 2729 of *LNCS*, pages 145–161. Springer, 2003.

[JKSS10a] K. Järvinen, V. Kolesnikov, A.-R. Sadeghi, and T. Schneider. Embedded SFE: Offloading server and network using hardware tokens. In *Financial Cryptography and Data Security (FC'10)*, volume 6052 of *LNCS*, pages 207–221. Springer, 2010.

[JKSS10b] K. Järvinen, V. Kolesnikov, A.-R. Sadeghi, and T. Schneider. Garbled circuits for leakage-resilience: Hardware implementation

Table 4: Comparison of the memory consumption.

Circuit	tot. number of gates	gates in memory	working set	FastGC [HEKM11]	Our Impl.
Original Hamming	7,163	7,163	1,800	28.5 MB	17 MB
Improved Hamming	5,362	5,362	1,800	26 MB	17 MB
Original AES	36,720	878	256	20.3 MB	16.9 MB
Improved AES	24,720	803	256	20.2 MB	18.7 MB
PRESENT	8,496	77	128	18.9 MB	18.6 MB
Fast Multiplication	57,072	57,072	1,074	74.4 MB	15 MB
Minimum	137,999,862	a) 137,999,862 b) 414 c) 69,000	a) 20,000,002 b) 104 c) 10,022	189 MB	a) 799.7 MB b) 18.4 MB c) 21.5 MB

and evaluation of one-time programs. In *Cryptographic Hardware and Embedded Systems (CHES'10)*, volume 6225 of *LNCS*, pages 383–397. Springer, 2010.

[JL09] S. Jarecki and X. Liu. Efficient oblivious pseudorandom function with applications to adaptive OT and secure computation of set intersection. In *Theory of Cryptography Conference (TCC'09)*, volume 5444 of *LNCS*, pages 577–594. Springer, 2009.

[Ker11] F. Kerschbaum. Automatically optimizing secure computation. In *ACM Computer and Communications Security (CCS'11)*, pages 703–714. ACM, 2011.

[KO62] A. A. Karatsuba and Y. Ofman. Multiplication of many-digital numbers by automatic computers. *SSSR Academy of Sciences*, 145:293–294, 1962.

[KS08] V. Kolesnikov and T. Schneider. Improved garbled circuit: Free XOR gates and applications. In *International Colloquium on Automata, Languages and Programming (ICALP'08)*, volume 5126 of *LNCS*, pages 486–498. Springer, 2008.

[KSS09] V. Kolesnikov, A.-R. Sadeghi, and T. Schneider. Improved garbled circuit building blocks and applications to auctions and computing minima. In *Cryptology and Network Security (CANS'09)*, LNCS. Springer, 2009.

[KSS12] B. Kreuter, A. Shelat, and C.-H. Shen. Billion-gate secure computation with malicious adversaries. In *USENIX Security Symposium (Security'12)*. USENIX, 2012.

[LP09a] Y. Lindell and B. Pinkas. A proof of Yao's protocol for secure two-party computation. *Journal of Cryptology*, 22(2):161–188, 2009.

[LP09b] Y. Lindell and B. Pinkas. Secure multiparty computation for privacy-preserving data mining. *Journal of Privacy and Confidentiality*, 1(1):59–98, 2009.

[Mal11] L. Malka. VMCrypt - modular software architecture for scalable secure computation. In *ACM Conference on Computer and Communications Security (CCS'11)*, pages 715–724. ACM, 2011.

[MLB12] B. Mood, L. Letaw, and K. Butler. Memory-efficient garbled circuit generation for mobile devices. In *Financial Cryptography and Data Security (FC'12)*, LNCS. Springer, 2012.

[MNPS04] D. Malkhi, N. Nisan, B. Pinkas, and Y. Sella. Fairplay — a secure two-party computation system. In *USENIX Security Symposium*, pages 287–302. USENIX, 2004.

[MZE12] W. Melicher, S. Zahur, and D. Evans. An intermediate language for garbled circuits. Poster at IEEE Symposium on Security and Privacy, 2012.

[NNOB12] J. B. Nielsen, P. S. Nordholt, C. Orlandi, and S. S. Burra. A new approach to practical active-secure two-party computation. In *Advances in Cryptology - CRYPTO'12*, volume 7417 of *LNCS*, pages 681–700. Springer, 2012.

[NP01] M. Naor and B. Pinkas. Efficient oblivious transfer protocols. In *ACM-SIAM Symposium On Discrete Algorithms (SODA'01)*, pages 448–457. Society for Industrial and Applied Mathematics, 2001.

[NPS99] M. Naor, B. Pinkas, and R. Sumner. Privacy preserving auctions and mechanism design. In *ACM Conference on Electronic Commerce*, pages 129–139. ACM, 1999.

[OPJM10] M. Osadchy, B. Pinkas, A. Jarrous, and B. Moskovich. SCiFI - a system for secure face identification. In *IEEE Symposium on Security and Privacy*, pages 239–254. IEEE, 2010.

[PSS09] A. Paus, A.-R. Sadeghi, and T. Schneider. Practical secure evaluation of semi-private functions. In *Applied Cryptography and Network Security (ACNS'09)*, volume 5536 of *LNCS*, pages 89–106. Springer, 2009.

[PSSW09] B. Pinkas, T. Schneider, Nigel P. Smart, and Stephen C. Williams. Secure two-party computation is practical. In *Advances in Cryptology - ASIACRYPT 2009*, volume 5912 of *LNCS*, pages 250–267. Springer, 2009.

[Sch12] T. Schneider. *Engineering Secure Two-Party Computation Protocols: Design, Optimization, and Applications of Efficient Secure Function Evaluation*. Springer, 2012.

[Yao86] A. C. Yao. How to generate and exchange secrets. In *Foundations of Computer Science (FOCS'86)*, pages 162–167. IEEE, 1986.

TabShots: Client-Side Detection of Tabnabbing Attacks

Philippe De Ryck, Nick Nikiforakis, Lieven Desmet, Wouter Joosen
iMinds-Distrinet, KU Leuven, 3001 Leuven, Belgium
Philippe.DeRyck@cs.kuleuven.be

ABSTRACT

As the web grows larger and larger and as the browser becomes the vehicle-of-choice for delivering many applications of daily use, the security and privacy of web users is under constant attack. Phishing is as prevalent as ever, with anti-phishing communities reporting thousands of new phishing campaigns each month. In 2010, tabnabbing, a variation of phishing, was introduced. In a tabnabbing attack, an innocuous-looking page, opened in a browser tab, disguises itself as the login page of a popular web application, when the user's focus is on a different tab. The attack exploits the trust of users for already opened pages and the user habit of long-lived browser tabs.

To combat this recent attack, we propose TabShots. TabShots is a browser extension that helps browsers and users to remember what each tab looked like, before the user changed tabs. Our system compares the appearance of each tab and highlights the parts that were changed, allowing the user to distinguish between legitimate changes and malicious masquerading. Using an experimental evaluation on the most popular sites of the Internet, we show that TabShots has no impact on 78% of these sites, and very little on another 19%. Thereby, TabShots effectively protects users against tabnabbing attacks without affecting their browsing habits and without breaking legitimate popular sites.

Categories and Subject Descriptors

K.6.5 [**Management of Computing and Information Systems**]: Security and Protection; H.3.5 [**Information Storage and Retrieval**]: Web-based services

Keywords

Tabnabbing, phishing, client-side protection

ASIA CCS'13, May 8–10, 2013, Hangzhou, China.

1. INTRODUCTION

Phishing, the process that involves an attacker tricking users into willingly surrendering their credentials, is as prevalent as ever. PhishTank, a volunteer-driven site for tracking phishing pages [13], in their latest publicly available report, reported a total of 22,851 valid phishing attempts just for July of 2012. In these attacks, an attacker targets the user and capitalizes on a user's inability of distinguishing a legitimate page from one that looks legitimate but is actually fraudulent. Phishing attacks can be conducted both on large and small scale, depending on an attacker's objectives. The latest publicized attack against the White House, involved the use of "spear phishing", a type of phishing that is targeting highly specific individuals and companies [9].

In 2010, Aza Raskin presented a new type of phishing attack which he called "tabnabbing" [14]. In tabnabbing, the user is lured into visiting a malicious site, which however looks innocuous. If a user keeps the attacker's site open and uses another tab of her browser to browse to a different website, the tabnabbing page takes advantage of the user's lack of focus (accessible through JavaScript as `window.onBlur`) to change its appearance (page title, favicon and page content) to look identical to the login screen of a popular site. According to Raskin, when a user returns back to the open tab, she has no reason to re-inspect the URL of the site rendered in it, since she already did that in the past. This type of phishing separates the visit of a site from the actual phishing attack and could, in theory, even trick users who would not fall victim to traditional phishing attacks.

In this paper we present TabShots, a countermeasure for detecting changes to a site when its tab is out of focus. TabShots allows a browser to "remember" what the tab looked like before it lost focus, and compare it with the appearance after regaining focus. More precisely, whenever a tab is fully loaded, TabShots records the favicon[1] and captures a screenshot of the visible tab. Whenever a user revisits a tab, a new capture is taken and compared to the previously stored one. If any changes are detected, the user is warned by adding a visual overlay on the current tab, showing exactly the content that was changed, assisting the user in distinguishing between legitimate changes and tabnabbing attacks. Our system is based on the user's visual perception of a site and not the HTML representation of it, allowing TabShots to withstand attacks that straightforwardly circumvent previously proposed, tabnabbing-detection systems. We implement TabShots as a Chrome extension and evaluate it against the top 1000 Alexa sites, showing that

[1]The small icon displayed in the tab's title space

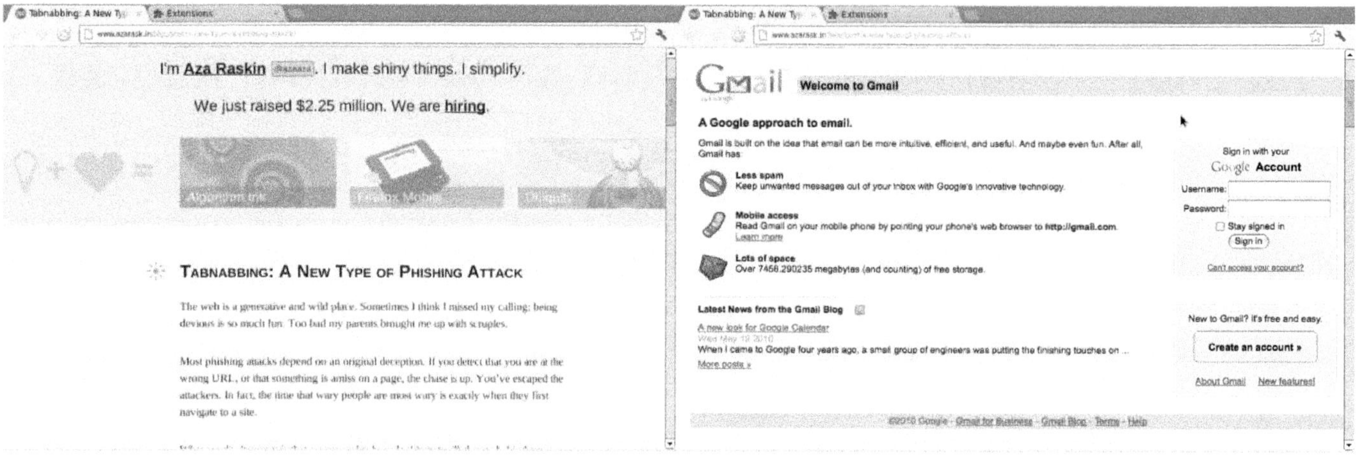

Figure 1: A seemingly innocuous page on the left performs a tabhabbing attack once the user switches focus, resulting in the page on the right [14].

78% of sites fall within a safe threshold of less than 5% changes, and an additional 19% fall within the threshold of less than 40% of changes. This means that TabShots effectively protects against tabnabbing attacks, without hindering a user's day-to-day browsing habits.

The rest of this paper is structured as follows: In Section 2 we first explore the original tabnabbing attack and then discuss possible variations taking advantages of the different implementations of the tabbing mechanism in popular browsers. In Section 3 we describe in detail the workings of TabShots and our implementation choices. In Section 4 we evaluate TabShots on security, performance and compatibility against the Alexa top 1000. In Section 5, we briefly describe how TabShots could be deployed on the server-side to create tabnabbing blacklists and expand protection to all users. In Section 6 we discuss the related work and conclude in Section 7.

2. BACKGROUND

2.1 Anatomy of a tabnabbing attack

Tabnabbing relies on the tab mechanism, which is common in all modern browsers. Users visit websites, but instead of navigating away from that website when they want to consume the content of a different website, they open a new tab, and use that tab instead. The old site remains open in the old tab, and many tabs can accumulate over time in a user's browser. A 2009 study of user's browsing habits revealed that users have an average of 3.2 tabs open in their browsers [6]. We expect that today, this number has increased, due to the sustained popularity of social networking sites and web applications that constantly update a user's page with new information. The latest features introduced by browsers attest to this popularity of multiple open tabs, since they give the user the ability to "pin" any given tab to the browser and treat it as a web application.

The steps of a tabnabbing attack as presented by Raskin [14] are the following:

1. An attacker convinces the user to visit a website under his control. This website appears to be an innocuous site that is not trying to fool the user into giving up her credentials. What the attacker must do, is convince the user to keep this tab open, and browse to a different website. This is easily achieved in a wide range of ways, for instance by providing an article that is both very interesting, but also too long to read in a single go, or some sort of free product that will be available in the near future. Directing the user away from the attacker's site is straightforward by adding the `target="_blank"` attribute to interesting hyperlinks, so that new links automatically open in a new tab or window.

2. JavaScript code running in the attacker's website is triggered when the current window has lost focus, by registering to the `window.onBlur` event handler.

3. The user keeps the attacker's website open and uses other tabs to surf the Internet.

4. The attacker realizes that his window is currently not in focus, and, after a possible delay of a few seconds in order to make sure that the user is busy consuming other content, changes the title, favicon and layout of the page to mimic the login screen of a web application, for instance the user's web mail or social networking site. The attacker can choose a default web application (like Gmail) under the assumption that most users have a Gmail account or can combine the tabnabbing attack with a history-revealing attack [8, 21], and present the login of a web application that he knows is visited in the past by the user. This process is also shown in Figure 1

5. At some point in the future, the user recognizes a tab with a familiar favicon (e.g. GMail) and unwittingly opens the attacker-controlled tab. At this point, the user is no longer checking the URL of the website, since it is a website that she opened in the past and thus "trusted". Given a convincing login screen, the user proceeds into typing her credentials in the given forms which are then transferred to the attacker, thus completing the tabnabbing attack.

Figure 2: Chrome keeps all tabs visible but shrinks the space alloted to each tab

Figure 3: After a number of tabs, Firefox hides older tabs in order to make space for the new ones

The main difference between tabnabbing and traditional phishing attacks is that the fake login form is decoupled from the visit of the malicious website. Thus, users who have been trained to spot phishing attacks by immediately checking the URL of the page they open, may fall victim to this variant of phishing. This "delayed maliciousness" can also be used to evade detection by any automated honeyclients which may be autonomously searching for phishing pages based on various heuristics [22]. If the honeyclient does not stay for long enough on the malicious page, or does not trigger the `window.onBlur` event, then the actual phishing page will never be shown and the attacker can avoid detection.

2.2 Overly Specific Detection

In the previous section, we described the anatomy of a tabnabbing attack, exactly as it was first presented by Raskin in 2010 [14]. According to Raskin, an attacker needs to change three things in order to conduct a successful tabnabbing attack: the page's title, the page's favicon and the page itself. Accordingly, currently known countermeasures depend on changes in these three properties, or include even more specific tabnabbing characteristics (more details in Section 6). This overly specific detection gives the attacker more flexibility to avoid detection.

One example of such flexibility is carrying out a tabnabbing attack without changing the title of the tab, simply by taking advantage of the tabbing behavior within a browser. While conducting our research, we noticed that different browsers behave differently when a user has many open tabs in one window. Figures 2 and 3 show how Chrome and Firefox handle many open tabs. Chrome, starts resizing the label of each tab, in an effort to keep all tabs visible. Here, one can notice that most of the title of each tab is hidden while favicons remain visible. On the other hand, Firefox starts hiding tabs which the user can access by clicking on the left arrow (circled in Figure 3). Moreover, Firefox preserves the title bar above the tabs, which Chrome dispenses in an effort to maximize the amount of space available for HTML.

In the case of Chrome, assuming that a user has many tabs open, the attacker can avoid the title change altogether, since it will likely not be visible to the user anyway.

In the next section, we present TabShots, which detects tabnabbing attacks using visual comparison. Since TabShots does not depend on fine-grained detection properties, we leave no room for an attacker to sneak through.

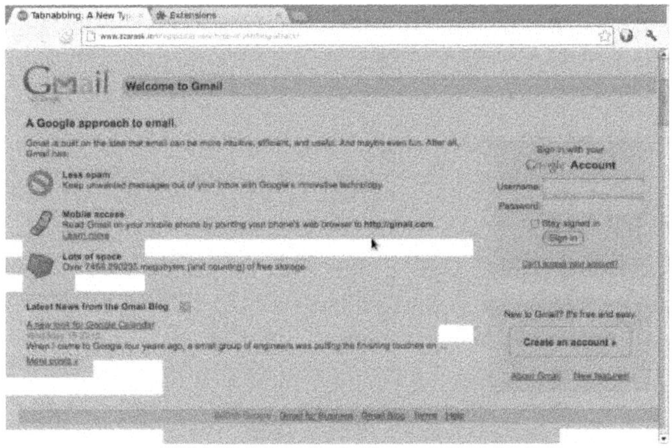

Figure 4: The overlay generated by TabShots for the attack from Figure 1. Here, only certain parts of the white background remained unchanged.

3. TabShots PROTOTYPE

3.1 Core idea

As discussed before, a successful tabnabbing attack depends on the user visiting a malicious page, shifting focus to a different tab and returning at some point, after which the malicious page has changed its looks to resemble a popular application's login form. In itself, a tabnabbing attack is extremely obvious to detect, since a convincing phishing page will differ from the previous content. Detection is however complicated by the tab being out of focus, and the user placing some trust in previously opened and visited tabs.

TabShots takes advantage of these obvious changes needed by a successful tabnabbing attack, by remembering what a tab looks like before it loses focus, and comparing that to what it looks like when it regains focus. Any changes that happened in the background will be detected, and communicated to the user by means of a colored overlay. This allows the user to decide for herself whether the changes are innocent (e.g. an incoming chat message) or malicious masquerading (e.g. a login form and GMail logo popping up). Figure 4 shows how TabShots detects the tabnabbing attack from Figure 1. This non-intrusive behavior guarantees compatibility with all existing sites, since changes are only highlighted and not blocked or prevented.

Our approach is purely built on the visible content of a tab, exactly as the user perceives it. This yields several advantages compared to techniques analyzing the structure and contents of a page. TabShots is invulnerable to HTML, CSS or JavaScript trickery, aimed at circumventing tabnabbing countermeasures (see Section 6), scrolling attacks or other obfuscation attacks.

3.2 Implementation details

TabShots is currently implemented as an extension for Google Chrome[2], but could easily be ported to other browsers supporting an extension system, provided they offer a reliable way to capture screenshots of tabs.

[2]A prototype of TabShots is available at `http://people.cs.kuleuven.be/~philippe.deryck/papers/asiaccs2013/`

In the following paragraphs, we discuss several implementation techniques and strategies for the major components of TabShots.

Capturing Tabs.

TabShots records the favicon and captures screenshots of the currently focused tab at regular intervals, keeping track of the latest version. This latest snapshot will be the basis for comparison when a tab regains focus. Capturing a screenshot of a tab in Google Chrome is trivial, since the browser offers an API call to capture the currently visible tab of a window. Capture data is stored as a data URL [10].

Capturing snapshots of a tab at regular intervals is a deliberate design decision, allowing TabShots to handle changes that happen in a tab while it is in focus. These changes typically occur in highly dynamic applications, such as Facebook or GMail, which often use AJAX techniques to dynamically update the contents of their pages. Ideally, a tab could be captured right before it loses focus, but since Google Chrome does not offer such an event, this feature cannot be implemented without a severe usability and performance penalty.

Comparing Tab Snapshots.

When a tab regains focus, TabShots needs to compare the current snapshot data with the stored data and detect any differences. Favicons are compared by source, and the screenshots are compared visually. Each screenshot is divided in a raster of fixed-size tiles (e.g., 10x10 pixels). Each tile is compared to its counterpart in the stored snapshot data. If the tiles do not match exactly, the area covered by it is marked as changed. The rastering and comparison algorithms are implemented using the recently introduced HTML5 canvas element, which offers extensive image manipulation capabilities.

One potential disadvantage of the screenshot analysis is the difficulty to detect a small change in a page that results in a visible shifting of contents (e.g. adding one message in front of a list). Such *false positives* may be addressed by a smarter comparison algorithm, that is able to detect movements within a screenshot.

The evaluation section (Section 4) discusses the chosen tile size and performance of the comparison algorithm in more detail.

Highlighting Differences.

Once the differences for a focused tab are calculated, TabShots injects an overlay into the page. This overlay is completely transparent, except for the differences, which are shown in semi-transparent red. The overlay is positioned in the top left corner and covers the entire visible part of the site. Setting the CSS directive *pointer-events: none* ensures that the overlay does not cause any unwanted interactions, and allows mouse and keyboard events to "fall through" the overlay onto the original content.

In order to detect a malicious page from actively trying to remove the overlay from the DOM, we implement a mutation event listener that is triggered when an element is removed. It then checks whether the overlay is still present and if not, immediately warns the user of this active malicious behavior.

Security Indicator.

In addition to the overlay of the changes on the current page, TabShots also has a browser toolbar icon, indicating the current status of the site. The icon's background color indicates how much of the site has changed, ranging from almost nothing ($< 10\%$, shown as green), over moderate ($< 40\%$, shown as yellow) to high ($> 40\%$, shown as red). Clicking on the icon shows a miniature view of the current tab combined with the overlay of detected changes. Having a security indicator as part of the browser environment ensures that even if a malicious page somehow manipulates or removes the overlay, the user still has a trustworthy notification mechanism.

The current notification mechanism is quite subtle, but follows other commonly accepted and implemented notification mechanisms, such as displaying a padlock when using a secure connection. If desired, the notification mechanism can be easily extended to something more visible, such as the warnings given in case of an invalid SSL certificate.

3.3 Alternative Design Decisions

During the design and development of TabShots, we considered different paths and options, leading to the outcome described here. For completion, we want to discuss two topics that drove the design and workings of TabShots in a bit more detail.

JavaScript-based Detection.

Instead of visually comparing screenshots, we might attempt to detect the malicious JavaScript code actually carrying out the tabnabbing attack. This is not a trivial task, since JavaScript's dynamic nature makes script analysis difficult. Furthermore, there are a multitude of ways to actually implement a tabnabbing attack. The attack example discussed earlier uses the `window.onBlur` event, but a tabnabbing attack is certainly not limited to only this event. Similarly, there are numerous ways to actually change the displayed content, ranging from the use of JavaScript to extensive use of available CSS techniques.

Regularly Capturing Tabs.

Currently, TabShots makes a capture of a tab at regular intervals, so it can compare the capture taken when the user returns to a fairly recent capture from before. Ideally, we would make a capture when the user leaves, and a capture when the user returns. Unfortunately, Chrome does not trigger an event when a user leaves a tab, only when a user focuses a new tab. At the moment this event is received, the new tab is already displayed. To take a screenshot of the tab that was just left, TabShots has to switch it back into display, take a capture and switch back to the new tab. Unfortunately, this cannot be implemented without very briefly revealing this process visually to the user, with a degraded user experience as a consequence.

4. EVALUATION

As discussed before, a tabnabbing attack takes place when a user leaves a innocuous-looking malicious tab unfocused. Tabnabbing is different from traditional phishing, since it exploits trust placed in a previously opened tab, whereas phishing simply tries to mislead the user.

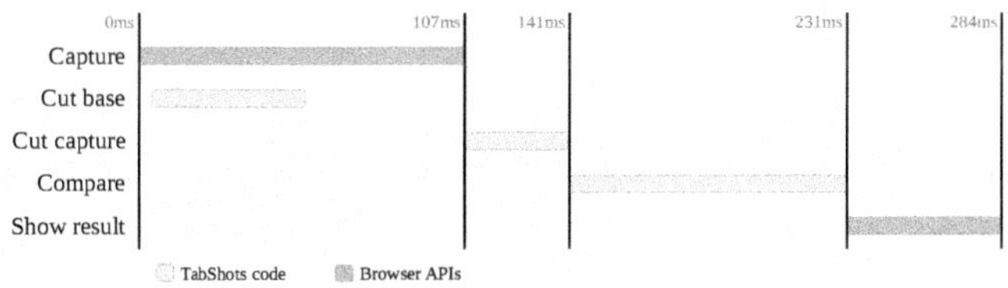

Figure 5: Breakdown of the average performance with a resolution of 1366x768.

Our evaluation of TabShots consists of three parts. First we discuss how TabShots effectively protects against all tabnabbing attacks. Second, we discuss the performance impact of TabShots. The third part elaborates on the setup and results of an experimental compatibility study using Alexa's 1000 most popular sites.

4.1 Security

The security guarantees offered by TabShots follow directly from its design. We recapitulate the three most important security properties here: (i) zero false negatives, (ii) user-friendly and clear overlay and (iii) secure toolbar indicator.

TabShots can not miss a tabnabbing attack by design, since it visually captures screenshots from a tab and compares them. In order for a tabnabbing attack to occur undetected, it has to ensure that the screenshots before and after losing focus are identical, meaning the page did not change while out of focus. This case is considered a classic phishing attack, and not a specific tabnabbing attack.

Second, TabShots injects an overlay of the focused tab, indicating which parts of the page have changed since its last focus. Using mutation events, TabShots detects if a malicious page actively tries to remove the overlay, and notifies the user with a strong security message.

Third, TabShots also adds an icon to the browser toolbar. Using a three-level color indication system, it notifies the user of how much a tab did change. The strength of this toolbar icon is that it runs in the context of the extension, and is completely out of reach to any page-specific code. This effectively prevents any manipulation by a malicious page.

4.2 Performance

In order to prevent tabnabbing attacks, TabShots must be capable of warning the user of any changes *before* she enters any sensitive information. Furthermore, since TabShots's algorithm is executed when a user switches tabs, it is crucial that there is no noticeable performance impact. The performance measurements and analysis of the main algorithm, discussed below, show that TabShots succeeds in quickly processing the captures and warning the user of any changes that occurred.

One important advantage of TabShots is that it fully operates in the background, without any blocking impact on any browser action or processing. When a user switches tabs, TabShots will perform the following steps:

1. Capture a screenshot of the newly focused tab
2. Cut the previously captured image of this tab (before it lost focus) into tiles
3. Cut the newly acquired screenshot into tiles
4. Compare the tiles of both screenshots and mark the differences
5. Inject the calculated overlay into the page and update the TabShots icon

For a browsing window with a resolution of 1366x768, the most common resolution at the time of this writing [18], TabShots is capable of performing these steps within an average time of 284ms after receiving the browser event fired by switching tabs. Fig. 5 shows a breakdown of this time into the steps mentioned before. Note that of these 284ms, 160ms are consumed by browser APIs, which are out of our control.

Currently, a large chunk of time is consumed by the comparison algorithm, which is a pixel-by-pixel comparison of each tile. The time used by this algorithm is strongly correlated to the number of changes within a page. If a difference between tiles is detected at the first pixel, there is no need to check the remaining pixels. Consequently, if a tabnabbing attack occurs, a lot of changes will be detected and TabShots's algorithm will perform even faster. Table 1 presents the number of milliseconds spent on comparison on our testing pages, where we use a div to change a certain percentage of a page, clearly showing the correlation between amount of changes and required processing time.

% changes	ms spent on comparison
0	126
25	86
50	60
75	32
100	4

Table 1: Correlation between amount of changes on a page and number of milliseconds consumed by the comparison algorithm.

Overall, one can see that TabShots is efficient enough to prevent tabnabbing attacks, before the user discloses her credentials to the phishing page and without a negative effect on the user's browsing experience. Moreover, if TabShots was to be implemented directly within the browser instead of through the browser's extension APIs, we expect that its overhead would be significantly lower.

4.3 Compatibility

Apart from the security guarantees offered by TabShots, its compatibility with existing sites is another important evaluation criterion. When using non-malicious web applications, the number of changes detected by TabShots, i.e. false positives, should be limited, even though the user can quickly determine whether a change is legitimate or not.

To determine the compatibility with current web applications, we ran TabShots on the top 1000 Alexa sites. Each site was loaded in a separate tab, and captured before and after it lost focus. These two captures were compared and analyzed for the number of changed blocks. Through our preliminary experimentation with TabShots, we discovered that a 10x10 tile-size strikes the desired balance between performance and precision. Smaller tiles would incur extra overhead, since as the number of tiles increase, so do the

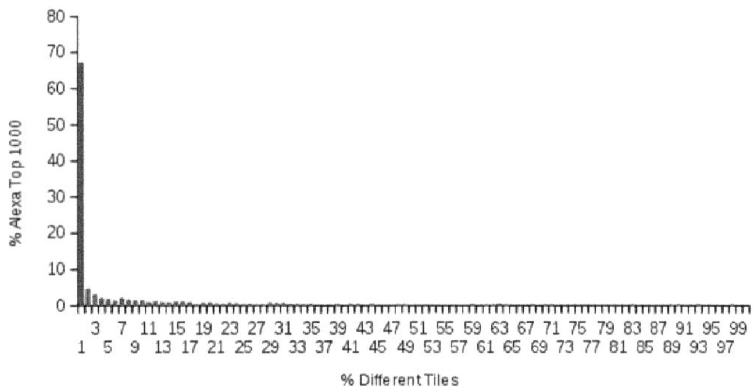

Figure 6: Compatibility analysis of the visual comparison algorithm with Alexa's top 1000 sites.

Domain	% of changed tiles
facebook.com	0.38
google.com	0.00
youtube.com	4.05
yahoo.com	5.31
baidu.com	0.00
wikipedia.org	0.73
live.com	2.65
twitter.com	2.91
qq.com	6.00
amazon.com	2.57
blogspot.com	0.32
linkedin.com	0.26
taobao.com	0.49
google.co.in	0.00
yahoo.co.jp	4.13
sina.com.cn	1.24
msn.com	23.22
google.com.hk	0.00
google.de	0.00
bing.com	0.00

Table 2: Compatibility analysis of the Alexa top 20 sites

checks between the old versions and the new ones, without a distinguishable improvement in pin-pointing the modified content.

Table 2 shows the results for the top 20 sites, and Fig. 6 shows a histogram of the entire top 1000, grouped by integer percentage values. The results show that 78% of sites fall within the safe threshold of less than 5% changed blocks, meaning there are no compatibility issues here. About 19% of sites have moderate changes, but still less than 40%. Manual verification shows that these changes are mainly caused by changing content such as image slideshows or dynamic advertisements. A typical example of an overlay of a dynamic advertisement is shown in Fig. 9. Finally, 3% of sites has more than 40% of changed blocks, which seem to be caused by changing background graphics. Fig. 7 and 8 respectively show the worst case scenario for the sites with moderate changes (less than 40%) and sites with heavy changes (more than 40%).

Note that even though certain sites have a high number of changed blocks, TabShots never interferes with a page,

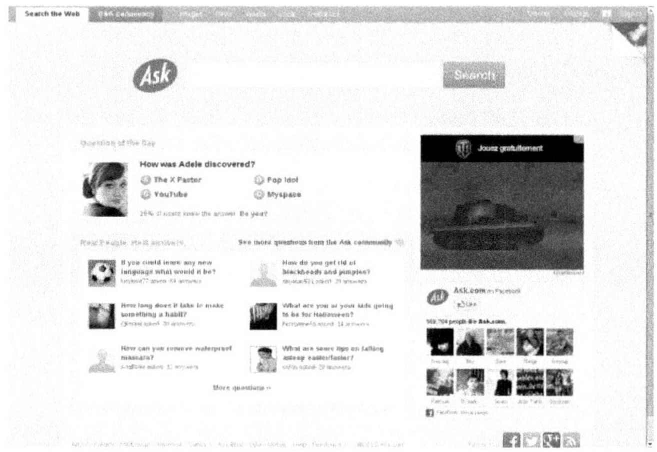

Figure 9: Screenshot of a typical dynamic advertisement being recognized by TabShots.

preventing any loss of functionality. If desired, a user can easily whitelist known trusted sites, to prevent needless overlaying of changed content. Additionally, a future extension of TabShots can incorporate a learning algorithm to identify dynamic parts of a site while the tab is in focus, which reduces the number of false positives.

The automated analysis gives a good idea of the impact on Alexa's top 1000, but is unfortunately not able to cover the authenticated parts of the sites. Therefore, we also tested the impact of TabShots on the daily use of several highly dynamic web applications, for example social networking applications (e.g. Facebook, Twitter) and webmail clients (e.g. GMail, Outlook Web Access). One noticeable effect is that the addition of a single element can cause a shifting of content within a page, which is currently flagged as a major change by the comparison algorithm. In future work, we can implement a comparison algorithm that detects such shifts and only marks the newly added content as a change.

5. BLACKLISTING TABNABBING

In the previous section, we described in detail the idea and implementation of TabShots. While this is sufficient for the protection of a user who has installed our browser extension, we deem it desirable to also protect users who are using different browsers or have not installed TabShots. This can be achieved through an optional server-side component which can aggregate information sent by individual browsers and, after validation, add the reported URLs in a blacklist. This server-side component would be the logical next step, to transition from a protection of a selected number of users (those with TabShots installed) to a more global protection, similar to Google's SafeBrowsing list of malicious sites [15] which is currently utilized by many modern browsers. In the rest of this section, we describe the possible workings of such a service.

In the stand-alone version of TabShots, once a user realizes that she is being targetted by a tabnabbing attack, she is instructed to simply navigate away from the malicious site without entering any private information. With a server-side component in place, the user can mark the current page as a "tabnabbing attack" through the UI of our extension.

Figure 7: Before and after shots of *americanexpress.com (#354)*, which has 38.93% of changed blocks, due to a background image that took longer to load.

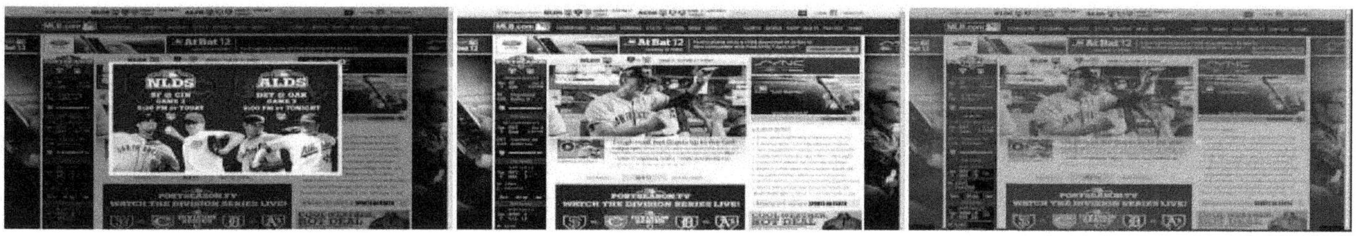

Figure 8: Before and after shots of *mlb.com (#355)*, which has 97.31% of changed blocks, due to an overlay that changed the intensity of the site to present an advertisement.

Once this happens, TabShots transfers to a server-side, data aggregator the following information:

1. The URL of the current page
2. The image of the page before the user switched tabs
3. The image of the page after the user switched tabs

The server-side aggregator has the responsibility of receiving reports from multiple users, filtering-out false reports and then adding the true positives on a blacklist. Filtering is necessary to stop attackers who wish to harm the reputation of the TabShots blacklist, by submitting legitimate sites that would then be automatically blocked. Our server-side service operates as follows:

For every previously unreported URL received by a user with TabShots installed, our service spawns an instrumented browser which visits the reported URL and captures a screenshot. Assuming that the current page is indeed performing tabnabbing, the malicious scripts will try to get information about their "visibility" through the `window.onBlur` event. Since our browsers are instrumented, we can trigger a `window.onBlur` event without requiring the actual presence of extra tabs. In the same way, any callbacks that the script registers using the `setTimeout` method, are immediately triggered, i.e., the malicious code is tricked into performing the tabnabbing attack, without the need of waiting. Once the callbacks are executed, our system takes another snapshot of the resulting page. The set of screenshots captured by the user is then compared with the set that was captured by our system. To account for changes in the pages due to advertisements and other legitimately-dynamic content, the screenshots are accepted if either the server-generated set is an identical copy of the user-generated one, or if they match over a certain configurable threshold (i.e. everything matches except certain dynamic areas).

Once the above process is complete, the URLs recognized as true positives are then sent to a human analyst who will verify that the resulting page is indeed a phishing page. A human analyst is necessary since our system cannot reason towards the maliciousness or legitimacy of the final changed page. Note, that human-assisted phishing verification is currently one of the most successful approaches, e.g., PhishTank [13], and its results are more trustworthy than any automated phishing-detection solution. The URLs that are marked as "tabnabbing", can then be added to a blacklist that browsers can subscribe to.

The users who report URLs that either never reach a human analyst (because the server-side screenshots did not match the user-provided ones), or reached a human analyst and were classified as "non-phishing" are logged, so that if they are found to consistently submit false positives, our system may adapt to ignore their future submissions.

Submitting screenshots to a third party service might be considered privacy-sensitive, so we take care to address these issues accordingly. Therefore, TabShots only submits a screenshot after explicit user approval. Additionally, the screenshot submission is only triggered after the user flagged a tabnabbing attack, so it is very likely that the captured site is malicious of nature, and does not contain any sensitive information.

6. RELATED WORK

Raskin was the first to present the tabnabbing attack in 2010 [14]. Others, presented variations of the attack, for instance redirecting the user through a meta-refresh tag to a new page, instead of changing the existing page with JavaScript [1], which would circumvent protections such as NoScript [12]. This attack however does not depend on user activity, but rather on a predefined timeout, by which the attacker hopes that the victim will have changed tabs and thus will not notice the changing web-site.

Unlu and Bicakci [20] proposed NoTabNab, a browser extension that monitors tabs in search for tabnabbing attacks. When a page loads, the extension records the title of the page, the URL, the favicon and several attributes of the topmost elements[3], as determined by the browser API call `document.elementFromPoint(x,y)`. While this approach is conceptually similar to ours, NoTabNab suffers from several issues that render it ineffective.

A first issue is that by capturing a tab when it is loaded, the extension will miss all content that is added dynamically (e.g. using AJAX) between loading a tab and actually switching to another tab. Second, the design of the detection mechanism offers an attacker several ways to evade it. For instance, an attacker can place all of the page's content in an iframe that spans the entire visible window. The `document.elementFromPoint` cannot "pierce" through the iframe, and will always return the iframe element, regardless of any changes that may happen inside the iframe. Another possible bypass, is through the overlay of a transparent element, that again stretches the entire page's content and allows clicks and interactions to "fall-through" to the actual phishing form under it. On the contrary, TabShots uses the screen-capturing API of the browser and is essentially using the same data as actually seen by the user. This design decision makes TabShots invulnerable to the aforementioned bypasses.

Suri et al. [19] propose the detection of tabnabbing through the use of tabnabbing "signatures". The authors claim that the combination of certain JavaScript APIs with HTML elements are tell-tale signs of a tabnabbing attack, and present two signatures, based on the presence of `onBlur`, `onFocus`, and other events within an iframe. Unfortunately, the authors make no attempt to characterize the false positives that their system would incur. Additionally, the presence of an iframe is by no means necessary for a tabnabbing attack. JavaScript code is capable to drastically change the appearance of a page through the addition and removal of styled HTML elements, thus allowing an attacker to bypass the authors' monitor. TabShots on the other hand, does not depend on anything other than the visual differences between the old and the new version of the tab, and thus will detect all visible changes, regardless of the technical means through which they are achieved.

While tabnabbing is a relatively new phishing technique, attackers have been trying to convince users to voluntarily give up their credentials for at least the last 17 years [3]. Several studies have been conducted, trying to identify why users fall victim to phishing attacks [5, 7] and various solutions have been suggested, such as the use of per-site "page-skinning" [4], security toolbars [23], images [2, 17], trusted password windows [16], use of past-activity knowledge [11] and automatic analysis of the content within a page [24]. Unfortunately, the problem is hard to address in a completely automatic way, and thus, the current deployed anti-phishing mechanisms in popular browsers are all black-list based [15]. The blacklists themselves are either generated automatically by automated crawlers, searching for phishing pages on the web [22] or are crowdsourced [13].

[3]When seeing a page as a stack of elements, the topmost elements potentially overlay other elements

7. CONCLUSION

Tabnabbing attacks are a type of phishing attacks where the attacker exploits the trust a user places in previously opened browser tabs, by making the malicious tab look like a legitimate login form of a known web application. This happens when the user is looking at another tab in the browser, making it very hard to detect and very easy to fall victim to.

Currently available countermeasures typically depend on several specific characteristics of a tabnabbing attack, and are easily bypassed or circumvented. Our countermeasure, TabShots, is the first to do a fully visual comparison, detecting any changes in an out-of-focus page and highlighting them, aiding the user in the decision whether to trust this page or not.

Our evaluation shows that TabShots protects users against potential tabnabbing attacks, with a minimal performance impact. Furthermore, an experimental evaluation using Alexa's top 1000 sites shows that 78% of these sites fall within the safe threshold of less than 5% changes in subsequent snapshots. This means that TabShots is fully compatible with these sites, and has very little impact on another 19%.

8. ACKNOWLEDGEMENTS

This research is partially funded by the Research Fund KU Leuven, IWT and the EU-funded FP7 projects NESSoS and Web- Sand.

With the financial support from the Prevention of and Fight against Crime Programme of the European Union European Commission – Directorate-General Home Affairs. This publication reflects the views only of the authors, and the European Commission cannot be held responsible for any use which may be made of the information contained therein.

9. REFERENCES

[1] E. Adler. Tabnabbing without JavaScript . `http://blog.eitanadler.com/2010/05/tabnabbing-without-javascript.html`.

[2] N. Agarwal, S. Renfro, and A. Bejar. Yahoo!'s Sign-in Seal and current anti-phishing solutions.

[3] AOL acts to thwart hackers. `http://simson.net/clips/1995/95.SJMN.AOL_Hackers.html`.

[4] R. Dhamija and J. D. Tygar. The battle against phishing: Dynamic security skins. In *Proceedings of the 2005 symposium on Usable privacy and security*, SOUPS '05, pages 77–88, New York, NY, USA, 2005. ACM.

[5] R. Dhamija, J. D. Tygar, and M. Hearst. Why phishing works. In *Proceedings of the SIGCHI conference on Human Factors in computing systems*, CHI '06, pages 581–590, New York, NY, USA, 2006. ACM.

[6] P. Dubroy. How many tabs do people use? (Now with real data!). `http://dubroy.com/blog/how-many-tabs-do-people-use-now-with-real-data/`.

[7] S. Egelman, L. F. Cranor, and J. Hong. You've been warned: an empirical study of the effectiveness of web browser phishing warnings. In *Proceedings of the SIGCHI conference on Human factors in computing systems*, CHI '08, pages 1065–1074, New York, NY, USA, 2008. ACM.

[8] D. Jang, R. Jhala, S. Lerner, and H. Shacham. An empirical study of privacy-violating information flows in JavaScript Web applications. In *Proceedings of CCS 2010*, pages 270–83. ACM Press, Oct. 2010.

[9] J. Leyden. Hackers break onto White House military network. `http://www.theregister.co.uk/2012/10/01/white_house_hack/`.

[10] L. Masinter. The "data" url scheme. 1998.

[11] N. Nikiforakis, A. Makridakis, E. Athanasopoulos, and E. P. Markatos. Alice, What Did You Do Last Time? Fighting Phishing Using Past Activity Tests. In *Proceedings of the 3rd European Conference on Computer Network Defense (EC2ND)*, volume 30, pages 107–117, 2009.

[12] NoScript - JavaScript/Java/Flash blocker for a safer Firefox experience! `http://noscript.net/`.

[13] PhishTank | Join the fight against phishing. `http://www.phishtank.com`.

[14] A. Raskin. Tabnabbing: A new type of phishing attack. `http://www.azarask.in/blog/post/a-new-type-of-phishing-attack/`.

[15] Safe Browsing API – Google Developers. `https://developers.google.com/safe-browsing/`.

[16] D. R. Sandler and D. S. Wallach. <input type="password">must die! In *Proceedings of W2SP 2008: Web 2.0 Security & Privacy 2008*, Oakland, CA, May 2008.

[17] SiteKey Security from Bank of America. `https://www.bankofamerica.com/privacy/online-mobile-banking-privacy/sitekey.go`.

[18] StatCounter. Screen resolution alert for web developers.

[19] R. K. Suri, D. S. Tomar, and D. R. Sahu. An approach to perceive tabnabbing attack. In *Internation Journal of Scientific & Technology Research*, volume 1, 2012.

[20] S. Unlu and K. Bicakci. Notabnab: Protection against the "tabnabbing attack". In *eCrime Researchers Summit (eCrime), 2010*, pages 1 –5, oct. 2010.

[21] Z. Weinberg, E. Y. Chen, P. R. Jayaraman, and C. Jackson. I still know what you visited last summer: Leaking browsing history via user interaction and side channel attacks. In *Proceedings of the 2011 IEEE Symposium on Security and Privacy*, SP '11, pages 147–161, 2011.

[22] L. Wenyin, G. Huang, L. Xiaoyue, Z. Min, and X. Deng. Detection of phishing webpages based on visual similarity. In *Special interest tracks and posters of the 14th international conference on World Wide Web*, WWW '05, pages 1060–1061, New York, NY, USA, 2005. ACM.

[23] M. Wu, R. C. Miller, and S. L. Garfinkel. Do security toolbars actually prevent phishing attacks? In *Proceedings of the SIGCHI conference on Human Factors in computing systems*, CHI '06, pages 601–610, New York, NY, USA, 2006. ACM.

[24] Y. Zhang, J. I. Hong, and L. F. Cranor. Cantina: a content-based approach to detecting phishing web sites. In *Proceedings of the 16th international conference on World Wide Web*, WWW '07, pages 639–648, New York, NY, USA, 2007. ACM.

Fuzzing the ActionScript Virtual Machine

Guanxing Wen, Yuqing Zhang, Qixu Liu, Dingning Yang
National Computer Network Intrusion Protection Center
School of Computer and Control, UCAS, Beijing 100049 PR China
wengx522@sina.com, {zhangyq, liuqixu}@ucas.ac.cn, yangdn@nipc.org.cn

ABSTRACT

Fuzz testing is an automated testing technique where random data is used as an input to software systems in order to reveal security bugs/vulnerabilities. Fuzzed inputs must be binaries embedded with compiled bytecodes when testing against ActionScript virtual machines (AVMs). The current fuzzing method for JavaScript-like virtual machines is very limited when applied to compiler-involved AVMs. The complete source code should be both grammatically and semantically valid to allow execution by first passing through the compiler. In this paper, we present *ScriptGene*, an algorithmic approach to overcome the additional complexity of generating valid ActionScript programs. First, nearly-valid code snippets are randomly generated, with some controls on instruction flow. Second, we present a novel mutation method where the former code snippets are lexically analyzed and mutated with runtime information of the AVM, which helps us to build context for undefined behaviours against compiler-check and produce a high code coverage. Accordingly, we have implemented and evaluated *ScriptGene* on three different versions of Adobe AVMs. Results demonstrate that *ScriptGene* not only covers almost all the blocks of the official test suite (Tamarin), but also is capable of nearly twice the code coverage. The discovery of six bugs missed by the official test suite demonstrates the effectiveness, validity and novelty of *ScriptGene*.

Categories and Subject Descriptors

D.2.5 [**Testing and Debugging**]: Testing tools

Keywords

Fuzz testing, Vulnerability discovery, ActionScript

1. INTRODUCTION

The theory of application virtual machines is well developed. An application virtual machine (VM) isolates the host OS from a single process. The VM is created when the associated process is started and destroyed when the associated process exits [27]. VM Implementations can be divided into two categories based on the code execution. JavaScript-like VMs accept source code directly, interpreting and executing code lines sequentially. ActionScript-like VMs accept only bytecodes generated by a compiler from source code. This additional compilation stage adds complexity to ActionScript-like VMs when compared to JavaScript-like VMs. Since many VMs are programmed using C/C++, it is not surprising that they contain bugs similar to other complex software systems.

Fuzz testing or fuzzing is a form of testing heavily used for finding security vulnerabilities in software. Since completely random strings generated by traditional fuzzing methods do not usually execute in VMs, grammar-based fuzzing is employed instead. Grammar-based whitebox approach of [17] has been used on a JavaScript VM. They presented a dynamic test-case generation algorithm, where symbolic execution directly generates grammar-based constraints, whose satisfiability is verified using a custom grammar-based constraint solver. The solver solves the altered constraints and directs code snippet generation to cover more execution paths of the VM. Unfortunately this approach is limited to JavaScript-like VMs, since feedback operates on the input directly. In comparison, ActionScript-like VMs only accept binaries embedded with bytecodes, which are the outputs of the compiler. Therefore, a hypothetical solver is unable to analyze and generate new pieces of code as inputs, rendering this approach ineffective. Grammar-based blackbox approach of [19] is yet another effective fuzzing method on VMs. Previously, this approach has also been restricted to testing JavaScript-like VMs. Both grammar-based whitebox and blackbox fuzzing methods focus only on the grammar, ignoring the substantial contributions of runtime class, property and API functions to the discovery of bugs in VMs.

To overcome the difficulties associated with fuzzing of ActionScript-like VMs, we have created *ScriptGene*. *ScriptGene* generates ActionScript (AS) source code, compiles and executes them on Adobe Flash ActionScript Virtual Machines (AVMs). As seen in figure 1, the left-hand side sequence of block diagrams, connected by downward arrows illustrates the *ScriptGene* code generation process. Prior to the code generation, a knowledge base is built by interpreting the grammar file of AS and the official test cases. Then, grammar-based blackbox methods generate nearly-valid code. The right-hand side sequence of the figure with upward arrows illustrates the mutation phase. A lexier marks all the identifiers in source code, pointing to functions, classes and variables with markers, followed by mutations and substitutions of markers with actual runtime class information. Lastly, the test case programs are created and sent to the target AVM.

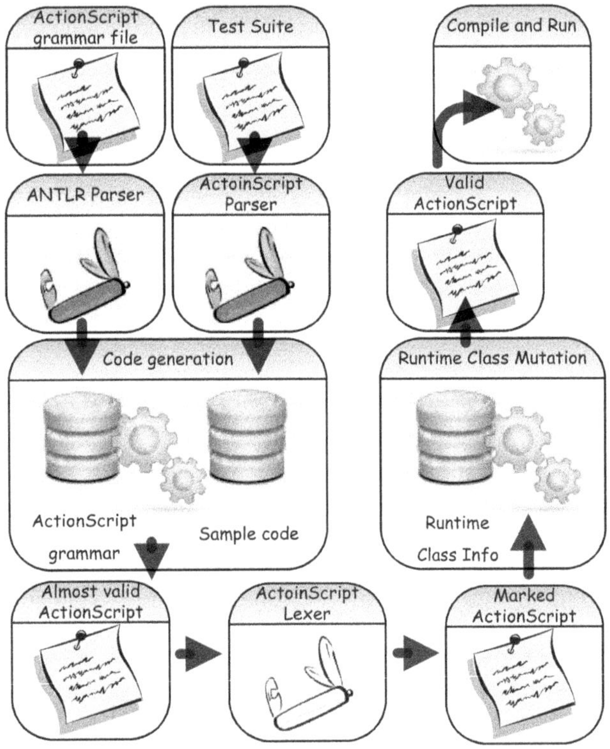

Figure 1: *ScriptGene* workflow.

In comparison with earlier approaches, *ScriptGene*'s test-generation techniques substantially advance the state of the art in fuzzing VMs by generating compilable and expressive random source code of high complexity, employing many AS language features. Compiler-validity is ensured by generating AS code that does not contain undefined classes, variables or functions; nor depend on random arrangements of the code structure. Currently, *ScriptGene* generates AS programs suitable for testing AVMs. Flexibility of our ideas underpinning *ScriptGene*, as well as our specific implementation should easily port to test other VMs. This paper summarizes the construction of *ScriptGene* and some of its current testing results.

Adobe Flash is a multimedia platform used to add animations to web pages and hybrid document formats such as PDFs and Office documents. Usually, Adobe Flash files with the ".SWF" (SWF) extension are treated as harmless files or as videos by the host OS. AS code inside SWFs are typically used to control animations and merely contain functions to operate natively on local resources (e.g., disk, process...). Thus, AVMs are running in trusted environments. If a SWF file triggers a vulnerability of the AVM and executes native code out of the trusted environment, the local resources could be compromised. Furthermore, Just-In-Time compilers (JIT) of AVMs can be used to breakthrough memory protection methods such as ASLR and DEP [16]. APT attack to RSA [15] in 2011 demonstrated the ramifications of AVM security vulnerabilities.

AVMs were chosen as the testing targets for two reasons. First and foremost, vulnerabilities in trusted environments are intrinsically harmful, more so than an equivalent language that operates on the local resources natively. Second, there appears to be a lack of existing grammar-based approaches focusing on ActionScript-like languages, where compilers are involved.

- Our first contribution is the implementation of a grammar-based blackbox method to fuzz test complier-involved AVMs. Our implementation has covered almost all the code blocks (>96%) of the official test cases.
- Our second contribution is the advancement of the state of the art in grammar-based blackbox test code generation, coupled with runtime class mutations. Our approach not only has allowed us to gain additional code coverage to nearly double that of the official test cases in our experiments, but also has discovered six previously undocumented bugs in three versions of AVMs [11] that were missed by the official test cases.

The remainder of this paper is organized as follows. Section 2 presents the technical details on *ScriptGene*, from the code generation phase to the runtime class mutation phase. Section 3 details the actual implementation and discusses our evaluations. Section 4 discusses related works and the state of the art in fuzz testing. Section 5 details the simplified strategy for *ScriptGene* and some of its limitations. Section 6 concludes the paper.

2. SCRIPTGENE

The basic concepts behind *ScriptGene* are in Section 2.1. The additional grammar controls are in Section 2.2. Our AS code generation algorithm is presented in Section 2.3. Extraction of runtime information for runtime class mutations is detailed in Section 2.4. Mutation of extracted information into code to create compilable source code is presented in Section 2.5.

2.1 Basic concepts and terminology

ActionScript and AVMs There are currently two versions of AS: AS2 and AS3, executed by AVM1 and AVM2. AS2 lacks many new features, such as 3D presentations and is less effective in code execution as compared to AS3. AS3 is chosen, since it is the version most frequently employed by current SWFs. AS3 based on ECMAScript4 standard [6] is an object-oriented programming language used to build SWFs. Unless otherwise noted, AS refers to AS3 and AVM refers to AVM2 hereafter.

Grammar files In order to generate and parse source code, the grammar file [7] written in ANTLR, containing the grammar standards of AS is utilized. ANTLR is a parser generator framework written by Parr and Quong [23]. Once the grammar file is established, the task of building a parser is similar to compiling source code. In addition to the AS parser, an ANTLR language parser parses the ANTLR format AS grammar file, since we wish to generate AS code snippets beyond merely parsing them. Since some features of ANTLR are only enabled under Java, Java is used to parse source code. Other parts of *ScriptGene* are programmed in Python, to leverage its string manipulation capabilities.

In total, there are three grammar files for *ScriptGene*: a grammar file used to build a parser for AS (G1), a grammar file used to guide the AS code generation process (G2) and a grammar file used to build a parser for ANTLR (G3). G2 is a modified version of G1, which will be explained in Section 2.2. G3 is the official ANTLR grammar file inside the example directory of its Java distributions.

Grammar rules in ANTLR format In ANTLR, AS grammar elements can be divided into three categories: rules, blocks and atoms. Rules comprise the main structures listed in an ANTLR format grammar file. Rules also contain sub-rules. Sub-rules are mixtures of blocks and atoms. Blocks consist of atoms. Atoms represent terminal and non-terminal nodes. Non-terminal nodes are other rules and terminal nodes are tokens or real characters inside a range. For the code generation phase, only the knowledge of node extensibility is required. If the node stands for a rule, it is extensible and ready to be extended with sub-rules, otherwise, no action is performed to this node until the end.

2.2 Adding control to grammar rules

The generated code should contain a diversity of grammar structures and runtime class interactions, while being easy to mutate and free of non-interesting code structure. Simultaneously, the generated code must be compilable. To satisfy these requirements, two controls are added to the production of AS code during opportune stages of *ScriptGene* before compilation. First control consists of modifications to the grammar file and the second control consists of adjustments to the context during the mutation phase (Section 2.5). Although adjustments of the grammar file are less flexible than that of the runtime class mutations Python code, the foremost occurrence of the grammar file in the *ScriptGene* generation process necessitates that it be given a higher adjustment priority. This allows the removal of root causes of potentially troublesome structures from the compiler. G2 is therefore created from G1 under a few controls (e.g., simplification, ignorance, modification...) of select rules. G2 is then transformed into a Python source file by the ANTLR parser.

Simplification rules ANTLR allows Java code to be inserted into grammar rules to analyze the complex context of the source code. However, this kind of Java code behaves as comments to the ANTLR language parser, and is ill-suited for generating code snippets. A clean-up simplification of this Java code is required to facilitate the process of parsing G2 into Python format.

Ignorance rules A rule such as *xmlPrimaryExpression* is used to parse XML expressions, which are uninteresting to us, since we focus on the grammar structures of the context and not on particular API functions. Code generation cycles will be wasted if these rules are left unmodified. Thus, these rules' sub-rules are replaced with short XML expression instances.

Modification rules A rule such as *functionBody* is rewritten with a one-time loop structure. All the functions generated will start from one-time for-loops. This allows most breaks and continues inside those functions compiler-legal statuses.

Expansion and replacement rules Consider the rule *tryStatement* as an example. The main part of this rule is embedded inside a block, whose expansion according to our algorithm can simply be a random pick from its sub-atoms (see Section 2.3.2). Initialization of *tryStatement* sub-rule selection will not work. Therefore, depth-first sub-rule selection is disabled in this situation. The rule is rewriten and the block inside the *tryStatement* is replaced with a newly defined rule *tryStatementBlock*. Initialization of sub-rule selection on *tryStatementBlock* will work and give the code generation phase a greater diversity.

```
1: tryStatement
   : TRY blockStatement
     ( catchClause+ finallyClause
     | catchClause+
     | finallyClause
     )
   ;
2: tryStatement
   : TRY blockStatement tryStatementBlock
   ;
   tryStatementBlock
   : catchClause+ finallyClause
   | catchClause+
   | finallyClause
   ;
```

Marking of rules related to functions and variable declarations This is done for the runtime class mutation phase, to reduce the complexity of lexical analysis subsequently. For example, if we attempt to expand the *variableDeclaration* rule during the code generation phase, the sophisticated variable expression will be very difficult for the mutation algorithm to control both the type and the value of the variable. Therefore, its sub-rules are replaced with markers such that we can directly alter the type and the value of the variable.

2.3 Grammar-based blackbox fuzzing

According to vulnerability reports of AVMs in the past 2 years [20], many vulnerabilities were caused by AS code containing interactions with runtime classes. Therefore, a main aspect of *ScriptGene* is to create opportunities for interactions between different runtime classes.

2.3.1 Key points of the design

There are three key points that should be taken into consideration while creating compiler-pass AS code to test a backend AVM.

Stricter environment For a JavaScript-like VM, even if the backend VM receives no executable code due to the inappropriate grammar context of the input code, it is still possible to test the frontend parser and trigger bugs in trusted environments. Unlike JavaScript code and its VM, AS code interfaces with the AVM via a compiler. Bugs of the parser (in the compiler) would not be a direct threat to the trusted environment. Accordingly, AS requires much stricter context than JavaScript, since any preceding line of invalid AS code will block the execution of the remaining code due to the compiler. I.e., AS should be global-valid while JavaScript needs to be only line-valid. Adding controls to the code generation phase and marking rules inside G2 (see Section 2.2) overcomes this difficulty.

Additional grammar structures in AVMs In order to compile and execute code on the AVM, some additional structures are needed in our AS code. Such structures are difficult to generate with random walks over grammar space, thus a brute force approach is inefficient. Therefore, code templates (see Section 2.5.2) are used to reconstruct the source code just one step before compilation.

Valid context Source files with proper structure and grammar still contain undefined variables, functions, classes, etc., and are thus still uncompilable. This is addressed by mutating the source code during the mutation phase (see Section 2.5) with real classes, API functions and predefined variables that are extracted from runtime libraries of the AVM.

2.3.2 *Generation of nearly-valid code snippets*

One possible approach for fuzzing AVMs based on grammar involves randomly generating source code that follow the grammar structure. However, naïve random walks over grammar rules are not guaranteed to terminate. Referring to the idea of *LangFuzz* [19], we bring an end to the random walk code generation process using instances of grammar rules learnt from known test suites. This is a good start, but in contrast to *LangFuzz*, it is not the final step before code execution. Furthermore, we enhance *LangFuzz*'s idea by incorporating both depth-first and breadth-first algorithms into *ScriptGene*, as described in algorithm 1. The

Algorithm 1 ActionScript code generation

```
Input:
    A decision of sub-rule selection, seednum;
Output:
    Nearly-valid code snippets, codearray;
Initialize codearray as [startrulenode];
Initialize the sub-rule selection based on seednum using
    a deterministic-random algorithm;
while (Number of active nodes and number of genera-
    tion cycles are still low) do
  for each node in codearray do
    if node is not an active node then
      Remain node unchanged;
    else
      Replace node with expanded nodes using
      breadth-first/depth-first algorithms of sub-rule s-
      elections;
    end if
  end for
end while
for each node in codearray do
  Replace node with rule instances of its type;
end for
return codearray;
```

AS code is generated on a per-cycle basis. Each cycle of generation expands the nodes inside the *codearray*. Iteration is used instead of recursion, since determination of possible code length and number of nodes are more convenient with iteration. These important parameters control the termination of the code generation phase.

Each level of expansion is done on active nodes. The type of the node determines if a node is active or not. In figures 2, 4 and 3, each circle represents a node, while the character inside states the content. A given node is a NULL atom if the node is blank (previous level of expansion has randomly chosen zero for the quantifiers or optional). The quantifiers ('*','+') and optional ('?') are stored without any simplifications and will be replaced by a random number in its valid range during code generation. This is also different from *LangFuzz* [19]. Before a level of expansion, the nodes in the *codearray* are scanned and a hierarchy is established for each cycle: rule->block->atom. During each cycle, only the highest hierarchy item is set active. This ordered expansion is more effective and intuitive when we need to examine the generation process. Both ordered and non-ordered expansion processes are illustrated in figures 2 and 4 for comparison. In a non-ordered expansion process, such as illustrated in figure 2, different structures will inevitably begin to embed each other (a block inside a block, an atom inside a block embedded further inside a rule, extended from an atom...). Additionally, expansion cycles may be wasted on extending structures that will eventually become NULL (the left branch of figure 2, for example). Therefore, the ordered process as depicted in figure 4 is our node expansion approach. Although figure 4 appears to

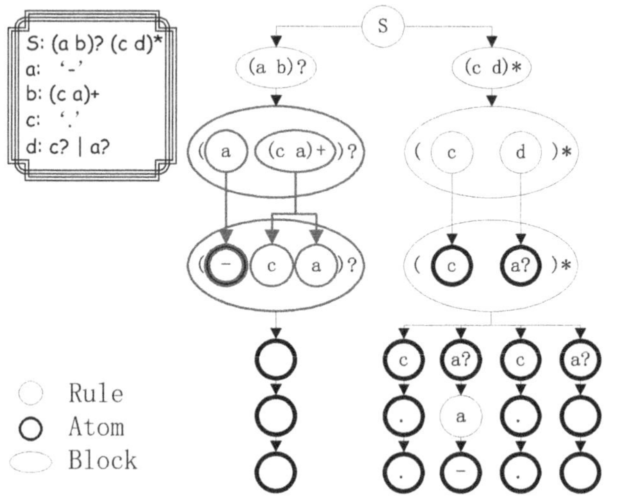

Figure 2: Node expansions in random order.

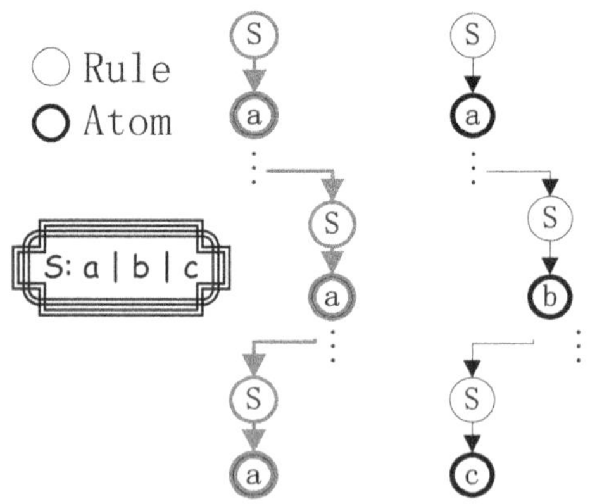

Figure 3: Left branch represents a depth-first sub-rule selection. Right branch represents a breadth-first sub-rule selection.

depict a breadth-first algorithm, it is not. Breadth-first or depth-first algorithms apply to the selection of sub-rules. Rules consist of sub-rules, separated by "|" in ANTLR such as "Rule S: a | b | c". Rule "S" can be selected as "a", "b" or "c". Our breadth-first method is a random selection among sub-rules, whereas our depth-first method holds on to the initial selection at the beginning of the code generation phase (see line 2 of algorithm 1). Figure 3 demonstrates the difference between breadth-first and depth-first algorithms. Our breadth-first method is similar to that used by [19]. Depth-first algorithm was added, based on results of our previous testings of the regular expression interpreter [31]. This addition enables us to test AVMs with nested grammar structures. The idea is akin to sending a "((((((expr))))))", to test the expression interpreter. Nested structures appear to be validated with less caution and return more bugs. Intuitively, to cover more grammar rules, breadth-first algorithms need longer code lengths, while depth-first algorithms need more generation cycles with different sub-rule initializations. To terminate the code generation process as in line

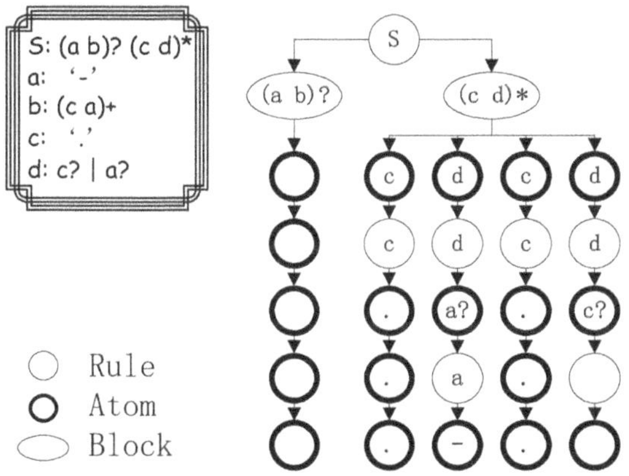

Figure 4: Node expansions following the rule, block, atom order.

13 of algorithm 1, we study code fragments (which are essentially examples for non-terminals in grammar rules and in our case, are instances of rules) from the AVM test suite in Tamarin [12], which is an open-source AVM project under Mozilla. The AS source code is taken from their acceptance test cases and dissected into pieces then categorized based on their rule types. A sample code fragment pool is built after parsing the test suite with the help of G1, in which there are several instances for each rule. The G1 grammar file originally belongs to a plug-in for Eclipse, intended to be part of a syntax highlighter function. It is imperfect when handling large assortments of code fragments. Since some parsed results may contain faults, the pool requires manual fixes. In addition, G2 is a modified version of G1, not all of the rules find their instances from this test suite. Thus, the missing ones are added manually.

2.4 Extraction of runtime information

After creating nearly-valid AS code snippets, all the identifiers pointing to labels, functions, variables and types are marked with the AS lexier. Some of these will be replaced with runtime classes, properties and API functions of the AVM, while others need modifications to suit the current context. A knowledge base is built for this purpose, containing the relevant runtime class information of the AVM.

Runtime class information can be obtained from mainly two sources. The first is the ActionsPanel_3.xml, found in the sub-directory of the SWF Integrated Development Environment (IDE), Adobe Flash Professional CS6 [4]. It contains sufficient information for the construction of expressions that declare variables to be instances of classes, call their member functions and modify their properties. However, it does not contain newly published APIs and classes.

The second one is the file playerglobal.swc, which accompanies newly released versions of Adobe Flash. This file contains the newly introduced APIs and runtime classes, and constitutes sufficient information for a compiler to determine if a class or function exists. By unpacking the playerglobal.swc file yields the files catalog.xml and library.swf. Inside catalog.xml are names of runtime classes, especially some undocumented API functions. Demonstrations of functions and properties are missing compared to ActionsPanel_3.xml. It is possible that they reside in library.swf in binary format. Instead of dealing with library.swf, the dynamic reflection mechanism of Flash is used to extract the missing information. Thus, *describeType* [1] is employed to restore the structure of the class using only its name.

After runtime information has been collected, verification is required for runtime class mutations, to ensure that they are compatible with our compilation environment. Our compiler asc.jar comes from Flex SDK [2], an open-source SDK designed to build Flash. Asc.jar performs compilation-checks by utilizing the information in playerglobal.abc. However, playerglobal.abc is not as up-to-date as playerglobal.swc. Therefore, the compilation and execution status of each class collected in the current AVM of interest is manually verified, to minimize the amount of non-executable test cases produced by the mutation phase.

To verify the compilation status of these runtime classes, an AS source file is built by importing all the classes collected, then called upon with *describeType* inside a "try-catch" clause. The classes that are not recognized (cause exceptions) by asc.jar and the target AVM are excluded, then the valid description output by *describeType* are recorded. For this purpose, a Python module is created that parses all the XML outputs of both *describeType* and ActionsPanel_3.xml, to build a runtime class pool in memory. This pool contains:

- name of runtime classes and their parent classes,
- name of member functions and the type of each parameter,
- name of dynamic and static properties of each class.

2.5 Runtime class mutation

This phase is a critical part of *ScriptGene*. Outputs of this phase are valid AS code that will be compiled into SWF files. The guiding idea of runtime class mutation is to bring more interactions between different classes and give a valid context to the code simultaneously. This is done in the following two steps.

2.5.1 *Step 1, mark identifiers*

We mark the identifiers pointing to names and types of functions, classes, catch variables, break variables, continue variables, labels and function-calls via lexical analysis. The lexier is based on G1, which recognizes all the identifiers. The identifiers are distinguished according to the characters of the former and latter tokens in the lexical sequence. For example, if a "while" clause is generated in code generation phase, then identifiers such as "DateCase" are neither API functions nor static classes and are not yet acceptable to the compiler. Subsequently, the nearly-valid AS source code will become a marked file with all the identifiers replaced by markers, for example:

```
1: while(new Object+=a,
   DateCase.setMilliseconds(newms)||c,
   i++||f||2);
2: while(new _varname_+=_varname_,
   _varname_._funcall_(_varname_)||_varname_,
   _varname_++||_varname_||2);
```

2.5.2 Step 2, replace markers

The input of this step is the output of the previous step, an AS source code file full of markers. Our goal is to mutate these markers with real classes, build a context for each variable that will be used and to correct lexical issues to compensate for G1. For example, marked AS code will replace "_userdefine_" in the following prebuilt template:

```
package{
    import flash.display.*;
    _import_
    _globalvardeclaration_
    _userdefine_
    public class Main extends MovieClip{
        public function Main():void{
            _functioncall_
        }
    }
}
```

When the mutation begins, the input will be a template file full of markers. All the classes, API functions, global variables and member properties will be enumerated from the runtime class information pool as built in Section 2.4. Every mutation cycle will commence by picking several classes from the runtime class information pool. Presently, we have adopted a limit of two classes per cycle. During this process, all markers in the template file will be replaced with meaningful code.

"_import_" will be replaced with "import" clause to include the class of our choice. "_globalvardeclaration_" will be replaced with declarations of class objects with "new" clauses. References are made to the runtime class information pool to check the existence of constructors. If so, a "new" clause is built for this class with appropriate parameters. Otherwise, the chosen class is a static class. Properties and member functions of such a class can be used directly without declaration.

Next, a variable pool and a function pool are initialized. The variable pool and the function pool stores the variable names and methods of runtime class objects of our choice respectively, including all of their properties and their parents'. These two pools will be referred to, when markers referencing variables and function-calls are encountered. In addition, there is one more global pool to record the classes that *ScriptGene* generates and their variables, functions and labels.

When *ScriptGene* scans the marked template file, it will encounter several kinds of markers. Markers referring to variables will be replaced as listed below.

- A variable pointing to the instance of a picked class.
- A property of the chosen class and its parent.
- A member function of the chosen class and its parent. Member functions can be seen as variables in AS under specific circumstances. Thus, they can be incorporated into grammar structures, such as assignment expressions.
- A property or member function of the chosen class in prototype form. This is added, since prototype is a way to inherit for this object-oriented program. We expect that it will lead to more interactions between different classes.

Under special circumstances, only pure variables are permitted (variables that are instances of classes), such as "_varname_++", "--_varname_". These situations are identified and replaced with only predefined instances of the chosen classes.

The function-call mutation is similar to the variable mutation. All the markers pointing to function-calls with member functions of chosen classes and their parents are replaced appropriately.

Other than the runtime class mutations, grammar structures in the marked template file also needs manual fixes. Some are fixed by altering G2 in Section 2.2, such as "continue" clauses. To accomplish this, we maintain a label pool to record recently declared labels. Once "continue" clauses are encountered that request for labels, the labels are substituted and appended. Although this process is not foolproof, it is necessary to enable a substantial portion of the code to compile. Other fixes on grammar structures are too trivial to be listed here and can be found on our website [11].

When all the variables and function-calls used in the code snippets have been declared, the instruction flow is altered to be logical and grammar errors are fixed. The "_functioncall_" markers are replaced by calls to member functions that *ScriptGene* has generated and mutated. Then, all the classes from the global pool are initialized and every function is called to enable our generated code to run in the target AVM. This completes the generation of compilable AS source code.

3. EVALUATION

Two experiments were conducted with *ScriptGene*. The first evaluates different generation and mutation strategies: breadth-first or depth-first during code generation phase and multiple or single templates during runtime class mutation phase.

The second compares our code coverage with the original Tamarin test suite. The results show that *ScriptGene* achieves a much better code coverage than Tamarin in all three versions of AVMs tested. A brief analysis of typical bugs found during the second experiment and details about how they were found will be given.

3.1 Testing conditions

The testing process consists of three parts: a generation phase, a runtime class mutation phase and an execution phase. The first part typically requires about half a day on a PC (equipped with an Intel Core Q9400 with 4GB of DDR3 RAM), to generate about ten thousand source code files with distinct sub-rule initializations and transform them to marked template files with lexical analysis. The mutation phase is then given a week of computation time. Millions of SWF files are produced and stored during this phase. Finally, all the SWF files are executed in the AVM.

To sequentially send test cases to the AVM via Adobe

Flash ActiveX Control, a wrapper was constructed and employed. When bugs occur that result in non-severe memory leaks, the AVM may not crash directly or immediately. Therefore, all test cases are sent by the wrapper in the same process, which amplifies memory leak effects on the AVM by accumulating multiple persistent memory leaks. This can either directly cause or hasten AVM crashes. Each SWF file is allowed at most one second of execution time before they are freed from memory.

Since we lack the source code of the AVMs tested, we resort to block coverage instead of line coverage to evaluate our fuzzing performance. To record the block coverage, we refer to the idea of PaiMei [24] and use PIDA to analyze the Adobe Flash ActiveX Control by recording the start and end addresses of each block inside. Basic block coverage is the same as statement coverage, except that the unit of code measured is each sequence of non-branching statements. A block is a basic element of the disassembled code in IDA Pro [8]. Once the start address is covered, all the assembled code in this block will be executed.

The debugger we have built is based on Pydbg (a subproject of PaiMei), to record block coverages and memory corruptions caused by the test cases. Once the wrapper begins loading the test cases, the debugger is attached and the breakpoints on basic blocks over Adobe Flash ActiveX Control are set. If any of the breakpoints are triggered, it is removed from the list for the remainder of the test. Subsequently, all the breakpoints that have been triggered (to record the total block coverage) during the execution phase are collected. If the AVM crashes due to some test cases, the debugger will record the current crash information and the test case file for further analysis.

There are two additional reasons for running the test cases in a single process. First, setting breakpoints on the wrapper requires no more than a few minutes, but are only done whenever the wrapper is restarted (e.g., such as due to crash-related wrapper terminations). Running the test cases in a single process instead of a new process for each test cases saves time spent on setting breakpoints. Second, in a multiple processes approach, it is possible to miss bugs that are only triggered together by multiple test cases. The root causes of these types of bugs are hard to locate in a multiple-test case, multi-processes scenario.

All the experiments are done through three versions of AVMs. Since many features of AS3 are not available until Adobe Flash 10, our first choice is version 10.0.45.2. This is the very first version of Adobe Flash 10. The second version, version 10.2.152.32 is chosen from the period when AVM's vulnerabilities began to attract security engineers' attention, around the time of CVE-2010-1297 [20]. The last version is the newest version available during our tests, version 11.3.300.265.

3.2 Comparison of code generation and mutation methodology combinations

In total, there are four code generation and mutation methodologies: breadth-first or depth-first methods for code generation, single or multiple templates methods for runtime class mutations. The first two methodologies that govern node-order expansions are given in Section 2.3.

The other methodologies regard to varying initial templates by choosing single or multiple templates during the mutation phase. For the single template methodology, an initialized selection of a single marked template file is used for the mutation phase. For the multiple templates methodology, each batch of mutations is done on a candidate selection, chosen randomly from all available marked template files. The multiple templates methodology is capable of more initialization choices of sub-rule selections when compared to the single template methodology. Four unique combinations of these methodologies are: breadth-first&single template (BS), breadth-first&multiple templates (BM), depth-first&single template (DS) and depth-first&multiple templates (DM).

Version	10.0.45.2	10.2.152.32	11.3.300.265
Blocks	15206	21819	34532
BM	4635(30.48%)	4330(19.85%)	5909(17.11%)
BS	3691(24.47%)	5152(23.61%)	5703(16.52%)
DM	3477(22.87%)	4660(21.36%)	7549(21.86%)
DS	2350(15.45%)	4353(19.95%)	4770(13.81%)

Table 1: Block coverages of different combinations of methodologies. The amount of covered blocks and the percentages of total blocks covered are tabulated.

Each combination is allotted a week of computation time for generation and mutations. The output of each is about $\sim 800,000$ SWF files. Then, an additional one to two weeks is required for execution, when the triggered breakpoints are recorded to calculate block coverage (see table 1).

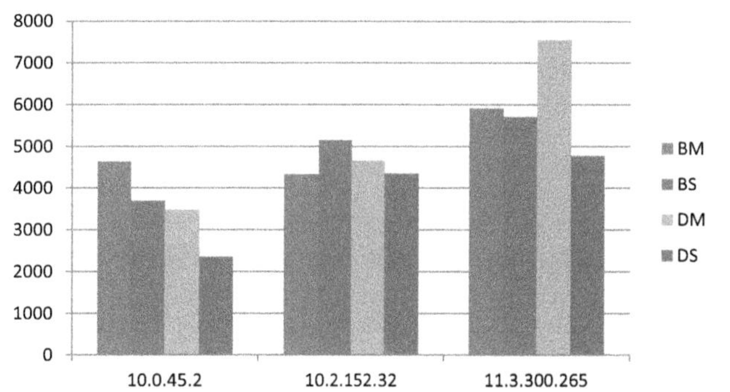

Figure 5: Block coverages under different versions of AVMs.

As seen from figure 5, the best combination in terms of coverage is not obvious, while the worst is the DS combination. During the tests on version 10.2.152.32, some test cases inside the BM combination have led to a memory bug of the AVM. Subsequently, the AVM entered a self-protection state that prevents further execution of a variety of SWF files. This explains the sudden fall of the BM combination's block coverage. For version 11.3.300.265, the DM combination has a greater increase in coverage as compared to the other combinations and the former results. This is because BM has randomly picked several marked template files containing more grammar structures than the other combinations. In fact, the DM and BM combinations are the most representative of ideal situations, where the diversity of initial templates is maximized, combined with extensive runtime class mutations. In reality, however, close approximations to this ideal results in a very resource-intensive process. Therefore, prior to fuzz testings, considerations should be made regarding the amount of computational resources available and choose methods accordingly.

These combinations are cross-compared in table 2. Each item contains three parts: blocks in common that have been covered during the tests and the proportion of these blocks to the total blocks. For example, the DM and DS cross-comparison under version 11.3.300.265 includes 4770 blocks that both combinations covered. These 4770 blocks represent 100% of the DS combination's coverage, and only 63.2% of the DM combination's coverage. This suggests that the D-S combination is less useful for testing version 11.3.300.265, as compared with the DM combination.

Version	10.0.45.2	10.2.152.32	11.3.300.265
BM,BS	3681 79.42%,99.73%	4098 94.64%,79.54%	5172 87.53%,90.69%
DM,DS	2329 66.98%,99.11%	3770 80.90%,86.61%	4770 63.19%,100.0%
BM,DM	3454 74.52%,99.34%	3959 91.43%,84.96%	5773 97.70%,76.47%
BS,DS	2322 62.91%,98.81%	4350 84.43%,99.93%	4769 83.62%,99.98%

Table 2: Cross-comparison between different combinations of methodologies.

To summarize, when the testing situation involves limited computational resources and/or time, the DS combination is inappropriate. The multiple templates methodologies seems to be more satisfactory than the single template methodologies. When the template methodology is fixed, the performance of the overall combination seems to depend on whether breadth-first or depth-first algorithms performs better on the particular AVM (e.g., the BM combination is better than the DM combination for version 10.0.45.2, and worse for version 11.3.300.265).

3.3 Comparisons between ScriptGene and Tamarin

The open source AVM implementation Tamarin contains acceptance test cases [3], which can be seen as the official test suite of AVMs and is employed as instances of grammar structures in the generation phase of *ScriptGene*. Block coverage comparisons between the test cases of Tamarin and *ScriptGene* should be a direct proof that *ScriptGene* is an effective approach to test AVMs. Normally, the test suite

Version	10.0.45.2	10.2.152.32	11.3.300.265
S	4667(30.69%)	5727(26.25%)	7697(22.29%)
T	2474(16.27%)	3351(15.36%)	3686(10.67%)
S,T	2408 51.60%,97.33%	3236 56.50%,96.57%	3622 47.06%,98.26%

Table 3: Block coverages and cross-comparison between *ScriptGene* (S) and Tamarin (T).

of Tamarin consists of AS source code and can only be built as ".abc" files that are compatible with the Tamarin Virtual Machine. However, thanks to Tamarin's support, a solution [10] has been created to compile SWF files with Tamarin-AS source code under cygwin [5]. There are 603 SWF files of Tamarin acceptance test cases. Using the same method as in Section 3.2 the block coverages of Tamarin test suite are recorded in the three versions of AVMs. The SWF test

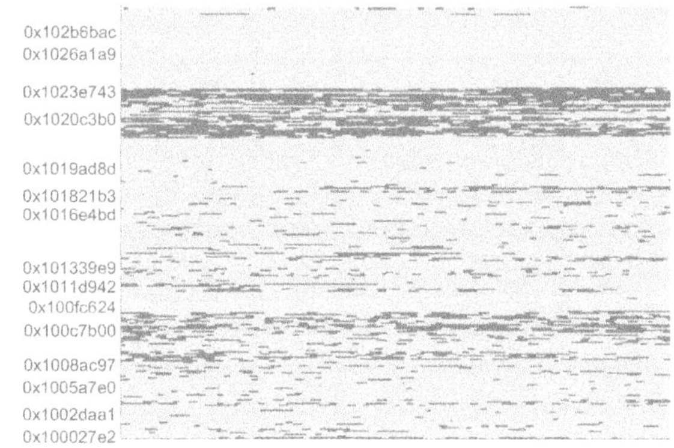

Figure 6: Block coverage of Tamarin.

cases generated by *ScriptGene* are the same as in the first experiment. Triggered blocks are merged among the different combinations to give us a net total block coverage of *ScriptGene*.

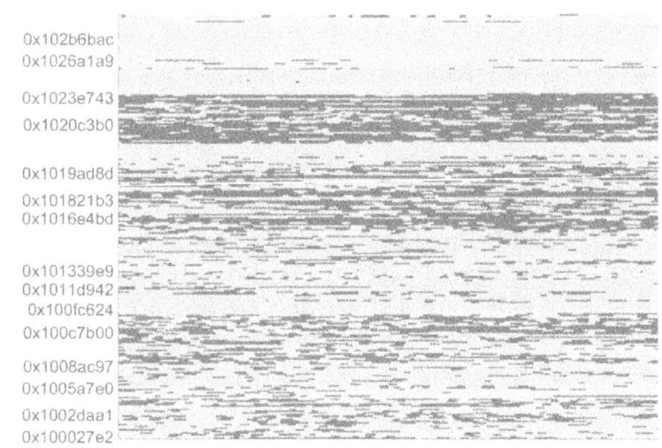

Figure 7: Block coverage of *ScriptGene*.

From table 3, it is seen that *ScriptGene* can cover most blocks (>96%) that are covered by Tamarin test cases in the three versions of AVMs. Vivid squares in figure 6 and figure 7 are used to demonstrate block coverages. Red strips represent covered blocks and dark yellow represents untouched blocks.

All the blocks are arranged according to their real memory addresses and lengths. The base address of Flash ActiveX DLL is 0x10000000, located at the bottom of the figures. Blocks with higher addresses are closer to the top. The figures look similar for the three versions tested. Therefore only one version is shown here. Figures 6,7,8 are the test results of version 10.0.45.2. The highest concentration of blocks covered by the Tamarin test suite lies near the top quarter of the memory space, which mainly deals with grammar and interpretation of the AS code. Blocks in lower addresses are those dealing with the SWF format and external resources of the operating system.

In figure 8, the blocks that was neither triggered by Tamarin nor *ScriptGene* are excluded. Remaining blocks are covered

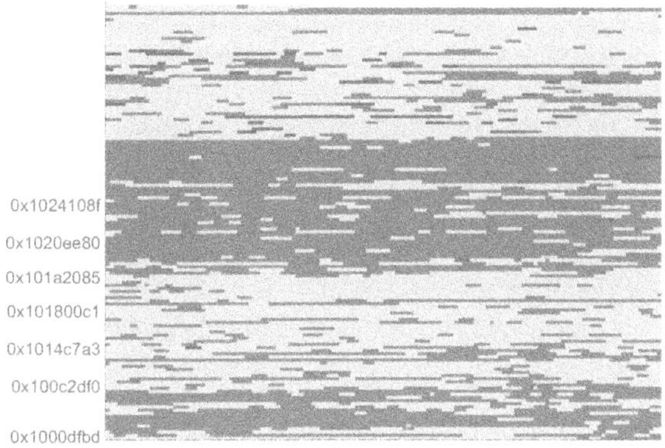

Figure 8: Cross-comparison between *ScriptGene* and Tamarin.

by only Tamarin (in red), only *ScriptGene* (in blue) or both Tamarin and *ScriptGene* (in light green). The improvement of block coverage by *ScriptGene* is seen to be very significant. The complexity of grammar structures produced by *ScriptGene* is certainly not less than that offered by the Tamarin test suite. In addition, with runtime class mutations, *ScriptGene* is able to create more interactions between various structures. We reason that both of these aspects have enabled us to achieve the higher block coverage.

3.4 Examples of bugs

Next, we would like to explain the generation process of the test cases that triggered some AVM bugs. These bugs were found by *ScriptGene* and missed by Tamarin. The original test cases can be found on our website [11]. Because our test cases contain many lines and each line consists of multiple structures of AS instructions, it is not immediately obvious which line or structure triggered a specific bug. Thus, for the following passages of this section, only the essential parts pertaining to each bug are maintained and simplified. *ScriptGene* found four bugs in version 10.0.45.2, one in version 10.2.152.32 and one in version 11.3.300.265. Although these bugs do not exist in the newest version of AVM at the time of writing, we have not yet found any published documentations on them. It is possible that Adobe adjusted the code of newer versions of AVMs, without reporting these bugs. Of the six bugs, three will be briefly detailed here, to illustrate the effectiveness of our strategies in code generation, runtime class mutation and testing.

3.4.1 Complex grammar structure generation

```
1: for each(var o in new <Object>
   [1,2,3,"hello",'out',"there",true,false,3.14159])
   {str+=o;}
2: for each(var _vardeclarationname_ in new<_varname_>
   [1,2,3,"hello",'out',"there",true,false,3.14159])
   {_varname_+=_varname_;}
3: for each(var var2:Number in new <var2>
   [1,2,3,"hello",'out',"there",true,false,3.14159])
   {var2+=var2;}
```

A bug leading to memory corruption is caused by malformed enumerations against the *vector* type. A *number* variable is made to point to a *string* under such code grammar, which leads to the AVM accessing unexpected memory. The ability to find this bug depends on the generation of valid code that are sufficiently grammatically-complex. As an example of the process of how this type of code snippet is generated, consider the following. First, the code generation phase outputs an AS code file containing a "for-loop" structure. Second, lexical analysis replaces the identifier with markers. Finally, in the mutation phase, "_vardeclarationname_" will be replaced with "varX:vartype" (where "X" is a number depending on the order of this pure variable and "vartype" will be chosen randomly from runtime class). "_varname_" will be replaced by a random pick from the variable pool (for instance, "varX").

3.4.2 Mutate with prototype

```
1: switch(actualmatch=string.match(pattern))
2: switch(_varname_=_varname_._funcall_(_varname_))
3: switch(
   var3.constructor=var3.append(
   flash.geom.Matrix3D.prototype.append
   ))
```

Member function *append* could cause memory corruptions when its parameter contains the prototype of this class. During the code generation phase of *ScriptGene*, a "switch" structure is built. Then, the identifier is marked and replaced by member functions, properties and instances of "flash.geom.Matrix3D" during runtime class mutations. This definitively supports the idea about bringing prototypes into the source code files, in order to allow more interactions between different classes as mentioned in Section 2.5.2.

3.4.3 Running tests cases in a single process

Another bug is found due to the method used to run the test cases. As mentioned before in Section 3.1, the wrapper of Flash ActiveX control loads the test cases sequentially in a single process.

In version 10.0.45.2, it was found that many access violations are triggered during the testing phase. Our debugger records the files that have caused the access violations. Attempts to reproduce the corruptions by sending the recorded test cases directly into the AVM results in no access violations. This implies that, some test cases prior to the recorded ones are probably also responsible for the access violations. Typically, this is around 2000-5000 test cases prior to the recorded test case. After several attempts to reproduce the corruptions, it is found that only the quantity of the test cases is important. If ~(30-50) test cases are loaded simultaneously with none being freed from memory, the corruption will be reproduced, irrespective of the content in the test cases.

Tests on version 11.3.300.265 found another bug similar to the bug of version 10.0.45.2. However, for this version, the required test cases increased to 40000-120000 to trigger the corruption. Our attempts to reproduce and analyze the bug with the same method used on version 10.0.45.2 was not fruitful, since the newer AVM takes counter-measures against unbounded memory allocations. When the AVM detects that the following operations can potentially exhaust the memory, it terminates the AVM immediately and produce a big exclamation mark in the Flash GUI. This protection mechanism complicates the simplification and distillation of the test cases. Thus, we are uncertain if these access violations are due to the same reason as in version 10.0.45.2. It is possible that these memory leak problems have existed in Flash for many versions. Each subsequent revision reduced the effect of memory leaks somewhat, possibly through counter-measures, such that newer versions gradually required more test cases to trigger the memory leaks.

4. RELATED WORK

Fuzz testing was introduced in 1972 by Purdom [25]. Purdom used a syntax-directed method to generate test sentences for a parser. He gave an efficient algorithm for generating short sentences from a context-free grammar such that each production of the grammar was used at least once and tested LR(1) parsers using this technique. It is one of the first attempts to automatically test a parser using its grammatical structure. In 1990, Miller et al. [21] were among the first to apply fuzz testing to real world applications. In their study, the authors used randomly generated program inputs to test various UNIX utilities. Since then, the technique of fuzz testing has been used in many different areas, such as protocol testing [13, 26], file format testing [28, 29], or mutations of valid inputs [22, 29].

The most relevant studies for this paper are recent ones on grammar-based fuzzing and test generations for compilers and interpreters. In 2011, Yang et al. [32] presented *CSmith*, a language-specific fuzzer operating on the C programming language grammar. *CSmith* is a pure generator-based fuzzer, generating C programs for testing compilers and is based on earlier works of the same authors and on the random C program generator published by Turner [30]. Drawing a parallel to our work, they have used the built-in grammar to create compilable programs. Furthermore, they introduced semantic rules during their generation process by using filter functions, which allow or disallow certain productions depending on the context. In contrast to our work, *ScriptGene* takes less control over the instruction flow (we only restrict the behaviour of breaks and continues) during the generation phase, and leave the compiler-pass task to the runtime class mutation phase. This approach allows us to generate a greater diversity of grammar structures during the generation phase.

In 2012, Holler at al. [19] proposed *LangFuzz*, a grammar-based fuzzing framework that has been proven to be versatile in discovering vulnerabilities in JavaScript engines. We especially adapted *LangFuzz*'s idea of the "Shortest Terminal String Algorithm" for *ScriptGene*. Similar to *LangFuzz*, we have used grammar instances extracted from a test suite of a certain language to terminate the code generation branch. In contrast, *LangFuzz* mainly deals with JavaScript that requires only line-validity, while *ScriptGene* must maintain full-text validation with the help of modified grammar rules and code templates. Furthermore, *ScriptGene* adds runtime class mutations, legalizes the code and covers more branches in tested AVMs simultaneously.

As shown by Godefroid et al. [17] in 2008, a grammar-based fuzzing framework that produces JavaScript test cases can increase coverage when linked to a constraint solver and coverage measurement tools. They present a dynamic test generation algorithm where symbolic execution directly generates grammar-based constraints whose satisfiability is verified using a custom grammar-based constraint solver. Unfortunately their work in its current form only suits well on JavaScript VMs, since the feedback adjusts the input directly. However, AVMs only accept bytecodes, which are the outputs of the compiler. This additional layer prevents the direct feedback, rendering the symbolic execution ineffective.

Fuzzing web browsers and their components is a promising field, particularly in the case of Adobe Flash, which are also used in some hybrid documents. To date, several approaches have been taken to find bugs/vulnerabilities in Flash.

Flash as a particular format of documents could be fuzzed using a variety of file format fuzzing frameworks such as Peach [9] and SPIKE [14]. One of the methods of file format fuzzing called Dumb Fuzz bitflips every bit in the entire file. Despite its name, Dumb Fuzz was found to be able to test AVMs to a certain extent. Since AS code is stored as bytecodes in a SWF, single bit change in these bytecodes would in general completely alter the original meaning of the AS code. Several vulnerabilities were confirmed to be found by Dumb Fuzz recently [20]. Most of these bugs are due to type confusion. Late in 2011, Google used large scale computing resources to fuzz AVMs, relying solely on Dumb Fuzz and found tens of bugs [18]. Nevertheless, this type of method is mutation based and highly dependant on the source of the parent SWF. The source for Google was 20T-B of downloaded SWF files. While 20TB cannot possibly include all the structures of AS, mutations on the bytecode level is a good way to explore additional code paths of AVMs by bypassing the compiler and alter the inputs directly, when computational resources permit.

There are other fuzzing approaches targeting Flash. However, most of them mainly focus on whether AS functions give enough validation to their parameters. In our previous work [31], we have attempted to fuzz the Regular Expression functions of AS. Using grammar-based mutation fuzzing, we have constructed much more sophisticated expressions to test the Regular Expression interpreter of AVMs, where several vulnerabilities were subsequently found.

5. DISCUSSION

Approximate grammars Other than the additional controls to the AS grammar rules as mentioned in Section 2.2, several other structures have been simplified. Package => Class => Function => Statement is essentially the generated code in a macro-view. Serialization of Package => Function, Class => Expression... is impossible due to the modifications of the grammar rules. Fixed serialization of code structures helps us keep track of the properties and functions and enables us to build them valid contexts. We need more domain knowledge of a particular language and its grammar, if we want to build a valid context for all complex grammar rules.

Resource limitations For the code generation phase, only several hours was needed to generate around ten thousand nearly-valid code snippets with different initializations of sub-rule selections. During each batch of runtime class mutations, only two of the classes from the pool were chosen. About ten days of mutation time yields only ~10% of the mutations possible under our current strategies. Considering all possible combinations and the depth of the grammar, these code snippets explore only a tiny subset of AVM execution paths, since the mutation phase requires far more time than the generation phase. This is practically inevitable with the blackbox fuzzing method. Additional information from whitebox analysis should be helpful to constrain the possibilities for future researches. Distributed computing environments such as cluster computing or cloud computing would be helpful for larger mutation diversities.

Finding new vulnerabilities We have used the compiler to generate complex bytecodes from AS source code, then execute the resulting files on the AVMs. Our approach is seen to be effective in achieving a good code coverage, as demonstrated in Section 3. However, we have found less bugs than expected, from our comparatively greater code coverage (compared to the Tamarin official test suite). Nevertheless, this is reasonable, since the discovery of bugs from random fuzzing is probabilistic. Evaluating a given fuzzing approach by the amount of bugs found is not appropriate, unless computational resources permit a large and statistically valid amount of runs. Therefore, code coverage and comparisons are better indicators of the effectiveness of *ScriptGene*. The discovery of bugs should be considered secondary to code coverage testing. In our future works, we would like to add bytecode mutations to the test cases produced by *ScriptGene*. Based on current code coverage, we deduce that bytecode mutations will be capable of identifying even more bugs by extending the code coverage. Bytecode mutations would also be a more direct method to test AVMs, since it bypasses the compiler-check.

6. CONCLUSION

ScriptGene is a novel approach to fuzz testing ActionScript Virtual Machines, where compilers are involved. We have extended and expanded the ideas of *LangFuzz* from JavaScript to ActionScript, to generate code snippets, then produced nearly-valid ActionScript code with additional controls. Finally, using runtime class mutations, we produce grammatically-complex compilable code that are rich in runtime class interactions to ultimately test a few AVMs. Our evaluation shows that our approach, *ScriptGene* explores deeper execution paths and is capable of nearly twice the code coverage compared to official tests (Tamarin). Our code coverage and discovery of unreported bugs found by *ScriptGene* in three different versions of AVMs demonstrate the effectiveness, validity and novelty of our approach.

7. ACKNOWLEDGMENTS

We thank Steven Ergong Zhang from University of Ottawa for his support and suggestions. This work is supported by National Natural Science Foundation of China (NSFC) under Grant 60970140, Beijing Natural Science Foundation (4122089) and China Postdoctoral Science Foundation Project (2011M500416, 2012T50152).

8. REFERENCES

[1] http://help.adobe.com/en_US/FlashPlatform/reference/actionscript/3/flash/utils/package.html#describeType().

[2] About flex sdk. http://sourceforge.net/adobe/flexsdk/wiki/About/.

[3] Actionscript acceptance tests. https://developer.mozilla.org/en-US/docs/Tamarin/Tamarin_Acceptance_Testing/Actionscript_Acceptance_Tests.

[4] Adobe flash professional cs6. http://www.adobe.com/en/products/flash.html.

[5] Cygwin. http://www.cygwin.com.

[6] Ecmascript. http://www.adobe.com/devnet/actionscript/articles/actionscript3_overview.html.

[7] Flex formatter. http://flexformatter.cvs.sourceforge.net/viewvc/flexformatter/ActionscriptInfoCollector/ASCollector.g3?view=log.

[8] Ida:about. http://www.hex-rays.com/products/ida/index.shtml.

[9] Peach fuzzing platform. http://peachfuzz.sourceforge.net/.

[10] Running tamarin acceptance tests. https://developer.mozilla.org/en-US/docs/Tamarin/Tamarin_Acceptance_Testing/Running_Tamarin_acceptance_tests.

[11] Scriptgene project. http://www.nipc.org.cn/project/ScriptGene.

[12] Tamarin. https://developer.mozilla.org/en-US/docs/Tamarin.

[13] D. Aitel. The advantages of block-based protocol analysis for security testing. *Immunity Inc., February*, 2002.

[14] D. Aitel. An introduction to spike, the fuzzer creation kit. *immunity inc. white paper*, 2004.

[15] B. Binde, R. McRee, and T. OąŕConnor. Assessing outbound traffic to uncover advanced persistent threat. *SANS Institute. Whitepaper*, 2011.

[16] D. Blazakis. Interpreter exploitation. In *Proceedings of the USENIX Workshop on Offensive Technologies*, 2010.

[17] P. Godefroid, A. Kiezun, and M. Levin. Grammar-based whitebox fuzzing. In *ACM SIGPLAN Notices*, volume 43, pages 206–215. ACM, 2008.

[18] Google. Fuzzing at scale, 2011. http://googleonlinesecurity.blogspot.nl/2011/08/fuzzing-at-scale.html.

[19] C. Holler, K. Herzig, and A. Zeller. Fuzzing with code fragments. In *Proceedings of the 21st USENIX conference on Security symposium*, Security'12, pages 38–38, Berkeley, CA, USA, 2012. USENIX Association.

[20] H. Li. Understanding and exploiting flash actionscript vulnerabilities, 2011.

[21] B. Miller, L. Fredriksen, and B. So. An empirical study of the reliability of unix utilities. *Communications of the ACM*, 33(12):32–44, 1990.

[22] P. Oehlert. Violating assumptions with fuzzing. *Security & Privacy, IEEE*, 3(2):58–62, 2005.

[23] T. Parr and R. Quong. Antlr: A predicated-ll (k) parser generator. *Software: Practice and Experience*, 25(7):789–810, 1995.

[24] pedramamini. Paimei. http://pedramamini.com/PaiMei/docs/.

[25] P. Purdom. A sentence generator for testing parsers. *BIT Numerical Mathematics*, 12(3):366–375, 1972.

[26] G. Shu, Y. Hsu, and D. Lee. Detecting communication protocol security flaws by formal fuzz testing and machine learning. *Formal Techniques for Networked*

and Distributed Systems–FORTE 2008, pages 299–304, 2008.

[27] J. Smith and R. Nair. *Virtual machines: versatile platforms for systems and processes.* Morgan Kaufmann, 2005.

[28] M. Sutton and A. Greene. The art of file format fuzzing. In *Blackhat USA Conference*, 2005.

[29] M. Sutton, A. Greene, and P. Amini. *Fuzzing: brute force vulnerabilty discovery.* Addison-Wesley Professional, 2007.

[30] B. TURNER. Random c program generator, 2007. `http://sites.google.com/site/brturn2/randomcprogramgenerator`.

[31] D. Yang, Y. Zhang, and Q. Liu. Blendfuzz: A model-based framework for fuzz testing programs with grammatical inputs. In *Trust, Security and Privacy in Computing and Communications (TrustCom), 2012 IEEE 11th International Conference on*, pages 1070–1076. IEEE, 2012.

[32] X. Yang, Y. Chen, E. Eide, and J. Regehr. Finding and understanding bugs in c compilers. *ACM SIGPLAN Notices*, 47(6):283–294, 2012.

Sensing-Enabled Channels for Hard-to-Detect Command and Control of Mobile Devices

Ragib Hasan
University of Alabama at Birmingham
Birmingham, Alabama 35294-1170
ragib@cis.uab.edu

Nitesh Saxena
University of Alabama at Birmingham
Birmingham, Alabama 35294-1170
saxena@cis.uab.edu

Tzipora Halevi
Polytechnic Institute of NYU
Brooklyn, NY 11201
tzipihalevi@yahoo.com

Shams Zawoad
University of Alabama at Birmingham
Birmingham, Alabama 5294-1170
zawoad@cis.uab.edu

Dustin Rinehart
University of Alabama at Birmingham
Birmingham, Alabama 35294-1170
drnehart@cis.uab.edu

ABSTRACT

The proliferation of mobile computing devices has enabled immense opportunities for everyday users. At the same time, however, this has opened up new, and perhaps more severe, possibilities for attacks. In this paper, we explore a novel generation of mobile malware that exploits the rich variety of sensors available on current mobile devices.

Two properties distinguish the proposed malware from the existing state-of-the-art. First, in addition to the misuse of the various traditional services available on modern mobile devices, this malware can be used for the purpose of targeted context-aware attacks. Second, this malware can be commanded and controlled over context-aware, out-of-band channels as opposed to a centralized infrastructure. These communication channels can be used to *quickly reach out to a large number of infected devices*, while offering a high degree of *undetectability*. In particular, unlike traditional network-based communication, the proposed sensing-enabled channels cannot be detected by monitoring the cellular or wireless communication networks. To demonstrate the feasibility of our proposed attack, we present different flavors of command and control channels based on acoustic, visual, magnetic and vibrational signaling. We further build and test a proof-of-concept Android application implementing many such channels.

Categories and Subject Descriptors

K.6.5 [**Security and Protection**]: Unauthorized access—*Malware* ; C.5.3 [**Microcomputers**]: Portable devices

Keywords

Mobile security, Mobile Malware, Command & Control, Mobile device sensors, Covert channel

ASIA CCS'13, May 8–10, 2013, Hangzhou, China.

1. INTRODUCTION

The mobile devices have become both "smart" as well as ubiquitous in the recent years. Today's mobile devices, such as smart phones or tablets, are equipped with a multitude of sensors, enabling them to detect their location, and learn the characteristics of their users and the surrounding environment. These rich capabilities have enabled many interesting applications and immense possibilities for ordinary users. However, at the same time, they have opened up the door towards new generation of mobile malware that can exploit the on-board sensors for malicious purposes. Ranging from eavesdropping over the phone call or user input [10,40,44], learning user's location [16] to snooping on the user's activities [34], mobile malware can gather sensitive information previously not available to traditional malware.

In this paper, we argue that the sensors present in mobile devices can also be used for out-of-band communication among malware infected devices as well as for targeted command and control. For example, the audio sensors present in a mobile phone can be used to trigger malware located in a specific physical region. Malware can also be triggered or commanded via audio/visual signaling transmitted through television or radio broadcasts. While triggering is currently possible via the cellular or wireless network channels, such messages can easily be detected by monitoring these communication channels either at the mobile phone or at the network gateways. Unlike the traditional command and control communication over a centralized infrastructure (such as a cellular network), out-of-band communication is *very hard to detect* and even harder to prevent. However, it can be still be used to *reach out to a large population* of infected mobile bots.

In addition to the misuse of the various traditional services available on modern mobile devices (such as phone calls or SMS/MMS), we posit that this malware can be used for the purpose of targeted context-aware attacks. For example, a malware that gets triggered in a movie theatre, via say a hidden audio signal embedded in a commercial, can be used for causing annoyance or even chaos; imagine, for instance, the infected devices in the theatre all playing a loud song or a siren suddenly.

Researchers have previously explored various forms of mobile malware and different means of malware communication channels including their detection and blocking. In particular, most of the work has focused on wireless or wired network based command

and control channels over the Internet, wireless, or cellular phone networks. The threat posed by the situational and environmental awareness of mobile malware and the use of the sensors to perform *hard-to-detect out-of-band communication* has not been studied.

1.1 Our Contributions

The contribution of this paper is threefold. *First*, we provide the first detailed study of environmental sensor-based covert channels in mobile malware. In particular, we present different flavors of out-of-band command and control channels based on acoustic, light, magnetic and vibrational signaling. Many of the proposed channels provide a means of *undetectable communication with a large number of malware infected devices* such as through the use of broadcast video or audio signals.

Second, to demonstrate the feasibility of our attacks, we build a proof-of-concept malware application using an off-the-shelf mobile phone on the Android platform implementing many of the proposed channels. We conduct several experiments to validate the effectiveness of these channels for command and control. Some of our experiments are conducted in real-life setting and further confirm the threat posed by the presented mobile malware.

Third, we sketch the possibility of building geographically localized attacks which can leverage upon the aforementioned out-of-band channels.

1.2 Scope and Ethical Aspects

The goal of this paper is to demonstrate the feasibility of out-of-band channels for command and control of the malware. As such, the scope of the paper is limited to the exploration, design and analysis of such channels. In particular, the emphasis is on sending undetectable triggers to infected devices over such channels. Developing models, either theoretical or experimental in nature, to estimate the latency and coverage of such channels (e.g., to determine how many infected devices can get triggered in a given time span) is beyond the scope of this work. However, we emphasize that many of our channels naturally provide rapid message delivery guarantees to a huge number of devices.

Although we are presenting essentially a new generation of attack against mobile devices, the purpose of this work is ethically sound and constructive. By pre-empting the design of this attack and possibly "staying ahead in the game" against the real attackers, our vision is to eventually come up with an effective defense against the envisioned attack. Due to the high level of incentive available to the attacker (high degree of undetectability), these attacks might very well be launched in the wild in the near future. By means of a publication on this topic, we hope to raise awareness about new threats, and motivate fellow researchers, device manufacturers and OS designers to build and deploy defenses before these attacks are launched in the wild. In fact, we also discuss potential approaches to defend against such malware, and aim to further extend them in the near future. Many recently published papers on precisely the same broad topic [8,13,19,26,37,42,46,49,50] (reviewed in Section 7) further support and justify this line of security research. Additionally, we are positive that the underlying communication channels designed as part of our research will come handy in other security applications in the future.

2. BACKGROUND AND THREAT MODEL

In this section, we discuss the motivation for out-of-band command and control, and the underlying threat model, and sketch the possibilities for localized context-aware attacks enabled by out-of-band channels.

2.1 Motivation: Why Use Sensors?

Many of today's mobile devices such as smart phones are equipped with optical, audio, vibration, and magnetic field sensors. Some of these sensors, such as accelerometers, are so sensitive that they have been repurposed for distributed sensing applications such as earthquake detection [11]. The optical sensor and the camera present in mobile phones are also becoming more sophisticated day by day. For example, the Apple iPhone 4GS contains an 8 mega pixel digital camera. Similarly, the microphones present in mobile phones are sensitive enough to pick up very subtle sounds from the surrounding environment. For instance, the iPhone 3g's built-in microphone can detect sound as low as 5 Hz to as high as 20 KHz [2]. Many devices besides smart phones, such as tablets and laptops, also come equipped with many of these sensors.

The sensors can serve as an appealing platform for out-of-band communication among malware infected devices as well as between the botmaster owner and infected devices. Unlike the traditional centralized means of communication for malicious purposes, such an out-of-band communication can remain *very hard to detect*, especially if covert and steganographic communication channels are used. However, it can still be used to communicate with a large number of devices, which can even span international borders. For example, messages sent embedded within the audio of a popular TV program can be delivered to a huge number of infected devices, whose users would be watching such a program.

Easy Detectability of Network-based Channels: For the botnet communication/triggering described above, a traditional network-based channel, e.g., a TCP/IP based channel, is easily detectable. TCP/IP-based triggers can be detected and/or blocked by a firewall or anti-malware software which monitors network packets. Applications constantly polling incoming packets, registering for push notifications, or accessing unknown web services can raise suspicion. Even triggers steganographically hidden in benign communication can still cause suspicion: why would an app run even benign protocols with random servers (other than known backends), assuming the botnet controller has not compromised legitimate servers?

Unique Advantages of Sensor-based Channels: The detection of out-of-band signals is complicated by the fact that the out-of-band trigger signal format can be free form. Traditional botnet command and control messages, on the other hand, travel over centralized networks obeying established protocols such as UDP or TCP/IP. However, the out-of-band covert channels can use arbitrary protocols to send the control and command messages. This makes the detection of such communication quite difficult in practice. Anti-malware software also does not know which medium is used by the botnet for covert communications: it must therefore monitor all sensors constantly. Also, accessing a sensor is not necessarily the "signature" of a malware: many legitimate applications need access to sensors for benign purposes.

Sensor-based trigger channels have other useful properties. For localized attacks, sensor channels are more effective than TCP/IP. In an area, not all phones may use the same network or have network connectivity, but most phones there share the same medium. Also, without knowing infected devices' IP addresses, the botnet controller has no way to trigger them (other than broadcasting/flooding entire network, which is infeasible), whereas in sensor-based channels, the controller does not have to know any addresses of infected devices.

2.2 System Model and Assumptions

Our system model is *no different* from the model employed in traditional command and control of malware. Namely, we assume that many mobile devices have already been corrupted with malware. Such corruption could take place, for example, when the user downloads an untrusted application – embedding the malware – from the application store of the service being used. However, to remain surreptitious, the malware on these devices will not activate or get triggered until indicated by the botmaster. These triggers will later be sent by the botmaster to (all or a subset of) infected devices over out-of-band channels. Such channels can also be used for the purpose of sending commands to the bots. However, in this paper, our primary emphasis is on triggers.

We assume that the (malware) application is allowed to run in the background and can access the device's on-board sensors without restriction. Indeed, as our prototype implementation shows, the Android platform supports such apps when using a microphone, a light sensor, and a magnetometer, as well as an accelerometer. Other operating systems, such as iOS, may not conform to this policy. However, it is also possible for the malware to attach itself to a benign app that needs constant access to sensors as a background process (e.g., a web search application that uses voice input).

Additionally, we assume that, with a very high probability, the mobile devices are switched on and are in close physical proximity of their users, either carried by the users in their pockets, purses or backpacks, or lying close to them. For example, while watching TV, a user's phone is placed next to him/her. Given that modern users heavily rely upon their mobile devices (especially phones), this is a valid assumption to make [17, 18, 24, 25, 29, 31, 33, 41, 45, 47, 48]. Under a rare circumstance that a phone is powered off or not close to the user, the messages can not delivered to that particular phone at that particular time. This would only degrade the overall reachability of the messages slightly.

Once activated by a trigger, the infected devices will carry out the attacks they are programmed for. Traditionally, these attacks could be used for the purpose of spam campaign, making illicit phone calls or sending SMS/MMS messages. In addition to these attacks, we envision another genre of attacks, which we discuss in the following subsection, that can be used for targeted and localized threats.

2.3 Localized Targeted Attacks

A localized mobile botnet consists of malware infected mobile devices physically present in a specific location. An attacker can launch localized attacks on or through these devices. For example, during a sports event, an attacker can trigger the malware-infected mobile devices in the sports arena. To do so, the attacker issues command and control messages broadcast via out-of-bound channels.

We outline following types of active context-aware attacks. Validating the feasibility of some of these attacks is beyond the scope of this paper, however.

Distributed Denial-of-Service Attacks: The infected mobile devices could be used to launch localized denial-of-service attacks on a certain network. For example, the infected devices present in an airport may be "commanded" to collectively bring down the airport's WiFi network.

Annoyance Attacks: The malware on the devices could be used to cause annoyance or even chaos in a public place. E.g., in a movie theater, an advertisement may be used to send covert triggers to all infected phones present in the theater; and then the malware on one phone may interact with other phones (say make phone calls to one another). The malware may also collectively play some loud music or sirens. Given that many people do not "silence" their phones in a theatre, such an attack can certainly cause a lot of annoyance.

Embarrassment Attacks: Selectively triggered malware could be used to cause embarrassment to the device's user or those present nearby. For instance, a person may be using her (infected) phone to project a presentation in a conference. As the person starts to speak, another infected phone in the room can trigger the phone malware (via some out-of-band channel), which would then project an embarrassing video onto the screen.

Safety Hazards: Public safety hazards are also within the purview of context-aware malware. As an example, the malware on the phone can be triggered when the infected phone is inside a driving car; the malware may then interact with the car's internal network and cause some serious problems. Similarly, a malware may get triggered inside a home/company and may then interfere with the home's wireless security system – perhaps dismantle it. This will clearly prompt the possibility of theft or burglary and may endanger the lives of the inhabitants.

Interference Attacks: Context-aware malware may be used to cause interference with the surrounding environment. For example, the infected mobile devices can selectively interfere with an aircraft radio system at the time of take-off or landing, or with the medical devices in a hospital [9, 30].

Distraction Attacks: Another viable and perhaps very interesting attack would be a "distraction" attack. Here, the malware aboard a user's mobile phone tries to distract the user while she is performing a security task (e.g., reading a security warning; pairing her devices, etc.). For example, the phone may play a ring tone or vibrate to distract the user. It would be interesting to see how users fair under such a distractive attack.

3. OUT-OF-BAND COMMAND AND CONTROL OVERVIEW

In this section, we provide an overview of different out-of-band channels using which a botmaster can trigger the infected devices. A detailed description of these channels and their various characteristics appears in the following section.

We divide these channels into two categories, one of which explicitly uses steganographic techniques to make the task of detection extremely difficult. The receiving side of these channels can be built using the following sensors on mobile devices:

- Audio sensor / microphone
- Camera / light sensor
- Magnetometer
- Accelerometer / vibration sensor

3.1 Steganographic Channels

A steganographic channel is one where the trigger signal is hidden inside another signal. This channel would usually involve the trigger being embedded inside a song or other audio. Trigger signal broadcasts using such a channel can be achieved either via audio signal embedding or live stream embedding, as discussed below.

Audio Signal Embedding: In this variant, the audio signal is embedded inside a recorded carrier audio which is later broadcast using TV or radio. Here, the attacker does not need to be physically present near the target. The trigger is embedded in an innocuous video or audio (e.g., a song or a TV program). When the program is broadcast through television or radio, any malware infected device in the physical proximity of a running television or radio will receive the signal. Other mechanisms, such as embedding the trigger signal inside music played by musical greeting cards, can also be used to spread the trigger.

Live Stream Embedding: In this variant, the carrier signal or audio, containing the trigger signal, is broadcast or played directly near the target mobile devices. The trigger signal can also be embedded in broadcast video or audio signals in a different manner. During live telecast of an event (such as a game or a speech), the attacker can play the trigger-embedded audio near the microphone used for the broadcast. The microphone can pick up the infected audio stream and thereby broadcast it to a large audience.

The attacker can also tap into the broadcast workplace music delivery networks and embed the trigger into the music. For example, background music services such as Muzak [38] deliver music to workplaces, hospitals and elevators, which may be leveraged for this purpose.

3.2 Non-Steganographic Channels

A non-steganographic triggering channel is one where the trigger signal is not hidden per se, but rather it is delivered directly.

Audio Patterns: The malware can be programmed to be triggered by a specific audio pattern (e.g., a song).

Ambient Light: The attacker can tap into the power supply of a building and cause rapid fluctuations in voltage, resulting in rapid but imperceptible flickers in the lights all across the building. A trigger message embedded via such flickers can be read by any infected phone in the building.

Magnetic Signalling: The attacker can induce a strong enough magnetic field and send the trigger by changing the strength of the field. Using this scheme, the attacker can hide a magnetic field inducing device in a crowded area, as an example, and trigger the devices whose users pass by the magnet.

Vibrational Signalling: Attackers can also use vibration channels. For example, messages can be encoded into vibrations produced by a subwoofer which can be read by nearby phone accelerometers.

4. CHANNEL CHARACTERISTICS

For transmitting command and control messages to mobile devices, we examined the properties of audio, light, magnetic field and vibrational channels overviewed in the previous section. In this section, we discuss the various properties of each of these channels including: *range*, *noise characteristics*, *adversarial control*, *coverage and reachability*, *latency*, *bandwidth*, and *steganographic capabilities*. A summary of this analysis is depicted in Table 1.

4.1 Audio Channel

An audio channel is composed of an audio signal, which is used to encode the trigger message. An audio channel can be steganographic or non-steganographic, as mentioned before. In the steganographic mode, the trigger signal is embedded in a song or other music/audio signal. In the non-steganographic mode, the audio signal is used directly.

The attacker can use a TV or a radio program, background music services, an internet TV program and musical greeting cards to transport its commands over the audio channels. When using a TV or radio program, the attacker would either need to manipulate an existing program (such as a popular show or a commercial) or insert a live audio stream at the recording of a live event (such as a sporting event). While the former capability may require the attacker to collude with an insider at the service itself, the latter capability only requires the attacker to hide a audio transmitter near the recording station. The attacker may also register its own commercial with the service within which it can insert the audio commands. Although this incurs a cost to the attacker, it offers the advantage of covert communication that can be broadcast to a large audience, possibly multiple times, in a day. Similarly, when using a musical service, the attacker would need to tamper with an ongoing music broadcast. This may again require collusion with an insider at the music service or hack into the inside network of the service.

In contrast to the traditional TV, radio or musical services, internet TV – especially P2P-TV [4, 5] – appears to be more lucrative for the attacker. Using such a channel, the attacker can simply send out its commands embedded within its *own program* (i.e., for which the attacker itself is the source of the streaming data). Since P2P-TV is completely decentralized and any user can become the source/receiver of a program, the attacker can very easily communicate with the infected devices through this mechanism. It is also possible for the attacker to insert its commands within an ongoing (popular) program as demonstrated in an attack [12] on a popular Chinese live streaming program called PPLive [4].

All of the above approaches provide broad coverage and reachability. The TV or radio programs can be used to communicate with all infected devices whose users are tuned to these programs (assuming that the devices are close to their users). The same applies to the Internet TV programs; here the users would be watching the programs on their computers. When using popular programs or live events in each of these settings, the attacker can ensure that its commands can be delivered to an extremely large number of infected devices across the globe. In the case of the workplace music setting, the commands can be received by the infected devices of all users present in the given building (including visitors). This may cover an enormous number of devices especially when a big hotel, library, or an airport, or perhaps many of them, is/are used as the attack target. With musical greeting cards, the commands can be delivered to any infected device whose user opens up or plays the cards. The reachability of this attack clearly depends upon the number of cards dispatched by the attacker. However, one can imagine a broad coverage especially during peak holiday seasons when users actively select/test greeting cards (and play them) at local stores. Those receiving these cards as gifts from others are also reachable because they will also be playing the music.

Another important measure of the effectiveness of command and control channels is their latency, i.e., the delay incurred in delivering the messages to infected devices. All of our settings introduce little latency and the delivery of messages can take place as soon as the users tune to the given program, or are present in a given building. Thus, the delay is only up to the broadcast of the program. With greeting cards, the delay is up to the point users open up and play the cards.

When looking at the audio channel properties themselves, the range and distortion might be two limiting factors. An audio channel must use audio signals that are strong enough to traverse the distance between the trigger transmitter and the mobile device. In addition, background noise often tends to drown the source audio signal. Luckily, in our attack setups, the desired communication range is short, less than a few feet at most. The Internet TV set-up is again the most promising in terms of the range since users' devices are expected to be only a few inches, or about a feet, away from their computers playing the program. Greeting cards also require similar ranges. TV programs, on the other hand, will usually require a communication range of several feet, given that the ideal TV viewing distance is several feet (depending upon the size of the TV screen) [6]. This, however, will be compensated by the powerful speakers of modern TV sets or their associated home theatre systems.

The sensitivity of mobile device microphones can also impact sound reception. Mobile phone microphone are reasonably sensitive (as discussed in Section 2.1), but most mobile operating systems perform noise cancellation of some form at the driver level. Hence, the audio received by a mobile phone may sometimes be not exactly the same frequency as the original sound. The audio channel can also be distorted if the target mobile device is inside pockets or purses.

Mode of Transport	Audio				Light	Magnet	Vibration
	TV / Radio Program	Music Services	P2P-TV Program	Musical Greeting Cards			
Targeted Devices	Any mobile device with a microphone (all phones, many laptops, tablets)			Any mobile device with a microphone and carried by the users (all phones)	Any device with a light sensor or camera (many phones, laptops or tablets)	Any device with a magnetometer (most smartphones)	Any device with a vibration sensor or an accelerometer (most smartphones, many laptops / tablets)
Attacker Control	•Insert audio at the recording of a live event •Manipulate an existing program or commercial •Register its own commercial	Manipulate an ongoing broadcast	• Stream its own program • Manipulate an ongoing program	Dispatch its own greeting cards	• Insert visual fluctuations in an existing TV program or commercial • Registers its own commercial • Tap into the power line	• Hide a magnet in a crowded area • Carry a magnet in a pocket or backpack; walk beside users in a crowded area	• Bring vibrating device close to the users • Induce a high bass sound in a TV or radio program
Coverage and Reachability	All users tuned to the program	All users in the building where the music is played	All users tuned to the program	All users who play /open the cards	• All users tuned to the TV broadcast • All users present in a given building	All users who pass by the magnet	All users who are in physical proximity of the vibration
Latency	Up to the telecast of the program	Up to the broadcast of the music	Up to the telecast of the program	Up to the propagation of the cards	Up to the TV telecast, or flickering of the lights	Up to the users' passing by the magnet	Up to the users' sensing vibrations
Steganographic?	Can be				Can be in the case of a TV broadcast	No, but imperceptible	Can be (imperceptible)
Desired Range	Several feet	Several feet	A feet or so	A feet or so	Several feet	Few centimeters	Several feet with strong subwoofers; else few centimeters
Presence of Noise	Noise is common				Noise not common	Little noise	Noise is common
Works when device is stowed?	Not always				No	Yes	Yes
Expected Bandwidth	Few bits per second						
Peer-based propagation?	No				No	No	Yes

Table 1: Comparison of different out-of-band channels for command and control

Furthermore, audio noise is fairly common and can complicate the task of audio decoding. Noise is quite common in public places.

The bandwidth of the command and control channel may also be important in some scenarios, especially if malware payloads or new malware programs are to be disseminated to the infected devices. The use of out-of-band channels in general, however, exhibits low bandwidths due their fundamental physical characteristics. The audio channel bandwidths are expected to be limited to only a few bits per second [32]. This is an obvious trade-off in using out-of-band command and control in contrast to the traditional channels – the former provide better undetectability while the latter possess better bandwidths. However, even with low bandwidth, the envisioned audio channels can definitely be used to rapidly send triggers to a large number of devices. In addition, to transmit larger messages or other data, the audio signalling could be spanned over the entire program (such as a game or a movie), or multiple programs, or multiple occurrences of the same program.

4.2 Light Channel

The light channel uses the ambient light sensors present in most smart phones and many laptops/tablets. These sensors are very sensitive to the ambient light. The intensity of most light sources depend on the voltage of the power supply. Thus, by changing the voltage, it is possible to modify the ambient light in an indoor location. An attacker using a light channel to broadcast the messages can tap into the power supply of the building and introduce fluctuations, by modifying the voltage or by introducing very short flickers. All rooms in the building will be affected under such an attack. Getting access to the power supply of a building can be a formidable obstacle for the adversary, although such attacks have previously been reported in the wild [36]. However, the payoff in this case is higher: by tapping into a single place, the adversary can cover all areas of a building. Another possibility is to embed the messages within the light variations into an existing or attacker chosen TV program or commercial, similar to some of the audio-based approaches.

Similar to audio channels, light channels also open up the possibility of a broad coverage. When using a TV broadcast, messages can reach all infected devices of users watching the broadcast (assuming the phones are near their users and not stowed inside pockets/purses). This is especially true when large screen TV sets are used such as in a pub or night club.

These channels also incur minor latencies. These are limited to the point the TV broadcast is screened or the lights are flickered in the building. In terms of undetectability, the flickering of the lights can be so quick that they are not perceptible by the humans present in the surroundings. The TV broadcast can be explicitly made steganographic similar in spirit to the audio based stego broadcast.

As far as the range of the light channel is concerned, several feet is easily workable. This clearly covers the indoor environment such as the scenario of watching a TV screen and fluctuating lights mounted on low ceilings.

The light channel, however, exhibits some limitations. First, to be able to receive a message via the light channel, the mobile device's light sensor must be unobstructed. If the mobile phone is kept inside a pocket or purse, the light channel can not be used. However, many computing devices other than mobile phones – such as tablets or laptops – also contain cameras or light sensors that can be used to measure the intensity of ambient light. The light channel can be effectively used on such devices as they are not usually stowed when powered on.

Another challenge with the light channel is the presence of noise. Ambient light sensors usually do not distinguish between light from different sources as long as it falls inside the visible light spectrum. Therefore, the noise in a light channel can be quite high if there are other frequently changing light sources (e.g., multiple TV sets displaying different programs in a pub) in the same indoor space. In contrast to the audio channel, however, the likelihood of the presence of visual background noise is much less, and thus the light channel is expected to be more robust in practice.

The light channel is expected to provide similar levels of bit rates to that of the audio channel for data transmission.

4.3 Magnetic Channel

The magnetic channel uses the magnetic field sensors (magnetometers or compass) present in most smart phones. Here, the attacker can place small electromagnets in a crowded area. The electromagnets can be controlled by a self-contained device, or via commands transmitted to it from the botmaster over a wireless or cellular network or text message [46]. The electromagnets are used to create a variable magnetic field. The trigger message can be encoded within the presence or absence, or a range of threshold values, of the magnetic field strength.

Magnetic fields dissipate quickly and are inversely proportional to the cube of the distance between the magnet and the device [15]. Thus, a challenge is to create a magnetic field strong enough to be detected above earth's natural magnetic field (the background field). Generation of such a magnetic field is significantly difficult and requires very high electric currents.[1]

However, the distance issue can be resolved by clever placement of the magnetic transmitter. For example, the transmitter can be placed in an elevator of a building or an entrance door frame of a popular building (a hotspot). In such a scenario, the mobile devices carried by people who walk past such transmitters can be triggered. It is also possible for the attacker to physically carry this magnetic device in a pocket or backpack and deliberately walk past users in a crowded area (such as a subway, mall or a sporting event).

Another scenario where the magnetic channel will prove beneficial to the adversary is the usage of Near Field Communication (NFC) phones. Here, the adversary can simply hide a magnet on the NFC readers used for payment transactions. As a user brings his/her phone close to the readers for making the payment, the phone will receive the message.

Compared to the audio and light channel, the reachability of the magnetic channel is a bit limited, and is restricted to triggering devices in the close vicinity of the magnetic source. However, it could still be leveraged to deliver messages to a significant number of devices especially in a crowded region or during rush hour. Its latency is up to the point users come in close physical proximity of the magnetic source.

A distinctive advantage of the magnetic channel over its counterparts is that it requires no line of sight and can work even when the devices are stowed inside pockets or purses. In fact, except of iron and steel, other materials have almost no effect on shielding magnetic fields. Another advantage of this channel is that it is least affected by the background noise (such as that introduced by other magnetic devices or the Earth's field) as long as a high enough detection threshold is used.

4.4 Vibrational Channel

In the vibration channel, the transmission is achieved by inducing vibrations which are then read by a vibration sensor or an accelerometer present on nearby devices. At first, it may seem difficult to induce vibrations to form such channels, but there are (at least) two possibilities for the attacker. One is to embed the messages as high bass sounds into a TV or radio program. Assuming that a subwoofer system is used as the audio transmitter connected to the TV or radio receiver (as in a home theatre or movie theatre system, for example), the associated vibrations can travel some distance and reach the nearby devices. The accelerometer on-board such devices can then decode the vibration patterns. This is especially true when using strong sub-woofer systems as well as buttkickers and vibrating seats already present in many home and movie theatre systems [1].

Another possibility is a peer-to-peer based gossiping channel whereby a vibrating mobile device transmits messages to a nearby device, which then vibrates in turn and transmits to its neighbors, and so on. Such a channel can be formed, for example, in a conference room where many people share a common surface (table) through which the vibrations can travel from one device to the other. This channel is quite feasible and inline with a recent work [34] which demonstrates that keystrokes on a laptop can be learned via an accelerometer on a nearby phone.

The subwoofer based TV broadcast channel provides the same level of coverage and latency as that of the TV based audio channels. The gossiping based channel, on the other hand, has limited reachability and relatively high latency due to its proximity requirements. However, it is highly undetectable due to its decentralized propagation. Both of the channels are also naturally imperceptible to the users. The TV based channels can be explicitly designed to be steganographic just like the audio.

The range of the TV based channel can be up to several feet, especially while using vibrating seats. The gossiping based channel is clearly restricted to a few centimeters. Both of these channels are affected by the background vibration noise, such as that caused by a subtle movement of the users themselves. They are also expected to provide low bandwidth, perhaps less than what is provided by the audio channels.

[1] According to the Biot Savart's Law [15], it will take a wire carrying a large amount of current (500 amperes) to generate a magnetic field strength of just 100 microtesla even from a distance of 1 m. A 500 ampere current will be impossible to induce; as a reference, a current of about 1 ampere can cause electrocution.

5. DESIGN, IMPLEMENTATION AND EXPERIMENTS

In the previous section, we discussed a wide variety of out-of-band command and control mechanisms. To demonstrate the feasibility of mobile malware triggered by covert channels, we have designed and implemented a selected set of these channels. These are listed below:

- Steganographic audio channel
- Direct (non-steganographic) audio channel
- Ambient light channel
- Magnetic-field channel

We have left out the feasibility analysis of vibrational channels from this paper, given that it is expected to exhibit more or less the same characteristics as the audio channel. This is an interesting item for future research, nevertheless.

5.1 Prototype Applications and Test Device

To evaluate the viability of sensor-based covert channels, we developed a set of applications on the Android 2.3.3 (Gingerbread) platform. Android provides support for different types of sensors needed as part of our channels, including the microphone, ambient light sensor and the magnetometer. We ran the prototype applications on an HTC Evo 4g smart phone. These applications ran in the background as Android services. The reported bandwidths in the following experiments are the maximum bandwidths achieved under experimental conditions.

5.2 Audio Channel Design

Direct Channel: When implementing the direct audio channel, we used frequency modulation to encode the messages. The audio signal was created using a 17 KHz carrier signal. The data transmission rate of this channel was 1 bit/second.

Steganographic Channel: Audio Signal Embedding: To develop this channel, we created an embedded signal based on replacing certain audio frequencies, similarly to the idea introduced by Gopalan et al [14]. We utilized two frequencies to encode our data: 1500 or 3000 Hz. If the embedded bit value is 0, then the leading frequency is 1500 Hz, otherwise, it is 3000 Hz. For each bit, the leading frequency power is set to 0.25 of the total frame power. The power of the other frequency is set to 0.001 of the total frame power. We divided the signal into frames of 0.25 sec each. Each frame is utilized to embed one bit. To obtain the frequency spectrum of the frame, we calculate its Fast Frequency Transform (FFT). The frame power is calculated as the summation of the square values of the absolute frequency coefficients.

We initially attempted to utilize only one frequency coefficient to embed each bit. However, this did not produce decodable results. To improve performance, instead of setting just one frequency component (closest to 1500 or 3000 Hz), the 3 frequency components closest to the leading frequency were set to the appropriate power. To detect the beginning of the signal, we added at the beginning of the audio a hail signal. The signal is a perfect sinus at 4500 Hz and lasts for 0.1 sec.

To decode the signal, we first produce a perfect hail signal of 0.1 sec and correlate it to the recorded signal using a 0.1 sec step to advance. We normalize both the hail signal and the inspected signal by setting the signal power of each segment to 1. We then calculate the correlation between the normalized hail signal and the inspected signal. If the correlation is above a certain threshold, we know we reached the region of interest. Once the region of interest has been reached, we further advance to the consecutive segments until the maximum correlation is achieved. At this point, we reduce the step size to 0.01 sec and examine the surrounding 0.1 sec around the found location until the updated maximum correlation is achieved.

At this stage, we decode the rest of the signal. We examine each 0.25 sec stream and calculate the FFT for it. We then calculate the power in the 3 frequencies closest to the frequencies of interest (1.5 KHz and 3 KHz). We divide the power of the first frequency coefficients set by the power of the second frequency coefficients set. If the resulting value is above a certain threshold, the value is marked as 0; otherwise, it is marked as 1.

Steganographic Channel: Live Stream Embedding: In this scenario, we add a signal which embeds a random bitstring to an unknown live stream. We create a high-frequency signal for each "mark" bit (a bit with value 1). We then play this signal together with the streaming signal and decode the result. We create a hail frequency with length of 0.1 sec and frequency of 4500 Hz at the beginning of the signal. For the "mark" bit, we utilize the 20K, 20.1K, 20.2K up to 20.5K frequencies. We create for each of these frequencies a perfect sinus and add all of them to create a combined signal of length 0.1 sec. To prevent "stretching" of the signal we further add a 0.1 sec of break after the mark. For the 0 bit, we leave a 0.2 seconds break. We play the resulting key audio stream together with the original stream. We then record the resulting audio and decode it.

To decode the signal, we first detect the hail frequency (in the same way as in the case of data embedded in a known signal). We then calculate the FFT for each 0.1 signal segment and add the sum of frequencies in the 20 - 20.5 KHz. We use a threshold on the sum to detect the value of each bit.

5.3 Audio Channel Experiments

5.3.1 *Direct Channel*

We conducted our experiments involving the non-steganographic or direct channel keeping in mind the various attack scenarios involving the audio channel: TV/radio broadcast, background music services, IP-TV program, and musical greeting cards (outlined in Section 4; summarized previously in Table 1). To this end, we decided to use a pair of low-end computer speakers as our audio transmitter. (A picture of our experimental set-up appears in Figure 4, moved to the appendix due to the space restrictions). This choice was made to simulate the transmission of audio from an average laptop (e.g., in the case of an IP-TV program), wall-mounted speakers (in the case of workplace music services), or even a musical greeting card. When sending audio via a TV set or a home theatre, much powerful speakers are used in practice. As our experiments below demonstrate, even with low-grade speakers and low volume levels, we were able to achieve reasonably long communication ranges.

Indoor Tests: We first tested the application inside a room (10 ft by 10 ft), with the phone placed on a desk. In this setting, we found that we can successfully send the trigger message to the phone without any data loss (i.e. 0% bit loss) from one side of the room to the other. This validates that the audio triggers can be effectively transmitted to phones, via the speakers of a desktop or laptop computer, in a small personal office setting.

Next, we tested the application in the hallway of a building. In this setting, there were a number of factors introducing environmental noise, such as people talking, kitchen appliances (hallway was alongside a kitchen), and air conditioners. We placed the speakers at one end of the hallway, and the phone at varying distances from the speakers. Despite the presence of "real" background noise, we were able to send the trigger message up to 55 feet distance (the hallway length) with no transmission errors or bit losses, even when the speaker volume was kept low.

To evaluate the effect of loud background noise, we next embedded the signal in a music video. We found that even this large noise introduced by the music video does not affect the transmission rate nor does it introduce any bit errors, up to the distance we tested (55 feet).

These experiments provide evidence that triggers can be easily dissipated via the speakers of a TV set even with large viewing distances, such as at a home, in a movie theatre or a pub with big TV screens. Given that these speakers under realistic conditions would be much powerful, the triggers can be sent, if needed, over much longer distances than the ones we tested. To conclude, the range of the audio channel can be much higher in practice, contrary to intuition.

Finally, we evaluated the scenario where the phone is kept inside a pocket or a purse. We found that the placement of the phone under two layers of clothing (a T-Shirt wrapped all over the phone, including its microphone) did not affect the reception of the trigger message from different distances. The phone was also able to receive the trigger even inside a purse when tested at a distance of up to 22 feet from the speaker. This suggests that audio triggers can be sent even when the phones are stowed inside users' pockets or purses. This ascertains specifically the possibility of transmitting triggers via musical cards in a store. Although we did not test the effect of keeping the phone inside small leather covers (which some people do), we believe that such materials might be able to completely shield the audio signals. In such cases, the triggers may not be received.

Outdoor Tests: We repeated our experiments outdoors in the presence of large environmental noise. We used the same set of low-end speakers to transmit the signal. Despite the large amount of noise due to passing by vehicles and people, the application was able to receive the trigger signal at a distance of 45 feet, with no bit errors. In case of the signal being embedded in a music video, the transmission range was reduced down to 25 feet.

This test indicates that the audio channel can be quite effective even in a realistic outdoor environment. Thus, it is not hard to imagine sending triggers to a large number of phones in a localized area such as a football stadium.

5.3.2 *Steganographic Channels*

For evaluating our steganographic channels, we further conducted some experiments in a office room setting. To test our algorithm for the embedded audio signal channel, we took a sentence from a wav file (from a James Bond movie), and used the encoding to insert a random 24-bit message in it. The audio was then played and recorded. The signal was decoded successfully and the correct key was detected. We ran the tests from close distance (using a laptop's built-in speaker). We also ran the tests from 3 ft away using a desktop with a speaker built into its monitor. We found the detection was successful from both distances. The sampling rate for the recording was 44.1 kHz. The transmission rate for this channel was 4 bits per second. The detection accuracy was reduced slightly when moving to a farther distance, achieving a 92% success rate at 6 ft (22 bits out of 24 bits successful) and 66% detection rate at 8 ft (16 bits out of 24 bits successful).

In case of the live stream channel, we encoded a random 128 bit long message. We played both the encoded message and a song in the background ("at last" by Etta James). Our tests showed that the entire message was decoded successfully. This channel had a bit rate of 5 bits per second. This was further tried from distances of 3, 6 and 8 ft, and the message was decoded successfully for all these distances.

The above result indicate that even steganographic channels can be effective at least in a non-noisy environment. Further tests are needed to evaluate these channels under larger amount of noise both indoor and outdoors.

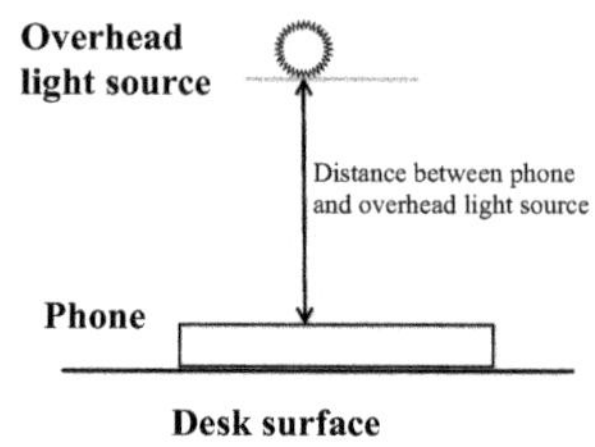

(a) Overhead lighting tests

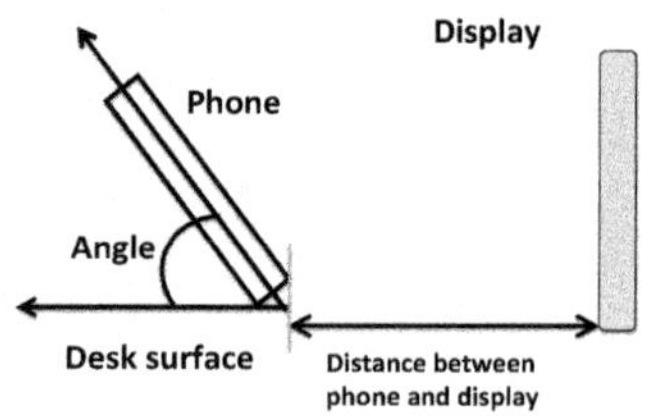

(b) Computer and TV lighting tests

Figure 1: Set-up for the light channel tests

5.4 Light Channel Design and Experiments

To evaluate the ambient light channel, we explored the effect of signals sent through two light sources: (1) overhead lighting, and (2) different types of computer and TV screens located near the phone. This captures the attacks whereby the triggers are sent via power fluctuations, or via TV/IP-TV programs. Figure 5 in the appendix shows a snapshot of our experimental set-up.

For this channel, we assumed that the mobile phone will be placed on top of a table. As mentioned earlier, we designed our malware application as a background service in this case. We note that the typical ambient light inside a room during daytime is between 120 and 240 lux units. At night, with a fluorescent light, the ambient light is about 120 lux units.

Overhead Lighting Tests: When placed on top of a desk, a mobile phone is usually about 6 feet away from the ceiling, which is where the light sources are commonly located. To simulate this scenario for the overhead light channel tests, the phone was placed flat on the desk in a office room, with the display facing up (the ambient light sensors of most smart phones are located at the top of their displays). Figure 1(a) depicts this set-up. We manually introduced small fluctuations in the light intensity of the overhead lights inside the room. Our experiments in this setting showed that by tweaking the light source, we were able to change the light intensity from 120 lux to 160 lux as detected by the phone's ambient light sensor. In this setting, we were able to send the trigger message to the phone successfully with 0% bit errors at a data rate of 0.5 bits per second.

Computer and TV Screen Tests: Next, we examined the feasibility of a signal being sent via a video played on a computer monitor or a TV. Figure 1(b) shows our experimental setup for this set of tests. Here, we varied the tilt angle between the phone and the desk, and measured the maximum range with 0% bit errors. The trigger signal was displayed on the monitor. To encode the messages, we programmatically increased and decreased the brightness of the screen.

The total amount of light emitted by a monitor depends on the size of the monitor. For our tests, we used a 17 inch laptop display, a

Angle (degree)	Bit Rate (bits/second)	Max Distance (inch)
90	1	3
90	0.5	65
60	0.5	16
45	0.75	5
45	0.5	11

Table 2: Results of the Laptop screen experiment

22 inch desktop LCD display, and a 48 inch LCD television. In case of the laptop and the TV, we performed the experiments at night with the overhead lights turned on, providing bright ambient light in the room. In case of the desktop screen, we performed the tests during the day (i.e., in presence of natural light) with a few overhead lights on.

Table 2 depicts the results we obtained when using the laptop screen. The angle between the phone and the surface was varied from 90 degree to 45 degree. We were able to sustain a bit rate of 0.5 bits/second at a distance of 65 inches or 5.5 feet with 0% bit loss when the phone is parallel to the laptop screen. As the tilt angle is lowered w.r.t. the desk surface, the range decreases. In summary, we achieved reasonable range even with a small screen monitor.

Table 3 shows the results of our tests when the desktop monitor was used. The results in this case are similar to the previous test with the laptop monitor.

Angle (degree)	Bit Rate (bits/second)	Max Distance (inch)
90	1	7
90	0.5	24
90	0.33	35
60	0.5	15
30	0.75	6
10	0.75	4

Table 3: Results of the 22 inch LCD desktop experiment

We found the best results (longest range) in the case of the 48 inch LCD TV. Results from our TV experiment are shown in Table 4. Even with the phone placed flat on a desk, we received the trigger message with 0% bit errors at a distance of 13 inches from the TV. We were even able to receive the signal at a distance of 100 inches (8.33 feet) when the phone was parallel to the TV screen. This clearly demonstrates the effectiveness of our light channel for sending triggers via TV broadcasts. Note that the normal TV viewing distances [6] are inline with the range we obtained based on our experiments.

5.5 Magnetic Channel Design and Experiments

For the magnetic-field channel, we built a programmatically controlled electro-magnet to encode the messages. At the receiving end, we retrieved the readings of the magnetic sensor of the phone via an application. The magnetometer built into the phone is a 3-axis vector magnetometer and thus provides the X, Y, Z components of the magnetic field relative to the spatial orientation of the phone, as shown in Figure 2. Choosing only one component of the magnetic field would require the device to be held in a particular orientation. Instead, a scalar value makes use of all three component values and allows the device to be in any orientation. We can simply derive this scalar value by using the formula for vector-to-scalar conversion: $\mathsf{scalar} = \sqrt{m_x^2 + m_y^2 + m_z^2}$, where x, y, z denote the X, Y, Z components of the magnetic field, respectively.

We first calculated the background magnetic field due to the Earth's magnetic core. We turned on the magnetic sensor on the phone and recorded the scalar values to measure the ambient magnetic field strength at various locations in the absence of a strong magnetic source in order to determine this baseline. The values we obtained fall within the range of 30 to 50 microtesla, which

Angle (degree)	Bit Rate (bits/second)	Max Distance (inch)
90	1	18
90	0.5	100
30	0.75	46
15	0.75	28
0	0.5	13

Table 4: Results of the 48 inch LCD TV experiment

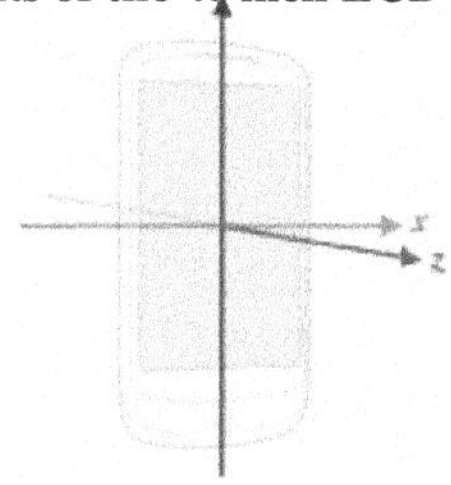

Figure 2: The reference axes across the phone

conforms to the fact that the Earth's magnetic field at the equator is 31 microtesla.

Next, we set out to test two different scenarios. First, when the (electro-)magnet is not covered, and second, when the magnet is covered by clothing, plastic and other material. This was done to study any potential effect on the magnetic field strength due to hiding the magnet on a door frame. For each scenario, we varied the orientation of the magnet with respect to the phone. A snapshot from our experiments is shown in Figure 5(a) in the appendix.

Figure 3(a) shows the various distances, between the phone and the magnet, at which the presence of the magnetic field can be *detected.* Notice that the magnetic field can be detected best when the magnet is oriented perpendicular to the front face of the phone (i.e., along the Z axis). In that orientation, we could detect the magnetic field at a maximum distance of 5 inches. However, at this distance, the magnetic field is not strong enough to send the trigger signal without bit errors. The field strength in this case hovers around the Earth's magnetic field, and fluctuates quite a bit.

Figure 3(b) depicts the maximum distances at which we were able to *send a signal* consistently without any bit errors. We were able to send a signal to the malware application at a maximum distance of 3.5 inches. At this distance, the magnetometer in the phone was able to detect a magnetic field of 60 microtesla from the electro-magnet, which is high enough to distinguish itself from the background/natural magnetic field.

We next repeated the above experiment by covering the electro-magnet with plastic. As depicted in Figures 3(c) and 3(d), the magnetic field is, however, virtually unchanged even in this case.

These results confirm that the magnetic channel is feasible in many scenarios where the magnet can be placed close to the phone without worrying about shielding effects.

5.6 Power Consumption

Finally, we investigated the power consumption of the sensors. We tested the power consumption by keeping the sensors running constantly for 10 minutes. We found that, in practice, sensors use a small amount of power. For example, the magnetometer, accelerometer, and light sensors cause virtually no change in battery consumption. Even the microphone consumed less than 1% of the battery capacity over regular consumption rate during a 10 minute period. Further details of sensor power usage can be found in Table 5.

The power consumption analysis shows that the malware application, even when accessing the sensors, does not cause significant power drain in practice. We point out that only the steganographic

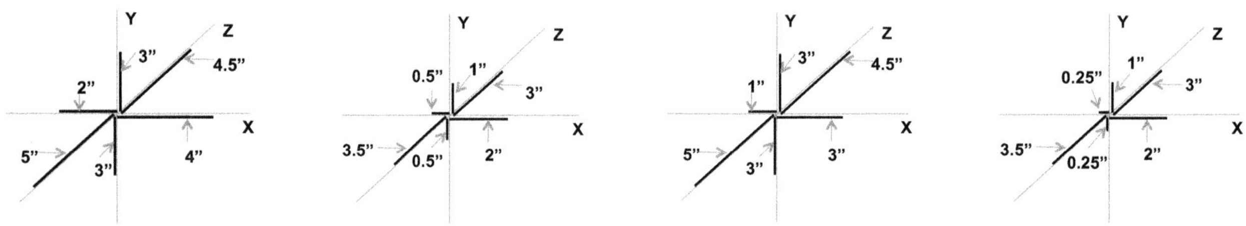

(a) Without cover (detection) (b) Without cover (signalling) (c) With cover (detection) (d) With cover (signalling)

Figure 3: Magnetic field distance tests. (a) and (c) show the test where presence of a magnetic field was detected by the malware. (b) and (d) show the test where we successfully sent the trigger to the malware with no bit errors. See Figure 2 for reference axes.

audio channel requires potentially expensive Fast Frequency Transform (FFT). The other channels do not use FFT. And while FFT does require computation resources, prior research shows it is feasible to reduce energy consumption [7]. Optimizations can parallelize FFT computation on ARM processors which are widely used in smartphones [35]. Another mechanism for evading power drain analysis is that, the attacker may send triggers only at certain times(e.g., once-a-day during telecast of a TV program where the trigger signal has been embedded). This significantly reduces likelihood of detection while still giving the attacker a periodic trigger window.

Sensor	Power consumption in 10 minutes (% of battery capacity)
No sensor	< 1
Microphone	2
Light sensor	1
Magnetometer	< 1
Accelerometer	< 1

Table 5: Power consumption of sensors.

6. DISCUSSION

6.1 Summary of Results and Further Analysis

Our experiments demonstrate the feasibility of sending command and control trigger messages to smart phones and similar devices over out-of-band covert channels.

Range: The direct audio channel exhibited the longest range – more than 55 feet indoors, 45 feet outdoors. It can also tolerate real-life background noise. We were able to achieve these communication ranges using only low-end PC speakers with minimal amplification and low-volume. In practice, we envision the audio channel to use better quality speakers or sound systems present in Televisions or other multimedia systems. The direct audio channel also has the advantage of working both indoors and outdoors. Our experiments also demonstrated that the audio reception works even if the phone will be kept in the user's pocket or a purse. However, we do not expect it to work when the phone is stowed inside leather pouches or inside other thick material. Therefore, our experiments provide evidence that the audio based command and control via TV or radio programs, background music services, Internet TV program and musical greeting cards is feasible. This is also true for our steganographic channels, although our current experiments with these channels were conducted in a typical office setting.

The ambient light channel has shorter range compared to the audio. However, it has the advantage of sending a trigger message simultaneously to all the phones in a targeted building. The light channel works best at night or in places with low illumination. During daytime, the presence of sunlight introduces a large amount of ambient light (a form of background noise), due to which receiving the messages over long distances might be difficult. However, when using large screen televisions, the messages can be relayed over reasonably long distances.

The magnetic channel, as expected, has the shortest range. This is due to the inherent property of magnetism – magnetic signals rapidly fade as they travel in the air. However, even this short range channel can be surreptitiously exploited in scenarios where the signal transmitter is hidden in the doorframe or on an elevator. The magnetic channel clearly would work in most, if not all, scenarios irrespective of the way the phones are stowed or carried by users.

Channel Bandwidth: A limitation of the sensor-based channels is their low-bandwidth. In our current implementation, we were able to achieve a maximum bandwidth of 5 bits per second with these channels. This limit is attributed to a number of different factors such as the limited sensitivity of mobile device sensors, the low sampling rate enforced by the Android OS and the need for undetectability of transmission. However, since command and control messages – especially the triggers – need not be too large, we believe that even this low bandwidth will be sufficient in most scenarios. To offset the low transmission rate, the messages can still be transmitted over large time spans, such as over an entire TV program or multiple occurrences of the same program. One possibility to improve the bandwidth is to employ multiple channels simultaneously. For example, the messages may be delivered in parallel over the audio channel as well a light channel, say while watching a TV. We believe the bandwidth limit to be inherent in the design of the sensing-enabled channels. This represents the obvious trade-off in using these channels for undetectable command and control when compared to other wired/wireless channels.

6.2 Possible Defense Mechanisms

Although it may be difficult to detect and prevent the out-of-band command and control presented in this paper, we recognize this to be an important problem worthy of further investigation. To this end, we suggest a few preliminary set of defenses.

An intrusion detection application running on the mobile phone itself can detect the sensor based signaling and prevent the malware application from receiving it by monitoring the sensor data stream. To this end, we envision two approaches. First, instead of giving the applications direct access to the sensors, a virtualization layer can be created between the sensors and the applications. The virtualization layer includes a monitor that constantly analyzes the sensor accesses by different applications as well as the sensor data streams to determine whether any malicious activity is taking place. The virtual sensor monitor may use machine learning techniques to determine the natural behavior of benign applications when run by a given user. The downside of this approach is that it may be rather heavyweight and requires the phone to monitor a large number of sensor readings. Second, in order to determine whether a malicious signal is present in the sensor readings, the phone can take random sensors samples and analyze them.

Another phone based defense could be based on the level of power consumed by applications that access sensors. A malicious application listening on to one or more of the sensors is likely to consume slightly more power compared to other applications (see Table 5). The feasibility of such a defense against malware is demonstrated in [23]. A related approach using memory footprints [21] can also be used to detect such stealthy mobile malware applications.

Finally, another line of defense is to monitor external media streams for malicious communication. For example, media streams such as radio, music, or video streams can be monitored by service providers prior to the transmission. These mechanisms can be effective against some of the attack scenarios presented in this work such as the use of TV or radio programs. However, it may be very difficult to detect the presence of our steganographic channels.

7. RELATED WORK

Most closely related to our work is the Bluetooth based command and control architecture proposed in [46]. In this approach, Bluetooth is used as the primary means of communication between the botmaster and infected devices. In particular, a hybrid approach is suggested whereby a set of seed nodes are first selected with which the botmaster communicates over the traditional network (such as SMS), and these seed nodes then relay the messages to other nodes in close physical proximity over Bluetooth. Based on public Bluetooth traces, [46] demonstrated that such an infrastructure of command and control is feasible. However, it exhibits high latencies in propagating messages. More specifically, such messages can reach about 70% of infected devices within 24 hours. Although some of our channels also exhibit similar latencies, many other channels offer an almost real-time delivery guarantees to an extremely large number of devices. Moreover, many of our channels do not at all rely upon a traditional infrastructure (such as SMS) and provide a means of undetectable communication, and some (such as audio) even possess communication range higher than that of Bluetooth.

HumaNet [8] is a similar architecture to that proposed in [46]. In contrast to [46], however, in this approach, the command and control messages are delivered only via phone to phone ad hoc communication (such as Bluetooth or WiFi ad hoc mode). This work also demonstrated the latencies similar to that of [46]. Compared to this work, the focus of our paper is on out-of-band communication of different flavors, including phone-to-phone communication in some cases.

Another related work is SkyNET [42] which makes use of compromised personal networks (such as personal WiFi) as a platform for command and control message dissemination. Other mobile botnets have been reported in scientific publications that make use of other services available on mobile devices for communication such as SMS/MMS [13, 19, 37, 49, 50] and available open WiFi networks [26]. Our work, on the other hand, utilizes out-of-band communication channels for command and control which are much harder to detect compared to the use of cellular or WiFi services.

The explicit use of steganography for the purpose of undetectable command and control, as employed in our stego audio channel, is also inline with Stegobot [39]. This is a botnet architecture that uses social networking based steganographic channels for command and control. In particular, image steganography is used to achieve this goal. Our steganographic audio channel, in contrast, embeds covert messages which are transmitted in real-time *over the air*, and not into static files or images.

Out-of-band communication in general has been employed in many security applications before. These include: proximity-based secure association of personal wireless devices (see surveys and usability studies [20, 22, 27, 28]) and user authentication [43]. Besides identifying a new application of out-of-band communication, i.e., undetected command and control, we have proposed many novel channels not previously explored.

8. CONCLUSION

The threat of mobile malware is rising rapidly. According to a recent report by Lookout Security [3], the Android phone users are becoming more and more vulnerable to malware. The same report states that, in several cases, legitimate applications were repackaged with malware code and distributed via the official Android app store or via updates/advertisements. Android and iPhone botnets have already been proposed [37, 49]; some have even appeared in the wild [49]. However, most of these mobile botnets used SMS or the Internet to build the command and control infrastructure – similar to traditional botnets, and are therefore detectable via traditional botnet defense mechanisms.

In this paper, we investigated the feasibility of sensing-enabled covert channels in mobile phones. Malware using such channels will be very difficult or impossible to detect using traditional means, because such the underlying command and control channels exploit non-network air-gaps to communicate. Our proof-of-concept prototype exemplifies this emerging problem – using off-the-shelf hardware and popular Android-based mobile phones, we were able to send surreptitious command and control messages without using any wireless or cellular networks. Our prototype malware application received the messages embedded in music, video, household lighting, or magnetic fields.

Malware with the capability of using such sensor-based covert channels can also open up new threats such as the creation of localized botnets and geo-targeted attacks, which we explored briefly in the paper.

Acknowledgment

This research was supported by a 2012 Google Faculty Research Award, the Office of Naval Research Grant #N000141210217, the Department of Homeland Security Grant #FA8750-12-2-0254, and the National Science Foundation under Grants #0937060, #1201927, and #1228236.

9. REFERENCES

[1] ButtKickers – Low Frequency Audio Transduceers. Available at http://www.thebuttkicker.com/.

[2] iPhone Microhone Frequency Response. Available at http://bit.ly/IDWfei/.

[3] Mobile Threat Report. Available at http://bit.ly/ImefoN/.

[4] PPLive. Available at http://www.synacast.com/en/.

[5] SopCast - Free P2P internet TV. Available at http://www.sopcast.org/.

[6] Suggested Viewing Distance for HDTV (Calculator). Available at http://hdinstallers.com/calculator.htm.

[7] M. Aboleaze and A. Elnaggar. Reducing memory references for FFT calculation. In *Proc. of the International Conference on Computer Design*, pages 26–28, 2006.

[8] A. J. Aviv, M. Sherr, M. Blaze, and J. M. Smith. Evading cellular data monitoring with human movement networks. In *Proc. of USENIX HotSec*, 2010.

[9] H. Bassen. Radiofrequency interference with medical devices. a technical information statement. *IEEE Engineering in Medicine and Biology*, 17(3):111–114, 1998.

[10] L. Cai and H. Chen. Touchlogger: inferring keystrokes on touch screen from smartphone motion. In *Proc. of USENIX HotSec*, 2011.

[11] E. Cochran, J. Lawrence, and C. Christensen. Quake-Catcher Network. Available at http://qcn.stanford.edu/, 2008.

[12] P. Dhungel, X. Hei, K. W. Ross, and N. Saxena. The pollution attack in p2p live video streaming: measurement results and defenses. In *Proc. of P2P-TV*, 2007.

[13] G. Geng, G. Xu, M. Zhang, Y. Yang, and G. Yang. An improved sms based heterogeneous mobile botnet model. In *Proc. of IEEE ICIA*, 2011.

[14] K. Gopalan and S. Wenndt. Audio steganography for covert data transmissions by imperceptible tone insertion. In *Communication systems and applications*, Available at `http://qcn.stanford.edu/`, 2004.

[15] D. J. Griffiths. *Introduction to Electrodynamics (Third Edition)*. Prentice Hall, 1999.

[16] J. Han, E. Owusu, T.-L. Nguyen, A. Perrig, and J. Zhang. ACComplice: Location Inference using Accelerometers on Smartphones. In *Proc. of COMSNETS*, Jan. 2012.

[17] Harris Interactive. Teenagers: A Generation Unplugged. Available at `http://bit.ly/IyH71E`, 2008.

[18] Harris Interactive. The Harris Poll - Cell Phone Usage Continues to Increase. Available at `http://bit.ly/IpgefF`, 2008.

[19] J. Hua and K. Sakurai. A sms-based mobile botnet using flooding algorithm. In *Proc. of WISTP*, 2011.

[20] I. Ion, M. Langheinrich, P. Kumaraguru, and S. Capkun. Influence of user perception, security needs, and social factors on device pairing method choices. In *Proc. of SOUPS*, 2010.

[21] M. Jakobsson and K. Johansson. Retroactive detection of malware with applications to mobile platforms. In *Proc. of USENIX HotSec*, 2010.

[22] R. Kainda, I. Flechais, and A. W. Roscoe. Two heads are better than one: Security and usability of device associations in group scenarios. In *Proc. of SOUPS*, pages 1–13, 2010.

[23] H. Kim, J. Smith, and K. G. Shin. Detecting energy-greedy anomalies and mobile malware variants. In *Proc. of MobiSys*. ACM, 2008.

[24] R. Kim. The World's a Cell-phone Stage. SFC, Available at `http://bit.ly/IpgSdf`, 2006.

[25] Knowledge Networks. New Study Shows Mobile Phones Merging New, Established Roles: Communicator, Shopping Aide, Entertainment and Research Hub. Available at `http://bit.ly/IDWyWv`, 2008.

[26] M. Knysz, X. Hu, Y. Zeng, and K. G. Shin. Can open wifi networks be lethal weapons for botnets? In *Proc. of INFOCOM*, to appear, 2012.

[27] A. Kobsa, R. Sonawalla, G. Tsudik, E. Uzun, and Y. Wang. Serial hook-ups: A comparative usability study of secure device pairing methods. In *Proc. of SOUPS*, 2009.

[28] A. Kumar, N. Saxena, G. Tsudik, and E. Uzun. Caveat emptor: A comparative study of secure device pairing methods. In *Proc. of PerCom*, 2009.

[29] D. Lee. College Student's Hand-phone Usage Culture Survey. In *University Culture Newspaper*, 2002.

[30] S. Lee, K. Fu, T. Kohno, B. Ransford, and W. Maisel. Clinically significant magnetic interference of implanted cardiac devices by portable headphones. *Heart rhythm : the official journal of the Heart Rhythm Society*, 6(10), October 2009.

[31] S. Lohr. As Cellphones Bulk Up, How Much Is Too Much? Available at `http://nyti.ms/IEk3hV`, 2005.

[32] C. V. Lopes and P. Q. Aguiar. Acoustic modems for ubiquitous computing. *IEEE Pervasive Computing, Mobile and Ubiquitous Systems*, 2(3):62–71, July-September 2003.

[33] Market Analysis and Consumer Research Organisation. Study of Mobile Phone Usage Among the Teenagers And Youth In Mumbai. Available at `http://bit.ly/JrjaCQ`, 2004.

[34] P. Marquardt, A. Verma, H. Carter, and P. Traynor. (sp)iPhone: decoding vibrations from nearby keyboards using mobile phone accelerometers. In *Proc. of ACM CCS*, 2011.

[35] P. Meerwald. KissFFT and ARM NEON. Online at `http://bit.ly/WtSjDI`, 2011.

[36] E. Mills. Attacking home automation networks over power lines. CNET, Available at `http://cnet.co/JCm8Ji`.

[37] C. Mulliner and J.-P. Seifert. Rise of the iBots: Owning a telco network. In *Proc. of MALWARE*, Oct. 2010.

[38] Muzak Inc. Music and More for any Businesses. Available at `http://www.muzak.com/`.

[39] S. Nagaraja, A. Houmansadr, P. Piyawongwisal, V. Singh, P. Agarwal, and N. Borisov. Stegobot: a covert social network botnet. In *Proc. of IH*, 2011.

[40] E. Owusu, J. Han, S. Das, A. Perrig, and J. Zhang. ACCessory: Keystroke Inference using Accelerometers on Smartphones. In *Proc. of HotMobile)*, Feb. 2012.

[41] W. K. Park. Mobile Phone Addiction. In *Mobile Communications: Re-negotiation of the Social Sphere*, 2006.

[42] T. Reed, J. Geis, and S. Dietrich. Skynet: a 3g-enabled mobile attack drone and stealth botmaster. In *Proc. of USENIX WOOT*, 2011.

[43] N. Saxena and J. H. Watt. Authentication technologies for the blind or visually impaired. In *Proc. of USENIX HotSec*, 2009.

[44] R. Schlegel, K. Zhang, X. yong Zhou, M. Intwala, A. Kapadia, and X. Wang. Soundcomber: A stealthy and context-aware sound trojan for smartphones. In *Proc. of NDSS*, 2011.

[45] A. N. Selian. Mobile Phones and Youth: A Look at the U.S. Student Market. Available at `http://bit.ly/ZnVMA4`, 2004.

[46] K. Singh, S. Sangal, N. Jain, P. Traynor, and W. Lee. Evaluating bluetooth as a medium for botnet command and control. In *Proc. of DIMVA*, 2010.

[47] P. L. Sunderland. Fieldwork and the Phone. 1999.

[48] T. A. Wikle. America's Cellular Telephone Obsession: New Geographies of Personal Communication. In *Journal of American and Comparative Cultures*, 2001.

[49] C. Xiang, F. Binxing, Y. Lihua, L. Xiaoyi, and Z. Tianning. Andbot: towards advanced mobile botnets. In *Proc. of USENIX LEET*, 2011.

[50] Y. Zeng, K. Shin, and X. Hu. Design of SMS commanded and controlled and P2P structured mobile botnets. In *Proc. of ACM WiSec*, 2012.

Appendix

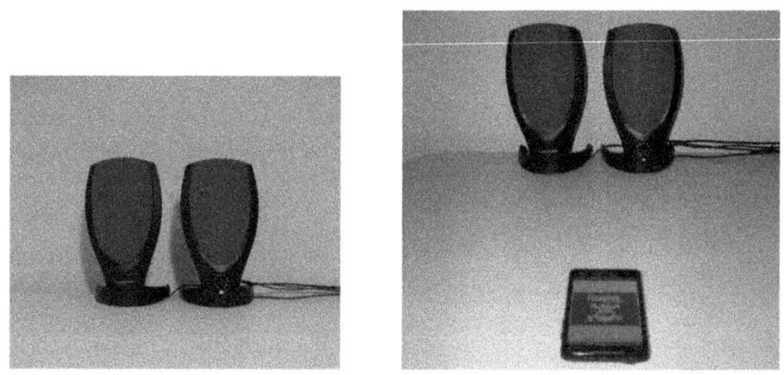

(a) We used low-end PC speakers to transmit the audio signal.

(b) A malware-infected phone can detect the trigger signal embedded in audio.

Figure 4: Equipment set-up for the audio channel

(a) A user is walking past the door with his phone inside the pocket, yet within close range of the transmitter.

(b) Phone placed flat on the desk, receiving signal from the overhead light.

Figure 5: Equipment set-up for the magnetic and light channel

For more images of our experimental setup, please visit `http://secret.cis.uab.edu`

LogicScope: Automatic Discovery of Logic Vulnerabilities within Web Applications

Xiaowei Li
Department of Electrical Engineering and Computer Science
Vanderbilt University
Nashville, TN USA 37203
xiaowei.li@vanderbilt.edu

Yuan Xue
Department of Electrical Engineering and Computer Science
Vanderbilt University
Nashville, TN USA 37203
yuan.xue@vanderbilt.edu

ABSTRACT

Logic flaws are an important class of vulnerabilities within web applications, which allow sensitive information and restrictive operations to be accessed at inappropriate application states. In this paper, we take a first step towards a systematic black-box approach to identifying logic vulnerabilities within web applications. We first construct a partial FSM over the expected input domain by collecting and analyzing the execution traces when users follow the navigation paths within the web application. Then, we test the application at each state by constructing unexpected input vectors and evaluating corresponding web responses. We implement a prototype system LogicScope and demonstrate its effectiveness using a set of real world web applications.

Categories and Subject Descriptors

K.6.5 [**Management of Computing and Information Systems**]: Security and Protection—*Unauthorized access*

General Terms

Design, Experimentation, Security

Keywords

Logic Vulnerability; Web Application Security; Finite State Machine

1. INTRODUCTION

Web applications have become one of the most prevalent technologies for information and service delivery over Internet. As they get increasingly complex to support sophisticated business functionalities, the vulnerabilities, as well as the attacks that target on those vulnerabilities, are both increasing. Vulnerabilities within web applications are commonly classified into two classes [3]. *Input validation vulnerabilities* are introduced by insufficient or erroneous sanitization of user inputs, which allows injection attacks (e.g., Cross-site Scripting, SQL injections), where malicious code is injected into web applications. The other class of vulnerabilities, which are referred to as *logic vulnerabilities*, (a.k.a, logic flaws) are caused by insufficient state checks, which allows sensitive information and restrictive operations to be accessed at inappropriate application states (i.e., state violation attacks). In contrast to input validation vulnerabilities, there have been very limited works that address logic vulnerabilities. The major challenge comes from the fact that application logic flaws are specific to the intended functionality of a particular web application. A real attack example is that in 2011, a huge amount of credit card information was leaked due to a logic flaw within the citigroup website [2].

Identifying logic flaws requires the knowledge of both the intended behavior (i.e., specification) of the application and the actual behavior as implemented in the application. Existing works (e.g., Waler [4], Sun et al. [12]) apply program analysis techniques over the application source code to identify the deviations of the implementation from the intended behavior. These techniques all require application source code for analysis or instrumentation, which makes them closely coupled with the application language (e.g., PHP, JSP), platform and implementation details. The correctness and accuracy of their derived intended behaviors are highly dependent on and thus limited by their capability of dealing with language-level details (e.g., if they can handle object-oriented code). Black-box analysis provides an appealing alternative that overcomes all the above limitations, but faces significant challenges, since it requires sufficient knowledge about the application logic under test without the application source code.

In this paper, we take a first step towards a systematic black-box approach to identifying a broad range of logic flaws within web applications. In particular, we model the logic of a web application using a finite state machine (FSM) and formalize logic vulnerabilities as the discrepancies between the intended FSM and the implementation FSM. The intended behavior of a web application can usually be observed when normal users follow the navigation paths within the web application [4][12]. In this case, we say the inputs (i.e., HTTP requests) from the users are *expected*. Over the expected inputs, the behavior of the intended FSM and the implementation FSM are consistent. The discrepancies are manifested over the application behaviors when unexpected inputs are fed into the application.

To identify such discrepancies, we first construct the in-

ASIA CCS'13, May 8–10, 2013, Hangzhou, China.

tended FSM as a partial FSM over the observed user inputs (i.e., expected input domain). Then, based on the inferred partial FSM, we test the application over unexpected inputs at each state to identify logic flaws. We present two methods for constructing test input vectors, corresponding to two commonly seen state violation attack vectors, i.e., parameter manipulation and forceful browsing attacks, and feed them into the application. We give an evaluation rule to determine whether the corresponding web response manifests a potential logic vulnerability. The reported potential logic vulnerabilities are evaluated manually are classified into real attack vectors and false positives.

Our contributions can be summarized as follows.

- We propose a systematic black-box approach to identifying logic flaws within web applications. In particular, we give a formalization of logic vulnerability based on the FSM model and present two test input generation methods. Our approach is independent of the application language, source code and implementation, thus can be scalable and general to handle a number of web applications and platforms.
- We implement a prototype system LogicScope and evaluate it using a set of real world web applications to demonstrate the effectiveness of LogicScope. LogicScope gives concrete attack vectors and evaluation evidences, which would greatly facilitate the confirmation and analysis of reported logic flaws.

The rest of paper is organized as follows. We present our system model and problem formalization in Section 2. Our approach and evaluation results are illustrated in Section 3 and 4, respectively. Section 5 discusses related works and Section 6 concludes the paper.

2. PROBLEM DESCRIPTION

In Figure 1, we present a small and vulnerable web application as the running example to illustrate how we formalize and identify logic flaws within web applications. This application uses two explicit session variables $_SESSION['privilege'] and $_SESSION['userid'] to remember the current user's access right and id. A user, who just logs in, will be redirected to index.php page. If the current user is an admin (i.e., $_SESSION['privilege'] is equal to "admin"), he/she is presented with links for adding new users, editing and deleting any of the registered users. If the current user is a regular user (i.e., $_SESSION['privilege'] is equal to "user"), he/she can only see the link for editing his/her own information.

We model a web application using a finite state machine (FSM) model (S, s_0, Σ, Λ, T, G), where S is the set of states, $s_0 \in S$ is the initial state, Σ is the set of input symbols (i.e., input alphabet), Λ is the set of output symbols (i.e., output alphabet), $T : S \times \Sigma \rightarrow S$ is the set of transition functions and $G : S \times \Sigma \rightarrow \Lambda$ is the set of output functions. T and G determine the next state and the output symbol, respectively, based on the current state and the input symbol.

To understand the logic flaws within a web application, we need to consider two FSMs associated with it: (1) the ideal FSM (denoted as F_{ideal}) models the intended (or expected) behavior of the web application without any security vulnerabilities; (2) the implementation FSM (denoted as F_{impl}) models the actual behavior of the web application

index.php
```
<?php
if ($_SESSION[`privilege`] == "admin") {
    echo "<a href = add_user.php>Add User </a>";
    $user_set = get_all_users();
    foreach ($user_set  as $user) {
        echo "<a href = edit_user.php?userid=" . $user . ">Edit User </a>";
        echo "<a href = delete_user.php?userid=" . $user . ">Delete User </a>";
    }
} else if ($_SESSION[`privilege`] == "user") {
    echo "<a href = edit_user.php?userid=" . $_SESSION[`userid`]  . ">Edit
User </a>";
}
?>
```

add_user.php
```
<?php
add_user();
?>
```

edit_user.php
```
<?php
edit_user($_GET[`userid`]);
?>
```

delete_user.php
```
<?php
if ($_SESSION[`privilege`] != "admin")
   die ("ACCESS DENIED!");
 delete_user($_GET[`userid`]);
?>
```

Figure 1: Example Application

as being implemented by the developer. If F_{impl} is equivalent to F_{ideal}, we say the implemented web application is secure as intended. If there exist discrepancies between F_{impl} and F_{ideal} and the discrepancies involve sensitive information or operations, we say the implemented web application has logic vulnerabilities.

As shown in Figure 2, the example application has three states: the user is not logged in (s_0), a regular user logs in (s_1) and an admin user logs in (s_2). Each input symbol $I \in \Sigma$ is an abstract representation of a web request, which consists of two parts: 1) the *key* (denoted by K) represents the syntax structure of the web request (e.g., K_1 = GET-edit_user.php : userid); 2) the *value* (denoted by V) represents the value domains of the parameters, which is related to the semantics of the web request (e.g., V_1 = [userid = $v_{constrained}$($_SESSION['userid'])] means that the userid parameter in the web request is always equal to the value of the session variable userid) (refer to Section 3.3 for details). $I_1 = K_1.V_1$ and $I_2 = K_1.V_2$ are two different input symbols due to the difference in their parameter value domains, although they have the same syntax structure. Similarly, each output symbol in Λ is an abstract representation of the web responses returned by the application to users.

The ideal FSM (F_{ideal}) for the application works as follows. At state s_1, since it is intended that the regular user can only edit his/her own information, when the regular user sends an input symbol $I_1 = K_1.V_1$, where the userid parameter is equal to the current user id, the application will respond with the edit_user page (output symbol O_1). When the regular user tries to edit other user's information (i.e., sending $I_2 = K_1.V_2$), delete a user (i.e., sending $I_3 = K_2.V$) or add a new user (i.e., sending $I_4 = K_3.V$), the application should reject such input symbols with an ACCESS_DENIED page (output symbol O_2).

However, the implemented application has two logic vulnerabilities, which are reflected as the discrepancies between F_{ideal} and F_{impl}, shown using the dashed lines with arrows in Figure 2. First, the edit_user.php page doesn't check if the userid parameter is equal to the current user id. Second, the add_user.php page doesn't check if the current user has the admin privilege, as the delete_user.php page does. These vulnerabilities allow two types of attacks. 1) *Parameter manipulation attack*: when input symbol I_2 is sent to the application at state s_1, output symbol O_1 is returned. This allows a regular user to edit other users' information. 2) *Forceful browsing attack*: when input symbol I_4 is sent to the application at state s_1, O_4 is returned, which allows a regular user to add new users.

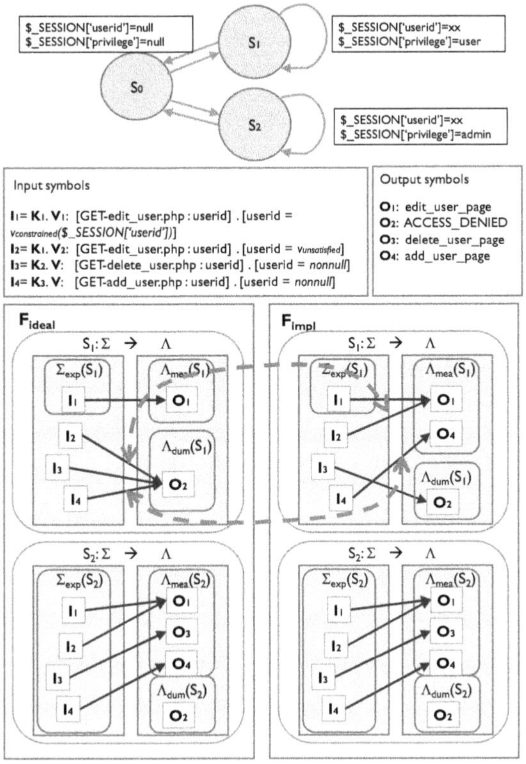

Figure 2: FSM Representation of Example Application

Since a user can send any web request to a web application, the input alphabet can be infinite and fed into the application at any state, resulting in corresponding output symbols. However, at a given state s, we observe that only a subset of input symbols are expected by the application (denoted as $\Sigma_{exp}(s)$) and processed to generate "meaningful" output symbols (i.e., $\Lambda_{mea}(s) = G(s, \Sigma_{exp}(s))$). The expected input symbols are the web requests that can be issued, if the user follows the navigation links of the web application and the meaningful output symbols are the web responses that provide the users with useful information. All the other input symbols, which are not expected at state s, should be rejected/mitigated by the application, resulting in "dummy" output symbols (i.e., $\Lambda_{dum}(s) = G(s, \Sigma - \Sigma_{exp}(s))$). A dummy output symbol means that the application responds the user with an error page or a redirection header pointing to a previously visited web page, without leaking any useful information. As shown in Figure 2, for state s_1, the expected input set is $\{I_1\}$, the meaningful output set is $\{O_1\}$ and the dummy output set is $\{O_2\}$. For state s_2, the expected input set is $\{I_1, I_2, I_3, I_4\}$ and the meaningful output set is $\{O_1, O_3, O_4\}$. A web application is expected to fully implement the intended functionality, which means the behaviors of F_{ideal} and F_{impl} over the expected input symbols should be consistent. However, the unexpected inputs may not be fully mitigated/rejected by F_{impl} as intended by F_{ideal}. Thus, we say a web application has a logic vulnerability at state s, if an input symbol, which is not expected at state s (called a malicious input and denoted by I_{mal}), is fed into the application, the application generates an output symbol that falls beyond the dummy output set (i.e., $G(s, I_{mal}) \in (\Lambda - \Lambda_{dum}(s))$).

3. APPROACH

3.1 High-level Overview

As stated in problem formulation, to identify logic vulnerabilities, we need to construct unexpected (malicious) inputs for each state and evaluate whether their outputs fall beyond the dummy output set. This is nontrivial since we have no knowledge about the entire input alphabet (and thus unexpected input) and the dummy output set at each state.

To approach this problem, we first construct a partial FSM over the expected input domain by observing the executions of the application when users follow the navigation paths provided by the application. To do so, we collect web requests, web responses and associated session variables during attack-free sessions and symbolize the traces through several steps: (1) State Construction (Section 3.2), in which we derive the set of application states S using collected session variables; (2) Input Symbolization (Section 3.3), in which we abstract concrete web requests into input symbols to profile the expected input domain at each state (i.e., $\Sigma_{exp}(s), \forall s \in S$); (3) Output symbolization, in which we leverage the same technique from BLOCK [7] to transform web responses (html pages) into output symbols, which allows us to obtain the mapping between the expected input and meaningful output symbols (i.e., $G(s, \Sigma_{exp}(s)) \rightarrow \Lambda_{mea}(s), \forall s \in S$). We also learn how the application transitions between states and the corresponding input symbols that trigger the transitions (i.e., $T : S \times \Sigma \rightarrow S$).

Then, we leverage the inferred partial FSM to construct unexpected inputs at each application state and test the application. In particular, we have to address two issues: (1) Test Input Generation: since we only have the knowledge of the expected input set given a state, how do we construct unexpected input symbols to test that state? We present two methods for generating test input vectors in Section 3.4. (2) Output Evaluation: since we only feed expected input symbols into the application, we have no knowledge about the dummy output set given a state. How do we determine whether the corresponding output symbol falls beyond the dummy output set, so as to report a potential logic vulnerability? We give the rule for evaluating output symbols in Section 3.5.

3.2 State Construction

A web application usually uses session variables to maintain the users' session state. Since a session variable can take infinite number of values, which results in infinite number of states, we first characterize the value domain for each session variable. We employ KS-test over the set of values collected for each session variable and categorize each variable into two types: (1) bounded, which means that the values of the variable are bounded to a finite set. For example, $_SESSION['privilege'] only assumes two values: user and admin, indicating two types of users. The value domain of bounded session variable consists of the set of observed values; (2) unbounded, which means the number of values the variable may take linearly increases with the number of samples. For example, $_SESSION['userid'] can take as many values as the number of users. The value domain of unbounded session variable consists of *null* and *nonnull*, where *null* means the variable doesn't exist.

Then, we construct the application state using the Cartesian product of the value domains of session variables, which we refer to as *state signature*, and identify the set of application states from the traces. One example state signature is [privilege = user] [userid = nonnull], which represents the application state when a regular user logs in. Theoretically, the application state space can be huge depending on the number of session variables and their value domains. In reality, we observe that the actual number of states is much smaller, since certain session variables are correlated with each other. For example, when a user logs in, both session variables are updated and there is no such state possible: [privilege = user] [userid = null].

3.3 Input Symbolization

Given a collection of web requests, we need to represent them with a finite number of input symbols. A web request consists of a HTTP method (we only consider GET and POST here), a URL path (the script file name in the case of PHP, e.g., index.php) and a set of input parameters. We symbolize each web request with a two-part structure *"Key.Value"*: (1) the *key* (denoted by K) represents the syntax structure of a web request, which is formed by the combination of HTTP method and URL path, referred to as the request key, with a set of parameters, whose order is sorted alphabetically, e.g., [GET-edit_user.php : userid]. (2) the *value* (denoted by V) represents the value domain of parameters, which indicates certain constraints over parameter values.

Since a parameter may assume an infinite number of values, we first profile each parameter and construct the "*Value*" part of the input symbol based on the value domains of all parameters, similar to how we build state signature. We take two steps to characterize the value domain of each parameter. First, since we observe that there may exist constraints between the parameter value and the specific application state, we profile each parameter at each state (i.e.,"local" profiling). We employ KS-test over the set of values collected for each parameter with the same request key at the same state and categorize the parameter into three types: (1) unbounded random variable (denoted as p_{ur}), which means the variable can take any random value. We represent its value domain with two abstract values, i.e., $\Theta(p_{ur}) = \{null, nonnull\}$, where Θ denotes the value domain. (2) unbounded constrained variable (denoted as p_{uc}), which means the variable can take an infinite number of values but is subject to certain constraints. Here, we only consider single state-related constraint, where the parameter value is always equal to a particular session variable (e.g., the parameter userid of request key GET-edit_user.php at state s_1 is always equal to $_SESSION['userid']). We represent its value domain with three values, i.e., $\Theta(p_{uc}) = \{null, v_{constrained}(sess), v_{unsatisfied}\}$, where $v_{constrained}$ denotes the value satisfying the constraint, *sess* denotes the session variable name and $v_{unsatisfied}$ denotes all the other values. (3) bounded variable (denoted as p_b), whose value domain is represented with the set of observed values plus two additional values: *null* and $v_{outofbound}$, where $v_{outofbound}$ denotes the values that are beyond the bounded set. In the second step, we combine the local views of the parameter value domains into a global view (i.e., "global" profiling). If the parameter has consistent value domain types over all the states, we keep its domain type and re-compute its value domain. For p_{ub}, its updated value domain is the set of values divided by multiple constraints, while in reality, this scenario rarely occurs. For p_b, additional values are added into its value domain. If the parameter has inconsistent value domain types over different states, we use the more restrictive domain type and further divide its value domain[1]. In the example application, the parameter userid of the request key GET-edit_user.php is constrained by $_SESSION['userid'] at state s_1, but inferred as an unbounded random variable at state s_2. Thus, its global value domain type is identified as unbounded constrained variable. In this way, we construct two atomic input symbols (i.e., $K_1.V_1$, $K_1.V_2$ as shown in Figure 2) at state s_2.

3.4 Test Input Generation

We present two methods for generating test input symbols at a given state s, as shown in Figure 3. Each method is designed to echo one type of attack vectors.

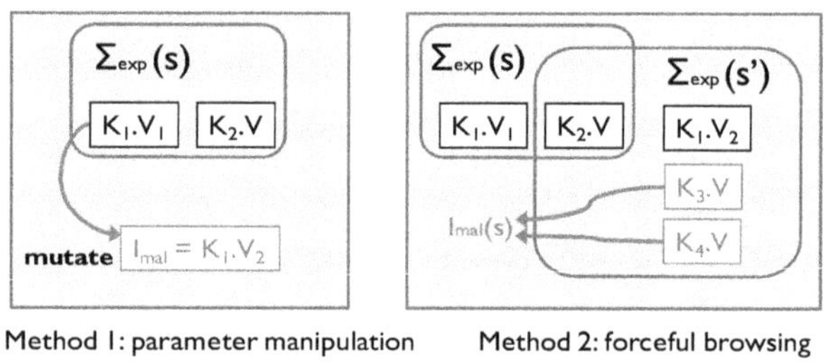

Figure 3: Test Input Generation

3.4.1 *Method 1 – Parameter Manipulation*

Given an expected input symbol $I = K.V \in \Sigma_{exp}(s)$ at state s, we directly mutate the input symbol, so that the mutated input symbol falls beyond the expected input set at state s, i.e., $I_{mal} = mutate(I) \notin \Sigma_{exp}(s)$, where *mutate* is a function over the input symbol. Specifically, the *mutate* function keeps the *"Key"* part of input symbol unchanged, but change the *"Value"* part by varying the values of one or several parameters. For an unbounded constrained variable, we may change its value from $v_{constrained}$ to $v_{unsatisfied}$; for an bounded variable, we may change its value to another value in the bounded set or $v_{outofbound}$. For example, as shown in Figure 3, the *"Value"* part of input symbol $K_1.V_1$ is mutated to be V_2 as a test input vector for state s. This method echoes the scenario of parameter manipulation attacks, where the attacker tampers the parameter values to violate the constraint between the web request and the current state.

3.4.2 *Method 2 – Forceful Browsing*

We leverage an expected input symbol from another state s', which falls beyond the expected input set of current state s, i.e., $I_{mal} \in \Sigma_{exp}(s') - \Sigma_{exp}(s)$. We select input symbols at state s' with a *"Key"* structure unobserved at state s as test

[1]The order of restrictiveness is defined as $p_b \gg p_{uc} \gg p_{ur}$, where $\gg$ means "is more restrictive than".

input vectors for state s. For example, as shown in Figure 3, the input symbols at state s' with *"Key"* K_3 and K_4 that are not observed at state s are used as test input vectors for state s. This method echoes the scenario of forceful browsing attacks, where the attacker provides a hidden sensitive link to the application that should not be accessible at current state.

3.5 Output Evaluation

Let the output symbol generated by the application after the test input vector I_{mal} is fed into the application at state s be O_{test}. The goal of the output evaluation is to determine whether O_{test} falls beyond the dummy output set (i.e., $O_{test} \notin \Lambda_{dum}(s)$). Since the dummy output set cannot be observed directly, here we apply the knowledge of the meaningful output set for evaluation. Based on the definition of dummy output, it is straightforward that there is no intersection between the meaningful and dummy output sets for any states. Thus, if the test output is the same as any meaningful output, we should report a potential logic vulnerability. Specifically, let I be the original input symbol from which I_{mal} is generated ($I_{mal} = mutate(I)$ in Method 1; $I_{mal} = I$ in Method 2), and O_{orig} be the output symbol generated by feeding the original input symbol I at the corresponding state (i.e., $O_{orig} = G(s, I)$ in Method 1; $O_{orig} = G(s', I)$ in Method 2, where s' is the state from which I is selected from). The output evaluation rule is:

if $O_{test} = O_{orig}$, *we report a potential logic vulnerability.*

If a test input vector can be generated from more than one input symbol (e.g., $K_1.V_3$ can be generated from both $K_1.V_1$ and $K_1.V_2$), which have different output symbols, the resulting output symbol will be compared to every possible output symbol.

Since our technique relies on the collected traces, which may not completely characterize the application behavior, we report those alerts as potential logic vulnerabilities, which require human efforts to analyze and confirm.

4. IMPLEMENTATION

We implement a prototype system LogicScope for identifying logic flaws within PHP web applications. Our technique only requires session information from the server side, which is usually maintained externally to the web application at either local files (e.g., at /var/lib/php5 for PHP) or in database table (e.g., django_session for Python). As shown Figure 4, LogicScope is composed of three major components, including *Trace Collector*, *Spec Analzyer* and *Testing Engine* and executed in three phases: trace collection, FSM inference and testing.

Trace Collector collects web traffic traces during attack-free sessions. We implement *Trace Collector* by enhancing WebScarab proxy with a *Session Inspector* module, which collects session variables through communicating with *Session Exporter* deployed on the server. To implement our technique for other platforms (e.g., JSP), *Session Exporter* is the only module that needs to be adapted.

Spec Analyzer is executed to derive both the partial FSM and the testing specification. Traces are first fed into *Symbolizer* and transformed into symbolized session logs, which are used for deriving the partial FSM (i.e., *FSM Analyzer*) and generating testing specification (i.e., *TestSpec Generator*), including a set of test input symbols for each state and corresponding output symbols for evaluation.

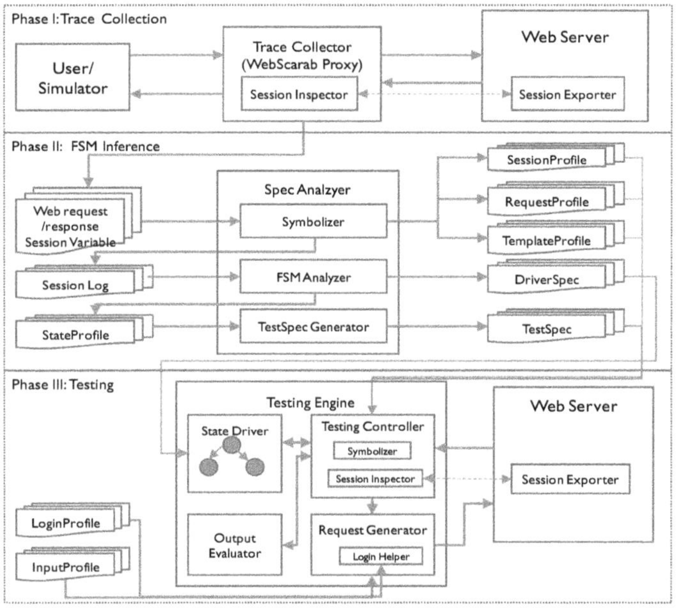

Figure 4: Prototype System Architecture

Testing Engine instantiates test input vectors into concrete web requests (i.e, *Request Generator*), feed them into the application and evaluate the corresponding web responses for vulnerability identification (i.e., *Output Evaluator*). In particular, *Testing Controller* is in charge of the entire testing procedure. *State Driver* computes the path leading to the target test state by feeding a sequence of input symbols to trigger the state transition step by step. When *Request Generator* instantiates input symbols, it is aware of the parameter value type and leverages pre-loaded InputProfile and *Login Helper* to generate meaningful parameter values.

5. EVALUATION

We select six real-world PHP web applications to evaluate our prototype system LogicScope. We deploy all web applications on a 2.13GHz Core 2 Linux server with 2GB RAM, running Ubuntu 10.10, Apache web server (version 2.2.16) and PHP (version 5.3.3). To facilitate trace collection, we developed user emulators for each application based on Selenium WebDriver. The user emulator will initiate a browser instance and automate the procedure of operating the web application, through performing a sequence of actions by following navigation links.

LogicScope first collects traces with the help of user emulators and performs specification inference. Table 1 shows the statistics of collected traces and inferred FSM, including the number of files, collected web requests, web responses, session variables, states, input and output symbols. Then, LogicScope launches the testing process against each application and also gives concrete attack vectors and evidences for manual inspection. We manually analyze the reported vulnerabilities and classify them into either true vulnerability or false positive. Table 2 shows the testing results, including the number of test input vectors generated by each method (*FB* for forceful browsing; *PM* for parameter manipulation), reported attack instances, real attack vectors and false positives (i.e., FP). We also compare the discovered vulnerabilities with the known vulnerabilities from public sources and report the false negatives (i.e., FN). In summary, we generate 233 test input vectors for all the applications and get 25 attack instances reported, among which, 18 are real attack vectors and 7 are false positives. Detailed results can be found in the technical report version [8].

Table 1: Summary of Traces and Inferred FSM (#)

Application	Files	Web req.	Web resp.	Session var.	State	Input symb.	Output symb.
Scarf	21	1348	1346	3	3	31	15
Wackopicko	52	2104	1650	1	2	18	7
Events Lister	37	1290	1287	2	2	25	11
Bloggit	24	2657	2645	1	2	18	23
openInvoice	25	1138	1083	5	5	73	11
OpenIT	25	1462	1453	5	5	16	6

Table 2: Summary of Testing Results (FB: forceful browsing, PM: parameter manipulation)

App	Method	Test input	Flagged attack	Real attack	FP	FN
Scarf	*FB*	5	1	0	1	0
	PM	44	9	8	1	0
Wackopicko	*FB*	5	2	2	0	1
	PM	16	0	0	0	0
Events Lister	*FB*	0	0	0	0	0
	PM	25	4	2	2	0
Bloggit	*FB*	6	2	0	2	0
	PM	20	0	0	0	1
openInvoice	*FB*	19	4	4	0	0
	PM	28	0	0	0	0
OpenIT	*FB*	3	3	2	1	0
	PM	62	0	0	0	0
Summary		233	25	18	7	2

6. RELATED WORKS

The key issue to address web application logic flaws is to infer the application logic specification. Waler [4] and Sun et al. [12] require application source code for analysis or instrumentation. In contrast, our approach is source code free. Second, Waler [4] can only identity violations of value-related invariants and NoTamper [1] only models the application logic behind form processing and validation. Our approach characterizes the application logic in a more general manner and can deal with a broader range of logic flaws. Furthermore, Swaddler [3], BLOCK [7] and SENTINEL [9] directly leverage the specification to detect attacks at runtime, while we have to construct testing cases for vulnerability exploitation in addition to specification inference.

Our work is also related to an array of fuzing testing works (FLAX [11], DART [5], SAGE [6], Kudzu [10]), which aim to achieve better coverage for discovering vulnerabilities through guided/directed test input generation. All of the above works require application source code, while our test input generation is based on a specification inferred in a black-box manner.

7. SUMMARY

In this paper, we present a systematic black-box approach to identifying logic flaws within web applications, implement and evaluate a prototype system LogicScope to demonstrate the effectiveness of our approach. We also would like to point out several limitations. First, LogicScope can not handle AJAX web applications. Second, LogicScope has limited capability in handling complex relationships/constraints within the database.

8. ACKNOWLEDGMENTS

This work was supported by NSF TRUST (The Team for Research in Ubiquitous Secure Technology) Science and Technology Center (CCF-0424422).

9. REFERENCES

[1] P. Bisht, T. Hinrichs, N. Skrupsky, R. Bobrowicz, and V. N. Venkatakrishnan. NoTamper: automatic blackbox detection of parameter tampering opportunities in web applications. In *CCS'10*, pages 607–618, 2010.

[2] Citigroup credit card information leakage in 2011. http://www.wired.com/threatlevel/2011/06/citibank-hacked/.

[3] M. Cova, D. Balzarotti, V. Felmetsger, and G. Vigna. Swaddler: An Approach for the Anomaly-based Detection of State Violations in Web Applications. In *RAID'07*, pages 63–86, 2007.

[4] V. Felmetsger, L. Cavedon, C. Kruegel, and G. Vigna. Toward Automated Detection of Logic Vulnerabilities in Web Applications. In *USENIX'10*, pages 143–160, 2010.

[5] P. Godefroid, N. Klarlund, and K. Sen. Dart: directed automated random testing. In *PLDI'05*, pages 213–223, 2005.

[6] P. Godefroid, M. Y. Levin, and D. A. Molnar. Automated whitebox fuzz testing. In *NDSS'08*, 2008.

[7] X. Li and Y. Xue. BLOCK: A Black-box Approach for Detection of State Violation Attacks Towards Web Applications. In *ACSAC'11*, pages 247–256, 2011.

[8] X. Li and Y. Xue. LogicScope: Automatic Discovery of Logic Vulnerabilities within Web Applications. Technical report, Vanderbilt University ISIS, 2012.

[9] X. Li, W. Yan, and Y. Xue. SENTINEL: securing database from logic flaws in web applications. In *CODASPY '12*, pages 25–36, 2012.

[10] P. Saxena, D. Akhawe, S. Hanna, F. Mao, S. McCamant, and D. Song. A symbolic execution framework for javascript. In *Oakland'10*, pages 513–528, 2010.

[11] P. Saxena, S. Hanna, P. Poosankam, and D. Song. FLAX: Systematic Discovery of Client-side Validation Vulnerabilities in Rich Web Applications. In *NDSS'10*, 2010.

[12] F. Sun, L. Xu, and Z. Su. Static Detection of Access Control Vulnerabilities in Web Applications. In *USENIX'11*, pages 11–11, 2011.

Protecting Function Pointers in Binary

Chao Zhang
LiST, Peking University
Beijing, China
chao.zhang@pku.edu.cn

Tao Wei*
LiST, Peking University
UC Berkeley
wei_tao@pku.edu.cn

Zhaofeng Chen
LiST, Peking University
Beijing, China
chenzhaofeng@pku.edu.cn

Lei Duan
LiST, Peking University
Beijing, China
lei_duan@pku.edu.cn

Stephen McCamant
University of Minnesota
USA
mccamant@cs.umn.edu

Laszlo Szekeres
Stony Brook University
USA
lszekeres@cs.stonybrook.edu

ABSTRACT

Function pointers have recently become an important attack vector for control-flow hijacking attacks. However, no protection mechanisms for function pointers have yet seen wide adoption. Methods proposed in the literature have high overheads, are not compatible with existing development process, or both. In this paper, we investigate several protection methods and propose a new method called FPGate (i.e., Function Pointer Gate). FPGate rewrites x86 binary executables and implements a novel method to overcome compatibility issues. All these protection methods are then evaluated and compared from the perspectives of performance and ease of deployment. Experiments show that FPGate achieves a good balance between performance, robustness and compatibility.

Categories and Subject Descriptors

D.4.6 [**Operating Systems**]: Security and Protection—*Access controls*; D.2.4 [**Software Engineering**]: Software/Program Verification—*Validation*

Keywords

Function Pointer Protection; Binary Rewriting

1. INTRODUCTION

Control-flow hijacking attacks [4, 16] have a long history, and corresponding protections [3, 17] have also existed for more than a decade, in an ongoing arms race. The current state of the art of protections pay much attention to the return addresses on the stack (i.e. targets of return instructions), but provides few protections against function pointers (i.e. targets of indirect call and jump instructions).

*Corresponding author.

And thus, function pointers have recently become an important attack vector. For instance, a buffer overflow or integer overflow vulnerability can be exploited to overwrite function pointers in the heap to hijack the control flow [6, 8]. Moreover, recent exploits [19] against use-after-free vulnerabilities replace objects' virtual function tables (i.e. `vtables`) to turn benign method calls into jumps to shellcode.

Representative protection schemes which protect function pointers from exploiting including PointGuard [7] and CFI (Control-Flow Integrity [1]). However, they require cooperation from developers or code-producers, or suffer from inefficiency and compatibility problems. We propose a new method called FPGate to achieve a balance.

First, FPGate needs no source-level information (e.g., PDB files or debug information needed by CFI). It utilizes relocation tables which are already required by ASLR (Address Space Layout Randomization [17]) in modern x86 binary executables, to disassemble binaries and identify all indirect transfer instructions and all valid jump targets.

Second, FPGate overcomes compatibility issues by encoding each valid function pointer into a pointer to a new *trampoline* chunk. Unlike other schemes, modules hardened by FPGate can inter-operate seamlessly with un-hardened ones.

Moreover, optimizations are introduced by FPGate to accelerate the run time execution, without any loss of security. As a result, FPGate introduces a negligible run time overhead, about 0.4% on SPECint2006.

FPGate fills most of the gap between existing lightweight protections on one hand, and CFI on the other. Combining FPGate with robust protections of return addresses is a sweet spot for security and usability. This combination provides a protection almost as strong as the original CFI. At the same time, it has low overhead and can be applied directly and progressively to a binary.

The key contributions of this paper are:

- We investigate several existing function pointer protection mechanisms, and point out their advantages and disadvantages.
- We propose a new efficient mechanism FPGate to protect function pointers in binary executables and provide incremental deployment.
- We compare FPGate with existing mechanisms in an experimental evaluation. The results show that FPGate achieves a good balance between performance, robustness and compatibility.

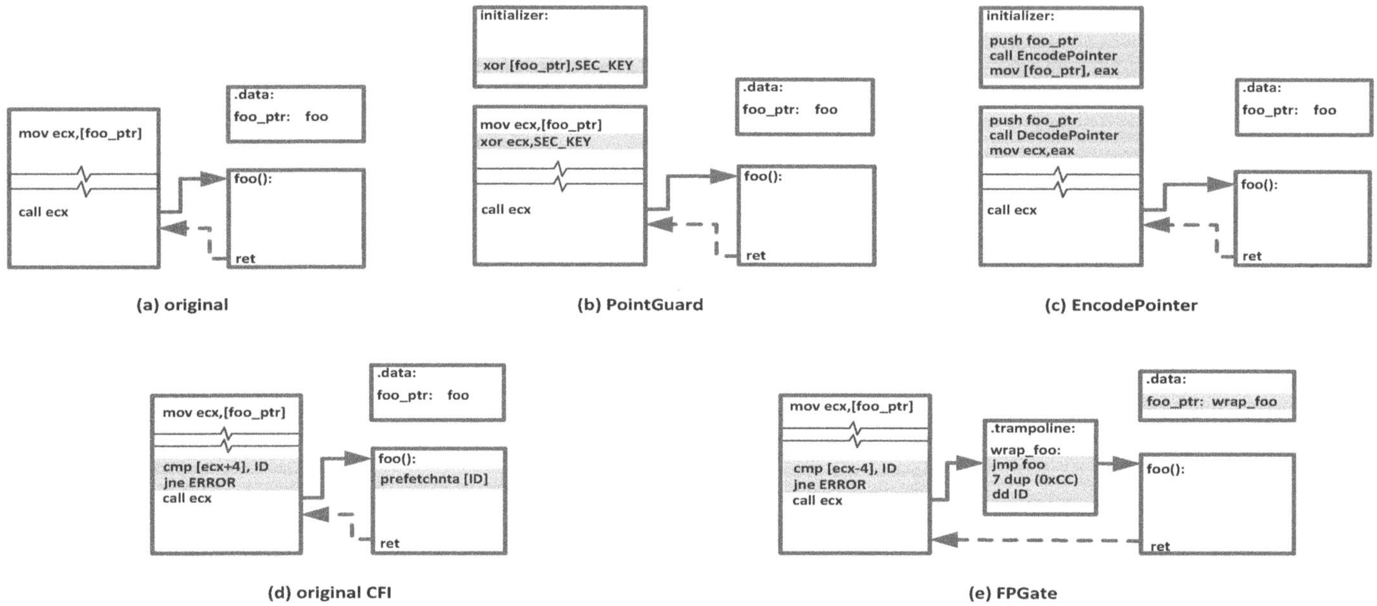

Figure 1: (a) An indirect call whose target is from memory. (b)-(e) Overview of existing function pointer protection schemes and FPGate.

2. STUDY OF EXISTING PROTECTIONS

As an important attack vector, the function pointer has drawn researchers' attentions for years. In existing protection techniques, legal function pointers are differentiated from illegal ones by encoding (or encrypting), by attaching identifiers (IDs), or by memory alignments. Several representative mechanisms are discussed here.

2.1 PointGuard

PointGuard [7] uses encryption to provide integrity for function pointers. A per-process secret key, i.e., `SEC_KEY`, is kept to encrypt and decrypt pointers and thus protect legal pointers from tampering. Before a function pointer is stored into memory, it is encrypted (i.e. XOR) with the secret key first. After the function pointer is read from memory, it is decrypted (i.e. XOR) and then stored in registers for use.

Figure 1(a) shows an indirect call example whose target is read from memory. After PointGuard is applied to this code snippet, the hardened version is shown in Figure 1(b). It is worth noting that a special initializer is instrumented by PointGuard to re-initialize all statically initialized pointers.

The integrity of function pointers hardened by PointGuard depends on the confidentiality of the secret key. However, the secret key is vulnerable to be stolen by attackers, such as through information leakage vulnerabilities. In addition, because the XOR operation is linear, an attacker may be able to manipulate an encoded pointer by overwriting only the lower bits of an address and launch an attack [2].

PointGuard works as a compiler extension and thus needs a target application's source code. Besides, with encryption and decryption, function pointers cannot flow between hardened modules and un-hardened ones.

2.2 EncodePointer

As PointGuard was never released, Microsoft implemented a similar approach beginning in Windows XP SP2. It provides two API routines that can be called at the discretion of the programmer, named `EncodePointer` and `DecodePointer` [13]. Figure 1(c) shows the hardened version using `EncodePointer` for the code snippet in Figure 1(a).

Rather than storing the secret key in user space like PointGuard, `EncodePointer` queries the system API `NtQueryInformationProcess` to acquire the secret key. It prevents a user-level attack from accessing the secret key directly, ameliorating the problem of information leakage vulnerabilities.

However, programmers have to decide which points to invoke `EncodePointer` and `DecodePointer` manually. In addition, each call to `EncodePointer/DecodePointer` queries the kernel, and introduces a significant run time overhead.

The requirement for source code and the compatibility problems between hardened and un-hardened modules in this protection scheme are analogous to those in PointGuard.

2.3 CFI

CFI [1] is a natural protection against control-flow hijacking attacks. It guarantees that all control-flow transfers in a program will be the ones intended in the original program (i.e., those represented in the compiler's control-flow graph).

In general, it inserts an ID (e.g., the `prefetchnta` instruction in Figure 1(d)) before each legal target (e.g., a function entry), and a check before each indirect transfer instruction. Before transferring to the computed target at runtime, the dynamic check validates whether the target's ID is correct.

CFI defeats a broad range of shellcode injection attacks, including sophisticated ROP (Return Oriented Programming [4]). However, despite its long history, CFI has not seen wide industrial adoption.

It is in part because CFI imposes a significant overhead. Even if only function pointers are protected by CFI (called as CFI-fp), it also introduces an overhead of 7.2%, as shown in Section 4.3. In addition, it requires source-level information (e.g., debug information) which are usually not available in COTS binary. Moreover, modules hardened by CFI cannot inter-operate seamlessly with un-hardened ones. For example, the instrumented runtime check will fail if the computed jump target falls into an un-hardened module.

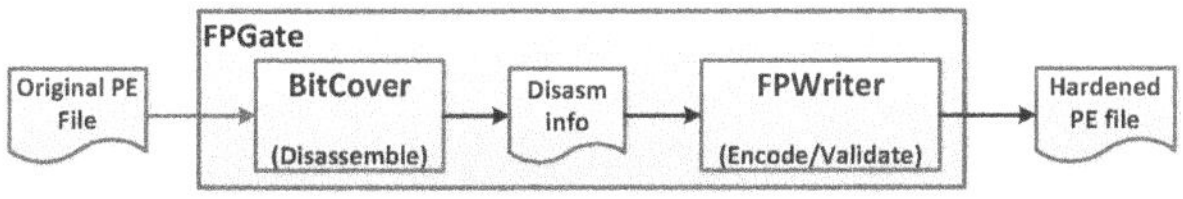

Figure 2: Architecture of FPGate

3. DESIGN & IMPLEMENTATION

3.1 Approach Overview

As PointGuard, FPGate encodes legal function pointers to differentiate them from illegal ones. Unlike PointGuard, pointers encoded by FPGate are valid code pointers, and can flow into un-hardened modules without compatibility issues.

On the other hand, a unique ID is attached to each encoded pointer for validating, as CFI. A novel code section, i.e., the *trampoline* section, is thus introduced for encoding pointers and holding IDs. Each valid function pointer is encoded into a pointer to a code chunk in the *trampoline*. Unlike CFI, most instructions' addresses will not change after rewriting. Meanwhile, with the *trampoline* layer, computed jump targets which fall into un-hardened modules can also be encoded to overcome compatibility issues.

The architecture of FPGate is shown in Figure 2. It consists of two core modules: BitCover and FPWriter. BitCover disassembles target binary and identifies all indirect call and jump instructions in addition to all valid jump targets. FPWriter then instruments runtime checks before indirect transfers, encodes valid jump targets with new code pointers, and creates the trampoline section. Figure 1(e) shows the code snippet which is rewritten by FPWriter.

3.2 Disassembling and Identifying

In general, it is challenging to disassemble an x86 PE [14] file correctly, because x86 is a CISC platform. However, we can take advantage of the fact that ASLR and DEP (Data Execution Prevention [3]) are widely adopted in Windows/x86 executables, particularly those whose developers care about security.

Combined with other policies, a custom disassembler BitCover [21] is built and can disassemble PE files generated by modern compilers. With BitCover, all indirect jumps or calls and their valid targets can be identified.

3.3 Validating Jump Targets

After identifying all indirect *call/jmp* instructions by BitCover, FPWriter instruments runtime checks before them.

As shown in Figure 1(e), a check which verifies the existence of a predefined ID around the jump target is inserted before the indirect call instruction (i.e., *call ecx*).

Similar as the CFI, the ID is carefully chosen in order to avoid conflicting with bytes in the application.

3.4 Encoding Legal Targets

Valid jump targets should be encoded to distinguish from invalid targets. According to our observation, legal jump targets fall into the following categories:

(A) Hard-coded function pointers in the executable file, such as those stored in the *vtables* and global variables.

(B) Entries in the import table.

(C) Function pointers generated at runtime by API, such as `GetProcAddress()`.

(D) Function pointers generated by `setjmp()`.

For function pointers of kind (A), they are simply encoded into a trampoline chunk pointer, as shown in Figure 1(e).

Considering the performance, hard-coded pointers which are directly used in `call/jmp` instructions, e.g., `call foo`, would not be encoded. Exception handlers used by the operating system are skipped encoding too. Moreover, code entries in switch tables [5] are not encoded. All these skipped pointers are protected by SafeSEH [12] or DEP, and cannot be tampered by attackers, and thus can be skipped safely.

The remaining cases B-D involve pointers from external modules, even un-encoded pointers, and thus cause compatibility issues. Unlike existing protections (e.g., CFI), FPGate can handle these issues with the extra *trampoline* layer.

3.5 Trampoline & Compatibility

FPWriter encodes each valid jump target into a pointer to a chunk in the trampoline. Each of these chunks holds a same ID and has 16 bytes. These chunks are continuous and can pass the validations instrumented before indirect jumps. As Figure 1(e) shows, the chunk will jump back to the original target to ensure applications behave correctly.

In the case of (B), function pointers are imported from external modules and will be updated when loading. As shown in Figure 3(a), a wrapper section (*.wrap*) is introduced to simulate the original IAT (Import Address Table, *.iat*). Whereas the IAT stores imported function pointers, the wrapper section stores pointers to trampoline chunks. Then, all references to the IAT are replaced with references to the wrapper section. Similarly, directly used imported function pointers, such as `call [imp_slot]`, are skipped encoding too in order to promote the performance.

For function pointers of kind (C), they are retrieved through API `GetProcAddress`, so they need be encoded at runtime. `GetProcAddress` is usually imported, and thus it can be processed in a same way as (B), except that the trampoline chunk for `GetProcAddress` does not jump back to original imported `GetProcAddress` directly. In fact, it jumps to a wrapper function for `GetProcAddress` (i.e. `gpa_wraper`), as shown in Figure 3(b). The wrapper function will call original `GetProcAddress`, and then stores the runtime retrieved function pointer into a preserved trampoline slot. And then the pointer to the preserved slot is returned as the new return value. As a result, the runtime retrieved function pointer is also encoded. It is worth noting that, the wrapper function has to set the preserved trampoline slots writable at runtime and turn it off when the wrapper function exits, in order to protect the preserved trampoline slots from tampering.

Function pointers of kind (D) are generated at runtime as well. The function `setjmp` retrieves its caller's (un-encoded) return address from the stack and saves it into a `jmp_buf` structure. Finally, this return address will be used by an indirect jump instruction in another API `longjmp` through querying the `jmp_buf` structure. We then replace each call instruction which invokes `setjmp` with two instructions: `PUSH ret_addr; JMP setjmp;`. The pushed return address `ret_addr` is a trampoline slot pointer, and it will jump back to the original return address, as shown in Figure 3(c).

As discussed above, the trampoline provides a demilitarized zone between hardened modules and un-hardened ones, and thus can help overcome the compatibility issues. This novel method can also be utilized and applied to other binary protection schemes, such as [21].

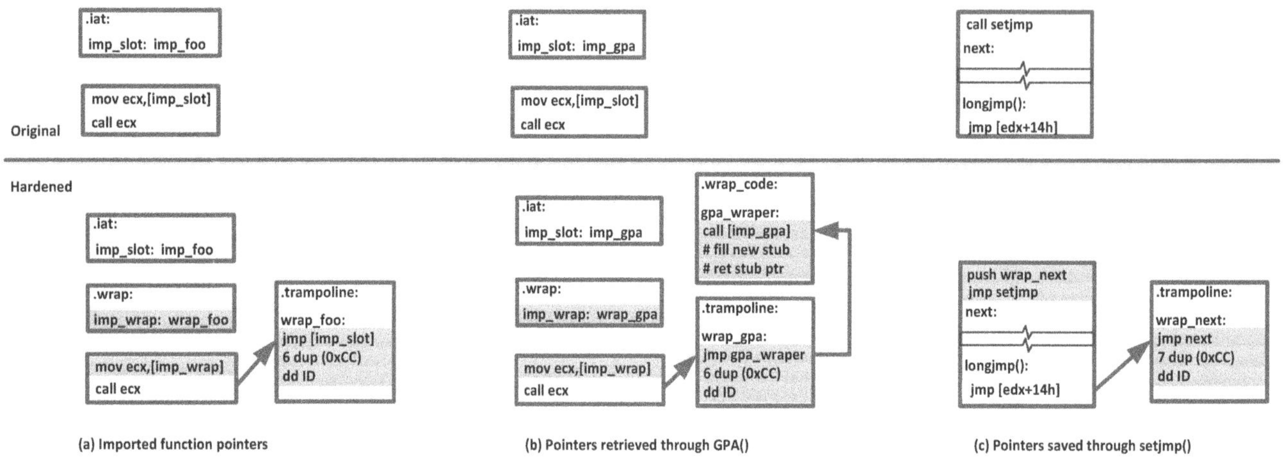

Figure 3: Special function pointers and FPGate's encoding countermeasures

4. EVALUATION

We implement a prototype of FPGate using C++ for the Windows platform targeting x86 PE executables. BitCover takes about 5K LOC and utilizes the Udis86 library (8K LOC, [18]) to parse x86 instructions, while FPWriter takes another 5K LOC and a custom PE file parser has 2K LOC.

We port PointGuard, EncodePointer and CFI discussed in Section 2 using binary rewriting, and then evaluate and compare their compatibility and performance with FPGate.

4.1 Correctness of FPGate

FPWriter encodes function pointers and instruments runtime checks before indirect `call/jmp` instructions in the executable. If there is an error in this rewriting process, the final hardened application will fail or crash at runtime. For example, if a function pointer fails to be identified by BitCover, then it will not be encoded by FPWriter, and finally cannot pass the instrumented runtime check.

Two groups of experiments are made to evaluate FPGate's correctness. First, FPGate is tested with 11 applications from the SPECint2006 benchmark [9]. Second, FPGate is tested with two real world browsers, Firefox 3.6.16 (FF3) and Internet Explorer 6 (IE6)[1]. In particular, the core modules `xul.dll` in FF3 and `mshtml.dll` in IE6 are hardened by FPGate, and other modules are left intact. These experiments show that hardened applications work correctly; i.e., they are semantically equivalent to original applications.

4.2 Compatibility Comparison

In this section, we discuss the compatibility of FPGate, and make a comparison with the other three protection schemes, i.e. PointGuard, EncodePointer and CFI-fp (a limited version of original CFI which protects only function pointers).

4.2.1 Compatible with Binary Applications

FPGate can be applied directly on binaries containing relocation tables, even without source-level information. For (seldom) binaries without relocation tables, FPGate work fine with the help of other tools or experts. For example, the commercial disassembler IDA Pro [10] can be easily utilized to generate a fake relocation table and help FPGate to be applied on these binaries.

On the other hand, original implementations of PointGuard and EncodePointer depend on source code, while the CFI-fp depends on debug information. However, in our porting, they are all ported successfully using binary rewriting.

4.2.2 Compatible with Un-hardened Modules

Function pointers may flow from hardened modules to un-hardened modules and vice-versa, and thus cause compatibility issues. With countermeasures presented in Section 3.5, modules hardened by FPGate can work seamlessly with un-hardened modules. The previous section which evaluates correctness of FPGate also proves this compatibility.

But for other three schemes, the compatibility issue is critical. Special efforts are need to fix this issue when porting.

In the case of PointGuard and EncodePointer, encoded function pointers are not valid pointers, i.e. they cannot be used directly in any un-hardened module. When porting, the binary has to be rewritten in such a way that each function pointer is decoded before it flows into external modules. For each specific application, lots of manual efforts are needed to identify and fix such pointer flows between modules.

On the other hand, function pointers from un-hardened modules may fail the runtime checks of the hardened modules. For imported function pointers, a bootstrap code chunk is inserted to encode them. For pointers retrieved by `GetProcAddress`, similar wrappers as FPGate are provided too. For `setjmp` and other cases, it is also handled like FPGate.

In the case of CFI-fp, compatibility issues occur when computed jump targets fall into un-hardened modules. In our porting, the error handler is set to be empty and will redirect the control flow to its original position.

In summary, FPGate has a good compatibility. PointGuard or EncodePointer needs a lot of manual effort to achieve a good compatibility. Compatibility issues of the original CFI implementation are hard to be fixed unless another layer (like the trampoline in FPGate) is introduced.

4.3 Performance Comparison

The performance experiments are made on a Windows 7 32-bit system, with an Intel Core2 CPU of 3.00GHz. The

[1]Core modules of newer IE cannot be replaced with custom modules because of the system protection. In addition, there are no public available exploits for newer browsers. As a result, we chose these two old softwares as a benchmark.

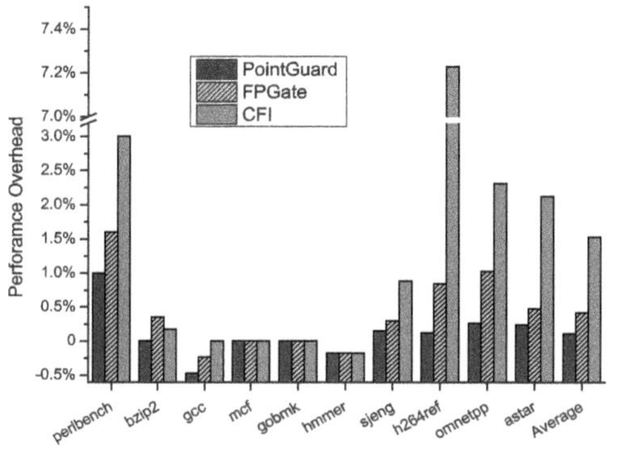

Figure 4: Overheads of PointGuard, FPGate and CFI-fp.

scripts in SPECint2006 are used to evaluate the original and hardened versions of applications' performance. And the average performance overheads are computed based on 9 trials.

Runtime overhead.

Figure 4 shows the performance overheads caused by FPGate, PointGuard and CFI-fp. The average/maximum overhead is about 0.11%/1% for PointGuard, about 0.42%/1.03% for FPGate, and about 1.53%/7.23% for CFI-fp. For most applications, the overhead brought by PointGuard is smallest, while that brought by CFI-fp is largest.

In addition, EncodePointer's performance is quite poor. Its average/maximum overhead is about 92%/750%, not shown in this figure due to the space limitation.

Statistics.

The upper half of Table 1 lists the count of modifications made by FPGate to the SPECint2006 applications, and the lower half for other real world applications.

More specifically, the columns under *Encoded pointers* in the table represent the count of pointers encoded by FPGate, including hard-coded function pointers, imported function pointers, pointers returned by `GetProcAddress` and pointers generated by `setjmp` functions.

The column under *Checks* records the count of indirect `call/jmp` instructions which are instrumented by FPGate, while the final column (i.e. the column under *Opt.*) collects the count of function pointers those are skipped encoding.

Performance Analysis.

PointGuard is fastest because it only needs one XOR instruction to encrypt and decrypt the jump targets. EncodePointer is slowest, because it queries the kernel heavily.

For the porting of CFI-fp, its overhead is about 1.5%, much faster than the original implementation in [1]. That is because this porting only protects function pointers but not return addresses.

But the CFI-fp is slower than FPGate, mainly due to the optimizations introduced by FPGate. As shown in the last column of Table 1, the count of function pointers skipped encoding is quite large, usually larger than the count of encoded function pointers. This optimization can greatly promote the performance of FPGate. However, this optimization cannot be deployed easily to CFI-fp. For example, for instruction *call foo*, FPGate can skip encoding the *foo* and thus introduce no overheads, but CFI cannot skip inserting an ID at the beginning of function *foo* because *foo* may be used as the target of another indirect jump.

In addition, the ID inserted by CFI-fp at the beginning of each jump target (e.g. function entries) is the **prefetchnta** instruction. This instruction has no side-effect but is extremely slow. For each direct (relative or absolute) jump or call instruction, the instrumented slow **prefetchnta** instructions at the beginning of the jump target will be always executed. As a result, CFI-fp is a little slower than FPGate.

Table 1: Modifications made by FPGate to applications

Benchmarks	Encoded pointers				Checks	Opt.
	#fp	#imp	#GPA	#setjmp	#indirect call/jmp	#skipped fp/import
SPECint2006						
400.perlbench	1,380	46	5	1	495	2,702
401.bzip2	73	20	3	0	123	290
403.gcc	3,501	26	3	1	774	10,254
429.mcf	73	20	3	0	105	238
445.gobmk	1,967	22	3	0	185	566
456.hmmer	117	23	3	0	125	429
458.sjeng	84	20	3	0	107	407
464.h264ref	185	20	3	0	465	369
471.omnetpp	4,798	30	3	1	1,853	1,065
473.astar	95	20	3	0	109	256
Browsers						
mshtml.dll (IE6)	1,526	139	21	0	10,452	29,557
xul.dll (FF3.6)	145,224	283	34	0	55,025	17,273

Table 2: Real World Exploit Samples Prevented by FPGate.

ID	App	Vul Type	Vul Module	Protected
CVE-2011-0065	FF 3.6	Use After Free	xul.dll	yes
CVE-2010-0249	IE 6	Use After Free	mshtml.dll	yes
CVE-2008-0348	coolplayer	Stack Overflow	core exe	yes
CVE-2010-5081	RM-MP3	Stack Overflow	core exe	yes
OSVDB-83362	urlhunter	Stack Overflow	core exe	yes
CVE-2007-1195	XM ftp	Format String	core exe	yes
OSVDB-82798	ComSndFTP	Format String	core exe	yes

4.4 Security Comparison

FPGate limits indirect instructions transfer only to valid targets with IDs. There is only one ID in FPGate, and thus it provides a weaker protection than CFI. However, due to the complexity of building a complete control flow graph, the count of unique IDs introduced by CFI for a given application is also small. And thus, the security gap between FPGate and CFI is not so much.

We chose 7 publicly available exploits from Metasploit [11] and ExploidDB [15] and tested them in a virtual machine running Windows XP SP3 within a separate experiment network. These exploits are tested against the original vulnerable applications and corresponding hardened applications. Table 2 shows the 7 vulnerabilities attacked by exploits we used. Results show that all these protection schemes can protect target browsers from attacking.

4.5 Miscellaneous

FPGate disassembles target binary and rewrites them. The static analysis time is positively correlated with the target file size and is usually small. As our experiments shows, it costs 63 seconds for `gcc` which is 1,200KB, and costs only 0.16 seconds for `mcf` which is 80KB.

FPGate instruments a trampoline section and allocates 16-bytes for each encoded function. The runtime memory overhead it introduced is positively correlated with the count of functions. For example, `perlbench` has about 700 functions, and the runtime memory overhead is about 10KB.

5. DISCUSSION

FPGate relies on relocation tables to disassemble binary and rewrite them. A recently published work REINS [20] utilizes IDA Pro to provide similar protection without relocation tables. However, IDA Pro is heuristic and suffers significant disassembly errors. REINS cannot identify all valid function pointers and thus maintain a lookup table to encode function pointers at runtime. A larger runtime overhead is then introduced. Moreover, if the computed jump targets fall into external modules and are retrieved through methods like `GetProcAddress`, REINS fails to protect the control flow, and causes compatibility issues.

FPGate uses only one ID and thus permit jumps to any valid target with an ID. This is vulnerable to jump-to-libc attacks. Attackers can hijack controls to valid jump targets and launch an attack. Further work is needed to defend against this kind of attack.

6. CONCLUSION

In this paper, we investigate several protection methods against function pointers, including PointGuard, EncodePointer and CFI. Inspired by them, a new approach called FPGate is proposed. FPGate limits all indirect calls and jumps to known valid targets. It can block various attacks against function pointers.

FPGate can be applied through binary rewriting on executables generated by modern compilers. It is compared with existing protections from the perspectives of performance, robustness, and ease of deployment. Results show that FPGate achieves a good balance between performance, robustness and compatibility.

Acknowledgments

This research was supported in part by the National Natural Science Foundation of China under the grant No. 61003216 and No. 61003217, the Chinese NDRC InfoSec Foundation under Grant No.[2010]3044, the National Science Foundation under Grant No. 0842695, 0831501 CT-L, CCF-0424422 and CNS-0831298, the Office of Naval Research under MURI Grant No. N000140911081, N000140710928 and award FA9550-09-1-0539, the AFOSR under grant FA9550-09-1-0539, and DARPA under award HR0011-12-2-005. Any opinions, findings, and conclusions or recommendations expressed in this material are those of the author(s) and do not necessarily reflect the views of the National Science Foundation.

7. REFERENCES

[1] M. Abadi, M. Budiu, U. Erlingsson, and J. Ligatti. Control-flow integrity: Principles, implementations, and applications. In *Proceedings of the 12th ACM Conference on Computer and Communications Security*, pages 340–353, 2005.

[2] S. Alexander. Defeating compiler-level buffer overflow protection. *The USENIX Magazine ;LOGIN*, 30(3):59–71, 2005.

[3] S. Andersen and V. Abella. Data execution prevention: Changes to functionality in microsoft windows xp service pack 2, part 3: Memory protection technologies. `http://technet.microsoft.com/en-us/library/bb457155.aspx`, 2004.

[4] E. Buchanan, R. Roemer, H. Shacham, and S. Savage. When good instructions go bad: generalizing return-oriented programming to risc. In *Proceedings of the 15th ACM conference on Computer and communications security*, pages 27–38, 2008.

[5] C. Cifuentes and M. Van Emmerik. Recovery of jump table case statements from binary code. In *Proceedings of the 7th International Workshop on Program Comprehension*, pages 192–199. IEEE, 1999.

[6] CORE Security Technologies Advisories. Core-2007-0219: Openbs's ipv6 mbufs remote kernel buffer overflow. `http://www.securityfocus.com/archive/1/462728/30/150/threaded`, 2007.

[7] C. Cowan, S. Beattie, J. Johansen, and P. Wagle. PointguardTM: protecting pointers from buffer overflow vulnerabilities. In *Proceedings of the 12th conference on USENIX Security Symposium*, volume 12, pages 91–104, 2003.

[8] M. Daniel, J. Honoroff, and C. Miller. Engineering heap overflow exploits with javascript. In *Proceedings of the 2nd conference on USENIX Workshop On Offensive Technologies*, 2008.

[9] J. L. Henning. Spec cpu2006 benchmark descriptions. *SIGARCH Comput. Archit. News*, 34:1–17, Sept. 2006.

[10] Hex-Rays SA. IDA Pro: a cross-platform multi-processor disassembler and debugger. `http://www.hex-rays.com/products/ida/index.shtml`.

[11] Metasploit Open Source Commitment. Metasploit penetration testing software & framework.

[12] Microsoft Visual Studio 2005. Image has safe exception handlers. `http://msdn.microsoft.com/en-us/library/9a89h429%28v=vs.80%29.aspx`.

[13] MSDN online library. EncodePointer function: provide another layer of protection for pointer values. `http://msdn.microsoft.com/en-us/library/bb432254(v=vs.85).aspx`.

[14] MSDN online library. Microsoft Portable Executable (PE) and Common Object File Format (COFF) Specification. `http://msdn.microsoft.com/en-us/windows/hardware/gg463119.aspx`.

[15] Offensive Security. Expoit database.

[16] A. One. Smashing the stack for fun and profit. *Phrack magazine*, 7:14–16, 1996.

[17] PaX Team. Pax address space layout randomization. `http://pax.grsecurity.net/docs/aslr.txt`, 2003.

[18] V. Thampi. Udis86 disassembler library for x86. `http://udis86.sourceforge.net/`.

[19] P. Vreugdenhil. Pwn2Own 2010 Windows 7 Internet Explorer 8 exploit. `http://vreugdenhilresearch.nl/Pwn2Own-2010-Windows7-InternetExplorer8.pdf`, 2010.

[20] R. Wartell, V. Mohan, K. Hamlen, and Z. Lin. Securing untrusted code via compiler-agnostic binary rewriting. In *Proceedings of the 28th Annual Computer Security Applications Conference (ACSAC'12)*, Orlando, FL, December 2012.

[21] C. Zhang, T. Wei, Z. Chen, L. Duan, L. Szekeres, S. McCamant, D. Song, and W. Zou. Practical control flow integrity & randomization for binary executables. In *Proceedings of the 34th IEEE Symposium on Security and Privacy*, San Francisco, CA, May 2013.

The (Un)Reliability of NVD Vulnerable Versions Data: an Empirical Experiment on Google Chrome Vulnerabilities [*]

Viet Hung Nguyen and Fabio Massacci
DISI, University of Trento, Italy
{viethung.nguyen, fabio.massacci}@unitn.it

ABSTRACT

NVD is one of the most popular databases used by researchers to conduct empirical research on data sets of vulnerabilities. Our recent analysis on Chrome vulnerability data reported by NVD has revealed an abnormally phenomenon in the data where almost vulnerabilities were originated from the first versions. This inspires our experiment to validate the reliability of the NVD vulnerable version data. In this experiment, we verify for each version of Chrome that NVD claims vulnerable is actually vulnerable. The experiment revealed several errors in the vulnerability data of Chrome. Furthermore, we have also analyzed how these errors might impact the conclusions of an empirical study on foundational vulnerability. Our results show that different conclusions could be obtained due to the data errors.

Categories and Subject Descriptors

H.4 [**Information Systems Applications**]: Miscellaneous

Keywords

Software security; vulnerability analysis; NVD reliability

1. INTRODUCTION

The last few years have seen a significant interest in empirical research on data sets of vulnerabilities. Public third-party vulnerability databases, *e.g.,* such as Bugtraq [12], ISS/XForce [3], National Vulnerability Database (NVD) [5], Open Source Vulnerability Database (OSVDB) [9], are mostly preferred by researchers due to their diversity, availability, and popularity. Among these, NVD is one of the most popular ones. The CVE-ID, *i.e.* the identifier of each NVD entry, is usually used as a common vulnerability identifier among other third-party data sources. In this type of research, the quality of data sources play a crucial role in empirical research on software vulnerabilities. If the data sources contain wrong data, any conclusion derived from these data sources may be potentially invalid.

Our research started from an abnormality in the data when we analyzed the NVD. According to our analysis of NVD data, all of vulnerabilities in Chrome v2–v12 were originated from version v1.0. To explain this, the following scenarios might occur: either yet more vulnerabilities in newer versions have not been detected, or there is a problem in the vulnerability data of Chrome, or both.

The analysis was based on an NVD data feature called *'vulnerable software and versions'* (or *vulnerable versions* for short). This feature remarks versions of particular applications that are vulnerable to the vulnerability described in the entry. For example, CVE-2008-7294 lists all Chrome versions before v3.0.195.24 in its *vulnerable versions*: this means that vulnerability affects Chrome v3.0 and all retrospective versions. According to an archive document[1], the information reported in this feature is *"obtained from various public and private sources. Much of this information is obtained (with permission) from CERT, Security Focus and ISSX-Force"*. Furthermore, our private communications with National Institute of Standards and Technology (NIST), host of NVD, and software vendors, have revealed a "paradox": NIST claimed vulnerable versions were taken from software vendors; whereas, software vendors claimed they did not know about this information. In other words, the original source of this feature is unknown to the public, and therefore its quality is unclear.

This raises a major threat to the validity of studies exploring this feature such as [4, 8, 10, 11, 14], and possibly others. We believe this may be a strong motivation to check for the reliability of NVD.

1.1 Contribution

The major contributions of this work are as follows:

- We present a replicable experiment to validate the reliability of "vulnerable software and versions" feature of NVD for Chrome. This experiment can be applied for other open source applications (*e.g.,* Firefox, Linux).
- We show that the error rates of vulnerabilities in Chrome versions are significant. The errors are both erroneously reporting vulnerabilities in past and future versions.

[*] This work is supported by the European Commission under the project EU-SEC-CP-SECONOMICS

[1] This page is removed, but can be accessed by url `http://web.archive.org/web/20021201184650/http://icat.nist.gov/icat_documentation.htm`

The rest of the paper is organized as follows. We present our research question and hypothesis (§2). After that we describe the validation method (§3) that we follow to conduct the experiment. Next, we report our result and perform analysis on collected data (§4). We also discuss threats to the validity and how to mitigate them (§5). Next we briefly review studies mostly related to our work (§6). Finally, we conclude our paper and discuss about the future work (§7).

2. RUNNING EXAMPLE AND RESEARCH QUESTION

We elaborate a running example on foundational vulnerabilities of Chrome to study the impact of the (un)reliability of NVD data. A *foundational vulnerability* [10] is one that was introduced in the very first version of a software (*i.e.* v1.0), but discovered later in newer versions. In theory, foundational vulnerabilities have higher chance to be exploited than others because they are exposed to attack longer than others. By finding these vulnerabilities in v1.0, attackers could use them to exploit recent versions (say, v20) at the release date. As the result, foundational vulnerabilities are a source for zero-day exploits.

By June 2012, NVD reported 539 vulnerabilities for 12 stable versions[2] of Chrome[3]. Out of these, 460 (85.3%) are reported as foundational. Figure 1 depicts the fraction of foundational vulnerabilities of Chrome. Clearly, each analyzed version of Chrome is rife with foundational vulnerabilities: 99.5% on average are foundational. We find unlikely that a lot of vulnerabilities were introduced in the first version, but none was introduced for the subsequent 11 versions. This motivates our research question as follows.

RQ1 *To what extent is the 'vulnerable versions' feature of the data reported by NVD truthworthy?*

To have such knowledge, for each pair of NVD entry and software version listed in the *vulnerable versions* data feature, we verify whether the NVD entry impacts the corresponding version or not. If it is not, this pair is an *error*. The ratio of the number of error and the number of pairs is the *error rate* which we use as an indicator for the unreliability. In many cases, a small error rate is acceptable. Depending on the type of study the acceptable threshold of errors may vary. Here we choose the threshold of 5% which is normally considered a threshold for statistical significance. We consider the error rate as significant if the median of error rates in individual versions is significantly greater than 5%. We test the median of error rates, rather than the mean because a previous study [4] has shown that vulnerabilities do not follow the normal distribution. Hence, we test the following hypothesis:

H1 *The median of error rates for vulnerabilities reported in Chrome versions is greater than* 5%.

We employ the non-parametric tests for the median to check for the significance.

[2]`http://omahaproxy.appspot.com/about`, visit on July 2012. This is a web application supported by Google team for tracking releases.

[3]We only consider 1+ year old versions as to allow their vulnerability data to mature.

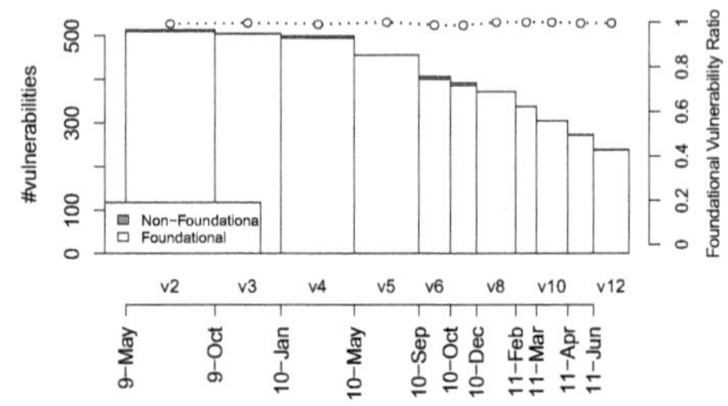

Above it is the stack bars represent the fraction of foundational and non-foundational vulnerabilities, below it is the release date of the versions.

Figure 1: Foundational vulnerabilities in Chrome.

3. VALIDATION METHOD

To verify 539 vulnerabilities for 12 versions of Chrome, we need to check 5,158 pairs[4] of vulnerability and version. Such huge amount of pairs is impossible for a manual verification. Additionally, the manual approach is not replicable. Thus, our proposed method is based on a repeatable and automatic approach [13] where security bug fixes are traced back to the code base to locate the vulnerable code responsible for vulnerabilities. Then we can determine whether a version claimed as vulnerable is actually vulnerable. Our method relies on the following assumptions.

ASS1 When developers commit a bug-fix, they denote the bug ID in the commit message.

ASS2 If the fragment of code responsible for a vulnerability is not there then the software is not vulnerable.

ASS3 If a vulnerability is fixed by only adding code to a vulnerable file, all versions containing the non-fixed revision of the vulnerable file are vulnerable.

By *vulnerable files* we mean the files developers changed to fix a vulnerability. Also, by *code responsible* for a vulnerability we mean the code that developers changed to fix the vulnerability. In some cases, the changed code might not be the vulnerable one, but it helped to remove the vulnerability, despite the original buggy code not being edited. For example, a vulnerability that could lead to SQL injection attack could be fixed by inserting a sanitizer around the source in another module. However, missing of such sanitizer does not mean the application is vulnerable to the same SQL injection attack. Even though the changed code is not buggy in some cases, we still abuse the concept and call it *code responsible*.

Figure 2 sketches the steps of the proposed method. The input of the process is a list of vulnerabilities and the output is list of vulnerabilities annotated with vulnerable versions. The details are as follows:

STEP 1 *Repository mining.* This step takes the list of vulnerabilities and the commit log (*i.e.* list of commits) generated by the repository to produce commits of security bug fixes. A commit of security bug-fixes is one that mentions a security bug ID[5] in its commit

[4]To check the reliability of a vulnerability affecting 12 versions of Chrome, we need to check 12 pairs of vulnerability and version.

[5]Bugs appear in the input list of vulnerabilities.

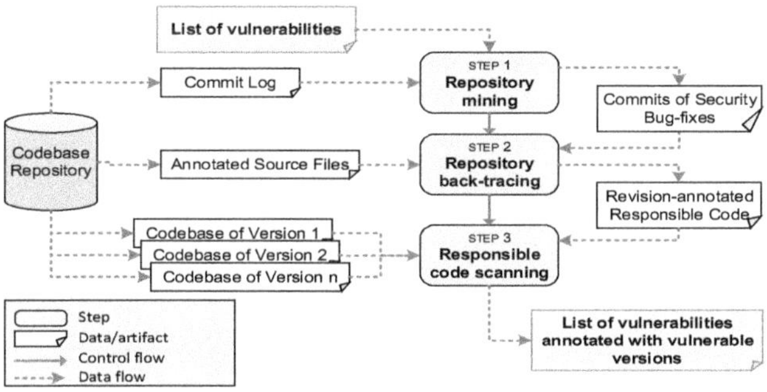

Figure 2: Validation method overview.

```
$ svn diff -c 95731 url_fixer_upper.cc
@@ -540,3 +540,6 @@
   bool is_file = true;
+  GURL gurl(trimmed);
+  if (gurl.is_valid() && ...)
+    is_file = false;
   FilePath full_path;
-  if (!ValidPathForFile(...) {
+  if (is_file && !ValidPathForFile(...) {
```

start line index and number of lines of the left, and the right revisions
added line preceded by a '+'
deleted line preceded by a '-'

Figure 3: An excerpt of the `diff` of two revisions.

message and in some special patterns. These patterns may be vary in different software. For Chrome, they are **BUG=n(,n)***, or **BUG=http://crbug.com/n** where n is the bug ID.

STEP 2 *Repository back-tracing* This step takes commits of security bug-fixes, and the annotated source files from the repository to produce revision-annotated responsible lines of code (LoC). For each source file f in each commit, let r_{fixed} be the revision of this commit. We compare revision r_{fixed} to revision $r_{fixed} - 1$ of file f using the **diff** command supported by the repository. The comparison output is in Unify Diff format, as exemplified by Figure 3, where we compare revision $r95730$ and $r95731$ of file **url_fixer_upper.cc**. By definition, responsible LoC appears in $r_{fixed} - 1$, but not in r_{fixed}. For instance, from Figure 3, the responsible LoC is $\{542\}$. We ignore trivial responsible LoC such as empty lines, or lines that contain only '{' or '}'.

Next, we execute **annotate** command for $r_{fixed} - 1$ of file f to obtain the revisions of responsible LoC. Figure 4 presents an excerpt of the annotated file **url_fixer_upper.cc**. We see that the revision of LoC 542 is $r15$.

There is a special case where the comparison between r_{fixed} and $r_{fixed} - 1$ contains no line preceded with the minus sign. It means developers fixed the vulnerability by adding code only (*e.g.,* security check). In this case we assume that all versions containing revision $r_{fixed} - 1$ and lower are vulnerable (see ASS3).

STEP 3 *Responsible code scanning.* This step looks for each revision-annotated responsible LoC in the code base of every version. If found, we append the corresponding version and the LoC to a list of version-annotated responsible LoC (see ASS2). From this list, we can identify vulnerable versions for each vulnerability.

Notice that there are *unverifiable* vulnerabilities for which the method can not verify the corresponding vulnerable versions. This could be due to a couple of reasons. First, there

```
$ svn annotate -c 95730 url_fixer_upper.cc
...
537:   15 initial.commit PrepareStringForFile...
538:   15 initial.commit
539:   15 initial.commit bool is_file = true;
541: 8536 estade@chromium.org FilePath full_path;
542:   15 initial.commit if (!ValidPathForFile(...)) {
543:   15 initial.commit  // Not a path as entered,
...
```

committed revision
committer

Figure 4: An excerpt of the annotation.

is no corresponding security bug for a vulnerability. Second, STEP 1 may not be always able to determine the commit for security bug-fixes of vulnerabilities.

Table 1 shows a few examples of Chrome vulnerabilities where we apply the method to verify their vulnerable versions. The two first columns indicate the input of the method where we have list of NVD entries and their corresponding bug. We additionally annotate the vulnerable versions reported by NVD for each entry next to the CVE-ID. The next columns show the outputs of the steps. The dash line indicates the data is not available. It means the corresponding NVD entry is not verifiable.

For a better understanding, we describe how the NVD entry 2011-2822 is verified as in Table 1. This vulnerability is reported to affect Chrome v1 up to v13. Its corresponding bug is 72492. In STEP 1, by scanning the log, the bug fix is found at revision $r95731$ of file **url_fixer_upper.cc**. In STEP 2, we diff revision $r95730$ and $r95731$ of this file (see Figure 3). The responsible LoC is determined as $\{542\}$. Then we annotate $r95730$ of the file to get the revision of the responsible LoC, which is $\{r15\}$ (see Figure 4). In STEP 3, we scan for this line in the code base of all versions, and found it in v1 to v13. Finally, we identify the vulnerable versions for this vulnerability, which are v1–v13.

In additional, we manually checked the verification output of some NVD entries (*e.g.,* 2011-1120, 2011-3087, and 2011-1124) for the correctness of the method. The detail validation of these entries is provided in [7]

4. RESULTS AND EXAMPLE REVISED

We apply the proposed method to verify vulnerabilities of major versions of Chrome from v1 to v12. By June 2012, NVD reported 539 entries that allegedly affect these versions of Chrome. Out of these, 503 entries have links to 552 security bugs in Chrome Issue Tracker in their *references* section. The method took 16 hours on a 3 x quad-core 2.83GHz Linux machine with 4GB of RAM to complete.

As the result, 167 NVD entries (31%) are *verifiable*, and 372 (69%) are unverifiable. Among the verifiable ones, 134 (81%) have errors, *i.e.* their verified vulnerable versions are different than reported ones. Among the unverifiable, 36(10%) do not have corresponding bugs, and for 336(90%) we could not locate their commits of security bug fixes. We have done a qualitative analysis on these entries and found that they are bugs in external projects used in Chrome, *e.g.,* WebKit – the HTML rendering engine, V8 – the java script engine, and so on. Therefore, their commits of bug fixes do not exists in the repository of Chrome. Later we will discuss how to work around this problem as a part of future work.

In the following, we analyze the difference of vulnerabilities in individual Chrome versions. Let cve be an NVD entry, and v be the version in analyzed, we define:

- $V(cve)$: is a set of reported vulnerable versions of cve.

Table 1: The execution of the method on few Chrome vulnerabilities.

INPUT		Output of STEP 1	Output of STEP 2	Output of STEP 3	
CVE-ID of NVD entry	Corresponding Bug	Commits for Bug-fixes	Revision-annotated responsible LoC	Version-annotated responsible LoC	Verified vulnerable versions
2011-2822 (v1–v13)	72492	url_fixer_upper.cc[1] ($r95731$)	$\langle r15, 542\rangle$	$\langle v1 - v13, 542\rangle$	v1–v13
2011-4080 (v1–v8)	68115	media_bench.cc[2] ($r70413$)	$\langle r26072, 352\rangle$, $\langle r53193, 353\rangle$	$\langle v3 - v8, 352\rangle$, $\langle v5 - v8, 353\rangle$	v3–v8
2012-1521 (v1–v18)	117110	–	–	–	–

[1]: `chrome/browser/net/url_fixer_upper.cc` [2]: `media/tools/media_bench/media_bench.cc`

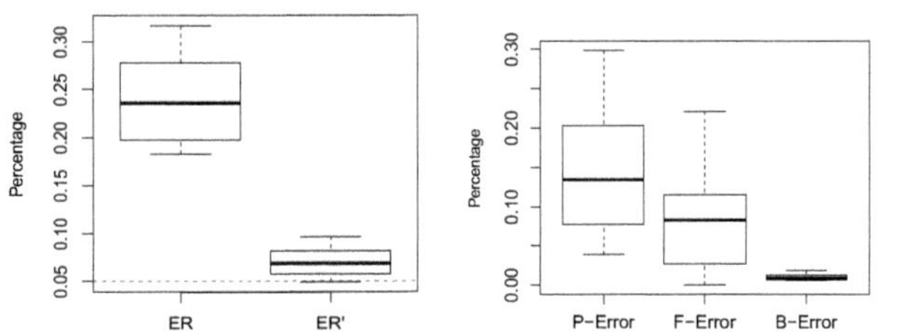

(a) Error Rate (b) Types of Error

In the whicker-box plot, the whickers represent the min and max value, the bold line in the middle is the median value, and the lower and upper part of the box are the quartile of the distribution. The blue dash line at 0.05 shows the threshold of error rate.

Figure 5: The errors in vulnerable version data of NVD entries for Chrome.

- $V'(cve)$: is a set of verified vulnerable versions of cve. If cve is unverifiable, $V'(cve) = \perp$.
- $verified(v) = \{cve | v \in V'(cve)\}$: is the cve which responsible code is detected in version v.
- $erroneous(v) = \{cve | v \in V(cve) \wedge v \notin V'(cve)\}$: is the set of verifiable cve which responsible code is not detected in version v.
- $unverifiable(v) = \{cve | v \in V(cve) \wedge V'(cve) = \perp\}$: is the set of unverifiable cve of version v.

For example, according to Table 1, $V(\text{2011-4080}) = \{v1 - v8\}$, $V'(\text{2011-4080}) = \{v3 - v8\}$. Then, 2011-4080 is a verified cve in v3 (*i.e.* 2011-4080 $\in verified(v3)$), but an erroneous cve in v1 (*i.e.* 2011-4080 $\in erroneous(v1)$).

By ignoring the unverifiable vulnerabilities, the error rate of a version v of Chrome is defined as the ratio of the number of erroneous vulnerabilities of v by the number of verifiable vulnerabilities of v, as shown in the following formula:

$$ER(v) = \frac{|erroneous(v)|}{|verified(v)| + |erroneous(v)|} \quad (1)$$

Being more optimistic, we assume all unverifiable vulnerabilities are all correct. The error rate is rewritten as follows:

$$ER'(v) = \frac{|erroneous(v)|}{|unverifiable(v)| + |verified(v)| + |erroneous(v)|} \quad (2)$$

Figure 5(a) shows the box plots for the distribution of the error rates in Chrome versions. In a box plot, the whickers represent the min and max values, the bold line in the middle is the median, and the lower and upper parts of the box are the quartiles of the distribution. According to the figure, since the median of both error rates ER, ER' are much greater than 5%. It remarkably denotes that the number of erroneous vulnerabilities is not negligible. This is confirmed

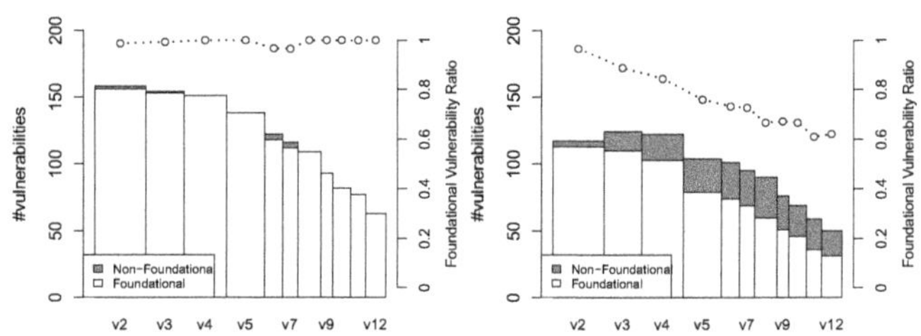

Figure 6: Verifiable (left) vs. Verified (right) vulnerabilities.

in the one-sided Wilcoxon rank-sign test where the null hypothesis is *"the median of error rates is 5%"*, and the alternative hypothesis is H1. The returned *p-value* for ER and ER' are almost zero ($2.44 \cdot 10^{-4}$, and $1.22 \cdot 10^{-4}$respectively). It means the error rates (both ER and ER') did not randomly happen and therefore are significantly greater than 5%.

We break down erroneous vulnerabilities into following categories:

- *stretched-past error* (*P-error*): is the set of erroneous vulnerabilities whose version v is older than all versions that the NVD entries are verified to impact to.

$$P\text{-}error(v) = \{cve \in erroneous(v) | v < \min(V'(cve))\}$$

- *future-version error* (*F-error*): is the set of erroneous vulnerabilities whose version v is newer than all versions that the NVD entries are verified to impact to.

$$F\text{-}error(v) = \{cve \in erroneous(v) | v > \max(V'(cve))\}$$

- *beta error* (*B-error*): is the set of erroneous vulnerabilities whose corresponding NVD entries only impact non-official versions, *i.e.* $V'(cve) = \emptyset$.

$$B\text{-}error(v) = \{cve \in erroneous(v) | V'(cve) = \emptyset\}$$

Similarly to (1), we calculate the *stretched-past error rate*, *future-error rate*, and *beta error rate*. Figure 5(b) reports the distributions of these rates. The *P-error* is slightly greater than *F-error*, and both of them are much greater than *B-error*. *B-error* seems to be negligible. We employ Wilcoxon rank-sum test to compare each pair of error categories. Since one category is compared with two other ones, the Bonferroni correction is applied *i.e.* the significance level is divided by 2: $\alpha = ^{0.05}/_2 = 0.025$. The test result confirms that both *P-error* and *F-error* are significantly greater than *B-error* since the returned *p-values* are less than α. The *p-value* $= 0.03 > \alpha$ of the comparison between *P-error* and *F-error* can be considered an evidence (even if not significant) that *P-error* is greater than *F-error*.

Hereafter we revise the running example about foundational vulnerability in Chrome. We assume that the same

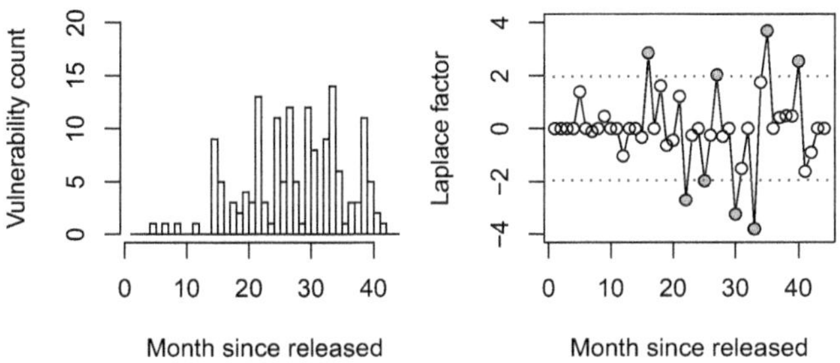

(a) Verifiable vulnerabilities

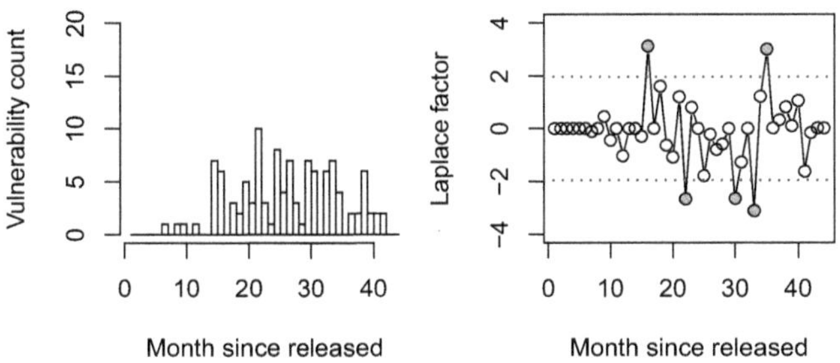

(b) Verified vulnerabilities

Figure 7: Trends in foundational vulnerability discovery.

ratio of errors would be applied in the unverifiable vulnerabilities. Therefore, in following analysis we study the impact of error in verifiable vulnerabilities when we study the trend of foundational vulnerability.

We have two data sets: Verifiable and Verified. The former is the set of verifiable vulnerabilities that we could verify by the proposed method. Their vulnerable versions are reported by NVD. The latter is also the same set, but their vulnerable versions are verified.

Figure 6 illustrates the fraction of foundational vulnerabilities in each version. Left is based on Verifiable, and right is based on Verified. In this figure, The circles denote the percentage of foundational vulnerabilities. By looking at the left picture, even though the absolute number of foundational vulnerabilities decreased, their fractions are almost unchanged. Additionally, as aforementioned, it is strange that vulnerabilities are only introduced in v1.0, but none are introduced in later versions. However, by looking at the right side, this phenomenon disappears. Moreover, there is a decreasing trend in the foundational vulnerabilities fractions from v2.0 to v12.0. We additionally employ the Wilcoxon rank-sum test on the absolute number and the fraction of foundational vulnerabilities in each version. The test results show that the difference between the left and the right is indeed not random since the returned *p-values* are almost zero (*i.e.* $3.82 \cdot 10^{-3}$, and $3.84 \cdot 10^{-3}$ respectively).

Furthermore, we replicate the analysis on the trend of foundational vulnerability discovery as described in [10]. Figure 7 exhibits the analysis result on Verifiable (Figure 7(a)) and Verified(Figure 7(b)). In the figure, left is the discovery rate of foundational vulnerabilities discovered monthly since the release date, right is the Laplace test for trend in monthly discovered foundational vulnerabilities. Two dotted horizonal lines at value 1.96 and -1.96 indicate the range such that if a value of Laplace factor is out of this range, it is significant evidence for either an increasing (> 1.96) or a decreasing (< -1.96) trend in the data. Such values are indicated as green (gray) circles. Again we see a clearly difference between two discovery rates between two data sets. The *p-value* of the Wilcoxon rank-sum test is almost zero (0.0008): this indicates that the difference is significant. We can also see the difference in the trend of discovery. By using Verifiable, we might observe several significant trends (both increasing and decreasing) of foundational vulnerability discovery. Some of these trends, however, disappear in Verified.

In short, our experiment provides evidence that the error in the *vulnerable versions* feature of NVD entries for Chrome is not negligible. Among the errors, NVD tends to commit more stretch-past error than others. It is one of the reasons for the abnormality that 99.5% vulnerabilities of Chrome are foundational. This error in NVD has significantly impact to the analysis of foundational vulnerabilities where different conclusions can be drawn.

5. THREAT TO VALIDITY

Construct validity includes threats affecting the approach by means of which we collect and verify vulnerabilities. Threats in this category come from the assumptions as follows.

By making the assumption ASS1, we delegate the completeness of our method to the responsibility of developers and the quality control of the software vendor. According to [2], there are two types of mistakes: the developers do not mention the bug ID in a bug-fix commit; and the developers mention a bug ID in a non-bug-fix commit. Also in [2], the authors showed that the latter is negligible, while the former does exist. To evaluate the impact of the latter mistakes, we have done a qualitative analysis on bug-fix reports, and we found that all analyzed bug-fix commits are actually bug fixes. As for the former mistakes, we check the completeness of the bug-fix commits for vulnerabilities. As discussed, we found a large portion of vulnerabilities for which we could not locate the bug-fix commits. Our qualitative analysis on these vulnerabilities reveals that they originate from external projects used in Chrome. We discuss a potential solution addressing this threat in future work.

The second assumption ASS2 is apparently syntactical and might not cover all the cases of bug fixes since it is extremely hard to automatically understand the root of vulnerabilities. The assumption also means that if a version contains at least one line of responsible code, this version is vulnerable. Together with the assumption ASS3, our method might overestimate the vulnerable versions by classifying safety code as buggy (error type I, false positive). However, since most Chrome vulnerabilities are reported as foundational, if we overestimate the vulnerable version, the reported *error rate* is a lower bound of the actual one.

Besides, a technical threat to construct validity may be the buggy implementation of the method. To minimize such problem, we employ multi-round test-and-fix approach where we ran the program on some vulnerabilities, then we manually checked the output, and fixed found bugs. We repeated this until no bug was found. Finally, we randomly checked the output again to ensure there was no mistake.

Internal validity concerns the causal relationship between the collected data and the conclusion. Our conclusions are based on statistical tests. These tests have their own assumptions. Choosing tests which assumptions are violated might end up with wrong conclusions. To reduce the risk we carefully analyzed the assumptions of the tests: for instance, we did not apply any tests with normality assumption since the distribution of vulnerabilities is not normal.

External validity is the extent to which our conclusion could be generalized to other applications. Our experiment is based on the vulnerability data of Chrome. So, to have a more generalized conclusion, a replication of this work on other applications should be done.

6. RELATED WORK

Śliwerski et al [13] proposed a technique that automatically locates fix-inducing changes. This technique first locates changes for bug fixes in the commit log, then determines earlier changes at these locations. These earlier changes are considered as the cause of the later fixes, and are called fix-inducing. This technique has been employed in several studies [13, 15] to construct bug-fix data sets. However, none of these studies mention how to address bug fixes which earlier changes could not be determined. These bug fixes were ignored and became a source of bias in their work.

Bird et al [2] conducted a study the level bias of techniques to locate bug fixes in code base. The authors have gathered a data set linking bugs and fixes in code base for five open source projects, and manual checked for the biases in their data set. They have found strong evidence of systematic bias in bug-fixes in their data set. Such bias might be also existed in other bug-fix data set, and could be a critical problem to any study relied on such biased data.

Antoniol et al [1] showed another kind of bias that the bug-fixes data set might suffer from. Many issues reported in many tracking system are not actual bug reports, but feature or improvement requests. Therefore, this might lead to inaccurate bug counts. However, such bias rarely happens for security bug reports. Furthermore, Nguyen et al [6], in an empirical study about bug-fix data sets, showed that the bias in linking bugs and fixes is the symptom of the software development process, not the issue of the used technique. Additionally, the linking bias has a stronger effect than the *bug-report-is-not-a-bug* bias.

7. CONCLUSION AND FUTURE WORK

In this paper we have conducted an experiment to verify the reliability of the *vulnerable versions* data of Chrome vulnerabilities reported by NVD. The experiment has revealed that the error in the *vulnerable versions* data is notable. Among verifiable vulnerabilities of individual Chrome version, approximately 25% of them are erroneous. If we assume that all unverifiable vulnerabilities are all correct, still more than 7% are erroneous in overall. We also demonstrated how these erroneous vulnerabilities could potentially impact the conclusion of foundational vulnerability study. Another study on the impact of erroneous vulnerabilities is further discussed in [7] . This experiment has shed a light into the (un)reliability of the NVD, and allows researchers to revisit the reliability of existing vulnerability databases.

However about two-third of Chrome vulnerabilities are unverifiable because they are vulnerabilities of the external projects used in Chrome. To be able to verify them, extra effort is required. First, we need to link the unverifiable vulnerabilities to the bug ID of the external projects. This could be done by parsing the Chrome bug report. Our qualitative study on several unverifiable vulnerability reports shows that all of them have links to bug reports of the external projects. Second, we apply the proposed method to identify vulnerable revisions of the external projects. Finally, we link these vulnerable revisions to the version of Chrome by looking at the repository of Chrome where the revisions of external projects used in individual versions of Chrome are mentioned. For example, Chrome v12.0 uses WebKit revision 80695, V8 revision 7138. A more detail discussion can be found in [7] .

Also as a part of future work, we plan to evaluate the robustness of the proposed method in identifying vulnerable revisions correctly. We also plan to repeat the experiment on other open source software like Firefox to have a better insight about the reliability of NVD.

8. REFERENCES

[1] G. Antoniol, K. Ayari, M. D. Penta, F. Khomh, and Y. Guhneuc. Is it a bug or an enhancement? a text-based approach to classify change requests. In *Proc. of CASCON'08*, pages 304–318, 2008.

[2] C. Bird, A. Bachmann, E. Aune, J. Duffy, A. Bernstein, V. Filkov, and P. Devanbu. Fair and balanced? bias in bug-fix datasets. pages 121–130. ACM, 2009.

[3] IBM. Internet Security System/X-Force, 2012. http://xforce.iss.net/.

[4] F. Massacci, S. Neuhaus, and V. H. Nguyen. After-life vulnerabilities: A study on firefox evolution, its vulnerabilities and fixes. In *Proc. of ESSoS'11*, 2011.

[5] National Institute of Standards and Technology. National Vulnerability Database, August 2012. http://web.nvd.nist.gov/.

[6] B. H. A. Nguyen, T.H.D.; Adams. A case study of bias in bug-fix datasets. In *Proc. of WCRE'10*, 2010.

[7] V. H. Nguyen and F. Massacci. The (un)reliability of nvd vulnerable versions data: an empirical experiment on google chrome vulnerabilities. *CoRR*, 2013. http://arxiv.org/abs/1302.4133.

[8] V. H. Nguyen and F. Massacci. An independent validation of vulnerability discovery models. In *Proc. of ASIACCS'12*, May 2012.

[9] OSVDB. The Open Source Vulnerability Database. http://www.osvdb.org.

[10] A. Ozment and S. E. Schechter. Milk or wine: Does software security improve with age? In *Proc. of USENIX'06*, 2006.

[11] E. Rescorla. Is finding security holes a good idea? *IEEE Sec. and Privacy*, 3(1):14–19, 2005.

[12] Security Focus. Bug Traq, 2012. http://www.securityfocus.com.

[13] J. Sliwerski, T. Zimmermann, and A. Zeller. When do changes induce fixes? In *Proc. of MSR'05*, pages 24–28, 2005.

[14] A. Younis, H. Joh, and Y. Malaiya. Modeling learningless vulnerability discovery using a folded distribution. In *Proc. of SAM'11*, pages 617–623, 2011.

[15] T. Zimmermann, R. Premraj, and A. Zeller. Predicting defects for eclipse. pages 9–15. IEEE Computer Society, 2007.

Horizon Extender: Long-term Preservation of Data Leakage Evidence in Web Traffic

David Gugelmann
ETH Zurich, Switzerland
gugelmann@tik.ee.ethz.ch

Dominik Schatzmann
ETH Zurich, Switzerland
schatzmann@tik.ee.ethz.ch

Vincent Lenders
armasuisse, Switzerland
vincent.lenders@armasuisse.ch

ABSTRACT

This paper presents Horizon Extender, a system for long-term preservation of data leakage evidence in enterprise networks. In contrast to classical network intrusion detection systems that keep only packet records of suspicious traffic (black-listing), Horizon Extender reduces the total size of captured network traces by filtering out all records that do not reveal potential evidence about leaked data (white-listing). Horizon Extender has been designed to exploit the inherent redundancy and adherence to protocol specification of general Web traffic. We show in a real-life network including more than 1000 active hosts that Horizon Extender is able to reduce the total HTTP volume by 99.8%, or the outgoing volume by 90.9% to 93.9%, while preserving sufficient evidence to recover retrospectively time, end point identity, and content of information leaked over the HTTP communication channel.

Categories and Subject Descriptors

C.2 [**COMPUTER-COMMUNICATION NETWORKS**]: General—*Security and protection (e.g., firewalls)*

Keywords

network security, data leakage, network forensics, incident investigation, aggregation.

1. INTRODUCTION

Digital forensics deals with collecting and preserving evidence for post mortem analysis of security incidents. Network traffic traces, including full payload content and aggregated network flows, prove to be very valuable for the forensic investigation of network-related data theft. For this reason, enterprises and government organizations increasingly deploy network forensics systems that preserve traces of network activity in their own network infrastructure.

Preserving complete network trace records for extended periods of time is becoming almost impossible because of ever increasing traffic volumes. That is why commercial network packet forensic collection systems focus their full payload collection on a limited time window, usually consisting of the last few days to weeks [1, 8]. However, since many real-world data theft incidents remain undetected for months or even years before being investigated, the relevant evidence is often missing. As an example, hackers have been suspected of having had deep access for nearly one decade to the IT systems of the former telecom giant Nortel Networks Corp. without being detected [10].

An existing approach to preserve network forensic evidence of security breaches for longer periods is to trigger packet collection when alarms are raised by a network intrusion detection system (NIDS). For example, *Time Machine* [14] or *Snort* [2] are systems that keep packet traces of selected flows that are identified as suspicious using predefined signatures. The drawback is that network evidence is only preserved for known threats, i.e., when signatures for the threats are known and activated a priori. However, it is well known that classical NIDS have significant false negatives rates, which in practice leads to missed relevant leakage evidence. Similarly, dynamically reducing storage space by packet sampling [15] leads to loss of evidence.

We introduce Horizon Extender, an alternative approach for reducing the total volume of collected traffic while preserving sufficient evidence of data leakage. Horizon Extender addresses the threat of a hacker or malicious insider extracting sensitive information from a network. He or she could try to steal data without being detected by hiding it in the noise of normal outbound traffic. Rather than searching for data leakage signatures at run-time, Horizon Extender aims to constrain the collected data to only records that may represent relevant evidence of leaked data.

Our work focuses specifically on HTTP traffic as this is the protocol accounting for a large portion of the overall traffic. Furthermore, many organizations, particularly those in which confidentiality is a high priority, will generally only allow outbound HTTP traffic, which is forced to go through a proxy server, and block everything else. Horizon Extender exploits the fact that HTTP traffic is highly redundant [3, 16] and mainly determined by the protocol specifications [5]. We exploit these two characteristics to reduce the total volume by filtering out forensically irrelevant HTTP fields and removing the redundancy of the remaining fields by means of deduplication. Yet, Horizon Extender preserves sufficient evidence to recover *time*, *host identities*, and *content* of a leakage incident, therefore covering three key parameters of interest in network forensics.

ASIA CCS'13, May 8–10, 2013, Hangzhou, China.

We have implemented Horizon Extender and evaluated its performance on HTTP network traffic obtained from an academic campus network with more than 1000 hosts over the period of seven days. Our main contribution is to show that Horizon Extender, which uses a combination of filtering, deduplication, data ordering, and compression, allows the reduction of the total HTTP volume required for archival by 99.8%, or by 90.9% to 93.9% of the outbound traffic, while preserving enough evidence for the investigation of data leakage incidents. The time horizon, in which data leakage evidence is retained, is therefore 11 to 16 times longer for the same storage budget when using Horizon Extender than when using raw HTTP log data.

Related work on the compression of HTTP traffic [3, 16] mainly focused on removing the redundancy in HTTP flows originating from servers. This paper focuses in contrast on HTTP client requests, i.e., the outbound HTTP traffic of monitored workstations. While server responses usually contain a message body that is several kilobytes in size, the average number of bytes sent with every client request is on the order of some hundred bytes. Repeated strings in HTTP requests are quite short, therefore current deduplication approaches that operate with block sizes on the order of several kilobytes to reduce the overhead caused by deduplication [12] are not efficient for HTTP requests. Like [8], Horizon Extender relies on various levels of aggregation; however, Horizon Extender not only uses time but also white lists and Web site popularity for selecting data for aggregation. In contrast to behavior-based outgoing HTTP traffic anomaly detection [4], our work applies behavior-based classification for identifying traffic that will likely be irrelevant for retrospective data leakage investigations. Our work is targeted at efficient preservation of information leaks in Web traffic, whereas a method to quantify information leaks is proposed in [5].

2. SETUP AND FORENSIC USE CASES

Network Setup. We consider a typical enterprise network setup. An internal network comprising client PCs is connected to the Internet through a dedicated HTTP proxy. The proxy blocks all traffic except HTTP and HTTPS Web traffic, which is intercepted and replicated to Horizon Extender for archival. The proxy provides clear text access to TLS-secured HTTP traffic so that Horizon Extender can handle HTTPS similar to HTTP.

Problem Statement. Although storage and computational costs have dropped steeply, archiving and processing fine-grained information on every HTTP stream on the network for long periods is currently impractical. For example, a router in our academic network observes an average of 2.4 TB/day of HTTP traffic. If we were to record every HTTP packet for a year, that would require about 870 TB of storage. Trying to store and process this volume of traffic would be massively expensive. In practice however, only a small portion of the traffic turns out to be relevant for forensic analysis of data leakage incidents. Nevertheless, it is difficult to know a priori what data is relevant as every HTTP request line may potentially be exploited to leak data. The key challenge is therefore how to decide a priori what data will be crucial when subsequently investigating an incident retrospectively. An additional important requirement in forensics is determinism. Aggregation and filtering of network records must be deterministic and reproducible over the entire processing chain. That is, it must be possible to tell whether a network record is missing because it did not happen or because the information was not retained. While deterministic processing is simple to achieve when archiving all network data in a first-in-first-out (FIFO) manner, this determinism is more challenging to guarantee for schemes that only record network traffic on behavior- and signature-based detection. Horizon Extender bridges the gap between traditional forensic tools that focus on determinism and completeness of the retrieved data [11] without concern for processing speed and storage space and anomaly detection based approaches [2, 13].

Attacker Model. We address the threat of a hacker or malicious insider extracting sensitive information from the internal network. He or she tries to steal data without being detected by hiding it in the noise of normal outbound HTTP traffic. For Web traffic, this often means stashing bytes in paths or header fields within seemingly benign requests. We assume that the attacker got root access on a set of client PCs within the internal network. The number of clients that the attacker may control is however limited to a maximum number (e.g., 10) or a fixed percentage (e.g., less than 10%) of the clients in the internal network. From these client PCs, he may therefore generate any traffic at any time. However, all leaked data must eventually pass through the HTTP proxy such that the attacker cannot escape the monitoring by Horizon Extender. Another assumption is that the attacker is not able to spoof IP addresses; this can be enforced by the IEEE 802.1X standard.

An attacker may further attempt to exploit hidden information channels. Since all connections go through the proxy server which terminates TCP connections, covert channels using fields in the TCP or IP header are not possible. However, an attacker may still establish a covert timing channel [7] by shaping the timing of requests, but the data leakage rate of such a channel is rather small. Since the problem of detecting and reducing the bandwidth of hidden information channels has already been studied [7], we will not discuss it further in the context of Horizon Extender. The scope of our work is therefore limited to attackers that leak information represented by legitimate content in the fields of the HTTP protocol. When the information is obfuscated (e.g. by encryption), Horizon Extender itself may not be able to decode the leaked content, but provides at the least an archived version of the obfuscated data.

Forensic Use Cases. The goal of Horizon Extender is to provide an administrator with the ability to analyze data leakage incidents retrospectively. Network forensic investigations are typically conducted on the *connection meta information* level or on the *payload* level. In particular, Horizon Extender focuses on the retention of network forensic evidence for the following generic purposes:

Attribution: Attribution deals with assigning leaked content, e.g., to attribute a known file that has been uploaded to a client and remote server identity.

Repudiation: Repudiation deals with providing assurance that hosts are innocent. That is to prove that particular hosts have not been involved in data leakage, e.g., by proving that a connection or upload did not take place.

Exploration: When a client has been identified as malicious by any external means, the collected evidence may serve to reconstruct the complete transfer history and to explore leakage of sensitive data.

3. TRAFFIC CLASSIFICATION

This section introduces the classification model on which Horizon Extender relies to classify HTTP sessions.

3.1 Trusted Service Model

Horizon Extender classifies traffic according to a trusted service model. The traffic is classified at the level of HTTP sessions. An HTTP session in HTTP/1.1 [9] is a sequence of network request-response transactions defined by methods like `GET`, `POST`, `HEAD`, etc. To decide which part of requests and responses should be archived, Horizon Extender differentiates two basic classes of sessions: sessions to *trusted* and *untrusted* Web services. Trusted Web services are trusted not to collude with malicious internal clients while untrusted services may potentially do so. When a service is trusted, the service is assumed to be well-behaving and the HTTP requests may only leak data over the reserved data fields of each request. For example, as shown in Figure 1, data leakage must happen over the request URI (line 1) or the message body (line 14) as reserved for uploading parameters. All other fields such as for example the `CONNECTION` or `COOKIE` fields are protocol specific and reserved for connection establishment and tracking. When the server is trusted and behaves according to the protocol specification, those fields do not pose a real data leakage threat, since they are internal connection state, only available to the hosting server, and not accessible to service users.

In contrast, a malicious service may collude with an internal client to leak data over any header fields of HTTP. It is therefore unsafe to assume that reserved fields will not be exploited by untrusted services for data leakage.

3.2 Popularity-assisted Classification

To classify HTTP sessions in *trusted* and *untrusted* services, Horizon Extender relies on a semi-automated behavior based classification scheme. Horizon Extender compiles a popularity ranking of accessed services by analyzing the percentage of internal hosts communicating with each service and displays the most popular services to the administrator for classification. This popularity-based classification has three major advantages: (1) the most popular services are likely to attract most traffic and therefore inherently cover a large portion of the total traffic volume; (2) popular Web services (e.g., Google, Facebook, etc) are more likely to be trustworthy[1]; (3) malicious manipulation of the service popularity is not trivial for an internal attacker since the popularity ranking is not volume based, thus a manipulation of the popularity may trivially be prevented by requiring a minimum number of distinct clients to define a service as popular. From the list of popular services, the administrator checks the hostname[2] of the Web services and classifies the services as trusted or untrusted. For trusted services, the administrator further categorizes the service as *interactive* or *non-interactive*. This distinction will help to further reduce the storage space requirements by discarding irrelevant content. Interactive services provide internal clients with the ability to download and upload content (e.g., Gmail, Facebook, Youtube, etc). Non-interactive services are pure download services where users retrieve content without being able to upload content (e.g., software updates hosted by CDNs or advertisement servers). While the HTTP upload data is relevant for interactive services, the session content of trusted non-interactive services is not relevant as these services do not pose a data leakage threat. For brevity, we will use the terms *trusted* for interactive trusted services and *white-listed* for non-interactive trusted services. This leads to the following three basic class definitions: ***untrusted, trusted, white-listed***.

[1]The fact that popular services are trusted does not mean that the service users are also trusted. It just states that the server is trusted to behave according to the HTTP specification.

[2]Services are classified by hostname because Web servers in the same domain might have different properties, e.g., static files included in an interactive Web site might be retrieved from servers in a dedicated subdomain that points to a CDN.

```
1:  POST /sec/wp-comments-post.php HTTP/1.1
2:  Host: blogs.example.com
3:  User-Agent: Mozilla/5.0 (Windows NT 6.1; rv:7.0.1)
[...]
8:  Connection: keep-alive
9:  Referer: http://blogs.example.com/sec/read-the-classics/
10: Cookie: __utma=116342964.432951434.1334764398.
    1331764393.1337764332.1; __utmb=116042265.4.15.1321764092;
11: Content-Type: application/x-www-form-urlencoded
12: Content-Length: 229
13:
14: author=john&email=john%40mail.example.com&url=&comment=
    A+comment+in+a+blog+might+be+used+to+leak+data&submit=Post
```

Figure 1: Sample `POST` request (lines truncated)

4. FORENSIC DATA STORAGE

At its core, Horizon Extender differentiates four aggregation levels for the representation of HTTP traffic records. The four aggregation levels are referred to as L1, L2, L3, and L4 with the higher the number the more aggregated are the network records contained in the aggregation level:

L1 (Raw HTTP): L1 represents a chronological raw trace of the complete HTTP requests as intercepted by the proxy. This data is further enriched with connection meta information including *time*, *source IP address*, *destination IP address*, *source port*, *destination port*, and *host*. *Host* is the hostname (DNS name) of the Web service. This level is used for short-term caching of incoming flows.

L2 (Compressed HTTP): At this aggregation level, L1 raw records are split into connection meta information records and reassembled data streams. The HTTP data streams are further split on newlines such that every line of outgoing HTTP traffic is transformed into one data block. Deduplication also operates on these data blocks, as explained in Sect. 5. This aggregation level is used for preserving flows classified as *untrusted*.

L3 (URI and message BODY): Level L3 uses the same meta information records and data structure for data blocks as level L2. To further save storage space, only data blocks containing request URIs and message bodies are retained. This aggregation level only captures data leakage that is conducted by submitting data over the message body or the request URI and is therefore used to store sessions classified as *trusted*.

L4 (Flow end points): In L4, only the meta information records of the HTTP flows are stored. This highly aggregated data representation allows the analysis of a client's remote communication end points and the amount of data that was sent and received. The HTTP content is however

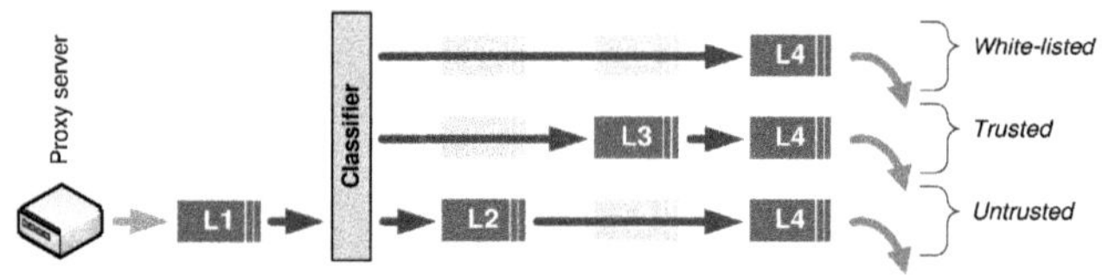

Figure 2: Data processing and archive states.

no longer available. This aggregation level is used to store the category of *white-listed* flows and old flows of the *trusted/untrusted* classes.

5. PROCESSING AND DEDUPLICATION

This section explains how Horizon Extender processes data being exported by the proxy server as visualized in Fig. 2.

Processing of Incoming Records. If approved traffic classification rules are already in place, the incoming L1 records are immediately classified and aggregated to the storage level assigned for the respective traffic category. After aggregation, records are deduplicated and cached until a time or size limit is reached. When the limit[3] is reached, the cached meta information records and data streams are ordered by client IP and remote service, compressed using gzip, and written to two separate files.

Deduplication is applied at L2 and L3 to remove redundancy in the traffic. Deduplication reduces storage size by replacing repeated byte sequences with a pointer to the previous storage location. Instead of using general content-aware block boundary calculation approaches like rolling hashes or Rabin fingerprints [6], Horizon Extender exploits that different HTTP fields are separated by newline and treats every line of the HTTP stream as one data block. Thus, the line number of a previous occurrence can be used as a pointer. Since the main source of redundancy in HTTP requests are static header fields that are sent with every request, this simple approach is efficient, especially in combination with ordering data by client IP and service and applying a file compression algorithm to the resulting deduplicated file. Ordering data blocks by client IP address and service improves the overall compression performance because data blocks of connections to the same service or from the same client often have byte strings in common (e.g., URIs of images located in the same directory). Thus, similar byte strings are moved close to each other such that they fall within the compression window of gzip.

Aggregation of Archived Data. When the storage budget assigned to records in the L2 and L3 aggregation levels is exhausted, Horizon Extender aggregates the oldest L2 and L3 records to the aggregation level L4, which means that corresponding payload data blocks are deleted. When the storage budget for aggregated L4 records is reached, Horizon Extender deletes the oldest connection meta information entries. Horizon Extender guarantees deterministic processing by grouping incoming network records by time and always moving complete groups of network records through the different archival states. Therefore, a group of network records is either in a state in which all relevant payload content is available given that services have been classified correctly or in a state in which only connection meta information is available for the investigation of data leakage.

[3]one day in the current implementation

6. EVALUATION

6.1 Methodology

To evaluate Horizon Extender, we collected HTTP traces at the border router of a university network. A selected subnetwork is used for the evaluation, covering different departments of the university. Our trace spans seven full days including Monday through Wednesday during one week in late 2012 and Thursday through Sunday on the following week. Based on the IPv4 address, we identified per day between 671 and 1888 regular internal clients issuing HTTP requests. The corresponding HTTP traffic on TCP port 80 sums up to 22.6 GB of outgoing and 1.1 TB of incoming data. We additionally classified clients as *active* on a daily basis if they communicated with more than 50 different Web servers per day, these were 112 to 1093 clients per day. Only active clients have been used for calculating the popularity ranking of Web services.

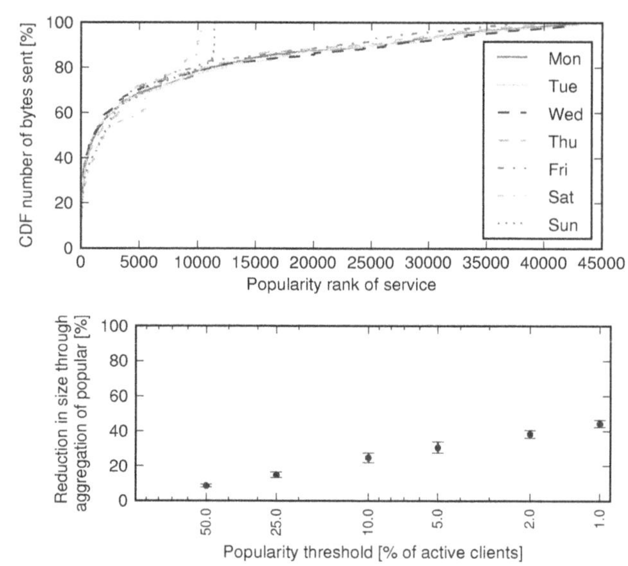

Figure 3: Bytes sent to HTTP services (top). Average reduction in size through aggregation of popular services from L2 to L3 (bottom).

6.2 Traffic Classification

Horizon Extender uses the popularity of Web services to compile traffic classification rules. In this section, we first discuss the correlation between popularity and traffic volume and show the impact of different popularity thresholds. Second, we discuss a list of services suggested to the administrator by Horizon Extender.

Analysis of Service Popularity. The cumulative distribution for the number of bytes sent to services vs. the popularity of services is shown in Fig. 3 (top) for every day. For the weekend days, there is a steep increase at the end of the curves. This is caused by individual clients uploading files to otherwise unpopular services.

Horizon Extender compiles a list of popular services by selecting all services whose popularity is higher than a certain threshold. The popularity is expressed as the percentage of active clients in the network visiting it. Fig. 3 (bottom) shows for different popularity thresholds by which percentage the average total volume of the data streams is reduced when aggregating corresponding connections from level L2 to L3. Because the popularity rank distribution (not shown) is heavy-tailed with the tail starting at around 10%, the volume savings are rising sharply once the popularity threshold is below 10%. Based on these numbers, we have set the popularity threshold for weekdays to 2% (around 20 clients).

Popular Service List. Next, we analyze how stable the list of popular services is. Fig. 4 shows how many new services appear every day on the list. We see that only few new services appear every day and that the number of new services is in general decreasing. Since the set of popular services is quite stable, Horizon Extender can use the same classification for multiple days to weeks with small degradation in compression performance. Table 1 presents the 15 most popular services for the *Mon* dataset in order to give the reader an intuition.

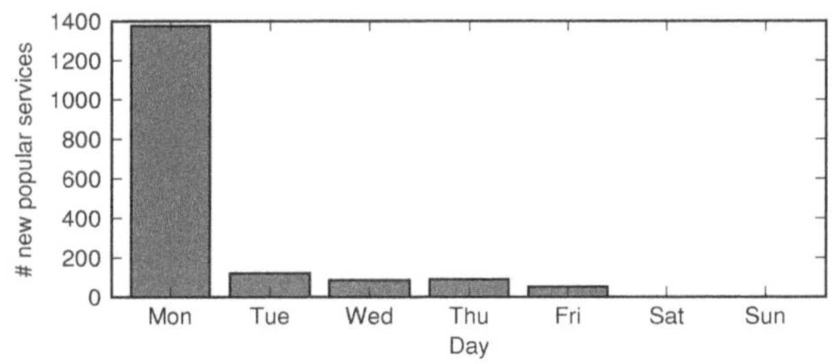

Figure 4: Number of new services appearing.

Host	Pop.	P1	P2	P3
www.google-analytics.com	87.7%	U	T	W
www.google.com	78.9%	U	T	T
ocsp.verisign.com	76.3%	U	T	W
www.facebook.com	75.8%	U	T	T
googleads.g.doubleclick.net	74.0%	U	T	W
www.google.ch	73.2%	U	T	T
platform.twitter.com	71.4%	U	T	W
s0.2mdn.net	70.6%	U	T	T
r.twimg.com	69.1%	U	T	T
p.twitter.com	68.9%	U	T	W
cdn.api.twitter.com	67.6%	U	T	T
connect.facebook.net	64.8%	U	T	W
pagead2.googlesyndication.com	64.4%	U	T	W
ocsp.thawte.com	64.1%	U	T	W
b.scorecardresearch.com	63.7%	U	T	W

Table 1: The 15 most popular services identified in the *Mon* dataset together with their classification according to different policies (U: untrusted, T: trusted, W: white-listed).

6.3 Effectiveness of Horizon Extender

Overall effectiveness. Next, we analyze the effectiveness of Horizon Extender for three administrative policies:
Policy P1 - no trust: An administrator behaving according to this policy does not trust any services. Therefore all services are classified as *untrusted (U)*.
Policy P2 - trust but no white list: The administrator closely follows Horizon Extenders suggestions and classifies all services as *trusted (T)* that Horizon Extender identifies using its popularity classification algorithm.
Policy P3 - trust and white list: All suggestions are classified as *trusted*, additionally the administrator goes through the 100 most popular services and white-lists *(W)* services, for which he thinks that they are non-interactive.

	Week days	Weekend	Overall
raw	100.0±0%	100.0±0%	100.0±0%
raw gzip	23.4±0.65%	30.0±2.17%	23.8±1.92%
P1 (L2)	8.4±0.73%	18.5±1.88%	9.1±2.75%
P2 (L2-L3)	5.8±0.79%	15.4±2.06%	6.5±2.63%
P3 (L2-L4)	5.4±0.79%	15.1±2.08%	6.1±2.66%
L4	0.3±0.01%	0.2±0.01%	0.3±0.01%

Table 2: Size of processed records compared to the size of raw records.

The list of popular services derived from the *Mon* dataset has been reviewed according to the three policies (see Tab. 1 for an excerpt) and has then been used to classify and accordingly aggregate connections; L2-L4 records have been deduplicated, ordered and compressed (using gzip) as discussed in Sect. 5, further meta information of L2 and L3 records has been aggregated for the administrative policies P2 and P3.

The results are shown in Tab. 2. Overall, the average size is reduced to 9.1% - 6.1% of the original outgoing data size, which is a factor of 2.6 to 3.9 better than only gzipping the raw data. Applying policy P2 instead of P1 reduces the total volume by approximately 29% and using P3 squeezes out another 6%. The difference between P2 and P3 is quite small because we only considered the top 100 services for white-listing and we assumed that there are no fully-trusted and therefore white-listed Web sites (e.g. a Web-based business management system) to which users upload content. The compression rate is worse at the weekend because file uploads account for a higher share of egress HTTP requests. Since block boundaries are set on newlines, which is not optimal for binary data, and because a file that is only seen once does not bring much redundancy, such data streams cannot be well deduplicated and Horizon Extender's compression ratio degrades towards the one of gzip.

In summary, Horizon Extender extends the time horizon for forensic investigations by a factor of 15 to 16 compared to raw log data and by a factor of 3.7 to 3.9 compared to compressed raw log data.

Analysis of Deduplication. Horizon Extender uses deduplication to efficiently store repeated data blocks. The impact of deduplication is illustrated in Fig. 5 for the *Mon* dataset. The upper plot shows the effect of deduplication on the total unique block count for selected request fields and the lower plot on total storage size required by the corresponding blocks. General header fields like `HOST` or `USER-AGENT` can be very well deduplicated. This is because they are contained in every request and usually stay constant for many requests. `COOKIE` blocks account for most volume. `GET URI` blocks account for most volume after deduplication since Web browsers usually cache requested objects and therefore do not send multiple requests for the same resource. In summary, deduplication reduces the total number of unique data blocks by 91.2% and the total file size or volume by 68.8%. Anand et al. reported a redundancy of 20% and 76% for outgoing HTTP traffic in a campus and large company network, respectively [3]. Our findings are therefore in line with what they reported for the large company network.

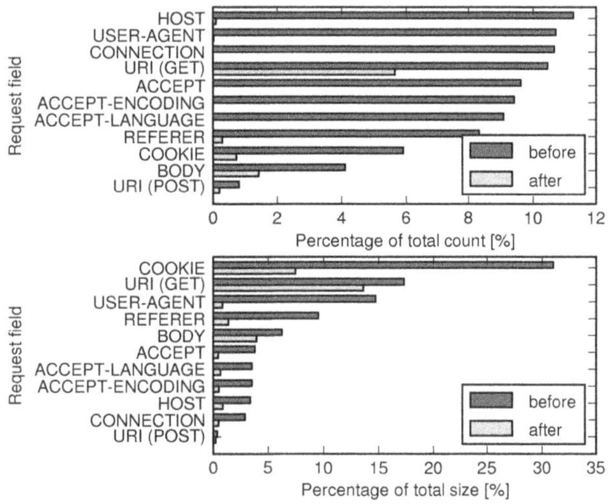

Figure 5: Impact of deduplication on total count (top) / total size (bottom).

7. DISCUSSION AND CONCLUSIONS

We have presented Horizon Extender, a network forensic collection system for HTTP traffic that aggregates records in contrast to traditional systems not merely according to their age but also according to their likelihood of being relevant for future security investigations. Horizon Extender relies on traffic classes to decide which parts of network records might be relevant for incident investigations. We have shown that the popularity of Web services is stable over several days to weeks and therefore efficient filters for traffic classification can be built with little manual configuration by analyzing connection patterns. By taking the different traffic classes into account and quickly aggregating and eventually discarding parts of network records that are not relevant for forensic investigations, Horizon Extender extends the time window in which information relevant for incident investigations is available by a factor of 15 to 16 in relation to the total outbound HTTP traffic volume.

Evasion in the context of Horizon Extender means that there is no evidence in Horizon Extender's logs for some data that has been leaked from a network. Covert timing channels have already been discussed in Sect. 2. Another attack is to deceive an administrator into marking a service as trusted even though the service is controlled by the attacker. However, assuming that the administrator updates the list of popular hosts only once a month, attackers must control multiple internal clients for more than a month, otherwise they cannot push the service to the list of popular hosts and there is still no guarantee that the administrator marks the service as trusted. Another attack would be to take over the servers of a popular service that has already been marked as trusted. However, since popular services generally are operated by large companies that monitor their services closely, this remains a challenging task.

8. ACKNOWLEDGEMENT

This work was partially supported by the Zurich Information Security Center. It represents the views of the authors.

9. REFERENCES

[1] Netwitness - Investigator. `http://netwitness.com/`. Last visited: 2012-12-03.

[2] Snort - Network intrusion prevention and detection system. `http://www.snort.org/`. Last visited: 2012-12-03.

[3] A. Anand, C. Muthukrishnan, A. Akella, and R. Ramjee. Redundancy in network traffic: findings and implications. In *Proc. of ACM SIGMETRICS '09*, 2009.

[4] K. Borders and A. Prakash. Web tap: detecting covert web traffic. In *Proc. of ACM CCS '04*, 2004.

[5] K. Borders and A. Prakash. Quantifying information leaks in outbound web traffic. In *Proc. of IEEE Symp. on Security and Privacy*, 2009.

[6] A. Broder. Some applications of Rabin's fingerprinting method. In *Sequences II: Methods in Communications, Security, and Computer Science*, pages 143–152. Springer Verlag, 1993.

[7] S. Cabuk, C. E. Brodley, and C. Shields. Ip covert timing channels: design and detection. In *Proc. of ACM CCS '04*, 2004.

[8] E. Cooke, A. Myrick, D. Rusek, and F. Jahanian. Resource-aware multi-format network security data storage. In *Proc. of ACM SIGCOMM workshop on LSAD '06*, 2006.

[9] R. Fielding, J. Gettys, J. Mogul, H. Frystyk, L. Masinter, P. Leach, and T. Berners-Lee. Hypertext transfer protocol – HTTP/1.1. RFC 2616, 1999.

[10] Financial Post. Nortel computers breached by hackers for over a decade: WSJ. `http://business.financialpost.com/2012/02/14/nortel-computers-breached-by-hackers-for-over-a-decade-wsj/`. Last visited: 2012-12-03.

[11] S. L. Garfinkel. Digital forensics research: The next 10 years. *Digital Investigation*, 7:S64–S73, Aug. 2010.

[12] F. Guo and P. Efstathopoulos. Building a high-performance deduplication system. In *Proc. of 2011 USENIX*, 2011.

[13] S. Kornexl, V. Paxson, H. Dreger, A. Feldmann, and R. Sommer. Building a time machine for efficient recording and retrieval of high-volume network traffic. In *Proc. of ACM SIGCOMM IMC '05*, 2005.

[14] G. Maier, R. Sommer, H. Dreger, A. Feldmann, V. Paxson, and F. Schneider. Enriching network security analysis with time travel. *SIGCOMM Comput. Commun. Rev.*, 38(4):183–194, Aug. 2008.

[15] A. Papadogiannakis, M. Polychronakis, and E. Markatos. RRDtrace: Long-term Raw Network Traffic Recording using Fixed-size Storage. In *2010 IEEE MASCOTS*, 2010.

[16] N. T. Spring and D. Wetherall. A protocol-independent technique for eliminating redundant network traffic. *ACM SIGCOMM Comput. Commun. Rev.*, 30(4):87–95, 2000.

Towards Fully Incremental Cryptographic Schemes

Kévin Atighehchi
ERISCS Research Group
Aix-Marseille University, France
kevin.atighehchi@univ-amu.fr

Traian Muntean
ERISCS Research Group
Aix-Marseille University, France
traian.muntean@univ-amu.fr

ABSTRACT

This paper focus on incremental cryptographic schemes that solve the *privacy* problem introduced by Bellare, Goldreich and Goldwasser. To our knowledge, none of the schemes designed so far provide simultaneously strong privacy guarantees and byte-wise incremental operations. We propose a new method that extends a block-wise incremental cryptographic scheme into a fully byte-wise incremental one while keeping good performances. This one insures the property of *perfect privacy* with the same average overhead for both the size of the cryptographic form and the number of operations to perform when applying the conjugate algorithm.

Categories and Subject Descriptors

E.3 [**Data**]: Data Encryption; D.4.6 [**Software**]: Security and Protection

Keywords

Incremental cryptography; Perfect privacy; Parallel cryptography

1. INTRODUCTION

Incremental cryptography, introduced in [1, 2, 3], is used to maintain up-to-date, at low computational cost, cryptographic transformations of files such as encryption, authenticated encryption, signature or fingerprint. More specifically, let us assume that we have a document D on which we have applied a cryptographic function $\mathcal{T}$ to generate $\mathcal{T}(D)$. Suppose that D is updated in D' via an update operation M. The goal is then to find an incremental update algorithm $\mathcal{I}$ which allows to compute the cryptographic transformation on D' from $\mathcal{T}(D)$ and M faster than if we would have had to compute this transformation on D' from scratch. Such a low computational efficiency for update operations is essential in areas such as virus protections [2], memory authentication [5], dictionary authentication [8], securing files systems and databases, or more generally files handling over networks of remote servers.

For incremental cryptographic schemes based on balanced tree, the run time of $\mathcal{I}$ is logarithmic in the number of blocks of D [2, 10, 5, 8]. In the ideal case, the run time of $\mathcal{I}$ is proportional with the amount of changes made in D to obtain D'. This is the case for schemes based on pair-wise chaining [7, 2, 4]. Schemes supporting efficient *insert* and *delete* operations may be based on pair-wise chaining or on dynamic data structures such as a signature scheme using a 2-3 Tree [2], a randomised version [10], or even a non-deterministic skip list [8]. Those based on randomised self-balancing data structures (Oblivious Tree [10] or non-deterministic skip list [8]) or on pair-wise chaining can ensure a perfect privacy property at each modification, that is, the updated cryptographic form is indistinguishable from an original one. Another interest of using a self-balancing data structure is its ability to ensure tamper-proof security [2, 10, 6] at quite low computational cost.

Perfectly private incremental cryptographic schemes designed so far are block-based incremental schemes. They can be impractical in general since they do not allow insertion or deletion of data of arbitrary lengths, an essential characteristic for workflow document editing. We propose here a new method that extends a block-wise incremental cryptographic scheme into a byte-wise one while keeping good performances. The method transforms byte-based updates to block-based updates and subsequently uses a block-based scheme as a black box. At each modification performed on the document, our approach insures the property of *perfect privacy* with the same average overhead for both the size of the cryptographic transformation and the number of operations to perform when applying the conjugate algorithm.

The paper is structured as follows. After having recalled definitions in Section 2, we exhibit some drawbacks of current incremental cryptographic schemes and we introduce a generic method to extend them into fully incremental schemes in Section 3. Then, security considerations are discussed in Section 4.

2. BACKGROUND AND DEFINITIONS

2.1 Cryptography and documents update

Cryptographic schemes take as input a document D which is divided into a sequence of fixed-size blocks $P_0, P_1, ..., P_n$. These schemes are defined thanks to a so-called mode of operation over these blocks. Thus, documents are viewed as strings over an alphabet $\sum = \{0,1\}^{8N}$ where N is the block-size in bytes.

Operations on documents. We recall here the block-wise space of modification $\mathcal{M}_B$, as defined for block-wise incremental cryptographic schemes. We borrow the notations defined in [4] and slightly redefine the main modification operations $M \in \mathcal{M}_B$ into the more general operations $M = (\text{blocks\_substitute}, i, j, \beta)$ which substitute block $i+1$ to block $j-1$ (included) by β, a string (possibly empty) constituted of an integral number of blocks.

ASIA CCS'13, May 8–10, 2013, Hangzhou, China.

Incremental cryptographic schemes. We recall here the very generic definition of an incremental cryptographic algorithm as presented by Bellare et *al* [2], with the same notations.

Definition 1. An incremental cryptographic scheme is specified by a 4-tuple of algorithms $\Pi = (\mathcal{K}, \mathcal{T}, \mathcal{I}, \mathcal{C})$ in which:

- $\mathcal{K}$, the key generation algorithm, is a probabilistic polynomial time algorithm that takes as input a security parameter k and returns a pair (K', K'') of keys called the transformation key and the conjugate key respectively.
- $\mathcal{T}$, the tranformation algorithm, is a probabilistic polynomial time (p.p.t.) algorithm that takes as input K' and document $D \in \sum^+$ and returns a cryptographic form $\mathcal{T}_{K'}(D)$.
- $\mathcal{I}$, the incremental update algorithm, is a p.p.t. algorithm that takes as input a key K', (a document D), modification operation $M \in \mathcal{M}_B$, and the cryptographic form t (related to D) and returns the modified cryptographic form t'.
- $\mathcal{C}$, the conjugate algorithm, is a deterministic polynomial time algorithm that takes as input a key K'' and a cryptographic form $t = \mathcal{T}_{K'}(D)$ and returns either document D or a special symbol $\perp$ to indicate that t is invalid.

These four algorithms of an incremental cryptographic scheme Π will be denoted, depending on the context, $\Pi = (\mathcal{K}, \mathcal{E}, \mathcal{I}, \mathcal{D})$ for an incremental encryption scheme and $\Pi = (\mathcal{K}, \mathcal{S}, \mathcal{I}, \mathcal{V})$ for an incremental signature scheme. We can distinguish several types of incremental cryptosystems: **(i) in the asymmetric case** : public key encryption schemes (digital signature schemes) where the decryption key K'' (resp. signature key K') is kept secret, while the encryption key K' (resp. verification key K'') is public; **(ii) in the symmetric case** : modes of encryption, symmetric signature schemes (Message Authentication Code), or even authenticated encryption schemes that combine encryption and authentication, where the secret key $K''(= K')$ is kept secret.

The above definition is very simplified. Regarding certain digital signatures *with appendix* or MACs (the list is certainly not exhaustive), the cryptographic form t should be interpreted cautiously as a couple of elements (D, t') where D is the document and t' is the appendix of the signature (or the tag). In the case of a signature scheme we will refer by t to the appendix of signature or the tag.

2.2 Security

Indistinguishability. Indistinguishibility measures the unability for an adversary to distinguish ciphertexts (and updates). We define an adversary $\mathcal{A}$ in a *find-then-guess* game: (i) ***"find" phase***: $\mathcal{A}$ makes queries to its encryption oracle $\mathcal{O}^{\mathcal{E}_{K'}}$ and update oracle $\mathcal{O}^{\mathcal{I}_{K'}}$ and eventually submits to the challenger either a pair of distinct chosen plaintexts (D_0, D_1) of same length, or a ciphertext C with a pair of modification operations (M_0, M_1) (of the same type and modifying the same amount at the same location); (ii) ***"guess" phase***: The challenger selects a bit $b \in \{0,1\}$ uniformly at random, encrypts the message D_b with $\mathcal{E}_{K'}$ or applies the update M_b on C with $\mathcal{I}_{K'}$, and the result is given to $\mathcal{A}$ which may then make more oracle queries. Finally $\mathcal{A}$ must output a guess for the value of b. We define the advantage of the adversary as twice the probability of a correct guess, minus 1/2. We say that Π is $(t, q_e, \mu, q_{inc}, \epsilon)$-secure in the sense of indistinguishability if, for any adversary $\mathcal{A}$ which runs in time t, making q_e queries to the encryption oracle $\mathcal{O}^{\mathcal{E}_{K'}}$ and q_i valid queries to the update oracle $\mathcal{O}^{\mathcal{I}_{K'}}$, the advantage of $\mathcal{A}$ is negligible. Note that in asymmetric case the encryption and incremental update oracles are replaced by the public key K'.

Unforgeability. Existential unforgeability measures the unability for an adversary to generate a pair (D, s) where D is a document that has not been signed by the signing oracle, nor issued from the update oracle, and s is a signature on that document. The adversary $\mathcal{A}$ has access to signature oracle $\mathcal{O}^{\mathcal{S}_{K'}}$ and incremental update oracle $\mathcal{O}^{\mathcal{I}_{K'}}$. We define the advantage of the adversary as the probability of a valid forgery, that is, for which $\mathcal{V}_{K''}(s) \neq \perp$. We say that Π is $(t, q_s, q_{inc}, \epsilon)$-secure in the sense of unforgeability if, for any adversary $\mathcal{A}$ which runs in time t, making q_s queries to the signing oracle $\mathcal{O}^{\mathcal{S}_{K'}}$ and q_i valid queries to the update oracle $\mathcal{O}^{\mathcal{I}_{K'}}$, the advantage of $\mathcal{A}$ is negligible.

Perfect privacy. An incremental cryptographic scheme is *oblivious* [10, 4, 2] if it produces updated cryptographic forms that are indistinguishable from original ones. Suppose that Alice sends a signature (or ciphertext) to Bob. The latter might be disappointed if he realizes that the signed (or ciphered) document has been obtained by an incremental update. This problem could arise with documents such as commitments, contracts whose cryptographic forms must not disclose any information at all about modifications, not even the fact that update operations have been performed.

3. BYTE-WISE INCREMENTAL CRYPTOGRAPHIC SCHEMES

If there exist incremental update cryptographic algorithms which support operations such as insertion, with a cost proportional to the number of blocks to insert, and deletion with a constant cost, this complexity is valid only when the resulting bytestring does not have problem of alignement, that is to say, when the size of the data being the subject of one of these two operations is a multiple of the block-size.

We illustrate solutions by taking as example a block-based incremental encryption scheme. A simple way to deal with such an incremental scheme is to use only one byte per block but this low payload implies an excessive cost for both the expansion of the ciphertext and the number of block-cipher evaluations. Another solution is to allow the insertion of a variable size of data without ensuring contiguity. Indeed, a plaintext block might be partially used before being ciphered, what would eliminate the need to recipher all following subsequent blocks since data are not re-aligned. Such a method is used by [9], in which a document is divided into parts whose sizes follow a bounded geometric distribution. The partitioned document is ciphered thanks to the authenticated encryption scheme RPC [4]. By using this mode, each variable-length block is encapsulated in the usable part (payload) of a RPC-block. The bound for the number of bytes is determined by the permissible payload. Nevertheless, this method is not without defects. First, the absence of standard padding for each encapsulated block renders this authenticated encryption not secure. Then, the upper bound for the distribution could be arbitrarily chosen so that we find a tradeoff between the expansion of the ciphertext, the total number of cipher/decipher evaluations for the encryption/decryption algorithm and the number of cipher evaluations for the update algorithm. Finally, the major defect is that the update algorithm does not take into account the readjustement of the distribution. This is even more drastic when the insertions are quite frequent and concern small amount of data.

An incremental variable-length block cryptographic scheme must ensure: (i) The obliviousness since statistical tests could reveal outlying regions in the sequence of variable-length parts and therefore the locations of insertions; (ii) Relatively to the size of the message, the conservation (on average) of the overhead for the cryptographic transformation size and the number of operations to perform in the conjugate algorithm.

Definition 2. An incremental variable-length block-based cryptographic scheme is perfectly private if and only if:

1. The distribution of blocks' lengths does not depend on information about modifications performed. Update operations are implemented so that this distribution is preserved.

2. The underlying incremental mode used to operate these blocks is itself perfectly private.

3.1 Variable-length block partition algorithm

The bytes string D is divided into variable-sized blocks $(B_i)_{i=0..n}$ whose corresponding lengths $(u_i)_{i=0..n}$ are generated from the same probability distribution ϕ on the set $\{1, \ldots, N\}$.

For convenience, we define a $Part$ operation which takes as input the probability distribution ϕ, the document D and outputs a partitioned form $\overline{D}$ (accompanied of the corresponding sequence of lengths):

1: Let S be the size of D and B the list of parts initially empty;
2: Generate a number a from the probability distribution ϕ and set $S = S - a$;
3: If $S > 0$ insert the part containing the next a bytes in B and go to step 2;
4: Otherwise, insert the part containing the next $S + a$ bytes in B and return B.

For instance, with a discrete uniform distribution $\mathcal{U}([1,\ N])$ and a fictitious block-size N of 4 bytes, a document of 19 bytes could be cut into parts $(B_i)_{i=0..6}$ of corresponding lengths $(2, 4, 3, 2, 3, 1, 4)$.

3.2 Operations on document

To obtain a fully incremental cryptosystem, we define a byte-wise space of modifications $\mathcal{M}_b$. The general fine-grained operations $M \in \mathcal{M}_b$ that we have to consider are $M = (\text{bytes\_substitute}, i, j, \delta)$ which substitute byte $i + 1$ to byte $j - 1$ (included) by δ, a bytestring of any length (possibly empty). Considering a n-byte string document $D = b_1 b_2 \ldots b_n$, a modification $(\text{bytes\_substitute}, i, j, \delta)$ can be interpreted as taking the string $b_1 b_2 \ldots b_{i-1} b_i b_j b_{j+1} \ldots b_{n-1} b_n$ and inserting δ just after the byte index i so that we obtain the new document $D' = b_1 b_2 \ldots b_{i-1} b_i \delta b_j b_{j+1} \ldots b_{n-1} b_n$.

Ideally, an update algorithm on inputs $\overline{D}$ and M returns an updated partition $\overline{D'}$ for the new document D'. Subsection 3.4 explains how are realized the update operations and give an example.

3.3 Notations

The main notations used are described in Table 1. $||$ is the concatenation operator. δ denotes the bytestring to insert. $|.|$ denotes the size of a data in bytes or the number of elements in a sequence (list). We specify a function $f(n) = \sum_{m=0}^{n} u_m$ that accumulates terms from u_0 to u_n. The sentinel indices $-\infty$ and $+\infty$ (used by convention either for a byte or a block) are necessary in the case of modification at the very beginning or the end of the document. We also need to specify a padding function *pad* taking as input a variable-length block B_k and padding it with a sufficient number of bits to obtain a block of size N. The choice of these padding bits depends on the use context. Thus, if its purpose is to be used in a signature scheme, *pad* assume an input block of size strictly lower than N and returns the padded block $B_k||10^{8(N-u_k)-1}$. However, if it has to be used in an encryption scheme, *pad* returns the padded block $B_k||0^{8(N-u_k)}$.

$P[a, b]$	substring of P from the byte a up to the byte b (included), $b \geq a \geq 1$
B_i	variable length block belonging to the partition of D
$\overline{D} = B = (B_i)$	list of B_i forming the partition
$u_i = \lvert B_i \rvert$	size of B_i, in number of bytes
$u_i \overset{\phi}{\leftarrow} \{1, \ldots, N\}$	u_i is drawn from the set $\{1, \ldots, N\}$ according to ϕ
$u = (u_i)$	list of variable lengths u_i
$\lvert B \rvert = \lvert u \rvert$	number of elements in the partition
B'_i	a new variable length block
$B' = (B'_i)$	list of B'_i forming the new subpartition
$u'_i = \lvert B'_i \rvert$	size of B'_i, in number of bytes
$u' = (u'_i)$	list of variable lengths u'_i
k_0, k_1	index k_0 of a block B_{k_0} after which the repartition starts, index k_1 of a block B_{k_1} before which the repartition terminates
$u_{-\infty}, u_{+\infty}$	sentinel lengths, valued at 0
$B_{-\infty}, B_{+\infty}$	sentinel blocks, designating an empty string

Table 1: Notations

3.4 Algorithmic approach

The challenge is the following, after a modification operation $M \in \mathcal{M}_b$, the resulting sequence of random draws (u_i) has to be as if each u_i (except the last one) has been picked up following the probability distribution ϕ.

In order to realize an insertion of a data δ (consider an insertion performed between two parts, for the sake of simplicity), the two first solutions that come to mind are the following: (i) partition δ in the same way as the document is partitioned, that is to say, in $\overline{\delta}$ and insert it at the good location in $\overline{D}$; (ii) repeat random draws until we find a sequence $(u_i)_{i=0...s}$ satisfying $\sum_{i=0}^{s} u_j = |\delta|$, and insert the partition $(B_0^\delta, B_1^\delta, \ldots, B_s^\delta)$ of δ at the good location in $\overline{D}$. None of these two approaches are satisfactory solutions since they lead to an observable bias after repeated insertions. The bad consequences are the absence of perfect privacy and even overheads (in space and time) since the partition is used as input to a block-based incremental scheme. Our approach proposes to ensure a (shift-)coupling of random walks, a random step in a walk corresponding to a random draw $u_i \sim \phi$. Thus, we will see that performing a modification while respecting the lengths distribution leads almost systematically to a repartitioning of an untouched but quite limited subpart of the document.

Case study of an insertion of δ at a byte located between the last byte of B_{k-1} and the last byte of B_k. Two cases may occur: (i) **case 1**: the insertion is between the last byte of B_{k-1} and the first byte of B_k; (ii) **case 2**: the insertion is between two bytes of B_k.

Let a random sequence $(u_i)_{i=0..n}$ be drawn i.i.d. from a discrete distribution. The problem is to generate a sequence of i.i.d. draws $(u'_i)_{i=k..l}$ from this distribution until we find a couple of indices (l, m) satisfying the following equality

$$\sum_{j=k}^{l} u'_j = \sum_{j=k}^{m} u_j + |\delta| \,. \qquad (1)$$

The resulting sequence of variable sized data blocks is then $B_0, \ldots, B_{k-1}, B'_k, \ldots, B'_l, B_{m+1}, \ldots, B_n$ where the subsequence of blocks $(B'_i)_{i=k..l}$ (of respective lengths $(u'_i)_{i=k..l}$) which contains the inserted data replaces the subsequence $(B_i)_{i=k..m}$. Let us see an example illustrating the **case 1** in Figure 1a: we have a 19 byte-sized document and we want to insert a 3 byte-sized data d right after the ninth byte. To achieve this, it is likely that the subsequence of variable-length blocks $(B_j)_{j\geq 3}$ need to be repartitioned according to our distribution law in a manner consistent with the equation 1 and the insertion of this new data. As can be seen, the first random draw $u'_3 = 1$ indicates that B'_3 can not receive the 3 bytes to insert, which brings us to the second random draw $u'_4 = 3$. If the sum of draws $u'_3 + u'_4 = 4$ allows for the insertion of d, one byte remains to be filled but this is not sufficient for the two following bytes of B_3, which brings us to the third draw $u'_5 = 2$. Once again, the sum of draws $u'_3 + u'_4 + u'_5 = 6$ allows for receiving the first five bytes $|d| + u_3$ of $d\|B_3$ but one byte remains unfilled and is insuficient for receiving the three bytes u_4 of B_4, which brings us to the fourth draw $u'_6 = 2$. Finally, the sum of draws $u'_3 + u'_4 + u'_5 + u'_6 = 8$ allows for receiving the data d and the content of the subsequence of blocks (B_3, B_4), which ends this partial repartition. Figure 1b illustrates the **case 2**. As one can see, this insertion could fragment B_1 and thus force us to include this block in the repartition. Finally this resulting partial repartition covers the data to insert and the content of the previous subpartition (B_1, B_2).

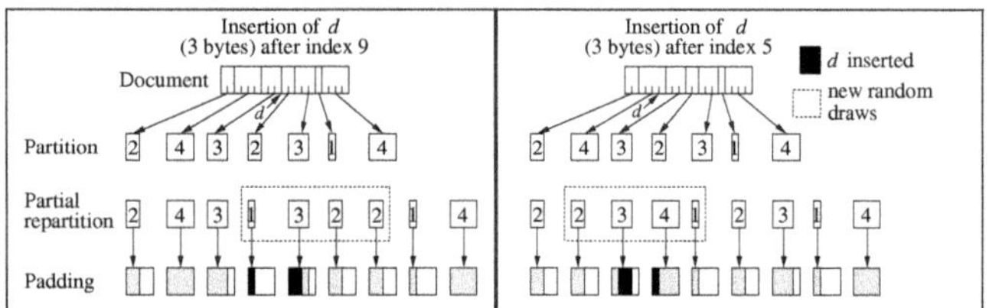

(a) Insertion between two parts B_2 and B_3

(b) Insertion inside a part B_1

Figure 1: Insertion of 3 bytes, $\forall i\ u_i \sim \mathcal{U}([1,\ N])$, $N = 4$

Algorithm 1 outputs the repartitioned subpart including the modification together with indices indicating the untouched parts. For instance, an output (k_0, k_1, B', u') means that the previous partition B is untouched from index 0 to k_0 (included) and from index k_1 (included) to $|u| - 1$. Consequently, the subpartition $(B_{k_0+1}, \ldots, B_{k_1-1})$ is replaced by $(B'_0, \ldots, B'_{|u'|-1})$. We describe also an algorithm *Top* that translates an operation in $\mathcal{M}_b$ into an operation in $\mathcal{M}_B$. This one is usefull for the generic construction presented in Subsection 3.5.

3.5 Generic construction

We define here a byte-wise incremental scheme by a 4-tuple of algorithms $\Xi = (SetUp,\ Transformation,\ IncUpdate,\ Conjugate)$. We write generic algorithms for the keys generation, tranformation and incremental update. As it does not seem possible to write a generic algorithm for the conjugate, we propose a version corresponding to the decryption algorithm of an encryption scheme and another one corresponding to the verification algorithm of a signature scheme. All these algorithms rely on the use of a block-based incremental cryptographic scheme Π which operates on strings over the alphabet $\sum = \{0,1\}^{8N}$.

A *SetUp* algorithm takes as input a security parameter k and returns the key pair generated by $\Pi.\mathcal{K}(k)$. If this one corresponds to a symmetric scheme then $K' = K''$, otherwise $K' \neq K''$. The algorithm 2a probabilistically partitions the document D in variable length blocks B_i, pads each blocks B_i in fixed length blocks P_i, concatenates them and apply the tranformation algorithm to the resulting string. Finally, the cryptographic form returned is (u, t). Note that for encryption schemes, the way of padding each block

---

Algorithm 1 BytesUpdate

Input: A probability distribution ϕ on the set $\{1, \ldots, N'\}$ where $N' \leq N$, the partition B of a document D, the list u of corresponding lengths, a byte-wise operation $M' = (\text{bytes\_substitute}, i, j, \delta)$.

1: $B' \leftarrow ()$; $u' \leftarrow ()$; $c \leftarrow |\delta|$;
2: **if** $i = -\infty$ **then**
3: $\quad k_0 \leftarrow -\infty$; go to step 10;
4: **if** $i = |D|$ **then**
5: $\quad k_0 \leftarrow |u| - 1$; $k_1 \leftarrow +\infty$; go to step 15;
6: $k_0 \leftarrow \text{argmin}_n \left(f(n) = \sum_{m=0}^{n} u_m | f(n) \geq i\right)$;
7: $l \leftarrow f(k_0) - i$;
8: **if** $l > 0$ **then**
9: $\quad \delta \leftarrow B_{k_0}[1, u_{k_0} - l]\|\delta$; $c \leftarrow c + u_{k_0} - l$; $k_0 \leftarrow k_0 - 1$;
10: $k_1 \leftarrow \text{argmin}_n \left(f(n) = \sum_{m=0}^{n} u_m | f(n) \geq j - 1\right)$;
11: $l \leftarrow f(k_1) - j + 1$;
12: **if** $l > 0$ **then**
13: $\quad \delta \leftarrow \delta\|B_{k_1}[u_{k_1} - l + 1, u_{k_1}]$; $c \leftarrow c + l$;
14: $k_1 \leftarrow k_1 + 1$
15: $x \overset{\phi}{\leftarrow} \{1, \ldots, N'\}$;
16: **if** $x = c$ **then**
17: $\quad u' \leftarrow (u', x)$; $B' \leftarrow (B', \delta)$; go to step 27;
18: **if** $x < c$ **then**
19: $\quad c \leftarrow c - x$; $u' \leftarrow (u', x)$; $B' \leftarrow (B', \delta[1, x])$;
20: $\quad \delta \leftarrow \delta[x + 1, c]$; go to step 15;
21: **else**$[x > c]$
22: $\quad$ **if** $k_1 = |u|$ **then**
23: $\quad\quad u' \leftarrow (u', c)$; $B' \leftarrow (B', \delta)$; $k_1 = +\infty$;
24: $\quad\quad$ go to step 27;
25: $\quad c \leftarrow c + u_{k_1}$; $\delta \leftarrow \delta\|B_{k_1}$; $k_1 \leftarrow k_1 + 1$;
26: $\quad$ go to step 16;
27: **return** (k_0, k_1, B', u');

Algorithm 2 Top

Input: A probability distribution ϕ, an operation $M' \in \mathcal{M}_b$, the partition B of a document D, the list u of corresponding lengths

Output: An operation $M \in \mathcal{M}_B$, the list of lengths u' of repartitioned parts

1: $P \leftarrow$ “”;
2: $(k_0, k_1, B', u') \leftarrow BytesUpdate(\phi, B, u, M')$;
3: **for** $i = 0$ to $|u'| - 1$ **do**
4: $\quad P_i \leftarrow pad(B'_i[1, u'_i])$; $P \leftarrow P\|P_i$;
5: $M \leftarrow (\text{blocks\_substitute}, k_0, k_1, P)$;
6: **return** (M, u');

---

does not matter contrary to signature schemes for which a standard padding is mandatory (for instance a sequence of bits starting by a single one and followed by consecutive zeroes). To avoid trivial forgeries, this standard padding has to appear whatever the size of the usable part is. That is why, in order to avoid repetition of our algorithms, we present them by assuming the use of the correct padding function and a probability distribution ϕ on a set $\{1, \ldots, N'\}$ where $N' < N$ in the case of a signature scheme or $N' = N$ in the case of an encryption scheme.

The algorithm 2b describes a decryption algorithm of an encryption scheme. It decrypts t and retreives the string of padded blocks, removes the padding of each blocks in order to obtain the variable

Input: A probability distribution ϕ, a document D, a key K'
$(u, \overline{D}) \leftarrow Part(\phi, D)$;
for $i = 0$ to $|u| - 1$ **do**
  $P_i \leftarrow pad(B_i)$;
$P \leftarrow P_0||P_1|| \ldots ||P_{|u|-1}$;
$t \leftarrow \Pi.\mathcal{T}(K', P)$;
return (u, t);

(a) Tranformation

Input: The cryptographic form (u, t), a key K''
$P \leftarrow \Pi.\mathcal{D}(K'', t)$;
for $i = 0$ to $|u| - 1$ **do**
  $B_i \leftarrow P_i[1, u_i]$;
$D \leftarrow B_0||B_1|| \ldots ||B_{|u|-1}$;
return D;

(b) Conjugate (Decryption of an encryption scheme)

Input: A document D, the cryptographic form (u, t), a key K''
$m \leftarrow 1$;
for $i = 0$ to $|u| - 1$ **do**
  $P_i \leftarrow pad(D[m, m+u_i-1])$;
  $m = m + u_i$;
$P \leftarrow P_0||P_1|| \ldots ||P_{|u|-1}$;
if $\Pi.\mathcal{V}(K'', P, t) = P$ **then**
  return D;
else
  return $\perp$;

(c) Conjugate (Verification of a signature scheme)

Input: A distribution ϕ, a cryptographic form (u, t), a modification operation $M' \in \mathcal{M}_b$, the partition $\overline{D}$, a key K'
$(M, u') \leftarrow Top(\phi, M', \overline{D}, u)$;
$(x_1, x_2, x_3, x_4) \leftarrow M$;
$t' \leftarrow \Pi.\mathcal{I}(K', t, M)$;
$v \leftarrow (u', (u_i)_{i=x_3..|u|-1})$;
$v \leftarrow ((u_i)_{i=0..x_2}, v)$;
return (v, t');

(d) IncUpdate

Figure 2: algorithms of Ξ

length blocks B_i and finally concatenates all the consecutive blocks B_i to reconstruct the document D.

The algorithm 2c describes a verification algorithm of a signature scheme. It divides D thanks to the predefined length partition u, pads each variable length blocks into fixed length blocks and concatenates all of these to obtain P. The validity of the signature t when compared to the string P is then checked-out. If this succeeds, the document D is returned, otherwise $\perp$ is returned.

The algorithm 2d describes the incremental update. It converts a byte-wise modification operation into a block-wise operation. The sequence of length u' which is retreived correponds to the partial repartition which takes into account the modification. Excluding indices x_2 and x_3 are then retreived: (*i*) x_2 corresponds to the index up to which the original partition u is unchanged; (*ii*) in the same way, x_3 corresponds to the index from which the partition is unchanged. The cryptographic form t is updated in t' following the block-wise operation. The length partition u is updated in v (between excluded indices x_2 and x_3). Then the updated cryptographic form returned is (v, t').

3.6 Efficiency analysis

Empirical analysis. The choice of the law ϕ and its parameters depends certainly on settings of the targeted application. At first glance, one could choose parameters that limit the expansion of the cryptographic form without addressing the effect on the update algorithm and the conjugate/tranformation algorithm complexities. To get an overview of the running time of an update, we have done simulations by assuming the use of a 16-byte block-based incremental scheme. Such use of block-size corresponds certainly to an incremental symmetric algorithm, such as for instance rECB [4] instanciated with AES. We are interested in estimation of performances of the extended block-based incremental algorithm. The statistical observations (Figures 3a, 3b) show the average number of blocks to insert in function of the size of the data to insert, due to the translation of an operation $M' \in \mathcal{M}_b$ into an operation $M \in \mathcal{M}_B$.

Three discrete distributions (uniform, binomial and geometric) may be suitable for the variable-length u_i. One can notice that for an insertion, the average running time of the update algorithm seems to be linear in the number of bytes to insert. Besides, as one might expect, the ranges of optimality are different depending on the chosen parameters. For example, when size u_i follows a binomial distribution, average results of Figure 3b shows that for inserts of less than 3 mega-bytes, the number of blocks inserted seems more interesting when P=0.75 compared to P=0.98.

Since the choice of the parameters of the probability distribution used for a given document is definitive, we must know the behavior of the application before making it, particularly regarding the statistics on the sizes of insertions if the goal is to optimize the computational resources. Indeed, for insertion concerning in average small sized data, a lower fill rate for the input blocks of the block-based incremental algorithm will correspond to a lower number of blocks inserted. Storage resources can also be optimized when we achieve a high fill rate, but at the counterpart of slower performances for insertions of small sized data. These storage ressources are always a concern, whatever the type of the cryptographic algorithm is. Indeed, more specifically for incremental encryption, a low fill rate corresponds to a large padding. Secondly, for incremental and tamper-proof signatures based on non-deterministic skip list or oblivious tree, the number of labeled nodes increases when the fill rate decreases. And finally, the list of variable lengths u takes more memory space when the fill rate is low.

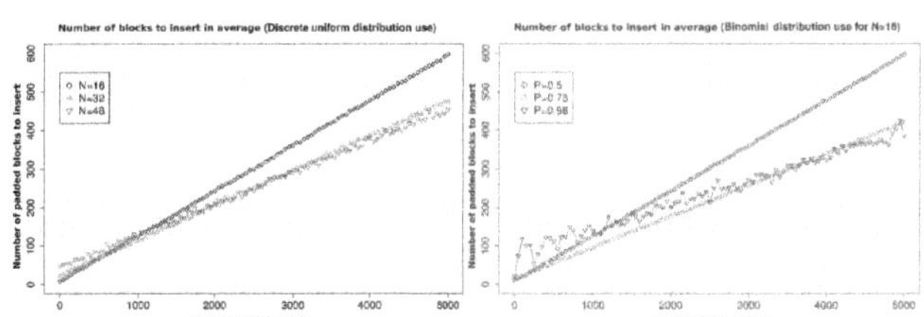

(a) $\forall\, i\; u_i \sim \mathcal{U}([1,\ N])$ (b) $\forall\, i\; u_i \sim \mathcal{B}(N-1, p) + 1$

Figure 3: Insertion cost

Modelization. More precisely, let $(X_i)_{i \geq 1}$ and $(Y_i)_{i \geq 1}$ be the sequences of i.i.d. random variables taking values in $\{1, \ldots, N'\}$ according to the distribution ϕ. We set the random walks S_n, T_m such that $S_n = \sum_{i=1}^{n} X_i$ and $T_m = \sum_{i=1}^{m} Y_i$ and the random subset Z of $\mathbb{N}^2$ such that $Z = \{(n, m) \in \mathbb{N}^2 \,; S_n - T_m = c\}$ where $c \in \mathbb{N}$. Note that c represents the number of contiguous bytes to insert. If (n, m) and (n', m') are distinct in Z then either $n < n'$ and $m < m'$ or $n' < n$ and $m' < m$. In other words $n \leq n'$ and $m' \leq m$ leads to $(n, m) = (n', m')$. Indeed, if $n \leq n'$ and $m' \leq m$ then $\sum_{i=1}^{n} X_i = \sum_{i=1}^{m} Y_i + c$ and $\sum_{i=1}^{n'} X_i = \sum_{i=1}^{m'} Y_i + c$ imply $\sum_{i=n+1}^{n'} X_i = -\sum_{i=m'+1}^{m} Y_i$. Therefore Z has the form $\{(n_k, m_k); k \in \mathbb{N}^*\}$ and the sequences (n_k) and (m_k) are strictly increasing. The question is then to determine the law of the couple of random variables (n_1, m_1).

Algorithmic analysis. Consider the following algorithm described below that takes as input an initial value c_0 for c. At first, we notice that we have two ways to terminate the algorithm, the traces of execution (5,1,2) or (9,2). We can reason about both the number n_1 of draws X and the number m_1 of draws Y.

1: $X \xleftarrow{\phi} \{1, \dots, N'\}$
2: **if** $X = c$ **then** stop
3: **if** $X < c$ **then**
4: $\quad c \leftarrow c - X$
5: $\quad$ go to step 1
6: **if** $X > c$ **then**
7: $\quad Y \xleftarrow{\phi} \{1, \dots, N'\}$
8: $\quad c \leftarrow c + Y$
9: $\quad$ go to step 2

We notice also the followings:

- If $c_0 > N'$, according to the Wald's equation [11] the average number of consecutive executions of the step 3 is upper bounded by $c_0/\mathbb{E}(X)$, after which we have $c \leq N'$.
- If $c_0 \leq N'$, whenever step 1 (or 7) is executed we have $c \leq N'$ and then a non-zero probability $P(x = c)$ (or $P(y = X - c)$ respectively) to terminate. Let us assume that ϕ is a uniform law. If we denote d_1 the total number of random draws ($d_1 = n_1 + m_1$) then $\mathbb{E}(n_1) \leq \mathbb{E}(d_1)$. It turns out that d_1 follows a geometric law of parameter $p = \frac{1}{N'}$, therefore $\mathbb{E}(n_1) \leq N'$. Upper bounds are possible for other cited distributions with a pessimistic approach.

Assuming a uniform distribution and $c_0 > N'$, let d_0 denote the number of random draws needed to satisfy the predicate $c \leq N'$ for the first time and d_1 the number of random draws needed to terminate the algorithm since the first satisfaction of this predicate. So, we can upper bound $\mathbb{E}(n_1)$ as follows:

$$\mathbb{E}(n_1) \leq \mathbb{E}(d_0) + \mathbb{E}(d_1) \leq \frac{2c_0}{1 + N'} + N'.$$

In more general terms, considering that an insertion can take place inside a part, we can upper bound the first moment (mean) of the distribution of n_1 by an affine function $\frac{2c_0}{1+N'} + N' + 2$. Therefore the average running time AT_{Ξ}^{inc} for inserting a bytestring δ in the extended scheme could be described as a function $AT_{\Xi}^{inc}(\delta, |D|) = (a\,|\delta| + b)T_{\Pi}^{inc}(1, |D|\,/m)$ where $T_{\Pi}^{inc}(1, |D|\,/m)$ is the running time for inserting one block in the block-based incremental scheme and $m = \mathbb{E}(X)$.

The average storage space required for the sequence u depends on the choice of the distribution and is upper bounded by $\frac{|D|+N'}{\mathbb{E}(X)} \left\lceil \frac{\log N'}{8} \right\rceil$. That represents, even for a basic distribution such as $\mathcal{U}([1,\ N'])$, a small storage overhead.

For the sake of simplicity and clarity of our subject, we have considered that blocks lengths are described in a simple sequence u. With such a sequence a byte index is reached in $\mathcal{O}(|u|)$ add operations. This could be felt for large size documents. The best solution is to partially accumulate the lengths and organize them into a self-balancing (and non-deterministic to conserve the perfect privacy property) data structure, in which case an access to a byte index, an update or an adding of a length value are all done in $\mathcal{O}(\log |u|)$ add operations.

4. SECURITY AND CONCLUSIONS

Whether we consider an updated partition or an original one, we see a sequence of lengths drawn from a discrete probability distribution, except the last one (due to the termination condition in the algorithm). In either case, this term depends on the document size and the summation of all previous drawn terms. Assuming that the underlying block-based incremental scheme is oblivious, the outputs of $\Xi.IncUpdate$ and $\Xi.Tranformation$ are still perfectly indistinguishable. Concerning indistinguishability of ciphertexts/updates (resp. unforgeability), the security of our extended block-based incremental encryption scheme (resp. signature scheme) can be reduced in a straightforward way to the security of the underlying block-based incremental encryption scheme (resp. signature scheme). Detailed proofs are avalaible in the full paper.

In this paper, we have shown that we can make an incremental byte-wise cryptographic scheme practical and perfectly private while conserving a good efficiency for many applications based on workflow of slight changes within documents. This opens doors to new schemes based on non-deterministic modes of operation which will be presented in a companion paper.

5. REFERENCES

[1] Mihir Bellare, Oded Goldreich, and Shafi Goldwasser. Incremental cryptography: The case of hashing and signing. In *CRYPTO*, 1994.

[2] Mihir Bellare, Oded Goldreich, and Shafi Goldwasser. Incremental cryptography and application to virus protection. In *Proc. of the 27th Ann. ACM Symp. on the Theory of Computing*, pages 45–56. ACM Press, 1995.

[3] Mihir Bellare and Daniele Micciancio. A new paradigm for collision-free hashing: incrementality at reduced cost. In *In Eurocrypt97*, pages 163–192. Springer-Verlag, 1997.

[4] Enrico Buonanno, Jonathan Katz, and Moti Yung. Incremental unforgeable encryption. In *FSE'01*, pages 109–124, 2001.

[5] Reouven Elbaz, David Champagne, Catherine H. Gebotys, Ruby B. Lee, Nachiketh R. Potlapally, and Lionel Torres. Hardware mechanisms for memory authentication: A survey of existing techniques and engines. *Transactions on Computational Science*, 4:1–22, 2009.

[6] Marc Fischlin. Lower bounds for the signature size of incremental schemes. In *In 38th Annual Symposium on Foundations of Computer Science*, pages 438–447, 1997.

[7] Bok-Min Goi, M. U. Siddiqi, and Hean-Teik Chuah. Incremental hash function based on pair chaining & modular arithmetic combining. In *INDOCRYPT*, pages 50–61, 2001.

[8] Michael T. Goodrich and Roberto Tamassia. Efficient authenticated dictionaries with skip lists and commutative hashing. Technical report, TECH. REP., JOHNS HOPKINS INFORMATION SECURITY INSTITUTE, 2001.

[9] Yan Huang and David Evans. Private editing using untrusted cloud services. In *ICDCS Workshops*, pages 263–272, 2011.

[10] Daniele Micciancio. Oblivious data structures: Applications to cryptography. In *In Proceedings of the 29th Annual ACM Symposium on the Theory of Computing*, pages 456–464. ACM Press, 1997.

[11] Abraham Wald. Some generalizations of the theory of cumulative sums of random variables. *The Annals of Mathematical Statistics*, 16(3):pp. 287–293, 1945.

Anonymous Attribute-Based Encryption Supporting Efficient Decryption Test

Yinghui Zhang
State Key Laboratory of Integrated Service Networks (ISN), Xidian University, Xi'an, P.R. China
yhzhaang@163.com

Xiaofeng Chen
State Key Laboratory of Integrated Service Networks (ISN), Xidian University, Xi'an, P.R. China
xfchen@xidian.edu.cn

Jin Li
School of Computer Science and Educational Software, Guangzhou University, Guangzhou, P.R. China
jinli71@gmail.com

Duncan S. Wong
Department of Computer Science, City University of Hong Kong, Hong Kong
duncan@cityu.edu.hk

Hui Li
State Key Laboratory of Integrated Service Networks (ISN), Xidian University, Xi'an, P.R. China
lihui@mail.xidian.edu.cn

ABSTRACT

Attribute-based encryption (ABE) has been widely studied recently to support fine-grained access control of shared data. Anonymous ABE, which is a relevant notion to ABE, further hides the receivers' attribute information in ciphertexts because many attributes are sensitive and related to the identity of eligible users. However, in existing anonymous ABE work, a user knows whether the attributes and the policy match or not only after repeating decryption attempts. And, the computation overhead of each decryption is high as the computational cost grows with the complexity of the access formula, which usually requires many pairings in most of the existing ABE schemes. As a result, this direct decryption method in anonymous ABE will suffer a severe efficiency drawback.

Aiming at tackling the challenge above, we propose a novel technique called match-then-decrypt, in which a *matching phase* is additionally introduced before the *decryption phase*. This technique works by computing special components in ciphertexts, which are used to perform the test that if the attribute private key matches the hidden attributes policy in ciphertexts without decryption. In our proposed construction, the computation cost of such a test is much less than one decryption operation. The proposed construction is proven to be secure. In addition, the results in simulation experiments indicate that the proposed solution is efficient and practical, which greatly improves the efficiency of decryption in anonymous ABE.

Categories and Subject Descriptors

E.3 [**Data Encryption**]: Public key cryptosystems

General Terms

Security

Keywords

Attribute-based encryption; Privacy; Attribute matching; Access control

1. INTRODUCTION

With the rapid development of cloud computing technology, more and more people have uploaded their various types of data into clouds either for ease of sharing or for cost saving. Naturally, people would like to make their private data only accessible to authorized users. In particular, differentiated data access is frequently required in the sense that users with different attributes or roles should be granted different level of access privileges.

Attribute-based encryption (ABE) is envisioned as a highly promising public key primitive for realizing scalable and fine-grained access control systems [15], where differential yet flexible access rights can be assigned to individual users. Especially, ciphertext-policy attribute-based encryption (CP-ABE) [1] puts access policy decisions in the hands of data owners. Though ABE can be directly applied to design secure access control, there is an increasing need to protect user privacy in access control systems. In order to address this problem, anonymous ABE was introduced in [8, 19] and further improved by [11, 12]. In anonymous CP-ABE, a user obtains his attribute secret key and if the attribute set associated with the secret key does not satisfy the access policy in the ciphertext, the user cannot decrypt and guess what access policy was specified by the data owner.

However, in existing anonymous ABE work, the user has to decrypt and decide whether his attributes satisfy the hidden policy in the ciphertext or not. Such a test should be repeated until a successful decryption or all of the possible tests have been considered. The decryption computation overhead in existing work is high as the computational cost grows with the complexity of the access formula, which usually requires many pairings in most of the existing ABE schemes. As a result, this direct decryption method in anonymous ABE will suffer a severe efficiency drawback. Therefore, it is desirable for users to efficiently decide before full decryption whether the hidden policy in a ciphertext matches his attributes. Currently, techniques for attribute matching detection in state-of-the-art CP-ABE schemes are implemented through repeated decryption and hence suffer efficiency limitations. In fact, most of the available detection techniques in CP-ABE schemes have to reveal the access policy in the ciphertext, which violates user privacy. In practice, each user may receive a large number of ciphertexts and such kind of decryption method involves a large computational overhead.

1.1 Our Contribution.

Our contribution are two-folds:

1. We introduce a new technique called match-then-decrypt into the decryption of anonymous ABE, in which a *matching phase* is added before the *decryption phase*. This technique works by computing special components in ciphertexts, which are used to perform the test that if the attribute private key matches the hidden attributes policy in ciphertexts without decryption. In our proposed construction, the computation cost of such a test is much less than one decryption operation.

2. We prove that the proposed construction is secure under the presented model. To be specific, our construction is proven to be selective ciphertext policy and chosen plaintext secure under the Decisional Bilinear Diffie-Hellman assumption and the Decisional Linear assumption. Experimental results show that the novel technique of match-then-decrypt can greatly improve the efficiency of decryption in anonymous ABE.

1.2 Related Work.

Since the introduction of ABE in implementing fine-grained access control systems [15], a plenty of researches have been done on flexible ABE schemes [1, 5, 6, 7, 8, 10, 11, 12, 13, 14, 17, 18, 19]. ABE comes in two flavors called key-policy ABE (KP-ABE) and ciphertext-policy ABE, which were both mentioned in [6]. The first KP-ABE construction [6] realized the monotonic access policy for key policies. To enable more flexible access policy, Ostrovsky et al. presented the first KP-ABE system that supports the expression of non-monotone formulas in key policies [14]. Bethencourt et al. proposed the first CP-ABE construction [1]. However, the construction [1] is only proved secure under the generic group model. To overcome this weakness, Cheung and Newport presented another construction [5] that is proved to be secure under the standard model. Later, Goyal et al. proposed a bounded CP-ABE scheme with expressive access structures and provable security under the standard model [7]. However, complexity of the construction is extremely high and can just serve as a theoretical feasibility.

All the above CP-ABE schemes suffer from a weakness that the access policy has to be revealed in the ciphertexts since decryptors must know how they should combine their secret key components for decryption. To further achieve recipient-anonymity, Kapadia et al. proposed a CP-ABE scheme [8], which can realize hidden ciphertext policies that are represented by AND of different attributes, but it is not collusion-resistant and needs an online semi-trusted server which may be a performance bottleneck in the system. Shi et al. proposed a predicate encryption scheme [16] that focuses on range queries over huge numbers. The security proof of [16] is based on a security notion weaker than ours, which is called match-revealing security in [16] and the number of attributes must be small because the decryption cost is exponential in the number of attributes. Boneh and Waters proposed a predicate encryption scheme based on the primitive called Hidden Vector Encryption [2]. The scheme in [2] can also realize the anonymous CP-ABE by using the opposite semantics of subset predicates. Furthermore, Katz, Sahai, and Waters proposed a novel predicate encryption scheme supporting inner product predicates [9]. However, their scheme is based on a special type of bilinear groups the order of which is a product of three (or two) large primes, and hence needs to deal with large group elements. Later, a more efficient anonymous CP-ABE scheme was constructed [11]. The security proof was given under the Decisional Bilinear Diffie-Hellman assumption and the Decision Linear assumption. Based on the same assumptions as [11], Jin et al. proposed an anonymous CP-ABE scheme [12] with shorter public parameters. There are also many works proposed to make further improvements on ABE [4, 12, 20, 21]. However, all these ABE schemes implement the attribute matching detection only after decryption and hence lose practicability due to large computation cost.

2. PRELIMINARIES

2.1 Bilinear Pairings

Let $\mathbb{G}$ and $\mathbb{G}_T$ be cyclic multiplicative groups of some large prime order p and we denote the identity of $\mathbb{G}_T$ as 1. We call $\hat{e}$ a bilinear pairing if $\hat{e} : \mathbb{G} \times \mathbb{G} \to \mathbb{G}_T$ is a map with the following properties:

1. Bilinear: $e(g^a, g^b) = e(g, g)^{ab}$ for all $a, b \in \mathbb{Z}_p$.
2. Non-degenerate: There exists $g_1, g_2 \in \mathbb{G}$ such that $e(g_1, g_2) \neq 1$.
3. Computable: There is an efficient algorithm to compute $e(g_1, g_2)$ for all $g_1, g_2 \in \mathbb{G}$.

2.2 Complexity Assumptions

The Decisional Bilinear Diffie-Hellman (DBDH) Assumption: Let $a, b, c, z \in_R \mathbb{Z}_p$ and $g \in_R \mathbb{G}$ be a generator. We say that the DBDH assumption holds in $\mathbb{G}$ if no probabilistic polynomial-time algorithm can distinguish the tuple $[g, g^a, g^b, g^c, \hat{e}(g, g)^{abc}]$ from the tuple $[g, g^a, g^b, g^c, g^z]$ with non-negligible advantage.

The Decision Linear (D-Linear) Assumption: Let $z_1, z_2, z_3, z_4, z \in_R \mathbb{Z}_p$ and $g \in_R \mathbb{G}$ be a generator. We say that the D-Linear assumption holds in $\mathbb{G}$ if no probabilistic polynomial-time algorithm can distinguish the tuple $[g, g^{z_1}, g^{z_2}, g^{z_1 z_3}, g^{z_2 z_4}, g^{z_3+z_4}]$ from the tuple $[g, g^{z_1}, g^{z_2}, g^{z_1 z_3}, g^{z_2 z_4}, g^z]$ with non-negligible advantage.

2.3 Access Structure

Intuitively, an access structure, namely a ciphertext policy in CP-ABE, is a rule W that returns either 0 or 1 given a set L of attributes. We say that L satisfies W if and only if W answers 1 on L. Usually, notation $L \models W$ is used to represent the fact that L satisfies W, and the case of L does not satisfy W is denoted by $L \not\models W$.

In our construction, we consider access structures consisting of a single AND-gate supporting multi-value attributes and wildcards. As a generalization of access structures in [5, 21], access structures in our construction are the same as those in [11, 12].

Formally, given an attribute list $L = [L_1, L_2, \cdots, L_n]$ and a ciphertext policy $W = [W_1, W_2, \cdots, W_n]$, $L \models W$ if $L_i = W_i$ or $W_i = *$ for all $1 \leq i \leq n$, and otherwise $L \not\models W$. Note that the wildcard $*$ in the ciphertext policy plays the role of "don't care" value.

3. DEFINITION AND SECURITY MODEL

3.1 Definition of Anonymous CP-ABE

An anonymous CP-ABE scheme consists of a suite of four algorithms: Setup, KeyGen, Encrypt, and Decrypt. These algorithms are specified as follows:

- $\mathsf{Setup}(1^\lambda) \rightarrow (PK, MK)$: The setup algorithm is run by the attribute center. On input a security parameter λ, it returns the system public key PK which is distributed to users, and the master key MK which is kept private.
- $\mathsf{KeyGen}(PK, MK, L) \rightarrow SK_L$: The key generation algorithm is run by the attribute center. On input the system public key PK, the master key MK and an attribute list L, it outputs SK_L as the attribute secret key associated with the attribute list L.
- $\mathsf{Encrypt}(PK, M, W) \rightarrow CT_W$: The encryption algorithm is run by the encryptor. On input the system public key PK, a message M and a ciphertext policy W, it generates a ciphertext CT_W as the encryption of M with respect to W.
- $\mathsf{Decrypt}(PK, CT_W, SK_L) \rightarrow M$ or $\perp$: The decryption algorithm involves two phases, that is, attribute matching detection and decryption phase. On input the system public key PK, a ciphertext CT_W of a message M under a ciphertext policy W, and a secret key SK_L associated with L, the ciphertext CT_W is tested and decrypted by a user with secret key SK_L as follows:
  1. Matching Phase: It returns $\perp$ to terminate decryption with overwhelming probability if the attribute list L does not satisfy the ciphertext policy W, that is, $L \not\models W$. Otherwise, $L \models W$, the Matching Phase ends by initiating the Decryption Phase.
  2. Decryption Phase: It returns the message M.

3.2 Security Model

We demonstrate security requirements for anonymous CP-ABE systems by modeling the capability of adversaries, and define corresponding security notions. The goals of an adversary in an anonymous CP-ABE system include extracting information of a plaintext from the ciphertext and distinguishing underlying access policies in ciphertexts, which can be integrated the following IND-sCP-CPA game involving an adversary $\mathcal{A}$ and a challenger $\mathcal{S}$.

1. Init: The adversary $\mathcal{A}$ commits to the challenge ciphertext policies W_0^*, W_1^*.
2. Setup: The challenger $\mathcal{S}$ chooses a sufficiently large security parameter λ, and runs the Setup algorithm to get a master key SK and the corresponding public key PK. It retains SK and gives PK to $\mathcal{A}$.
3. Phase 1: In addition to hash queries, the adversary $\mathcal{A}$ issues a polynomially bounded number of queries to the following key generation oracle:
   - KeyGen oracle $\mathcal{O}_{KeyGen}$: The adversary $\mathcal{A}$ submits an attribute list L, if $(L \models W_0^* \wedge L \models W_1^*)$ or $(L \not\models W_0^* \wedge L \not\models W_1^*)$, the challenger $\mathcal{S}$ gives $\mathcal{A}$ the secret key SK_L. Otherwise, it outputs $\perp$.
4. Challenge: Once $\mathcal{A}$ decides that Phase 1 is over, it outputs two equal length messages M_0, M_1 from the message space, on which it wishes to be challenged with respect to W_0^* and W_1^*. It is required that $M_0 = M_1$ if any secret key on L satisfying $L \models W_0^* \wedge L \models W_1^*$ has been queried. The challenger $\mathcal{S}$ randomly chooses a bit $\nu \in \{0, 1\}$, computes $CT_{W_\nu^*} = \mathsf{Encrypt}(PK, M_\nu, W_\nu^*)$ and sends $CT_{W_\nu^*}$ to $\mathcal{A}$.
5. Phase 2: The same as Phase 1.
6. Guess: The adversary $\mathcal{A}$ outputs a guess bit $\nu' \in \{0, 1\}$ and wins the game if $\nu' = \nu$.

The advantage of an adversary $\mathcal{A}$ in the IND-sCP-CPA game is defined as $\mathsf{Adv}_{\mathsf{CP\text{-}ABE}}^{\mathsf{IND\text{-}sCP\text{-}CPA}}(\mathcal{A}) = \left|\Pr[\nu' = \nu] - \frac{1}{2}\right|$.

Definition 1. *An anonymous CP-ABE scheme is said to be IND-sCP-CPA secure if no probabilistic polynomial-time adversary can break the IND-sCP-CPA game with non-negligible advantage.*

4. THE PROPOSED CONSTRUCTION

In this section, we present a CPA-secure anonymous CP-ABE scheme, which achieves the protection of user privacy and enables efficient attribute matching detection.

4.1 Construction

- $\mathsf{Setup}(1^\lambda)$: Let $\mathbb{G}, \mathbb{G}_T$ be cyclic multiplicative groups of prime order p, and $\hat{e} : \mathbb{G} \times \mathbb{G} \rightarrow \mathbb{G}_T$ be a bilinear map. Define a hash function $H : \{0,1\}^* \rightarrow \mathbb{G}$. Assume there are n attributes in universe and the universal attribute set is $\mathcal{U} = \{\omega_1, \omega_2, \cdots, \omega_n\}$. And, each attribute has multiple values, where $S_i = \{v_{i,1}, v_{i,2}, \cdots, v_{i,n_i}\}$ is the multi-value set for ω_i and $|S_i| = n_i$. The attribute center chooses $y \in_R \mathbb{Z}_p$, $g_1, g_2 \in_R \mathbb{G}$. Next the attribute center computes $Y = \hat{e}(g_1, g_2)^y$. The system public key is published as $PK = \langle g, g_1, g_2, Y \rangle$. The master key is $MK = \langle y \rangle$.
- $\mathsf{KeyGen}(PK, MK, L)$: Let $L = [L_1, L_2, \cdots, L_n]$ be the attribute list for the user who obtains the corresponding attribute secret key. The attribute center chooses $r_1, r_2, \cdots, r_{n-1} \in_R \mathbb{Z}_p$ and computes $r_n = y - \sum_{i=1}^{n-1} r_i \mod p$. Also, the attribute center chooses $r \in_R \mathbb{Z}_p$ and $\{\hat{r}_i, \lambda_i, \widehat{\lambda}_i \in_R \mathbb{Z}_p\}_{1 \leq i \leq n}$, sets $\hat{r} = \sum_{i=1}^{n} \hat{r}_i$, and computes $\left[\widehat{D}_0, D_{\Delta,0}\right] = [g_2^{y-\hat{r}}, g_1^r]$. For $1 \leq i \leq n$, the attribute center computes
$$\left[D_{\Delta,i}, D_{i,0}, D_{i,1}, \widehat{D}_{i,0}, \widehat{D}_{i,1}\right] =$$
$$\left[g_2^{\hat{r}_i} H(i||v_{i,k_i})^r, g_2^{\lambda_i}, g_1^{r_i} H(0||i||v_{i,k_i})^{\lambda_i}, g_1^{\widehat{\lambda}_i}, g_2^{r_i} H(1||i||v_{i,k_i})^{\widehat{\lambda}_i}\right],$$
where $L_i = v_{i,k_i}$. The secret key is
$$SK_L = \langle \widehat{D}_0, D_{\Delta,0}, \{D_{\Delta,i}, D_{i,0}, D_{i,1}, \widehat{D}_{i,0}, \widehat{D}_{i,1}\}_{1 \leq i \leq n} \rangle.$$
- $\mathsf{Encrypt}(PK, M, W)$: To encrypt a message $M \in \mathbb{G}_T$ under a ciphertext policy $W = [W_1, W_2, \cdots, W_n]$, an encryptor chooses $s, s', s'' \in_R \mathbb{Z}_p$, and computes $\widetilde{C} = MY^s$, $C_\Delta = \hat{e}(g, g)^s Y^{s'}$, $C_0 = g^s$, $\widehat{C}_0 = g_1^{s'}$, $C_1 = g_2^{s''}$, $\widehat{C}_1 = g_1^{s-s''}$. Then for $1 \leq i \leq n$ and $1 \leq t \leq n_i$, the encryptor computes $\left[C_{i,t,\Delta}, C_{i,t,0}, \widehat{C}_{i,t,0}\right]$ as follows:

1. If $v_{i,t} \in W_i$, then $\left[C_{i,t,\Delta}, C_{i,t,0}, \widehat{C}_{i,t,0}\right] =$

$$\left[H(i||v_{i,t})^{s'}, H(0||i||v_{i,t})^{s''}, H(1||i||v_{i,t})^{s-s''}\right].$$

2. If $v_{i,t} \notin W_i$, then $\left[C_{i,t,\Delta}, C_{i,t,0}, \widehat{C}_{i,t,0}\right]$ are random elements in $\mathbb{G}$.

Then the ciphertext of M with respect to W is

$$CT_W = \langle C_\Delta, C_0, \widehat{C}_0, \widetilde{C}, C_1, \widehat{C}_1, \{\{C_{i,t,\Delta}, C_{i,t,0}, \widehat{C}_{i,t,0}\}_{1 \le t \le n_i}\}_{1 \le i \le n}\rangle.$$

- Decrypt(PK, CT_W, SK_L): The ciphertext CT_W is tested and decrypted by a user with secret key SK_L as follows:

  1. Matching Phase: The user checks whether $L \models W$ in terms of the following Equality (1). To be specific, $L \models W$ if and only if Equality (1) holds:

  $$\frac{C_\Delta}{\hat{e}(g, C_0)} = \frac{\hat{e}\left(\widehat{C}_0, \widehat{D}_0 \prod_{i=1}^n D_{\Delta,i}\right)}{\hat{e}\left(\prod_{i=1}^n C_{i,t,\Delta}, D_{\Delta,0}\right)}, \quad (1)$$

  where $L_i = v_{i,t}$. If $L \not\models W$, it returns $\perp$. Otherwise, it initiates the Decryption Phase if $L \models W$.

  2. Decryption Phase: The user decrypts and computes M as follows:

  $$M = \frac{\widetilde{C} \prod_{i=1}^n \hat{e}(C_{i,t,0}, D_{i,0})\, \hat{e}\left(\widehat{C}_{i,t,0}, \widehat{D}_{i,0}\right)}{\prod_{i=1}^n \hat{e}(C_1, D_{i,1})\, \hat{e}\left(\widehat{C}_1, \widehat{D}_{i,1}\right)},$$

  where $L_i = v_{i,t}$.

4.2 Consistency of the Proposed Construction

For $1 \le i \le n$, suppose $L_i = v_{i,t}$. We first show that Equality (1) holds, which means the attribute matching detection is valid in decryption of the proposed construction. Indeed,

$$\begin{aligned}
&\frac{\hat{e}\left(\widehat{C}_0, \widehat{D}_0 \prod_{i=1}^n D_{\Delta,i}\right)}{\hat{e}\left(\prod_{i=1}^n C_{i,t,\Delta}, D_{\Delta,0}\right)} \\
=\ & \frac{\hat{e}\left(g_1^{s'}, g_2^{y-\hat{r}} \prod_{i=1}^n \left(g_2^{\hat{r}_i} H(i||v_{i,t})^{r}\right)\right)}{\hat{e}\left(\prod_{i=1}^n H(i||v_{i,t})^{s'}, g_1^{r}\right)} \\
=\ & \hat{e}\left(g_1^{s'}, g_2^{y-\hat{r}} \left(\prod_{i=1}^n g_2^{\hat{r}_i}\right)\right) = \hat{e}\left(g_1^{s'}, g_2^{y-\hat{r}} g_2^{\hat{r}}\right) \\
=\ & \hat{e}\left(g_1^{s'}, g_2^{y}\right) = \hat{e}(g_1, g_2)^{ys'} \\
=\ & \frac{C_\Delta}{\hat{e}(g, C_0)}.
\end{aligned}$$

On the other hand, the message can be successfully recovered as follows:

$$\begin{aligned}
&\frac{\widetilde{C} \prod_{i=1}^n \hat{e}(C_{i,t,0}, D_{i,0})\, \hat{e}\left(\widehat{C}_{i,t,0}, \widehat{D}_{i,0}\right)}{\prod_{i=1}^n \hat{e}(C_1, D_{i,1})\, \hat{e}\left(\widehat{C}_1, \widehat{D}_{i,1}\right)} \\
=\ & \frac{\widetilde{C} \prod_{i=1}^n \hat{e}\left(H(0||i||v_{i,t})^{s''}, g_2^{\lambda_i}\right) \hat{e}\left(H(1||i||v_{i,t})^{s-s''}, g_1^{\hat{\lambda}_i}\right)}{\prod_{i=1}^n \hat{e}\left(g_2^{s''}, g_1^{r_i} H(0||i||v_{i,t})^{\lambda_i}\right) \hat{e}\left(g_1^{s-s''}, g_2^{r_i} H(1||i||v_{i,t})^{\hat{\lambda}_i}\right)} \\
=\ & \frac{MY^{s}}{\prod_{i=1}^n \hat{e}\left(g_2^{s''}, g_1^{r_i}\right) \hat{e}\left(g_1^{s-s''}, g_2^{r_i}\right)} \\
=\ & \frac{M\hat{e}(g_1, g_2)^{ys}}{\prod_{i=1}^n \hat{e}(g_1, g_2)^{sr_i}} = \frac{M\hat{e}(g_1, g_2)^{ys}}{\hat{e}(g_1, g_2)^{ys}} = M.
\end{aligned}$$

4.3 Security Results

THEOREM 1. *If there exists a polynomial-time adversary $\mathcal{A}$ that can attack the proposed construction in the IND-sCP-CPA model with advantage ϵ, then ϵ is negligible and $\epsilon \le \epsilon_{DBDH} + n\epsilon_{DL}$, where ϵ_{DBDH} and ϵ_{DL} respectively denotes the advantage of a distinguisher of a DBDH challenge and a D-Linear challenge, and n represents the total number of attributes in universe.*

PROOF. (Sketch) Suppose that the adversary $\mathcal{A}$ commits to the challenge ciphertext policies $W_0^* = [W_{0,1}^*, W_{0,2}^*, \cdots, W_{0,n}^*]$ and $W_1^* = [W_{1,1}^*, W_{1,2}^*, \cdots, W_{1,n}^*]$ at the beginning of the game. Based on the IND-sCP-CPA security model, we use a sequence of hybrid games to prove that $\mathcal{A}$ cannot win the original security game denoted by $\mathbf{G}$ with non-negligible probability. We begin by slightly modifying the game $\mathbf{G}$ into a game $\mathbf{G}_0$. The definitions of games $\mathbf{G}$ and $\mathbf{G}_0$ are the same except the generation of the challenge ciphertext. In $\mathbf{G}_0$, to be precise, if the adversary did not obtain the secret key SK_L with respect to the attribute list L satisfying the condition of $[L \models W_0^* \wedge L \models W_1^*]$, then the challenge ciphertext component $\widetilde{C}$ is a random element in $\mathbb{G}_T$ regardless of the random coin, while the rest of the challenge ciphertext are generated in a normal way. On the other hand, if the adversary obtained the secret key SK_L whose associated attribute list L is such that $[L \models W_0^* \wedge L \models W_1^*]$, then the challenge ciphertext is generated correctly. In this case, we have $\mathbf{G} = \mathbf{G}_0$. In any case, we have

LEMMA 1. *Under the DBDH assumption, the difference between advantages of $\mathcal{A}$ in game $\mathbf{G}$ and game $\mathbf{G}_0$ is negligible in the security parameter λ. To be specific, we have $|\Pr[\mathcal{E}] - \Pr[\mathcal{E}_0]| \le \epsilon_{DBDH}$.*

Subsequently, we modify the game $\mathbf{G}_0$ by changing the ciphertext components $\{\{C_{i,t,\Delta}, C_{i,t,0}, \widehat{C}_{i,t,0}\}_{1 \le t \le n_i}\}_{1 \le i \le n}$ and define a sequence of games as follows. For $v_{i,t}$ such that $(v_{i,t} \in W_{0,i}^* \wedge v_{i,t} \in W_{1,i}^*)$ or $(v_{i,t} \notin W_{0,i}^* \wedge v_{i,t} \notin W_{1,i}^*)$, the components $\{C_{i,t,\Delta}, C_{i,t,0}, \widehat{C}_{i,t,0}\}$ are generated as in the real game through the sequence of all the games. In the case that there exists $v_{i,t}$ such that $(v_{i,t} \in W_{0,i}^* \wedge v_{i,t} \notin W_{1,i}^*)$ or $(v_{i,t} \notin W_{0,i}^* \wedge v_{i,t} \in W_{1,i}^*)$, the components $\{C_{i,t,\Delta}, C_{i,t,0}, \widehat{C}_{i,t,0}\}$ generated normally in game $\mathbf{G}_{l-1}$ are replaced with the random elements from the group $\mathbb{G}$ in the new modified game $\mathbf{G}_l$ regardless of the random coin. The process is repeated until there is no component $v_{i,t}$ satisfying $(v_{i,t} \in W_{0,i}^* \wedge v_{i,t} \notin W_{1,i}^*)$ or $(v_{i,t} \notin W_{0,i}^* \wedge v_{i,t} \in W_{1,i}^*)$. In the last game of the sequence, the advantage of the adversary $\mathcal{A}$ is zero since that $\mathcal{A}$ is given a ciphertext chosen from the same distribution regardless of the random coin. By replacing the well-formed ciphertext components in $\mathbf{G}_{l-1}$ with the random elements from the group $\mathbb{G}$ in $\mathbf{G}_l$ in this way, we can embed a D-Linear challenge into the ciphertext such that a distinguisher of $\mathbf{G}_{l-1}$ and $\mathbf{G}_l$ leads to a distinguisher of the D-Linear challenge. To be specific, we have

LEMMA 2. *Under the D-Linear assumption, the difference between advantages of $\mathcal{A}$ in game $\mathbf{G}_{l-1}$ and game $\mathbf{G}_l$ is negligible in the security parameter λ. Specifically, we have $|\Pr[\mathcal{E}_{l-1}] - \Pr[\mathcal{E}_l]| \le \epsilon_{DL}$ for $1 \le l \le l_{max}$.*

Suppose that the above games sequentially compose $\{\mathbf{G}, \mathbf{G}_0, \mathbf{G}_1, \cdots, \mathbf{G}_{l_{max}}\}$, where $\mathbf{G}$ is the original attack game and $\mathbf{G}_{l_{max}}$ is the last one which gives no advantage to $\mathcal{A}$. Let $\mathcal{E}$ be the event that $\nu' = \nu$ in the original game $\mathbf{G}$ and $\mathcal{E}_l$ be the event that

$\nu' = \nu$ in $\mathbf{G}_l$ for $0 \leq l \leq l_{max}$. Then, we have $\epsilon = \left|\Pr[\mathcal{E}] - \frac{1}{2}\right| = |\Pr[\mathcal{E}] - \Pr[\mathcal{E}_{l_{max}}]|$. From the triangle inequality, it follows that

$$\begin{aligned}\epsilon &\leq |\Pr[\mathcal{E}] - \Pr[\mathcal{E}_0]| + \sum_{l=1}^{l_{max}} |\Pr[\mathcal{E}_{l-1}] - \Pr[\mathcal{E}_l]| \\ &\leq \epsilon_{\mathrm{DBDH}} + n\epsilon_{\mathrm{DL}}.\end{aligned}$$

Then it is obvious that ϵ is negligible, and hence the proposed scheme is IND-sCP-CPA secure under the DBDH assumption and the D-Linear assumption. ■

REMARK 1. *We can apply the Canetti-Halevi-Katz technique [3] to obtain a selective ciphertext policy and chosen ciphertext secure extension based on strongly existentially unforgeable one-time signatures.*

5. PERFORMANCE COMPARISON

In this section, we compare the security and efficiency performance of the proposed scheme with some existing CP-ABE schemes [1, 5, 11, 12].

Table 1 mainly presents the security comparison with respect to the complexity assumption, the security model and anonymity. Note that anonymity can be applied to protect users' privacy. From the security comparison in Table 1, it is clear that only Nishide et al's scheme [11], Li et al's scheme [12] and ours achieve the goal of protecting users' privacy and they have the same access policy. It is important to note that only the proposed construction enjoys the desirable property of match-then-decrypt, and hence can be used to alleviate the computation burden of users in the attribute-based setting.

On the other hand, it is indispensable for receivers to perform attribute matching detection (*i.e.*, decryption test) in attribute-based setting. Considering the protection of user privacy, we only compare the cost of attribute matching detection of schemes [11, 12] with that of ours in Figure 1. Our experiment is simulated on a LINUX machine with Intel Core 2 processors running at 2.40 GHz and 2G memory. The order p of groups $\mathbb{G}$ and $\mathbb{G}_T$ is set as a prime of length 160 bits. It is obvious that the proposed scheme is significantly more efficient than schemes [11, 12]. The reason is that the cost of attribute matching detection is mainly determined by the number of fundamental cryptographic operations, especially the most expensive *pairing* operation. And, the number of *pairing* operations for attribute matching detection is constant in the proposed scheme and linearly grows with n in schemes [11, 12].

6. CONCLUSION

The notion of anonymous ABE can be applied to hide the receivers' attribute information in ciphertexts. However, the decryption computation overhead is high as the computational cost grows with the complexity of the access formula. Aiming at improving the decryption efficiency, in this paper, we introduced a new technique called match-then-decrypt into the decryption of anonymous ABE, in which a *matching phase* is added before the *decryption phase*. This technique performs the test that if the attribute private key matches the hidden attributes policy in ciphertexts without decryption. In the proposed construction, the computation cost of such a test is much less than one decryption operation. Our construction is proven to be secure. In particular, experimental results show that the proposed solution is efficient and practical, and that the novel technique of match-then-decrypt can greatly improve the efficiency of decryption in anonymous ABE.

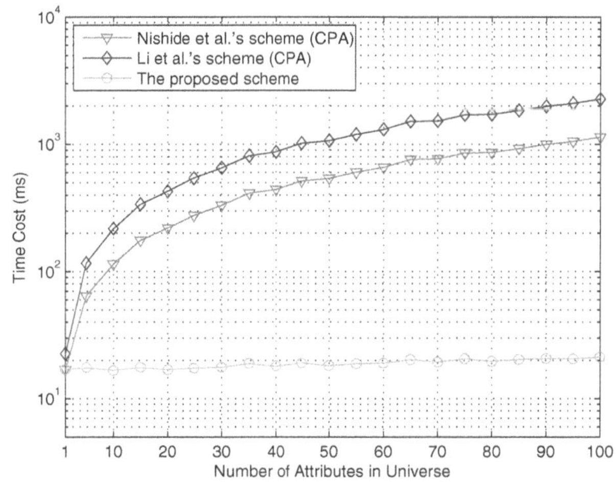

Figure 1: Comparison of cost for attribute matching detection.

7. ACKNOWLEDGMENTS

We are grateful to the anonymous referees for their invaluable suggestions. This work is supported by the National Natural Science Foundation of China (Nos.61272455, 61272457, and 61100224), China 111 Project (No.B08038), the Program for Changjiang Scholars and Innovative Research Team in University (IRT1078), and the Fundamental Research Funds for the Central Universities (Nos.K50511010001 and JY10000901034). Particularly, this work is supported by the Graduate Student Innovation Fund of Xidian University (Research on key security technologies of large-scale data sharing in cloud computing).

8. REFERENCES

[1] J. Bethencourt, A. Sahai, and B. Waters. Ciphertext-policy attribute-based encryption. In *Proc. of SP'07*, Oakland, California, USA, May 20-23, pages 321–334, 2007.

[2] D. Boneh and B. Waters. Conjunctive, subset, and range queries on encrypted data. In *Proc. of TCC'07*, KNAW Trippenhuis Amsterdam, The Netherlands, volume 4392 of *LNCS*, pages 535–554. Springer Berlin-Heidelberg, 2007.

[3] R. Canetti, S. Halevi, and J. Katz. Chosen-ciphertext security from identity-based encryption. In *Proc. of EUROCRYPT'04*, volume 3027 of *LNCS*, pages 207–222. Springer Berlin-Heidelberg, 2004.

[4] M. Chase and S. S. Chow. Improving privacy and security in multi-authority attribute-based encryption. In *Proc. of CCS'09*, New York, NY, USA, pages 121–130. ACM Press. 2009.

[5] L. Cheung and C. Newport. Provably secure ciphertext policy abe. In *Proc. of CCS'07*, New York, NY, USA, pages 456–465. ACM Press. 2007.

[6] V. Goyal, O. Pandey, A. Sahai, and B. Waters. Attribute-based encryption for fine-grained access control of encrypted data. In *Proc. of CCS'06*, New York, NY, USA, pages 89–98. ACM Press. 2006.

[7] V. Goyal, A. Jain, O. Pandey, and A. Sahai. Bounded ciphertext policy attribute based encryption. In *Proc. of*

Table 1: Security Comparison of CP-ABE Schemes

Schemes	Expressiveness	Complexity assumption	Security model	Anonymity
BSW [1]	Type 1 [a]	Generic group model	CPA	×
CN [5]	Type 2 [b]	DBDH	CPA	×
NYO [11]	Type 3 [c]	DBDH, D-Linear	CPA	✓
LRZW [12]	Type 3	DBDH, D-Linear	CPA	✓
The proposed scheme	Type 3	DBDH, D-Linear	CPA	✓

[a] Tree-based structure.
[b] AND-gates on positive and negative attributes with wildcards.
[c] AND-gates on multi-valued attributes with wildcards.

ICALP'08, volume 5126 of *LNCS*, pages 579–591. Springer Berlin-Heidelberg, 2008.

[8] A. Kapadia, P. P. Tsang, and S. W. Smith. Attribute-based publishing with hidden credentials and hidden policies. In *Proc. of NDSS'07*, San Diego, California, USA, February 28-March 2, pages 179–192. The Internet Society, 2007.

[9] J. Katz, A. Sahai, and B. Waters. Predicate encryption supporting disjunctions, polynomial equations, and inner products. In *Proc. of EUROCRYPT'08*, volume 4965 of *LNCS*, pages 146–162. Springer Berlin-Heidelberg, 2008.

[10] A. Lewko, T. Okamoto, A. Sahai, K. Takashima, and B. Waters. Fully secure functional encryption: Attribute-based encryption and (hierarchical) inner product encryption. In *Proc. of EUROCRYPT'10*, volume 6110 of *LNCS*, pages 62–91. Springer Berlin-Heidelberg, 2010.

[11] T. Nishide, K. Yoneyama, and K. Ohta. Abe with partially hidden encryptor-specified access structure. In *Proc. of ACNS'08*, New York, USA, June 3-6, volume 5037 of *LNCS*, pages 111–129. Springer Berlin-Heidelberg, 2008.

[12] J. Li, K. Ren, B. Zhu, and Z. Wan. Privacy-aware attribute-based encryption with user accountability. In *Proc. of ISC'09*, Pisa, Italy, September 7-9, volume 5735 of *LNCS*, pages 347–362. Springer Berlin-Heidelberg, 2009.

[13] T. Okamoto and K. Takashima. Fully secure functional encryption with general relations from the decisional linear assumption. In *Proc. of CRYPTO'10*, volume 6223 of *LNCS*, pages 191–208. Springer Berlin-Heidelberg, 2010.

[14] R. Ostrovsky, A. Sahai, and B. Waters. Attribute-based encryption with non-monotonic access structures. In *Proc. of CCS'07*, New York, pages 195–203. ACM Press. 2007.

[15] A. Sahai and B. Waters. Fuzzy identity-based encryption. In *Proc. of EUROCRYPT'05*, volume 3494 of *LNCS*, pages 557–557. Springer Berlin-Heidelberg, 2005.

[16] E. Shi, J. Bethencourt, T.-H. Chan, D. Song, and A. Perrig. Multi-dimensional range query over encrypted data. In *Proc. of SP'07*, pages 350–364. 2007.

[17] B. Waters. Ciphertext-policy attribute-based encryption: An expressive, efficient, and provably secure realization. In *Proc. of PKC'11*, volume 6571 of *LNCS*, pages 53–70. Springer Berlin-Heidelberg, 2011.

[18] S. Yamada, N. Attrapadung, G. Hanaoka, and N. Kunihiro. Generic constructions for chosen-ciphertext secure attribute based encryption. In *Proc. of PKC'11*, volume 6571 of *LNCS*, pages 71–89. Springer Berlin-Heidelberg, 2011.

[19] S. Yu, K. Ren, and W. Lou. Attribute-based content distribution with hidden policy. In *Proc. of NPSec'08*, Orlando, Florida, USA, pages 39–44. 2008.

[20] S. Yu, K. Ren, W. Lou, and J. Li. Defending against key abuse attacks in kp-abe enabled broadcast systems. In *Proc. of Securecomm'09*, Athens, Greece, September 14-17, volume 19 of *LNCS*, pages 311–329. Springer Berlin-Heidelberg, 2009.

[21] S. Yu, C. Wang, K. Ren, and W. Lou. Attribute based data sharing with attribute revocation. In *Proc. of ASIACCS'10*, New York, NY, USA, pages 261–270. ACM Press. 2010.

A Group Signature Scheme with Unbounded Message-Dependent Opening

Kazuma Ohara
UEC[*] and AIST[†]
Tokyo and Ibaraki, Japan
k-ohara@uec.ac.jp

Yusuke Sakai[‡]
UEC and AIST
Tokyo and Ibaraki, Japan
yusuke.sakai@uec.ac.jp

Keita Emura
NICT[§]
Tokyo, Japan
k-emura@nict.go.jp

Goichiro Hanaoka
AIST
Ibaraki, Japan
hanaoka-goichiro@aist.go.jp

ABSTRACT

Group signature with message-dependent opening (GS-MDO) is a kind of group signature in which only the signers who have created group signatures on problematic messages will be identified. In the previous GS-MDO scheme, however, the number of problematic messages is bounded owing to a limitation of the Groth-Sahai proofs. In this paper, we propose the first GS-MDO scheme with the unbounded-MDO functionality in the random oracle model. Our unbounded GS-MDO scheme is based on the short group signature scheme proposed by Boneh, Boyen, and Shacham and the Boneh-Franklin identity-based encryption scheme. To combine these building blocks and to achieve CCA-anonymity, we also construct a special type of multiple encryption. This technique yields an efficient construction compared with the previous bounded GS-MDO scheme: the signature of our scheme contains about 16 group elements (3630 bits), whereas that of the previous scheme has about 450 group elements (75820 bits).

Categories and Subject Descriptors

D.4.6 [**Management of Computing and Information Systems**]: Security and Protection—*authentication*; E.3 [**Data**]: Data Encryption—*public key cryptosystems*

[*]UEC: University of Electro-Communications

[†]AIST: National Institute of Advanced Industrial Science and Technology

[‡]The second author is supported by a JSPS Fellowship for Young Scientists.

[§]NICT: National Institute of Information and Communications Technology

General Terms

Theory, Security

Keywords

Group Signature; Unbounded Message-Dependent Opening; Random Oracle Model

1. INTRODUCTION

Group signature is a kind of digital signature, proposed by Chaum and van Heyst [9]. Each signer belongs to a group and can create a group signature, which can be verified whether it was created by a group member, without identifying the actual signer. If it is necessary to identify the signer of a group signature (e.g., a problem that may be caused by the group signature), only the authority called the "opener" can identify the corresponding signer of the group signature. However, since the opener can identify the signers without any restriction, it seems that the opener's power is unfairly strong. For example, there is a possibility that a non-problematic user will be identified.

To decentralize the power of the opener, Sakai et al. proposed a model for group signature with a new capability, which is called group signature with message-dependent opening (GS-MDO) [15], where a new authority called the "admitter" is introduced. If a group signature is found on a problematic message, the admitter issues a token corresponding to this message, and the opener can identify the signer of this group signature only when using both the given token and the opener's secret key. One of the significant functionalities of GS-MDO is that no interaction between the opener and the admitter is required. In other words, tokens can merely be published by the admitter, and once the opener receives the token, he is able to open the signatures without any further interaction with any other entity. Furthermore, once the admitter releases a token for a problematic message, the opener can open all signatures on that message.

An application of GS-MDO is an anonymous bulletin board system, which allows users to post their comments without revealing their own identity. In the case of disputes, which may be caused by posting inappropriate comments (e.g., leaking some personal information of others, a crime notice, and so on), GS-MDO comes into effect: the admitter indicates messages that should be prohibited, and the opener can identify only the writers who have posted inappropriate comments on the board. Another application of GS-MDO is an anonymous auction. To bid anonymously, each bidder produces a group signature on his bidding price. After checking the validity of the group signatures, the admitter issues a token only to open the signatures on the highest bid, thereby determining the winner(s). One benefit of this system (compared with applying the conventional group signature) is that no loser(s) of this auction can be identified by the opener.

Sakai et al. also presented a generic construction of GS-MDO from a tag-based key encapsulation mechanism (tag-based KEM), digital signature, identity-based KEM (IB-KEM), and non-interactive zero knowledge (NIZK) proof. Moreover, they gave an instantiation of the generic construction using Shacham's decision-linear (DLIN) Cramer-Shoup encryption scheme (actually a tag-based KEM variant thereof), the Abe-Haralambiev-Ohkubo structure-preserving signature [2, 1], k-resilient a DLIN variant of the Heng-Kurosawa IB-KEM [14]), and the Groth-Sahai proof system [13], which is used as an instantiation of the NIZK proof in the standard model.

One of the weaknesses of the specific construction by Sakai et al. is that it only achieves "k-bounded" security, in the sense that the number of tokens the admitter can issue must be determined when the scheme is set up. This stems from the (in)compatibility of the known identity-based encryption (IBE) schemes and the Groth-Sahai proof system. As the implication result by Sakai et al. suggests, using IBE as a building block to construct a GS-MDO scheme is indispensable. Recall a widely used approach for constructing group signature, in which each group member is assigned a digital signature and generate group signature by encrypting that digital signature and prove the validity of encrypted signature by using an NIZK proof. It is also crucial to use an non-interactive *zero-knowledge* proof system to ensure "well-formedness" of the encrypted certificate, which serves as a group signature. Unfortunately, the Groth-Sahai proof system provides the zero-knowledge property only for a restricted type of theorem. Basically, when a theorem involves equations between elements of the target group $\mathbb{G}_T$, the zero-knowledge property is not guaranteed, which is why the instantiation by Sakai et al. used a k-resilient scheme to instantiate IB-KEM instead of an ordinary IBE scheme.

Such a k-bounded limitation seems to be unavoidable. All known IBE schemes (of the discrete-log type) depend on pairing-type assumptions, and more particularly contain target-group elements in their ciphertexts. Hence the "well-formedness" of an IBE ciphertext is described as equations between target-group elements, for which the Groth-Sahai proof system does not provide the zero-knowledge property. Considering the fact that the Groth-Sahai proof system is currently the only choice for an efficient instantiation of a zero-knowledge proof system, we have no other way of providing a proof of this type.

Of course if we want an unbounded GS-MDO scheme, it is easily possible by applying general NIZK techniques (as in the BMW/BSZ constructions [4, 5]). However, it might be hard to achieve an efficient instantiation of the known NIZK proofs for general NP-languages (The same holds even in the random oracle model).

From the above considerations, it is difficult to construct efficient unbounded GS-MDO schemes in the standard model. Therefore, applying random oracles seems to be a reasonable solution.

Our Contribution.

In this paper, we propose the first unbounded GS-MDO scheme in the random oracle model. The proposed scheme is based on the Boneh-Boyen-Shacham (BBS) group signature [7], which is one of the most efficient group signature schemes in the random oracle model. The opening procedure is implemented by linear encryption, and a user's certificate is implemented by the Boneh-Boyen short signature [6]. The functionality of MDO is realized by adopting the Boneh-Franklin (BF) IBE [8]. In order to combine the short group signature and the BF IBE, we replace the linear encryption with a certain type of 2-out-of-2 multiple encryption.

Note that the BBS scheme satisfies only CPA-anonymity, the security game of which does not allow the adversary to access the opening oracle. We make a further improvement to the above approach to achieve CCA-anonymity. This improvement is carried out by changing the "linear" part of the multiple encryption to a kind of double encryption similar to the Naor-Yung construction and adding a validity check component that ensures "well-formedness" of the ciphertext, which could be reminiscent of the Cramer-Shoup encryption scheme [11, 12].

For efficiency reasons, to realize the proof, we do not follow the construction of the Cramer-Shoup scheme directly, but instead use the Fiat-Shamir heuristics. This strategy yields a more efficient construction compared with that obtained when directly using another DLIN-based CCA-secure public-key encryption (PKE) is directly used, e.g., Shacham's DLIN-variant of the Cramer-Shoup PKE scheme [16].

The proposed scheme simultaneously achieves a higher degree of efficiency and security than the previous scheme by Sakai et al. The signature contains 16 group elements (3630 bits for 80-bit security), whereas that of the previous scheme contains about 450 elements (about 76000 bits for the same security level). Furthermore, the proposed scheme allows the admitter to issue an *unbounded* number of tokens, which is not achieved by the previous efficient construction. See Section 4 for a detailed comparison.

2. PRELIMINARIES

In this section, we present a formal model of the GS-MDO proposed by Sakai et al. in [15], and computational assumptions for our proposed scheme.

2.1 Group Signature with Message-dependent Opening

GS-MDO is an extension of group signature, which allows members of the group to sign a message anonymously. In addition, as described in the introduction, there are two authorities in GS-MDO, the opener and the admitter. The admitter is able to issue a token that is specific to a message. The opener is able to identify the signer of a signature on a message for which a token from the admitter is available.

A GS-MDO scheme consists of five probabilistic polynomial-time algorithms $(\mathsf{GKg}, \mathsf{GSig}, \mathsf{GVf}, \mathsf{Td}, \mathsf{Open})$. GKg takes as inputs $(1^\lambda, 1^n)$ where λ is a security parameter and n is the number of group members, and outputs $(gpk, ok, ak,$

$(gsk_i)_{1 \le i \le n})$ where gpk is a group public key, ok is an opening key for the opener, ak is the message specification key for the admitter, and gsk_i is a group signing key for each group member i. GSig takes as inputs (gpk, i, gsk_i, M) where M is a message, and outputs a group signature σ. GVf takes as inputs (gpk, M, σ) and outputs $\top$ or $\perp$. Td takes as inputs (gpk, ak, M) and outputs the token t_M for M. Open takes as inputs $(gpk, ok, M, \sigma, t_M)$ and outputs $i \in \{1, 2, \ldots, n\}$ or $\perp$.

Note that in the case where the number of tokens issued by the admitter is bounded by k (like the Sakai et al.'s scheme [15]), the group key generation algorithm GKg takes as input $(1^\lambda, 1^n, 1^k)$.

2.2 The Security Requirements of GS-MDO

Sakai et al. defined the following three requirements for a GS-MDO (the definition of these properties is based on Bellare, Micciancio, and Warinschi's security model [4]).

Opener Anonymity.

The opener should be unable to identify the signer of any group signature without cooperation with the admitter, even if some group members are corrupted. This requirement is formalized by the following game. Note that, in order to model anonymity against the opener, we give the opening key to the adversary in this game.

DEFINITION 1. *We say that a GS-MDO scheme* $\Pi = (\mathsf{GKg}, \mathsf{GSig}, \mathsf{GVf}, \mathsf{Td}, \mathsf{Open})$ *has opener anonymity if for all probabilistic polynomial-time adversaries* $\mathcal{A}$*, the success probability of* $\mathcal{A}$ *in the following game between a challenger is negligible in the security parameter* λ.

Setup. *The challenger runs* $\mathsf{GKg}(1^\lambda, 1^n)$ *and obtain* $(gpk, ok, ak, (gsk_i)_{1 \le i \le n})$. *Then the challenger sends* $(gpk, ok, (gsk_i)_{1 \le i \le n})$ *to* $\mathcal{A}$.

Token Query I. $\mathcal{A}$ *is allowed to interact with a token oracle. For a token query for* M*, the challenger runs* $\mathsf{Td}(gpk, ak, M)$ *to obtain* t_M *and return* t_M *to* $\mathcal{A}$.

Challenge. *At some point* $\mathcal{A}$ *requests a challenge for* $i_0, i_1 \in \{1, \ldots, n\}$ *and a message* M^*. *The challenger chooses a random bit* b*, and return* $\mathsf{GSig}(gpk, gsk_{i_b}, M^*)$. *In this phase* $\mathcal{A}$ *is forbidden to submit* M^* *which is already queried in Token Query I.*

Token Query II. $\mathcal{A}$ *continues to query tokens. In this phase* $\mathcal{A}$ *is forbidden to query* M^* *which is submitted in challenge phase.*

Guess. $\mathcal{A}$ *outputs a bit* b'. *The advantage of* $\mathcal{A}$ *is defined by the absolute difference between the probability that* b' *is equal to* b *and* $1/2$.

Admitter Anonymity.

The admitter should be unable to identify the signer of any group signature without cooperation with the opener, even if some group members are corrupted. This requirement is formalized by the following game. Note that, in order to model anonymity against the admitter, we give the message specification key to the adversary in this game.

DEFINITION 2. *We say that a GS-MDO scheme* $\Pi = (\mathsf{GKg}, \mathsf{GSig}, \mathsf{GVf}, \mathsf{Td}, \mathsf{Open})$ *has admitter anonymity if for all probabilistic polynomial-time adversaries* $\mathcal{A}$*, the success probability of* $\mathcal{A}$ *in the following game between a challenger is negligible in the security parameter* λ.

Setup. *The challenger runs* $\mathsf{GKg}(1^\lambda, 1^n)$ *and obtain* $(gpk, ok, ak, (gsk_i)_{1 \le i \le n})$. *Then the challenger sends* $(gpk, ak, (gsk_i)_{1 \le i \le n})$ *to* $\mathcal{A}$.

Open Query I. $\mathcal{A}$ *is allowed to interact with an open oracle. For an open query for* (M, σ)*, the challenger runs* $\mathsf{Td}(gpk, ak, M)$ *to obtain* t_M *and return* $\mathsf{Open}(gpk, ok, M, \sigma, t_M)$ *to* $\mathcal{A}$.

Challenge. *At some point* $\mathcal{A}$ *requests a challenge for* $i_0, i_1 \in \{1, \ldots, n\}$ *and a message* M^*. *The challenger chooses a random bit* b*, and return* $\mathsf{GSig}(gpk, gsk_{i_b}, M^*)$.

Open Query II. $\mathcal{A}$ *continues to query tokens. In this phase* $\mathcal{A}$ *is forbidden to query* σ^* *which is submitted in challenge phase.*

Guess. $\mathcal{A}$ *outputs a bit* b'. *The advantage of* $\mathcal{A}$ *is defined by the absolute difference between the probability that* b' *is equal to* b *and* $1/2$.

Traceability.

Even the opener and the admitter collude, they should not be able to produce any forged signature or untraceable signature. This requirement is formalized by the following definition.

DEFINITION 3. *We say that a GS-MDO scheme* $\Pi = (\mathsf{GKg}, \mathsf{GSig}, \mathsf{GVf}, \mathsf{Td}, \mathsf{Open})$ *has traceability if for all probabilistic polynomial-time adversaries* $\mathcal{A}$*, the success probability of* $\mathcal{A}$ *in the following game between a challenger is negligible in the security parameter* λ.

Setup. *The challenger runs* $\mathsf{GKg}(1^\lambda, 1^n)$ *and obtain* $(gpk, ok, ak, (gsk_i)_{1 \le i \le n})$. *Then the challenger sends* $(gpk, ok, ak, (gsk_i)_{1 \le i \le n})$ *to* $\mathcal{A}$.

Private Key Query. $\mathcal{A}$ *is allowed to interact with a private key oracle. For a private key query for* i*, the challenger returns* gsk_i *to* $\mathcal{A}$.

Signing Query. $\mathcal{A}$ *is allowed to interact with signing oracle. For a signing query for* (i, M)*, the challenger returns* $\mathsf{GSig}(gpk, i, gsk_i, M)$ *to* $\mathcal{A}$.

Forge. $\mathcal{A}$ *outputs a message-signature pair* (M^*, σ^*). *The adversary wins the game if* $\mathsf{GVf}(gpk, M^*, \sigma^*) = \top$ *and one of the following conditions (a) and (b) holds: (a)* $\mathsf{Open}(gpk, ok, M^*, \sigma^*, \mathsf{Td}(gpk, ak, M^*)) = \perp$*, or (b)* $\mathsf{Open}(gpk, ok, M^*, \sigma^*, \mathsf{Td}(gpk, ak, M^*)) = i^* \neq \perp$ *and both the signing key of the user* i^* *and a signature on* (i^*, M^*) *are never queried to the above oracles. The advantage of* $\mathcal{A}$ *is defined by the probability that* $\mathcal{A}$ *wins the game.*

2.3 The Computational Assumptions

Let $\mathcal{G}$ be a probabilistic polynomial-time algorithm that takes a security parameter 1^λ as input and generates a parameter $(p, \mathbb{G}, \mathbb{G}_T, e, g)$ of *bilinear groups*, where p is a λ-bit prime, $\mathbb{G}$ and $\mathbb{G}_T$ are groups of order p, e is a bilinear map from $\mathbb{G} \times \mathbb{G}$ to $\mathbb{G}_T$, and g is a generator of $\mathbb{G}$. We then describe several computational assumptions on which the proposed scheme is based.

The q-strong Diffie-Hellman Assumption.

Let $(p, \mathbb{G}, \mathbb{G}_T, e, g) \xleftarrow{\$} \mathcal{G}(1^\lambda)$, $\gamma \xleftarrow{\$} \mathbb{Z}$ and $A_i \leftarrow g^{\gamma^i}$ for $i = 0, \ldots, q$. The q-strong Diffie-Hellman problem in $\mathbb{G}$ is stated as follows: given $(g, (A_i)_{0 \le i \le q})$, output $(c, g^{1/(\gamma+c)})$ where $c \in \mathbb{Z}_p^*$. The advantage of an algorithm $\mathcal{A}$ against the q-strong Diffie-Hellman problem is defined as

$$\mathrm{Adv}_{\mathcal{A}}^{q\text{-SDH}}(\lambda) = \Pr[\mathcal{A}(g, (A_i)_{0 \le i \le q}) = (c, g^{1/(\gamma+c)}) \wedge c \in \mathbb{Z}_p].$$

We say that the q-strong Diffie-Hellman assumption holds if $\mathrm{Adv}_{\mathcal{A}}^{q\text{-SDH}}(\lambda)$ is negligible in λ for any probabilistic polynomial-time algorithm $\mathcal{A}$.

The Decision Linear Assumption.

Let u, v, $h \xleftarrow{\$} \mathbb{G}$, α, β, $r \xleftarrow{\$} \mathbb{Z}_p$ and $g_1 \leftarrow u^\alpha, g_2 \leftarrow v^\beta$. The decision linear problem in $\mathbb{G}$ is stated as follows: given $(u, v, h, u^\alpha, v^\beta, z)$, output 1 if $z = h^{\alpha+\beta}$, otherwise 0 if $z = h^r$. The advantage of an algorithm $\mathcal{A}$ against the decision linear problem is defined as

$$\mathrm{Adv}_{\mathcal{A}}^{\mathrm{DLIN}}(\lambda) = |\Pr[\mathcal{A}(u, v, h, u^\alpha, v^\beta, z) = 1 \mid z = h^{\alpha+\beta}] - \Pr[\mathcal{A}(u, v, h, u^\alpha, v^\beta, z) = 1 \mid z = h^r]|.$$

We say that the decision linear assumption holds if $\mathrm{Adv}_{\mathcal{A}}^{\mathrm{DLIN}}(\lambda)$ is negligible in λ for any probabilistic polynomial-time algorithm $\mathcal{A}$.

The Decision Bilinear Diffie-Hellman Assumption.

Let $(p, \mathbb{G}, \mathbb{G}_T, e, g) \xleftarrow{\$} \mathcal{G}(1^\lambda)$ and a, b, c, $r \xleftarrow{\$} \mathbb{Z}_p$. The decision linear problem in $(\mathbb{G}, \mathbb{G}_T)$ is stated as follows: given (g, g^a, g^b, g^c, z), output 1 if $z = e(g, g)^{abc}$, otherwise 0 if $z = e(g, g)^r$. The Advantage of an algorithm $\mathcal{A}$ against the decision bilinear Diffie-Hellman problem is defined as

$$\mathrm{Adv}_{\mathcal{A}}^{\mathrm{DBDH}}(\lambda) = |\Pr[\mathcal{A}(g, g^a, g^b, g^c, z) = 1 \mid z = e(g, g)^{abc}] - \Pr[\mathcal{A}(g, g^a, g^b, g^c, z) = 1 \mid z = e(g, g)^r]|.$$

We say that the decision bilinear Diffie-Hellman assumption holds if $\mathrm{Adv}_{\mathcal{A}}^{\mathrm{DBDH}}(\lambda)$ is negligible in λ for any probabilistic polynomial-time algorithm $\mathcal{A}$.

3. THE PROPOSED SCHEME

In this section we describe the proposed scheme and its security.

Underlying Idea.

Our proposed scheme is based on the BBS group signature, in which each group member is provided with a (Boneh-Boyen) signature [6], to certify the membership of the owner. The group signature consists of two parts: the first part is the linear encryption of the certificate, while the second part is the "signature of knowledge" of the encrypted certificate. The decryption key for the linear encryption, which is held by the opener, is used to revoke the anonymity of any group signature.

We extend the BBS group signature scheme by replacing the linear encryption with a certain type of 2-out-of-2 multiple encryption of ordinary PKE and IBE. The multiple encryption is designed to ensure that only when *both* the PKE and the IBE are decrypted, can the entire ciphertext be decrypted. Such multiple-encryption can be accomplished using a simple 2-out-of-2 secret sharing.

This feature enables us to realize the MDO functionality. If the opener only possesses the decryption key of the PKE and the admitter holds the master secret of the IBE, the decryption key of the IBE (under a certain ID) can serve as the message-dependent token. The multiple encryption ensures that the opener cannot identify the originator of a signature even if the opener has the decryption key of the PKE scheme. Furthermore, if the opener receives the token, which is merely a derived decryption key of the IBE, the opener, using both his own decryption key and the message-specific decryption key received from the admitter, is able to decrypt the ciphertext included in the group signature, thereby identifying the originator of the signature.

Our Construction.

$\mathsf{GKg}(1^\lambda, 1^n)$. The proposed scheme uses two hash functions $H_1\colon \{0,1\}^* \to \mathbb{G}$ and $H_2\colon \{0,1\}^* \to \mathbb{Z}_p$. They are modeled as random oracles in the security analysis. Given a security parameter 1^λ, the algorithm runs $\mathcal{G}(1^\lambda)$ to generate a parameter of bilinear group $(p, \mathbb{G}, \mathbb{G}_T, e, g)$. Then the algorithm selects a random element u, v, $h \xleftarrow{\$} \mathbb{G} \setminus \{1\}$ and random integers ξ_1, ξ_2, ξ_3, ζ, $\gamma \xleftarrow{\$} \mathbb{Z}_p$, sets $g_1 \leftarrow u^{\xi_1} h^{\xi_3}$, $g_2 \leftarrow v^{\xi_2} h^{\xi_3}$, $y \leftarrow g^\zeta$, and $w \leftarrow g^\gamma$. The algorithm then select $x_i \xleftarrow{\$} \mathbb{Z}_p$ and sets $A_i \leftarrow g^{1/(\gamma+x_i)}$ for each user i $(1 \le i \le n)$. Finally the algorithm outputs the group public key $gpk \leftarrow (p, \mathbb{G}, \mathbb{G}_T, e, g, u, v, h, g_1, g_2, y, w, H_1, H_2)$, the message-specification key $ak \leftarrow \zeta$, the opening key $ok \leftarrow (\xi_1, \xi_2, \xi_3, (e(A_i, g))_{1 \le i \le n})$, and the users' signing keys $(gsk_i)_{1 \le i \le n} \leftarrow (A_i, x_i)_{1 \le i \le n}$.

$\mathsf{GSig}(gpk, i, gsk_i, M)$. Given an input (gpk, i, gsk_i, M), the algorithm generates a group signature as follows: choose random α, β, ρ, $\eta \xleftarrow{\$} \mathbb{Z}_p$, computes

$$(T_1, T_2, T_3, T_4) \leftarrow (u^\alpha, v^\beta, h^{\alpha+\beta}, g_1^\alpha g_2^\beta A_i g^\eta)$$

and

$$(T_5, T_6) \leftarrow (g^\rho, e(y, H_1(M))^\rho e(g, g)^{-\eta}).$$

Then choose random r_α, r_β, r_ρ, r_η, r_x, $r_{\alpha x}$, $r_{\beta x}$, $r_{\rho x}$, $r_{\eta x} \xleftarrow{\$} \mathbb{Z}_p$, compute

$$\begin{aligned}
R_1 &\leftarrow u^{r_\alpha},\\
R_2 &\leftarrow v^{r_\beta},\\
R_3 &\leftarrow h^{r_\alpha + r_\beta},\\
R_4 &\leftarrow e(T_4, g)^{r_x} e(g_1, w)^{-r_\alpha} e(g_1, g)^{-r_{\alpha x}}\\
&\quad \cdot e(g_2, w)^{-r_\beta} e(g_2, g)^{-r_{\beta x}} e(g, w)^{-r_\eta} e(g, g)^{-r_{\eta x}},\\
R_5 &\leftarrow g^{r_\rho},\\
R_6 &\leftarrow e(y, H_1(M))^{r_\rho} e(g, g)^{-r_\eta},\\
R_7 &\leftarrow T_1^{r_x} u^{-r_{\alpha x}},\\
R_8 &\leftarrow T_2^{r_x} v^{-r_{\beta x}},\\
R_9 &\leftarrow T_5^{r_x} g^{-r_{\rho x}},\\
R_{10} &\leftarrow T_6^{r_x} e(y, H_1(M))^{-r_{\rho x}} e(g, g)^{r_{\eta x}},
\end{aligned}$$

compute $c \leftarrow H_2(M, T_1, \ldots, T_6, R_1, \ldots, R_{10})$, and further computes

$$\begin{aligned}
s_\alpha &\leftarrow r_\alpha + c\alpha, \\
s_\beta &\leftarrow r_\beta + c\beta, \\
s_\rho &\leftarrow r_\rho + c\rho, \\
s_\eta &\leftarrow r_\eta + c\eta, \\
s_x &\leftarrow r_x + cx_i, \\
s_{\alpha x} &\leftarrow r_{\alpha x} + c\alpha x_i, \\
s_{\beta x} &\leftarrow r_{\beta x} + c\beta x_i, \\
s_{\rho x} &\leftarrow r_{\rho x} + c\rho x_i, \\
s_{\eta x} &\leftarrow r_{\eta x} + c\eta x_i.
\end{aligned}$$

Finally let σ be

$$(T_1, \ldots, T_6, c, s_\alpha, s_\beta, s_\rho, s_\eta, s_x, s_{\alpha x}, s_{\beta x}, s_{\rho x}, s_{\eta x})$$

and output σ as the group signature.

Remark 1. The above $R_1, \ldots, R_{10}$, c, s_α, s_β, s_ρ, s_η, s_x, $s_{\alpha x}$, $s_{\beta x}$, $s_{\rho x}$, and $s_{\eta x}$ come from a Schnorr-type protocol that proves knowledge α, β, ρ, η, and x satisfying the equations

$$\begin{aligned}
T_1 &= u^\alpha, \\
T_2 &= v^\beta, \\
T_3 &= h^{\alpha+\beta}, \\
e(g, g) &= e(T_4 {g_1}^{-\alpha} {g_2}^{-\beta} g^{-\eta}, wg^x), \\
T_5 &= g^\rho, \\
T_6 &= e(y, H_1(M))^\rho e(g, g)^{-\eta}.
\end{aligned}$$

More concretely, introducing four intermediate variables δ_1, δ_2, δ_3, and δ_4 (which are intended to be $\delta_1 = \alpha x$, $\delta_2 = \beta x$, $\delta_3 = \rho x$, and $\delta_4 = \eta x$), the underlying protocol proves knowledge α, β, ρ, η, x, δ_1, δ_2, δ_3, and δ_4 satisfying the equations

$$\begin{aligned}
T_1 &= u^\alpha, \\
T_2 &= v^\beta, \\
T_3 &= h^{\alpha+\beta}, \\
e(g, g)/e(T_4, w) &= e(T_4, g)^x e(g_1, w)^{-\alpha} e(g_1, g)^{-\delta_1} \\
&\quad \cdot e(g_2, w)^{-\beta} e(g_2, g)^{-\delta_2} \\
&\quad \cdot e(g, w)^{-\eta} e(g, g)^{-\delta_4}, \\
T_5 &= g^\rho, \\
T_6 &= e(y, H_1(M))^\rho e(g, g)^{-\eta}, \\
1 &= T_1^x u^{-\delta_1}, \\
1 &= T_2^x v^{-\delta_2}, \\
1 &= T_5^x g^{-\delta_3}, \\
1 &= T_6^x e(y, H_1(M))^{-\delta_3} e(g, g)^{\delta_4}.
\end{aligned}$$

$\mathsf{GVf}(gpk, M, \sigma)$. Given gpk, M and σ, the algorithm verifies the signature as follows. The algorithm computes R'_1, R'_2, R'_3, R'_4, R'_5, R'_6, R'_7, R'_8, R'_9, and R'_{10} by letting

$$\begin{aligned}
R'_1 &\leftarrow u^{s_\alpha} T_1^{-c}, \\
R'_2 &\leftarrow v^{s_\beta} T_2^{-c}, \\
R'_3 &\leftarrow h^{s_\alpha + s_\beta} T_3^{-c}, \\
R'_4 &\leftarrow e(T_4, g)^{s_x} e(g_1, w)^{-s_\alpha} e(g_1, g)^{-s_{\alpha x}} \\
&\quad \cdot e(g_2, w)^{-s_\beta} e(g_2, g)^{-s_{\beta x}} \\
&\quad \cdot e(g, w)^{-s_\eta} e(g, g)^{-s_{\eta x}} \\
&\quad \cdot (e(g, g)/e(T_4, w))^{-c}, \\
R'_5 &\leftarrow y^{s_\rho} T_5^{-c}, \\
R'_6 &\leftarrow e(y, H_1(M))^{s_\rho} e(g, g)^{-s_\eta} T_6^{-c}, \\
R'_7 &\leftarrow T_1^{s_x} u^{-s_{\alpha x}}, \\
R'_8 &\leftarrow T_2^{s_x} v^{-s_{\beta x}}, \\
R'_9 &\leftarrow T_5^{s_x} g^{-s_{\rho x}}, \\
R'_{10} &\leftarrow T_6^{s_x} e(y, H_1(M))^{-s_{\rho x}} e(g, g)^{s_{\eta x}}.
\end{aligned}$$

Then the algorithm verifies whether the equation

$$c = H_2(M, T_1, \ldots, T_6, R'_1, \ldots, R'_{10})$$

holds. If the equation holds, the algorithm outputs $\top$, otherwise outputs $\perp$.

$\mathsf{Td}(gpk, ak, M)$. Given gpk, $ak = \zeta$, and M, the algorithm generates a token t_M as $t_M \leftarrow H_1(M)^\zeta$ and outputs t_M.

$\mathsf{Open}(gpk, ok, M, \sigma, t_M)$. Given gpk, ok, M, σ, and t_m, the algorithm first verifies the signature using the algorithm GVf. If the signature is invalid, the algorithm outputs $\perp$. If the signature is valid, the algorithm then find i which satisfies the equation

$$e\left(\frac{T_4}{T_1^{\xi_1} T_2^{\xi_2} T_3^{\xi_3}}, g\right) \cdot \frac{T_6}{e(T_5, t_M)} = e(A_i, g).$$

When such i exists, the algorithm outputs i, otherwise outputs $\perp$.

The security of our proposed scheme is proved as follows.

THEOREM 1. *If the decision bilinear Diffie-Hellman assumption holds, the proposed construction has opener anonymity in the random oracle model.*

THEOREM 2. *If the decision linear assumption holds, the proposed construction has admitter anonymity in the random oracle model.*

THEOREM 3. *If the q-strong Diffie-Hellman assumption holds, the proposed construction has traceability in the random oracle model.*

For the details of the proof of these theorems, see the full version of this paper.

4. COMPARISON

Finally, we present a brief comparison of the proposed construction and some related constructions in Table 1. Concretely we compare the proposed scheme with the instantiation presented by Sakai et al. [15] and the BBS group signature scheme, on which the proposed construction is based.

As shown in Table 1, compared with the scheme by Sakai et al., the proposed scheme realizes improvements in two aspects: the first is removing the a priori upper bound on the number of tokens the admitter can issue, while the second is substantially reducing the signature size. For 80-bit security,

Table 1: **Performance comparison, where DLIN, DBDH, q-SDH, and SFP respectively stand for the decision linear assumption, the decision bilinear Diffie-Hellman assumption, the q-strong Diffie-Hellman assumption, and the simultaneous flexible pairing assumption [3]. The bit size for the estimation is given in [10].**

	Signature length # of $[\mathbb{G},\mathbb{Z}_p,\mathbb{G}_T]$-elements	Message-dependent Opening	Assumption	w/o RO	Anonymity
Proposed	[5,10,1] (3630 bits)	unbounded	DLIN, DBDH, q-SDH	-	CCA
Sakai et al. [15]	[446,0,0] (75820 bits)[1]	k-bounded	DLIN, SFP	✓	CCA
BBS [7]	[3,6,0] (1530 bits)	-	DLIN, q-SDH	-	CPA

[1]This scheme needs an additional (strongly unforgeable) one-time signature scheme, and therefore the total number of group elements becomes slightly larger, and additional complexity assumptions may be required.

the signature length of our scheme is 3630 bits, while those of the schemes by Sakai et al. [15] and Boneh et al. [7] are 75820 bits and 1530 bits (the bit size for the estimation is given in [10]), respectively. These two improvements are achieved at the cost of using random oracles, as indicated in the "w/o RO" column in the table.

Compared with the BBS scheme, we believe that the proposed scheme achieves MDO at a relatively reasonable cost given the increase in signature size. As shown in the table, the signature size of the proposed scheme is almost twice as long as that of the BBS scheme, which we believe to be reasonable. We also note that the proposed scheme achieves CCA-anonymity, which guarantees a remarkably higher level of security than the CPA-anonymity achieved by the BBS scheme.

5. ACKNOWLEDGMENTS

The authors are grateful to Hovav Shacham for pointing out the possibility of using random oracles to achieve unbounded security. The authors would like to thank Shin-Akarui-Angou-Benkyou-Kai for the valuable discussions and comments.

6. REFERENCES

[1] M. Abe, G. Fuchsbauer, J. Groth, K. Haralambiev, and M. Ohkubo. Structure-preserving signatures and commitments to group elements. In T. Rabin, editor, *CRYPTO 2010*, volume 6223 of *LNCS*, pages 209–236. Springer, Heidelberg, 2010.

[2] M. Abe, K. Haralambiev, and M. Ohkubo. Signing on elements in bilinear groups for modular protocol design. Cryptology ePrint Archive, Report 2010/133, 2010. `http://eprint.iacr.org/`.

[3] M. Abe, K. Haralambiev, and M. Ohkubo. Group to group commitments do not shrink. In D. Pointcheval and T. Johansson, editors, *EUROCRYPT 2012*, volume 7237 of *LNCS*, pages 301–317. Springer, Heidelberg, 2012.

[4] M. Bellare, D. Micciancio, and B. Warinschi. Foundations of group signatures: Formal definitions, simplified requirements, and a construction based on general assumptions. In E. Biham, editor, *EUROCRYPT 2003*, volume 2656 of *LNCS*, pages 644–644. Springer, Heidelberg, 2003.

[5] M. Bellare, H. Shi, and C. Zhang. Foundations of group signatures: The case of dynamic groups. In A. Menezes, editor, *CT-RSA 2005*, volume 3376 of *LNCS*, pages 136–153. Springer, Heidelberg, 2005.

[6] D. Boneh and X. Boyen. Short signatures without random oracles and the SDH assumption in bilinear groups. *J. Cryptol.*, 21:149–177, 2008.

[7] D. Boneh, X. Boyen, and H. Shacham. Short group signatures. In M. Franklin, editor, *CRYPTO 2004*, volume 3152 of *LNCS*, pages 227–242. Springer, Heidelberg, 2004.

[8] D. Boneh and M. Franklin. Identity-based encryption from the Weil pairing. *SIAM J. Comput.*, 32(3):586–615, 2003.

[9] D. Chaum and E. van Heyst. Group signatures. In D. W. Davies, editor, *EUROCRYPT '91*, volume 547 of *LNCS*, pages 257–265. Springer, Heidelberg, 1991.

[10] J. Chen, H. W. Lim, S. Ling, H. Wang, and H. Wee. Shorter IBE and signatures via asymmetric pairings. Cryptology ePrint Archive, Report 2012/224, 2012. `http://eprint.iacr.org/`.

[11] R. Cramer and V. Shoup. A practical public key cryptosystem provably secure against adaptive chosen ciphertext attack. In H. Krawczyk, editor, *CRYPTO 1998*, volume 1462 of *LNCS*, pages 13–25. Springer, Heidelberg, 1998.

[12] R. Cramer and V. Shoup. Design and analysis of practical public-key encryption schemes secure against adaptive chosen ciphertext attack. *SIAM Journal on Computing*, 33(1):167–226, 2003.

[13] J. Groth and A. Sahai. Efficient non-interactive proof systems for bilinear groups. In N. Smart, editor, *EUROCRYPT 2008*, volume 4965 of *LNCS*, pages 415–432. Springer, Heidelberg, 2008.

[14] S.-H. Heng and K. Kurosawa. k-resilient identity-based encryption in the standard model. In T. Okamoto, editor, *CT-RSA 2004*, volume 2964 of *LNCS*, pages 67–80. Springer, Heidelberg, 2004.

[15] Y. Sakai, K. Emura, G. Hanaoka, Y. Kawai, T. Matsuda, and K. Omote. Group signatures with message-dependent opening. In *Pairing 2012*, pages 270–294, 2012.

[16] H. Shacham. A Cramer-Shoup encryption scheme from the linear assumption and from progressively weaker linear variants. Cryptology ePrint Archive, Report 2007/074, 2007. `http://eprint.iacr.org/`.

Attribute-based Fine-Grained Access Control with Efficient Revocation in Cloud Storage Systems

Kan Yang
Dept. of Computer Science
City University of Hong Kong
Hong Kong SAR
kanyang3@student.cityu.edu.hk

Xiaohua Jia
Dept. of Computer Science
City University of Hong Kong
Hong Kong SAR
csjia@cityu.edu.hk

Kui Ren
University at Buffalo
State University of New York
NY, USA
kuiren@buffalo.edu

ABSTRACT

A cloud storage service allows data owner to outsource their data to the cloud and through which provide the data access to the users. Because the cloud server and the data owner are not in the same trust domain, the semi-trusted cloud server cannot be relied to enforce the access policy. To address this challenge, traditional methods usually require the data owner to encrypt the data and deliver decryption keys to authorized users. These methods, however, normally involve complicated key management and high overhead on data owner. In this paper, we design an access control framework for cloud storage systems that achieves fine-grained access control based on an adapted Ciphertext-Policy Attribute-based Encryption (CP-ABE) approach. In the proposed scheme, an efficient attribute revocation method is proposed to cope with the dynamic changes of users' access privileges in large-scale systems. The analysis shows that the proposed access control scheme is provably secure in the random oracle model and efficient to be applied into practice.

Categories and Subject Descriptors

D.4.6 [**Security and Protection**]: Access controls, Cryptographic controls

General Terms

Security, Theory

Keywords

Access Control, Cloud Storage, Attribute Revocation, CP-ABE

1. INTRODUCTION

Cloud storage is an important service of cloud computing [4], which allows data owners to host their data in the cloud that provides "24/7/365" data access to the users (data consumers). Because the cloud storage service separates the roles of the data owner from the data service provider, and the data owner does not interact with the user directly for providing data access service, the data access control becomes a challenging issue for cloud storage systems. Existing methods [6] usually delegate data access control to a trusted server and let it be in charge of defining and enforcing access policies. However, the cloud server cannot be fully trusted by data owners, since the cloud server may give data access to unauthorized users to make more profit (e.g., the competitor of a company). Thus, traditional server-based data access control methods are no longer suitable for cloud storage systems. The aim of this paper is to study the data access control issue in cloud storage systems, where the data owner is in charge of defining and enforcing the access policy.

The Ciphertext-Policy Attribute-based Encryption (CP-ABE) [1, 7] is regarded as one of the most suitable technologies for data access control in cloud storage systems, because it gives the data owner more direct control on access policies and the policy checking occurs "inside the cryptography". However, due to the *attribute revocation problem*, it is very costly to apply the CP-ABE approach to control the data access in cloud storage systems. To address this problem, there are two requirements: 1) *Backward Security* The revoked user (whose attribute is revoked) cannot decrypt any new published ciphertext which requires the revoked attribute to decrypt. 2) *Forward Security* The newly joined user who has sufficient attributes is still able to decrypt the ciphertexts which were published before it joined the system. Existing attribute revocation methods proposed for CP-ABE systems usually rely on the fully trusted server, thus they cannot be applied into the cloud storage systems. The attribute revocation is still an open problem in the design of attribute-based data access control schemes for cloud storage systems.

In this paper, we study the data security issues in cloud storage systems and propose a fine-grained data access control scheme, where the server is not required to be fully trusted and data owners are not required to be online all the time.

The main contributions of this work are summarized as follows.

1) We design an efficient revocable CP-ABE scheme as the underling techniques to achieve fine-grained data access control for cloud storage systems. In the proposed scheme, the access policy is defined and enforced by data owners rather than by cloud server.

2) We propose a secure and efficient attribute revocation method to cope with the dynamic change of users' access privileges in large-scale systems. It achieves both forward security and backward security, and incurs less computation cost and communication overhead. Moreover, the revocation is conducted efficiently on attribute level rather than on user level.

3) We further provide the security analysis and performance analysis to show that our scheme is secure in the random oracle model and efficient to be applied into practice.

ASIA CCS'13, May 8–10, 2013, Hangzhou, China.

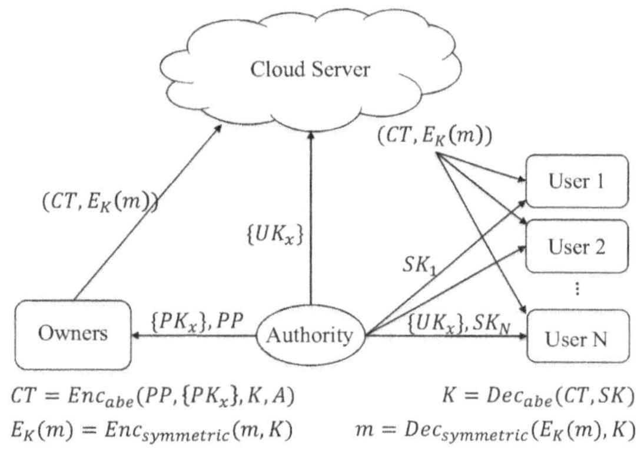

Figure 1: System Model of Access Control in Cloud Storage

The remaining of this paper is organized as follows. Section 2 describes the definition of system model and security model. In Section 3, we describe the attribute-based fine-grained access control scheme and the efficient attribute revocation method. Section 4 gives the analysis of our scheme in terms of security and efficiency. In Section 5, we give the related work on data access control and attribute revocation in ABE system. Finally, the conclusion is given in Section 6.

2. SYSTEM AND SECURITY MODEL

2.1 System Model

We consider an access control system for cloud storage service, as described in Figure 1. There are four entities in the system: Authority, Data owners (owners), Cloud server (server) and Data consumers (users).

The authority is responsible for entitling/revoking/re-granting attributes to/from/to users according to their role or identity in the system. It assigns secret keys to users when they are entitled attributes and maintains a version number of each attribute. When an attribute is revoked, the authority will update the version number of the revoked attribute, and generate a corresponding update key. The update key is used to update secret keys for non-revoked users [1] and the ciphertexts on the cloud servers.

The owners determine the access policies and encrypt their data under the policies before hosting them in the cloud (For simplicity, the data here means the content key[2]).

The cloud server stores the owners' data and provides data access service to users. But the server does not engage in the data access control. Instead, we assume the ciphertext may be accessed by all the legal users in the system. But, the *access control happens inside the cryptography*. That is only the users who possess eligible attributes (satisfying the access policy) can decrypt the ciphertext.

Each user is entitled a set of attributes according to its roles or identity in the system. However, the user's attribute set may dynamically change due to the role changed of the user in the system. For example, when a user is degraded from the manager to the normal worker, some of its attributes should be revoked, while sometimes the revoked attribute need to be re-granted to the user. The user can decrypt the ciphertext only when he/she has sufficient attributes satisfying the access policy associated with the ciphertext.

[1]We use *non-revoked users* to denote those users who hold the revoked attribute but not been revoked.

[2]In practical, the data is encrypted with a content key by using symmetric encryption method, and the content key is encrypted by using CP-ABE.

2.2 Data Access Control Framework

The framework of the data access control is defined as follows.

DEFINITION 1 (ACCESS CONTROL SCHEME). *An access control scheme is a collection of the following algorithms:* Setup, SKeyGen, Encrypt, Decrypt, UKeyGen, SKUpdate *and* CTUpdate, *where* UKeyGen, SKUpdate *and* CTUpdate *are used for attribute revocation.*

Setup$(1^{\lambda}) \rightarrow (MK, PP, \{PK_x\}, \{VK_x\})$. The setup algorithm is run by the authority. It takes no input other than the implicit security parameter λ. It outputs a master key MK, the public parameters PP, the set of all the public attribute keys $\{PK_x\}$ and the set of all attribute version keys $\{VK_x\}$.

SKeyGen$(MK, S, \{VK_x\}_{x\in S}) \rightarrow SK$. The secret key generation algorithm is run by the authority. It takes as inputs the master key MK, a set of attributes S that describes the secret key, and the corresponding set of attribute version keys $\{VK_x\}_{x\in S}$. It outputs the user's secret key SK.

Encrypt$(PP, \{PK_x\}, m, \mathbb{A}) \rightarrow CT$. The encryption algorithm is run by the data owner. It takes as inputs the public parameters PP, the set of public attribute key $\{PK_x\}$, a message m and an access structure $\mathbb{A}$ over the universe of attributes. The algorithm will encrypt m such that only a user who possesses a set of attributes satisfying the access structure will be able to decrypt the message. It outputs a ciphertext CT.

Decrypt$(CT, SK) \rightarrow m$. The decryption algorithm is run by the user. It takes as inputs the ciphertext CT which contains an access structure $\mathbb{A}$ and the secret key SK for a set of attributes S. If the set of attributes S satisfies the access structure $\mathbb{A}$, then the algorithm will decrypt the ciphertext and return a message m.

UKeyGen$(MK, VK_{x'}) \rightarrow (\widetilde{VK}_{x'}, UK_{x'})$. The update key generation algorithm is run by the authority. It takes as inputs the master key MK and the current version key $VK_{x'}$ of the revoked attribute x'. It outputs a new version key $\widetilde{VK}_{x'}$ of the revoked attribute x' and an update key $UK_{x'}$.

SKUpdate$(MK, SK, UK_{x'}) \rightarrow \widetilde{SK}$. The secret key update algorithm is run by the authority. It takes as inputs the master key MK, the current secret key SK and the update key $UK_{x'}$ of the revoked attribute x'. It outputs a new secret key $\widetilde{SK}$.

CTUpdate$(CT, UK_{x'}) \rightarrow \widetilde{CT}$. The ciphertext update algorithm is run by the cloud server. It takes as inputs the ciphertext CT and the update key $UK_{x'}$. It outputs a new ciphertext $\widetilde{CT}$.

2.3 Security Model

In cloud storage systems, we assume that: 1) The server may give access permission to the users who do not have access permission. 2) The server is curious but honest. It is curious about the content of the encrypted data or the received message, but will execute correctly the task assigned by the authority. 3) The users, however, are dishonest and may collude to obtain unauthorized access to data.

We now describe the security model for CP-ABE systems by the following game between a challenger and an adversary as follows.

Setup. The challenger runs the Setup algorithm and gives the public parameters PP, and the public keys PK to the adversary.

Phase 1. The adversary is given oracle access to secret keys SK that corresponding to sets of attributes $S_1, S_2, \cdots, S_{q1}$ and the update keys UK.

Challenge. The adversary submits two equal length messages M_0 and M_1. In addition, the adversary gives a challenge access structure $\mathbb{A}^*$ such that none of the secret keys and updated keys queried from Phase 1 satisfy the access structure. The challenger flips a random coin b, and encrypts M_b under the access structure $\mathbb{A}^*$. Then, the ciphertext CT^* is given to the adversary.

Phase 2. Phase 1 is repeated with the restrictions: 1) none of sets of attributes $S_{q1+1}, \cdots, S_q$ satisfy the access structure corresponding to the challenge; 2) none of the updated secret keys $\widetilde{SK}$ (generated by the queried SK and update keys UK) can decrypt the challenge ciphertext.

Guess. The adversary outputs a guess b' of b.

The advantage of an adversary $\mathcal{A}$ in this game is defined as $Pr[b' = b] - 1/2$. This security model can easily be extended to handle chosen-ciphertext attacks by allowing for decryption queries in Phase 1 and Phase 2.

DEFINITION 2. *A revocable CP-ABE scheme is secure if all polynomial time adversaries have at most a negligible advantage in the above game.*

3. ATTRIBUTE-BASED ACCESS CONTROL WITH EFFICIENT REVOCATION

In this section, we first give an overview of our method and then propose the detailed construction of access control scheme. After that, we describe our attribute revocation method to cope with the dynamic changes of users' attributes in large-scale storage systems.

3.1 Overview

To achieve fine-grained access control, the owner first divides the data into several components according to the logic granularities and encrypts each data component with different content keys by using symmetric encryption techniques. Then, the owner applies our proposed CP-ABE method to encrypt each content key, such that only the user whose attributes satisfy the access structure in the ciphertext can decrypt the content keys. Users with different attributes can decrypt different number of content keys and thus obtain different granularities of information from the same data.

Since the constructions of the existing CP-ABE schemes are not suitable for attribute revocation, we cannot directly apply them as the underlying techniques to design the data access control scheme. Thus, we first design a new underlying CP-ABE scheme that supports attribute revocation, where each attribute is assigned with a version number. When an attribute is revoked from a user, the authority generates a new version key and an update key for this revoked attribute. With the update key, all the non-revoked users can update their secret keys (Backward Security). By using the update key, the components associated with the revoked attribute in the ciphertext can also be updated to the current version. To improve the efficiency, we delegate the workload of ciphertext update to the server by using the proxy re-encryption method, such that the newly joined user is also able to decrypt the previous published data, which are encrypted with the previous public keys (Forward Security).

3.2 Scheme Construction

Let $\mathbb{G}$ and $\mathbb{G}_T$ be the multiplicative groups with the same prime order p and $e : \mathbb{G} \times \mathbb{G} \rightarrow \mathbb{G}_T$ be the bilinear map. Let g be the generator of $\mathbb{G}$. Let $H : \{0,1\}^* \rightarrow \mathbb{G}$ be a hash function such that the security will be modeled in the random oracle.

The construction of our access control scheme consists of four phases: *System Initialization, Key Generation by Authority, Data Encryption by Owners* and *Data Decryption by Users*.

Phase 1: System Initialization

CT_1	$E_{K_1}(m_1)$	$\cdots$	$\cdots$	CT_n	$E_{K_n}(m_n)$

Figure 2: Data Format on Cloud Server

The authority initializes the system by running the Setup algorithm. It randomly chooses $\alpha, \beta, \gamma, a \in \mathbb{Z}_p$ as the master key $MK = (\alpha, \beta, \gamma, a)$. Then, it generates the public parameters PP as

$$PP = (\ g,\ g^a,\ g^{1/\beta},\ g^\beta,\ e(g,g)^\alpha\ )$$

For each attribute x, the authority generates a random number $v_x \in \mathbb{Z}_p$ as the initial attribute version number $VK_x = v_x$ and then applies it to generate a public attribute key PK_x as

$$PK_x = (\ PK_{1,x} = H(x)^{v_x},\ PK_{2,x} = H(x)^{v_x \gamma}\ ).$$

All the public parameters PP and the public attribute keys $\{PK_x\}$ are published on the public bulletin board of the authority, such that all the owners in the system can freely get them.

Phase 2: Secret Key Generation for Users

When a user joins the system, the authority first assigns a set of attributes S to this user according to its role or identity. Then, it generates the secret key SK for this user by running the secret key generation algorithm SKeyGen. It takes as inputs the master key MK, a set of attributes S that describes the secret key, and the corresponding set of attribute version keys $\{VK_x\}_{x \in S}$. It then chooses a random number $t \in \mathbb{Z}_p$ and generates the user's secret key as

$$SK = (\ K = g^{\frac{\alpha}{\beta}} \cdot g^{\frac{at}{\beta}}, L = g^t, \forall x \in S : K_x = g^{t\beta^2} \cdot H(x)^{v_x t \beta}\ ).$$

The authority then sends SK to the user via a secure channel.

Phase 3: Data Encryption by Owners

Before outsourcing data M to the cloud servers, the owner processes the data as follows. 1) It first divides the data into several data components as $M = \{m_1, \cdots, m_n\}$ according to the logic granularities;2) It encrypts each data component m_i with different content keys $k_i (i = 1, \cdots, n)$ by using the symmetric encryption techniques; 3) For each content key $k_i (i = 1, \cdots, n)$, the owner defines the access structure $\mathcal{M}$ over the universe of attributes $\mathbb{S}$ and then encrypts k_i under this access structure by running the encryption algorithm Encrypt.

The encryption algorithm Encrypt can be constructed as follows. It takes as inputs the public parameters PP, a set of public attribute key $\{PK_x\}$, a content key k and a LSSS access structure $(\mathcal{M}, \rho)$. Let $\mathcal{M}$ be a $l \times n$ matrix, where l denotes the number of attributes involved in the encryption. The function ρ associates rows of $\mathcal{M}$ to attributes. It first chooses a random encryption exponent $s \in \mathbb{Z}_p$ and a random vector $\vec{v} = (s, y_2, \cdots, y_n) \in \mathbb{Z}_p^n$, where $y_2, \cdots, y_n$ are used to share the encryption exponent s. For $i = 1$ to l, it computes $\lambda_i = \vec{v} \cdot \mathcal{M}_i$, where $\mathcal{M}_i$ is the vector corresponding to the i-th row of $\mathcal{M}$. Then, it randomly chooses $r_1, r_2, \cdots, r_l \in \mathbb{Z}_p$ and computes the ciphertext as

$$CT = (\ C = ke(g,g)^{\alpha s},\ C' = g^{\beta s},\ C_i = g^{a\lambda_i} (g^\beta)^{-r_i} H(\rho(i))^{-r_i v_{\rho(i)}},$$
$$D_{1,i} = H(\rho(i))^{v_{\rho(i)} r_i \gamma},\ D_{2,i} = g^{\frac{r_i}{\beta}}\ (i = 1, \cdots, l)\ ).$$

The owner then uploads the encrypted data to the server in the format as described in Figure 2.

Phase 4: Data Decryption by Users

Upon receiving the data from the server, the user runs the decryption algorithm Decrypt to obtain the corresponding content keys and uses them to further decrypt data components. Only the attributes that the user possesses satisfy the access structure defined in the ciphertext CT, the user can get the data component successfully. Users with different attributes will be able to decrypt different number of data components, such that they can get different granularities of information from the same data.

The decryption algorithm Decrypt is constructed as follows. It takes as inputs a ciphertext CT attached with the access structure $(\mathcal{M},\rho)$ and the secret key for a set of attributes S. Suppose that the user's attribute set S satisfies the access structure and let $I \subset \{1,2,\cdots,l\}$ be defined as $I=\{i:\rho(i)\in S\}$. Then, it chooses a set of constants $\{w_i \in \mathbb{Z}_p\}_{i\in I}$ and reconstructs the encryption exponent as $s=\sum_{i\in I} w_i\lambda_i$ if $\{\lambda_i\}$ are valid shares of the secret s according to $\mathcal{M}$. The decryption algorithm first computes

$$\frac{e(C',K)}{\prod_{i\in I}(e(C_i,L)e(D_{2,i},K_{\rho(i)}))^{w_i}} = e(g,g)^{\alpha s}. \quad (1)$$

It can then decrypt the content key as $k=C/e(g,g)^{\alpha s}$. The user then uses the content keys to further decrypt the data.

3.3 Efficient Attribute Revocation for CP-ABE

In large-scale data storage systems, the users' access privileges may dynamically change. For example, when a user is leaving the system, it loses the access privileges of all the data in the cloud, which is called *User Revocation*. When a user is degraded in the system, it only loses part access privilege as some attributes should be removed from it, which is called the *Attribute Revocation*.

Suppose an attribute x' is revoked from a user μ. Our revocation method includes three phases: *Update Key Generation by Authority*, *Secret Key Update for non-revoked Users* and *Ciphertext Update by Cloud Server*.

Phase 1: Update Key Generation by Authority

When there is an attribute revocation, the authority runs the update key generation algorithm $\textbf{UKeyGen}(MK, VK_{x'}) \rightarrow (\widetilde{VK}_{x'}, UK_{x'})$. It takes the master key MK and the current version key $VK_{x'}$ of the revoked attribute x' as inputs. It generates a new attribute version key $\widetilde{VK}_{x'}$ by randomly choosing a number $\tilde{v}_{x'} \in \mathbb{Z}_p(\tilde{v}_{x'} \neq v_{x'})$. Then, the authority computes the update key as

$$UK_{x'} = (\ UK_{1,x'} = \frac{\tilde{v}_{x'}}{v_{x'}},\ UK_{2,x'} = \frac{v_{x'}-\tilde{v}_{x'}}{v_{x'}\gamma}\ ).$$

It outputs a new version key $\widetilde{VK}_{x'}$ of the attribute x' and an update key $UK_{x'}$ that can be used for updating the secret keys of non-revoked users and the ciphertexts that are associated with the revoked attribute x'. Then, the authority sends the update key $UK_{x'}$ to the cloud server (for ciphertext updating) via secure channels.

The authority also updates the public attribute key of the revoked attribute x' as $\widetilde{PK}_{x'} = (\ \widetilde{PK}_{1,x'} = H(x')^{\tilde{v}_{x'}},\ \widetilde{PK}_{2,x'} = H(x')^{\tilde{v}_{x'}\gamma}\ )$. After that, the authority broadcasts a message to all the owners that the public attribute key of the revoked attribute x' is updated. Then, all the owners can obtain the new public attribute key of the revoked attribute from the public bulletin board of the authority.

Phase 2: Secret Key Update for Non-revoked Users

Each non-revoked user submits two components $L=g^t$ and $K_{x'}$ of the secret key SK to the authority. Upon receiving these components, the authority runs the SKUpdate to compute a new component $\tilde{K}_{x'}$ associated with the revoked attribute x' as

$$\tilde{K}_{x'} = (K_{x'}/L^{\beta^2})^{UK_{1,x'}} \cdot L^{\beta^2} = g^{t\beta^2}\cdot H(x')^{\tilde{v}_{x'}t\beta}.$$

Then, it returns the new component $\tilde{K}_{x'}$ to the non-revoked user. The user's secret key is updated by replacing the component $K_{x'}$ associated with the revoked attribute x' with the new one $\tilde{K}_{x'}$:

$$\widetilde{SK} = (\ K,\ L,\ \tilde{K}_{x'},\ \forall x \in S\backslash\{x'\}: K_x\ ).$$

Note that only the component associated with the revoked attribute x' in the secret key needs to be updated, while all the other components are kept unchanged.

Phase 3: Ciphertext Update by Cloud Server

To ensure that the newly joined user who has sufficient attributes can still decrypt those previous data which are published before it joined the system, all the ciphertexts associated with the revoked attribute are required to be updated to the latest version. Intuitively, the ciphertext update should be done by data owners, which will incur a heavy overhead on the data owner. To improve the efficiency, we move the workload of ciphertext update from data owners to the cloud server, such that it can eliminate the huge communication overhead between data owners and cloud server, and the heavy computation cost on data owners. The ciphertext update is conducted by using proxy re-encryption method, which means that the server does not need to decrypt the ciphertext before updating.

Upon receiving the update key UK_x from the authority. The cloud server runs the ciphertext update algorithm CTUpdate to update the ciphertext associated with the revoked attribute x'. It takes as inputs the ciphertext CT and the update key $UK_{x'}$. It updates the ciphertext associated with x' as

$$\begin{aligned}\widetilde{CT} = (\ & \tilde{C}=C,\ \tilde{C}'=C',\ \forall i=1\ to\ l:\ \tilde{D}_{2,i}=D_{2,i},\\ & if\ \rho(i)\neq x':\ \tilde{C}_i = C_i,\ \tilde{D}_{1,i}=D_{1,i},\\ & if\ \rho(i)=x':\ \tilde{C}_i = C_i\cdot(D_{1,i})^{UK_{2,x'}},\ \tilde{D}_{1,i}=(D_{1,i})^{UK_{1,x'}}\ )\end{aligned}$$

It is obvious that our scheme only requires to update those components associated with the revoked attribute in the ciphertext, while the other components are not changed. In this way, our scheme can greatly improve the efficiency of attribute revocation.

The ciphertext update can not only guarantee the forward security of the attribute revocation, but also can reduce the storage overhead on the users (i.e., all the users need to hold only the latest secret key, rather than to keep records on all the previous secret keys). The cloud server in our system is required to be semi-trusted. Even when the cloud server is not semi-trusted in some circumstance, which means that the server will not update the ciphertexts correctly. The forward security cannot be guaranteed, but our system can still achieve the backward security.

4. ANALYSIS OF OUR PROPOSED ACCESS CONTROL SCHEME

4.1 Security Analysis

We conclude the security analysis as the following Theorems:

THEOREM 1. *When the decisional q-parallel BDHE assumption holds, no polynomial time adversary can selectively break our system with a challenge matrix of size $l^*\times n^*$, where $n^*\leq q$.*

PROOF. Suppose we have an adversary $\mathcal{A}$ with non-negligible advantage $\varepsilon = Adv_{\mathcal{A}}$ in the selective security game against our construction and suppose it chooses a challenge matrix M^* with the dimension at most q columns. Under the constraint that none of the updated secret keys $\widetilde{SK}$ (generated by both the queried secret keys SKs and update keys UKs) can decrypt the challenge ciphertext, we can build a simulator $\mathcal{B}$ that plays the decisional q-parallel BDHE problem with non-negligible advantage. The detailed proof will be shown in the full version of our work. □

THEOREM 2. *Our proposed access control scheme is secure against the unauthorized access.*

Table 1: Comparison of Storage Overhead

Entity	Our Scheme	[3]
Authority	$(4+n_a)\cdot\lvert p\rvert$	$2\lvert p\rvert$
Owner	$(2+n_a)\lvert g\rvert+\lvert g_T\rvert$	$2\lvert g\rvert+\lvert g_T\rvert$
Server	$\lvert g_T\rvert+(3l+1)\lvert g\rvert$	$2\lvert g_T\rvert+(3l+3)\lvert g\rvert+\frac{l\cdot\lvert n_u\rvert\cdot\lvert p\rvert}{2}$
User	$(2+n_{a,i})\cdot\lvert g\rvert$	$(2n_{a,i}+1)\lvert g\rvert+\log(n_u+1)\lvert p\rvert$

PROOF. From the definition of the unauthorized access, there are two scenarios: 1) Users who do not have sufficient attributes satisfying the access structure may try to access and decrypt the data. 2) When one or some attributes of the user are revoked, the user may still try to access the data with his/her previous secret key.

For the first scenario, the users who do not have sufficient attributes cannot decrypt the ciphertext by using their own secret keys. We also consider the collusion attack from multiple users, in our scheme, the user's secret key is generated with a random number, such that they may not be the same even if the users have the same set of attributes. Thus, they cannot collude their secret keys together to decrypt the ciphertext.

For the second scenario, suppose one attribute is revoked from a user, the authority will choose another version key to generate the update key and sends it to the server for updating all the ciphertexts associated with the revoked attribute, such that the ciphertexts are associated with the latest version key of the revoked attributes. Due to the different values of the version key in the ciphertext, the revoked user is not able to use the previous secret key to decrypt the ciphertext. □

4.2 Performance Analysis

We give the analysis of our scheme by comparing with [3] in terms of storage overhead, communication cost and computation efficiency. Let $|p|$ be the size of elements in $\mathbb{Z}_p$. Let $|g|$ and $|g_T|$ be the element size in $\mathbb{G}$ and $\mathbb{G}_T$ respectively. Let n_a and n_u denote the total number of attributes and users in the system respectively. Let $n_{a,i}$ denote the number of attributes the user i possesses and let l denote the number of attributes associated with the ciphertext.

4.2.1 Storage Overhead

Table 2 shows the comparison of storage overhead on each entity in the system. The main storage overhead on the authority comes from the master key in [3]. Besides the master key, in our scheme, the authority needs to hold a version key for each attribute. Both the public parameters and the public attribute keys contribute the storage overhead on the owner in our scheme, which is linear to the total number of attributes in the system. Although the data is stored on the server in the format as shown in Figure 2, we do not consider the storage overhead caused by the encrypted data, which are the same in both our scheme and [3]. Our scheme only requires the server to store the ciphertext, while the server in [3] needs to store both the message head and the ciphertext which is also linear to the number of users in the system. The storage overhead on each user in our scheme is associated with the number of attributes it possesses, while in [3] the storage overhead on each user is not only linear to the number of attributes it possesses but also linear to the number of users in the system. Usually, the number of users are much larger than the number of attributes in the system, which means that our scheme incurs less storage overhead.

4.2.2 Communication Cost

As illustrated in Table 3, the communication cost in the system is mainly caused by the keys and ciphertexts. In our scheme, the communication cost between the authority and the user comes from both the user's secret keys and the update keys, while in [3] only the secret key contributes the communication cost between the authority and the user. The communication cost between the authority and the owner mainly comes from the public keys. In our scheme, when there is an attribute revocation, the owner needs to get the latest public attribute key of the revoked attributes, which also contributes the communication between the authority and the owner.

Table 2: Comparison of Communication Cost

Communication Cost between	Our Scheme	[3]
Auth.&User	$4\lvert g\rvert+n_{a,i}\lvert g\rvert$	$\lvert g\rvert+2n_{a,i}\lvert g\rvert$
Auth.&Owner	$2\lvert g\rvert+\lvert g_T\rvert+n_a\lvert g\rvert$	$2\lvert g\rvert+\lvert g_T\rvert$
Server&User	$\lvert g_T\rvert+(3l+1)\lvert g\rvert$	$\lvert g_T\rvert+(2l+1)\lvert g\rvert+$ $(l\cdot\lvert n_u\rvert/2+\log(n_u+1))\lvert p\rvert$
Server&Owner	$\lvert G_T\rvert+(3l+1)\cdot\lvert G\rvert$	$(l+1)\lvert G_T\rvert+2l\lvert G\rvert$

In our scheme, the communication cost between the server and the user comes from the ciphertext. But in [3], besides the ciphertext, the message head (which contains the path keys) also contributes the communication cost between the server and the users, which is linear with the number of all the users in the system. Thus, our scheme incurs less communication cost between the server and the user than [3]. The ciphertext contributes the main communication cost between the server and the owner. Because the size of ciphertext in our scheme is much smaller than the one in [3], the communication cost between the sever and the owner is much less than the one in [3].

4.2.3 Computation Efficiency

We implement our scheme and [3] on a Linux system with an Intel Core 2 Duo CPU at 3.16GHz and 4.00GB RAM. The code uses the Pairing-Based Cryptography (PBC) library version 0.5.12 to implement the schemes. We use a symmetric elliptic curve α-curve, where the base field size is 512-bit and the embedding degree is 2. The α-curve has a 160-bit group order, which means p is a 160-bit length prime. The size of the plaintext is set to be 1 KByte. All the simulation results are the mean of 20 trials.

We compare the computation efficiency between our scheme and [3] in terms of encryption, decryption and re-encryption [3]. From the Figure 3(a), we can see that the time of encryption is linear with the total number of attributes in the system. The encryption phase in our scheme is more efficient than the one in [3]. That is because, in [3], the owner first encrypts the data by using the CP-ABE scheme and sends the ciphertext to the server. Upon receiving the ciphertext from the owner, the server will re-encrypt the ciphertext with a randomly generated encryption exponent. Then, the server encrypts this exponent with a set of attribute group keys by using the broadcast encryption approach. Correspondingly, in the phase of decryption, the user should first decrypt the exponent with its own path key and uses it to decrypt the data together with the secret key. In our scheme, however, the user only needs to use the secret key to decrypt the data, which is more efficient than the [3] as illustrated in the Figure 3(b).

During the attribute revocation, our scheme only requires to update those components associated with the revoked attribute of the ciphertext, while the [3] should re-encrypt all the components of the ciphertext. Besides, the re-encryption in [3] should generate a new

[3]Note that we do not consider the computation of symmetric encryption for data components since they are the same in both our scheme and [3].

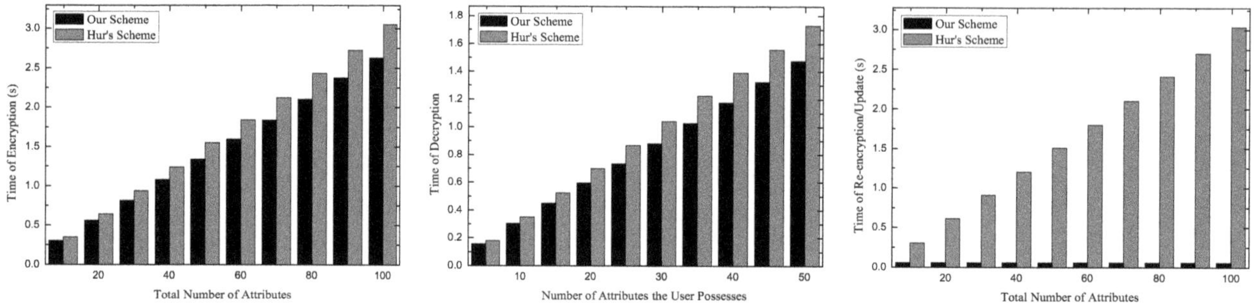

(a) Encryption (b) Decryption (c) Re-encryption

Figure 3: Computation Cost Comparison

encryption exponent and encrypt this new exponent with the new set of attribute group key by using broadcast encryption approach. Thus, as illustrated in Figure 3(c), the time of re-encryption phase in [3] is linear with the total number of attributes, while the time of ciphertext update in our scheme is constant to the number of revoked attributes.

5. RELATED WORK

The attribute-based encryption (ABE) technique [1,2,5,7] is regarded as one of the most suitable technologies for data access control in cloud storage systems.There are two complementary forms of ABE, Key-Policy ABE (KP-ABE) [2] and Ciphertext-Policy ABE (CP-ABE) [1,7]. In KP-ABE, attributes are used to describe the encrypted data and access policies over these attributes are built into user's secret keys; while in CP-ABE, attributes are used to describe the user's attributes and the access policies over these attributes are attached to the encrypted data.

In [8], the authors proposed a fine-grained data access control scheme based on the KP-ABE approach [2]. In their scheme, the data owner encrypts the data with a content key and then encrypt the content key by using the KP-ABE technique. The data owner assigns the access structure and the corresponding secret key to users by encrypting them with the user's public key and stores it on the server. However, their scheme requires the data owner to always be online for user joining, which is not appropriate in cloud storage systems. Some access control schemes are proposed based on CP-ABE [1,3], since CP-ABE is considered to be more suitable for data access control in cloud storage systems than KP-ABE. It allows data owners to define an access structure on attributes and encrypt the data under this access structure, such that data owners can define the attributes that the user needs to possess in order to decrypt the ciphertext. However, the revocation issue in CP-ABE is still an open problem.

To deal with the attribute revocation issue in ABE system, Yu *et al.* [9] proposed an attribute revocation method for CP-ABE, but they require the server to decide which users can update their secret keys according to the revoked user identity list, such that the server is required to be fully trusted. Hur *et al.* [3] also proposed an attribute revocation scheme in CP-ABE by allowing the server to re-encrypt the ciphertext with a set of attribute group keys. It can conduct the access right revocation on attribute level rather than on user level. During the attribute revocation, the server needs to change the attribute group key for the attribute which is affected by the membership change and re-encrypts the ciphertext with the new set of group attribute keys. This may incur high computation cost on the server. Also the server should be fully trusted. However, the server in cloud storage systems cannot be trusted and thus [3] cannot be applied in our problem. Therefore, the attribute revocation is still an open problem in attribute-based data access control.

6. CONCLUSION

In this paper, we proposed a fine-grained data access control scheme based on CP-ABE approach, where the owner was in charge of defining and enforcing the access policy. We also proposed an efficient attribute revocation method for CP-ABE, which can greatly reduce the cost of attribute revocation. Although this work is specific to cloud storage systems, but it is true that an untrusted remote storage system is an application of the work.

7. ACKNOWLEDGMENT

This work is supported by Research Grants Council of Hong Kong [Project No. CityU 114112] and in part by US National Science Foundation under grants CNS-1262277 and CNS-1116939.

8. REFERENCES

[1] J. Bethencourt, A. Sahai, and B. Waters. Ciphertext-policy attribute-based encryption. In *S&P'07*, pages 321–334. IEEE Computer Society, 2007.

[2] V. Goyal, O. Pandey, A. Sahai, and B. Waters. Attribute-based encryption for fine-grained access control of encrypted data. In *CCS'06*, pages 89–98. ACM, 2006.

[3] J. Hur and D. K. Noh. Attribute-based access control with efficient revocation in data outsourcing systems. *IEEE Trans. Parallel Distrib. Syst.*, 22(7):1214–1221, 2011.

[4] P. Mell and T. Grance. The NIST definition of cloud computing. Technical report, National Institute of Standards and Technology, 2009.

[5] A. Sahai and B. Waters. Fuzzy identity-based encryption. In *EUROCRYPT'05*, pages 457–473. Springer, 2005.

[6] K. Sohr, M. Drouineaud, G.-J. Ahn, and M. Gogolla. Analyzing and managing role-based access control policies. *IEEE Trans. Knowl. Data Eng.*, 20(7):924–939, 2008.

[7] B. Waters. Ciphertext-policy attribute-based encryption: An expressive, efficient, and provably secure realization. In *PKC'11*, pages 53–70. Springer, 2011.

[8] S. Yu, C. Wang, K. Ren, and W. Lou. Achieving secure, scalable, and fine-grained data access control in cloud computing. In *INFOCOM'10*, pages 534–542. IEEE, 2010.

[9] S. Yu, C. Wang, K. Ren, and W. Lou. Attribute based data sharing with attribute revocation. In *ASIACCS'10*, pages 261–270. ACM, 2010.

Covert Computation
Hiding Code in Code for Obfuscation Purposes

Sebastian Schrittwieser*, Stefan Katzenbeisser‡, Peter Kieseberg†, Markus Huber†, Manuel Leithner†, Martin Mulazzani†, Edgar Weippl†

Vienna University of Technology*
sebastian.schrittwieser@tuwien.ac.at

Darmstadt University of Technology‡
skatzenbeisser@acm.org

SBA Research†
{pkieseberg, mhuber, mleithner, mmulazzani, eweippl}@sba-research.org

ABSTRACT

As malicious software gets increasingly sophisticated and resilient to detection, new concepts for the identification of malicious behavior are developed by academia and industry alike. While today's malware detectors primarily focus on syntactical analysis (i.e., signatures of malware samples), the concept of semantic-aware malware detection has recently been proposed. Here, the classification is based on models that represent the underlying machine and map the effects of instructions on the hardware. In this paper, we demonstrate the incompleteness of these models and highlight the threat of malware, which exploits the gap between model and machine to stay undetectable. To this end, we introduce a novel concept we call *covert computation*, which implements functionality in side effects of microprocessors. For instance, the flags register can be used to calculate basic arithmetical and logical operations. Our paper shows how this technique could be used by malware authors to hide malicious code in a harmless-looking program. Furthermore, we demonstrate the resilience of *covert computation* against semantic-aware malware scanners.

Categories and Subject Descriptors

C.2.0 [**Computer-Communication Networks**]: General-Security and protection

Keywords

code obfuscation; side effects; malware detection

1. INTRODUCTION

Malware detection is an important research problem in computer security that strives to spot malicious routines in software. In recent years, the threat of malware, viruses, spyware, and trojans has dramatically increased and resulted in a cat-and-mouse game between malware authors and developers of anti-malware software. The ultimate goal of malware detectors (commonly known as virus scanners) is to determine whether a program includes malicious routines or not. Since the early days of malware defense, this was done by matching a signature of known malware against the software to be analyzed [2]. With the increasing amount of malicious software, the extraction of signatures from malware samples as the sole detection technique became inefficient as well as insufficient. Over time, other identification methods were developed and used in combination with malware signatures, which are still widely used by anti-malware software [9]. Heuristic-based malware detection identifies malicious code by statistically analyzing its structure and behavior without depending on prior knowledge of the malware. However, this approach suffers from false positives as well as false negatives as decisions are based on statistical models for maliciousness. Furthermore, the rise of concepts such as polymorphism and metamorphism lead to an entirely new class of malware, which is resistant against signature-based detectors, as these focus exclusively on malware syntax and ignore malware semantics. The idea of semantic-aware malware detection was introduced by Christodorescu et al. in 2005 [3]. Their approach is based on the definition of templates for malicious behavior and is more resistant to simple obfuscation techniques such as *garbage insertions* [5] and *equivalent instruction replacement* [7] as the semantics of the code are analyzed.

In this paper, we demonstrate that today's static malware detection approaches ignore fundamental knowledge of the underlying hardware and thus are ineffective against our novel *covert computation* obfuscation method. Static analysis techniques for binary code are based on a specific machine model in order to understand the functionality of the analyzed program. Such a model describes how code is interpreted by the machine, i.e., how a specific instruction influences the state of the microprocessor. Based on the entirety of effects that a sequence of instructions has on the model, its maliciousness is evaluated by the malware detection software. This model, however, is a simplified, abstract representation of the real machine. This simplification poses a problem for model-based code analysis, as abstract models are not strong enough to entirely simulate the effects of

ASIA CCS'13, May 8–10, 2013, Hangzhou, China.

the code running on real hardware. It is possible to refine machine models and make them more expressive, but this results in a tradeoff between correctness and complexity. In a malware detection context, the complexity of testing whether a given code matches a model for malicious behavior has to be low enough for the problem to be decidable in real time. The fundamental dilemma of static malware detection is that, on the one hand, code can be made arbitrarily complex with acceptable performance losses, while, on the other hand, a model strong enough to perform a complete evaluation of code semantics would reach an impractical level of complexity for real-life applications.

As main contributions of this paper we introduce a novel approach for code obfuscation called *covert computation*, based upon side effects in today's microprocessor architectures. We further show feasibility of our concept based on instruction side effects in the flags register as well as `LOOP` and string instructions. We finally demonstrate how our approach fundamentally raises the bar for semantic-aware code analysis.

2. RELATED WORK

The use of code obfuscation to prevent reverse engineering of any given software is a well-studied field [4, 17]. A formal concept of code obfuscation has been defined by Barak et al. [1]. Although this work shows that a universal obfuscator for any type of software does not exist and perfectly secure software obfuscation is not possible, various types of code obfuscation are still used by today's malware to "raise the bar" for detection.

Various malware obfuscation approaches presented in the literature follow the concept of polymorphism [13], which hides malicious code by packing or encrypting it as data that cannot be interpreted by the analysis machine. Thus, an unpacking routine has to be used to turn this data back into machine-interpretable code. A number of approaches have been suggested to defeat this obfuscation technique, such as detecting malicious code with model checking [11] or symbolic execution [6]. Today's malware detection systems evaluate the maliciousness of a program based on structural and behavioral patterns [10]. Christodorescu et al. [3] first introduced the concept of semantic-aware malware detection. In their paper, the authors define formal semantics for the maliciousness of programs and a semantic-aware matching algorithm for malware detection based on them. Templates for malicious behavior are defined and matched against the potential malware. If both have the same effect on memory, the binary is identified as malicious. This approach can deal with simple forms of obfuscation but does not recognize instruction replacements based on patterns completely. The concept was further formalized by Preda et al. in 2007 [14] and 2008 [15].

Recently, Wu et al. [18] introduced the concept of mimimorphism, which aims at obfuscating malicious code against both static and statistical detection systems. In the area of code obfuscation, various approaches for hiding a program's semantics can be found. In recent literature on code obfuscation, several authors have proposed the removal of instruction patterns in order to increase de-obfuscation complexity. Recent work by De Sutter et al. [7] on avoiding characteristic instruction patterns normalizes the distribution of instructions used in a program by replacing rare ones with semantically equivalent blocks of more frequently used instructions with equivalent blocks of less frequently used instructions. The drawback to this approach is, however, that an equal distribution of instructions used by a program is statistically unlikely and therefore easily detectable for a code analyst. Furthermore, semantic-aware detection approaches such as described in [3] can implement the replacement patterns in their templates for malicious behavior.

Giacobazzi [8] first theoretically discussed the idea of making code analysis more difficult by forcing the detection system to become incomplete. However, no practical approach of this idea was given in the paper. Moser et al. [12] discussed the question whether static analysis alone allows reliable malware detection.

3. APPROACH

All semantic-aware malware identification techniques follow the same basic approach. The classification of maliciousness is based on a model of the underlying system (i.e., the microprocessor), which describes how a specific instruction modifies the system's state. The quality and completeness of the model are crucial for a high identification rate. An incomplete model is not able to map all effects that an instruction has on the hardware and thus cannot evaluate its impact on the system and the system's state after executing the instruction. Current models for malware detection are focused on the instruction layer but do not fully map all effects of an instruction on the model. Today's microprocessors are highly complex systems with hundreds of different instructions that influence the processor's state.

In general terms, there are two types of models for us to consider. First, a human analyst defines his or her own model of the machine when trying to understand the meaning of a program's code. Given that the analyst knows the purpose of a specific instruction, he or she can perceive the code's meaning on a semantic layer and draw conclusions concerning the functionality of a sequence of instructions. However, it is drastically more difficult to keep the program's entire state, which is modified constantly while executing instructions, in mind. Thus, the human model of a machine can be described as a very basic semantic representation. The second model that has to be considered is the one of an automatic analysis tool, which makes a decision regarding the maliciousness of a program based on predefined templates and patterns. If side effects are not implemented in this model, its impact cannot be evaluated and is missed by the analysis tool.

In order to hide the implementation of a specific functionality of a program, we identified the possibility of implementing it based on features of the processor that are not described by its model and thus not evaluated. Analyzing such code on the semantic layer would not identify the hidden functionality as it is not contained in the basic semantics of a sequence of instructions, but in some deeper abstraction layer that is not included in the model.

3.1 Side Effects

In computer science, side effects in general describe any persistent modification of a program's state or its environment after executing a basic block, which is a straight block of instructions with one entry and one exit point. This includes setting a global variable, writing to the file system or accessing auxiliary equipment. Side effects are a fundamental prerequisite for a program's ability to interact with the user, the underlying computing system or other programs. Depending on the programming paradigm, side effects are more or less frequently used in programming languages. While in functional programming languages, such as Haskell, side effects do not exist or are restricted to a minimum, imperative programming (e.g., C, Java, and many others) makes use of side effects more frequently.

While most side effects are intended by the developer and are an integral part of the functionality (e.g., writing data to the file system or interacting with the outside world over a network connection), some side effects modify states of the program or the outside world without the developer's direct knowledge. The developer has a mental model of instructions and its effects, which might not cover the entire functionality of the instruction. The same applies to a code analyst and even, in a similar form, to machine-based malware detection systems. Code is analyzed based on expectations of what functionality a specific instruction implements. This can be the mental model of the human code analyst or the hardware model that is implemented in the automatic malware analysis system. Consider the x86 instruction `ADD EAX, EBX`. The core functionality, which is expected by the analyst, is the calculation of the sum of the two operands `EAX` and `EBX`. Thus, the new state of the machine after the execution of the instruction includes a modified register `EAX` that now holds the sum of the registers `EAX` and `EBX`. However, there is another register that was influenced by the `ADD` instruction. Within the flags register, several bits could have been modified by the instruction, depending on the result of the operation. For example, the Zero flag is set to 1 if the result of the operation is 0.

3.2 Using side effects to hide functionality

Usually, side effects are avoided at an early stage of software development (e.g., developers are discouraged from using global variables heavily), as they could influence the program's state in a way that was not considered by the developer and cause an unpredictable malfunction of the program.

In a security context, we have identified side effects as excellent vehicles for hiding malicious functionality inside arbitrary program code. The idea of injecting malicious functionality by making program code look harmless is not new. Winning examples of the *Underhanded C Contest*[1], an annual contest for writing innocent-looking C code implementing covert malicious behavior, use very subtle techniques for hosting hidden functionality. In contrast to that contest, which is based on the high-level language C, we describe covert functionality that is tied very closely to the underlying hardware, based on side effects in the microprocessor and implemented at assembly level.

3.3 Flags

Flags in microprocessors are status bits that are used for indicating states and conditions of different operations performed by the microprocessor. For example, most of today's architectures such as x86 implement a Zero flag, which is set when the result of an arithmetic or logical operation is equal to zero. The value of a flag bit is then used for conditional jumps and is therefore responsible for modifying the control flow of a program's execution.

[1]http://underhanded.xcott.com (Last accessed February 12th 2013)

In the context of side effects within programs, flags are the target of side effects. In x86 as well as in many other microprocessor architectures, most arithmetic and logical operations influence at least one bit in the flags register. Flags can therefore be seen as global variables that are permanently modified by instructions. The concept of *covert computation* uses flags for basic operations in such a way that the flags are directly used for performing the calculation and storing intermediate results by using conditional jumps. In the following, we show how flags can be used as intermediate storage for the calculation of logical operations. This concept can easily be adopted for arithmetic operations.

In a bitwise logical operation the calculation of the four different possible combinations (00/01/10/11) can be represented by two conditional jumps. The two input operands are stored in two arbitrary flags. In a first step, the value of the first flag is determined by implementing a conditional jump, which evaluates this specific flag. In x86, for example, the Zero flag can be evaluated with a `JZ` (jump if zero) or a `JNZ` (jump if not zero) instruction. For each case (jump was performed or not performed), the second phase again implements a conditional jump, thus resulting in four possible outgoing control flows, each representing one possible result of the bitwise operation of the values stored in the two flags. Figure 1 illustrates the approach. The two source values 1 and 0 are stored in the Zero and the Carry flag. Two conditional jumps are implemented to calculate the logical `XOR`. The first one (`JZ`) evaluates the value stored in the Zero flag. For both cases – the jump is either performed or not – a second conditional jump (`JC`), which evaluates the carry flag, is implemented. The control flow of the software now follows the calculation of a logical XOR, and at each of the four possible end points of the control flow graph, code can be implemented that stores the result of the calculation to some output register.

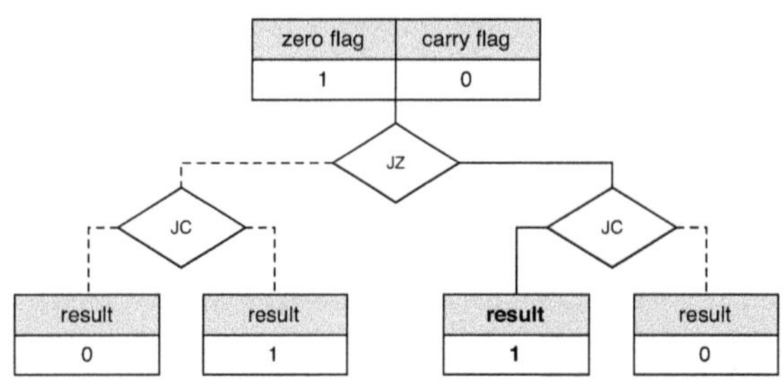

Figure 1: Calculation of XOR using the Zero and the Carry flag.

This concept can be extended to the full length of 32-bit operands very easily by repeating the two jumps for each bitwise operation. Other logical operations can be represented as side effects analogous to the `XOR` replacement pattern.

3.4 Other side effects in the x86 architecture

Many instructions of the x86 architecture modify the state of the processor in a way that is not the primary functionality of the instruction. In the following, we exemplarily describe side effects of the `LOOP` as well as string instructions and explain how side effects can be used to emulate other instructions.

LOOP instruction. The LOOP instruction in x86 behaves as follows: The value of the counter register (CX/ECX) is decremented by one. If it contains 0 after this operation, the loop terminates and execution continues past the LOOP instruction. Otherwise, a short jump to the relative offset specified as the operand to LOOP is taken. While the obvious behavior of the LOOP instruction is the repeated execution of one or more instructions, it also can be perform other, not so obvious functionality. One of the most trivial ways to repurpose this instruction's behavior is by using it as a short JMP or conditional jump. This can be achieved by ensuring that ECX does not contain 0 (or writing a value unequal to 0 to ECX if the given condition is met) and instead of using a JMP instruction, writing LOOP <label>. However, there are far more sophisticated ways to emulate other given instructions using LOOP. In Listing 1, the standard SUB instruction is represented using a combination of LOOP and XCHG.

```
SUB EAX,200

  ⇓

MOV ECX,200
XCHG EAX,ECX
LOOP -1
```

Listing 1: SUB with LOOP instruction

At first glance, this loop may seem entirely pointless, simply exchanging the values of EAX and ECX at every iteration and decreasing the value of ECX. However, because ECX constantly witches between containing the actual loop counter and the value of EAX, the latter is actually decremented by the former until either of them reaches the value 0. Note that this sequence of commands will not work as intended if EAX contains a value between 0 and 0x200. A variety of other instructions can be emulated using LOOP, e.g., MOV between registers (by moving the target value into ECX and incrementing the target register's value at every loop iteration).

String instructions. Another source of side effects in the x86 architecture are the string instructions. MOVS, SCAS, CMPS, STOS, and LODS instructions are intended to operate on continuous blocks of memory instead of single bytes. Because of the fact that these instructions modify the registers ESI, EDI, and ECX it is possible to emulate ADD, SUB, INC, and DEC instruction. An example of a replacement pattern for the INC instruction is given in Listing 2. The replacement pattern will clobber the value of ESI and is only applicable if the value of EAX points to a memory location that is accessible to the program.

```
INC EAX

  ⇓

XCHG EAX, ESI
LODS
XCHG EAX, ESI
```

Listing 2: Arithmetic operations with string instructions.

4. EVALUATION

In this section we discuss the effectiveness of the concept of *covert computation* for obfuscation purposes. We first considered assessing its resilience against commercial malware detectors by using real malware samples that were modified to implement some of their functionality in side effects. However, as pointed out in [12], this type of evaluation would be of doubtful value. The detection engines of today's virus scanners are mainly signature-based, which means that modifying the binary code would most likely destroy the signature. It would then come as no surprise to have a detection rate that was lower than the one for the original binaries. As this effect can be simply tracked down to the modification of the signature and not to the concept of covert functionality in the code, it would heavily restrict the significance of the evaluation. Therefore, we decided to focus our evaluation on the concept of semantic-aware malware detection. We performed a theoretical analysis to evaluate the resilience of our approach against Christodorescu et al.'s [3] approach. For semantic-aware malware detection, the binary program is disassembled and brought to an architecture-independent intermediate representation, which is matched against templates describing malicious behavior. In order to be able to detect basic obfuscation methods like *register reassignment* or *instruction reordering* (e.g., by inserting jumps in the control flow graph), so-called def-use chains are utilized. Furthermore, a value-preservation oracle is implemented for detecting NOP instructions as well as NOP fragments.

Normalization of intermediate representation. The approach introduced by Christodorescu et al. [3] is based on IDAPro for decompilation of the program to be analyzed. By generating an intermediate representation (IR), semantically equivalent instruction replacement patterns such as INC EAX, ADD EAX, 1, and SUB EAX, -1 are normalized with semantically disjoint operations and can then be matched against the generic template, which describes malicious behavior.

Semantics detection. Since the general problem of deciding whether one program is an obfuscated form of another program is closely related to the halting problem, which in general is undecidable [16], the presented algorithm uses the following strategy to match the program to the template: The algorithm tries to match (unify) each template node to a node in the program. In case two matching nodes are found, the def-use relationships in the template are evaluated with respect to the program code. If they hold true in the actual program, the program fragment matches the template.

Value preservation and NOP detection. The goal of this analysis step lies in the detection of NOP operations, i.e., program fragments that do not change the values of the watched variables. The following strategies were implemented by the authors in [3]: (i) Matching instructions against a library of known NOP commands and NOP fragments, (ii) symbolic execution of the code sequence with randomized initial states, as well as (iii) two different theorem provers.

Resilience against the approach. As outlined by the authors, the semantic-aware malware detection approach is

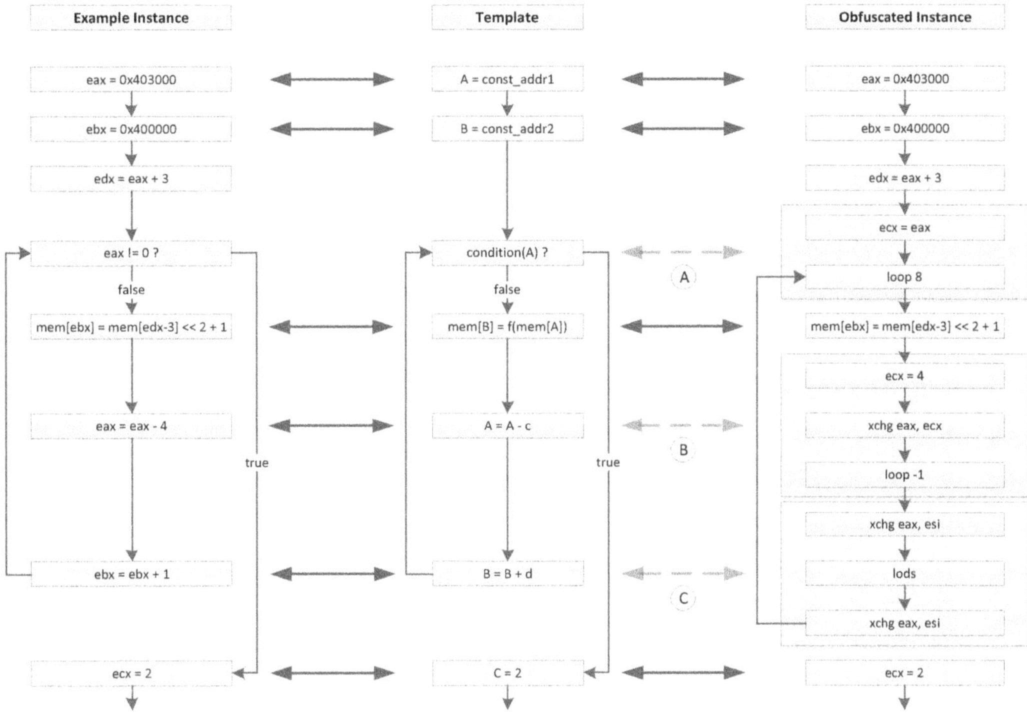

Figure 2: Resilience against semantic-aware malware detection.

able to detect *instruction reordering* and *register reassignment* as well as a *garbage insertion*. Furthermore, with respect to the underlying instruction replacement engine, a limited set of *replaced instructions* can be detected. However, this approach is not able to detect obfuscation techniques using *equivalent functionality* or *reordered memory access*. In Figure 2 we give an example of a code fragment (left) that is matched to a template (center) and an obfuscated form (right) of the same fragment. The obfuscation steps applied are flagged with the letters (A) and (B). Note that for reasons of simplicity, `JMP` instructions have been omitted from the illustration.

Since our obfuscation technique does not work by inserting `NOP` fragments, the direct detection and removal of `NOP` elements has no impact on our approach. Nevertheless, we use these mechanisms in the course of the matching algorithm in order to check for value preservation. The semantic detection relies heavily on the algorithm applying local unification by trying to find bindings of program nodes to template nodes. It is important to note that the bindings may differ at different program points, i.e., one variable in the template may be bound to different registers in the program, and the binding is therefore not consistent. The idea behind this approach lies in the possibility to detect register reassignments. In order to eliminate inconsistent matches that cannot be solved using register reassignment, a mechanism based on def-use chains and value-preservation (using `NOP` detection) is applied.

The local unification used to generate the set of candidate matches that is then reduced using def-use chains and value preservation is limited by several restrictions. The following two are the most important ones with respect to our obfuscation method: (i) If operators are used in a template node, the node can only be unified with program nodes containing the same operators and (ii) symbolic constants in template nodes can only be unified with program constants. The obfuscation pattern (B) in Figure 2 violates restrictions (i) and (ii) as, e.g., the simple "+"-function is replaced by a `MOV` instruction followed by looping an `XCHG` instruction. The same holds true for obfuscation pattern (C). In case of obfuscation pattern (A) even the control flow graph was changed as the explicit jump instruction following the condition as well as the condition itself are replaced by an assignment and a `LOOP` instruction. Thus, the local unification engine is not able to match these program fragments to the respective template fragments. In order to generate the set of match candidates, the local unification procedure must be able to match program nodes with template nodes, relying on the IR-engine to detect semantically identical program nodes and to convert them into the same intermediate representation. However, authors state that "[...] same operation [...] has to appear in the program for that node to match.". For example, an arithmetic left shift (`eax = eax << 1`) would not match a multiplication by 2 (`x = x * 2`) despite these instructions being semantically equivalent. Therefore, we can safely conclude that replacements with side effects as proposed in our concept would not match in the local unification as they do not use the same operations as the original code for implementing a specific functionality.

One could argue that once the concept of *covert computation* is publicly known, malware detectors could simply improve the hardware models on which the instruction replacement engine is based to be able to identify malicious behaviors implemented in side effects. While in theory, every single aspect of the hardware could be mapped to the machine model, we strongly believe that this is an unrealistic assumption as increasing the level of detail and completeness of the model is costly and reduces its practical applicability in real-life malware detection scenarios, where the decision on maliciousness has to be made in real time. A more complex model also increases the complexity of the evaluation, so the model has to be kept as general as possible, preventing completeness in semantic-aware program analysis. Today's virus scanners as well as semantic-aware malware detection concepts are not even able to cover the entire semantics of side effects-free code. Following the original argument of the possibility of a complete model, mapping these semantics should have been even more trivial. The second important aspect is diversity. Christodorescu et al. [3] argue that a malware author would have to "devise multiple equivalent, yet distinct, implementations of the same computation, to evade detection". With *covert computation* we have shown that side effects in the microprocessor can be used to achieve exactly this requirement.

5. CONCLUSION

In this paper, we have shown that the complexity of today's microprocessors, which support a myriad of different instructions, can be exploited to hide functionality in a program's code as small code portions. In the context of malware, we demonstrated that existing concepts for semantic-aware malware detection systems are not able to analyze the semantics of code that is implemented in side effects and argue that the indirectly shown incompleteness of machine models for semantic-aware malware detection raises the threat of malware that exploits exactly this knowledge gap. A successful implementation of the obvious mitigation strategy – improving the models – is doubtful, as, while it might be possible to map the side effects of a specific instruction in the model, it would complex to evaluate the impact of the side effects of an entire sequence of instructions. Models are abstract representations of the real world, which implies that some information is lost in its development. The increasing complexity of the real world makes it hard to design models that are strong enough to entirely simulate the effects of the code running on real hardware. The obfuscator has an important advantage over code analysts as he or she can make the code arbitrarily complex. In contrast, analysts have to keep their models simple to avoid an impractical level of complexity for testing whether a given code matches a model for malicious behavior. Thus, the feasibility of implementing a complete hardware model for malware detection is an unrealistic assumption.

Acknowledgments

The research was funded under Grant 826461 (FIT-IT) and COMET K1 by the FFG – Austrian Research Promotion Agency.

6. REFERENCES

[1] B. Barak, O. Goldreich, R. Impagliazzo, S. Rudich, A. Sahai, S. Vadhan, and K. Yang. On the (im) possibility of obfuscating programs. In *Advances in Cryptology—Crypto 2001*, pages 1–18. Springer, 2001.

[2] M. Christodorescu and S. Jha. Testing malware detectors. *ACM SIGSOFT Software Engineering Notes*, 29(4):34–44, 2004.

[3] M. Christodorescu, S. Jha, S. Seshia, D. Song, and R. Bryant. Semantics-aware malware detection. In *Security and Privacy, 2005 IEEE Symposium on*, pages 32–46. IEEE, 2005.

[4] C. Collberg and C. Thomborson. Watermarking, tamper-proofing, and obfuscation-tools for software protection. *Software Engineering, IEEE Transactions on*, 28(8):735–746, 2002.

[5] C. Collberg, C. Thomborson, and D. Low. A taxonomy of obfuscating transformations. Technical report, Department of Computer Science, The University of Auckland, New Zealand, 1997.

[6] J. Crandall, Z. Su, S. Wu, and F. Chong. On deriving unknown vulnerabilities from zero-day polymorphic and metamorphic worm exploits. In *Proceedings of the 12th ACM conference on Computer and communications security*, pages 235–248. ACM, 2005.

[7] B. De Sutter, B. Anckaert, J. Geiregat, D. Chanet, and K. De Bosschere. Instruction set limitation in support of software diversity. *Information Security and Cryptology–ICISC 2008*, pages 152–165, 2009.

[8] R. Giacobazzi. Hiding information in completeness holes: New perspectives in code obfuscation and watermarking. In *Software Engineering and Formal Methods, 2008. SEFM'08. Sixth IEEE International Conference on*, pages 7–18. IEEE, 2008.

[9] K. Griffin, S. Schneider, X. Hu, and T. Chiueh. Automatic generation of string signatures for malware detection. In *Recent Advances in Intrusion Detection*, pages 101–120. Springer, 2009.

[10] N. Idika and A. Mathur. A survey of malware detection techniques. *Purdue University*, page 48, 2007.

[11] J. Kinder, S. Katzenbeisser, C. Schallhart, and H. Veith. Detecting malicious code by model checking. *Detection of Intrusions and Malware, and Vulnerability Assessment*, pages 514–515, 2005.

[12] A. Moser, C. Kruegel, and E. Kirda. Limits of static analysis for malware detection. In *Computer Security Applications Conference, 2007. ACSAC 2007. Twenty-Third Annual*, pages 421–430. IEEE, 2007.

[13] C. Nachenberg. Computer virus-coevolution. *Communications of the ACM*, 50(1):46–51, 1997.

[14] M. Preda, M. Christodorescu, S. Jha, and S. Debray. A semantics-based approach to malware detection. In *ACM SIGPLAN Notices*, volume 42, pages 377–388. ACM, 2007.

[15] M. Preda, M. Christodorescu, S. Jha, and S. Debray. A semantics-based approach to malware detection. *ACM Transactions on Programming Languages and Systems (TOPLAS)*, 30(5):25:1–25:53, 2008.

[16] A. Turing. On computable numbers, with an application to the entscheidungsproblem. In *Proceedings of the London Mathematical Society*, volume 42, pages 230–265, 1936.

[17] S. Udupa, S. Debray, and M. Madou. Deobfuscation: Reverse engineering obfuscated code. In *Reverse Engineering, 12th Working Conference on*, pages 10–pp. IEEE, 2005.

[18] Z. Wu, S. Gianvecchio, M. Xie, and H. Wang. Mimimorphism: a new approach to binary code obfuscation. In *Proceedings of the 17th ACM conference on Computer and communications security*, pages 536–546. ACM, 2010.

Proof of Plaintext Knowledge for Code-Based Public-Key Encryption Revisited

Rong Hu
Graduate School of Mathematics, Kyushu University
744 Motooka Nishi-ku
Fukuoka 819-0395 Japan
r-hu@math.kyushu-u.ac.jp

Kirill Morozov
Institute of Mathematics for Industry, Kyushu University
744 Motooka Nishi-ku
Fukuoka 819-0395 Japan
morozov@imi.kyushu-u.ac.jp

Tsuyoshi Takagi
Institute of Mathematics for Industry, Kyushu University
744 Motooka Nishi-ku
Fukuoka 819-0395 Japan
takagi@imi.kyushu-u.ac.jp

ABSTRACT

In a recent paper at Asiacrypt'2012, Jain et al point out that Véron code-based identification scheme is not perfect zero-knowledge. In particular, this creates a gap in security arguments of proof of plaintext knowledge (PPK) and verifiable encryption for the McEliece public key encryption (PKE) proposed by Morozov and Takagi at ACISP'2012. We fix the latter result by showing that PPK for the code-based Niederreiter and McEliece PKE's can be constructed using Stern zero-knowledge identification scheme, which is unaffected by the above mentioned problem. Since code-based verifiable encryption uses PPK as a main ingredient, our proposal presents a fix for the McEliece verifiable encryption as well. In addition, we present the Niederreiter verifiable encryption.

Categories and Subject Descriptors

F.m [**Theory of Computation**]: Miscellaneous—*zero-knowledge proofs*; E.3 [**Data**]: Data Encryption—*public key cryptosystems*; E.4 [**Data**]: Coding and Information Theory—*error correcting codes*

General Terms

Security, Theory

Keywords

Proof of plaintext knowledge, verifiable encryption, Niederreiter cryptosystem, McEliece cryptosystem, zero-knowledge proof

1. INTRODUCTION

Postquantum cryptography encompasses schemes which remain secure even against attacks using quantum computers. Two important candidates for the postquantum public key encryption are code-based public key encryption (PKE) schemes by McEliece [25] and Niederreiter [27]. Their security is based on hardness of decoding, which is a well-studied cryptographic assumption [7, 31, 15, 8, 17, 9, 3].

It is important to develop code-based cryptographic protocols in order to introduce a wide spectrum of secure services which cryptography can offer. Quite a few of such protocol already exists such as identification or digital signatures – see [15, 12] for surveys, and also [29] for related results.

In this paper, we focus on *proof of plaintext knowledge (PPK)* for code-based PKE. Suppose that a prover P encrypted the plaintext m into ciphertext c on the public key pk. Now, PPK allows P to convince a verifier V, who does not have the secret key, that P knows m. We consider zero-knowledge (ZK) proofs which means that PPK does not leak any information on m to V.

Verifiable encryption with respect to some binary relation R on the plaintexts is a zero-knowledge proof on public inputs pk, c, and δ that allows a prover P to convince a verifier V that c is a ciphertext of m under pk such that $(m, \delta) \in R$.

1.1 Related Works

Proof of Plaintext Knowledge. PPK were first introduced by Aumann and Rabin [1] in the generic case of any PKE, and then investigated by Katz [22], who presented efficient PPK for RSA, Rabin, ElGamal and Paillier cryptosystems using Sigma-protocols. The first PPK in postquantum setting for the lattice-based Ajtai-Dwork PKE was presented by Goldwasser and Kharchenko [20], and then a number of works for lattice-based systems followed [36, 37, 5]. It is worth noting that the works by Xagawa et al [36] and Xagawa and Tanaka [37] use a modification of Stern code-based ZK identification scheme [34]. The latter scheme was also used by Kobara et al [23] for a similar purpose in code-based oblivious transfer.

Recently, Morozov and Takagi [26] presented PPK and verifiable encryption for the McEliece PKE using code-based Véron identification scheme [35], which is, in a sense, the dual of Stern scheme [34]. However, Jain et al [21] pointed out a gap in the proof of zero-knowledge property of Véron's scheme. This also created a gap in the proofs of both primitives in [26].

It is also worth noting that Jain et al [21] presented efficient commitments and ZK proofs based on a variant of the learning parity with noise (LPN) problem (for details about this problem, see e.g. [29]).

Verifiable Encryption. Verifiable encryption was first introduced by Stadler [33] as a tool for publicly verifiable secret sharing, and later generalized by Asokan et al [2] and applied to fair exchange of digital signatures. See the works by Camenisch and Damgård [10] and by Camenisch and Shoup [11] for further developments on this topic. Verifiable encryption for Ajtai-Dwork PKE was presented in [20].

1.2 Our Contribution and Discussion

In this work, we show that the basic idea presented in [26] is valid by introducing a (zero-knowledge) PPK for Niederreiter and McEliece PKE using Stern ZK identification scheme [34]. We suggest secure parameter sets and estimate the performance of our proposal. In particular, for 80-bit equivalent security, our protocol requires 9.1 kilobytes of communication, in 28 rounds, which seems to be practical.

One can imagine various applications, where the receiver of the ciphertext might want to ensure in advance that the sender knows the plaintext inside it. For instance, in case of an auction, when the encrypted bid is sent, an adversary might intercept it and re-send it on his own behalf, in attempt to bring the auction to a draw. Clearly, deployment of PPK would prevent the above attack. One might argue that an authentication mechanism such as digital signature might be a cheaper solution, however it depends on a particular application, since PPK does *not* bind the bidder to the bid. Therefore, the interactive nature of PPK would allow the bidder to authenticate the bid, but deny it later for privacy reasons.

It is worth noting that our result implies interactive code-based IND-CCA1 PKE in the standard model [18, 19, 22].

Throughout our work, we assume the presence of the public key infrastructure (or any other key authentication mechanism), which ensures validity of receiver's public key to be used in PPK.

As in [26], our protocol works also in the standard model, as any commitment scheme can be used. In particular, one may use (efficient) computationally hiding and statistically binding commitment scheme based on hardness of syndrome decoding from [14] or [21].

The type of zero-knowledge which we obtain, being statistical or computational, depends solely on the employed commitment scheme. In case of [14, 21], which are perfectly binding and computationally hiding, the ZK property holds in computational sense.

2. PRELIMINARIES

We borrow some parts of the presentation in this section from [26].

Let us first fix some notation. Let J be an ordered subset as follows: $\{j_i, \dots, j_m\} = J \subseteq \{1, \dots, n\}$, then we denote a vector $(x_{j_1}, \dots, x_{j_m}) \in \mathbb{F}_2^m$ by x_J. Similarly, we denote by M_J the submatrix of a $(k \times n)$ matrix M consisting of the columns which correspond to the indexes of J. A concatenation of matrices $X \in \mathbb{F}_2^{k \times n_0}$ and $Y \in \mathbb{F}_2^{k \times n_1}$ is written as $(X|Y) \in \mathbb{F}_2^{k \times (n_0+n_1)}$, and for $k=1$ this will denote concatenation of vectors. We denote by $x \xleftarrow{\$} \mathcal{X}$ a uniformly random selection of an element from its domain $\mathcal{X}$. A set of $(n \times n)$ permutation matrices is denoted by $\mathcal{S}_n$.

We denote by $\langle A(a), B(b)\rangle(c)$ a random variable representing the output of a Turing machine B following an execution of an interactive two-party protocol between a Turing machine A with private input a and a Turing machine B with private input b on a joint input c, where A and B have uniformly distributed random tapes. If a party, say A, has no input, then we omit it by writing just A (instead of $A(a)$) in the above notation.

In our two-party protocols, we will denote an honest prover by P and an honest verifier by V, while a dishonest party will be denoted by $\widetilde{\mathsf{P}}$ and $\widetilde{\mathsf{V}}$, respectively.

We call a function $\epsilon(n)$ *negligible in some parameter* n, if $\epsilon(n) = 2^{-\omega(\log n)}$. We call a probability $1-\epsilon(n)$ *overwhelming*, when $\epsilon(n)$ is negligible. Occasionally, we may omit mentioning of the security parameter. In these cases, by saying that a quantity is negligible (overwhelming), we mean that it is *negligible (overwhelming) in the security parameter.* The Hamming weight of $x \in \mathbb{F}_2^n$ is denoted as $w_H(x)$.

2.1 Security Assumptions

DEFINITION 1. *Syndrome Decoding (SD) Problem.*
Input: $H \xleftarrow{\$} \mathbb{F}_2^{(n-k)\times n}$, $y \xleftarrow{\$} \mathbb{F}_2^{n-k}$ *and* $0 < t \in \mathbb{N}$.
Output: $s \in \mathbb{F}_2^n$ *such that* $w_H(s) \le t$, $Hs^T = y$.

This problem was shown to be NP-complete by Berlekamp et al [7]. Its equivalent dual version can be formulated as follows.

DEFINITION 2. *General Decoding (G-SD) Problem.*
Input: $G \xleftarrow{\$} \mathbb{F}_2^{k\times n}$, $y \xleftarrow{\$} \mathbb{F}_2^n$ *and* $0 < t \in \mathbb{N}$.
Output: $x \in \mathbb{F}_2^k$, $e \in \mathbb{F}_2^n$ *s.t.* $w_H(e) \le t$, $xG \oplus e = y$.

The following three problems use the quantities defined in the next subsection. No polynomial-time algorithm is known for these problems [15, 17, 9].

DEFINITION 3. *Niederreiter Problem.*
Input: A public key of Niederreiter cryptosystem (H^{pub}, t), *where* $H^{pub} \in \mathbb{F}_2^{(n-k)\times n}$, $0 < t \in \mathbb{N}$; *and a Niederreiter ciphertext* $c \in \mathbb{F}_2^n$.
Output: $m \in \mathbb{F}_2^k$ *such that* $d_H(H, M, P, s) = e$.

DEFINITION 4. *McEliece Problem.*
Input: A public key of McEliece cryptosystem (G^{pub}, t), *where* $G^{pub} \in \mathbb{F}_2^{k\times n}$, $0 < t \in \mathbb{N}$; *and a McEliece ciphertext* $c \in \mathbb{F}_2^n$.
Output: $m \in \mathbb{F}_2^k$ *such that* $d_H(mG^{pub}, c) = t$.

DEFINITION 5. *Goppa Code Distinguishing Problem.*
Input: $H \in \mathbb{F}_2^{(n-k)\times n}$.
Decide: If H *is a parity check matrix of an* (n,k) *irreducible Goppa code, or of a random* (n,k)*-code?*

2.2 Niederreiter Cryptosystem

For a survey on the material presented in this and in the next subsections, we refer the reader to the paper by Engelbert et al [15].

The Niederreiter PKE consists of the following triplet of algorithms $(\mathcal{K}, \mathcal{E}, \mathcal{D})$ with system parameters $n, t \in \mathbb{N}$:

• Key generation algorithm $\mathcal{K}$: On input n, t, generate the following matrices:

– $H \in \mathbb{F}_2^{(n-k)\times n}$ – the parity check matrix of an irreducible binary Goppa code which can correct up to t errors. Its decoding algorithm is denoted as $\mathsf{Dec}_{\mathcal{H}}$.

– $M \in \mathbb{F}_2^{(n-k)\times(n-k)}$ – a random non-singular matrix.

– $P \in \mathbb{F}_2^{n\times n}$ – a random permutation matrix.

– $H^{pub} = MHP \in \mathbb{F}_2^{(n-k)\times n}$.

Output the public key $pk = (H^{pub}, t)$ and the secret key $sk = (M, H, P, \mathsf{Dec}_{\mathcal{H}})$.

• Encryption algorithm $\mathcal{E}$: On input a plaintext $m \in \mathbb{F}_2^n$ such that $w_H(m) = t$, and the public key pk, output the ciphertext $c = H^{pub}m^T$.

• Decryption algorithm $\mathcal{D}$: On input a ciphertext c and the secret key sk, calculate:

– $M^{-1}c = (HP)m^T$.

– Since $(HP)m^T = H(Pm^T)$, use the syndrome decoding algorithm $\mathsf{Dec}_{\mathcal{H}}$ to recover Pm^T.

–Output $m^T = P^{-1}Pm^T$.

It is easy to check correctness of the decryption algorithm: After in the first step of decryption, we obtain a syndrome of the permuted plaintext Pm^T. Since the the decoding algorithm $\mathsf{Dec}_{\mathcal{H}}$ is known, it is easy to recover the plaintext.

We note that the plaintext space of Niederreiter PKE is the set of weight-t binary vectors. For representation of arbitrary binary vectors (of an appropriate length) as valid plaintexts, see the work by Cover [13] and its improvements by Sendrier [32].

2.3 McEliece Cryptosystem

The McEliece PKE consists of the following triplet of algorithms $(\mathcal{K}, \mathcal{E}, \mathcal{D})$ with system parameters $n, t \in \mathbb{N}$:

• Key generation algorithm $\mathcal{K}$: On input n, t, generate the following matrices:

– $G \in \mathbb{F}_2^{k\times n}$ – the generator matrix of an irreducible binary Goppa code correcting up to t errors. Its decoding algorithm is denoted as $\mathsf{Dec}_{\mathcal{G}}$.

– $S \in \mathbb{F}_2^{k\times k}$ – a random non-singular matrix.

– $P \in \mathbb{F}_2^{n\times n}$ – a random permutation matrix (of size n).

– $G^{pub} = SGP \in \mathbb{F}_2^{k\times n}$.

Output the public key $pk = (G^{pub}, t)$ and the secret key $sk = (S, G, P, \mathsf{Dec}_{\mathcal{G}})$.

• Encryption algorithm $\mathcal{E}$: On input a plaintext $m \in \mathbb{F}_2^k$ and the public key pk, choose a vector $e \in \mathbb{F}_2^n$ of weight t at random, and output the ciphertext $c = mG^{pub} \oplus e$.

• Decryption algorithm $\mathcal{D}$: On input c and the secret key sk, calculate:

– $cP^{-1} = mSG \oplus eP^{-1}$.

– $mSG = \mathsf{Dec}_{\mathcal{G}}(cP^{-1})$.

– Let $J \subseteq \{1, \ldots, n\}$ be s.t. G_J is invertible.

Output $m = (mSG)_J(G_J)^{-1}S^{-1}$.

It is easy to check that the decryption algorithm correctly recovers the plaintext: Since in the first step of decryption, the permuted error vector eP^{-1} is again of weight t, the decoding algorithm $\mathsf{Dec}_{\mathcal{G}}$ successfully corrects these errors in the next step.

2.4 Proof of Plaintext Knowledge

We use the definition of [22]. For a PKE scheme $(\mathcal{K}, \mathcal{E}, \mathcal{D})$, denote by $c = \mathcal{E}_{pk}(m; R)$ a ciphertext of a plaintext m under public key pk using randomness R. We will call (m, R) a *witness* to the decryption of c under pk. Informally, in a PPK protocol, a sender P proves to a receiver V the knowledge of a witness to the decryption for some ciphertext c under the known public key pk.

DEFINITION 6. *Let* $\Pi = (\mathsf{P}, \mathsf{V})$ *be a tuple of PPT algorithms.* Π *is a* proof of plaintext knowledge *for encryption scheme* $(\mathcal{K}, \mathcal{E}, \mathcal{D})$ *if the following conditions hold:*

(Completeness) *For all pk output by* $\mathcal{K}(1^n)$ *and all c with witness w to the decryption of c under pk, we have that* $\langle \mathsf{P}(w), \mathsf{V}\rangle(pk, c) = 1$. *(When* V *outputs 1 we say it* accepts.*)*

(Soundness) *For all pk output by* $\mathcal{K}(1^n)$, *all c produced under pk, and for any* $\widetilde{\mathsf{P}}$, *we have that* $\Pr[\langle \widetilde{\mathsf{P}}, \mathsf{V}\rangle(pk, c) = 1]$ *is negligible.*

(Zero-knowledge) *There exists a PPT Turing machine* $\mathcal{SIM}$ *(called a* simulator*) such that, for all pk output by* $\mathcal{K}(1^n)$, *all PPT* $\widetilde{\mathsf{V}}$, *and all w, the following distributions are indistinguishable:*

$$\{c = \mathcal{E}_{pk}(m; R) : \langle \mathsf{P}(w), \widetilde{\mathsf{V}}\rangle(pk, c)\},$$

$$\{c = \mathcal{E}_{pk}(m; R) : \langle \mathcal{SIM}, \widetilde{\mathsf{V}}\rangle(pk, c)\},$$

in case of statistical, *respectively* computational *indistinguishability, we call the property* statistical, *respectively* computational *zero-knowledge (ZK).*

3. PPK FOR NIEDERREITER PKE

We construct the proof of plaintext knowledge for Niederreiter encryption using Stern ZK identification scheme [34]. We take the Niederreiter public key, i.e. a permuted and scrambled parity-check matrix of an irreducible binary Goppa code correcting up to t errors as common data. The plaintext is used as witness and the ciphertext is used as public identification. Since the Niederreiter PKE is deterministic, we consider the string R representing randomness in Definition 6 as empty.

We observe that the security proof of Stern's scheme [34] does not use the fact the common code is random. Of importance are only the facts that it has a particular minimum distance, which is provided by construction of the Niederreiter public key, and that the decoding problem for it is hard. The latter is ensured by the hardness of Niederreiter Problem (see Section 2.1) which we assume to hold.

Witness: $m \in \mathbb{F}_2^n$, $w_H(m) = t$, where the parameters n and t are described in Section 2.2.

Common data: (H^{pub}, t) s.t. $H^{pub} \in \mathbb{F}_2^{(n-k)\times n}$ – the Niederreiter public key, and $c = H^{pub}m^T$ – the Niederreiter ciphertext.

Protocol 1 (Niederreiter PPK).

1. P computes $y \stackrel{\$}{\leftarrow} \mathbb{F}_2^n$ and $\pi \stackrel{\$}{\leftarrow} \mathcal{S}_n$ and sends three commitments:
   - $C_1 = Com(\pi, H^{pub}y^T)$,
   - $C_2 = Com(y\pi)$,
   - $C_3 = Com((y + m)\pi)$.
2. V sends $b \stackrel{\$}{\leftarrow} \{0, 1, 2\}$.
3. V performs the following checks and rejects, if any check fails:
   - If $b = 0$,
     - P sends y, π and opens C_1 and C_2.
     - V directly checks validity of the opened values.

– If $b = 1$,

– P sends $y + m, \pi$ and opens C_1 and C_3.

– V checks validity by computing the following: $H^{pub}y^T = H^{pub}(y+m)^T + c$, and then verifying that the opening of C_1 is $(\pi, H^{pub}y^T + c)$, and that the opening of C_3 is $(y+m)\pi$.

– If $b = 2$,

– P sends $y\pi, m\pi$ and opens C_2 and C_3.

– V checks validity of opened values by verifying that C_2 opens to $y\pi$, C_3 opens to $y\pi + m\pi$, and that $w_H(m\pi) = t$.

Denote a protocol consisting of r independent iterations of Protocol 1 by $\mathsf{PPK}(H^{pub}, c; m)$, with some appropriately chosen r.

THEOREM 1. *Protocol $\mathsf{PPK}(H^{pub}, c; m)$ is a proof of plaintext knowledge for the Niederreiter public key encryption according to Definition 6 assuming hardness of Niederreiter Problem, in the standard model.*

PROOF. We generally follow the proof of [34], but for the proof of soundness we use the argument from [35], since it is shorter[1].

Completeness. It is easy to check that P who knows the plaintext can answer all of three challenges correctly. This implies that $\langle \mathsf{P}(w), \mathsf{V}\rangle(pk, c) = 1$.

Soundness.

LEMMA 1. *If V accepts $\widetilde{\mathsf{P}}$'s proof with probability at least $(\frac{2}{3})^r + \epsilon$, then there exists a PPT algorithm WE which, with overwhelming probability, computes a witness m.*

The proof will appear in the full version of this paper. We only note that the machine WE, constructed in the proof, finds a valid witness, hereby contradicting the hardness of the Niederreiter problem, unless the binding property of the commitment is violated. Therefore, for a cheating prover $\widetilde{\mathsf{P}}$, we must have $\Pr[\langle \widetilde{\mathsf{P}}, V\rangle(pk, c) = 1] \leq (2/3)^r + \epsilon$, which is negligible in n and r.

We emphasize that the proof does not require any additional assumptions on the Niederreiter public key (such as for instance, its indistinguishability from a random matrix), except for those made in the statement of the Niederreiter problem.

Zero-knowledge. This property guarantees that the execution of the protocol does not leak any information (in computational, or in statistical sense) to the cheating polynomial-time verifier $\widetilde{\mathsf{V}}$, who might decide on an arbitrary strategy for choosing her challenges.

LEMMA 2. *Protocol 1 is computational (respectively statistical) zero-knowledge according to Definition 6, if the used commitment scheme is computationally (respectively unconditionally) hiding.*

The proof will appear in the full version of this paper.

The above two lemmas now conclude the proof of Theorem 1. □

[1]Remember that the gap in the proof of [35] pointed out in [21] concerned only the proof of zero-knowledge property.

4. PERFORMANCE EVALUATION

In this section, we estimate security and performance of the proposed scheme. We chose the parameters of the Niederreiter scheme according to the estimation by Peters [30] for the running time of the information set decoding algorithm. We first choose the code length n (as a power of 2, for convenience) and $k = n - t\log_2 n$ (again for convenience), then we find the smallest value of t which provides us with 80 or 128 bits of equivalent security. We fix the soundness failure probability in our PPK to be at most 2^{-16}, since this value is a minimal requirement in the ISO/IEC-9798-5 standard for the zero-knowledge techniques for entity authentication. Remembering that the soundness failure probability is 2/3 in each round, we will need 28 rounds in total.

We consider the original Niederreiter public key encryption, with binary irreducible Goppa codes used for generation of public keys, without any optimizations. Then, the size of the public key is $(n-k)n$ bits, the ciphertext size is n bits.

In order to keep estimation simple, we assume the random oracle model, and construct commitments using idealized hash functions $h : \{0,1\}^* \to \{0,1\}^{l_c}$, taking $l_c = 160$. In order to commit to a value x (we will think of a binary representation of x), P will simply compute $h(x)$. Then, V's checks in our protocol will be performed as in [34] by computing the corresponding hash values. For instance, we will compute $C_1 = h(\pi, H^{pub}y^T)$, and when $b = 0$, V will use the values π and $H^{pub}y^T$ received from P to compute $h(\pi, H^{pub}y^T)$ and check that it is equal to C_1 received in Step 1.

We assume that the representations of y and π, respectively, are generated using pseudorandom generators with seed length $l_s = 128$. Sending seeds instead of actual values will allows us to save on communication.

The expected communication cost of our protocol is:

$$r\big(3l_c + 2 + (3l_s + 3n)/3\big).$$

we call it "expected" because in the thirm term, we average over the sizes of responses to the challenge b.

Prover's expected amount of computation, which is also an upper bound on that amount for the verifier, is as follows:

$$r\big((2n-1)(n-k) + n\big) \text{ binary operations.}$$

This is just a rough estimate of the total computation cost, since the costs of computing permutations, commitments, and pseudorandom generation – which are implementation-specific – are not included.

Our proposed recommended parameters and the resulting costs are summarized in Table 1.

We can see that the communication cost of the proposed protocol appears to be within the practical feasibility range. For instance, for the equivalent 80 bit security, it is 9.1 kilobytes. For comparison, it is 6.9 times smaller than the (non-optimized) public key size of the Niederreiter PKE, which is equal to 62.3 kilobytes for the relevant parameters.

5. PPK FOR MCELIECE PKE

One natural way to construct the proof of plaintext knowledge for the McEliece PKE [25] would be to replace the flawed Véron ZK identification scheme [35] with the ZK proofs of Jain et al [21] in the construction of [26].

A potential problem is that the scheme of [21] is based of a variant of LPN problem. This, in turn, will require us

Equivalent security (bits)	80	128
Code length n	2048	4096
Code dimension k	1806	3676
Weight of error vector t	22	35
Public key size (Kbytes)	62.3	132.3
Communication (Kbytes)	9.1	16.1
Prover's computation (operations over $\mathbb{F}_2$)	$2^{24.73}$	$2^{26.52}$

Table 1: Parameters and Performance of the Proposed PPK Protocol.

to make an additional Goppa Code Distinguishing assumption (see Sec. 2.1) on pseudorandomness of the McEliece public key. Although this assumption has not yet been disproved directly, a distinguisher for *high rate* Goppa codes was presented by Faugère et al in [16]. Therefore, we prefer to avoid this assumption whenever possible – in particular, when constructing our PPK.[2]

Our approach is to use the equivalence between McEliece and Niederreiter PKE's observed by Li et al [24]. In order to construct PPK for the McEliece ciphertext, one will first compute an "equivalent" plaintext, along with an "equivalent" public key of the Niederreiter PKE, and then use Protocol 1. In fact the equivalent public key may be pre-computed and distributed the same way as the original McEliece public key, since the equivalence is easy to verify.

In details, suppose that the McEliece ciphertext is $c = mG^{pub} + e$, where m, G^{pub} and e are computed as described in Section 2.3. Let the equivalent Niederreiter public key H^{pub} be chosen a basis of the kernel of G^{pub}, i.e.

$$H^{pub}(G^{pub})^T = \mathbf{0}, \tag{1}$$

where $\mathbf{0}$ is the zero-vector of the appropriate length. In other words, H^{pub} is a parity check matrix of the linear code generated by G^{pub}. The equivalent Niederreiter ciphertext c_N is computed as a syndrome of the ciphertext c, using H^{pub} as a parity check matrix: $c_N = H^{pub}c^T = H^{pub}(mG^{pub})^T + H^{pub}e^T = H^{pub}e^T$.

Let us now briefly argue that our proposal indeed works. First of all, it is easy to check that H^{pub} is chosen a basis of the kernel of G^{pub} since Equation (1) and the ranks of these matrices are easy to verify. Then, note that multiplication of H^{pub} by $(c')^T$, where c' is an arbitrary n-bit vector, will phase out any codewords of the code generated by G^{pub} from c'. Now the obtained expression is of the form $H^{pub}(e')^T$, $e' = c' + mG^{pub}$ for some $m \in \mathbb{F}_2^k$. If $w_H(e') = t$, then c' is a valid McEliece ciphertext, but also e' is a valid witness for Protocol 1. Otherwise, c' is not a valid McEliece ciphertext, but also e' is not a valid witness, and a prover must fail the PPK by Theorem 1.

Note that so far we only explained that Protocol 1 works as a ZK proof of validity for the McEliece PKE. However, the prover, who knows an error vector of appropriate weight, can compute the plaintext simply by solving an overdefined system of linear equations as $c + e = mG^{pub}$, where G^{pub} is public. Therefore, this is also a proof of plaintext knowledge.

In Definition 6, we would consider the string m representing the plaintext as empty.

[2]At the same time, we admit that deployment of the ZK proofs of [21] seems to be a prospective way for enhancing the code-based verifiable encryption of [26].

Trivially, the same PPK will work for the IND-CPA variant of the McEliece PKE [28], where the plaintext is padded with uniform randomness.

6. VERIFIABLE ENCRYPTION

Informally, verifiable encryption with respect to some binary relation R on the plaintexts is a ZK proof on public inputs pk, c, and δ that allows P to convince V that c is a ciphertext of m under pk such that $(m, \delta) \in R$. The simplest example is the equality relation $R_{eq} = \{(m, m')|m = m'\}$, i.e. that a given ciphertext c is an encryption of a given plaintext m under public key pk. We refer the reader to [26] and references therein for a formal definition and related results.

The work [26] presents the McEliece verifiable encryption with respect to the equality relation. In other words, P convinces V that a given message m is contained in the Randomized McEliece encryption, see Section 2.3 for description.

Informally, their protocol works as follows: First, P runs PPK on the input ciphertext $c = (r|m)G^{pub} + e = rG_0^{pub} + mG_1^{pub} + e$, where $(G^{pub})^T = ((G_0^{pub})^T|(G_1^{pub})^T)$, $G_0^{pub} \in \mathbb{F}_2^{k_0 \times n}$ and $G_1^{pub} \in \mathbb{F}_2^{k_1 \times n}$ are the sub-matrices of G^{pub} corresponding to randomness and plaintext, respectively. Secondly, both players compute $c' = c + mG_1^{pub} = rG_0^{pub} + e$, hereby canceling out the plaintext, and then run PPK with c' as ciphertext and G_0^{pub} as public key.

Now, we can see that Protocol 1 can be used as PPK in the above construction. Hereby, we fix the McEliece verifiable encryption scheme of [26].

It was noted in [26], that although the resulting PPK is zero-knowledge, the fact that V learns m (together with the fact that c is a valid McEliece ciphertext) implies that $\widetilde{\mathsf{V}}$ can now produce a valid encryption $c_a = rG_0^{pub} + e + m_a G_1^{pub}$ for an arbitrary $m_a \in \mathbb{F}_2^k$. Note that although the Randomized McEliece PKE is clearly malleable, the above attack is not feasible for $\widetilde{\mathsf{V}}$ prior to the protocol execution. This is not a problem of the protocol, but rather the property of the Randomized McEliece encryption. Therefore, some authentication technique must be applied in order to avoid this attack.

The construction of Niederreiter verifiable encryption using PPK is similar. We deter its presentation to the full version of this paper.

7. ACKNOWLEDGMENTS

The authors would like to thank the anonymous reviewers of AsiaCCS 2013 for pointing out some errata. Rong Hu was supported by China Scholarship Council. Kirill Morozov was supported by a *kakenhi* Grant-in-Aid for Young Scientists (B) 24700013 from Japan Society for the Promotion of Science.

8. REFERENCES

[1] Y. Aumann and M.O. Rabin. A Proof of Plaintext Knowledge Protocol and Applications. Manuscript. June, 2001. Available as slides from 1998 IACR Distinguished Lecture by M.O. Rabin: `http://www.iacr.org/publications/dl/rabin98/rabin98slides.ps`.

[2] N. Asokan, V. Shoup, M. Waidner. Optimistic Fair Exchange of Digital Signatures (Extended Abstract). EUROCRYPT 1998: 591-606.

[3] A. Becker, A. Joux, A. May, A. Meurer: Decoding Random Binary Linear Codes in $2^{n/20}$: How $1 + 1 = 0$ Improves Information Set Decoding. EUROCRYPT 2012: 520-536.

[4] Mihir Bellare, Phillip Rogaway: Random Oracles are Practical: A Paradigm for Designing Efficient Protocols. ACM Conference on Computer and Communications Security 1993: 62-73.

[5] R. Bendlin, I. Damgård. Threshold Decryption and Zero-Knowledge Proofs for Lattice-Based Cryptosystems. TCC 2010: 201-218.

[6] D.J. Bernstein, T. Lange, C. Peters. Wild McEliece. Selected Areas in Cryptography 2010: 143-158.

[7] E. Berlekamp, R. McEliece, and H. van Tilborg. On the inherent intractability of certain coding problems, IEEE Trans. on Inf. Theory 24, 1978: 384-386.

[8] D.J. Bernstein: Grover vs. McEliece. PQCrypto 2010: 73-80.

[9] D.J. Bernstein, T. Lange and C. Peters. Smaller Decoding Exponents: Ball-Collision Decoding. CRYPTO 2011: 743-760.

[10] J. Camenisch, I. Damgård. Verifiable Encryption, Group Encryption, and Their Applications to Separable Group Signatures and Signature Sharing Schemes. ASIACRYPT 2000: 331-345.

[11] J. Camenisch, V. Shoup: Practical Verifiable Encryption and Decryption of Discrete Logarithms. CRYPTO 2003: 126-144.

[12] Pierre-Louis Cayrel, Mohammed Meziani: Post-quantum Cryptography: Code-Based Signatures. AST/UCMA/ISA/ACN 2010: 82-99.

[13] T. Cover. Enumerative source encoding. IEEE Transactions on Information Theory, 19(1):73-77, January 1973.

[14] R. Dowsley, J. van de Graaf, J. Müller-Quade, A.C.A. Nascimento: Oblivious Transfer Based on the McEliece Assumptions. ICITS 2008: 107-117.

[15] D. Engelbert, R. Overbeck and A. Schmidt: A Summary of McEliece-Type Cryptosystems and their Security, Journal of Mathematical Cryptology, vol. 1, Walter de Gruyter, 2007: 151-199.

[16] Jean-Charles Faugère, Ayoub Gauthier-Umaña, Valérie Otmani, Ludovic Perret, and Jean-Pierre Tillich. A Distinguisher for High Rate McEliece Cryptosystems. Information Theory Workshop (ITW), 2011: 282-286.

[17] M. Finiasz, N. Sendrier: Security Bounds for the Design of Code-Based Cryptosystems. ASIACRYPT 2009: 88-105.

[18] Z. Galil, S. Haber, M. Yung: Symmetric Public-Key Encryption. CRYPTO 1985: 128-137.

[19] O. Goldreich: Foundations of Cryptography I: Basic Tools. Cambridge University Press, 2001.

[20] S. Goldwasser, D. Kharchenko: Proof of Plaintext Knowledge for the Ajtai-Dwork Cryptosystem. TCC 2005: 529-555.

[21] A. Jain, S. Krenn, K. Pietrzak, A. Tentes: Commitments and Efficient Zero-Knowledge Proofs from Learning Parity with Noise. ASIACRYPT 2012: 663-680.

[22] J. Katz: Efficient and Non-malleable Proofs of Plaintext Knowledge and Applications. EUROCRYPT 2003: 211-228.

[23] K. Kobara, K. Morozov, R. Overbeck: Coding-Based Oblivious Transfer. MMICS 2008: 142-156.

[24] Y. X. Li, R. H. Deng, and X. M. Wang: The Equivalence of McEliece's and Niederreiter's Public-Key Cryptosystems. IEEE Trans. Inform. Theory 40, 1994: 271-273.

[25] R.J. McEliece: A Public-Key Cryptosystem Based on Algebraic Coding Theory. Deep Space Network Progress Rep., 1978.

[26] K. Morozov, T. Takagi: Zero-Knowledge Protocols for the McEliece Encryption. ACISP 2012: 180-193.

[27] H. Niederreiter: Knapsack-type Cryptosystems and Algebraic Coding Theory. Prob. of Control and Inf. Theory, 15(2), 1986: 159-166.

[28] R. Nojima, H. Imai, K. Kobara, K. Morozov. Semantic security for the McEliece cryptosystem without random oracles. Des. Codes Cryptography 49(1-3), 2008: 289-305.

[29] K. Pietrzak: Cryptography from Learning Parity with Noise. SOFSEM 2012: 99-114.

[30] C. Peters: Information-Set Decoding for Linear Codes over F_q. PQCrypto 2010: 81-94.

[31] N. Sendrier. On the security of the McEliece public-key cryptosystem. Information, Coding and Mathematics – Proceedings of Workshop honoring Prof. Bob McEliece on his 60th birthday, Kluwer, 2002: 141-163.

[32] N. Sendrier: Encoding information into constant weight codewords, ISIT 2005: 435-438.

[33] M. Stadler: Publicly Verifiable Secret Sharing. EUROCRYPT 1996: 190-199.

[34] J. Stern: A new paradigm for public key identification. IEEE Trans. Inform. Theory 42(6), 1996: 1757-1768. A conference version: J. Stern: A New Identification Scheme Based on Syndrome Decoding. CRYPTO 1993: 13-21.

[35] P. Véron. Improved identification schemes based on error-correcting codes. Appl. Algebra Eng. Commun. Comput. 8(1), 1996: 57-69.

[36] K. Xagawa, A. Kawachi, K. Tanaka. Proof of Plaintext Knowledge for the Regev Cryptosystems. Tech.rep. C-236, Tokyo Inst. of Technology, 2007.

[37] K. Xagawa, K. Tanaka. Zero-Knowledge Protocols for NTRU: Application to Identification and Proof of Plaintext Knowledge. ProvSec 2009: 198-213.

An Efficient and Probabilistic Secure Bit-Decomposition

Bharath K. Samanthula, Hu Chun, and Wei Jiang
Department of Computer Science, Missouri S&T
500 West 15th Street, Rolla, Missouri 65409
{bspq8, chwrc, wjiang}@mst.edu

ABSTRACT

Many secure data analysis tasks, such as secure clustering and classification, require efficient mechanisms to convert the intermediate encrypted integers into the corresponding encryptions of bits. The existing bit-decomposition algorithms either do not offer sufficient security or are computationally inefficient. In order to provide better security as well as to improve efficiency, we propose a novel probabilistic-based secure bit-decomposition protocol for values encrypted using public key additive homomorphic encryption schemes. The proposed protocol guarantees security as per the semi-honest security definition of secure multi-party computation (MPC) and is also very efficient compared to the existing method. Our protocol always returns the correct result, however, it is probabilistic in the sense that the correct result can be generated in the first run itself with very high probability. The computation time of the proposed protocol grows linearly with the input domain size in bits. We theoretically analyze the complexity of the proposed protocol with the existing method in detail.

Categories and Subject Descriptors

K.6.5 [**Management of Computing and Information Systems**]: Security and Protection (D.4.6)

General Terms

Algorithms, Security

Keywords

Security; Encryption; Bit-decomposition

1. INTRODUCTION

Statistical data analysis is an essential task in many data mining and business intelligence applications [10]. However, when the data come from multiple parties and where user privacy is a big concern, we need to perform the data analysis task in a privacy-preserving manner. The data analysis task becomes even more challenging when the data is in encrypted form which is quite common in outsourced databases.

Though the goal of this paper is not to investigate the existing privacy preserving data mining tasks over the encrypted data, we focus on specific underlying secure operations used in these tasks. In particular, performing secure comparison on the intermediate encrypted results (i.e., given the encryptions of two integers x and y in the cloud, we need to securely evaluate whether x is greater than or equal to y) is an essential operation in many data analysis tasks such as clustering and classification. Nevertheless, most of the existing secure comparison methods [2, 5] require encryptions of the individual bits of the corresponding encrypted value.

At first, it seems simply operating on the encrypted bits from the beginning of data analysis process is a better approach. However, this will lead to inefficient protocols considering the amount of computation involved in massive data analysis tasks. Therefore, many secure data analysis tasks require efficient mechanisms to convert the intermediate encrypted integers into the corresponding encryptions of bits. The process of securely converting an encrypted value of x into encryptions of the individual bits of x is termed as secure bit-decomposition (SBD) [1, 4, 14]. SBD acts as an important primitive in various secure multi-party computation (MPC) [6, 17, 18] protocols such as secure comparison, public modulo and private exponentiation on encrypted integers [7, 9].

Existing SBD methods are based on either secret sharing schemes [15] or threshold homomorphic cryptosystem [3] under MPC framework. We emphasize that SBD methods which are based on secret sharing schemes require at least three parties; therefore, they are not applicable to our problem domain (i.e., semi-honest two-party setting). Whereas, the SBD method proposed in [14], which is based on the (threshold) Paillier homomorphic cryptosystem, can easily be adopted to the two-party case. However, there are two main disadvantages regarding the SBD protocol proposed in [14]:

- The round and communication complexities are independent of the domain size of x (in bits), for $0 \leq x < 2^m$ and $m < K$, where K is the Paillier encryption key size [12]. Instead, they depend on the value of K. Therefore, the case where $m < K$, which is true for most of the practical applications, has no significant benefit with respect to both round and communication complexities. We emphasize that the protocol in [14], for smaller domain values (i.e., $0 \leq x < 2^m$), is not secure since the intermediate decrypted values are not uniformly random in $\mathbb{Z}_N$. On the other hand, the general solution (referred to as the BITREP gate) proposed in [14], for $x \in \mathbb{Z}_N$, is secure but it requires $O(K)$ rounds and also the communication complexity is bounded by $O(K^2)$ in bits;
- The BITREP gate [14] is not efficient. Therefore, using it as a building block in the relevant secure data analysis tasks may reduce the overall throughput.

To overcome the above disadvantages of BITREP gate, we pro-

ASIA CCS'13, May 8–10, 2013, Hangzhou, China.

pose an efficient (both in terms of computation and communication) SBD protocol that also requires less number of communication rounds between the two parties in comparison to BITREP. The proposed protocol is very efficient and probabilistic in the number of iterations. However, we show that, with very high probability, our protocol gives the correct result in the first run itself.

1.1 Problem Statement

In our problem setting, we consider two semi-honest (also referred to as honest-but-curious) parties Alice and Bob [6]. We assume that Alice generates a Paillier[1] public/secret key pair (pk, sk) and broadcasts the public key pk to Bob.

Let $\langle E, D\rangle$ be the encryption and decryption functions associated with the public/secret key pair $\langle pk, sk\rangle$. Without loss of generality, assume that Bob holds the Paillier encrypted value $E(x)$, where $0 \leq x < 2^m$ (here m is referred to as the domain size of x in bits). We explicitly assume that x is not known to Alice and Bob. Suppose $\langle x_0, \ldots, x_{m-1}\rangle$ denotes the binary representation of x where x_0 and x_{m-1} are the least and most significant bits respectively. The goal of this paper is to convert encryption of x into the encryptions of the individual bits of x without disclosing any information regarding x to both Alice and Bob. We refer to such a process as secure bit-decomposition (SBD). More formally, we define the SBD protocol as follows:

$$\text{SBD}(E(x)) \rightarrow \langle E(x_0), \ldots, E(x_{m-1})\rangle \quad (1)$$

At the end of the SBD protocol, the values $E(x_0), \ldots, E(x_{m-1})$ are known only to Bob and nothing is revealed to Alice. Note that since SBD protocol is used as a sub-routine in many secure applications, leaking either the value of x or any of the bit values (x_i's) to either Alice or Bob may not be allowed.

Also, we observe that the proposed protocol can be directly used for (Paillier) threshold based setting [3], where the secret key sk is shared between Alice and Bob, without affecting the complexities.

1.2 Our Contributions

This paper presents an efficient (in terms of both computation and communication costs) and probabilistic SBD protocol. The main contributions of this paper are summarized as follows.

(a). **Security -** The proposed protocol provides perfect security as per the standard semi-honest security definition of secure multi-party computation (MPC) framework [6, 17, 18].

(b). **Efficiency -** Our protocol is efficient compared to the solution in [14] both in terms of computation and communication.

(c). **Round Complexity -** The proposed SBD protocol requires exactly m number of communication rounds between Alice and Bob, where $0 \leq x < 2^m$. In general, when $m < K$ which is true in most of the practical applications, the round complexity of our protocol is improved by a factor of $\frac{K}{m}$ over [14].

(d). **Accuracy -** Since the proposed protocol is probabilistic in nature, we include a checking step to securely verify whether the result is correct or not. That is, the steps involved in the proposed protocol can be run in an iterative fashion until the result is correct. Nevertheless, for most of the practical applications, we show that the probability for our protocol to produce the correct result in the first run itself is statistically indistinguishable from "1". More details are given in Section 4.

[1]We emphasize that any additive homomorphic encryption scheme can be utilized in the proposed protocol, however, to be consistent with the previous work, in this paper we use Paillier cryptosystem.

The rest of the paper is organized as follows. In Section 2, we discuss the previous works related to our problem domain. Section 3 explains some definitions and properties as a background knowledge which are used throughout the paper. Then, we provide a detailed description of the proposed SBD protocol along with a running example in Section 4. Also, apart from the security analysis, we theoretically analyze the correctness of the proposed protocol and compare its complexities with those of existing work. We conclude the paper with future work in Section 5.

2. RELATED WORK

2.1 Security Definition

In this paper, the terms privacy and security are interchangeable. In general, privacy and security are closely related to the amount of information disclosed during the execution of a protocol. There are many ways to define information disclosure. This paper adopts the well-known security definition of semi-honest (also referred to as honest-but-curious) model from the field of secure multi-party computation (MPC) [6, 17, 18]. Informally speaking, a protocol Π privately computes function f (known to both parties) if whatever can be obtained from a party's view of a (semi-honest) execution could be essentially obtained from the input and output available to the party. We refer the reader to [6] for detailed security definitions.

2.2 Existing SBD Algorithms

As mentioned in Section 1, the existing SBD algorithms are based on either the secret sharing scheme [15] (where the secret $x \in \mathbb{Z}_p$ is shared among n parties and p is a prime) or the cryptographic setting [3] (Paillier encrypted value of x) under the MPC framework [6, 17, 18].

2.2.1 SBD algorithms under secret sharing

The SBD problem was first introduced in [1] under secret sharing setting assuming the value of x is much smaller than p. However, as mentioned in [4, 14], their approach follows a more complex procedure by involving an addition circuit for securely adding n numbers bitwise; therefore, leading to inefficient solution. Plus, their approach is only passively secure and round complexity is also bounded by $O(\log_2 n + \log_2 m)$ (non-constant-rounds); where n (≥ 3) is the number of parties and m is the (upper-bound) number of bits required to represent x.

Damgård et al. [4] proposed the first constant round solution to the SBD problem using linear secret sharing scheme such as Shamir's scheme [15]. However, as mentioned in [13, 16], their method is still expensive as it requires $O(l \log_2 l)$ secure multiplications, where $l = \lceil \log_2 p \rceil$. Also, their method requires 38 communication rounds. Following the work from [4], Nishide and Ohta [11] improved the round and communication complexity by a constant factor. Their method requires 25 rounds and computation complexity is still bounded by $O(l \log_2 l)$ secure multiplications. Toft [16] has proposed an almost linear and constant rounds solution to the SBD problem, under secret sharing, by reducing the bit-decomposition problem to a postfix comparison.

2.2.2 SBD algorithms under Paillier cryptosystem

As an independent work, most related to ours, Schoenmakers and Tuyls [14] considered multi-party threshold (Paillier) homomorphic cryptosystem [3] and solved the problem of converting a given Paillier encrypted integer x (under $\mathbb{Z}_N$) into Paillier encryptions of the individual bits of x (referred to as BITREP gate). We emphasize that the BITREP gate can directly be applied to a two-party case. However, though their method is secure, the compu-

Algorithm 1 Binary(x) $\rightarrow \langle x_0, \ldots, x_{m-1} \rangle$

Require: A positive decimal integer x, where $0 \leq x < 2^m$

1: $i \leftarrow 0$
2: **while** $i \neq m$ **do**
3: $\quad x_i \leftarrow x \bmod 2$
4: $\quad x \leftarrow \lfloor \frac{x}{2} \rfloor$ {observe that x is updated to current quotient q_i}
5: $\quad i \leftarrow i + 1$
6: **end while**

tation and communication costs of BITREP gate are still high and also the round complexity is bounded by $O(K)$.

In a recent work, Reisted and Toft [13] proposed the first constant-round linear solution for the SBD problem under arbitrary setting. However, we observe that their protocol is still inefficient (requiring $O(K)$ secure multiplications). In addition, which is also the main disadvantage, the security of their protocol is not perfect because it uses an LSB gate of [14] which is not secure as a sub-routine. Therefore, for the rest of this paper, we compare our proposed solution with the BITREP (since it is the only existing secure solution, to the best of our knowledge, under Paillier cryptosystem).

3. PRELIMINARIES

3.1 Standard Binary Conversion Method

Our protocol uses standard binary conversion algorithm as a baseline. Let x be an integer such that $0 \leq x < 2^m$. The overall steps involved in the standard binary conversion method are highlighted in Algorithm 1. Briefly, we first divide x by 2. The remainder 0 or 1 (i.e., $x \mod 2$) will be the bit in question and then x is replaced by the quotient (denoted by q_0, where $q_0 = \lfloor \frac{x}{2} \rfloor$). This process is repeated until m iterations.

3.2 Paillier Cryptosystem

The Paillier cryptosystem is an additive homomorphic and probabilistic asymmetric encryption scheme whose security is based on the Decisional Composite Residuosity Assumption [12]. Let E be the encryption function with public key pk given by (N, g) and D be the decryption function with secret key sk given by a trapdoor function λ (that is, the knowledge of the factors of N). Here, $N = p * q$ of bit length K (security parameter usually of 1024 bits); where p and q are large primes of similar bit length, and generator $g \in \mathbb{Z}^*_{N^2}$. For any given $y, z \in \mathbb{Z}_N$, the Paillier cryptosystem exhibits the following properties:

a. **Homomorphic Addition -** $E(y+z) \leftarrow E(y) * E(z) \bmod N^2$;

b. **Homomorphic Multiplication -** $E(z * y) \leftarrow E(y)^z \bmod N^2$;

c. **Semantic Security -** The encryption scheme is semantically secure [6]. Briefly, given a set of ciphertexts, an adversary cannot deduce any additional information about the plaintext.

4. THE PROPOSED PROTOCOL

In this section, we present our new probabilistic SBD protocol (denoted as SBD_p) based on the standard binary conversion algorithm. In addition, we analyze its security and correctness, and also theoretically compare the complexity of SBD_p with the existing secure bit-decomposition protocol (i.e, BITREP gate [14]).

As mentioned earlier, we assume that Bob holds a Paillier encryption of integer x i.e., $E(x)$ such that x is not known to both Alice and Bob, where $0 \leq x < 2^m$. The goal of SBD_p is to compute $E(x_0), \ldots, E(x_{m-1})$ without revealing either the value of x

Algorithm 2 $\text{SBD}_\text{p}(E(x)) \rightarrow \langle E(x_0), \ldots, E(x_{m-1}) \rangle$

Require: Bob has Paillier encrypted value $E(x)$, where x is not known to both parties and $0 \leq x < 2^m$; (Note: The public key (g, N) is known to both Alice and Bob whereas the secret key sk is known only to Alice)

1: $l \leftarrow 2^{-1} \mod N$
2: $T \leftarrow E(x)$
3: **for** $i = 0 \rightarrow m - 1$ **do**
4: $\quad E(x_i) \leftarrow$ Encrypted_LSB(T, i)
5: $\quad Z \leftarrow T * E(x_i)^{N-1} \bmod N^2$
$\quad$ {update T with the encrypted value of q_i}
6: $\quad T \leftarrow Z^l \bmod N^2$
7: **end for**
8: $\gamma \leftarrow$ SVR($E(x), \langle E(x_0), \ldots, E(x_{m-1}) \rangle$)
9: **if** $\gamma = 1$ **then**
10: $\quad$ return
11: **else**
12: $\quad$ go to Step 2
13: **end if**

or any of x_i's to Alice and Bob. At the end of the SBD_p protocol, only Bob knows the encryptions of x_i's, for $0 \leq i \leq m-1$. The proposed SBD_p protocol mainly involves the following two stages:

- **Stage 1** - Computing Encrypted Bits of x:
  In this Stage, Alice and Bob (with input $E(x)$) jointly compute encryptions of the individual bits of x in an iterative fashion. At the end of this stage, only Bob knows $E(x_i)$'s, for $0 \leq i \leq m-1$. Other than the output, during this process, nothing is revealed to Alice and Bob.
- **Stage 2** - Secure Verification of Result:
  Following from Stage 1, Alice and Bob securely verify whether the result from Stage 1 is correct or not. Only the output of the verification stage is revealed to Alice and Bob. If the result from Stage 1 is correct, then both parties terminate the protocol. Otherwise, the SBD_p protocol is repeated.

The overall steps involved in SBD_p are highlighted in Algorithm 2. Now, we explain the steps involved in each stage in detail.

Stage 1 - Computing Encrypted Bits of x: Initially, Bob computes l as the multiplicative inverse of 2 modulo N and assigns $E(x)$ to T. Then, in an iterative fashion, Bob and Alice jointly compute the encrypted value $E(x_i)$, for $0 \leq i \leq m-1$, by updating T based on the standard binary conversion given in Algorithm 1. Briefly, in the first iteration, Bob computes $E(x_0)$ (the encryption of least-significant bit of x) using T. Then he updates T with the encryption of quotient q_0 (which is equivalent to $\lfloor \frac{x}{2} \rfloor$). This process is continued until m iterations. Note that in iteration i, encryption of bit x_i (i.e., $E(x_i)$) is revealed only to Bob, for $0 \leq i \leq m-1$. Before explaining the steps of Stage 1, we first discuss its basic idea which follows from Observations 1 and 2.

OBSERVATION 1. *For any given x, let $y = x + r \mod N$, where r is a random number from Z_N. Here the relation between y and r depends on whether $x + r \mod N$ leads to an overflow (i.e., $x + r$ is greater than N) or not. We observe that y is always greater than r if there is no overflow. Similarly, in case of overflow, y is always less than r.*

The above observation is clear from the fact that $y = x + r$ if there is no overflow. On the other hand, if there is an overflow, then $y = x + r - N$.

Algorithm 3 Encrypted_LSB$(T, i) \rightarrow E(x_i)$

Require: Bob has T from current iteration i

1: Bob:

(a). $Y \leftarrow T * E(r) \bmod N^2$, where r is random in $\mathbb{Z}_N$

(b). Send Y to Alice

2: Alice:

(a). Receive Y from Bob

(b). $y \leftarrow D(Y)$

(c). **if** y is even **then** $\alpha \leftarrow E(0)$
else $\alpha \leftarrow E(1)$

(d). Send α to Bob

3: Bob:

(a). Receive α from Alice

(b). **if** r is even **then** $E(x_i) \leftarrow \alpha$
else $E(x_i) \leftarrow E(1) * \alpha^{N-1} \bmod N^2$

(c). return $E(x_i)$

OBSERVATION 2. *For any given $y = x + r \bmod N$, where N is odd, the following property regarding the least significant bit of x (i.e., x_0) always hold:*

$$x_0 = \begin{cases} \lambda_1 \oplus \lambda_2 & \text{if } r \text{ is even} \\ 1 - (\lambda_1 \oplus \lambda_2) & \text{otherwise} \end{cases}$$

Here λ_1 denotes whether an overflow occurs or not, and λ_2 denotes whether y is odd or not. That is, $\lambda_1 = 1$ if $r > y$ (i.e., overflow), and 0 otherwise. Similarly, $\lambda_2 = 1$ if y is odd, and 0 otherwise. Observe that $1 - (\lambda_1 \oplus \lambda_2)$ is the negation of bit $\lambda_1 \oplus \lambda_2$. Also note that N is always odd in the Paillier cryptosystem.

EXAMPLE 1. *Consider an even integer x, i.e., $x_0 = 0$. Without loss of generality, suppose r be an odd integer such that $x + r \bmod N$ causes an overflow. Then, $y = x + r \bmod N$ is always an even integer. That is $\lambda_2 = 0$. Also, $\lambda_1 = 1$ since there is an overflow. As r is odd, following from Observation 2, we have* $1 - (\lambda_1 \oplus \lambda_2) = 1 - (1 \oplus 0) = 0 = x_0$. □

By using the properties of Observation 2, Stage 1 computes the encryption of least significant bit of T by iteratively updating T to the encryption of corresponding quotient. The main steps involved in Stage 1 are shown as steps 2 to 7 in Algorithm 2. Stage 1 utilizes a sub-protocol (denoted by Encrypted_LSB) to compute the encryption of least significant bit from current T. The overall steps involved in the Encrypted_LSB protocol are shown in Algorithm 3.

To start with, in the first iteration, Bob randomizes the value of T (note that initially $T = E(x)$) by computing $Y = T * E(r) \bmod N^2$, and sends Y to Alice[2]. Upon receiving, Alice decrypts it to get $y = D(Y)$, where $y = x + r \bmod N$. If y is odd, Alice computes $\alpha = E(1)$, else she computes $\alpha = E(0)$, and sends it to Bob. Observe that $\alpha = E(\lambda_2)$. Also it is possible to securely compute $E(\lambda_1)$ using existing secure integer comparison protocol [8] on y (known only to Alice) and r (known only to Bob). However, though this construction is better than BITREP gate, we observe that the efficiency gains are not that significant. Therefore, in our SBD$_\text{p}$ protocol we decided to use a probabilistic assumption. That is, Bob simply assumes $\lambda_1 = 0$ (no overflow). We emphasize that though Bob assumes no overflow, $y = x + r \bmod N$ can still have overflow which depends on the actual values of x and r. In the later parts of this section, we show that for many practical applications, the above probabilistic assumption is reasonable.

[2] In Paillier's cryptosystem, ciphertext multiplications are followed by modulo N^2 operation so that the resulting ciphertext is in $\mathbb{Z}_{N^2}$.

Algorithm 4 SVR$(E(x), \langle E(x_0), \ldots, E(x_{m-1})\rangle) \rightarrow \gamma$

Require: Bob has $E(x)$ and $\langle E(x_0), \ldots, E(x_{m-1})\rangle$

1: Bob:

(a). $U \leftarrow \prod_{i=0}^{m-1}(E(x_i))^{2^i} \bmod N^2$

(b). $V \leftarrow U * E(x)^{N-1} \bmod N^2$

(c). $W \leftarrow V^{r'} \bmod N^2$, where r' is random in $\mathbb{Z}_N$

(d). Send W to Alice

2: Alice:

(a). Receive W from Bob

(b). **if** $D(W) = 0$ **then** $\gamma \leftarrow 1$
else $\gamma \leftarrow 0$

(c). Send γ to Bob

Once Bob receives α from Alice, he computes $E(x_i)$ depending on whether r is even or odd as below.

- If r is even, then $E(x_0) = \alpha = E(\lambda_2)$.
- Else $E(x_0) = E(1) * \alpha^{N-1} \bmod N^2 = E(1 - \lambda_2)$

Since λ_1 is assumed to be 0, following from Observation 2, we have $\lambda_1 \oplus \lambda_2 = \lambda_2$. Also, note that $N - 1$ is equivalent to -1 under $\mathbb{Z}_N$. After this, Bob updates T to $E(q_0) = E(\lfloor \frac{x}{2} \rfloor)$ by performing following homomorphic additions:

- $Z = T * E(x_0)^{N-1} \bmod N^2 = E(x - x_0)$
- $T = Z^l \bmod N^2 = E((x - x_0) * 2^{-1}) = E(q_0)$

where $l = 2^{-1} \bmod N$. The main observation is that $x - x_0$ is always a multiple of 2; therefore, $(x - x_0) * 2^{-1}$ always gives the correct quotient under $\mathbb{Z}_N$. The above process is continued iteratively such that in iteration i, Bob computes $E(x_i)$ and updates T to $E(q_i)$, for $0 \leq i \leq m - 1$.

Stage 2 - Secure Verification of Result: Since Bob assumes $\lambda_1 = 0$ during each iteration of Stage 1, there is a possibility for the wrong outcome. Though this probability is very low for many practical applications, as will be shown later, it is better to provide a mechanism to mitigate this effect. Along this direction, Stage 2 securely verifies whether the result of Stage 1 is correct and takes appropriate steps to compute the correct result (only in the case of wrong result from Stage 1).

The overall steps involved in Stage 2 of SBD$_\text{p}$ are highlighted as steps 8 to 13 in Algorithm 2. As shown in step 8 of Algorithm 2, Stage 2 utilizes a sub-protocol that performs the secure verification of the result (denoted by SVR). The main steps involved in SVR are given in Algorithm 4. Initially, Bob computes the encrypted integer corresponding to the encrypted bits computed in Stage 1 and performs the following homomorphic operations.

- $U = E(x_0 + \ldots + 2^{m-1} * x_{m-1})$
- $V = E(x_0 + \ldots + 2^{m-1} * x_{m-1} - x)$
- $W = V^{r'} = E(r' * (x_0 + \ldots + 2^{m-1} * x_{m-1} - x))$

Table 1: Complexity comparison of existing Paillier based SBD algorithm with our work

Method	Rounds	Computations	Communications
BITREP gate [14]	$O(K)$	$21K+2$ encryptions and $179K$ exponentiations	$O(K^2)$ bits
This paper	$O(m)$	$3m+1$ encryptions and $4m+2$ exponentiations	$O(m*K)$ bits

After this, Bob sends W to Alice. Upon receiving W, Alice decrypts it and sets γ to 1 (denoting the result from Stage 1 is correct) if $D(W)=0$, else γ is set to 0 (denoting the result is incorrect), and sends γ to Bob. Observe that $\gamma=0$ iff $x=x_0+\ldots+2^{m-1}*x_{m-1}$, i.e., result of Stage 1 is correct. Note that $D(W)$ is either 0 or random; therefore, the above process does not reveal any information about x and x_i's to Alice. Finally, if $\gamma=1$, Bob terminates the SBD_p protocol implying the result of Stage 1 is correct. Else, Bob initiates Stage 1 of SBD_p (i.e., go to Step 2 of Algorithm 2) and the above process is repeated until the result is correct.

EXAMPLE 2. *Suppose $x=5$ and $m=3$. We show various intermediate results during the execution of Stage 1 in the proposed protocol. Initially T is set to $E(5)$. We assume that r is even and there is no overflow. Note that r is different in each iteration.*

Iteration 0:

$$\begin{aligned} y &= 5+r \mod N = an\ odd\ integer \\ E(x_0) &= \alpha = E(1) \\ T &= E((5-1)*2^{-1}) = E(2) \end{aligned}$$

Iteration 1:

$$\begin{aligned} y &= 2+r \mod N = an\ even\ integer \\ E(x_1) &= \alpha = E(0) \\ T &= E((2-0)*2^{-1}) = E(1) \end{aligned}$$

Iteration 2:

$$\begin{aligned} y &= 1+r \mod N = an\ odd\ integer \\ E(x_2) &= \alpha = E(1) \\ T &= E((1-1)*2^{-1}) = E(0) \end{aligned}$$

The output of Stage 1 is $\langle E(1), E(0), E(1)\rangle$. □

4.1 Correctness

We emphasize that the final encrypted bits computed from the SBD_p protocol are always correct. This is because of the secure verification step (i.e., Stage 2) in SBD_pwhich makes sure that the protocol terminates only if the output of Stage 1 is correct. It is possible for the SBD_p protocol to generate the correct output after many runs of Stage 1 & 2 (which depends on x and random r chosen in each iteration of Stage 1). Therefore, a natural question to ask is whether there is an upper bound on the number of runs in SBD_p to produce the correct output. However, the proposed protocol produces the correct output in the first run of Stage 1 and 2 for many practical applications. We theoretically prove this below.

In general, many secure data analysis tasks such as secure clustering and classification has m much smaller than K. During Stage 1, in each iteration, r can take any value in $\mathbb{Z}_N$. We observe that if $r\in[N-2^m, N)$, only then the corresponding computed (encrypted) bit can be wrong (due to overflow). That is, the number of possible values of r that can give rise to error are 2^m. Since we have N number of possible values for r, the probability for producing wrong bit is $\frac{2^m}{N}\approx\frac{1}{2^{K-m}}$. Therefore, the probability for computing the encryption of x_i correctly is approximately $1-\frac{1}{2^{K-m}}$. This probability remains same for all the bits since r is chosen independently in each iteration. Hence, the probability for Stage 1 to compute the correct output is given by:

$$\left(1-\frac{1}{2^{K-m}}\right)^m \approx e^{-\frac{m}{2^{K-m}}}$$

In general, for many real-world applications, m can be at most 100. Therefore, for 1024-bit key size (i.e., $K=1024$), the probability for SBD_p to produce correct output in the first run is approximately $e^{-\frac{100}{2^{924}}}\approx 1$. Hence, for practical domain values of x, with a probability of almost 1, the SBD_p protocol gives the correct output in the first run itself. Based on the above analysis, it is clear that Stage 2 is not required in practice. We emphasize that even in the extreme case such as $m=950$, the probability for SBD_p to produce correct output in the first run is $e^{-\frac{950}{2^{74}}}\approx 1$.

4.2 Security Analysis

In the SBD_p protocol, the communication between Alice and Bob occurs both in Stage 1 and 2. During Stage 1, in each ith iteration, Bob randomizes current T (from step 1(a) of Algorithm 3) and sends it to Alice. Therefore, the decryption of Y by Alice (from step 2(b) of Algorithm 3) yields a random value y which is uniformly distributed in $\mathbb{Z}_N$. Thus, no information about x is revealed to Alice. Also, depending on whether y is even or odd, Alice sends $\alpha=E(0)$ or $E(1)$ to Bob. Here, the ciphertext α is uniformly random in $\mathbb{Z}_{N^2}$ and does not provide any information regarding the corresponding plaintext to Bob due to the semantic security of Paillier cryptosystem [12]. Note that, after computing $E(x_{i-1})$ in the ith iteration, the process of updating T is done locally by Bob, for $1\leq i\leq m$. Hence, it is clear that no information related to x is revealed to Alice and Bob during Stage 1 of SBD_p.

During Stage 2, Bob sends $W=V^{r'} \mod N^2$ to Alice (step 1(d) of Algorithm 4), where r' is a random number in $\mathbb{Z}_N$ known only to Bob. Since the decryption of W by Alice gives either 0 or a uniformly random value in $\mathbb{Z}_N$, no information is revealed to Alice and Bob except whether the output of Stage 1 is correct or not. Therefore, Stage 2 is secure under the assumption (which is also practical) that the output γ can be revealed to both parties. We emphasize that revealing γ does not leak any information regarding x to Alice and Bob. In addition, the intermediate results which are passed as inputs from Stage 1 to Stage 2 are pseudo-random. Hence, by composition theorem [6], the sequential composition of the two stages lead to a secure protocol.

4.3 Complexity Analysis

The comparison results of our approach with the BITREP gate are as shown in Table 1. As mentioned in [14], the BITREP gate utilizes an addition and subtraction circuit whose round complexities are bounded by $O(K)$, where K is the encryption key size. Also, the overall number of computations involved in the BITREP gate are roughly $21K+2$ encryptions and $179K$ exponentiations under the assumption that Paillier's encryption and decryption operations take similar amount of time. The communication complexities of addition and subtraction circuits, as mentioned in [14], are bounded by $O(K^2)$ bits. Therefore, the overall communication complexity of BITREP gate is bounded by $O(K^2)$ bits.

On the other hand, for the SBD_p protocol, Stage 1 involves m number of iterations, where each iteration requires one round of communication between Alice and Bob. Whereas, Stage 2 requires one round of communication between Alice and Bob. As shown earlier, for many practical applications, the probability of getting the correct result in the first instantiation of the SBD_p protocol is very high. Therefore, we assume that Stage 1 and 2 are run only once. Thus, the round complexity of SBD_p is bounded by $O(m)$.

During Stage 1 of SBD_p, in each iteration, Bob has to randomize current T which requires one encryption and one homomorphic addition operations. Alice further performs one decryption and one encryption operations (at step 2(b) and 2(c) of Algorithm 3). Also, if r is odd, Bob has to perform one exponentiation and one homomorphic addition operations (at step 3(b) of Algorithm 3). In addition, Bob has to perform two exponentiation and one homomorphic addition operations while updating T in each iteration. We emphasize that the values of $l = 2^{-1} \mod N$ and $E(1)$ (step 3(b) of Algorithm 3) are pre-computed and can be re-used in each iteration. Therefore, Stage 1 requires $3m$ encryptions (assuming that encryption and decryption under Paillier take almost similar time), $3m$ exponentiation, and $3m$ homomorphic addition operations. In general, the cost of $3m$ homomorphic additions is negligible compared to the cost of $3m$ exponentiations. Therefore, the computation cost of Stage 1 is bounded by $3m$ encryptions and $3m$ exponentiations.

Furthermore, Stage 2 of SBD_p requires one decryption, $m + 2$ exponentiation, and m homomorphic addition operations. Hence, the total computation cost of SBD_p (one instantiation) is roughly $3m + 1$ encryptions and $4m + 2$ exponentiations.

The communication complexity of our protocol entirely depends on Stage 1. During Stage 1, in each iteration, Bob sends encrypted value Y to Alice. Also, Alice sends an encrypted value α to Bob. Since we have m iterations, the total communication complexity of SBD_p is bounded by $O(m * K)$ bits. Based on the above analysis, when $m < K$, it is clear that the complexities of SBD_p are significantly much better than those of BITREP. However, when $m = K$ (i.e., $x \in \mathbb{Z}_N$), the communication and round complexities of our protocol are the same as in BITREP. Nevertheless, in either case, SBD_p is more efficient than BITREP as shown in Table 1.

5. CONCLUSION AND FUTURE WORK

Secure bit-decomposition (SBD) acts as an important primitive in various secure computation operations such as comparison and public modulo. The existing SBD protocol is inefficient and also requires a number of communication rounds bounded by $O(K)$, where K is the encryption key size. Along this direction, this paper presented a new efficient and probabilistic SBD protocol, referred to as SBD_p, to securely convert a Paillier encrypted value of x, where $0 \leq x < 2^m$, into encryptions of the individual bits of x.

The proposed protocol always generates the correct result. However, we theoretically showed that the probability for SBD_p to generate the correct output in the first run itself is almost 1 for many practical applications. In addition, we analyzed the complexities of the proposed protocol with the existing secure approach (i.e., BITREP gate [14]). We emphasize that, apart from the efficiency gain, the round and communication complexities of our protocol are improved by a factor of $\frac{K}{m}$ compared to BITREP. As a future work, we will focus on extending our protocol to multi-party case.

Acknowledgment

This material is based upon work supported by the Office of Naval Research under Award No. N000141110256 and NSF under award No. CNS-1011984.

6. REFERENCES

[1] J. Algesheimer, J. Camenisch, and V. Shoup. Efficient computation modulo a shared secret with application to the generation of shared safe-prime products. In *CRYPTO*, pages 417–432. Springer-Verlag, 2002.

[2] I. F. Blake and V. Kolesnikov. One-round secure comparison of integers. *Journal of Mathematical Cryptology*, 3(1):37–68, May 2009.

[3] R. Cramer, I. Damgård, and J. B. Nielsen. Multiparty computation from threshold homomorphic encryption. In *EUROCRYPT*, pages 280–299. Springer-Verlag, 2001.

[4] I. Damgård, M. Fitzi, E. Kiltz, J. B. Nielsen, and T. Toft. Unconditionally secure constant-rounds multi-party computation for equality, comparison, bits and exponentiation. In *Theory of Cryptography Conference (TCC)*, volume 3876, pages 285–304. Springer, 2006.

[5] M. Fischlin. A cost-effective pay-per-multiplication comparison method for millionaires. In *Proceedings of the Conference on Topics in Cryptology: The Cryptographer's Track at RSA*, pages 457–472. Springer-Verlag, 2001.

[6] O. Goldreich. *The Foundations of Cryptography*, volume 2. Cambridge University Press, 2004.

[7] J. Guajardo, B. Mennink, and B. Schoenmakers. Modulo reduction for paillier encryptions and application to secure statistical analysis. In *Financial Cryptography and Data Security*, volume 6052, pages 375–382. Springer, 2010.

[8] A. E. Nergiz, M. E. Nergiz, T. Pedersen, and C. Clifton. Practical and secure integer comparison and interval check. In *Proceedings of Second International Conference on Social Computing*, pages 791–799. IEEE Computer Society, 2010.

[9] C. Ning and Q. Xu. Constant-rounds, linear multi-party computation for exponentiation and modulo reduction with perfect security. In *ASIACRYPT*, volume 7073 of *LNCS*, pages 572–589. Springer, 2011.

[10] R. Nisbet, J. Elder, and G. Miner. *Handbook of Statistical Analysis and Data Mining Applications*. Academic Press, 2009.

[11] T. Nishide and K. Ohta. Multiparty computation for interval, equality, and comparison without bit-decomposition protocol. In *PKC*, pages 343–360. Springer-Verlag, 2007.

[12] P. Paillier. Public key cryptosystems based on composite degree residuosity classes. In *Advances in Cryptology - Eurocrypt '99*, pages 223–238. Springer-Verlag, 1999.

[13] T. Reistad and T. Toft. Linear, constant-rounds bit-decomposition. In *ICISC*, pages 245–257. Springer-Verlag, 2010.

[14] B. Schoenmakers and P. Tuyls. Efficient binary conversion for paillier encrypted values. In *EUROCRYPT*, pages 522–537. Springer-Verlag, 2006.

[15] A. Shamir. How to share a secret. *Communications of the ACM*, 22(11):612–613, nov 1979.

[16] T. Toft. Constant-rounds, almost-linear bit-decomposition of secret shared values. In *Proceedings of the The Cryptographers' Track at the RSA Conference on Topics in Cryptology*, pages 357–371. Springer-Verlag, 2009.

[17] A. C. Yao. Protocols for secure computations. In *Symposium on Foundations of Computer Science*, pages 160–164, Washington, DC, USA, 1982. IEEE Computer Society.

[18] A. C. Yao. How to generate and exchange secrets. In *Symposium on Foundations of Computer Science*, pages 162–167. IEEE Computer Society, 1986.

Defining Verifiability in e-Auction Protocols

Jannik Dreier
Université Grenoble 1, CNRS
jannik.dreier@imag.fr

Hugo Jonker
University of Luxembourg
hugo.jonker@uni.lu

Pascal Lafourcade
Université Grenoble 1, CNRS
pascal.lafourcade@imag.fr

ABSTRACT

An electronic auction protocol will only be used by those who trust that it operates correctly. Therefore, e-auction protocols must be *verifiable*: seller, buyer and losing bidders must all be able to determine that the result was correct. We pose that the importance of verifiability for e-auctions necessitates a formal analysis. Consequently, we identify notions of verifiability for each stakeholder. We formalize these and then use the developed framework to study the verifiability of two examples, the protocols due to Curtis et al. and Brandt, identifying several issues.

Categories and Subject Descriptors

C.2.2 [**Network Protocols**]: Protocol verification; K.4.4 [**Electronic Commerce**]: Security

Keywords

Verifiability; e-Auction; Formal Methods; Protocol Analysis

1. INTRODUCTION

Auctions provide sellers and buyers with a way to exchange goods for a mutually acceptable price. Unlike a marketplace, where the sellers compete with each other, auctions are a seller's market where *buyers* bid against each other over the goods for sale. There are many different types of auctions, varying how to determine winner and price. For example, in an English auction, buyers bid publicly against each other and the highest bid wins (e.g. [16]). A Vickrey auction is rather similar, except that the winning buyer pays the price of the second-highest bid (see e.g. [10, 14]). Conversely, in a Dutch auction, the price starts too high – the auctioneer keeps lowering the price until a buyer claims the good for that price (e.g. [17]). Sealed-bid auctions are auctions that run in one round: in the bidding phase, each buyer submits one sealed bid, which are then simultaneously opened in the opening phase. Sealed-bid auctions can be used to implement auctions where the bidder with the highest bid wins (e.g. [4, 15, 1, 5]), but also Vickrey-style auctions, where the winner pays the second-highest price (e.g. [14]). In this paper, we focus on verifiability of sealed-bid auctions (though our results are applicable to English and, in a limited sense, Dutch auctions).

Auctions involve the following stakeholders:

- *Bidders*: prospective buyers, who want to pay as little as possible.
- *Seller*: a seller, striving to sell for as much as possible.
- *Auctioneer*: the party organizing the auction.

As is readily apparent, the interests of the various stakeholders are opposed. Buyers are in competition with each other for the goods on sale, sellers are in competition with buyers for the price of the goods, the auctioneer may profit directly from overvalued sale price (which provides an incentive to collude with the seller), but a reputation for undervalued sale prices will ensure many repeat customers (which provides incentive to collude with buyers). Consider e.g. the case where there are several bids for the same price. In such a case, an auctioneer might prefer the most "active" bidder instead of the normal tie-breaking rules, and so favor frequent customers over occasional ones.

There are thus no impartial parties to oversee the correctness of the process to determine selling price and winner. For this reason, an auction system must provide some form of *verifiability* for each involved party – irrespective of how the auction process is run and the winner is determined.

Auction verifiability is easy to achieve in isolation, as happens in English "shout-out" auctions. However, maintaining verifiability while ensuring other properties (non-repudiation, privacy, etc.) is far harder. Too often, newly proposed auction protocols proudly show how they achieve these other properties, while only acknowledging the requirement for verifiability in passing. Typically, verifiability is subsequently claimed without providing any formal proof, e.g., [3, 6]. To address this, we propose a generic formal framework applicable independent of the type of auction. The framework consists of formal tests of verifiability.

Contribution. The main contribution of this work is to identify a set of scheme-independent definitions, which, taken together, cover verifiability of auctions. To this end, we focus on the bidders (distinguishing verifiability for losing bidders from the winning bidder) and the seller. We present this framework as a set of formal verifiability tests. Moreover, we investigate the auction protocols by Curtis et al. [6] and by Brandt [3]. Both protocols claim verifiability, yet we identify issues with each.

ASIA CCS'13, May 8–10, 2013, Hangzhou, China.

Related work. There are relatively few formal analyses of auction protocols. Dong et al. [7] study privacy properties of the protocol by Abe and Suzuki [1] in the applied π-calculus. More recently, Dreier et al. [8] used the applied π-calculus to formalize several properties (privacy, non-repudiation, non-cancellation and fairness, but not verifiability) for auction protocols, and studied (and found problems with) two auction protocols. Besides these, verifiability in auctions has (to the best of our knowledge) only been studied for particular schemes. However, in the field of voting several more generic definitions of verifiability have emerged, and we look there for inspiration.

In voting, the property of *individual verifiability* – a voter can verify that her vote counts correctly for the result – has been a well-established notion since the field's inception [9, 2, 18, 11]. Sako and Killian [18] introduced the concept of *universal verifiability*: the property that any observer may verify (using only public information) the correctness of the result. Kremer et al. [12] introduced the notion of *eligibility verifiability*: the property whereby any observer may verify, using only public information, that the set of cast votes from which the result is determined originates only from eligible voters, and each eligible voter cast at most one vote. Finally, Küsters et al. [13] introduced the notion of *accountability*: when verifiability fails, it is possible to identify the person responsible for the failure.

While the intuition behind these notions carries (to some degree) over to auctions, we do note, that unlike voting, auctions involve *competing* parties – an illegal bid (e.g., by the seller) may increase the winning price, while not changing the winner. Hence, a lack of verifiability in auctions can compromise fairness. In addition, winner/price determination can be quite complex depending on the type of auction, e.g. for second-price or multi-good auctions. Thus, verification of voting systems does not translate directly to verification of auction systems.

Outline. In Section 2, we describe our modeling of auctions. In Section 3, we formalize the verifiability definitions, taking into account the point of view of the seller and the bidders. In Section 4, we then apply our model to two examples and exhibit problems in both cases.

2. MODELING AUCTION PROTOCOLS

We consider a set of bidders $\mathcal{B}$ and a seller S. We do not model other parties as only bidders and the seller verify the execution of the protocol.

Bids are of type *Bid* (in the simplest case just a price). When being submitted the bids might be encrypted or anonymized to ensure privacy, hence we use the type *EBid* for such bids. We assume that there is a public list $\mathbb{L}$ of length n and type *List*(*EBid*) of all submitted bids, for example a bulletin board. To define the soundness of the verification tests we need a mapping between both types, i.e. a function $getPrice\colon EBid \mapsto Bid$ that gives the bid for an encrypted bid. This function does not need to be computable for any party, as it is only used in the soundness definition.

Bidders have to register at some point, or are otherwise authenticated when bidding, in order to be able to obtain their goods once the auction has ended. This could for example be implemented using signatures, authentication tokens, MACs etc. Therefore we require a function $isReg\colon EBid \mapsto \{true, false\}$ that returns *true* if a bid was submitted by a registered bidder, and not modified – this integrity protection is necessary to prevent manipulation of bids.

In addition, we require a public function that - given a list of bids - computes the index of the winning bid within the list of all bids: $win\colon List(Bid) \mapsto Index$. This might simply be the index of the maximal bid among all bids, but there may be more complex operations to determine this index depending on the type of auction or to deal with ties (i.e. several maximal bids).

Lastly, we assume that the variable *winBid* of type *Index* refers to the index of the announced winning bid at the end of the auction, and that each bidder has a variable *myBid* of type *Index* that refers to the index of his bid in $\mathbb{L}$.

Note that for a list l we write $l[i]$ to denote the i-th element of the list starting with 1, and $Indices(l)$ to denote the set of indices of l, i.e. $\{1, \dots, n\}$ if l contains n elements.

DEFINITION 1. *An auction protocol is a tuple* ($\mathcal{B}$, S, $\mathbb{L}$, *getPrice*, *isReg*, *win*, *winBid*) *where*

- $\mathcal{B}$ *is the set of bidders,*
- S *is the seller,*
- $\mathbb{L}$ *is a list of all submitted bids,*
- $getPrice\colon EBid \mapsto Bid$ *is a function mapping submitted bids to bids,*
- $isReg\colon EBid \mapsto \{true, false\}$ *is a function that returns true if a bid was submitted by a registered bidder,*
- $win\colon List(Bid) \mapsto Index$ *is a function that returns the index of the winning bid,*
- *winBid is a variable referring to the index of the winning bid at the end of the auction.*

3. DEFINING VERIFIABILITY

In this section, we formally define verifiability for auction protocols. In the first part we consider only first-price auctions. Thereafter we generalize the definitions to account for second-price, multi-price, and other types of auctions.

3.1 First-Price Auctions

To understand which verifications are needed, we start by discussing three different stakeholder's perspectives:

- A *losing bidder* wants to be convinced that he actually lost, i.e. that:
  - the winning bid was actually superior to his bid (as defined by the *win* function), and
  - that the winning bid was submitted by another bidder (preventing both seller and auctioneer from maliciously adding or manipulating bids to influence the final price).
- A *winning bidder* wants to check that:
  - he actually submitted the winning bid,
  - the final price is correctly computed,
  - all other bids originated from bidders, and
  - no bid was modified.

Together, these verification checks ensure that the winning bidder is indeed the correct winner, for the correct price. Moreover, the last two checks ensure that the auction process was only influenced by legitimate bidders – neither seller nor auctioneer influenced the process.

- The *seller* wants to verify that:
  - the announced winner is correct, and
  - the winning price is correct,

  in particular if the outcome of the auction was not determined publicly (e.g. privately by the auctioneer, or using distributed computations among the bidders).

To execute these verifications, we introduce the notion of *Verification Tests*.

We define a *Verification Test* as an efficient terminating algorithm that takes as input the data visible to a participant of an auction protocol and returns a Boolean value.

We deliberately do not specify more details at this point as they will depend on the underlying protocol model. Such a test could be a logical formula (whose size is polynomial in the input) in a symbolic model or a polynomial-time Turing-machine in a computational model. Obviously there can be different tests for different participants (e.g. for bidders and the seller), since they may have different views of the protocol execution.

We define verifiability as follows.

DEFINITION 2 (VERIFIABILITY - 1ST-PRICE AUCTIONS). *An auction protocol* $(\mathcal{B}, S, \mathbb{L}, getPrice, isReg, win, winBid)$ *ensures* Verifiability *if we have Verification Tests* rv_s, rv_w, ov_l, ov_w, ov_s *respecting the following soundness conditions:*

1. *Registration and Integrity Verifiability (RV):*
   - *Anyone can verify that all bids on the list were submitted by registered bidders:*
     $rv_s = true \implies \forall b \in \mathbb{L}\colon isReg(b) = true$
   - *Anyone can verify that the winning bid is one of the submitted bids:*
     $rv_w = true \implies winBid \in Indices(\mathbb{L})$
2. *Outcome Verifiability (OV):*
   - *A losing bidder can verify that his bid was not the winning bid:*
     $ov_l = true \implies myBid \neq win(getPrice(\mathbb{L}))$
   - *A winning bidder can verify that his bid was the winning bid:*
     $ov_w = true \implies myBid = win(getPrice(\mathbb{L}))$
   - *The seller can verify that the winning bid is actually the highest submitted bid:*
     $ov_s = true \implies winBid = win(getPrice(\mathbb{L}))$

as well as the following completeness condition:

- *If all participants follow the protocol correctly, the above tests succeed (i.e., the implications hold in the opposite direction,* $\Longleftarrow$*, as well).*

where – with abuse of notation – we write $getPrice(\mathbb{L})$ *for* $getPrice(\mathbb{L}[1]), \ldots, getPrice(\mathbb{L}[n])$.

Consider the perspective of a losing bidder: He can verify that his bid was not the winning bid (ov_l), and that the winning bid was among the ones submitted by registered bidders, which were also not modified (rv_s and rv_w). Similarly a winning bidder can check that his bid was actually the winning bid (ov_w), and that the other bids were submitted by other bidders and not modified (rv_s). Lastly, the seller can also check that the bids using for computing the winner were submitted only by registered bidders (rv_s and rv_w), and that the outcome was correct (ov_s). Hence these tests cover all the verifications discussed above.

In the case of soundness, we require the conditions to hold even in the presence of malicious participants (since the tests should check if they did their work correctly), whereas in the case of completeness we only consider honest participants. This is necessary as otherwise e.g. a dishonest auctioneer could announce the correct result, but publish incorrect evidence. Hence the verification tests fail although the outcome is correct, but this acceptable since the auctioneer did not "work correctly" in the sense that he deviated from the protocol specification.

Definition 2 can be applied to sealed-bid auctions, where all bids are submitted in a private way, as well as English auctions, where the price increases with each publicly announced bid. These latter are verified by applying the verification tests after each price increase.

Example.

Consider a simple auction system where all bidders publish their (not encrypted and not signed) bids on a bulletin board, and at the end of the bidding phase the auctioneer announces the winner. In this case there is a simple test for rv_w: anyone can simply test if the winning bid is one of published ones. However there is no test for rv_s since bids are not authenticated. If we require bidders to sign their bids before publishing them, we also have a simple test for rv_s: verifying the signatures.

It is clear that we have simple tests for ov_l, ov_w and ov_s since everybody can compute the winner on the public list of unencrypted bids. This however means that the protocol ensures no privacy, and no fairness since a bidder can chose his price depending on the previously submitted bids. If we add encryption for the bids to address this shortcoming, the situation becomes more complex and the auctioneer has to prove that he actually computed the winner correctly, for example using zero-knowledge proofs.

3.2 Other Types of Auctions

Our definition can be extended to other auctions, including second-price auctions, more general $(M+1)$st-price auctions, and even bulk-good auctions that have multiple winners at different prices. The price in these types of auctions may also depend on the other submitted bids – not only on the winning bid. To deal with this, we enrich our model of an auction protocol with a type *Price*. The function *win* now returns lists of winners and prices $win\colon List(Bid) \mapsto List(Index) \times List(Price)$. We also assume that there are two variables *winPrice* and *myPrice* instantiated as the announced list of winning prices and the price announced to a winning bidder respectively. Similarly *winBid* is now instantiated as a list of indices of bids.

For such auctions, registration verifiability does not change, but winner(s) and seller also want to verify the price they pay to prevent a malicious party from increasing price(s).

DEFINITION 3 (GENERALIZED VERIFIABILITY). *An auction protocol* $(\mathcal{B}, S, \mathbb{L}, getPrice, isReg, win, winBid, winPrice)$ *ensures* Verifiability *if we have Verification Tests* rv_s, rv_w, ov_l, ov_w, ov_s *respecting the following conditions:*

1. *Soundness:*

   (a) *Registration and Integrity Verifiability (RV):*

   - *Anyone can verify that all bids on the list were submitted by registered bidders:* $rv_s = true \implies \forall b \in \mathbb{L}: isReg(b) = true$
   - *Anyone can verify that the winning bids are among the submitted bids:* $rv_w = true \implies$ $\boldsymbol{\forall b \in winBid}: b \in Indices(\mathbb{L})$

   (b) *Outcome Verifiability (OV):* **Let** $\boldsymbol{(indexes, prices) = win(getPrice(\mathbb{L}))}$.

   - *A losing bidder can verify that his bid was not the winning bids:* $ov_l = true \implies \boldsymbol{myBid \notin indexes}$
   - *A winning bidder can verify that his bid was among the winning bids, and that his price is correct:* $ov_w = true \implies \boldsymbol{\exists i}: \boldsymbol{(myBid = indexes[i]} \wedge \boldsymbol{myPrice = prices[i])}$
   - *The seller can verify that the list of winners and the winning prices are correctly determined:* $ov_s = true \implies$ $\boldsymbol{(winBid = indexes \wedge winPrice = prices)}$

2. *Completeness: If all participants follow the protocol correctly, the above tests succeed (i.e., the implications hold in the opposite direction,* $\Longleftarrow$*, as well).*

where – with abuse of notation – we write $getPrice(\mathbb{L})$ *for* $getPrice(\mathbb{L}[1]), \ldots, getPrice(\mathbb{L}[n])$. *Differences to Def. 2 are marked in* **bold**.

Note that e.g. in the case of a second-price auction verifying the price, for example in test ov_w, may implicitly include some more registration verification, namely checking that the second-highest bid was actually submitted by a bidder. Otherwise a malicious seller could add a higher second-highest bid or manipulate the existing one to achieve a higher selling price. This is however included in our model as the function *win* only works on the list $\mathbb{L}$, hence adding another bid later on to manipulate the bidding price violates the test, and adding or manipulating a bid in $\mathbb{L}$ violates rv_s.

4. CASE STUDIES

In this section, we discuss two case studies: The protocols by Curtis et al. [6] and Brandt [3].

4.1 Protocol by Curtis et al. [6]

The protocol by Curtis et al. [6] was designed to support any type of sealed-bid auction while guaranteeing fairness, privacy, verifiability and non-repudiation.

4.1.1 Informal Description

The main idea of the protocol is the following: using a public-key infrastructure (PKI), the bidders register with a trusted Registration Authority (RA), who issues pseudonyms that will then be used for submitting bids to the Seller (S). The seller eventually receives all bids in clear and can hence apply any possible *win* function. However, he can only link bids to pseudonyms, not to bidders. The protocol is split into three phases: Registration, Bidding, and Winner determination.

- *Registration:* Each bidder sends his identity, a hash of his bidding price b_i and a signature of $h(b_i)$ to the RA. The RA checks the identity and the signature using the PKI, and replies with an encrypted and signed message containing a newly generated pseudonym p and the hashed bid $h(b_i)$.
- *Bidding:* The RA generates a new symmetric key k. Each bidder will send $c = Enc_{pk_S}(b_i)$, his bid b_i encrypted with the seller's public key, and a signature of c, together with his pseudonym to the RA. The RA will reply with a signature on c, and encrypts the bidders message, together with the hashed bid $h(b_i)$ from phase one, using the symmetric key k. This encrypted message is then send to the seller.
- *Winner determination:* After all bids have been submitted, the RA will reveal the symmetric key k to the seller. The seller can then decrypt the bids, verify the correctness of the hash and determine the winner. To identify the winner using the pseudonym he can ask the RA to reveal the true identity.

4.1.2 Formal Model

We have the set of bidders $\mathcal{B}$ and a seller S. We do not need to specify the type of bids *Bid* since the protocol supports any type of bids. The bids are published when the auctioneer reveals the symmetric key, i.e. $\mathbb{L}$ contains bids of the following type: $(Pseudo \times PEnc(Bid) \times Hash)$, where $Pseudo$ is the type of pseudonyms, $PEnc$ is a public-key encryption and $Hash$ are hash values. The function *getPrice* will simply decrypt the encrypted bid (the second entry of the tuple). The function *isReg* will return *true* if and only if the hash value is correct, the pseudonym was actually attributed by the RA and the bid was submitted correctly signed by the bidder with this pseudonym. The protocol is independent of the used auction mechanism and hence does not define *win*. The seller will simply decrypt all bids and can then apply any function *win*. He will publish the winning price and the winning bidders pseudonym, and *winBid* will denote the index of the bid containing this pseudonym.

4.1.3 Analysis

Since the seller does the winner determination on his own, there is a simple test for ov_s: He can check his own computations. As the computation of the winner is not specified in order to support any type of auction, we cannot give tests for ov_l and ov_w – they would have to be designed as a function of the used auction algorithm. Yet there is also a test for rv_w: Checking if the pseudonym appears in the list of bids.

However, the messages from the RA to the seller are not authenticated, hence there can be no suitable tests for rv_s

once the (encrypted) bids are revealed. Even if they were authenticated, this still requires trusting the RA, since there is no way to verify if a pseudonym actually corresponds to a bidder. This also shows a simple attack: the RA can create a new pseudonym and submit a bid under this pseudonym, which may allow him to manipulate the auction outcome.

Curtis et al. explicitly state that the RA needs to be trusted. However, they themselves propose a property "robustness" which states that an auction protocol should be able to handle corrupt behavior. Clearly, their protocol is not robust to dishonest behavior by the RA, and verifiability of their protocol requires trust in the RA. We argue that basing verifiability tests on such a trust assumption at least partly contradicts the main point of verifiability, which is to eliminate such trust assumptions by providing evidence that a trusted party actually behaved honestly.

4.2 Protocol by Brandt [3]

The protocol by Brandt [3] realizes a first-price sealed-bid auction and was designed to ensure full privacy in a completely distributed way. It exploits the homomorphic properties of a distributed El-Gamal Encryption scheme for a secure multi-party computation of the winner.

4.2.1 Informal Description

The participating bidders and the seller communicate using a bulletin board, i.e. an append-only memory accessible for everybody. The bids are encoded as bit-vectors where each entry corresponds to a price. The protocol then uses linear algebra operations on the bid vectors to compute a function f_i, which returns a vector containing one entry "1" if bidder i submitted the highest bid, and different numbers ($\neq 1$) otherwise. To be able to compute this function in a completely distributed way, and to guarantee that no coalition of malicious bidders can break privacy, these computations are performed on the encrypted bids using homomorphic properties of a distributed El-Gamal encryption.

In a nutshell, the protocol realizes the following steps:

1. Firstly, the distributed key is generated: each bidder chooses his part of the secret key and publishes the corresponding part of the public key on the bulletin board.

2. Each bidder then computes the joint public key, encrypts his bid using this key and publishes it on the bulletin board.

3. Then function f_i is calculated for every bidder i using the homomorphic property of the encryption scheme.

4. The outcome of this computation (n encrypted values) are published on the bulletin board, and each bidder partly decrypts each value using his secret key.

5. These shares are sent to the seller, who can combine all to obtain the result (i.e. all f_i). He publishes part of the shares such that each bidder j can only compute his f_j to see if he won or lost (using his knowledge and the published shares), but not the other f_i.

4.2.2 Formal Model

We have a set of bidders $\mathcal{B}$ and a seller S. The list of all submitted bids $\mathbb{L}$ is published on the bulletin board. The function *getPrice*(C) decrypts the bid using the joint private key. The function *win* returns the index of the highest bid submitted, in case of ties the one submitted by the bidder with the smallest index. The protocol has two particularities: Firstly there is no registration (and hence no meaningful function *isReg*), and secondly the winner is not publicly announced – only the winning bidder and the seller know at the end who won. We can still assume that *winBid* gives the index of the winning bid, although only the seller and the winning bidder have access to it. We assume that there is a "magical" function *isReg* that can check if a bid was submitted by a registered bidder, however the absence of registration and authentication means that we cannot implement it.

4.2.3 Analysis

The protocol includes no authentication or registration, hence there is no suitable test for rv_s. An attacker may hence submit bids on behalf of a bidder, which cannot be detected using a verification test. Yet using the values published on the bulletin board everybody can check if the values used for the computation were the previously submitted bids, and as the winning index will be among them, we have a test for rv_w.

The author claims that the protocol is verifiable as the parties have to provide zero-knowledge proofs for their computations, however there are two problems.

Firstly a winning bidder cannot verify if he actually won. To achieve privacy, the protocol hides all outputs of f_i except for the entry containing "1". This is done by exponentiation of all entries x_i of the return vector x with random values, i.e. by calculating $x_i^{\sum_j r_j}$. If x_i is one, this will still return one, but a random value for any other value of x. Yet these random values r_j may add up to zero (mod q), hence the returned value will be $x_i^0 = 1$ and the bidder will conclude that he won ($x_i = 1$), although he actually lost ($x_i \neq 1$). Hence simply verifying the proofs is not sufficient – such a test ov_w would not be sound. For the same reason the seller might observe two or more "1"-values even though all proofs are correct, and will be unable to decide which bidder actually won. He could even exploit such a situation to his advantage: He can simply tell both bidders that they won and take money from both, although there is only one good to sell. If the bidders do not exchange additional data there is no way for them to discover that something went wrong, since the seller is the only party having access to all values. The probability of the random values adding up to zero is low, yet this means that there are cases where the verifiability tests are not sound.

Secondly the author does not exactly specify the proofs that have to be provided in the joint decryption phase. If the bidders only prove that they use the same private key on all decryptions (and not also that it is the one they used to generate their public key), they may use a wrong one. This will lead to a wrong decryption where with very high probability no value is "1", as they will be random. Hence all bidders will think that they lost, thus allowing a malicious bidder to block the whole auction, as no winner is determined. Hence, if we assume that ov_l consists in verifying the proofs, a bidder trying to verify that he lost using the proofs might perform the verification successfully, although the result is incorrect and he actually won – since he would have observed a "1" if the vector had been correctly decrypted. This problem can be addressed by requiring the bidders to also prove that they used the same private key as in the key generation phase.

5. CONCLUSION

In this work, we identified the types of verifiability necessary for the stakeholders in auctions. We then formalized these requirements in a protocol-independent way, resulting in tests $rv_s, rv_w, ov_l, ov_w, ov_s$, which together constitute a general verifiability framework for auction protocols.

We illustrated the use of the proposed tests by two case studies, that analyzed the auction protocols by Curtis et al. [6] and by Brandt [3], respectively. The protocol by Curtis et al. is correct only for a trusted Registration authority – which runs contrary to the point of verification: that the authorities no longer need to be trusted. Brandt's protocol does not have sound verifiability tests: it is technically possible for a losing bidder to conclude he won. Moreover, it may also be possible for a bidder to prevent anyone from winning by using a wrong decryption key. To prevent this, bidders must prove that the private key matches the previously announced public key. Additionally, the protocol does not provide sufficient registration and authentication mechanisms to allow registration verifiability.

Future work. We are currently working on a full application of these definitions to various auction protocols in both the symbolic model and the computational model.

Looking further ahead, we are interested in the full relationship between fairness and verifiability in auctions. As illustrated, there exist verifiability requirements without which violations of fairness may occur. The exact relationship between fairness and verifiability however is an open question.

6. ACKNOWLEDGMENTS

This work was partly supported by the ANR project ProSe (decision ANR-2010-VERS-004-01).

7. REFERENCES

[1] M. Abe and K. Suzuki. Receipt-free sealed-bid auction. In *Proc. 5th Conference on Information Security*, volume 2433 of *LNCS*, pages 191–199. Springer, 2002.

[2] J. Benaloh and D. Tuinstra. Receipt-free secret-ballot elections (extended abstract). In *Proc. 26th Symposium on Theory of Computing*, STOC '94, pages 544–553, New York, NY, USA, 1994. ACM.

[3] F. Brandt. How to obtain full privacy in auctions. *International Journal of Information Security*, 5:201–216, 2006.

[4] C. Cachin. Efficient private bidding and auctions with an oblivious third party. In *Proc. 6th Conference on Computer and Communications Security (CCS'99)*, pages 120–127. ACM Press, 1999.

[5] X. Chen, B. Lee, and K. Kim. Receipt-free electronic auction schemes using homomorphic encryption. In *Proc. 6th Conference on Information Security and Cryptology*, volume 2971 of *LNCS*, pages 259–273. Springer, 2003.

[6] B. Curtis, J. Pieprzyk, and J. Seruga. An efficient eAuction protocol. In *Proc. 7th Conference on Availability, Reliability and Security (ARES'07)*, pages 417–421. IEEE Computer Society, 2007.

[7] N. Dong, H. L. Jonker, and J. Pang. Analysis of a receipt-free auction protocol in the applied pi calculus. In *Proc. 7th Workshop on Formal Aspects in Security and Trust (FAST'10)*, volume 6561 of *LNCS*, pages 223–238. Springer-Verlag, 2011.

[8] J. Dreier, P. Lafourcade, and Y. Lakhnech. Formal verification of e-auction protocols. In *Principles of Security and Trust (POST)*, 2013. To appear.

[9] A. Fujioka, T. Okamoto, and K. Ohta. A practical secret voting scheme for large scale elections. In J. Seberry and Y. Zheng, editors, *Advances in Cryptology – AUSCRYPT '92*, volume 718 of *LNCS*, pages 244–251. Springer Berlin / Heidelberg, 1992.

[10] M. Harkavy, J. D. Tygar, and H. Kikuchi. Electronic auctions with private bids. In *Proc. 3rd USENIX Workshop on Electronic Commerce*. Usenix, 1998.

[11] M. Hirt and K. Sako. Efficient receipt-free voting based on homomorphic encryption. In *Proc. 19th Annual Conference on the Theory andApplications of Cryptographic Techniques: Advances inCryptology (EUROCRYPT)*, volume 1807 of *LNCS*, pages 539–556. Springer, 2000.

[12] S. Kremer, M. D. Ryan, and B. Smyth. Election verifiability in electronic voting protocols. In *Proc. 15th European Symposium on Research in Computer Security (ESORICS 2010)*, volume 6345 of *LNCS*, pages 389–404. Springer, 2010.

[13] R. Küsters, T. Truderung, and A. Vogt. Accountability: definition and relationship to verifiability. In *Proc. 17th Conference on Computer and Communications Security (CCS'10)*, CCS '10, pages 526–535. ACM, 2010.

[14] H. Lipmaa, N. Asokan, and V. Niemi. Secure vickrey auctions without threshold trust. In *Proc. 6th Conference on Financial Cryptography*, volume 2357 of *LNCS*, pages 87–101. Springer, 2003.

[15] M. Naor, B. Pinkas, and R. Sumner. Privacy preserving auctions and mechanism design. In *Proc. 1st Conference on Electronic Commerce*, pages 129–139. ACM Press, 1999.

[16] K. Omote and A. Miyaji. A practical english auction with one-time registration. In *Proc. 6th Australasian Conference on Information Security and Privacy*, volume 2119 of *LNCS*, pages 221–234. Springer, 2001.

[17] T. E. Rockoff and M. Groves. Design of an internet-based system for remote dutch auctions. *Internet Research*, 5:10–16, 1995.

[18] K. Sako and J. Kilian. Receipt-free mix-type voting scheme - A practicalsolution to the implementation of a voting booth. In *Proc. 14th Conference on the Theory andApplications of Cryptographic Techniques: Advances inCryptology (EUROCRYPT'95)*, volume 921 of *LNCS*, pages 393–403. Springer, 1995.

Verifiable and Private Top-k Monitoring

Xuhua Ding
School of Information Systems
Singapore Management University
xhding@smu.edu.sg

HweeHwa Pang
School of Information Systems
Singapore Management University
hhpang@smu.edu.sg

Junzuo Lai [*]
Department of Computer Science,
Jinan University, China
junzuolai@smu.edu.sg

ABSTRACT

In a data streaming model, records or documents are pushed from a data owner, via untrusted third-party servers, to a large number of users with matching interests. The match in interest is calculated from the correlation between each pair of document and user query. For scalability and availability reasons, this calculation is delegated to the servers, which gives rise to the need to protect the privacy of the documents and user queries. In addition, the users need to guard against the eventuality of a server distorting the correlation score of the documents to manipulate which documents are highlighted to certain users.

In this paper, we address the aforementioned privacy and verifiability challenges. We introduce the first cryptographic scheme which concurrently safeguards the privacy of the documents and user queries in such a data streaming model, while enabling users to verify the correlation scores obtained. We provide techniques to bound the computation demand in decrypting the correlation scores, and we demonstrate the overall practicality of the scheme through experiments with real data.

Categories and Subject Descriptors

H.2.4 [**Systems**]: Query processing; H.2.7 [**Database Administration**]: Security, integrity, and protection

Keywords

Vector product, correlation computation, verifiability, privacy

1. INTRODUCTION

Data streaming applications like Web access analysis, profile-driven marketing, environment sensing, stock trading and online bidding are now commonplace, driven by the widespread adoption of mobile devices and RFID technology. Such applications can generate high volume of data, which are often streamed to an intermediary for query processing and analysis to produce aggregate results for end-user consumption. As the intermediary may not be trusted by the data owner/generator or the end-users, the data and queries need to be protected.

To motivate our data streaming setting, suppose that Figure 1 depicts the system model of a surveillance application. There are three parties in the model – a data Owner, one or more Servers, and many Users. The Owner operates various security checkpoints of a country or sensitive installation. Here, a picture of the face is taken of each visitor passing through a checkpoint. From the picture, a feature vector of the relative position, size and shape of the eyes, nose, jaw, etc. are automatically extracted. The original picture and feature vector form a document $\mathbf{d}$ that is streamed to a shared Server. One of the Users is an intelligence agency which is monitoring a list of subjects. The intelligence agency is likely to have pictures of each subject; the feature vector extracted from each picture is registered with the Server as a standing query $\mathbf{q}$. The Server computes the correlation score between $\mathbf{d}$ and $\mathbf{q}$, and returns the score along with the document identifier to the agency. The k documents that best match the query are displayed on the agency's alert screen. Since the Server is administered by another agency or an outsourced service provider whose operators may not have appropriate security clearances, the following system and security requirements become necessary:

- *Query privacy.* The feature values in $\mathbf{q}$ must be known only to the particular User who issued it. The reason is while the original picture cannot be reconstructed directly from $\mathbf{q}$, an adversary could extract feature vectors from his own pictures and attempt to match them against $\mathbf{q}$ to discover what the intelligence agency is interested in. Furthermore, the actual correlation score between $\mathbf{d}$ and $\mathbf{q}$ should be available only to the issuing User. Neither the Owner who possesses $\mathbf{d}$, the Server that computes the score nor other users can be allowed to deduce $\mathbf{q}$.
- *Document privacy.* The feature values in $\mathbf{d}$ must not be revealed to the Server, so as to prevent an adversary from matching $\mathbf{d}$ against his own pictures. To safeguard the rights of the visitors who have their pictures taken, the Owner may selectively restrict the features in $\mathbf{d}$ that various Users are allowed to query over.
- *Verifiability.* Without direct access to $\mathbf{d}$, the User who issued $\mathbf{q}$ would want to verify the correlation score between $\mathbf{d}$ and $\mathbf{q}$. This is to guard against an adversary at the Server manipulating the score of $\mathbf{d}$ to elude detection by the User.
- *Result ranking.* In order to identify the k most promising document matches, the User needs to be able to retrieve the actual correlation score for each $\mathbf{d}$.

The system and security requirements of our data streaming model exceed the capability of existing schemes in the literature. In the

[*]The author carried out this work at the Singapore Management University.

ASIA CCS'13, May 8–10, 2013, Hangzhou, China.

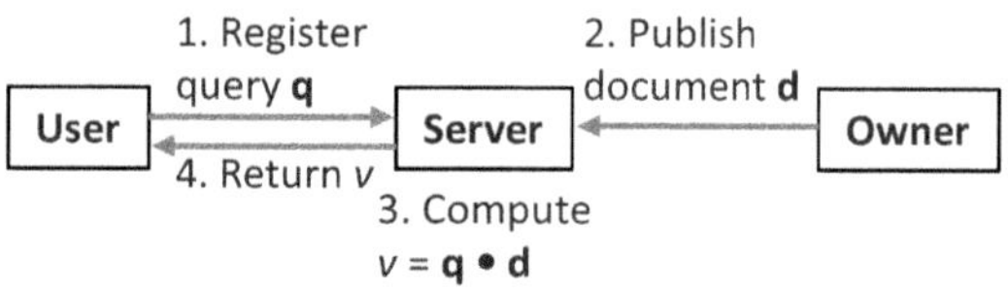

Figure 1: Data Streaming Model

data stream security literature, verifiability and privacy have been addressed separately. The former includes [12] which allows users to verify the arrival, removal and update of data falling within a selection range on an attribute, and [7] for authenticating selection-aggregation queries over an attribute of interest. Privacy in data streaming has been achieved through randomization [1] and anonymization [2]. In cryptography, the most relevant line of work is predicate encryption that supports the inner product of two vectors, e.g., [6]. However, predicate encryption schemes are not designed for our data streaming model in which the document vector $\mathbf{d}$ and the query vector $\mathbf{q}$ are generated by different parties and must be kept secret from each other.

CONTRIBUTIONS In this paper, we propose a Cryptographic Verifiable and Private Monitoring (CVPM) scheme to enable our data streaming model in Figure 1 which satisfies the aforementioned security requirements. In CVPM, each User encrypts her standing query vectors to safeguard their privacy; the encrypted query vectors are lodged with the Server. Whenever the Owner has a document to release, he generates for it an encrypted document vector which is distributed to the Server. Upon receiving a new encrypted document vector $\mathbf{d}$, the Server computes the protected correlation coefficient $v = \mathbf{q} \cdot \mathbf{d}$ for each query $\mathbf{q}$, and outputs v to the issuing User. The User then deciphers the output to obtain and verify the correlation score.

To the best of our knowledge, our scheme is the first for the data streaming model that concurrently achieves verifiability and privacy protection. Specifically, it ensures that the document vectors cannot be exploited by the Server or the User, other than for computing correlation scores. In addition, neither the Owner nor the Server gains an advantage in compromising the privacy of the user queries. Finally, the Server cannot tamper with the document correlation scores without being detected by the User. Through extensive experiments involving real datasets, we also demonstrate the practicality of CVPM for a broad spectrum of applications.

2. RELATED WORK

Data stream security has been studied extensively in various research communities. Table 1 summarizes the characteristics of the most relevant ones with respect to the system and security requirements identified in the Introduction.

In [8], Lindner and Meier discuss general security concerns in architecting a data stream management system. To safeguard the privacy of data streams, [1] proposes to inject randomized noise. The condensation scheme in [2] to achieve anonymization supports incremental update, and is applicable to data streams. [3] and [11] introduce protocols for a server to search a stream for files that contain given query keywords; the protocols protect the privacy of the query, but not the data stream.

Verifiability has been addressed by a different group of studies. These include [12] which allows users to verify the arrival, removal and update of data falling within a selection range on an attribute, [7] for authenticating selection-aggregation queries over an attribute of interest, and [10] for verifying the output of aggregation functions like MAX and SUM. None of the above schemes

Symbol	Meaning
$\mathbf{d}$	Document vector
k_d	Bit length of each coordinate in $\mathbf{d}$
$\mathbf{q}$	Query vector
k_q	Bit length of each coordinate in $\mathbf{q}$
m	Dimensionality of $\mathbf{d}$ and $\mathbf{q}$
m_q	Number of coordinates that the User specified in $\mathbf{q}$
v	$v = \mathbf{q} \cdot \mathbf{d}$ is the score of $\mathbf{d}$ given $\mathbf{q}$
n	$n = p_1 p_2$ where p_1, p_2 are distinct, large prime numbers
$\mathbb{G}, \mathbb{G}_T$	Multiplicative groups of order n
$\hat{e}$	Bilinear map $\hat{e} : \mathbb{G} \times \mathbb{G} \rightarrow \mathbb{G}_T$
$\mathbb{G}_1$	Subgroup of $\mathbb{G}$, of order p_1
$\mathbb{G}_2$	Subgroup of $\mathbb{G}$, of order p_2
$H(.)$	A one-way hash function

Table 2: Notation

support the correlation computation that we need. Moreover, they provide either privacy or verifiability, but not both.

Among existing cryptographic protocols, the ones that support inner product, which is what the common correlation coefficients entail, are most relevant to our problem setting. The rest of this section focuses on those schemes and explains why they do not meet our requirements.

In the predicate encryption scheme proposed in [6], an entity possessing a secret key token associated with a vector $\mathbf{x}$ can decrypt a public key encryption ciphertext associated with another vector $\mathbf{y}$ on the condition that $\mathbf{x} \cdot \mathbf{y} = 0$. The scheme is not suitable for our data streaming model because it does not protect the privacy of $\mathbf{x}$ against the token holder (the Server in our context), who can infer $\mathbf{x}$ by choosing any $\mathbf{y}$ and producing a ciphertext with the public key. Note that in our model, we require the privacy of both vectors against the Server.

The above privacy issue is addressed by the symmetric-key predicate encryption scheme in [13]. Here, the generation of both the secret token for $\mathbf{x}$ and the ciphertext associated with $\mathbf{y}$ require secret inputs; this prevents the token holder from generating ciphertext for his chosen $\mathbf{y}$. However, this scheme requires a one-to-one mapping between the secrets used for encrypting $\mathbf{x}$ and $\mathbf{y}$, whereas our data streaming model requires a one-to-many mapping as the Owner's data stream serves multiple Users simultaneously.

Our problem setting is also related to privacy-preserving scalar product schemes such as [5, 15]. These schemes are built on an interactive protocol in which two parties, each holding a secret input, co-compute the scalar product without revealing their inputs to each other. They do not permit the computation to be carried out by an untrusted intermediary (the Server) though, a key requirement of our data streaming model. There is also no provision to check whether the scalar product is computed correctly.

3. CRYPTOGRAPHIC VERIFIABLE AND PRIVATE MONITORING SCHEME

In this section, we introduce our CVPM scheme for the data streaming model. Table 2 summarizes the frequently used notations, which will be explained as they are used.

3.1 Overview

As shown in Figure 1, our data streaming model comprises a set of Users who issue standing queries $\mathbf{q}$, an Owner who generates a stream of documents $\mathbf{d}$, and one or more Servers that match $\mathbf{q}$ with $\mathbf{d}$. The Server may also enforce access policies on

Scheme	Query Privacy	Document Privacy	Verifiability	Result Ranking
Random noise injection [1]	No	Yes	No	No
Data condensation [2]	No	Yes	No	No
Classifier training on data stream [14]	No	Yes	No	No
Keyword search on file stream [3], [11]	Yes	No	No	No
Authenticate streamed data [12]	No	No	Yes	No
Authenticate selection-aggregation queries [7]	No	No	Yes	No
Authenticate aggregation functions [10]	No	No	Yes	No

Table 1: Properties of Existing Security Schemes for Data Streaming Model

which features in $\mathbf{d}$ can be queried by particular users. Our CVPM scheme is made up of six algorithms that are executed by the parties in the model: `Setup`, `UserReg`, `QueryGen`, `DataGen`, `ServPro` and `UserDec`.

Owner first runs `Setup` to initialize the cryptographic setting shared by all participants and generate its own secret key. To join the system, a User first registers with the Owner. If the User satisfies the admission policy, the Owner executes `UserReg` to (i) produce a secret for the User to encrypt her queries, and (ii) produce a secret that enables the Server to process the User's queries subsequently. Based on her interest, the User runs `QueryGen` to generate the encryption $\mathbf{Q}_u$ of her query $\mathbf{q}$ and deposits $\mathbf{Q}_u$ with the Server. No two users share the same secret. This ensures that the encrypted queries are user-specific; in other words, identical queries from two users would lead to different encrypted queries.

At runtime, the Owner generates a stream of documents as well as the corresponding document vectors. For each newly published document vector $\mathbf{d}$, the Owner runs `DataGen` to produce an encrypted document vector $\mathbf{D}$ for the Server; $\mathbf{D}$ is common to *all* the Users in the system. The Server then executes `ServPro` for each registered query $\mathbf{Q}_u$ and returns the encrypted correlation score v to the corresponding User.

On receiving the response, the User runs `UserDec` with her secret to recover v. We provide constraints that bound the search space for v, so that its decryption is computationally feasible. A User cannot use her secret to decrypt correctly the encrypted scores intended for other users.

Adversarial model & security objectives The adversary could be a Server that executes the protocol honestly but is curious to know the document features and user queries, or cheats in computing the document scores. The adversary could also be an unauthorized party who managed to intercept and issue messages in the system. We assume the adversary is rational in that it does not run denial-of-service attacks. The security objectives include query privacy, document privacy and verifiability, as defined in the Introduction.

3.2 Solution Construction

Setup The Owner carries out the following initialization steps. It first generates two cyclic groups $\mathbb{G}$ and $\mathbb{G}_T$, both of order $n = p_1 p_2$ where p_1, p_2 are distinct prime numbers, and g is a generator of $\mathbb{G}$. Furthermore, there exists a non-degenerate and efficient bilinear mapping $\hat{e} : \mathbb{G} \times \mathbb{G} \to \mathbb{G}_T$. Hence, $\mathbb{G} = \langle g \rangle$ and $\mathbb{G}_T = \langle \hat{e}(g, g) \rangle$. The Owner computes $g_1 = g^{p_2}$, $g_2 = g^{p_1}$, and $h_1 = \hat{e}(g_1, g_1)$, $h_2 = \hat{e}(g_2, g_2)$. Let $\mathbb{G}_1 = \langle g_1 \rangle$ and $\mathbb{G}_2 = \langle g_2 \rangle$ denote two subgroups in $\mathbb{G}$ of order p_1 and p_2, respectively. The Owner then selects x randomly from $\mathbb{Z}_n$, denoted as $x \in_R \mathbb{Z}_n$. The secret key of the Owner is $sk = (g_1, h_1, x)$. It publishes $(\mathbb{G}, g, n, g_2, h_2)$ to enable the Server and Users to perform group operations in $\mathbb{G}$.

User Registration The Owner executes Algorithm `UserReg` described in Figure 2 to initialize the User's state with $\{A_u, h_u, K_u\}$, and to update the Server with B_u.

> Algorithm: `UserReg`
> Input: sk; Output: $\{B_u\}$ to Server, $\{A_u, h_u, K_u\}$ to User
>
> 1. Select $a_u, b_u \in_R \mathbb{Z}_n$ such that $a_u + b_u = x \bmod n$.
> 2. Compute $A_u = g^{a_u}$ and $B_u = g^{b_u}$.
> 3. Select $k_u \in_R \mathbb{Z}_n$; compute $h_u = h_1^{k_u}$ and $K_u = g_1^{k_u}$.
> 4. Securely send B_u to the Server; securely send $\{A_u, h_u, K_u\}$ to the User.

Figure 2: User Registration by the Owner

Query submission With $\{A_u, h_u, K_u\}$ from the Owner, the User initializes her standing query request $\mathbf{q} = \{q_i\}_{i=1}^m$ where $q_i \in [0, 2^{k_q}) \subseteq \mathbb{Z}_{p_1}$, according to her interests; k_q is the bit length of each coordinate in $\mathbf{q}$. She deposits the encryption $\mathbf{Q}_u$ of $\mathbf{q}$ with the Server as a standing query over the data stream. The algorithm for the User's query generation is described in Figure 3.

> Algorithm: `QueryGen`
> Input: K_u, $\mathbf{q} = \{q_i\}_{i=1}^m$; Output: $\mathbf{Q}_u$ to Server, $\mathbf{r}$ to User
>
> 1. Choose a random vector $\mathbf{r} = \{r_i\}_{i=1}^m$ with $r_i \in \mathbb{Z}_n$.
> 2. Compute vector $\mathbf{Q}_u = \{Q_i\}_{i=1}^m$ where $Q_i = K_u^{q_i} g_2^{r_i}$ for every $i \in [1, m]$.
> 3. Securely send $\mathbf{Q}_u$ to the Server.

Figure 3: Query Generation by Users

Data encryption For each document vector $\mathbf{d} = \{d_i\}_{i=1}^m$ where $d_i \in [0, 2^{k_d}) \subseteq \mathbb{Z}_{p_1}$ and k_d is the bit length of the coordinates in $\mathbf{d}$, the Owner prepares an encrypted document vector $\mathbf{D}$ by running Algorithm `DataGen` described in Figure 4. $\mathbf{D}$ is the common document ciphertext for all the users in the system. In the algorithm, the Owner encrypts each d_i with a random number t_i derived from a fresh seed r, and $H(.)$ is a secure one-way hash function.

> Algorithm: `DataGen`
> Input: sk, $\mathbf{d} = \{d_i\}_{i=1}^m$; Output: $\{\mathbf{D}, C\}$ to Server
>
> 1. Choose $r \in_R \mathbb{Z}_n$ and compute $C = g^r$.
> 2. Set $t_i = H(i, \hat{e}(g^r, g^x))$ for every $i \in [1, m]$.
> 3. Compute $\mathbf{D} = \{D_i\}_{i=1}^m$ where $D_i = g_1^{d_i} g_2^{t_i}$.
> 4. Send $\{\mathbf{D}, C\}$ to the Server.

Figure 4: Data Generation by the Owner

Server processing On receiving encrypted document vector $\mathbf{D}$, the Server executes Algorithm `ServPro` shown in Figure 5 for all the users in its service domain. Given registered standing query $\mathbf{Q}_u$ from a User, the algorithm first computes a ciphertext of the correlation score for the User, i.e., an encrypted form of the inner product of $\mathbf{d}$ and $\mathbf{q}$. Then, it computes a partial result to aid the User in deciphering the correlation score.

User decryption The User executes Algorithm `UserDec` in Figure 6 to get the final correlation score v. With each coordinate in

Algorithm: ServPro
Input: $\{\mathbf{D}, C\}$, $\{\mathbf{Q_u}, B_u\}$; Output: $\{W, C, C_1\}$ to User

For every standing query $\mathbf{Q_u}$, perform the following steps:

1. Compute $w_i = \hat{e}(D_i, Q_i)\ \forall 1 \le i \le m$; then compute $W = \prod_{i=1}^{m} w_i$.
2. Compute $C_1 = \hat{e}(C, B_u)$.
3. Send $\{W, C, C_1\}$ to the User.

Figure 5: Correlation Computation by the Server

$\mathbf{d}$ and $\mathbf{q}$ having bit length k_d and k_q respectively, v is in the range of $[0, 2^{k_d+k_q} \cdot m_q)$ where $m_q \le m$ is the number of coordinates that the User specified in $\mathbf{q}$. However, the User may predefine a threshold t to filter out documents with low correlation scores, which signify that they are too dissimilar to the User's interest and can be discarded without examination. This limits the search for v to a narrower range $[t, 2^{k_d+k_q} \cdot m_q)$ in Step 5. Note that t is chosen by and known only to the User.

Furthermore, when the document streaming rate is high, the User may wish to monitor only the top-k documents within a sliding window (of the most recent documents, or of documents that arrived within the last several seconds). The correlation score v_k of the k-th highest ranking document provides another threshold to limit the search for v. Therefore, the User only needs to search for the score v of the new document within a bounded range:

$$v \in [\max\{t, v_k\}, 2^{k_d+k_q} \cdot m_q) \qquad (1)$$

The constraints in the formula enable the User to avoid a full discrete logarithm computation for v, by exploiting secret application-specific values t, v_k and m_q which only the User knows. If v is found, the User may request for the actual document from the Owner; otherwise, the User ignores the document.

Algorithm: UserDec
Input: $\{W, C, C_1\}$, $\{\mathbf{r}, h_u, A_u\}$; Output: v

1. Compute $C_2 = \hat{e}(C, A_u)$ and $s = C_1 C_2$.
2. For $1 \le i \le m$, compute $t_i = H(i, s)$.
3. Compute $R = \sum_{i=1}^{m} r_i t_i$.
4. Compute $W' = W / h_2^R$.
5. Check whether there exists $v \in [\max\{t, v_k\}, 2^{k_d+k_q} \cdot m_q)$ such that $h_u^v = W'$, t is a relevance threshold, v_k is the score of the current k-th result document, and m_q is the number of coordinates used in $\mathbf{q}$.

Figure 6: User Decryption

Where the Owner has specified access policies, W in Algorithm ServPro and R in Algorithm UserDec would aggregate over only features that the User is authorized to query.

Finally, if any input parameter to UserDec has been tampered with, with overwhelming probability R would not match, so W' would have order n and there exists no v that satisfies $h_u^v = W'$.

3.3 Decryption of Document Score

The UserDec algorithm requires a discrete logarithm in Step 5 to get the document score v. Although general discrete logarithm is considered to be intractable for a large cyclic group, it is feasible for the User to find v within the restricted range in Formula 1 with t, v_k and m_q which are known only to her. If the range for v is narrow, the User can pre-generate a look-up table for v. If not, the User executes the baby-step giant-step procedure, with space and time complexities $O(\sqrt{2^{k_d+k_q} \cdot m_q})$, to find v. In our target applications, typically the number of features that Users specify in queries satisfies $m_q \ll m$ and is within a hundred or two. Moreover, k_d and k_q are also small. Thus, we expect the time and space complexity of the baby-step giant-step procedure to be tolerable. Additionally, the optimization opportunities described below often allow the discrete logarithm computation to terminate early. We will validate the practicality of the optimized UserDec algorithm through experiments in Section 4.

3.3.1 *Optimizing Document Score Decryption*

There exist several algorithms for solving the discrete logarithm problem, such as the baby-step giant-step algorithm, the Pohlig-Hellman algorithm and the index-calculus algorithm [9]. Although the latter two are superior in asymptotic time complexity, they are difficult to optimize to take advantage of the constraints expressed in Formula 1. Instead, we pick the baby-step giant-step algorithm for the User's discrete logarithm computation. The optimization works as follows to find $v \in [\max(t, v_k), 2^{k_d+k_q} \cdot m_q)$ satisfying $h_u^v = W'$ in the UserDec algorithm.

Let $\tau = \left\lceil \sqrt{2^{k_d+k_q} \cdot m_q} \right\rceil$ and $v = c_1 \cdot \tau + c_0$ for some $0 \le c_1 < \tau$ and $0 \le c_0 < \tau$. Thus, $h_u^{-c_1 \tau} W' = h_u^{c_0}$. The User creates beforehand a lookup table for $\langle h_u^{c_0}, c_0 \rangle$ with $h_u^{c_0}$ as search key, for $0 \le c_0 < \tau$. For each v, the User iteratively checks whether $h_u^{-c_1 \tau} W'$ exists in the lookup table, decrementing c_1 from $\tau - 1$ down towards $\lfloor \max(t, v_k)/\tau \rfloor$. The procedure can be suspended after any iteration; as long as c_1 is kept, the procedure can be resumed subsequently.

Suppose that the User has a set of top-k documents in place when he obtains the answer W' for a new document $\mathbf{d}$. He needs to evaluate immediately whether its score v falls within $[\max(t, v_k), 2^{k_d+k_q} \cdot m_q)$. If so, $\mathbf{d}$ replaces one of the earlier documents in the top-k result. If not, the User has established an upper bound $\overline{v} = v_k$ for v. The User also knows the earliest that $\mathbf{d}$ may be considered for the top-k result again is when one of the current top-k documents expires and v_k changes.

Now suppose that in between document arrivals, the User has several documents $\mathbf{d}_i$ for which the upper bound score $\overline{v}_i = v_k$. Instead of recovering the actual score v_i for each of those documents in turn, the User will lower their $\overline{v}_i$'s uniformly (by decrementing c_1 in the computation of their document scores in unison). This allows the User to fully decrypt the higher document scores first, and discover the more relevant candidates that are likely to replace the next expiring document in the top-k result.

3.3.2 *Computing Document Score in Memory Constrained User Devices*

The optimization in Section 3.3.1 limits the search for the document score as expressed in Formula 1, and is particularly helpful when $2^{k_d+k_q} \cdot m_q$ is large. However, the size of the lookup table remains at $\lceil \sqrt{2^{k_d+k_q} \cdot m_q} \rceil$ entries, which may exceed the available memory in user devices that are severely resource-constrained. We now discuss how our extended baby-step giant-step procedure trades computation for memory space in finding $v \in [\max(t, v_k), 2^{k_d+k_q} \cdot m_q)$ such that $h_u^v \equiv W'$ for given h_u and W'.

The User begins by allocating a lookup table within the available memory space. Let τ denote the number of entries in the table. We represent v as a polynomial $v = c_\alpha \tau^\alpha + c_{\alpha-1} \tau^{\alpha-1} + \ldots + c_0$, where $\alpha = \lfloor \log_\tau (2^{k_d+k_q} \cdot m_q) \rfloor$ and $c_i \in [0, \tau)\ \forall i \in [0, \alpha]$. Thus, $h_u^{-(c_\alpha \tau^{\alpha-1} + \ldots + c_1)\tau} W' = h_u^{c_0}$. As before, the User populates the lookup table for $\langle h_u^{c_0}, c_0 \rangle$ with $h_u^{c_0}$ as search key, for $0 \le c_0 < \tau$. Following that, the User iteratively checks whether $h_u^{-i\tau} W'$ exists in the lookup table, decrementing i from $\tau^\alpha - 1$ towards $\lfloor \max(t, v_k)/\tau \rfloor$.

3.4 Compressing Document & Query Vectors

In certain applications, the document vector $\mathbf{d}$ and query vector $\mathbf{q}$ are expected to contain mostly zero coordinates. One such example is the vector space model for text retrieval [16], where each feature corresponds to a dictionary word. Since most of the documents and queries contain only a very small subset of the dictionary words, only a small number of coordinates in the document and query vectors are non-zero. In this situation, the Owner may leave out some of the zero coordinates in $\mathbf{d}$, so that the Server can exclude them from query processing. This reduces the server processing overhead, at the expense of exposing the dropped coordinates in $\mathbf{d}$. Likewise, the User may drop some of the zero coordinates in $\mathbf{q}$. The concrete scheme is as follows.

For each $\mathbf{d}$, the Owner creates a certified Bloom filter [4] on the index position of the non-zero coordinates in $\mathbf{d}$. After that, those zero coordinates whose index gets a false positive on the Bloom filter must also stay in $\mathbf{d}$; the remaining zero coordinates whose index gets a negative on the Bloom filter may be omitted.

As proposed in [4], a Bloom filter is designed to support membership checks on a set $\mathcal{F}$ of f key values, $\mathcal{F} = \{F_1, F_2, \ldots, F_f\}$. To construct a Bloom filter with ϕ bits, we choose ρ independent hash functions $H_1, H_2, \ldots, H_\rho$, each with a range of $[1, \phi]$. For each value $F_i \in \mathcal{F}$, the filter bits at positions $H_1(F_i)$, $H_2(F_i)$, ..., $H_\rho(F_i)$ are set to 1. To check whether a given F is in $\mathcal{F}$, we examine the bits at $H_1(F), H_2(F), \ldots, H_\rho(F)$. If any of the bits is 0, F cannot possibly be in $\mathcal{F}$; otherwise there is a high probability that F is in $\mathcal{F}$. In other words, the Bloom filter admits controlled false positive rates but no false negatives. The false positive rate is

$$FP = \left(1 - \left(1 - \frac{1}{\phi}\right)^{\rho f}\right)^{\rho} \approx \left(1 - e^{-\rho f/\phi}\right)^{\rho} \qquad (2)$$

Mathematically, FP is minimized for $\rho = \phi \ln 2/f$. Since ρ must be an integer, we will use $\rho = \lfloor \phi \ln 2/f \rfloor$. Given the value of f and the target FP rate, we can thus set ρ and ϕ accordingly.

Applying the Bloom filter to our setting, f corresponds to the number of non-zero coordinates in $\mathbf{d}$, and the number of false positives is $FP \times (m - f)$. Thus, the Server sees $f + FP \times (m - f)$ remaining coordinates in $\mathbf{d}$. Together with the parameters ϕ and ν, the Server can deduce the value of f, i.e., the actual number of non-zero coordinates among the remaining coordinates. If we want to prevent that, then the index positions of some of the zero coordinates should go into the Bloom filter too, so that f yields only a loose upper bound on the number of non-zero coordinates.

Likewise for $\mathbf{q}$, the User may randomly drop some of the zero coordinates. The remaining coordinates, along with their index positions, are submitted to the Server.

In processing a document $\mathbf{d}$ against a query $\mathbf{q}$, the Server evaluates the inner product $\mathbf{q} \cdot \mathbf{d}$ only over the common coordinates in $\mathbf{q}$ and $\mathbf{d}$ that are accessible to each User. Along with the answer W', the Server returns the index positions $\mathcal{I}$ that are in $\mathbf{q}$ but have been omitted from $\mathbf{d}$, as well as the certified Bloom filter. The user can verify $\mathcal{I}$ against the Bloom filter, which by construction will not produce false positives on the indices in $\mathcal{I}$.

4. EMPIRICAL VALIDATION

In this section, we evaluate the overall practicality of our CVPM scheme, and the effectiveness of the optimization techniques introduced in Section 3.3.1 in particular.

4.1 Experiment Set-Up

Datasets: To ensure that our observations are generalizable, we have run CVPM on several datasets. Here we report on two real datasets from the UCI KDD Archive[1]. The datasets are picked because they vary from each other in key properties that stress different algorithms in CVPM. The first, Corel Image, contains feature vectors extracted from 68,040 photo images. Each feature vector is a point in 32-dimensional HSV color space, so $m = 32$. We discretize every dimension into 2^8 integer values. The feature vectors contain many zero coordinates, and the correlation between feature vectors are generally low. The dataset relates closely with the main motivating application in Section 1.

The second dataset, US Census, is a discretized version of part of the data collected in the 1990 U.S. census. The dataset includes 2,458,285 records (vectors), each with $m = 68$ attribute values. All the attributes are in the range [0, 20]. The correlation between vectors are high, relative to that in the Corel Image data. The data in this collection essentially are user profiles, hence they simulate the user profile matching application described in the Introduction.

Methodology: For both datasets, we extract 100 vectors randomly to be user queries. The rest of the vectors are shuffled and fed into the document stream from the Owner. The arrival rate is λ documents/minute. The User maintains a sliding window of the documents that arrived in the last w hour, and the k documents with the highest correlation scores in this window constitute the query result. For each experiment setting, the performance measures are averaged across the 100 queries.

We implemented the CVPM scheme in C language, on the PBC cryptography library from Stanford University[2]. The experiments are run on a MacBook Pro with 2.8 GHz Intel Core i7 CPU and 8 GB of main memory.

Metrics: Our primary performance metrics include: (a) the average time taken by the Owner in executing the `DataGen` algorithm on a new document; (b) the average time taken by the Server to execute the `ServPro` algorithm for each document-query pair; and (c) the average time taken by the User to run the `UserDec` algorithm and update her top-k result upon a document arrival, with and without the optimization described in Section 3.3.1.

4.2 Corel Image Experiment

For the Corel Image data, we set the sliding window w to one hour and vary λ from 1 to 30 documents/minute. Referring to Formula 1, the threshold t is set to 0, so pruning of the search space for the correlation score v of a new document relies solely on v_k, the correlation score of the k-th result document. Figure 7(a) plots the average user execution time per query for $k = 10, 20, 40$ and 60. The figure also gives the execution time of 'No Opt', which disables the optimization techniques in Section 3.3.1.

With a larger k, v_k becomes lower. If the correlation score v of an arriving document is below v_k, the User will have to execute more iterations in the baby-step giant-step algorithm before she can conclude that the new document is not (yet) eligible for the top-k result. This explains why a larger k lengthens the execution time. For a fixed k setting, the execution time drops as λ increases. The reason is that the sliding window accumulates more documents, in the process pushing up the k-th highest correlation score and narrowing the search space for v. The execution time is bounded from below by the cost of the bilinear pairing operation $\hat{e}(.,.)$ (indicated by the dotted horizontal line). In the worst case, the User executes the baby-step giant-step algorithm until v is discovered, so the execution time is bounded from above by the 'No Opt' method.

The execution time incurred by the Owner and Server are summarized in Table 3. The cost of the `DataGen` algorithm, incurred

[1] http://kdd.ics.uci.edu/databases/
[2] http://crypto.stanford.edu/pbc/

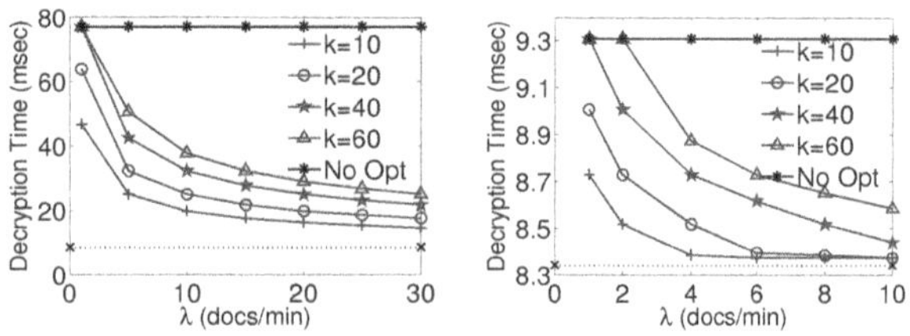

(a) Corel Image Data (b) US Census Data

Dataset	Owner	Server
Corel Image	217.80	275.56
US Census	449.82	576.19

Table 3: Owner and Server Processing Times (msec)

once for every new document, is just over 200 msec; this cost is dominated by the exponentiation operations in Step 3 of the algorithm. The cost of the `ServPro` algorithm is incurred once for every document-query pair and stands at 275 msec; this cost is dominated by the bilinear pairing operations in Step 1 of the algorithm. These performance levels can be maintained for higher data rates by employing multiple CPUs.

4.3 US Census Experiment

Turning to the US Census data, we again set the sliding window w at one hour. Here λ varies from 1 to 10 documents/minute. The average user execution time per query is summarized in Figure 7(b). The qualitative behavior of CVPM observed here is the same as in the previous experiment. Quantitatively, the execution times are lower now because the narrower feature domains (i.e., smaller k_d and k_q values in Formula 1) constrain the search space for the correlation score v of arriving documents.

Finally, the corresponding execution time of the Owner and Server are reported in Table 3. Both are higher than in the previous experiment, on account of the larger number of features in this dataset – $m = 68$ – compared to $m = 32$ in the Corel Image data.

4.4 Summary of Experiment Results

Besides the Corel Image and US Census data, we have also experimented with several other datasets like the Insurance Company Benchmark, also available from the UCI KDD Archive. The common observations across the experiments are as follows:

- The optimization arising from Formula 1 is particularly effective with high document rate λ, large sliding window w or low k settings, as they tend to raise the correlation score v_k of the k-th result document.
- The User cost is sensitive to the number of features and their domain sizes. If there are many features with large domains, it may be necessary to bucketize the feature values; effectively, this induces a coarser resolution on the features.
- Where the feature vectors have high dimensionality, it may be necessary to parallelize the computation at the Owner and Server. This is straightforward to implement, as the computation on different coordinates in the document vector can be carried out independently.
- CVPM delivers acceptable performance for many practical application settings, with sub-second execution time for every party in the data streaming model.

5. CONCLUSION

In this paper, we formulate the security requirements in a data streaming model. The model employs an untrusted server to compute the correlation scores between documents streamed from the data owner, and the standing queries issued by users. The correlation computation translates to an inner product of the respective document and query vectors. We present the first cryptographic scheme that concurrently safeguards the privacy of the documents and queries, while enabling users to verify the correctness of the correlation scores received. Through extensive experiments, we demonstrate that the proposed scheme achieves practical execution time for a wide spectrum of application settings.

Acknowledgements

HweeHwa Pang is supported by Research Grant 12-C220-SMU-004 from the Singapore Management University.

6. REFERENCES

[1] C. C. Aggarwal. On randomization, public information and the curse of dimensionality. In *Proc. of ICDE*, 2007.
[2] R. Agrawal and R. Srikant. Privacy-preserving data mining. In *Proc. of ACM SIGMOD*, 2000.
[3] J. Bethencourt, D. X. Song, and B. Waters. New constructions and practical applications for private stream searching. In *Proc. of IEEE S& P*, 2006.
[4] B. Bloom. Space/Time Trade-Offs in Hash Coding with Allowable Errors. *Communications of the ACM*, 13(7):422–426, July 1970.
[5] B. Goethals, S. Laur, H. Lipmaa, and T. Mielikainen. On private scalar product computation for privacy-preserving data mining. In *Proc. of ICISC*, 2004.
[6] J. Katz, A. Sahai, and B. Waters. Predicate encryption supporting disjunctions, polynomial equations, and inner products. In *EUROCRYPT*, pages 146–162, April 2008.
[7] F. Li, K. Yi, M. Hadjieleftheriou, and G. Kollios. Proof-infused streams: Enabling authentication of sliding window queries on streams. In *Proc. of VLDB*, pages 147–158, September 2007.
[8] W. Lindner and J. Meier. Towards a secure data stream management system. In *Trends in Enterprise Application Architecture*, pages 114–128, August 2005.
[9] A. J. Menezes, P. C. van Oorschot, and S. A. Vanstone. *Handbook of Applied Cryptography*. CRC Press, 1997.
[10] S. Nath, H. Yu, and H. Chan. Secure outsourced aggregation via one-way chains. In *Proc. of ACM SIGMOD*, 2009.
[11] R. Ostrovsky and W. E. S. III. Private searching on streaming data. In *Advances in Cryptology (CRYPTO)*, 2005.
[12] S. Papadopoulos, Y. Yang, and D. Papadias. CADS: Continuous authentication on data streams. In *Proc. of VLDB*, pages 135–146, September 2007.
[13] E. Shen, E. Shi, and B. Waters. Predicate privacy in encryption systems. In *Proc. of TCC*, 2009.
[14] Y. Xu, K. Wang, A. W.-C. Fu, R. She, and J. Pei. Privacy-preserving data stream classification. In *Privacy-Preserving Data Mining*. Springer, 2008.
[15] Z. Yang, R. Wright, and H. Subramaniam. Experimental analysis of a privacy-preserving scalar protocol. *International Journal of Computer Systems Science and Engineering*, 21(1):47–52, January 2006.
[16] J. Zobel and A. Moffat. Inverted Files for Text Search Engine. *ACM Computing Surveys*, 38(2):6, July 2006.

Author Index

www.ingramcontent.com/pod-product-compliance
Ingram Content Group UK Ltd.
Pitfield, Milton Keynes, MK11 3LW, UK
UKHW051416290926
15471UKWH00009B/1639

9 781450 322805